MW01472610

Fundamentals of Power Semiconductor Devices

B. Jayant Baliga

Fundamentals of Power Semiconductor Devices

Springer

B. Jayant Baliga
Power Semiconductor Research Center
North Carolina State University
1010 Main Campus Drive
Raleigh, NC 27695-7924
USA

ISBN 978-0-387-47313-0 e-ISBN 978-0-387-47314-7

Library of Congress Control Number: 2008923040

Printed on acid-free paper.

9 8 7 6 5 4 3 2 1

springer.com

Dedication

The author would like to dedicate this book to his wife, Pratima, for her unwavering support throughout his career devoted to the enhancement of the performance and understanding of power semiconductor devices.

Preface

Today the semiconductor business exceeds $200 billion with about 10% of the revenue derived from power semiconductor devices and smart power integrated circuits. Power semiconductor devices are recognized as a key component for all power electronic systems. It is estimated that at least 50% of the electricity used in the world is controlled by power devices. With the widespread use of electronics in the consumer, industrial, medical, and transportation sectors, power devices have a major impact on the economy because they determine the cost and efficiency of systems. After the initial replacement of vacuum tubes by solid-state devices in the 1950s, semiconductor power devices have taken a dominant role with silicon serving as the base material. These developments have been referred to as the *Second Electronic Revolution*.

Bipolar power devices, such as bipolar transistors and thyristors, were first developed in the 1950s. Because of the many advantages of semiconductor devices compared with vacuum tubes, there was a constant demand for increasing the power ratings of these devices. Their power rating and switching frequency increased with advancements in the understanding of the operating physics, the availability of larger diameter, high resistivity silicon wafers, and the introduction of more advanced lithography capability. During the next 20 years, the technology for the bipolar devices reached a high degree of maturity. By the 1970s, bipolar power transistors with current handling capability of hundreds of amperes and voltage blocking capability of over 500 V became available. More remarkably, technology was developed capable of manufacturing an individual power thyristor from an entire 4-inch diameter silicon wafer with voltage rating over 5,000 V.

My involvement with power semiconductor devices began in 1974 when I was hired by the General Electric Company at their corporate research and development center to start a new group to work on this technology. At that time, I had just completed my Ph.D. degree at Rensselaer Polytechnic Institute by performing research on a novel method for the growth of epitaxial layers of compound semiconductors.[1–4] Although I wanted to explore this approach after joining the semiconductor industry, I was unable to secure a position at any of the major research laboratories due to a lack of interest in this unproven growth technology. Ironically, the OMCVD epitaxial growth process that I pioneered with Professor Ghandhi has now become the most commonly used method for the growth of high quality compound semiconductor layers for applications such as lasers, LEDs, and microwave transistors.

My first assignment at GE was to develop improved processes for the fabrication of high voltage thyristors used in their power distribution business. Since the thyristors were used for high voltage DC transmission and electric locomotive drives, the emphasis was on increasing the voltage rating and current handling capability. The ability to use neutron transmutation doping to produce high resistivity n-type silicon with improved uniformity across large diameter wafers became of interest at this time. I was fortunate in making some of the critical contributions to annealing the damage caused to the silicon lattice during neutron irradiation making this process commercially viable.[5] This enabled increasing the blocking voltage of thyristors to over 5,000 V while being able to handle over 2,000 A of current in a single device.

Meanwhile, bipolar power transistors were being developed with the goal of increasing the switching frequency in medium power systems. Unfortunately, the current gain of bipolar transistors was found to be low when it was designed for high voltage operation at high current density. The popular solution to this problem, using the Darlington configuration, had the disadvantage of increasing the on-state voltage drop resulting in an increase in the power dissipation. In addition to the large control currents required for bipolar transistors, they suffered from poor safe-operating-area due to second breakdown failure modes. These issues produced a cumbersome design, with snubber networks, that raised the cost and degraded the efficiency of the power control system.

In the 1970s, the power MOSFET product was first introduced by International Rectifier Corporation. Although initially hailed as a replacement for all bipolar power devices due to its high input impedance and fast switching speed, the power MOSFET has successfully cornered the market for low voltage (<100 V) and high switching speed (>100 kHz) applications but failed to make serious inroads in the high voltage arena. This is because the on-state resistance of power MOSFETs increases very rapidly with increase in the breakdown voltage. The resulting high conduction losses, even when using larger more expensive die, degrade the overall system efficiency.

In recognition of these issues, I proposed two new thrusts in 1979 for the power device field. The first was based upon the merging of MOS and bipolar device physics to create a new category of power devices.[6] My most successful innovation among MOS-bipolar devices has been the insulated gate bipolar transistor (IGBT). Soon after commercial introduction in the early 1980s, the IGBT was adopted for all medium power electronic applications. Today, it is manufactured by more than a dozen companies around the world for consumer, industrial, medical, and other applications that benefit society. The triumph of the IGBT is associated with its huge power gain, high input impedance, wide safe operating area, and a switching speed that can be tailored for applications depending upon their operating frequency.

The second approach that I suggested in 1979 for enhancing the performance of power devices was to replace silicon with wide bandgap semiconductors. The basis for this approach was an equation that I derived relating the on-resistance of the drift region in unipolar power devices to the basic properties of the semiconductor material. This equation has since been referred to as Baliga's figure of merit (BFOM). In addition to the expected reduction in the on-state resistance with higher carrier mobility, the equation predicts a reduction in on-resistance as the inverse of the cube of the breakdown electric field strength of the semiconductor material.

The first attempt to develop wide-bandgap-semiconductor-based power devices was undertaken at the General Electric Corporate Research and Development Center, Schenectady, NY, under my direction. The goal was to leverage a 13-fold reduction in specific on-resistance for the drift region predicted by the BFOM for gallium arsenide. A team of ten scientists was assembled to tackle the difficult problems of the growth of high resistivity epitaxial layers, the fabrication of low resistivity ohmic contacts, low leakage Schottky contacts, and the passivation of the GaAs surface. This led to an enhanced understanding of the breakdown strength[7] for GaAs and the successful fabrication of high performance Schottky rectifiers[8] and MESFETs.[9] Experimental verification of the basic thesis of the analysis represented by BFOM was therefore demonstrated during this period. Commercial GaAs-based Schottky rectifier products were subsequently introduced in the market by several companies.

In the later half of the 1980s, the technology for the growth of silicon carbide was developed at North Carolina State University (NCSU) with the culmination of commercial availability of wafers from CREE Research Corporation. Although data on the impact ionization coefficients of SiC were not available, early reports on the breakdown voltage of diodes enabled estimation of the breakdown electric field strength. Using these numbers in the BFOM predicted an impressive 100–200-fold reduction in the specific on-resistance of the drift region for SiC-based unipolar devices. In 1988, I joined NCSU and subsequently founded the Power Semiconductor Research Center (PSRC) – an industrial consortium – with the objective of exploring ideas to enhance power device performance. Within the first year of the inception of the program, SiC Schottky barrier rectifiers with breakdown voltage of 400 V were successfully fabricated with on-state voltage drop of about 1 V and no reverse recovery transients.[10] By improving the edge termination of these diodes, the breakdown voltage was found to increase to 1,000 V. With the availability of epitaxial SiC material with lower doping concentrations, SiC Schottky rectifiers with breakdown voltages over 2.5 kV have been fabricated at PSRC.[11] These results have motivated many other groups around the world to develop SiC-based power rectifiers. In this regard, it has been my privilege to assist in the establishment of national programs to fund research on silicon carbide technology in the United States, Japan, and Switzerland–Sweden. Meanwhile, accurate measurements of the impact ionization coefficients for 6H-SiC and 4H-SiC in defect-free regions were performed at PSRC using an electron beam excitation method.[12] Using these coefficients, a BFOM of over 1,000 is predicted for SiC, providing even greater motivation to develop power devices from this material.

Although the fabrication of high performance, high voltage Schottky rectifiers has been relatively straightforward, the development of a suitable silicon carbide MOSFET structure has been problematic. The existing silicon power D-MOSFET and U-MOSFET structures do not directly translate to suitable structures in silicon carbide. The interface between SiC and silicon dioxide, as a gate dielectric, needed extensive investigation due to the large density of traps that prevent the formation of high conductivity inversion layers. Even after overcoming this hurdle, the much higher electric field in the silicon dioxide when compared with silicon devices, resulting from the much larger electric field in the underlying SiC, leads to reliability problems. Fortunately, a structural innovation called the ACCUFET, to overcome both of these problems, was proposed and demonstrated at PSRC.[13] In this structure, a buried P^+ region is used to shield the gate region from the high electric field within the SiC drift region. This concept is applicable to devices that utilize either accumulation channels or inversion channels. Devices with low specific on-resistance have been demonstrated at PSRC using both 6H-SiC and 4H-SiC with epitaxial material capable of supporting over 5,000 V.[14] This device structure has been subsequently emulated by several groups around the world.

The availability of power semiconductor devices with high input impedance has encouraged the development of integrated control circuits. In general, the integration of the control circuit is preferred over the discrete counterpart due to reduced manufacturing costs at high volumes and improved reliability from a reduction of the interconnects. Since the complexity of including additional circuitry to an IC is relatively small, the incorporation of protective features such as over-temperature, over-current, and over-voltage has become cost effective. In addition, the chips can contain encode/decode CMOS circuitry to interface with a central microprocessor or computer in the system for control and diagnostic purposes. This technology is commonly referred to as *Smart Power Technology*.[15]

The advent of smart power technology portends a *Second Electronic Revolution*. In contrast to the integrated circuits for information processing, this technology enables efficient control of power and energy. These technologies can therefore be regarded as complementary, similar to the brain and muscles in the human body. Smart power technology is having an enormous impact on society. The widespread use of power semiconductor devices in consumer, industrial, transportation, and medical applications brings greater mobility and comfort to billions of people around the world. Our ability to improve the efficiency for the control of electric power results in the conservation of fossil fuels, which in turn provides reduction of environmental pollution.

Due to these developments, it is anticipated that there will be an increasing need for technologists trained in the discipline of designing and manufacturing power semiconductor devices. This textbook provides the knowledge in a tutorial format suitable for self-study or in a graduate/senior level university course. In comparison with my previous textbooks[16,17] (which have gone out of print), this book provides a more detailed description of the operating physics of power devices. Analytical expressions have been rigorously derived using the fundamental semiconductor Poisson's, continuity, and conduction equations. The electrical characteristics of all the power devices discussed in this book can be computed using these analytical solutions as shown by typical examples provided in each section. Due to increasing interest in the utilization of wide bandgap semiconductors for power devices, the book includes the analysis of silicon carbide structures. To corroborate the validity of the analytical formulations, I have included the results of two-dimensional numerical simulations using MEDICI[18] in each section of the book. The simulation results are also used to elucidate further the physics and point out two-dimensional effects whenever relevant.

In Chap. 1, a broad introduction to potential applications for power devices is provided. The electrical characteristics for ideal power rectifiers and transistors are then defined and compared with those for typical devices. Chapter 2 provides the transport properties of silicon and silicon carbide that have relevance to the analysis and performance of power device structures. Chapter 3 discusses breakdown voltage, which is the most unique distinguishing characteristic for power devices, together with edge termination structures. This analysis is pertinent to all the device structures discussed in subsequent chapters of the book.

Chapter 4 provides a detailed analysis of the Schottky rectifier structure. On-state current flow via thermionic emission is described followed by the impact of image force barrier lowering on the reverse leakage current. These phenomena influence the selection of the barrier height to optimize the power losses as described in the chapter. The influence of the tunneling current component is also included in this chapter due to its importance for silicon carbide Schottky rectifiers.

Chapter 5 describes the physics of operation of high voltage P–i–N rectifiers. The theory for both low-level and high-level injection conditions during on-state current flow is developed in detail. The impact of this on the reverse recovery phenomenon during turn-off is then analyzed. The influence of end region recombination, carrier–carrier scattering, and auger recombination are included in the analysis.

In Chap. 6, an extensive discussion of the operating principles and design considerations is provided for the power metal–oxide–semiconductor field effect transistor (MOSFET) structure. The influence of the parasitic bipolar transistor on the blocking voltage is described together with methods for its suppression. The basic physics of creating channels in the MOSFET structure is then developed. The concepts of threshold voltage, transconductance, and specific on-resistance are described. Various components of the on-state resistance are analyzed and optimization procedures are provided. Both the commercially available DMOS and UMOS structures are analyzed here. The modification of the physics required to produce a superlinear transfer characteristic is included due to its relevance for RF and audio applications. A detailed analysis of the device capacitances is then provided for use in the analysis of the switching behavior. Analysis of the gate charge is included here because of its common use in comparing device designs. The switching characteristics of the power MOSFET are then related to its capacitance, including the impact of the Miller effect. This is followed by discussion of the safe-operating-area, the integral body diode, high temperature characteristics, and complementary (p-channel) devices. A brief description of the process flow for the D-MOSFET and U-MOSFET structures is given in the chapter for completeness. The last portion of the chapter focuses on silicon carbide technology with the options of the Baliga–Pair configuration, the shielded planar structure, and the shielded trench-gate structure described in detail.

Chapter 7 is devoted to bipolar power transistors. The basic theory for current transport and gain in an N–P–N transistor is first developed followed by a discussion of issues relevant to power transistors. The various breakdown modes of the bipolar transistors are then explained. The physics governing the current gain of the bipolar transistor is extensively analyzed including high-level injection effects, the current-induced base, and emitter current crowding. The output characteristics for the bipolar transistor are then described with analysis of the saturation region, the quasisaturation mode, and the output resistance. This is followed by analysis of the switching characteristics. The influence of stored charge on the switching behavior of the bipolar transistor is described in detail during both the turn-on and turn-off transients. Issues dealing with second breakdown are then considered followed by ways to improve the current gain by using the Darlington configuration.

The physics of operation of the power thyristor is considered in Chap. 8. The impact of the four layer structure on the forward and reverse blocking capability is first analyzed including the use of cathode shorts. The on-state characteristics for the thyristor are then shown to approach those for a P–i–N rectifier. The gate triggering and holding currents are related to the cathode short design. Under switching characteristics, the turn-on physics is discussed with description of the involute design, the amplifying gate, and light-activated gate structures. The commutated switching behavior is also analyzed together with a discussion of voltage transients. The basic principles of the gate turn-off (GTO) thyristor are then described with analytical models for the storage, voltage-rise and current-fall times. The chapter concludes with the description of triacs, which are commonly used for AC power control.

The insulated gate bipolar transistor (IGBT) is discussed in depth in Chap. 9. The benefits of controlling bipolar current transport in a wide base P–N–P transistor using a MOS channel are explained. The design of both reverse blocking (symmetric) and unidirectional blocking (asymmetric) structures is considered here. The on-state characteristics of the IGBT are then extensively analyzed including the impact of high-level injection in the wide-base region and the finite injection efficiency of the collector junction. The discussion includes not only the basic symmetric IGBT structure but also the asymmetric structure and the transparent emitter structure. The utilization of lifetime control is compared with changes to the N-buffer-layer and P^+ collector doping concentrations. After developing the current saturation model for the IGBT structure, the output characteristics for the three types of IGBT structure are derived. The impact of the stored charge on the switching behavior of the device is then analyzed for the case of no-load, resistive-load, and inductive-load conditions for each of the three types of structures. The optimization of the power losses in the IGBT structure is then performed, allowing comparison of the three types of structures. The next section of the chapter describes the complementary (p-channel) IGBT structure. This is followed by an extensive discussion of methods for suppression of the parasitic thyristor in the IGBT due to its importance for designing stable devices. The next section on the safe-operating-area includes analysis of the FBSOA, RBSOA, and SCSOA. The trench-gate IGBT structure is then demonstrated to produce lower on-state voltage drop. This is followed by discussion of scaling up the voltage rating for the IGBT and its excellent characteristics for high ambient temperatures. Various methods for improving the switching speed of the IGBT structure and optimizing its cell structure are then discussed. The chapter concludes with the description of the reverse conducting IGBT structure.

The final chapter (Chap. 10) provides the basis for the comparison of various power devices from an applications viewpoint. A typical motor drive case is selected to demonstrate the reduction of power losses by optimization of the on-state and switching characteristics of the devices. The importance of reducing the reverse recovery current in power rectifiers is highlighted here.

Throughout the book, emphasis is placed on deriving simple analytical expressions that describe the underlying physics and enable representation of the device electrical characteristics. This treatment is invaluable for teaching a course on power devices because it allows the operating principles and concepts to be conveyed with quantitative analysis. The analytical approach used in the book based on physical insight will provide a good foundation for the reader. The results of two-dimensional numerical simulations have been included to supplement and reinforce the concepts. Due to space limitations, only the basic power device structures have been included in this book. Advanced structures will be covered in monographs to be subsequently published. I am hopeful that this book will be widely used for the teaching of courses on solid-state devices and that it will become an essential reference for the power device industry well into the future.

Raleigh, NC

B. Jayant Baliga

References

[1] B.J. Baliga and S.K. Ghandhi, "Heteroepitaxial InAs Grown on GaAs from Triethylindium and Arsine", Journal of the Electrochemical Society, Vol. 121, pp 1642–1650, 1974.

[2] B.J. Baliga and S.K. Ghandhi, "Growth and Properties of Heteroepitaxial GaInAs Alloys Grown on GaAs Substrates from Trimethylgallium, Triethylindium and Arsine", Journal of the Electrochemical Society, Vol. 122, pp 683–687, 1975.

[3] B.J. Baliga and S.K. Ghandhi, "The Preparation and Properties of Tin Oxide Films Formed by the Oxidation of Trimethytin", Journal of the Electrochemical Society, Vol. 123, pp 941–944, 1976.

[4] B.J. Baliga and S.K. Ghandhi, "Preparation and Properties of Zinc Oxide Films Grown by the Oxidation of Diethylzinc", Journal of the Electrochemical Society, Vol. 128, pp 558–561, 1981.

[5] B.J. Baliga et al., "Defect Levels Controlling the Behavior of Neutron Transmutation Doped Silicon during Annealing", NTD Conference, April 1987.

[6] B.J. Baliga, "Evolution of MOS-Bipolar Power Semiconductor Technology", Proceedings IEEE, pp 409–418, 1988.

[7] B.J. Baliga et al., "Breakdown Characteristics of Gallium Arsenide", IEEE Electron Device Letters, Vol. EDL-2, pp 302–304, 1981.

[8] B.J. Baliga et al., "Gallium Arsenide Schottky Power Rectifiers", IEEE Transactions on Electron Devices, Vol. ED-32, pp 1130–1134, 1985.

[9] P.M. Campbell et al., "Trapezoidal-Groove Schottky-Gate Vertical-Channel GaAs FET", IEEE Electron Device Letters, Vol. EDL-6, pp 304–306, 1985.

[10] M. Bhatnagar, P.K. McLarty, and B.J. Baliga, "Silicon-Carbide High-Voltage (400 V) Schottky Barrier Diodes", IEEE Electron Device Letters, Vol. EDL-13, pp 501–503, 1992.

[11] R.K. Chilukuri and B.J. Baliga, "High Voltage Ni/4H-SiC Schottky Rectifiers", IEEE International Symposium on Power Semiconductor Devices and ICs, pp 161–164, 1999.

[12] R. Raghunathan and B.J. Baliga, "Temperature dependence of Hole Impact Ionization Coefficients in 4H and 6H-SiC", Solid State Electronics, Vol. 43, pp 199–211, 1999.

[13] P.M. Shenoy and B.J. Baliga, "High Voltage Planar 6H-SiC ACCUFET", International Conference on Silicon Carbide, III-Nitrides, and Related Materials, Abstract Tu3b-3, pp 158–159, 1997.

[14] R.K. Chilukuri and B.J. Baliga, PSRC Technical Report TR-00-007, May 2000.

[15] B.J. Baliga, "Smart Power Technology: An Elephantine Opportunity", Invited Plenary Talk, IEEE International Electron Devices Meeting, pp 3–6, 1990.

[16] B.J. Baliga, "Modern Power Devices", Wiley, New York, 1987.

[17] B.J. Baliga, "Power Semiconductor Devices", PWS, Boston, MA, 1996.

[18] MEDICI TCAD Simulator, Avanti Corporation, Fremont, CA 94538.

Contents

Chapter 1

Introduction

Modern society is increasingly dependent upon electrical appliances for comfort, transportation, and healthcare, motivating great advances in power generation, power distribution and power management technologies. These advancements owe their allegiance to enhancements in the performance of power devices that regulate the flow of electricity. After the displacement of vacuum tubes by solid state devices in the 1950s, the industry relied upon silicon bipolar devices, such as bipolar power transistors and thyristors. Although the ratings of these devices grew rapidly to serve an ever broader system need, their fundamental limitations in terms of the cumbersome control and protection circuitry led to bulky and costly solutions. The advent of MOS technology for digital electronics enabled the creation of a new class of devices in the 1970s for power switching applications as well. These silicon power MOSFETs have found extensive use in high frequency applications with relatively low operating voltages (below 100 V). The merger of MOS and bipolar physics enabled the creation of yet another class of devices in the 1980s. The most successful innovation in this class of devices has been the insulated gate bipolar transistor (IGBT). The high power density, simple interface, and ruggedness of the IGBT have made it the technology of choice for all medium and high power applications, with perhaps the exception of high voltage DC transmission systems. Even the last remaining bastion for the conventional power thyristors is threatened by the incorporation of MOS-gated structures.

Power devices are required for systems that operate over a broad spectrum of power levels and frequencies. In Fig. 1.1, the applications for power devices are shown as a function of operating frequency. High power systems, such as HVDC power distribution and locomotive drives, requiring the control of megawatts of power operate at relatively low frequencies. As the operating frequency increases, the power ratings decrease for the devices, with typical microwave devices handling about 100 W. All these applications are served by silicon devices. Thyristors are

B.J. Baliga, *Fundamentals of Power Semiconductor Devices*, doi: 10.1007/978-0-387-47314-7_1,

favored for the low frequency, high power applications, IGBTs for the medium frequency and power applications, and power MOSFETs for the high frequency applications.

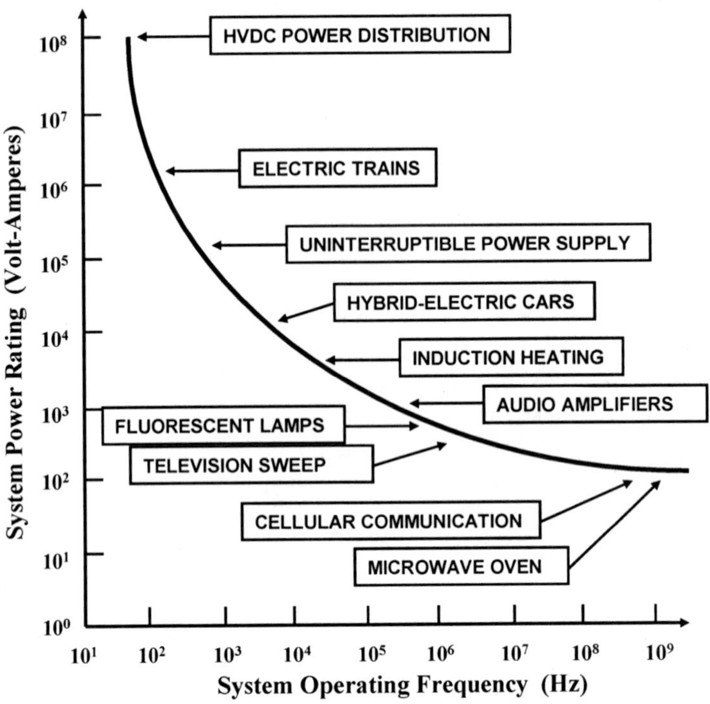

Fig. 1.1 Applications for power devices

Another approach to classification of applications for power devices is in terms of their current and voltage handling requirements, as shown in Fig. 1.2. On the high power end of the chart, thyristors are available that can individually handle over 6,000 V and 2,000 A, enabling the control of over 10 MW of power by a single monolithic device. These devices are suitable for the HVDC power transmission and locomotive drive (traction) applications. For the broad range of systems that require operating voltages between 300 and 3,000 V with significant current handling capability, the IGBT has been found to be the optimum solution. When the current requirements fall below 1 A, it is feasible to integrate multiple devices on a single monolithic chip to provide greater functionality for systems such as telecommunications and display drives. However, when the current exceeds a few amperes, it is more cost effective to use discrete power MOSFETs with appropriate control ICs to serve applications such as automotive electronics and switch mode power supplies. Consequently, no single device structure exists at this time that is suitable for serving all the applications, leaving plenty of room for further innovations.

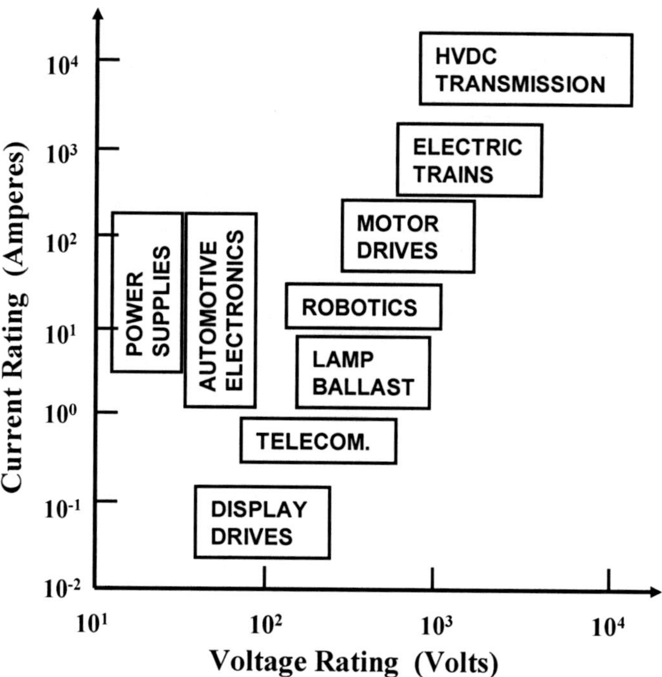

Fig. 1.2 System ratings for power devices

1.1 Ideal and Typical Power Switching Waveforms

An ideal power device must be capable of controlling the flow of power to loads with zero power dissipation. The loads encountered in systems may be inductive in nature (such as motors and solenoids), resistive in nature (such as heaters and lamp filaments), or capacitive in nature (such as transducers and LCD displays). Most often, the power delivered to a load is controlled by turning-on a power device on a periodic basis to generate pulses of current that can be regulated by a control circuit. The ideal waveforms for the power delivered through a power switch are shown in Fig. 1.3. During each switching cycle, the switch remains on for a time upto t_{ON} and maintains an off-state for the remainder of the period T. For an ideal power switch, the voltage drop during the on-state is zero, resulting in no power dissipation. Similarly, during the off-state, the (leakage) current in the ideal power switch is zero, resulting in no power dissipation. In addition, it is assumed that the power switch makes the transition between the on-state and off-state instantaneously, resulting in no power loss as well.

The waveforms observed with typical power switches produce power dissipation during the on-state, off-state, as well as during the switching transients. As shown in Fig. 1.4, typical power devices exhibit a voltage drop (V_F) in the on-state, which results in a power dissipation in the on-state given by:

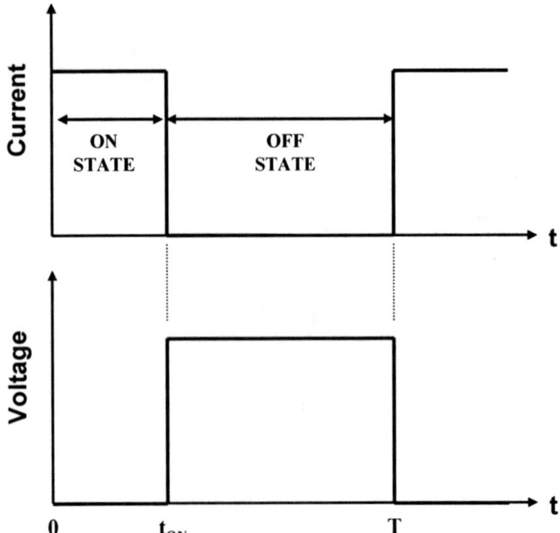

Fig. 1.3 Ideal switching waveforms for power delivery

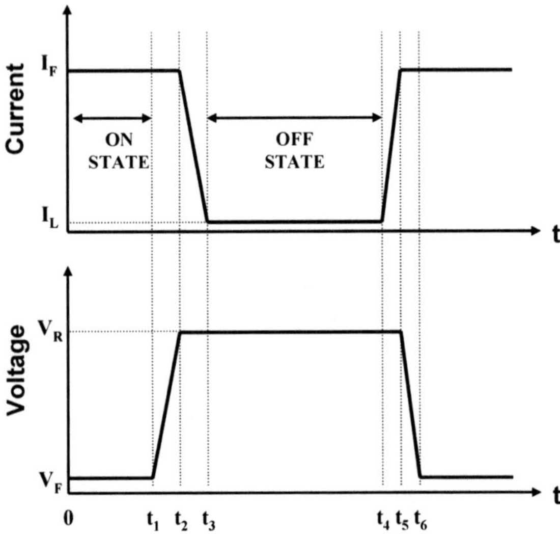

Fig. 1.4 Typical switching waveforms for power delivery

$$P_L(\mathrm{on}) = \delta I_{\mathrm{F}} V_{\mathrm{F}} \tag{1.1}$$

where I_{F} is the on-state current. In this expression, δ is referred to as the duty cycle given by:

$$\delta = t_1 / T \tag{1.2}$$

where T is the time period (the reciprocal of the operating frequency).

Power dissipation also occurs in the off-state given by

$$P_L(\text{off}) = (1-\delta) I_L V_R \tag{1.3}$$

where I_L is the leakage current exhibited by the device in its off-state due to supporting a reverse bias (V_R). This expression holds true if the switching times are small compared with the time period. If this assumption is not valid, the power dissipation can be obtained by using:

$$P_L(\text{off}) = (t_4 - t_3) I_L V_R / T \tag{1.4}$$

Most often, the power dissipation in the off-state is small when compared with the other components and can be neglected. However, this does not hold true at elevated temperatures and when Schottky contacts are supporting the reverse voltage.

The power dissipation that occurs during switching must be treated separately for the turn-off transient and the turn-on transient. During the turn-off transient for typical loads that are inductive in nature, the voltage across the switch increases rapidly to the DC-supply voltage, followed by a decrease in the current flowing through the switch. For the linearized waveforms shown in Fig. 1.4, the power loss during the turn-off transient can be calculated using:

$$P_L(\text{turnoff}) = 0.5 (t_3 - t_1) I_F V_R f \tag{1.5}$$

where f is the operating frequency. In a similar manner, the power loss during the turn-on transient can be calculated using:

$$P_L(\text{turnon}) = 0.5 (t_6 - t_4) I_F V_R f \tag{1.6}$$

The total power dissipation incurred in the switch is obtained by combining these terms:

$$P_L(\text{total}) = P_L(\text{on}) + P_L(\text{off}) + P_L(\text{turnoff}) + P_L(\text{turnon}) \tag{1.7}$$

At low operating frequencies, the on-state power loss is usually dominant, making it desirable to develop power switches with low on-state voltage drops. At high operating frequencies, the switching power losses are usually dominant, making it desirable to develop power switches with fast switching speeds or small transition times. Unfortunately, it is usually necessary to perform a trade off between minimizing the on-state and switching power losses in most designs. As power switch technology advances, the total power loss for the optimized design continues to reduce, providing enhancements to the efficiency of power systems.

1.2 Ideal and Typical Power Device Characteristics

As discussed in the previous section, silicon power devices have served the industry for well over five decades but cannot be considered to have ideal device characteristics. In general, power electronic circuits require both rectifiers to control the direction of current flow and power switches to regulate the duration of current flow. Neither of these components exhibits the ideal characteristics that are required in power circuits to prevent power dissipation.

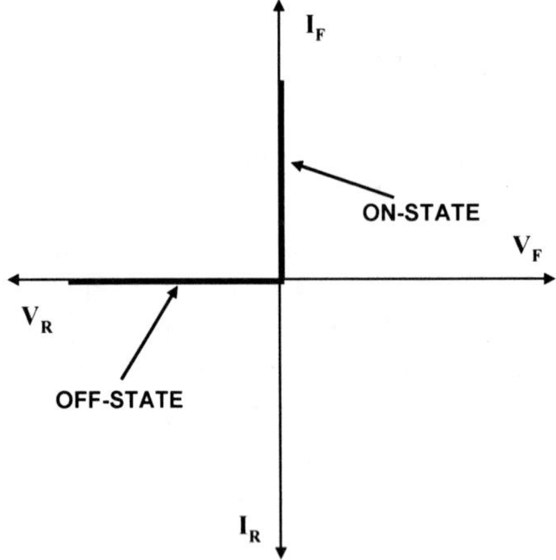

Fig. 1.5 Characteristics of an ideal power rectifier

An ideal power rectifier should exhibit the current–voltage (*i–v*) characteristics shown in Fig. 1.5. In the forward conduction mode, the first quadrant of operation in the figure, it should be able to carry any amount of current with zero on-state voltage drop. In the reverse blocking mode, the third quadrant of operation in the figure, it should be able to hold off any value of voltage with zero leakage current. Further, the ideal rectifier should be able to switch between the on-state and the off-state, with zero switching time.

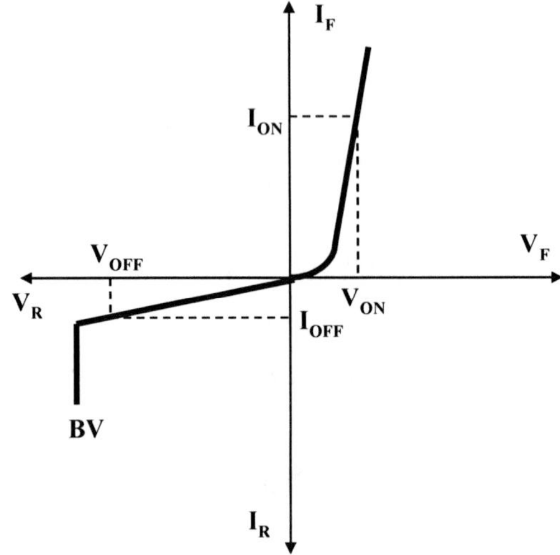

Fig. 1.6 Characteristics of a typical power rectifier

Actual silicon power rectifiers exhibit the *i–v* characteristics illustrated in Fig. 1.6. They have a finite voltage drop (V_{ON}) when carrying current on the on-state, leading to "conduction" power loss. They also have a finite leakage current (I_{OFF}) when blocking voltage in the off-state, creating off-state power loss. In addition, the doping concentration and thickness of the drift region of the silicon device must be carefully chosen with a design target for the breakdown voltage (BV). Moreover, the power dissipation in power devices increases when their voltage rating is increased because of an increase in the on-state voltage drop.

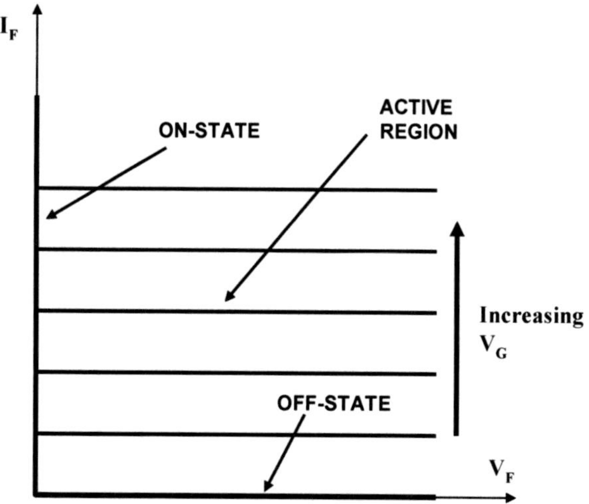

Fig. 1.7 Characteristics of an ideal transistor

The *i–v* characteristics of an ideal power switch are illustrated in Fig. 1.7. As in the case of the ideal rectifier, the ideal transistor conducts current in the on-state with zero voltage drop and blocks voltage in the off-state with zero leakage current. In addition, the ideal device can operate with a high current and voltage in the active region, with the saturated forward current in this mode controlled by the applied gate bias. The spacing between the characteristics in the active region is uniform for an ideal transistor indicating a gain that is independent of the forward current and voltage.

The *i–v* characteristics of a typical power switch are illustrated in Fig. 1.8. This device exhibits a finite resistance when carrying current in the on-state as well as a finite leakage current while operating in the off-state (not shown in the figure because its value is much lower than the on-state current levels). The breakdown voltage of a typical transistor is also finite as indicated in the figure with "BV". The typical transistor can operate with a high current and voltage in the active region. This current is controlled by the base current for a bipolar transistor while it is determined by a gate voltage for a MOSFET or IGBT (as indicated in the figure). It is preferable to have gate voltage controlled characteristics because the drive circuit can be integrated to reduce its cost. The spacing between the characteristics in the active region is nonuniform for a typical transistor with a square-law behavior for devices operating with channel pinch-off in the current saturation mode. Recently, devices operating under a new super-linear mode have been proposed and demonstrated for wireless base-station applications.[1] These devices exhibit an equal spacing between the saturated drain current characteristics as the gate voltage is increased. This is an ideal behavior when the transistor is used for the amplification of audio, video, or cellular signals because it eliminates signal distortion that occurs with the characteristics shown in Fig. 1.8.

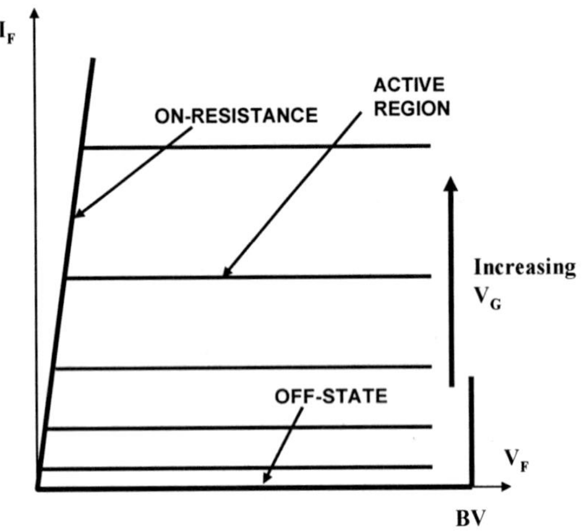

Fig. 1.8 Characteristics of a typical transistor

1.3 Unipolar Power Devices

Bipolar power devices operate with the injection of minority carriers during on-state current flow. These carriers must be removed when switching the device from the on-state to the off-state. This is accomplished by either charge removal via the gate drive current or via the electron-hole recombination process. These processes introduce significant power losses that degrade the power management efficiency. It is therefore preferable to utilize unipolar current conduction in a power device. The commonly used unipolar power diode structure is the Schottky rectifier that utilizes a metal-semiconductor barrier to produce current rectification. The high voltage Schottky rectifier structure also contains a drift region, as shown in Fig. 1.9, which is designed to support the reverse blocking voltage. The resistance of the drift region increases rapidly with increasing blocking voltage capability, as discussed later in this chapter. Silicon Schottky rectifiers are commercially available with blocking voltages of up to 100 V. Beyond this value, the on-state voltage drop of silicon Schottky rectifiers becomes too large for practical applications. Silicon P–i–N rectifiers are favored for designs with larger BVs due to their low on-state voltage drop despite slower switching properties. Silicon carbide Schottky rectifiers have much lower drift region resistance, enabling design of very high voltage devices with low on-state voltage drop and excellent switching characteristics.

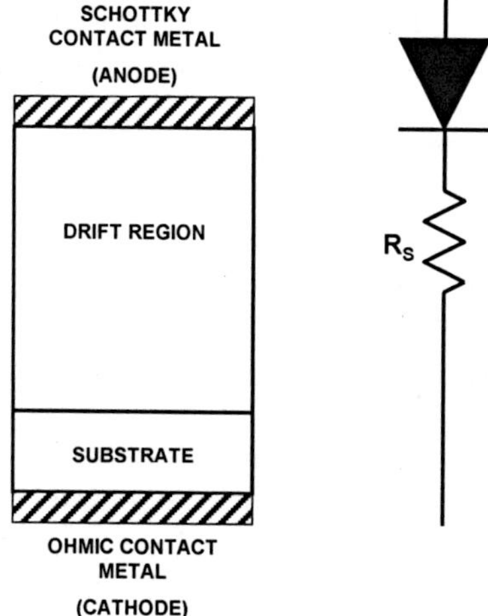

Fig. 1.9 The power Schottky rectifier structure and its equivalent circuit

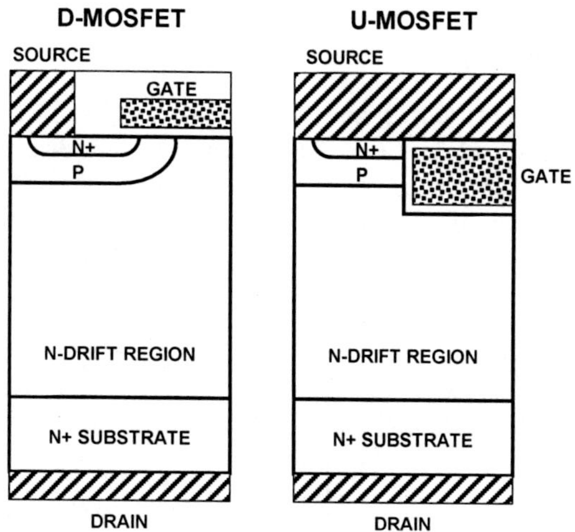

Fig. 1.10 Silicon power MOSFET structures

The most commonly used unipolar power transistor is the silicon power metal-oxide-semiconductor field-effect-transistor or MOSFET. Although other structures such as JFETs or SITs have been explored,[2] they have not been popular for power electronic applications because of their normally-on behavior. The commercially available silicon power MOSFETs are based upon the structures shown in Fig. 1.10. The D-MOSFET was first commercially introduced in the

1970s and contains a "planar-gate" structure. The P-base and the N^+ source regions are self-aligned to the edge of the polysilicon gate electrode by using ion-implantation of boron and phosphorus, followed by their respective drive-in thermal cycles. The n-type channel is defined by the difference in the lateral extension of the junctions under the gate electrode. The device supports positive voltage applied to the drain across the P-base/N-drift region junction. The voltage blocking capability is determined by the doping and thickness of the N-drift region. Although low voltage (<100 V) silicon power MOSFETs have low on-resistances, the drift region resistance increases rapidly with increasing blocking voltage, limiting the performance of silicon power MOSFETs to below 200 V. It is common-place to use the IGBT for higher voltage applications because of the smaller on-state power dissipation.

The silicon U-MOSFET structure became commercially available in the 1990s. It has a gate structure embedded within a trench etched into the silicon surface. The N-type channel is formed on the side-wall of the trench at the surface of the P-base region. The channel length is determined by the difference in vertical extension of the P-base and N^+ source regions as controlled by the ion-implant energies and drive times for the dopants. The silicon U-MOSFET structure was developed to reduce the on-state resistance by the elimination of the JFET component within the D-MOSFET structure.

1.4 Bipolar Power Devices

The commonly available silicon power bipolar devices are the bipolar transistor and the gate turn-off thyristor or GTO. These devices were originally developed in the 1950s and widely used for power switching applications until the 1970s when the availability of silicon power MOSFETs and IGBTs supplanted them. The structures of the bipolar transistor and GTO are shown in Fig. 1.11. In both devices,

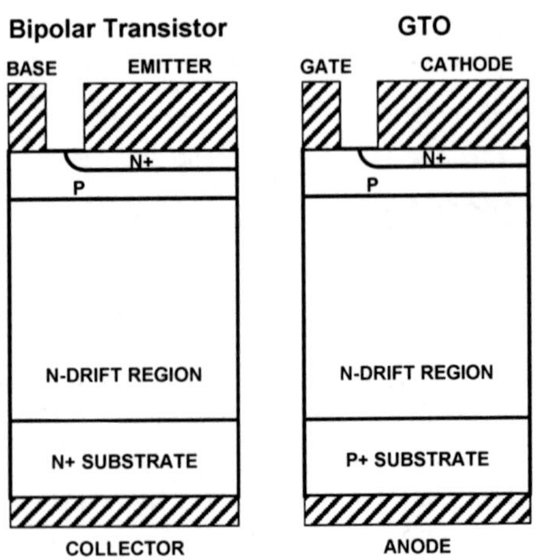

Fig. 1.11 Silicon bipolar device structures

injection of minority carriers into the drift region modulates its conductivity, reducing the on-state voltage drop. However, this charge must be subsequently removed during switching, resulting in high turn-off power losses. These devices require a large control (base or gate) current, which must be implemented with discrete components, leading to an expensive bulky system design. To address this short-coming, MOS-bipolar structures were proposed and developed in the 1980s.

1.5 MOS-Bipolar Power Devices

The most widely used silicon high voltage (>300 V) device for power switching applications is the *IGBT*, which was developed in the 1980s by combining the physics of operation of bipolar transistors and MOSFETs.[3] The structure of the IGBT is deceptively similar to that for the power MOSFET as shown in Fig. 1.12 if viewed as a mere replacement of the N^+ substrate with a P^+ substrate. However, this substitution creates a four-layer parasitic thyristor, which was initially considered a show-stopper because its latch-up results in destructive failure due to loss of gate control. Fortunately the parasitic thyristor can be suppressed by the addition of the P^+ region within the cell while retaining the benefits of the P^+ substrate for the injection of minority carriers into the N-drift region, resulting in greatly reducing its resistance. This has enabled the development of high voltage IGBT products with high current carrying capability.

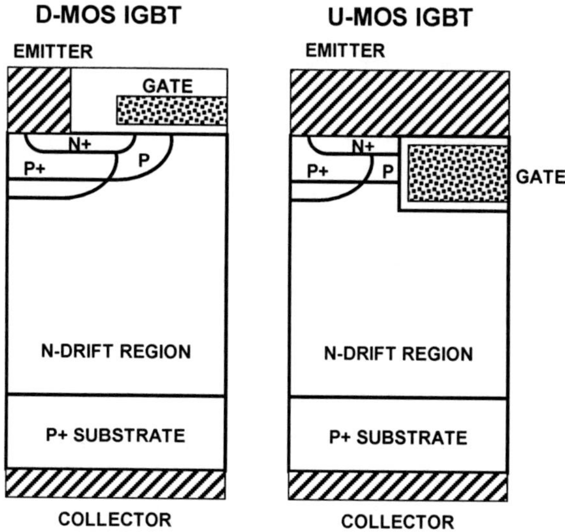

Fig. 1.12 IGBT device structures

The on-state voltage drop for the IGBT increases with increasing voltage blocking capability due to the necessity to widen the N-drift region. To improve upon the on-state voltage drop while maintaining the advantages of small control currents, many MOS-gated thyristor structures have been explored. The first structure to garner attention was the MOS-controlled thyristor (MCT),[4,5] which is illustrated in Fig. 1.13 with an N-type drift region. A reduced on-state voltage drop when compared with the IGBT is obtained by allowing the four-layer thyristor structure to latch-up. The thyristor is switched off by the application of a negative gate bias to form a current path that shunts the N^+ emitter/P-base junction such that the gain of the upper N–P–N transistor is sufficiently diminished to prevent the regenerative action within the thyristor. One disadvantage of the structure shown in Fig. 1.13 is the complexity of forming the P-channel MOSFET within the P-base region. Another issue that has hindered its application is the lack of current saturation capability. Without gate controlled current saturation, it becomes necessary to incorporate snubbers with the device adding cost, complexity, and power losses to the system.

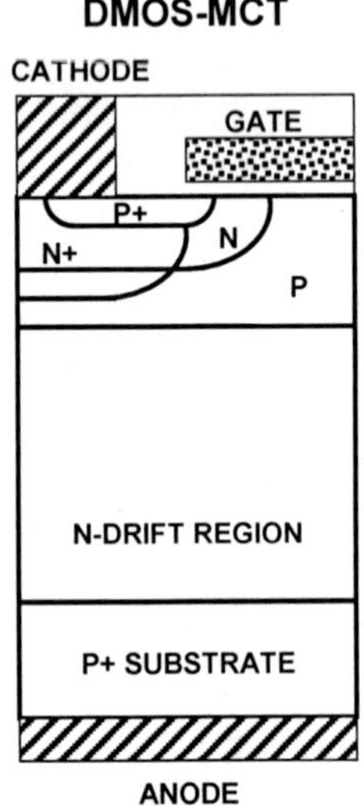

Fig. 1.13 MOS-controlled thyristor (MCT) structure

One solution to the above issues is the base resistance controlled thyristor (BRT)[6] structure shown in Fig. 1.14. In this structure, a diverter region is integrated adjacent to the four-layer thyristor. The application of a positive bias to the gate allows turning on the thyristor, while the application of a negative bias turns off the thyristor by raising its holding current above the operating current level. The process for fabrication of this structure is compatible with that used for manufacturing IGBTs while its on-state voltage drop is similar to that for the MCT structure.

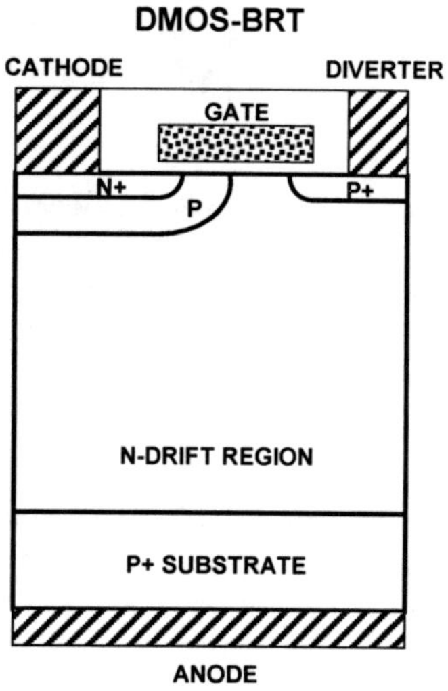

DMOS-BRT
CATHODE
DIVERTER
GATE
N+
P
P+
N-DRIFT REGION
P+ SUBSTRATE
ANODE

Fig. 1.14 Base resistance controlled thyristor (BRT) structure

D-MOS EST
CATHODE
GATE
GATE
N+
P+
P
N+ Cathode
P
N-DRIFT REGION
P+ SUBSTRATE
ANODE

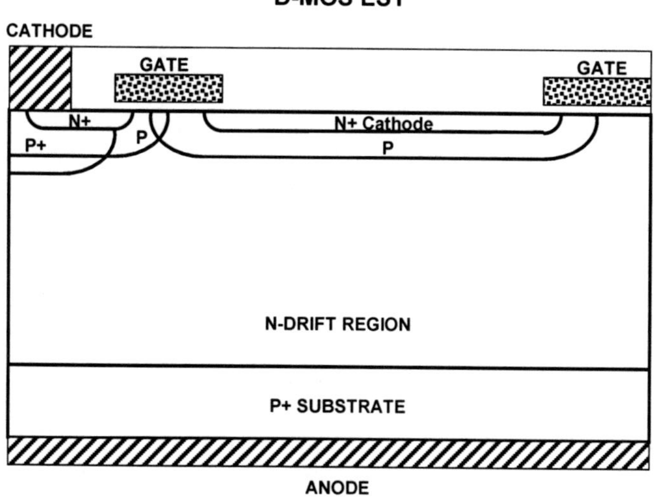

Fig. 1.15 Emitter switched thyristor (EST) structure

The MCT and BRT structure do not allow for saturated current flow under gate control, a key feature of IGBTs that is extensively utilized in power electronic circuits. This capability has been successfully incorporated in the emitter switched thyristor (EST)[7] structure shown in Fig. 1.15. The EST structure contains a four-layer thyristor without direct contact to the N^+ cathode region. The current flowing into the N^+ cathode region is constrained to flow via an n-channel MOSFET to the cathode contact. This provides complete gate controlled current saturation capability. However, the on-state voltage drop is slightly larger than that for the MCT or BRT due to the additional voltage drop in the channel of the MOSFET.

1.6 Ideal Drift Region for Unipolar Power Devices

The unipolar power devices discussed above contain a drift region, which is designed to support the blocking voltage. The properties (doping concentration and thickness) of the *ideal drift region* can be analyzed by assuming an abrupt junction profile with high doping concentration on one side and a low uniform doping concentration on the other side, while neglecting any junction curvature effects by assuming a parallel-plane configuration. The resistance of the ideal drift region can then be related to the basic properties of the semiconductor material.[8]

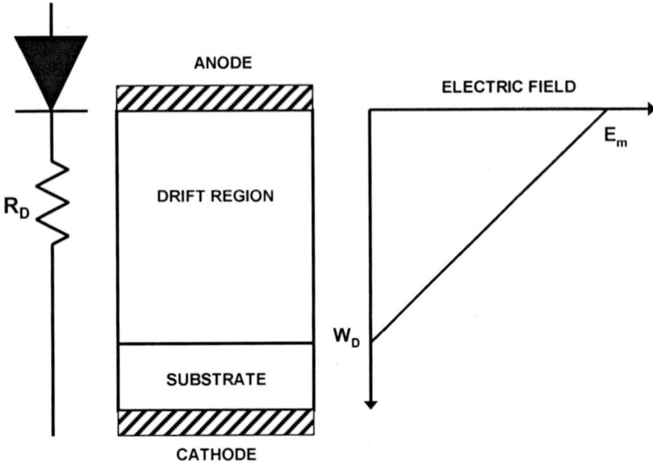

Fig. 1.16 The ideal drift region and its electric field distribution

The solution of Poisson's equation leads to a triangular electric field distribution, as shown in Fig. 1.16, within a uniformly doped drift region with the slope of the field profile being determined by the doping concentration. The maximum voltage that can be supported by the drift region is determined by the maximum electric field (E_m) reaching the critical electric field (E_c) for breakdown for the semiconductor material. The critical electric field for breakdown and the doping concentration then determine the maximum depletion width (W_D).

The specific resistance (resistance per unit area) of the ideal drift region is given by:

$$R_{on.sp} = \left(\frac{W_D}{q\mu_n N_D} \right) \tag{1.8}$$

Since this resistance was initially considered to be the lowest value achievable with silicon devices, it has historically been referred to as the *ideal specific on-resistance of the drift region.* More recent introduction of the charge-coupling concept, described later in this chapter, has enabled reducing the drift region resistance of silicon devices to below the values predicted by this equation. The depletion width under breakdown conditions is given by:

$$W_{\mathrm{D}} = \frac{2\mathrm{BV}}{E_{\mathrm{C}}} \tag{1.9}$$

where BV is the desired breakdown voltage. The doping concentration in the drift region required to obtain this BV is given by:

$$N_{\mathrm{D}} = \frac{\varepsilon_{\mathrm{S}} E_{\mathrm{C}}^2}{2q\mathrm{BV}} \tag{1.10}$$

Combining these relationships, the specific resistance of the ideal drift region is obtained:

$$R_{\text{on-ideal}} = \frac{4\mathrm{BV}^2}{\varepsilon_{\mathrm{S}} \mu_n E_{\mathrm{C}}^3} \tag{1.11}$$

The denominator of this equation ($\varepsilon_{\mathrm{S}} \mu_n E_{\mathrm{C}}^3$) is commonly referred to as *Baliga's figure of merit for power devices*. It is an indicator of the impact of the semiconductor material properties on the resistance of the drift region. The dependence of the drift region resistance on the mobility (assumed to be for electrons here because in general they have higher mobility values than for holes) of the carriers favors semiconductors such as gallium arsenide and indium arsenide. However, the much stronger (cubic) dependence of the on-resistance on the critical electric field for breakdown favors wide band gap semiconductors such as silicon carbide.[9] The critical electric field for breakdown is determined by the impact ionization coefficients for holes and electrons in semiconductors as discussed in the next chapter.

As an example, the change in the specific on-resistance for the drift region with critical electric field and mobility is shown in Fig. 1.17 for the case of a BV of 1,000 V. The location of the properties for silicon, gallium arsenide, and silicon carbide are shown in the figure by the points. The improvement in drift region resistance for GaAs in comparison with silicon is largely due to its much greater mobility for electrons. The improvement in drift region resistance for SiC in comparison with silicon is largely due to its much larger critical electric field for breakdown. On the basis of these considerations, excellent high voltage Schottky rectifiers were developed from GaAs in the 1980s[10] and from silicon carbide in the 1990s.[11] Interest in the development of power devices from wide-band-gap semiconductors, including gallium nitride, continues to grow.

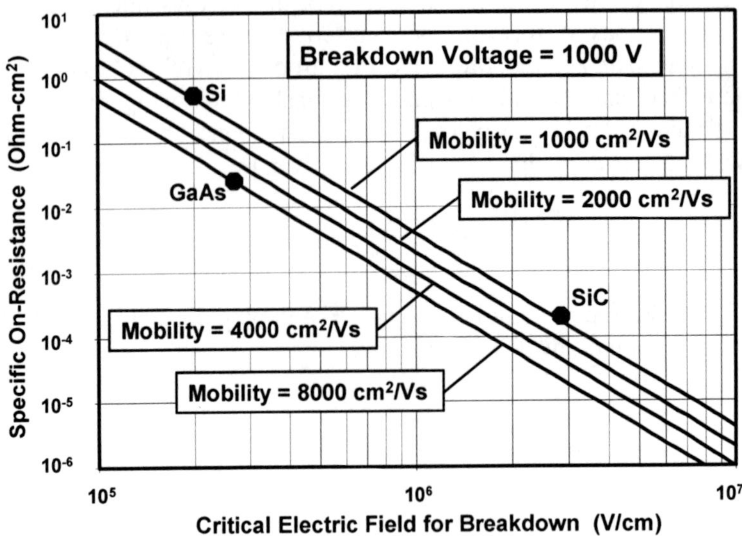

Fig. 1.17 Specific on-resistance of the *ideal drift region*

1.7 Charge-Coupled Structures: Ideal Specific On-Resistance

The depletion region extends in one-dimension from a junction or Schottky contact during the blocking mode for the conventional structure discussed in the previous section. In the charge-coupled structure, the voltage blocking capability is enhanced by the extension of depletion layers in two-dimensions. This effect is created by the formation of a horizontal Schottky contact on the top surface as illustrated in Fig. 1.18, which promotes the extension of a depletion region along the vertical or *y*-direction. Concurrently, the presence of the vertical P–N junction created by the alternate N- and P-type regions promotes the extension of a depletion region along the horizontal or *x*-direction. These depletion regions conspire to produce a two-dimensional charge coupling in the N-drift region, which alters the electric field profile.

The optimization of the charge-coupled structure requires proper choice of the doping concentration and thickness of the N- and P-type regions. It has been found that the highest BV occurs when the charge in these regions is given by:

$$Q_{\text{optimum}} = 2qN_{\text{D}}W_{\text{N}} = \varepsilon_{\text{S}}E_{\text{C}} \tag{1.12}$$

where q is the charge of an electron (1.6×10^{-19} C), N_{D} is the doping concentration of the N-type drift region, W_{N} is half the width of the N-type drift region as shown in Fig. 1.18, ε_{S} is the dielectric constant of the semiconductor, and E_{C} is the critical electric field for breakdown in the semiconductor. For silicon, the optimum charge

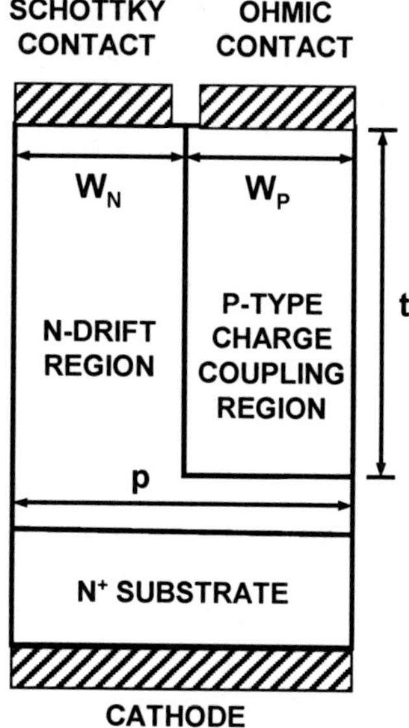

Fig. 1.18 Basic charge coupled Schottky diode structure using P-N junctions

is found to be 3.11×10^{-7} C cm^{-2} based upon a critical electric field of 3×10^{5} V cm^{-1}. The optimum charge is often represented as a dopant density per unit area, in which case it takes a value of about 2×10^{12} cm^{-2} for silicon. A slightly lower value for the doping concentration in the drift region may be warranted as discussed later in this section of the chapter.

The specific on-resistance for the drift region in the charge-coupled structures is given by:

$$R_{\mathrm{D,sp}} = \rho_{\mathrm{D}}\left(\frac{p}{W_{\mathrm{N}}}\right)t \tag{1.13}$$

where ρ_{D} is the resistivity of the N-type drift region, t is the trench depth, and p is the cell pitch. Here, the uniform electric field is assumed to be produced only along the trench where the charge coupling occurs and the resistance of the remaining portion of the N-drift region is neglected. Using the relationship between the resistivity and the doping concentration, this equation can be written as:

$$R_{\mathrm{D,sp}} = \frac{tp}{q\mu_{\mathrm{N}} N_{\mathrm{D}} W_{\mathrm{N}}} \tag{1.14}$$

Combining this expression with (1.12):

$$R_{\mathrm{D,sp}} = \frac{2tp}{\mu_{\mathrm{N}} Q_{\mathrm{optimum}}} \tag{1.15}$$

If the electric field along the trench, at the on-set of breakdown in the charge-coupled device structure, is assumed to be uniform at a value equal to the critical electric field of the semiconductor:

$$t = \frac{\mathrm{BV}}{E_{\mathrm{C}}} \tag{1.16}$$

Using this expression, as well as the second part of (1.12), in (1.15) yields:

$$R_{\mathrm{D,sp}} = \frac{2\mathrm{BV}p}{\mu_{\mathrm{N}}\varepsilon_{\mathrm{S}}E_{\mathrm{C}}^{2}} \tag{1.17}$$

This is a fundamental expression for the *ideal specific on-resistance of vertical charge-coupled devices*. By comparison of this expression with that for the one-dimensional case (see (1.11), it can be observed that the specific on-resistance for the charge-coupled devices increases linearly with the BV unlike the more rapid quadratic rate for the conventional drift region. In addition, it is worth pointing out that the specific on-resistance for the drift region in the charge-coupled structure can be reduced by decreasing the pitch. This occurs because the doping concentration in the drift region increases when the pitch is reduced so as to maintain the same optimum charge. The larger doping concentration reduces the resistivity and hence the specific on-resistance. However, the analysis of the specific on-resistance for the drift region in charge-coupled device structures must be tempered by several considerations. First, it must be recognized that the mobility will become smaller when the doping concentration becomes larger. Second, the critical electric field for breakdown becomes smaller for the charge-coupled structures because the high electric field in the drift region extends over a larger distance, producing enhanced impact ionization. If a critical electric field for breakdown in the drift region for charge-coupled structures is reduced to 2×10^{5} V cm^{-1}, an optimum charge of 2.07×10^{-7} C cm^{-2}, with a corresponding dopant density of about 1.3×10^{12} cm^{-2}, is more appropriate for silicon.

In designing the drift region for charge-coupled structures, it is important to recognize that, unlike in the conventional one-dimensional case, the doping concentration of the drift region is dictated by the cell pitch and not the BV. The BV in the charge-coupled structure is determined solely by the depth of the trench used to provide the charge coupling effect and is independent of the doping concentration of the N-drift region. In the case of silicon charge-coupled devices, the doping concentration for the N-type drift region is provided in Fig. 1.19 for the case of equal widths for the N-type and P-type charge coupling regions. For a typical cell pitch of 1 μm, the doping concentration in the N-type drift region is about 1.3×10^{16} cm^{-3} when a critical electric field of 2×10^{5} V cm^{-1} is assumed.

It is interesting to compare the ideal specific on-resistance for the drift region in the silicon charge-coupled structures to that for the one-dimensional parallel-plane case. This comparison is done in Fig. 1.20 using three values for the

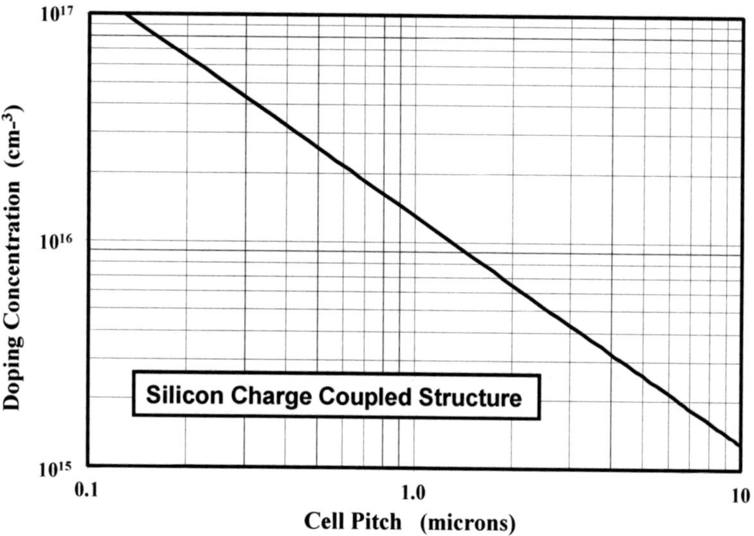

Fig. 1.19 Drift region doping concentration for the charge coupled structure

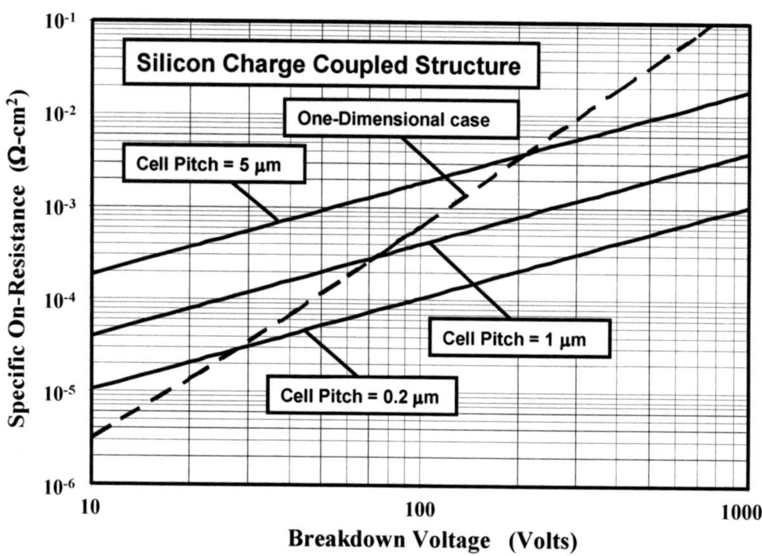

Fig. 1.20 Ideal specific on-resistance for the charge coupled structure (solid lines: charge coupled structures; dashed line: one dimensional case)

cell pitch in the case of the charge-coupled structures. The doping concentration in the N-drift region increases when the cell pitch is reduced from 5 μm to 0.2 μm, as already shown in Fig. 1.19, leading to a decrease in the specific on-resistance. The resulting reduction of the mobility with increasing doping concentration was included during the calculation of the specific on-resistance in Fig. 1.20. There is a cross-over in the specific on-resistance for the two types of structures. For the cell pitch of 1 μm, the cross-over occurs at a BV of about 70 V. The cross-over moves to a BV of about 200 V when the cell pitch is increased to 5 μm, and to about 30 V if a smaller cell pitch of 0.2 μm is used. Consequently, the charge-coupled structure is more attractive for reducing the specific on-resistance when the cell pitch is smaller. This entails a more complex process technology with higher attendant costs.

As a particular example, consider the case of silicon devices designed to support 200 V. In the case of the conventional structure with a one-dimensional junction, the specific on-resistance of the drift region is found to be 3.4 mΩ cm^2 if a critical electric field for breakdown of 3×10^5 V cm^{-1} is used. In contrast, the specific on-resistance for the drift region of the charge-coupled structure with a cell pitch of 1 μm is found to be only 0.86 mΩ-cm^2 if a critical electric field for breakdown of 2×10^5 V cm^{-1} is used. In this calculation, a bulk mobility of 1,220 cm^2 V^{-1} s^{-1} was used corresponding to a doping concentration of 1.3×10^{16} cm^{-3} in the N-type portion of the drift region. In this example, the drift region for the charge-coupled structure would have a thickness of 10 μm when compared with 12.5 μm needed in the conventional structure.

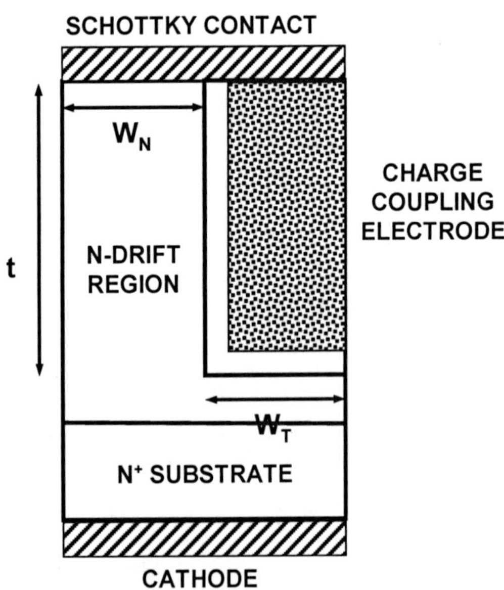

Fig. 1.21 Basic charge coupled Schottky diode structure using MOS structure

An alternate method for creating a charge-coupled structure, which is simpler from a fabrication viewpoint, is illustrated in Fig. 1.21. A uniform electric field in the drift region can be created with this structure by using a graded doping profile.[12] These devices have been demonstrated to provide significant reduction of the on-state voltage drop for Schottky rectifiers designed to block 30–200 V with low leakage currents.[13] A detailed discussion of charge-coupled devices is beyond the scope of this book.

1.8 Summary

The desired characteristics for power semiconductor rectifiers and transistors have been reviewed in this chapter. The characteristics of typical silicon power devices have been compared with those for the ideal case. Various unipolar and bipolar power device structures have been briefly introduced here. Only the basic power device structures are discussed in this book because of space constraints.

Problems

1.1 Define the current and voltage ratings of power devices for typical automotive applications.

1.2 Define the current and voltage ratings of power devices for typical computer power supply applications.

1.3 Define the current and voltage ratings of power devices for typical motor control applications.

1.4 Define the current and voltage ratings of power devices for typical electric locomotive applications.

1.5 What are the characteristics of an ideal power device?

1.6 Describe the characteristics of an ideal power rectifier.

1.7 How do the characteristics of actual power rectifiers defer from those for the ideal device?

1.8 Describe the characteristics of an ideal power transistor.

1.9 How do the characteristics of actual power transistors defer from those for the ideal device?

1.10 Why are unipolar power device structures more attractive for applications than bipolar power devices?

1.11 Calculate Baliga's figure-of-merit for a semiconductor with an electron mobility of 2,000 $cm^2\ V^{-1}\ s^{-1}$ and critical breakdown electric field strength of 5×10^5 V cm^{-1}.

1.12 Determine Baliga's figure-of-merit for gallium nitride assuming it has an electron mobility of 900 $cm^2\ V^{-1}\ s^{-1}$ and critical breakdown electric field strength of 3.3×10^6 V cm^{-1}.

References

[1] B.J. Baliga, "Silicon RF Power Devices," World Scientific, Singapore, 2005.

[2] B.J. Baliga, "Modern Power Devices," Wiley, New York, 1987.

[3] B.J. Baliga, "Evolution of MOS-Bipolar Power Semiconductor Technology," Proceedings of the IEEE, pp. 409–418, 1988.

[4] V.A.K. Temple, "MOS Controlled Thyristors (MCTs)," IEEE International Electron Devices Meeting, Abstract 10.7, pp. 282–285, 1984.

[5] M. Stoisiek and H. Strack, "MOS-GTO – A Turn-off Thyristor with MOS Controlled Emitter Shorts," IEEE International Electron Devices Meeting, Abstract 6.5, pp. 158–161, 1985.

[6] M. Nandakumar, et al., "A New MOS-Gated Power Thyristor Structure with Turn-off achieved by Controlling the Base Resistance," IEEE Electron Device Letters, Vol. EDL-12, pp. 227–229, 1991.

[7] B.J. Baliga, "The MOS-Gated Emitter Switched Thyristor," IEEE Electron Device Letters, Vol. EDL-11, pp. 75–77, 1990.

[8] B.J. Baliga, "Semiconductors for High Voltage Vertical Channel Field Effect Transistors," Journal of Applied Physics, Vol. 53, pp. 1759–1764, 1982.

[9] B.J. Baliga, "Silicon Carbide Power Devices," World Scientific, Singapore, 2006.

[10] B.J. Baliga, et al., "Gallium Arsenide Schottky Power Rectifiers," IEEE Transactions on Electron Devices, Vol. ED-32, pp. 1130–1134, 1985.

[11] M. Bhatnagar, P.M. McLarty, and B.J. Baliga, "Silicon Carbide High Voltage (400V) Schottky Barrier Diodes," IEEE Electron Device Letters, Vol. EDL-13, pp. 501–503, 1992.

[12] B.J. Baliga, "Schottky Barrier Rectifiers and Method of Forming Same," U.S. Patent 5,612,567, Issued 18 March 1997.

[13] S. Mahalingam and B.J. Baliga, "The Graded Doped Trench MOS Barrier Schottky Rectifier," Solid State Electronics, Vol. 43, pp. 1–9, 1999.

Chapter 2

Material Properties and Transport Physics

At present, most power semiconductor devices are manufactured using silicon as the base material. Both unipolar and bipolar devices have been successfully developed from silicon to serve a very broad range of applications. As shown in the previous chapter, the resistance of the drift region in power devices can be drastically reduced by replacing silicon with wide band gap semiconductors. Although some effort was undertaken in the 1980s to develop power devices from gallium arsenide, interest in this technology has dwindled because it is much more promising to make the power devices from silicon carbide. For this reason, the relevant properties of silicon carbide are included with those of silicon in this chapter for the purposes of comparison and design. Analytical expressions that can model these properties are provided to facilitate the analysis of power device structures.

Although the fundamental material properties of the semiconductor govern the operating characteristics of power devices, processing techniques used to control these properties impose technological constraints that are equally important for obtaining the desired device characteristics. For this reason, the chapter includes a discussion of current technology, such a neutron transmutation doping (NTD) for controlling the resistivity and electron irradiation for controlling the minority carrier lifetime. These technologies have been specifically developed for manufacturing power devices and are almost exclusively used for these types of structures.

2.1 Fundamental Properties

The fundamental properties of silicon[1] and silicon carbide[2–4] are summarized and compared in this section. Only the properties for the 4H poly-type of silicon carbide have been included here because its properties are superior to those of the

B.J. Baliga, *Fundamentals of Power Semiconductor Devices*, doi: 10.1007/978-0-387-47314-7_2,

other poly-types of silicon carbide. The basic properties of relevance to power devices are the energy band gap, the impact ionization coefficients, the dielectric constant, the thermal conductivity, the electron affinity, and the carrier mobility. Since the intrinsic carrier concentration and the built-in potential can be extracted by using this information, their values have also been provided for silicon and silicon carbide in this section.

Properties	Silicon	4H-SiC
Energy Band Gap (eV)	**1.11**	**3.26**
Relative Dielectric Constant	**11.7**	**9.7**
Thermal Conductivity (W/cm-K)	**1.5**	**3.7**
Electron Affinity (eV)	**4.05**	**3.7**
Density of States Conduction Band (cm^{-3})	**2.80×10^{19}**	**1.23×10^{19}**
Density of States Valence Band (cm^{-3})	**1.04×10^{19}**	**4.58×10^{18}**

Fig. 2.1 Fundamental material properties

The fundamental material parameters for silicon and 4H-SiC are summarized in Fig. 2.1 for comparison purposes. It is worth noting that the energy band gap for 4H-SiC is three times larger than that for silicon. This results in a much lower intrinsic carrier concentration at any given temperature, as well as much smaller impact ionization coefficients at any given electric field, for this semiconductor when compared with silicon, as discussed in more detail in the next section. The dielectric constant and the density of states are comparable for both these semiconductors. The thermal conductivity for 4H silicon carbide is two to three times that for silicon, enabling superior heat extraction from devices. The difference in the electron affinity between the semiconductors appears small but is quite significant when determining the characteristics of Schottky barrier contacts.

2.1.1 Intrinsic Carrier Concentration

The intrinsic carrier concentration is determined by the thermal generation of electron-hole pairs across the energy band gap of a semiconductor. Its value can be calculated by using the energy band gap (E_G) and the density of states in the conduction (N_C) and valence (N_V) bands:

$$n_i = \sqrt{np} = \sqrt{N_C N_V} e^{-E_G/2kT} \tag{2.1}$$

where k is Boltzmann's constant (1.38×10^{-23} J K^{-1}) and T is the absolute temperature. For silicon, the intrinsic carrier concentration is given by:

$$n_i = 3.87 \times 10^{16} T^{3/2} e^{-(7.02 \times 10^3)/T} \tag{2.2}$$

while that for 4H-SiC is given by:

$$n_i = 1.70 \times 10^{16} T^{3/2} e^{-(2.08 \times 10^4)/T} \tag{2.3}$$

Using these equations, the intrinsic carrier concentration can be calculated as a function of temperature. The intrinsic carrier concentrations for silicon and 4H-SiC are plotted in Figs. 2.2 and 2.3.

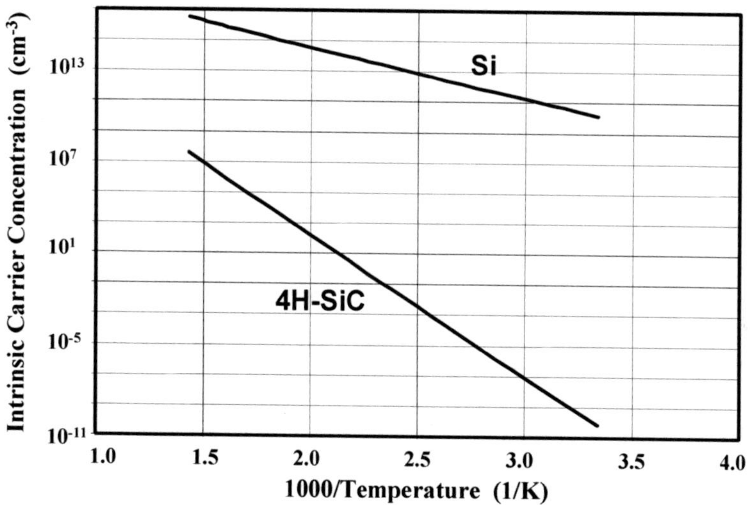

Fig. 2.2 Intrinsic carrier concentration

The traditional method for making this plot is shown in Fig. 2.2 by using an inverse temperature scale for the x-axis. The additional figure (Fig. 2.3) is provided here because it is easier to relate the value of the intrinsic carrier concentration to the temperature, with its values directly plotted on the x-axis. It is obvious that the intrinsic carrier concentration for silicon carbide is far smaller than that for silicon due to the large difference in band gap energy. At room temperature (300 K), the intrinsic carrier concentration for silicon is 1.4×10^{10} cm^{-3} while that for 4H-SiC is only 6.7×10^{-11} cm^{-3}. This indicates that the bulk generation current is negligible for the determination of the leakage current in silicon carbide devices but not for silicon structures. Surface generation currents may become dominant in 4H-SiC devices because of the presence of states within the band gap.

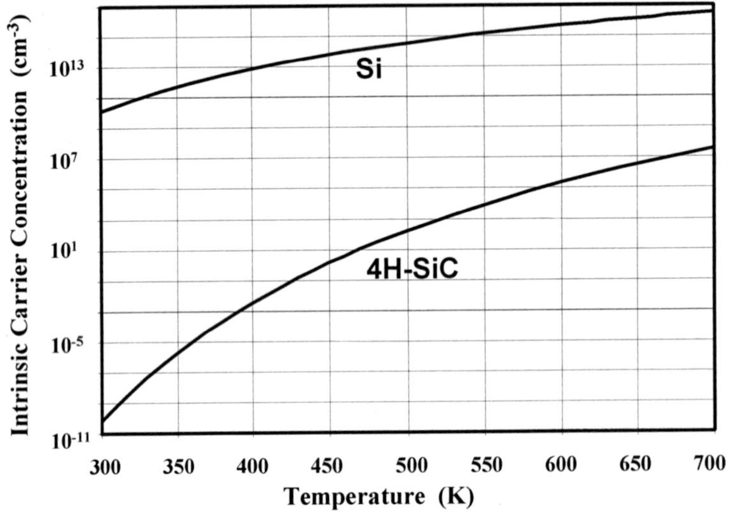

Fig. 2.3 Intrinsic carrier concentration

For silicon, the intrinsic carrier concentration becomes equal to a typical doping concentration of 1×10^{15} cm^{-3} at a relatively low temperature of 540 K or 267°C. In contrast, the intrinsic carrier concentration for 4H-SiC is only 3.9×10^{7} cm^{-3} even at 700 K or 427°C. The development of mesoplasmas has been associated with the intrinsic carrier concentration becoming comparable to the doping concentration.[5] Mesoplasmas create current filaments that have very high current density, leading to destructive failure in semiconductors. This is obviously much less likely in silicon carbide.

2.1.2 Band Gap Narrowing

The energy band diagram for silicon at low doping concentrations takes the form illustrated in Fig. 2.4(a), with the density of states varying as the square root of energy away from the band edges. The donor and acceptor levels are located at discrete positions within the band gap and are clearly separated from the conduction and valence band edges. The well-defined separation between the conduction and valence band edges is called the *energy band gap* (E_G). The energy band gap for silicon and 4H silicon carbide is provided in Fig. 2.1.

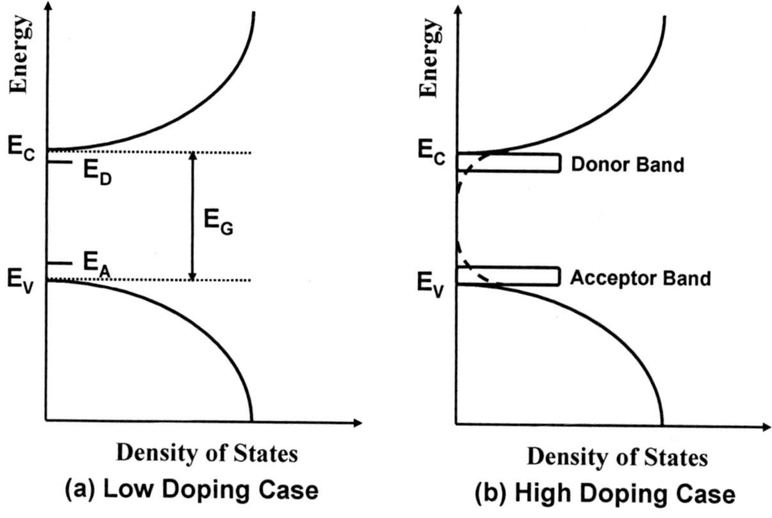

Fig. 2.4 Energy band diagrams at (a) low and (b) high doping concentrations

At high doping levels in the semiconductor, the band structure is altered by three effects. First, as the impurity density becomes large, the spacing between the individual impurity atoms becomes small. The interaction between adjacent impurity atoms leads to a splitting of the impurity levels into an impurity band as illustrated in Fig. 2.4(b). This phenomenon is observed when the doping concentration exceeds 10^{18} cm^{-3}. Second, the conduction and valence band edges no longer exhibit a parabolic shape. The statistical distribution of the dopant atoms in the lattice introduces point-by-point differences in local doping concentration and lattice potential, leading to disorder. This results in the formation of band tails as illustrated in Fig. 2.4(b) by the dashed lines. This phenomenon is observed when the doping concentration approaches 10^{21} cm^{-3}. Third, the interaction between the free carriers and more than one impurity atom leads to a modification of the density of states at the band edges. This phenomenon is called *rigid band gap narrowing*.

The formation of the impurity bands, the band tails, and the rigid band gap narrowing are majority carrier effects related to the reduced dopant atom spacing within the semiconductor lattice. In addition to this, the electrostatic interaction of the minority carriers and the high concentration of majority carriers lead to a reduction in the thermal energy required to create an electron–hole pair. This phenomenon is also a rigid band gap reduction that does not distort the energy dependence of the density of states. The rigid band gap reduction arising from the screening of the minority carriers by the high concentration of majority carriers can be derived by using Poisson's equation[6]:

$$\frac{1}{r^2}\frac{\mathrm{d}}{\mathrm{d}r}\left(r^2\frac{\mathrm{d}V}{\mathrm{d}r}\right)=-\frac{q}{\varepsilon_S}\Delta n \tag{2.4}$$

where V is the electrostatic potential of the screened carriers at a radial distance r. The excess electron concentration Δn is given by Boltzmann's equation:

$$\Delta n = n_0(\mathrm{e}^{qV/kT}-1) \tag{2.5}$$

For the case of low excess electron concentrations:

$$\Delta n = \left(\frac{qn_0V}{kT}\right) \tag{2.6}$$

Solution of (2.4) using the excess electron concentration given by (2.6) yields:

$$V(r)=\frac{q}{4\pi\varepsilon_S}\left(\frac{\mathrm{e}^{-r/r_S}}{r}\right) \tag{2.7}$$

where the screening radius r_S is given by:

$$r_S=\sqrt{\frac{\varepsilon_S kT}{q^2 n_0}} \tag{2.8}$$

The electric field distribution for the screened Coulombic potential can be obtained by the integration of (2.7):

$$E(r)=\frac{q}{4\pi\varepsilon_S r}\left(\frac{1}{r}-\frac{1}{r_S}\right)\mathrm{e}^{-r/r_S} \tag{2.9}$$

In comparison, the unscreened electric field distribution at low doping concentrations is given by:

$$E(r)=\frac{q}{4\pi\varepsilon_S r^2} \tag{2.10}$$

The reduction of the band gap is given by the difference in the electrostatic energy between the screened and unscreened cases:

$$\Delta E_G=\frac{\varepsilon_S}{2}\int_0^\infty\left[E_0^2(r)-E^2(r)\right]\mathrm{d}r \tag{2.11}$$

The solution of this integral yields:

$$\Delta E_G=\frac{3q^2}{16\pi\varepsilon_S}\sqrt{\frac{q^2 n_0}{\varepsilon_S kT}} \tag{2.12}$$

In the case of low injection levels and under the assumption that all the dopant atoms are ionized:

$$n_0 = N_D^+ = N_D \tag{2.13}$$

where N_D is the doping concentration. Consequently,

$$\Delta E_G = \frac{3q^2}{16\pi\varepsilon_S}\sqrt{\frac{q^2 N_D}{\varepsilon_S kT}} \tag{2.14}$$

This expression, valid for small changes in the band gap, indicates that the band gap narrowing is proportional to the square root of the doping concentration. Using the appropriate parameters for silicon:

$$\Delta E_G = 22.5\times 10^{-3}\sqrt{\frac{N_I}{10^{18}}} \tag{2.15}$$

where N_I is the impurity (donor or acceptor) concentration responsible for the reduction in the band gap. The decrease of the energy gap for silicon at room temperature due to high doping concentrations as predicted by the above theory is shown in Fig. 2.5. This theoretical curve has been found to provide a good fit with the measured data up to a doping concentration of 10^{19} cm^{-3}, indicating that the screening of minority carriers by the majority carriers is a dominant effect responsible for band gap narrowing in silicon.[7–10]

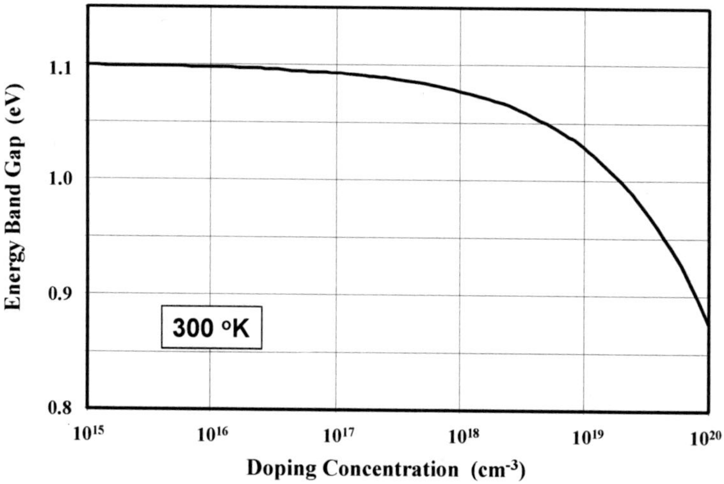

Fig. 2.5 Band gap narrowing in silicon at high doping concentrations

The band gap narrowing phenomenon has a strong impact upon the product of the equilibrium concentrations of electrons and holes in the heavily doped semiconductor material. The p–n product can be represented by an *effective intrinsic carrier concentration* in a manner similar to that for lightly doped semiconductors:

$$(np) = n_{ie}^2 \tag{2.16}$$

The larger values for the p–n product arising from band gap narrowing are related to the smaller energy band gap by:

$$n_{ie}^2 = n_i^2 \, e^{q\Delta E_G / kT} \tag{2.17}$$

where ΔE_G is the reduction of the band gap due to the combined effects of impurity band formation, band tailing, and screening of minority carriers. By using (2.15), the effective intrinsic carrier concentration at room temperature for silicon is given by:

$$n_{ie} = 1.4 \times 10^{10} \exp\left(0.433\sqrt{\frac{N_I}{10^{18}}}\right) \tag{2.18}$$

where N_I is the doping concentration in cm^{-3}. The intrinsic carrier concentration predicted by this equation increases by more than one order of magnitude when the doping concentration exceeds 3×10^{19} cm^{-3}, as shown in Fig. 2.6. These high doping concentrations are commonly encountered in the diffused end regions of power devices. The performance of bipolar power devices is affected by the band gap narrowing phenomenon because the injection of minority carriers into the end regions is enhanced by the larger intrinsic carrier concentration.

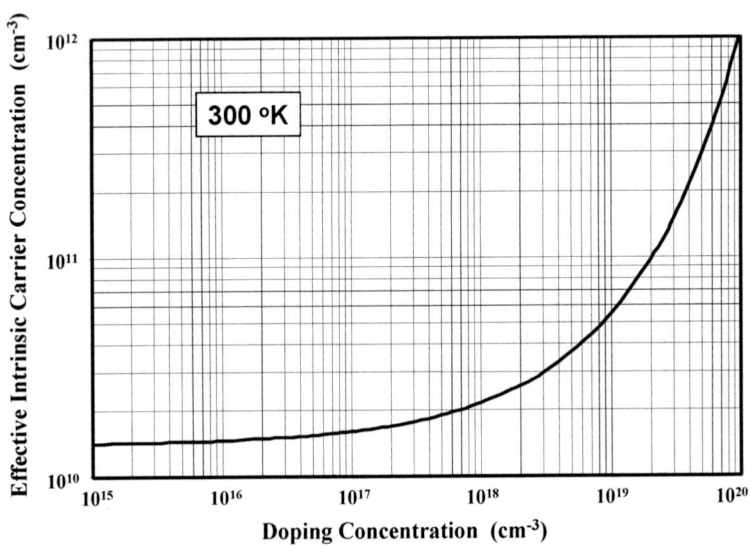

Fig. 2.6 Effective intrinsic carrier concentration in silicon at high doping levels

2.1.3 Built-in Potential

The built-in potential of P–N junctions can play an important role in determining the operation and design of power semiconductor devices. As an example, the built-in potential determines the zero-bias depletion width, which is important for the calculation of the on-resistance in power D-MOSFETs. The built-in voltage is given by:

$$V_{bi} = \frac{kT}{q}\ln\left(\frac{N_A^- N_D^+}{n_i^2}\right) \tag{2.19}$$

where N_A^- and N_D^+ are the ionized impurity concentrations on the two sides of an abrupt P–N junction. For silicon, their values are equal to the doping concentration because of the small dopant ionization energy levels.

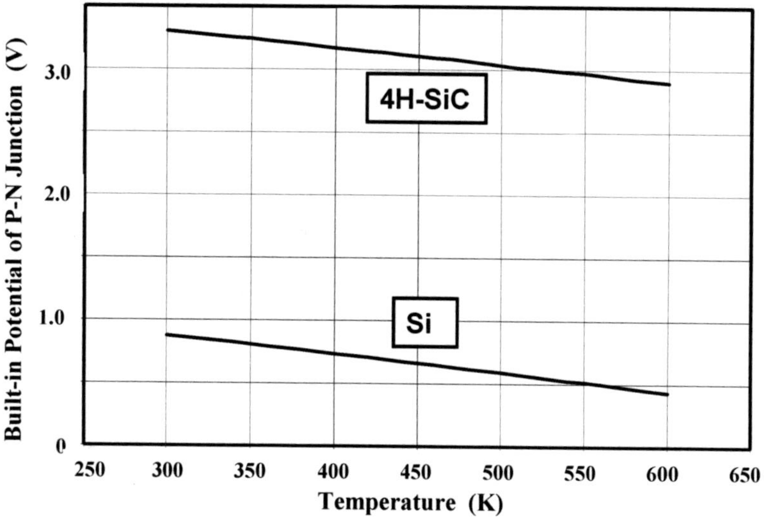

Fig. 2.7 Built-in potential for a P-N junction in Si and 4H-SiC

The calculated built-in potential for silicon and 4H-SiC is shown in Fig. 2.7 as a function of temperature for the typical operating range of power devices. In making the plots, the product ($N_A^- N_D^+$) was assumed to be 10^{35} cm^{-6}. This would be applicable for a typical doping concentration of 1×10^{19} cm^{-3} on the heavily doped side of the junction and 1×10^{16} cm^{-3} on the lightly doped side of the junction. The built-in potential for silicon carbide is much larger than that for silicon because of the far smaller values for the intrinsic carrier concentration. This can be a disadvantage because of the larger zero bias depletion width for silicon carbide. As an example, the larger zero bias depletion width consumes space within the D-MOSFET cell structure, increasing the on-resistance by constricting the area through which current can flow. In contrast, the larger built-in potential for SiC and its associated larger zero bias depletion width can be taken advantage of in making innovative device structures, such as the ACCUFET (see Chap. 6), that are tailored to the unique properties of SiC.

2.1.4 Zero-Bias Depletion Width

The calculated zero bias depletion widths for silicon and 4H-SiC are shown in Fig. 2.8 as a function of the doping concentration on the lightly doped side of an abrupt P–N junction. For both silicon and silicon carbide, it can be seen that the zero bias depletion width can be substantial in size at low doping concentrations, making it important to take this into account during device design and analysis. The zero bias depletion width for SiC is much (about 2×) larger than that for silicon for the same doping concentration. However, it is worth noting that the doping concentration in SiC is much (about 100×) greater for a given breakdown voltage than for silicon as discussed later in the book. Consequently, the zero-bias depletion width for 4H-SiC devices is smaller than that for silicon devices with the same breakdown voltage.

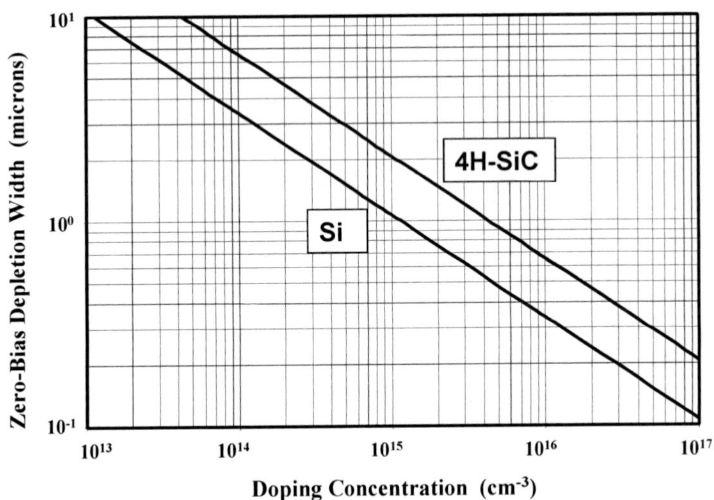

Fig. 2.8 Zero-bias depletion width in Si and 4H-SiC

2.1.5 Impact Ionization Coefficients

The *impact ionization coefficient for holes* (α_p) is defined as the number of electron–hole pairs created by a hole traversing 1 cm through the depletion layer along the direction of the electric field. Similarly, the *impact ionization coefficient for electrons* (α_n) is defined as the number of electron–hole pairs created by an electron traversing 1 cm through the depletion layer along the direction of the electric field. The impact ionization coefficients for semiconductors are dictated by Chynoweth's Law[11,12]:

$$\alpha = a\,\mathrm{e}^{-b/E} \tag{2.20}$$

where E is the electric field component in the direction of current flow. The parameters a and b are constants that depend upon the semiconductor material and the temperature. For silicon, the impact ionization rates have been measured as a function of electric field and temperature.[13,14] Based upon these extensive measurements, the parameters for silicon have been found to be $a_n = 7 \times 10^5$ cm^{-1} and $b_n = 1.23 \times 10^6$ V cm^{-1} for electrons; $a_p = 1.6 \times 10^6$ cm^{-1} and $b_p = 2 \times 10^6$ V cm^{-1} for holes for electric fields ranging between 1.75×10^5 and 6×10^5 V cm^{-1}.

The impact ionization coefficients for silicon carbide have been recently measured as a function of temperature by using an electron beam excitation method.[15] This method allowed the extraction of impact ionization rates in defect-free regions of the material by isolating diodes containing defects with an EBIC (electron beam induced current) image. This was important because substantially enhanced impact ionization rates were discovered when the measurements were conducted at defect sites.[16] For defect-free material, the extracted values for the impact ionization coefficient parameters for holes in 4H-SiC were found to be:

$$a(\text{4H-SiC}) = 6.46 \times 10^6 - 1.07 \times 10^4 T \tag{2.21}$$

with

$$b(\text{4H-SiC}) = 1.75 \times 10^7 \tag{2.22}$$

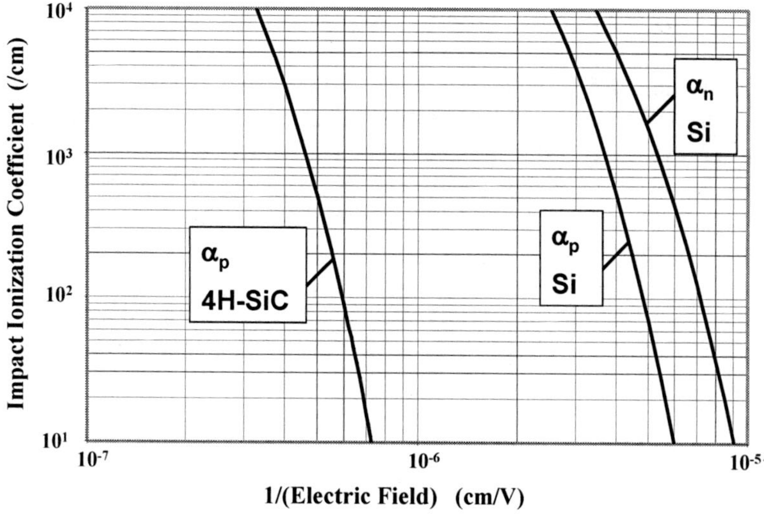

Fig. 2.9 Impact ionization coefficients for silicon and 4H-SiC

The impact ionization coefficients for 4H-SiC can be compared with those for silicon in Figs. 2.9 and 2.10. It is traditional to graph the ionization coefficients as the reciprocal of the electric field as shown in Fig. 2.9. However, it is more convenient to use the graph shown in Fig. 2.10 for extracting the ionization coefficients at known values for the electric field. In both semiconductors, there is a very strong dependence of the ionization coefficients upon the electric field. Consequently, the breakdown voltage of power devices can be severely reduced by the presence of a high localized electric field within the structure.

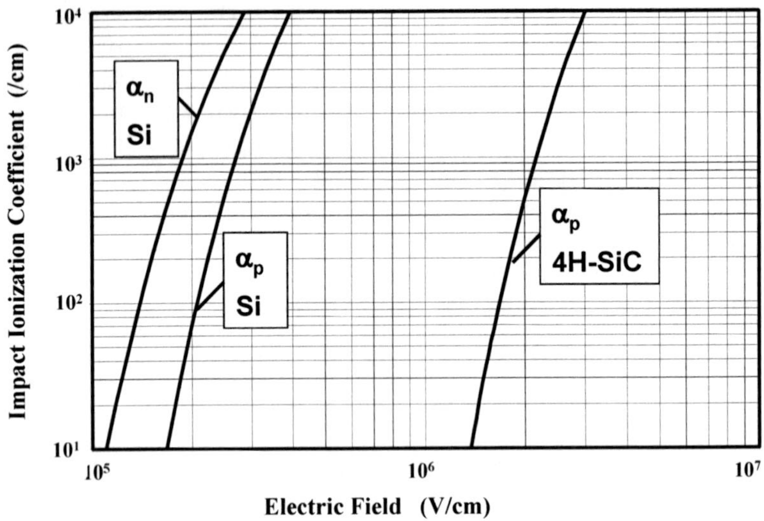

Fig. 2.10 Impact ionization coefficients for silicon and 4H-SiC

From the above figures, it can be seen that the onset of significant generation of carriers by impact ionization occurs at much larger electric fields in 4H-SiC when compared with silicon. As a consequence, breakdown in 4H-SiC devices occurs when the electric fields are in the range of $2–3 \times 10^6$ V cm^{-1} – an order of magnitude larger than that for silicon. This implies a much larger critical electric field E_c for breakdown for 4H-SiC, resulting in a big increase in the *Baliga's figure of merit* (see Chap. 1). The critical electric field for breakdown is discussed in more detail in the next chapter. It is worth pointing out that the impact ionization coefficient for electrons is much larger than that for holes in silicon while the opposite holds true for 4H-SiC.

2.1.6 Carrier Mobility

The resistivity (ρ) of a semiconductor region is given by:

$$\rho = \frac{1}{q\mu N} \tag{2.23}$$

where μ by the mobility, which is a function of the carrier type (electrons or holes), the doping concentration (N), and the lattice temperature (T). A good understanding of the mobility for the carriers is therefore necessary to model current transport in the semiconductor regions within devices.

In semiconductors, carriers are accelerated by the presence of an electric field and achieve an average velocity determined by the carrier scattering processes. Silicon is an indirect band gap semiconductor in which electron transport, under the influence of an applied electric field, occurs in six equivalent conduction band minima while the transport of holes occurs at two degenerate subbands. The inequality between the mobility of electrons and holes in silicon arises because of differences between the shapes of the conduction and valence band minima. As the free carriers are transported along the direction of the electric field, their velocity increases until they undergo scattering. In the bulk, the scattering can occur either by interaction with the lattice or at ionized donor and acceptor atoms. Consequently, the mobility is dependent upon the lattice temperature and the ionized impurity concentration. When the free carriers are transported near the semiconductor surface, additional scattering is observed, which decreases the mobility to below the bulk values. In bipolar power devices, a high concentration of holes and electrons can be simultaneously injected into the base region during on-state current flow. The probability for mutual scattering between electrons and holes increases under these conditions, reducing the effective mobility.

The *mobility* (μ) is defined as the proportionality constant relating the average carrier velocity (v_D) to the electric field (E):

$$v_D = \mu E \tag{2.24}$$

This expression is valid at low electric fields. However, when the electric field exceeds 10^4 V cm^{-1}, as commonly encountered in power devices, the velocity is no longer found to increase in proportion to the electric field. In fact, the velocity approaches a constant value known as the *saturated drift velocity*. The dependence of the mobility upon the electric field strength must be accounted for during the analysis of the electrical characteristics of power devices.

Temperature Dependence

At low doping concentrations in the semiconductor region, the scattering of free carriers occurs predominantly by interaction with the lattice vibrations. This *lattice scattering phenomenon* can occur by means of either acoustical phonons or optical phonons. Optical phonon scattering is important at high temperatures and electric fields while acoustical phonon scattering is dominant at low electric fields. In addition, at around room temperature, intervalley scattering mechanisms become important. For high purity, lightly doped, silicon, it has been experimentally determined that acoustical phonon scattering is dominant at temperatures below 50 K and the mobility varies as $T^{-3/2}$, where T is the absolute temperature. At around room temperature, intervalley scattering comes into effect and the mobility for silicon is given by:

$$\mu_{\mathrm{n}} = 1360\left(\frac{T}{300}\right)^{-2.42} \tag{2.25}$$

for electrons[17] and:

$$\mu_{\mathrm{p}} = 495\left(\frac{T}{300}\right)^{-2.20} \tag{2.26}$$

for holes.[18] This variation of the mobility for electrons and holes in silicon is shown in Fig. 2.11 for the case of low doping concentrations (below 10^{15} cm^{-3}) in the semiconductor. There is a rapid reduction of the mobility with increasing temperature, with a reduction by a factor of 2 when the temperature reaches 400 K. Since the doping concentration in the drift region of high voltage silicon devices is below 10^{15} cm^{-3}, it is important to include the impact of the decrease in the mobility with temperature when analyzing the device characteristics because power devices operate at above room temperature due to power dissipation within the structure.

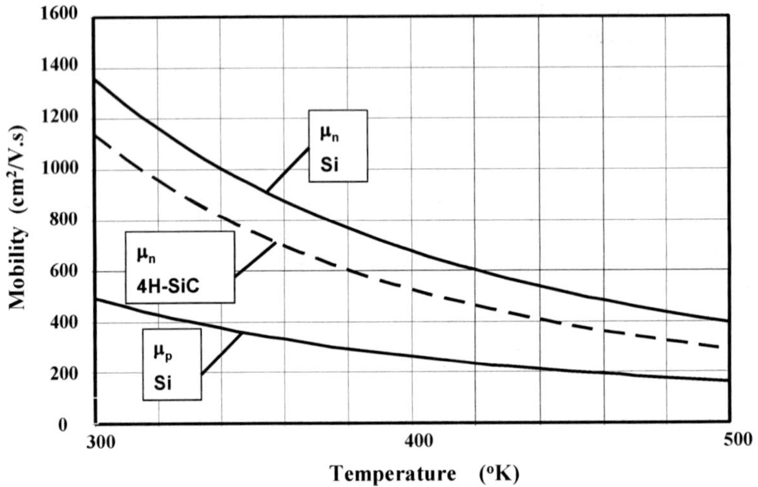

Fig. 2.11 Temperature dependence of the mobility for electrons and holes in silicon and 4H-SiC

For 4H-SiC, the temperature dependence of the mobility at low doping concentrations can be modeled by[4]:

$$\mu_{\mathrm{n}}(\text{4H-SiC}) = 1140\left(\frac{T}{300}\right)^{-2.70} \tag{2.27}$$

This variation is also shown in Fig. 2.11 by the dashed line. Only the mobility for electrons in 4H-SiC is discussed here because it is preferable to make unipolar devices from silicon carbide with n-type drift regions.

Doping Dependence

The presence of ionized donor or acceptor atoms in the semiconductor lattice contributes to a reduction in the mobility due to the additional Coulombic scattering of the free carriers. The impact of ionized impurity scattering is the largest at low temperatures because of the diminishing influence of lattice scattering. As the doping concentration is increased, the impact of ionized impurity scattering becomes larger, resulting in a reduction of the mobility.

For silicon, the ionized impurity scattering effects are small at room temperature for doping concentrations below 10^{16} cm^{-3}. Under these conditions, the mobility for electrons and holes can be considered to be independent of the doping concentration, with a magnitude of 1,360 and 495 cm^2 V^{-1} s^{-1}, respectively. At larger doping levels, the measured data[19] for the mobility of electrons in silicon at room temperature as a function of the doping concentration can be modeled by:

$$\mu_{\mathrm{n}}(\mathrm{Si}) = \frac{5.10\times10^{18} + 92N_{\mathrm{D}}^{0.91}}{3.75\times10^{15} + N_{\mathrm{D}}^{0.91}} \tag{2.28}$$

where N_{D} is the donor concentration per cubic centimeter. When the doping concentration becomes greater than 10^{19} cm^{-3}, the mobility for electrons in silicon approaches a constant value of about 90 cm^2 V^{-1} s^{-1}. This behavior is graphed in Fig. 2.12.

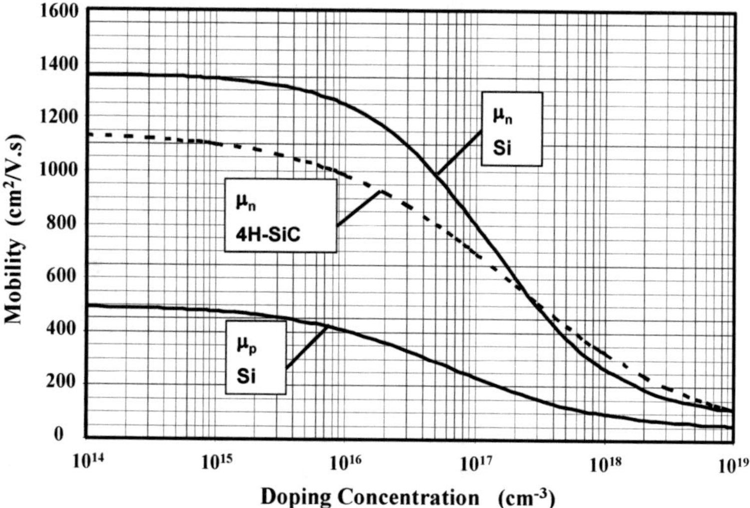

Fig. 2.12 Bulk mobility for electrons and holes in silicon and 4H-SiC

In the same manner, the measured data for the mobility of holes in silicon at room temperature as a function of the doping concentration can be modeled by:

$$\mu_{\mathrm{p}}(\mathrm{Si}) = \frac{2.90 \times 10^{15} + 47.7 N_{\mathrm{A}}^{0.76}}{5.86 \times 10^{12} + N_{\mathrm{A}}^{0.76}} \tag{2.29}$$

where N_A is the acceptor concentration per cubic centimeter. When the doping concentration becomes greater than 10^{19} cm^{-3}, the mobility for the holes in silicon approaches a constant value of about 50 cm^2 V^{-1} s^{-1}. This behavior is also graphed in Fig. 2.12.

For silicon carbide, the mobility of electrons at room temperature as a function of the doping concentration can be modeled by[3]:

$$\mu_{\mathrm{n}}(\mathrm{4H\text{-}SiC}) = \frac{4.05 \times 10^{13} + 20 N_{\mathrm{D}}^{0.61}}{3.55 \times 10^{10} + N_{\mathrm{D}}^{0.61}} \tag{2.30}$$

This behavior of the electron mobility in 4H-SiC at room temperature is also plotted in Fig. 2.12 as a function of doping concentration. The dependence of the electron mobility on doping concentration in silicon carbide has been theoretically modeled by taking into account acoustic and polar optical phonon scattering as well as intervalley scattering.[20]

Electric Field Dependence

The mobility for electrons and holes in silicon is independent of the magnitude of the electric field only when the electric field is below 10^3 V cm^{-1}. In this regime of operation, the velocity for the free carriers can be assumed to increase linearly with increasing electric field. At electric fields above 10^3 V cm^{-1}, the velocity for electrons and holes in silicon has been found to increase sub-linearly with increasing electric field, as shown in Fig. 2.13, until it reaches a saturated value. The time of flight technique[21] has been used to measure the velocity of electrons and holes in silicon at various values for the electric field and at various ambient temperatures. Based upon this empirical data, analytical expressions can be derived relating the drift velocity to the electric field.

For the analysis of power devices, it is convenient to define an average mobility as the ratio of the drift velocity to the electric field in the semiconductor. At low doping concentrations in silicon, the average mobility for electrons can be related to the electric field using:

$$\mu_{\mathrm{n}}^{\mathrm{av}} = \frac{9.85 \times 10^{6}}{\left(1.04 \times 10^{5} + E^{1.3}\right)^{0.77}} \tag{2.31}$$

Similarly, at low doping concentrations in silicon, the average mobility for holes can be related to the electric field using:

$$\mu_{p}^{av} = \frac{8.91 \times 10^{6}}{\left(1.41 \times 10^{5} + E^{1.2}\right)^{0.83}} \tag{2.32}$$

At electric fields below 10^3 V cm^{-1}, the mobility for electrons and holes has a constant value and becomes inversely proportional to the electric field when it exceeds 10^5 V cm^{-1}.

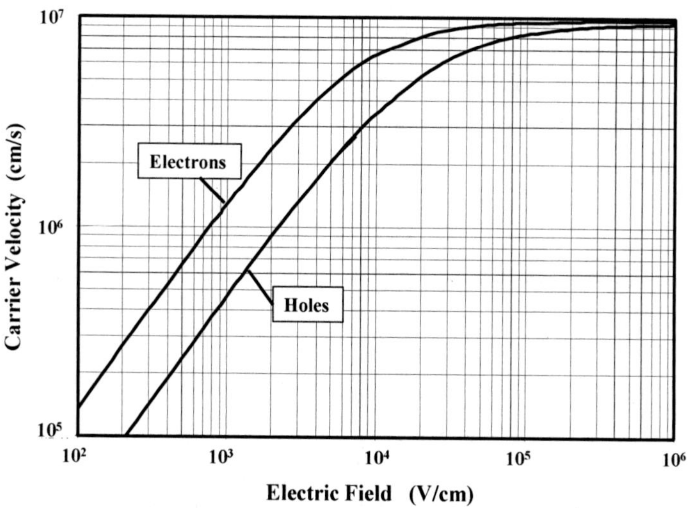

Fig. 2.13 Velocity for electrons and holes in silicon

When the electric field becomes high, the second term in the denominator for the above expressions becomes dominant and the average mobility becomes inversely proportional to the electric field strength. This implies that the velocity for the free carriers has become constant at a value given by the numerator in the expressions. This phenomenon is called *drift velocity saturation.* Just as the mobility is dependent upon the ambient temperature, the saturated drift velocity has also been found to be temperature dependent. For the case of electrons transported along the $\langle 111 \rangle$ axis in silicon:

$$v_{\text{sat,n}} = 1.434 \times 10^{9} T^{-0.87} \tag{2.33}$$

where T is the absolute temperature in Kelvin. Similarly, for the case of holes transported along the <111> axis in silicon:

$$v_{\text{sat,p}} = 1.624 \times 10^{8} T^{-0.52} \tag{2.34}$$

where T is the absolute temperature in Kelvin. Since power devices are usually rated to operate between −25°C and 150°C, the decrease in the saturated drift velocity with increasing temperature predicted by these equations is plotted in Fig. 2.14 over a slightly broader range of temperatures.

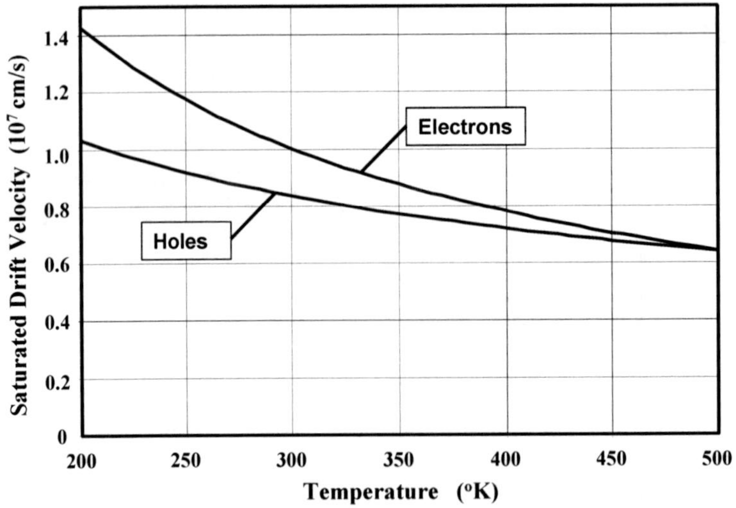

Fig. 2.14 Saturated drift velocity for electrons and holes in silicon

It is worth pointing out that the saturated drift velocity for electrons is slightly larger than that for holes in silicon. However, the difference becomes smaller as the temperature increases. These results are applicable for lightly doped silicon. Monte Carlo calculations of the effect of ionized impurity atoms upon high field transport in silicon indicate that, in addition to the decrease in the mobility, the saturated drift velocity for electrons and holes also decreases when the doping concentration exceeds 10^{18} cm^{-3}. Since the drift regions in power devices, where the high electric fields are prevalent, are doped well below this level, the use of the saturated drift velocity as described earlier is adequate for the analysis of their characteristics.

The earlier discussion relates to the transport of free carriers along the <111> direction in the silicon lattice. The mobility of electrons and holes in silicon at low electric fields is independent of the orientation of the electric field with respect to the crystal lattice. However, anisotropic transport of the free carriers is observed at high electric fields.[22] This anisotropy is caused by the ellipsoidal shape of the six conduction band valleys and the warped nature of the valence sub-bands. This anisotropic behavior for transport of carriers in silicon has generally been ignored during the analysis of silicon power devices.

At low doping concentrations in silicon carbide, the average mobility for electrons can be related to the electric field using[4]:

$$\mu_{\mathrm{n}}^{\mathrm{av}} = \frac{\mu_0}{\left[1+(\mu_0 E / v_{\mathrm{sat,n}})^2\right]^{0.5}} \tag{2.35}$$

where μ_0 is the mobility for electrons in 4H-SiC at low electric fields (1,140 $cm^2 V^{-1} s^{-1}$) and $v_{sat,n}$ is the saturated drift velocity for electrons in 4H-SiC, which has a value of 2×10^7 cm s^{-1}. This behavior of the average mobility for electrons in 4H-SiC is plotted in Fig. 2.15. As expected, at electric fields below 10^3 V cm^{-1}, the mobility for electrons has a constant value and becomes inversely proportional to the electric field when it exceeds 10^5 V cm^{-1}. As in the case of silicon, when the electric field becomes high, the second term in the denominator for the above expression becomes dominant and the average mobility becomes inversely proportional to the electric field strength.

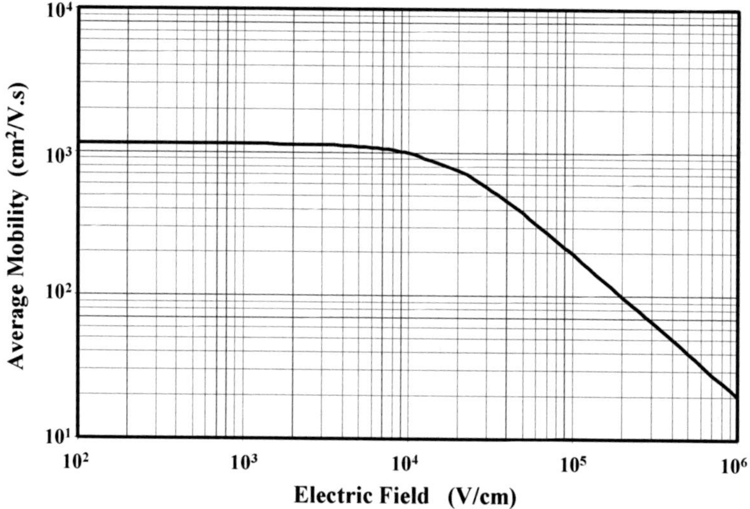

Fig. 2.15 Average mobility for electrons in 4H-SiC

Injection Level Dependence

In highly doped regions of semiconductors, majority carrier transport by drift is dominant because of the low minority carrier density. However, in lightly doped regions of semiconductor devices, current transport occurs by a combination of drift due to the presence of an electric field and diffusion due to the presence of concentration gradients for the free carriers. Within the drift region of bipolar power devices, the transport of both electrons and holes (i.e. both majority and minority carriers) contributes to the current flow.

If a low electric field is assumed to be present, the continuity equation that governs the transport of minority carriers in an n-type semiconductor can be expressed as:

$$\frac{\partial p'}{\partial t} = -\frac{p'}{\tau_a} + D_a \frac{\partial^2 p'}{\partial x^2} \tag{2.36}$$

where p' is the excess hole concentration, and D_a and τ_a are defined as the *ambipolar diffusion coefficient* and the *ambipolar lifetime*, respectively. The ambipolar diffusion coefficient is a function of the free carrier concentration:

$$D_a = \frac{(n+p) D_n D_p}{(nD_n + pD_p)} \tag{2.37}$$

where n and p are the electron and hole concentrations, and D_n and D_p are the electron and hole diffusion constants, respectively. These diffusion constants can be obtained from the respective mobility values by using the Einstein relationship for nondegenerate semiconductors:

$$D = \frac{kT}{q} \mu \tag{2.38}$$

where k is Boltzmann's constant and T is the absolute temperature. At low injection levels and for doping levels of several orders of magnitude above the intrinsic carrier concentration, the ambipolar diffusion coefficient is equal to the minority carrier diffusion constant. This does not hold true at high injection levels that occur in bipolar power devices during current transport with high on-state current density.

At low injection levels, when the density of minority carriers is far smaller than the majority carrier concentration, the mobility responsible for carrier transport is controlled by either phonon or ionized impurity scattering, as already discussed. However, at high injection levels in the lightly doped base regions of bipolar power devices during on-state current flow, the density of both electrons and holes becomes approximately equal in order to preserve charge neutrality. Since the density of both carriers is simultaneously large, the probability for the mutual Coulombic interaction of the free carriers about a common center of mass becomes significant, resulting in a decrease in the mobility of both carriers. The diffusion coefficient and the mobility decrease as the injected carrier concentration increases because of this *carrier–carrier scattering phenomenon.* If the carrier–carrier scattering phenomenon is treated like ionized impurity scattering, the mobility can be shown to vary inversely with the injected carrier concentration when the injection level exceeds 10^{17} cm^{-3}. Using Matthiessen's rule:

$$\frac{1}{\mu} = \frac{1}{\mu_0} + \frac{n \ln\left(1 + 4.54 \times 10^{11} n^{-0.667}\right)}{1.428 \times 10^{20}} \tag{2.39}$$

where μ is the mobility at a carrier density of "n" per cubic centimeter, and μ_0 is the majority carrier mobility (i.e. either μ_n or μ_p). In deriving this expression, it has been assumed that high-level injection conditions prevail so that the concentration of electrons (n) is equal to the concentration of holes (p). The decrease in the mobility for electrons and holes in silicon with increasing injected carrier concentration is plotted in Fig. 2.16. This decrease in the mobility has to be accounted for during the modeling of the forward voltage drop and surge current handling capability of bipolar power devices at high current densities.

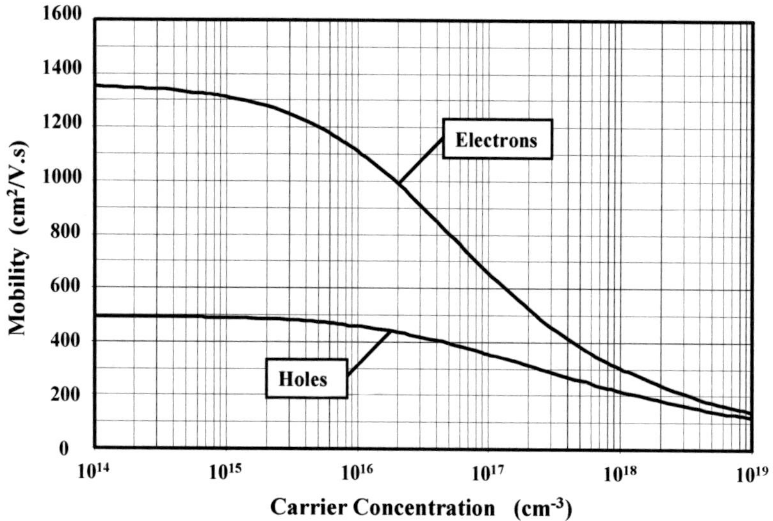

Fig. 2.16 Carrier–carrier scattering limited mobility for electrons and holes in silicon

Surface Scattering Dependence

Current flow in power MOSFETs and IGBTs occurs by the transport of carriers close to the oxide–semiconductor interface. In the case of an n-channel MOSFET, on-state current flow is induced by the formation of an inversion layer on the P-base region by the application of a gate voltage to create a strong electric field normal to the semiconductor region. The conductivity of the inversion layer depends upon the number of free carriers induced by the gate bias in the inversion layer and their mobility along the surface. Current transport through the channel is achieved by the application of a transverse electric field with the current limited by the transport properties close to the semiconductor surface.

At the high electric fields required across the gate oxide to create the inversion layer, the free carriers in the channel are confined to a thin layer located close to the surface. At low temperatures, the motion of the carriers within the inversion layer is quantized in a direction perpendicular to the surface. At room temperature, the kinetic energy of the electrons spreads their wave-function among the quantum levels and the electron transport can be treated using classical statistics.

A pictorial representation of the oxide-semiconductor interface is shown in Fig. 2.17 with a rough interface between the layers. In power MOSFETs and IGBTs, an inversion layer consisting of electrons is created near the semiconductor surface because of the applied electric field (E_N) normal to the semiconductor surface. The electrons in the inversion layer have a Gaussian distribution as illustrated on the right hand side of the figure, and are located close to the surface, typically in the range of 100 Å. In addition to Coulombic scattering due to ionized acceptors, the electrons in the inversion layer undergo several other scattering processes: (a) surface phonon scattering due to lattice vibrations, (b) additional Coulombic scattering due to the interface state charge and the fixed oxide charge, and (c) surface roughness scattering due to the deviation of the surface from a specular interface. As in the case of carrier transport in the bulk, phonon scattering is important at high temperatures when the thermal energy in the lattice becomes large. The influence of Coulombic scattering due to the interface state charge and the fixed oxide charge is important in the case of lightly inverted surfaces and when the processing conditions produce high interface state and fixed oxide charge densities. The surface roughness scattering is predominant under strong inversion conditions because the electric field normal to the surface must be increased in order to produce a high inversion layer concentration. The high normal electric field increases the velocity of the carriers towards the surface and brings the inversion layer charge distribution closer to the surface. Both these factors enhance carrier scattering at the rough interface with the oxide as illustrated in the figure. The surface roughness scattering is sensitive to processes that influence the smoothness of the surface. A quantitative understanding of the carrier mobility in inversion and accumulation layers is of importance for the proper modeling and design of power MOS-gated devices.

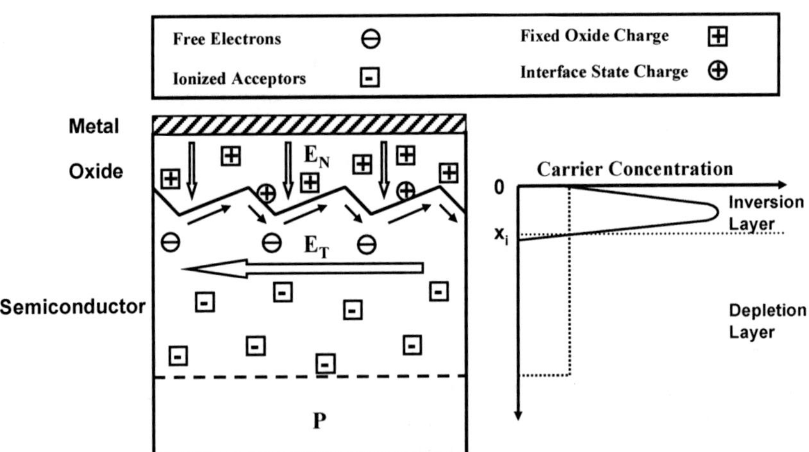

Fig. 2.17 Current transport at the interface between the oxide and the semiconductor in a MOSFET structure

The dependence of the surface mobility upon various parameters is discussed below. To facilitate this discussion, it is useful to define an *effective mobility* (μ_e) for carriers in the inversion or accumulation layers:

$$\mu_e = \frac{\int_0^{x_i} \mu(x)n(x)\mathrm{d}x}{\int_0^{x_i} n(x)\mathrm{d}x} \tag{2.40}$$

where $\mu(x)$ and $n(x)$ are the local mobility and free carrier concentration in the inversion layer at a depth x from the oxide-semiconductor interface and x_i is the thickness of the inversion layer. Defined in this manner, the effective mobility is a measure of the conductance of the inversion layer and is, therefore, eminently suitable for the calculation of the channel resistance in MOS-gated devices. In fact, determination of the effective mobility is usually performed by using lateral MOSFETs specifically designed to extract this parameter. In doing these measurements, special attention must be given to the surface orientation, the direction of the current flow vector, and the surface charge density because they influence the magnitude of the effective mobility. In addition, charge trapping has been found to have a strong impact on the inversion layer mobility is silicon carbide. Special Hall mobility structures have been used to extract the effective mobility in this case.

The surface mobility has been found to be a strong function of the electric field (E_N) normal to the surface rather than the inversion layer concentration, which is dependent upon the background doping level and the applied gate bias. As illustrated in Fig. 2.17, the inversion layer charge is distributed near the surface with a Gaussian profile. The electric field experienced by the free carriers in the inversion layer varies from the peak electric field at the oxide-semiconductor interface to that at the depletion layer interface. For the same peak surface electric field, the average electric field in the inversion layer will vary with background doping level. To describe the effect of the electric field strength normal to the surface upon the effective mobility, it is appropriate to use the average electric field rather than the peak surface electric field. This *effective electric field* (E_{eff}) is given by:

$$E_{eff} = \frac{1}{\varepsilon_S}\left(\frac{1}{2}Q_{inv} + Q_B\right) \tag{2.41}$$

where Q_{inv} is the inversion layer charge density per square centimeter and Q_B is the bulk depletion layer charge density per square centimeter. For the same peak surface electric field, the effective electric field will become smaller as the background doping level decreases.

The effective mobility in an inversion layer decreases with increasing electric field as discussed earlier. A typical variation of the effective mobility with increasing electric field normal to the surface is shown in Fig. 2.18. In general, it has been found that the effective mobility increases very rapidly with increasing inversion layer concentration under weak inversion conditions as indicated by the dashed line in the figure. As the inversion layer concentration increases, the effective mobility has a peak value and then decreases. The increase in the effective mobility at low inversion layer concentrations has been related to Coulombic scattering at charged interface states. If the surface state density is adequately reduced by suitable process conditions, the effective mobility remains independent of the inversion layer concentration in the weak inversion regime.[23] When the electric field normal to the surface exceeds 10^4 V cm^{-1}, the effective mobility begins to decrease with increasing inversion layer charge. The enhancement of the surface scattering by the larger normal electric field is responsible for this degradation.

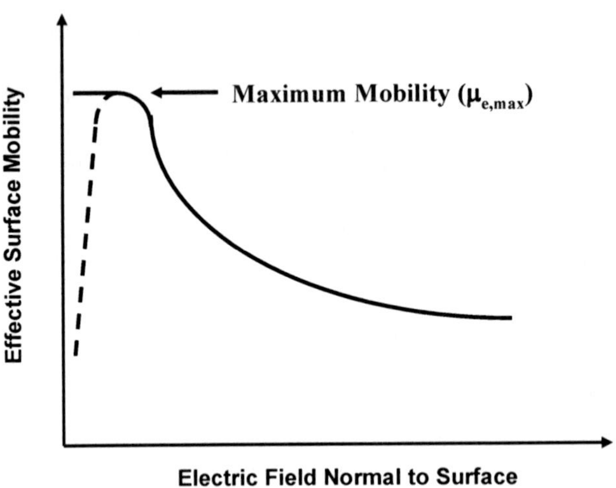

Fig. 2.18 Effective inversion layer mobility

For electrons in silicon inversion layers, the effective mobility can be described as a function of the effective electric field by:

$$\mu_e = \mu_{e,max} \left(\frac{E_C}{E_{eff}} \right)^{\beta} \tag{2.42}$$

where $\mu_{e,max}$ is the maximum effective mobility for a given background doping level and fixed oxide charge density. The parameters E_c and β are empirical constants that are dependent upon the fixed oxide charge density and the background doping level. These constants have been measured for the commonly used oxidation techniques for silicon device fabrication employing steam or dry oxygen.[24]

The parameter β is a measure of the rate of degradation of the effective mobility with increasing electric field normal to the surface. It has a strong influence on the on-resistance and transconductance of power MOS-gated devices. Based upon empirical extraction:

$$\beta_{wet} = 0.341 - 5.44 \times 10^{-3} \log(N_A) \tag{2.43}$$

for wet oxidation conditions and

$$\beta_{\text{dry}} = 0.313 - 6.05 \times 10^{-3} \log(N_{\text{A}}) \tag{2.44}$$

for dry oxidation conditions. The difference between the wet and dry oxidation conditions is believed to be related to the slower growth rate of dry oxide resulting in a smoother interface between the oxide and the semiconductor. Since the surface roughness scattering becomes dominant at high electric fields, dry oxidation produces superior effective mobility under strong inversion conditions. Under typical power device processing conditions, $\beta = 0.25$, i.e. the effective mobility decreases inversely as the fourth power of the effective electric field.

The constant E_{c} is dependent upon the fixed charge density and the background doping level. Based upon empirical extraction:

$$E_{\text{c}} = 2 \times 10^{-4} N_{\text{A}}^{0.25} A \text{e}^{BQ_{\text{f}}} \tag{2.45}$$

where N_{A} is the background doping concentration and Q_{f} is the fixed charge density. The parameters A and B are dependent upon the oxide growth process conditions. For wet oxidation conditions, $A = 2.79 \times 10^4$ V cm^{-1} and $B = 8.96 \times 10^{-2}$ cm^2, while for dry oxidation conditions $A = 2.61 \times 10^4$ V cm^{-1} and $B = 1.3 \times 10^{-1}$ cm^2. The fixed charge density Q_{f} is in units of 10^{11} cm^{-2} in this expression. The influence of the substrate doping concentration is small, whereas the fixed charge density has a strong influence on the effective mobility over a broad range of electric field strength.

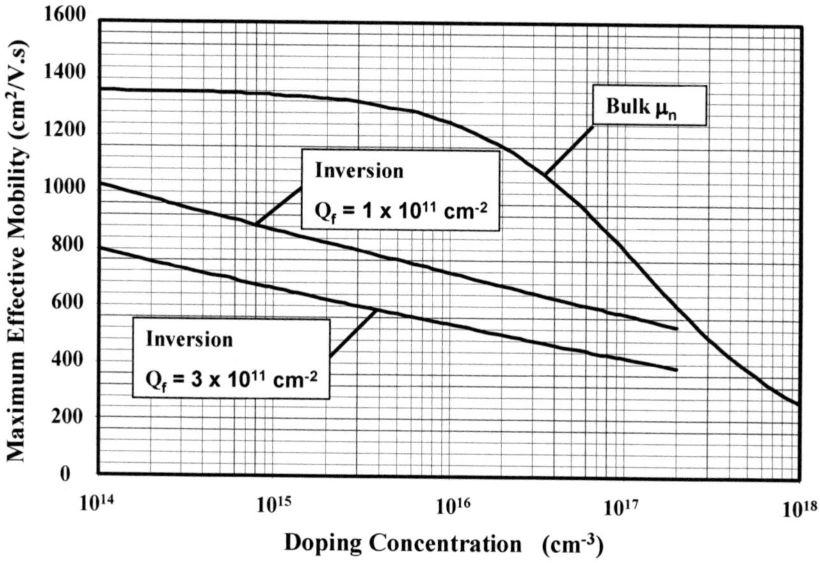

Fig. 2.19 Effective inversion layer mobility for electrons in silicon

The maximum effective inversion layer mobility for electrons in silicon has been measured as a function of the substrate doping concentration.[24] The observed behavior is plotted in Fig. 2.19 for two cases of fixed oxide charge together with the bulk mobility for comparison. It can be seen that the effective mobility decreases logarithmically with increasing doping concentration in the substrate. This decrease is not due to enhanced ionized impurity scattering. Since the inversion layer thickness is on the order of 100 Å in thickness, the number of ionized impurities is less than 10^{10} per square centimeter for doping levels up to 10^{16} cm^{-3}. This is an order of magnitude smaller than the typical fixed oxide charge density, making ionized impurity scattering ineffective. The observed decrease in the effective mobility with increasing doping concentration arises primarily from the higher electric fields, which must be applied normal to the surface to produce surface inversion.

The measured data indicates that the maximum inversion layer mobility is 75–85% of the bulk value for substrate doping ranging from 10^{14} to 10^{17} cm^{-3} if the fixed oxide charge is small. If the processing conditions produce a large fixed oxide charge, the maximum effective mobility can be severely degraded. This behavior of the maximum inversion layer mobility can be empirically modeled by:

$$\mu_{e,\max} = \frac{\mu_0}{1+\alpha\, Q_f} \tag{2.46}$$

where the parameters μ_0 and α are dependent upon the background doping concentration (N_A) per inverse cubic centimeters as given by:

$$\mu_0 = 3490 - 164\log(N_A) \tag{2.47}$$

and

$$\alpha = -0.104 + 0.0193\log(N_A) \tag{2.48}$$

The plots in Fig. 2.19 take into account this dependence of the inversion layer mobility on background doping concentration. If the interface charge density is significant, it should be added to the fixed charge (Q_f) when computing the inversion layer mobility using (2.46).

The effective mobility for holes in inversion layers[25,26] has also been measured in silicon but in less detail than for electrons. The observed decrease in the effective mobility for holes upon the inversion layer charge is shown in Fig. 2.20. This variation can be described by (2.42) when the inversion layer concentration exceeds 5×10^{11} cm^{-2} by using $E_c = 3 \times 10^4$ V cm^{-1} and $\beta = 0.28$. The dependence of these parameters on the doping concentration and fixed charge has not been studied in detail.

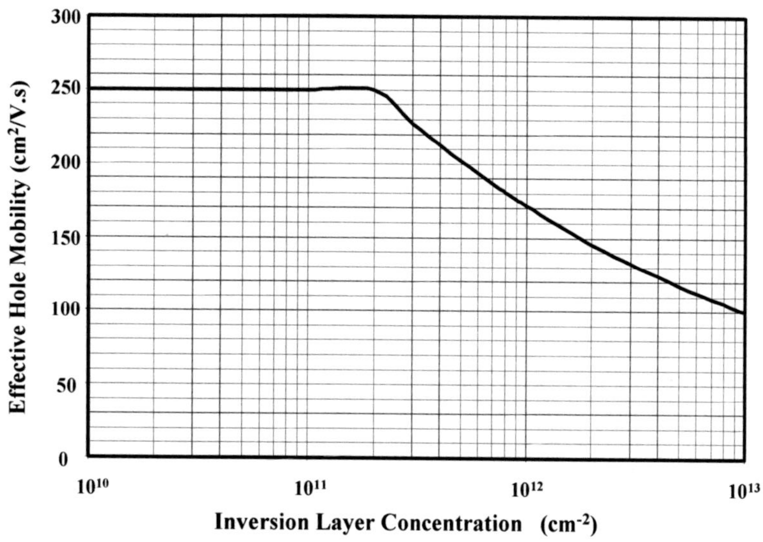

Fig. 2.20 Effective inversion layer mobility for holes in silicon

The effective mobility in inversion layers in silicon has also been found to be anisotropic. Its value depends not only upon the orientation of the plane of the surface on which the inversion layer is formed but also upon the direction of the current flow vector within this surface. Measurements[27,28] of the effective mobility for electrons on many surface orientations of silicon have demonstrated that:

$$\mu_e(100) > \mu_e(111) > \mu_e(110) \tag{2.49}$$

while for holes on silicon surfaces

$$\mu_p(110) > \mu_p(111) > \mu_p(100) \tag{2.50}$$

This variation of the effective mobility upon the crystal orientation has been correlated with the anisotropy of the conductivity effective mass in silicon. Based upon these observations, it is typical to manufacture n-channel power MOSFETs and IGBTs using (100) oriented silicon as the starting material.

Intraplanar anisotropy[24,27] of the effective mobility for electrons in inversion layers has also been reported for silicon. For the same substrate doping concentration and fixed oxide charge density, the effective mobility along the [100] direction has been found to be 30% larger than that along the [110] direction. The ratio of the effective mobility remains constant with increasing effective electric field normal to the surface, indicating that surface roughness is isotropic and that the difference between the mobilities arises from the anisotropy of the conductivity effective mass in silicon. Similar studies have been undertaken for silicon carbide.[29] The results show only a weak dependence on the orientation because the high interface state density tends to mask the orientation dependence.

The above results are applicable for surfaces prepared by chemi-mechanical polishing processes that are standard practice in the industry. They are relevant to the design and analysis of planar gate power MOSFET and IGBT structures. In recent years, a larger channel density has been achieved for these devices by locating the MOS-gate electrode within trenches etched into the silicon surface. The effective mobility for electrons in inversion layers formed along the trench sidewalls is usually lower than that on the polished planar surface because of enhanced surface roughness. The roughness of the surface depends upon the etch chemistry and can be reduced by decreasing the etch rate. The results of measurements[30] of the effective mobility on trench sidewalls have demonstrated that mobilities comparable to those on the planar surface can be obtained if proper precautions are taken to ensure smoothness of the trench sidewalls followed by sacrificial oxidation steps to remove the surface damage created by the reactive-ion-etch process used to fabricate the trenches.

An accumulation layer is also utilized to improve current transport in power MOSFET and IGBT structures. The accumulation layer is formed when the electric field normal to the semiconductor surface attracts majority carriers to the surface. The carriers located in the accumulation layer are distributed further from the surface than in the case of inversion layers, reducing the impact of surface roughness scattering. The effective mobility can therefore be expected to be larger in accumulation layers than that for the same type of carriers in an inversion layer. The measured effective mobility for electrons in silicon accumulation layers[24] has been reported to be about 80% of the bulk mobility. The accumulation mobility also decreases with increasing electric field normal to the surface as observed for inversion layers and exhibits anisotropy as well. The anisotropy of the effective mobility for electrons in silicon has been ascribed to the anisotropy of the effective mass for electrons in silicon.

As in the case of current transport in the bulk, the velocity for the carriers within inversion layers can also undergo saturation when the transverse electric field becomes large. The saturated drift velocity in a silicon inversion layer is a function of the surface orientation and has been found to be lower than that observed in the bulk. In the case for holes, the largest saturated drift velocity is observed for the (100) orientation.[31] In the case of electrons, the saturated drift velocity has been measured as a function of temperature for various surface orientations.[32] The measured data can be fitted by using:

$$v^{\mathrm{i}}_{\mathrm{sat,n}} = v_0 - \delta T \tag{2.51}$$

where v_0 is the saturated drift velocity at 0 K and T is the absolute temperature in Kelvin. The values for v_0 for the different orientations are as follows: 8.9×10^6 cm s^{-1} for the (100) orientation, 7.4×10^6 cm s^{-1} for the (111) orientation, and 6.2×10^6 cm s^{-1} for the (110) orientation. A value of 7×10^3 cm s^{-1} K^{-1} for the constant δ provides a good fit to the measured data for all the three orientations.

2.2 Resistivity

In the absence of the dopants, the intrinsic resistivity is determined by the electrons and holes created by the thermal generation process. The resistivity of a semiconductor region can be controlled by the addition of dopants to the crystal lattice. In the case of the starting material, the resistivity of n-type silicon is usually controlled by the addition of phosphorus and the resistivity of p-type silicon is usually controlled by the addition of boron during the growth of the crystal.

2.2.1 Intrinsic Resistivity

An intrinsic semiconductor region contains an equal concentration of electrons and holes produced by the thermal generation process. The intrinsic resistivity (ρ_i) of the semiconductor region is then governed by the transport of both carriers as given by:

$$\rho_i = \frac{1}{qn_i\left(\mu_n + \mu_p\right)} \tag{2.52}$$

where μ_n and μ_p are the mobility for electrons and holes, and n_i is the intrinsic carrier concentration. At room temperature, the intrinsic resistivity for silicon is 2.5×10^5 Ω cm. At elevated temperatures, the intrinsic resistivity is rapidly reduced by the increase in the intrinsic carrier concentration. By using (2.2) for the intrinsic carrier concentration:

$$\rho_i = 1.75 \times 10^{-7}\, T^{0.8}\, e^{7020/T} \tag{2.53}$$

where T is the absolute temperature in Kelvin.

2.2.2 Extrinsic Resistivity

The resistivity of a semiconductor region can be altered by the addition of controlled amounts of impurities. The addition of impurity elements from the fifth column of the periodic table to the silicon lattice produces n-type material with a preponderance of electrons, and the addition of impurity elements from the third column of the periodic table to the silicon lattice produces p-type material with a preponderance of holes. The most commonly used n-type dopants during the fabrication of power devices are phosphorus and arsenic as in the case of integrated circuits for signal processing. The most commonly used p-type dopant for integrated circuits is boron. Although boron is also used for the fabrication of power devices, such as power MOSFETs and IGBTs, aluminum and gallium are commonly used for the fabrication of power thyristors and GTOs due to their larger diffusion coefficients.

The diffusion coefficients for gallium and aluminum are compared with that for boron in Fig. 2.21 for the temperature range used for the fabrication of power device structures. In modern power MOSFETs and IGBTs, the dopants are not driven by large distances after ion-implantation of the dopants. This favors the use of boron for the formation of the P-base regions within these structures. However, in the case of power thyristors designed to support very high voltages, it is advantageous to form deep p-type regions with a graded doping profile to enable supporting some of the voltage in the diffused side of the junction. Such deep, graded junctions are also favorable for the formation of beveled edge terminations with breakdown voltages approaching the ideal parallel-plane breakdown voltage. Consequently, it is commonplace to use gallium and aluminum as p-type dopants for the P-base regions within thyristors and GTOs. The diffusion coefficient for aluminum in silicon is a factor of 5 times larger than that for boron at typical temperatures. Thus, a deep junction with depth of 100 μm can be achieved using aluminum diffusion at 1,200°C in 98 h (4 days). In contrast, it would require 556 h (23 days) to obtain the same junction depth by using boron diffusion at the same temperature. A considerable savings in the processing time is obviously feasible by using aluminum instead of boron in this case. Another advantage of using aluminum in place of boron is that its ionic radius is comparable to that for silicon, resulting in minimizing misfit strain in the lattice. Superior junctions can be formed using sealed-tube diffusion of aluminum to obtain a low surface concentration with a highly graded doping profile.

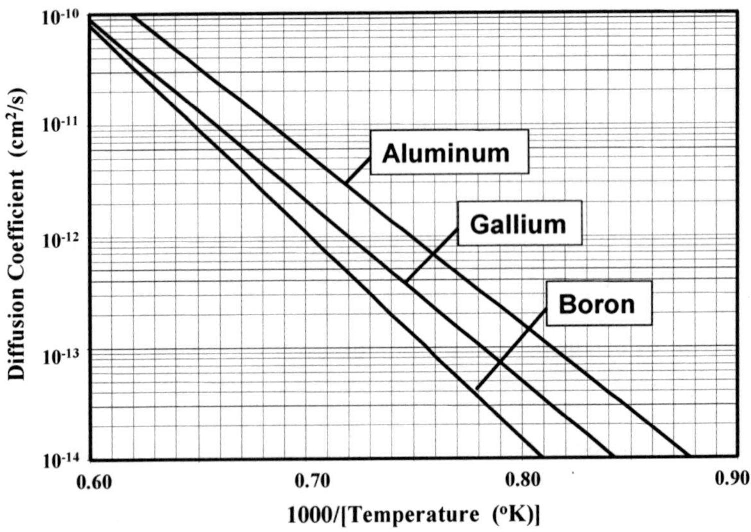

Fig. 2.21 Dopant diffusion coefficients in silicon

The extrinsic resistivity of silicon doped with phosphorus has been measured[33] over a broad range of doping concentrations. Similar measurements[34] have also been conducted for P-type silicon doped with boron. This resistivity data is shown in Fig. 2.22 as a function of the doping concentration. The resistivity of P-type silicon is larger than that for N-type silicon at any given doping concentration due to the smaller mobility for holes in silicon.

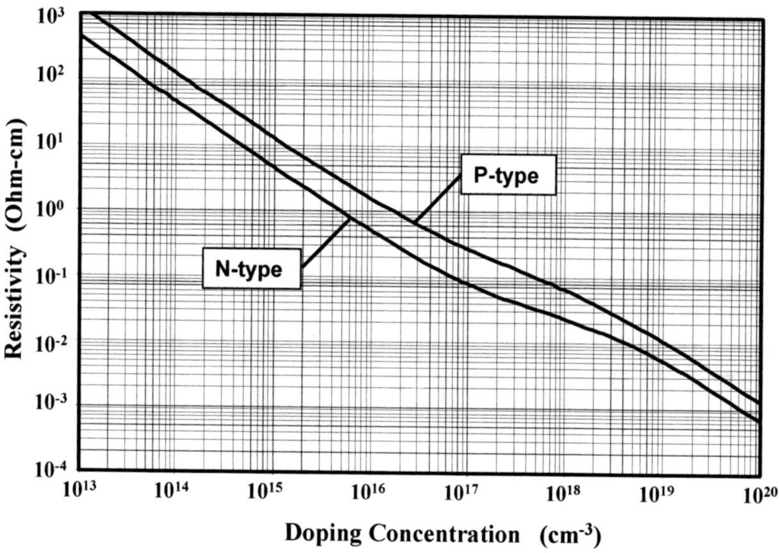

Fig. 2.22 Resistivity of N-type and P-type silicon

Using the relationship between doping concentration and mobility for electrons in silicon given by (2.28), the resistivity for N-type silicon can be related to the doping concentration:

$$\rho_n(\text{Si}) = \frac{3.75\times10^{15} + N_D^{0.91}}{1.47\times10^{-17}N_D^{1.91} + 8.15\times10^{-1}N_D} \tag{2.54}$$

where N_D is the donor concentration per cubic centimeter When the doping concentration becomes less than 10^{15} cm^{-3}, the mobility for electrons in silicon approaches a constant value of about 1,360 cm^2 V^{-1} s^{-1}. The resistivity then becomes inversely proportional to the doping concentration:

$$\rho_n(\text{Si}) = \frac{4.6\times10^{15}}{N_D} \tag{2.55}$$

This expression is suitable for computing the values for high resistivity silicon with low doping concentrations, such as the N-base regions of high voltage thyristors and GTOs. When the doping concentration becomes greater than 10^{19} cm^{-3}, the mobility for electrons in silicon again approaches a constant value of about

90 cm^2 V^{-1} s^{-1}. The resistivity then also becomes inversely proportional to the doping concentration:

$$\rho_n(\mathrm{Si}) = \frac{6.94\times 10^{16}}{N_D} \tag{2.56}$$

In a similar manner, by using the relationship between doping concentration and mobility for holes in silicon given by (2.29), the resistivity for P-type silicon can be related to the doping concentration:

$$\rho_p(\mathrm{Si}) = \frac{5.86\times 10^{12} + N_A^{0.76}}{7.63\times 10^{-18} N_A^{1.76} + 4.64\times 10^{-4} N_A} \tag{2.57}$$

where N_A is the acceptor concentration per cubic centimeter. When the doping concentration becomes less than 10^{15} cm^{-3}, the mobility for holes in silicon approaches a constant value of about 495 cm^2 V^{-1} s^{-1}. The resistivity then becomes inversely proportional to the doping concentration:

$$\rho_p(\mathrm{Si}) = \frac{1.26\times 10^{16}}{N_A} \tag{2.58}$$

This expression is suitable for computing the values for high resistivity silicon with low doping concentrations, such as the drift regions of high voltage p-channel power MOSFETs. When the doping concentration becomes greater than 10^{19} cm^{-3}, the mobility for holes in silicon again approaches a constant value of about 50 cm^2 V^{-1} s^{-1}. The resistivity then also becomes inversely proportional to the doping concentration:

$$\rho_p(\mathrm{Si}) = \frac{1.25\times 10^{17}}{N_A} \tag{2.59}$$

The resistivity for the starting material required for the fabrication of power devices with breakdown voltages above 100 V exceeds 1 Ω cm for N-type silicon and 3 Ω cm for P-type silicon. The starting material used for the fabrication of power devices is preferably bulk material because of its lower defect density when compared with epitaxial material. When the breakdown voltage exceeds 2,000 V, the thickness of the drift region is sufficient to allow utilization of bulk material with diffusion performed on the both sides of the wafer to form the semiconductor regions in thyristors, GTOs and IGBTs. When the breakdown voltage is below 1,500 V, the thickness of the drift region becomes too small to allow handling the thin wafers through the manufacturing process without breakage problems. Epitaxial starting material is therefore preferred for power MOSFETs and IGBTs rated at lower voltages.

2.2.3 Neutron Transmutation Doping

In the case of thyristors with blocking voltage capability exceeding 2,000 V, bulk N-type material is used as the starting material. Since a low concentration of oxygen is required to avoid the formation of precipitates that can adversely impact the breakdown characteristics, float zone (FZ) silicon is preferred over Czochralski (CZ) silicon. These high voltage devices must also be capable of handling very high currents as indicated for the applications discussed in Chap. 1. It is typical to fabricate thyristors and GTOs out of an entire 3- or 4-inch-diameter silicon wafer. Variations in the resistivity across the wafer can have a strong impact on the breakdown voltage distribution. In devices that contain back-to-back junctions, such as thyristors, the breakdown voltage is limited by open-base transistor breakdown. The device breakdown voltage can then become limited by reach-through effects at the higher resistivity regions and avalanche breakdown effects at the lower resistivity regions. Large variations in the resistivity require making design compromises, which enlarge the thickness of the drift region, leading to an increase in the on-state voltage drop and power losses.

The inhomogeneous resistivity in silicon wafers grown using the float-zone process occurs in the form of striations that are attributed to local fluctuations in the growth rate caused by temperature variations. These changes in the growth rate induce variations in the dopant incorporation in the crystal because the effective segregation coefficient is determined by the growth rate. Advanced crystal growth methods have enabled reducing the resistivity variation to the range of +15% to −15%. Even superior staring material can be obtained by using the NTD process.

The NTD process[35] is based upon the conversion of the Si^{30} isotope into phosphorus atoms by the absorption of thermal neutrons. Naturally occurring silicon consists of three isotopes: the Si^{28} isotope has an abundance of 92.27% in the lattice; the Si^{29} isotope has an abundance of 4.68% in the lattice; and the Si^{30} isotope has an abundance of 3.05% in the lattice. These isotopes are uniformly distributed within the silicon lattice. The NTD process takes place in a nuclear reactor by the absorption of thermal neutrons. The capture of thermal neutrons by all three of the naturally occurring isotopes occurs in proportion to their abundance in the lattice and their capture cross-sections. The absorption of a neutron by the isotopes leads to the following nuclear reactions:

$$\mathrm{Si}^{28}(n,\gamma) \quad \mathrm{Si}^{29} \tag{2.60}$$

$$\mathrm{Si}^{29}(n,\gamma) \quad \mathrm{Si}^{30} \tag{2.61}$$

$$\mathrm{Si}^{30}(n,\gamma) \quad \mathrm{Si}^{31} \rightarrow P^{31} + \beta \tag{2.62}$$

The cross-sections for the absorption of thermal neutrons for these reactions are 0.08, 0.28, and 0.13 barns (one barn equals 10^{-24} cm^2), respectively. The first two reactions merely alter the isotope ratios within the silicon crystal. Since the doping induced by the NTD process is small when compared with the density of silicon atoms, this alteration can be neglected. The third reaction creates the phosphorus atoms that can be utilized to produce n-type doping in the silicon crystal.

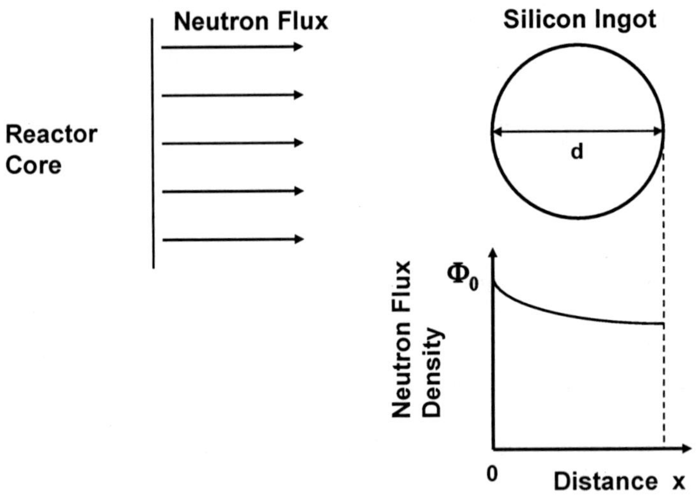

Fig. 2.23 Neutron transmutation doping of a silicon Ingot

The phosphorus donor concentration created by the NTD process is given by:

$$C_{\text{donor}} = 2.06 \times 10^{-4}\, \Phi t \tag{2.63}$$

where Φ is the thermal neutron flux density and t is the irradiation time. Based upon the homogeneous distribution of the Si^{30} isotope in the silicon crystal, a homogeneous distribution of phosphorus can be created if a uniform flux of neutrons is ensured within all portions of the silicon crystal during the irradiation. However, the uniformity of the resistivity within the silicon crystal after irradiation is compromised by the abruption of the neutrons as they pass through the crystal and the presence of any inhomogeneously distributed dopants in the starting material.

During the neutron transmutation process, the silicon ingot is placed inside a nuclear reactor at a location containing a uniform flux of thermal neutrons emanating from the core of the reactor as illustrated in Fig. 2.23. If the silicon ingot is assumed to be surrounded by a medium with the same thermal neutron absorption properties as silicon, the influence of the absorption of the thermal neutrons by the silicon can be modeled by

$$\Phi = \Phi_0\, \mathrm{e}^{-x/b} \tag{2.64}$$

where Φ is the thermal neutron flux density at a position x in the ingot as indicated in Fig. 2.23 and Φ_0 is the thermal neutron flux density at $x = 0$ (at the edge of the ingot). The constant b is the decay length (19 cm) for the absorption of thermal neutrons in silicon. For an ingot of diameter d, a phosphorus concentration proportional to Φ_0 is created at one edge of the ingot while a lower concentration (proportional to $\Phi_0\, e^{-d/b}$) is created at its opposite edge. This produces an inhomogeneous distribution of phosphorus by the NTD process, leading to a resistivity variation given by ±tanh $(d/2b)$.

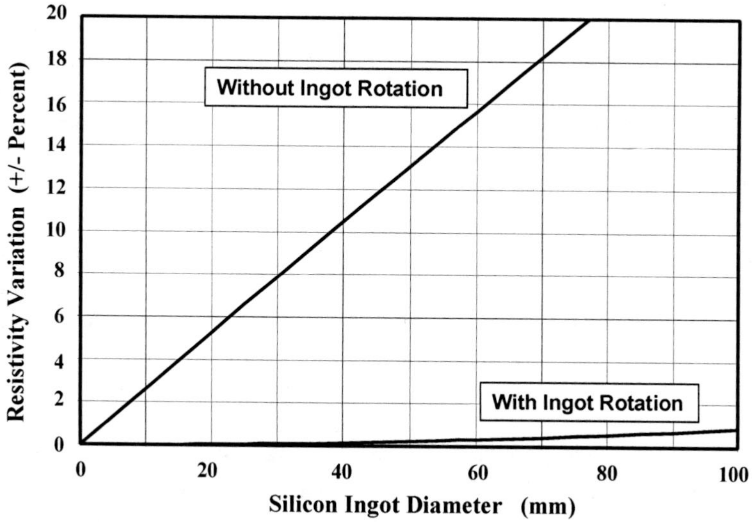

Fig. 2.24 Resistivity variation produced by the NTD process in a silicon Ingot

The variation of the resistivity across the diameter of the silicon ingot due to the absorption of the thermal neutrons is shown in Fig. 2.24. It can be seen that the absorption of the thermal neutrons can lead to a significant variation in the resistivity when the ingot diameter exceeds 20 mm. Modern high current thyristors are manufactured using wafers with diameters of upto 4 in. (100 mm). A resistivity variation of over ±20% in these wafers with the NTD process would exceed that achievable with the FZ process, making this technology unacceptable. This problem can be overcome by the rotation of the silicon ingots during the neutron irradiation. If it is assumed that the ingot undergoes many complete rotations during the irradiation time in the reactor, the maximum resistivity will occur at the axis of the ingot, with the lowest resistivity at the periphery. The ratio of the neutron dose at the periphery to the dose at the axis of the cylindrical ingot is given by[36]:

$$\frac{\Phi(\text{periphery})}{\Phi(\text{center})} = 1 + \frac{1}{16}\left(\frac{d}{b}\right)^2 \tag{2.65}$$

This produces a maximum resistivity variation of $\pm[32(b/d)^2 - 1]^{-1}$ across the crystal. This variation is also shown in Fig. 2.24 for comparison with the case of irradiation without rotation of the ingot. A significant improvement in the homogeneity of the doping within the crystal is produced with ingot rotation, leading to a resistivity variation of less than 2% even for wafers with diameters of 4 in. In actual practice, the silicon ingot is surrounded by heavy water or graphite in the reactor during the neutron irradiation. This tends to increase the decay length, which results in even superior uniformity of the doping than predicted by the above expressions.

As mentioned earlier, the presence of inhomogeneously distributed dopants in the starting material can have a strong impact on the resistivity variation in the silicon wafer after the NTD process. Because of the high segregation coefficient for boron in silicon, a more homogeneous doping can be achieved in P-type starting material than in N-type starting material. However, it is worth pointing out that for the same initial resistivity, the background doping concentration for P-type material is larger than that in N-type starting material because of the lower mobility for holes in silicon. This becomes a consideration during the choice of the starting material because a lower initial dopant concentration is preferable. Both P-type and N-type wafers have been used for the manufacturing of the neutron transmutation doped silicon. In both cases, the concentration of the phosphorus atoms created by the NTD process should be much larger than the initial doping concentration in order to take full advantage of the improved resistivity variation inherent to the NTD process.

If the initial starting material has a resistivity variation of $\pm\alpha\%$, then the resistivity variation after the NTD process will be given by $\pm\alpha(\rho_f/\rho_i)$ for N-type starting material and by $\pm\alpha(\mu_n/\mu_p)(\rho_f/\rho_i)$ for P-type starting material. Here, ρ_f and ρ_i are the final (or post) irradiation and initial (or pre) irradiation resistivity, respectively. Typical values of the preirradiation resistivity variation are ±15%. In this case, if the resistivity variation after the NTD process must be less than ±3%, N-type starting material must have a resistivity of more than 5 times the postirradiation value, and P-type starting material must have a resistivity of more than 15 times the postirradiation value. P-type silicon can be grown with resistivity over 5,000 Ω cm, enabling the production of very homogeneously doped silicon with resistivity as high as 200 Ω cm by using the NTD process.

When first proposed, the acceptance of the NTD process for the manufacturing of very homogeneously doped material for power device applications was delayed by the relatively low lifetime in this material when compared with the FZ starting material because the neutron transmutation process is accompanied by severe lattice damage. The lattice damage arises from the displacement of silicon atoms because of (a) Gamma-recoil during the transmutation reaction, (b) crystal irradiation by the energetic electrons (β particles) produced by the nuclear reaction (see (2.62)), and (c) bombardment by the fast neutrons produced in the nuclear reactors. These processes introduce deep levels in the silicon band gap, which act as recombination centers. The influence of fast neutron damage, which is more difficult to remove by annealing, can be mitigated by locating the ingot in portions of the nuclear reactor where the fast neutron flux is low. It is also common practice to clad the ingot in cadmium, which can selectively absorb the fast neutrons. In addition, the absorption of thermal neutrons by the phosphorus atoms generated by the NTD process can produce sulfur, and the absorption of fast neutrons by the silicon can produce magnesium resulting in additional deep levels in the band gap. Fortunately, the deep levels produced by the crystal damage could be eliminated by annealing the silicon ingot at appropriate temperatures.[37] The concentrations of sulfur and magnesium produced during the manufacturing of high resistivity silicon were also found to have a negligible impact on the minority carrier lifetime.

Because of the crystal damage produced by the above mechanisms, the resistivity of the silicon after irradiation is over 10^5 Ω cm. Annealing the silicon ingot at between 750 and 800°C has been found to be sufficient for the removal of the damage so that the resistivity is determined by the phosphorus concentration produced by the NTD process.[37] The minority carrier lifetime is increased to the range of 1 ms, which is sufficient for the manufacturing of high voltage thyristors. These postirradiation processes have enabled improving the quality of the silicon wafers to a level where the NTD process has become widely accepted as the starting material for manufacturing power thyristors.

2.3 Recombination Lifetime

A continuous balance between the generation and recombination of electron–hole pairs occurs in semiconductors operating under thermal equilibrium conditions. The creation of excess carriers by an external stimulus disturbs this equilibrium. When the external stimulus is removed, the excess carrier density decays and the carrier concentrations are eventually returned to the equilibrium values. The rate of recovery is governed by the minority carrier lifetime. The recovery to equilibrium conditions occurs because of several simultaneous processes: (a) recombination occurring because of an electron dropping directly from the conduction band into the valence band; (b) recombination occurring because of an electron dropping from the conduction band and a hole dropping from the valence band into a recombination level located within the energy band gap; and (c) recombination occurring because of electrons from the conduction band and holes from the valence band dropping into surface traps.

The transitions that occur during the recombination processes are schematically illustrated in Fig. 2.25. During recombination, the energy of the carriers is dissipated by one of several mechanisms: (1) the emission of a photon (referred to as *radiative recombination*); (2) the distribution of the energy into the lattice in the form of phonons (referred to as *multi-phonon recombination*); and (3) the transmission of the energy to a third particle, which can be either an electron or a hole (referred to as *Auger recombination*). All these processes simultaneously assist in the recovery of the carrier density to its equilibrium value.

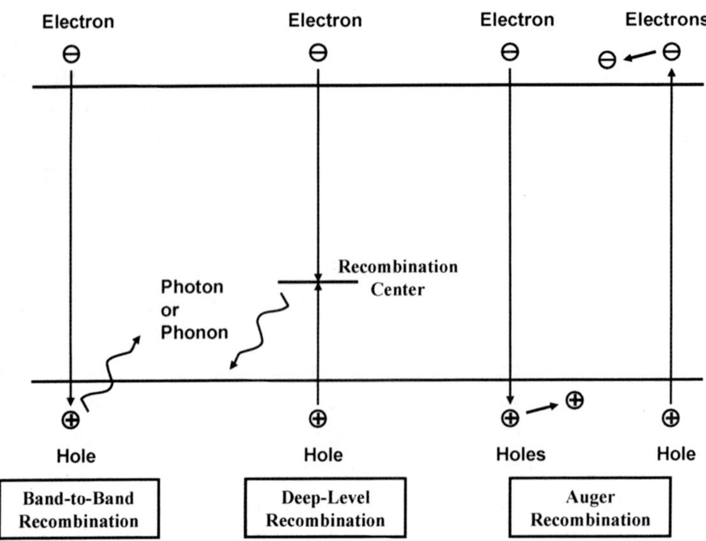

Fig. 2.25 Recombination processes in a semiconductor

In a semiconductor such as silicon with an indirect band gap structure, the probability for direct transitions from the conduction band to the valence band is small. Consequently, the radiative recombination lifetime for silicon is on the order of 1 s. In comparison, the density of recombination centers is sufficiently high, even in high purity silicon used to fabricate power devices, so as to reduce the lifetime associated with recombination via the deep levels to less than 100 μs. This recombination process, therefore, proceeds much more rapidly and is predominant under most device operating conditions. The Auger recombination process becomes important in the determination of the minority carrier lifetime in very heavily doped regions and during very high injection levels in the drift regions of bipolar power devices.

2.3.1 Shockley–Read–Hall Recombination

The statistics for the recombination of electrons and holes in semiconductors was first treated independently by Shockley and Read[38] and by Hall.[39] Their analysis shows that the rate of recombination (U) in the steady state via a single deep level recombination center is given by:

$$U = \frac{\delta n\, p_0 + \delta p\, n_0 + \delta n \delta p}{\tau_{p0}(n_0 + \delta n + n_1) + \tau_{n0}(p_0 + \delta p + p_1)} \tag{2.66}$$

where δn and δp are the excess electron and hole concentrations, respectively, created by the external stimulus; n_0 and p_0 are the equilibrium concentrations for the electrons and holes, respectively; τ_{p0} and τ_{p0} are the minority carrier lifetimes in heavily doped N-type and P-type silicon, respectively; and n_1 and p_1 are the equilibrium electron and hole density, respectively, when the Fermi level position coincides with the recombination level position in the band gap. These concentrations are given by:

$$n_1 = N_C\, e^{(E_r - E_C)/kT} = n_i\, e^{(E_r - E_i)/kT} \tag{2.67}$$

and

$$p_1 = N_V\, e^{(E_V - E_r)/kT} = n_i\, e^{(E_i - E_r)/kT} \tag{2.68}$$

where N_C and N_V are the density of states in the conduction and valence bands, respectively; E_C, E_r, and E_V are the conduction band, recombination level, and the valence band energy levels, respectively; k is Boltzmann's constant; and T is the absolute temperature.

The recombination lifetime is defined as:

$$\tau = \frac{\delta n}{U} = \tau_{p0}\left(\frac{n_0 + n_1 + \delta n}{n_0 + p_0 + \delta n}\right) + \tau_{n0}\left(\frac{p_0 + p_1 + \delta n}{n_0 + p_0 + \delta n}\right) \tag{2.69}$$

In deriving this relationship from (2.66), the excess electron (δn) and hole (δp) concentrations are assumed to be equal in order for space charge neutrality to prevail. Since silicon power devices are usually made from N-type starting material, the rest of the treatment in this section will be confined to N-type silicon. Analogous equations can be derived for P-type silicon. In the case of N-type silicon, the electron density (n_0) is much larger than the hole density (p_0). Taking this into account together with the expressions in (2.67) and (2.68) and assuming that the density of states in the conduction and valence bands are equal, it can be shown that:

$$\frac{\tau}{\tau_{p0}} = \left[1 + \frac{1}{(1+h)} e^{(E_r - E_F)/kT}\right] + \zeta\left[\frac{h}{(1+h)} + \frac{1}{(1+h)} e^{(2E_i - E_r - E_F)/kT}\right] \tag{2.70}$$

where E_i and E_F are the intrinsic and Fermi energy level positions, respectively. In this equation, ζ is the ratio of the minority carrier lifetime in heavily doped P-type to that in heavily doped N-type silicon:

$$\zeta = \frac{\tau_{n0}}{\tau_{p0}} \tag{2.71}$$

and h is the normalized injection level given by:

$$h = \frac{\delta n}{n_0} \tag{2.72}$$

The minority carrier lifetime in heavily doped N-type and P-type material is dependent upon the capture cross-sections for holes (C_p) and electrons (C_n) at the recombination center:

$$\tau_{n0} = \frac{1}{C_n N_r} = \frac{1}{v_{Tn} \sigma_{cn} N_r} \tag{2.73}$$

and

$$\tau_{p0} = \frac{1}{C_p N_r} = \frac{1}{v_{Tp} \sigma_{cp} N_r} \tag{2.74}$$

where v_{Tn} and v_{Tp} are the thermal velocities for electrons and holes, respectively; σ_{cn} and σ_{cp} are the capture cross-sections for electrons and holes, respectively, at the recombination center; and N_r is the density of the recombination centers. Using these expressions:

$$\zeta = \frac{\tau_{n0}}{\tau_{p0}} = \frac{v_{Tp} \sigma_{cp}}{v_{Tn} \sigma_{cn}} = 0.827 \frac{\sigma_{cp}}{\sigma_{cn}} \tag{2.75}$$

it can be seen that the parameter ζ is independent of the recombination center concentration and is a weak function of temperature because the capture cross-sections do not vary significantly with temperature.

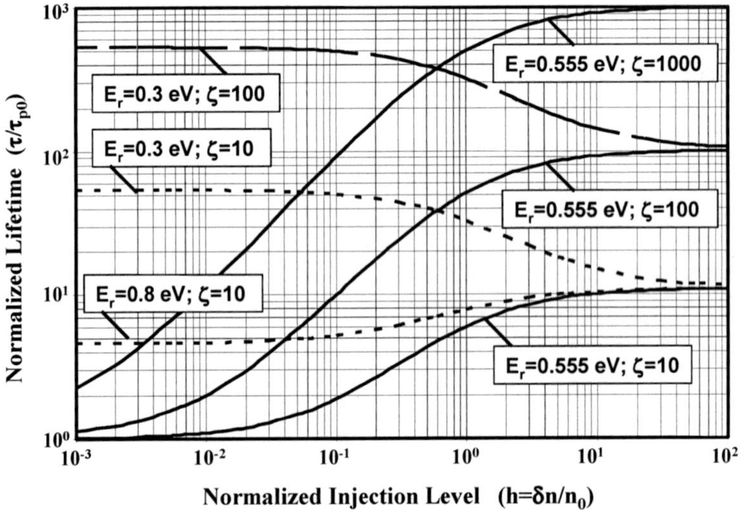

Fig. 2.26 Minority carrier lifetime in silicon

Well-developed techniques, such as deep-level transient spectroscopy, are available for the extraction of the recombination center density, the capture cross-sections, and the position of the deep level in the energy band gap. Using this data, it is possible to compute the minority carrier lifetime pertinent to any operating conditions in the semiconductor regions. As an example, the variation of the minority carrier lifetime with injection level is shown in Fig. 2.26 for N-type silicon with a doping concentration of 5×10^{13} cm^{-3} at room temperature. In this figure, a range of values for the recombination center location (E_r) and the parameter ζ have been considered.

At low injection levels ($h << 1$), the minority carrier lifetime asymptotes to a constant value, called the *low-level lifetime* (τ_{LL}), which is independent of the injection level. Using this condition in (2.70), the normalized low-level lifetime is given by:

$$\frac{\tau_{LL}}{\tau_{p0}} = \left[1 + e^{(E_r - E_F)/kT}\right] + \zeta\left[e^{(2E_i - E_r - E_F)/kT}\right] \qquad (2.76)$$

From this expression and Fig. 2.26, it can be observed that the low-level lifetime is dependent upon the recombination level position and the parameter ζ. For example, the low-level lifetime becomes larger when the recombination center moves away from either side of the mid-gap (0.555 eV) position. This is due to a reduction in the probability for capturing electrons when the recombination level is below mid-gap and a reduction in the probability for capturing holes when the recombination level is above mid-gap. When the recombination center is located below mid-gap (case of $E_r = 0.3$ eV in the figure), the low-level lifetime becomes dependent on the magnitude of the parameter ζ as well.

When the injection level becomes large ($h >> 1$), the lifetime asymptotically approaches a constant value referred to as the *high-level lifetime* (τ_{HL}). Using this condition in (2.70), the normalized high-level lifetime is given by:

$$\frac{\tau_{HL}}{\tau_{p0}} = 1 + \zeta \qquad (2.77)$$

From this expression, it can be seen that the high-level lifetime is not dependent on the position of the recombination center resulting in the lifetime approaching the same value at high injection levels in Fig. 2.26 for the cases with the same value for the parameter ζ. It is worth pointing out the lifetime can either increase or decrease with increasing injection level. As discussed later in this section, it is preferable to have an increase in the lifetime with increasing injection level, together with a large value for the parameter ζ, for bipolar power devices.

2.3.2 Low-Level Lifetime

The low-level lifetime depends upon the position of the recombination center in the energy gap and the relative capture cross-sections for holes and electrons as described by (2.76). To elucidate this dependence, an example of the variation of the low-level lifetime with recombination level position is provided in Fig. 2.27 for the case of N-type silicon with doping concentration of 5×10^{13} cm^{-3}. The low-level lifetime has its smallest value when the recombination center lies close to the mid-gap position. As the recombination center position is shifted towards the conduction band edge, the probability for capturing holes decreases, resulting in a larger value for the low-level lifetime. Similarly, the probability for capturing electrons decreases when the recombination center position is shifted towards the valence band edge, resulting in a larger value for the low-level lifetime.

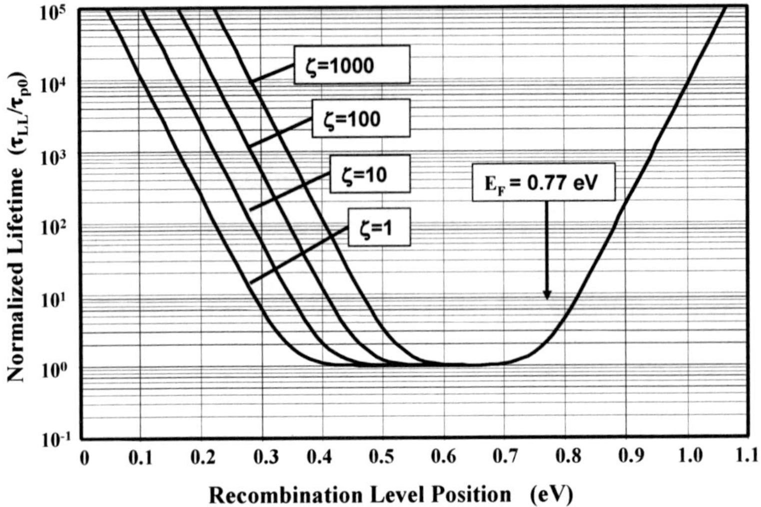

Fig. 2.27 Low-level lifetime in silicon

When the position of the recombination center is above the mid-gap location, the second term in the square brackets in (2.76) becomes small and the low-level lifetime becomes independent of the parameter ζ. In this case, the low-level lifetime increases rapidly when the recombination level position exceeds the Fermi level position. If the recombination center is located below the mid-gap position, the second term in square brackets in (2.76) becomes dominant and the low-level lifetime becomes dependent of the parameter ζ. To achieve high switching speed in bipolar power devices, it is necessary to have a rapid removal of the excess injected carriers within their drift regions by the recombination process. This requires a small value for the low-level lifetime. It can be seen from Fig. 2.27 that this is achievable by locating the recombination center near the mid-gap position. These results also indicate that in the presence of multiple recombination centers, the one located close to mid-gap will have the strongest influence on the recombination process.

2.3.3 Space-Charge Generation Lifetime

Power devices must be designed to support high voltages with small current flow in order to minimize the power dissipation in the blocking mode of operation. Although an ideal device should exhibit zero current flow in this mode of operation, in practice a finite current, referred to as the *leakage current*, is always observed. The leakage current arises either from the diffusion of free carriers, which are generated within a minority carrier diffusion length from the edge of the depletion region (referred to as the *diffusion current*) or from carriers generated within the depletion layer (referred to as the *space charge generation current*). In the case of silicon, the generation current is much larger in magnitude than the diffusion current at room temperature. At high temperatures, the diffusion current can become comparable to the generation current.

The leakage current associated with the generation of carriers within the depletion region, called the *space-charge generation current* (I_{SC}), can be derived under the assumption of a uniform generation rate within the depletion region with all the generated carriers swept out of the depletion region by the strong prevailing electric field[40]:

$$I_{SC} = qAW_D U_G \tag{2.78}$$

where q is the charge of an electron, A is the area of the P–N junction, W_D is the width of the depletion region, and U_G is the generation rate for carriers in the depletion region. The generation rate for carriers is given by:

$$U_G = \frac{n_i}{\left(\tau_{p0}\dfrac{n_1}{n_i} + \tau_{n0}\dfrac{p_1}{n_i}\right)} \tag{2.79}$$

The denominator of this expression is referred to as the *space-charge generation lifetime* (τ_{SC}). By using (2.67) and (2.68), it can be shown that:

$$\tau_{SC} = \tau_{p0}\left[e^{(E_r - E_i)/kT} + \zeta\, e^{(E_i - E_r)/kT}\right] \tag{2.80}$$

The space-charge generation current can then be obtained as:

$$I_{SC} = qAW_D \frac{n_i}{\tau_{SC}} \tag{2.81}$$

A large value for the space-charge generation lifetime is desirable in order to reduce the space-charge generation leakage current.

As in the case of the low-level lifetime, the space-charge generation lifetime is a function of the position of the recombination center in the band gap and the parameter ζ. An example of the variation of the space-charge generation lifetime with the position of the recombination center is shown in Fig. 2.28 for the case of a doping concentration of 5×10^{13} cm^{-3} in silicon. The lowest value for the space-charge generation lifetime, and consequently the highest leakage currents, occur when the recombination center is located close to the mid-gap position. From this viewpoint, it is desirable for the recombination center to be located away from mid-gap towards one of the band edges. These results also indicate that in the presence of multiple recombination centers, the one located close to mid-gap will have the strongest influence on the leakage current because of the generation of carriers within depletion regions.

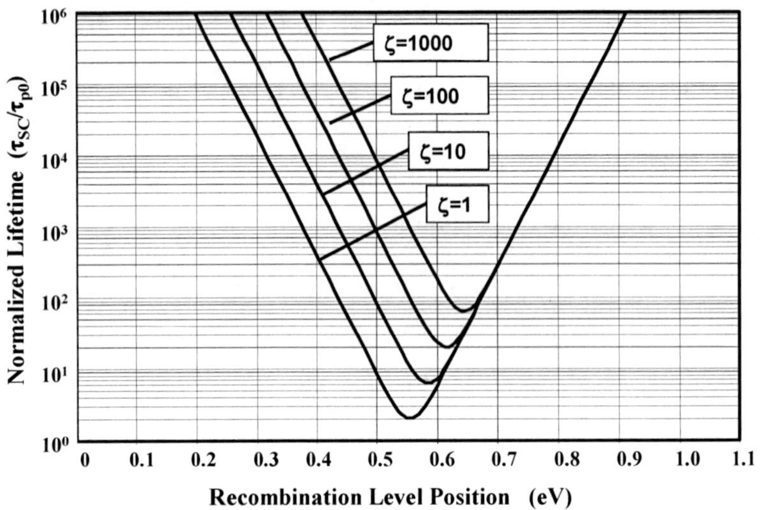

Fig. 2.28 Space-charge generation lifetime in silicon

The above analysis assumes that the leakage current is proportional to the width of the depletion layer because the generation of carriers occurs uniformly through the region. Experimental measurements on silicon devices indicate that the uniform generation of carriers occurs over only a fraction of the width of the depletion region.[41] The deviation is strongest when the applied reverse bias is small across the P–N junction and becomes small at larger reverse bias voltages. This phenomenon is usually neglected during the analysis of power devices.

2.3.4 Recombination Level Optimization

In the case of unipolar power devices, it is unnecessary to reduce the lifetime because of the absence of minority carrier injection during current conduction. Consequently, it is common practice to maintain a clean process to maximize the lifetime in order to reduce the leakage current. However, in the case of bipolar power devices that are dependent upon the substantial injection of minority carriers to reduce the resistance to on-state current flow, it becomes necessary to reduce the lifetime to achieve a fast switching speed because the injected carriers must be rapidly removed to reinstate the blocking mode.

In the previous sections, it was demonstrated that the low-level and space-charge generation lifetimes are dependent upon the position of the recombination center in the band gap as well as the parameter ζ, which is determined by the capture cross-sections for holes and electrons at the center. Empirical studies performed with the introduction of various elements into silicon, as well as its irradiation with high energy particles, have revealed a plethora of deep levels within the band gap. In principle, any of these deep levels could be utilized to reduce the lifetime to the desired value by controlling their concentration. However, some deep level impurities have been observed to provide superior power device performance.[42] This motivated the theoretical optimization of the deep level properties to obtain the best performance in bipolar power devices.[43]

Low-Level Injection Devices

In the case of bipolar power devices that operate with low-level injection conditions during current conduction, the injected minority carrier concentration is well below the doping concentration of the drift region. Here, it is desirable to obtain a small low-level lifetime in order to speed up the removal of the excess minority carriers during the process of switching the device from the on-state to the off-state. At the same time, it is desirable to maintain a large space-charge generation lifetime in order to minimize the leakage current. As demonstrated by the plots in Fig. 2.27, it is necessary to locate the recombination center close to mid-gap to achieve a small low-level lifetime. Unfortunately, as shown in Fig. 2.28, it is necessary to locate the recombination center away from mid-gap to achieve a large space-charge generation lifetime. To resolve this conflict, an optimum location for the recombination center can be derived by taking the ratio of the space-charge generation lifetime to the low-level lifetime. By using (2.76) and (2.80):

$$\frac{\tau_{SC}}{\tau_{LL}} = \frac{e^{(E_r - E_i)/kT} + \zeta\, e^{(E_i - E_r)/kT}}{1 + e^{(E_r - E_F)/kT} + \zeta\, e^{(2E_i - E_F - E_r)/kT}} \tag{2.82}$$

To maximize the performance of bipolar power devices operating in the on-state with low injection levels, it is necessary to choose the recombination center properties so that the above relationship is maximized.

To illustrate the optimization procedure, an example of the variation of the (τ_{SC}/τ_{LL}) ratio with recombination center location is shown in Fig. 2.29 for the case of silicon with doping concentration of 5×10^{13} cm^{-3}. The capture cross-section parameter ζ was assumed to have a value of 100 when making these plots. It is apparent that the ratio has its highest value when the recombination center is located near the conduction or valence band edges. When the temperature is increased from 300 to 400 K, the same trend is observed. However, the value of the ratio becomes smaller by nearly an order to magnitude, indicating that it is more difficult to minimize the leakage current at elevated temperatures when performing the optimization.

It can be observed in Fig. 2.29 that the (τ_{SC}/τ_{LL}) ratio remains high over a range of values for the recombination center location as long as it is located close to the band edges. This range is a function of the doping concentration in the silicon and the capture cross-section parameter ζ. If the range is defined to extend to 10% below the maximum value for the lifetime ratio, as indicated by the arrows in Fig. 2.29, a relationship between the optimum position of the recombination center and the doping concentration can be established.

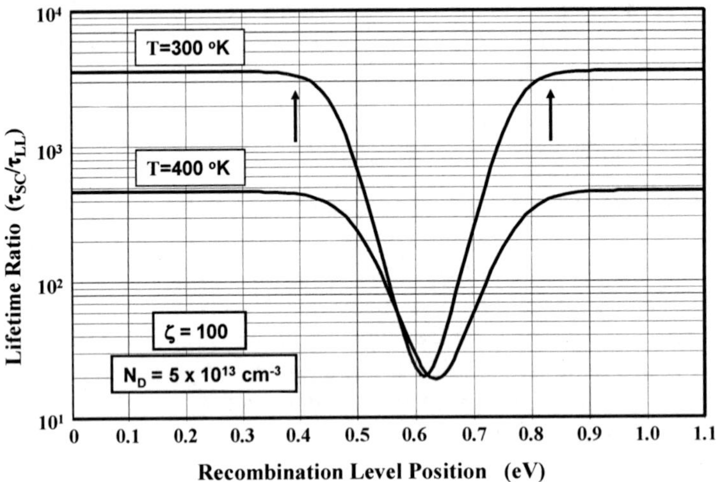

Fig. 2.29 Optimization of recombination center for silicon devices operating under low-level injection conditions in the on-state

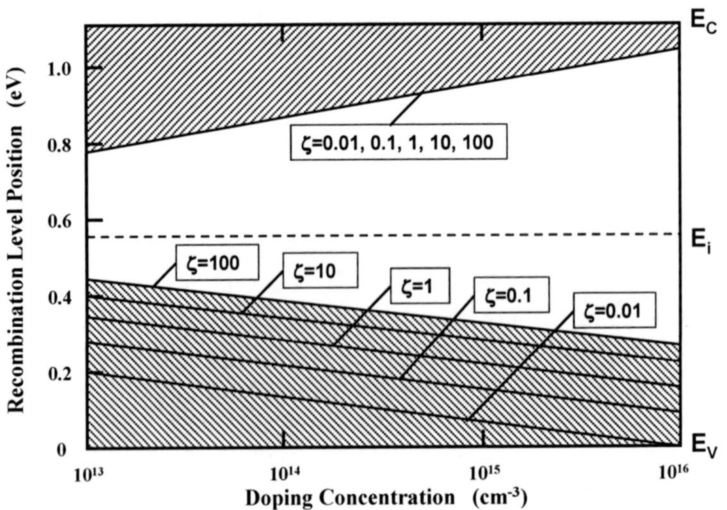

Fig. 2.30 Preferred locations for the recombination center for silicon devices operating under low-level injection conditions in the on-state

For the case of N-type silicon, the preferred locations of the recombination center within the band gap are indicated in Fig. 2.30 by the shaded regions. The shaded area in the figure located above mid-gap applies to all values of the capture cross-section parameter ζ. When the recombination center is located below mid-gap, a broader range of positions is indicated for larger values of the capture cross-section parameter ζ. For example, for a doping concentration of 1×10^{14} cm^{-3}, the recombination center must be located within 0.15 eV from the valence band edge if $\zeta = 0.01$ but can be located at upto 0.4 eV from the valence band edge if $\zeta = 100$. Further, it is worth pointing out that a broader range of positions for the recombination center is available for lightly doped silicon, providing more latitude during the selection of the lifetime controlling impurity for devices with larger breakdown voltages.

From Fig. 2.29, it can be seen that the maximum value for the (τ_{SC}/τ_{LL}) ratio occurs when the recombination center is located close to either the conduction or valence band edges. In both cases, the maximum value is a function of the doping concentration, but does not depend upon the capture cross-section parameter ζ. From (2.82) it can be shown that the maximum value for the (τ_{SC}/τ_{LL}) ratio is given by:

$$\frac{\tau_{SC}}{\tau_{LL}} = e^{(E_F - E_i)/kT} = \frac{n_0}{n_i} \tag{2.83}$$

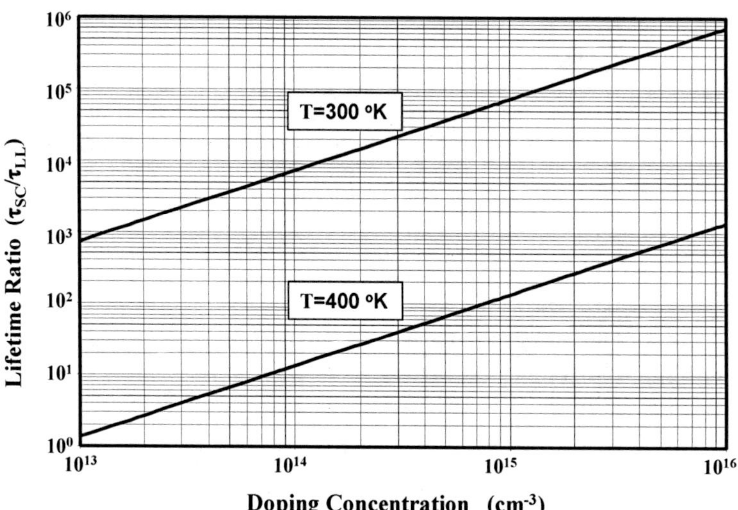

Fig. 2.31 Maximum achievable (τ_{SC}/τ_{LL}) ratio for silicon devices operating under low-level injection conditions in the on-state

The variation of the maximum achievable value for the (τ_{SC}/τ_{LL}) ratio with doping concentration is shown in Fig. 2.31 at 300 and 400 K. It is apparent that the

(τ_{SC}/τ_{LL}) ratio becomes smaller with decreasing doping concentration and increasing temperature. Consequently, it becomes more difficult to achieve a reduction of the low-level lifetime in high resistivity silicon, without a substantial increase in the leakage current, when compared with low resistivity silicon. This problem becomes aggravated as the temperature increases, a common problem due to self-heating in power devices.

High-Level Injection Devices

Most bipolar power devices are designed to operate with high-level injection conditions during current conduction so that the injected minority carrier concentration is well above the doping concentration of the drift region. This reduces the resistance of the drift region, resulting in a smaller on-state voltage drop and power loss. Here, it is desirable to obtain a large value for the high-level lifetime in order to maximize the concentration of the injected minority carriers and hence the conductivity of the drift region during current flow. At the same time, it is desirable to minimize the low-level lifetime to facilitate the rapid removal of the excess minority carriers, resulting in turning off the current in a short time to reduce power dissipation during the switching transients.

The optimization of the recombination center properties for power devices operating under high injection conditions during on-state current flow can be performed by taking the ratio of the high-level lifetime to the low-level lifetime.[43] By using (2.76) and (2.77):

$$\frac{\tau_{HL}}{\tau_{LL}} = \frac{1+\zeta}{1+e^{(E_r-E_F)/kT}+\zeta e^{(2E_i-E_F-E_r)/kT}} \tag{2.84}$$

This (τ_{HL}/τ_{LL}) ratio can be maximized with respect to the recombination level position (E_r) by setting:

$$\frac{d}{dE_r}\left(\frac{\tau_{HL}}{\tau_{LL}}\right) = 0 \tag{2.85}$$

and solving for E_r. This produces the first optimization criterion for power devices operating with high-level injection:

$$E_r = E_i + \frac{kT}{2}\ln(\zeta) \tag{2.86}$$

The optimum recombination center position is found to be independent of the doping concentration in the drift region and purely a function of the capture cross-section parameter ζ and the temperature. The optimum location for the recombination center lies above mid-gap when the cross-section parameter ζ is greater than 1 and below mid-gap when its value is less than 1, as shown by the plots in Fig. 2.32. There is a slight shift in the optimum position with temperature.

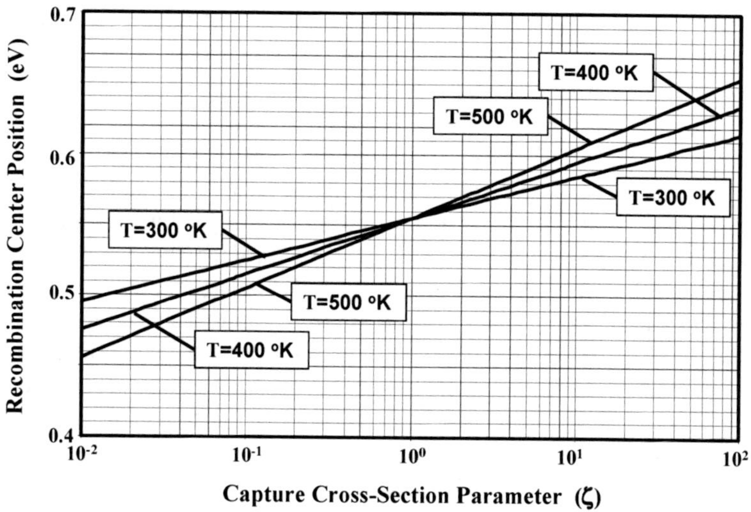

Fig. 2.32 Optimum location for the recombination center in silicon devices operating under high-level injection conditions in the on-state

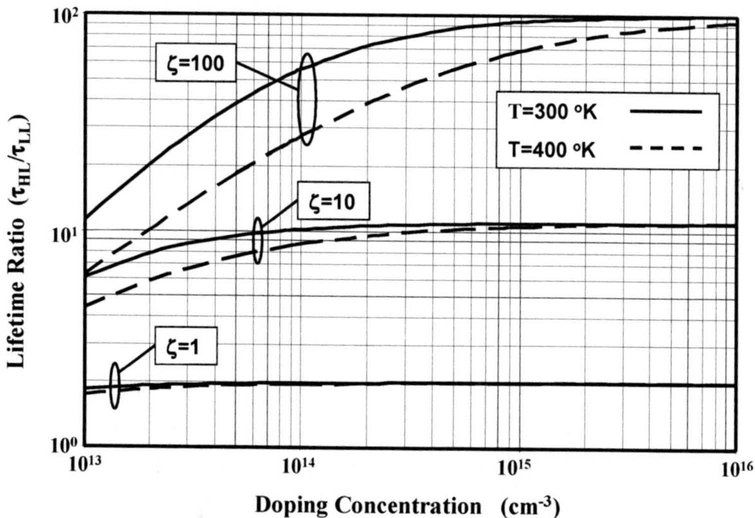

Fig. 2.33 Optimization of the recombination center in silicon devices operating under high-level injection conditions in the on-state

To achieve the highest degree of conductivity modulation of the drift region in power devices operating under high injection levels while maintaining a fast switching speed, it is necessary to obtain a large absolute value for the (τ_{HL}/τ_{LL}) ratio. For N-type silicon, this occurs when the capture cross-section parameter ζ is large. It becomes more difficult to achieve large values for the (τ_{HL}/τ_{LL}) ratio at lower doping concentrations for the drift region and at higher temperatures. This can be illustrated by plotting the (τ_{HL}/τ_{LL}) ratio as a function of doping concentration as shown in Fig. 2.33 for several values for the capture cross-section parameter ζ at 300 and 400 K for the case of a recombination center located at 0.45 eV above the valence band edge. The temperature dependence of the (τ_{HL}/τ_{LL}) ratio arises from the variation of the low-level lifetime with temperature.

For power devices operating under high injection levels during current conduction, it is desirable to achieve a rapid transition in the lifetime from low-level to high-level injection conditions. This maximizes the range of injection levels over which the high-level lifetime remains large so that the on-state voltage drop is low and simultaneously provides a small lifetime at low injection levels to speed up the turn-off process during the switching transient. The rate of change in the lifetime from low to high injection levels can be maximized by choosing a recombination center position such that:

$$\frac{\mathrm{d}}{\mathrm{d}E_{\mathrm{r}}}\left(\frac{\mathrm{d}\tau}{\mathrm{d}h}\right)=0 \tag{2.87}$$

where τ is the lifetime described by (2.70) and h is the normalized injection level. The solution to this equation happens to be identical to that given in (2.86). The optimum recombination center positions are therefore found to lie close to the center of the energy band gap as shown in Fig. 2.32. Unfortunately, such recombination centers also enhance the space-charge generation current. Consequently, the optimum recombination center location for the high-level injection case conflicts with the need to maintain low leakage currents in power devices.

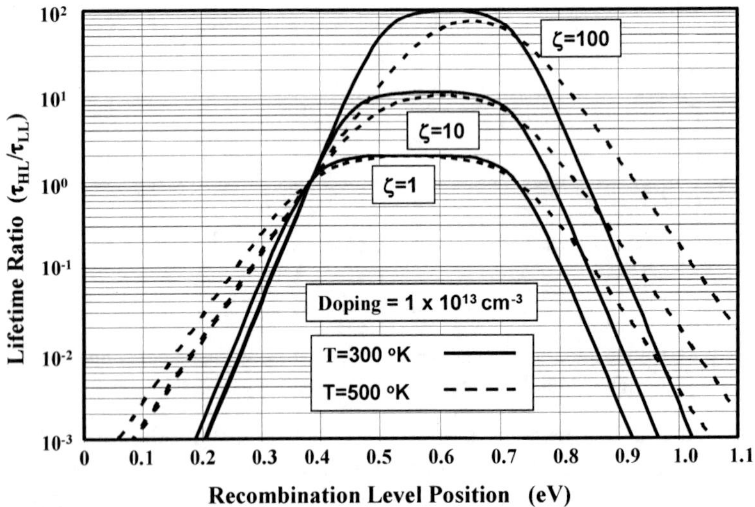

Fig. 2.34 Optimization of the recombination center in silicon devices operating under high-level injection conditions in the on-state

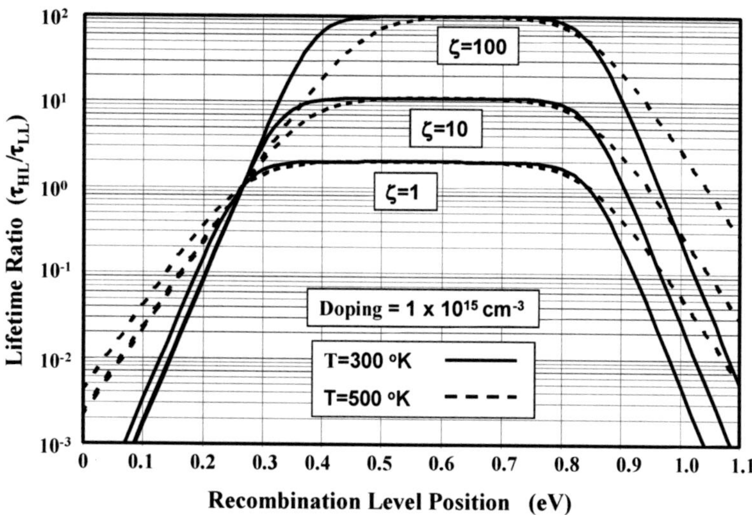

Fig. 2.35 Optimization of the recombination center in silicon devices operating under high-level injection conditions in the on-state

It is possible to resolve this conflict by a more detailed examination of the variation of the (τ_{HL}/τ_{LL}) ratio with the position of the recombination center. The variation of the (τ_{HL}/τ_{LL}) ratio with the position of the recombination center is shown in Figs. 2.34 and 2.35 for two examples of drift region doping concentrations of 1×10^{13} cm^{-3} and 1×10^{15} cm^{-3} for the case of three values for the capture cross-section parameter ζ. In the case of the higher doping level in the drift region, the (τ_{HL}/τ_{LL}) ratio exhibits a broad maxima for all three values of the capture cross-section parameter ζ. Consequently, the recombination center can be located away from the mid-gap position without suffering a significant reduction in the value for the (τ_{HL}/τ_{LL}) ratio. The leakage current can be reduced by this approach without significant impact on the on-state voltage drop. This behavior is also observed for the case of the low doping concentration at 300 K but a broad maximum is not observed at 500 K for the larger values for the capture cross-section parameter ζ.

An optimum recombination center position that will provide a large value for (τ_{HL}/τ_{LL}) ratio while minimizing the leakage current can be chosen by selecting a location where the ratio is 10% below its maximum value for those instances when a broad maxima is observed and at the peak for the ratio in those instances when a sharp maxima is observed. The optimum recombination center position obtained with this approach is shown in Fig. 2.36 as a function of the doping concentration in the drift region. It is worth pointing out that although two optimum locations for the recombination center can be defined on either side of the maxima in the above figures, only the position above the mid-gap position has been chosen because this produces the desired increase in the lifetime with injection level. The flat lines at 500 K observed at lower doping concentrations are associated with cases where there is a sharp peak in the variation of the (τ_{HL}/τ_{LL}) ratio. It can be observed that the optimum location for the recombination center moves closer to the mid-gap position as the doping concentration in the drift region is reduced. This indicates that it is more difficult to achieve a low leakage current in devices designed to support higher voltages.

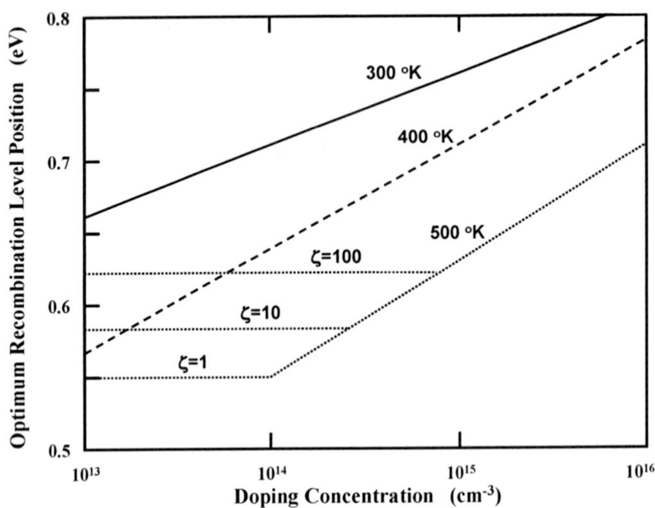

Fig. 2.36 Optimum location for the recombination center in silicon devices operating under high-level injection conditions in the on-state

Technological Constraints

The introduction of deep levels in the energy gap due to the presence of recombination centers can result in changes in the resistivity of the drift region. When the recombination center concentration approaches the doping concentration, electrons are transferred from the donor into the deep level instead of into the conduction band. The decrease in the net free electron concentration produces an increase in the resistivity, a phenomenon referred to as *compensation*. Changes in the resistivity of the drift region can produce a reduction in the breakdown voltage of power devices. It is therefore desirable to minimize the compensation effect by using the lowest concentration of recombination centers that can produce the desired reduction in the minority carrier lifetime. The minority carrier lifetime is related to the concentration of the recombination centers via the relationships in (2.73) and (2.74). The same value for the minority carrier lifetime can be achieved with a smaller value for the recombination center concentration if its capture cross-section for holes and electrons is larger.

From a practical manufacturing standpoint, it is also desirable that the same process for the introduction of the recombination centers be utilized for silicon drift regions over a broad range of resistivity. This enables the manufacturing of a variety of power products with a single lifetime control process. From this viewpoint, the recombination center properties should be chosen so that the (τ_{HL}/τ_{LL}) ratio remains high over a broad range of doping concentration levels.

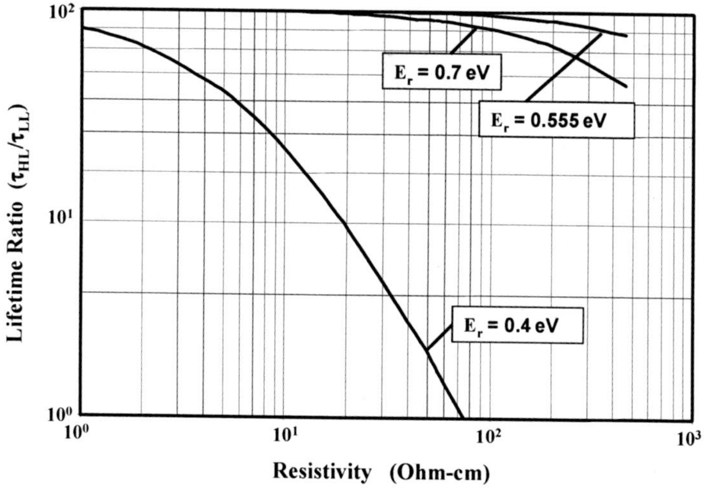

Fig. 2.37 Technological considerations for the recombination center in silicon devices operating under high-level injection conditions in the on-state

The change in the (τ_{HL}/τ_{LL}) ratio with variation of the resistivity for an N-type silicon region is shown in Fig. 2.37 for the case of a cross-section parameter ζ value of 100. In this plot, three cases for the position of the recombination center have been shown for comparison. It is obvious that there is a large reduction in the (τ_{HL}/τ_{LL}) ratio when the recombination center is located below the mid-gap position. The least reduction of the (τ_{HL}/τ_{LL}) ratio is observed when the recombination center is located at the mid-gap position. However, this is detrimental to achieving low space-charge generation currents in the blocking mode. It is therefore preferable to locate the recombination center slightly above mid-gap (as shown by the case with energy level of 0.7 eV), consistent with the optimum recombination center position discussed in the previous sections.

2.3.5 Lifetime Control

The manufacturing of power semiconductor devices is performed using pristine processing conditions to minimize the introduction of recombination centers. The as-fabricated devices have slow turn-off speed because of the long time required for the removal of the minority carriers by the recombination process. To satisfy the performance requirements of various power electronic systems, the turn-off speed of the devices is enhanced by the introduction of recombination centers into the lattice in a controlled manner. A larger quantity of recombination centers must be introduced into devices used for applications operating at higher frequencies to reduce switching losses.

Two fundamental approaches have been developed for the control of lifetime in power devices. The first method is based upon the thermal diffusion of an impurity that exhibits deep levels in the silicon band gap. The second method entails the creation of lattice damage in the form of vacancies and interstitial atoms in the silicon crystal by bombardment with high energy particles. Historically, the diffusion based process was developed well before the particle irradiation process. A large number of impurities have been found to produce deep levels in the silicon band gap.[44] Although many of these impurities are potential candidates for reducing the lifetime in silicon power devices, only gold and platinum have been extensively used for commercial devices. Similarly, although the lattice damage could be achieved by using any energetic particle, only electron and Co^{60}-Gamma beam radiation have been utilized for commercial devices because of their optimal range for absorption in silicon leading to a uniform lifetime reduction with depth.

The diffusion of gold and platinum in silicon occurs much more rapidly than dopant atoms, such as boron and phosphorus. Consequently, the introduction of the deep level impurities is performed after fabrication of the device structure but just prior to the application of the metallization. To ensure a homogeneous distribution of the deep level impurity within the wafer, it is customary to perform the diffusion at a temperature between 800 and 900°C. The concentration of the deep level impurity in the silicon lattice is adjusted by choosing the exact diffusion temperature because the solid solubility of the impurity depends upon it. A higher diffusion temperature produces a larger concentration of the deep level impurity, resulting in a lower lifetime. The wafer temperature must be rapidly reduced by quenching after the diffusion time to lock in the impurities within the lattice.

The diffusion temperature for deep level impurities exceeds the eutectic point between silicon and the aluminum metallization used for power devices. Consequently, the metal can be applied only after the introduction of the deep level impurity in the diffusion method. This implies that the devices cannot be tested prior to performing the lifetime control process. Further, small variations in the diffusion temperature can produce a large variation in the device characteristics, resulting in a poor distribution of device parameters during manufacturing. These problems can be overcome by utilizing the high energy particle radiation process.

The bombardment of silicon with high energy particles produces damage by the displacement of silicon atoms from their lattice sites. The displacement of silicon atoms from the lattice sites creates a vacancy as well as a silicon atom in the interstitial position. It has been found that the vacancy in the silicon lattice is highly mobile at room temperature. Consequently, the defects that remain after irradiation are composed of complexes of the vacancy with impurity atoms and two adjacent vacancy sites in the lattice (called a *divacancy*). In the high resistivity regions of silicon bipolar power devices, where the minority carrier recombination processes determine the switching speed, the divacancy defect has been identified to be the dominant recombination site for both p-type and n-type regions.[45,46]

For controlling the lifetime in bipolar power devices, both MeV electron irradiation[42] and Co^{60}-Gamma irradiation[47] have been shown to be practical processing techniques. There are many advantages to using these irradiation techniques: (a) the irradiation can be performed at room temperature after complete device fabrication (including metallization), allowing the initial testing of the device characteristics as well as the characteristics immediately following the irradiation; (b) the lifetime can be precisely controlled by accurately monitoring the radiation dose; (c) a tighter distribution in the device characteristics can be achieved because of improved control over the density of the recombination centers; (d) the device characteristics can be trimmed by performing the irradiation in steps with electrical testing after each step; (e) the irradiation damage can be removed by annealing the devices at 400°C, allowing the recovery of devices that may have received an overdose during the irradiation; and (f) the irradiation process is cleaner when compared with the diffusion of impurities at high temperatures, avoiding common problems with the contamination of equipment with deep level impurities such as gold.

The concentration of the recombination centers created by electron irradiation is proportional to the total flux of the high energy electrons that bombard the silicon wafer per unit area. The minority carrier lifetime is dependent on the recombination centers produced by the irradiation as well as the presence of any recombination centers prior to the radiation. The minority carrier lifetime after electron irradiation is given by:

$$\frac{1}{\tau_f} = \frac{1}{\tau_i} + K\phi \tag{2.88}$$

where τ_i is the initial lifetime prior to electron irradiation and ϕ is the total fluence of electrons per square centimeter. The constant K depends upon the energy of the electrons used to create the lattice damage. For the case of 3 MeV electrons, the constant K has a value of 10^{-8} cm^2 s^{-1}. The electron irradiation dose can be accurately monitored by the measurement of the beam current. If the initial minority carrier lifetime is large, the postirradiation lifetime can be precisely controlled. In addition, the final lifetime becomes uniform across the area of even large devices, such as thyristors and GTOs manufactured using an entire 100-mm wafer. During the fabrication of these devices, special gettering methods are employed to achieve a high lifetime in the as-fabricated devices to obtain better control and uniformity of the lifetime after electron irradiation.

Although the electron irradiation technique offers many benefits, lifetime control using gold diffusion is still used because superior on-state current conduction characteristics are observed for this method. In addition, the defects produced by electron irradiation have been found to dissociate at elevated temperatures, producing concerns regarding the long-term stability of the power device products. However, annealing studies[45] have demonstrated that, although the divacancy defect does dissociate at 300°C, the lifetime does not change appreciably because of the creation of a new recombination center. Thus, electron irradiation does not result in long-term instability under typical operating conditions for power devices.

The introduction of recombination centers by doping silicon with gold or platinum and by the electron irradiation process produces multiple deep levels in the energy band gap. The position of the deep levels is provided in Fig. 2.38 for these three cases. The diffusion of gold into silicon produces one acceptor and one donor level in the energy gap.[48] In the case of platinum, four deep levels have been identified in the energy gap,[49] while three deep levels have been observed in N-type silicon[45] in the case of electron irradiation. As explained in the previous section, the deep level closest to the mid-gap position has the strongest impact upon the minority carrier recombination process. To calculate the lifetime, it is necessary to know the capture cross-section for holes and electrons at these dominant deep levels. This data is provided in Fig. 2.39 for the three cases.

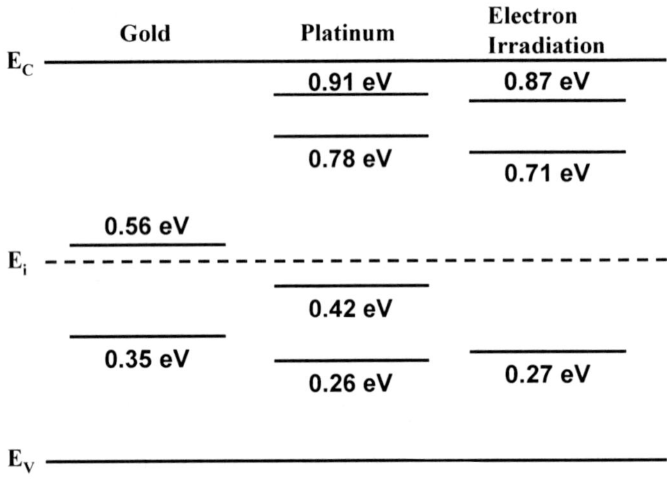

Fig. 2.38 Deep-levels produced by recombination centers in silicon

Using the data provided in the table, the (τ_{HL}/τ_{LL}) ratio can be calculated for each of the dominant levels for the three cases. The variation of the (τ_{HL}/τ_{LL}) ratio is plotted in Fig. 2.40 as a function of the resistivity for N-type silicon. It can be observed that the variation is much smaller for the case of gold and electron irradiation when compared with platinum, making these techniques more attractive for lifetime control in products with a wide range of breakdown voltages. It can also be seen that the ratio is much larger for gold when compared with electron irradiation, indicating that the use of gold will provide superior on-state characteristics in bipolar power devices. This has been experimentally confirmed for silicon P-i-N rectifiers.[42]

Impurity	Dominant Energy Level Position (eV)	Capture Cross-Section for Holes (cm^2)	Capture Cross-Section for Electrons (cm^2)	Capture Cross-Section Parameter (ζ)
Gold	0.56	6.08×10^{-15}	7.21×10^{-17}	69.7
Platinum	0.42	2.70×10^{-12}	3.20×10^{-14}	69.8
Electron Irradiation	0.71	8.66×10^{-16}	1.61×10^{-16}	4.42

Fig. 2.39 Properties of the dominant deep-levels produced by recombination centers in silicon

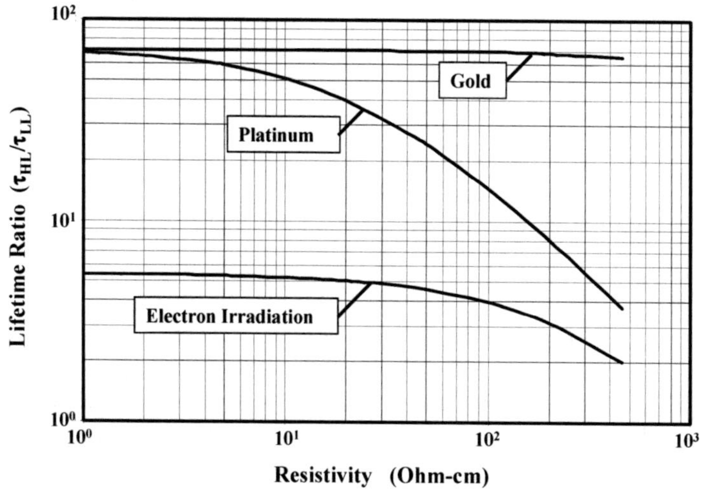

Fig. 2.40 High-level to low-level lifetime ratio for recombination centers in silicon

It is also important to consider the impact of lifetime control performed using the above three cases on the leakage current produced by space-charge generation. The (τ_{SC}/τ_{LL}) ratio is plotted in Fig. 2.41 for the three cases as a function of temperature. At room temperature, the lowest ratio is observed for gold because the recombination center lies close to the mid-gap position. The larger values for the (τ_{SC}/τ_{LL}) ratio for the case of platinum and electron irradiation indicate that devices fabricated with these methods for lifetime control will exhibit lower leakage currents. This has indeed been observed in power rectifiers fabricated using the three lifetime control processes.[42] Based upon these theoretical considerations, electron irradiation is superior to gold for lifetime control while maintaining a low leakage current. Further, the previously mentioned advantages of electron irradiations from a practical manufacturing stand point make it an attractive technology.

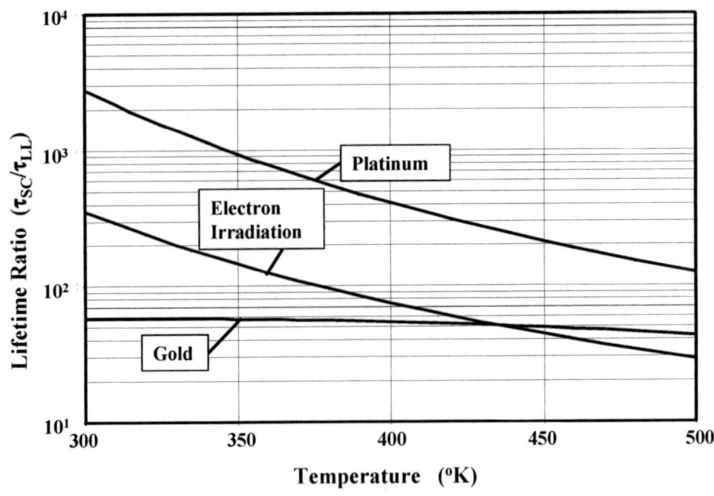

Fig. 2.41 Space-charge generation to low-level lifetime ratio for recombination centers in silicon

2.3.6 Auger Recombination

In the heavily doped N-type and P-type regions within power devices, another mechanism facilitates the recombination of the injected minority carriers. This process, called Auger recombination, occurs by the transfer of the energy and momentum released by the recombination of an electron–hole pair to a third mobile particle – electrons in the case of heavily doped N-type material and holes in the case of heavily doped P-type material. This recombination mechanism can also become important in the lightly doped drift regions within power devices when the injection level becomes very high, such as during on-state current flow under surge conditions. The simultaneous presence of a high density of both electrons and holes increases the probability for the transfer of energy to a third mobile particle in this case. Theoretical analyses[50,51] of the Auger recombination phenomenon in indirect gap semiconductors, such as silicon, indicate that the process can occur with or without phonon assistance, and may involve either direct band-to-band transitions or transitions via deep levels.

The Auger recombination process has been experimentally characterized by observing the decay of excess minority carriers at low injection levels in heavily doped N-type and P-type silicon,[52] as well as by observing the decay of excess free carriers at high injection levels in lightly doped silicon.[53] In the case of heavily doped N-type silicon, since the Auger process transfers the energy and momentum due to recombination to an electron in the conduction band, Auger recombination lifetime is given by:

$$\tau_{\mathrm{A}}^{\mathrm{N}} = \frac{1}{C_{\mathrm{AN}} n^2} = \frac{1}{2.8 \times 10^{-31} n^2} \tag{2.89}$$

where n is the concentration of electrons. In the case of heavily doped P-type silicon, since the Auger process transfers the energy and momentum because of recombination to a hole in the valence band, Auger recombination lifetime is given by:

$$\tau_{\mathrm{A}}^{\mathrm{P}} = \frac{1}{C_{\mathrm{AP}} p^2} = \frac{1}{1.0 \times 10^{-31} p^2} \tag{2.90}$$

where p is the concentration of holes. In the case of Auger recombination occurring at high injection levels with a large density of holes and electrons, the Auger recombination lifetime is given by:

$$\tau_{\mathrm{A}}^{\Delta n} = \frac{1}{C_{\mathrm{A}\Delta\mathrm{n}} \Delta n^2} = \frac{1}{3.4 \times 10^{-31} \Delta n^2} \tag{2.91}$$

where Δn is the excess carrier concentration. The Auger coefficients (C_{A}) are not the same for the three cases because of the difference in the structure of the conduction and valence bands in the energy-momentum space.

The Auger recombination lifetime calculated using the above expressions is plotted in Fig. 2.42. The plots are shown for doping levels upto 1×10^{20} cm^{-3} for the case of heavily doped regions because this is common practice while fabricating power devices. The extremely low Auger lifetime values for doping levels above 1×10^{19} cm^{-3} have an impact on the diffusion length given by:

$$L = \sqrt{D\tau} \tag{2.92}$$

where D is the diffusion coefficient given by the Einstein relationship:

$$D = \frac{kT}{q}\mu \tag{2.93}$$

where μ is the mobility for the minority carriers. A reduction in the lifetime produces enhanced recombination currents in the end regions of power devices that can degrade performance as discussed later in the book.

The Auger lifetime for the case of high-level injection in lightly doped regions is even lower than that in heavily doped regions because the Auger processes requiring two electrons and a hole and two holes and an electron can occur simultaneously because of the large density for both carriers. This plot of the Auger lifetime is shown for carrier densities of upto 5×10^{18} cm^{-3}, which can occur at very high on-state current densities in power rectifiers operating at surge current conditions.

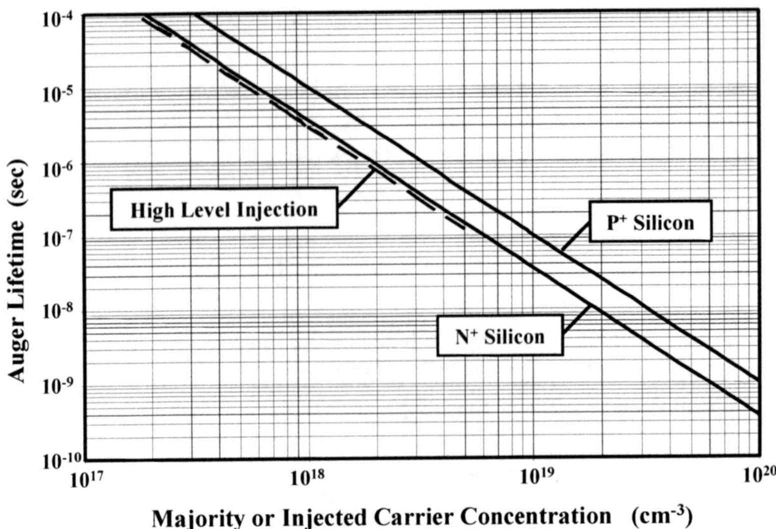

Fig. 2.42 Auger lifetime in silicon

The Auger recombination process occurs simultaneously with the Shockley–Read–Hall recombination process discussed in the previous section. The effective recombination lifetime is then given by:

$$\frac{1}{\tau_{\text{eff}}} = \frac{1}{\tau_{\text{SRH}}} + \frac{1}{\tau_{\text{A}}} \tag{2.94}$$

where τ_{SRH} and τ_{A} are the lifetimes due to the Shockley–Read–Hall and the Auger recombination processes, respectively. In the case of high voltage rectifiers and thyristors, the devices are manufactured with low defect densities to reduce the on-state voltage drop. The impact of the Auger recombination process can be substantial in this case because of the large value for the Shockley–Read–Hall lifetime. This tends to increase the on-state voltage drop for power rectifiers and thyristors under surge current conditions as discussed later in the book.

2.4 Ohmic Contacts

Contacts to power devices are made by using metal films deposited on the surface of its semiconductor layers. A common requirement is to make ohmic contacts to N-type and P-type regions. The resistance of the contacts must be made as small as possible in order to minimize the on-state voltage drop and power dissipation during current conduction. This can be achieved by using a metal-semiconductor contact with low barrier height and a high doping concentration in the semiconductor to promote tunneling current across the contact. For metal–semiconductor contacts with high doping level in the semiconductor, the contact resistance determined by the tunneling process[54] is dependent upon the barrier height (ϕ_{bn}) and the doping concentration (N_D) as given by:

$$R_c = \exp\left[\frac{2\sqrt{\varepsilon_S m^*}}{h}\left(\frac{\phi_{bn}}{\sqrt{N_D}}\right)\right] \tag{2.95}$$

where ε_S is the dielectric constant of the semiconductor, m^* is the effective mass for electrons, and h is Plank's constant. Typically, specific contact resistances of less than $1 \times 10^{-5}\ \Omega\ \text{cm}^2$ must be achieved in power device structures.

The specific contact resistance calculated using the above formula is plotted in Fig. 2.43 as a function of the doping concentration using the barrier height as a parameter. The barrier heights of metal contacts to silicon are in the range of 0.5–0.7 eV. From the figure, it can be concluded that a specific contact resistance of $1 \times 10^{-5}\ \Omega\ \text{cm}^2$ can be obtained for a doping concentration of $5 \times 10^{19}\ \text{cm}^{-3}$ if the barrier height is 0.6 eV. Such high surface doping concentrations can be achieved for N-type silicon by using ion-implantation of phosphorus. Similar values are achievable for contacts to P-type silicon by using aluminum as the metal and raising the surface doping concentration to about $5 \times 10^{19}\ \text{cm}^{-3}$ by ion-implantation of boron.

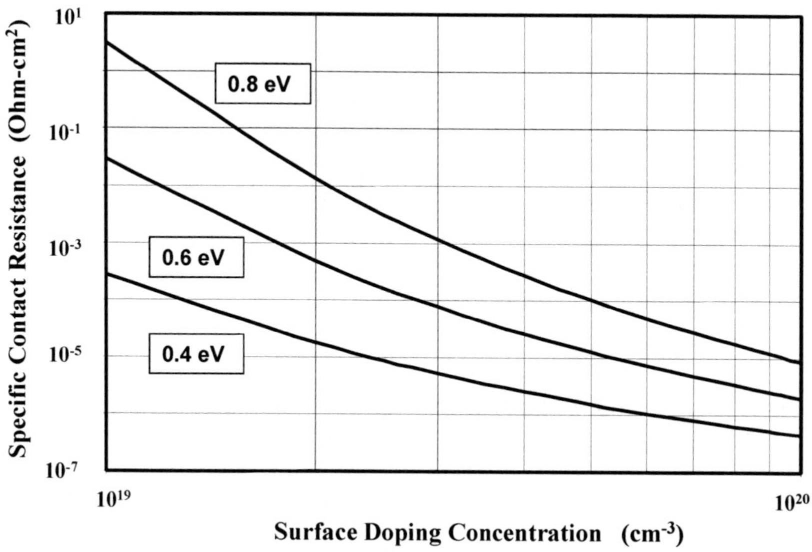

Fig. 2.43 Specific resistance at metal-semiconductor contacts

2.5 Summary

The properties of silicon and silicon carbide relevant to power devices have been reviewed in this chapter. Among semiconductor properties that are particularly important to the analysis and operation of power devices are the impact ionization coefficients, which determine the breakdown voltage capability. The carrier mobility, in the bulk and at inversion layers formed at the oxide interface, is another important parameter that is required for the analysis of unipolar and MOS-bipolar structures. For bipolar power devices, the minority carrier lifetime is a critical parameter that determines the switching speed as well as the on-state voltage drop. Since most commercially available power devices are manufactured using silicon as the base material, these properties have been treated in detail in this chapter. The basic properties required for the analysis of unipolar silicon carbide power device structures has also been provided in this chapter. For a more detailed description of the properties for silicon carbide relevant to the analysis of power devices, the reader is referred to information available in the literature.[55]

The basic properties of silicon and silicon carbide that have been described in this chapter will be extensively utilized for the analysis of the electrical characteristics of the power rectifiers and transistors in subsequent sections of the book. A thorough knowledge of these properties, including the impact of high doping concentrations and elevated temperatures, is essential to predicting the device electrical characteristics in an accurate manner.

Problems

2.1 Determine the intrinsic carrier concentration for silicon at 300, 400, and 500 K.

2.2 Determine the intrinsic carrier concentration for 4H-SiC at 300, 400, and 500 K.

2.3 Calculate the band-gap narrowing in silicon at a doping concentration of 1×10^{19} cm^{-3}.

2.4 Calculate the intrinsic carrier concentration in silicon at 300 K for a region with doping concentration of 1×10^{19} cm^{-3}.

2.5 Calculate the built-in potential for silicon at 300, 400, and 500 K using a doping concentration of 1×10^{19} cm^{-3} on the P-side and 1×10^{16} cm^{-3} on the N-side of the junction.

2.6 Calculate the built-in potential for 4H-SiC at 300, 400, and 500 K using a doping concentration of 1×10^{19} cm^{-3} on the P-side and 1×10^{16} cm^{-3} on the N-side of the junction.

2.7 Determine the impact ionization coefficients for electrons and holes in silicon at an electric field of 2×10^{5} V cm^{-1}. What is the ratio of the values?

2.8 Determine the impact ionization coefficients for holes in 4H-SiC at an electric field of 2×10^{6} V cm^{-1}.

2.9 Determine the mobility for electrons and holes in silicon at 300, 400, and 500 K for a doping concentration of 1×10^{14} cm^{-3}. What is the ratio of the values at each temperature?

2.10 Determine the mobility for electrons in 4H-SiC at 300, 400, and 500 K for a doping concentration of 1×10^{14} cm^{-3}.

2.11 Determine the mobility for electrons and holes in silicon at 300 K for doping concentrations of 1×10^{15}, 1×10^{16}, 1×10^{17}, 1×10^{18}, and 1×10^{19} cm^{-3}.

2.12 Determine the mobility for electrons in 4H-SiC at 300 K for doping concentrations of 1×10^{15}, 1×10^{16}, 1×10^{17}, 1×10^{18}, and 1×10^{19} cm^{-3}.

2.13 Calculate the velocity and average mobility for electrons in silicon at 300 K for electric field values of 1×10^{2}, 1×10^{3}, 1×10^{4}, and 1×10^{5} V cm^{-1}.

2.14 Calculate the velocity and average mobility for electrons in 4H-SiC at 300 K for electric field values of 1×10^{2}, 1×10^{3}, 1×10^{4}, 1×10^{5}, and 1×10^{6} V cm^{-1}.

2.15 Determine the mobility for electrons and holes in silicon at 300 K for injected concentrations of 1×10^{15}, 1×10^{16}, 1×10^{17}, 1×10^{18}, and 1×10^{19} cm^{-3}. Assume that high-level injection conditions prevail in the material.

2.16 Determine the intrinsic resistivity for silicon and 4H-SiC at 300 K.

2.17 Determine the resistivity for N- and P-type silicon at 300 K for doping concentrations of 1×10^{13}, 1×10^{14}, 1×10^{15}, 1×10^{16}, 1×10^{17}, 1×10^{18}, and 1×10^{19} cm^{-3}.

2.18 Determine the resistivity for N-type 4H-SiC at 300 K for doping concentrations of 1×10^{13}, 1×10^{14}, 1×10^{15}, 1×10^{16}, 1×10^{17}, 1×10^{18}, and 1×10^{19} cm^{-3}.

2.19 Determine the low-level lifetime at 300 K in N-type silicon with doping concentration of 1×10^{14} cm^{-3} for the cases of recombination centers located at an energy of 0.3, 0.5, 0.7, and 0.9 eV from the valence band. Assume a value of 1 μs for τ_{p0} and 10 for the ζ parameter.

2.20 Determine the space-charge-generation lifetime at 300 K in N-type silicon with doping concentration of 1×10^{14} cm^{-3} for the cases of recombination centers located at an energy of 0.3, 0.5, 0.7, and 0.9 eV from the valence band. Assume a value of 1 μs for τ_{p0} and 10 for the ζ parameter.

2.21 Determine the high-level lifetime at 300 K in N-type silicon with doping concentration of 1×10^{14} cm^{-3} for the cases of recombination centers located at an energy of 0.3, 0.5, 0.7, and 0.9 eV from the valence band. Assume a value of 1 μs for τ_{p0} and 10 for the ζ parameter.

2.22 Determine the Auger recombination lifetime at 300 K in N- and P-type silicon with doping concentration of 1×10^{19} cm^{-3}.

References

[1] S.M. Sze, "Physics of Semiconductor Devices," Wiley, New York, 1981.
[2] G.L. Harris, "Properties of Silicon Carbide," IEE Inspec, 1995.
[3] M. Ruff, H. Mitlehner, and R. Helbig, "SiC Devices: Physics and Numerical Simulations," IEEE Transactions on Electron Devices, Vol. ED-41, pp. 1040–1954, 1994.
[4] N.G. Wright, et al., "Electrothermal Simulation of 4H-SiC Power Devices," Silicon Carbide, III-Nitrides, and Related Materials – 1997, Material Science Forum, Vol. 264, pp. 917–920, 1998.
[5] S.K. Ghandhi, "Semiconductor Power Devices," Wiley, New York, 1977.
[6] H.P.D. Lanyon and R.A. Turf, "Bandgap Narrowing in Heavily Doped Silicon," IEEE International Electron Devices Meeting, Abstract 13.3, pp. 316–319, 1978.
[7] J.W. Slotboom and H.C. DeGraf, "Measurements of Bandgap Narrowing in Silicon Bipolar Transistors," Solid State Electronics, Vol. 19, pp. 857–862, 1976.

[8] A.W. Wieder, "Arsenic Emitter Effects," IEEE International Electron Devices Meeting, Abstract 18.7, pp. 460–462, 1978.
[9] R. Mertens and R.J. Van Overstraeten, "Measurement of the Minority Carrier Transport Parameters in Heavily Doped Silicon," IEEE International Electron Devices Meeting, Abstract 13.4, pp. 320–323, 1978.
[10] G.E. Possin, M.S. Adler, and B.J. Baliga, "Measurements of the p–n Product in Heavily Doped Epitaxial Emitters," IEEE Transactions on Electron Devices, Vol. ED-31, pp. 3–17, 1984.
[11] A.G. Chynoweth, "Ionization Rates for Electrons and Holes in Silicon," Physical Review, Vol. 109, pp.1537–1545, 1958.
[12] A.G. Chynoweth, "Uniform Silicon P–N Junctions II: Ionization rates for Electrons," Journal of Applied Physics, Vol. 31, pp 1161–1165, 1960.
[13] C.R. Crowell and S.M. Sze, "Temperature Dependence of Avalanche Multiplication in Semiconductors," Applied Physics Letters, Vol. 9, pp 242–244, 1966.
[14] R. Van Overstraeten and H. De Man, "Measurement of the Ionization Rates in Diffused Silicon P–N Junctions," Solid State Electronics, Vol. 13, pp. 583–590, 1970.
[15] R. Raghunathan and B.J. Baliga, "Temperature Dependence of Hole Impact Ionization Coefficients in 4H and 6H SiC," Solid State Electronics, Vol. 43, pp. 199–211, 1999.
[16] R. Raghunathan and B.J. Baliga, "Role of Defects in Producing Negative Temperature Dependence of Breakdown Voltage in SiC," Applied Physics Letters, Vol. 72, pp. 3196–3198, 1998.
[17] C. Canali, et al., "Electron Drift Velocity in Silicon," Physical Review, Vol. B12, pp. 2265–2284, 1975.
[18] G. Ottaviani, et al., "Hole Drift Velocity in Silicon," Physical Review, Vol. B12, pp. 3318–3329, 1975.
[19] C. Jacobini, et al., "A Review of Some Charge Transport Properties of Silicon," Solid State Electronics, Vol. 20, pp. 77–89, 1977.
[20] H. Iwata and K.M. Itoh, "Theoretical Calculation of the Electron Hall Mobility in n-type 4H- and 6H-SiC," Silicon Carbide and Related Materials – 1999, Materials Science Forum, Vol. 338–342, pp. 879–884, 2000.
[21] C. Canali, et al., "Electron and Hole Drift Velocity Measurements in Silicon," IEEE Transactions on Electron Devices, Vol. ED-22, pp. 1045–1047, 1975.
[22] C. Jacobini, et al., "A Review of Some Charge Transport Properties of Silicon," Solid State Electronics, Vol. 20, pp. 77–89, 1977.
[23] A.A. Guzev, G.L. Kurishev, and S.P. Sinista, "Scattering Mechanisms in Inversion Channels of MIS Structures in Silicon," Physica Status Solidi, Vol. A14, pp. 41–50, 1972.
[24] S.C. Sun and J.D. Plummer, "Electron Mobility in Inversion and Accumulation Layers on Thermally Oxidized Silicon Surfaces," IEEE Transactions on Electron Devices, Vol. ED-27, pp. 1497–1508, 1980.

[25] N.J. Murphy, et al., "Carrier Mobility in Silicon MOST's," Solid State Electronics, Vol. 12, pp. 775–786, 1969.
[26] O. Leistiko, et al., "Electron and Hole Mobilities in Inversion Layers on Thermally Oxidized Silicon Surfaces," IEEE Transactions on Electron Devices, Vol. ED-12, pp. 248–254, 1965.
[27] T. Sato, et al., "Mobility Anisotropy of Electrons in Inversion Layers on Oxidized Silicon Surfaces," Physical Review, Vol. B-4, pp. 1950–1960, 1971.
[28] A. Ohwada, et al., "Effect of Crystal Orientation upon Electron Mobility at the Si/SiO_2 Interface," Japanese Journal of Applied Physics, Vol. 8, pp. 629–630, 1969.
[29] S. Sridevan and B.J. Baliga, "Phonon Scattering Limited Mobility in SiC Inversion Layers," PSRC Technical Report, TR-98-03.
[30] T. Syau and B.J. Baliga, "Mobility Study on RIE Etched Silicon Surfaces Using SF_6/O_2 Gas Etchants," IEEE Transactions on Electron Devices, Vol. 40, pp. 1997–2005, 1993.
[31] T. Sato, et al., "Drift Velocity Saturation of Holes in Silicon Inversion Layers," Journal of Physical Society of Japan, Vol. 31, pp. 1846–1847, 1971.
[32] F.F. Fang and A.B. Fowler, "Hot Electron Effects and Saturation Velocities in Silicon Inversion Layers," Journal of Applied Physics, Vol. 41, pp. 1825–1831, 1970.
[33] F. Mousty, P. Osoja, and L. Passari, "Relationship between Resistivity and Phosphorus Concentration in Silicon," Journal of Applied Physics, Vol. 45, pp. 4576–4580, 1974.
[34] D.M. Caughey and R.F. Thomas, "Carrier Mobilities in Silicon Empirically Related to Doping and Field," Proceedings of the IEEE, Vol. 55, pp. 2192–2193, 1967.
[35] M. Tanenbaum and A.D. Mills, "Preparation of Uniform Resistivity N-type Silicon by Nuclear Transmutation," Journal of the Electrochemical Society, Vol. 108, pp. 171–176, 1961.
[36] H.M. Janus and O. Malmos, "Application of Thermal Neutron Irradiation for Large Scale Production of Homogeneous Phosphorus Doping of Float Zone Silicon," IEEE Transactions on Electron Devices, Vol. ED-23, pp. 797–802, 1976.
[37] B.J. Baliga and A.O. Evwaraye, "Defect Levels Controlling the Behavior of Neutron Transmutation Doped Silicon during Annealing," in 'Neutron Transmutation Doping in Semiconductors', Plenum Press, New York, 1979.
[38] W. Shockley and W.T. Read, "Statistics of the Recombination of Holes and Electrons," Physical Review, Vol. 87, pp. 835–842, 1952.
[39] R.N. Hall, "Electron-Hole Recombination in Germanium," Physical Review, Vol. 87, pp. 387–388, 1952.
[40] C.T. Sah, R.N. Noyce, and W. Shockley, "Carrier Generation and Recombination in P–N Junctions and P–N Junction Characteristics," Proceedings of the IRE, Vol. 45, pp. 1228–1243, 1957.

[41] P.U. Calzolari and S. Graffi, "A Theoretical Investigation of the Generation Current in Silicon P–N Junctions under Reverse Bias," Solid State Electronics, Vol. 15, pp. 1003–1011, 1972.

[42] B.J. Baliga and E. Sun, "Comparison of Gold, Platinum, and Electron Irradiation for Controlling Lifetime in Power Rectifiers," IEEE Transactions on Electron Devices, Vol. ED-24, pp. 1103–1108, 1977.

[43] B.J. Baliga and S. Krishna, "Optimization of Recombination Levels and Their Capture Cross-sections in Power Rectfiers and Thyristors," Solid State Electronics, Vol. 20, pp. 225–232, 1977.

[44] A.G. Milnes, "Deep Impurities in Semiconductors," Wiley, New York, 1973.

[45] A.O. Evwaraye and B.J. Baliga, "The Dominant Recombination Centers in Electron Irradiated Semiconductor Devices," Journal of the Electrochemical Society, Vol. 124, pp. 913–916, 1977.

[46] B.J. Baliga and A.O. Evwaraye, "Correlation of Lifetime with Recombination Centers in Electron Irradiated P-type Silicon," Journal of the Electrochemical Society, Vol. 130, pp. 1916–1918, 1983.

[47] R.O. Carlson, Y.S. Sun, and H.B. Assalit, "Lifetime Control in Silicon Power Devices by Electron and Gamma Irradiation," IEEE Transactions on Electron Devices, Vol. ED-24, pp. 1103–1108, 1977.

[48] J.M. Fairfield and B.V. Gokhale, "Gold as a Recombination Center in Silicon," Solid State Electronics, Vol. 8, pp. 685–691, 1965.

[49] K.P. Lisiak and A.G. Milnes, "Platinum as a Lifetime Control Deep Impurity in Silicon," Journal of Applied Physics, Vol. 46, pp. 5229–5235, 1975.

[50] L. Huldt, "Band-to-Band Auger Recombination in Indirect Gap Semiconductors," Physica Status Solidi, Vol. A8, pp. 173–187, 1971.

[51] A. Huag, "Carrier Density Dependence of Auger Recombination," Solid State Electronics, Vol. 21, pp. 1281–1284, 1978.

[52] J. Dziewior and W. Schmid, "Auger Coefficients for Highly Doped and Highly Excited Silicon," Applied Physics Letters, Vol. 31, pp. 346–348, 1977.

[53] K.G. Svantesson and N.G. Nilson, "Measurement of Auger Recombination in Silicon by Laser Excitation," Solid State Electronics, Vol. 31, pp. 1603–1608, 1978.

[54] S.M. Sze, "Physics of Semiconductor Devices," Wiley, New York, p. 304, 1981.

[55] B.J. Baliga, "Silicon Carbide Power Devices," World Scientific, Singapore, 2006.

Chapter 3

Breakdown Voltage

The most unique feature of power semiconductor devices is their ability to withstand high voltages. In transistors designed for microprocessors and semiconductor memories, the pressure to reduce their size to integrate more devices on a monolithic chip has resulted in a reduction in their operating voltage. In contrast, the desire to control larger power levels in motor drive and power distribution systems has encouraged the development of power devices with larger breakdown voltages. Typical applications for power devices were illustrated in Fig. 1.2. Depending upon the application, the breakdown voltage of devices can range from 20 to 30 V for voltage regulator modules (power supplies) used to deliver power to microprocessors in personal computers and servers to over 5,000 V for devices used in power transmission networks.

In a semiconductor, the ability to support high voltages without the onset of significant current flow is limited by the avalanche breakdown phenomenon, which is dependent on the electric field distribution within the structure. High electric fields can be created within the interior of power devices as well as at their edges. The design optimization of power devices must be performed to meet the breakdown voltage requirements for the application while minimizing the on-state voltage drop, so that the power dissipation is reduced.

Power devices are designed to support high voltages within a depletion layer formed across either a P–N junction, a metal–semiconductor (Schottky barrier) contact, or a metal–oxide–semiconductor (MOS) interface. Any electrons or holes – that enter the depletion layer either due to the space-charge generation phenomenon or by diffusion from adjacent quasineutral regions – are swept out by the electric field produced in the region by the applied voltage. As the applied voltage is increased, the electric field in the depletion region increases, resulting in acceleration of the mobile carriers to higher velocities. In the case of silicon, the mobile carriers attain a saturated drift velocity of 1×10^7 cm s^{-1} when the electric

B.J. Baliga, *Fundamentals of Power Semiconductor Devices*, doi: 10.1007/978-0-387-47314-7_3,

field exceeds 1×10^5 V cm^{-1} as discussed in Chap. 2. With further increase in the electric field, the mobile carriers gain sufficient kinetic energy from the electric field, so that their interaction with the lattice atoms produces the excitation of electrons from the valence band into the conduction band. The generation of electron–hole pairs due to energy acquired from the electric field in the semiconductor is referred to as the *impact ionization.* Since the electron–hole pairs created by impact ionization also undergo acceleration by the electric field in the depletion region, they participate in the creation of further pairs of electrons and holes. Consequently, impact ionization is a multiplicative phenomenon, which produces a cascade of mobile carriers being transported through the depletion region leading to a significant current flow through it. Since the device is unable to sustain the application of higher voltages due to a rapid increase in the current, it is considered to undergo *avalanche breakdown*. Thus, avalanche breakdown limits the maximum operating voltage for power devices.

In this chapter, the physics of avalanche breakdown is analyzed in relation to the properties of the semiconductor region that is supporting the voltage. After treating the one-dimensional junction, the edge terminations for power devices are described. Power devices require special edge terminations due to their finite area. The electric field at the edges usually becomes larger than in the middle of the device leading to a reduction of the breakdown voltage. Significant effort has been undertaken to develop a good understanding of the electric field enhancement at the edges, and methods have been proposed to mitigate the increase in the electric field. Various edge termination approaches are discussed in detail in this chapter because of their importance to maximizing the performance of power devices.

3.1 Avalanche Breakdown

The maximum voltage that can be supported by a power device before the onset of significant current flow is limited by the avalanche breakdown phenomenon. In power devices, the voltage is supported across depletion regions. As discussed in Chap. 2, mobile carriers are accelerated in the presence of a high electric field until they gain sufficient energy to create hole–electron pairs upon collision with the lattice atoms. This impact ionization process determines the current flowing through the depletion region in the presence of a large electric field. An impact ionization coefficient was defined in Chap. 2 as the number of electron–hole pairs created by a mobile carrier traversing 1 cm through the depletion region along the direction of the electric field. The impact ionization coefficients for electrons and holes are a strong function of the magnitude of the electric field as shown in Fig. 2.10.

3.1.1 Power Law Approximations for the Impact Ionization Coefficients

It is convenient to use a power law, referred to as the *Fulop's approximation*[1]:

$$\alpha_F(\text{Si}) = 1.8 \times 10^{-35} E^7 \tag{3.1}$$

for the impact ionization coefficients even though they actually increase exponentially with increasing electric field, when performing analytical derivations pertinent to the performance of silicon power devices. The impact ionization coefficient obtained by using this approximation is shown in Fig. 3.1 by the dashed line together with the impact ionization coefficient for electrons in silicon as governed by the Chynoweth's law (shown by the solid line). In the same manner, it is convenient to use the *Baliga's power law approximation*[2] for the impact ionization coefficients for 4H-SiC for analytical derivations:

$$\alpha_{\mathrm{B}}(\text{4H-SiC}) = 3.9 \times 10^{-42} E^{7}. \tag{3.2}$$

The impact ionization coefficient obtained by using this approximation is also shown in Fig. 3.1 by the dashed line together with the impact ionization coefficient for holes in 4H-SiC as governed by the Chynoweth's law (shown by a solid line).

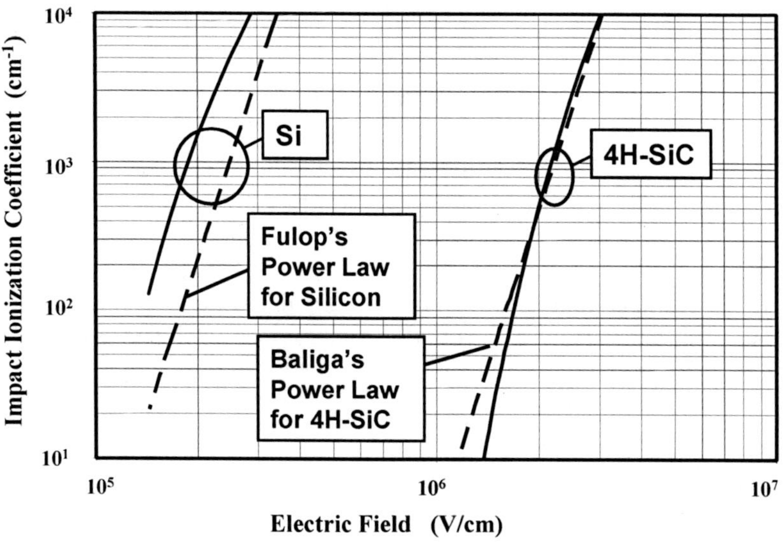

Fig. 3.1 Power law approximations for the impact ionization coefficients in silicon and 4H-SiC

During numerical simulations of power device structures, it is customary to use the Chynoweth's formula for the impact ionization coefficients. However, the power law approximations are valuable for obtaining analytical solutions for the breakdown voltage of planar one-dimensional junctions and the influence of various edge terminations on the breakdown voltage. These analytical solutions provide insight into the physics determining the breakdown phenomenon, enabling the design of improved device structures.

3.1.2 Multiplication Coefficient

The avalanche breakdown condition is defined by the impact ionization rate becoming infinite. To analyze this, consider a one-dimensional reverse-biased N^+/P junction with a depletion region extending primarily in the P-region. If an electron–hole pair is generated at a distance x from the junction, the hole will be swept toward the contact to the P-region, while the electron is simultaneously swept toward the junction with the N^+ region. If the electric field in the depletion region is large, these carriers will be accelerated until they gain sufficient energy to create electron–hole pairs during collisions with the lattice atoms. Based upon the definitions for the impact ionization coefficients, the hole will create $[\alpha_p\, dx]$ electron–hole pairs when traversing a distance dx through the depletion region. Simultaneously, the electron will create $[\alpha_n\, dx]$ electron–hole pairs when traversing a distance dx through the depletion region. The total number of electron–hole pairs created in the depletion region due to a single electron–hole pair initially generated at a distance x from the junction is given by[3,4]

$$M(x) = 1 + \int_0^x \alpha_n M(x)\mathrm{d}x + \int_x^W \alpha_p M(x)\mathrm{d}x, \tag{3.3}$$

where W is the width of the depletion layer. A solution for this equation is given by

$$M(x) = M(0)\exp\left[\int_0^x (\alpha_n - \alpha_p)\mathrm{d}x\right], \tag{3.4}$$

where $M(0)$ is the total number of electron–hole pairs at the edge of the depletion region. Using this expression in (3.3) with $x = 0$ provides a solution for $M(0)$:

$$M(0) = \left\{1 - \int_0^W \alpha_p \exp\left[\int_0^x (\alpha_n - \alpha_p)\mathrm{d}x\right]\mathrm{d}x\right\}^{-1}. \tag{3.5}$$

Using this expression in (3.4) gives

$$M(x) = \frac{\exp\left[\int_0^x (\alpha_n - \alpha_p)\mathrm{d}x\right]}{1 - \int_0^W \alpha_p \exp\left[\int_0^x (\alpha_n - \alpha_p)\mathrm{d}x\right]\mathrm{d}x}. \tag{3.6}$$

This expression for $M(x)$, referred to as the *multiplication coefficient*, allows calculation of the total number of electron–holes pairs created as a result of the generation of a single electron–hole pair at a distance x from the junction if the electric field distribution along the impact ionization path is known. The avalanche breakdown condition, defined to occur when the total number of electron–hole pairs generated within the depletion region approaches infinity, corresponds to M becoming equal to infinity. This condition is attained by setting the denominator of (3.6) to zero:

$$\left\{\int_0^W \alpha_{\mathrm{p}} \exp\left[\int_0^x (\alpha_{\mathrm{n}} - \alpha_{\mathrm{p}})\mathrm{d}x\right]\mathrm{d}x\right\} = 1. \tag{3.7}$$

The expression on the left-hand side of (3.7) is known as the *ionization integral.* During the analysis of avalanche breakdown in power devices, it is common practice to find the voltage at which the ionization integral becomes equal to unity. If the impact ionization coefficients for electrons and holes are assumed to be equal, the avalanche breakdown condition can be written as

$$\int_0^W \alpha\,\mathrm{d}x = 1. \tag{3.8}$$

This approach to the determination of the breakdown voltage is valid for power rectifiers and MOSFETs where the current flowing through the depletion region is not amplified. In devices, such as thyristors and IGBTs, the current flowing through the depletion region becomes amplified by the gain of the internal transistors. In these cases, it becomes necessary to solve for the multiplication coefficient instead of using the ionization integral.

The multiplication coefficient for a high-voltage $\mathrm{P^+/N}$ diode is given by[5]

$$M_{\mathrm{p}} = \frac{1}{1-(V/\mathrm{BV})^6}, \tag{3.9}$$

where V is the applied reverse bias voltage and BV is the breakdown voltage, while that for an $\mathrm{N^+/P}$ diode is given by

$$M_{\mathrm{n}} = \frac{1}{1-(V/\mathrm{BV})^4}. \tag{3.10}$$

Thus, the reverse current for a $\mathrm{P^+/N}$ diode approaches infinity at a faster rate with increasing voltage than for an $\mathrm{N^+/P}$ diode. This has been related to the diffusion current due to holes from the N-region in the $\mathrm{P^+/N}$ diode.

3.2 Abrupt One-Dimensional Diode

Power devices are designed to support high voltages across a depletion layer formed at either a P–N junction, a metal–semiconductor (Schottky barrier) contact, or a metal–oxide–semiconductor (MOS) interface. The onset of the avalanche breakdown condition can be analyzed for all these cases, by assuming that the voltage is supported across only one side of the structure. This holds true for an abrupt P–N junction with a very high doping concentration on one side when compared with the other side. In junctions formed with a shallow depth and a high surface concentration with a lightly doped underlying region of opposite conductivity type, the depletion region extends primarily in the lightly doped region allowing their treatment as abrupt junctions.

The analysis of a one-dimensional abrupt junction can be used to understand the design of the drift region within power devices. The case of a P^+/N junction is illustrated in Fig. 3.2 where the P^+ side is assumed to be very highly doped, so that the electric field supported within it can be neglected. When this junction is reverse biased by the application of a positive bias to the N-region, a depletion region is formed in the N-region together with the generation of a strong electric field within it that supports the voltage. The Poisson's equation for the N-region is then given by

$$\frac{d^2V}{dx^2} = -\frac{dE}{dx} = -\frac{Q(x)}{\varepsilon_S} = -\frac{qN_D}{\varepsilon_S}, \tag{3.11}$$

where $Q(x)$ is the charge within the depletion region due to the presence of ionized donors, ε_S is the dielectric constant for the semiconductor, q is the electron charge, and N_D is the donor concentration in the uniformly doped N-region.

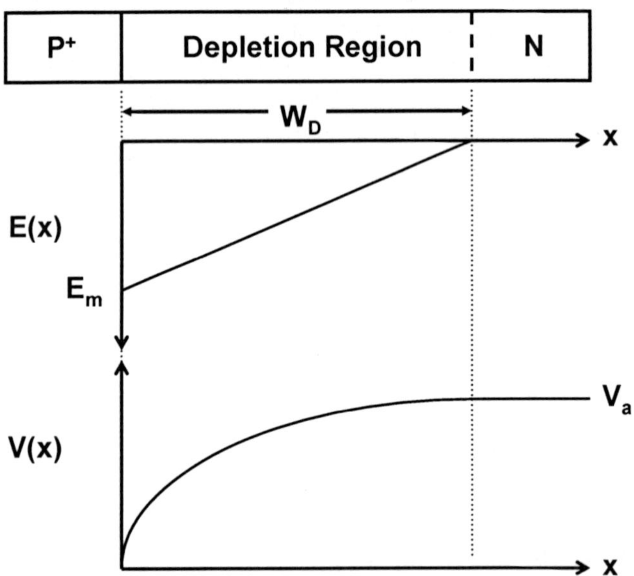

Fig. 3.2 Electric field and potential distribution for an abrupt parallel-plane P^+/N junction

Integration of (3.11) with the boundary condition that the electric field must go to zero at the edge of the depletion region (i.e., at $x = W_D$) provides the electric field distribution:

$$E(x) = -\frac{qN_D}{\varepsilon_S}(W_D - x). \tag{3.12}$$

The electric field has a maximum value of E_m at the P^+/N junction ($x = 0$) and decreases linearly to zero at $x = W_D$. Integration of the electric field distribution through the depletion region provides the potential distribution:

$$V(x) = \frac{qN_D}{\varepsilon_S}\left(W_D x - \frac{x^2}{2}\right). \tag{3.13}$$

This equation is obtained by using the boundary condition that the potential is zero at $x = 0$ within the P^+ region. The potential varies quadratically as illustrated in the figure. The thickness of the depletion region (W_D) can be related to the applied reverse bias (V_a) by using the boundary condition:

$$V(W_D) = V_a, \tag{3.14}$$

$$W_D = \sqrt{\frac{2\varepsilon_S V_a}{qN_D}}. \tag{3.15}$$

Using these equations, the maximum electric field at the junction can be obtained:

$$E_m = \sqrt{\frac{2qN_D V_a}{\varepsilon_S}}. \tag{3.16}$$

When the applied bias increases, the maximum electric field approaches values at which significant impact ionization begins to occur. The breakdown voltage is determined by the ionization integral becoming equal to unity:

$$\int_0^W \alpha\, dx = 1, \tag{3.17}$$

where α is the impact ionization coefficient discussed in Chap. 2. To obtain a closed-form solution for the breakdown voltage, it is convenient to use the power law for the impact ionization coefficient in place of Chynoweth's law. Substituting the Fulop's power law into (3.17) with the electric field distribution given by (3.12), analytical solutions for the breakdown voltage and the corresponding maximum depletion layer width can be derived for silicon:

$$BV_{PP}(\text{Si}) = 5.34 \times 10^{13} N_D^{-3/4} \tag{3.18}$$

and

$$W_{PP}(\text{Si}) = 2.67 \times 10^{10} N_D^{-7/8}. \tag{3.19}$$

In a similar manner, substituting the Baliga's power law into (3.17) with the electric field distribution given by (3.12), analytical solutions for the breakdown voltage and the corresponding maximum depletion layer width can be derived for 4H-SiC:

$$\mathrm{BV_{PP}(4H\text{-}SiC)} = 3.0 \times 10^{15} N_D^{-3/4} \tag{3.20}$$

and

$$W_{\mathrm{PP}}(\mathrm{4H\text{-}SiC}) = 1.82 \times 10^{11} N_D^{-7/8}. \tag{3.21}$$

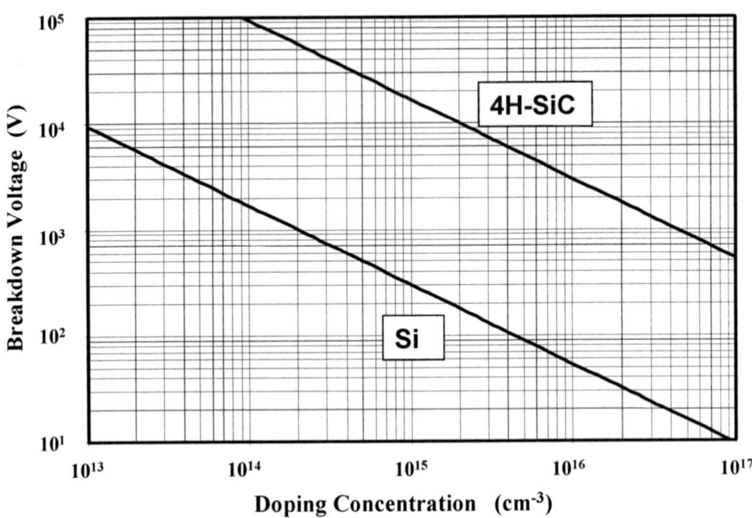

Fig. 3.3 Breakdown voltage for abrupt parallel-plane junctions in Si and 4H-SiC

The breakdown voltage is plotted in Fig. 3.3 as a function of the doping concentration on the lightly doped side of the junction. It can be seen that the breakdown voltage decreases with increasing doping concentration. It is worth pointing out that it is possible to support a much larger voltage in 4H-SiC when compared with silicon for any given doping concentration. The ratio of the breakdown voltage in 4H-SiC to that in silicon for the same doping concentration is found to be 56.2. It is also obvious from this figure that for a given breakdown voltage, it is possible to use a much higher doping concentration in the drift region for 4H-SiC devices when compared with silicon devices. The ratio of the doping concentration in the drift region for a 4H-SiC device to that for a silicon device with the same breakdown voltage is found to be 200.

The maximum depletion width reached at the onset of breakdown is shown in Fig. 3.4 for silicon and 4H-SiC. It can be seen that the thickness of the lightly doped side of the junction must be increased to support larger voltages. For the same doping concentration, the maximum depletion width in 4H-SiC is 6.8 times larger than that in silicon because it can sustain a much larger electric field. However, for a given breakdown voltage, the depletion width in 4H-SiC is smaller than for a silicon device because of the much larger doping concentration in the drift region. This smaller depletion width, in conjunction with the far larger doping concentration, results in an enormous reduction in the specific on-resistance of the drift region in 4H-SiC when compared with silicon.

The onset of the avalanche breakdown for an abrupt parallel-plane junction, as defined by the above equations, is accompanied by a maximum electric field at the junction referred to as the *critical electric field* for breakdown.

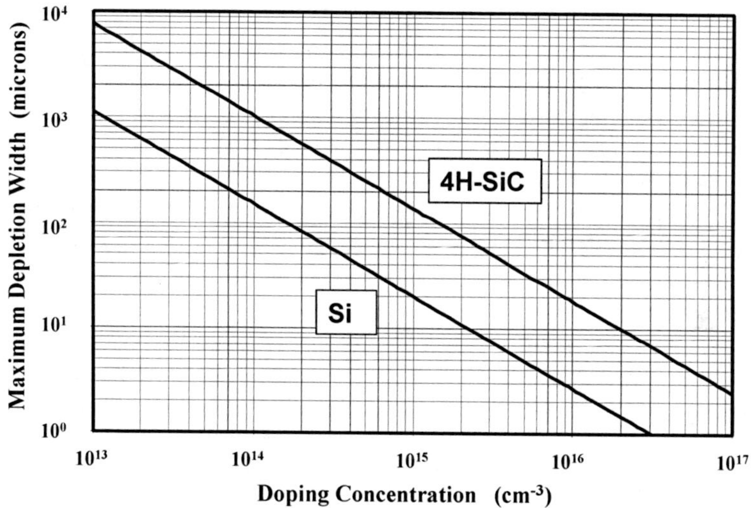

Fig. 3.4 Maximum depletion width at breakdown in Si and 4H-SiC

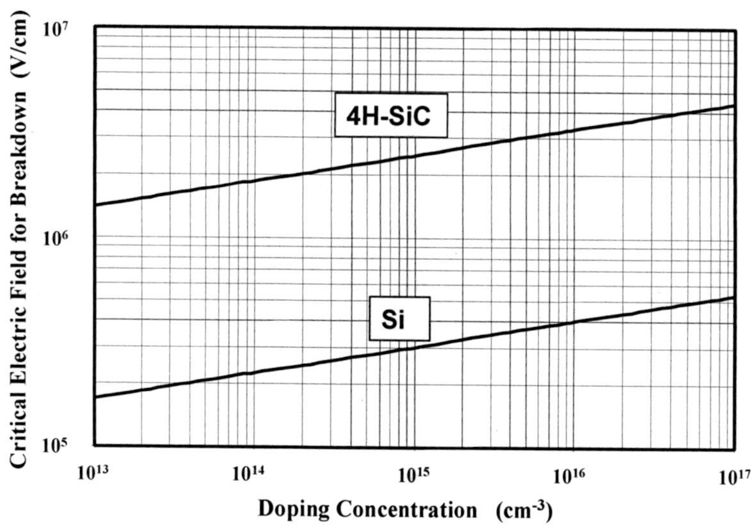

Fig. 3.5 Critical electric field for breakdown in Si and 4H-SiC

Combining (3.16) and (3.18), the critical electric field for breakdown in silicon is given by

$$E_C(\mathrm{Si}) = 4010 N_D^{1/8}, \tag{3.22}$$

while that for 4H-SiC is given by

$$E_C(\text{4H-SiC}) = 3.3\times10^4 N_D^{1/8}. \tag{3.23}$$

The critical electric field for 4H-SiC can be compared with that for silicon in Fig. 3.5. In both cases, the critical electric field is a weak function of the doping concentration. For the same doping concentration, the critical electric field in 4H-SiC is 8.2 times larger than in silicon. The larger critical electric field in 4H-SiC results in a much larger *Baliga's Figure of Merit* (see Chap. 1).

The critical electric field is a useful parameter for identifying the onset of avalanche breakdown in power device structures. Due to the very strong dependence of the impact ionization coefficients on the electric field strength, avalanche breakdown can be usually assumed to occur when the electric field within any local region of a power device approaches the critical electric field. However, it is important to note that this provides only an indication of the onset of breakdown and the exact breakdown voltage must be determined by extracting the ionization integral. This is particularly true for devices where the electric field deviates from the triangular shape pertinent to an abrupt parallel-plane junction.

3.3 Ideal Specific On-Resistance

The specific on-resistance of the drift region is related to the breakdown voltage by (1.11) which is repeated here for discussion:

$$R_{\mathrm{on,sp}} = \frac{4\mathrm{BV}^2}{\varepsilon_S \mu_n E_C^3}. \tag{3.24}$$

An accurate modeling of the specific on-resistance requires taking into account the dependence of the critical electric field and mobility on the doping concentration, which varies as the breakdown voltage is changed. It is possible to do this by computing the doping concentration for achieving a given breakdown voltage and then using the equations for the depletion width and mobility as a function of doping concentration to obtain the specific on-resistance:

$$R_{\mathrm{on,sp}} = \frac{W_{\mathrm{PP}}}{q\mu_n N_D}. \tag{3.25}$$

The specific on-resistance projected for the drift region in 4H-SiC devices by using the above method is compared with that for silicon devices in Fig. 3.6. The values for 4H-SiC are about 2,000 times smaller than for silicon devices for the same breakdown voltage. This has encouraged the development of unipolar power devices,[2] such as Schottky rectifiers and MOSFETs, from 4H-SiC.

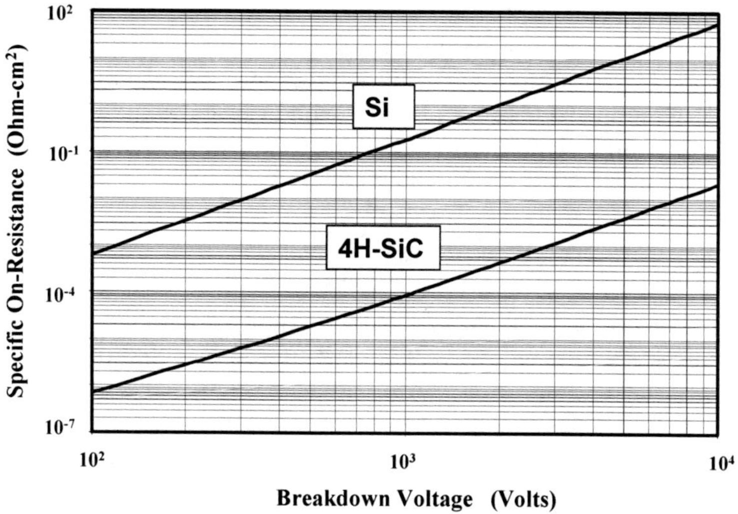

Fig. 3.6 Specific on-resistance of drift regions in 4H-SiC and silicon

3.4 Abrupt Punch-Through Diode

In the case of some power devices, such as P-i-N rectifiers, the resistance of the drift region is greatly reduced during on-state current flow by the injection of a large concentration of minority carriers. In these cases, the doping concentration of the drift region does not determine the resistance to the on-state current flow. Consequently, it is preferable to use a thinner depletion region with a reduced

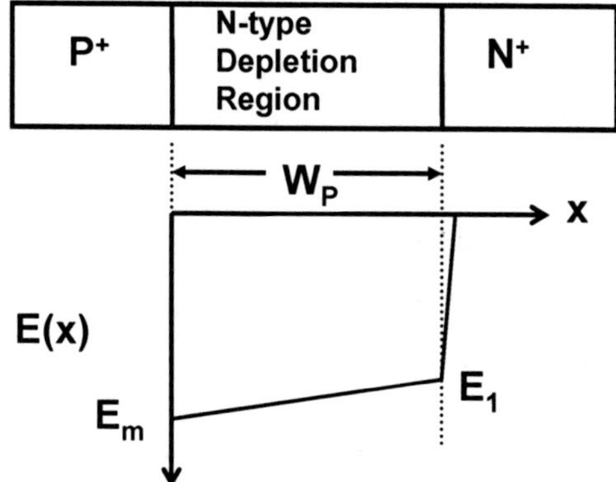

Fig. 3.7 The punch-through design for a P-i-N rectifier

doping concentration to support the voltage. This configuration for the drift region is called the *punch-through design*.

The electric field distribution for the punch-through design is shown in Fig. 3.7. In comparison with the triangular electric field distribution shown in Fig. 3.2, the electric field for the punch-through design takes a trapezoidal shape. The electric field varies more gradually through the drift region due to its lower doping concentration and then very rapidly with distance within the N^+ end region due to its very high doping concentration. The electric field at the interface between the drift region and the N^+ end region is given by

$$E_1 = E_m - \frac{qN_{DP}}{\varepsilon_S} W_P, \tag{3.26}$$

where E_m is the maximum electric field at the junction, N_{DP} is the doping concentration in the N-type drift region, and W_P is the width of the N-type drift region.

The voltage supported by the punch-through diode is given by

$$V_{PT} = \left(\frac{E_m + E_1}{2} \right) W_P \tag{3.27}$$

if the small voltage supported within the N^+ end region is neglected. The punch-through diode undergoes avalanche breakdown when the maximum electric field (E_m) becomes equal to the critical electric field (E_C) for breakdown. Using this condition in (3.27) together with the field distribution in (3.26), the breakdown voltage for the punch-through diode is given by

$$BV_{PT} = E_C W_P - \frac{qN_{DP} W_P^2}{2\varepsilon_S}. \tag{3.28}$$

The breakdown voltages calculated using this relationship are shown in Fig. 3.8 for silicon punch-through diodes with various thicknesses for the drift region. In performing these calculations, the change in the critical electric field with doping concentration was taken into account. For any doping concentration for the drift region, the breakdown voltage for the punch-through diode is reduced due to the truncation of the electric field at the N^+ end region. The breakdown voltage becomes smaller as the thickness of the drift region is reduced. From the point of view of designing the drift region for a P-i-N rectifier, it is possible to obtain a breakdown voltage of 1,000 V with a drift region thickness of about 50 μm. In contrast, a drift region thickness of 80 μm would be required in the nonpunch-through case. This reduced drift region thickness with the punch-through design is beneficial not only for reducing the on-state voltage drop but also for reducing the stored charge and consequently the reverse recovery power loss as discussed later in the book.

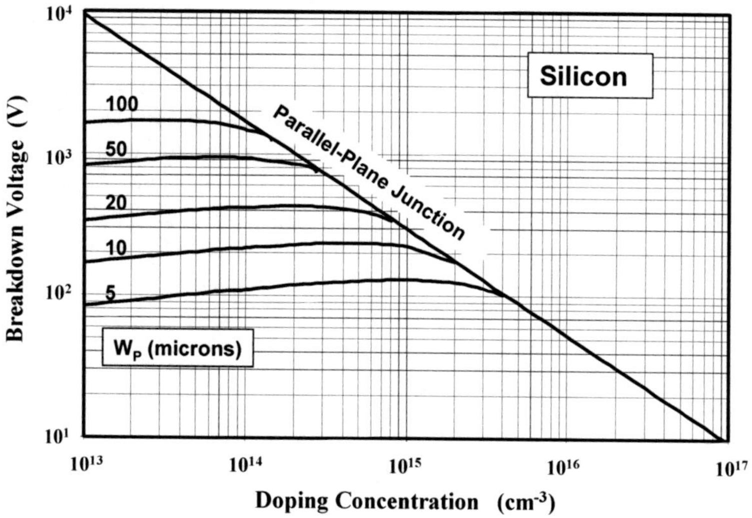

Fig. 3.8 Breakdown voltages for the silicon P-i-N diodes with punch-through design

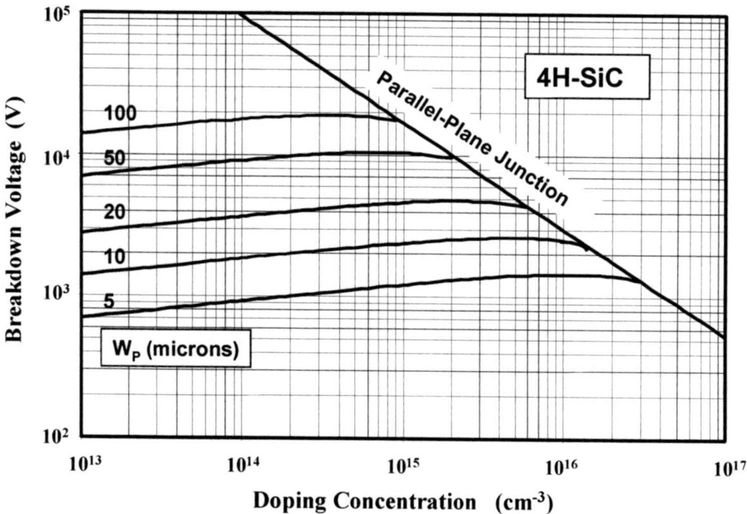

Fig. 3.9 Breakdown voltages for the 4H-SiC P-i-N diodes with punch-through design

A similar analysis for the breakdown voltages can be performed for punch-through diodes fabricated from 4H-SiC with various thicknesses for the drift region (Fig. 3.9). In performing these calculations, the change in the critical electric field with doping concentration, as described by (3.23), must be taken into account. In comparison with silicon punch-through diodes, a much higher (~10 times) doping concentration can be used in the drift region for 4H-SiC to achieve the punch-through design with a given thickness for the drift region. From the point of view of designing the drift region for a P-i-N rectifier, it is possible to obtain a breakdown voltage of 10,000 V with a drift region thickness of about 50 μm in 4H-SiC. In contrast, a drift region thickness of 80 μm would be required in the nonpunch-through case. This reduced drift region thickness with the punch-through design is beneficial for reducing the on-state voltage drop. However, the minority carrier lifetime in 4H-SiC has been found to be low resulting in poor conductivity modulation of the drift region. It is therefore advisable to maintain a high doping concentration in the drift region for P-i-N rectifiers fabricated from 4H-SiC.

3.5 Linearly Graded Junction Diode

Power devices fabricated using junctions with high surface doping concentration and shallow thickness tend to behave like the abrupt junction diodes that were discussed in the previous sections. Power devices, such as thyristors, that are designed to support very high voltages (above 2,000 V) rely upon junctions with low surface concentration and large depth to enhance the blocking voltage capability. In addition, power devices with low (<50 V) blocking voltages, such as low-voltage power MOSFETs, require drift regions with relatively high doping concentrations that are comparable with the doping level on the diffused side of the

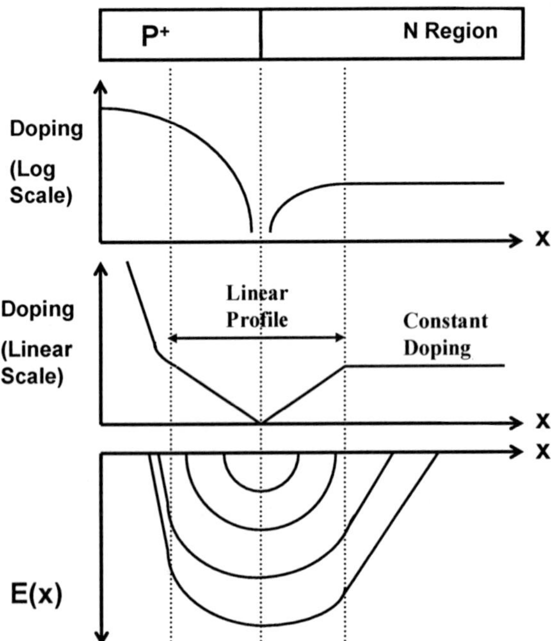

Fig. 3.10 The diffused junction diode

junction. A significant fraction of the reverse bias voltage is supported within the diffused side of the junction in this case as well.

These types of junctions can be analyzed by assuming a linearly graded doping profile in the vicinity of the junction. A typical doping profile for a diffused junction diode is illustrated in Fig. 3.10. For diffused junctions, it is customary to plot the profile with the doping concentration displayed using a logarithmic scale as shown in the upper part of the figure. Due to the compensation of the N-region by the P-type dopant in the vicinity of the junction, the profile has a linear net doping distribution as illustrated in the lower portion of the figure. The diffused junction can therefore be treated as a combination of a linearly graded junction and a uniformly doped junction.

If the linear doping grading is sufficiently steep, the maximum electric field at the junction can reach the critical electric field with the depletion region confined to this portion of the doping profile. The linearly graded junction is illustrated in Fig. 3.11 together with the electric field and potential distributions. Note that the depletion region extends to both side of the metallurgical junction by a distance W. With a positive voltage applied to the N-region, the junction becomes reverse biased with a net negative charge on the P-side due to the ionized acceptors having a greater concentration than the donors, while a net positive charge develops on the N-side due to the ionized donors having a greater concentration than the acceptors. The concentration of the net charge varies linearly with distance with a grade constant G.

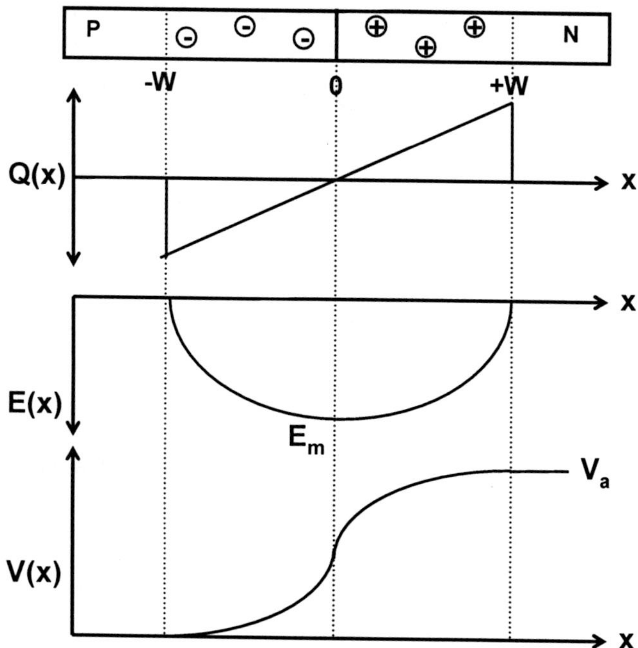

Fig. 3.11 The linearly graded junction diode

The breakdown voltage of this linearly graded junction can be analyzed by using the following charge distribution profile in Poisson's equation:

$$Q(x) = qGx. \tag{3.29}$$

Applying this charge distribution to the Poisson's equation gives

$$\frac{d^2V}{dx} = -\frac{dE}{dx} = -\frac{Q(x)}{\varepsilon_S} = \frac{-qGx}{\varepsilon_S}. \tag{3.30}$$

Integration of this equation with the boundary condition that the electric field must be zero at the edge of the depletion region ($x = W$) provides the electric field distribution:

$$E(x) = \frac{qG}{2\varepsilon_S}(x^2 - W^2). \tag{3.31}$$

The electric field varies parabolically with distance with its maximum value at the junction given by

$$E_m = \frac{qGW^2}{2\varepsilon_S}. \tag{3.32}$$

Integration of the electric field distribution through the depletion region with the boundary conditions that the potential is zero at $x = -W$ on the P-side of the junction yields

$$V(x) = \frac{qG}{\varepsilon_S}\left(\frac{W^3}{3} + \frac{W^2x}{2} - \frac{x^3}{6}\right). \tag{3.33}$$

This voltage distribution is shown at the bottom of Fig. 3.11. The depletion layer width (W) on both sides of the junction can be obtained by using the boundary condition that the voltage on the N-side of the junction is equal to the applied bias (V_a):

$$W = \left(\frac{3\varepsilon_S V_a}{qG}\right)^{1/3}. \tag{3.34}$$

In the case of devices, such as power MOSFETs, the extension of the depletion layer on the diffused side of the junction can lead to reach-thorough breakdown at well below the avalanche breakdown voltage. The depletion width calculated by using (3.34), based upon approximation of the diffused junction by a linearly graded junction, provides an analytical approach to designing the width of the P-base region.

A closed-form analytical solution for the breakdown voltage of the linearly graded junction can be obtained by determination of the voltage at which the impact ionization integral becomes equal to unity. Using the ionization integral given by (3.17) with Fulop's approximation for the impact ionization coefficients and the electric field distribution given by (3.31),

$$\int_{-W}^{W} 1.8\times10^{-35}\left[\frac{qG}{2\varepsilon_S}(x^2-W^2)\right]^7 \mathrm{d}x = 1. \tag{3.35}$$

The solution for this equation provides the depletion width at the point of breakdown for the linearly graded junction:

$$W_{CL} = 9.1\times10^5 G^{-7/15}. \tag{3.36}$$

Using this depletion width in (3.34), the breakdown voltage for the linearly graded junction is found to be given by

$$BV_L = 9.2\times10^9 G^{-2/5}. \tag{3.37}$$

As illustrated by the electric field distribution in Fig. 3.10, the diffused junction diode usually behaves as a combination of a linearly graded junction and an abrupt junction with uniform doping on the lightly doped side. The extension of the depletion region into the diffused side of the junction enhances the breakdown voltage to above that derived earlier for the abrupt parallel-plane junction because of the additional voltage that is supported on the diffused side of the junction. This can be taken advantage of during the design of low-voltage (<30 V) power MOSFETs.

3.6 Edge Terminations

All semiconductor devices have a finite size, which is achieved by sawing through the wafers to produce the chips that go into packages. The sawing of wafers, performed by using diamond-coated blades, produces severe damage to the crystal. In the case of power devices, if the sawing is performed through the junction that must support a high voltage, the crystal damage creates a high leakage current that degrades the breakdown voltage and its stability with respect to time. This problem can be addressed by using special junction terminations around the edges of the power devices, so that the depletion regions of the high-voltage junctions do not intersect with the saw lanes where the damage is located. Another approach that can be used to control and preserve a high breakdown voltage is by shaping the surface of the edges of the device. The earliest method for shaping the edges was by mesa etching. Subsequently, the beveling of the edges of wafers was found to be very effective in preserving the breakdown voltage of high-voltage power rectifiers and thyristors. With the widespread availability of ion implantation for the fabrication of power devices in the 1980s, the use of a lightly doped zone at the edges of junctions has been found to be effective in achieving high breakdown voltages. These methods for enhancing the breakdown voltage of power devices are discussed in this section.

3.6.1 Planar Junction Termination

A cornerstone of modern semiconductor devices is the planar junction formed by the diffusion of impurities through a window in a silicon dioxide mask grown on the silicon surface. Consider a rectangular window etched in a silicon dioxide masking layer on an N-type silicon substrate as illustrated in the upper part of Fig. 3.12. The dopant can be diffused into the silicon surface exposed within the window by dopants introduced via the vapor phase or by low energy ion implantation. It is customary to thermally drive the dopant atoms into the silicon at elevated temperatures to produce junction depths appropriate for power devices. During this drive-in process, the dopants migrate vertically downward within the diffusion window to produce a parallel-plane junction. However, at the edges of diffusion window, the dopants migrate laterally under the silicon dioxide while being driven downward. If the lateral diffusion is assumed to be equal to the junction depth, this process produces a cylindrical-shaped junction at the edges of the diffusion window as illustrated at the bottom of the figure. The breakdown voltage of the planar junction is reduced by the presence of this junction curvature.[6]

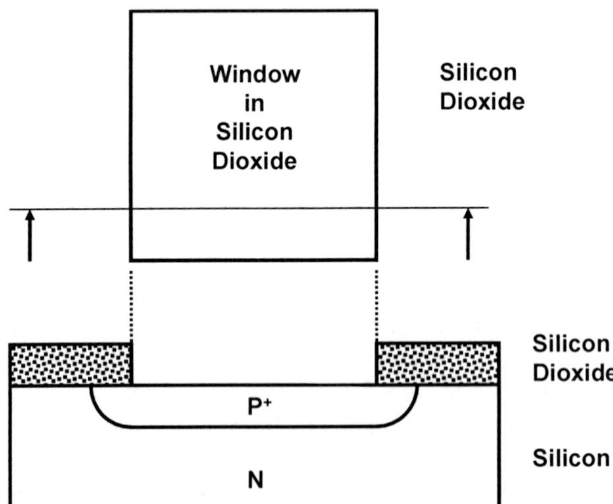

Fig. 3.12 The planar junction obtained by diffusion through a window in a silicon dioxide mask

At the corners of the rectangular diffusion window, a sharp point is formed. The dopants are driven in three dimensions away from this corner, producing a junction which is one-eight of a spherical surface. An even greater electric field enhancement occurs due to the presence of these spherical junctions at the four corners of the rectangular window. The breakdown voltage at these locations is even lower than that at the edges of the window where the cylindrical junctions are located. Lateral diffusion in silicon under a silicon dioxide mask has been found to be 85% of the vertical depth. These junctions can be approximated as either cylindrical or spherical junctions to derive analytical solutions to the breakdown voltage.

Cylindrical Junction

A cross section of the cylindrical junction is shown in Fig. 3.13 with junction depth of r_J. If the P^+ diffused side of the junction is assumed to have a very high doping concentration when compared with the N-type substrate, the depletion layer will extend only on the N-side to a depth r_D as shown in the figure. The analysis of the breakdown of this structure has been performed by numerical simulations.[7] The breakdown voltage for the cylindrical junction can also be analytically obtained by solving Poisson's equation in cylindrical coordinates[8]:

$$\frac{1}{r}\frac{\mathrm{d}}{\mathrm{d}r}\left(r\frac{\mathrm{d}V}{\mathrm{d}r}\right)=-\frac{1}{r}\frac{\mathrm{d}}{\mathrm{d}r}(rE)=-\frac{Q(r)}{\varepsilon_S}=-\frac{qN_D}{\varepsilon_S}, \tag{3.38}$$

where the potential $V(r)$ and electric field $E(r)$ are defined along the radius vector r extending into the depletion region as shown in Fig. 3.13. Integration of this equation with the boundary condition that the electric field must be zero at the depletion region boundary (r_D) in the N-type region provides the electric field distribution:

$$E(r)=\frac{qN_D}{2\varepsilon_S}\left(\frac{r^2-r_D^2}{r}\right). \tag{3.39}$$

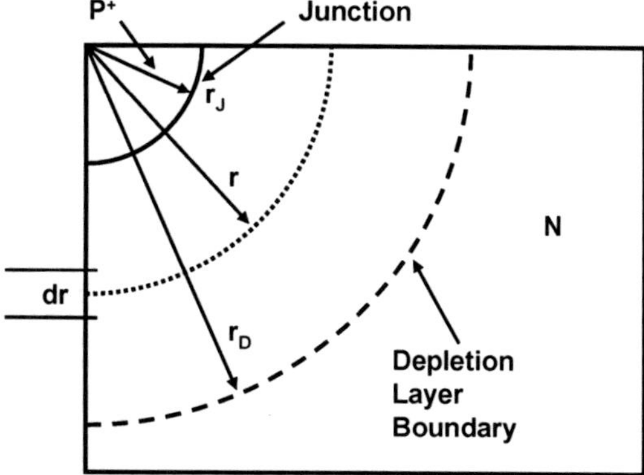

Fig. 3.13 The cylindrical junction

As in the case of the parallel-plane junction, the maximum value for the electric field occurs at the metallurgical junction located at $r = r_J$:

$$E_{\text{m,CYL}}(r_{\text{J}}) = \frac{qN_{\text{D}}}{2\varepsilon_{\text{S}}}\left(\frac{r_{\text{J}}^2 - r_{\text{D}}^2}{r_{\text{J}}}\right). \tag{3.40}$$

The maximum electric field generated in the cylindrical junction is significantly larger than that observed in the parallel-plane case. This can be demonstrated by considering the case of a cylindrical junction whose radius of curvature (r_{J}) is small when compared with the depletion width and consequently the radius r_{D}. Since r_{J} is much smaller than r_{D} in this case, the maximum electric field is given by

$$E_{\text{m,CYL}}(r_{\text{J}}) = -\frac{qN_{\text{D}}r_{\text{D}}^2}{2\varepsilon_{\text{S}}r_{\text{J}}}. \tag{3.41}$$

The maximum electric field for the parallel-plane junction is given by (3.12) with $x = 0$:

$$E_{\text{m,PP}} = -\frac{qN_{\text{D}}W_{\text{D}}}{\varepsilon_{\text{S}}}. \tag{3.42}$$

If the depletion widths for the two cases are assumed to be approximately equal at the same reverse bias voltage (i.e., $r_{\text{D}} = W_{\text{D}}$), then the enhancement of the electric field at the cylindrical junction can be obtained by taking the ratio of the maximum electric field for the two cases:

$$\frac{E_{\text{m,CYL}}}{E_{\text{m,PP}}} = \frac{r_{\text{D}}}{2r_{\text{J}}}. \tag{3.43}$$

From this equation, it can be concluded that, for shallow junctions with small radii of curvature, the maximum electric field is significantly larger than for the parallel-plane case. For example, if the junction depth is 1 μm and the depletion region has a thickness of 30 μm, the maximum electric field at the cylindrical junction will be 15 times larger than that for the parallel-plane case. Since impact ionization is a very strong function of the electric field, avalanche breakdown will occur at a lower voltage for the cylindrical junction than for a parallel-plane junction with the same doping concentration on the lightly doped side.

The potential distribution for the cylindrical junction, obtained by integration of the electric field distribution, is given by

$$V(r) = \frac{qN_{\text{D}}}{2\varepsilon_{\text{S}}}\left[\left(\frac{r^2 - r_{\text{J}}^2}{2}\right) + r_{\text{D}}^2 \ln\left(\frac{r}{r_{\text{J}}}\right)\right]. \tag{3.44}$$

The width of the depletion layer for the cylindrical junction can be obtained by using the boundary conditions for the voltage, namely, the voltage being equal to zero on the highly doped side and V_{a} in the lightly doped side. The breakdown voltage for the cylindrical junction can be obtained by performing the ionization integral using the electric field distribution given by (3.39). To obtain a closed-form solution for the ionization integral, it is convenient to make the approximation that the electric field varies inversely with distance from the junction:

$$E(r) = -\frac{qN_D}{2\varepsilon_S}\frac{r_D^2}{r} = -\frac{K_C}{r}. \tag{3.45}$$

The electric field distribution obtained by using this hyperbolic approximation is compared with that given by (3.39) in Fig. 3.14 for the case of $r_J = 0.1r_D$. The hyperbolic approximation provides a good fit to the exact case in the vicinity of the junction where the electric field is large. Since the impact ionization coefficients are a very strong function of the electric field, the approximation is satisfactory for the evaluation of the ionization integral. It is worth pointing out that the hyperbolic approximation implies that the electric field distribution extends to an infinite distance from the junction. Consequently, the ionization integral must also be performed to infinity when using the hyperbolic approximation for the electric field.

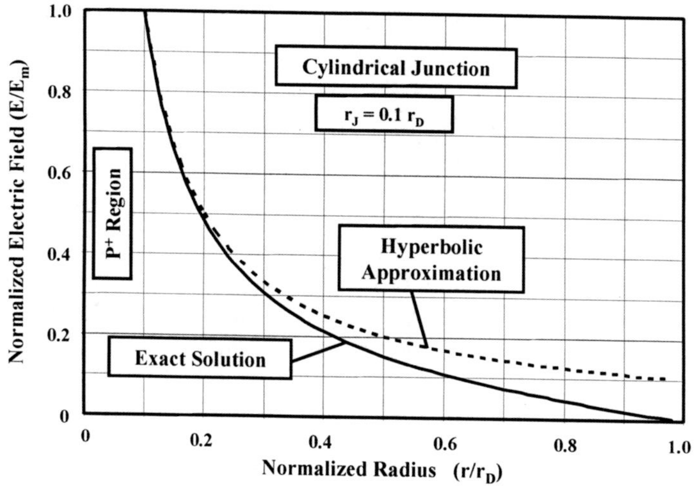

Fig. 3.14 Comparison of the electric field distribution for the hyperbolic approximation with the exact case for a cylindrical junction

Evaluation of the ionization integral using the hyperbolic variation of the electric field together with Fulop's law for the ionization coefficient yields a solution for the breakdown condition for the cylindrical junction:

$$K_C = \left(\frac{6r_J^6}{1.8\times10^{-35}}\right)^{1/7} = \frac{qN_D r_D^2}{2\varepsilon_S}. \tag{3.46}$$

By combining this condition for breakdown with (3.41) for the maximum electric field, the critical electric field for breakdown at the cylindrical junction is obtained:

$$E_{C,CYL} = \left(\frac{3.25 \times 10^{35}}{r_J} \right)^{1/7}. \tag{3.47}$$

The critical electric field for breakdown in the case of cylindrical junctions can be compared with the critical electric field for the parallel-plane junction by taking their ratio:

$$\frac{E_{C,CYL}}{E_{C,PP}} = \left(\frac{3W_{PP}}{4r_J} \right)^{1/7}. \tag{3.48}$$

In deriving this relationship, the critical electric field ($E_{C,PP}$) for the parallel-plane case was related to the depletion width (W_{PP}) by using (3.19) and (3.22). Since the radius of curvature of the junction is assumed to be small when compared with the depletion layer thickness, the above relationship indicates that the critical electric field for breakdown in the cylindrical junction is larger than that for the parallel-plane junction. This difference is associated with the high electric field being located over a shorter distance in the case of the cylindrical junction when compared with the parallel-plane junction.

The breakdown voltage for the cylindrical junction can be obtained by using $r = r_D$ in (3.44) with the value for r_D defined by the breakdown condition as governed by (3.46). To generalize the solution to represent a wide variety of junctions with different doping concentrations on the lightly doped side, it is convenient to normalize the breakdown voltage of the cylindrical junction to that for the parallel-plane case. To obtain this generalized solution, it is also convenient to normalize the radius of curvature of the junction to the depletion layer thickness at breakdown for the parallel-plane junction. This methodology provides the normalized breakdown voltage for the cylindrical junction:

$$\frac{BV_{CYL}}{BV_{PP}} = \frac{1}{2}\left[\left(\frac{r_J}{W_{PP}} \right)^2 + 2\left(\frac{r_J}{W_{PP}} \right)^{6/7} \right] \ln\left[1 + 2\left(\frac{W_{PP}}{r_J} \right)^{8/7} \right] - \left(\frac{r_J}{W_{PP}} \right)^{6/7}. \tag{3.49}$$

It is worth pointing out that this relationship was derived under the assumption that the radius of curvature of the junction is small when compared with the depletion layer thickness (i.e., $r_J / W_{PP} \ll 1$). Since power devices are usually fabricated with shallow junctions when compared with the large depletion layer widths required to support high voltages, (3.49) is usually valid for their analysis.

The normalized breakdown voltage for cylindrical junctions as predicted by (3.49) is plotted in Fig. 3.15 as a function of the normalized radius of curvature. This graph is valid for junctions fabricated from any doping concentration on the lightly doped side as long as the radius of curvature is small when compared with the depletion layer width at breakdown for the parallel-plane junction with the same doping concentration on the lightly doped side. It can be seen that the breakdown voltage for the cylindrical junction increases when the radius of curvature (or junction depth) is increased. Shallower junction can be used for power devices with lower breakdown voltages due to the relatively high doping concentration and small depletion layer widths in the drift region. Power devices with high breakdown voltages require large junction depths to reduce the degradation of the breakdown voltage due to junction curvature. As an example, if a junction depth of 3 μm is used with a background doping concentration of 1×10^{15} cm^{-3} corresponding to a depletion layer width of 20 μm, the normalized breakdown voltage is 40% of the breakdown voltage for the parallel-plane case. However, if the doping concentration of the drift region is reduced to 4×10^{14} cm^{-3} corresponding to a depletion layer width of 60 μm, the normalized breakdown voltage is reduced to only 25% of the breakdown voltage for the parallel-plane case. In practice, it is impractical to obtain a normalized radius of curvature of more than 0.4, making it difficult to raise the normalized breakdown voltage for the cylindrical junction to above 50% of the parallel-plane case. It is common practice to incorporate floating field rings and field plates to enhance the breakdown voltage of cylindrical junctions as discussed later in this chapter.

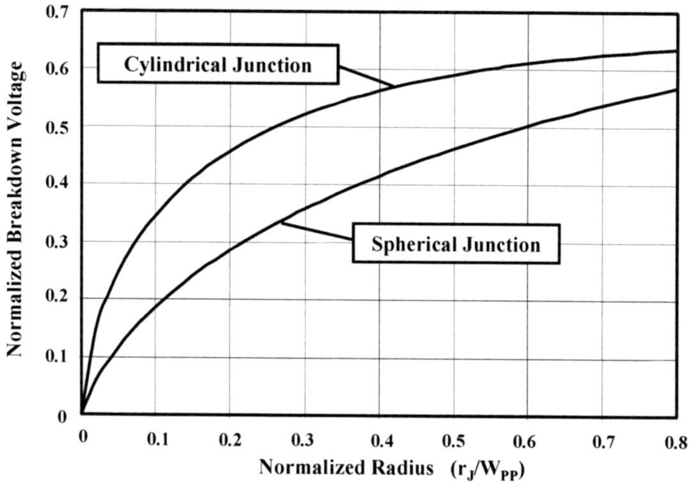

Fig. 3.15 Breakdown voltages of cylindrical and spherical junctions normalized to the parallel-plane case

Simulation Example

To gain further insight into the physics of operation for the cylindrical junction, the results of two-dimensional numerical simulations are provided in this section for the case of a drift region with doping concentration of 3.8×10^{14} cm^{-3}. The simulations were performed by using impact ionization coefficients reported in the literature for silicon.[9] At this doping concentration, the parallel-plane breakdown voltage was found to be 520 V with a depletion region thickness of 41 μm for the case of a P^+ region with a surface doping concentration of 1×10^{20} cm^{-3} and depth of 5 μm. The breakdown voltage obtained with the simulations is smaller than that (620 V) predicted by the analytical formulation in (3.18), because Fulop's law underestimates the magnitude for the impact ionization coefficients as can be seen in Fig. 3.1. The electric field at the junction was found to be 2.35×10^5 V cm^{-1} at breakdown. This field is also lower than the critical electric field for breakdown predicted by (3.22) due to the underestimation of the magnitude for the impact ionization coefficient by Fulop's law.

The breakdown voltages of planar junctions with cylindrical curvature were obtained by performing simulations with various junction depths ranging from 2 to 20 μm. The breakdown voltages obtained with the two-dimensional numerical

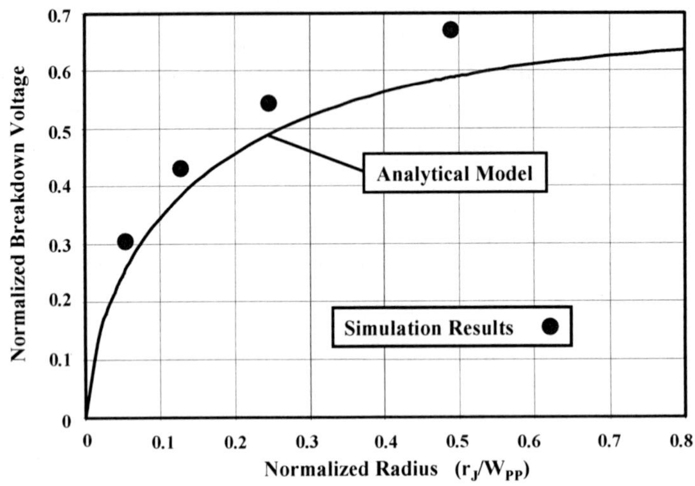

Fig. 3.16 Breakdown voltages of cylindrical junctions: comparison of simulated results with analytical model

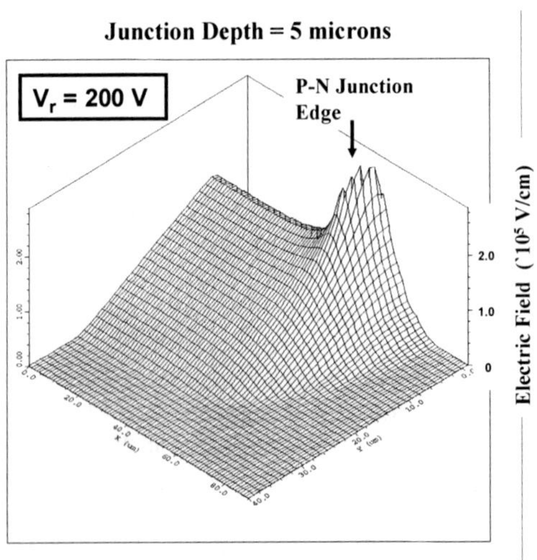

Fig. 3.17 Electric field distribution in a cylindrical junction

simulations for the various junction depths were normalized to the breakdown voltage of the parallel-plane junction as determined by the simulations. The normalized breakdown voltage for these cylindrical junctions is plotted in Fig. 3.16 as a function of the normalized radius of curvature together with the analytically derived curve previously shown in Fig. 3.15. A good agreement between the results of the simulations and the analytical formulation is observed at small junction depths, confirming that the analytical model provides a reasonable approach for the analysis of the breakdown voltage for cylindrical junctions. However, the analytical model underestimates the breakdown voltage by more than 10% when the normalized junction depth is larger than 0.4.

A three-dimensional view of the electric field at the cylindrical junction is shown in Fig. 3.17 for the case of a junction depth of 5 μm at a reverse bias of 200 V. It can be seen that there is an enhancement in the electric field at the

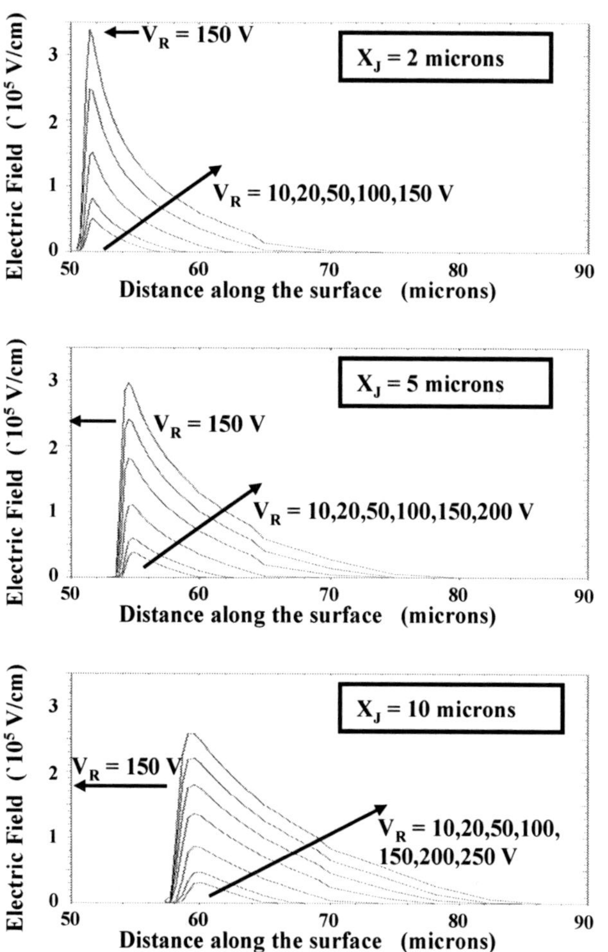

Fig. 3.18 Electric field profiles in cylindrical junctions with various junction depths

corner of the junction with much larger electric fields in this location when compared with the middle of the junction. The degree of enhancement of the electric field depends upon the radius of curvature of the junction. This is illustrated in Fig. 3.18 by comparing the electric field profile at various reverse bias voltages for junctions with different depths. It can be observed that a larger electric field develops at the cylindrical junction at the same magnitude for the reverse bias voltage (see arrows at V_R = 150 V for the three cases) when its junction depth is smaller. This is responsible for a reduction in the breakdown voltage.

Spherical Junction

It is common practice to fabricate devices in integrated circuits by using rectangular windows, as illustrated in Fig. 3.12, because this simplifies the layout of the chip. As already discussed, a spherical junction is formed at each of the four corners of the rectangular diffusion window. A cross-sectional view of this spherical junction is identical to that illustrated in Fig. 3.13 for the cylindrical junction. However, the electric field is enhanced even further than for the cylindrical junction, because the field lines approach a point in three dimensions for the spherical junction while they approach a line from two dimensions for the cylindrical junction. This difference in behavior can be analyzed by performing the solution for Poisson's equation[8] in spherical coordinates:

$$\frac{1}{r^2}\frac{\mathrm{d}}{\mathrm{d}r}\left(r^2\frac{\mathrm{d}V}{\mathrm{d}r}\right) = -\frac{1}{r^2}\frac{\mathrm{d}}{\mathrm{d}r}(r^2 E) = -\frac{Q(r)}{\varepsilon_S} = -\frac{qN_D}{\varepsilon_S}, \tag{3.50}$$

where the potential $V(r)$ and electric field $E(r)$ are defined along the radius vector r extending into the depletion region as shown in Fig. 3.13. Integration of this equation with the boundary condition that the electric field must be zero at the depletion region boundary (r_D) in the N-type region provides the electric field distribution:

$$E(r) = \frac{qN_D}{3\varepsilon_S}\left(\frac{r^3 - r_D^3}{r^2}\right). \tag{3.51}$$

The maximum value for the electric field for the spherical junction also occurs at the metallurgical junction located at $r = r_J$:

$$E_{m,SP}(r_J) = \frac{qN_D}{3\varepsilon_S}\left(\frac{r_J^3 - r_D^3}{r_J^2}\right). \tag{3.52}$$

The maximum electric field generated in the spherical junction is not only significantly larger than that observed in the parallel-plane case but also exceeds that generated in the cylindrical junction. This can be demonstrated by considering the case of a spherical junction whose radius of curvature (r_J) is small when compared with the depletion width and consequently the radius r_D. Since r_J is much smaller than r_D in this case, the maximum electric field is given by

$$E_{\mathrm{m,SP}}(r_{\mathrm{J}}) = -\frac{qN_{\mathrm{D}}r_{\mathrm{D}}^{3}}{3\varepsilon_{\mathrm{S}}r_{\mathrm{J}}^{2}}. \tag{3.53}$$

If the depletion widths for the spherical and cylindrical cases are assumed to be approximately equal at the same reverse bias voltage (i.e., $r_{\mathrm{D}} = W_{\mathrm{D}}$), then the enhancement of the electric field at the spherical junction over that at the cylindrical junction can be obtained by taking the ratio of the maximum electric field for the two cases:

$$\frac{E_{\mathrm{m,SP}}}{E_{\mathrm{m,CYL}}} = \frac{2r_{\mathrm{D}}}{3r_{\mathrm{J}}}. \tag{3.54}$$

From this equation, it can be concluded that, for shallow junctions with small radii of curvature, the maximum electric field generated at the sharp corners of diffusion windows is significantly larger than along the sides of the window. For example, if the junction depth is 1 μm and the depletion region has a thickness of 30 μm, the maximum electric field at the spherical junction will be 20 times larger than that at the cylindrical junction. Since impact ionization is a very strong function of the electric field, avalanche breakdown will occur at the corners of the rectangular diffusion window at a lower voltage than at the edges or middle of the diffused region.

The potential distribution for the spherical junction, obtained by integration of the electric field distribution, is given by

$$V(r) = \frac{qN_{\mathrm{D}}}{3\varepsilon_{\mathrm{S}}}\left[\left(\frac{r^{2} - r_{\mathrm{J}}^{2}}{2}\right) + r_{\mathrm{D}}^{3}\left(\frac{1}{r} - \frac{1}{r_{\mathrm{J}}}\right)\right]. \tag{3.55}$$

The width of the depletion layer for the spherical junction can be obtained by using the boundary conditions for the voltage, namely, the voltage being equal to zero on the highly doped side and V_{a} in the lightly doped side. The breakdown voltage for the spherical junction can be obtained by performing the ionization integral using the electric field distribution given by (3.51). To obtain a closed-form solution for the ionization integral, it is convenient to make the approximation that the electric field varies inversely as the square of the distance from the junction:

$$E(r) = -\frac{qN_{\mathrm{D}}}{3\varepsilon_{\mathrm{S}}}\frac{r_{\mathrm{D}}^{3}}{r^{2}} = -\frac{K_{\mathrm{S}}}{r^{2}}. \tag{3.56}$$

The electric field distribution obtained by using this inverse square approximation is compared with that given by (3.51) in Fig. 3.19 for the case of $r_{\mathrm{J}} = 0.1r_{\mathrm{D}}$. The inverse square approximation provides a very good fit to the exact case especially in the vicinity of the junction where the electric field is large. Since the impact ionization coefficients are a very strong function of the electric field, the approximation is satisfactory for the evaluation of the ionization integral. It is worth pointing out that this approximation implies that the electric field distribution extends to an infinite distance from the junction. Consequently, the ionization integral must also be performed to infinity when using the inverse square approximation for the electric field.

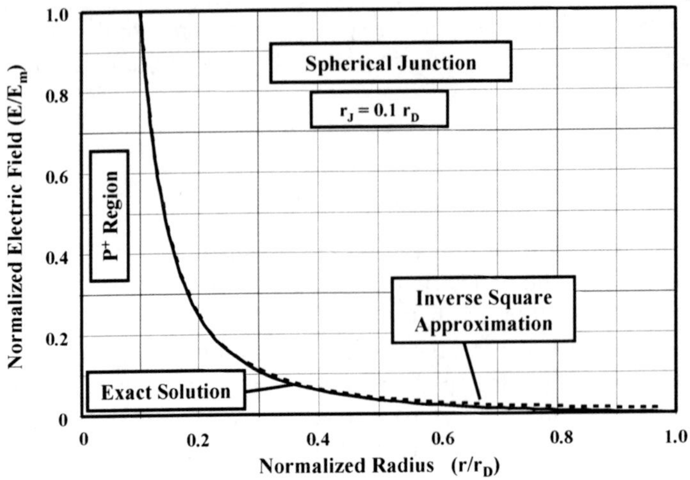

Fig. 3.19 Comparison of the electric field distribution for the inverse square approximation with the exact case for a spherical junction

Evaluation of the ionization integral using the inverse square variation of the electric field together with Fulop's law for the ionization coefficient yields a solution for the breakdown condition for the cylindrical junction:

$$K = \left(\frac{13 r_J^{13}}{1.8 \times 10^{-35}}\right)^{1/7} = \frac{q N_D r_D^3}{3 \varepsilon_S}. \tag{3.57}$$

By combining this condition for breakdown with (3.53) for the maximum electric field, the critical electric field for breakdown at the spherical junction is obtained:

$$E_{C,SP} = \left(\frac{7.2 \times 10^{35}}{r_J}\right)^{1/7}. \tag{3.58}$$

The critical electric field for breakdown in the case of cylindrical junctions can be compared with the critical electric field for the parallel-plane junction by taking their ratio:

$$\frac{E_{C,SP}}{E_{C,PP}} = \left(\frac{13 W_{PP}}{8 r_J}\right)^{1/7}. \tag{3.59}$$

In deriving this relationship, the critical electric field ($E_{C,PP}$) for the parallel-plane case was related to the depletion width (W_{PP}) by using (3.19) and (3.22). Since the radius of curvature of the junction is assumed to be small when compared with the depletion layer thickness, the above relationship indicates that the critical electric field for breakdown in the spherical junction is larger than that for the parallel-plane junction. This difference is associated with the high electric field being located over a shorter distance in the case of the spherical junction when compared with the parallel-plane junction.

The breakdown voltage for the spherical junction can be obtained by using $r = r_D$ in (3.55) with the value for r_D defined by the breakdown condition as governed by (3.57). To generalize the solution to represent a wide variety of junctions with different doping concentrations on the lightly doped side, it is convenient to normalize the breakdown voltage of the spherical junction to that for the parallel-plane case. To obtain this generalized solution, it is also convenient to normalize the radius of curvature of the junction to the depletion layer thickness at breakdown for the parallel-plane junction. This methodology provides the normalized breakdown voltage for the spherical junction:

$$\frac{\mathrm{BV_{SP}}}{\mathrm{BV_{PP}}} = \left(\frac{r_J}{W_{PP}}\right)^2 + 2.14\left(\frac{r_J}{W_{PP}}\right)^{6/7} - \left[\left(\frac{r_J}{W_{PP}}\right)^3 + 3\left(\frac{r_J}{W_{PP}}\right)^{13/7}\right]^{2/3}. \tag{3.60}$$

It is worth pointing out that this relationship was derived under the assumption that the radius of curvature of the junction is small when compared with the depletion layer thickness (i.e., $r_J / W_{PP} \ll 1$). Since power devices are usually fabricated with shallow junctions when compared with the large depletion layer widths required to support high voltages, (3.60) is usually valid for their analysis.

The normalized breakdown voltage for spherical junctions as predicted by (3.60) is also plotted in Fig. 3.15 as a function of the normalized radius of curvature. This graph is valid for junctions fabricated from any doping concentration on the lightly doped side as long as the radius of curvature is small when compared with the depletion layer width at breakdown for the parallel-plane junction with the same doping concentration on the lightly doped side. It can be seen that the breakdown voltage for the spherical junction increases when the radius of curvature (or junction depth) is increased.

The breakdown voltage at the spherical junctions formed at the corners of the rectangular diffusion window is substantially lower than the breakdown voltage at the cylindrical junctions formed at the straight edges of the window. For instance, at a normalized radius of curvature of 0.2, the breakdown voltage is reduced from 46 to 28% of the parallel-plane value. This is detrimental to achieving low on-state voltage drops in power devices, because the doping concentration must be reduced and thickness of the drift region must be increased to achieve the desired breakdown voltage in the presence of the spherical junction curvature. This problem can be overcome by rounding the corners of the diffusion window. It is common practice to design the windows of high-voltage power devices with the corners rounded with a radius that is at least twice the thickness of the depletion region (W_{PP}) for the parallel-plane junction at breakdown for the drift region.

3.6.2 Planar Junction with Floating Field Ring

An elegant method for improving the breakdown voltage of planar junctions is by surrounding the diffusion window with a floating field ring.[10] This can be implemented by opening a diffusion window for the floating field ring simultaneously with the main junction with no additional process steps. A top view of the structure is shown in Fig. 3.20 together with a cross section. Note that there is no metal contact made to the floating field ring allowing it to attain a potential intermediate to the voltage applied to the cathode. Although the depth of the floating field ring could be greater or smaller than the main junction, in practice the floating field ring is invariably fabricated at the same time as the main junction giving it the same depth. The analysis in this section will therefore be confined to this case.

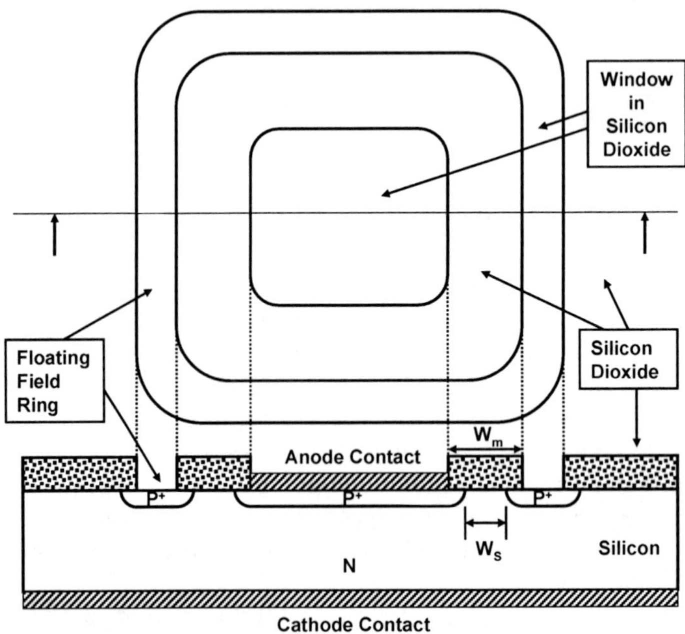

Fig. 3.20 The planar junction with floating field ring

In order for the floating field ring to perturb the electric field at the main junction, it must be located within the depletion width of the main junction. The optimum location of the floating field ring has been determined by numerical simulations.[11] If the spacing of the floating ring is too close to the main junction, its potential becomes close to that of the main junction and the breakdown voltage is not substantially improved, because a high electric field develops at the floating field ring. On the other hand, if the floating field ring is placed too far from the main junction, it has minimal effect on the electric field at the main junction, resulting in insubstantial improvement in the breakdown voltage. It is necessary to place the floating field ring at an optimal spacing to provide an improvement in the breakdown voltage. It is worth pointing out that the spacing on the mask for the design of the floating field ring termination must take into account the lateral diffusion of the junctions. Once the optimum spacing of W_S is obtained, the spacing on the mask is given by

$$W_m = W_S + 2x_J. \tag{3.61}$$

The potential assumed by the floating field ring can be analytically determined under the assumption that the presence of the floating field ring does not perturb the extension of the depletion region from the main junction. If the floating ring is assumed to have a small width, then its potential can be assumed to equal to the potential within the depletion region at a distance W_S from the junction. Using the potential distribution for the parallel-plane junction given by (3.13) with $x = W_S$ and a depletion width of W_D,

$$V_{FFR} = \frac{qN_D}{\varepsilon_S}\left(W_D W_S - \frac{W_S^2}{2}\right). \tag{3.62}$$

Replacing W_D by using (3.15),

$$V_{FFR} = \sqrt{\frac{2qN_D W_S^2 V_a}{\varepsilon_S}} - \frac{qN_D W_S^2}{2\varepsilon_S}. \tag{3.63}$$

This equation indicates that the potential of the floating field ring will increase as the square root of the applied bias to the cathode. It is worth pointing out that (3.63) is valid only when the depletion layer width of the main junction exceeds the field ring spacing. When the applied reverse bias is insufficient to allow the depletion layer from the main junction to overlap the floating field ring, the potential of the floating field ring is equal to the reverse bias applied to the cathode.

The potential of the floating field ring calculated by using (3.63) is shown in Fig. 3.21 for various examples of the field ring spacing together with the case of a very large field ring spacing where the field ring potential is equal to the applied bias to the cathode. For the smallest spacing of 3 μm, the potential at the field ring increases in proportion to the applied bias up to about 10 V and then increases at a more gradual rate after the depletion layer from the main junction extends to the floating field ring. When the field ring spacing is increased to 5, 10, and 15 μm, this transition occurs at a larger applied bias of about 20, 40, and 100 V, respectively. This has a strong influence on the electric field at the edges of the main junction and floating field ring junction.

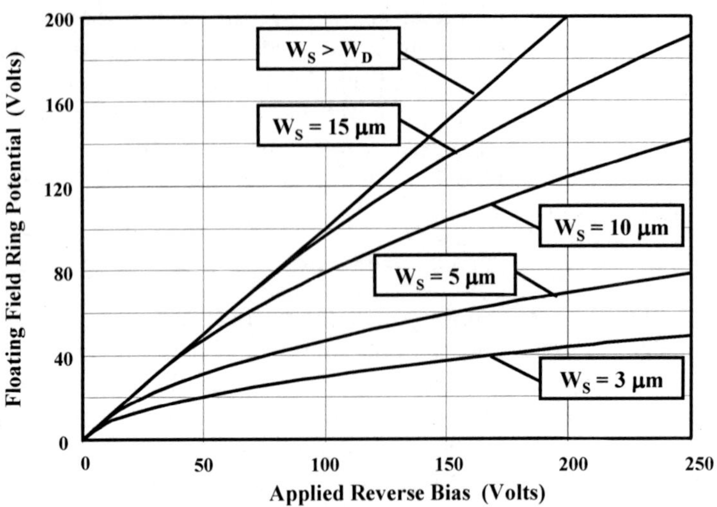

Fig. 3.21 Floating field ring potential for various field ring spaces

As discussed earlier, the floating field ring must be located at an optimal position to maximize the breakdown voltage. With an optimum spacing (W_S), the electric field at the main junction and floating field ring junction simultaneously becomes equal to the critical electric field for breakdown. Analysis of the breakdown voltage for the floating field ring termination[12] can be performed under the assumption that the electric field at the main junction is determined by the difference in voltage between the main junction and the floating field ring, while the electric field at the floating field ring junction is governed by the cylindrical junction. Under these assumptions, the potential of the floating field ring (V_{FFR}) must be equal to the breakdown voltage of the cylindrical junction with the same radius of curvature and doping concentration in the drift region:

$$\frac{V_{FFR}}{BV_{PP}} = \frac{BV_{CYL}}{BV_{PP}}. \tag{3.64}$$

The difference in the voltage between the main junction (V_M) and the floating field ring (V_{FFR}) can be obtained by using the voltage distribution for a cylindrical junction given by (3.44) with $r = W_S$:

$$(V_M - V_{FFR}) = \frac{qN_D}{2\varepsilon_S}\left[\left(\frac{r_J^2 - W_S^2}{2}\right) + r_D^2 \ln\left(\frac{W_S}{r_J}\right)\right]. \tag{3.65}$$

Under breakdown, the potential of the main junction (V_M) becomes equal to the breakdown voltage of the floating field ring termination (BV_{FFR}). Making use of (3.63), (3.65), and the basic relationship for the parallel-plane junction,

$$BV_{PP} = \frac{1}{2} E_C W_{PP} = \frac{qN_D}{2\varepsilon_S} W_{PP}^2 \tag{3.66}$$

to normalize the solution, it can be shown that

$$\left(\frac{BV_{FFR} - BV_{CYL}}{BV_{PP}} \right) = \frac{1}{2}\left(\frac{r_J}{W_{PP}} \right)^2 - 0.96\left(\frac{r_J}{W_{PP}} \right)^{6/7} + 1.92\left(\frac{r_J}{W_{PP}} \right)^{6/7} \ln\left[1.386\left(\frac{W_{PP}}{r_J} \right)^{4/7} \right]. \tag{3.67}$$

It is worth pointing out that this expression is valid under the assumptions that (a) the radius of curvature for the junction is small when compared with the depletion layer width at breakdown for the parallel-plane junction and (b) the floating field ring is located at an optimal spacing from the main junction. The normalized breakdown voltage calculated using this analytical formulation is compared with the breakdown voltage for the cylindrical junction in Fig. 3.22.

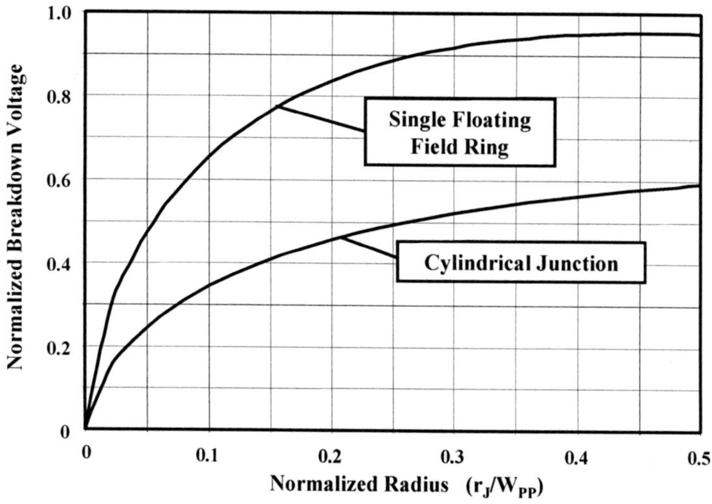

Fig. 3.22 Breakdown voltages for cylindrical junctions with a single floating field ring

From Fig. 3.22, it can be observed that the breakdown voltage can be approximately doubled by the addition of a floating field ring to the cylindrical junction. This provides a powerful method for increasing the breakdown voltage with no additional processing steps because the floating field ring can be formed simultaneously with the main junction. This design requires precise location of the floating field ring at the optimum spacing from the main junction. An analytical solution for the optimal spacing[12] can be derived by solving for W_S in (3.65) with V_M equal to BV_{FFR} and V_{FFR} equal to BV_{CYL} and eliminating N_D by using (3.66):

$$W_S^2 - 2\sqrt{\frac{BV_{FFR}}{BV_{PP}}} W_{PP} W_S + \left(\frac{BV_{CYL}}{BV_{PP}}\right) W_{PP}^2 = 0. \tag{3.68}$$

This quadratic equation provides an elegant solution for the optimum field ring spacing, in terms of the breakdown voltage of the cylindrical and floating field ring cases when the spacing is normalized to the depletion width of the parallel-plane junction at breakdown:

$$\frac{W_S}{W_{PP}} = \sqrt{\frac{BV_{FFR}}{BV_{PP}}} - \sqrt{\left(\frac{BV_{FFR}}{BV_{PP}}\right) - \left(\frac{BV_{CYL}}{BV_{PP}}\right)}. \tag{3.69}$$

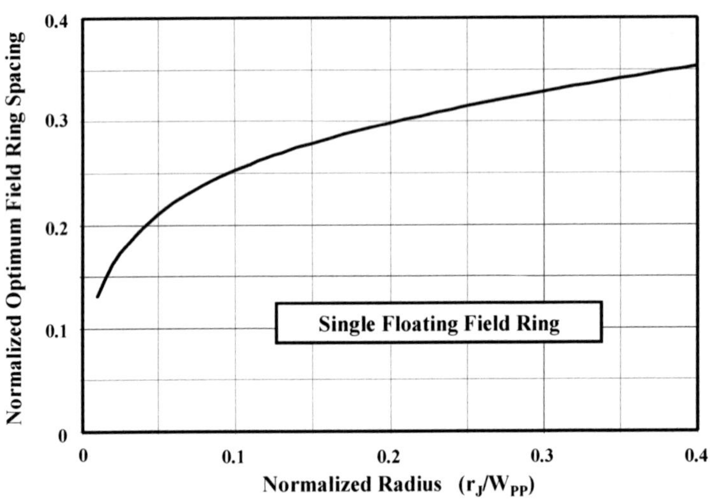

Fig. 3.23 Optimum spacing for a single floating field ring

The normalized optimum spacing for the floating ring is a function of the ratio of the radius of curvature of the junction to the depletion layer width at breakdown for the parallel-plane junction, because the normalized breakdown voltages in (3.69) depend on this ratio. A plot for the normalized optimum spacing for the floating field ring is provided in Fig. 3.23. The optimum spacing becomes larger with increasing normalized radius of curvature for the junction. The optimum spacing is in the range of 0.15–0.35 times the depletion layer width for the parallel-plane junction at breakdown. For a doping concentration of 3.8×10^{14} cm^{-3} (previously used in the context of cylindrical junctions) and a junction depth of 5 μm, a normalized radius of curvature of 0.11 is obtained because the depletion layer width at breakdown is 47 μm. Using this value, an optimum spacing for the floating field ring is found to be 12 μm by using the analytical solution corresponding to a normalized value of 0.258.

Simulation Example

To gain further insight into the operation of the cylindrical junction with floating field ring, the results of two-dimensional numerical simulations are provided in this section for the case of a drift region with doping concentration of 3.8×10^{14} cm^{-3}. At this doping concentration, the parallel-plane breakdown voltage was found to be 520 V with a depletion region thickness of 41 μm for the case of a P$^+$ region with a surface doping concentration of 1×10^{20} cm^{-3} and depth of 5 μm as discussed earlier. The simulations of the floating field ring termination were performed with various spacing between the main junction and the floating field ring. In all cases, the main junction had a width of 50 μm and the width of the window for the floating field ring was maintained at 20 μm.

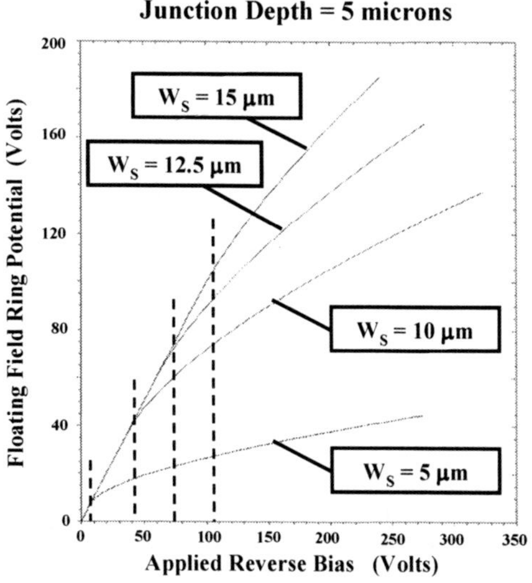

Fig. 3.24 Floating field ring potential determined by numerical simulations

The potential of the floating field ring was monitored during the simulations as a function of the applied reverse bias to the main junction. As expected, the floating ring potential increases with increasing applied reverse bias as shown in Fig. 3.24. The potential of the floating field ring remains equal to the applied bias up to the corresponding vertical dashed line marked in the figure for each value for the spacing. This potential is very well predicted by the analytical solutions shown in Fig. 3.21. The potential increases more gradually, as described by the analytical model, when the applied bias extends the depletion region of the main junctions beyond the spacing of the floating field ring. The predictions of the analytical model, as shown in Fig. 3.21 in this domain of operation, are in good agreement with the simulation results for spacing above 5 μm providing credence to the model even though it was based upon simple one-dimensional considerations.

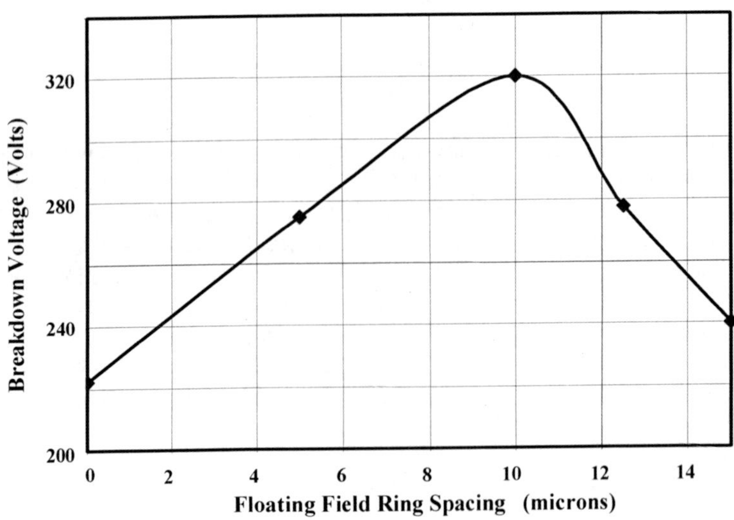

Fig. 3.25 Breakdown voltages of cylindrical junctions with a single floating field ring

The breakdown voltages of the cylindrical junction with floating field rings were obtained by performing simulations with various floating field ring spacing ranging up to 15 μm. The breakdown voltage was found to go through a maximum value as the floating field ring spacing was increased as shown in Fig. 3.25. The maximum breakdown voltage was found to be 320 V, which is 62% of the parallel-plane breakdown voltage. The normalized breakdown voltage (66% of the parallel-plane breakdown voltage) predicted by the analytical model for this case of normalized radius of curvature of 0.11 is in good agreement with the simulated value.

The optimum spacing for the floating field ring obtained from the simulations is 10 μm, which is 24% of the depletion layer width for the parallel-plane junction. The analytical model predicts an optimum spacing of 0.26 times the depletion layer width for the parallel-plane junction. Consequently, the analytical model provides an excellent tool for choice of the placement for the floating field ring. This spacing can then be further refined by performing two-dimensional numerical simulations.

To gain further insight into the impact of an optimum placement of the floating field ring, the electric field distribution is shown in Fig. 3.26 for three values for the field ring spacing. In each case, the electric field has been plotted along the surface for various reverse bias voltages applied to the cathode. In the case of the smallest spacing of 5 μm, the electric field increases more rapidly at the edge of the floating field ring located at 90 μm. This results in breakdown occurring at the edge of the floating field ring at a lower reverse bias voltage. In the case of the largest spacing of 15 μm, the electric field increases more rapidly at the edge of the main junction located at 55 μm. This results in breakdown occurring at the main junction at a lower reverse bias voltage. For the optimum floating field ring spacing of 10 μm, the maximum electric field at the edge of the field ring is equal to the maximum electric field developed at the main junction. In this case, the

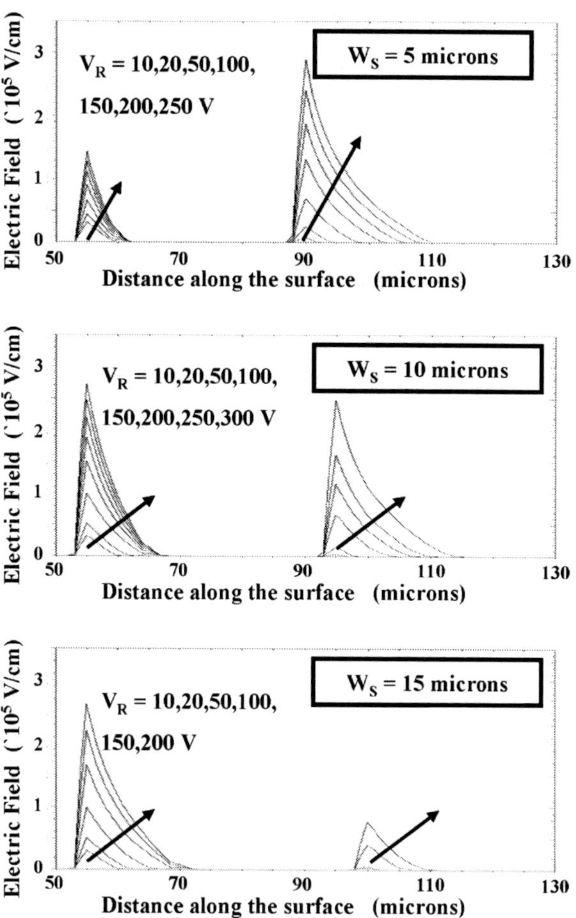

Fig. 3.26 Electric field profiles in cylindrical junctions with single floating field ring with various spacing

breakdown voltage is maximized with breakdown occurring simultaneously at the edge of the floating field ring and the edge of the main junction.

A three-dimensional view of the electric field distribution is shown at a reverse bias of 300 V in Fig. 3.27 for the case of a cylindrical junction with a single floating field ring located at 10 μm from the main junction. It can be observed that the electric fields at the edges of the main junction and the field ring are equal in magnitude, indicating that the floating field ring is located at an optimum distance from the main junction. However, the electric fields at these edges are much larger than in the parallel-plane portion of the main junction. This enhancement of the electric field is responsible for the breakdown voltage for even the optimum floating field ring design being less than the parallel-plane breakdown voltage.

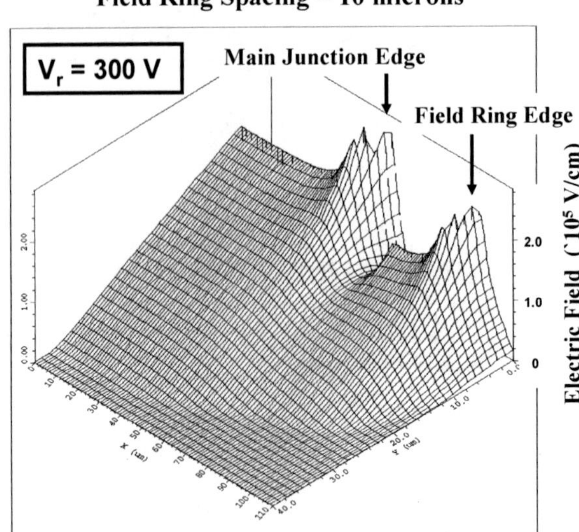

Fig. 3.27 Three-dimensional view of the electric field distribution for a cylindrical junction with a single floating field ring

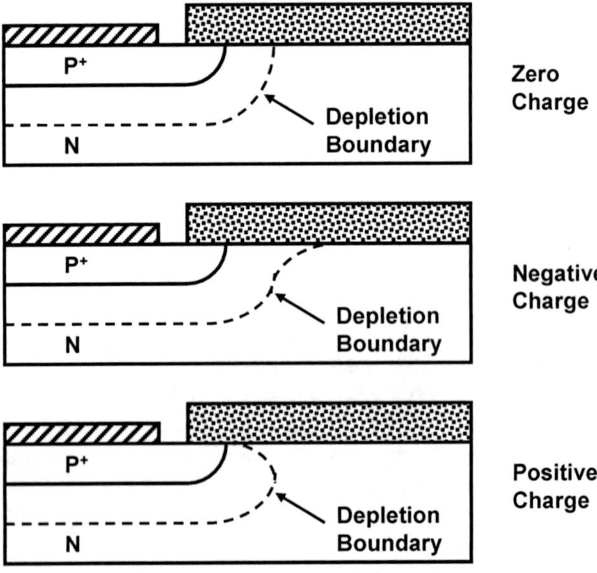

Fig. 3.28 Impact of fixed oxide charge on the depletion boundary for a cylindrical junction

The breakdown voltage of the edge termination with floating field rings has been found to be sensitive to the presence of fixed oxide charge in the passivation layer between the main junction and the floating field ring. This can be understood by considering the influence of the oxide charge on the depletion layer under the passivation oxide as schematically illustrated in Fig. 3.28. The presence of a negative charge in the oxide compensates the positive charge at the ionized donors, producing an extension of the depletion layer along the surface. The presence of a positive charge has the opposite effect on the depletion layer at the surface. In the case of a cylindrical junction, the presence of a positive charge in the oxide, which occurs in thermally grown oxide layers on silicon, enhances the junction curvature and electric field resulting in a reduction of the breakdown voltage. In the case of the edge termination with a floating field ring, the presence of the positive charge in the oxide perturbs the potential acquired by the floating field ring. This disturbs the electric field distribution for an optimally spaced floating field ring designed without taking this charge into account.

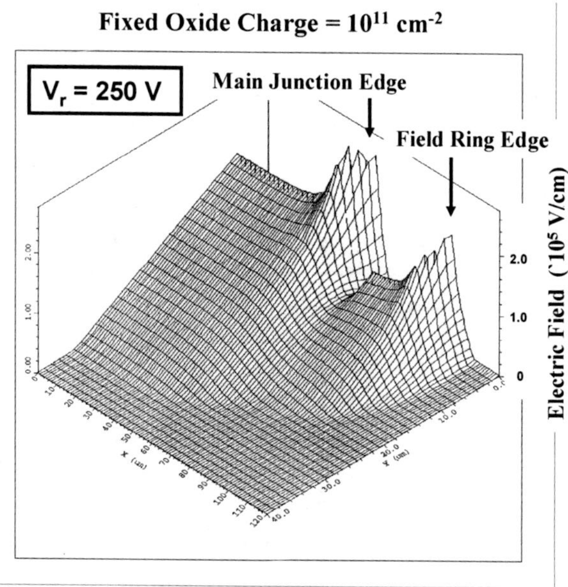

Fig. 3.29 Three-dimensional view of the electric field distribution for a cylindrical junction with a single floating field ring

As an example of the impact of the fixed oxide charge on the floating field ring termination, numerical simulations for the case of floating field ring spacing of 10 μm were repeated after inclusion of an oxide charge of 1×10^{11} cm^{-2}. The breakdown voltage obtained by the simulations was 256 V, which is considerably lower than before inclusion of the oxide charge. This can be understood by examination of the electric field distribution within the structure as shown in Fig. 3.29 at a reverse bias of 250 V. It can be seen that the electric field at the edge of the main junction has been enhanced in relation to the electric field at the edge of the floating field ring due to the positive charge in the oxide.

In principle, it is possible to account for the presence of the oxide charge during the optimization of the design of the location of the floating field ring. Unfortunately, the charge in the thermally grown field oxide during the fabrication of power devices can vary from wafer to wafer and even across a wafer by as much as 10^{11} cm^{-2}. The results of the simulations demonstrate that the breakdown voltage for an edge termination with single floating field ring can be significantly reduced due to the presence of this charge. This can produce a wide distribution in the breakdown voltages of the power devices which is detrimental to getting a high yield during their manufacturing. This problem can be mitigated by using multiple floating field rings.

Floating Field Ring Width

From the point of view of saving space occupied by the edge termination at the periphery of the power devices, it is advantageous to reduce the width of the floating field ring as much as possible. However, it has been found that the effectiveness of the floating field ring in terms of improving the breakdown voltage of the cylindrical junction is compromised if its width becomes too small. This is schematically illustrated in Fig. 3.30, where the depletion layer is shown for a narrow and wide floating field ring. From this standpoint, it is necessary to make the width of the floating field ring at least equal to the depletion width (W_{PP}) of the parallel-plane junction at breakdown.

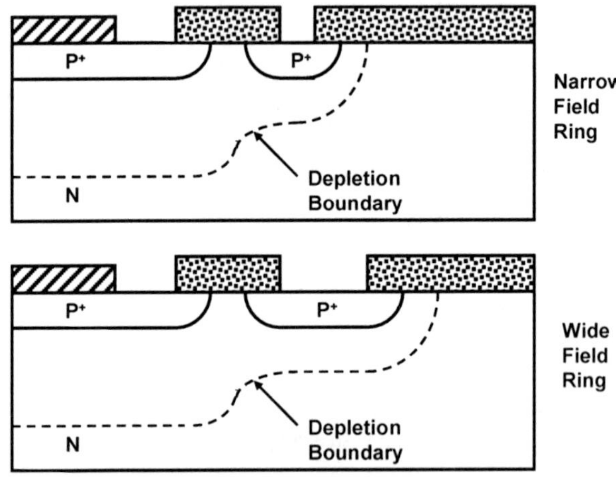

Fig. 3.30 Impact of the width of the floating field ring

3.6.3 Planar Junction with Multiple Floating Field Rings

In the case of devices that are designed to support less than 50 V, the depletion width is relatively small and a single floating field ring is usually sufficient to provide a breakdown voltage close to that for the parallel-plane junction even when the junction depth is less than 5 μm. However, as the breakdown voltage of the device (and consequently the depletion width) increases, the breakdown voltage for a cylindrical junction with a single floating field ring becomes much lower than that for the parallel-plane junction. This problem can be overcome by the placement of multiple floating field rings around the main junction.

The electric field developed at the edge of the single floating field ring can be reduced by the addition of another floating field ring that surrounds it. This favors a closer spacing for the first floating field ring from the main junction than the optimum spacing with a single floating field ring. This approach can be applied to each additional floating field ring around the main junction. In addition, the width of the floating field rings can be reduced when they are placed away from the main junction because the depletion width is smaller under them.

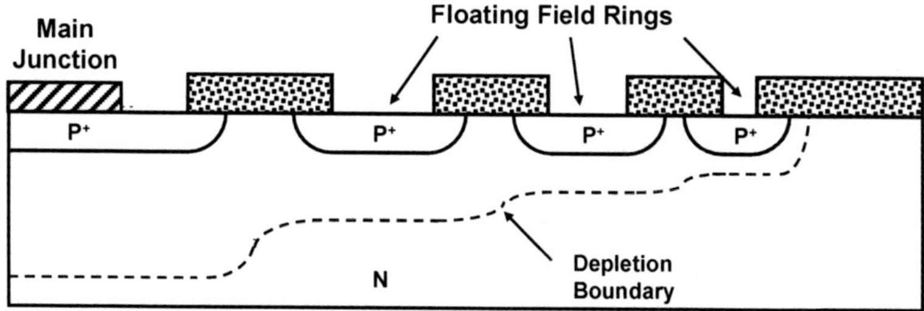

Fig. 3.31 Multiple floating field ring termination with graded spacing and width

Two approaches to designing the edge termination with multiple floating field rings have evolved. In the first approach shown in Fig. 3.31, both the floating field ring width and the spacing between the field rings are reduced with increasing distance from the main junction. This produces a gradual variation of the depletion width at the termination which is favorable for reducing electric fields. In this approach, the width of the outer rings is reduced in proportion to the underlying depletion layer width because this reduces the space occupied by the termination. The optimization of the spacing between the rings requires a precise knowledge of the charge in the field oxide. With an optimum design, the electric field at the outer edge of all the field rings is equal, so that avalanche breakdown occurs simultaneously at these locations.

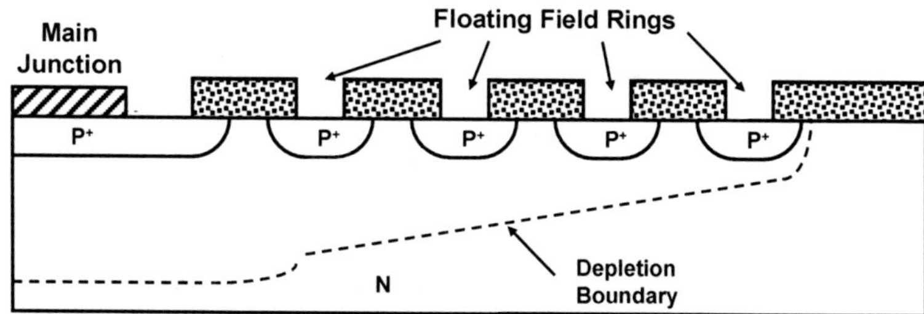

Fig. 3.32 Multiple floating field ring termination with equal spacing and width

In the second approach shown in Fig. 3.32, all the floating field rings are equally spaced and their widths are made equal. This allows accommodating more floating field rings within a given space on the edge of the chip. This produces a finer gradation of the depletion region at the edge resulting in reducing the electric field. The presence of more floating field rings is believed to reduce the impact of variation in the oxide charge.

In principle, the use of multiple floating field rings allows increasing the breakdown voltage of planar junctions arbitrarily close to the parallel-plane breakdown voltage by the addition of a very large number of rings. In practice, there is a diminishing benefit in terms of increasing the breakdown voltage from the addition of floating rings, while more space is occupied by the edge termination resulting in a larger die size and cost. Although planar edge terminations with up to ten floating field rings have been reported, it is usually practical to use only up to three floating field rings to enhance the breakdown voltage.

3.6.4 Planar Junction with Field Plate

A planar junction with a metal field plate located at its edge over the field oxide is illustrated in Fig. 3.33. The electric field at the edge of a planar junction can be modulated by the application of a bias voltage to the metal.[13] With no bias voltage applied to the field plate, the depletion region boundary has the form indicated by case A for the cylindrical junction. When a positive bias is applied to the field plate with respect to the N-type substrate, it attracts electrons toward the surface. This shrinks the extension of the depletion layer along the surface as indicated by case B. This will enhance the electric field at the junction producing a reduction in the breakdown voltage. On the other hand, if a negative bias is applied to the field plate, it repels electrons away from the surface. This will produce an expansion of the depletion region at the surface as indicated by case C. This will result in a reduction of the electric field at the junction leading to an increase in the breakdown voltage.

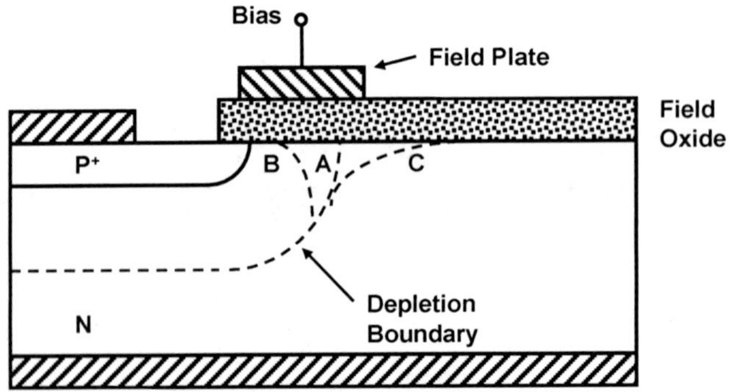

Fig. 3.33 Planar junction with biased field plate over the field oxide

It has been found[13] that the breakdown voltage of the diode with the field plate (BV_{FP}) is related to the magnitude of the negative bias (V_{FP}) applied to the field plate:

$$BV_{FP} = mV_{FP} + K, \tag{3.70}$$

where m and K are constants. The value for m is close to unity especially for smaller field oxide thicknesses.

With the application of a sufficient potential to the field plate, it is possible to approach the breakdown voltage of a parallel-plane junction. However, it is not practical to provide a separate bias to a field plate in the case of discrete power devices because of the additional package terminal as well as the cost associated with the bias circuit. An alternative approach is to form the field plate by extending the contact metal for the P^+ region over the field oxide at the edge of the junction as shown in Fig. 3.34. In this case, the application of a negative voltage to the P^+ region to reverse bias the P–N junction also provides a negative bias to the field plate. This produces an expansion of the depletion region along the surface as illustrated in the figure. The resulting reduction in the electric field at the cylindrical junction (at point A) will increase the breakdown voltage. However, a high electric field can be produced at the edge of the field plate at point B. This can result in premature breakdown at this location which can reduce the breakdown voltage.

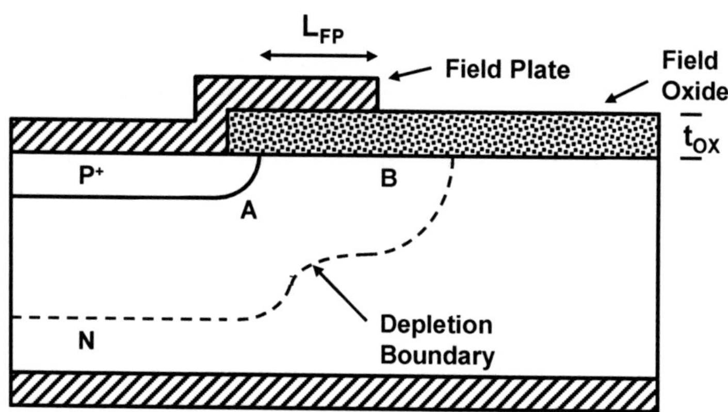

Fig. 3.34 Planar junction with metal field plate over the field oxide

Analysis of the breakdown at the edge of the field plate can be performed by treating the edge of the field plate as a cylindrical junction with the oxide under the field plate serving as the highly doped side of the junction. The difference in the dielectric constants for silicon dioxide and silicon must be taken into account when making this analogy. In accordance with Gauss' law, the electric field in the semiconductor is related to the electric field in the oxide in proportion to their permittivity. Based upon this, the junction depth corresponding to an oxide thickness of t_{OX} is given by

$$x_J = \left(\frac{\varepsilon_{Si}}{\varepsilon_{OX}}\right) t_{OX} \approx 3t_{OX}. \tag{3.71}$$

The breakdown voltage at the field plate for a field oxide thickness of 1 μm would be equivalent to the breakdown voltage of a cylindrical junction with a depth of 3 μm. Using this junction depth, the breakdown voltage can be obtained by using the analytical formulations developed for cylindrical junctions. However, it is important to avoid sharp corners at the field plate to prevent degradation of the breakdown voltage to that for a spherical junction with the same junction depth. For the case of a doping concentration of 3.8×10^{14} cm^{-3} in the N-type substrate, the breakdown voltage at the edge of the field plate can be determined to be 174 V using the cylindrical junction analysis. This value is pessimistic as shown by the results of two-dimensional numerical simulations.

Simulation Example

To gain further insight into the operation of the cylindrical junction with a field plate, the results of two-dimensional numerical simulations are provided in this section for the case of a drift region with doping concentration of 3.8×10^{14} cm^{-3}. At this doping concentration, the parallel-plane breakdown voltage was found to be 520 V with a depletion region thickness of 41 μm for the case of a P^+ region with a surface doping concentration of 1×10^{20} cm^{-3} and depth of 5 μm. The simulations of the field plate termination were performed with various lengths (L_{FP}) of the field plate and different field oxide thickness (t_{OX}). Note that the length of the field plate is defined as its extension from the edge of the junction (see Fig. 3.34) and not the edge of the diffusion window. In all cases, the main junction had a width of 50 μm.

In the absence of the field plate, the breakdown voltage of the cylindrical junction was found to be 220 V. For the case of a field oxide thickness of 1 μm, the breakdown voltage was found to increase upon the addition of the field plate. The increase in the breakdown voltage is dependent upon the length of the field plate as shown in Fig. 3.35. It can be seen that the extension of the field plate beyond 15 μm will not produce enhancement of the breakdown voltage for this case. The addition of the field plate provides a 40% improvement in the breakdown voltage over that for the cylindrical junction. This is comparable with the improvement in breakdown voltage obtained by using a single optimally spaced field ring.

A three-dimensional view of the electric field distribution within the silicon is shown in Fig. 3.36 for the case of cylindrical junction with a field plate at a reverse bias of 300 V. The field plate had a length of 10 μm and the field oxide thickness was 1 μm. The electric field in the semiconductor reaches a magnitude of about 3×10^5 V cm^{-1} at a much larger voltage when compared with the cylindrical junction (see Fig. 3.17). The electric field at the edge of the field plate can be seen to be slightly larger than at the edge of the cylindrical junction.

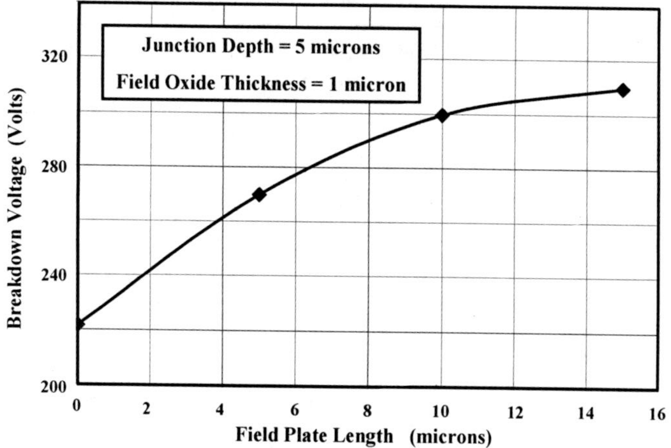

Fig. 3.35 Breakdown voltages of cylindrical junctions with a field plate

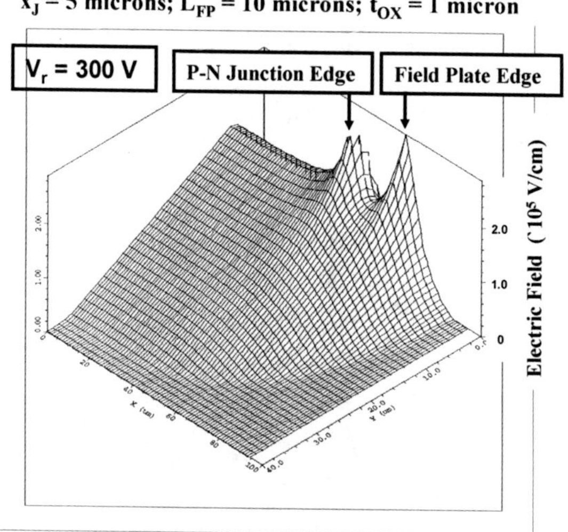

Fig. 3.36 Electric field distribution in a cylindrical junction with a field plate

The breakdown voltage of the cylindrical junction with the field plate termination is also dependent upon the thickness of the field oxide. This is demonstrated in Fig. 3.37 for the case of a junction depth of 5 μm and a field plate length of 10 μm. A field oxide thickness of 1 μm is necessary to take full advantage of the incorporation of the field plate. The breakdown voltage degrades when the field oxide thickness is reduced below this value. In the case of an oxide thickness of 0.25 μm, the breakdown voltage is reduced to 205 V which is lower than that for the cylindrical junction. This reduction is due to the enhanced electric field at the edge of the field plate as shown in Fig. 3.38. It can be seen that this electric field exceeds that at the junction resulting in avalanche breakdown being initiated at the edge of the field plate. Although this is consistent with the predictions of the analytical model, the magnitude of the breakdown voltage is significantly larger.

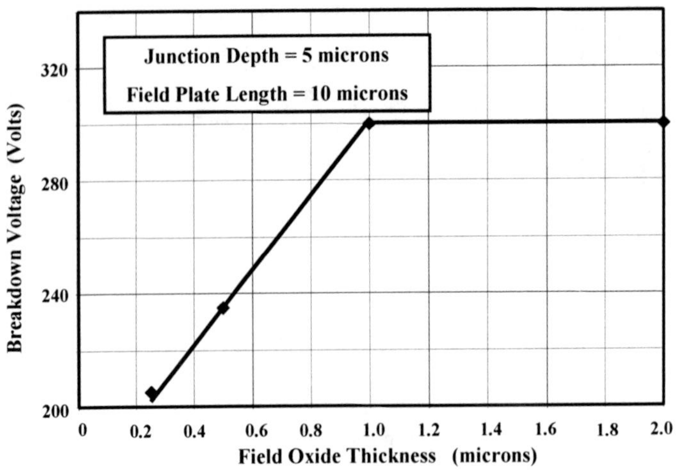

Fig. 3.37 Breakdown voltages of cylindrical junctions with a field plate

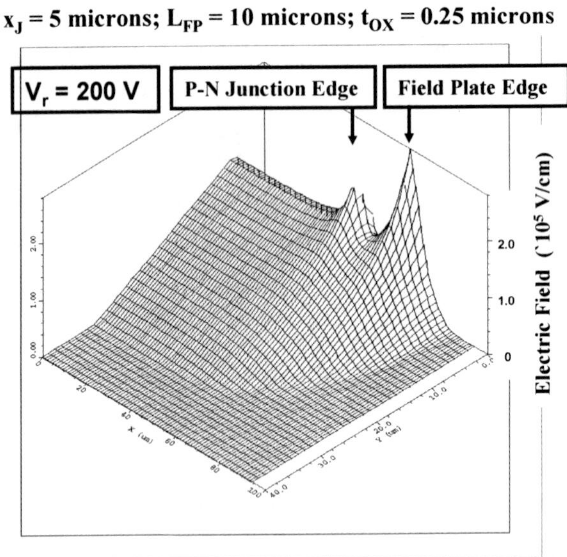

Fig. 3.38 Electric field distribution in a cylindrical junction with a field plate

The effectiveness of the field plate in improving the breakdown voltage of cylindrical junctions is also dependent upon the junction depth. The field plate has a stronger impact on the electric field for shallower junction depths due to its proximity. As an example, when the junction depth was reduced from 5 to 2 μm while maintaining the same field oxide thickness of 1 μm and field plate length of 10 μm, the breakdown voltage obtained by the numerical simulations was found to be 280 V. In this case, the breakdown voltage is enhanced from 160 V for the cylindrical junction by nearly a factor of 2 times.

3.6.5 Planar Junction with Field Plates and Field Rings

A popular edge termination design for power devices combines the effectiveness of the multiple floating field rings with field plates. As shown in Fig. 3.39, the field plates are designed to extend over the space between the junctions. The field plates reduce the electric field at the edges of all the junctions and also prevent mobile ions from entering the field oxide. Mobile ions can be introduced at the surface of power devices during the packaging operations. It has been found that the presence of mobile ions can produce instabilities (called *walkout*) in the breakdown voltage because they are redistributed by the electric field.

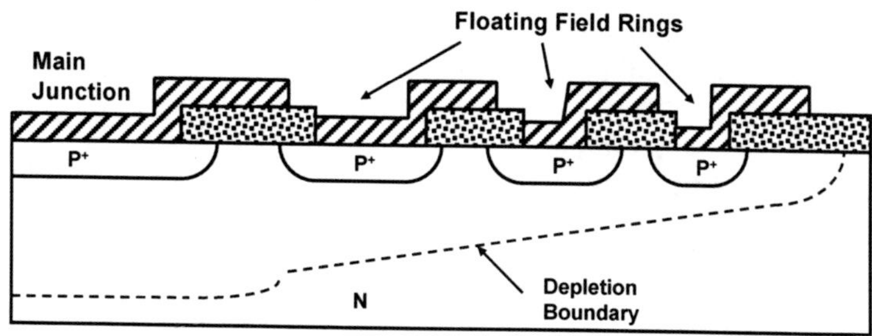

Fig. 3.39 Multiple floating field ring termination with field plates

The floating field ring termination is the most practical solution for improving the breakdown voltage of power devices with voltage ratings up to 1,500 V. These devices include power rectifiers, power MOSFETs, and IGBTs. Until the development of this approach, it was common practice to etch the region between devices on a wafer to expose the junction followed by passivation using dielectrics deposited on the etched surface. This method, called *mesa termination*, was difficult to control resulting in wide variability in the breakdown voltage. The evolution of the planar junction termination with floating field rings brought about a major improvement in the manufacturability of power devices.

3.6.6 Bevel Edge Terminations

Power rectifiers and thyristors with voltage ratings above 2,000 V are required for high power systems such as power distribution networks. Due to the high power levels encountered in these applications, the current rating for the devices is also very large (see Fig. 1.2). Consequently, these types of devices are manufactured with an entire wafer serving as a single device. In addition, to improve the breakdown voltage capability, it is advantageous to prepare deep diffusions with low surface concentration because this creates a graded doping profile at the junction. A substantial portion of the applied reverse bias can then be supported within the diffused region which enhances the breakdown voltage.

The diffusion coefficient for boron is too low even at 1,200°C (see Fig. 2.21) to enable the fabrication of junctions with depths of 50–100 μm. It is necessary to utilize aluminum and gallium as P-type dopants to prepare such deep junctions with low surface concentrations by performing the diffusion at 1,200°C in a sealed tube. The masking of such diffusions was not considered to be possible until the 1970s making it impossible to fabricate planar junctions. A process for the fabrication of planar junctions was eventually developed by using silicon nitride as the masking layer.[14] Meanwhile, techniques were developed for reducing the electric field at the edges of the wafer by using a bevel. The process of beveling consists of removal of silicon at the edges of the wafer at a precisely controlled angle. The beveling of the edges has been demonstrated to enhance the breakdown voltage by reducing the electric field at the edges when compared with cutting the wafer orthogonal to the surface.

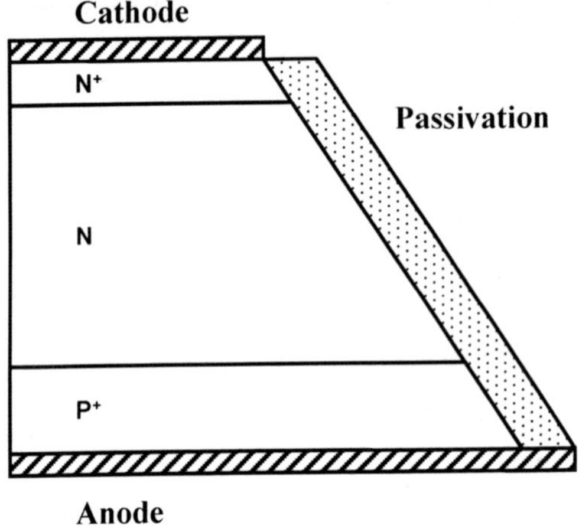

Fig. 3.40 Power rectifier with positive bevel edge termination

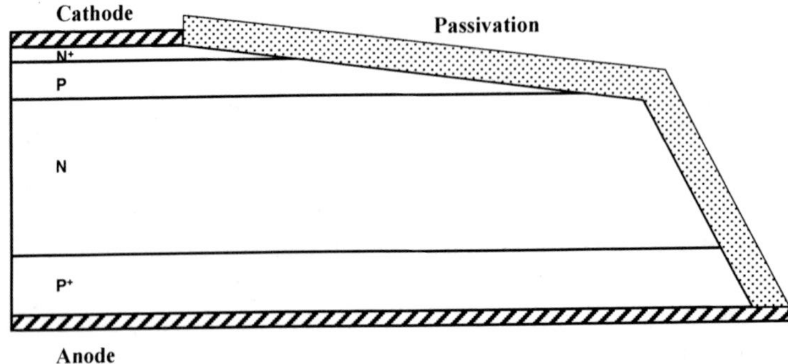

Fig. 3.41 Power thyristor with negative bevel edge termination at the upper high voltage junction

Two configurations for the bevel edge termination have been found to be successful for applications in power rectifiers and thyristors. The positive bevel configuration, shown in Fig. 3.40, is preferable for the termination of the single high-voltage junction within high-voltage power rectifiers. It is also utilized to terminate the anode junction in high-voltage thyristors to obtain a stable reverse-blocking capability. In general, a positive bevel angle is defined as one where more material is removed from the edge when progressing from the heavily doped side to the lightly doped side of the P–N junction.

The negative bevel termination, shown in Fig. 3.41, is used for the termination of the forward-blocking junction in thyristors. In general, a negative bevel angle is defined as one where more material is removed from the edge when progressing from the lightly doped side to the heavily doped side of the P–N junction. As illustrated in the figure, a relatively shallow angle is required to obtain stable performance with the negative bevel termination. These terminations are discussed in more detail below.

Positive Bevel

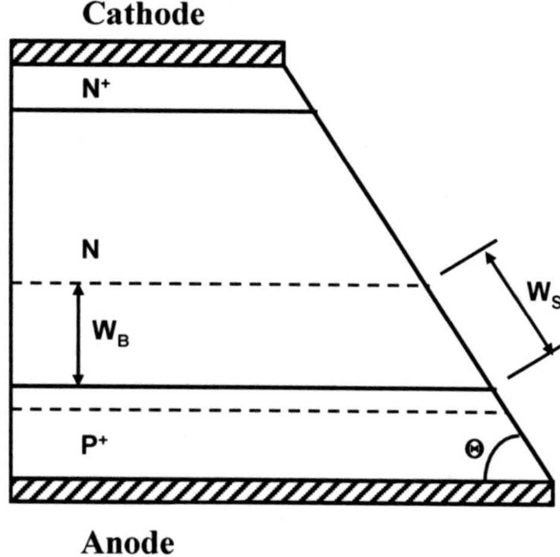

Fig. 3.42 Positive bevel edge termination: model A

The positive bevel edge termination with a bevel angle θ is illustrated in Fig. 3.42 with a simple model for the positions of the depletion regions on both sides of the P–N junction shown by the dashed lines. In this model A, the depletion regions are assumed to retain a flat shape all the way to the bevel edge. The width of the depletion region along the beveled surface (W_S) can be related to the depletion width of the P–N junction in the bulk (W_B) by

$$W_S = \frac{W_B}{\sin(\theta)}. \tag{3.72}$$

Since the same voltage is being supported across the P–N junction in the bulk and at the beveled surface, the maximum electric field at the surface of a positive bevel (E_{mPB}) is related to the maximum electric field in the bulk (E_{mB}) by

$$E_{mPB} = E_{mB}\left(\frac{W_B}{W_S}\right) = E_{mB}\sin(\theta). \tag{3.73}$$

This simple model overestimates the magnitude of the electric field at the surface of the positive bevel because it does not take into account the impact of the removal of the charge due to the bevel.

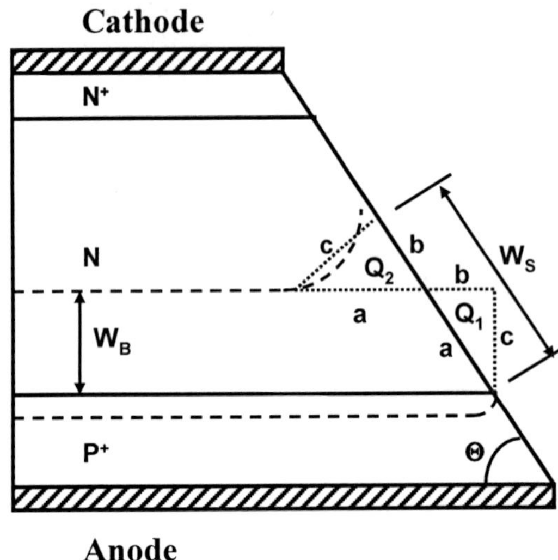

Fig. 3.43 Positive bevel edge termination: model B

The influence of the bevel on the depletion region at the edges is included in model B. A positive bevel edge termination is illustrated in Fig. 3.43 for this case with the positions of the depletion regions on both sides of the P–N junction shown by the dashed lines. To maintain charge balance between the P- and N-sides of the junction, the depletion region in the P-type region is reduced at the edge until it is pinned at the junction. At the same time, the depletion region on the N-type region expands at the edge to compensate for the removal of the charge (labeled Q_1 in the figure). The expansion of the depletion region in the N-type region near the edge can be represented by a right-angled triangle with an area Q_2 which is equal to the removed charge Q_1.

The width of the depletion region along the beveled surface (W_S) can be related to the depletion width of the P–N junction in the bulk (W_B). The hypotenuse (labeled a in the figure) of the two right-angled triangles for the charges Q_1 and Q_2 must be equal because the triangles have equal area and share a common angle.

The other sides are also related as indicated by the letters *b* and *c* in the figure. Since the dimension *c* is equal to the depletion width in the bulk (W_B),

$$a = \frac{W_B}{\sin(\theta)} \tag{3.74}$$

and

$$b = \frac{W_B}{\tan(\theta)}. \tag{3.75}$$

The width of the depletion region along the beveled surface is then given by

$$W_S = a + b = W_B\left(\frac{1}{\sin(\theta)} + \frac{1}{\tan(\theta)}\right). \tag{3.76}$$

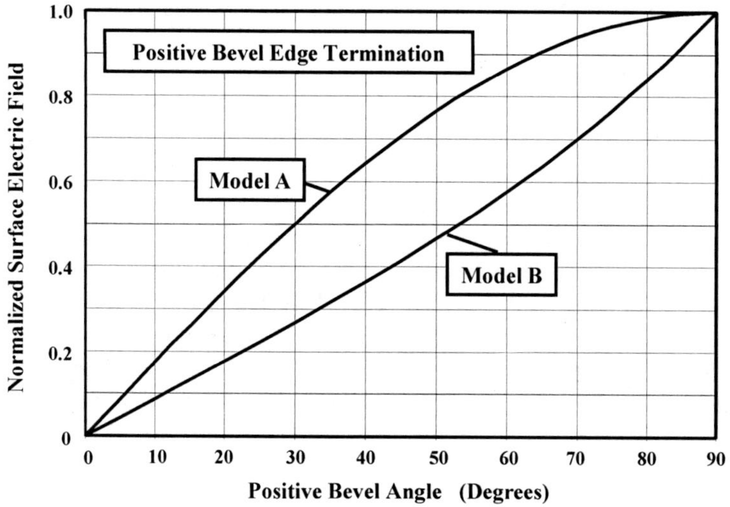

Fig. 3.44 Surface electric field in the positive bevel edge termination

Since the same voltage is being supported across the P–N junction in the bulk and at the beveled surface, the maximum electric field at the surface of a positive bevel (E_{mPB}) is given by

$$E_{mPB} = E_{mB}\left(\frac{W_B}{W_S}\right) = E_{mB}\left(\frac{\sin(\theta)}{1+\cos(\theta)}\right). \tag{3.77}$$

The variation of the normalized surface electric field to the bulk value predicted by this equation is shown in Fig. 3.44 as a function of the positive bevel angle together with the predictions of the simple model A. It can be seen that a significant reduction of the surface electric field can be achieved by using the positive bevel. With a positive bevel angle of 45°, the surface electric field is about 40% of the bulk value. It has been found that surface breakdown can occur at lower electric fields than in the bulk due to the presence of imperfections. However, a reduction in the electric field by 40% is usually sufficient to shift the breakdown from the surface to the bulk. This ensures stable operation while maximizing the breakdown voltage to that for a parallel-plane junction. It is worth pointing out that the positive bevel edge termination is the only technique that has been found to ensure that the surface electric field is well below that in the bulk of the device.

Two-dimensional numerical simulations[15] of the positively beveled junction have confirmed the reduction of the electric field at the surface. The maximum surface electric field has been extracted for a variety of positive bevel angles using simulations with different background doping concentration for the N-type region and various doping profiles for the P-type region.[16] The normalized surface electric field obtained from the simulations has a remarkably similar behavior (shown in Fig. 3.44 with model B) to that obtained with the simple analytical method described above. These results indicate that the optimum positive bevel angle is in the range of 30°–60°. Larger angles are not recommended because of the enhanced surface electric field while smaller angles lead to wasted space on the edge of the wafer.

The positive bevel angle for high-voltage power devices, such as power rectifiers and thyristors, is constructed by first attaching the wafer to a molybdenum heat sink. A nozzle is used to bombard the edge of the wafer (while it is rotated) with an abrasive powder (called *grit*). The angle of the nozzle in relation to the wafer surface determines the positive bevel angle. The damage caused to the silicon crystal by the grit blasting must be removed prior to the passivation of the surface. This technique is commonly used to terminate the high-voltage blocking junction in power rectifiers and the reverse-blocking junction in high-voltage power thyristors.

Simulation Example

To gain further insight into the operation of the positive bevel junction termination, the results of two-dimensional numerical simulations are provided in this section for the case of a drift region with doping concentration of 5×10^{13} cm^{-3}. At this doping concentration, the parallel-plane breakdown voltage was found to be 3,000 V with a depletion region thickness of 300 μm for the case of a P$^+$ region with a surface doping concentration of 1×10^{19} cm^{-3} and depth of 50 μm. These values are representative of the anode junction in high-voltage thyristors to provide its reverse-blocking capability. The simulations of the positive bevel termination were performed with various bevel angles ranging from 15° to 90°. All the structures had an oxide layer over the bevel surface as the passivation.

The reduction of the maximum surface electric field obtained from the numerical simulations is shown in Fig. 3.45 for the various bevel angles. Although the surface electric field is slightly larger than predicted by the analytical model B, the results of the simulations clearly demonstrate the benefits of using a positive bevel angle as a termination for high-voltage, large area devices. The reduction of the surface electric field predicted by model B is more than that indicated by the simulations, because the model does not account for the extra charge created by the positive bevel within the diffused P^+ side of the junction. Based upon the results of the simulations, a positive bevel angle of 45° is recommended and commonly used for power devices.

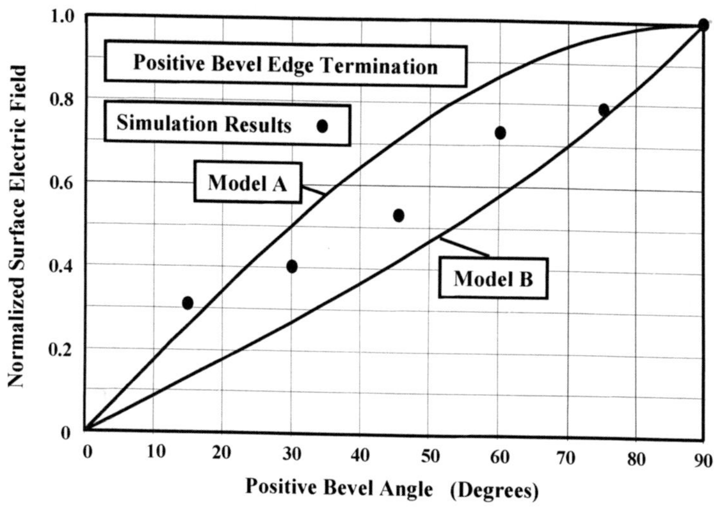

Fig. 3.45 Maximum surface electric field with a positive bevel termination

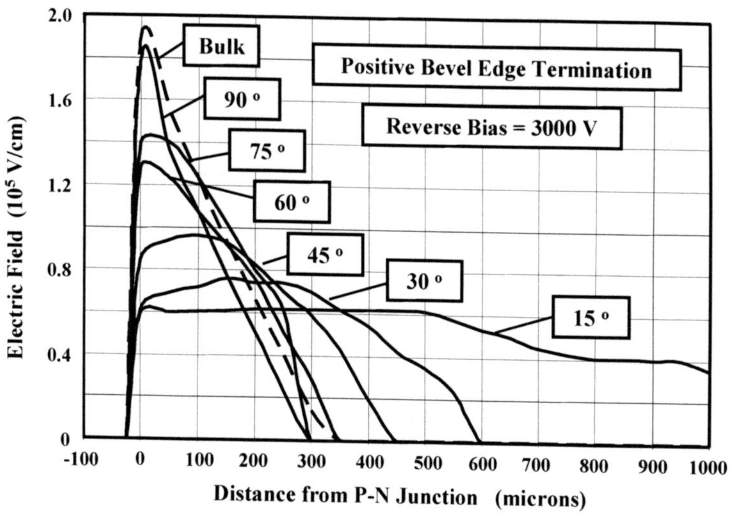

Fig. 3.46 Electric field profiles in a positive bevel termination

The electric field profiles along the surface of the bevel obtained from the simulations are compared with the electric field profile in the bulk as shown in Fig. 3.46 (dashed line). As expected, the electric field in the bulk has a maxima located at the junction with a magnitude of 1.9×10^5 V cm^{-1} at a reverse bias of 3,000 V. The electric field profile at the surface is similar in nature for a bevel angle of 90°. However, when the positive bevel angle is reduced, the peak of the surface electric field shifts away from the junction. Since the reverse bias voltage is supported over a larger distance along the surface, the magnitude of the maximum electric field is also smaller than that in the bulk. This redistribution of the electric field is responsible for suppressing surface breakdown in the case of the positive bevel termination, ensuring that the P–N junction is able to support the parallel-plane breakdown voltage.

Negative Bevel

A negative bevel is defined as one in which the area decreases when proceeding from the lightly doped side to the highly doped side of the P–N junction. Since the negative bevel removes more charge from the P^+ side of the junction than the N-side, the depletion region expands on the P-side and contracts on the N-side at the surface as illustrated in Fig. 3.47. The expansion of the depletion region on the P-side is relatively small due to the high doping concentration of the diffused region. Consequently, the width of the depletion region at the surface (W_S) is smaller than the depletion width in the bulk (W_B) for a negatively beveled junction. This implies that the electric field at the surface will be larger than in the bulk for this termination leading to unstable surface breakdown. It is, therefore, counterintuitive to make use of negative bevel edges in power devices.

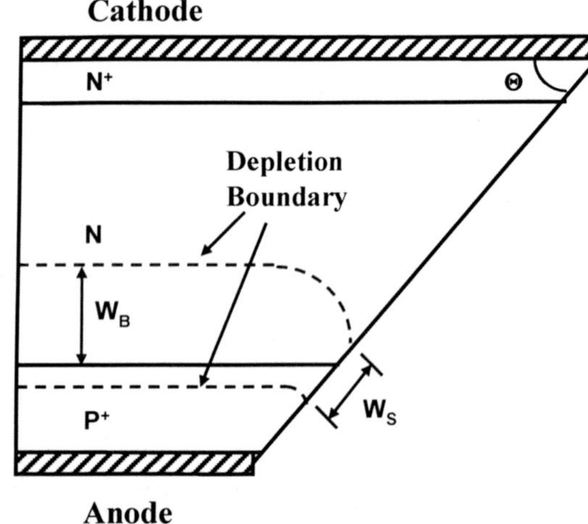

Fig. 3.47 Negative bevel edge termination

A reduction of the surface electric field can be obtained by using a negative bevel angle of sufficiently small value in combination with a highly graded diffused side of the junction. The doping concentration of the P-base region in high-voltage power thyristors has to be reduced to obtain a sufficiently large current gain for the inherent N–P–N transistor within the four layer structure. This provides an opportunity to use a negative bevel to terminate the forward-blocking junction in these devices as illustrated in Fig. 3.41.

To analyze the electric field at the surface of a negative bevel junction, consider the P–N junction shown in Fig. 3.48 with the depletion region boundaries indicated by the dashed lines. The removal of more charge on the heavily doped side of the junction produces a reduction of the depletion layer width on the lightly doped N-side of the junction at the edges. For a shallow bevel angle, it has been reported[17] that the depletion region gets pinned at the junction located at the bevel surface. At the same time, the depletion region extends along the surface on the diffused side of the junction to compensate for the charge removed by the bevel. This extension of the depletion region on the diffused side of the junction is strongly dependent on the doping profile. A low concentration gradient for the P-region favors a greater extension of the depletion region resulting in a reduction of the surface electric field. Such shallow gradients can be produced by forming very deep junction with low surface concentrations using aluminum as the dopant.

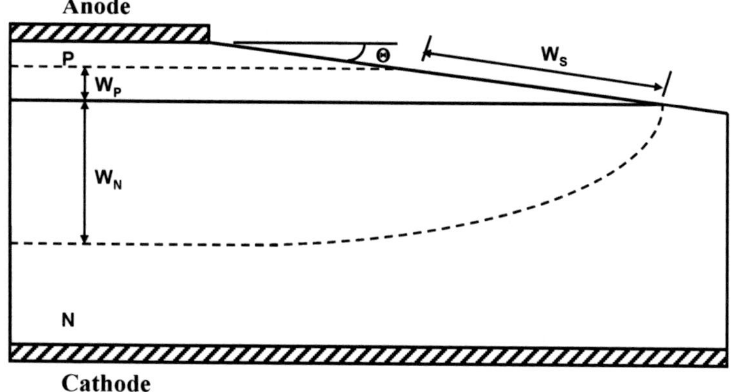

Fig. 3.48 Negative bevel edge termination with graded doping profile on the diffused side of the junction

For the model as depicted in Fig. 3.48, the extension of the depletion region along the surface (W_S) is given by

$$W_S = \frac{W_P}{\sin(\theta)}, \tag{3.78}$$

where W_P is the depletion width on the diffused side of the junction. Since the same voltage is being supported across the P–N junction in the bulk and at the beveled surface, the maximum electric field at the surface of a negative bevel (E_{mNB}) is given by

$$E_{mNB} = E_{mB}\left(\frac{W_N}{W_S}\right) = E_{mB}\frac{W_N}{W_P}\sin(\theta). \tag{3.79}$$

The variation of the normalized surface electric field to the bulk value predicted by this equation is shown in Fig. 3.49 as a function of the negative bevel angle for the case of various values of W_N/W_P. It can be seen that a significant reduction of the surface electric field can be achieved by using a very shallow negative bevel angle. For instance, the surface electric field can be reduced to 50% of the bulk value for a negative bevel angle of 2.5° when W_N/W_P is 10. Such shallow bevel angles can be fabricated by using special wafer lapping equipment that allows precise adjustment of the angle between the wafer surface and the polishing pad. A space of 1–2 mm is consumed by the negative bevel at the perimeter of the wafer resulting in a small loss in the active area for current conduction.

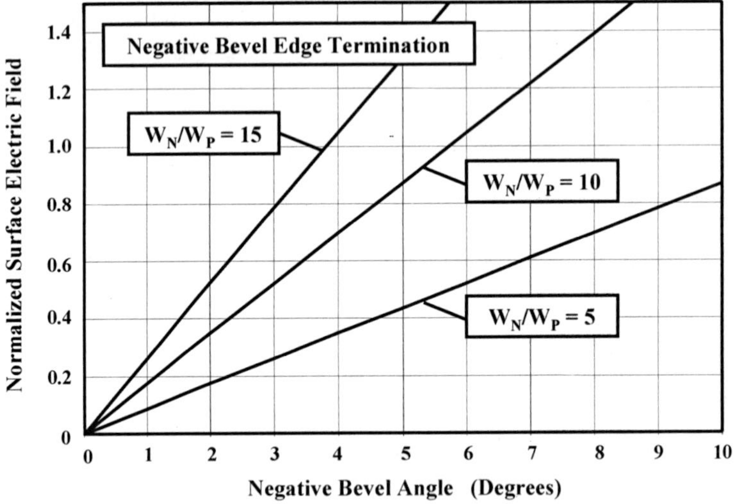

Fig. 3.49 Reduction of the maximum surface electric field with a negative bevel termination

It has been experimentally observed that the breakdown voltage of a junction terminated with a negative bevel is lower than that predicted for the parallel-plane junction. This was discovered to be due to the presence of an enhanced electric field in the bulk in the vicinity of the bevel surface.[17] Consequently, although the breakdown voltage can be stabilized by the reduction of the surface electric field with a negative bevel, its magnitude is below that of the parallel-plane junction.

Negative bevels are only used for devices, such as thyristors, with back-to-back junctions, to enable the termination of the opposing high-voltage junctions.

Simulation Example

To gain further insight into the operation of the negative bevel junction termination, the results of two-dimensional numerical simulations are provided in this section for the case of a drift region with doping concentration of 5×10^{13} cm^{-3}. At this doping concentration, the breakdown voltage for the parallel-plane junction was found to be 3,000 V with a depletion region thickness of 300 μm on the lightly doped side. A P–N junction with a highly graded P-region was achieved by using a surface doping concentration of 1×10^{17} cm^{-3} with a depth of 100 μm. These values are representative of the P-base region in high-voltage thyristors to provide its forward-blocking capability. The simulations of the negative bevel termination were performed with various bevel angles ranging from 2.5° to 10°. All the structures had an oxide layer over the bevel surface as the passivation. The depletion layer width on the P-side of the junction was found to be about 30 μm. Thus, the simulated structure has a W_N/W_P ratio of 10.

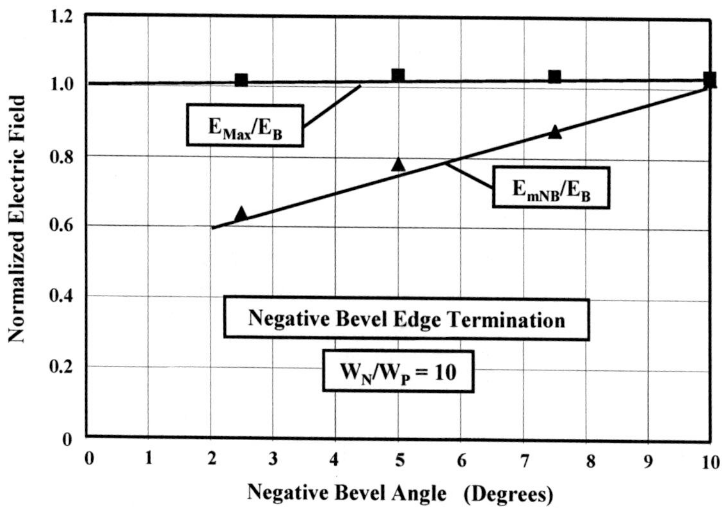

Fig. 3.50 Surface electric field reduction with a negative bevel termination

The reduction of the maximum surface electric field obtained from the numerical simulations is shown in Fig. 3.50 for the various bevel angles. The surface electric field predicted by the simple analytical model is in the range of values obtained using the simulations. Despite the reduction of the surface electric field at the bevel for the shallow angles, the breakdown voltage was found to be reduced to about 85% of the parallel-plane value. This reduction is due to an enhanced electric field in the vicinity of the negative bevel. As an example, the electric field profile is shown in Fig. 3.51 along the vertical direction at a location where the surface electric field has its maximum value. It can be seen that the maximum electric field in the bulk (E_{Max}) is larger than at the surface of the negative bevel (E_{mNB}). This maximum electric field is slightly larger than the maximum electric field observed in the bulk parallel-plane portion of the P–N junction (E_B). This phenomenon is responsible for the reduction of the breakdown voltage with a negative bevel termination.

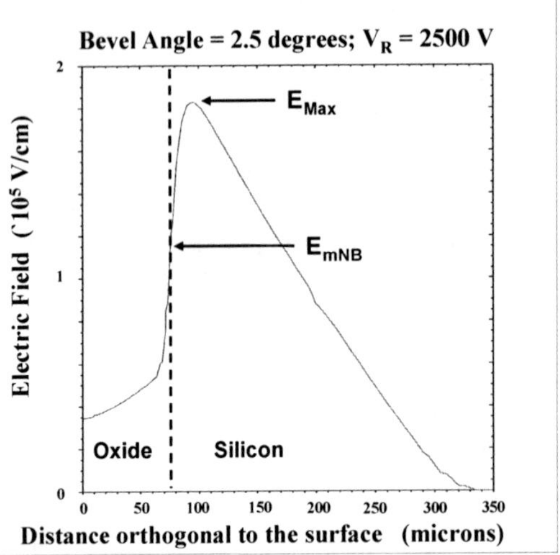

Fig. 3.51 Electric field profile in a negative bevel termination

3.6.7 Etch Terminations

One of the earliest methods for the fabrication of multiple high-voltage devices on a single wafer relied upon etching a moat around the reverse-blocking junction of each device as illustrated in Fig. 3.52. A variety of masking materials, such as black-wax, photoresist, and metal, were utilized with etch solutions containing a mixture of nitric, hydrofluoric, and sulfuric acid. Although widely used for the manufacturing of products in the 1950s and 1960s, this approach fell out of favor due to the lack of control over the shape and depth of the moat. The deep contour of the moat was also a major problem for passivation of the junction.

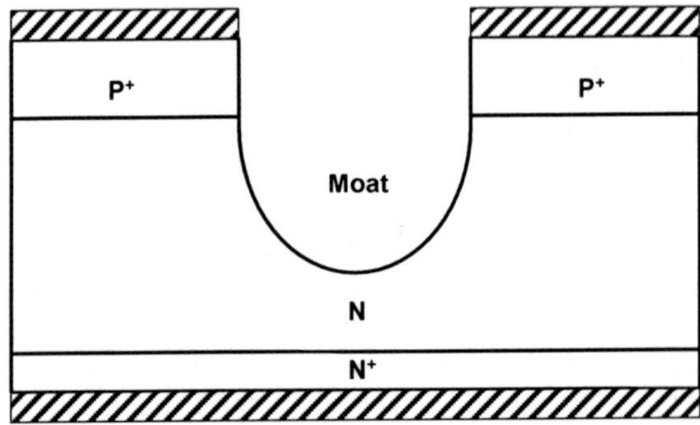

Fig. 3.52 Moat etch termination

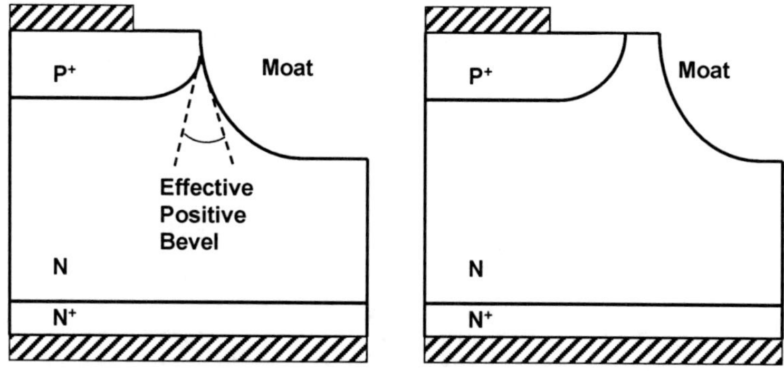

Fig. 3.53 Moat etch termination

It can be seen from the illustration of the moat etch termination in Fig. 3.52 that the surface at the edges of the junction approximates a 90° bevel angle. This results in a relatively high surface electric field at the termination producing poor stability in the breakdown voltage. This situation can be improved upon by creating a planar junction with the moat located to create an effective positive bevel. Two examples of this approach are illustrated in Fig. 3.53. In the first case, shown on the left-hand side, the moat intersects the planar junction producing a local positive bevel. This has been demonstrated to reduce the surface electric field.[18] The second approach, shown on the right-hand side, places the moat within the depletion region extending from the P–N junction. This is also equivalent to a positive bevel based upon its definition of more material being removed from the lightly doped side of the junction than the heavily doped side. Although this method for edge termination is superior to the moat etch shown in Fig. 3.52, it is not commonly used in modern power devices because the wet chemical etching of silicon has fallen out of favor in fabrication facilities.

3.6.8 Junction Termination Extension

The bevel edge terminations rely upon the selective removal of charge from the two sides of the P–N junction at the edges. A complementary approach for altering the surface electric field at the edges is based upon selectively adding charge to the junction. This can be done by ion implantation of a P-type region at the edge of a planar P^+ diffusion as shown in Fig. 3.54. This P-type region has been named the *junction termination extension* (JTE) region.[19] The charge within the P-type region can be precisely adjusted with the ion implant dose, providing better control and uniformity over the charge at the edges of the junction than with the bevel terminations while retaining a planar surface. This is advantageous for manufacturing multiple small power devices on a single wafer.

The charge within the JTE region must be precisely controlled to maximize the breakdown voltage. If the charge is small, it has little impact on the

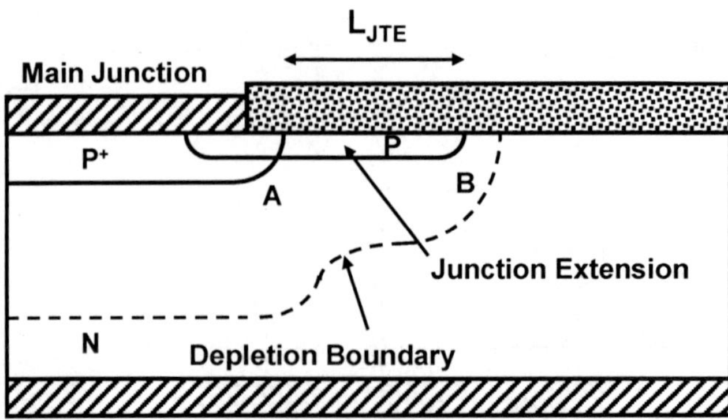

Fig. 3.54 Junction termination extension

electric field distribution and the maximum electric field will occur at point A as in the case of the unterminated planar junction. This will result in a breakdown voltage limited by the cylindrical junction curvature as discussed earlier in this chapter. If the charge in the JTE is high, it will merely serve as an extension of the junction to point B with a smaller radius of curvature. This will result in a reduction of the breakdown voltage due to the enhanced curvature at the cylindrical junction located at point B. To reduce the electric field at the main junction at point A without encountering breakdown at point B at low reverse bias voltages, the charge in the JTE region must be such that it is completely depleted by the reverse bias.

For a homogeneously doped region, the charge within the depletion region is related to the doping concentration and maximum electric field by

$$Q = \int_0^W qN_{\mathrm{A}}\mathrm{d}x = qN_{\mathrm{A}}W_{\mathrm{D}} = \varepsilon_{\mathrm{S}}E_{\mathrm{m}}. \tag{3.80}$$

The maximum electric field at the junction becomes equal to the critical electric field when the junction reaches its breakdown voltage. Under these conditions,

$$Q_{\mathrm{OPT}} = \varepsilon_{\mathrm{S}}E_{\mathrm{C}} \approx 2.07\times10^{-9}\,\mathrm{C\,cm^{-2}} \tag{3.81}$$

if a critical electric field of 2×10^5 V cm^{-1} is assumed for silicon. The corresponding dopant dose is 1.3×10^{12} cm^{-2} in the JTE region. If the dose for the P-type implant is chosen to produce this charge, the JTE region will become completely depleted at the breakdown condition. Since the entire JTE region is depleted, the electric field is distributed along the surface over the length (L_{JTE}) of the junction termination region. If the length of the JTE region is chosen to be much longer than the depletion width of the main junction, the surface electric field can be reduced to below that for a parallel-plane junction. Consequently, the breakdown voltage of this termination can approach that of an ideal parallel-plane junction.

In practice, the breakdown voltage for the JTE has been found to be strongly dependent upon the charge of the ion-implanted JTE region. Although ion implant doses can be precisely metered, the charge in the JTE region can vary due to the dopant segregation during annealing of the implant and the growth of the oxide layer for the passivation of the surface. In addition, the electric field distribution in the JTE region can be perturbed by the fixed oxide charge which can vary across the wafer. This can result in a variation of the breakdown voltage across the wafer as well as unstable behavior if mobile ions are introduced into the passivation during packaging of the die.

An elegant approach to improving the performance of the JTE is by using multiple zones of ion-implanted regions at the edge of the main junction.[20] By using three zones with a charge reduction by a factor of 2 times when proceeding from the main junction toward the exterior, breakdown voltages of over 90% of the parallel-plane value have been reported. Although these three JTE zones could be fabricated by using three masking and implant steps, an elegant and less expensive approach uses a single mask with a variable window size when proceeding from the main junction to the exterior as illustrated in Fig. 3.55. The JTE dose decreases when proceeding away from the main junction because of the smaller effective doping concentration in the silicon in spite of using a single ion implantation step. The JTE zones can also be merged together by using a suitable annealing step to diffuse the dopant sideways between the windows.

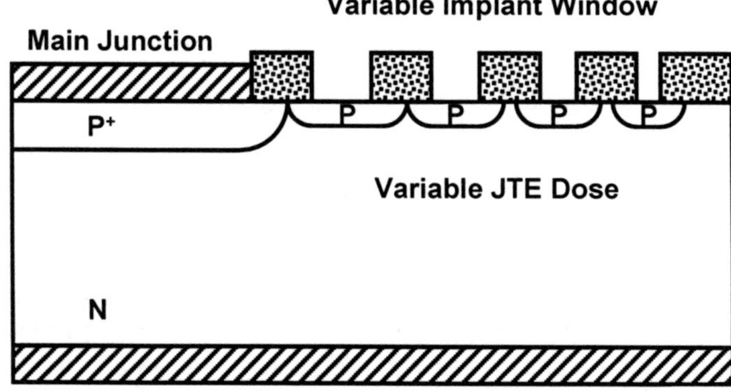

Fig. 3.55 Junction termination extension with variable lateral doping

Simulation Example

To gain further insight into the operation of the JTE, the results of two-dimensional numerical simulations are provided in this section for the case of a drift region with doping concentration of 3.8×10^{14} cm^{-3}. This doping concentration corresponds to that used earlier for the cylindrical junction and floating field ring examples. The breakdown voltage for the parallel-plane junction in this case was found to be 520 V with a depletion region thickness of 41 μm on the lightly doped side. The JTE region was created using a Gaussian doping profile with various P-type doses.

The breakdown voltages obtained using the two-dimensional numerical simulations are plotted in Fig. 3.56 as a function of the dose for the P-type dopant in the JTE region. These breakdown voltages were observed for the case of a JTE region length of 40 μm, which is equal to the width of the depletion region at breakdown for the parallel-plane junction. It can be seen that the breakdown voltage has a maximum value of about 90% of the parallel-plane junction at a JTE region dose of 1.3×10^{12} cm^{-2}. This optimum dose for the JTE region observed with the simulations is in remarkably good agreement with the optimum dose predicted by (3.81).

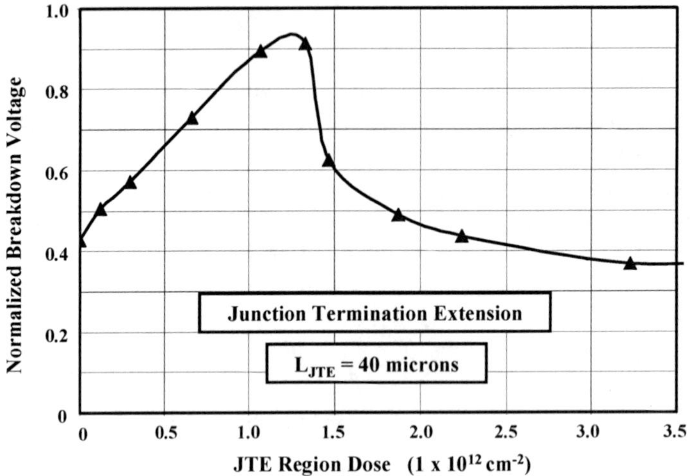

Fig. 3.56 Breakdown voltage obtained with the junction termination extension for a length of 40 μm

The change in the breakdown voltage with increasing dose in the JTE region can be understood by examination of the electric field profile within the JTE region. At low values for the dose, the electric field along the surface within the JTE region exhibits a maximum value at the edge of the main junction as shown in Fig. 3.57 for the case of dose of 0.3×10^{12} cm^{-2}. For high values of the dose, the electric field along the surface within the JTE region exhibits a maximum value at the edge of the JTE region as shown in Fig. 3.57 for the case of dose of 1.5×10^{12} cm^{-2}. When a dose of 1.1×10^{12} cm^{-2} is used, the electric field becomes relatively flat within the JTE region. This results in the highest breakdown voltage occurring at the optimum dose of 1.3×10^{12} cm^{-2}.

The length of the JTE region must also be sufficient to obtain the full benefits of reduction of the electric field along the surface at the edge of the planar junction. To illustrate this, the results of two-dimensional numerical simulations for the JTE are shown in Fig. 3.58 for the case of a JTE dose of 1.1×10^{12} cm^{-2}. A substantial increase in the breakdown voltage over that for the cylindrical junction is obtained when the length of the JTE region extends to the width of the depletion region for the parallel-plane junction at breakdown. An increase in the length of the JTE region much beyond this length does not produce further improvement in the breakdown voltage while consuming space at the edge of the device.

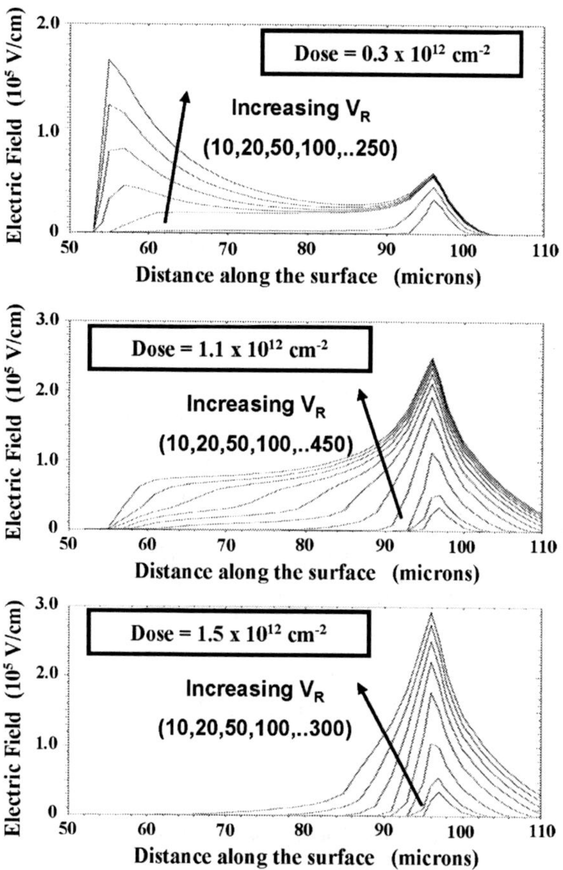

Fig. 3.57 Electric field profiles within the JTE region for three values of the JTE dose

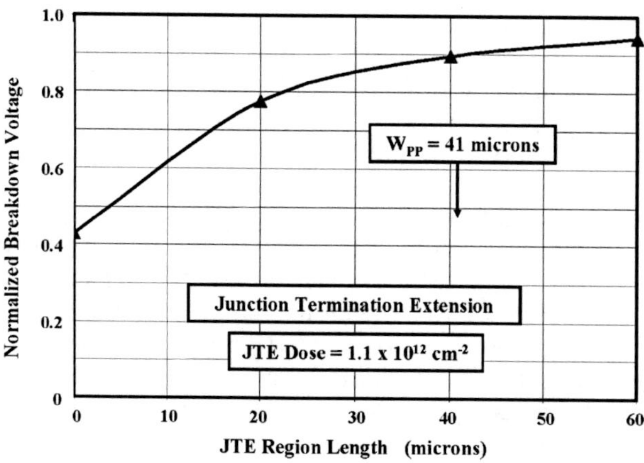

Fig. 3.58 Breakdown voltage obtained with the junction termination extension

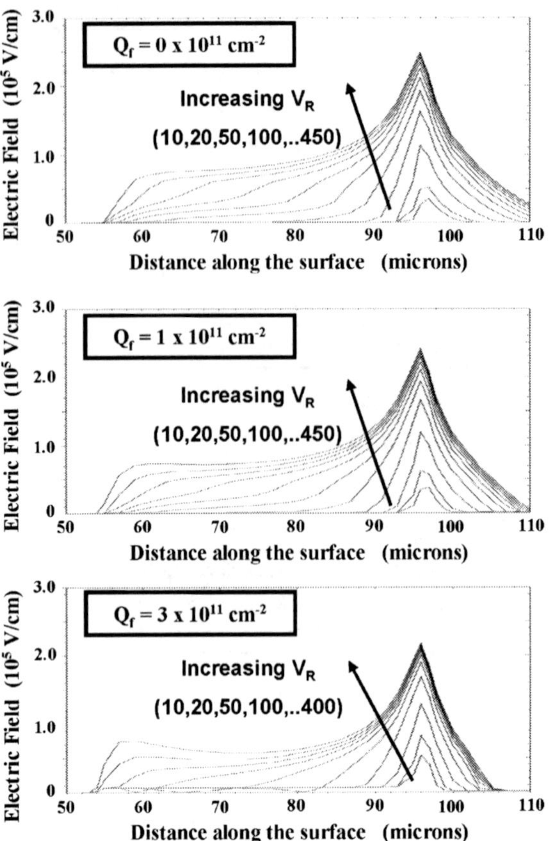

Fig. 3.59 Impact of fixed oxide charge on the electric field profiles within the JTE region

As discussed earlier, the breakdown voltage for the JTE is sensitive to the presence of charge above the JTE region. The optimum charge in the JTE region has been shown to be 1.3×10^{12} cm^{-2}. Unfortunately, the fixed oxide charge in thermally grown oxide over the silicon surface can be of this order of magnitude. If an optimum dose is used for the JTE region, the presence of the fixed oxide charge alters the electric field distribution along the surface at the JTE region producing a reduction of the breakdown voltage. To illustrate the influence of any surface charge on the breakdown voltage of the JTE, two-dimensional numerical simulations were conducted for the case of the JTE structure with a dose of 1.1×10^{12} cm^{-2} and length of 40 μm with fixed oxide charge of 1×10^{11} cm^{-2}. The breakdown voltage was found to be reduced from 465 V (89% of the breakdown voltage for the parallel-plane junction) without the fixed oxide charge to 450 V (86% of the breakdown voltage for the parallel-plane junction) with the fixed oxide charge. When the charge was increased to 3×10^{11} cm^{-2}, the breakdown voltage was reduced to 416 V (80% of the breakdown voltage for the parallel-plane junction).

These reductions of the breakdown voltage in the presence of the fixed oxide charge can be correlated with changes in the electric field distribution along the surface in the JTE region. The electric field profile is not significantly modified by the presence of a fixed oxide charge of 1×10^{11} cm^{-2} as shown in Fig. 3.59. However, when the charge is increased to 3×10^{11} cm^{-2}, the electric field at the edge of the JTE region is enhanced leading to the observed reduction of the breakdown voltage. Based upon these results, it can be concluded that the fixed oxide charge should be reduced to less than 1×10^{11} cm^{-2} during the passivation of the JTE.

3.7 Open-Base Transistor Breakdown

In the previous sections of this chapter, the breakdown voltage of the P–N junction has been analyzed including the impact of enhanced electric fields at their terminations. These results have relevance to power devices with single blocking junctions, such as power rectifiers and power MOSFETs. In many other power devices, such as power thyristors and IGBTs, the structures contain back-to-back P–N junctions. The maximum voltage that can be supported by these structures becomes limited by *open-base transistor breakdown*. In this situation, the current generated by impact ionization is amplified by the gain of the bipolar transistor.[21]

An open-base P–N–P transistor is illustrated in Fig. 3.60 with a positive bias applied to electrode on the right-hand side. The applied voltage produces a forward bias across the junction J_2 while junction J_1 becomes reverse biased. The reverse-biased junction J_1 supports the voltage with the development of a depletion

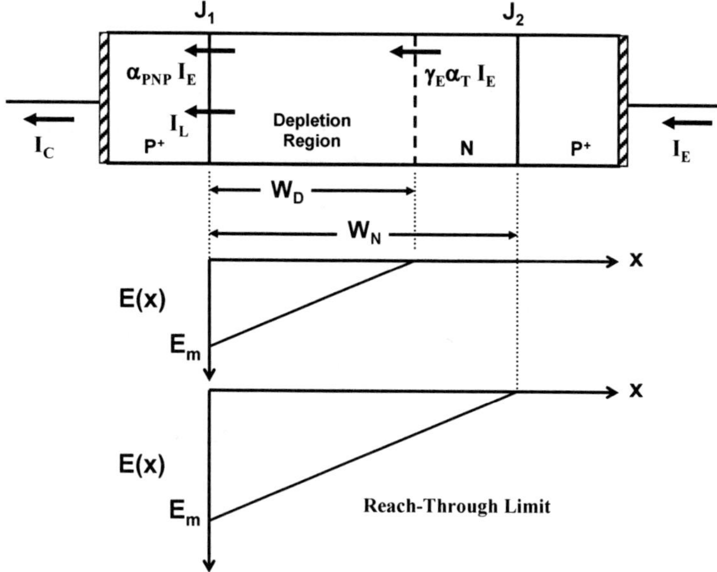

Fig. 3.60 Open-base transistor breakdown analysis

region in the N-base region. The boundary of the depletion region is indicated by the dashed line. The electric field profile for this case is shown in the middle of the figure. If the width (W_N) of the N-base region is large, breakdown will occur when the maximum electric field (E_m) becomes equal to the critical electric field for breakdown for the semiconductor. The breakdown voltage for this case is the same as that of the parallel-plane junction. This corresponds to the multiplication coefficient becoming infinitely large.

When the width (W_N) of the N-base region becomes shorter than the depletion width for breakdown for the parallel-plane junction, the breakdown voltage for the P–N–P transistor is reduced due to the reach-through effect. Open-base transistor breakdown is precipitated by the injection of holes from the forward-biased junction J_2 with the current flow amplified by the gain of the bipolar transistor. The current due to the injected holes is indicated by the arrow labeled ($\gamma_E \alpha_T I_E$) at the depletion boundary and ($\alpha_{PNP} I_E$) at the collector junction J_1. Here, the emitter injection efficiency (γ_E) is close to unity and the base transport factor (α_T) is less than unity, while the common emitter current gain (α_{PNP}) of the P–N–P transistor becomes larger than unity due to the onset of carrier multiplication at high bias voltages. In addition, the current due to the generation of carriers in the depletion region and the neutral region is indicated by the arrow labeled I_L in the figure. Using Kirchhoff's law,

$$I_C = \alpha_{PNP} I_E + I_L = I_E, \tag{3.82}$$

leading to

$$I_C = I_E = \frac{I_L}{(1 - \alpha_{PNP})}. \tag{3.83}$$

Based upon this equation, it can be concluded that the current will become large when the current gain of the transistor approaches unity. The criterion for breakdown for the open-base transistor can therefore be written as

$$\alpha_{PNP} = \gamma_E \alpha_T M = 1, \tag{3.84}$$

where M is the multiplication coefficient. These terms are discussed in detail in the chapter on bipolar power transistors. Due to the low doping concentration in the N-base region in symmetric blocking devices like thyristors and IGBTs to enable the support of high voltages, the injection efficiency can be assumed to be equal to unity. The base transport factor is less than unity as determined by the undepleted base width ($W_N - W_D$) and the minority carrier diffusion length (L_p):

$$\alpha_T = \cosh^{-1}\left[\frac{W_N - W_D}{L_p}\right], \tag{3.85}$$

where W_D is the width of the depletion layer. The depletion region width is related to the applied reverse bias voltage by (3.15). The multiplication coefficient is also a function of the applied reverse bias as given by (3.9) for the case of a P^+/N diode. The breakdown voltage for the open-base transistor can therefore be determined by the evaluation of the current gain as a function of the applied reverse bias to determine the voltage at which it becomes equal to unity.

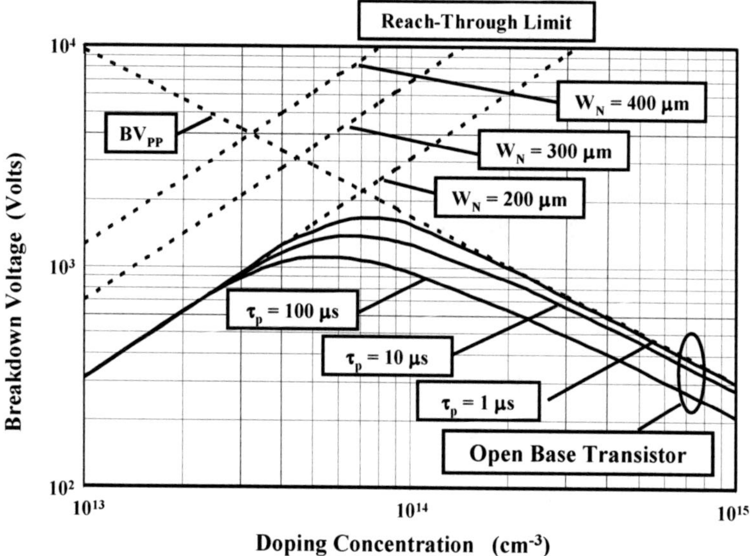

Fig. 3.61 Open-base transistor breakdown voltage

As an example, the breakdown voltage for the open-base transistor is plotted in Fig. 3.61 for the case of an N-base width of 200 μm. Three values for the minority carrier (hole) lifetime are taken into consideration. In addition, the boundaries defined by the pure avalanche breakdown (BV_{PP}) and the reach-through breakdown are included for comparison. The reach-through limit is defined as the voltage at which the depletion region width becomes equal to the width of the N-base region. This voltage is given by

$$BV_{RT} = \frac{qN_D W_N^2}{2\varepsilon_S}, \tag{3.86}$$

where N_D is the doping concentration in the N-base region. It is worth pointing out that the reach-through breakdown limit does not take into consideration avalanche multiplication at the reverse-biased junction. Instead, it is assumed that when the depletion region extends through the entire N-base region, any further applied bias produces the injection of minority carriers from the forward-biased junction leading to the onset of high-current flow.

From the figure, it can be observed that, on the one hand, when the doping concentration in the N-base region is low (less than 3×10^{13} cm^{-3} in the example), the open-base transistor breakdown voltage is limited by the reach-through phenomenon. On the other hand, when the doping concentration of the N-base region is high (more than 3×10^{14} cm^{-3} in the example with a low minority carrier lifetime), the open-base transistor breakdown voltage is limited by the avalanche multiplication phenomenon. The highest breakdown voltage is observed at a doping concentration of about 7×10^{13} cm^{-3} for this case with an N-base width of 200 μm. The open-base transistor breakdown voltage is always lower than the avalanche breakdown voltage with a greater reduction for cases with larger values for the minority carrier lifetime. For a minority carrier lifetime of 1 μs, the highest open-base breakdown voltage is 1,670 V at an N-base doping concentration of 7×10^{13} cm^{-3} when compared with an avalanche breakdown voltage of 2,200 V at this doping concentration. This reduction must be taken into account during the design of devices, such as thyristors, which must exhibit both high forward- and reverse-blocking capability.

Simulation Example

To gain further insight into the operation of the open-base transistor, the results of two-dimensional numerical simulations are provided in this section for the case of an N-base width of 200 μm. The breakdown voltages obtained using the two-dimensional numerical simulations are plotted in Fig. 3.62 as a function of the doping concentration in the N-base region for the case of a minority carrier lifetime of 10 μs. The open-base transistor breakdown voltages predicted by the analytical model are in excellent agreement with the results of the simulations providing confidence in the model. The maximum breakdown voltage was found to be 1,420 V at a doping concentration of 8×10^{13} cm^{-3} with the simulations when compared with 1,380 V obtained using the analytical model at a doping concentration of 7×10^{13} cm^{-3}. When the minority carrier lifetime was decreased to 1 μs in the

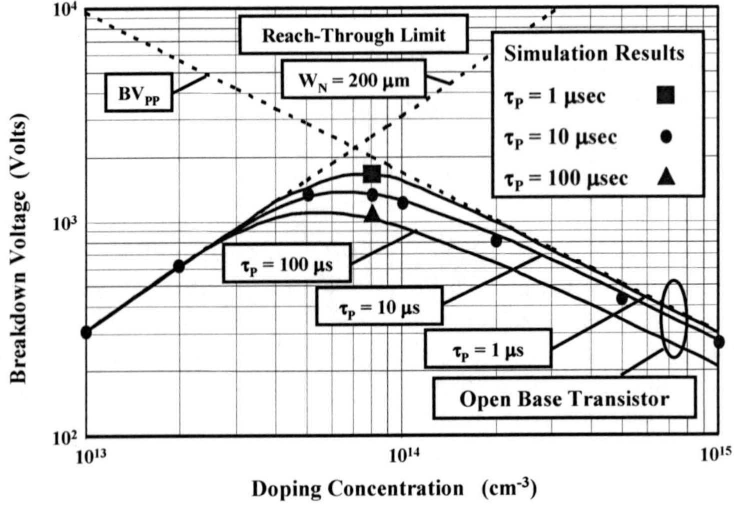

Fig. 3.62 Open-base transistor breakdown voltage for an N-base width of 200 μm

simulations, the breakdown voltage was found to increase to 1,600 V at a doping concentration of 8×10^{13} cm^{-3} when compared with 1,670 V obtained using the analytical model. When the minority carrier lifetime was increased to 100 μs in the simulations, the breakdown voltage was found to decrease to 1,190 V at a doping concentration of 8×10^{13} cm^{-3} when compared with 1,110 V obtained using the analytical model.

3.7.1 Composite Bevel Termination

Power devices designed to support high voltages in the first and third quadrant of operation contain an open-base transistor structure with two back-to-back junctions. It is necessary to provide an edge termination for both of these junctions simultaneously. One approach to achieving a reduction of the surface electric field at both the junctions is by combining a positive bevel with a negative bevel as illustrated in Fig. 3.63. The depletion region boundary for the case of a positive bias applied to the anode, for operation in the first quadrant, is indicated in the figure by the dotted lines. In this case, the upper P–N junction J_1 becomes reverse biased leading to the extension of the depletion region downward toward junction J_2. This results in open-base transistor breakdown in the bulk region of the thyristor, which is smaller than the breakdown voltage at the negative bevel termination. In contrast, the depletion region boundary for the case of a negative bias applied to the anode, for operation in the third quadrant, is indicated in the figure by the dashed lines. In this case, the lower P–N junction J_2 becomes reverse biased leading to the extension of the depletion region upward toward junction J_1. This also results in open-base transistor breakdown in the bulk region of the thyristor, which is smaller than the breakdown voltage at the positive bevel termination. The design of the thyristor structure must take this into consideration.

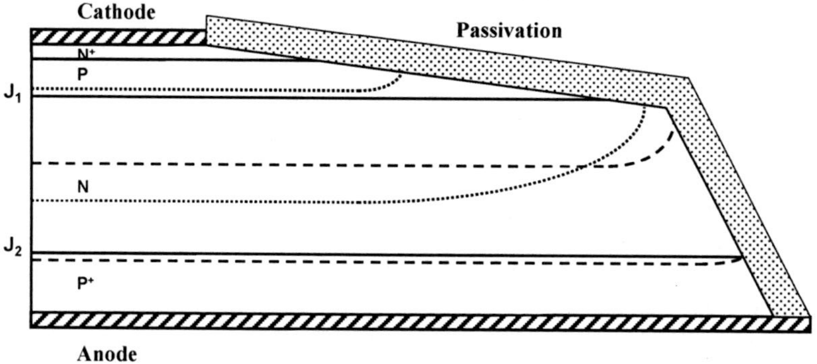

Fig. 3.63 Composite positive/negative bevel termination

3.7.2 Double-Positive Bevel Termination

Another approach to providing the edge termination for devices, such as thyristors designed to support high voltages in the first and third quadrant of operation with two back-to-back junctions, is by using a double-positive bevel. This concept relies on creating a local positive bevel for both of the high-voltage junctions. Two shapes for the double-positive bevel have been explored. The V-shape shown on the left-hand side of Fig. 3.64 can be produced by grit blasting at various angles to the edge of the wafer. The rounded shape shown on the right-hand side of the figure can be produced by using slurry on a wire that contacts the wafer edge while the wafer is rotated about its center.

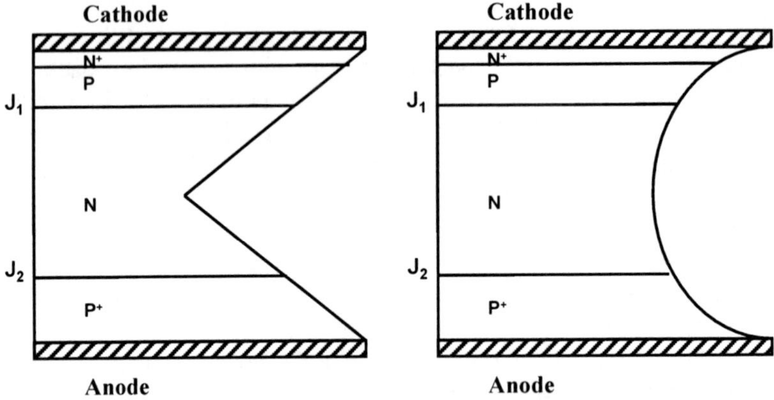

Fig. 3.64 Double-positive bevel terminations

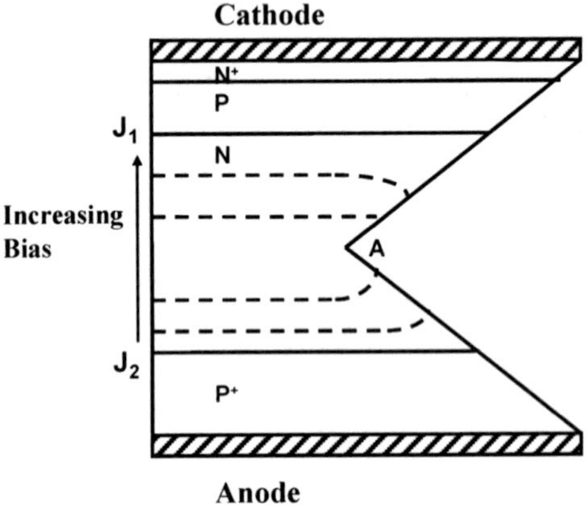

Fig. 3.65 Depletion layer extension in the double-positive bevel termination

The depletion region boundary for the case of a negative bias applied to the anode, for operation in the third quadrant, is indicated in Fig. 3.65 by the dashed lines for the V-shaped double-positive bevel termination. In this case, the lower P–N junction J_2 becomes reverse biased leading to the extension of the depletion region upward toward junction J_1. At lower reverse bias voltages, the edge termination behaves as a positive bevel termination with the depletion region expanding at the edges. This reduces the electric field at the bevel surface. However, with increasing reverse bias, the depletion region eventually extends past the corner of the bevel (point A). Beyond this voltage, the angle of the bevel is reversed producing a local negative bevel. This alters the depletion region shape as indicated in the figure. In spite of this change in the bevel angle, the double-positive bevel termination has been found[22] to suppress the surface electric field enabling breakdown to occur in the bulk. Two manufacturing difficulties that had to be overcome while developing this edge termination were to avoid breakage of the thin wafer edges created by the bevel and the passivation of the surface due to its concave topography.

Simulation Example

To gain further insight into the operation of the double-positive bevel termination, the results of two-dimensional numerical simulations are provided in this section for the case of the V-shaped edge termination. The structure had an N-base width of 300 μm with a doping concentration of 5×10^{13} cm^{-3}. The diffused P$^+$ region had a junction depth of 50 μm. When a minority carrier lifetime of 1 μs was used, the open-base transistor breakdown voltage for the parallel-plane structure was found to be just above 2,500 V for both polarities of the bias voltage. The double-bevel termination had a bevel angle of 45° in both directions. As expected, the breakdown voltages obtained for the double-bevel termination with both polarities of bias were found to be the same as that for the parallel-plane junction.

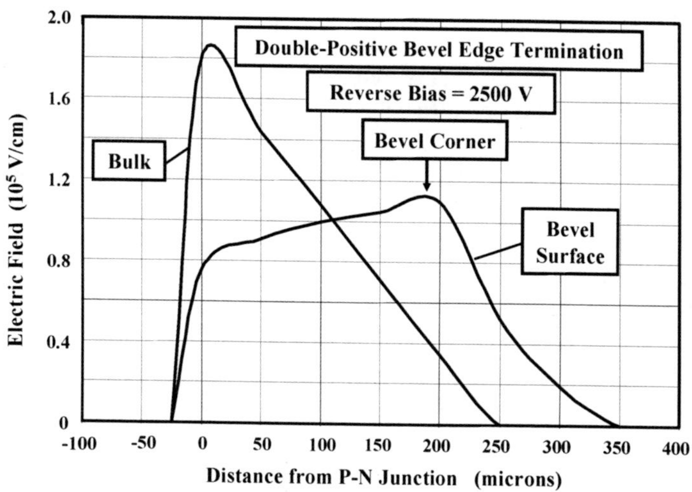

Fig. 3.66 Electric field profiles within the double-positive bevel termination

The electric field distribution within the double-positive bevel termination is shown in Fig. 3.66 at a bias of 2,500 V. The electric field in the bulk exhibits the triangular shape with a maximum value of 1.85×10^5 V cm^{-1} located at the junction. The electric field along the bevel edge has a much lower value of 0.85×10^5 V cm^{-1} at the junction. The maximum electric field occurs in the vicinity of the corner of the bevel because of the change in the bevel angle from a positive value to a negative value. However, this electric field is also much lower than in the bulk, ensuring a breakdown voltage equal to that for the open-base transistor with parallel-plane junctions. A much less reduction of the surface electric field has been reported for the rounded bevel termination.[16]

3.8 Surface Passivation

The leakage current and breakdown voltage of power devices can be compromised by poor surface passivation. The presence of mobile charge close to the semiconductor surface can lead to alterations of the electric fields at the edges during device operation producing changes in the breakdown voltage. Inadequate surface preparation of the silicon surface can lead to the presence of defects, such as dislocations, that adversely impact the leakage current due to a high density of deep levels in the band gap. These defects can also initiate premature breakdown due to localized enhancements of the electric field.

For very high-voltage power devices, such as thyristors with bevel edge terminations, the commonly used surface passivation is with rubberized coatings or organic polymers.[23] The surface damage produced by the beveling process must be first removed by chemical etching of the surface, immediately followed by coating with silicone rubber or polyimide layers. The passivation layer is then cured. The devices are enclosed in a hermetically sealed package to minimize the presence of mobile ions and moisture.

For planar devices, it is commonplace to use silicon dioxide as the passivation layer at the edges. However, sodium and potassium ions are known to migrate through the oxide creating instability in the breakdown voltage. This can be prevented by covering the oxide with a silicon nitride or oxynitride film using plasma-enhanced chemical vapor deposition.[24] Another approach used for planar power devices is to utilize semi-insulating polycrystalline silicon (SIPOS) films.[25] The resistivity of these films can be controlled by adjusting their oxygen content. Power bipolar transistors with breakdown voltages as high as 10,000 V have been fabricated using the SIPOS passivation method.

3.9 Summary

A relatively high breakdown voltage is the most distinguishing feature for a power device. This chapter has provided the criteria for the design of the breakdown voltage for typical P–N junction diodes that are representative of the internal structure of power devices. In practical structures, the breakdown voltage can be drastically reduced by the enhancement of the electric field at the edges of the devices. Various methods to suppress this electric field enhancement have been analyzed in this chapter. For devices with areas that are a small fraction of the wafer area, the most attractive edge terminations utilize planar junctions. For high-current devices fabricated using an entire wafer, it is possible to bevel the edge to reduce the electric field and ensure bulk breakdown.

Problems

3.1 Compare the impact ionization coefficient obtained using Fulop's approximation with that for electrons and holes in silicon at an electric field of 2×10^5 V cm^{-1}.

3.2 Compare the impact ionization coefficient obtained using Baliga's approximation with that for holes in 4H-SiC at an electric field of 2×10^6 V cm^{-1}.

3.3 Calculate the parallel-plane breakdown voltage for silicon abrupt P–N junctions at drift region doping concentrations of 1×10^{13}, 1×10^{14}, 1×10^{15}, and 1×10^{16} cm^{-3}.

3.4 Calculate the parallel-plane breakdown voltage for 4H-SiC abrupt P–N junctions at drift region doping concentrations of 1×10^{14}, 1×10^{15}, 1×10^{16}, and 1×10^{17} cm^{-3}.

3.5 Calculate the maximum depletion layer width at breakdown for silicon abrupt P–N junctions at drift region doping concentrations of 1×10^{13}, 1×10^{14}, 1×10^{15}, and 1×10^{16} cm^{-3}.

3.6 Calculate the maximum depletion layer width at breakdown for 4H-SiC abrupt P–N junctions at drift region doping concentrations of 1×10^{14}, 1×10^{15}, 1×10^{16}, and 1×10^{17} cm^{-3}.

3.7 Compare the critical electric field at breakdown for silicon and 4H-SiC abrupt P–N junctions with the same breakdown voltage of 1,000 V.

3.8 Compare the ideal specific on-resistance for the drift region in silicon and 4H-SiC devices with the same breakdown voltage of 1,000 V.

3.9 Calculate the width of the drift region for a silicon punch-through diode to achieve a breakdown voltage of 1,000 V if the drift region doping concentration is 2×10^{13} cm^{-3}.

3.10 Calculate the width of the drift region for a 4H-SiC punch-through diode to achieve a breakdown voltage of 1,000 V if the drift region doping concentration is 1×10^{15} cm^{-3}.

3.11 Calculate the breakdown voltage for a cylindrical junction termination with a depth of 3 μm for a silicon drift region with doping concentration of 1×10^{14} cm^{-3}.

3.12 Calculate the breakdown voltage for a spherical junction termination with a depth of 3 μm for a silicon drift region with doping concentration of 1×10^{14} cm^{-3}.

3.13 Calculate the breakdown voltage for a junction termination using the single optimally located floating field ring with a depth of 3 μm for a silicon drift region with doping concentration of 1×10^{14} cm^{-3}.

3.14 Determine the spacing for the single optimally located floating field ring in Problem 3.13. What is the mask dimension required for this design?

3.15 Determine the normalized surface electric field for a positive bevel termination with an angle of 45°.

3.16 Determine the normalized surface electric field for a negative bevel termination with an angle of 3° if the ratio of the depletion layer widths on the lightly doped side to the heavily doped side of the junction is 10.

3.17 Determine the optimum charge for the junction termination extension in a silicon device.

3.18 Determine the optimum charge for the junction termination extension in a 4H-SiC device.

3.19 Calculate the breakdown voltage for an open-base silicon transistor with a drift region doping concentration of 5×10^{13} cm^{-3} and thickness of 300 μm if the low-level lifetime is 10 μs. Compare this value with the avalanche breakdown voltage and the reach-through breakdown voltage.

3.20 Determine the impact of changing the drift region doping concentration to 2×10^{13} cm^{-3} in Problem 3.19.

References

[1] W. Fulop, "Calculation of Avalanche Breakdown of Silicon P–N Junctions", Solid-State Electronics, Vol. 10, pp. 39–43, 1967.

[2] B.J. Baliga, "Silicon Carbide Power Devices", World Scientific, Singapore, 2006.

[3] R.J. McIntyre, "Multiplication Noise in Uniform Avalanche Diodes", IEEE Transactions on Electron Devices, Vol. ED-13, pp. 164–168, 1966.

[4] S.K. Ghandhi, "Semiconductor Power Devices", p. 39, Wiley, New York, 1977.
[5] N.R. Howard, "Avalanche Multiplication in Silicon Junctions", Journal of Electronics and Control, Vol. 13, pp. 537–544, 1962.
[6] S.M. Sze and G. Gibbons, "Effect of Junction Curvature on Breakdown Voltage in Semiconductors", Solid-State Electronics, Vol. 9, pp. 831–845, 1966.
[7] V.A.K. Temple and M.S. Adler, "Calculation of the Diffusion Curvature Related Avalanche Breakdown in High Voltage Planar P–N Junctions", IEEE Transactions on Electron Devices, Vol. ED-22, pp. 910–916, 1975.
[8] B.J. Baliga and S.K. Ghandhi, "Analytical Solutions for the Breakdown Voltage of Abrupt Cylindrical and Spherical Junctions", Solid-State Electronics, Vol. 19, pp. 739–744, 1976.
[9] R. Van Overstraeten and H. DeMan, "Measurements of the Ionization Rates in Diffused Silicon P–N Junctions", Solid-State Electronics, Vol. 13, pp. 583–608, 1970.
[10] Y.C. Koa and E.D. Wolley, "High Voltage Planar P–N Junctions", Proceeding of the IEEE, Vol. 55, pp. 1409–1414, 1967.
[11] M.S. Adler et al., "Theory and Breakdown Voltage of Planar Devices with a Single Field Limiting Ring", IEEE Transactions on Electron Devices, Vol. ED-24, pp. 107–113, 1977.
[12] B.J. Baliga, "Closed Form Analytical Solutions for the Breakdown Voltage of Planar Junctions Terminated with a Single Floating Field Ring", Solid-State Electronics, Vol. 33, pp. 485–488, 1990.
[13] A.S. Grove, O. Leistiko, and W.W. Hooper, "Effect of Surface Fields on the Breakdown Voltage of Planar Silicon P–N Junctions", IEEE Transactions on Electron Devices, Vol. ED-14, pp. 157–162, 1967.
[14] B.J. Baliga, "Deep Planar Gallium and Aluminum Diffusions in Silicon", Journal of the Electrochemical Society, Vol. 126, pp. 292–296, 1979.
[15] J. Cornu, "Field Distribution Near the Surface of Beveled P–N Junctions of High Voltage Devices", IEEE Transactions on Electron Devices, Vol. ED-20, pp. 347–352, 1973.
[16] M.S. Adler and V.A.K. Temple, "Maximum Surface and Bulk Electric Fields at Breakdown for Planar and Beveled Devices", IEEE Transactions on Electron Devices, Vol. ED-25, pp. 1266–1270, 1978.
[17] M.S. Adler and V.A.K. Temple, "A General Method for Predicting the Avalanche Breakdown Voltage of Negative Beveled Devices", IEEE Transactions on Electron Devices, Vol. ED-23, pp. 956–960, 1976.
[18] V.A.K. Temple, B.J. Baliga, and M.S. Adler, "The Planar Junction Etch for High Voltage and Low Surface Fields in Planar Devices", IEEE Transactions on Electron Devices, Vol. ED-24, pp. 1304–1310, 1977.
[19] V.A.K. Temple, "Junction Termination Extension: A New Technique for Increasing Avalanche Breakdown Voltage and Controlling Surface Electric Fields at P–N Junctions", IEEE International Electron Devices Meeting, Abstract 20.4, pp. 423–426, 1977.

[20] R. Stengle and U. Gosele, "Variation of Lateral Doping – A New Concept to Avoid High Voltage Breakdown of Planar Junctions", IEEE International Electron Devices Meeting, Abstract 6.4, pp. 154–157, 1985.
[21] A. Herlet, "The Maximum Blocking Capability of Silicon Thyristors", Solid-State Electronics, Vol. 8, pp. 655–671, 1965.
[22] J. Cornu, S. Schweitzer, and O. Kuhn, "Double Positive Beveling: A better Edge Contour for High-Voltage Devices", IEEE Transactions on Electron Devices, Vol. 21, pp. 181–184, 1974.
[23] R.R. Verderber et al., "Passivation of High Power Rectifiers", IEEE Transactions on Electron Devices, Vol. ED-17, pp. 797–799, 1970.
[24] R.E. Blaha and W.R. Fahrner, "Passivation of High Breakdown Voltage P–N–P Structures by Thermal Oxidation", Journal of the Electrochemical Society, Vol. 123, pp. 515–518, 1976.
[25] T. Matsushita et al., "Highly Reliable High Voltage Transistors by Use of the SIPOS Process", IEEE Transactions on Electron Devices, Vol. ED-23, pp. 826–830, 1976.

Chapter 4

Schottky Rectifiers

A Schottky rectifier is formed by making an electrically nonlinear contact between a metal and the semiconductor drift region. The Schottky rectifier is an attractive unipolar device for power electronic applications due to its relatively low on-state voltage drop and its fast switching behavior. It has been widely used in power supply circuits with low operating voltages due to the availability of excellent devices based upon silicon technology. In the case of silicon, the maximum breakdown voltage of Schottky rectifiers has been limited by the increase in the resistance of the drift region. Commercially available devices are generally rated at breakdown voltages of less than 100 V. Novel silicon structures that utilize the charge-coupling concept have allowed extending the breakdown voltage to the 200 V range.[1,2]

Many applications described in Chap. 1 require fast switching rectifiers with low on-state voltage drop that can also support over 500 V. The much lower resistance of the drift region for silicon carbide enables development of such Schottky rectifiers with very high breakdown voltages.[3] These devices not only offer fast switching speed but also eliminate the large reverse recovery current observed in high-voltage silicon P-i-N rectifiers. This reduces switching losses not only in the rectifier but also in the IGBTs used within the power circuits.[4]

In this chapter, the basic structure of the power Schottky rectifier is first introduced to define its constituent elements. This chapter then provides a discussion of the basic principles of operation of the metal–semiconductor contact. The current transport mechanisms that are pertinent to power devices are elucidated for both the forward and reverse mode of operation. In the first quadrant of operation, the thermionic emission process is dominant for power Schottky rectifiers. In the third quadrant of operation, the influence of Schottky barrier lowering has a strong impact on the leakage current for silicon devices. In the case of silicon carbide devices, the influence of tunneling current must also be taken into account when performing the analysis of the reverse leakage current.

B.J. Baliga, *Fundamentals of Power Semiconductor Devices*, doi: 10.1007/978-0-387-47314-7_4,

The tradeoff between reducing power dissipation in the on-state and the off-state for Schottky rectifiers is also analyzed in this chapter. This tradeoff requires taking into account the maximum operating temperature for the application. The power dissipation in the Schottky rectifier is shown to depend upon the barrier height as well as the duty cycle.

4.1 Power Schottky Rectifier Structure

The basic one-dimensional structure of the metal–semiconductor or Schottky rectifier structure is shown in Fig. 4.1 together with electric field profile under reverse bias operation. The applied voltage is supported by the drift region with a triangular electric field distribution if the drift region doping is uniform. The maximum electric field occurs at the metal contact. The device undergoes breakdown when this field becomes equal to the critical electric field for the semiconductor.

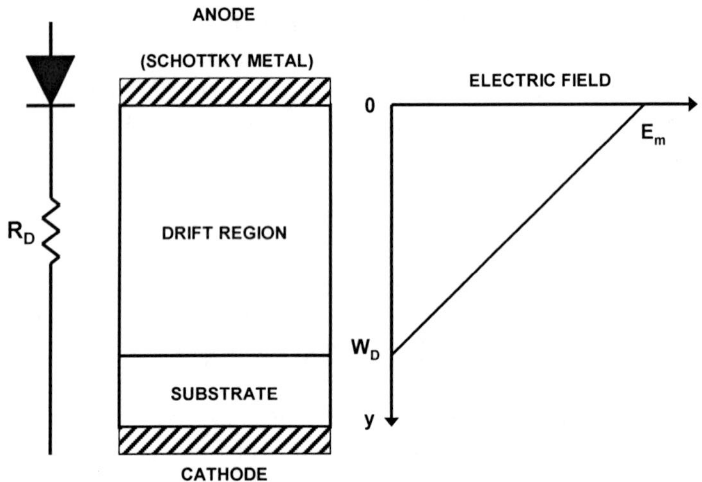

Fig. 4.1 Electric field distribution in a Schottky rectifier

When a negative bias is applied to the cathode, current flow occurs in the Schottky rectifier by the transport of electrons over the metal–semiconductor contact and through the drift region as well as the substrate. The on-state voltage drop is determined by the voltage drop across the metal–semiconductor interface and the ohmic voltage drop in the resistance of the drift region, the substrate, and its ohmic contact.

At typical on-state operating current density levels, the current transport is dominated by majority carriers. Consequently, there is insignificant minority carrier stored charge within the drift region in the power Schottky rectifier. This enables switching the Schottky rectifier from the on-state to the reverse-blocking off-state in a rapid manner by establishing a depletion region within the drift region. The fast switching capability of the Schottky rectifier enables operation at high frequencies with low power losses, making this device popular for high frequency switch-mode power supply applications. With the advent of high-voltage Schottky rectifiers based upon silicon carbide, they are expected to be utilized in motor control applications as well.

4.2 Metal–Semiconductor Contact

Nonlinear current transport across a metal–semiconductor contact has been known for a long time. The potential barrier responsible for this behavior was ascribed to the presence of a stable space-charge layer by Walter Schottky in 1938. In this section, the principles for the formation of a rectifying contact between a metal and an N-type semiconductor region are described. This enables relating the Schottky barrier height between the metal and the semiconductor to their fundamental properties.

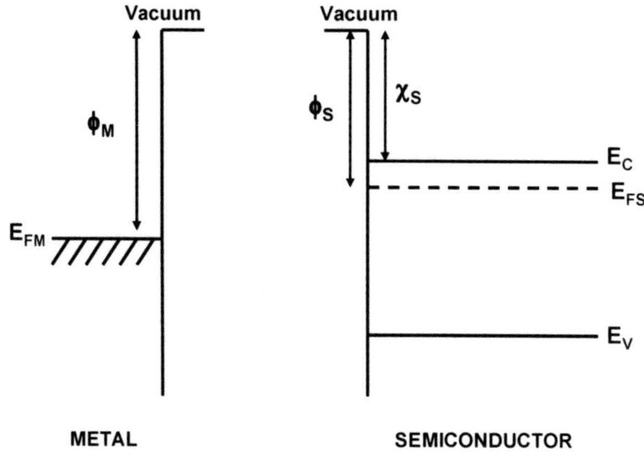

Fig. 4.2 Energy band diagram for a metal and a semiconductor in isolation

The energy band diagram for a metal and an N-type semiconductor is shown in Fig. 4.2 when they are isolated from each other. In general, the position of the Fermi level in the metal and the semiconductor will have different energy values. In the example shown in the figure, the Fermi level in the semiconductor lies above the Fermi level for the metal. The work function for the metal (Φ_M) is defined as the energy required to move an electron from the Fermi level position in the metal (E_{FM}) to a state of rest in free space outside the surface of the metal. In the same manner, the work function for the semiconductor (Φ_S) is defined as the energy required to move an electron from the Fermi level position in the semiconductor (E_{FS}) to a state of rest in free space outside the surface of the semiconductor. Since no electrons are located at the Fermi level position in the semiconductor, it is useful to define an electron affinity for the semiconductor (χ_S)

as the energy required to move an electron from the bottom of the conduction band in the semiconductor (E_C) to a state of rest in free space outside the surface of the semiconductor. The work function and electron affinity for the semiconductor are related by

$$\Phi_S = \chi_S + (E_C - E_{FS}). \tag{4.1}$$

The potential difference between the Fermi level in the semiconductor and the Fermi level in the metal is called the *contact potential* (V_C) which is given by

$$qV_C = (E_{FS} - E_{FM}) = \Phi_M - \Phi_S = \Phi_M - (\chi_S + E_C - E_{FS}). \tag{4.2}$$

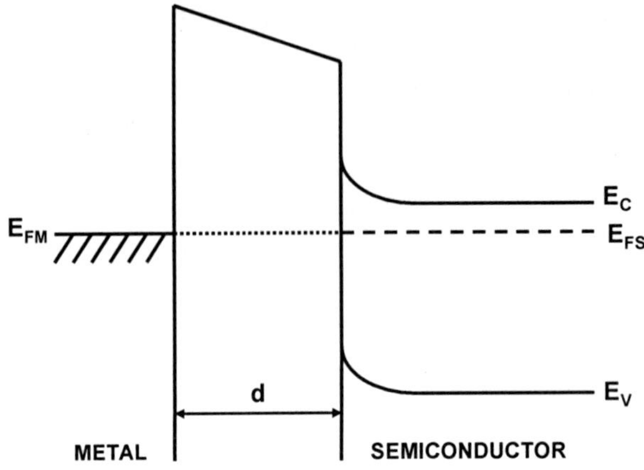

Fig. 4.3 Energy band diagram for a metal and a semiconductor after making an electrical connection between them

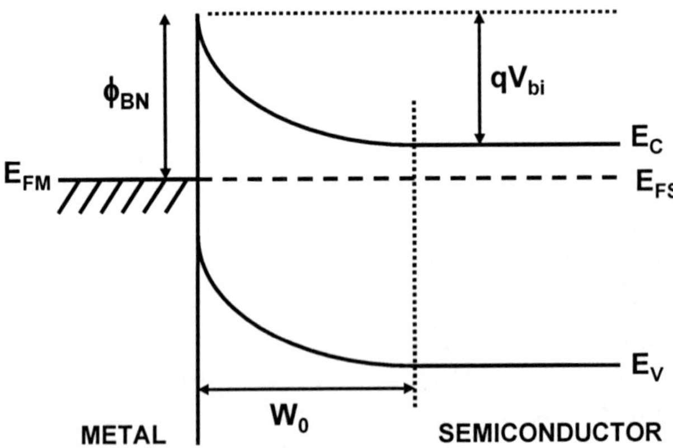

Fig. 4.4 Energy band diagram for a metal–semiconductor junction after making an intimate contact of their surfaces

When an electrical connection is provided between the metal and the semiconductor, electrons are transferred from the semiconductor to the metal due to their greater energy until thermal equilibrium is established. This transfer of electrons creates a negative charge in the metal and a positive charge within a depletion region formed at the semiconductor surface. The resulting band structure is illustrated in Fig. 4.3 for the case of a separation *d* between the metal and the semiconductor surfaces. When the metal and the semiconductor surfaces are brought into contact by reducing the separation *d* to zero, the band structure for the metal–semiconductor contact is obtained as illustrated in Fig. 4.4. The entire contact potential is now supported within the depletion region formed at the surface of the semiconductor. This voltage is therefore also referred to as the *built-in potential* (V_{bi}) of the metal–semiconductor contact.

The Schottky barrier height (Φ_{BN}) is related to the built-in potential by

$$\Phi_{BN} = qV_{bi} + (E_C - E_{FS}). \tag{4.3}$$

Another useful relationship for obtaining the Schottky barrier height is

$$\Phi_{BN} = \Phi_M - \chi_S, \tag{4.4}$$

because these properties for the materials are known. The built-in potential creates a zero-bias depletion region within the semiconductor whose width is given by

$$W_0 = \sqrt{\frac{2\varepsilon_S V_{bi}}{qN_D}}. \tag{4.5}$$

4.3 Forward Conduction

Current flow across the metal–semiconductor junction can be produced by the application of a negative bias to the N-type semiconductor region. This produces a shift in the energy band structure as illustrated in Fig. 4.5. Current flow across the interface then occurs mainly due to majority carriers – electrons for the case of an N-type semiconductor. The current transport across the contact can take place via four basic processes[5] that are schematically shown in the figure:

1. The transport of electrons from the semiconductor into the metal over the potential barrier (referred to as the *thermionic emission current*)
2. The transport of electrons by quantum mechanical tunneling through the potential barrier (referred to as the *tunneling current*)
3. The transport of electrons and holes into the depletion region followed by their recombination (referred to as the *recombination current*)
4. The transport of holes from the metal into the neutral region of the semiconductor followed by recombination (referred to as the *minority carrier current*)

In the case of power rectifiers, the doping concentration in the semiconductor must be relatively low to support the reverse bias (or blocking) voltage. This spreads the depletion region over a substantial distance. Consequently, the potential barrier is

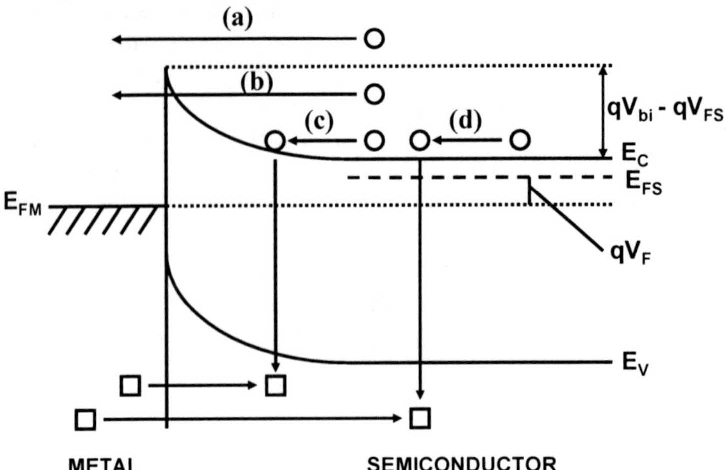

Fig. 4.5 Energy band diagram for a metal–semiconductor junction after the application of a forward bias voltage (electrons are shown as *circles* and holes are shown as *squares*)

not sharp enough to allow substantial current via the tunneling process. The recombination current in the space-charge region is observable only at very low on-state current levels. The current transport due to the injection of holes is usually negligible unless the Schottky barrier height is large. In power Schottky rectifiers, the barrier height is intentionally reduced to lower the on-state voltage drop making the minority carrier current small. Consequently, the current flow via the thermionic emission process is the dominant current transport mechanism in silicon and silicon carbide Schottky power rectifiers.

In the case of high mobility semiconductors, such as silicon, gallium arsenide, and silicon carbide, and for power rectifiers with low doping concentrations in the semiconductor, the thermionic emission theory can be used to describe the current flow across the Schottky barrier interface[6]:

$$J = AT^2 \mathrm{e}^{-(q\Phi_{\mathrm{BN}}/kT)}[\mathrm{e}^{(qV/kT)} - 1], \tag{4.6}$$

where A is the effective Richardson's constant, T is the absolute temperature, k is the Boltzmann's constant, and V is the applied bias. An effective Richardson's constant of 110, 140, and 146 A cm^{-2} K^{-2} can be used for n-type silicon,[6] gallium arsenide,[6] and 4H silicon carbide,[3] respectively. This expression, based upon the superimposition of the current flux from the metal and the semiconductor[7] which balance out at zero bias, holds true for both positive and negative voltages applied to the metal contact.

When a forward bias is applied (positive values for V in (4.6)), the first term in the square brackets of the equation becomes dominant allowing calculation of the forward current density:

$$J = AT^2 e^{-(q\Phi_{BN}/kT)} e^{(qV_{FS}/kT)}, \tag{4.7}$$

where V_{FS} is the forward voltage drop across the Schottky contact. In the case of power Schottky rectifiers, a thick lightly doped drift region must be placed below the Schottky contact as illustrated in Fig. 4.1 to allow supporting the reverse-blocking voltage. A resistive voltage drop (V_R) occurs across this drift region which increases the on-state voltage drop of the power Schottky rectifier beyond V_{FS}. In case of current transport by the thermionic emission process, there is no modulation of the resistance of the drift region because minority carrier injection is neglected. Due to the small thickness (typically less than 50 μm) of the drift region for power Schottky diodes, it is grown on top of a heavily doped N^+ substrate as a handle during processing and packaging of the devices. The resistance contributed by the substrate (R_{SUB}) must be included in the analysis because it can be comparable with that of the drift region especially for silicon carbide devices. In addition, the resistance of the ohmic contact (R_{CONT}) to the cathode may make a substantial contribution to the on-state voltage drop.

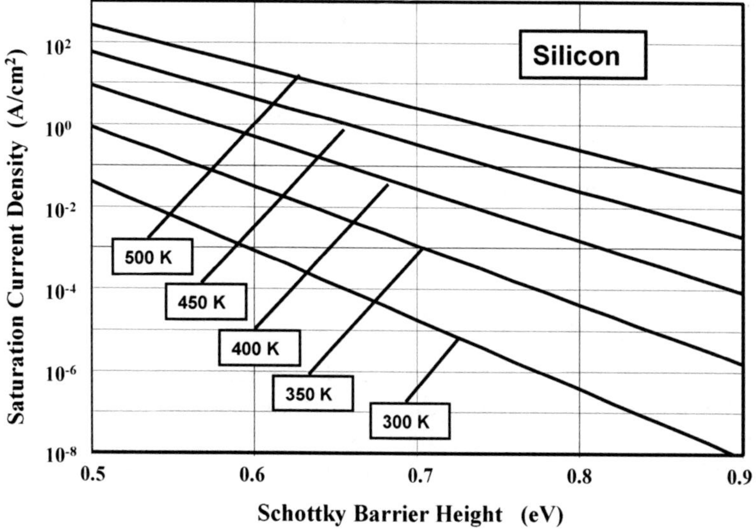

Fig. 4.6 Saturation current density for silicon Schottky barrier rectifiers

The on-state voltage drop (V_F) for the power Schottky rectifier, after including the resistive voltage drop, is given by

$$V_F = V_{FS} + V_R = \frac{kT}{q} \ln\left(\frac{J_F}{J_S}\right) + R_{S,SP} J_F, \tag{4.8}$$

where J_F is the forward (on-state) current density, J_S is the saturation current density, and $R_{S,SP}$ is the total series-specific resistance. In this expression, the saturation current is given by

$$J_{\mathrm{S}} = AT^2 \mathrm{e}^{-(q\Phi_{\mathrm{BN}}/kT)} \tag{4.9}$$

and the total series-specific resistance is given by

$$R_{\mathrm{S,SP}} = R_{\mathrm{D,SP}} + R_{\mathrm{SUB}} + R_{\mathrm{CONT}}. \tag{4.10}$$

The saturation current is a strong function of the Schottky barrier height and the temperature as shown in Fig. 4.6 for silicon devices. (A corresponding plot for 4H-SiC is provided in reference 3 for the range of barrier heights typical for this material.) The barrier heights chosen for this plot are in the range for typical metal contacts with silicon. The saturation current density increases with increasing temperature and reduction of the barrier height. This has an influence not only on the on-state voltage drop but also on an even greater impact on the reverse leakage current as discussed in Sect. 4.4.

As discussed in Chap. 1, the specific on-resistance of the drift region is given by

$$R_{\text{on-ideal}} = \frac{4\mathrm{BV}^2}{\varepsilon_{\mathrm{S}} \mu_{\mathrm{n}} E_{\mathrm{C}}^3}. \tag{4.11}$$

The specific on-resistance of the drift region for 4H-SiC is approximately 2,000 times smaller than for silicon devices for the same breakdown voltage as shown earlier in Fig. 3.6. Their values are given by

$$R_{\mathrm{D,SP}} = R_{\text{on-ideal}}(\mathrm{Si}) = 5.93 \times 10^{-9} \mathrm{BV}^{2.5} \tag{4.12}$$

and

$$R_{\mathrm{D,SP}} = R_{\text{on-ideal}}(\text{4H-SiC}) = 2.97 \times 10^{-12} \mathrm{BV}^{2.5}. \tag{4.13}$$

In addition, it is important to include the resistance associated with the thick, highly doped N^+ substrate because this is comparable with that for the drift region in some instances. The specific resistance of the N^+ substrate can be determined by taking the product of its resistivity and thickness. For silicon, N^+ substrates with resistivity of 1 mΩ cm are available. If the thickness of the substrate is 200 μm, the specific resistance contributed by the N^+ substrate is $2 \times 10^{-5}\ \Omega\ \mathrm{cm}^2$. For silicon carbide, the available resistivity of the N^+ substrates is substantially larger. For the available substrates with a typical resistivity of 0.02 Ω cm and thickness of 200 μm, the substrate contribution is $4 \times 10^{-4}\ \Omega\ \mathrm{cm}^2$. The specific resistance of the ohmic contact to the N^+ substrate can be reduced to less than $1 \times 10^{-6}\ \Omega\ \mathrm{cm}^2$ with adequate attention to increasing the doping concentration at the contact and by using ohmic contact metals with low barrier heights as discussed in Sect. 2.4.

The calculated forward conduction characteristics for silicon Schottky rectifiers are shown in Fig. 4.7 for various breakdown voltages. For this figure, a Schottky barrier height of 0.7 eV was chosen because this is a typical value used in actual power devices. It can be seen that the series resistance of the drift region does not adversely impact the on-state voltage drop for the device with a breakdown voltage of 50 V at a nominal on-state current density of 100 A cm^{-2}. However, this resistance becomes significant when the breakdown voltage exceeds 100 V, limiting the application of silicon Schottky rectifiers to systems, such as switch-mode power supply circuits, operating at voltages below 100 V.

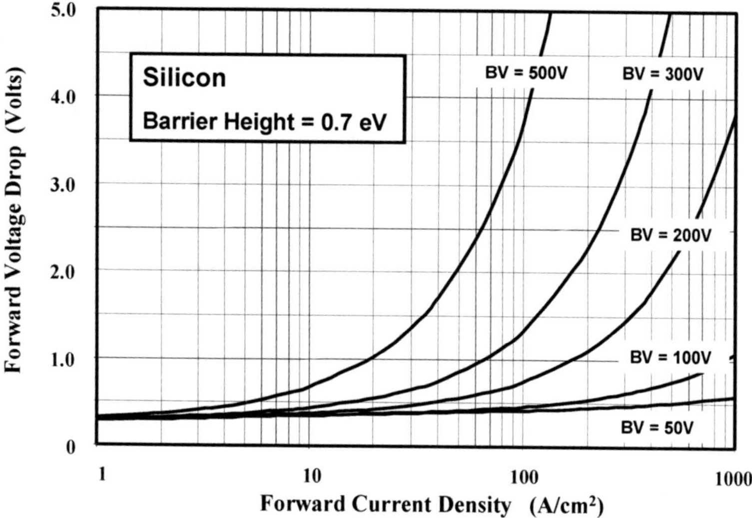

Fig. 4.7 Forward characteristics of silicon Schottky rectifiers

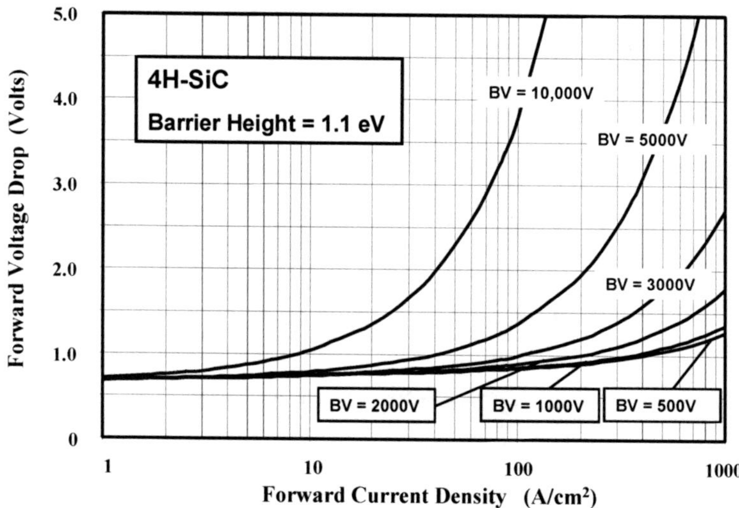

Fig. 4.8 Forward characteristics of 4H-SiC Schottky rectifiers

The significantly smaller resistance of the drift region enables scaling of the breakdown voltage of silicon carbide Schottky rectifiers to much larger voltages typical of medium and high power electronic systems, such as those used for motor control. The forward characteristics of high-voltage 4H-SiC Schottky rectifiers are shown in Fig. 4.8 for the case of a Schottky barrier height of 1.1 eV. The N^+ substrate resistance used for these calculations was $4 \times 10^{-4}\ \Omega\ cm^2$. It can be seen that the drift region resistance does not produce a significant increase in on-state voltage drop until the breakdown voltage exceeds 3,000 V. From these results, it can be concluded that silicon carbide Schottky rectifiers are excellent companion diodes for medium and high power electronic systems that utilize insulated gate bipolar transistors (IGBTs). Their fast switching speed and absence of reverse recovery current can reduce power losses and improve the efficiency in motor control applications.[4]

The choice of the Schottky barrier height has an impact on the on-state voltage drop. To illustrate this, the calculated forward conduction characteristics for silicon Schottky rectifiers are shown in Fig. 4.9 for various Schottky barrier heights. For this figure, a breakdown voltage of 50 V was chosen because this is a typical value for power devices. It can be seen that an increase in the on-state voltage drop occurs in proportion to the magnitude of the Schottky barrier height. It is therefore attractive to use a low Schottky barrier height for power rectifiers to reduce the on-state voltage drop.

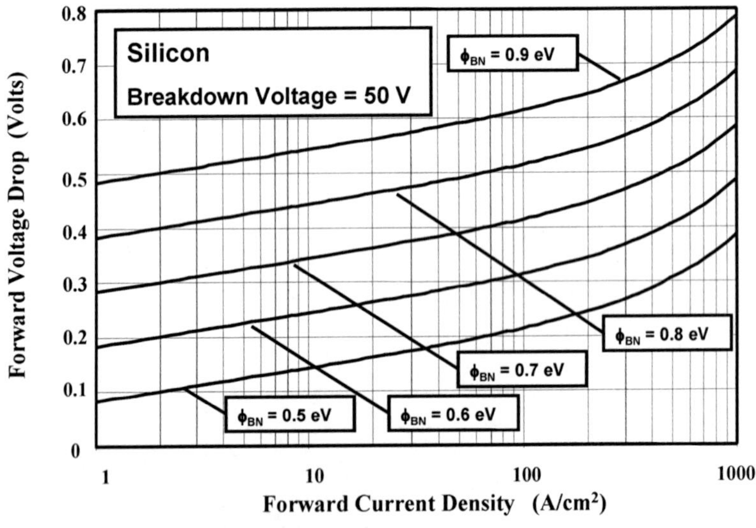

Fig. 4.9 Forward characteristics of silicon Schottky rectifiers

As discussed above, the on-state voltage drop for silicon Schottky power rectifiers designed to support low voltages is determined mainly by the voltage drop across the metal–semiconductor contact. By using (4.8) and (4.9) and neglecting the resistive voltage drop, the on-state voltage drop is given by

$$V_F = \Phi_{BN} + \frac{kT}{q}\ln\left(\frac{J_F}{AT^2}\right). \tag{4.14}$$

Since the logarithmic term in this expression has a negative value, the forward voltage drop for the Schottky diode decreases with increasing temperature. Examples of the variation of the on-state voltage drop for Schottky rectifiers with temperature are shown in Fig. 4.10 for various cases of the Schottky barrier height. The observed decrease in the on-state voltage drop with temperature is favorable for reducing power losses but can cause current localization within devices.

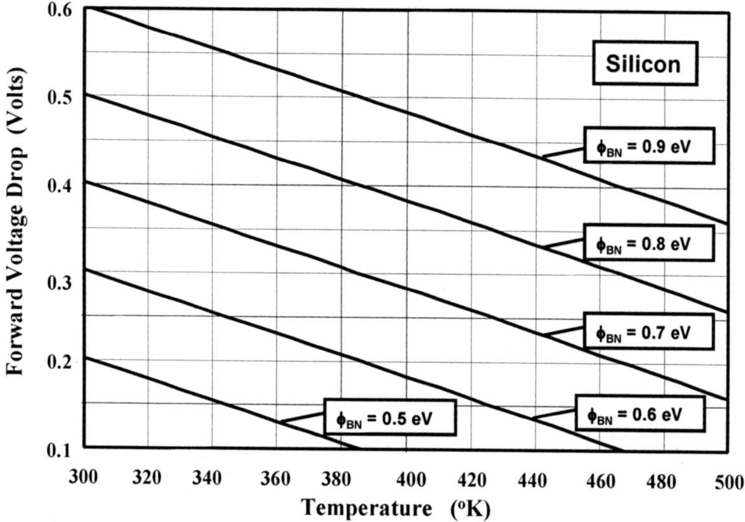

Fig. 4.10 Temperature dependence of the on-state voltage drop for silicon Schottky rectifiers

Simulation Example

To gain further insight into the physics of operation for the Schottky rectifier, the results of one-dimensional numerical simulations are provided in this section for the case of a device with a breakdown voltage of 50 V. This is a typical breakdown voltage for commercially available silicon power Schottky rectifiers. This breakdown voltage was obtained with a drift region with doping concentration of 8×10^{15} cm^{-3} and thickness of 3 μm. As discussed earlier, the doping concentration was lower during the simulations when compared with the analytical model because Fulop's law underestimates the impact ionization coefficients for silicon. The lower doping concentration results in a larger specific on-resistance for the drift region which increases the on-state voltage drop. The forward conduction characteristics were obtained by sweeping the cathode voltage in the negative direction. Simulations were performed for various values for the Schottky barrier height to examine the impact on the injection of minority carriers (holes) into the drift region. The minority carrier lifetimes (τ_{p0} and τ_{n0}) were assigned a value of 10 μs during these simulations.

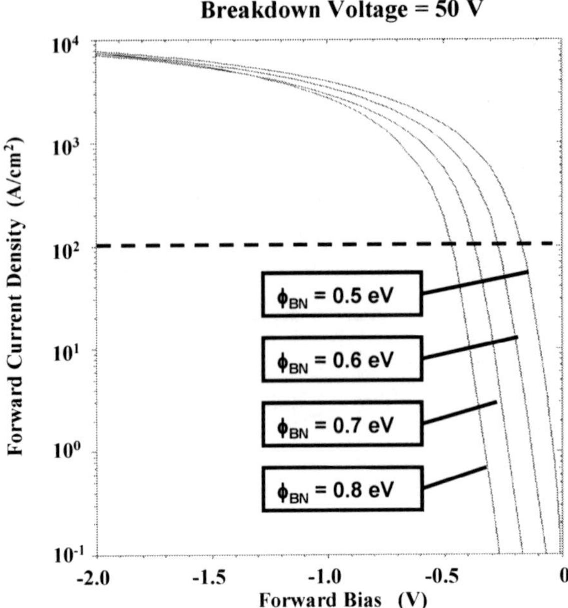

Fig. 4.11 Forward conduction characteristics for silicon Schottky rectifiers with breakdown voltage of 50 V

The forward characteristics obtained with the simulations for these Schottky rectifiers are shown in Fig. 4.11 for various Schottky barrier heights. These characteristics are in excellent agreement with those shown in Fig. 4.9 based upon the analytical model. For example, the on-state voltage drop obtained at a current density of 100 A cm^{-2} is 0.41 V for both the simulation and the analytical case when the barrier height is 0.7 eV. The analytical model is therefore sufficient for the analysis of the silicon Schottky rectifier.

The on-state current flow in the Schottky rectifier was discussed earlier with the aid of Fig. 4.5. It was pointed out that one of the current flow mechanisms is by minority carrier injection into the drift region. The simulations allow analysis of this contribution by examination of the hole concentration in the drift region. Since the injection process is known to be sensitive to the barrier height,[8] the hole concentration in the drift region is shown in Fig. 4.12 for various barrier heights. In all cases, the on-state voltage drop was chosen as 0.5 V, which is the typical operating value for silicon rectifiers because it produces an on-state current density of about 100 A cm^{-2}. It can be observed that the injected hole concentration increases with increasing barrier height as expected. However, even for the largest barrier height of 0.8 eV, the injected hole density is less than 10^{13} cm^{-3} in the drift region, which is 1,000 times smaller than the doping concentration. This confirms that the on-state operation of the silicon Schottky rectifier can be performed while neglecting the injection of minority carriers. It also confirms that the stored charge in the silicon Schottky rectifier is small allowing rapid switching from the on-state to the off-state during circuit operation.

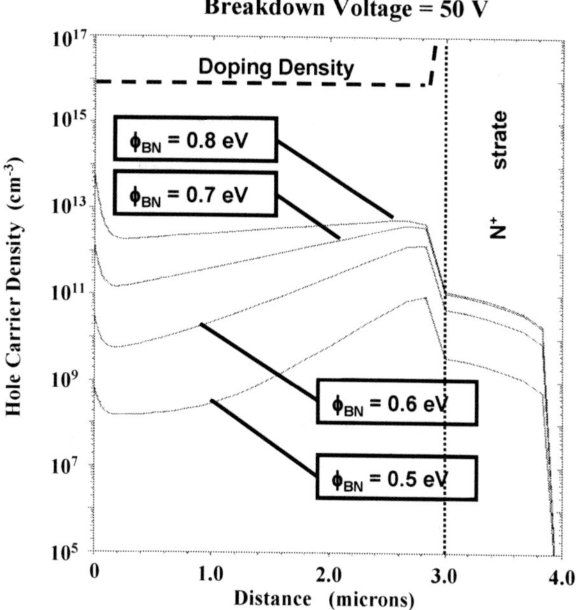

Fig. 4.12 Injected hole concentration for silicon Schottky rectifiers at an on-state voltage drop of 0.5 V

4.4 Reverse Blocking

When a reverse bias is applied to the Schottky rectifier, the voltage is supported across the drift region with the maximum electric field located at the metal–semiconductor contact as shown in Fig. 4.1. The energy band diagram corresponding to this condition is illustrated in Fig. 4.13. Since no voltage can be supported within the metal, the reverse-blocking capability of the Schottky rectifier is governed by the physics for the abrupt P–N junction that was discussed in Chap. 3. If a parallel-plane breakdown voltage is assumed, the drift region doping and width for a silicon device are given by

$$N_D = 2 \times 10^{18} (BV_{PP})^{-4/3} \tag{4.15}$$

and

$$W_D = 2.58 \times 10^{-6} (BV_{PP})^{7/6}. \tag{4.16}$$

In the case of actual power Schottky rectifiers, the breakdown voltage is constrained by breakdown at the edges. Edge terminations that have been used to raise the breakdown voltage of Schottky rectifiers close to the parallel-plane value are discussed in Sect. 4.10.

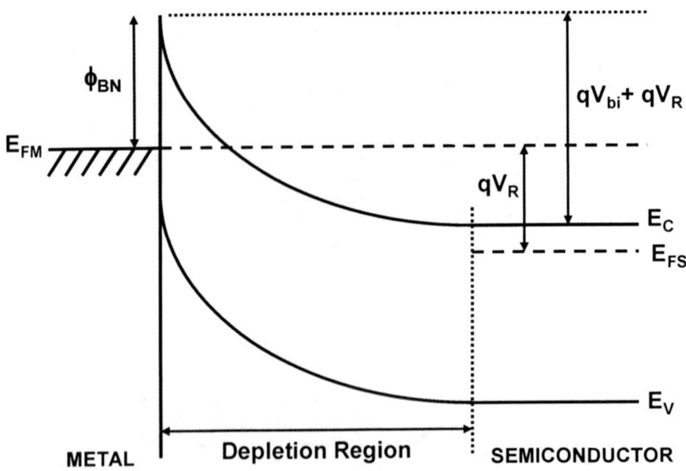

Fig. 4.13 Energy band diagram for a metal–semiconductor junction after the application of a reverse bias voltage

4.4.1 Leakage Current

The leakage current for Schottky rectifiers is comprised of three components:

1. Space-charge generation current arising from the depletion region
2. Diffusion current arising from carrier generation in the neutral region
3. Thermionic emission current across the metal–semiconductor contact

Due to the relatively small barrier height utilized in silicon Schottky rectifiers, the thermionic emission component is dominant. The leakage current for the Schottky rectifier can be obtained by using (4.6) and substituting a negative bias of magnitude V_R applied to the diode:

$$J_L = AT^2 e^{-(q\Phi_{BN}/kT)}[e^{-(qV_R/kT)} - 1]. \tag{4.17}$$

Since the typical reverse bias voltages (V_R) are much greater than the thermal energy (kT/q), the exponential term in the square brackets becomes very small under reverse-blocking conditions. Consequently, the leakage current is determined by the saturation current:

$$J_L = -AT^2 e^{-(q\Phi_{BN}/kT)} = -J_S. \tag{4.18}$$

As previously discussed with reference to Fig. 4.6 for the saturation current, the leakage current due to the thermionic emission process is a strong function of the Schottky barrier height and the temperature. To reduce the leakage current and minimize power dissipation in the blocking state, a large Schottky barrier height is required. Further, a very rapid increase in leakage current occurs with increasing temperature as shown in Fig. 4.14. If the power dissipation due to the leakage current becomes dominant, the resulting increase in the device temperature produces a positive feedback mechanism, which can lead to unstable operation of the Schottky rectifier due to thermal runaway. This destructive failure mechanism for power Schottky rectifiers must be avoided by sufficiently increasing the Schottky barrier height even though this increases the on-state voltage drop. A larger Schottky barrier height is warranted for power Schottky rectifiers that must operate at higher ambient temperatures. This tradeoff is discussed in more detail later in this chapter.

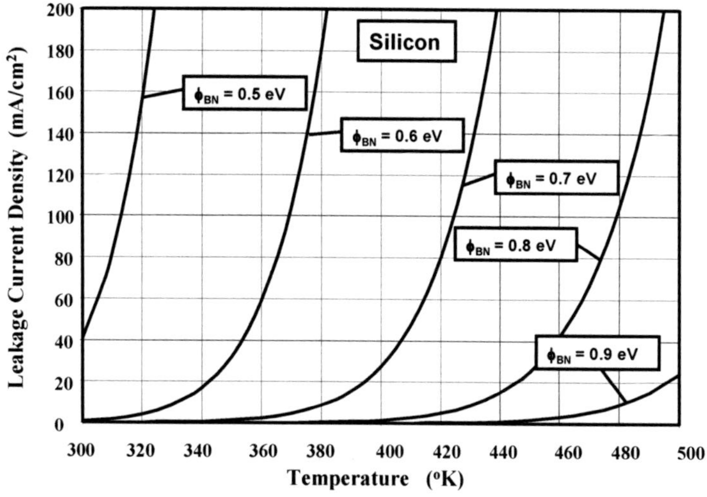

Fig. 4.14 Temperature dependence of the leakage current for silicon Schottky rectifiers

4.4.2 Schottky Barrier Lowering

Based upon the above analysis, the leakage current of the Schottky rectifier should be independent of the magnitude of the applied reverse bias voltage. However, actual power Schottky rectifiers exhibit a significant increase in the leakage current with increasing reverse bias voltage. This increase in the leakage current is far greater than the space-charge generation current within the expanding depletion region with increasing reverse bias voltage.

Under reverse-blocking operation, it has been found that there is a reduction of the Schottky barrier height due to the image force lowering phenomenon.[9] To analyze this phenomenon, consider the energy band diagram for the metal–semiconductor contact shown in Fig. 4.15. When an electron in the semiconductor approaches the metal at a distance x from the interface, a positive mirror image charge of the same magnitude occurs in the metal at a distance $-x$ from the interface. This produces an electrostatic force on the electron given by

$$F(x) = \frac{q^2}{4\pi\varepsilon_S (2x)^2}. \tag{4.19}$$

This attractive force between the particles creates a negative potential energy for the electron inside the semiconductor, which is the work done to move the electron from position x to infinity. The corresponding image force potential (V_I) is given by

$$qV_I = \int_x^{\infty} F(x)\mathrm{d}x = -\frac{q^2}{16\pi\varepsilon_S}\int_x^{\infty}\frac{\mathrm{d}x}{x^2} = -\frac{q^2}{16\pi\varepsilon_S x}. \tag{4.20}$$

The negative image force potential combines with the positive potential due to the Schottky barrier, producing a maximum at a distance X_M from the interface. At this location, the image force potential is equal to the potential drop across the depletion region due to the prevailing electric field indicated by the arrow in the figure. Since the maximum is located close to the interface, it can be assumed that the electric field at this location is approximately equal to the maximum electric field (E_M) at the Schottky contact. Equating the image force potential (V_I) at location X_M to the potential drop ($E_M X_M$) in the depletion region gives

$$\frac{q}{16\pi\varepsilon_S X_M} = E_M X_M. \tag{4.21}$$

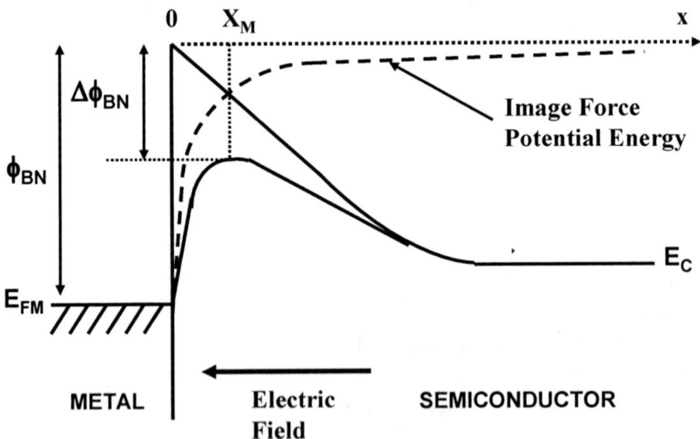

Fig. 4.15 Band diagram illustrating the image force lowering of the Schottky barrier height

The reduction of the barrier height due to the image force lowering phenomenon, indicated as $\Delta\Phi_{BN}$ in Fig. 4.15, is then given by

$$\Delta\Phi_{BN} = 2E_M X_M. \tag{4.22}$$

Using (4.21) to eliminate X_M, the barrier lowering is found to be determined by the maximum electric field (E_M) at the metal–semiconductor interface:

$$\Delta\Phi_{BN} = \sqrt{\frac{qE_M}{4\pi\varepsilon_S}}. \quad (4.23)$$

For a one-dimensional structure, the maximum electric field is related to the applied reverse bias voltage (V_R) by

$$E_M = \sqrt{\frac{2qN_D}{\varepsilon_S}(V_R + V_{bi})}. \quad (4.24)$$

As an example, the reduction of the barrier height for a silicon Schottky rectifier, for the case of a drift region doping concentration of 1×10^{16} cm^{-3}, is shown in Fig. 4.16. The reduction of the Schottky barrier height is 0.065 eV at the maximum reverse bias voltage. Although this change in barrier height may appear to be small, it can lead to a substantial increase in the leakage current with increasing reverse bias voltage.

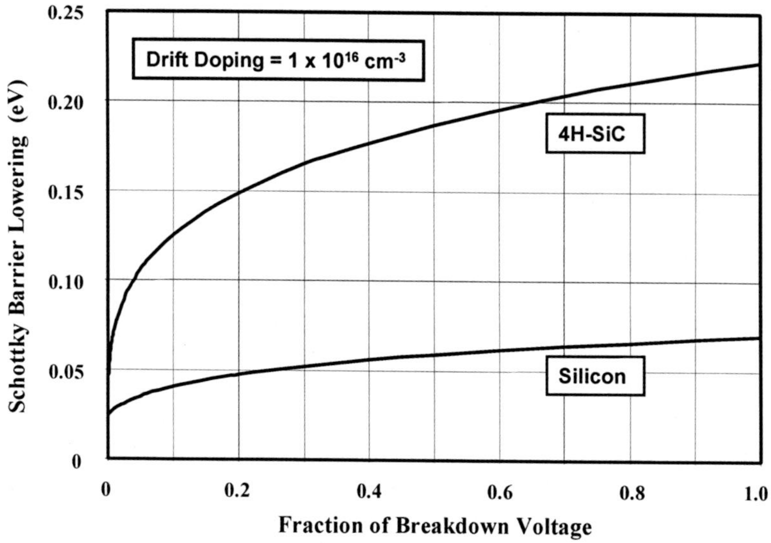

Fig. 4.16 Schottky barrier lowering for silicon and 4H-SiC Schottky rectifiers

The leakage current for the Schottky rectifier including the effect of Schottky barrier lowering is given by

$$J_L = -AT^2 e^{-q(\Phi_{BN} - \Delta\Phi_{BN})/kT}. \quad (4.25)$$

The leakage currents calculated with and without the Schottky barrier lowering effect are compared for the case of a silicon device with a breakdown voltage of 50 V in Fig. 4.17. In making these plots, the leakage current due to space-charge generation was neglected because it is much smaller than the leakage current across the metal–semiconductor contact. It can be seen that the leakage current is enhanced by a factor of 5 times due to the barrier lowering phenomenon as the reverse voltage increases and approaches the breakdown voltage.

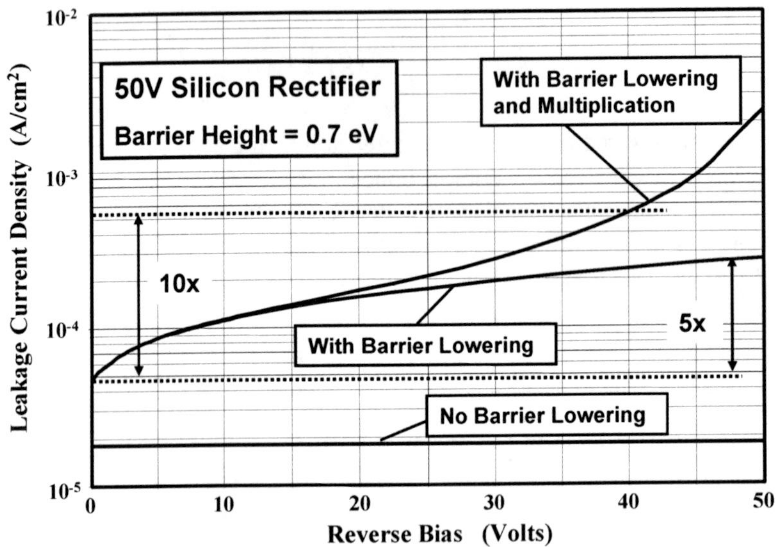

Fig. 4.17 Leakage current density for a 50-V silicon Schottky rectifier

4.4.3 Prebreakdown Avalanche Multiplication

The actual reverse leakage current for silicon Schottky rectifiers has been found to increase by an even greater degree than predicted by the Schottky barrier lowering phenomenon. This increase in leakage current can be accounted for by including the effect of prebreakdown avalanche multiplication of the large number of free carriers being transported through the Schottky rectifier structure at the high electric fields associated with reverse bias voltages close to the breakdown voltage.[10] This impact ionization process can be treated as a purely electron-initiated process due to the relatively large thermionic emission current across the metal–semiconductor contact. The total number of electrons that reach the edge of the depletion region will be larger than those crossing the metal–semiconductor contact by a factor M_n, which is the electron multiplication factor. An analytical expression for the electron multiplication factor can be derived by using a power series approximation for the impact ionization coefficients α_n and α_p:

$$\alpha_n = 6.6 \times 10^{-24} E^{4.93} \qquad (4.26)$$

and

$$\alpha_{\mathrm{p}} = 2.3 \times 10^{-24} E^{4.93} = 0.344 \alpha_{\mathrm{n}}. \tag{4.27}$$

It has been assumed that the ratio of the impact ionization coefficients for electrons and holes remains independent of the electric field. The multiplication coefficient (M_{n}) is then determined from the maximum electric field (E_{M}) at the metal–semiconductor contact:

$$M_{\mathrm{n}} = \{1 - 1.52[1 - \exp(-7.22 \times 10^{-25} E_{\mathrm{m}}^{4.93} W_{\mathrm{D}})]\}^{-1}, \tag{4.28}$$

where W_{D} is the depletion layer width. The leakage current density for a silicon Schottky rectifier with drift region doping concentration of 1×10^{16} cm^{-3} is shown in Fig. 4.17 after including the influence of the prebreakdown multiplication coefficient. The effect of including the multiplication coefficient is apparent at high voltages when the electric field approaches the critical electric field for breakdown. The leakage currents obtained, after including the effects of Schottky barrier lowering and prebreakdown multiplication, are consistent with the characteristics of commercially available silicon devices, which exhibit an order of magnitude increase in leakage current from low reverse bias voltages to the rated voltage (about 80% of the breakdown voltage).

4.4.4 Silicon Carbide Rectifiers

Since the low specific on-resistance of the drift region in silicon carbide devices is associated with the much larger electric field in the material before the onset of impact ionization, the Schottky barrier lowering in silicon carbide rectifiers can be expected to be significantly larger than in silicon devices. For the case of a drift region doping level of 1×10^{16} cm^{-3}, the barrier lowering is found to be three times larger in silicon carbide at the corresponding breakdown voltage as shown in Fig. 4.16. In preparing this graph, the reverse voltage was normalized to the breakdown voltage because of the different breakdown voltages for the silicon (50 V) and silicon carbide devices (3,000 V).

The enhanced Schottky barrier lowering in silicon carbide devices leads to a more rapid increase in leakage current with increasing reverse bias as shown in Fig. 4.18. The leakage current is predicted by this model to increase by about three orders of magnitude when the reverse voltage approaches the breakdown voltage. The observed increase in leakage current with applied reverse bias voltage for high-voltage silicon carbide Schottky rectifiers is much greater than can be accounted for with the Schottky barrier lowering model[11–13] despite the much large barrier lowering effect. The experimentally observed increase in leakage current is about six orders of magnitude with increase in reverse bias voltage.

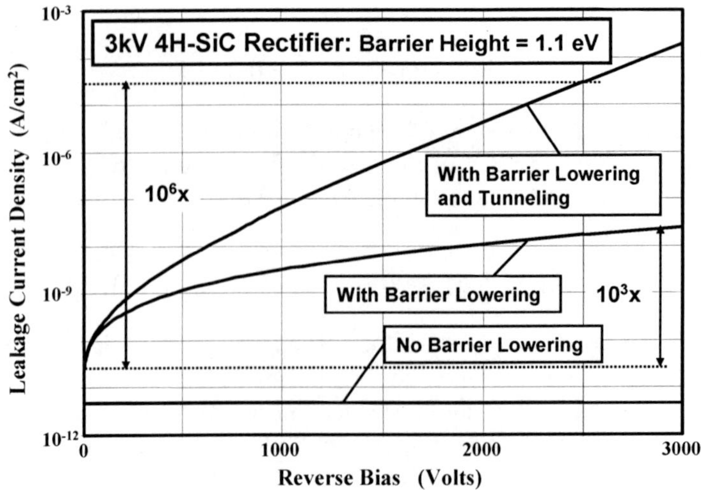

Fig. 4.18 Leakage current density for a 3-kV 4H-SiC Schottky rectifier

Tunneling Current

To explain the more rapid increase in leakage current observed in silicon carbide Schottky rectifiers, it is necessary to include the field emission (or tunneling) component of the leakage current.[14] The thermionic field emission model for the tunneling current leads to a barrier lowering effect proportional to the square of the electric field at the metal–semiconductor interface. When combined with the thermionic emission model, the leakage current density can be written as

$$J_S = AT^2 \exp\left(-\frac{q\Phi_{BN}}{kT}\right)\exp\left(\frac{q\Delta\Phi_{BN}}{kT}\right)\exp(C_T E_M^2), \tag{4.29}$$

where C_T is a tunneling coefficient. A tunneling coefficient of 8×10^{-13} cm^2 V^{-2} was found to yield an increase in leakage current by six orders of magnitude as shown in Fig. 4.18 consistent with the experimental observations. Thus, the inclusion of the tunneling model enhances the leakage current by another three orders of magnitude beyond that due to the Schottky barrier lowering phenomenon.

As discussed above, the leakage current in silicon carbide Schottky rectifiers increases much more rapidly with reverse voltage than in silicon devices. Fortunately, larger barrier heights can be utilized in silicon carbide devices when compared with silicon devices to reduce the absolute magnitude of the leakage current density, because an on-state voltage drop of 1–1.5 V is acceptable for such high-voltage structures. This enables maintaining an acceptable level of power dissipation in the reverse-blocking mode. For example, in the case of the 3-kV 4H-SiC Schottky diode discussed above, the reverse power dissipation at room temperature is less than 1 W cm^{-2} when compared with an on-state power dissipation of 100 W cm^{-2}. The expected increase in leakage current with temperature must of course be taken into account to ensure that the reverse power dissipation remains below the on-state power dissipation for stable operation. The leakage current can be suppressed by shielding the Schottky contact[3] using the junction barrier-controlled Schottky (JBS) rectifier structure[15] originally proposed for silicon devices.

4.5 Device Capacitance

The reverse-blocking voltage is supported across a depletion region in the power Schottky rectifier as shown in Fig. 4.13. The thickness of the depletion region (W_D) is related to the applied reverse bias voltage (V_R) by

$$W_D = \sqrt{\frac{2\varepsilon_S}{qN_D}(V_R + V_{bi})}, \tag{4.30}$$

where V_{bi} is the built-in voltage. The specific capacitance (capacitance per unit area) associated with this depletion region is given by

$$C_{SBD,SP} = \frac{\varepsilon_S}{W_D}, \tag{4.31}$$

where ε_S is the dielectric constant of the semiconductor.

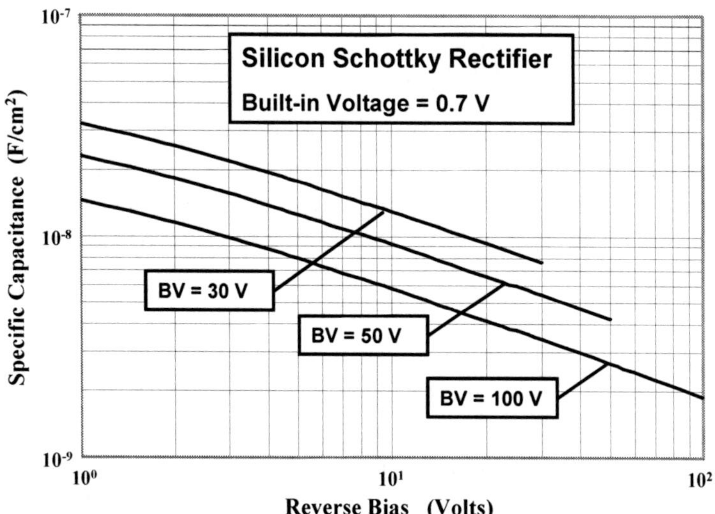

Fig. 4.19 Specific capacitance for silicon Schottky rectifiers

The specific capacitance calculated using the above relationships is plotted in Fig. 4.19 for silicon Schottky rectifiers with breakdown voltages of 30, 50, and 100 V. For these devices, the built-in voltage was assumed to be 0.7 V. The specific capacitance decreases with increasing reverse bias voltage due to the expansion of the depletion region into the drift region. At any given reverse bias voltage, the specific capacitance is smaller for the higher breakdown voltage structure due to the smaller doping concentration in the drift region. A typical value for the specific capacitance for silicon Schottky rectifiers is 10^{-8} F cm^{-2}.

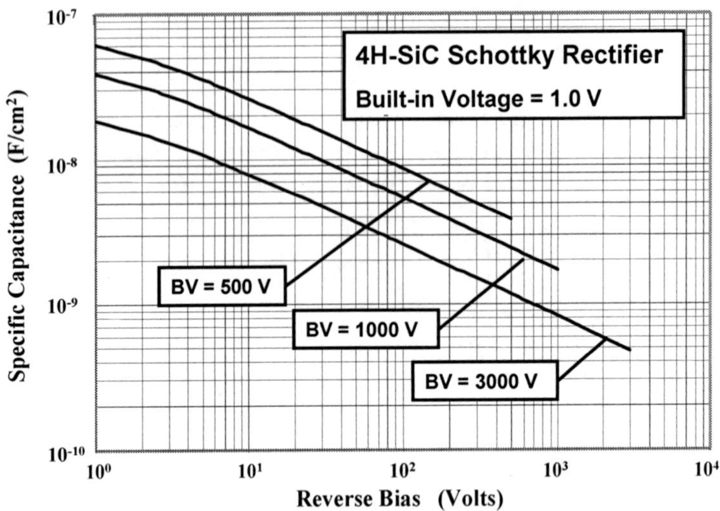

Fig. 4.20 Specific capacitance for 4H-SiC Schottky rectifiers

The specific capacitance for the 4H-SiC Schottky rectifiers calculated using the above relationships is plotted in Fig. 4.20. In this case, a slightly larger built-in potential of 1.0 V was used due to the larger Schottky barrier heights utilized in silicon carbide devices. Larger values for the breakdown voltages for the silicon carbide Schottky diodes, of interest for their applications, were selected for this figure. The behavior of the specific capacitance is similar to that for silicon devices. A typical value for the specific capacitance for these higher breakdown voltage 4H-SiC Schottky rectifiers is also 10^{-8} F cm^{-2}.

4.6 Thermal Considerations

The Schottky power rectifier is used to control the direction of current flow in power circuits, such as switch-mode power supplies, operating at high frequencies. The device operates for a part of the cycle in the on-state and the off-state for the rest of the cycle with rapid switching transients between these modes. A typical set of current–voltage waveforms for the Schottky rectifier are shown in Fig. 4.21 under the assumption that the switching intervals can be neglected. As discussed earlier in this chapter, the minority carrier stored charge is very small in the silicon Schottky rectifier allowing the device to rapidly switch between the on-state and the off-state. In practical circuit boards, care must be taken to ensure that current ringing due to the stray inductance of the board and the diode capacitance is minimized.

The power dissipation incurred in the power Schottky rectifier can be calculated by adding the power loss during the on-state with the power dissipated in the off-state.[16] Unlike most power devices, the off-state power loss in the Schottky rectifier becomes significant, especially at elevated temperatures, due to the relatively large leakage current. If the power loss due to the leakage current becomes greater than the power loss due to the on-state current flow, the Schottky rectifier can undergo thermal runaway leading to destructive failure.

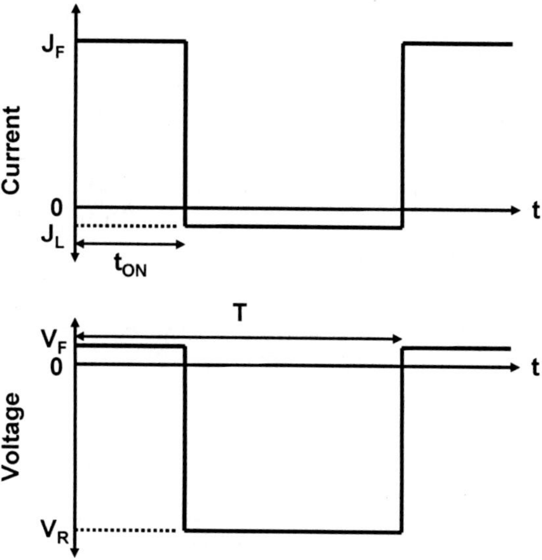

Fig. 4.21 Typical switching waveforms for a Schottky rectifier

As shown in Fig. 4.21, typical Schottky rectifiers exhibit a voltage drop (V_F) during current conduction in the forward direction. This results in power dissipation per unit area in the on-state given by

$$P_L(\text{on}) = \delta J_F V_F, \tag{4.32}$$

where J_F is the on-state current density. In this expression, δ is referred to as the *duty cycle* given by

$$\delta = t_{ON} / T, \tag{4.33}$$

where t_{ON} is the on-state duration and T is the time period (the reciprocal of the operating frequency). The on-state power dissipation decreases with increasing temperature because the on-state voltage drop decreases as shown in Fig. 4.10.

The power dissipation per unit area in the off-state is given by

$$P_L(\text{off}) = (1-\delta) J_L V_R, \tag{4.34}$$

where J_L is the leakage current exhibited by the device in its off-state due to supporting a reverse bias (V_R). The power dissipation in the off-state increases rapidly with temperature due to an increase in the leakage current as shown in Fig. 4.14.

The total power dissipation incurred in the diode is obtained by combining these terms:

$$P_L(\text{total}) = P_L(\text{on}) + P_L(\text{off}). \tag{4.35}$$

As the temperature of the diode is increased from room temperature, the on-state power dissipation decreases, resulting in a reduction of the total power dissipation because the leakage current is small. However, the leakage current increases rapidly at high temperatures, resulting in an increase in the power dissipation with temperature. Consequently, the power dissipation in the Schottky rectifier goes through a minimum with increasing temperature as illustrated in Fig. 4.22 for the case of a device with breakdown voltage of 50 V. A duty cycle of 50%, a Schottky barrier height of 0.7 eV, a reverse bias voltage of 30 V, and an on-state current density of 100 A cm^{-2} were chosen for this example.

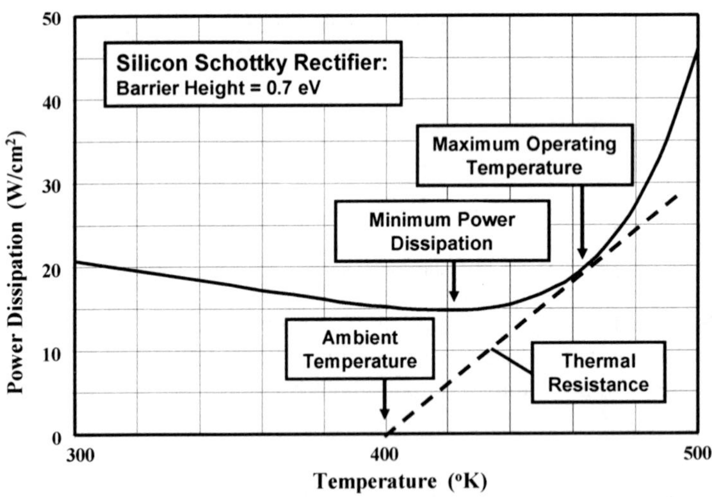

Fig. 4.22 Typical power dissipation for a silicon Schottky rectifier

The maximum stable operating temperature for the Schottky rectifier is limited by the thermal impedance of the package and heat sink. If a tangent is drawn from the ambient temperature to the power dissipation curve as shown in Fig. 4.22, the maximum stable operating temperature is obtained as shown in the figure. Although stable operation is theoretically predicted below this temperature point, it is prudent to keep the maximum operating temperature below the point of minimum power dissipation indicated in the figure.

The Schottky barrier height has a strong influence on the maximum operating temperature and the minimum power dissipation. This is illustrated in Fig. 4.23 for the case of a silicon Schottky rectifier with a breakdown voltage of

50 V. Here, a duty cycle of 50% was chosen with a reverse bias voltage of 30 V and an on-state current density of 100 A cm^{-2} for these diodes. As the barrier height is increased from 0.5 to 0.9 eV, the temperature at which the minimum power dissipation occurs shifts from 300 K to above 500 K. Thus, it becomes necessary to use a larger Schottky barrier height when designing silicon Schottky rectifiers for high-temperature operation. It can be observed from the figure that this is accompanied by an increase in the power dissipation within the rectifier.

The maximum operating temperature is also dependent on the duty cycle. This is illustrated in Fig. 4.24 for the case of 50-V silicon Schottky rectifiers with a barrier height of 0.7 eV. Here, a reverse bias voltage of 30 V and an on-state current

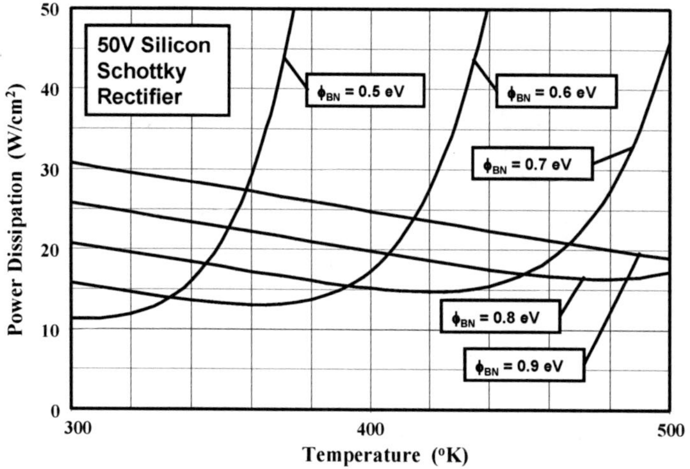

Fig. 4.23 Power dissipation for silicon Schottky rectifiers

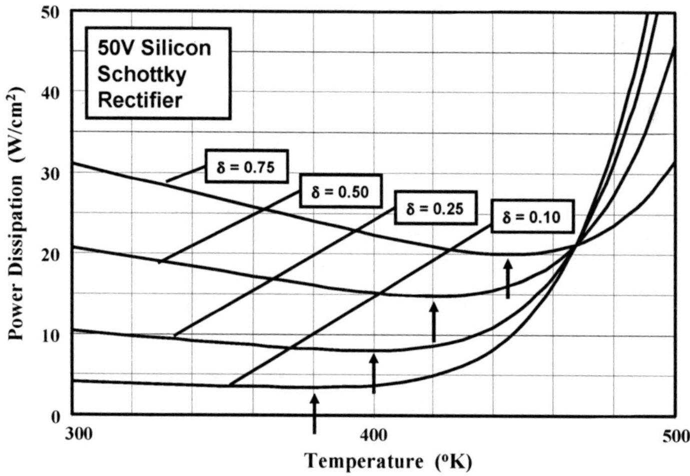

Fig. 4.24 Power dissipation for silicon Schottky rectifiers

density of 100 A cm^{-2} were assumed for these diodes. It can be seen that at room temperature, the power dissipation is reduced for smaller duty cycles because the on-state power dissipation is dominant. However, the temperature at which the minimum power dissipation occurs, indicated by the arrows in the figure, is also smaller for smaller duty cycles. It becomes necessary to raise the barrier for low duty cycle operation to enable operation at high temperatures leading to an increase in the power dissipation.

It is interesting to note that the power dissipation curves all cross one another at the same temperature. This implies that there is a temperature at which the power dissipation becomes independent of the duty cycle. This temperature can be determined by using (4.35) with

$$\frac{\mathrm{d}P_{\mathrm{L}}}{\mathrm{d}\delta} = 0. \tag{4.36}$$

By using (4.14) and (4.18) for the on-state voltage drop and the leakage current, this condition is defined by

$$J_{\mathrm{F}}\left[\Phi_{\mathrm{BN}} + \frac{kT}{q}\ln\left(\frac{J_{\mathrm{F}}}{AT^2}\right)\right] - AT^2 \mathrm{e}^{-(q\Phi_{\mathrm{BN}}/kT)} V_{\mathrm{R}} = 0. \tag{4.37}$$

Iterative solution of this equation with the previously defined device parameters yields a temperature of 466 K, which is in agreement with the plots in Fig. 4.24.

4.7 Fundamental Tradeoff Analysis

As demonstrated in Sect. 4.6, it is necessary to adjust the barrier height to minimize the power losses for applications. Low barrier heights should be chosen for power Schottky rectifiers that are intended for applications with large duty cycles where the power losses due to forward conduction are dominant. However, the leakage current increases, resulting in a low maximum operating temperature if a small Schottky barrier height is chosen. Similarly, large barrier heights are required for Schottky rectifiers used in applications with high reverse bias stress and elevated ambient temperatures. It is therefore necessary to make a tradeoff between reducing the forward voltage drop and minimizing the leakage current by appropriate choice of the barrier height.

In the case of low-voltage (<50 V) Schottky rectifiers, it is possible to neglect the influence of the series resistance on the on-state voltage drop. In this case, the on-state voltage drop is given by

$$V_{\mathrm{F}} = \Phi_{\mathrm{BN}} + \frac{kT}{q}\ln\left(\frac{J_{\mathrm{F}}}{AT^2}\right). \tag{4.38}$$

If the impact of Schottky barrier lowering and prebreakdown avalanche multiplication is neglected, then the leakage current for the Schottky rectifier is equal to the saturation current density:

$$J_L = J_S = AT^2 e^{-(q\Phi_{BN}/kT)}. \tag{4.39}$$

A tradeoff relationship that is useful during the design of Schottky power rectifiers can be derived by combining the above equations with the elimination of the barrier height:

$$J_L = J_F e^{-(qV_F/kT)}. \tag{4.40}$$

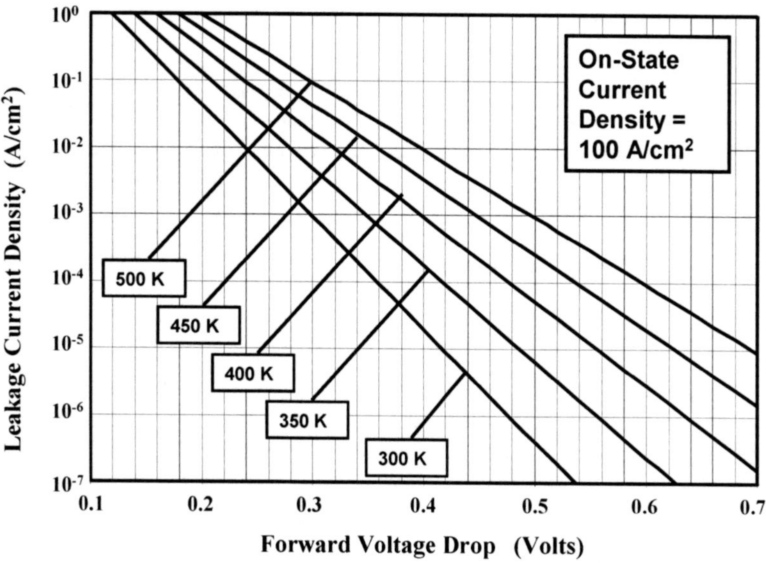

Fig. 4.25 Fundamental tradeoff curves for Schottky rectifiers

The calculated tradeoff curves for Schottky rectifiers operating at various junction temperatures are shown in Fig. 4.25. It is worth pointing out that these tradeoff curves are fundamental in nature because they are independent of the semiconductor material. This implies that the performance of silicon Schottky rectifiers with low breakdown voltages cannot be improved by replacement with other semiconductors, such as gallium arsenide or silicon carbide. Thus, the plots in Fig. 4.25 establish the minimum expected leakage current in Schottky rectifiers for any given on-state voltage drop and operating temperature. For example, the tradeoff curve for 400 K indicates that the minimum leakage current for a Schottky rectifier will be 1 mA cm^{-2} if the barrier height is chosen to obtain an on-state voltage drop of 0.4 V. The performance of actual devices can be expected to be worse than this due to the impact of the series resistance on increasing the on-state voltage drop and the influence of Schottky barrier lowering and prebreakdown avalanche multiplication on increasing the leakage current.

4.8 Device Technology

A variety of metals have been used to manufacture silicon Schottky barrier rectifiers. As previously pointed out, it is necessary to use a metal with low barrier height to reduce the power dissipation. This is satisfactory if the operating ambient temperature for the diode is low. As the ambient temperature increases, it becomes necessary to use metals with larger barrier heights to suppress the leakage current. The Schottky barrier height depends upon the work function of the metal. The work functions and corresponding barrier heights[17] are tabulated in Fig. 4.26 for cleaved N-type silicon surfaces with metal deposited in an ultrahigh vacuum environment. As expected, the barrier height increases with increasing work function for the metal. These values for the barrier height are consistent with an electron affinity of about 4.0 eV for silicon.

Metal	Cr	W	Mo	Pt
Work Function (eV)	4.50	4.60	4.60	5.30
Barrier Height (eV)	0.57	0.61	0.59	0.81

Fig. 4.26 Work functions and Schottky barrier heights for metals on N-type silicon

When the metal–silicon interface is subjected to an anneal process at elevated temperatures, the metal reacts with the silicon producing a metal silicide which has a different work function. The measured barrier heights for various metal silicides on N-type silicon are provided in Fig. 4.27. Platinum silicide is commonly used for power Schottky rectifiers that must be designed to operate at high temperatures.

Metal-Silicide	$CrSi_2$	WSi_2	$MoSi_2$	$PtSi_2$
Barrier Height (eV)	0.57	0.65	0.55	0.78

Fig. 4.27 Schottky barrier heights for metal silicides on N-type silicon

4.9 Barrier Height Adjustment

The barrier height is usually decided by the choice of the metal when manufacturing power Schottky rectifiers. An alternative approach to adjusting the barrier height is by employing a shallow ion implant at the surface of the semiconductor. The addition of a surface layer, whose thickness is less than the electron mean free path of about 100 Å, with a carefully controlled dose can change the effective barrier height between the metal and the semiconductor.[18] For an N-type drift region, the addition of an N-type surface layer will reduce the effective barrier height while it will be increased by the incorporation of a P-type surface layer. This method is attractive because it allows the selection of the metal based upon the metallurgical properties of the interface with the semiconductor for stable operation while simultaneously tailoring the barrier height using the ion-implanted layer for the intended application.

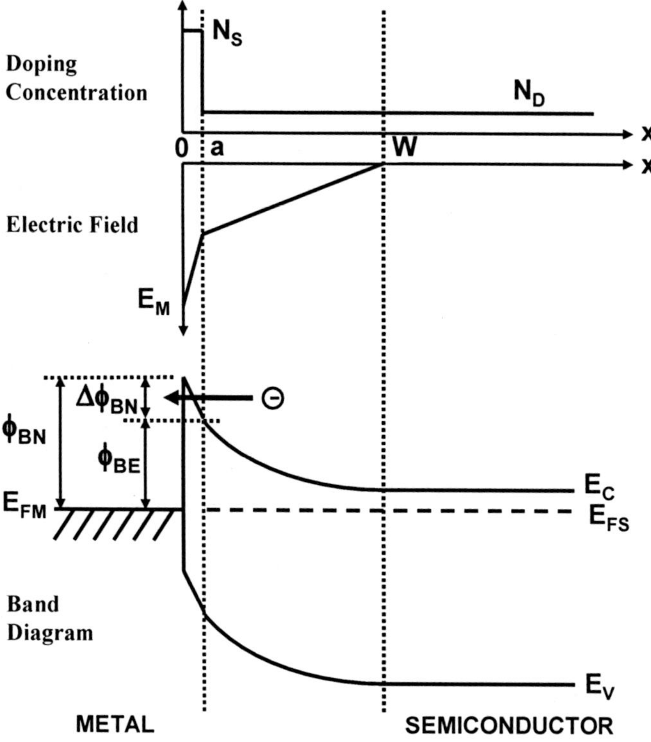

Fig. 4.28 Metal–semiconductor contact with a thin, highly doped, N-type surface layer

One method for the optimization of the barrier height for power Schottky rectifiers is by starting with a large barrier height and then reducing the barrier by ion implantation of an N-type surface layer into the N-type drift region. To analyze the reduction of the barrier height, the doping profile for this case is shown in Fig. 4.28 under the assumption of a uniform concentration in the surface layer and the drift region. The electric field profile for this doping profile is also shown in the figure. The slope of the electric field profile in the zone from $x = 0$ to $x = a$ is determined by the larger doping concentration of the surface layer, while that in the drift region from $x = a$ to $x = W$ is determined by its lower doping concentration. It can be seen from the band diagram in Fig. 4.28 that a narrow potential barrier is formed at the metal–semiconductor interface. As indicated in the figure, electrons can pass through this barrier creating a tunneling current. This additional current component can be analyzed as a reduction in the barrier height, producing an enhancement of the thermionic emission current.

The electric field profile obtained by solving Poisson's equation with the doping profile shown in Fig. 4.28 is given by the following equations:

$$E(x) = -E_{\mathrm{M}} + \frac{qN_{\mathrm{S}}x}{\varepsilon_{\mathrm{S}}} \tag{4.41}$$

from $x = 0$ to $x = a$ and

$$E(x) = -\frac{qN_{\mathrm{D}}}{\varepsilon_{\mathrm{S}}}(W - x) \tag{4.42}$$

from $x = a$ to $x = W$. The maximum electric field (E_{M}) at the metal–semiconductor contact is given by

$$E_{\mathrm{M}} = \frac{q}{\varepsilon_{\mathrm{S}}}[N_{\mathrm{S}}a + N_{\mathrm{D}}(W - a)]. \tag{4.43}$$

The reduction of the Schottky barrier height can be obtained by substitution of this expression into (4.23). If the dose ($N_{\mathrm{S}}a$) of the ion implant to create the N-type surface layer is much greater than the charge in the depletion region at zero bias, the reduction of the barrier height is given by

$$\Delta\Phi_{\mathrm{BN}} = \frac{q}{\varepsilon_{\mathrm{S}}}\sqrt{\frac{aN_{\mathrm{S}}}{4\pi}}. \tag{4.44}$$

The effective barrier height (Φ_{BE}) can be calculated by subtracting this barrier reduction from the metal–semiconductor barrier height (Φ_{BN}).

The reduction of the Schottky barrier height that can be achieved with a shallow N-type ion implant is shown in Fig. 4.29. It is possible to obtain a barrier reduction in the range of 0.05–0.15 eV by using a dose of between 10^{12} and 10^{13} cm^{-2}. This has been experimentally confirmed by using antimony implanted into an N-type drift region.[19] Antimony was chosen as the N-type dopant because of its large mass and low diffusion coefficient in silicon. The large atomic mass for antimony ensures that the dopant is located close to the silicon surface for ion implant energies of 5–10 keV, the lowest energies available in commercial ion implanters. Its low diffusion coefficient ensures that the dopant does not get re-distributed during the postimplant annealing step to activate the dopant and remove the implant damage. It has been found that the damage from ion implantation can also alter the barrier height as well as contribute to leakage current by the generation of current at the metal–semiconductor interface via deep levels in the band gap. Consequently, appropriate annealing steps are required after the ion implant to produce good Schottky diode characteristics.

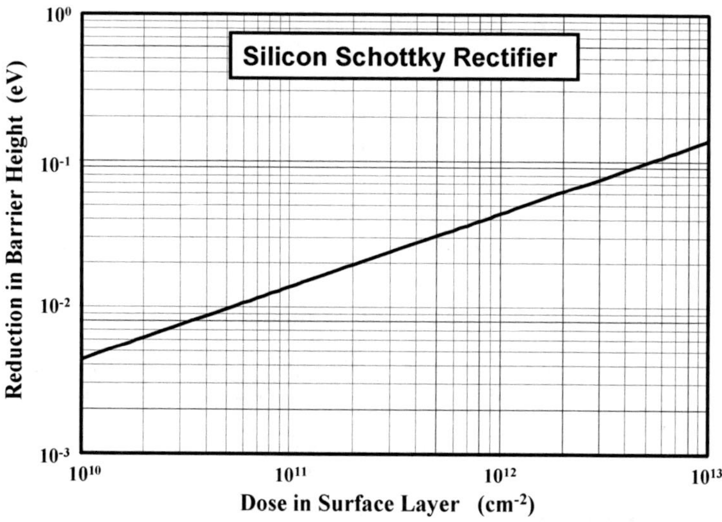

Fig. 4.29 Reduction of the Schottky barrier height with a thin, highly doped, N-type surface layer

4.10 Edge Terminations

The power Schottky barrier rectifier can be fabricated without the need for the additional processing steps that are required for creating P–N junctions. A Schottky rectifier with a metal field plate structure is illustrated in Fig. 4.30a with thermally grown silicon dioxide as the passivation at the edges. With this edge termination, a high electric field develops at the edges of metal (at point A). This not only degrades the breakdown voltage but also contributes to enhancement of the leakage current due to Schottky barrier lowering. The extension of Schottky contact metal over the oxide reduces the electric field at point A as discussed for field plates in Chap. 3. Care must be taken to design the field plate termination, so that premature breakdown is not initiated at point B.

An improved field plate structure, illustrated in Fig. 4.30b, can be created by using the local oxidation of silicon (LOCOS) process. In the LOCOS process, a tapered field oxide is grown at the edges by utilizing a silicon nitride mask in the active area of the diode. This produces a tapered oxide at the edges due to the formation of the "birds-beak" effect, as illustrated in the figure, which assists in reducing the electric field at Schottky metal contact.

A more commonly used approach for providing the edge termination for silicon Schottky power rectifiers is by incorporation of a P^+ guard ring, as illustrated in Fig. 4.30c, even though this entails extra processing steps. The guard ring overlaps the edges of the Schottky contact metal completely screening it from high electric fields. The breakdown voltage of this edge termination is the same as that of the cylindrical junction discussed in Chap. 3 if the corners of the diode are sufficiently rounded to avoid formation of spherical junctions. The presence of the P^+ guard ring creates a P–N junction in parallel with the Schottky diode. If the Schottky rectifier is designed to operate with an on-state voltage drop below 0.6 V, as is typical for silicon Schottky rectifiers with low breakdown voltages, the P–N junction does not get sufficiently forward biased to inject minority carriers into the drift region under normal operating conditions. This preserves the fast switching properties of the Schottky rectifiers essential for their application in high frequency power circuits. Under surge current levels, where the diode is subjected to a very high on-state current density, the injection from the P–N junction is beneficial for reducing the on-state voltage drop and power dissipation.

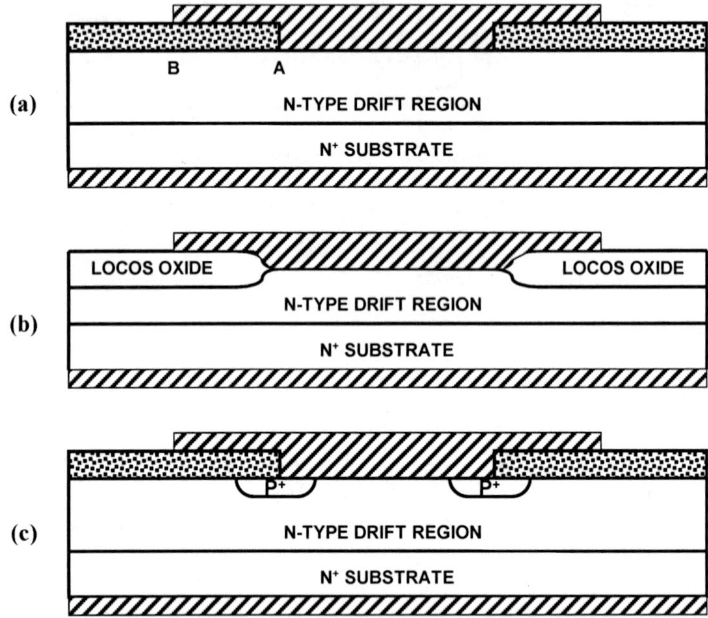

Fig. 4.30 Edge terminations for Schottky barrier rectifiers

4.11 Summary

The physics of operation of the Schottky rectifier has been described in this chapter. For power devices with relatively high breakdown voltages, the dominant current conduction mechanism is by the thermionic emission process. This process governs the fundamental relationship between the on-state voltage drop and the leakage current for power Schottky rectifiers. For power Schottky rectifiers, it is necessary to include the impact of the series resistance of the drift region on the on-state voltage drop. This series resistance limits the performance of silicon rectifiers to a breakdown voltage of less than 200 V. In addition, the Schottky barrier lowering and prebreakdown avalanche multiplication must be taken into consideration when analyzing the leakage current for silicon devices, because they can enhance the leakage current by an order of magnitude for high reverse bias voltages. In the case of silicon carbide Schottky rectifiers, it is possible to extend the breakdown voltage to at least 3,000 V due to the much smaller resistance in the drift region. However, the reverse leakage current in silicon carbide devices is significantly enhanced by the tunneling current at high reverse bias voltages.

The leakage current in silicon and silicon carbide Schottky rectifiers can be suppressed by using the JBS rectifier structure.[1–3] A detailed analysis of these structures is beyond the scope of this textbook due to space limitations.

Problems

4.1 Calculate the barrier height for a Schottky contact to silicon made using a metal with a work function of 4.6 eV.

4.2 Calculate the specific resistance for the ideal drift region for a silicon Schottky barrier rectifier designed to block 100 V.

4.3 Calculate the on-state voltage drop for a silicon Schottky barrier rectifier designed to block 100 V under the following assumptions (a) parallel-plane breakdown voltage, (b) on-state current density of 100 A cm^{-2}, (c) barrier height of 0.8 eV, (d) operation at room temperature (300 K), and (e) zero substrate and ohmic contact resistance. Provide the voltage drop across the Schottky barrier and the drift region.

4.4 Calculate the specific resistance for the ideal drift region for a silicon Schottky barrier rectifier designed to block 1,000 V.

4.5 Calculate the on-state voltage drop for a silicon Schottky barrier rectifier designed to block 1,000 V under the following assumptions (a) parallel-plane breakdown voltage, (b) on-state current density of 100 A cm^{-2}, (c) barrier height of 0.8 eV, (d) operation at room temperature (300 K), and (e) zero substrate and ohmic contact resistance. Provide the voltage drop across the Schottky barrier and the drift region.

4.6 Calculate the specific resistance for the ideal drift region for a 4H-SiC Schottky barrier rectifier designed to block 1,000 V.

4.7 Calculate the on-state voltage drop for a 4H-SiC Schottky barrier rectifier designed to block 1,000 V under the following assumptions (a) parallel-plane breakdown voltage, (b) on-state current density of 100 A cm^{-2}, (c) barrier height of 1.1 eV, (d) operation at room temperature (300 K), and

(e) zero substrate and ohmic contact resistance. Provide the voltage drop across the Schottky barrier and the drift region.

4.8 A silicon Schottky barrier rectifier is designed to block 100 V.
(a) Calculate the leakage current density without Schottky barrier lowering.
(b) Calculate the leakage current density with Schottky barrier lowering.
(c) What is the barrier reduction in eV due to the image force?
Use the following assumptions (a) parallel-plane breakdown voltage, (b) reverse bias voltage of 80 V, (c) barrier height of 0.8 eV, (d) no impact ionization, and (e) no generation or diffusion current.

4.9 A 4H-SiC Schottky barrier rectifier is designed to block 1,000 V.
(a) Calculate the leakage current density without Schottky barrier lowering and tunneling.
(b) Calculate the leakage current density with Schottky barrier lowering but without tunneling.
(c) Calculate the leakage current density with Schottky barrier lowering and tunneling.
(d) What is the barrier reduction in eV due to the image force?
Use the following assumptions (a) parallel-plane breakdown voltage, (b) reverse bias voltage of 800 V, (c) barrier height of 1.1 eV, (d) no impact ionization, and (e) no generation or diffusion current.

4.10 Calculate the specific capacitance for a silicon Schottky barrier rectifier designed to block 100 V at reverse bias voltages of 10, 20, 40, and 80 V.

4.11 Calculate the specific capacitance for a 4H-SiC Schottky barrier rectifier designed to block 1,000 V at reverse bias voltages of 100, 200, 400, and 800 V.

4.12 Calculate the power dissipation for a silicon Schottky barrier rectifier designed to block 100 V at 300, 350, 400, 450, and 500 K. Use the following assumptions (a) parallel-plane breakdown voltage, (b) reverse bias voltage of 80 V, (c) barrier height of 0.8 eV, (d) duty cycle of 50, and (e) on-state current density of 100 A cm^{-2}. Estimate the temperature at which minimum power dissipation is observed.

References

[1] B.J. Baliga, "Schottky Barrier Rectifiers and Methods of Forming the Same", U.S. Patent 5,612,567, March 18, 1997.

[2] B.J. Baliga, "The Future of Power Semiconductor Technology", Proceedings of the IEEE, Vol. 89, pp. 822–832, 2001.

[3] B.J. Baliga, "Silicon Carbide Power Devices", World Scientific, Singapore, 2005.

[4] B.J. Baliga, "Power Semiconductor Devices for Variable Frequency Drives", Proceedings of the IEEE, Vol. 82, pp. 1112–1122, 1994.
[5] C.R. Crowell and S.M. Sze, "Current Transport in Metal–Semiconductor Barriers", Solid-State Electronics, Vol. 9, pp. 1035–1048, 1966.
[6] S.M. Sze, "Physics of Semiconductor Devices", pp. 254–258, 2nd Edition, Wiley, New York, 1981.
[7] H.A. Bethe, "Theory of the Boundary Layer of Crystal Rectifiers", MIT Radiation Laboratory Report, Vol. 43, p. 12, 1942.
[8] S.M. Sze, "Physics of Semiconductor Devices", pp. 265–270, 2nd Edition, Wiley, New York, 1981.
[9] E.H. Rhoderick and R.H. Williams, "Metal–Semiconductor Contacts", pp. 35–38, 2nd Edition, Oxford Science, Oxford, 1988.
[10] S.L. Tu and B.J. Baliga, "On the Reverse Blocking Characteristics of Schottky Power Diodes", IEEE Transactions on Electron Devices, Vol. 39, pp. 2813–2814, 1992.
[11] M. Bhatnagar, P.K. McLarty, and B.J. Baliga, "Silicon-Carbide High-Voltage (400 V) Schottky Barrier Diodes", IEEE Electron Device Letters, Vol. 13, pp. 501–503, 1992.
[12] F. Dahlquist et al., "A 2.8 kV, Forward Drop JBS Diode with Low Leakage", Silicon Carbide and Related Materials – 1999, Material Science Forum, Vol. 338–342, pp. 1179–1182, 2000.
[13] Y. Sugawara, K. Asano, and R. Saito, "3.6 kV 4H-SiC JBS Diodes with Low Ron", Silicon Carbide and Related Materials – 1999, Material Science Forum, Vol. 338–342, pp. 1183–1186, 2000.
[14] T. Hatakeyama and T. Shinohe, "Reverse Characteristics of a 4H-SiC Schottky Barrier Diode", Silicon Carbide and Related Materials – 2001, Material Science Forum, Vol. 389–393, pp. 1169–1172, 2002.
[15] B.J. Baliga, "The Pinch-Rectifier: A Low Forward Drop, High Speed Power Diode", IEEE Electron Device Letters, Vol. 5, pp. 194–196, 1984.
[16] D.J. Page, "Theoretical Performance of the Schottky Barrier Power Rectifier", Solid-State Electronics, Vol. 15, pp. 505–515, 1972.
[17] E.H. Rhoderick and R.H. Williams, "Metal–Semiconductor Contacts", pp. 48–55, 2nd Edition, Oxford Science, Oxford, 1988.
[18] J.M. Shannon, "Reducing the Effective Height of a Schottky Barrier Using Low Energy Ion Implantation", Applied Physics Letters, Vol. 24, pp. 369–371, 1974.
[19] S. Ashok and B.J. Baliga, "Effect of Antimony Ion Implantation on Al–Silicon Schottky Diode Characteristics", Journal of Applied Physics, Vol. 56, pp. 1237–1239, 1984.

Chapter 5

P-i-N Rectifiers

Power rectifiers are needed in power electronic circuits to control the direction for current flow. In the case of silicon Schottky barrier rectifiers, it was demonstrated in Chap. 4 that their on-state voltage drop becomes large when the device is designed to support more than 200 V in the reverse-blocking mode. Power device applications, such as motor control, require rectifiers with blocking voltages ranging from 300 to 5,000 V. Silicon P-i-N rectifiers have been developed for these high-voltage applications.

In a P-i-N rectifier, the reverse-blocking voltage is supported across a depletion region formed with a P–N junction structure. The voltage is primarily supported within the N-type drift region with the properties of the P-type region optimized for good on-state current flow. In Chap. 3, it was established that any given reverse-blocking voltage can be supported across a thinner drift region by utilizing the punch-through design. It is beneficial to use a low doping concentration for the N-type drift region in this design. It is therefore referred to as an *i-region* (implying that the drift region is intrinsic in nature).

The silicon P-i-N rectifiers that are designed to support large voltages rely upon the high-level injection of minority carriers into the drift region. This phenomenon greatly reduces the resistance of the thick, very lightly doped drift region necessary to support high voltages in silicon. Consequently, the on-state current flow is not constrained by the low doping concentration in the drift region. A reduction of the thickness of the drift region, by utilizing the punch-through design, is beneficial for decreasing the on-state voltage drop.

In the case of silicon carbide rectifiers, it was demonstrated in Chap. 3 that the drift region doping level is relatively large and its thickness is much smaller than for silicon devices to achieve very high breakdown voltages. This enables the design of 4H-SiC-based Schottky rectifiers with reverse-blocking capability of up to at least 3,000 V with low on-state voltage drop. Based upon the inherent fast

B.J. Baliga, *Fundamentals of Power Semiconductor Devices*, doi: 10.1007/978-0-387-47314-7_5,

switching capability of Schottky rectifiers, it is anticipated that silicon carbide-based Schottky rectifiers will displace silicon P-i-N rectifiers for applications with reverse-blocking capability of up to 3,000 V.[1] However, this displacement will require further progress with reducing the cost of silicon carbide technology. Meanwhile, silicon P-i-N rectifiers will continue to play an important role in applications.

5.1 One-Dimensional Structure

The basic one-dimensional P-i-N rectifier structure is illustrated in Fig. 5.1. As previously discussed in Chap. 3, the thickness of the drift region must be designed to sustain the reverse breakdown voltage under the assumption that the depletion region punches through the drift region to the N^+ region. The voltage drop across the structure during on-state current flow depends upon the voltage drop across this drift region in addition to the voltage drop across the P–N junction.

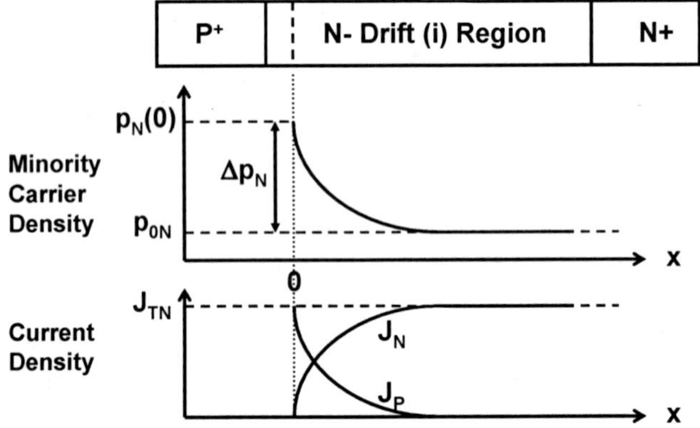

Fig. 5.1 Carrier distribution and current flow under low-level injection conditions for a P-i-N rectifier

The on-state current flow in the P-i-N rectifier is governed by three current transport mechanisms:

1. At very low-current levels, the current transport is dominated by the recombination process within the space-charge layer of the P–N junction (referred to as the *recombination current*).
2. At low-current levels, the current transport is dominated by the diffusion of minority carriers injected into the drift region (referred to as the *diffusion current*).
3. At high-current levels, the current transport is dictated by the presence of a high concentration of both electrons and holes in the drift region (referred to as the *high-level injection current*).

These current transport phenomena are discussed in this section. The impact of current flow due to injection into the heavily doped P^+ and N^+ regions (often referred to as the *end regions*) is then taken into account. The role of Auger recombination on the effective lifetime for the carriers in the drift region at high injection levels must also be considered to accurately account for the behavior of the on-state characteristics of P-i-N rectifiers at very high (surge) current levels.

5.1.1 Recombination Current

When the current flowing across the P–N junction is very small, the current transport is dominated by the recombination of carriers within the depletion region of the junction. For an applied bias V_a across the diode, the minority carrier concentration on the P-side of the junction is given by the Shockley boundary condition (also called the "Law of the Junction"):

$$n_P(0) = \frac{n_i^2}{N_A} e^{qV_a/kT}, \tag{5.1}$$

where n_i is the intrinsic carrier concentration, N_A is the acceptor doping concentration, q is the electron charge, k is a Boltzmann's constant, and T is the absolute temperature. Consequently, the *p–n* product on the P-side of the junction is given by

$$p_P(0)n_P(0) = n_i^2 e^{qV_a/kT}, \tag{5.2}$$

because the majority carrier concentration [$p_P(0)$] is equal to the doping concentration (N_A). The *p–n* product on the N-side of the P–N junction is also given by the same expression based upon a similar analysis for this region. This allows making the assumption that the *p–n* product must be given by the same equation throughout the depletion region.

The recombination rate governed by the Shockley–Read–Hall theory for a single recombination center is given by[2]

$$U = \frac{pn - n_i^2}{\tau_{p0}[(n+n_1)+\zeta(p+p_1)]}, \tag{5.3}$$

where ζ is the ratio of the minority carrier lifetime in heavily doped P-type material (τ_{n0}) to the minority carrier lifetime in heavily doped N-type material (τ_{p0}); n_1 and p_1 are, respectively, the electron and hole concentrations if the Fermi level is located at the recombination center. These concentrations are equal to the intrinsic carrier concentration (n_i) if the recombination center is located at the midgap position. From (5.2), under the assumption that the electron and hole concentrations are equal within the depletion region:

$$n = p = n_i e^{qV_a/2kT}. \tag{5.4}$$

Using this concentration for the carriers in (5.3),

$$U = \frac{n_i^2 e^{qV_a/kT} - n_i^2}{\tau_{p0}[(n_i e^{qV_a/2kT} + n_i) + \zeta(n_i e^{qV_a/2kT} + n_i)]}. \tag{5.5}$$

Since the applied bias is much greater than the thermal voltage (kT/q), this equation can be written as

$$U = \frac{n_i}{\tau_{p0}[1+\zeta]} e^{qV_a/2kT}. \tag{5.6}$$

From (2.80) with the recombination center located at the midgap position,

$$\tau_{SC} = \tau_{p0}[1+\zeta]. \tag{5.7}$$

Thus, the recombination rate under forward current flow is given by

$$U = \frac{n_i}{\tau_{SC}} e^{qV_a/2kT}. \tag{5.8}$$

The forward current density produced by this recombination process within the depletion region of width W_D is then given by

$$J_{FR} = qUW_D = \frac{qn_i W_D}{\tau_{SC}} e^{qV_a/2kT}. \tag{5.9}$$

At very low on-state current levels, the forward current density in P-i-N rectifiers has been observed to follow this dependence on the on-state voltage drop.

5.1.2 Low-Level Injection Current

When the forward bias across the P-i-N rectifier is increased, the current flow becomes dominated by the diffusion current associated with the injection of minority carriers into the neutral regions on either side of the P–N junction. The profile for the injected minority carriers on the N-side of the junction is illustrated in Fig. 5.1. The concentration of holes at the edge of the depletion region is related to the applied forward bias by the “Law of the Junction” (also referred to as the Shockley boundary conditions) derived under Boltzmann’s quasiequilibrium assumptions:

$$p_N(0) = p_{0N} e^{qV_a/kT}, \tag{5.10}$$

where p_{0N} is the hole concentration in equilibrium on the N-side of the junction. Note that the x-axis is defined to begin at the edge of the depletion region on the N-side of the junction. The excess minority carrier density (Δp_N) at the edge of the depletion region is then given by

$$\Delta p_N = p_N(0) - p_{0N} = p_{0N}(e^{qV_a/kT} - 1). \tag{5.11}$$

The injected carriers diffuse away from the edge of the depletion region with a characteristic length (referred to as the *minority carrier diffusion length* (L_p)), producing an exponential reduction of the excess minority carrier density:

$$\delta p_N(x) = \Delta p_N e^{-x/L_p}. \tag{5.12}$$

At the edge of the depletion region, the total current flow due to injection into the N-side of the junction is entirely due to the diffusion current[3] given by

$$J_{TN} = J_p(0) = qD_p\left(\frac{d\delta p_N}{dx}\right)_{x=0}, \tag{5.13}$$

where D_p is the diffusion coefficient for holes. Using (5.11) and (5.12) to describe the excess minority carrier distribution,

$$J_{TN} = \frac{qD_p p_{0N}}{L_p}(e^{qV_a/kT} - 1). \tag{5.14}$$

As indicated in the lower part of Fig. 5.1, this minority carrier diffusion current decreases when proceeding away from the junction. The same total current flow is maintained by a small electric field developed in the neutral region to create a majority carrier drift component. This majority carrier current increases when proceeding away from the junction as shown in Fig. 5.1.

During forward bias, injection of minority carriers (electrons) also occurs into the P-side of the junction. Based upon an exponential decay of this carrier concentration in the P-side of the junction when proceeding away from the edge of the depletion region with a characteristic length L_n, the total current flow due to injection on the P-side of the junction can be shown to be given by

$$J_{TP} = \frac{qD_n n_{0P}}{L_n}(e^{qV_a/kT} - 1), \tag{5.15}$$

where D_n is the diffusion coefficient for electrons and n_{0P} is the minority carrier concentration in equilibrium on the P-side of the junction. Power P-i-N rectifiers are always designed with a high doping concentration (N_A) on the P-side of the junction in comparison with the doping concentration (N_D) on the N-side, which serves as the drift region. Consequently,

$$p_{0N} = \frac{n_i^2}{N_D} \gg n_{0P} = \frac{n_i^2}{N_A}. \tag{5.16}$$

Based upon this difference in the minority carrier density on the two sides of the junction, the current (J_{TP}) due to injection into the P-side of the junction can be assumed to be negligible when compared with the current (J_{TN}) due to injection into the N-side of the junction. The total current flow across the P–N junction under low-level injection conditions is therefore given by (5.14). It is worth pointing out that the rate of increase in current with forward bias voltage predicted by this equation is more rapid than that for the recombination current.

In deriving the above relationship, the width of the drift region was assumed to be very large when compared with the diffusion length for holes. If this does not hold true, the minority carrier distribution in the drift region is altered because its concentration is reduced to the minority carrier density in the N^+ region at the boundary between the drift region and the N^+ region. Since the doping concentration in the N^+ region is large, the minority carrier concentration can be assumed to be zero at the boundary. Using this boundary condition, it can be shown[4] that the current flow due to injection into the N-drift region is given by

$$J_{\mathrm{TN}} = \frac{qD_{\mathrm{p}}p_{0\mathrm{N}}}{L_{\mathrm{p}}\tanh(W_{\mathrm{N}}/L_{\mathrm{p}})}(\mathrm{e}^{qV_{\mathrm{a}}/kT} - 1), \tag{5.17}$$

where W_N is the width of the N-drift region.

5.1.3 High-Level Injection Current

The N-drift region in the P-i-N rectifier must be lightly doped to support high voltages in the reverse-blocking mode. When the on-state voltage drop increases, the injected minority carrier concentration also increases in accordance with (5.10). Ultimately, the concentration of the minority carriers exceeds the background doping concentration (N_D) in the drift region. This is defined as *high-level injection*. When the injected hole concentration in the drift region becomes much

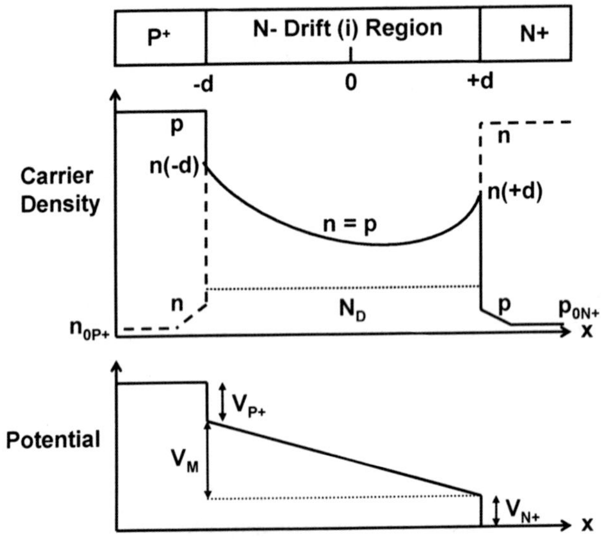

Fig. 5.2 Carrier and potential distribution under high-level injection conditions for a P-i-N rectifier

greater than the background doping concentration, charge neutrality requires that the concentrations for electrons and holes become equal:

$$n(x) = p(x). \tag{5.18}$$

The large concentration of free carriers reduces the resistance of the drift region. This phenomenon is referred to as the *conductivity modulation* of the drift region. Conductivity modulation of the drift region is beneficial for allowing the transport of a high-current density through lightly doped drift regions with a low on-state voltage drop.

The carrier distribution within the drift region $n(x)$ can be obtained by solving the continuity equations for the N-region[5–7]:

$$\frac{\partial n}{\partial t} = -\frac{n}{\tau_{HL}} + D_n \frac{\partial^2 n}{\partial x^2} + \mu_n \frac{\partial}{\partial x}(nE), \tag{5.19}$$

$$\frac{\partial p}{\partial t} = -\frac{p}{\tau_{HL}} + D_p \frac{\partial^2 p}{\partial x^2} - \mu_p \frac{\partial}{\partial x}(pE), \tag{5.20}$$

where D_n and D_p are the diffusion coefficients for electrons and holes, respectively, and τ_{HL} is the high-level lifetime in the drift region. Combining these equations after multiplying (5.19) by $(\mu_p p)$ and (5.20) by $(\mu_n n)$ gives

$$\frac{\partial n}{\partial t} = -\frac{n}{\tau_{HL}} + \left(\frac{\mu_p p D_n + \mu_n n D_p}{\mu_p p + \mu_n n}\right)\frac{\partial^2 n}{\partial x^2}. \tag{5.21}$$

In deriving this equation, it has been assumed that the transport of carriers due to the electric field can be neglected when compared with the current due to the diffusion of the carriers. The Einstein relationship between the diffusion coefficient and the mobility gives

$$D = \frac{kT}{q}\mu. \tag{5.22}$$

Since the carrier density for electrons and holes is equal in accordance with (5.18), (5.21) can be written under steady-state conditions as

$$\frac{\partial n}{\partial t} = 0 = -\frac{n}{\tau_{HL}} + D_a \frac{\partial^2 n}{\partial x^2}, \tag{5.23}$$

where τ_{HL} is the high-level lifetime in the drift region and D_a is the ambipolar diffusion coefficient given by

$$D_a = \frac{p+n}{\dfrac{p}{D_n} + \dfrac{n}{D_p}} = \frac{2D_n D_p}{D_n + D_p} \tag{5.24}$$

due to charge neutrality (see (5.18)). The general solution for the carrier concentration governed by (5.23) is given by

$$n(x) = A\cosh\left(\frac{x}{L_a}\right) + B\sinh\left(\frac{x}{L_a}\right), \tag{5.25}$$

with the constants A and B determined by the boundary conditions for the N-drift region. The parameter L_a in this equation, referred to as the *ambipolar diffusion length*, is given by

$$L_a = \sqrt{D_a \tau_{HL}}. \tag{5.26}$$

At the junction between the N-drift region and the N$^+$ cathode region (located at $x = +d$ in Fig. 5.1), the total current flow occurs exclusively by electron transport:

$$J_T = J_n(+d) \tag{5.27}$$

and

$$J_p(+d) = 0. \tag{5.28}$$

Similarly, at the junction between the N-drift region and the P$^+$ anode region (located at $x = -d$ in Fig. 5.1), the total current flow occurs exclusively by hole transport:

$$J_T = J_p(-d) \tag{5.29}$$

and

$$J_n(-d) = 0. \tag{5.30}$$

Using (5.28), the hole current due to drift and diffusion can be written as

$$J_p(+d) = q\mu_p p(+d)E(+d) - qD_p\left(\frac{dp}{dx}\right)_{x=+d} = 0. \tag{5.31}$$

Combining this equation with (5.18) and the Einstein relationship,

$$E(+d) = \frac{kT}{qn(+d)}\left(\frac{dn}{dx}\right)_{x=+d}. \tag{5.32}$$

Equation (5.27) for the total current flow due to electron transport at this boundary can be written as

$$J_T = q\mu_n n(+d)E(+d) + qD_n\left(\frac{dn}{dx}\right)_{x=+d}. \tag{5.33}$$

Using (5.32) for the electric field $E(+d)$,

$$J_{\mathrm{T}} = 2qD_{\mathrm{n}}\left(\frac{\mathrm{d}n}{\mathrm{d}x}\right)_{x=+d}. \tag{5.34}$$

In the same manner,

$$J_{\mathrm{T}} = 2qD_{\mathrm{p}}\left(\frac{\mathrm{d}n}{\mathrm{d}x}\right)_{x=-d}. \tag{5.35}$$

Based upon these equations,

$$\left(\frac{\mathrm{d}n}{\mathrm{d}x}\right)_{x=+d} = \frac{J_{\mathrm{T}}}{2qD_{\mathrm{n}}} < \left(\frac{\mathrm{d}n}{\mathrm{d}x}\right)_{x=-d} = \frac{J_{\mathrm{T}}}{2qD_{\mathrm{p}}}. \tag{5.36}$$

This indicates that the slope of the carrier concentration is smaller at the cathode side when compared with the anode side.

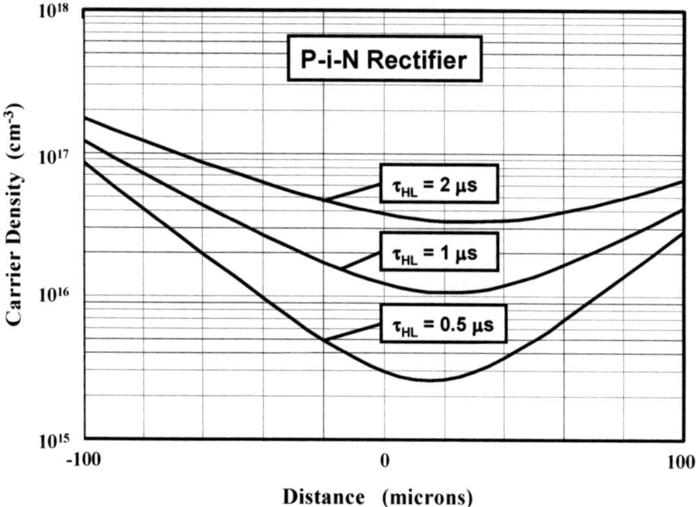

Fig. 5.3 Carrier distribution under high-level injection conditions for a P-i-N rectifier with various high-level lifetime values

The above boundary conditions can be used to obtain the constants A and B in (5.25) resulting in

$$n(x) = p(x) = \frac{\tau_{\mathrm{HL}} J_{\mathrm{T}}}{2qL_{\mathrm{a}}}\left[\frac{\cosh(x/L_{\mathrm{a}})}{\sinh(d/L_{\mathrm{a}})} - \frac{\sinh(x/L_{\mathrm{a}})}{2\cosh(d/L_{\mathrm{a}})}\right]. \tag{5.37}$$

The catenary carrier distribution described by this equation was schematically illustrated in Fig. 5.2. As a particular example, the carrier distributions calculated by using (5.37) are shown in Fig. 5.3 for the case of three values for the high-level lifetime for a diode with drift region thickness of 200 μm. The largest concentrations for the electrons and holes in the drift region occur at its boundary with the P^+ and N^+ end regions. The droop in the carrier density toward the center of the drift region is determined by the ambipolar diffusion length. A larger droop in concentration occurs with the smallest diffusion length, and a smaller average carrier concentration is observed when the lifetime is reduced.

The reduction of the average carrier density injected into the drift region with reduction of the lifetime can be deduced from charge control considerations. Under steady-state conditions, the current flow in the P-i-N rectifier can be related to sustaining the recombination of holes and electrons within the drift region if the recombination within the end regions is neglected. Consequently,

$$J_{\mathrm{T}} = \int_{-d}^{+d} qR\,\mathrm{d}x, \tag{5.38}$$

where R is the recombination rate given by

$$R = \frac{n(x)}{\tau_{\mathrm{HL}}}. \tag{5.39}$$

Using an average carrier density (n_{a}) within the drift region, these equations can be combined to yield

$$J_{\mathrm{T}} = \frac{2qn_{\mathrm{a}}d}{\tau_{\mathrm{HL}}}. \tag{5.40}$$

The average carrier density in the drift region is then given by

$$n_{\mathrm{a}} = \frac{J_{\mathrm{T}}\tau_{\mathrm{HL}}}{2qd}. \tag{5.41}$$

From this relationship, it can be concluded that the average carrier density in the drift region will increase with the on-state current density and decrease with reduction of the lifetime. This behavior is exhibited by the carrier distribution in Fig. 5.3. For the case of an on-state current density of 100 A cm^{-2} and a drift region thickness ($2d$) of 200 μm with a high-level lifetime of 1 μs, the average carrier concentration obtained by using (5.41) is 3×10^{16} cm^{-3}, which is consistent with the carrier distribution shown in Fig. 5.3.

The specific resistance of the drift region can be calculated from the average carrier density with the acknowledgement that both electrons and holes are available for current transport:

$$R_{\mathrm{i,SP}} = \frac{2d}{q(\mu_{\mathrm{n}} + \mu_{\mathrm{p}})n_{\mathrm{a}}}. \tag{5.42}$$

Using (5.41) for the average carrier density,

$$R_{\mathrm{i,SP}} = \frac{4d^2}{(\mu_{\mathrm{n}} + \mu_{\mathrm{p}})J_{\mathrm{T}}\tau_{\mathrm{HL}}}. \tag{5.43}$$

The voltage drop across the drift region (middle region) is then given by

$$V_{\mathrm{M}} = J_{\mathrm{T}}R_{\mathrm{i,SP}} = \frac{4d^2}{(\mu_{\mathrm{n}} + \mu_{\mathrm{p}})\tau_{\mathrm{HL}}}. \tag{5.44}$$

From this equation, it can be concluded that the voltage drop across the drift region is independent of the current density flowing through it. This unusual behavior occurs due to the presence of a high concentration of minority carriers, contrary to Ohm's law for drift regions without the conductivity modulation. Thus, the conductivity modulation phenomenon at high injection levels enables maintaining a low-voltage drop across the drift region, which is extremely beneficial for obtaining a low on-state voltage drop in power P-i-N rectifiers.

A more accurate analysis of the voltage drop across the drift region can be performed by integration of the electric field. The electric field in the drift region can be obtained from the carrier distribution given by (5.37). The hole and electron currents flowing in the drift region are given by

$$J_{\mathrm{p}} = q\mu_{\mathrm{p}}\left(pE - \frac{kT}{q}\frac{\mathrm{d}p}{\mathrm{d}x}\right) \tag{5.45}$$

and

$$J_{\mathrm{n}} = q\mu_{\mathrm{n}}\left(nE + \frac{kT}{q}\frac{\mathrm{d}n}{\mathrm{d}x}\right). \tag{5.46}$$

The total current at any location in the drift region is constant and given by

$$J_{\mathrm{T}} = J_{\mathrm{p}} + J_{\mathrm{n}}. \tag{5.47}$$

Combining these relationships,

$$E(x) = \frac{J_{\mathrm{T}}}{q(\mu_{\mathrm{n}} + \mu_{\mathrm{p}})n} - \frac{kT}{2qn}\frac{\mathrm{d}n}{\mathrm{d}x}. \tag{5.48}$$

Here, the charge neutrality condition $n(x) = p(x)$ was also utilized. The first term in (5.48) takes into account the ohmic voltage drop due to current flow through the drift region. The second term in (5.48) is associated with the asymmetrical carrier gradient produced by the difference in the mobility for electrons and holes.

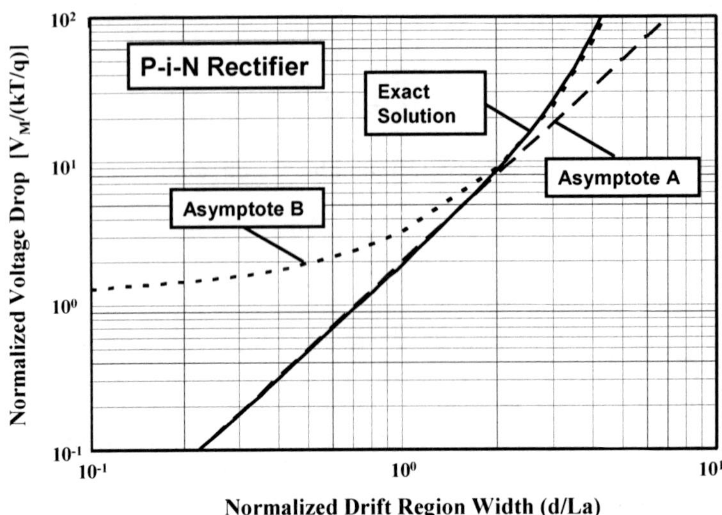

Fig. 5.4 Voltage drop in the drift (middle) region of a P-i-N rectifier

The integration of the electric field distribution given in (5.48) yields the voltage drop across the drift (or middle) region[5–7]:

$$\frac{V_\mathrm{M}}{kT/q} = \left\{ \begin{array}{l} \dfrac{8b}{(b+1)^2} \dfrac{\sinh(d/L_\mathrm{a})}{\sqrt{1-B^2\tanh^2(d/L_\mathrm{a})}} \\ \times \arctan\left[\sqrt{1-B^2\tanh^2(d/L_\mathrm{a})}\sinh(d/L_\mathrm{a})\right] \end{array} \right\} \tag{5.49}$$

$$+B\ln\left[\frac{1+B\tanh^2(d/L_\mathrm{a})}{1-B\tanh^2(d/L_\mathrm{a})}\right],$$

where $b = (\mu_\mathrm{n}/\mu_\mathrm{p})$ and $B = (\mu_\mathrm{n} - \mu_\mathrm{p})/(\mu_\mathrm{n} + \mu_\mathrm{p})$. This complex equation can be approximated by using two asymptotes as illustrated in Fig. 5.4. For d/L_a ratios of up to 2, the asymptote A given by

$$V_\mathrm{M} = \frac{2kT}{q}\left(\frac{d}{L_\mathrm{a}}\right)^2 \tag{5.50}$$

provides a good fit. For d/L_a ratios of greater than 2, the asymptote B given by

$$V_\mathrm{M} = \frac{3\pi kT}{8q}\mathrm{e}^{(d/L_\mathrm{a})} \tag{5.51}$$

provides a good fit. As discussed earlier in connection with (5.44), all these terms for the voltage drop across the drift region are independent of the on-state current density. The voltage drop across the drift region increases rapidly with increasing d/L_a ratio. When this ratio is 0.1, the middle region voltage drop is only 0.5 mV. It increases to a voltage drop of about 50 mV for a d/L_a ratio of unity and becomes 0.7 V when the d/L_a ratio increases to 3. Thus, the increase in the voltage drop across the middle region degrades the on-state voltage drop when the lifetime is reduced to enhance the switching speed.

The on-state voltage drop in the P-i-N rectifier consists of the voltage drop across the P^+/N junction, the middle region, and the N/N^+ interface. The voltage drop across the P^+/N junction can be determined from the injected minority carrier density:

$$p(-d) = p_{0N} e^{qV_{P+}/kT}, \tag{5.52}$$

where p_{0N} is the minority carrier density in the N-type drift region in equilibrium and V_{P+} is the voltage drop across the P^+/N junction. Relating the minority carrier concentration in equilibrium to the doping level N_D in the drift region,

$$V_{P+} = \frac{kT}{q} \ln\left[\frac{p(-d) N_D}{n_i^2}\right]. \tag{5.53}$$

Similarly, applying the "Law of the Junction" on the cathode side,

$$n(+d) = n_{0N} e^{qV_{N+}/kT}, \tag{5.54}$$

where n_{0N} is the majority carrier density in the N-type drift region in equilibrium and V_{N+} is the voltage drop across the N^+/N junction. Since the majority carrier concentration in equilibrium is equal to the doping level N_D in the drift region,

$$V_{N+} = \frac{kT}{q} \ln\left[\frac{n(+d)}{N_D}\right]. \tag{5.55}$$

The voltage drop associated with the two end regions is therefore given by

$$V_{P+} + V_{N+} = \frac{kT}{q} \ln\left[\frac{n(+d) n(-d)}{n_i^2}\right]. \tag{5.56}$$

In deriving this expression, the charge neutrality condition $n(x) = p(x)$ under high-level injection was assumed.

The voltage drop across the end regions has been combined with the voltage drop across the middle region to derive a relationship between the on-state current density (J_T) and the total on-state voltage drop (V_{ON}) for the P-i-N rectifier[5–7]:

$$J_T = \frac{2qD_a n_i}{d} F\left(\frac{d}{L_a}\right) e^{qV_{ON}/2kT}, \tag{5.57}$$

where

$$F\left(\frac{d}{L_a}\right) = \frac{(d/L_a)\tanh(d/L_a)}{\sqrt{1-0.25\tanh^4(d/L_a)}} e^{-qV_M/2kT}. \tag{5.58}$$

From (5.57), it is apparent that, for a fixed on-state current density, the on-state voltage drop will be smaller if the function $F(d/L_a)$ is large.

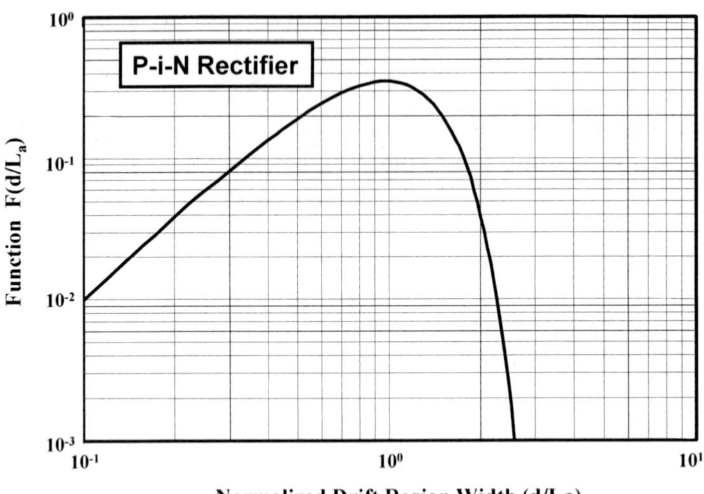

Fig. 5.5 Function $F(d/L_a)$ for a P-i-N rectifier

The variation of the function $F(d/L_a)$ with increasing (d/L_a) ratio is plotted in Fig. 5.5. It can be seen that this function has a maximum value when $d/L_a = 1$. Thus, to minimize the on-state voltage drop, the lifetime should be adjusted until the diffusion length is equal to one-half of the width of the drift region. It is worth pointing out that the function $F(d/L_a)$ decreases very rapidly when the (d/L_a) ratio increases beyond a value of 3. This leads to a very rapid increase in the on-state voltage drop when the diffusion length is less than one-sixth of the drift region width.

The on-state voltage drop for a P-i-N rectifier can be derived from (5.57) for a device with its half-width (d) determined by the punch-through breakdown voltage capability:

$$V_{ON} = \frac{2kT}{q}\ln\left[\frac{J_T d}{2qD_a n_i F(d/L_a)}\right]. \tag{5.59}$$

The calculated on-state voltage drop for a silicon P-i-N rectifier with a drift region width of 200 μm is shown in Fig. 5.6 at an on-state current density of 100 A cm^{-2}. As expected, the on-state voltage drop exhibits a minimum at a (d/L_a) ratio of unity and increases rapidly when the (d/L_a) ratio exceeds 3.

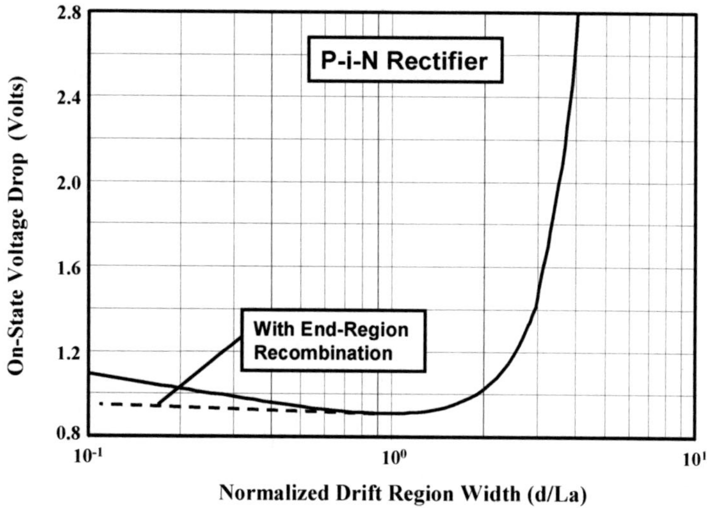

Fig. 5.6 On-state voltage drop for a P-i-N rectifier

5.1.4 Injection into the End Regions

The high-level injection conditions within the drift region described above determine the on-state voltage drop at typical current densities in the range of 100–200 A cm^{-2}. At even greater on-state current densities, the injection of minority carriers into the end regions must be taken into account.[8,9] The total current flow must therefore accommodate not only the recombination of carriers in the drift region but also the recombination of carriers in the end regions. Thus,

$$J_{\text{T}} = J_{\text{P+}} + J_{\text{M}} + J_{\text{N+}}. \tag{5.60}$$

Consequently, the current density associated with the middle region (J_{M}) is no longer equal to the total on-state current density (J_{M}) as assumed in Sect. 5.1.3 but has a smaller value. This reduces the injection level in the drift region corresponding to any given total current density resulting in an increase in the voltage drop across the middle region.

Due to the high doping concentrations in the end regions, the injected minority carrier density in these regions is well below the majority carrier density even during operation at very high on-state current densities. The current corresponding to the end regions can therefore be analyzed using low-level injection theory under the assumption of a uniform doping concentration in these regions. Based upon (5.17),

$$J_{P+} = \frac{qD_{nP+}n_{0P+}}{L_{nP+}\tanh(W_{P+}/L_{nP+})} e^{qV_{P+}/kT} = J_{SP+}e^{qV_{P+}/kT}, \tag{5.61}$$

where W_{P+} is the width of the P^+ region, L_{nP+} is the minority carrier diffusion length in the P^+ region, D_{nP+} is the minority carrier diffusion coefficient in the P^+ region, n_{0P+} is the minority carrier concentration in the P^+ region, and V_{P+} is the voltage drop at the P^+/N junction. In a similar manner,

$$J_{N+} = \frac{qD_{pN+}p_{0N+}}{L_{pN+}\tanh(W_{N+}/L_{pN+})} e^{qV_{N+}/kT} = J_{SN+}e^{qV_{N+}/kT}, \tag{5.62}$$

where W_{N+} is the width of the N^+ region, L_{pN+} is the minority carrier diffusion length in the N^+ region, D_{pN+} is the minority carrier diffusion coefficient in the N^+ region, p_{0N+} is the minority carrier concentration in the N^+ region, and V_{N+} is the voltage drop at the N^+/N interface. In (5.61) and (5.62), J_{SP+} and J_{SN+} are referred to as the *saturation current densities* for the heavily doped P^+ anode and N^+ cathode regions, respectively. They are a measure of the quality of the end regions as determined by their doping profiles and processing conditions. Typical values for the saturation current density are in the range from 1×10^{-13} to 4×10^{-13} A cm^{-2} for silicon devices.

The injected carrier concentrations on the two sides of the P^+/N junction are related under quasiequilibrium conditions by

$$p_{P+}(-d)n_{P+}(-d) = p(-d)n(-d). \tag{5.63}$$

Under low-level injection conditions within the P^+ anode region,

$$p_{P+}(-d) = p_{0P+} \tag{5.64}$$

and

$$n_{P+}(-d) = n_{0P+}e^{qV_{P+}/kT}. \tag{5.65}$$

Using these relationships in (5.63),

$$p(-d)n(-d) = p_{0P+}n_{0P+}e^{qV_{P+}/kT} = n_{ieP+}^2 e^{qV_{P+}/kT}, \tag{5.66}$$

where n_{ieP+} is the effective intrinsic carrier concentration in the P^+ anode region including the influence of band-gap narrowing as discussed in Chap. 2. Due to charge neutrality considerations $p(-d) = n(-d)$, leading to

$$e^{qV_{P+}/kT} = \left[\frac{n(-d)}{n_{ieP+}}\right]^2. \tag{5.67}$$

Using this expression in (5.61),

$$J_{P+} = J_{SP+}\left[\frac{n(-d)}{n_{ieP+}}\right]^2. \tag{5.68}$$

A similar derivation performed for the N$^+$ cathode side yields

$$J_{N+} = J_{SN+}\left[\frac{n(+d)}{n_{ieN+}}\right]^2, \tag{5.69}$$

where n_{ieN+} is the effective intrinsic carrier concentration in the N$^+$ cathode region including the influence of band-gap narrowing. From these equations, it can be concluded that the carrier concentration in the drift region will increase as the square root of the current density if the end region recombination becomes dominant. Under these circumstances, the middle region voltage drop is no longer independent of the current density resulting in an increase in the total on-state voltage drop. The influence of end region recombination can be mitigated by optimization of the doping profile to minimize the saturation current densities.

5.1.5 Carrier–Carrier Scattering Effect

The current flow through the drift region in the P-i-N rectifier is accomplished with the injection of a high concentration of holes and electrons. As discussed in Chap. 2, the presence of a high concentration of both electrons and holes increases the probability for mutual Coulombic interaction about a common center of mass. This carrier–carrier scattering phenomenon produces a reduction of the mobility for both electrons and holes as shown in Fig. 2.16. The concomitant reduction of the diffusion length produces a decrease in the carrier concentration in the central portion of the drift region. This phenomenon increases the on-state voltage drop of the P-i-N rectifier, especially at surge current levels.

5.1.6 Auger Recombination Effect

At high on-state current densities, the carrier concentration in the drift region becomes sufficiently large to favor the Auger recombination process. As discussed in Chap. 2, Auger recombination occurs by the transfer of energy due to the recombination of an electron–hole pair to a third mobile particle. The rate of recombination is then given by

$$R = \frac{n(x)}{\tau_{HL}} + C_{A\Delta n}[n(x)]^2, \tag{5.70}$$

where $C_{A\Delta n}$ is the Auger coefficient for the high-level injection case. This change in the recombination rate can be taken into account by rewriting (5.23):

$$D_{\mathrm{a}}\frac{\partial^2 n}{\partial x^2}=\frac{n(x)}{\tau_{\mathrm{HL}}}+C_{\mathrm{A\Delta n}}[n(x)]^2. \tag{5.71}$$

A solution for this differential equation is

$$n(x)=n(x_0)\left[cn\left(\frac{x-x_0}{L}\Big/ m\right)\right]^{-1}, \tag{5.72}$$

where $cn(u/m)$ is a Jacobian elliptic function of argument u. In this equation, x_0 is the value for x at which dn/dx = 0. The parameter m is given by

$$m=\frac{0.5C_{\mathrm{A\Delta n}}\tau_{\mathrm{HL}}[n(x_0)]^2+1}{C_{\mathrm{A\Delta n}}\tau_{\mathrm{HL}}[n(x_0)]^2+1} \tag{5.73}$$

and

$$L=\sqrt{D_{\mathrm{a}}\tau_{\mathrm{HL}}(2m-1)} \tag{5.74}$$

is the modified ambipolar diffusion length. The onset of Auger recombination in the drift region depresses the carrier concentration in the central portion, leading to an increase in the on-state voltage drop at very high-current densities. Auger recombination can also be a factor when determining the current flow into the end regions. The high doping concentrations in the P^+ anode and N^+ cathode regions favor the Auger recombination process due to the large majority carrier density. This reduces the minority carrier diffusion length in the end regions which enhances the end region current levels.

When only the recombination within the drift region is taken into account, it is found that the on-state voltage drop goes through a minimum when the (d/L_a) ratio becomes equal to unity as shown in Fig. 5.6. However, with the inclusion of end region recombination, including the band-gap narrowing and Auger recombination effects, the forward drop becomes nearly independent of the (d/L_a) ratio when its value is below unity. This is illustrated in Fig. 5.6 by the dashed line. During the fabrication of P-i-N rectifiers, it is important to optimize the doping profile for the end regions. Although a high doping concentration favors high injection efficiency, the on-set of band-gap narrowing and Auger recombination can counteract the benefits of higher doping levels. This has led to the speculation that an optimum doping profile exists for minimizing the adverse impact of re-combination in the end regions. However, the measured saturation current densities for end regions fabricated using a variety of techniques (including alloying and diffusion) indicate a weak dependence of the saturated current density upon the processing conditions and the doping profile.

5.1.7 Forward Conduction Characteristics

The analysis of current flow in a P-i-N rectifier in the previous sections indicates that the relationship between the current density and the voltage drop across the rectifier depends upon the injection level. At very low-current levels, space-charge generation controls the current flow with the current proportional to $(qV_{ON}/2kT)$. When the current is controlled by minority carrier injection into the drift region with the minority carrier concentration well below the background doping concentration, the current flow occurs by diffusion under low-level injection conditions. The current flow is then proportional to (qV_{ON}/kT). With further increase in the forward current density, the injected carrier density in the drift region exceeds the background doping concentration leading to high-level injection conditions. The current flow then once again becomes proportional to $(qV_{ON}/2kT)$. In this mode of operation, the injected carrier concentration in the drift region increases in proportion to the current density resulting in a constant voltage drop across the drift region. At even larger on-state current densities, the influence of the recombination in the end regions reduces the injected carrier density in the drift region. This produces a more rapid increase in the on-state voltage drop. This

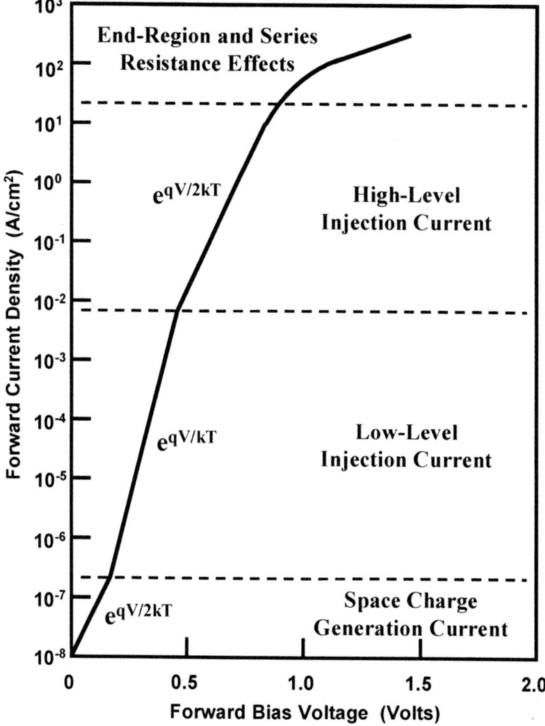

Fig. 5.7 On-state characteristics for a P-i-N rectifier

overall behavior is captured in Fig. 5.7 where a typical on-state characteristic is shown for a P-i-N rectifier under the various modes of operation.

Simulation Example

To gain further insight into the physics of operation for the P-i-N rectifier, the results of two-dimensional numerical simulations are provided in this section for a structure designed for supporting 3,000 V. For this case, a drift region with doping concentration of 4.6×10^{13} cm^{-3} was used with a thickness of 300 μm. The P$^+$ and N$^+$ end regions had a surface concentration of 1×10^{19} cm^{-3} and a depth of about 5 μm. The on-state characteristics were obtained for various values for the lifetime (τ_{p0} and τ_{n0}). In all cases, it was assumed that $\tau_{p0} = \tau_{n0}$. The influence of band-gap narrowing, Auger recombination, and carrier–carrier scattering was included during the numerical simulations.

The on-state characteristics obtained from the numerical simulations are shown in Fig. 5.8 for the case of lifetime (τ_{p0} and τ_{n0}) of 10 μs in the drift region. Several distinct regimes of operation are apparent in the shape of the characteristics. At current densities ranging between 10^{-7} and 10^{-3} A cm^{-2}, the device operates in the low-level injection regime. Here, the slope of the *i*–*v* characteristic exhibits the expected (qV_{ON}/kT) behavior with the forward voltage drop increasing at the rate of 60 mV per decade of increase in the on-state current density. At larger

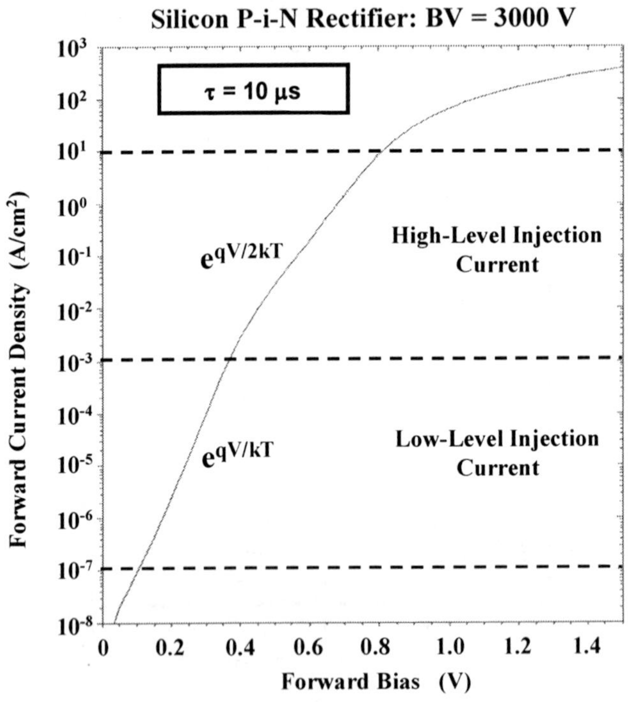

Fig. 5.8 On-state characteristics for a 3,000-V P-i-N rectifier

current densities ranging between 10^{-3} and 10^{1} A cm^{-2}, the device operates in the high-level injection regime. Here, the slope of the *i–v* characteristic exhibits the expected $(qV_{ON}/2kT)$ behavior with the forward voltage drop increasing at the rate of 120 mV per decade of increase in the on-state current density. This behavior validates the analytical theory described in the previous sections for current conduction in the P-i-N rectifier.

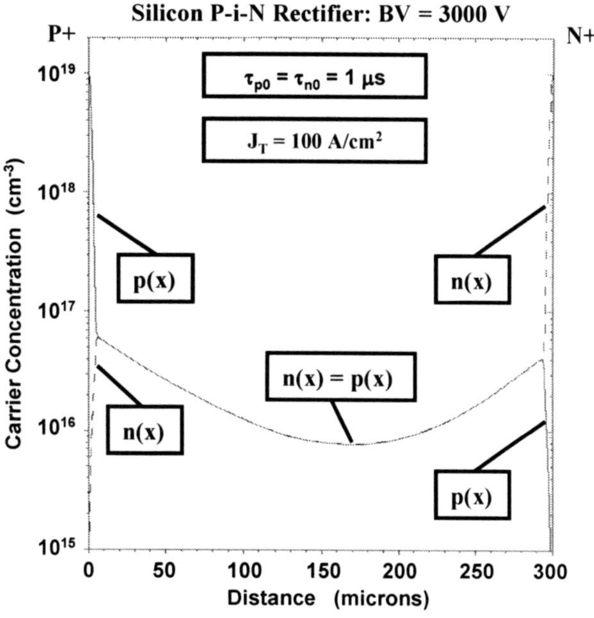

Fig. 5.9 Carrier distribution within a 3,000-V P-i-N rectifier

The carrier distribution within the P-i-N rectifier is shown in Fig. 5.9 for the case of a lifetime (τ_{p0} and τ_{n0}) of 1 μs in the drift region at an on-state current density of 100 A cm^{-2}. Here, the hole concentration is shown with a solid line while the electron concentration is shown by a dashed line. It can be seen that high-level injection conditions prevail in the drift region because the injected carrier concentration is far greater than the background doping concentration. The hole and electron concentrations are equal in magnitude throughout the drift region and exhibit the expected catenary shape derived in the previous sections. The average carrier concentration (4×10^{16} cm^{-3}) calculated by using (5.41) is in good agreement with the carrier density obtained from the simulations.

The impact of changing the lifetime in the 3,000-V structure is shown in Fig. 5.10. It can be seen that the on-state voltage drop is low for a lifetime of 100 μs. This is consistent with the small value of about 0.3 for the (d/L_a) ratio. When the lifetime is reduced to 10 μs, the on-state voltage drop increases only slightly because the (d/L_a) ratio is still close to unity. However, the on-state voltage drop increases substantially when the lifetime is reduced to 1 μs because the (d/L_a) ratio has become significantly greater than 1. The on-state voltage drop obtained with the simulations is close to that obtained using the analytical model, as shown in Fig. 5.11 by the square symbols, providing further credence to the model. In addition, the results obtained by varying the lifetime in a 1,000-V P-i-N rectifier with an N-drift region width of 60 μm are shown by the triangular symbols.

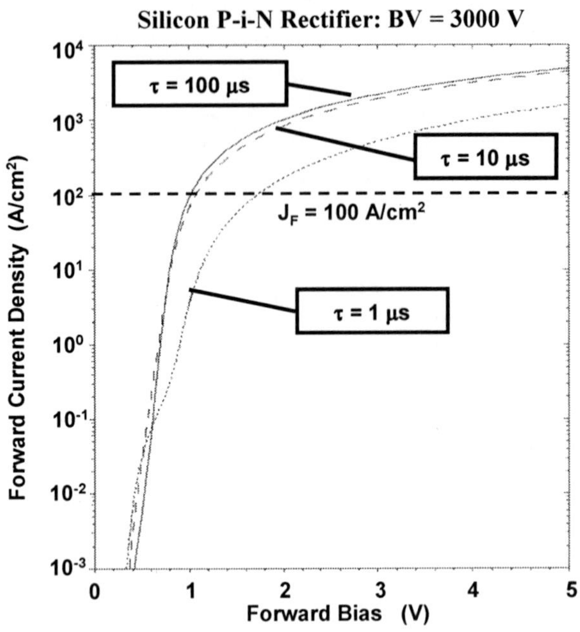

Fig. 5.10 On-state characteristics for 3,000-V P-i-N rectifiers

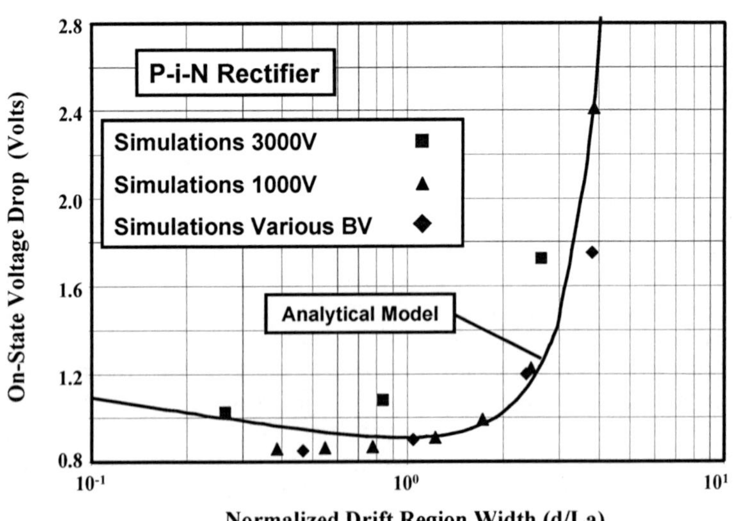

Fig. 5.11 On-state voltage drop for P-i-N rectifiers

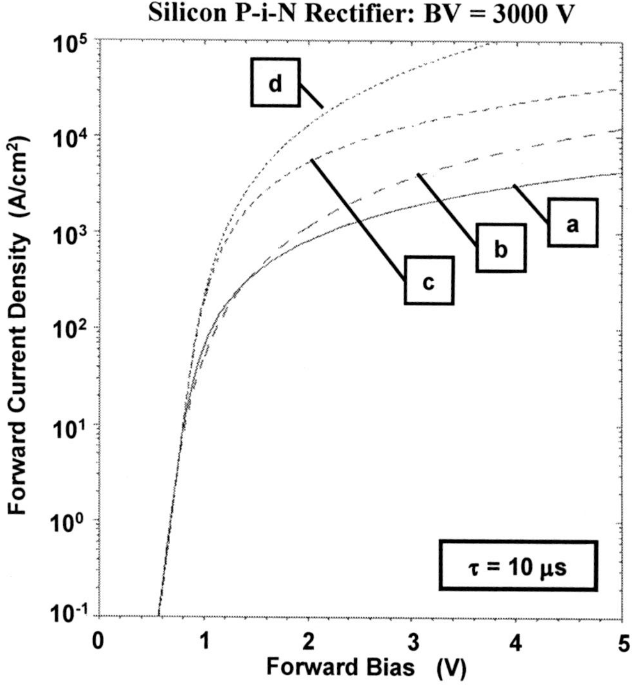

Fig. 5.12 On-state characteristics for 3,000-V P-i-N rectifiers. Models: (*a*) all; (*b*) without carrier–carrier scattering; (*c*) without carrier–carrier scattering and band-gap narrowing; and (*d*) without carrier–carrier scattering, band-gap narrowing, and Auger recombination

The numerical simulations provide a convenient tool for examination of the impact of the various physical processes that influence the on-state *i–v* characteristics for P-i-N rectifiers. In Fig. 5.12, the *i–v* characteristics are shown for the 3,000-V P-i-N rectifier when various models are progressively removed during the simulations. It can be seen that the carrier–carrier scattering phenomenon alters the shape of the characteristics due to its impact on carrier distribution in the drift region. The impact of the removal of the band-gap narrowing phenomenon is especially strong resulting in a reduction of the on-state voltage drop at surge current levels (current density of 10^3 A cm^{-2} and greater). Auger recombination has its greatest impact at current densities in excess of 10^4 A cm^{-2}.

The net influence of the carrier–carrier scattering, band-gap narrowing, and Auger recombination phenomena on the free carrier distribution within the drift region is shown in Fig. 5.13. A lifetime of 10 μs was used in the drift region during the simulations. As discussed in the previous sections, these phenomena enhance the recombination current in the end regions producing a reduction of the carrier concentration in the drift region. It can be observed from the simulations that the inclusion of the models results in depressing the free carrier concentration in the central portion of the drift region by a factor of about 2 times.

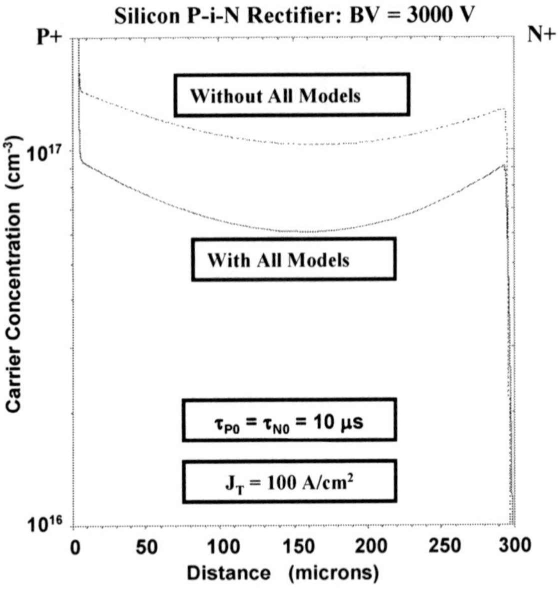

Fig. 5.13 Influence of carrier–carrier scattering, band-gap narrowing, and Auger recombination on the free carrier distribution within a 3,000-V P-i-N rectifier

To increase the blocking voltage capability of the P-i-N rectifier, it is necessary to increase the thickness of the drift region and reduce its doping concentration. The on-state voltage drop of the P-i-N rectifier can be expected to increase with increasing blocking voltage capability if the lifetime in the drift region is held constant. This is illustrated in Fig. 5.14 for the case of a lifetime (τ_{p0} and τ_{n0}) of 1 μs when the cathode was biased in the negative direction to forward bias the diode. Here, the on-state voltage drop is increasing with blocking voltage rating due to the increase in the width ($2d$) of the drift region while the diffusion length (L_a) is held constant. The drift region widths used during these simulations were 36, 81, 183, 294, 530, and 1,200 μm for the blocking voltages of 500, 1,000, 2,000, 3,000, 5,000, and 10,000 V, respectively. The on-state voltage drop begins to increase very rapidly when the blocking voltage exceeds 3,000 V because the (d/L_a) ratio becomes much larger than unity at this point. The on-state voltage drops for these P-i-N rectifiers are also shown in Fig. 5.11 with the diamond symbols. The good agreement with the analytical model provides further confirmation of its ability to accurately predict the behavior of P-i-N rectifiers.

The *i–v* characteristics shown in Fig. 5.14 indicate that the lifetime is too short for obtaining conductivity modulation for the case of a 10-kV P-i-N rectifier when a lifetime of 1 μs is prevailing in the drift region. Conductivity modulation of the drift region can be achieved for this diode if the lifetime is increased to above 10 μs as shown in Fig. 5.15. With adequate conductivity modulation with a lifetime

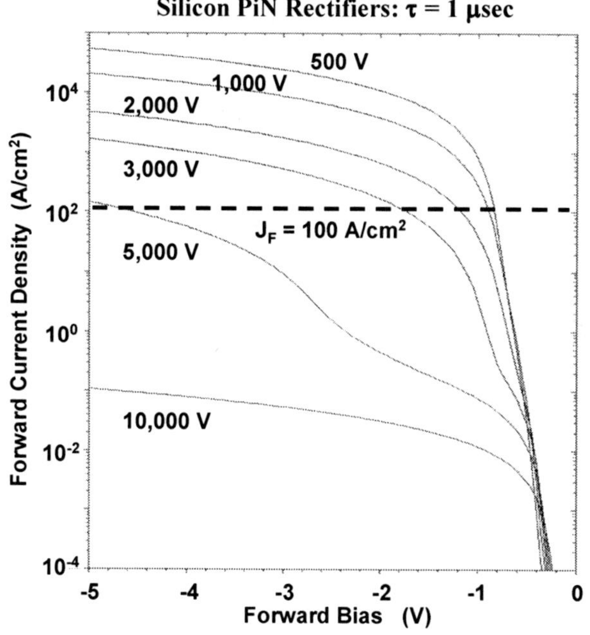

Fig. 5.14 Forward conduction characteristics of silicon P-i-N rectifiers

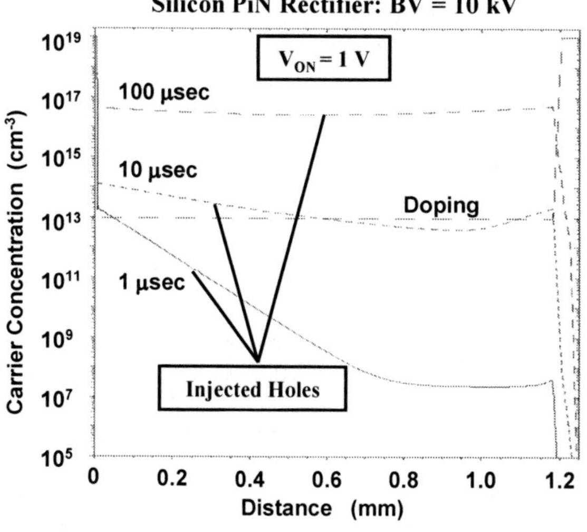

Fig. 5.15 Conductivity modulation within 10-kV silicon P-i-N rectifiers

of 100 μs, the on-state characteristics are greatly improved as shown in Fig. 5.16, resulting in an on-state voltage drop of 1.8 V at a forward current density of 100 A cm^{-2}.

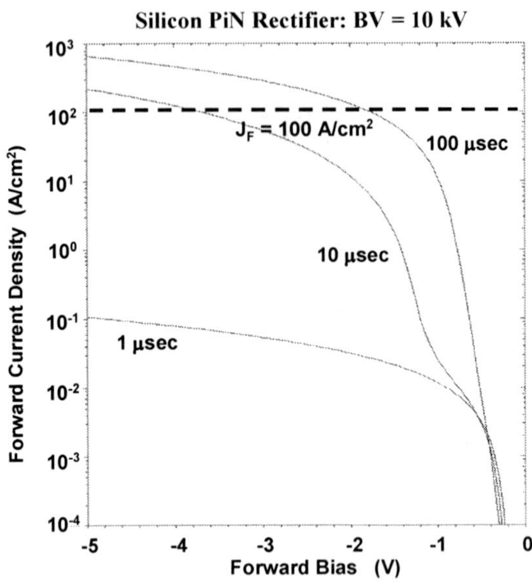

Fig. 5.16 Forward conduction characteristics of 10-kV silicon P-i-N rectifiers

During operation in power circuits, the junction temperature in the P-i-N rectifiers increases due to power dissipation. It is therefore important to evaluate the influence of the temperature upon the *i–v* characteristics in the forward conduction mode of operation. As an example, these characteristics are shown in Fig. 5.17 for the case of the 3,000-V P-i-N rectifier with a lifetime (τ_{p0} and τ_{n0}) of 10 μs in the drift region. The on-state voltage drop at a forward current density of 100 A cm^{-2} is observed to reduce slightly with temperature. This is due to a reduction of the voltage drop at the junctions. Unfortunately, this behavior favors the development of "hot spots" within devices where the current density can become large. However, the positive temperature coefficient for the on-state voltage drop for current densities above 300 A cm^{-2} indicates that stable operation is possible with only moderate nonuniformities in the current distribution within silicon P-i-N rectifiers.

The free carrier concentration within the drift region of the 3,000-V P-i-N rectifier obtained with the numerical simulations is shown in Fig. 5.18. This behavior is consistent with the analytical model due to a reduction in the diffusion length with temperature. The resulting increase in the stored charge within the drift region prolongs the reverse recovery during turn-off and increases the power losses in applications for P-i-N rectifiers.

Fig. 5.17 Forward conduction characteristics of a silicon P-i-N rectifier

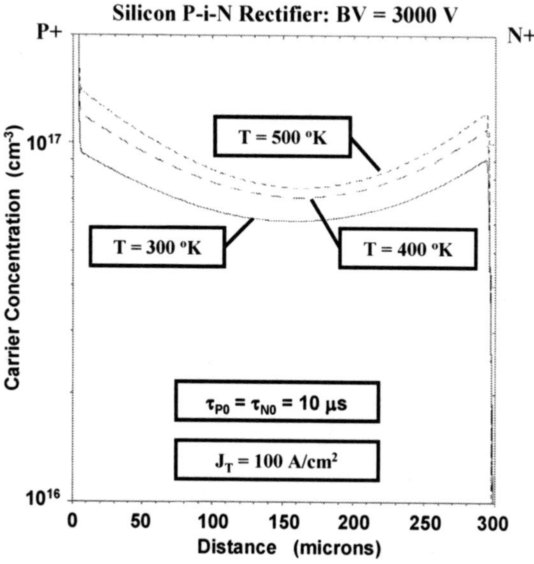

Fig. 5.18 Increase in the free carrier concentration within a 3,000-V P-i-N rectifier with increasing temperature

5.2 Silicon Carbide P-i-N Rectifiers

Due to the much larger electric field that can be supported in silicon carbide, the width of the drift region is much smaller than that for the corresponding silicon device with the same breakdown voltage. This implies that the stored charge in the silicon carbide P-i-N rectifier will be much smaller than for the silicon device providing an improvement in the switching behavior. Unfortunately, the improved switching performance is accompanied by a substantial increase in the on-state voltage drop associated with the larger energy band gap for silicon carbide.

The physics of operation of the silicon carbide P-i-N rectifier is the same as that described in the previous sections. However, the parameters for the silicon carbide device defer from the silicon device. This has a strong impact on the voltage drop associated with the junctions. It was previously demonstrated that the junction voltage drop is given by

$$V_{P+} + V_{N+} = \frac{kT}{q} \ln\left[\frac{n(+d)n(-d)}{n_i^2}\right]. \tag{5.75}$$

Although the injected carrier concentrations $n(+d)$ and $n(-d)$ can be assumed to be similar in magnitude to those in a silicon P-i-N rectifier, the intrinsic carrier concentration for 4H-SiC is only 6.7×10^{-11} cm^{-3} at 300 K, due to its larger energy band gap, when compared with 1.4×10^{10} m^{-3} for silicon. This produces an increase in the junction voltage drop from 0.82 V for the silicon diode to 3.24 V for the 4H-SiC diode if the free carrier concentration in the drift region is assumed to be 1×10^{17} cm^{-3}. The power dissipation in the 4H-SiC P-i-N rectifier is therefore four times greater than in the silicon device. The expected improvement in the switching behavior is mitigated by the large on-state power loss. Consequently, it is preferable to develop silicon carbide Schottky diodes for voltage ratings of up to 5,000 V and P-i-N diodes for voltage ratings above 10,000 V.

Simulation Example

To illustrate the reduction in the stored charge within silicon carbide P-i-N rectifiers, consider the case of a 4H-SiC P-i-N rectifier designed to support 10,000 V. The thickness of the drift region for this device is only 80 μm when compared with 1,200 μm required for the silicon structure. The drift region also has relatively high doping level of 2×10^{15} cm^{-3}. Due to the smaller thickness, good conductivity modulation of the drift region is observed even for a very small lifetime (τ_{p0} and τ_{n0}) value of 100 ns as shown in Fig. 5.19. However, poor conductivity modulation occurs for a typical lifetime (τ_{p0} and τ_{n0}) value of 10 ns in the drift region observed in 4H-SiC. Methods for improving the minority carrier lifetime in 4H-SiC are required to assure good diode characteristics. The plot in Fig. 5.19 was obtained at a forward bias of 4 V because of the larger junction potential for 4H-SiC as discussed above. Thus, the improved switching performance is obtained at the significant disadvantage of high on-state power loss.

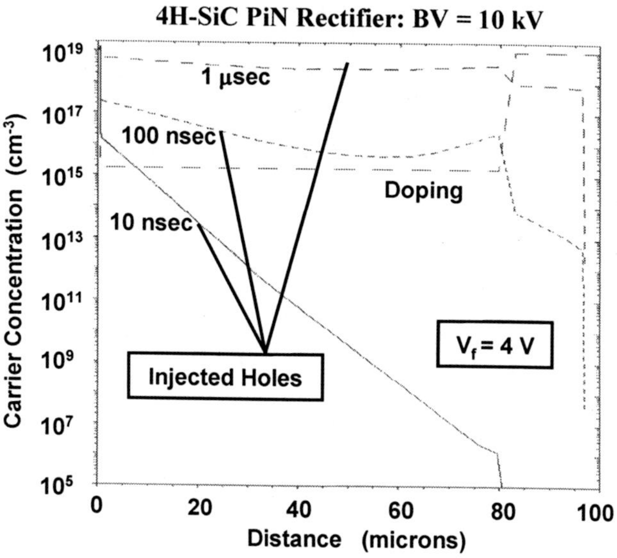

Fig. 5.19 Conductivity modulation within 10-kV 4H-SiC P-i-N rectifiers

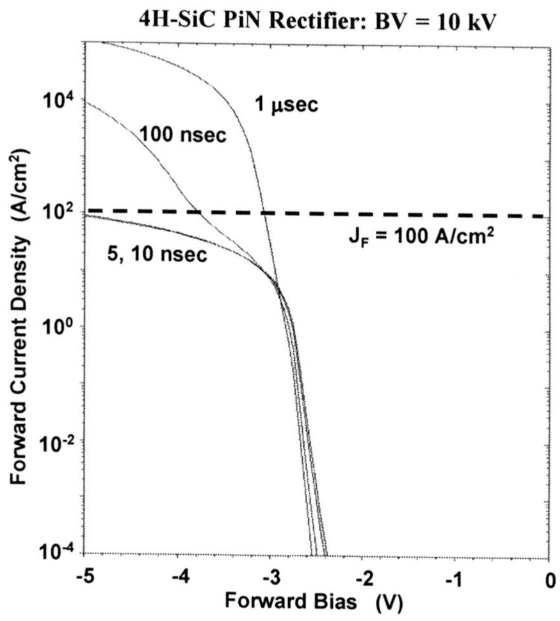

Fig. 5.20 Forward conduction characteristics of 10-kV 4H-SiC P-i-N rectifiers

The forward *i–v* characteristics for the 10-kV 4H-SiC P-i-N rectifiers obtained from the numerical simulations are shown in Fig. 5.20 for various lifetime values. It can be observed that the on-state voltage drop is determined by the unmodulated resistance of the drift region when the lifetime is at 5 or 10 ns. When the lifetime is increased to 100 ns, the conductivity modulation of the drift region reduces the on-state voltage drop. The results of numerical simulations of the forward characteristics of 4H-SiC P-i-N rectifiers with various breakdown voltage capabilities are shown in Fig. 5.21 for the case of a lifetime of 5 ns. The influence of conductivity modulation of the drift region can be observed even for the case of a breakdown voltage of 5,000 V. However, the on-state voltage drop for all the devices is relatively high (more than 3 V) when compared with silicon devices. For this reason, Schottky barrier rectifiers fabricated from 4H-SiC offer much superior performance in terms of both the on-state voltage drop and switching behavior until the blocking voltage exceeds 5,000 V.

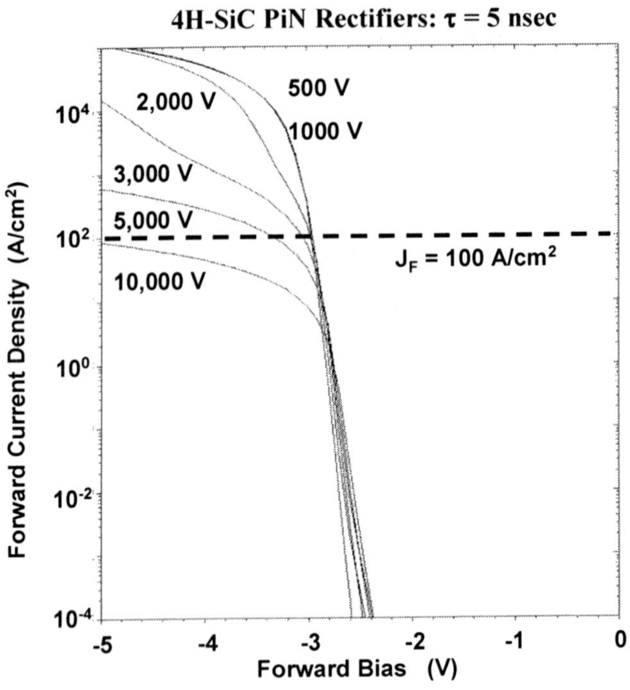

Fig. 5.21 Forward conduction characteristics of 4H-SiC P-i-N rectifiers

5.3 Reverse Blocking

The reverse-blocking voltage capability of the P-i-N rectifier is determined by the punch-through electric field distribution profile as previously described in Sect. 3.4. The punch-through design enables reduction of the thickness of the drift region, which is beneficial for reducing the on-state voltage drop as discussed in the previous sections. Since the doping concentration of the drift region is small in the punch-through design, the drift region becomes completely depleted at a relatively low reverse bias voltage given by

$$V_{PT} = \frac{qN_D(2d)^2}{2\varepsilon_S}. \tag{5.76}$$

Beyond this voltage, the depletion region volume for the P-i-N rectifier remains independent of the reverse bias voltage under the assumption that the end regions are heavily doped.

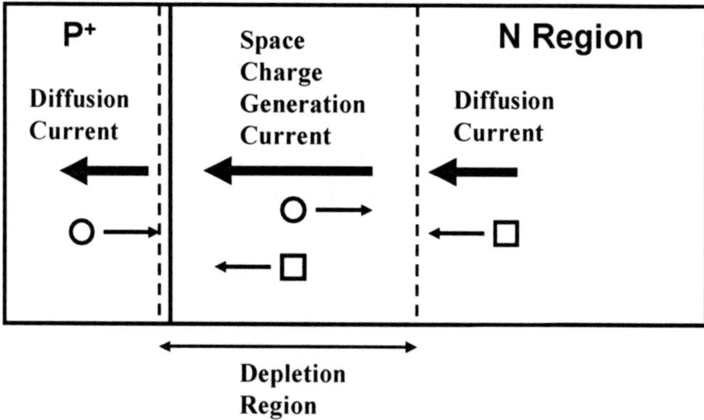

Fig. 5.22 Leakage current components in a P–N junction (electrons: *circles*, holes: *squares*)

The leakage current for a reverse-biased P–N junction is produced by a combination of the space-charge generation current and the diffusion current as illustrated in Fig. 5.22. Any electron–hole pairs generated within the depletion region are swept out as indicated in the figure due to the prevailing electric field. The generation of electron–hole pairs in a semiconductor is governed by the generation rate:

$$U = \frac{n_i}{\tau_{SC}}, \tag{5.77}$$

where τ_{SC} is the space-charge generation lifetime discussed in Chap. 2. This generation occurs throughout the volume of the space-charge region. Consequently, the space-charge generation current is given by

$$J_{SC} = \frac{qW_D n_i}{\tau_{SC}}, \tag{5.78}$$

where W_D is the width of the depletion region.

An additional component of the leakage current is associated with the generation of electron–hole pairs in the neutral regions. Any minority carriers generated far from the junction will recombine without contributing to the current flow. However, any minority carriers generated in the proximity of the junction can diffuse to the depletion region boundary and get swept to the opposite side of the junction by the electric field. An expression for the diffusion current was already derived in Sect. 5.1.2 for current flow across a P–N junction:

$$J_{\mathrm{TN}} = \frac{qD_{\mathrm{p}}p_{0\mathrm{N}}}{L_{\mathrm{p}}}(\mathrm{e}^{qV_{\mathrm{a}}/kT} - 1) \tag{5.79}$$

for the hole current component in the N-region and

$$J_{\mathrm{TP}} = \frac{qD_{\mathrm{n}}n_{0\mathrm{P}}}{L_{\mathrm{n}}}(\mathrm{e}^{qV_{\mathrm{a}}/kT} - 1) \tag{5.80}$$

for the electron component in the P-region. Since a large negative bias is applied under reverse-blocking conditions, the diffusion components of the leakage current are given by

$$J_{\mathrm{LN}} = \frac{qD_{\mathrm{p}}p_{0\mathrm{N}}}{L_{\mathrm{p}}} = \frac{qD_{\mathrm{p}}n_{\mathrm{i}}^2}{L_{\mathrm{p}}N_{\mathrm{D}}} \tag{5.81}$$

and

$$J_{\mathrm{LP}} = \frac{qD_{\mathrm{n}}n_{0\mathrm{P}}}{L_{\mathrm{n}}} = \frac{qD_{\mathrm{n}}n_{\mathrm{i}}^2}{L_{\mathrm{n}}N_{\mathrm{A}}}. \tag{5.82}$$

In the case of the P-i-N rectifier, the entire drift region is depleted at a relatively small reverse bias voltage due to its low doping concentration. The space-charge generation current can therefore be assumed to arise from the entire width ($2d$) of the drift region while the diffusion currents are generated in the P^+ and N^+ end regions. The total leakage current for the P-i-N rectifier is then given by

$$J_{\mathrm{LT}} = \frac{qD_{\mathrm{n}}n_{\mathrm{i}}^2}{L_{\mathrm{n}}N_{\mathrm{AP+}}} + \frac{q(2d)n_{\mathrm{i}}}{\tau_{\mathrm{SC}}} + \frac{qD_{\mathrm{p}}n_{\mathrm{i}}^2}{L_{\mathrm{p}}N_{\mathrm{DN+}}}. \tag{5.83}$$

The influence of heavy doping effects on the intrinsic carrier concentration and the diffusion lengths can enhance the leakage current arising from the end regions. However, the leakage current in the P-i-N rectifier, under reverse-blocking conditions, is determined primarily by the space-charge generation current. The contributions due to the diffusion currents from the end regions become comparable with the space-charge generation current only at elevated temperatures.

As an example, consider the case of a P-i-N rectifier with a drift region width ($2d$) of 100 μm and a space-charge generation lifetime of 10 μs. The leakage current components calculated using (5.83) are plotted in Fig. 5.23 between 300 and 500 K under the assumption that the P^+ and N^+ end regions have a doping concentration of $1 \times 10^{19}\ \mathrm{cm}^{-3}$ and minority carrier lifetime of 1 ns. It is obvious that the leakage current due to the space-charge generation process is dominant over this temperature range. It will therefore determine the total leakage current density.

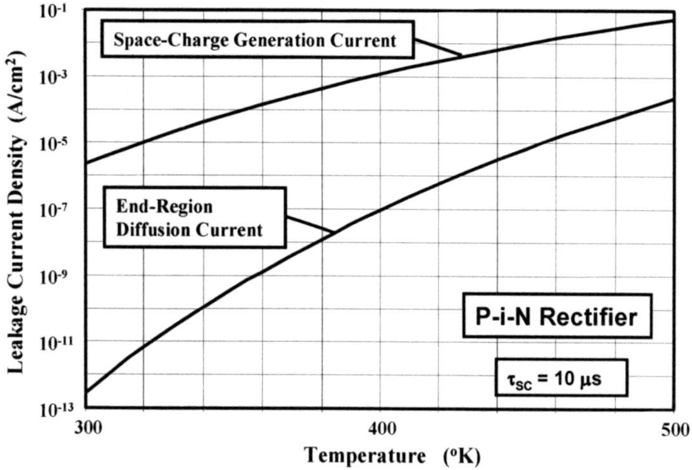

Fig. 5.23 Leakage current components in a silicon P-i-N rectifier

Simulation Example

To validate the above model for the leakage current in the P-i-N rectifier, the results of numerical simulations on the 3,000-V silicon P-i-N rectifier structure, whose forward characteristics were discussed in Sect. 5.2, are described here. A reverse bias of 1,000 V was chosen to completely deplete the drift region. The leakage current was extracted over a temperature range of 300–500 K. A lifetime (τ_{p0} and τ_{n0}) value of 10 μs was used in the drift region. The end regions had a

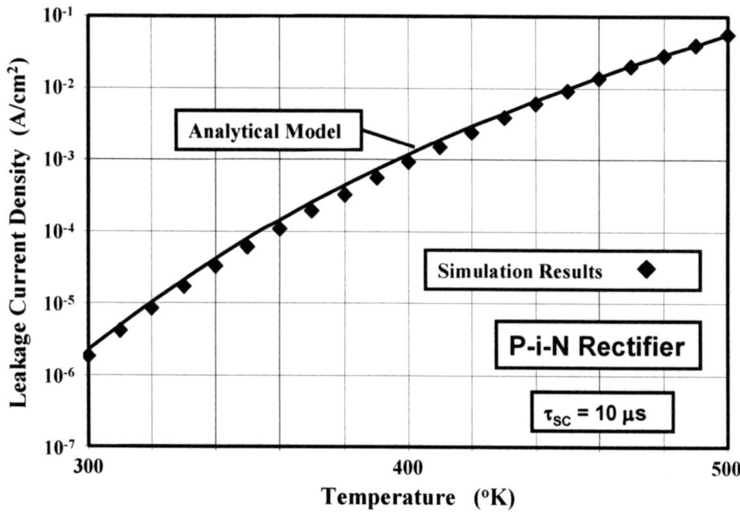

Fig. 5.24 Leakage current in a 3-kV silicon P-i-N rectifier

Gaussian doping profile with a surface concentration of 1×10^{19} cm^{-3} and depth of about 10 μm. A leakage current density obtained using the simulations is compared with that calculated using the model in Fig. 5.24. The excellent agreement between the calculated values and those obtained with the numerical simulations provides confirmation that the analytical model can be utilized for the analysis of the leakage current in P-i-N rectifiers.

5.4 Switching Performance

Power rectifiers control the direction of current flow in circuits used in various power conditioning applications. They operate for part of the time in the on-state when the bias applied to the anode is positive and for the rest of the time in the blocking state when the bias applied to the anode is negative. During each operating cycle, the diode must be rapidly switched between these states to minimize power losses. The transition of the diode from the blocking state to the on-state is accompanied by an overshoot in the anode voltage which increases the power dissipation. This phenomenon is referred to as the *forward recovery*. Even greater power losses are incurred when the diode switches from the on-state to the reverse-blocking state. The stored charge within the drift region of the power rectifier must be extracted before it is able to support high voltages. This produces a large reverse current for a short time duration. This phenomenon is referred to as the *reverse recovery*.

5.4.1 Forward Recovery

When a P-i-N rectifier is switched from the reverse-blocking mode to the forward conduction mode, the voltage drop across the diode in the forward direction can be substantially larger than the on-state voltage drop under steady-state operation. This is especially true if the anode current is increased at a rapid rate. Unlike in the steady-state case that was previously analyzed in Sect. 5.1.3, the drift region does not get modulated in proportion to the current density in the transient case due to the finite rate for the diffusion of minority carriers. Consequently, the voltage drop across the middle region of the P-i-N rectifier is no longer independent of the current density. If the current increases at a rapid rate, a portion of the drift region can remain without conductivity modulation. Since this portion of the drift region will have a high resistance due to its low doping concentration, the voltage drop across the diode is much greater than under steady-state operation.

The increase in the forward voltage drop when the P-i-N rectifier is turned on with a high ramp rate for the anode current can be analyzed by solving for the injected carrier distribution as a function of both time and position in the drift region. The excess majority carrier density is of interest here to determine the conductivity modulation of the drift region. The excess majority carrier concentration (δn) injected into the drift region is governed by the continuity equation:

$$\frac{\partial \delta n}{\partial t} = \frac{1}{q}\frac{\partial J_{\rm n}}{\partial x} - \frac{\partial n}{\tau_{\rm n}}. \tag{5.84}$$

If the current flow is dominated by diffusion, the electron current density is given by

$$J_{\rm n} = qD_{\rm n}\frac{\partial \delta n}{\partial x}, \tag{5.85}$$

leading to the diffusion equation[10] for electrons:

$$\frac{\partial \delta n}{\partial t} = D_{\rm n}\frac{\partial^2 \delta n}{\partial x^2} - \frac{\partial n}{\tau_{\rm n}}. \tag{5.86}$$

Since the forward voltage overshoot occurs only under high ramp rates for the current within a time duration which is small compared with the recombination lifetime, the recombination process can be neglected leading to

$$\frac{\partial \delta n(x,t)}{\partial t} = D_{\rm n}\frac{\partial^2 \delta n(x,t)}{\partial x^2}. \tag{5.87}$$

This equation governs the distribution of the excess electrons in the drift region as a function of both space and time. The solution for the excess electron concentration is of the form

$$\delta n(x,t) = A(t){\rm e}^{-\left(x/\sqrt{4D_{\rm n}t}\right)}, \tag{5.88}$$

where the term $A(t)$ is determined by the current density, which is increasing as a function of time. The current density at the edge of the P–N junction (at $x = 0$) under high-level injection conditions is given by (see derivation for (5.34))

$$J = J_{\rm n}(0) = 2qD_{\rm n}\left(\frac{{\rm d}\delta n}{{\rm d}x}\right)_{x=0}. \tag{5.89}$$

Using the carrier distribution in (5.88),

$$J = 2qD_{\rm n}\frac{A(t)}{\sqrt{4D_{\rm n}t}} = qA(t)\sqrt{\frac{D_{\rm n}}{t}}. \tag{5.90}$$

During the forward recovery transient, the current density increases at a constant rate (a). Thus,

$$J = at. \tag{5.91}$$

From these equations, the coefficient $A(t)$ is obtained:

$$A(t) = \frac{at^{3/2}}{q\sqrt{D_n}}. \tag{5.92}$$

This provides the excess electron concentration at the edge of the P–N junction, which is increasing as a function of time due to the increase in the current density. The excess electron concentration then decays away from the P–N junction with a diffusion length given by

$$L(t) = \sqrt{4D_n t}. \tag{5.93}$$

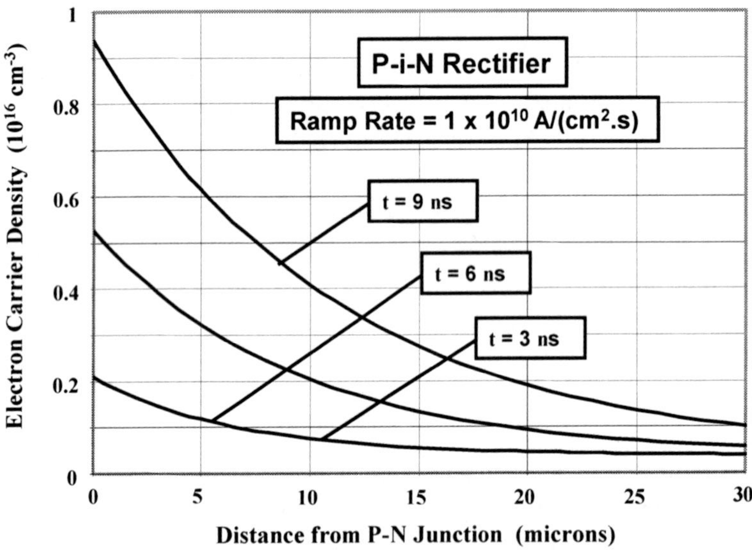

Fig. 5.25 Electron concentration in a silicon P-i-N rectifier during the forward recovery process

The evolution of the excess electron concentration within the drift region is then given by

$$\delta n(x,t) = \frac{at^{3/2}}{q\sqrt{D_n}} e^{-\left(x/\sqrt{4D_n t}\right)}. \tag{5.94}$$

The total electron concentration within the drift region can then be obtained by adding the electron concentration due to the donor dopant atoms to the excess electron concentration:

$$n(x,t) = \delta n(x,t) + N_D = \frac{at^{3/2}}{q\sqrt{D_n}} e^{-\left(x/\sqrt{4D_n t}\right)} + N_D. \tag{5.95}$$

This total electron concentration is plotted in Fig. 5.25 for three instances of time in the case of a ramp rate of 1×10^{10} A cm^{-2} s^{-1}. This P-i-N rectifier structure had a drift region thickness of 60 μm with a doping concentration of 5×10^{13} cm^{-3}. As time progresses from 3 to 6 to 9 ns, the electron concentration at the P–N junction ($x = 0$) increases due to the increase in the current density from 30 to 60 to 90 A cm^{-2}. Concurrently, the electrons are distributed further into the drift region by the diffusion process.

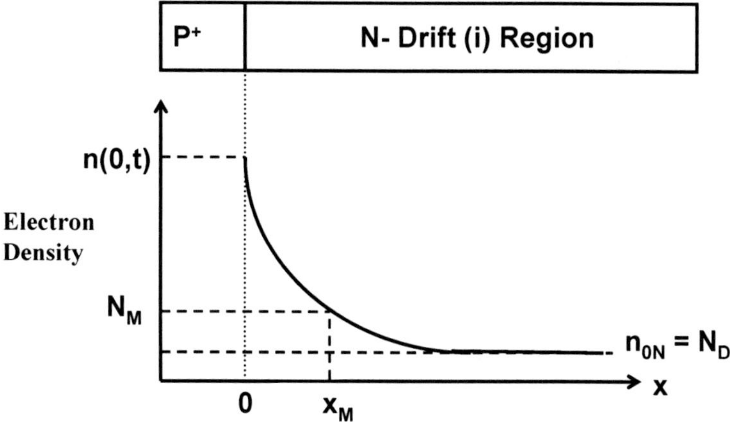

Fig. 5.26 Electron concentration in a silicon P-i-N rectifier during the forward recovery process

The voltage drop across the drift region can be calculated by taking the product of the current density at any time and the resistance of the drift region. Based upon the above electron concentration distribution, it can be concluded that the drift region resistance reduces with time because a larger proportion of the drift region becomes conductivity modulated as the current ramps up. To analyze the reduction of the resistance of the drift region, consider a segment at distance x from the P–N junction with a small thickness dx. The resistivity at this location is given by

$$\mathrm{d}R = \rho(x)\mathrm{d}x = \frac{\mathrm{d}x}{q\mu_{\mathrm{n}}(x,t)n(x,t)}, \tag{5.96}$$

where the electron mobility is a function of position and time because the electron concentration depends on these parameters. (It is necessary to account for carrier–carrier scattering for the determination of the mobility due to the relatively large concentration of both minority and majority carriers.) The resistance of the drift region during the forward recovery transient is then given by

$$R_D(t) = \int_0^{W_D} dR = \int_0^{W_D} \frac{dx}{q\mu_n(x,t)n(x,t)}. \tag{5.97}$$

A simple analytical solution for this resistance cannot be derived due to the dependence of the mobility on the electron carrier concentration.

An alternative approach for the analysis of the drift region resistance is based upon defining a "modulation concentration" (N_M) as indicated in Fig. 5.26 and assuming that the resistance of the drift region with electron concentration above this value is negligible when compared with rest of the drift region, where no modulation is assumed to occur. The conductivity-modulated portion of the drift region then has a distance x_M as shown in the figure. This distance can be obtained by using (5.95):

$$x_M(t) = \sqrt{4D_n t}\ \ln\left[\frac{A(t)+N_D}{N_M}\right]. \tag{5.98}$$

The expansion of this conductivity-modulated region with time during the ramp up of the current is shown in Fig. 5.27 for the three cases of the ramp rates when a value of 2×10^{14} cm^{-3} was used for the "modulation concentration" (N_M). In each case, the conductivity-modulated region is formed after a time delay during which the current increases to the level required to achieve high-level injection in the drift region. This time delay is therefore shorter for the faster ramp rates. The conductivity-modulated region then grows as shown in the figure. A faster rate of growth occurs with increasing ramp rates.

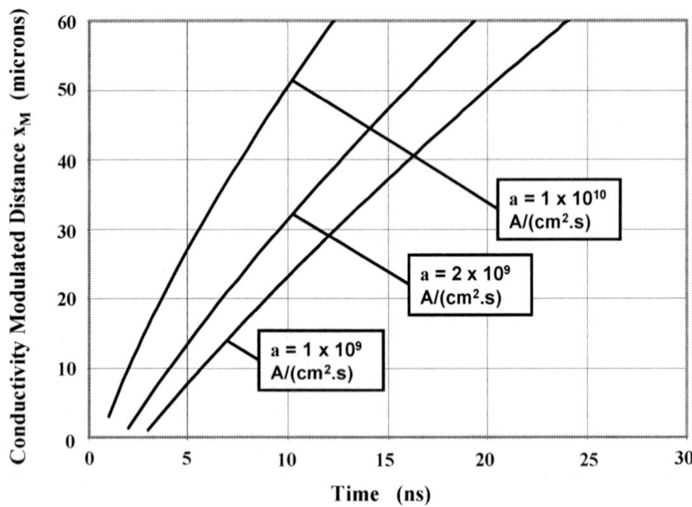

Fig. 5.27 Growth of the conductivity-modulated region during the forward recovery process

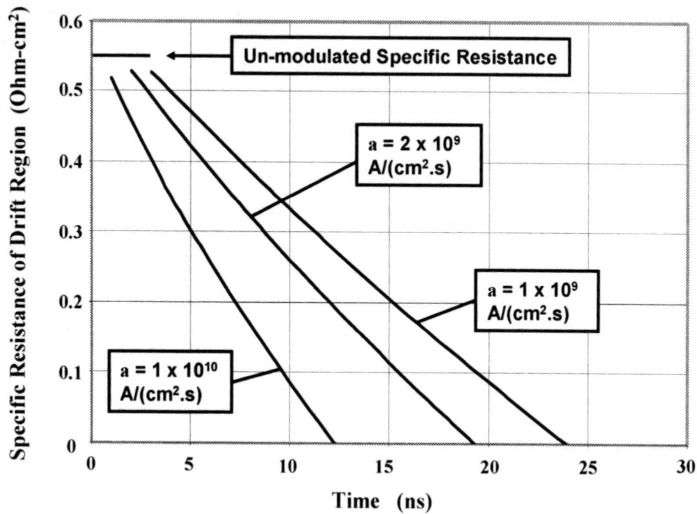

Fig. 5.28 Decrease of the drift region resistance during the forward recovery process

The resistance of the drift region decreases, as shown in Fig. 5.28 for the three cases of the ramp rate, as the modulated portion expands with time. In this structure, the drift region had a width of 60 μm with a doping concentration of 5×10^{13} cm^{-3} resulting in an unmodulated specific resistance of 0.55 Ω cm^2. The voltage drop across the drift region [$v_D(t)$] can be obtained by multiplying the specific resistance of the unmodulated region and the current density pertaining to each time instant:

$$v_D(t) = R_D(t)J_F(t). \tag{5.99}$$

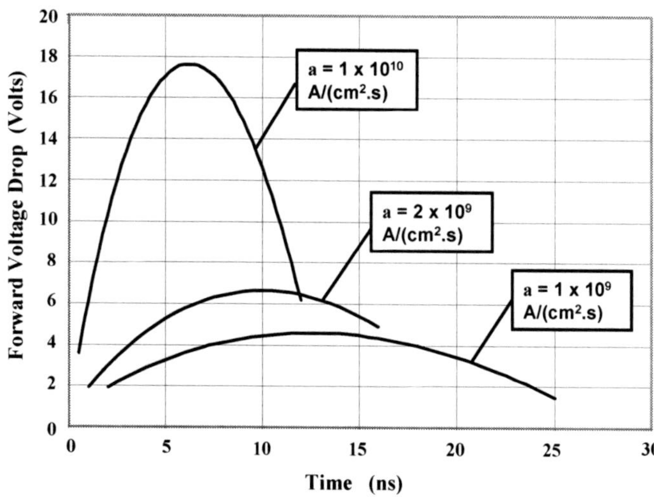

Fig. 5.29 Voltage drop across the P-i-N rectifier during the forward recovery process

The total voltage drop across the P-i-N rectifier during the forward recovery transient consists of the junction voltage drop plus the voltage drop across the unmodulated portion of the drift region. The evolution of the forward voltage drop across the P-i-N rectifier with time during the transient as described by the analytical model is shown in Fig. 5.29. The portion of the transient below 1 V is not shown because the model is based upon high-level injection in the drift region. The maximum forward voltage drop (or overshoot voltage) can be observed to become larger with increasing rate of rise of the current density. For this rectifier structure, the maximum forward voltage drop increases from 4.6 to 6.6 to 17.4 V as the ramp rate increases from 1×10^9 to 2×10^9 to 1×10^{10} A cm^{-2} s^{-1}. The time at which the maximum forward voltage drop is observed for each of these cases decreases from 13.5 to 10 to 6.5 ns. According to the model, the voltage overshoot is not a function of the minority carrier lifetime.

Simulation Example

To validate the above model for the forward recovery transient in the P-i-N rectifier, the results of numerical simulations on a 1,000-V silicon P-i-N rectifier structure are described here. The structure had a drift region thickness of 60 μm with a doping concentration of 5×10^{13} cm^{-3}. The lifetime (τ_{p0} and τ_{n0}) in the drift region was 1 μs. The cathode current was ramped from zero to a steady-state value of 100 A cm^{-2} using ramp rates of 1×10^9, 2×10^9, and 1×10^{10} A cm^{-2} s^{-1}. The voltage drop across the rectifier during the transient is shown in Fig. 5.30 for the three cases.

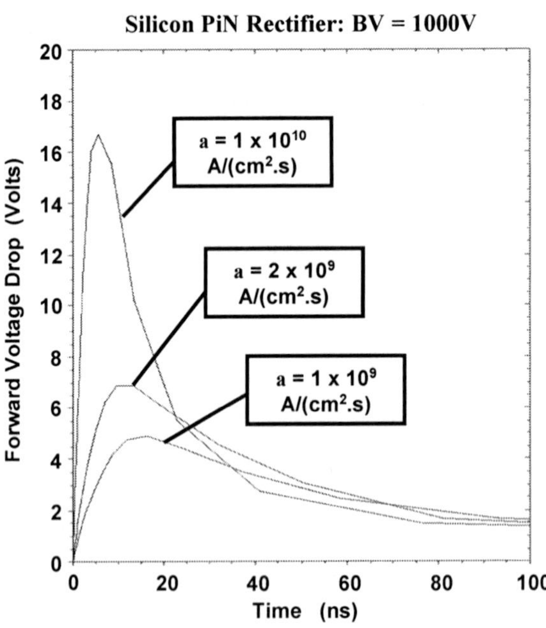

Fig. 5.30 Forward voltage overshoot in a 1,000-V silicon P-i-N rectifier

The maximum forward voltage drops observed for the ramp rates of 1×10^9, 2×10^9, and 1×10^{10} A cm^{-2} s^{-1} are 4.9, 6.9, and 16.7 V, respectively. The peak in the voltage overshoot occurs at 13, 10, and 6 ns for the three cases. The predictions of the analytical model are in excellent agreement with these values providing validation of the model.

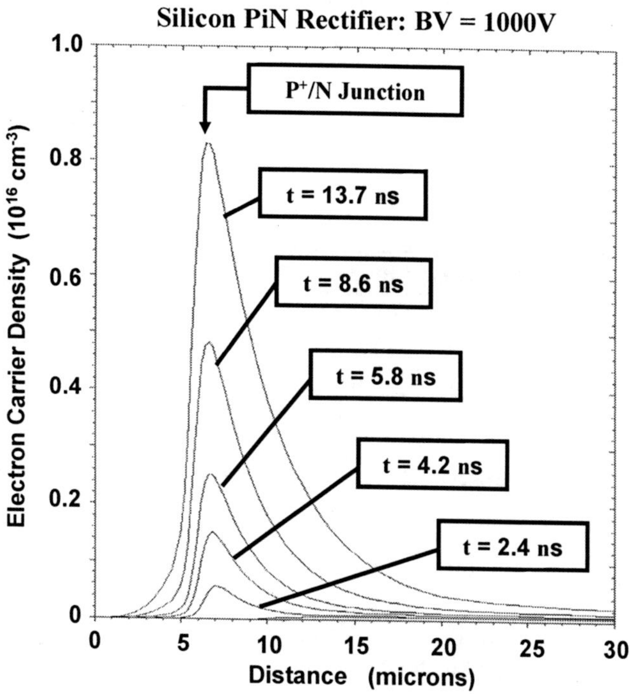

Fig. 5.31 Carrier distributions in a 1,000-V silicon P-i-N rectifier during the forward recovery process

The electron carrier concentration in the drift region can be extracted from the transient simulations at various points in time. As an example, the electron concentration profiles are shown in Fig. 5.31 for the case of a ramp rate of 1×10^{10} A cm^{-2} s^{-1}. The electron concentration exhibits an exponential distribution when proceeding away from the P–N junction with the carrier concentration increasing with time. As time progresses during the turn-on transient, the electron density modulates a greater portion of the drift region. The electron distribution predicted by the analytical model (see Fig. 5.25) is consistent with that observed with the numerical simulations.

The numerical simulations of the forward recovery process for the above 1,000-V P-i-N rectifier structure were repeated with a lifetime of 10 μs in the drift region. The resulting waveform for the forward voltage overshoot is shown in Fig. 5.32. It is obvious that there is no difference between the voltage overshoot for a minority carrier lifetime of 1 and 10 μs. The predictions of the analytical model are consistent with this behavior.

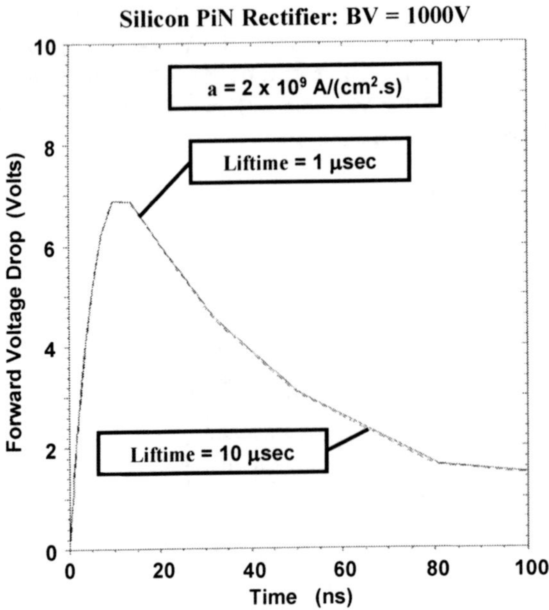

Fig. 5.32 Forward voltage overshoot in a 1,000-V silicon P-i-N rectifier (*solid line*: 1 μs, *dashed line*: 10 μs)

5.4.2 Reverse Recovery

The presence of a large concentration of free carriers in the drift region during on-state current flow is responsible for the low on-state voltage drop of high-voltage silicon P-i-N rectifiers. To switch the diode from its on-state mode to the reverse-blocking mode, it is necessary to remove these free carriers to enable the formation of a depletion region that can support a high electric field. The process of switching the P-i-N rectifier from the on-state to the blocking state is referred to as the *reverse recovery*.

When the voltage applied to the anode of the P-i-N rectifier is reversed from positive to negative, the current does not monotonically reduce to zero. If the reversal in the voltage is performed with a circuit comprising a voltage source and a series resistance, a constant reverse current is observed immediately after the voltage changes from its positive value to a negative value.[11] This current persists until the stored charge is sufficiently removed to allow the P–N junction to support the voltage by the formation of a depletion layer. This reverse recovery process pertains to a *resistive load.*

In power electronic circuits, it is commonplace to use power rectifiers with an *inductive load.* In this case, the current reduces at a constant ramp rate (a) as illustrated in Fig. 5.33 until the diode is able to support voltage. Consequently, a large *peak reverse recovery current* (J_{PR}) occurs due to the stored charge followed

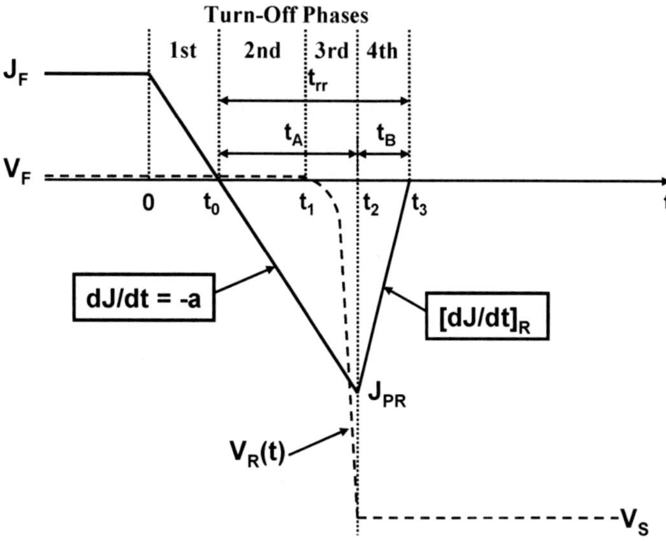

Fig. 5.33 Anode current and voltage waveforms for the P-i-N rectifier during the reverse recovery process

by the reduction of the current to zero. The power rectifier remains in its forward-biased mode with a low on-state voltage drop until time t_1. The voltage across the diode then rapidly increases to the supply voltage with the rectifier operating in its reverse bias mode. The current flowing through the rectifier in the reverse direction reaches a maximum value (J_{PR}) at time t_2 when the reverse voltage becomes equal to the reverse bias supply voltage (V_S).

The simultaneous presence of a high current and voltage produces large instantaneous power dissipation in the power rectifier. The peak reverse recovery current also flows through the power switch that is controlling the switching event. This increases the power losses in the transistor. In the case of typical motor control PWM circuits that utilize IGBTs as power switches, a large reverse recovery current can trigger latch-up failure that can destroy both the transistor and the rectifier. It is therefore desirable to reduce the magnitude of the peak reverse recovery current and the time duration of the recovery transient. This time duration is referred to as the *reverse recovery time* (t_{rr}). The waveforms and power losses for a typical motor control circuit are discussed in the concluding chapter of this book.

An analytical model for the reverse recovery process for the turn-off of a P-i-N rectifier under a constant rate of change of the current (*current ramp rate*) can be created by assuming that the concentration of the free carriers in the drift region can be linearized as illustrated in Fig. 5.34. This approach was first proposed for the step-recovery process.[12] It is extended to the ramp-recovery process here. As shown in the figure, the catenary carrier distribution established by the on-state current flow is approximated by an average value in the middle of the drift region and a linearly varying portion with a concentration of $n(-d)$ at $x = 0$

to the average concentration of n_a at a distance $x = b$. These carrier concentrations can be obtained from (5.37) and (5.41):

$$n(-d) = \frac{\tau_{HL} J_F}{2qL_a}\left[\frac{\cosh(-d/L_a)}{\sinh(d/L_a)} - \frac{\sinh(-d/L_a)}{2\cosh(d/L_a)}\right] \tag{5.100}$$

and

$$n_a = \frac{J_F \tau_{HL}}{2qd}, \tag{5.101}$$

where J_F is the forward (or on-state) current density.

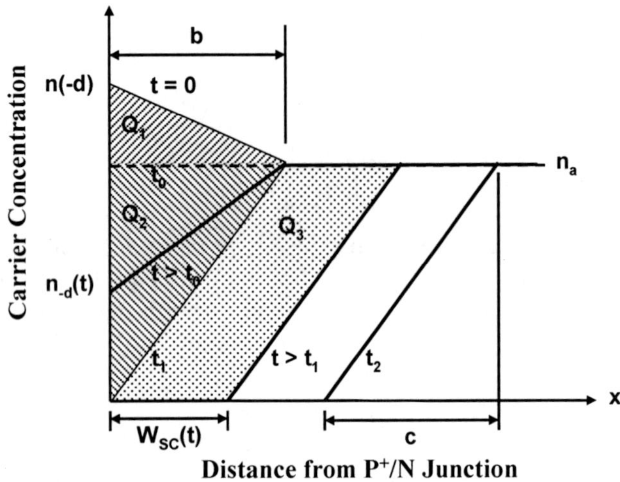

Fig. 5.34 Carrier distribution profiles in the P-i-N rectifier during the reverse recovery process

The current flowing through the rectifier at any time during the turn-off transient is determined by the rate of diffusion of the carriers at the P^+/N junction boundary as described earlier during the discussion of on-state operation in Sect. 5.1.3:

$$J_F = 2qD_a\left(\frac{dn}{dx}\right)_{x=-d}. \tag{5.102}$$

In this equation, the ambipolar diffusion coefficient is used to take into account carrier–carrier scattering effects.

In the first phase of the turn-off process, the current density in the P-i-N rectifier changes from the on-state current density (J_F) to zero at time t_0. The distance b in Fig. 5.34 can be obtained by relating the charge Q_1 removed during the first phase to the current flow. At the end of the first phase, the carrier profile becomes flat at time t_0, as indicated by the dashed line in Fig. 5.34, because the current is zero at this time. The change in the stored charge within the drift region during the first phase can then be obtained from the cross-hatched area, indicated by Q_1, in the figure:

$$Q_1 = \frac{qb}{2}[n(-d) - n_a]. \tag{5.103}$$

This charge can be related the current flow during the turn-off transient from $t = 0$ to $t = t_0$:

$$Q_1 = \int_0^{t_0} J(t)\mathrm{d}t = \int_0^{t_0} (J_F - at)\mathrm{d}t = J_F t_0 - \frac{a t_0^2}{2}. \tag{5.104}$$

The time t_0 at which the current crosses zero is given by

$$t_0 = \frac{J_F}{a}. \tag{5.105}$$

Combining the above relationships,

$$b = \frac{J_F^2}{qa[n(-d) - n_a]}. \tag{5.106}$$

The carrier concentration $n(-d)$ in (5.100) can be written as

$$n(-d) = \frac{J_F \tau_{HL}}{2qL_a} K, \tag{5.107}$$

where

$$K = \left[\frac{\cosh(-d/L_a)}{\sinh(d/L_a)} - \frac{\sinh(-d/L_a)}{2\cosh(d/L_a)}\right]. \tag{5.108}$$

Using (5.107), in conjunction with (5.101), in (5.106),

$$b = \frac{2dL_a J_F}{a\tau_{HL}(Kd - L_a)}. \tag{5.109}$$

The distance b can therefore be calculated from the device parameters (d and τ_{HL}), the on-state current density, and the ramp rate a.

The second phase of the turn-off process occurs from the time t_0 at which the current crosses zero up to the time t_1 when the P^+/N junction can begin to support voltage. The carrier profile at time t_1 is shown in Fig. 5.34 as extending from a zero concentration at the junction (located at $x = 0$) and the average concentration n_a at a distance b from the junction. After time t_1, a depletion region forms at the P^+/N junction with the zero carrier concentration at some distance away from the junction. The time t_1 can be obtained by analysis of the charge removal during the turn-off transient from $t = t_0$ to $t = t_1$. In Fig. 5.34, the charge removed during this time interval is indicated by the cross-hatched area marked Q_2. This area is given by

$$Q_2 = \frac{1}{2} q n_a b. \tag{5.110}$$

This charge can be related the current flow during the turn-off transient from $t = t_0$ to $t = t_1$:

$$Q_2 = \int_{t_0}^{t_1} J(t)\mathrm{d}t = \int_{t_0}^{t_1} (at)\mathrm{d}t = \frac{a}{2}(t_1^2 - t_0^2). \tag{5.111}$$

Using (5.105) for t_0,

$$Q_2 = \frac{a}{2}\left(t_1^2 - \frac{J_F^2}{a^2}\right). \tag{5.112}$$

Combining (5.110) and (5.112),

$$t_1 = \sqrt{\frac{q n_a b}{a} + \frac{J_F^2}{a^2}}. \tag{5.113}$$

Making use of the (5.101) for the average carrier concentration n_a and (5.106) for the distance b,

$$t_1 = \frac{J_F}{a}\sqrt{\frac{L_a}{Kd - L_a} + 1}. \tag{5.114}$$

Based upon this expression, the end of the second phase occurs earlier when the ramp rate is increased. This accelerates the point at which the rectifier can begin to support a reverse bias voltage.

During the entire time from $t = 0$ until time $t = t_1$, the P^+/N junction within the P-i-N rectifier remains forward biased because the minority carrier density in the drift region at the junction [$p(-d, t)$] is above the equilibrium minority carrier concentration (p_{0N}). Under the assumptions of high-level injection conditions in the drift region, the minority carrier density [$p(-d, t)$] is equal to the majority carrier density [$n(-d, t)$] that is illustrated in Fig. 5.34. Based upon (5.102), the current density at any point in time is given by

$$J(t) = 2qD_a\left(\frac{\mathrm{d}n}{\mathrm{d}x}\right)_{x=-d} = 2qD_a\frac{[n(-d,t) - n_a]}{b}. \tag{5.115}$$

The carrier concentration in the drift region at the junction is therefore related to the current density by

$$p(-d,t) = n(-d,t) = n_a + \frac{J(t)b}{2qD_a} = n_a + \frac{(J_F - at)b}{2qD_a}. \tag{5.116}$$

This expression is valid for both positive and negative values for the current density during the turn-off transient until time t_1. The voltage drop across the forward-biased junction during this time interval can be obtained using the Boltzmann's relationship:

$$V_F(t) = \frac{kT}{q}\ln\left[\frac{p(-d)}{p_{0N}}\right] = \frac{kT}{q}\ln\left[\frac{(J_F - at)b}{2qD_a p_{0N}} + \frac{n_a}{p_{0N}}\right]. \tag{5.117}$$

This expression describes the change in the voltage drop across the P-i-N rectifier during the turn-off transient until it is able to support a reverse bias voltage.

During the third phase of the turn-off transient, the P-i-N rectifier begins to support an increasing voltage. This requires the formation of a space-charge region $W_{SC}(t)$ at the P^+/N junction that expands with time as illustrated in Fig. 5.34. The expansion of the space-charge region is achieved by further extraction of the stored charge in the drift region, resulting in the reverse current continuing to increase after time t_1. The growth of the reverse bias voltage across the P-i-N rectifier can be analytically modeled under the assumption that the sweep out of the stored charge is occurring at an approximately constant current. In this case, the slope of the carrier distribution profile remains constant as shown in Fig. 5.34.

In Fig. 5.34, the charge removed at a time t, after the P^+/N junction is reverse biased at time t_1, is indicated by the shaded area marked Q_3. The area of this parallelogram is given by

$$Q_3 = qn_a W_{SC}(t). \tag{5.118}$$

This charge can be related the current flow during the turn-off transient from time t_1 to time t:

$$Q_3 = \int_{t_1}^{t} J(t)\mathrm{d}t = \int_{t_1}^{t} (J_F - at)\mathrm{d}t = J_F(t - t_1) - \frac{a}{2}(t^2 - t_1^2). \tag{5.119}$$

Combining these relationships for the charge Q_3 provides an expression for the growth of the space-charge region as a function of time:

$$W_{SC}(t) = \frac{a}{2qn_a}(t^2 - t_1^2) - \frac{J_F}{qn_a}(t - t_1). \tag{5.120}$$

The voltage supported across this space-charge region can be obtained by solving the Poisson's equation:

$$\frac{\mathrm{d}^2V}{\mathrm{d}x^2} = -\frac{\mathrm{d}E}{\mathrm{d}x} = -\frac{Q(x)}{\varepsilon_S}, \tag{5.121}$$

where $Q(x)$ is the charge in the space-charge region. Unlike the blocking mode of operation, where the charge in the depletion region consists of the ionized donor charge, during the turn-off process an additional charge is contributed by the large reverse current flow. This charge is due to the holes that are transiting through the space-charge region due to the removal of the stored charge. Since the electric field within the space-charge region is large, it can be assumed that these holes are moving at the saturated drift velocity ($v_{sat,p}$). The concentration of the holes within the space-charge region is then related to the current density (J_R) by

$$p(t) = \frac{J_R(t)}{qv_{sat,p}} = \frac{at - J_F}{qv_{sat,p}}. \tag{5.122}$$

The voltage supported by the space-charge region is then given by

$$V_R(t) = \frac{q[N_D + p(t)]}{2\varepsilon_S}[W_{SC}(t)]^2. \tag{5.123}$$

This expression, in conjunction with (5.120) for the expansion of the space-charge width, indicates a rapid rise in the voltage supported by the P-i-N rectifier after time t_1. The end of the third phase occurs when the reverse bias across the P-i-N rectifier becomes equal to the supply voltage (V_S). Using this value in (5.123) together with (5.120), the time t_2 (and hence J_{PR}) can be obtained.

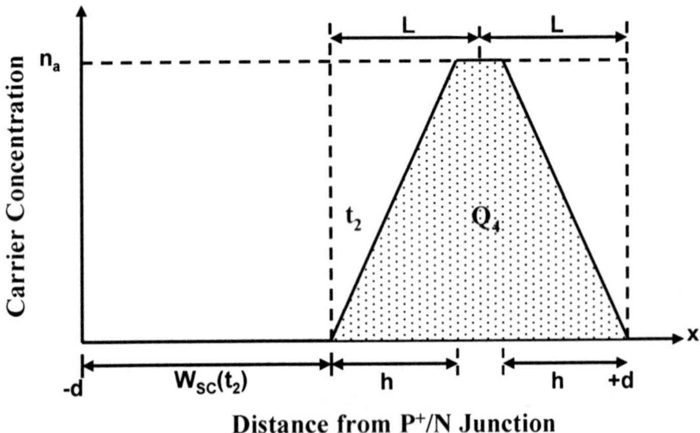

Fig. 5.35 Stored charge within the P-i-N rectifier after the end of the third phase

During the fourth phase of the turn-off process, the reverse current rapidly reduces at approximately a constant rate as illustrated in Fig. 5.33 while the voltage supported by the P-i-N rectifier remains constant at the supply voltage. The stored charge within the drift region after the end of the third phase is illustrated in Fig. 5.35 by the shaded area marked Q_4. At the end of the third phase of the turn-off process ($t = t_2$), the peak reverse recovery current J_{PR} is flowing through the structure. This current can be related to the free carrier profile by

$$J_{PR} = 2qD_a \frac{dn}{dx} = 2qD_a \frac{n_a}{h}, \tag{5.124}$$

where the dimension h is shown in Fig. 5.35. Using this equation,

$$h = \frac{2qD_a n_a}{J_{PR}}. \tag{5.125}$$

The stored charge remaining in the drift region at time t_2 is then given by

$$Q_4 = qn_a[2d - W_{SC}(t_2) - h]. \tag{5.126}$$

This charge must be removed during the fourth phase of the turn-off process. During the fourth phase, the current reduces to zero at an approximately constant rate, indicated as $[dJ/dt]_R$ in Fig. 5.33, over a time period t_B extending from time t_2 to t_3. The charge removed due to the current flow during this time is given by

$$Q_R = \frac{1}{2} J_{PR} t_B. \tag{5.127}$$

This can be equated to the charge left in the drift region at the end of the third phase if recombination during this time is neglected due to the short duration of this time interval relative to the minority carrier lifetime. The time interval (t_B) for the reduction of the reverse current is then obtained:

$$t_B = (t_3 - t_2) = \frac{2qn_a}{J_{PR}}[2d - W_{SC}(t_2) - h]. \tag{5.128}$$

The reverse ramp rate is then given by dividing the peak reverse recovery current by this time interval:

$$\left[\frac{dJ}{dt}\right]_R = \frac{J_{PR}}{t_B} = \frac{J_{PR}^2}{2qn_a[2d - W_{SC}(t_2) - h]}. \tag{5.129}$$

A smaller value for the reverse $[di/dt]$ is desirable to reduce voltages developed across stray inductances in the circuit. These voltages cause an increase in the voltage supported by all the devices in the circuit making it necessary to enhance their breakdown voltages. This is detrimental to system performance due to an overall increase in power dissipation in the semiconductor components.

The utility of the analytical model can be illustrated by performing the analysis of the reverse recovery for a specific P-i-N rectifier structure. Consider the case of a P-i-N rectifier designed to support 1,000 V with a drift region thickness of 60 μm and doping concentration of 5×10^{13} cm^{-3} that was previously analyzed in Sect. 5.4.1 for the forward recovery process. The reverse recovery process in this structure is analyzed here using the analytical solutions for various ramp rates, lifetime values, and reverse supply voltages. In all cases, the reverse recovery is assumed to begin with on-state operation at a current density of 100 A cm^{-2}.

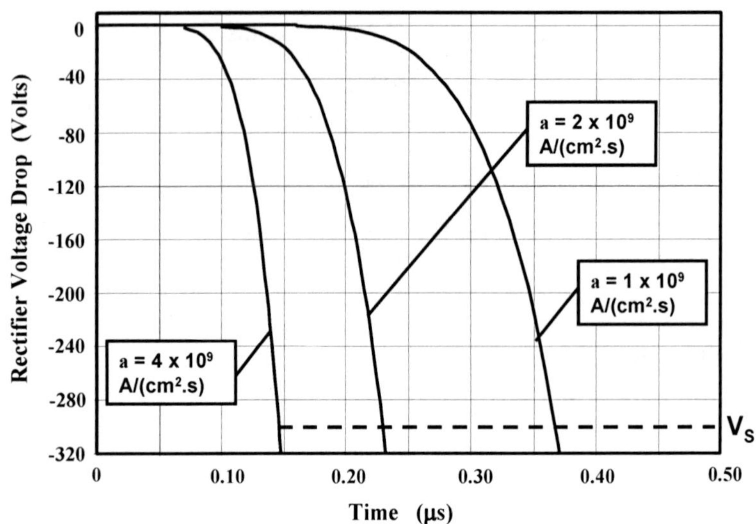

Fig. 5.36 Analytically calculated voltage waveforms for a 1,000-V P-i-N rectifier during the reverse recovery process using various ramp rates

The voltage waveforms calculated using the analytical solutions are shown in Fig. 5.36 for the case of a high-level lifetime of 0.5 μs. With this lifetime, the average free carrier concentration in the drift region was found to be 5.2×10^{16} cm^{-3} for this structure at the on-state current density of 100 A cm^{-2}. According to the analytical model (see (5.114)), the time t_1 at which the junction becomes reverse biased increases from 51 to 80 to 134 ns as the ramp rate decreases from 4×10^9 to 2×10^9 to 1×10^9 A cm^{-2} s^{-1}. Before this time, the voltage across the rectifier is slightly positive with a value given by (5.117). The voltage then increases rapidly and reaches 300 V (indicated by the dashed line in the figure) at 145, 230, and 370 ns, respectively, for the three cases. This point in the voltage waveforms defines the end of the third phase.

The peak reverse recovery current occurs at the end of the third phase. The peak reverse recovery current densities predicted by the analytical model are 480, 360, and 270 A cm^{-2}, respectively, for the three cases of the ramp rate as can be observed in Fig. 5.37 which shows the current waveforms obtained using the analytical model. After the third phase, the reverse current reduces to zero at a constant rate. The time duration (t_B), during which the reverse current reduces to zero, becomes smaller with a reduction in the ramp rate. The values for t_B predicted by the analytical model decrease from 70 to 60 to 28 ns as the ramp rate decreases from 4×10^9 to 2×10^9 to 1×10^9 A cm^{-2} s^{-1}. These combinations of the peak reverse recovery current and period t_B produce a reverse [d*J*/d*t*] ranging from 6 to 9.5×10^9 A cm^{-2} s^{-1}.

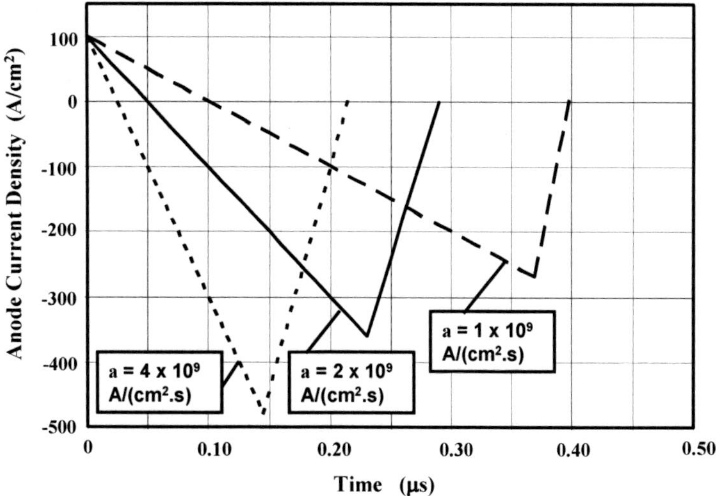

Fig. 5.37 Analytically calculated current waveforms for a 1,000-V P-i-N rectifier during the reverse recovery process using various ramp rates

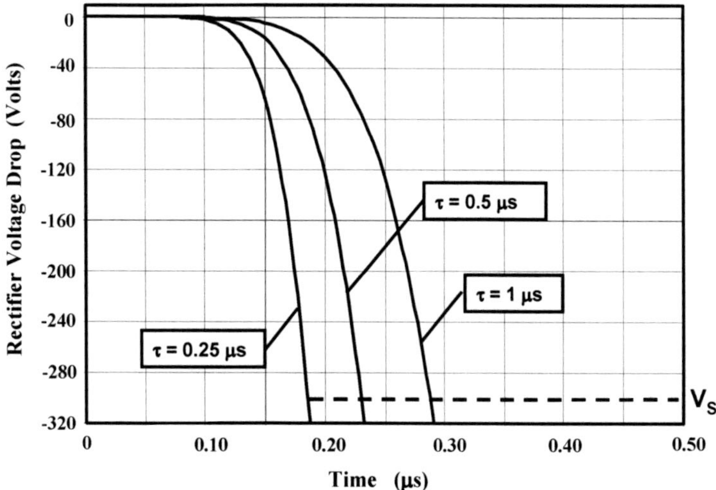

Fig. 5.38 Analytically calculated voltage waveforms for a 1,000-V P-i-N rectifier during the reverse recovery process for various lifetime values

The analytical model can also be utilized to examine the influence of the minority carrier lifetime on the reverse recovery process. Consider the case of the same 1,000-V P-i-N rectifier structure switched off from an on-state current density of 100 A cm^{-2} at a ramp rate of 2×10^9 A cm^{-2} s^{-1}. The voltage waveforms predicted by the analytical model for the reverse recovery process with lifetime values of 0.25, 0.5, and 1 μs in the drift region are shown in Fig. 5.38. The model predicts no change in the time t_1 for the end of the first phase and a faster rate of increase in the anode voltage during the second phase when the lifetime is reduced.

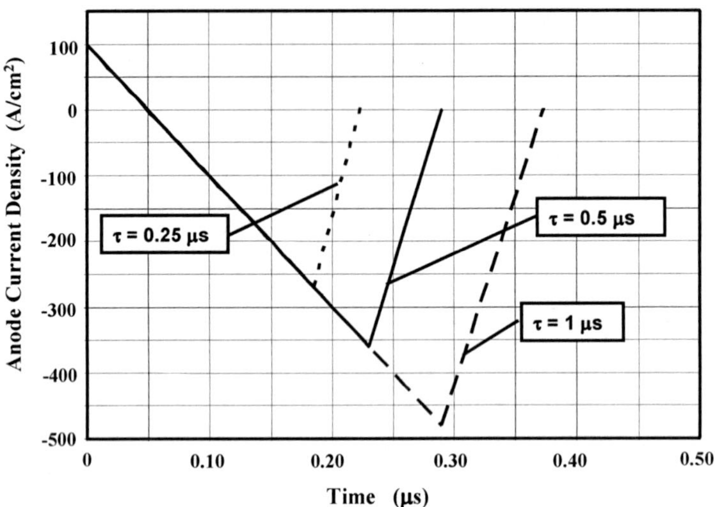

Fig. 5.39 Analytically calculated current waveforms for a 1,000-V P-i-N rectifier during the reverse recovery process for various lifetime values

The current flow during the reverse recovery process is shown in Fig. 5.39 for the case of the three lifetime values. The peak reverse recovery current density predicted by the analytical model reduces from 480 to 360 to 270 A cm^{-2}, respectively, when the lifetime is reduced from 1 to 0.5 to 0.25 μs as can be observed from Fig. 5.39. The time duration (t_B) for the fourth phase, during which the reverse current reduces to zero, also becomes smaller with a reduction in the lifetime. The values for t_B predicted by the analytical model are 38, 60, and 84 ns for lifetime values of 0.25, 0.5, and 1 μs, respectively. These combinations of the peak reverse recovery current and period t_B produce a reverse [dJ/dt] ranging from 5.7 to 7×10^9 A cm^{-2} s^{-1}.

The analytical model also enables analysis of the impact of changing the reverse recovery voltage on the reverse recovery process. Consider the case of the same 1,000-V P-i-N rectifier structure switched off from an on-state current density of 100 A cm^{-2} at a ramp rate of 2×10^9 A cm^{-2} s^{-1}. The voltage waveform predicted by the analytical model is shown in Fig. 5.40 with dashed lines indicating the point at which the reverse bias voltage reaches values of 90, 300, and 600 V during the reverse recovery transient. As the reverse voltage is increased, it takes a longer time interval to produce the wider space-charge region that is needed to support the voltage. This is accompanied by a larger value for the peak reverse recovery current as shown in Fig. 5.41. The larger space-charge region, formed at larger reverse bias voltages, removes a greater fraction of the stored charge as well. This produces a substantial reduction of the period t_B resulting in very high reverse [dJ/dt] as observed in the figure.

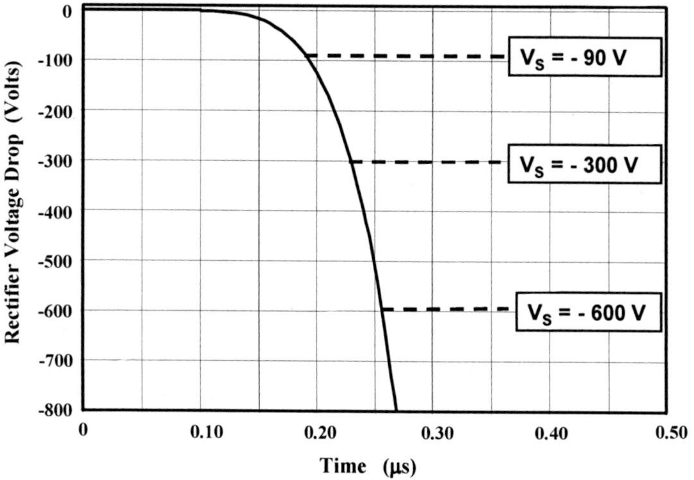

Fig. 5.40 Analytically calculated voltage waveforms for a 1,000-V P-i-N rectifier during the reverse recovery process for various supply voltages

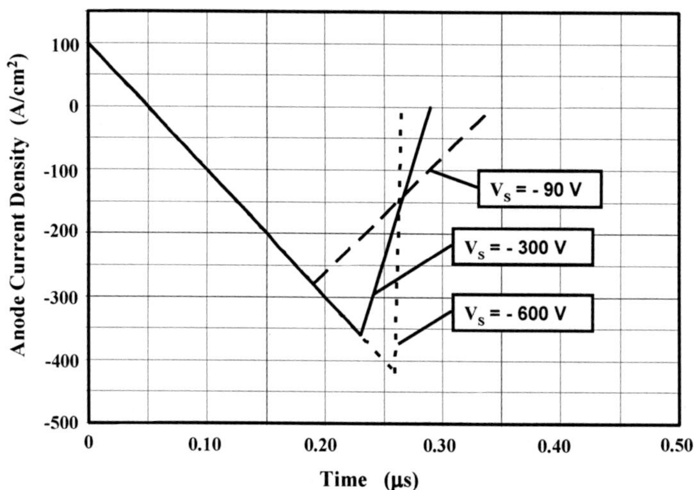

Fig. 5.41 Analytically calculated current waveforms for a 1,000-V P-i-N rectifier during the reverse recovery process for various supply voltages

The peak reverse recovery current density predicted by the analytical model increases from 280 to 360 to 420 A cm^{-2} when the voltage is increased from 90 to 300 to 600 V, respectively, as can be observed from Fig. 5.41. The time duration (t_B) for the fourth phase, during which the reverse current reduces to zero, also becomes smaller with an increase in the reverse voltage. The values for t_B predicted by the analytical model are 153, 60, and only 1 ns for reverse voltages of 90, 300, and 600 V, respectively. The t_B value for the 600-V case indicates that almost all the stored charge has been extracted by the extension of the space-charge region at this large reverse voltage. These combinations of the peak reverse recovery current and period t_B produce a drastic increase in the reverse [dJ/dt] ranging from 1.8 to 6 to 420 × 10^9 A cm^{-2} s^{-1} when the reverse voltage is increased from 90 to 300 to 600 V, respectively.

Simulation Example

To validate the above model for the reverse recovery transient in the P-i-N rectifier, the results of numerical simulations on a 1,000-V silicon P-i-N rectifier structure are described here. The structure had a drift region thickness of 60 μm with a doping concentration of 5 × 10^{13} cm^{-3}. The cathode current was ramped from 100 A cm^{-2} in the on-state using various values of negative ramp rates. In addition, the impact of changing the lifetime and the reverse supply voltage was examined for comparison with the analytical model.

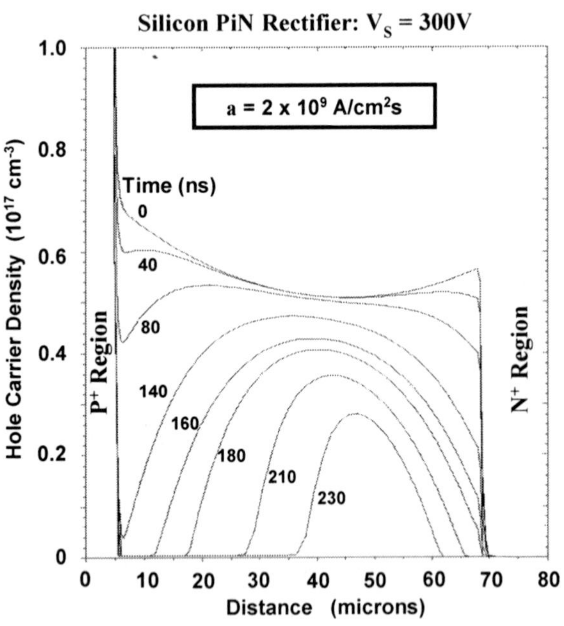

Fig. 5.42 Carrier distribution in a 1,000-V silicon P-i-N rectifier during phases 1–3 of the reverse recovery transient

First, consider the case of varying the negative ramp rate from 1 × 10^9 to 2 × 10^9 to 4 × 10^9 A cm^{-2} s^{-1}. For these cases, a lifetime (τ_{p0} and τ_{n0}) value of 1 μs was used during the numerical simulations. The average carrier concentration in the drift region under steady-state conditions with an on-state current density of 100 A cm^{-2} was found to be about 5×10^{16} cm^{-3} as shown in Fig. 5.42. This value is obtained by using a high-level lifetime of 0.5 μs in (5.101), indicating that end region recombination currents are significant in this structure. The carrier concentration profile exhibits a zero slope at time $t = 40$ ns, corresponding to time $t = t_0$ in the analytical model (see Fig. 5.34). The slope of the carrier profile then becomes positive, as shown for the time $t = 80$ ns. During this time, the carrier concentration at the junction is well above the equilibrium value, indicating that the P$^+$/N junction is still forward biased. At time $t = 140$ ns, the carrier concentration at the junction becomes close to zero, corresponding to the time $t = t_1$ in the analytical model (see Fig. 5.33). The value for t_1 obtained using the analytical model is about 120 ns in good agreement with the simulations.

The carrier profiles for subsequent time instances of 160, 180, 210, and 230 ns are also shown in Fig. 5.42. It can be observed that the depletion region expands from the P$^+$/N junction during this time interval. The analytical model predicts a depletion region width of 38 μm when the reverse bias voltage reaches 300 V, in excellent agreement with the simulations. The analytical model also predicts the end of phase 3 at time $t_2 = 230$ ns in very good agreement with the simulations. Consequently, the peak reverse recovery current predicted by the model also agrees with the simulations.

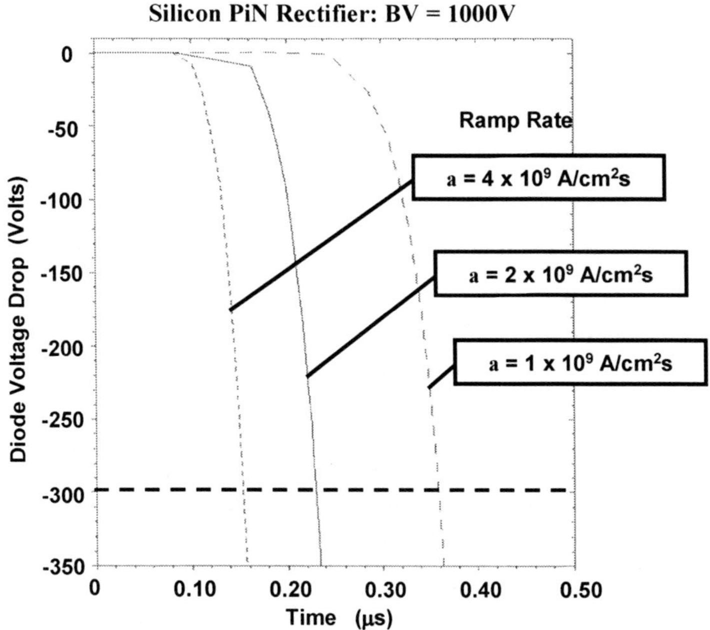

Fig. 5.43 Voltage waveforms for a 1,000-V silicon P-i-N rectifier during the reverse recovery transient with various ramp rates

The diode voltage and current waveforms obtained with the aid of the numerical simulations are shown in Figs. 5.43 and 5.44, respectively. These waveforms have the same features predicted by the analytical model (see Figs. 5.36 and 5.37). The peak reverse currents obtained using the model are in good agreement with those observed in the simulations. However, the transient observed with the numerical simulations during phase 4 occurs with a constant ramp rate followed by a more abrupt reduction of the current. The ramp rate observed with the simulations is in the range between 7 and 9×10^9 A cm^{-2} s^{-1} as predicted by the analytical model.

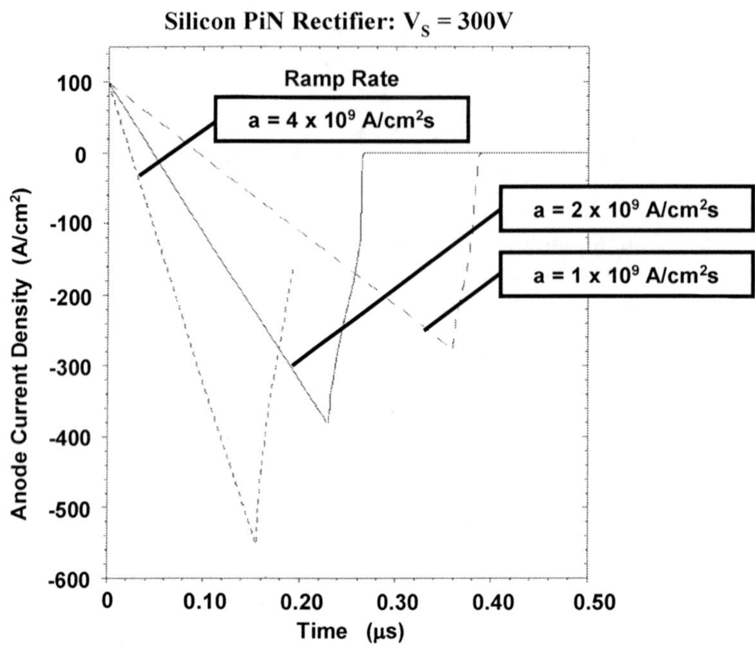

Fig. 5.44 Current waveforms for a 1,000-V silicon P-i-N rectifier during the reverse recovery transient with various ramp rates

During the fourth phase of the turn-off process, the remaining free carriers in the drift region are removed by further extension of the depletion region as the current ramps down to zero. The reduction of the reverse current is accompanied by a reduction in the concentration of holes within the space-charge region as described by (5.122). This reduces the net positive charge in the space-charge region allowing its expansion in spite of the constant reverse voltage across the diode. Since the P-i-N rectifier is designed with a punch-through architecture, the space-charge region eventually expands through the entire drift region removing all the stored charge. The removal of the stored charge, observed with the numerical simulations, is shown in Fig. 5.45.

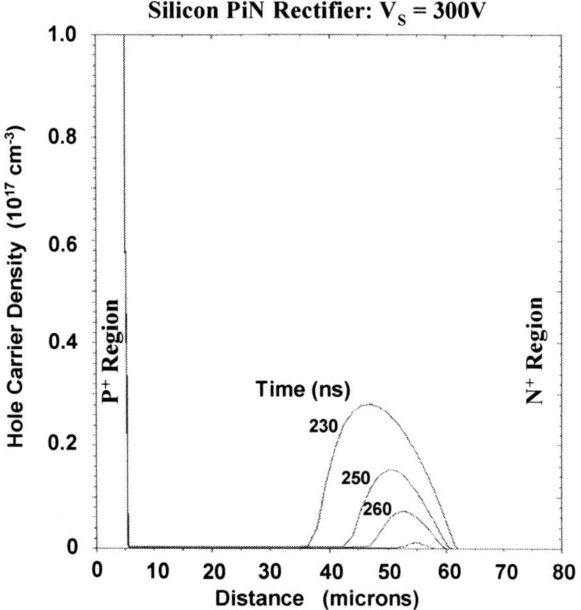

Fig. 5.45 Carrier distribution in a 1,000-V silicon P-i-N rectifier during phase 4 of the reverse recovery transient

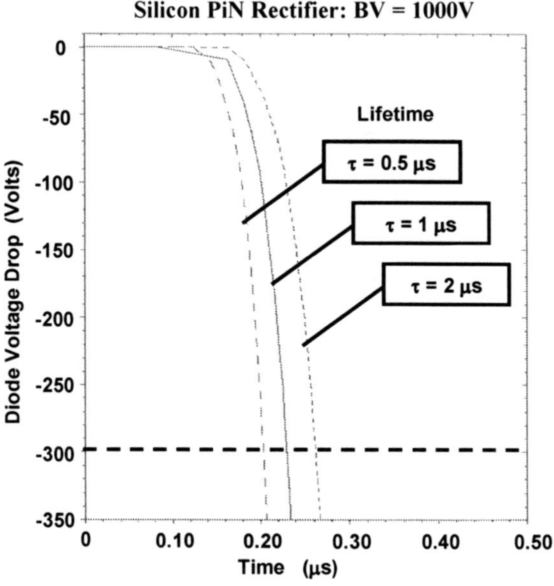

Fig. 5.46 Voltage waveforms for a 1,000-V silicon P-i-N rectifier during the reverse recovery transient with various lifetimes

The validity of the analytical model can also be examined by observation of the impact of changes in the lifetime on the reverse recovery process. To illustrate this, the lifetime was increased and reduced by a factor of 2 times while performing the reverse recovery at ramp rate of 2×10^9 A cm^{-2} s^{-1}. The diode voltage and current waveforms obtained using the numerical simulations are shown in Figs. 5.46 and 5.47, respectively.

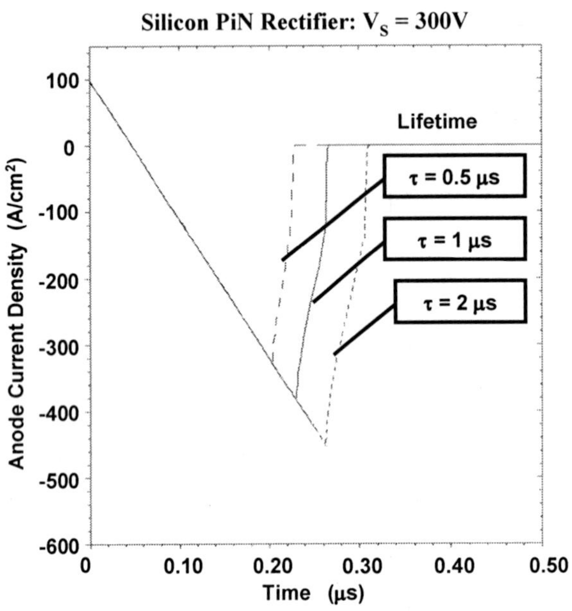

Fig. 5.47 Current waveforms for a 1,000-V silicon P-i-N rectifier during the reverse recovery transient with various lifetimes

The waveforms shown in the above figures have the same features as the waveforms obtained using the analytical model (see Figs. 5.38 and 5.39). There is no change in the time t_1 for the end of the second phase while the time taken for the voltage to increase to 300 V reduces with a reduction in the lifetime. The peak reverse recovery current decreases with a reduction of the lifetime as predicted by the model. The time duration for the fourth phase also increases with increasing lifetime as predicted by the model, resulting in a slightly smaller reverse [d*i*/d*t*]. It can be concluded that the analytical model provides an accurate description of the impact of changes in the lifetime on the reverse recovery process.

As a further validation of the model, numerical simulations of the 1,000-V P-i-N rectifier structure were performed with various reverse bias supply voltages. The voltage waveform is shown in Fig. 5.48 for the case of a ramp rate of 2×10^9 A cm^{-2} s^{-1} and lifetime of 1 μs. The time taken to arrive at the increasing reverse bias voltages is consistent with the prediction of the analytical model (see Fig. 5.40). Consequently, the peak reverse recovery currents predicted by the analytical model are also in agreement with those obtained from the numerical simulations. These waveforms are shown in Fig. 5.49.

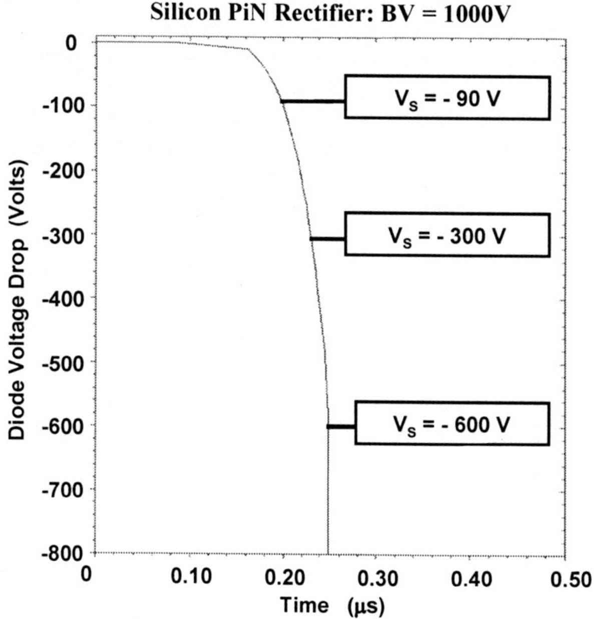

Fig. 5.48 Voltage waveforms for a 1,000-V silicon P-i-N rectifier during the reverse recovery transient with various supply voltages

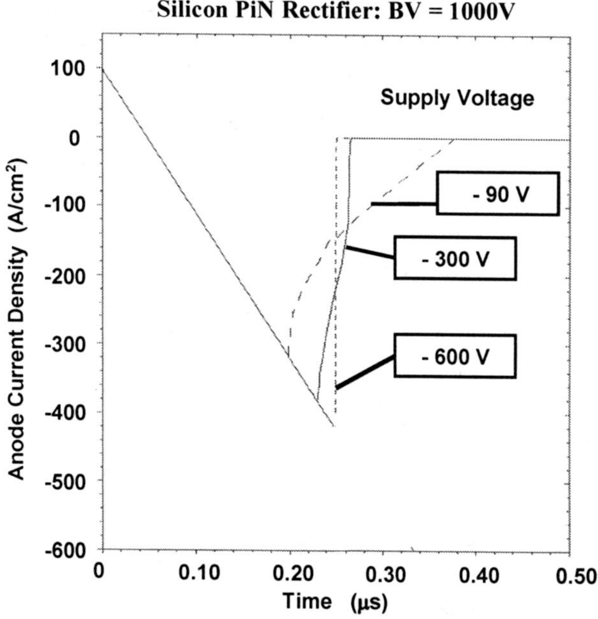

Fig. 5.49 Current waveforms for a 1,000-V silicon P-i-N rectifier during the reverse recovery transient with various supply voltages

The current waveforms observed with the numerical simulations have the same features predicted by the analytical model (see Fig. 5.41). When the reverse voltage is increased to 600 V, the fourth phase of the reverse recovery occurs with an abrupt reduction of the reverse current as predicted by the analytical model. The peak reverse currents obtained using the model are also in good agreement with those observed in the simulations. The extremely high reverse [d*i*/d*t*] associated with an abrupt drop in the reverse recovery current is a problem in power circuits where it produces large voltage spikes across any stray inductances that are in series with the diode.

The above simulation results provide validation for the analytical model developed in this section and give additional insight into the carrier distribution and transients. Based upon the excellent match between the predictions of the analytical model and the simulations, it can be concluded that the model is able to account for the all four phases of the reverse recovery process and predict the proper dependence of the reverse recovery current on the ramp rate, the lifetime, and the reverse supply voltage.

5.5 P-i-N Rectifier Structure with Buffer Layer

In Sect. 5.4.2, it was found that the reverse recovery for the P-i-N rectifier had a high rate of change in the reverse recovery current during the fourth phase of the turn-off process. The high rate of change in the reverse recovery current has been referred to as the *snappy recovery*, which can lead to high transient voltages developed across the stray inductance in the diode circuit path. One approach to mitigating this effect is by the incorporation of a *buffer layer* in the drift region (Fig. 5.50). The buffer layer is a more heavily doped N-type region added at the cathode side of the drift region. Its doping concentration is chosen to be sufficiently large, so that it cannot be depleted by the reverse bias voltage applied to the diode during the reverse recovery process. At the same time, its doping concentration is chosen to be low enough to allow conductivity modulation resulting in some stored

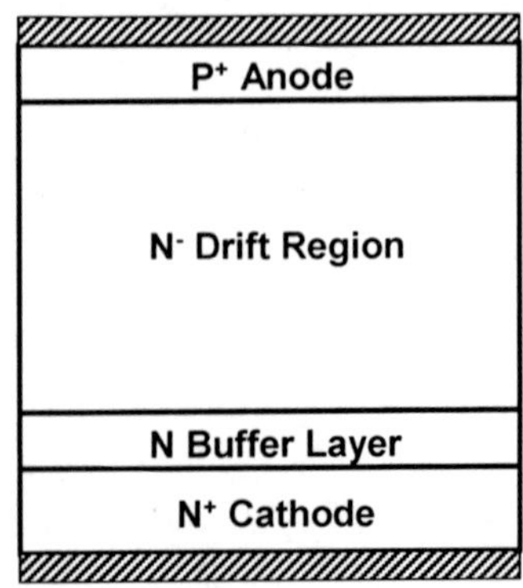

Fig. 5.50 P-i-N rectifier structure with buffer layer

charge within the buffer layer. During the fourth phase of the reverse recovery process, this stored charge is not rapidly removed by the applied bias producing a smaller rate of change in the reverse current.

This structure provides a relatively small improvement in the reverse [di/dt] while increasing the on-state voltage drop due to the increase in the thickness of the drift region. The added complexity of growing a drift region with two different doping concentrations makes it an unattractive approach as well. A greater reduction of the reverse [di/dt], with a *soft recovery* characteristic, can be achieved by utilizing the nonpunch-through structure. This approach is discussed in Sect. 5.6.

5.6 Nonpunch-Through P-i-N Rectifier Structure

In Sect. 5.4, it was demonstrated that the punch-through structure is preferable for a P-i-N rectifier to reduce its on-state voltage drop because of a smaller width for the drift region. However, it was found that this results in complete removal of the stored charge during the first three phases of the reverse recovery process. Consequently, a very high [di/dt] is produced during the fourth phase of the turn-off process which is detrimental to operation in power circuits. One method to avoiding this problem is to use a nonpunch-through architecture for the drift region. In this approach, the drift region doping concentration and thickness are chosen to achieve ideal parallel-plane breakdown voltage at the desired voltage rating for the rectifier. The drift region is then completely depleted only if the applied reverse supply voltage is equal to the breakdown voltage. Since the rectifier is usually utilized in circuits with lower supply voltages, a portion of the drift region remains undepleted during the entire reverse recovery process. The stored charge in the undepleted portion of the drift region is then removed by recombination. This produces a significant reduction in the reverse [di/dt] during the reverse recovery process.

The reverse recovery process for the nonpunch-through structure is identical to that for the punch-through structure for the first three phases. However, unlike the punch-through case, the stored charge cannot be removed by the extension of the space-charge region through the entire drift region. Instead, the current reduces as governed by the diffusion of the excess holes remaining in the drift region toward the edge of the space-charge region in a time frame that is much shorter than the recombination lifetime. Since the voltage is supported across the space-charge region during this process, the electric field can be assumed to be small during the diffusion of the excess carriers in the undepleted region. The drift component of the current in the continuity equation for the excess carriers can therefore be neglected.

The excess carrier concentration [$\delta p(x, t)$] during the fourth phase is then governed by the following equation:

$$\frac{\partial \delta p(x,t)}{\partial t} = D_{\mathrm{p}} \frac{\partial^2 \delta p(x,t)}{\partial x^2}. \tag{5.130}$$

A solution for this excess charge distribution is of the form

$$\delta p(x,t) = A\mathrm{e}^{-x^2/4D_{\mathrm{p}}t} + B\mathrm{e}^{x^2/4D_{\mathrm{p}}t}, \tag{5.131}$$

where A and B are constants determined by the boundary conditions. During the fourth phase of the reverse recovery process, the excess carrier concentration is zero at the edge of the space-charge region and has a maximum value at the center of the undepleted region as shown in Fig. 5.35. If the distance between the center of the undepleted region and the edge of the space-charge region is defined as

$$L = \frac{1}{2}[2d - W_{\mathrm{SC}}(t_2)], \tag{5.132}$$

then the maximum excess concentration lies at

$$x_{\mathrm{M}} = W_{\mathrm{SC}}(t_2) + L = \frac{1}{2}[2d + W_{\mathrm{SC}}(t_2)]. \tag{5.133}$$

The carrier concentration at the maximum value is given by

$$\delta p(x_{\mathrm{M}},t) = \left(\frac{\Delta p}{4D_{\mathrm{p}}t^3}\right), \tag{5.134}$$

where the term Δp can be obtained using the excess carrier concentration at $x = x_{\mathrm{M}}$ at the beginning of the fourth phase. If the excess concentration at this location at the beginning of the fourth phase (time $t = t_2$) is assumed to be equal to average carrier concentration n_{a}, then

$$\Delta p = 4D_{\mathrm{p}}t_2^3 n_{\mathrm{a}} \tag{5.135}$$

and

$$\delta p(x_{\mathrm{M}},t_2) = \left(\frac{t_2}{t}\right)^3 n_{\mathrm{a}}. \tag{5.136}$$

Using this boundary condition together with

$$\delta p[W_{\mathrm{SC}}(t_2),t] = 0 \tag{5.137}$$

yields the free carrier distribution in the undepleted region:

$$\delta p(x,t) = n_{\mathrm{a}}\left(\frac{t_2}{t}\right)^3 \left\{\mathrm{e}^{\Delta x^2/4D_{\mathrm{p}}t} - \mathrm{e}^{L^2/4D_{\mathrm{p}}t}\left[\frac{\sinh(\Delta x^2/4D_{\mathrm{p}}t)}{\cosh(L^2/4D_{\mathrm{p}}t)}\right]\right\}, \tag{5.138}$$

where Δx is measured from the position $x = x_M$ and time t extends beyond the end of the third phase at $t = t_2$.

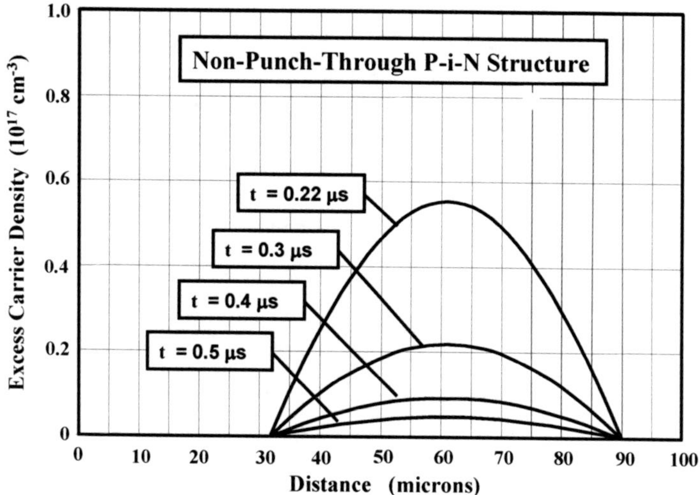

Fig. 5.51 Analytically calculated decay of stored charge during the fourth phase for the nonpunch-through structure

The excess charge distribution during the fourth phase of the turn-off process, as predicted by the analytical model, is shown in Fig. 5.51 for a nonpunch-through P-i-N rectifier structure designed to support 1,000 V. The structure had an N-type drift region with a doping concentration of 2×10^{14} cm^{-3} and a thickness of 90 µm. The distribution was calculated for the case of a reverse supply voltage of 300 V. The depletion width at the end of the third phase is 32 µm resulting in an undepleted region of about 60 µm. The peak of the excess carrier distribution resides at about 60 µm from the P^+/N-drift junction. The excess carriers are extracted by diffusion to the edge of the depletion region within about 0.4 µs after phase 3. This time frame is shorter than the high-level recombination lifetime of 0.8 µs within the drift region for this structure.

The reverse recovery current for the nonpunch-through structure during the fourth phase in the turn-off process can be obtained from the free carrier distribution profile:

$$J_R(t) = -qD_p \left.\frac{\delta p}{\delta x}\right|_{x=-L} . \tag{5.139}$$

By using the excess charge distribution in (5.138) in this equation, the reverse recovery current is obtained:

$$J_{\mathrm{R}}(t) = \frac{qn_{\mathrm{a}}L}{2t}\left(\frac{t_2}{t}\right)^3 e^{L^2/4D_{\mathrm{p}}t}\left[1 + \frac{\cosh(L^2/4D_{\mathrm{p}}t)}{\sinh(L^2/4D_{\mathrm{p}}t)}\right]. \tag{5.140}$$

The calculated reverse recovery current waveforms for various cases of the high-level lifetime in the drift region are shown in Fig. 5.52. It can be seen that the peak reverse recovery current becomes smaller when the lifetime is reduced. The current exhibits an exponential decay after the occurrence of the peak reverse current. This behavior is different from that observed for the punch-through structure.

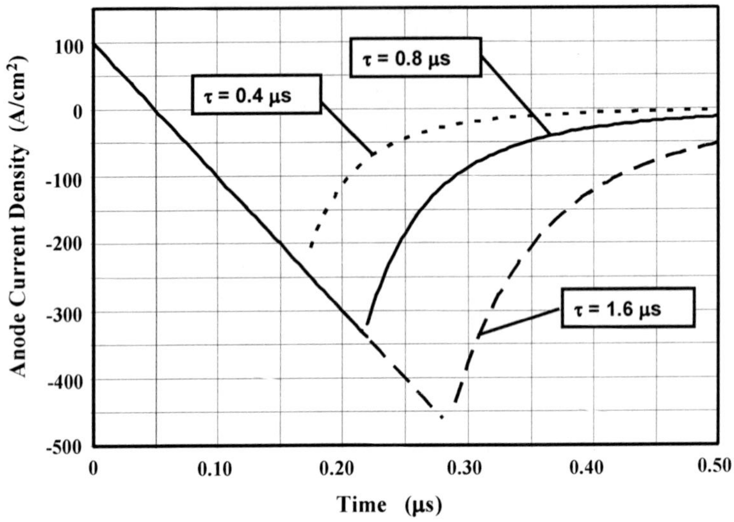

Fig. 5.52 Analytically calculated reverse recovery current waveforms for the nonpunch-through structure

Simulation Example

To validate the above model for the reverse recovery transient in the nonpunch-through P-i-N rectifier structure, the results of numerical simulations on a 1,000-V silicon P-i-N rectifier are described here. The structure had a drift region thickness of 85 μm with a doping concentration of 2×10^{14} cm^{-3}. The breakdown voltage obtained for this case by using numerical simulations was found to be just above 1,000 V. The on-state voltage drop was found to be 0.89 V at a current density of 100 A cm^{-2} for the case of a lifetime (τ_{p0} and τ_{n0}) of 1 μs, which is slightly larger than the 0.86 V observed for the punch-through structure.

For comparison of the reverse recovery transient with the punch-through structure, the cathode current was ramped down from an on-state current density of 100 A cm^{-2} using negative ramp rate of 2×10^9 A cm^{-2} s^{-1}. The resulting reverse recovery waveforms for the current flowing through the punch-through and nonpunch-through rectifier structures are shown in Fig. 5.53 for the case of a lifetime value of 1 μs. The reverse current exhibits the same peak reverse recovery current in both cases, as expected from the analytical model, because the first three phases of the turn-off process are identical. During the fourth phase, the initial reduction of current occurs at a similar rate for both structures. However, the abrupt reduction of the current with a very high reverse [d*i*/d*t*] in the punch-through structure is replaced by an exponential gradual reduction in current for the nonpunch-through case. This behavior is consistent with the model for the carrier distribution developed above.

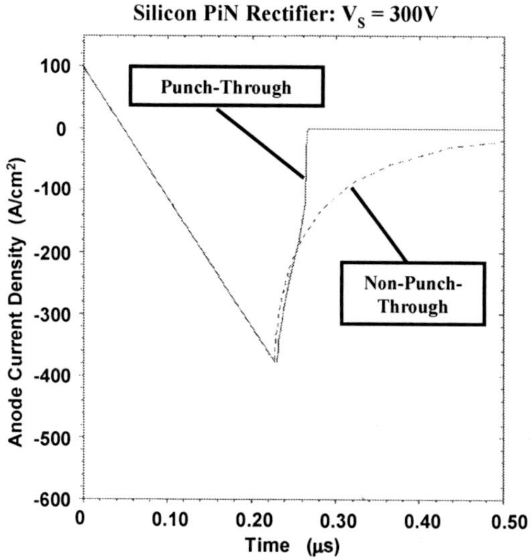

Fig. 5.53 Comparison of the current waveforms for 1,000-V silicon nonpunch-through and punch-through P-i-N rectifiers during the reverse recovery transient

The carrier distribution profiles during the first three phases of the reverse recovery process are shown in Fig. 5.54 for the nonpunch-through structure. These carrier profiles are similar to those in Fig. 5.42 for the punch-through structure. However, due to the larger doping concentration in the drift region, the space-charge region extends to only 33 μm when compared with 38 μm for the punch-through case. A larger amount of stored charge is retained within the nonpunch-through structure at the end of the third phase. The reverse voltage reaches 300 V at the same time (230 ns) for the nonpunch-through case resulting in the same peak reverse recovery current for both structures.

During the fourth phase of the turn-off process, the remaining free carriers in the drift region are removed by diffusion of the stored charge toward the edge of the space-charge region. The removal of the stored charge during the fourth phase, observed with the numerical simulations, is shown in Fig. 5.55. The analytical model described above is consistent with the behavior shown in this figure. Note that free carriers are present in the drift region for a much longer time frame when compared with the punch-through structure. This is responsible for the gradual (exponential) reduction of the reverse recovery current for the nonpunch-through structure.

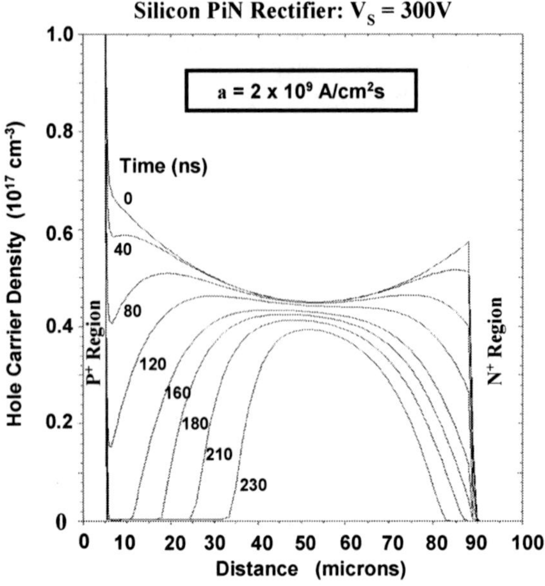

Fig. 5.54 Carrier distribution in a 1,000-V silicon nonpunch-through P-i-N rectifier during phases 1–3 of the reverse recovery transient

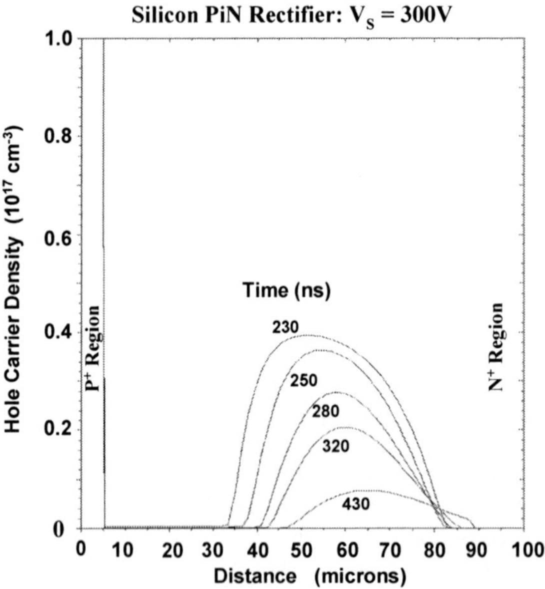

Fig. 5.55 Carrier distribution in a 1,000-V silicon nonpunch-through P-i-N rectifier during phase 4 of the reverse recovery transient

The validity of the analytical model can also be examined by observation of the impact of changes to the lifetime on the reverse recovery process. To illustrate this, the lifetime was increased and reduced by a factor of 2 times while performing the reverse recovery at ramp rate of 2×10^9 A cm^{-2} s^{-1}. The current waveforms obtained using the numerical simulations are shown in Fig. 5.56. The waveforms shown in the figure have the same features as the waveforms obtained using the analytical model (see Fig. 5.52). The peak reverse recovery current decreases with a reduction of the lifetime as predicted by the model. The time duration for the fourth phase also increases with increasing lifetime as predicted by the model, resulting in a slightly smaller reverse [d*i*/d*t*]. It can therefore be concluded that the analytical model provides an accurate description of the impact of changes in the lifetime on the reverse recovery process.

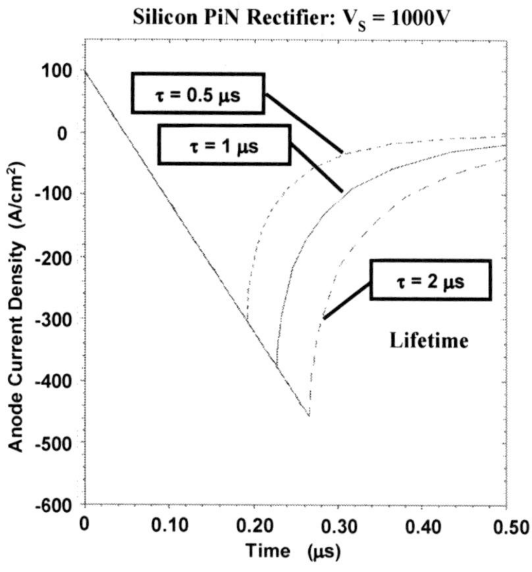

Fig. 5.56 Current waveforms for a 1,000-V silicon nonpunch-through P-i-N rectifiers during the reverse recovery transient with various lifetimes

The above simulation results provide validation for the analytical model developed in this section and give additional insight into the carrier distribution and transients. Based upon the excellent match between the predictions of the analytical model and the simulations, it can be concluded that the model is able to account for the all four phases of the reverse recovery process and predict the proper dependence of the reverse recovery current on the ramp rate, the lifetime, and the reverse supply voltage. The nonpunch-through structure is found to offer an improvement in the reverse recovery [d*i*/d*t*] with a *soft recovery* behavior. This is desirable to reduce spikes in the voltages in the power circuit. However, the longer recovery time produces an increase in the power dissipation reducing the efficiency for power conversion.

5.7 P-i-N Rectifier Tradeoff Curves

In the previous sections, it was demonstrated that the peak reverse recovery current and the turn-off time can be reduced by reducing the minority carrier lifetime in the drift region of the P-i-N rectifier structure. This enables reduction of the power losses during the switching transient. However, the on-state voltage drop in a P-i-N rectifier increases when the minority carrier lifetime is reduced, which produces an increase in the power dissipation during on-state current flow. For power system applications, it is desirable to reduce the total power dissipation produced in the rectifiers. This also reduces the heat generated within the power devices maintaining a lower junction temperature, which is desirable to prevent thermal runaway and reliability problems. To minimize the power dissipation, it is commonplace to perform a tradeoff between on-state and switching power losses for power P-i-N rectifiers by developing tradeoff curves.

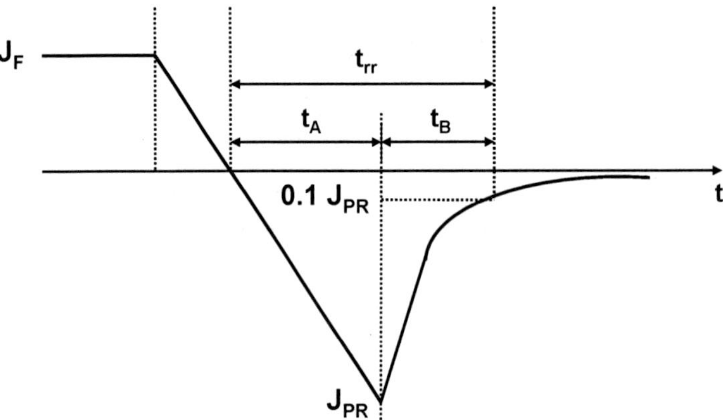

Fig. 5.57 Typical reverse recovery current waveform for the P-i-N rectifier structure defining the reverse recovery turn-off time

One type of the tradeoff curve for a power P-i-N rectifier can be generated by plotting the on-state voltage drop against the reverse recovery turn-off time. A turn-off waveform that includes a nonlinear portion during the fourth phase is illustrated in Fig. 5.57. The turn-off time (t_{rr}) is defined as the time taken for the reverse current to reduce to 10% of the peak reverse recovery current (J_{PR}) after the current crosses zero. This time can be extracted from the measured characteristics of devices using automated test equipment. In addition, it is common practice to extract the times t_A and t_B as defined on the waveform in Fig. 5.57. A larger $[t_A/t_B]$ ratio is considered to be desirable to mitigate the voltage spikes created in power circuits by a large $[di/dt]$ in the reverse direction.

The tradeoff curve between the on-state voltage drop and the reverse recovery time obtained by using the analytical models is shown in Fig. 5.58 by the solid line. The results obtained using two-dimensional numerical simulations for the punch-through structure are also shown in the figure for comparison. It can be observed that there is good agreement between them until the lifetime is reduced below 0.05 μs to achieve a reverse recovery time of less than 0.1 μs. The reverse recovery time predicted by the analytical models is larger than observed with the simulations because free carrier recombination was neglected during the reverse recovery process. This assumption is not valid when the recombination lifetime is reduced to below 0.05 μs.

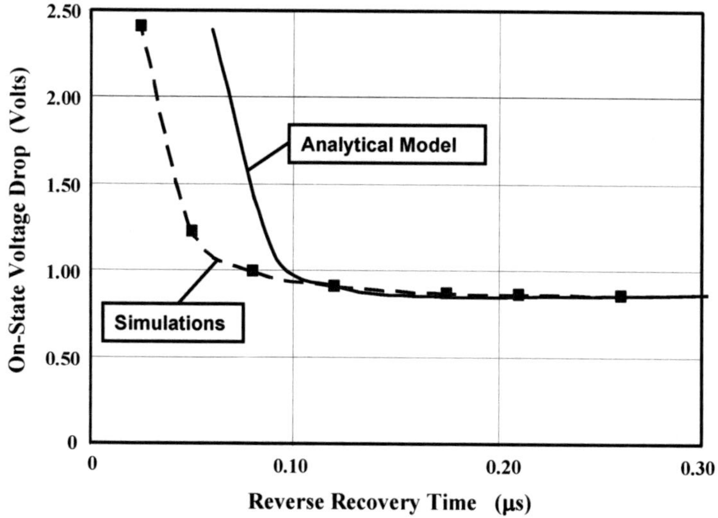

Fig. 5.58 Tradeoff curve for the punch-through P-i-N rectifier structure

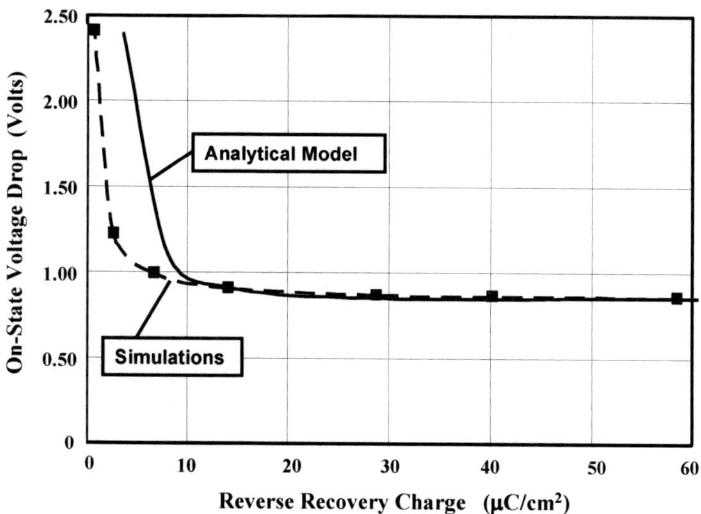

Fig. 5.59 Tradeoff curve for the punch-through P-i-N rectifier structure

Another common practice for displaying the tradeoff in power losses in power P-i-N rectifiers is by plotting the on-state voltage drop against the reverse recovery charge (Q_{rr}). The reverse recovery charge can be extracted by integration of the turn-off waveform for the current. This graph obtained with the analytical model and the simulations is shown in Fig. 5.59 for the punch-through structure.

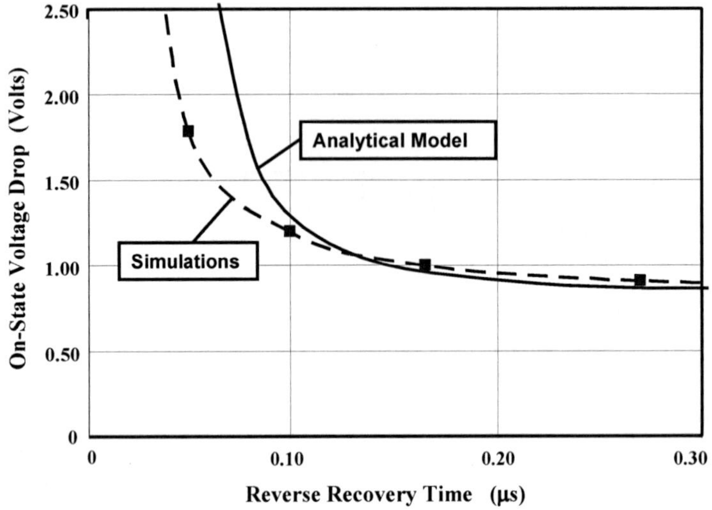

Fig. 5.60 Tradeoff curve for the nonpunch-through P-i-N rectifier structure

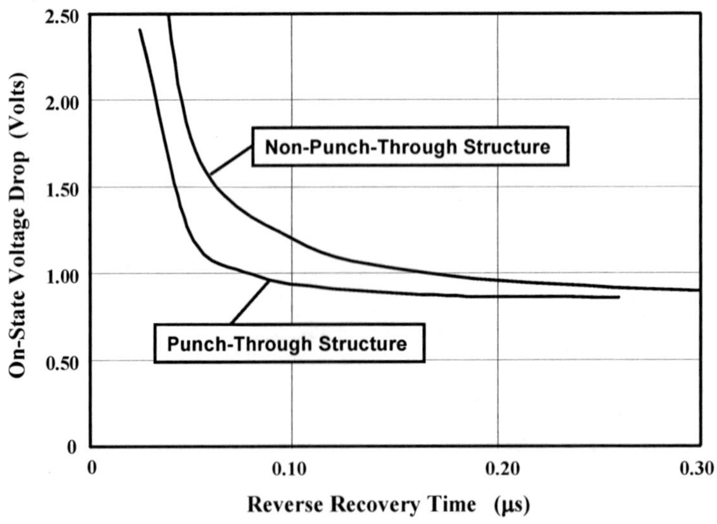

Fig. 5.61 Comparison of the tradeoff curves for the punch-through and nonpunch-through P-i-N rectifier structures

A similar analysis can be preformed for the nonpunch-through P-i-N rectifier structure. The resulting tradeoff curve between the on-state voltage drop and the reverse recovery turn-off time is shown in Fig. 5.60. The values obtained by using the analytical model for the nonpunch-through structure are compared with those obtained from the two-dimensional numerical simulations. The analytical model works in a satisfactory manner until the recombination lifetime is reduced to a value comparable with the reverse recovery time.

A comparison between the tradeoff curves for the punch-through and nonpunch-through structures is provided in Fig. 5.61. From this figure, it is apparent that the tradeoff curve for the punch-through design is located below that for the nonpunch-through design. This implies that the reverse recovery time is shorter for the punch-through design for any given on-state voltage drop. For example, if the on-state voltage drop is 1 V, the reverse recovery time for the nonpunch-through structure is 0.16 μs compared with only 0.08 μs for the punch-through design. This implies a reduction in the turn-off power loss by about a factor of 2 by utilization of the punch-through structure.

Lifetime Control

The adjustment of the recombination lifetime in the drift region can be performed by using a variety of techniques that introduce deep level centers within the band gap. As described in detail in Sect. 2.3.5, the position of the deep level within the band gap and the capture cross sections for electrons and holes are unique to each process. The most commonly used approaches in the power semiconductor industry have been by the diffusion of gold and platinum or by electron irradiation.

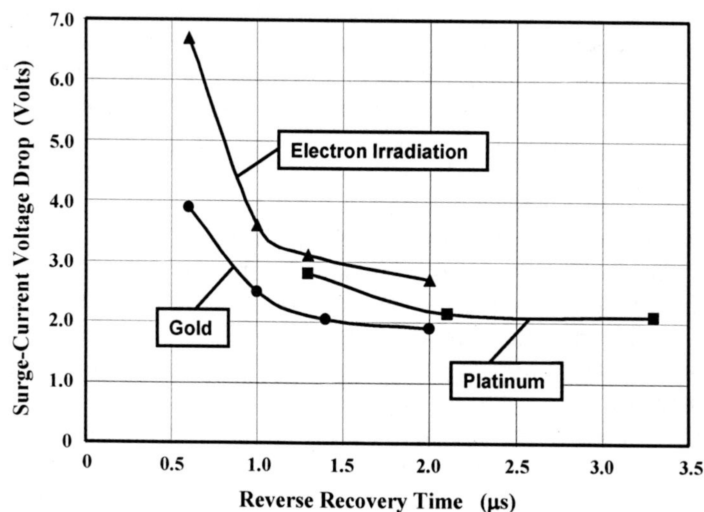

Fig. 5.62 Tradeoff curves for the different methods for controlling lifetime in the P-i-N rectifier

The experimentally measured tradeoff curves[13] obtained for P-i-N rectifiers built using 60–90 Ω cm N-type silicon with breakdown voltage capability of 2,500 V are shown in Fig. 5.62. In this work, the forward voltage drop was measured under surge current levels to emphasize the differences between the methods for lifetime control. It can be seen from the figure that the best tradeoff curve is exhibited by the gold doping method. These results are consistent with the analysis provided in Sect. 2.3.5. In spite of the superior tradeoff curve for gold doping, the power semiconductor industry has migrated to electron irradiation because of the lower leakage current and its manufacturing convenience, as described in Sect. 2.3.5.

5.8 Summary

The physics of operation of the P-i-N rectifier has been analyzed in this chapter. Analytical expressions have been derived for the on-state and blocking state, as well as the forward and reverse recovery transients. At low-current levels, current flow in this diode occurs by the familiar diffusion current transport phenomenon. However, at higher current levels, the injected minority carrier density in the drift region exceeds the relatively low doping concentration required to achieve high breakdown voltages. This high-level injection in the drift region modulates its conductivity producing a reduction in the on-state voltage drop. If recombination in the drift region is dominant, the voltage drop across the drift region becomes independent of the on-state current density. These phenomena allow operation of the silicon P-i-N rectifier with an on-state voltage drop of only 1 V making it very attractive for power electronic applications.

The P-i-N rectifier can support a large voltage in the reverse-blocking mode by appropriate choice of the doping concentration and thickness of the drift region. A punch-through design is favored because it reduces the thickness of the drift region. A narrower drift region contains a smaller amount of stored charge during on-state operation enabling faster turn-off.

The switching of the P-i-N rectifier from the on-state to the reverse-blocking state is accompanied by a significant current flow in the reverse direction. This reverse current produces large power dissipation in the rectifier as well as the power switches in power converter circuits. The reverse recovery current and the reverse recovery turn-off time can be reduced by reduction of the recombination lifetime in the drift region. Since this is accompanied by an increase in the on-state voltage drop, it is customary to perform a tradeoff analysis to minimize the overall power dissipation.

Silicon carbide-based P-i-N rectifiers have a much narrower drift region thickness when compared with silicon devices due to the higher critical electric field for breakdown. This favors a faster switching speed with reduced reverse recovery current. However, the larger band gap for silicon carbide produces an on-state voltage drop that is four times larger than for the silicon rectifiers. For this reason, silicon carbide P-i-N rectifiers are of interest only when the blocking voltage capability exceeds 10,000 V.

Problems

5.1 Determine the on-state current density for a silicon P-i-N rectifier at which the average injected carrier concentration becomes five times the doping concentration of 5×10^{13} cm^{-3} in a drift region with a width of 200 μm. The high-level lifetime in the drift region is 1 μs. Neglect end region recombination.

5.2 Plot the distribution of the injected carrier concentration as a function of distance within the drift region for the above P-i-N rectifier at an on-state current density of 100 A cm^{-2}.

5.3 Obtain the on-state voltage drop for the above P-i-N rectifier at an on-state current density of 100 A cm^{-2}.

5.4 Obtain the on-state voltage drop for the above P-i-N rectifier at an on-state current density of 100 A cm^{-2} if the high-level lifetime is increased by a factor of 2 times.

5.5 Obtain the on-state voltage drop for the above P-i-N rectifier at an on-state current density of 100 A cm^{-2} if the high-level lifetime is decreased by a factor of 2 times.

5.6 Determine the reverse breakdown voltage of the P-i-N rectifier in Problem 5.1.

5.7 What is the drift region thickness to achieve the same reverse breakdown voltage for a 4H-SiC P-i-N rectifier?

5.8 Obtain the on-state voltage drop for the 4H-SiC P-i-N rectifier in Problem 5.7 at an on-state current density of 100 A cm^{-2} if the high-level lifetime in the drift region is 0.1 μs.

5.9 Determine the leakage currents for the silicon P-i-N rectifier in Problem 5.1 at 300, 400, and 500 K if the space-charge generation and the low-level lifetimes are equal to the high-level lifetime. Assume that the entire drift region is depleted.

5.10 A silicon P-i-N rectifier has a drift region with doping concentration of 5×10^{13} cm^{-3} and thickness of 60 μm. The diode is turned on using a current ramp rate of 2×10^9 A cm^{-2} s^{-1}. Determine the time at which the peak occurs in the voltage overshoot. Determine the values for the conductivity modulation distance and the specific resistance of the drift region at this time. What is the maximum overshoot voltage?

5.11 The P-i-N rectifier in Problem 5.10 undergoes reverse recovery with a current ramp rate of 2×10^9 A cm^{-2} s^{-1} from an initial on-state current density of 100 A cm^{-2}. Determine the time taken for the current to cross zero. What is the time taken for the reverse voltage to reach a supply voltage of 300 V? Use a high-level lifetime of 0.5 μs in the drift region.

5.12 For the P-i-N rectifier in Problem 5.11, determine the peak reverse recovery current density. What is the space-charge layer width at this time?

5.13 For the P-i-N rectifier in Problem 5.11, determine the time t_B for the reverse current to decay to zero. What is the reverse [dJ/dt] for this rectifier?

5.14 What are the advantages of replacing the punch-through P-i-N rectifier design in Problem 5.10 with a nonpunch-through design? Define the drift region doping concentration and thickness for the nonpunch-through design to achieve the same reverse-blocking voltage as the punch-through structure.

References

[1] B.J. Baliga, "Silicon Carbide Power Devices", World Scientific, Singapore, 2005.
[2] S.M. Sze, "Physics of Semiconductor Devices", 2nd Edition, pp. 35–38, Wiley, New York, 1981.
[3] B.G. Streetman and S.K. Banerjee, "Solid State Electronic Devices", 6th Edition, pp. 184–193, Prentice Hall, Englewood Cliffs, NJ, 2006.
[4] S.K. Ghandhi, "Semiconductor Power Devices", p. 96, Wiley, New York, 1977.
[5] S.K. Ghandhi, "Semiconductor Power Devices", pp. 112–128, Wiley, New York, 1977.
[6] R.N. Hall, "Power Rectifiers and Transistors", Proceedings of the IRE, Vol. 40, pp. 1512–1518, 1952.
[7] H. Benda and E. Spenke, "Reverse Recovery Processes in Silicon Power Rectifiers", Proceedings of the IEEE, Vol. 55, pp. 1331–1354, 1967.
[8] N.R. Howard and G.W. Johnson, "PIN Silicon Diodes at High Forward Current Densities", Solid-State Electronics, Vol. 8, pp. 275–284, 1965.
[9] A. Herlet, "The Forward Characteristics of Silicon Power Rectifiers at High Current Densities", Solid-State Electronics, Vol. 11, pp. 717–742, 1968.

---

[10] B.G. Streetman and S.K. Banerjee, “Solid State Electronic Devices”, 6th Edition, p. 141, Prentice Hall, Englewood Cliffs, NJ, 2006.
[11] B.G. Streetman and S.K. Banerjee, “Solid State Electronic Devices”, 6th Edition, pp. 208–211, Prentice Hall, Englewood Cliffs, NJ, 2006.
[12] S.K. Ghandhi, “Semiconductor Power Devices”, pp. 9, 130–135, Wiley, New York, 1977.
[13] B.J. Baliga and E. Sun, “Comparison of Gold, Platinum, and Electron Irradiation for Controlling Lifetime in Power Rectifiers”, IEEE Transactions on Electron Devices, Vol. ED-24, pp. 685–688, 1977.

Chapter 6

Power MOSFETs

The vertical power metal–oxide–semiconductor field effect transistor (MOSFET) structure was developed in the mid-1970s to obtain improved performance when compared with the existing power bipolar transistors.[1] One of the major issues with the power bipolar transistor structure was its low-current gain when designed to support high voltages. In addition, power bipolar transistors could not be operated at high frequencies due to the large storage time related to the injected charge in their drift regions and were prone to destructive failure during hard switching in applications with inductive loads. The replacement of these current-controlled devices with a voltage-controlled device was attractive from an application's viewpoint. The high input impedance of the metal–oxide–semiconductor (MOS)-gate structure simplified the drive circuit requirements when compared with bipolar transistors being used at that time. In addition, their superior switching speed opened new applications operating in the 10–50 kHz frequency domain. Today, power MOSFETs are the most commonly used power switches in applications where the operating voltages are below 200 V.

The vertical power MOSFETs were initially considered to be ideal power switches due to their high input impedance and fast switching speed. However, their power-handling capability was constrained by the internal resistance within the structure between the drain and source electrodes. The power dissipated due to the ohmic voltage drop in the internal resistance limited the current-handling capability of the power MOSFETs as well as the efficiency of the power circuits in which they were utilized.

The first commercially successful power MOSFETs were developed by using the double-diffusion process in the mid-1970s. In these VD-MOSFET structures, the channel was formed by controlling the depth of two junctions. This enabled formation of short channel length structures without resorting to expensive high-resolution lithography. To increase the current-handling capability of the VD-

B.J. Baliga, *Fundamentals of Power Semiconductor Devices*, doi: 10.1007/978-0-387-47314-7_6,

MOSFETs, their internal resistance was substantially reduced from the mid-1970s to the mid-1980s by the use of enhanced design rules. However, it became apparent that the VD-MOSFET structure contained basic parasitic resistances that impeded further progress with improving the internal resistance.

In the 1990s, an alternate device structure was developed based upon using trench technology that had evolved for dynamic random access memory (DRAM) applications. The trench-gate or U-MOSFET structure offered the opportunity to reduce the internal resistance of the power MOSFET closer to the ideal value by elimination of a JFET region within the VD-MOSFET structure. The optimization of this structure also enabled increasing the operating frequency for power MOSFETs to the 1-MHz range.

In this chapter, the basic operating principles of the power MOSFET structure are discussed in detail. In the power MOSFET structure, the current flow between the drain and source electrodes is controlled by the formation of an inversion layer induced by the bias applied to the gate electrode. The physics of operation of the basic MOS structure is therefore described at the beginning of the chapter. Using this background, basic relationships for the internal resistance are then developed for the VD-MOSFET and U-MOSFET structures.

Although the power MOSFET does not require any gate drive current during steady-state operation due to the presence of the gate insulator, a substantial gate current is required to charge the input gate capacitance during the turn-on and turn-off event for each operating cycle. The switching power losses can supersede the on-state power loss when the power MOSFET is operated at high frequencies in applications such as in switch-mode power supplies. For this reason, analysis of the gate capacitance for the power MOSFET structure is provided in this chapter. In addition, the basic switching transients during turn-on and turn-off are described using analytical formulations.

The results of two-dimensional numerical simulations of typical cell designs are provided in this chapter to elucidate the operating principles for the power MOSFET structures. The results of the simulations also enable validation of the analytical models that are developed for the analysis of the power MOSFET structures. Although the exact values for the doping profiles, the gate oxide thickness, and channel length used in these simulations may differ from those used by any particular manufacturer, the basic structures for the VD-MOSFET and U-MOSFET devices used in the industry are similar to those described in this chapter. Consequently, the models developed here are generally application to all power MOSFET structures.

6.1 Ideal Specific On-Resistance

As discussed above, the performance of power MOSFETs is restricted by the internal resistance. It is useful to ascertain the minimum value for the internal resistance for a power MOSFET structure that is capable of supporting a desired blocking voltage. In the idealized case, it is assumed that the structure can support the blocking voltage with no degradation due to the edge termination or local electric field enhancement within the cell structure. In addition, all the resistances within the device structure, apart from that of the drift region, are considered to be parasitic resistances that have been reduced to zero.

The electric field distribution was illustrated in Chap. 1 (see Fig. 1.16) for the case of an abrupt parallel-plane junction. Based upon this electric field distribution, an equation was derived for the resistance per unit area for the drift region:

$$R_{\text{on-sp,ideal}} = \frac{4\text{BV}_{\text{PP}}^2}{\varepsilon_{\text{S}}\mu E_{\text{C}}^3}, \tag{6.1}$$

where BV_{PP} is the parallel-plane breakdown voltage, ε_{S} is the dielectric constant for the semiconductor, μ is the mobility, and E_{C} is the critical electric field for breakdown. This resistance is referred to as the *ideal specific on-resistance*.

In the case of silicon power MOSFETs, the doping concentration in the drift region is sufficiently low, so that the mobility can be assumed to be constant. For the case of an N-type drift region that is utilized for n-channel power MOSFETs, the ideal specific on-resistance is given by

$$R_{\text{on-sp,ideal}}(\text{n-channel}) = 5.93\times 10^{-9}\,\text{BV}_{\text{PP}}^{2.5} \tag{6.2}$$

after taking into account the change in critical electric field with doping concentration. In a similar manner, for the case of a P-type drift region that is utilized for p-channel power MOSFETs, the ideal specific on-resistance is given by

$$R_{\text{on-sp,ideal}}(\text{p-channel}) = 1.63\times 10^{-8}\,\text{BV}_{\text{PP}}^{2.5} \tag{6.3}$$

after taking into account the difference in the mobility for electrons and holes.

The ideal specific on-resistance for silicon calculated by using (6.1) is shown in Fig. 6.1. In performing these calculations, the change in the critical electric field strength and the mobility with doping concentration associated with each breakdown voltage was taken into account. The ideal specific on-resistance for P-type silicon is approximately three times larger than that for N-type silicon due to the difference between the mobility for holes and electrons. For this reason, the n-channel power MOSFET is preferred over the p-channel power MOSFET for most applications. A preponderance of power MOSFETs that are commercially available are consequently n-channel structures. However, p-channel devices are required in power circuits that utilize complementary devices and for battery charger circuits.

The ideal specific on-resistance is considered to be the lowest resistance per unit area that can be achieved for a power MOSFET designed with a chosen blocking voltage. A comparison of the actual internal specific on-resistance of any particular cell architecture with the ideal specific on-resistance provides a measure of the quality of the cell design. Progress with power MOSFET cell architectures and improvements in the cell design rules have enabled reduction of the specific on-resistance closer to the ideal value with successive generations of devices.

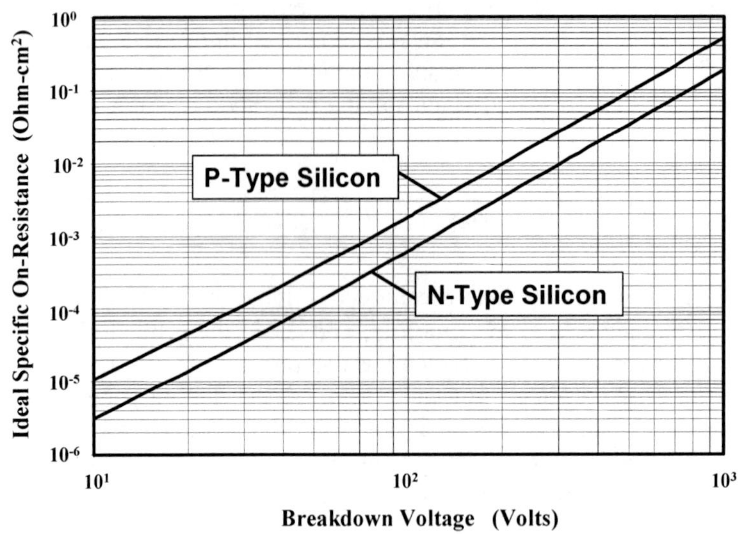

Fig. 6.1 Ideal specific on-resistance for power MOSFET structures

6.2 Device Cell Structure and Operation

The basic operation of the MOSFET entails the formation of a conductive channel at the surface of the semiconductor below an insulator by the application of a voltage to a gate electrode. The first silicon MOSFET structure fabricated by using a thermally grown gate oxide was reported in 1960.[2,3] These lateral structures were not designed to support high voltages or handle high-current levels. In the 1970s, it was recognized that a vertical architecture is required to handle high voltages and currents to produce a power device. In a lateral device structure, the drain and source electrodes must be interdigitated. This can only be achieved by using thin metal electrodes which have poor current-handling capability. In a vertical device structure, the two high current-carrying electrodes can be located on opposite sides of the wafer. The vertical structure enables the use of thick source and drain electrodes avoiding transport of the current through thin metal fingers. In addition, the potential distribution within the vertical structure is more favorable for supporting high voltages.

In all the vertical power MOSFET structures, the N^+ source and drain regions are separated by a P-base region. The P–N junction formed between the P-base region and the drain is used to support high voltages by utilizing a lightly doped drift region. To suppress the N–P–N parasitic transistor formed within the power MOSFET structure, the P–N junction formed between the N^+ source region and the P-base region is short circuited by overlapping the source metal across this junction. This also allows the source electrode to act as the substrate contact for the MOS structure formed between the gate and the P-base region.

In the power MOSFET structure, there is no injection of minority carriers during on-state current flow. Consequently, the device can be rapidly switched from the on-state to the off-state by removal of the gate bias to extinguish the channel. The speed with which the channel can be removed under gate control is dictated by the input capacitance. A more rapid switching can be achieved by increasing the gate current during the switching intervals.

Historically, the first power MOSFET structures were fabricated by using a V-groove process. Due to instabilities in these structures, a planar structure based upon the double-diffusion process was subsequently developed that rapidly supplanted the V-MOSFET structure. Most recently, a U-MOSFET structure was developed by using a trench-gate structure. The U-MOSFET structure has enabled significant reduction of the resistance of power MOSFETs allowing it to gain in popularity for commercial applications.

6.2.1 The V-MOSFET Structure

The first high-voltage power MOSFET structure was developed by using a V-groove etching process[4] during the 1970s. A cross section of this V-MOSFET structure is illustrated in Fig. 6.2. The N^+ source and drain regions in the vertical power MOSFET structure are separated by a P-base region, resulting in the formation of two P–N junctions labeled J_1 and J_2 in the figure. A V-groove is formed at the upper surface that penetrates through both the junctions. The gate electrode is placed inside the V-groove after creating a gate oxide on its surface, preferably by thermal oxidation of the silicon. Without the application of a gate

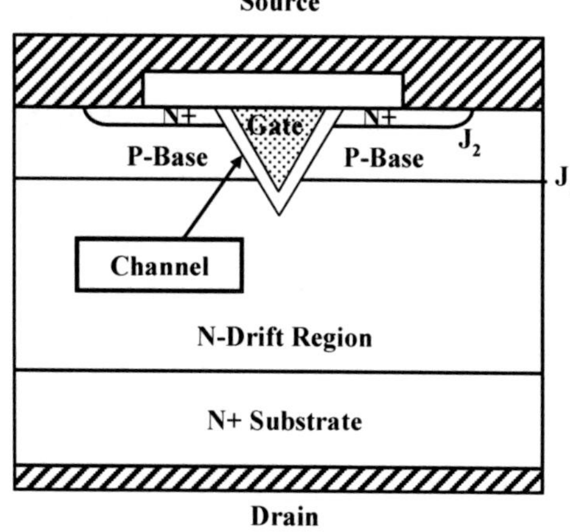

Fig. 6.2 V-groove MOSFET structure

bias, junction J_1 becomes reverse biased when a positive bias is applied to the drain electrode. A high voltage can be supported under these conditions by appropriate choice of the doping concentration and thickness of the N-drift region. The second junction J_2 is short circuited by overlapping the source electrode over the junction as illustrated in the figure to suppress the parasitic bipolar transistor. With proper design considerations, the breakdown voltage approaches that for a P–N diode.

Current flow between the drain and source electrodes of the V-MOSFET structure can be induced by the formation of a channel at the surface of the P-base region below the gate oxide. A positive bias applied to the gate electrode attracts electrons to the semiconductor surface under the gate oxide. These electrons provide a path for current flow between the source and the drain. The maximum current-carrying capability is determined by the internal resistance within the structure. A smaller resistance can be achieved by using smaller dimensions in the cell structure to increase the channel density.

The V-MOSFET structure fell out of favor because of manufacturing difficulties. The V-groove was formed by using a potassium hydroxide-based etch for silicon, which exhibits different etch rates for various silicon surface orientations. It was found that the potassium from the etch solutions contaminates the gate oxide, producing instabilities during long-term operation of the V-MOSFET structure. In addition, the sharp corner at the bottom of the V-groove was found to degrade the breakdown voltage.

6.2.2 The VD-MOSFET Structure

A cross section of the basic cell structure for the vertical-diffused (VD) MOSFET is illustrated in Fig. 6.3. This device structure is fabricated by starting with an N-type epitaxial layer grown on a heavily doped N^+ substrate. The channel is formed

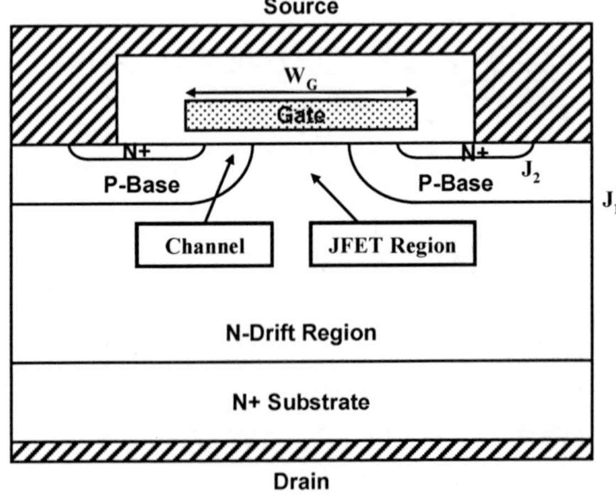

Fig. 6.3 VD-MOSFET structure

by the difference in lateral extension of the P-base and N^+ source regions produced by their diffusion cycles. Both regions are self-aligned to the left-hand side and right-hand side of the gate region during ion implantation to introduce the respective dopants. A refractory gate electrode, such as polysilicon, is required to allow diffusion of the dopants under the gate electrode at elevated temperatures.

Without the application of a gate bias, a high voltage can be supported in the VD-MOSFET structure when a positive bias is applied to the drain. In this case, junction J_1 formed between the P-base region and the N-drift region becomes reverse biased. The voltage is supported mainly within the thick lightly doped N-drift region. Drain current flow in the VD-MOSFET structure is induced by the application of a positive bias to the gate electrode. This produces an inversion layer at the surface of the P-base region under the gate electrode. This inversion layer channel provides a path for transport of electrons from the source to the drain when a positive drain voltage is applied.

After transport from the source region through the channel, the electrons enter the N-drift region at the upper surface of the device structure. They are then transported through a relatively narrow JFET region located between the adjacent P-base regions within the VD-MOSFET structure. The constriction of the current flow through the JFET region substantially increases the internal resistance of the VD-MOSFET structure. A careful optimization of the gate width (W_G) is required, as discussed later in this chapter, to minimize the internal resistance for this structure. In addition, it is customary to enhance the doping concentration in the JFET region to reduce the resistance to current flow through this portion of the device structure.

After being transported through the JFET region, the electrons enter the N-drift region below junction J_1. The current spreads from the relatively narrow JFET region to the entire width of the cell cross section. This nonuniform current distribution within the drift region enhances its resistance, making the internal resistance of the VD-MOSFET structure larger than the ideal specific on-resistance of the drift region. The large internal resistance for the VD-MOSFET structure provided motivation for the development of the trench-gate power MOSFET structure in the 1990s.

6.2.3 The U-MOSFET Structure

During the late 1980s, the technology for etching trenches in silicon became available due to its application for making charge storage capacitors within DRAM chips. This process was adapted by the power semiconductor industry to develop the trench-gate or U-MOSFET structure. As shown in Fig. 6.4, the trench extends from the upper surface of the structure through the N^+ source and P-base regions into the N-drift region. The gate electrode is placed within the trench after the formation of the gate oxide by thermal oxidation of the bottom and sidewalls.

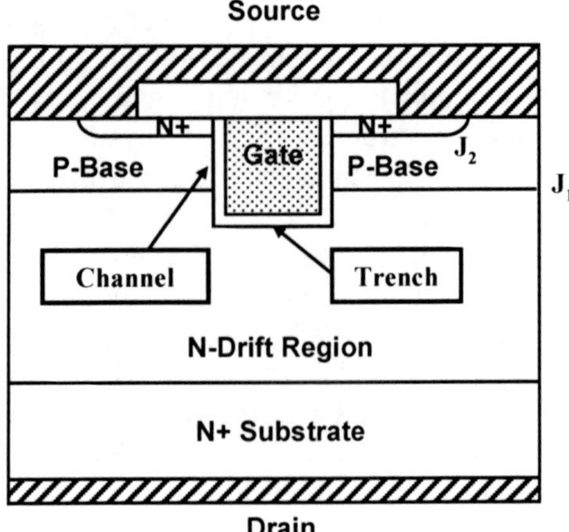

Fig. 6.4 U-MOSFET structure

Without the application of a gate bias, a high voltage can be supported in the U-MOSFET structure when a positive bias is applied to the drain. In this case, junction J_1 formed between the P-base region and the N-drift region becomes reverse biased. The voltage is supported mainly within the thick lightly doped N-drift region. Since the gate is at zero potential during the blocking mode of operation, a high electric field is also developed across the gate oxide. To avoid reliability problems arising from the enhanced electric field in the gate oxide at the trench corners, it is customary to round the bottom of the trench.

Drain current flow in the U-MOSFET structure is induced by the application of a positive bias to the gate electrode. This produces an inversion layer channel at the surface of the P-base region along the vertical sidewalls of the trench. This inversion layer channel provides a path for transport of electrons from the source to the drain when a positive drain voltage is applied. After transport from the source region through the channel, the electrons enter the N-drift region at the bottom of the trenches. The current then spreads to the entire width of the cell cross section. Consequently, there is no JFET region in the U-MOSFET structure, enabling a significant reduction of the internal resistance when compared with the VD-MOSFET structure. The reduced internal resistance for the U-MOSFET structure provided motivation for the development of these devices in the 1990s.

6.3 Basic Device Characteristics

The power MOSFET structure contains two back-to-back junctions created between the N^+ source, P-base, and N-type drain regions. In principle, it could support voltage in both the first and third quadrant of operation. However, the junction J_2 between the N^+ source and P-base regions is invariably short circuited to suppress the parasitic N–P–N transistor. Consequently, the power MOSFET can only support a high voltage in the first quadrant of operation. In the third quadrant, it behaves like a forward-biased diode.

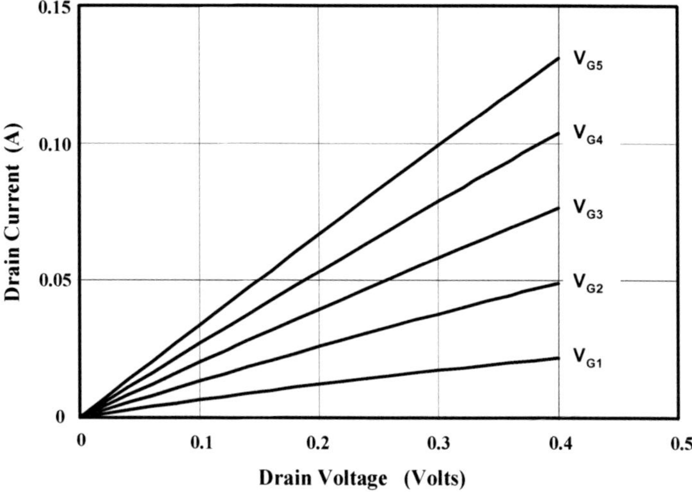

Fig. 6.5 Power MOSFET characteristics in the first quadrant at low drain bias voltages

In the first quadrant of operation, the power MOSFET can carry high-current levels if a gate bias is applied to create a channel to connect the N^+ source and N-type drain regions. At low drain bias voltages, the *i–v* characteristics for the power MOSFET resemble those for a resistor whose value can be modulated by the gate bias. This behavior is illustrated in Fig. 6.5 for various gate bias voltages. These characteristics are observed at low gate bias voltages when the channel resistance for the power MOSFET is much larger than the resistance of the drift region. At high gate bias voltages, the drift region resistance can become dominant. Under these conditions, the resistance of the power MOSFET will no longer reduce with increasing gate bias.

When the drain bias voltage approaches and exceeds the gate bias voltage, the current in the power MOSFET saturates as illustrated in Fig. 6.6. The gate voltage-controlled current saturation in the power MOSFET is a useful behavior when switching inductive loads. It allows limiting the current flow by the device during switching transients. Since the device simultaneously supports a high voltage, considerable power dissipation occurs during this mode of operation. The power MOSFET can sustain this power dissipation as long as the junction temperature rise due to the heating is kept below 200°C.

A useful parameter that is commonly used for describing the operation of the power MOSFET structure is its transconductance. The *transconductance* is defined as the rate of change in the drain current with incremental gate voltage:

$$g_m = \frac{\Delta I_D}{\Delta V_G} = \frac{I_{D4} - I_{D3}}{V_{G4} - V_{G3}}, \tag{6.4}$$

where the currents and voltages are indicated in the figure. A large transconductance is desirable to obtain a high drain current with a small gate bias voltage. In addition, the switching speed of the power MOSFET improves with increasing transconductance.

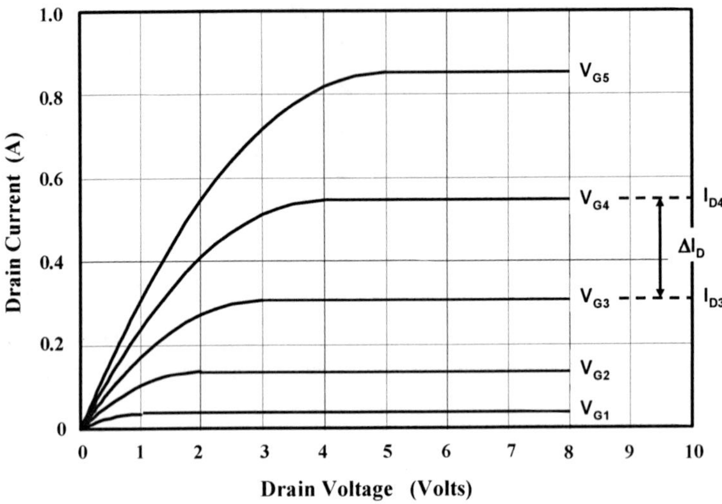

Fig. 6.6 Power MOSFET characteristics in the first quadrant at high drain bias voltages

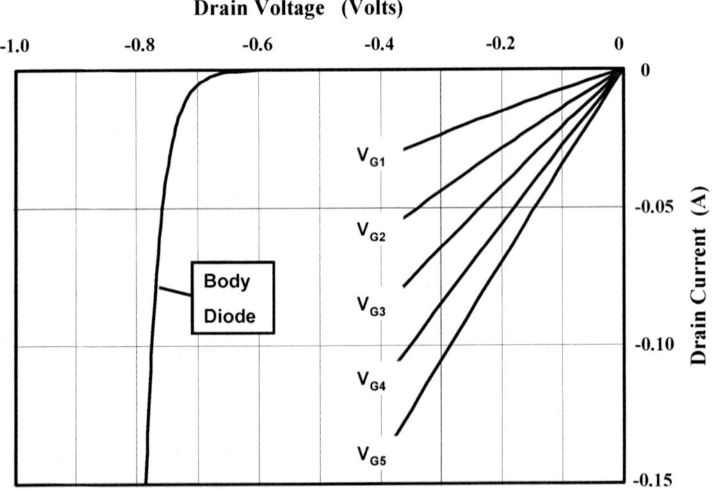

Fig. 6.7 Power MOSFET characteristics in the third quadrant at low drain bias voltages

When a negative bias is applied to the drain terminal of the power MOSFET structure, the junction J_1 between the P-base region and the N-drift region becomes forward biased. Current flow between the drain and source electrodes can now occur because the source electrode is also connected to the P-base region in the power MOSFET structures to suppress the parasitic N–P–N transistor. This is referred to as the current flow through the *body diode* of the power MOSFET. The body diode current flows when the drain bias voltage exceeds approximately 0.7 V in magnitude in the negative direction as shown in Fig. 6.7. It is possible to induce larger drain current flow at low negative drain bias voltages by the application of a gate bias. This current flow occurs via the channel in the power MOSFET structure whose resistance determines the on-state voltage drop in the third quadrant. This is referred to as the *synchronous rectification* mode because the MOSFET is behaving like a rectifier by selectively applying the gate bias only when the drain voltage is negative during circuit operation. The utilization of the power MOSFET as a synchronous rectifier has become very popular for increasing the efficiency of switch-mode power supplies.

6.4 Blocking Voltage

The power MOSFET structure is capable of supporting a high voltage in the first quadrant of operation when the drain bias voltage is positive. During operation in the blocking mode, the gate electrode is shorted to the source electrode by the external gate bias circuit. The application of a positive drain bias voltage produces a reverse bias across junction J_1 between the P-base region and the N-drift region. Most of the applied voltage is supported across the N-drift region. The doping concentration of donors in the N-epitaxial drift region and its thickness is chosen to attain the desired breakdown voltage. In devices designed to support low voltages (<50 V), the doping concentration of the P-base region is comparable with the doping concentration of the N-drift region leading to a graded junction profile. Consequently, a fraction of the applied drain voltage is supported across a depletion region formed in the P-base region. It is preferable to make the depth of the P-base region smaller to reduce the channel length in the power MOSFET structure. However, the design of the P-base region must take into account reduction of the breakdown voltage due to the depletion region in the P-base region reaching through to the N^+ source region.

6.4.1 Impact of Edge Termination

In practical devices, the maximum blocking voltage (BV) of the power MOSFET is invariably decided by the edge termination that surrounds the device cell structure. The most commonly used edge termination for VD-MOSFETs is based upon floating field rings and field plates. The enhanced electric field at the edges limits the breakdown voltage to about 80% of the parallel-plane breakdown voltage (BV_{PP}):

$$BV_{PP} = \left(\frac{BV}{0.8}\right). \quad (6.5)$$

Consequently, the doping and thickness of the N-drift region must be chosen to achieve a parallel-plane breakdown voltage, which is 25% larger than the blocking voltage for the device:

$$N_D = \left(\frac{5.34 \times 10^{13}}{BV_{PP}}\right)^{4/3}. \quad (6.6)$$

A common design error that can occur is to make the thickness of the drift region below the P–N junction equal to the depletion width for the ideal parallel-plane junction with the above doping concentration. In actuality, the maximum depletion width is limited to that associated with the blocking voltage of the structure. The thickness of the drift region required below the P–N junction is therefore less than the depletion width for the ideal parallel-plane junction with the above doping concentration and is given by

$$t = W_D(BV) = \sqrt{\frac{2\varepsilon_S BV}{qN_D}}. \quad (6.7)$$

The resistance of the drift region can be reduced by using this thickness rather than the maximum depletion width corresponding to the doping concentration given by (6.6).

6.4.2 Impact of Graded Doping Profile

For power MOSFET structures with low (<50 V) breakdown voltages, the doping concentration of the drift region is comparable with the doping concentration of the P-base region. As discussed later in this chapter, the maximum doping concentration (N_{PS}) of the P-base region at the silicon surface must be chosen to obtain a threshold voltage for the power MOSFET in the range of 1–2 V. This produces a graded doping profile for the junction J_1 between the P-base region and the N-drift region as illustrated in Fig. 6.8. In this figure, the concentrations of the donors and acceptors are shown by the solid lines while the dashed lines represent the net doping concentration after taking into account compensation near the junctions.

The electric field developed across junction J_1 is also illustrated in Fig. 6.8. Due to the graded doping profile, the electric field extends on both sides of junction J_1. The electric field in the P-base region supports a portion of the applied positive drain voltage. This implies that the same breakdown voltage can be achieved with a larger doping concentration and a smaller thickness for the N-drift region. This improvement can be translated to increasing the breakdown voltage at the edge termination if the P-base region is used at the edges of the power MOSFET structure. A reduction of the resistance for the power MOSFET structure can be achieved by taking into account the voltage supported within the P-base region. An improvement in the specific on-resistance of 20% can be achieved by taking into account the graded doping profile.

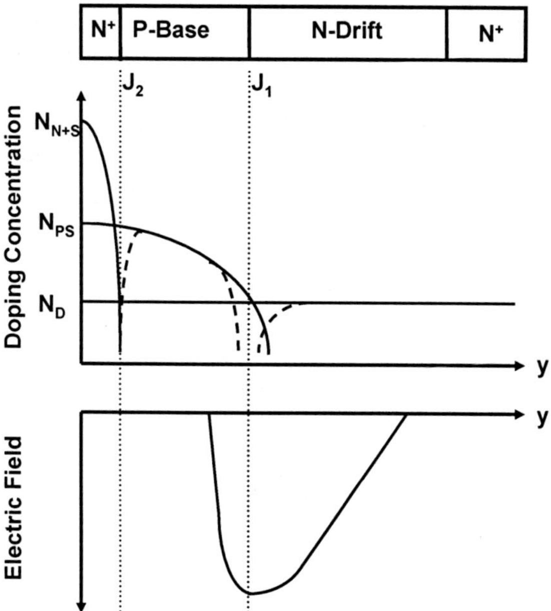

Fig. 6.8 Doping profile and electric field distribution for the power MOSFET structure

6.4.3 Impact of Parasitic Bipolar Transistor

The power MOSFET structure contains a parasitic bipolar transistor formed between the N^+ source region, the P-base region, and the N-drift region. Since the P-base region has a narrow width to obtain a shorter channel length for the power MOSFET structure, the inherent current gain (β) of the N–P–N transistor is large. Consequently, the open-base breakdown voltage (BV_{CEO}) of the N–P–N transistor is much lower than the base-collector breakdown voltage (BV_{CBO}):

$$BV_{CEO} = \frac{BV_{CBO}}{\beta^{1/6}}. \tag{6.8}$$

A reduction of the breakdown voltage can be prevented by short circuiting the emitter and base of the N–P–N transistor to suppress the current gain. For this reason, the source electrode is designed to overlap junction J_2 formed between the N^+ source and P-base regions in all the power MOSFET structures. This short circuits the emitter and base of the parasitic N–P–N transistor and simultaneously provides a contact to the substrate of the MOS structure for creating the inversion channel.

A reduction of the breakdown voltage can occur in the power MOSFET structure in spite of short circuiting the N^+ source and P-base regions when the thickness of the P-base region is reduced. As illustrated in Fig. 6.8, the electric field extends into the P-base region during the blocking mode. When the width of the P-base region is reduced, the depletion region formed within the P-base region can exceed its thickness prior to the on-set of breakdown due to impact ionization. This phenomenon is called *reach-through breakdown*. To avoid reach-through limited breakdown, the thickness of the P-base region must exceed the depletion region width in the P-base region under avalanche breakdown conditions.

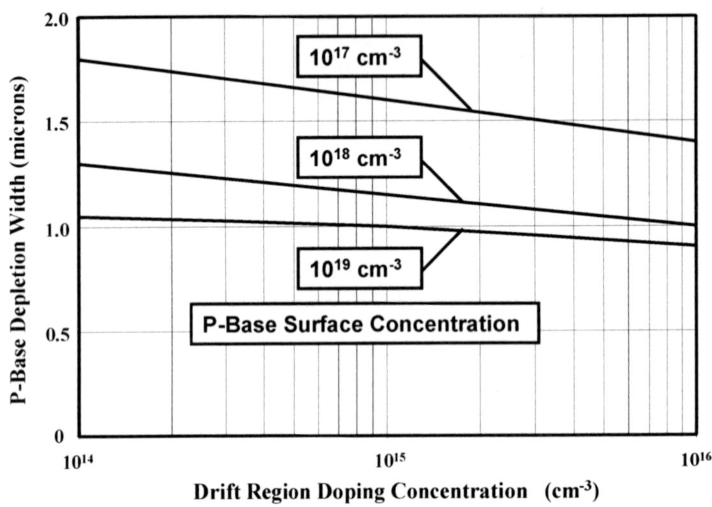

Fig. 6.9 Depletion layer width within the P-base region of the power MOSFET structure

The depletion layer thickness in the P-base region at the on-set of avalanche breakdown can be extracted by numerical simulations of a P–N junction with the appropriate doping profile. In a typical power VD-MOSFET structure, the depth of the P-base region is about 2.5 μm. The depletion region width in the P-base region was obtained for this junction depth for various surface doping concentrations for the P-base region. The doping concentration for the N-drift region was varied between 1×10^{14} and 1×10^{16} cm^{-3} to represent power MOSFET structures with a wide range of breakdown voltages. The results are provided in Fig. 6.9. They indicate that the base width for the power MOSFET structure must exceed 1 μm to avoid reach-through limited breakdown. This in turn limits the channel length of the power MOSFET structure to greater than 1 μm. It is worth pointing out that the depletion width in the P-base region increases when the N-drift region doping concentration is reduced. This implies that the minimum channel length is larger for power MOSFETs designed to support larger blocking voltages. This is acceptable because the channel resistance is only a small fraction of the drift region resistance for high-voltage power MOSFETs as shown later in this chapter.

6.4.4 Impact of Cell Pitch

Optimization of the power VD-MOSFET structure requires adjustment of the pitch to obtain the minimum possible specific on-resistance as discussed later in this chapter. The variation of the pitch is performed while maintaining a minimum size for the polysilicon window in the power MOSFET structure. The smallest polysilicon window is determined by photolithographic and alignment design rules. As the cell pitch is increased, the width of the polysilicon (W_G in Fig. 6.3) also increases. This has an impact on the breakdown voltage of the cell structure.

The construction of the VD-MOSFET structure by using the double-diffusion process creates planar junctions in each of the device cells. For a linear cell topology, a cylindrical junction is produced within each cell. As discussed in Chap. 3, cylindrical junctions have reduced breakdown voltages due to electric field crowding at the junction curvature. Fortunately, the junction curvature effect is mitigated by the presence of the gate electrode, which acts as a field plate because the gate is shorted by the external circuit to the source during operation of the power MOSFET in the blocking mode. The MOS structure between the P-base regions operates in the deep depletion mode because the minority carriers are removed via the reverse-biased P–N junctions.

In spite of the field plate action, the breakdown voltage of the VD-MOSFET cell structure is a function of the width (W_G) of the gate electrode.[5] A small gate width brings the junctions closer together reducing the influence of the junction curvature, which results in an improvement in the breakdown voltage. However, a small gate width should be used with caution because a drastic increase in the on-resistance can occur. As the gate width is enlarged, the breakdown voltage does not continue to degrade because of the field plate action provided by the gate structure.

Simulation Example

To illustrate the influence of the planar junction curvature on the breakdown within the VD-MOSFET structure, it is instructive to use the results of two-dimensional numerical simulations of a particular device structure. Consider the case of a structure designed with blocking voltage capability of 35 V corresponding to a power MOSFET with a rating of 30 V. For an edge termination that limits the breakdown voltage to 80% of the parallel-plane value, the parallel-plane breakdown voltage is found to be 43.75 V. The doping concentration in the N-drift region to achieve this parallel-plane breakdown voltage is 1.3×10^{16} cm^{-3} according to (6.6). To prevent reach-through problems, the VD-MOSFET structure was designed with a channel length of 1.6 μm by using a P-base junction depth of 2.4 μm and an N^+ source depth of 0.8 μm along the surface under the gate.

The breakdown voltages for the VD-MOSFET structure obtained by using two-dimensional numerical simulations are shown in Fig. 6.10. For these simulations, the cell pitch was varied by changing the width of the gate electrode while maintaining a polysilicon window of 8 μm. It can be observed that the breakdown voltage increases when the cell pitch is reduced. At large values for the cell pitch, the breakdown voltage reaches a plateau at 43 V. As expected, the cell breakdown voltage exceeds the breakdown voltage at the edge termination. Note that the breakdown voltages obtained in the numerical simulations exceed the parallel-plane breakdown voltage calculated by using (6.6) because of the additional voltage supported within the P-base region.

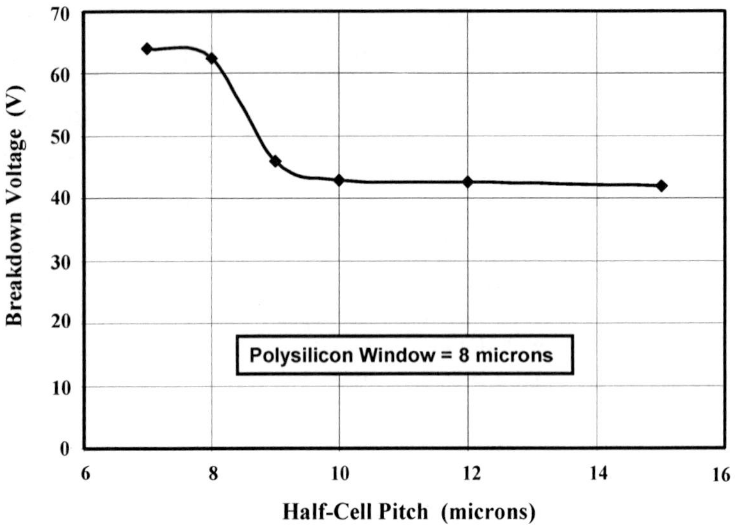

Fig. 6.10 Breakdown voltages for the power VD-MOSFET structure

The reason for the improvement in the breakdown voltage with smaller cell pitch can be elucidated by examination of the potential distribution in the VD-MOSFET structure. The potential distribution obtained from the numerical simulations for the case of a half-cell pitch of 15 μm is shown in Fig. 6.11. The potential contours are displayed at 5-V intervals. It can be observed that a depletion region forms below the P–N junction as well as under the gate electrode. This demonstrates that the MOS structure operates under deep depletion conditions. Although the gate structure acts like a field plate for the P–N junction, a crowding of the potential contours is observed at location "A" in the figure. This reduces the breakdown voltage to below that for the parallel-plane junction. When the cell pitch is reduced, the crowding of the potential contours is reduced as demonstrated in Fig. 6.12 for the case of a half-cell pitch of 8 μm. The P–N junctions of the adjacent cells in the VD-MOSFET structure are now sufficiently close to reduce the electric field at the junction. This allows the structure with the smaller cell pitch to support a larger voltage as shown in Fig. 6.10.

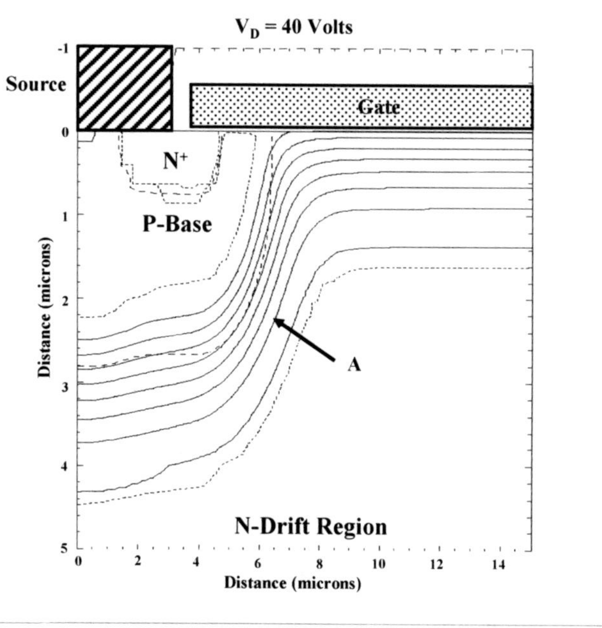

Fig. 6.11 Potential contours for the power VD-MOSFET structure with half-cell pitch of 15 μm

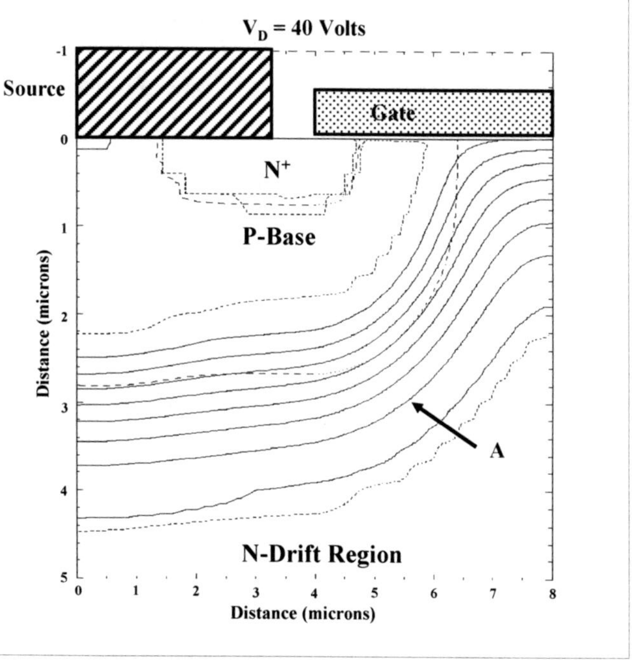

Fig. 6.12 Potential contours for the power VD-MOSFET structure with half-cell pitch of 8 μm

6.4.5 Impact of Gate Shape

The VD-MOSFET cell structure contains a planar-gate topology where the gate electrode is located on the flat upper surface of the semiconductor. Although electric field enhancement occurs at the planar junctions in this structure as discussed in Sect. 6.4.4, there is no electric field enhancement at the gate electrode because the edges of the gate electrodes overlap the highly doped N^+ source regions. The edges of the gate electrode are screened from the applied drain potential by the presence of the N^+ source regions. In addition, the presence of the P–N junctions below the gate region can screen it from the drain bias when the gate width is reduced. In these designs, the P–N junctions act like the gate regions of a JFET structure producing a potential barrier below the gate structure.

In the case of the V-groove power MOSFET structure, a sharp corner is formed at the bottom of the groove under the gate electrode. In order for the channel to extend from the N^+ source to the N-drift region, the tip of the groove must extend through junction J_1 into the drift region. During operation in the blocking mode, the highest electric field occurs at junction J_1. Consequently, the sharp tip of the V-groove is located in the vicinity of the highest electric fields within the structure. The presence of a sharp point within the depletion region enhances the electric field leading to a reduction in the breakdown voltage. Since this high electric field is located at the edge of the channel, electrons that gain energy from the electric field are launched into the gate oxide in the vicinity of the channel. The electrons launched into the oxide are trapped within the oxide leading to a shift in the threshold voltage of the MOSFET. The V-groove power MOSFET structure is therefore prone to *hot electron instability* problems. For these reasons, the V-groove architecture for the power MOSFET is no longer used to manufacture commercial devices.

A high electric field can also be generated at the corners of the trenches in the U-MOSFET structure. In this structure, the gate must extend from the upper surface through junction J_1 to form a channel extending from the N^+ source region to the N-drift region. As in the case of the V-groove structure, the bottom corners of trench are located in close proximity to the voltage supporting junction J_1. The enhancement of the electric field at the trench corners leads to a reduction of the breakdown voltage and hot electron instability problems. These problems can be overcome by rounding the bottom of the trench and adding a deep P^+ region to shield the corners of the trench.[6] Although effective for improving the breakdown voltage, the presence of the deep P^+ regions introduces a JFET region within the U-MOSFET structure which can degrade its on-resistance.

Simulation Example

To illustrate the influence of the trench corner on the electric field distribution within the U-MOSFET structure, it is instructive to use the results of two-dimensional numerical simulations of a particular device structure. Consider the case of a structure designed with blocking voltage capability of 35 V corresponding to a power MOSFET with a rating of 30 V. For an edge termination that limits the breakdown voltage to 80% of the parallel-plane value, the parallel-plane breakdown voltage is found to be 43.75 V. The doping concentration in the N-drift region to achieve this parallel-plane breakdown voltage is 1.3×10^{16} cm^{-3} according to (6.6). To prevent reach-through problems, the U-MOSFET structure was designed with a channel length of 1.0 μm by using a P-base junction depth of 1.5 μm and an N$^+$ source depth of 0.5 μm. The doping profiles for this structure will be discussed in more detail later in this chapter. The U-MOSFET structure was designed with a trench width of 1 μm, a mesa width of 4 μm, and a trench depth of 2 μm.

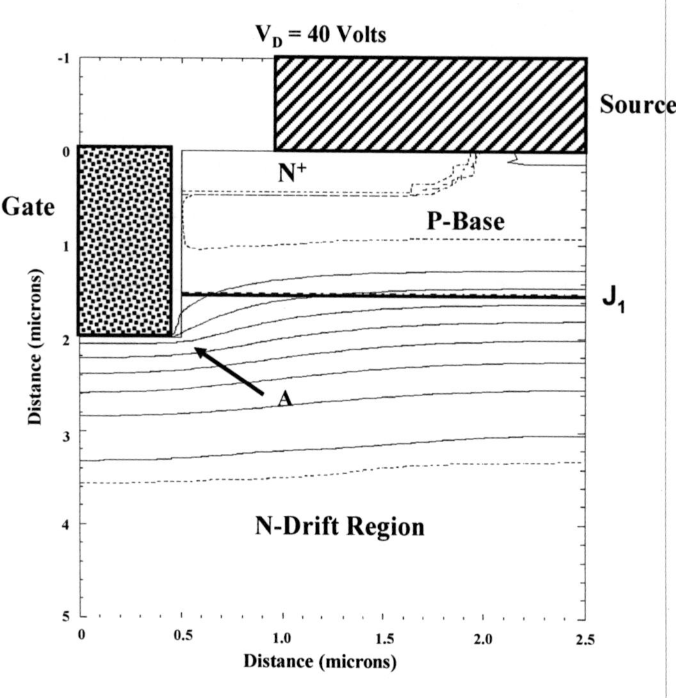

Fig. 6.13 Potential contours within the power U-MOSFET structure

The breakdown voltage for the U-MOSFET structure obtained by using two-dimensional numerical simulations was found to be 44.7 V which is consistent with the design goals for a 30-V rating. The potential contours at a drain bias of 40 V are displayed in Fig. 6.13 when the gate bias is zero. It can be observed that the potential lines are more closely spaced around the trench corner (location A in the figure) indicating an enhanced electric field.

A quantitative assessment of the increase in the electric field at the trench corner can be obtained from Fig. 6.14. Here, the electric field profile taken along the vertical surface of the trench (solid line) at a drain bias of 40 V is compared with the electric field profile taken at the center of the mesa (dashed line). The electric field at the center of the mesa exhibits the behavior expected for a P–N junction with a graded doping profile in the proximity of the junction. The maximum electric field in this location is 2.8×10^5 V cm^{-1}. In contrast, the electric field at the trench corner (at a depth of 2 μm from the surface) is nearly 7×10^5 V cm^{-1}. The enhanced electric field near the trench corners can lead to the generation of hot electrons that can be injected into the gate oxide. These electrons become trapped in the gate oxide producing a shift in the threshold voltage. It is customary to round the bottom of the trenches to avoid this problem.

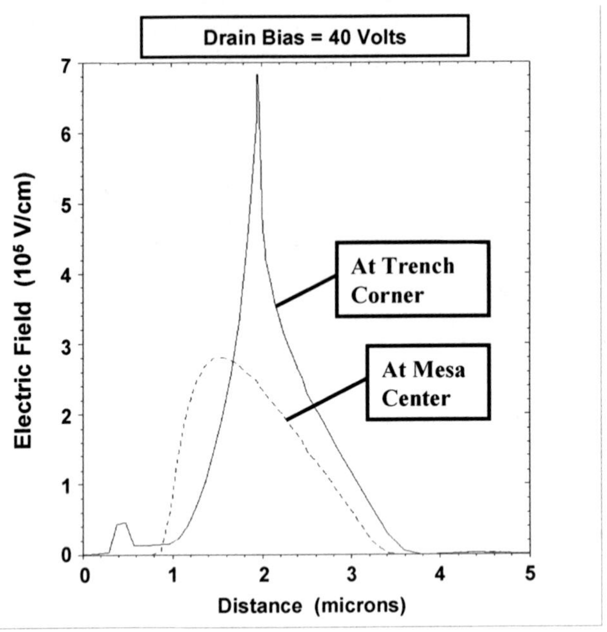

Fig. 6.14 Electric field profiles in the power U-MOSFET structure

6.4.6 Impact of Cell Surface Topology

As indicated earlier in this section, the breakdown voltage for the power VD-MOSFET structure is determined by the edge termination design. For an edge termination utilizing planar junctions with floating field rings and field plates, a breakdown voltage of 80% of the ideal parallel-plane value is achievable. With the use of the junction termination extension (JTE) technique, it is possible to raise the breakdown voltage at the edge termination close to that for the ideal parallel-plane junction. With this improvement at the edges of the device, the breakdown shifts to the power MOSFET cell structure. Consequently, any electric field crowding inside the power MOSFET cell structure can lead to degradation in the breakdown voltage. The breakdown voltage of the power MOSFET structure then becomes dependent on the surface topology of the cell layout.

Many topologies for the cell design at the top surface of power MOSFET structures were proposed and examined from the point of view of reducing the on-resistance[7] during the early stages of development of these devices. The impact of the cell topology on the breakdown voltage and on-resistance was subsequently analyzed after improvements to the edge termination were achieved.[8] Various surface topologies for the power VD-MOSFET structure are illustrated in Fig. 6.15. In the illustrations, the polysilicon is shown by the shaded area.

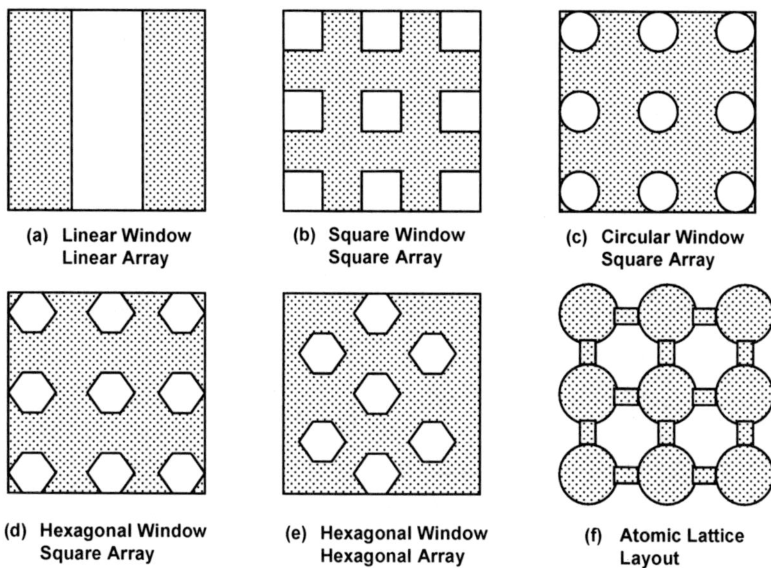

Fig. 6.15 Cell design surface topologies for power VD-MOSFET structures

In the case of the linear cell topology shown in Fig. 6.15a, a cylindrical junction is formed at the edges of the polysilicon window as previously shown in the cell cross section in Fig. 6.3. The breakdown voltage of the power MOSFET cell is then expected to be determined by the enhanced electric field at cylindrical junctions as discussed in Chap. 3. However, the linear cell topology must be terminated at the ends of the fingers to interconnect the polysilicon regions and obtain a common gate contact. It is customary to perform this termination of the fingers with a rectangular shape due to the ease of the layout during design. A rectangular polysilicon window creates sharp corners at the ends of the finger similar to those shown in Fig. 6.15b for the square-shaped polysilicon windows. Spherical junctions are formed at these sharp corners with much lower breakdown voltage than the cylindrical junctions as discussed in Chap. 3.

The formation of spherical junctions can be avoided by creating round polysilicon windows as illustrated in Fig. 6.15c for the case of a square array. If the radius of the polysilicon window is large compared with the junction depth of the P-base region, the breakdown voltage can approach that for the linear cell topology. The circular polysilicon windows can also be placed in a hexagonal array. However, the layout of devices is based upon utilization of rectangular pixels, which do not allow formation of perfectly circular shapes. It is therefore preferable to utilize hexagonal windows. This type of surface topology is illustrated in Fig. 6.15d for a square array and in Fig. 6.15e for a hexagonal array.

An improvement in breakdown voltage when compared with the cylindrical junction can be achieved with the atomic lattice layout (ALL) design, which is illustrated in Fig. 6.15f. In this design, the polysilicon is constructed in the form of circular regions which are interconnected with small bars of polysilicon. When the P-base region diffuses under the polysilicon, it creates a *saddle junction*. The saddle junction has the opposite curvature to that for a cylindrical junction, which ameliorates the electric field crowding and increases the breakdown voltage. This allows using a higher doping concentration in the drift region for the ALL design, producing a reduction of the specific on-resistance in power MOSFET structures. The largest improvement in the specific on-resistance obtained with the ALL design is in the case of power MOSFET structures designed to support high voltages because the drift region resistance is the dominant component.

6.5 Forward Conduction Characteristics

On-state current flow can be induced in the n-channel power MOSFET structure by the application of a positive gate bias to create a channel extending from the N^+ source region to the N-drift region. Current conduction occurs by the transport of electrons from the source to the drain via the channel and the drift region. At drain bias voltages that are small when compared with the gate bias voltage, the characteristics of the power MOSFET resemble those of a resistor whose magnitude can be controlled by the gate bias. This is referred to as the *linear region* of operation. When the drain bias voltages become comparable with the gate bias voltage, the resistance to current flow increases because the drain bias counteracts the gate bias voltage. Eventually, the drain current saturates at larger drain bias voltages. This is referred to as the *saturation region* of operation. The saturated drain current is also a function of the gate bias voltage.

The *i–v* characteristics for the power MOSFET structure are governed by the behavior of the channel region as a function of the gate and drain bias voltages. The channel properties are determined by the MOS stack that comprises the gate structure. The physics of operation of the MOS stack is therefore analyzed in the first portion of this section of the chapter. Using this analysis, analytical expression for the channel resistance can be derived as a function of the gate and drain bias. These solutions are similar to those provided in other textbooks that describe the operation of MOSFET structures.[9,10] In the case of the power MOSFET structure, it is necessary to include the impact of current flow through the drift region due to its relatively large resistance.

The low on-state resistance of the power MOSFET structure is its most important attribute for power electronic applications. In conjunction with the inherent fast switching capability, this property of power MOSFET structures enables maintaining low power losses in applications. A detailed analysis of the on-resistance is therefore provided in this section. The impact of the device cell topology must be taken into account during the computation of the on-resistance.

This includes the surface layout of the cell structure, and whether planar- or trench-gate architectures are utilized.

6.5.1 MOS Interface Physics

The conductivity of the channel in the power MOSFET structure depends upon the density of free carriers and their mobility. The charge density in the channel is governed by the gate bias. The transport mobility for these free carriers is reduced when compared with transport in the bulk of the semiconductor due to the additional surface scattering phenomena that were discussed in Chap. 2. To obtain the charge density in the semiconductor, a one-dimensional MOS structure will be first considered with an ideal dielectric layer. The ideal dielectric layer is assumed to prevent the transport of charge between the gate and the semiconductor and contains no charge within it. The treatment will be performed for a P-type semiconductor which represents the base region for n-channel power MOSFET structures.

Flat Band Conditions

The energy band diagram for the ideal MOS structure with a P-type semiconductor region is shown in Fig. 6.16 for the case of no bias applied to the gate electrode. An ideal MOS structure is defined as one that has (a) an insulator with infinite resistivity, (b) charge only in the metal and the semiconductor, and (c) work function of the metal equal to that for the semiconductor. Under the last assumption, the Fermi levels for the metal (E_{FM}) and the semiconductor (E_{FS}) have the same energy. Consequently, there is no transfer of charge between the metal

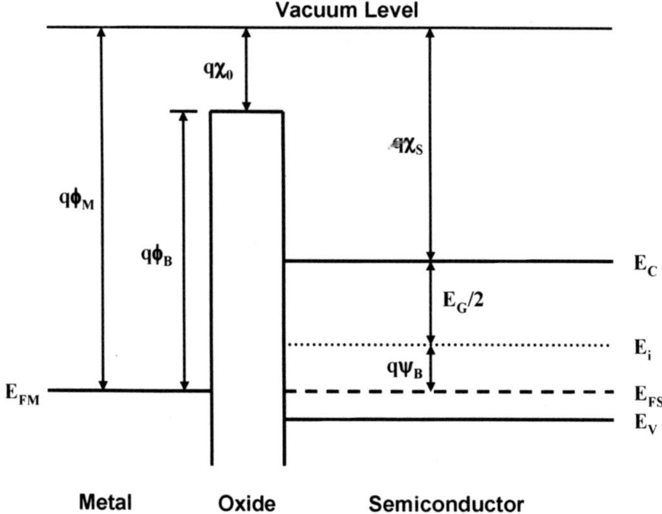

Fig. 6.16 MOS structure operating under flat band conditions

and the semiconductor in the absence of a gate bias, leading to the *flat band* conditions illustrated in the figure.

From the figure, it can be observed that

$$q\phi_{\mathrm{M}} = q\chi_{\mathrm{S}} + \frac{E_{\mathrm{G}}}{2} + q\psi_{\mathrm{B}} = q\phi_{\mathrm{B}} + q\chi_{\mathrm{O}}, \tag{6.9}$$

where ϕ_M is the work function for the metal, χ_S is the electron affinity for the semiconductor, E_G is the energy band gap for the semiconductor, ψ_B is the potential difference between the intrinsic and Fermi levels in the bulk of the semiconductor, ϕ_B is the barrier height between the metal and the oxide, and χ_O is the electron affinity for the oxide. The position of the Fermi level in the semiconductor can be obtained by using

$$\psi_{\mathrm{B}} = \left(\frac{E_{\mathrm{i}} - E_{\mathrm{FS}}}{q}\right) = \frac{kT}{q}\ln\left(\frac{p_0}{n_{\mathrm{i}}}\right), \tag{6.10}$$

where E_i is the position of the intrinsic level, n_i is the intrinsic concentration, and p_0 is the hole concentration which can be assumed to be equal to the doping concentration (N_A) for silicon. For a typical doping concentration of 1×10^{17} cm^{-3} in the P-base region for silicon power MOSFET structures, the Fermi level is located at 0.41 eV below the intrinsic level in the semiconductor at room temperature.

Accumulation Conditions

When a negative bias is applied to the metal electrode in the MOS structure, the negative charge developed in the metal attracts the positively charged holes in the semiconductor toward the interface between the semiconductor and the oxide. The presence of excess majority carriers at the surface of the semiconductor is referred to as *accumulation*. The enhanced concentration of holes at the surface of the semiconductor is accompanied by a reduction of energy difference between the Fermi level and the valence band as illustrated in Fig. 6.17. A small band bending at the surface is sufficient to induce a large density of holes, indicated by the circular symbol with positive charge, at the interface between the oxide and the semiconductor. Consequently, the excess charge in the semiconductor is located close to the interface with the oxide, leading to all the applied bias on the metal being supported across the oxide. The charge in the semiconductor then increases in proportion to the applied bias to the metal electrode. This situation is applicable to the MOS regions within power MOSFET structures where the gate overlaps the drift region.

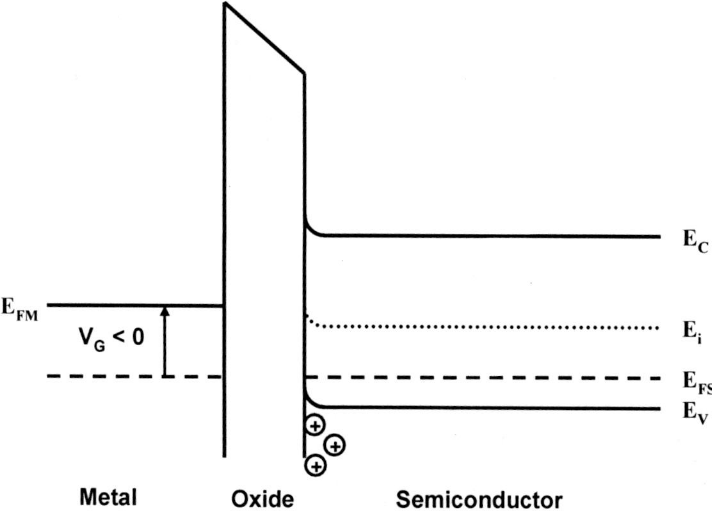

Fig. 6.17 MOS structure operating under accumulation conditions

Depletion Conditions

When a positive bias is applied to the metal electrode in the MOS structure, the positive charge developed in the metal repels the positively charged holes in the semiconductor away from the interface between the semiconductor and the oxide. At small positive bias voltages, the semiconductor at the oxide interface becomes depleted leading to the energy band bending illustrated in Fig. 6.18. The ionized acceptor atoms in the depletion region are indicated by the square symbols with negative charge. In this case, the Fermi level in the semiconductor is located

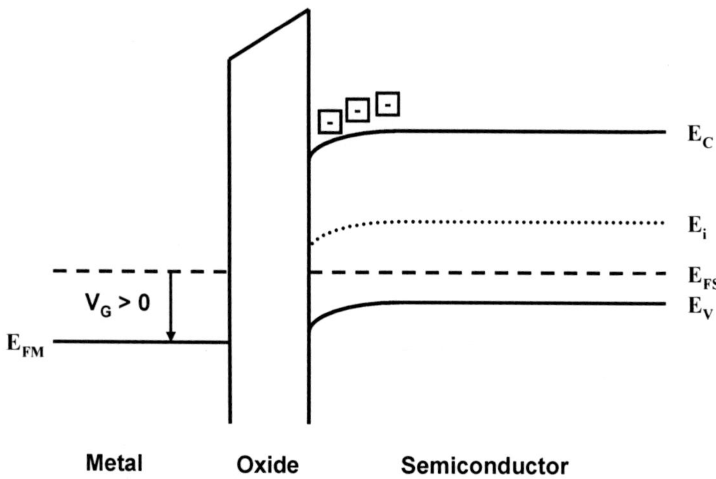

Fig. 6.18 MOS structure operating under depletion conditions

further away from the valence band near the oxide when compared with the bulk. The applied bias to the metal electrode is shared between the oxide and the semiconductor. The width of the depletion region formed in the semiconductor depends upon its doping concentration and the surface potential. Since there are no mobile charge carriers resident in the semiconductor under the oxide, these conditions are applicable for the P-base region of the n-channel power MOSFET structure at positive gate bias voltages before the formation of the channel. The gate bias voltage must be below the threshold voltage of the power MOSFET structure for operation under depletion conditions.

Inversion Conditions

As the positive bias applied to the metal electrode in the MOS structure is increased, the band bending also increases until the intrinsic level crosses the Fermi level. The semiconductor surface now has the properties of an N-type semiconductor because the Fermi level lies above the intrinsic level. The free electrons created within this *inversion region* can be utilized to form the channel in the power MOSFET structure. The band structure corresponding to these bias conditions is illustrated in Fig. 6.19. As long as the Fermi level remains close to the intrinsic level in the semiconductor near the oxide interface, the concentration of electrons is small. This condition is referred to as *weak inversion*. When the positive bias applied to the metal electrode is sufficiently large to bend the bands such that the density of electrons at the oxide–semiconductor interface exceeds the density of the majority carriers in the bulk of the semiconductor, the condition is referred to as *strong inversion*. The channel of power MOSFET structures must be operated under strong inversion conditions to reduce its resistance by the application of a sufficiently large gate bias voltage.

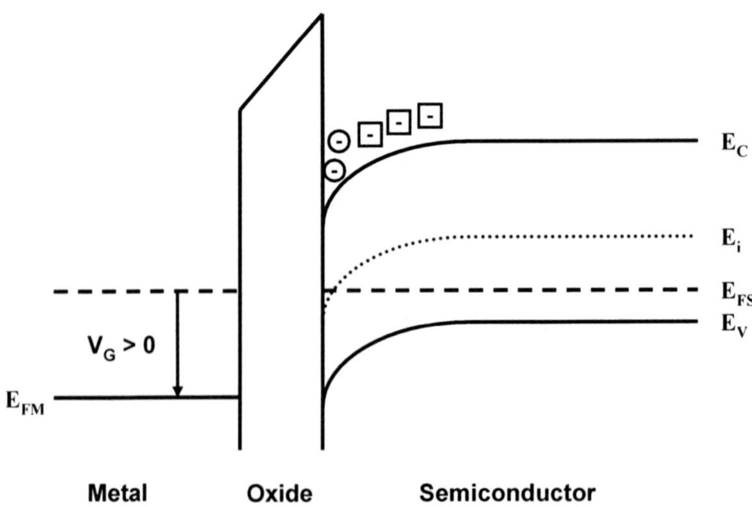

Fig. 6.19 MOS structure operating under inversion conditions

6.5.2 MOS Surface Charge Analysis

The conductivity of the channel in the power MOSFET structure is determined by the mobile charge density created by the gate bias as well as the mobility for the carriers near the oxide–semiconductor interface. The impact of surface scattering on the carrier mobility was described in Chap. 2. Depending upon the strength of the electric field normal to the semiconductor surface, the mobility can be degraded by a factor of 2–3 times when compared with the mobility in the bulk. To compute the charge available in the channel for current conduction, it is necessary to analyze the potential distribution in the MOS structure.[11]

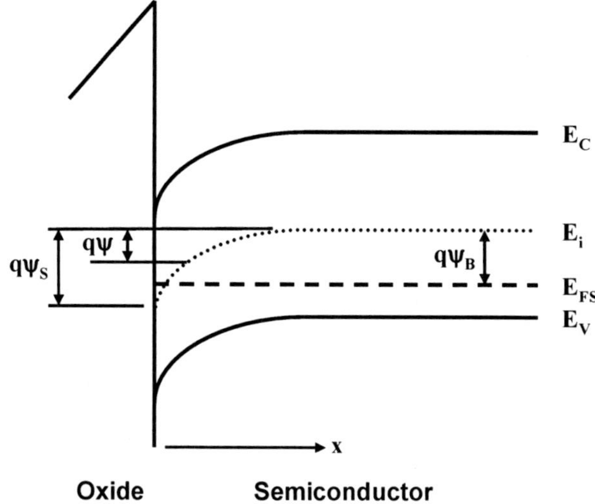

Fig. 6.20 Potential within the MOS structure operating under inversion conditions

The potential distribution within the MOS structure is shown in Fig. 6.20 for the case of a positive bias applied to the gate electrode. The potential distribution [$\psi(x)$] can be derived by solving Poisson's equation:

$$\frac{\mathrm{d}^2\psi}{\mathrm{d}x^2} = -\frac{\rho(x)}{\varepsilon_S}, \tag{6.11}$$

with the charge density $\rho(x)$ given by

$$\rho(x) = q(N_D^+ - N_A^- + p_P - n_P), \tag{6.12}$$

where N_D^+ and N_A^- are the ionized donor and acceptor concentrations, respectively. The electron and hole concentrations at a depth x in the semiconductor are related to the potential [$\psi(x)$] at that point:

$$n_P = n_{P0}\mathrm{e}^{q\psi/kT}, \tag{6.13}$$

$$p_{\mathrm{P}} = p_{\mathrm{P0}} \mathrm{e}^{-q\psi/kT}, \tag{6.14}$$

where n_{P0} and p_{P0} are the concentrations for electrons and holes in the bulk of the semiconductor in equilibrium. Since charge neutrality exists in the bulk of the semiconductor far removed from the oxide interface,

$$N_{\mathrm{D}}^{+} - N_{\mathrm{A}}^{-} = n_{\mathrm{P0}} - p_{\mathrm{P0}}. \tag{6.15}$$

Combining these relationships, the Poisson's equation can be rewritten as

$$\frac{\mathrm{d}^2\psi}{\mathrm{d}x^2} = -\frac{q}{\varepsilon_{\mathrm{S}}}[p_{\mathrm{P0}}(\mathrm{e}^{q\psi/kT} - 1) - n_{\mathrm{P0}}(\mathrm{e}^{q\psi/kT} - 1)]. \tag{6.16}$$

The electric field distribution can be obtained by integration of the potential distribution:

$$E(x) = -\frac{\mathrm{d}\psi}{\mathrm{d}x} = \frac{\sqrt{2}kT}{qL_{\mathrm{D}}} F\left(\frac{q\psi}{kT}, \frac{n_{\mathrm{P0}}}{p_{\mathrm{P0}}}\right), \tag{6.17}$$

where

$$F\left(\frac{q\psi}{kT}, \frac{n_{\mathrm{P0}}}{p_{\mathrm{P0}}}\right) = \left\{\left[\mathrm{e}^{-q\psi/kT} + \left(\frac{q\psi}{kT}\right) - 1\right] + \frac{n_{\mathrm{P0}}}{p_{\mathrm{P0}}}\left[\mathrm{e}^{q\psi/kT} - \left(\frac{q\psi}{kT}\right) - 1\right]\right\}^{1/2}. \tag{6.18}$$

The term L_{D} in (6.17) is given by

$$L_{\mathrm{D}} = \sqrt{\frac{kT\varepsilon_{\mathrm{S}}}{q^2 p_{\mathrm{P0}}}}. \tag{6.19}$$

It is referred to as the *extrinsic Debye length for holes*.

The total space charge per unit area (Q_{S}) within the semiconductor is related to the electric field at the semiconductor surface (E_{S}) by Gauss's law:

$$Q_{\mathrm{S}} = -\varepsilon_{\mathrm{S}} E_{\mathrm{S}}. \tag{6.20}$$

The electric field at the semiconductor surface can be obtained from (6.17) by using the surface potential (ψ_{S}). Combining this with (6.20),

$$Q_{\mathrm{S}} = \frac{\sqrt{2}\varepsilon_{\mathrm{S}} kT}{qL_{\mathrm{D}}} F\left(\frac{q\psi_{\mathrm{S}}}{kT}, \frac{n_{\mathrm{P0}}}{p_{\mathrm{P0}}}\right). \tag{6.21}$$

This is a generally applicable equation for computation of the total charge per unit area within the semiconductor under both positive and negative gate bias conditions. It can therefore be used to determine the total charge per unit area within the semiconductor under accumulation, depletion, and inversion conditions.

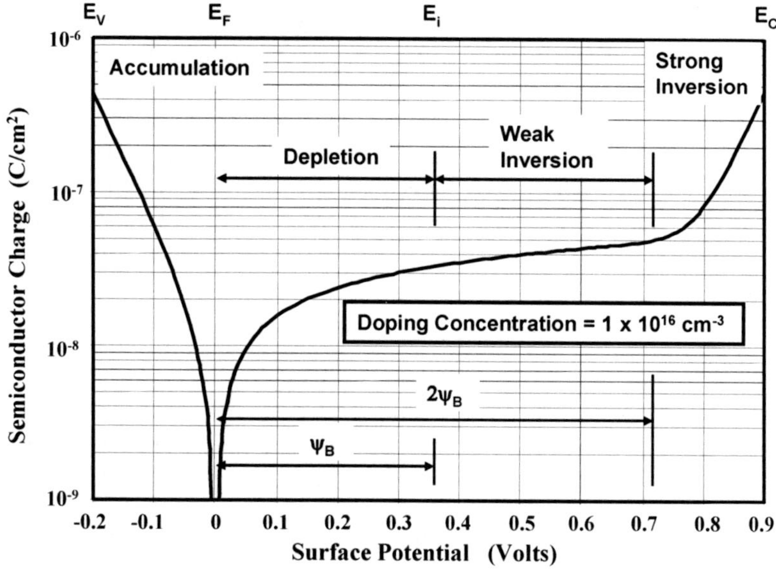

Fig. 6.21 Total charge per unit area within the semiconductor for an MOS structure

The total charge in the semiconductor is shown in Fig. 6.21 for the case of a P-type semiconductor region with doping concentration of 1×10^{16} cm^{-3}. For this doping concentration, the Fermi level is located at 0.206 eV from the valence band edge. The total charge in the semiconductor is zero when the surface potential is zero as expected for flat band conditions. For negative values for the surface potential pertinent to the application of a negative bias to the metal electrode in the MOS structure, the charge increases rapidly creating a positively charged accumulation layer. When the surface potential has a negative value, the first term in (6.18) becomes dominant allowing the approximation:

$$Q_{\mathrm{S}}(\text{accumulation}) = \frac{\sqrt{2}\varepsilon_{\mathrm{S}}kT}{qL_{\mathrm{D}}} \mathrm{e}^{-q\psi_{\mathrm{S}}/2kT}. \tag{6.22}$$

This expression indicates an exponential increase in the charge with surface potential. A large charge in the accumulation layer is therefore produced if a negative bias is applied to the MOS structure with a P-type semiconductor region. In the case of n-channel power MOSFET structures, accumulation layers are formed in portions where the gate structure overlaps the N-drift region when a positive gate bias is applied to create the inversion layer over the P-base region. The charge in the accumulation layer is formed with only a small amount of band bending and is confined to a depth of only 200 Å from the oxide–semiconductor interface. Due to the large charge at the oxide–semiconductor interface, the electric field is confined to the oxide layer within the MOS structure under these bias conditions.

The charge increases relatively slowly for small positive values of the surface potential pertinent to the application of a small positive bias to the metal electrode in the MOS structure. As long as the surface potential (ψ_S) is smaller than the bulk potential (ψ_B), the intrinsic level at the surface does not cross the Fermi level. The surface then remains P-type with a depletion layer formed in the semiconductor below the oxide. The total charge per unit area in the depletion layer can be obtained by using (6.21). In the depletion mode of operation, the second term in (6.18) becomes dominant allowing the approximation:

$$Q_S(\text{depletion}) = \frac{\varepsilon_S}{L_D}\sqrt{\frac{2kT}{q}\psi_S}. \tag{6.23}$$

This expression indicates that the total charge in the semiconductor increases gradually as the square root of the surface potential as can be observed in Fig. 6.21. By using (6.19) for the extrinsic Debye length, (6.23) can be rewritten as

$$Q_S(\text{depletion}) = \sqrt{2q\varepsilon_S p_{P0}\psi_S}. \tag{6.24}$$

When the positive bias applied to the gate electrode is increased, the band bending at the semiconductor increases until the intrinsic level at the surface crosses the Fermi level. At this point, the density of electrons begins to exceed the density of holes at the surface. This creates the weak inversion domain of operation. The total charge in the semiconductor continues to increase relatively slowly as the square root of the surface potential under these conditions:

$$Q_S(\text{weak inversion}) = \sqrt{2q\varepsilon_S p_{P0}\psi_S}. \tag{6.25}$$

At large positive bias voltages applied to the gate electrode, the band bending at the semiconductor surface becomes sufficient to make the surface potential greater than twice the bulk potential. Although the ratio of the minority carrier density to the majority carrier density (n_{P0}/p_{P0}) is small (only 2×10^{-12} for the example with P-type doping density of 1×10^{16} cm^{-3}), the fourth term in (6.18) becomes dominant under these conditions. The total charge in the semiconductor then increases rapidly with increase in surface potential:

$$Q_S(\text{strong inversion}) = \frac{\sqrt{2}\varepsilon_S kT}{qL_D}\sqrt{\frac{n_{P0}}{p_{P0}}}e^{q\psi_S/2kT}. \tag{6.26}$$

By using (6.19) for the extrinsic Debye length, (6.26) can be rewritten as

$$Q_S(\text{strong inversion}) = \sqrt{2\varepsilon_S kTn_{P0}}e^{q\psi_S/2kT}. \tag{6.27}$$

From this expression, it can be concluded that the charge in the semiconductor increases exponentially with increasing surface potential as shown in Fig. 6.21. This charge is located within a depth of only 50 Å of the oxide–semiconductor interface. The large amount of mobile charge created with a positive bias applied to the gate electrode under these strong inversion conditions can be utilized to form the channel in power MOSFET structures.

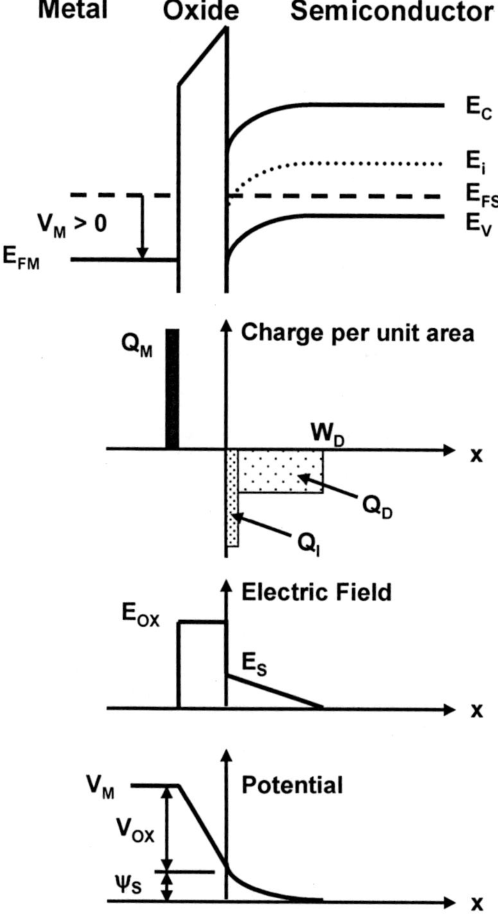

Fig. 6.22 Charge, electric field, and potential distributions within the MOS structure

The charge, electric field, and potential distributions under inversion conditions are illustrated in Fig. 6.22. The inversion layer charge (Q_I) is confined to a very small distance from the surface with the depletion layer charge extending much further into the semiconductor. The bias voltage applied to the metal electrode in the MOS structure is shared between the oxide and the semiconductor. The electric field varies linearly with distance in the semiconductor due to its uniform doping concentration. Once the MOS structure enters the strong inversion regime of operation, all the incremental gate bias is supported across the gate oxide due to the increase in inversion layer charge at the oxide–semiconductor interface.

Consequently, the depletion width in the semiconductor remains constant after the on-set of strong inversion.

6.5.3 Maximum Depletion Width

As discussed in Sect. 6.5.2, a depletion region forms in the semiconductor due to the band bending under the application of a positive bias to the metal electrode of an MOS structure with a P-type semiconductor region. The depletion region width increases with increasing positive bias applied to the metal electrode until the on-set of strong inversion conditions. Any further increase in the voltage applied to the metal electrode produces an increase in the charge within the inversion region once strong inversion conditions are reached. Consequently, the depletion region for an MOS structure has a maximum value defined by the on-set of strong inversion conditions.

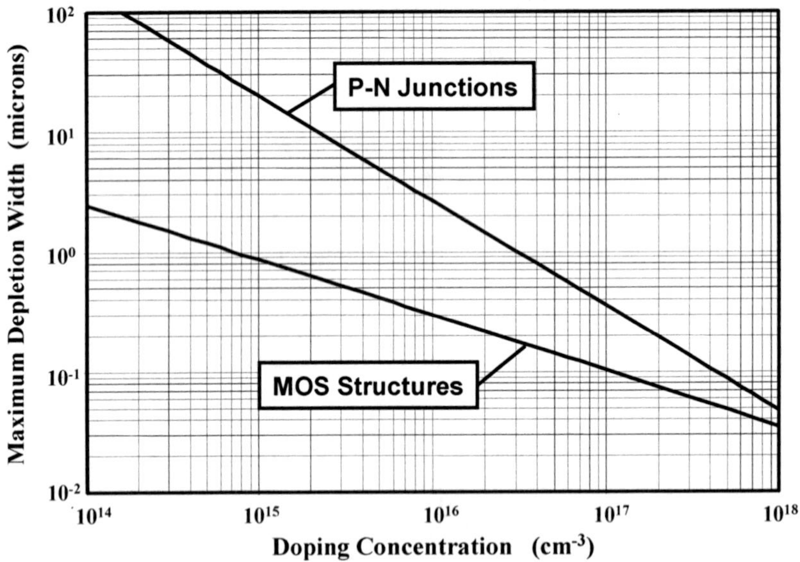

Fig. 6.23 Maximum depletion layer width for silicon MOS structures

The strong inversion condition occurs when the band bending is sufficient to make the surface potential (ψ_S) equal to twice the magnitude of the bulk potential (ψ_B). This potential can therefore be used to compute the maximum depletion layer width for the MOS structure:

$$W_M = \sqrt{\frac{2\varepsilon_S}{qN_A}(2\psi_B)}. \tag{6.28}$$

Using (6.10) for the bulk potential in this equation, with the concentration for holes in equilibrium set equal to the doping concentration of the P-type semiconductor region, yields

$$W_{\mathrm{M}} = \sqrt{\frac{4\varepsilon_{\mathrm{S}}kT}{q^{2}N_{\mathrm{A}}}\ln\left(\frac{N_{\mathrm{A}}}{n_{\mathrm{i}}}\right)}. \tag{6.29}$$

The maximum depletion width for silicon MOS structures is provided in Fig. 6.23 for a broad range of doping concentrations in the semiconductor. The maximum depletion width for P–N junctions as limited by avalanche breakdown is also shown in the figure for comparison. The maximum depletion width for MOS structures is much smaller than for the P–N junction due to the formation of the inversion layer. For typical doping concentrations in the range of $1–5 \times 10^{17}$ cm^{-3} for the base regions of power MOSFET structures, the maximum depletion width for the MOS region is less than 0.1 μm. This value is much smaller than the thickness of the P-base region, allowing the MOS depletion phenomenon to be neglected during the design of the power MOSFET channel.

6.5.4 Threshold Voltage

In the previous sections, it was demonstrated that the application of a positive bias to the metal electrode in the MOS structure with a P-type semiconductor substrate first produces a depletion region. The formation of an inversion layer occurs only after exceeding a positive bias voltage sufficient to bend the bands, so that the surface potential (ψ_{S}) becomes equal to the bulk potential (ψ_{B}). Even at this bias voltage, the MOS structure operates in the weak inversion domain with a relatively low concentration of carriers in the inversion layer as shown in Fig. 6.21. The concentration of carriers is too low in the weak inversion domain of operation to allow significant current flow through the channel in a power MOSFET structure. Once the MOS structure enters the strong inversion domain of operation, the carrier density in the inversion layer becomes sufficient to allow the conduction of current through the channel in power MOSFET structures.

The voltage applied to the metal electrode of the MOS structure is shared between the oxide and the semiconductor as illustrated in Fig. 6.22. Using the potential at the semiconductor surface (ψ_{S}),

$$V_{\mathrm{M}} = V_{\mathrm{OX}} + \psi_{\mathrm{S}}, \tag{6.30}$$

where V_{OX} is the voltage supported across the oxide. The voltage supported by the oxide can be related to the total charge in the semiconductor (Q_{S}) by

$$V_{\mathrm{OX}} = E_{\mathrm{OX}}t_{\mathrm{OX}} = \frac{Q_{\mathrm{S}}}{\varepsilon_{\mathrm{OX}}}t_{\mathrm{OX}} = \frac{Q_{\mathrm{S}}}{C_{\mathrm{OX}}}, \tag{6.31}$$

where C_{OX} is the specific capacitance of the oxide. The *threshold voltage* (V_{TH}) is defined as the voltage applied to the metal electrode to enter the strong inversion domain of operation. When the semiconductor surface enters the strong inversion mode of operation, the surface potential (ψ_S) is equal to twice the bulk potential (ψ_B) as shown in Fig. 6.21. Using this value for ψ_S in (6.30) together with (6.31) for the voltage supported by the oxide,

$$V_{TH} = \frac{Q_S}{C_{OX}} + 2\psi_B. \tag{6.32}$$

Applying (6.27) for the total semiconductor charge per unit area under strong inversion conditions with the surface potential equal to twice the bulk potential,

$$V_{TH} = \frac{\sqrt{2\varepsilon_S n_{P0}}}{C_{OX}} e^{q\psi_B/kT} + 2\psi_B. \tag{6.33}$$

The bulk potential (ψ_B) can be related to the doping concentration in the semiconductor by using (6.10), leading to

$$V_{TH} = \frac{\sqrt{4\varepsilon_S kTN_A \ln(N_A/n_i)}}{C_{OX}} + \frac{2kT}{q}\ln\left(\frac{N_A}{n_i}\right). \tag{6.34}$$

The first term is usually dominant in this equation for the threshold voltage of power MOSFET structures. After substituting for the specific oxide capacitance, the threshold voltage is then given by

$$V_{TH} = \frac{t_{OX}}{\varepsilon_{OX}}\sqrt{4\varepsilon_S kTN_A \ln\left(\frac{N_A}{n_i}\right)}. \tag{6.35}$$

Based upon this equation, the threshold voltage increases linearly with increasing oxide thickness and approximately as the square root of the doping concentration in the semiconductor. This knowledge can be used to improve the design of power MOSFET and IGBT structures as discussed later.

The threshold voltage calculated by using (6.34) for silicon MOS structures with P-type substrates is shown in Fig. 6.24 for gate oxide thicknesses ranging from 100 to 10,000 Å. The doping concentration in the semiconductor substrate is used as a parametric variable to allow application of the results to a wide variety of devices. At low doping concentrations (e.g., 1×10^{15} cm^{-3}), the first term in (6.34) becomes small resulting in a weak dependence of the threshold voltage on the oxide thickness. At large doping concentrations (e.g., 1×10^{18} cm^{-3}) and oxide thicknesses, the first term in (6.34) becomes dominant resulting in a strong dependence of the threshold voltage on the oxide thickness as described by (6.35). From the figure, it can be concluded that a threshold voltage of about 2 V will be obtained at a doping concentration of 1×10^{17} cm^{-3} for n-channel power MOSFET structures with typical gate oxide thickness in the range of 500–1,000 Å. The threshold voltage for p-channel, MOSFET structures is also represented by the plots in Fig. 6.24 but the threshold voltages have a negative value.

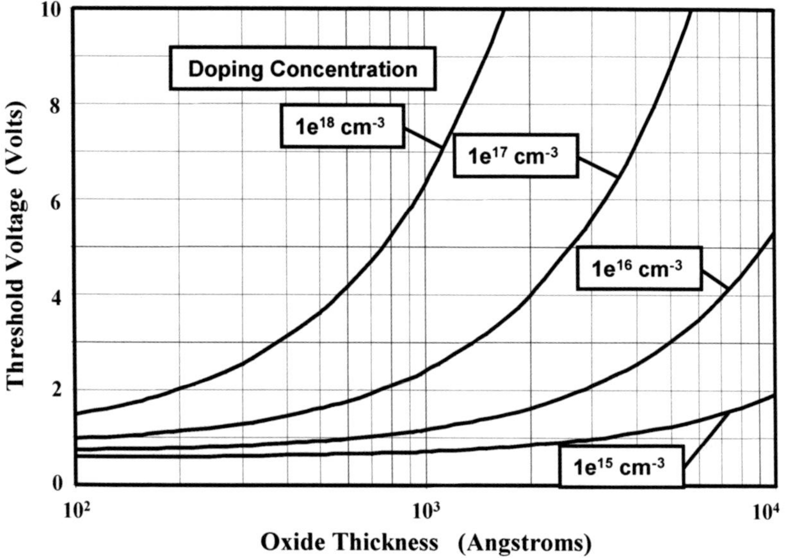

Fig. 6.24 Threshold voltage for silicon MOS structures

The plots for the threshold voltage in Fig. 6.24 are based upon the assumption of a uniform doping concentration in the semiconductor both along the channel and perpendicular to the semiconductor surface. The actual doping profile in the channel of power MOSFET structures varies as illustrated in Fig. 6.8. In this situation, the threshold voltage is determined by the largest compensated doping concentration along the channel, which occurs in the P-base region near its junction with the N^+ source region. Due to the small depth of the depletion region at this doping concentration in the P-base region, the one-dimensional analysis for the threshold voltage developed in this section can be applied to determine the threshold voltage for power MOSFET structures. The threshold voltage for actual power MOSFET structures is altered due to (a) the unequal work function for the gate electrode and the semiconductor and (b) the presence of charge in the oxide. The impact of these phenomena on n- and p-channel power MOSFET structures is discussed below.

Work Function Difference

If the metal and semiconductor do not have equal work functions, transfer of charge occurs between them to establish equilibrium conditions. An example for the case of a P-type semiconductor is shown in Fig. 6.25 when the work function of the metal is smaller than that for the semiconductor, namely, for the case of a negative work function difference (ϕ_{MS}) between the metal and the semiconductor. The transfer of charge from the semiconductor produces a depletion region in the semiconductor at the oxide interface.

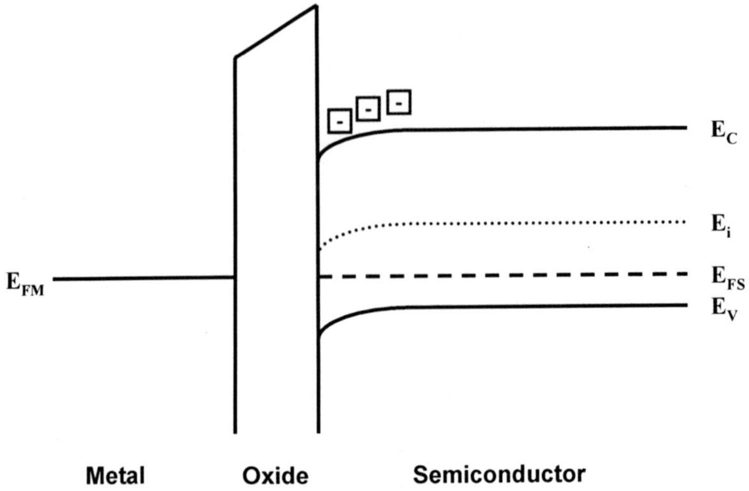

Fig. 6.25 Band diagram for an MOS structure with negative work function difference

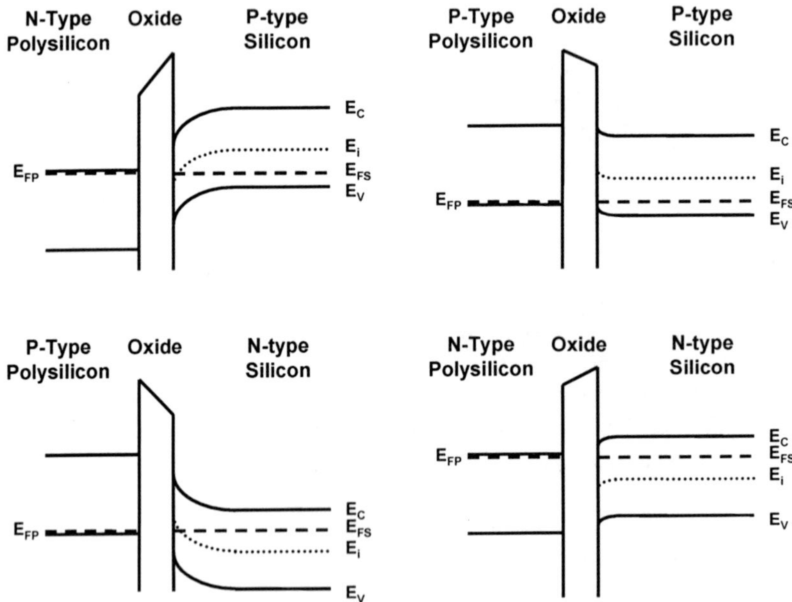

Fig. 6.26 Band diagrams for silicon MOS structures with heavily doped polysilicon gate regions

The difference in work function between the metal and the semiconductor is given by (see Fig. 6.16)

$$q\phi_{MS} = q\phi_B + q\chi_O - (q\chi_S + E_i + q\psi_B). \quad (6.36)$$

It is common practice to fabricate the silicon power MOSFET structures using a polysilicon gate electrode. The VD-MOSFET structure is fabricated by ion implantation of the P-base and N^+ source regions with the gate edge acting as the masking material. The utilization of polysilicon provides a refractory gate material that can withstand the high temperatures required during driving the dopants into the silicon in the vertical and lateral directions after the ion implantation. This process allows formation of submicron channels without high-resolution lithography. The polysilicon gate can be doped with either N-type (usually phosphorus) or P-type (usually boron) impurities. The choice of the dopant type alters the work function for the gate material.

The energy band diagrams for the case of silicon MOS structures with heavily doped polysilicon gate electrodes are illustrated in Fig. 6.26. For the case of an N-type polysilicon gate electrode and P-type silicon substrate, the work function difference (ϕ_{MS}) is negative resulting in the formation of a depletion region in the semiconductor. This reduces the threshold voltage for the corresponding n-channel power MOSFET structure. The reduction of threshold voltage can be compensated by increasing the doping concentration in the P-base region, which is desirable to suppress reach-through breakdown and shorten the channel length. In contrast, an accumulation layer forms in the P-type silicon substrate when heavily doped P-type polysilicon is used as the gate electrode. Consequently, n-channel power MOSFET structures are fabricated using heavily doped N-type polysilicon as the gate electrode. In the case of p-channel power MOSFET structures, it is customary to use heavily doped P-type polysilicon as the gate electrode for the same reasons as illustrated by the energy band diagrams in the lower portion of Fig. 6.26.

The work function difference calculated for the four possible MOS structures is provided in Fig. 6.27. For a typical doping concentration of between 1 and 5×10^{17} cm^{-3} for the base region in power MOSFET structures, a work function difference of about 1 V is obtained by using N^+ polysilicon for n-channel devices and P^+ polysilicon for p-channel devices. This potential difference can be utilized to enhance the performance of power MOSFET structures.

The impact of changing the gate electrode from N^+ polysilicon to P^+ polysilicon on the threshold voltage is shown in Fig. 6.28 for the case of n-channel power MOSFET structures with P-type base regions. The case with no work function difference between the gate electrode and the semiconductor is indicated by the dashed line for reference. There is shift in the threshold voltage by about 1 V in the negative direction if N^+ polysilicon is used instead of P^+ polysilicon for the gate electrode material. A threshold voltage of 2 V can be achieved with a doping concentration of 8×10^{16} cm^{-3} for the case of N^+ polysilicon as the gate electrode when compared with a doping concentration of 2×10^{16} cm^{-3} for the case of P^+ polysilicon. This increase in doping concentration enables reduction of the channel length and suppression of the parasitic bipolar transistor in the power MOSFET structure. It is worth pointing out that the plot shown in Fig. 6.28 can also be utilized for the analysis of p-channel power MOSFET structures if the voltage polarity for the threshold voltage is reversed. In this case, a higher doping concentration for the N-base region is obtained by using P^+ polysilicon as the gate electrode material to achieve the same magnitude for the threshold voltage.

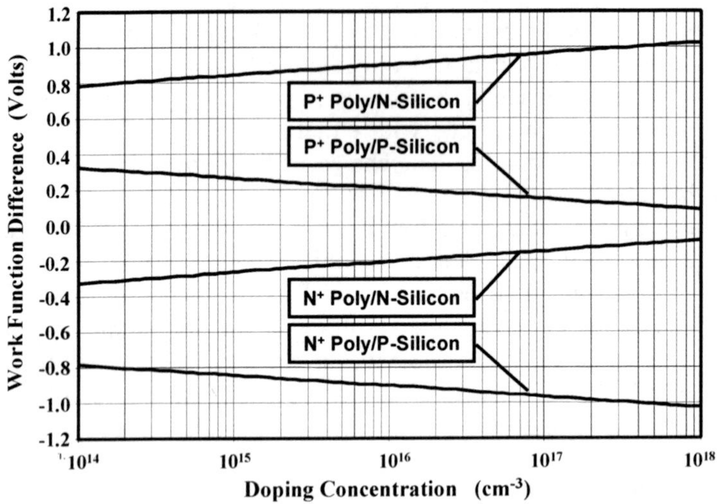

Fig. 6.27 Work function difference for silicon MOS structures with heavily doped polysilicon gate regions

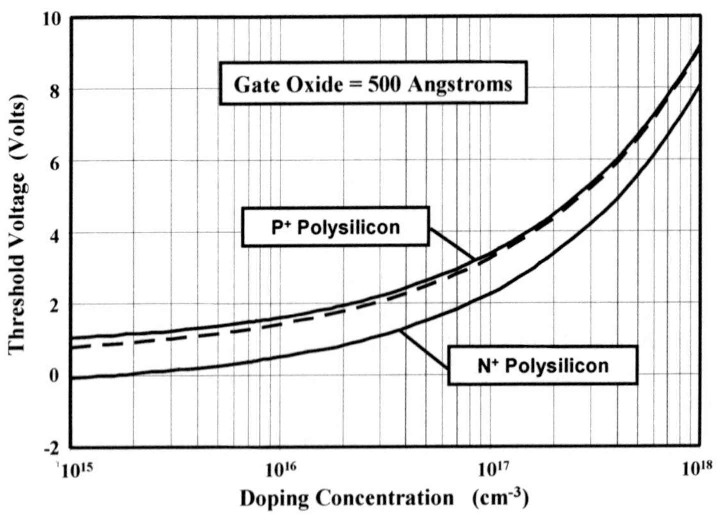

Fig. 6.28 Threshold voltage for silicon n-channel power MOSFET structures with heavily doped polysilicon gate regions

Oxide Charge

The growth of silicon dioxide on the surface of silicon by thermal oxidation is a widely used method for producing the gate oxide for CMOS and power devices. It has been observed that the thermally grown oxide contains a variety of positive charges that can be classified as (a) mobile ion charge, (b) fixed oxide charge, (c) trapped oxide charge, and (d) interface state charge. The location of these charges within the MOS sandwich is schematically illustrated in Fig. 6.29.

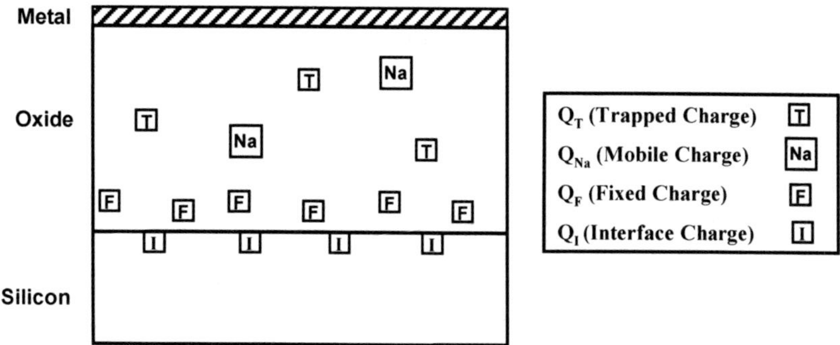

Fig. 6.29 Charge in the oxide grown on silicon surfaces

The presence of mobile ion charge (labeled as Q_{Na} in the figure) was discovered during the early stages of the development of MOS technology. These charges were found to drift through the oxide layer under the presence of an electric field at relatively low temperatures (150–200°C).[12] The mobile ions were found to be associated with the incorporation of either sodium or potassium atoms and identified to have positive charge. One of the reasons for the failure of the V-groove technology for power MOSFET structures was the utilization of a potassium hydroxide-based etchant solution for preparing the V-groove. The residual potassium in the gate oxide introduced a large concentration of mobile potassium ions in the gate oxide. Even in the case of the planar CMOS and power MOSFET structures, sodium contamination was found to be introduced from the solvents used to clean the wafers. The mobile ions have a stronger influence on the semiconductor when located closer to the oxide interface. As the mobile ions migrate through the oxide during device operation, they produce an undesirable shift in the threshold voltage with time. Improvements in the quality of the cleaning solutions and clean-room practices have eliminated this problem.

The oxide grown on silicon surfaces also contains a trapped oxide charge (labeled as Q_T in the figure) due to imperfections. In addition, there are fixed oxide charges (labeled as Q_F in the figure) located close to the silicon surface produced by the incomplete oxidation of the silicon. Although located close to the silicon surface, the fixed oxide charges are not in communication with the silicon and cannot be charged and discharged during normal device operating conditions. The density of the fixed oxide charge depends upon the surface orientation of the silicon and the oxidation conditions. Since a smaller density of fixed oxide charge can be obtained by using the {100} surface of silicon, this is the preferred orientation for wafers used to manufacture the planar VD-MOSFET structure.

In addition, the sudden termination of the semiconductor lattice at the oxide–silicon interface introduces positively charged interface state charges (labeled as Q_I in the figure). These charges are in communication with the silicon and can be charged and discharged during normal device operation. Their density also depends upon the surface orientation of the silicon and the oxidation conditions. Since a smaller density of interface state oxide charge can be obtained by using the {100} surface of silicon, this is the preferred orientation for wafers used to manufacture the planar VD-MOSFET structure.

Under carefully controlled process conditions, it is possible to reduce the density of the fixed and interface state charges to below 2×10^{10} cm^{-2}. However, in practical power MOSFET structures, the charge density can be much greater with values ranging up to 5×10^{11} cm^{-2}. This occurs due to the high-temperature drive-in processes that are required to achieve the desired junction depths subsequent to the growth of the gate oxide. It is therefore necessary to account for the presence of the positive charge in the gate oxide when determining the threshold voltage. A convenient approach for taking the oxide charge into account is to assume that there is an effective total positive charge in the oxide (Q_{OX}) located at the metal–oxide interface as illustrated in Fig. 6.30. The presence of the positive charge produces band bending at the semiconductor surface even without the application of a bias to the metal electrode. The band bending for the case of a P-type semiconductor creates a depletion region as shown in the figure. This reduces the threshold voltage for the MOS structure.

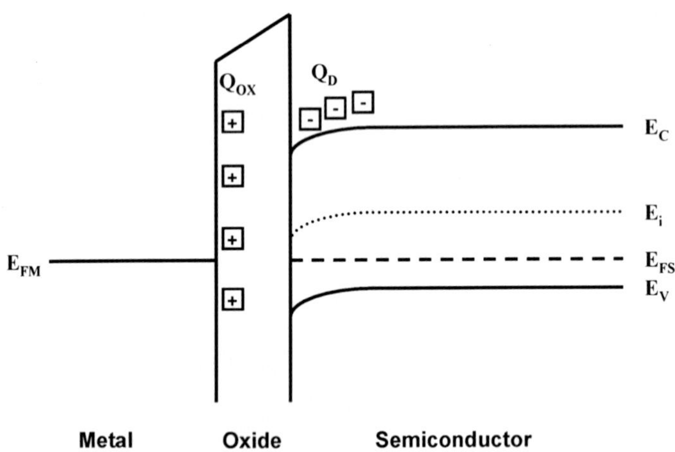

Fig. 6.30 Energy band diagram for an MOS structure in the presence of positive charge in the oxide

The impact of the positive charge in the oxide on the threshold voltage for power MOSFET structures can be conveniently performed by assuming that all the charge is located at the metal–oxide interface. Under this assumption, the threshold voltage is given by

$$V_{\mathrm{TH}} = \frac{\sqrt{4\varepsilon_{\mathrm{S}} kTN_{\mathrm{A}} \ln(N_{\mathrm{A}}/n_{\mathrm{i}})}}{C_{\mathrm{OX}}} + \frac{2kT}{q}\ln\left(\frac{N_{\mathrm{A}}}{n_{\mathrm{i}}}\right) - \frac{Q_{\mathrm{OX}}}{C_{\mathrm{OX}}}, \tag{6.37}$$

where the total *effective charge* in the oxide (Q_{OX}) includes the mobile ion charge (Q_{Na}), the trapped oxide charge (Q_{T}), the fixed oxide charge (Q_{F}), and the interface state charge (Q_{I}) after taking into account the fact that these charges are distributed throughout the oxide.

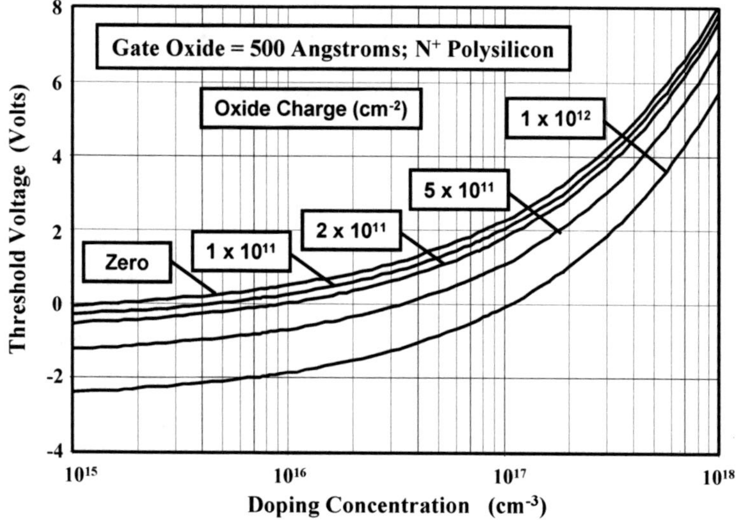

Fig. 6.31 Threshold voltage shift for silicon n-channel power MOSFET structures due to oxide charge

The threshold voltages calculated for a typical case of a gate oxide thickness of 500 Å are provided in Fig. 6.31 for various values of the oxide charge. It can be observed that the threshold voltage is shifted in the negative direction by about 1 V due to the presence of an oxide charge of 5×10^{11} cm^{-2} in magnitude. This is desirable from the point of view of increasing the doping concentration for the P-base region. For instance, the doping concentration for the P-base region, required to obtain a threshold voltage of 2 V, is increased from 8×10^{16} to 1.8×10^{17} cm^{-3} with an oxide charge of 5×10^{11} cm^{-2}. However, it is necessary to ensure a uniform concentration for the oxide charge across the wafer and from wafer to wafer during manufacturing to take advantage of the positive oxide charge.

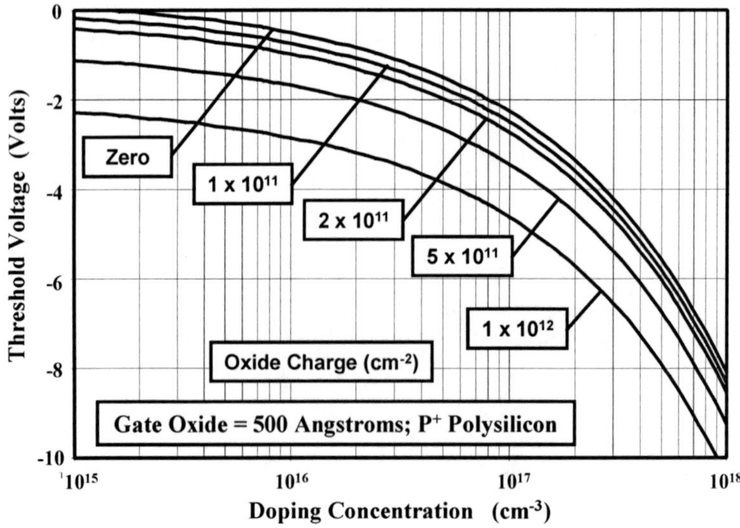

Fig. 6.32 Threshold voltage shift for silicon p-channel power MOSFET structures due to oxide charge

In the case of p-channel power MOSFET structures, the positive oxide charge shifts the threshold voltage in the negative direction as shown in Fig. 6.32. This undesirable increase in the threshold voltage must be compensated by reducing the doping concentration for the N-base region. For instance, the doping concentration for the N-base region, required to obtain a threshold voltage of −2 V, is decreased from 8×10^{16} to 2×10^{16} cm^{-3} with an oxide charge of 5×10^{11} cm^{-2}. The low N-base doping concentration will make the MOSFET structure prone to reach-through breakdown problems unless the thickness of the N-base region is made very large. The long channel length for such structures, in conjunction with the lower mobility for holes in silicon, makes the on-resistance for the p-channel MOSFET structure very high. Consequently, in the case of p-channel power MOSFET structures, it is imperative to reduce the oxide charge as much as possible during device fabrication.

Field Oxide

The power MOSFET structure must be constructed with a large number of cells that are interconnected by the source and gate electrodes on the upper surface. The source electrode is typically fabricated using a thick layer of aluminum because it covers a large area of the top surface while interconnecting the device cells. In contrast, the gate electrode is located below the source electrode as illustrated in the cross sections for the power MOSFET structures (see Fig. 6.3). The polysilicon layer used as the gate electrode has a substantial resistance (typically 30 Ω sq^{-1}). It is necessary to periodically make a metal contact to the polysilicon to form a gate bus that is brought to a gate contact pad on one location on the chip surface.

During the design of the gate bus architecture, it is necessary to ensure that an inversion layer is not formed below this region when the gate bias is applied to the power MOSFET structure to switch it on. This can be accomplished by employing a thick field oxide layer below the gate bus to increase the threshold voltage. From Fig. 6.24, it can be concluded that a field oxide thickness greater than 6,000 Å is sufficient to raise the threshold voltage above 10 V. The same thick field oxide layer is usually also utilized for the formation of the edge termination for the power MOSFET structure.

6.5.5 Channel Resistance

In the power MOSFET structure, the on-state current flow is established by the formation of an n-channel region that connects the N^+ source region with the N-drift region. The electrical properties of the channel determine the on-state resistance and the output characteristics of the device. The electrical characteristics of the channel can be analyzed by considering the basic lateral n-channel MOSFET structure shown in Fig. 6.33. This structure consists of N^+ source and drain regions formed in a P-base region. The contact to the P-base region (or substrate for the lateral MOSFET structure) is achieved by overlapping the source contact metal with the junction J_2 between the N^+ source and P-base regions.

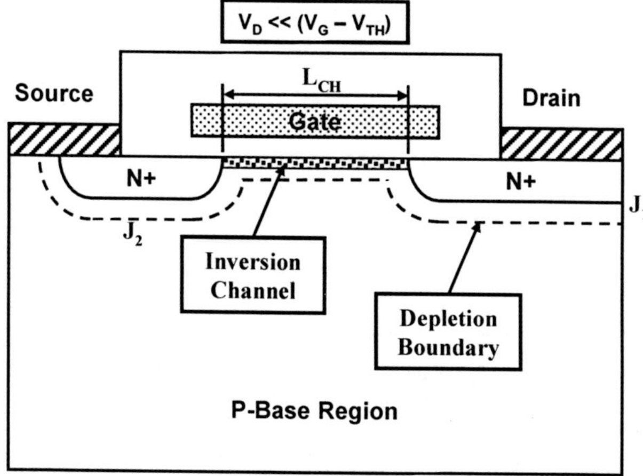

Fig. 6.33 Lateral MOSFET structure representing the channel region under very small drain bias voltages

The junction J_1 between the N^+ drain and P-base regions becomes reverse biased when a positive bias is applied to the drain electrode. Without a gate bias, no current flow transpires between the source and the drain of the lateral MOSFET structure with the voltage supported by junction J_1. However, if a positive bias is applied to the gate electrode above the threshold voltage for the MOS structure, a strong inversion layer forms below the gate to create the channel. The charge in the channel under the gate is given by

$$Q_n = C_{OX}(V_G - V_{TH}), \tag{6.38}$$

because the gate voltage in excess of the threshold voltage creates the strong inversion layer charge located at the oxide–semiconductor interface. If the drain bias voltage is assumed to be small when compared with the gate bias voltage, the charge in the inversion layer is uniform throughout the channel. This is indicated in the figure as a channel with constant thickness along the entire channel. The resistance of the channel is then given by

$$R_{CH} = \frac{L_{CH}}{Z\mu_{ni}C_{OX}(V_G - V_{TH})}, \tag{6.39}$$

where L_{CH} is the channel length as indicated in the figure, Z is the channel width orthogonal to the cross section in the figure, and μ_{ni} is the inversion layer mobility for electrons. The inversion layer mobility was discussed in detail in Chap. 2.

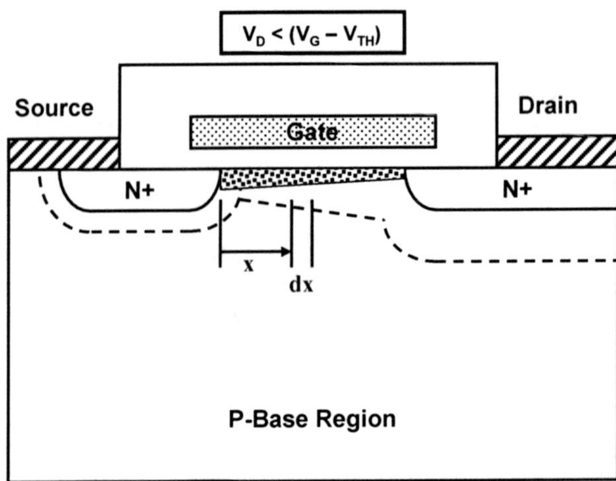

Fig. 6.34 Lateral MOSFET structure representing the channel region with drain bias comparable with the gate bias voltage

When the current flowing through the channel increases, a finite voltage drop occurs along the channel due to its finite resistance. The positive voltage developed in the channel due to the current flow opposes the applied gate bias. Consequently, the charge formed in the inversion layer becomes smaller at the drain end of the channel when compared with the charge induced at the source end of the channel. This is illustrated in Fig. 6.34 by a smaller thickness for the channel region on the drain side when compared with the source end.

When the drain current increases, the drain bias voltage also increases until it eventually becomes equal to the difference between the gate bias voltage and the threshold voltage. At this drain bias voltage, there is no longer sufficient voltage difference between the gate and the semiconductor to produce a strong inversion layer at the drain end of the channel. This is schematically illustrated in Fig. 6.35 as a channel with zero thickness on the drain end. This situation is referred to as *channel pinch-off*. The drain current saturates under these bias conditions at a value determined by the gate bias voltage.

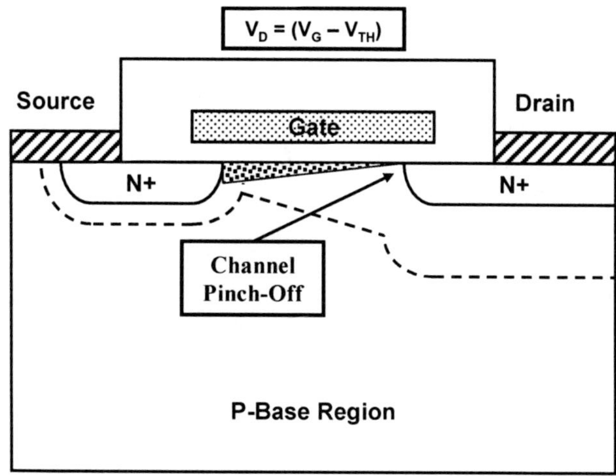

Fig. 6.35 Lateral MOSFET structure representing the channel region under channel pinch-off conditions

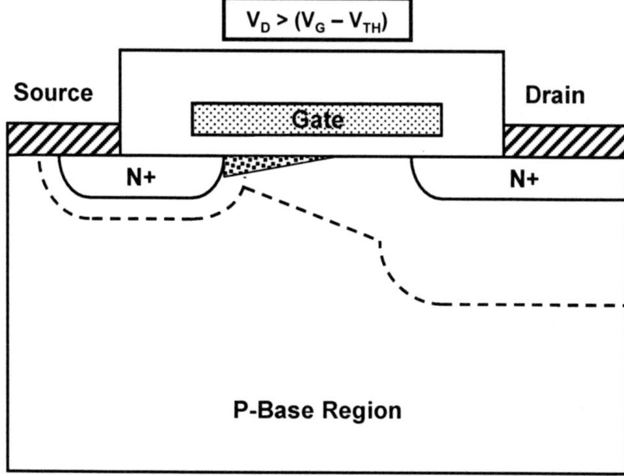

Fig. 6.36 Lateral MOSFET structure representing the channel region for drain bias voltages well above channel pinch-off conditions

If the drain voltage is increased beyond the value required to achieve channel pinch-off conditions, the drain current remains constant. The additional drain voltage is supported along the channel with a depletion region formed between the edge of the inversion layer and the N^+ drain region as illustrated in Fig. 6.36. Although the inversion layer no longer extends along the entire length of the channel, drain current flow can still occur because electrons are transported from the edge of the inversion layer through the depletion region by the prevailing longitudinal electrical field. Under these operating conditions, the effective channel length is reduced resulting in a finite output resistance for the MOSFET as discussed later in this chapter.

The resistance of the inversion layer that forms the channel for the lateral MOSFET structure is a function of the gate and drain bias voltages. An analytical treatment of the channel resistance can be performed under the following assumptions (a) an ideal gate MOS structure with no charge transport through the gate dielectric, (b) a free carrier mobility in the inversion layer that is independent of the transverse and longitudinal electric fields, (c) a uniformly doped base region, (d) current transport through the channel exclusively by the drift phenomenon, (e) negligible leakage current in the substrate, and (f) a longitudinal electric field along the surface much smaller when compared with the transverse electric field created by the gate bias. The last assumption is referred to as the *gradual channel approximation.*

Consider an elemental segment of the channel of width dx located at a distance x from the N^+ source region as illustrated in Fig. 6.34. The resistance of the elemental segment is determined by the local inversion layer charge density [$Q_n(x)$]:

$$\mathrm{d}R = \frac{\mathrm{d}x}{Z\mu_{ni}Q_n(x)}, \tag{6.40}$$

where μ_{ni} is the inversion layer mobility. The charge in the inversion layer at this location is determined by the gate bias voltage and the local potential $V(x)$ in the channel:

$$Q_n(x) = C_{OX}[V_G - V_{TH} - V(x)]. \tag{6.41}$$

The voltage drop across the elemental segment due to the channel current flow is given by

$$\mathrm{d}V = I_D \mathrm{d}R, \tag{6.42}$$

because the same drain current (I_D) must prevail through the entire channel. Combining these expressions and integrating along the channel,

$$\int_0^{L_{CH}} I_D \mathrm{d}x = Z\mu_{ni}C_{OX}\int_0^{V_D}(V_G - V_{TH} - V)\mathrm{d}V, \tag{6.43}$$

leading to

$$I_D = \frac{Z\mu_{ni}C_{OX}}{2L_{CH}}[2(V_G - V_{TH})V_D - V_D^2]. \tag{6.44}$$

This equation describes the *i–v* characteristics for the channel in the lateral MOSFET structure until the drain voltage reaches the pinch-off voltage for the MOSFET, which is given by

$$V_P = V_G - V_{TH}. \tag{6.45}$$

When the drain bias voltage is very small when compared with the gate voltage, the second term in (6.44) becomes negligible and the drain current increases linearly with increasing drain bias voltage:

$$I_D = \frac{Z\mu_{ni}C_{OX}}{L_{CH}}[(V_G - V_{TH})V_D]. \tag{6.46}$$

In the *linear regime of operation*, a channel resistance can be defined:

$$R_{CH} = \frac{V_D}{I_D} = \frac{L_{CH}}{Z\mu_{ni}C_{OX}(V_G - V_{TH})}. \tag{6.47}$$

The same expression was derived previously by assuming a uniform inversion layer charge for the entire channel. From this expression, it can be concluded that the channel resistance can be decreased by increasing the gate bias voltage. Although this behavior is observed in MOSFET structures, the reduction of the resistance becomes limited by a reduction of the mobility with increasing gate bias due to the concomitant increase in the electric field normal to the semiconductor surface. In addition, the resistance for power MOSFET structures becomes limited by other components that are discussed later in this chapter.

The transconductance for the MOSFET structure in the linear regime of operation is given by

$$g_{mL} = \frac{dI_D}{dV_G} = \frac{Z\mu_{ni}C_{OX}}{L_{CH}}V_D. \tag{6.48}$$

It can be concluded from this expression that the transconductance is independent of the gate bias voltage and proportional to the drain bias voltage. This behavior is illustrated in Fig. 6.5 for a typical MOSFET structure.

When the drain bias voltage is made comparable with the gate bias voltage, the drain current increases sublinearly, as predicted by (6.44), due to a reduction of the inversion layer charge at the drain end of the channel. Eventually, the drain voltage becomes equal to the pinch-off voltage leading to the saturation of the drain current. Substituting the pinch-off voltage as given by (6.45) into (6.44) provides an expression for the saturated drain current:

$$I_{D,sat} = \frac{Z\mu_{ni}C_{OX}}{2L_{CH}}(V_G - V_{TH})^2. \tag{6.49}$$

Based upon this equation, the saturated drain current increases as the square of the gate bias voltage. This is commonly referred to as the *square-law characteristic* for MOSFET structures. In this expression, the drain current is independent of the drain bias voltage, indicating that the output impedance for the MOSFET structure is infinite in magnitude.

The transconductance for the MOSFET structure in the saturated current regime of operation is given by

$$g_{\mathrm{mS}} = \frac{\mathrm{d}I_{\mathrm{D}}}{\mathrm{d}V_{\mathrm{G}}} = \frac{Z\mu_{\mathrm{ni}}C_{\mathrm{OX}}}{L_{\mathrm{CH}}}(V_{\mathrm{G}} - V_{\mathrm{TH}}). \tag{6.50}$$

It can be concluded from this expression that the transconductance is proportional to the gate bias voltage and independent of the drain bias voltage. The dependence of the transconductance on the gate bias voltage is a fundamental nonlinearity inherent in MOSFET structures that operated with current saturation due to channel pinch-off. Recently, a new mode of operation, discussed in Sect. 6.11, that produces a transconductance that is independent of the gate bias voltage has been discovered for power MOSFET structures.[13] These superlinear MOSFET structures are ideal for amplification of analog signals in audio and cellular applications.

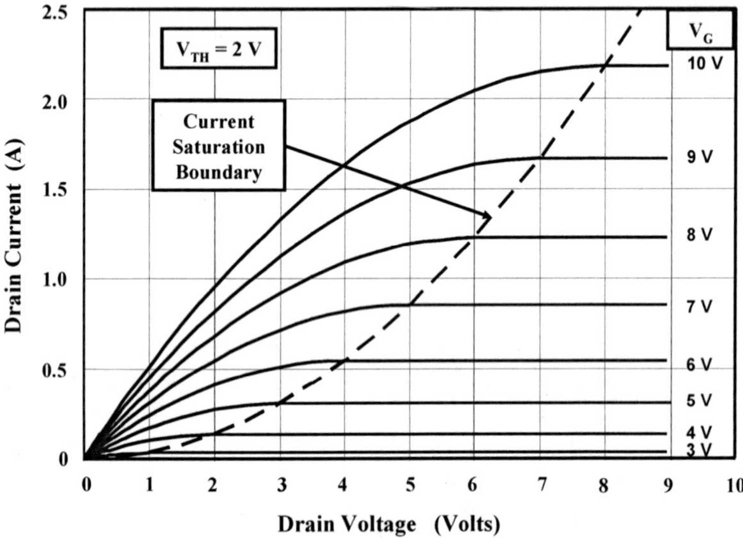

Fig. 6.37 Output characteristics for a MOSFET structure operating under channel pinch-off conditions

The *i*–*v* characteristics for an MOSFET structure operating under channel pinch-off controlled conditions are provided in Fig. 6.37. In this structure, a channel width (Z) of 1 cm and channel length (L_{CH}) of 1 μm were assumed with a gate oxide thickness of 500 Å and an inversion layer mobility of 200 cm^2 V^{-1} s^{-1}.

The structure has a threshold voltage of 2 V. The linear regime of operation can be observed at drain bias voltages below 1 V. The drain current reaches its saturated value at the boundary defined by the dashed line. For each gate bias voltage, the drain current becomes constant when the drain current exceeds the pinch-off voltage. In the saturated current regime of operation, the drain current increases in a nonlinear manner as defined by the square-law relationship between the drain current and gate bias voltage.

6.6 Power VD-MOSFET On-Resistance

The on-resistance (R_{ON}) for a power MOSFET structure is defined as the total resistance to current flow between the drain and source electrodes when a gate bias is applied to turn on the device. The on-resistance limits the maximum current-handling capability of the power MOSFET structure. The power dissipated in the power MOSFET structure during on-state operation is given by

$$P_{D} = I_{D}V_{D} = I_{D}^{2}R_{ON}, \tag{6.51}$$

which can also be expressed on a unit area basis:

$$\frac{P_{D}}{A} = P_{DA} = J_{D}^{2}R_{ON,SP}, \tag{6.52}$$

where A is the active area of the device, J_D is the on-state drain current density, and $R_{ON,SP}$ is the specific on-resistance of the power MOSFET structure. The power dissipation per unit area is limited by the allowable maximum junction temperature (T_{JM}) based upon reliability considerations. The increase in temperature above the ambient value (T_A) is determined by the thermal impedance of the package (R_θ). Combining these relationships, the maximum current density as limited by continuous on-state operation is given by

$$J_{DM} = \sqrt{\frac{T_{JM} - T_{A}}{R_{ON,SP}R_{\theta}}}. \tag{6.53}$$

From this expression, it can be concluded that the current-handling capability for the power MOSFET structure can be increased by reducing its specific on-resistance. This has been one of the main objectives of the power semiconductor development community by using a combination of structural innovations and improvements to the process technology.

The power VD-MOSFET structure is shown in Fig. 6.38 with its internal resistance components. In addition to the channel resistance (R_{CH}) that was discussed in Sect. 6.5.5, there are seven resistances that must be analyzed to obtain the total

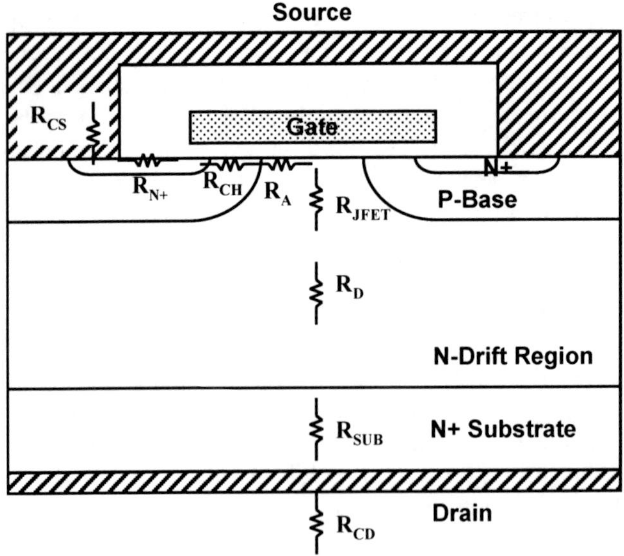

Fig. 6.38 Power VD-MOSFET structure with its internal resistances

on-resistance between the source and drain electrodes when the device is turned on. It is customary to analyze not only the resistance for a particular cell design but also the specific resistance for each of the components by multiplying the cell resistance with the cell area. The total on-resistance for the power MOSFET structure is obtained by the addition of all the resistances because they are considered to be in series in the current path between the source and drain electrodes:

$$R_{ON} = R_{CS} + R_{N+} + R_{CH} + R_{A} + R_{JFET} + R_{D} + R_{SUB} + R_{CD}. \qquad (6.54)$$

Each of the resistances within the power VD-MOSFET structure is analyzed below.

A cross section of the power VD-MOSFET structure is illustrated in Fig. 6.39 with various dimensions that can be used for the analysis of the on-resistance components. Here, W_{Cell} is the pitch for the linear cell geometry analyzed in this section, W_G is the width of the gate electrode, W_{PW} is the width of the polysilicon window, W_C is the width of the contact window to the N^+ source and P-base regions, and W_S is the width of the photoresist mask used during the N^+ source ion implantation. In addition, the current flow pattern in this device structure is indicated by the shaded area. In this illustration, it is assumed that the current spreads at a 45° angle in the drift region and then becomes uniformly distributed when it enters the N^+ substrate.

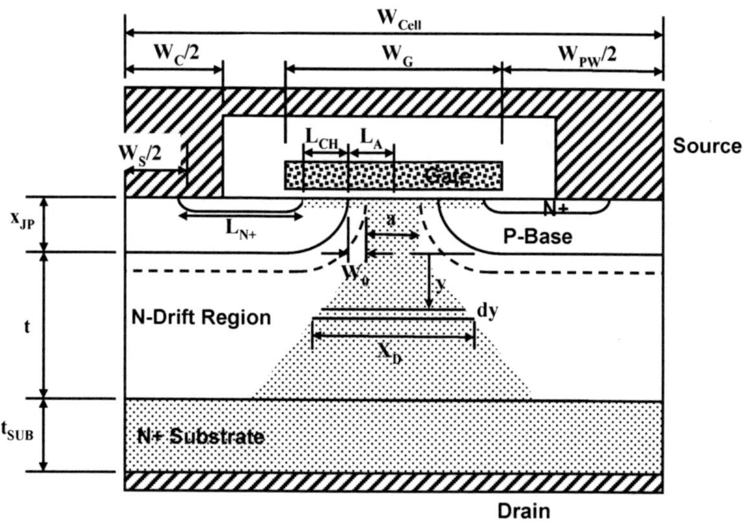

Fig. 6.39 Power VD-MOSFET structure with current flow (model A) used for analysis of its internal resistances

6.6.1 Source Contact Resistance

The contact to the N^+ source region is made over a relatively small area in the VD-MOSFET structure because the size of the window in the polysilicon must be minimized to obtain the lowest possible specific on-resistance as demonstrated later in this section. Further, the area for the contact is limited by the need to simultaneously contact the P-base region with the source electrode to suppress the parasitic N–P–N bipolar transistor. During device fabrication, the P-base region is defined by the polysilicon edges within the VD-MOSFET cell structure. One edge of the N^+ source region is also defined by the same polysilicon edge to control the channel length by the relative diffusion depth of the N^+ source and P-base regions under the gate electrode. The other end of the N^+ source region is defined by using a photoresist boundary. The contact between the source metal electrode and the N^+ source region occurs only from this boundary to the edge of the contact window, which is defined by another photolithographic masking step.

For computation of the contact resistance between the N^+ source region and its electrode, it is necessary to determine the contact area for the source region. The contact area for the N^+ source region is determined by the difference in width of the contact window (W_C) and the N^+ source ion implant window (W_S). The contact resistance to each of the N^+ source regions within the power VD-MOSFET cell structure can be obtained by dividing the specific contact resistance (ρ_C), which is determined by the work function of the contact metal and the doping concentration at the surface of the N^+ region as discussed in Chap. 2, with the contact area:

$$R_{CS} = \frac{2\rho_C}{Z(W_C - W_S)}, \tag{6.55}$$

where Z is the length of the cell in the orthogonal direction to the cross section shown in the figure. The specific contact resistance for the source regions within the power VD-MOSFET structure can then be obtained by multiplying the source contact resistance with the cell area ($W_{Cell}Z$), while recognizing that there are two source regions within each cross section shown in the figure:

$$R_{CS,SP} = \rho_C \frac{W_{Cell}}{W_C - W_S}. \tag{6.56}$$

This expression indicates that the contact resistance to the N^+ source regions can be significantly enhanced when the design rules are reduced to shrink the size of the polysilicon window.

The cell width is related to the polysilicon gate width (W_G) and the polysilicon window width (W_{PW}):

$$W_{Cell} = W_G + W_{PW}. \tag{6.57}$$

To illustrate the amplification of the contact resistance to the source region, consider a conservative power VD-MOSFET structural design with a polysilicon gate width (W_G) of 10 μm, a polysilicon window width (W_{PW}) of 10 μm, a source contact window width (W_C) of 8 μm, and a source boundary width (W_S) of 4 μm. In this design, the width of the contact to the N^+ source region is only 2 μm while the cell width (W_{Cell}) is 20 μm. Consequently, the contact resistance to the source gets amplified by a factor of 5 times. If the design rules are altered to reduce the polysilicon window (W_{PW}) to 5 μm by reducing the source contact window width (W_C) to 4 μm and a source boundary width (W_S) to 3 μm, the width of the contact to the N^+ source region is only 1 μm while the cell width (W_{Cell}) is 15 μm. Consequently, the contact resistance to the source gets amplified by a factor of 7.5 times. It is necessary to ensure that the specific contact resistance (ρ_C) is less than $1 \times 10^{-5}\ \Omega\ \mathrm{cm}^2$, by increasing the surface doping concentration of the N^+ region above $5 \times 10^{19}\ \mathrm{cm}^{-3}$, to avoid a strong impact on the total on-resistance for the power MOSFET structure. It is also common practice to use a low barrier height contact metal such as titanium or titanium silicide to reduce the specific contact resistance.

6.6.2 Source Region Resistance

Upon entering the N^+ source region from the contact, the current must flow along the source region until it reaches the channel. The resistance contributed by the source region is determined by the sheet resistance of the N^+ diffusion (ρ_{SQN+}) and its length (L_{N+}):

$$R_{\text{SN+}} = \rho_{\text{SQN+}} \frac{L_{\text{N+}}}{Z}. \tag{6.58}$$

The length of the N^+ source region is determined by the windows, shown in Fig. 6.39, used during the fabrication of the cell:

$$L_{\text{N+}} = \frac{W_{\text{PW}} - W_{\text{S}}}{2} + 2x_{\text{JN+}}, \tag{6.59}$$

where $x_{\text{JN+}}$ is the depth of the source region.

The specific resistance contributed by the current flow through the N^+ source region is obtained by multiplying its resistance with the cell area and taking into account the presence of two source regions within each cell:

$$R_{\text{SN+,SP}} = \frac{\rho_{\text{SQN+}} L_{\text{N+}} W_{\text{Cell}}}{2}. \tag{6.60}$$

The specific resistance of the N^+ source region is usually negligible in the power VD-MOSFET structure. In the case of the conservative power VD-MOSFET design with cell pitch of 20 μm, the length of the N^+ source region is 5 μm if the junction depth of the source region is 1 μm. For a typical sheet resistance for the N^+ source region of 10 Ω sq^{-1}, the specific resistance contributed by current flow through the source region is only 0.01 mΩ cm^2.

6.6.3 Channel Resistance

In Sect. 6.5.5, a lateral MOSFET structure was analyzed to obtain an expression for the channel resistance:

$$R_{\text{CH}} = \frac{L_{\text{CH}}}{Z\mu_{\text{ni}} C_{\text{OX}} (V_{\text{G}} - V_{\text{TH}})}, \tag{6.61}$$

where L_{CH} is the channel length, μ_{ni} is the inversion layer mobility, C_{OX} is the specific capacitance of the gate oxide, V_{G} is the gate bias voltage, and V_{TH} is the threshold voltage. In the case of the power VD-MOSFET structure, the channel length is defined by the difference in the depth of the P-base and N^+ source junctions below the gate as indicated in Fig. 6.39:

$$L_{\text{CH}} = x_{\text{JP}} - x_{\text{N+}}, \tag{6.62}$$

where x_{JP} is the junction depth of the P-base region and $x_{\text{N+}}$ is the junction depth of the N^+ source region. These junction depths are usually defined for doping profiles taken orthogonal to the upper surface of the semiconductor. In the power VD-MOSFET structure, the pertinent junction depths are along the surface of the semiconductor below the gate oxide. These depths can be smaller than the vertical junction depth by a factor of 80%.[14] In addition, it is common practice to enhance the doping concentration in the JFET region located between the P-base regions to reduce its resistance as discussed later in this section. The enhanced doping usually has its largest value at the surface of the semiconductor. This tends to reduce the junction depth (x_{JP}) of the P-base region, making the channel length shorter than that given by (6.62).

The specific on-resistance contributed by the channel in the power VD-MOSFET structure can be obtained from (6.61) by multiplying the cell resistance with the cell area after accounting for the fact that there are two channels within each cross section of the structure that feed the current from the source into the JFET region:

$$R_{CH,SP} = \frac{L_{CH} W_{Cell}}{2 \mu_{ni} C_{OX} (V_G - V_{TH})}. \tag{6.63}$$

For the conservative power MOSFET design with a cell width of 20 μm, the specific resistance contributed by the channel at a gate bias of 10 V is 0.92 mΩ cm^2 if the channel length is 1 μm and the gate oxide thickness is 500 Å. When the cell pitch is reduced to 15 μm with the more aggressive design rules, the specific resistance contributed by the channel at a gate bias of 10 V is reduced to 0.69 mΩ cm^2. The reduction in the specific on-resistance contributed by the channel is due to the increase in the channel density, which is defined as the channel width per unit area. The channel resistance can be reduced by decreasing the gate oxide thickness but this is accompanied by a proportionate increase in the input capacitance, which can slow down the switching speed of the power MOSFET structure.

6.6.4 Accumulation Resistance

In the power MOSFET structure, the current flowing through the inversion channel enters the drift region at the edge of the P-base junction. The current then spreads from the edge of the P-base junction into the JFET region. The current spreading phenomenon is aided by the formation of an accumulation layer in the semiconductor below the gate oxide due to the positive gate bias applied to turn on the device. For an analytical treatment, it is convenient to first calculate the resistance for current flow via the accumulation layer from the edge of the P-base region to the center of the gate. This distance L_A, shown in Fig. 6.39, is given by

$$L_A = \left(\frac{W_G}{2} - x_{JP} \right). \tag{6.64}$$

The resistance of the accumulation layer between the edge of the P-base junction and the center of the gate electrode is then given by

$$R_A = \frac{L_A}{Z \mu_{nA} C_{OX} (V_G - V_{TH})}. \tag{6.65}$$

The specific on-resistance contributed by the accumulation layer in the power VD-MOSFET structure can be obtained from (6.65) by multiplying the above resistance with the cell area after accounting for the fact that there are two accumulation layer paths within each cross section of the structure that feed the current from the two source regions into the JFET region:

$$R_{A,SP} = K_A \frac{(W_G - 2x_{JP})W_{Cell}}{4\mu_{nA} C_{OX}(V_G - V_{TH})}. \tag{6.66}$$

In writing this expression, a coefficient K_A has been introduced to account for the current spreading from the accumulation layer into the JFET region. A typical value for this coefficient is 0.6 based upon the current flow observed from numerical simulations of power VD-MOSFET structures. The threshold voltage in this expression is for the on-set of formation of the accumulation layer. Although this has a lower value than the threshold voltage previously derived for the formation of the inversion layer, it is convenient to use the same threshold voltage when performing the analytical computations.

For the conservative power MOSFET design with a cell width of 20 μm, the specific resistance contributed by the accumulation layer at a gate bias of 10 V is 0.66 mΩ cm^2 if the P-base junction depth (x_{JP}) is 2 μm and the gate oxide thickness is 500 Å. When the cell pitch is reduced to 15 μm with the more aggressive design rules, the specific resistance contributed by the accumulation layer at a gate bias of 10 V is reduced to 0.50 mΩ cm^2. The accumulation layer resistance can be reduced by decreasing the gate width (W_G). However, this increases the JFET and drift region resistances as discussed below.

6.6.5 JFET Resistance

The electrons entering from the channel into the drift region are distributed into the JFET region via the accumulation layer formed under the gate electrode. The spreading of current in this region was accounted for by using a constant K_A of 0.6 for the accumulation layer resistance. Consequently, the current flow through the JFET region can be treated with a uniform current density. In the power VD-MOSFET structure, the cross-sectional area for the JFET region increases with distance below the semiconductor surface due to the planar shape of the P-base junction. However, to simplify the analysis, a uniform cross section for the current flow with a width a will be assumed for the JFET region as illustrated by the shaded area in Fig. 6.39. The width of the current flow is related to the device structural parameters:

$$a = W_G - 2x_{JP} - 2W_0, \tag{6.67}$$

where W_0 is the zero-bias depletion width for the JFET region. The depletion region boundary is indicated in Fig. 6.39 with the dashed lines. In the model, it is assumed that no current can flow through the depleted region because all the free carriers have been swept out by the prevailing electric field across the junction. The zero-bias depletion width (W_0) in the JFET region can be computed by using the doping concentrations on both sides of the junction:

$$W_0 = \sqrt{\frac{2\varepsilon_S N_A V_{bi}}{qN_{DJ}(N_A + N_{DJ})}}, \tag{6.68}$$

where N_A is the doping concentration in the P-base region and N_{DJ} is the doping concentration in the JFET region. In (6.68), the built-in potential is also related to the doping concentrations on both sides of the junction:

$$V_{bi} = \frac{kT}{q}\ln\left(\frac{N_A N_{DJ}}{n_i^2}\right). \tag{6.69}$$

In practical devices, the P-base region is diffused into the N-drift region producing a graded doping profile as shown in Fig. 6.8. However, these expressions based upon assuming a uniform doping concentration for the P-base region are adequate for analytical computations. Further, it is common practice to enhance the doping concentration for the JFET region above that for the drift region. It is then appropriate to use the enhanced doping concentration (N_{DJ}) of the JFET region in the above expressions.

If the JFET region is assumed to extend to the bottom the P-base junction, the resistance of the JFET region can be obtained by using

$$R_{JFET} = \frac{\rho_{JFET} x_{JP}}{Z_a} = \frac{\rho_{JFET} x_{JP}}{Z(W_G - 2x_P - 2W_0)}, \tag{6.70}$$

where ρ_{JFET} is the resistivity of the JFET region given by

$$\rho_{JFET} = \frac{1}{q\mu_n N_{DJ}}, \tag{6.71}$$

where μ_n is the bulk mobility appropriate to the doping level of the JFET region.

The specific on-resistance contributed by the JFET region in the power VD-MOSFET structure can be obtained by multiplying the above resistance with the cell area:

$$R_{JFET,SP} = \frac{\rho_{JFET} x_{JP} W_{Cell}}{W_G - 2x_P - 2W_0}. \tag{6.72}$$

For the conservative power MOSFET design with a cell width of 20 μm, the specific resistance contributed by the JFET region is 0.192 mΩ cm^2 if the P-base junction depth (x_{JP}) is 2 μm and resistivity of the JFET region is 0.267 Ω cm based upon a JFET doping concentration of 2×10^{16} cm^{-3}. When the cell pitch is reduced to 15 μm with the more aggressive design rules, the specific resistance contributed by the JFET region is reduced to 0.144 mΩ cm^2. The JFET region resistance can be reduced by increasing the gate width (W_G). However, this increases the channel and accumulation layer resistances.

6.6.6 Drift Region Resistance

The resistance contributed by the drift region in the power VD-MOSFET structure is enhanced well above that for the ideal drift region discussed earlier in this chapter due to current spreading from the JFET region. The cross-sectional area for the current flow in the drift region increases from the width a of the JFET region as illustrated in Fig. 6.39 by the shaded area. Several models for this current spreading have been proposed in the literature.[7,15]

The first model (model A) for the resistance of the drift region developed in this section is based upon the current flow pattern shown in Fig. 6.39. In this pattern, it is assumed that the cross-section width (X_D) for current flow increases at a 45° angle from the JFET region. It is also assumed that the width of the polysilicon window (W_{PW}) in conjunction with the depth of the P-base region (x_{JP}) does not allow the current flow paths to merge when the current reaches the N$^+$ substrate. Under these conditions, the width of the cross-sectional area for current flow at a depth y below the JFET region, as indicated in the figure, is given by

$$X_D = a + 2y. \tag{6.73}$$

The resistance of an elemental segment of thickness dy at this depth below the JFET region is then given by

$$\mathrm{d}R_D = \frac{\rho_D \mathrm{d}y}{ZX_D} = \frac{\rho_D \mathrm{d}y}{Z(a+2y)}. \tag{6.74}$$

The resistance contributed by the drift region can be obtained by integration of the above elemental resistance from $y = 0$ to $y = t$:

$$R_D = \frac{\rho_D}{2Z} \ln\left[\frac{a+2t}{a}\right]. \tag{6.75}$$

The specific on-resistance contributed by the drift region in the power VD-MOSFET structure can be obtained by multiplying the above resistance with the cell area:

$$R_{D,SP} = \frac{\rho_D W_{Cell}}{2} \ln\left[\frac{a+2t}{a}\right]. \tag{6.76}$$

For the conservative power MOSFET design with a cell width of 20 μm, the specific resistance contributed by the drift region is 0.337 mΩ cm^2 if the P-base junction depth (x_{JP}) is 2 μm and resistivity of the drift region is 0.50 Ω cm based upon a doping concentration of 1×10^{16} cm^{-3}. When the cell pitch is reduced to

15 μm with the more aggressive design rules, the specific resistance contributed by the drift region is reduced to 0.255 mΩ cm^2. The drift region resistance can be reduced by increasing the gate width (W_G). However, this increases the channel and accumulation layer resistances. In these designs, the current spreads to a width of 10.4 μm before reaching the N$^+$ substrate, so that there is no overlap of the current flow paths within cell width of either 20 or 15 μm.

An alternate model (model B) for the current distribution in the drift region is illustrated in Fig. 6.40 with the shaded area. In this model, the current is assumed to spread from the JFET region to the entire cell width when it reaches the N$^+$ substrate. Under these conditions, the width of the cross-sectional area for current flow at a depth y below the JFET region, as indicated in the figure, is given by

$$X_D = a + \frac{W_{Cell} - a}{t} y. \tag{6.77}$$

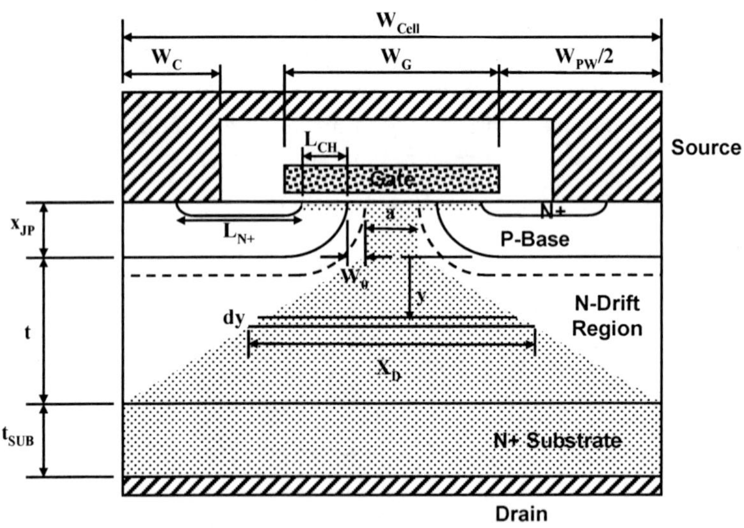

Fig. 6.40 Power VD-MOSFET structure with current distribution (model B) in the drift region

The resistance of an elemental segment of thickness dy at this depth below the JFET region is then given by

$$dR_D = \frac{\rho_D dy}{Z X_D} = \frac{\rho_D t\, dy}{Z[at + (W_{Cell} - a)y]}. \tag{6.78}$$

The resistance contributed by the drift region can be obtained by integration of the above elemental resistance from $y = 0$ and $y = t$:

$$R_D = \frac{\rho_D t}{Z(W_{Cell} - a)} \ln\left[\frac{W_{Cell}}{a}\right]. \tag{6.79}$$

The specific on-resistance contributed by the drift region in the power VD-MOSFET structure can be obtained by multiplying the above resistance with the cell area:

$$R_{\mathrm{D,SP}} = \frac{\rho_{\mathrm{D}} t W_{\mathrm{Cell}}}{W_{\mathrm{Cell}} - a} \ln\left[\frac{W_{\mathrm{Cell}}}{a}\right]. \tag{6.80}$$

In the special case where $(W_{\mathrm{Cell}} - a)/2$ becomes equal to t, the current spreads at a 45° angle in this model resulting in the same expression for the specific on-resistance for the drift region given by (6.76).

For the conservative power MOSFET design with a cell width of 20 μm, the specific resistance contributed by the drift region calculated by using this model is 0.237 mΩ cm^2 if the P-base junction depth (x_{JP}) is 2 μm and resistivity of the drift region is 0.50 Ω cm based upon a doping concentration of 1×10^{16} cm^{-3}. When the cell pitch is reduced to 15 μm with the more aggressive design rules, the specific resistance contributed by the drift region is reduced to 0.211 mΩ cm^2. Model B predicts a smaller value for the drift region resistance when compared with model A with the 45° distribution angle due to greater current spreading in the drift region.

To increase the blocking voltage capability of the power VD-MOSFET structure, it is necessary to reduce the doping concentration and increase the thickness of the N-drift region to support the voltage. The analytical model for the drift region resistance with the 45° spreading angle can be extended to this situation by taking into account the overlap of the current flow paths as illustrated in Fig. 6.41. The resistance of the drift region is now determined by two portions: a first portion with a cross-sectional area that increases with the depth and a second portion with a uniform cross-sectional area for the current flow.

Under these conditions, the width of the cross-sectional area for the first portion of the current path at a depth y below the JFET region is given by

$$X_{\mathrm{D}} = a + 2y. \tag{6.81}$$

The resistance of an elemental segment of thickness dy at this depth below the JFET region is then given by

$$\mathrm{d}R_{\mathrm{D}} = \frac{\rho_{\mathrm{D}} \mathrm{d}y}{Z X_{\mathrm{D}}} = \frac{\rho_{\mathrm{D}} \mathrm{d}y}{Z(a + 2y)}. \tag{6.82}$$

The resistance contributed by the first portion of the drift region can be obtained by integration of the above elemental resistance for the segment over the limits $y = 0$ and $y = (W_{\mathrm{Cell}} - a)/2$:

$$R_{\mathrm{D1}} = \frac{\rho_{\mathrm{D}}}{2Z} \ln\left[\frac{W_{\mathrm{Cell}}}{a}\right]. \tag{6.83}$$

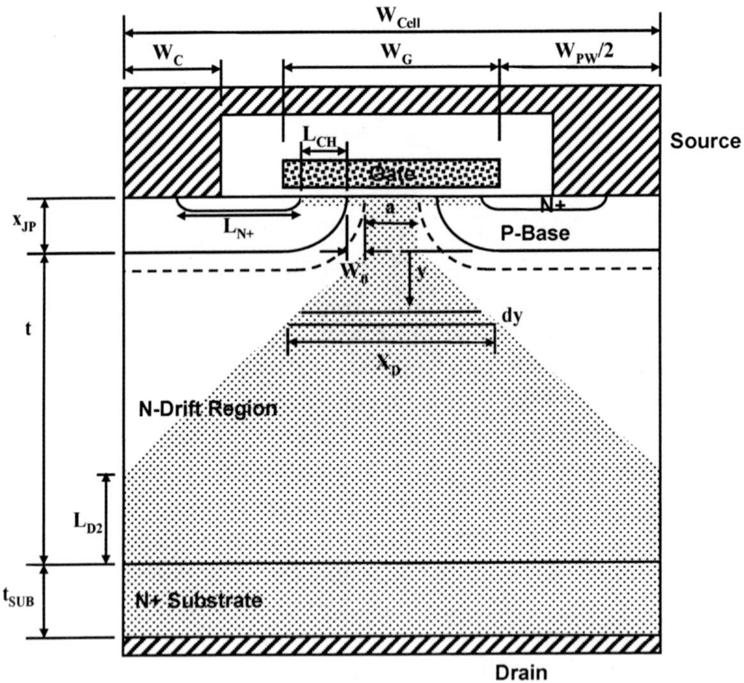

Fig. 6.41 Power VD-MOSFET structure with overlapping current distribution paths (model C) in the drift region

The resistance contributed by the second portion of the drift region is determined by current flowing through a uniform cross-sectional width (W_{Cell}). The distance (L_{D2}, shown in Fig. 6.41) over which the cross-sectional area is uniform is given by

$$L_{\mathrm{D2}} = t + \frac{a}{2} - \frac{W_{\mathrm{Cell}}}{2}. \tag{6.84}$$

The resistance of this portion of the drift region is then given by

$$R_{\mathrm{D2}} = \frac{\rho_{\mathrm{D}}}{ZW_{\mathrm{Cell}}}\left(t + \frac{a}{2} - \frac{W_{\mathrm{Cell}}}{2}\right). \tag{6.85}$$

The specific on-resistance contributed by the drift region in the high-voltage power VD-MOSFET structure (using model C) can be obtained by multiplying the above two resistances with the cell area:

$$R_{\mathrm{D,SP}} = \frac{\rho_{\mathrm{D}} W_{\mathrm{Cell}}}{2} \ln\left[\frac{W_{\mathrm{Cell}}}{a}\right] + \rho_{\mathrm{D}}\left(t + \frac{a}{2} - \frac{W_{\mathrm{Cell}}}{2}\right). \tag{6.86}$$

Consider a power VD-MOSFET structure constructed using a drift region with doping concentration of 1×10^{15} cm^{-3}. This structure is capable of supporting a parallel-plane breakdown voltage of 300 V with a depletion width of 20 μm. For the conservative power MOSFET design with a cell width of 20 μm, the specific resistance contributed by the drift region calculated by using the above model is 11.9 mΩ cm^2 if the P-base junction depth (x_{JP}) is 2 μm and resistivity of the drift region is 4.65 Ω cm based upon the above doping concentration. When the cell pitch is reduced to 15 μm with the more aggressive design rules, the specific resistance contributed by the drift region is reduced to 10.57 mΩ cm^2. The improvement in the specific on-resistance obtained with the smaller design rules is not as large as in the case of the device with lower breakdown voltage.

6.6.7 N$^+$ Substrate Resistance

When the current reaches the bottom of the N-drift region, it is very quickly distributed throughout the heavily doped N$^+$ substrate. The current flow through the substrate can therefore be assumed to occur with a uniform cross-sectional area. Under this assumption, the specific resistance contributed by the N$^+$ substrate is given by

$$R_{\text{SUB,SP}} = \rho_{\text{SUB}} t_{\text{SUB}}, \tag{6.87}$$

where ρ_{SUB} and t_{SUB} are the resistivity and thickness of the N$^+$ substrate, respectively. When processing the wafers used to manufacture power MOSFET structures, it is necessary to use a substrate with a thickness of 500 μm on which the thin (2–50 μm) epitaxial drift region is grown to avoid breakage during handling of the wafers in the processing equipment. The lowest available resistivity for a typical phosphorus-doped silicon wafer is 0.003 Ω cm. The specific resistance contributed by this wafer is 0.15 mΩ cm^2. Although this value is acceptable for power MOSFET structures designed to support high (>200 V) voltages, it makes a large contribution to power MOSFET structures designed to support low (<50 V) voltages. Consequently, it is common practice to reduce the thickness of the wafer just before application of the drain electrode metal. A substrate contribution of 0.06 mΩ cm^2 can be obtained by reducing the substrate thickness to 200 μm. This is an acceptable value for most power MOSFET products. Power MOSFET products have been manufactured with wafer thickness reduced to only 50 μm. In addition, special wafer doping techniques have evolved that allow reduction of the resistivity of the N$^+$ substrate to the 0.001 Ω cm range allowing the specific resistance contributed by the substrate to be reduced to 0.02 mΩ cm^2.

6.6.8 Drain Contact Resistance

Before entering the drain electrode, the current flows through the contact resistance between the drain metal and the N$^+$ substrate. Due to the uniform current flow at the drain contact, its resistance is not amplified unlike in the case of the source contact. It is therefore satisfactory to achieve a contact resistivity of $1 \times 10^{-5}\ \Omega\ \text{cm}^2$. This can be obtained by using a titanium contact layer to achieve a low barrier height, followed by a coating of nickel and silver. The nickel acts as a barrier between the titanium and the silver. The silver layer is ideal for mounting the chips to the package by using solders.

6.6.9 Total On-Resistance

The total specific on-resistance for the power VD-MOSFET structure can be computed by adding all the above eight components. For the case of the conservative power VD-MOSFET design with a cell pitch (W_{Cell}) of 20 μm to support 50 V, the total specific on-resistance is $2.24\ \text{m}\Omega\ \text{cm}^2$. The contributions from each of the eight components of the on-resistance are summarized in Fig. 6.42 together with their percentage contributions. By comparing the numbers, it can be observed that the channel and accumulation layer resistances are the largest components. The ideal specific on-resistance for a device capable of supporting 50 V obtained by using (6.2) is $0.105\ \text{m}\Omega\ \text{cm}^2$. Consequently, the specific on-resistance for this design of the power VD-MOSFET structure is 20 times larger than the ideal case. The specific on-resistance for the VD-MOSFET structure can be brought closer to the ideal value by using smaller design rules and optimization of the cell structure as discussed in Sect. 6.7.

Resistance	Value (mOhm-cm^2)	Percentage Contribution
Source Contact ($R_{CS,SP}$)	0.05	2.2
Source ($R_{N+,SP}$)	0.01	0.4
Channel ($R_{CH,SP}$)	0.92	41.0
Accumulation ($R_{A,SP}$)	0.66	29.5
JFET ($R_{JFET,SP}$)	0.19	8.5
Drift ($R_{D,SP}$)	0.34	15.2
Substrate ($R_{SUB,SP}$)	0.06	2.7
Drain Contact ($R_{DS,SP}$)	0.01	0.4
Total ($R_{T,SP}$)	2.24	100

Fig. 6.42 On-resistance components within the 50-V power VD-MOSFET structure with 20-μm cell pitch

Simulation Example

To validate the above model for the on-resistance of the power VD-MOSFET structure, the results of two-dimensional numerical simulations on a structure designed with a 30-V rating are described here. The structure had a drift region thickness of 3 μm with a doping concentration of 1.3×10^{16} cm^{-3} located below the P-base region. The P-base region and N^{+} source regions have depths of 3 and 1 μm, respectively. The doping concentration in the JFET region was enhanced by using an additional N-type doping concentration of 1×10^{16} cm^{-3} with a depth of 3 μm. A three-dimensional view of the doping distribution is shown in Fig. 6.43 from the left-hand edge of the structure to the center of the polysilicon gate region. The N^{+} source and P-base regions are aligned to the gate edge, which is located at 4 μm from the left-hand side. The structure also includes a P^{+} region located at the left-hand side to suppress the bipolar transistor. The enhancement of the doping in the N-drift region near the surface is due to the additional JFET doping.

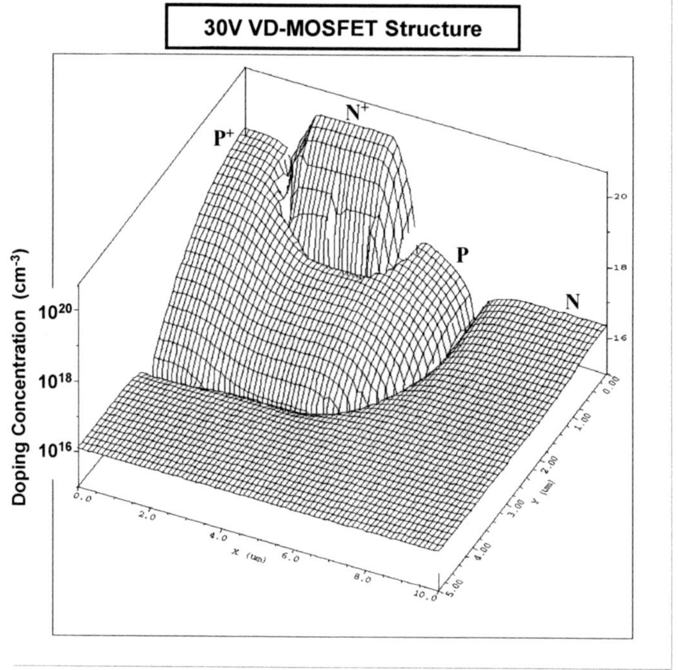

Fig. 6.43 Doping distribution for the VD-MOSFET structure

The lateral doping profile taken along the surface under the gate electrode is shown in Fig. 6.44. From the profile, it can be observed that the doping concentration of the JFET region has been increased to 2.1×10^{16} cm^{-3} due to the additional doping. This reduced the lateral extension of the P-base region to 2.4 μm when compared with a vertical depth of 3 μm. The N^{+} source region has a lateral depth of 0.8 μm leading to a channel length of 1.6 μm. As discussed earlier in this chapter, this channel length is sufficient to prevent reach-through breakdown providing a cell breakdown voltage of 43 V. The maximum compensated P-type doping concentration in the channel is 1.5×10^{17} cm^{-3}. In conjunction with a gate oxide thickness of 500 Å, a threshold voltage of 3.5 V was obtained for this doping concentration.

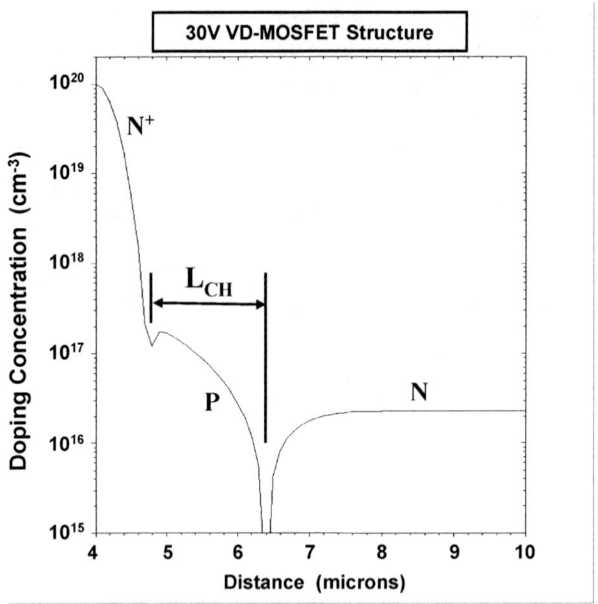

Fig. 6.44 Channel doping profile for the VD-MOSFET structure

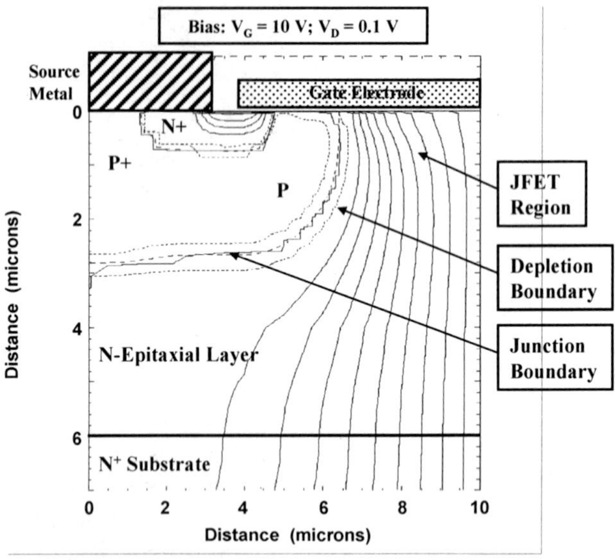

Fig. 6.45 Current flow distribution pattern in the VD-MOSFET structure

To analyze the current flow pattern in the power VD-MOSFET structure, the above device structure was turned on by the application of a gate bias of 10 V. The on-state current flow pattern at a small drain bias of 0.1 V is shown in Fig. 6.45. In the figure, the depletion layer boundary is shown by the dotted lines and the junction boundary is delineated by the dashed line. The depletion layer width (W_0) in the JFET region is 0.24 μm in good agreement with the value computed by using the analytical model. It can be observed that the current flows from the channel and distributes into the JFET region via the accumulation layer. Within the JFET region, the cross-sectional area is approximately constant with a width ($a/2$) of 3.4 μm. From the JFET region width of 3.4 μm, the current spreads to a width of 6.5 μm at a depth of 6 μm when it reaches the N^+ substrate. Consequently, the width of the cross section increases by 3 μm when the current travels a distance of 3 μm from the bottom of the P-base junction. This confirms that the current spreads at a 45° angle as assumed in model A.

From the transfer characteristics at a drain bias of 0.1 V, a threshold voltage of 3.5 V was extracted. The total specific on-resistance obtained from the numerical simulations for the above power VD-MOSFET structure is 1.60 mΩ cm^2 at a gate bias of 10 V. To compare this with the values predicted by the model, it is necessary to extract the mobility in the inversion and accumulation layers within the simulated VD-MOSFET structure. The channel and accumulation layer mobility were extracted by simulation of a long channel lateral MOSFET structure with the same gate oxide thickness. The inversion layer mobility was found to be 450 cm^2 V^{-1} s^{-1} while that for the accumulation layer was found to be 1,000 cm^2 V^{-1} s^{-1} at a gate bias of 10 V. When these values were used in model A with the dimensions and doping concentrations for this VD-MOSFET structure, the calculated value for the specific on-resistance is 1.58 mΩ cm^2 at a gate bias of 10 V. This confirms the validity of the model with a 45° spreading angle for the current flow in the drift region. This model is therefore preferable to model B where the current is assumed to spread over the entire width of the cell.

6.7 Power VD-MOSFET Cell Optimization

In Sect. 6.6, it was demonstrated that the JFET and drift region specific resistance contributions can be reduced by increasing the width of the gate electrode. Unfortunately, the specific resistance contributions from the channel and accumulation regions increase when the gate width is increased. Consequently, it is necessary to optimize the width (W_G) of the gate electrode to obtain the lowest possible specific on-resistance.[7,15]

6.7.1 Optimization of Gate Electrode Width

The optimization procedure for minimizing the specific on-resistance of the power VD-MOSFET structure can be illustrated by using an example. Consider the case of power VD-MOSFET structure designed for a blocking capability of 30 V that was discussed in the simulation example in Sect. 6.6. The specific on-resistance calculated by using model A for this case is provided in Fig. 6.46 when the gate width is increased from 4 to 18 μm while maintaining the same polysilicon window size of 8 μm. A gate bias of 10 V was used with a threshold voltage of 3.5 V for this plot.

The analytical model predicts a minimum in the total specific on-resistance for a gate electrode width of 9.5 μm with a value of 1.54 mΩ cm^2, which is a factor of 53 times larger than the ideal specific on-resistance for a breakdown voltage of 30 V. There is a relatively broad minimum with low specific on-resistances for structures with gate electrode widths ranging from 8 to 12 μm. However, the specific on-resistance increases very rapidly when the gate electrode width is made less than 6 μm. This occurs because approximately 5.3 μm is consumed under the gate electrode by the lateral P-base diffusion and the zero-bias depletion width, resulting in a drastic increase in the resistance contributed by the JFET and drift regions. The results obtained from the numerical simulations of power VD-MOSFET structures with various gate electrode widths are also included in the figure. The agreement between the values obtained from the numerical simulations and the values calculated by using model A with a wide range of gate electrode widths provides credence to the model.

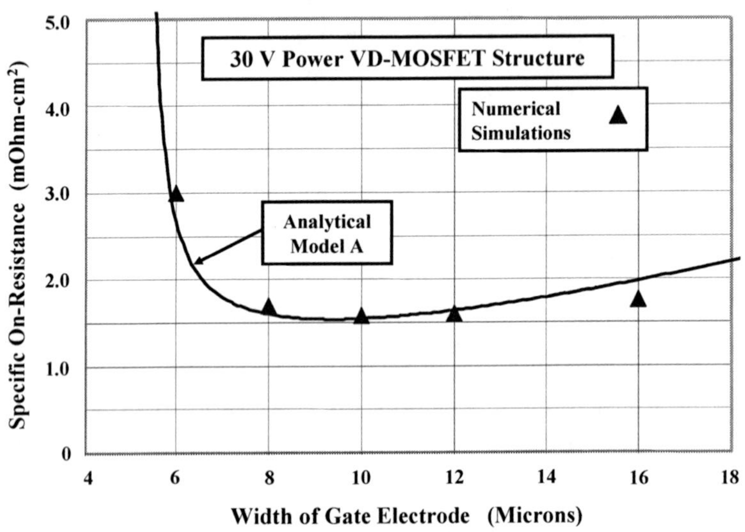

Fig. 6.46 Optimization of the power VD-MOSFET linear cell structure to obtain the minimum specific on-resistance

All the major components of the internal resistance within the power VD-MOSFET structure are dependent on the cell pitch. Only the substrate and drain contact resistances do not depend upon this design parameter. It is important to have a good understanding of the relative contributions of the various major components to the specific on-resistance during the optimization procedure. As an example, the four major contributors to the specific on-resistance for the power

VD-MOSFET structure designed for operation at 30 V are provided in Fig. 6.47 as a function of the gate electrode width. From these plots, the drastic increase in the JFET and drift region contributions at gate electrode widths of less than 6 μm is strikingly apparent. It is also apparent that the channel and accumulation layer contributions become dominant once the gate electrode width exceeds 10 μm. The minimum specific on-resistance occurs at a gate width of 9.5 μm. For this optimum design, the channel contribution is the largest component of the specific on-resistance.

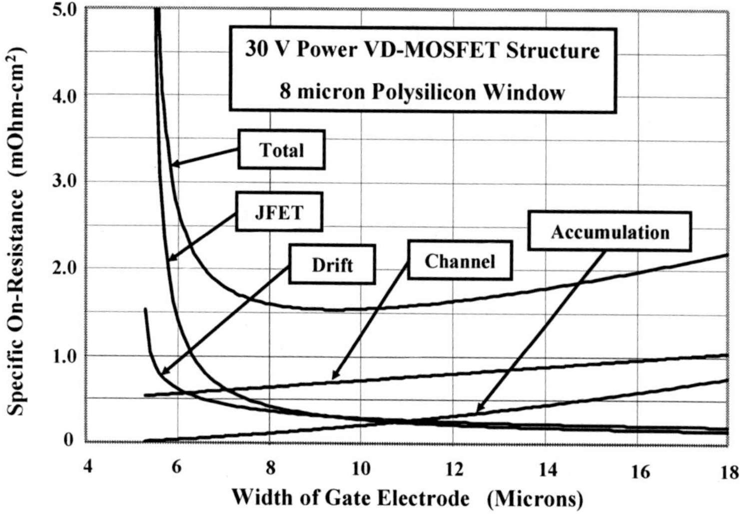

Fig. 6.47 Internal resistance components within the 30-V power VD-MOSFET linear cell structure

6.7.2 Impact of Breakdown Voltage

The breakdown voltage capability for the power VD-MOSFET structure can be increased by using a higher resistivity drift region with a larger thickness. Consequently, the resistance contributed by the drift region increases with increasing breakdown voltage. The model A developed in Sect. 6.7.1 can be utilized to analyze the change in the relative contributions of the various components of the internal resistance when the breakdown voltage is increased.

The four major components of the specific on-resistance for the power VD-MOSFET structure capable of operating at 100 V are shown in Fig. 6.48 together with the total specific on-resistance. The parallel-plane breakdown voltage for this device was assumed to be 125 V due to the impact of the edge termination. The rest of the structural parameters are the same as those of the 30-V device structure discussed in Sect. 6.7.1. It is apparent from the figure that the minimum total specific on-resistance has increased to 3 mΩ cm^2 at an optimum gate electrode width of 11.5 μm. The total specific on-resistance of the power VD-

MOSFET structure is a factor of 5 times larger than the ideal specific on-resistance of 0.593 mΩ cm^2.

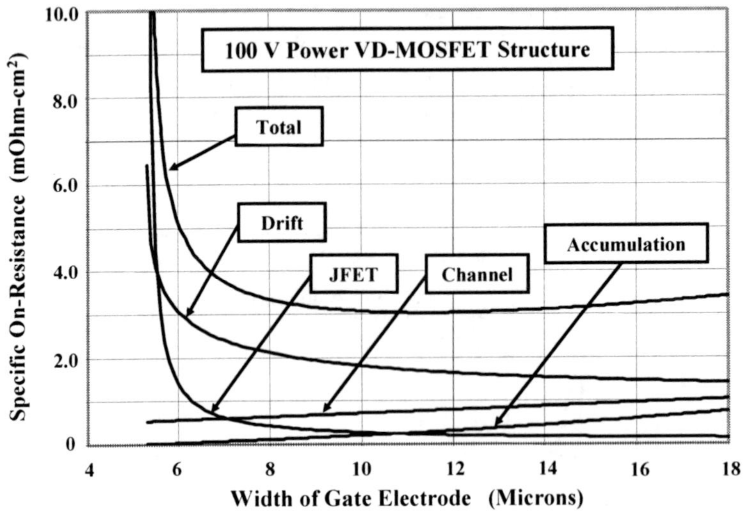

Fig. 6.48 Internal resistance components within the 100-V power VD-MOSFET linear cell structure

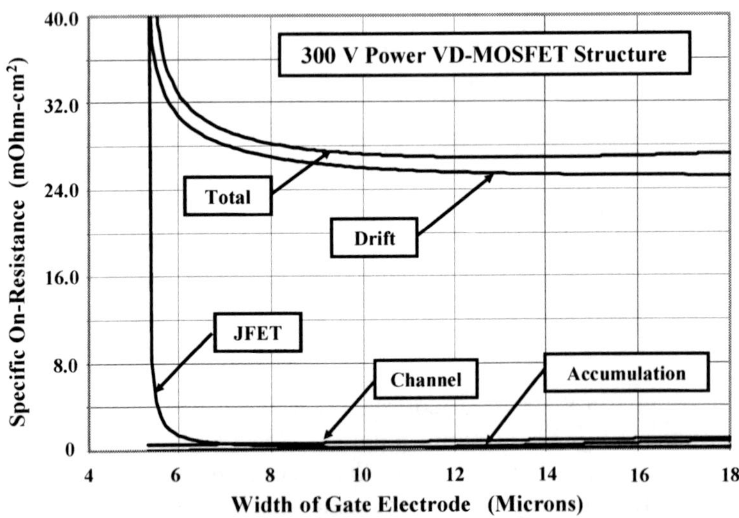

Fig. 6.49 Internal resistance components within the 300-V power VD-MOSFET linear cell structure

The four major components of the specific on-resistance for the power VD-MOSFET structure are shown in Fig. 6.49 together with the total specific on-resistance when the operating voltage is extended to 300 V. In this case, the current flow paths begin to merge under the polysilicon window. It is therefore necessary to use model C instead of model A for computation of the resistances. The parallel-plane breakdown voltage for this device was assumed to be 375 V due to the impact of the edge termination. The rest of the structural parameters are the same as those of the 30-V device structure discussed in Sect. 6.7.1. From the figure, it can be observed that all the other components are negligible when compared with the drift region resistance contribution. It is also apparent that the minimum total specific on-resistance has increased to 26.9 mΩ cm^2 at an optimum gate electrode width of 13.2 μm. The total specific on-resistance of the power VD-MOSFET structure is a factor of 2.9 times larger than the ideal specific on-resistance of 9.24 mΩ cm^2.

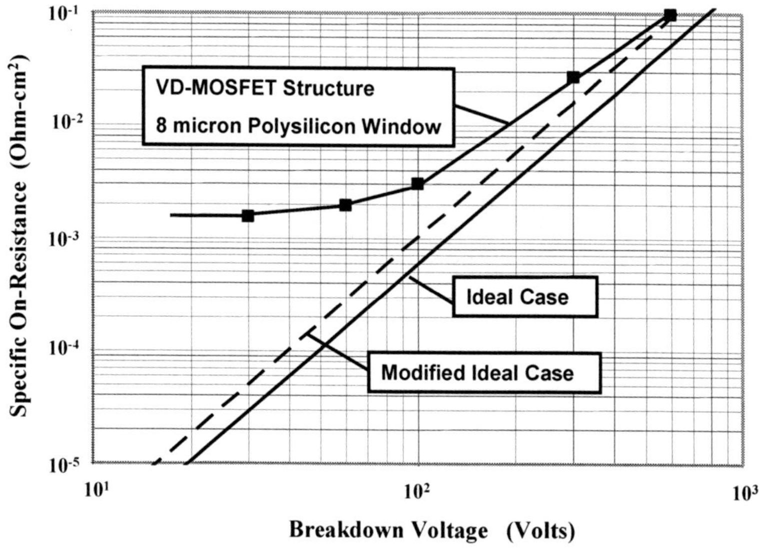

Fig. 6.50 Specific on-resistance of the power VD-MOSFET linear cell structure

From the above analysis of the power VD-MOSFET structures with different operating voltage capability, it can be seen that the minimum specific on-resistance increases and the optimum gate electrode width shifts to a larger value when the blocking voltage is increased. As the blocking voltage capability increases, the drift region contribution begins to dominate making the specific on-resistance of the power VD-MOSFET structure closer to the ideal specific on-resistance. These changes can be observed in Fig. 6.50 where the specific on-resistance for the VD-MOSFET structures is compared with the ideal specific on-resistance of the drift region. It can be observed that the specific on-resistance asymptotes to the ideal specific on-resistance at high breakdown voltages and to a constant value at low breakdown voltages. The asymptote at low breakdown voltages is created because the channel resistance becomes dominant in the power VD-MOSFET structure. In the figure, a modified ideal specific on-resistance line is also shown by the dashed lines. In this case, the ideal specific on-resistance was computed after taking into account an 80% reduction of the breakdown voltage due to the edge termination (as also assumed for the power VD-MOSFET structure). Its specific on-resistance is very close to the modified ideal specific on-resistance at high voltages, indicating that improvements in the edge termination for high-voltage power MOSFET structures can improve their performance.

It is common practice in the industry to use the planar VD-MOSFET architecture for high-voltage devices because of the mature manufacturing technology. This structure also has superior ruggedness under adverse operating conditions such as switching inductive loads with the simultaneous presence of high voltage and current. However, at lower blocking voltage levels, the power VD-MOSFET architecture yields a specific on-resistance much larger than the ideal case. This situation can be improved upon by using higher resolution design rules as discussed in Sect. 6.7.3.

6.7.3 Impact of Design Rules

In the power VD-MOSFET structure, it is desirable to reduce the size of the polysilicon window to improve the current distribution within the drift region. As indicated by the current flow pattern in Fig. 6.39 for a device with relatively low breakdown voltage, a dead space is created below the center of the polysilicon window due to the inability of the current to spread from the JFET region into this portion of the drift region. Consequently, the specific on-resistance contributed by the drift region is reduced when the polysilicon window is reduced. In addition, a reduction of the polysilicon window decreases the cell pitch. A smaller cell pitch reduces the contributions to the specific on-resistance from the channel, accumulation, and JFET regions.

A reduction of the size of the polysilicon window in the power VD-MOSFET structure can be achieved by resorting to smaller design rules for the various masking layers used to fabricate the device. Since smaller design rules require investment in photolithographic technology with higher resolution, the power semiconductor industry has made gradual progress toward this goal over the last several decades. The size of the polysilicon window (W_{PW}) is determined by the nesting of the contact window (W_C) and the N^+ source definition mask (W_S). As the contact window size approaches the size of the polysilicon window, the probability of gate-to-source short-circuits gets enhanced because of the proximity of the source metal and the polysilicon edge. The problem for power VD-MOSFET structures is exacerbated by the extremely long channel widths (Z), which can exceed several meters in magnitude, for high-current devices. Depending on the alignment tolerances and the shape of the dielectric layer covering the gate electrode at its edges, a severe loss in the yield due to the gate–source short circuits can occur prohibiting the manufacturing of devices.

The size of the source definition window (W_S) is constrained by the need to make a contact to the P-base region with the source electrode in this location of the power VD-MOSFET structure. If the source definition window is too small, the lateral diffusion of the N^+ source regions can overlap under this region resulting in a poor contact to the P-base region, which can drastically diminish the dynamic performance of the device due to creation of an open-base N–P–N transistor. If the contact window size (W_C) is made close to that of the source definition window, the size of the contact to the N^+ source region can become too small producing an increase in the specific on-resistance.

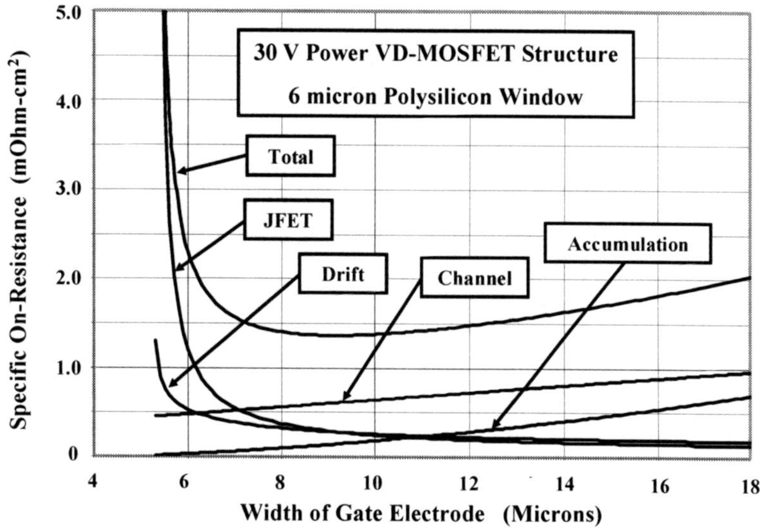

Fig. 6.51 Impact of polysilicon window size on the optimization of a 30-V power VD-MOSFET linear cell structure

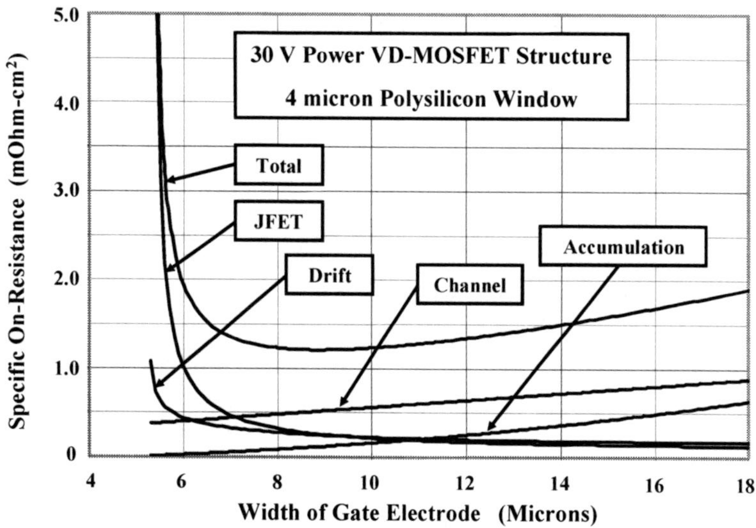

Fig. 6.52 Impact of polysilicon window size on the optimization of a 30-V power VD-MOSFET linear cell structure

The change in the optimization curves for the power VD-MOSFET structure with the size of the polysilicon window can be analyzed by using model A for the specific on-resistance. The results obtained for the case of a power VD-MOSFET structure with 30-V operating capability are provided in Figs. 6.51 and 6.52 by reducing the polysilicon window to 6 and 4 μm, respectively. It can be observed from these figures that the total specific on-resistance reduces when the polysilicon window is made smaller. In this particular example, the specific on-resistance is reduced from 1.54 to 1.36 to 1.21 mΩ cm^2 as the polysilicon window is reduced from 8 to 6 to 4 μm. This indicates that an approximately 10% enhancement in performance can be achieved with each of the successive iterations in reducing the polysilicon window size. It is also worth pointing out that the optimum size of the polysilicon electrode decreases when the polysilicon window size is reduced. In this particular example, the optimum polysilicon electrode width is reduced from 9.5 to 9.1 to 8.8 μm as the polysilicon window is reduced from 8 to 6 to 4 μm.

6.7.4 Impact of Cell Topology

In the previous sections, a linear cell topology was used during the analysis of the specific on-resistance. Many alternative cell topologies can be conceived for the power VD-MOSFET structure as illustrated in Fig. 6.15. The cross section of the device structure is the same for all of these topologies. However, the channel density varies depending upon the shape of the surface topology. The specific on-resistance for each topology can be derived by using the current flow pattern in Fig. 6.39 and accounting for the channel length (Z) within each unit cell. The hexagonal window design can be approximated as a circular cell. As discussed for the linear cell topology, the contact resistances and the N$^+$ source resistance can be neglected when compared with the other components of the internal resistance in the power VD-MOSFET structure.

Square Polysilicon Window in a Square Cell Array

The square cell array can be treated by using a unit cell with a width of W_{Cell} in Fig. 6.39. The area of the cell is then $(W_{\mathrm{Cell}})^2$. Within this unit cell space, the channel has a width of $4(W_{\mathrm{PW}} + 2x_{\mathrm{JN+}})$, where $x_{\mathrm{JN+}}$ is the depth of the N$^+$ source diffusion. The channel resistance is then given by

$$R_{\mathrm{CH}} = \frac{L_{\mathrm{CH}}}{4(W_{\mathrm{PW}} + 2x_{\mathrm{JN+}})\mu_{\mathrm{ni}} C_{\mathrm{OX}} (V_{\mathrm{G}} - V_{\mathrm{TH}})}, \tag{6.88}$$

where L_{CH} is the channel length, μ_{ni} is the inversion layer mobility, C_{OX} is the specific capacitance of the gate oxide, V_{G} is the gate bias voltage, and V_{TH} is the threshold voltage. The specific on-resistance contributed by the channel can be obtained from (6.88) by multiplying the cell resistance with the cell area:

$$R_{\text{CH,SP}} = \frac{L_{\text{CH}} W_{\text{Cell}}^2}{4(W_{\text{PW}} + 2x_{\text{JN+}})\mu_{\text{ni}} C_{\text{OX}} (V_{\text{G}} - V_{\text{TH}})}. \tag{6.89}$$

For the conservative power VD-MOSFET design with a cell width of 20 μm, polysilicon window of 10 μm, and N^+ source region depth of 1 μm, the specific resistance contributed by the channel at a gate bias of 10 V is 0.76 mΩ cm^2 if the channel length is 1 μm and the gate oxide thickness is 500 Å. This value is smaller than that obtained for the linear cell topology.

For the square array topology, the resistance of the accumulation layer between the edge of the P-base junction and the center of the gate electrode can be derived by using a distance L_A given by (6.64) with a periphery given by $4(W_{\text{PW}} + 2x_{\text{JP}})$, where x_{JP} is the depth of the P-base diffusion:

$$R_{\text{A}} = \frac{L_{\text{A}}}{4(W_{\text{G}} + 2x_{\text{JP}})\mu_{\text{nA}} C_{\text{OX}} (V_{\text{G}} - V_{\text{TH}})}. \tag{6.90}$$

The specific on-resistance contributed by the accumulation layer in the square array can be obtained from (6.90) by multiplying the above resistance with the cell area:

$$R_{\text{A,SP}} = K_{\text{A}} \frac{(W_{\text{G}} - 2x_{\text{JP}}) W_{\text{Cell}}^2}{8(W_{\text{PW}} + 2x_{\text{JP}})\mu_{\text{nA}} C_{\text{OX}} (V_{\text{G}} - V_{\text{TH}})}. \tag{6.91}$$

For the conservative power MOSFET design with a cell width of 20 μm, the specific resistance contributed by the accumulation layer in the square array at a gate bias of 10 V is 0.47 mΩ cm^2 if the P-base junction depth (x_{JP}) is 2 μm and the gate oxide thickness is 500 Å. This value is slightly smaller than that obtained for the linear cell topology.

In the square array, the cross-sectional area for the JFET region can be assumed to have a uniform cross-sectional width $a/2$ (with a given by (6.67) as in the linear cell topology), and the periphery for the JFET region can be taken to be approximately equal to four times the cell pitch. If the JFET region is assumed to extend to the bottom the P-base junction, the resistance of the JFET region can then be obtained by using

$$R_{\text{JFET}} = \frac{\rho_{\text{JFET}} x_{\text{JP}}}{2W_{\text{Cell}} (W_{\text{G}} - 2x_{\text{P}} - 2W_0)}. \tag{6.92}$$

The specific on-resistance contributed by the JFET region in the square cell array can be obtained by multiplying the above resistance with the cell area:

$$R_{\text{JFET,SP}} = \frac{\rho_{\text{JFET}} x_{\text{JP}} W_{\text{Cell}}}{2(W_{\text{G}} - 2x_{\text{P}} - 2W_0)}. \tag{6.93}$$

For the conservative power MOSFET design with a cell width of 20 μm, the specific resistance contributed by the JFET region in the square array design is 0.096 mΩ cm^2 if the P-base junction depth (x_{JP}) is 2 μm and resistivity of the JFET region is 0.267 Ω cm based upon a JFET doping concentration of 2×10^{16} cm^{-3}. This value is smaller than that obtained with the linear cell topology.

The resistance contributed by the drift region in the square cell array can be analyzed by using model A for the current flow pattern. The periphery for the current flow in the square array can be assumed to be approximately four times the cell pitch. The resistance contributed by the drift region in the square array is then given by

$$R_D = \frac{\rho_D}{4W_{Cell}} \ln\left[\frac{a+2t}{a}\right]. \tag{6.94}$$

The specific on-resistance contributed by the drift region in the square array can be obtained by multiplying the above resistance with the cell area:

$$R_{D,SP} = \frac{\rho_D W_{Cell}}{4} \ln\left[\frac{a+2t}{a}\right]. \tag{6.95}$$

For the conservative power MOSFET design with a cell width of 20 μm, the specific resistance contributed by the drift region for the square array is 0.168 mΩ cm^2 if the P-base junction depth (x_{JP}) is 2 μm and resistivity of the drift region is 0.50 Ω cm based upon a doping concentration of 1×10^{16} cm^{-3}. This value is much smaller than that obtained with the linear cell topology.

The specific resistance contributed by the N$^+$ substrate in the square array is the same as that for the linear cell topology:

$$R_{SUB,SP} = \rho_{SUB} t_{SUB}, \tag{6.96}$$

where ρ_{SUB} and t_{SUB} are the resistivity and thickness of the N$^+$ substrate, respectively. The specific resistance contributed by the substrate can be assumed to be 0.06 mΩ cm^2. The total specific on-resistance for the power VD-MOSFET structure with square cell topology can be computed by adding all the above components. For the case of the conservative power VD-MOSFET structure designed with a cell pitch (W_{Cell}) of 20 μm to support 50 V, the total specific on-resistance for the square array is 1.56 mΩ cm^2, which is 30% smaller than that for the linear cell topology.

Hexagonal Window in a Hexagonal Cell Array

The hexagonal cell array with hexagonal windows can be analyzed by approximating these features as circular in shape with a diameter of W_{Cell} in the cross section shown in Fig. 6.39. The area of the cell is then $(\pi/4)(W_{Cell})^2$. Within this unit cell space, the channel has a width of $\pi(W_{PW} + 2x_{JN+})$, where x_{JN+} is the depth of the N$^+$ source diffusion. The channel resistance is then given by

$$R_{CH} = \frac{L_{CH}}{\pi(W_{PW} + 2x_{JN+})\mu_{ni}C_{OX}(V_G - V_{TH})}, \tag{6.97}$$

where L_{CH} is the channel length, μ_{ni} is the inversion layer mobility, C_{OX} is the specific capacitance of the gate oxide, V_G is the gate bias voltage, and V_{TH} is the threshold voltage. The specific on-resistance contributed by the channel in the power VD-MOSFET structure can be obtained from (6.97) by multiplying the cell resistance with the cell area:

$$R_{CH,SP} = \frac{L_{CH}W_{Cell}^2}{4(W_{PW} + 2x_{JN+})\mu_{ni}C_{OX}(V_G - V_{TH})}, \tag{6.98}$$

which is identical to that for the square array. For the conservative power MOSFET design with a cell diameter of 20 μm, polysilicon window of 10 μm, and N$^+$ source region depth of 1 μm, the specific resistance contributed by the channel at a gate bias of 10 V is 0.76 mΩ cm^2 if the channel length is 1 μm and the gate oxide thickness is 500 Å. This value is slightly smaller than that obtained for the linear cell topology.

For the circular array topology, the resistance of the accumulation layer between the edge of the P-base junction and the center of the gate electrode can be derived by using a distance L_A given by (6.64) with a periphery given by $\pi(W_{PW} + 2x_{JP})$, where x_{JP} is the depth of the P-base diffusion:

$$R_A = \frac{L_A}{\pi(W_G + 2x_{JP})\mu_{nA}C_{OX}(V_G - V_{TH})}. \tag{6.99}$$

The specific on-resistance contributed by the accumulation layer in the circular array can be obtained from (6.99) by multiplying the above resistance with the cell area:

$$R_{A,SP} = K_A \frac{(W_G - 2x_{JP})W_{Cell}^2}{8(W_{PW} + 2x_{JP})\mu_{nA}C_{OX}(V_G - V_{TH})}, \tag{6.100}$$

which is identical to that for the square array. For the conservative power MOSFET design with a cell diameter of 20 μm, the specific resistance contributed by the accumulation layer in the circular array at a gate bias of 10 V is 0.47 mΩ cm^2 if the P-base junction depth (x_{JP}) is 2 μm and the gate oxide thickness is 500 Å. This value is slightly smaller than that obtained for the linear cell topology.

In the circular array, the cross-sectional area for the JFET region can be assumed to have a uniform cross-sectional width $a/2$ (with a given by (6.67) as in the linear cell topology), and the periphery for the JFET region can be taken to be approximately equal to π times the cell diameter. If the JFET region is assumed to extend to the bottom the P-base junction, the resistance of the JFET region can then be obtained by using

$$R_{JFET} = \frac{2\rho_{JFET} x_{JP}}{\pi W_{Cell}(W_G - 2x_P - 2W_0)}. \tag{6.101}$$

The specific on-resistance contributed by the JFET region in the circular cell array can be obtained by multiplying the above resistance with the cell area:

$$R_{JFET,SP} = \frac{\rho_{JFET} x_{JP} W_{Cell}}{2(W_G - 2x_P - 2W_0)}, \tag{6.102}$$

which is identical to that for the square array. For the conservative power MOSFET design with a cell diameter of 20 μm, the specific resistance contributed by the JFET region in the circular array design is 0.096 mΩ cm^2 if the P-base junction depth (x_{JP}) is 2 μm and resistivity of the JFET region is 0.267 Ω cm based upon a JFET doping concentration of 2×10^{16} cm^{-3}. This value is smaller than that obtained with the linear cell topology.

The resistance contributed by the drift region in the circular cell array can be analyzed by using model A for the current flow pattern. The periphery for the current flow in the circular array can be assumed to be approximately π times the cell diameter. The resistance contributed by the drift region in the circular array is then given by

$$R_D = \frac{\rho_D}{\pi W_{Cell}} \ln\left[\frac{a+2t}{a}\right]. \tag{6.103}$$

The specific on-resistance contributed by the drift region in the circular array can be obtained by multiplying the above resistance with the cell area:

$$R_{D,SP} = \frac{\rho_D W_{Cell}}{4} \ln\left[\frac{a+2t}{a}\right], \tag{6.104}$$

which is identical to that for the square array. For the conservative power MOSFET design with a cell diameter of 20 μm, the specific resistance contributed by the drift region for the circular array is 0.168 mΩ cm^2 if the P-base junction depth (x_{JP}) is 2 μm and resistivity of the drift region is 0.50 Ω cm based upon a doping concentration of 1×10^{16} cm^{-3}. This value is much smaller than that obtained with the linear cell topology.

The specific resistance contributed by the N$^+$ substrate in the circular array is the same as that for the linear cell topology:

$$R_{SUB,SP} = \rho_{SUB} t_{SUB}, \tag{6.105}$$

where ρ_{SUB} and t_{SUB} are the resistivity and thickness of the N^+ substrate, respectively. The specific resistance contributed by the substrate can be assumed to be 0.06 mΩ cm^2. Since all the specific on-resistance components for the circular (hexagonal) cell topology are identical to those for the square cell topology, the total specific on-resistance for the power VD-MOSFET structure with circular (hexagonal) cell topology is also 1.56 mΩ cm^2, which is smaller than that for the linear cell topology.

Atomic Lattice Layout Cell Array

The ALL cell topology can be analyzed by using the same procedure as the other topologies described above with the cross section shown in Fig. 6.39. The area of the ALL cell is $(W_{Cell})^2$. However, the cell width is given by

$$W_{Cell} = W_G + W_B, \tag{6.106}$$

where W_G is the diameter of the circular polysilicon islands and W_B is the width of the bars connecting the polysilicon islands. These bars can be shorter than the polysilicon windows in the previous designs because the contact to the N^+ source and P-base regions is made at the four corners of the cells.

Within the ALL unit cell space, the channel has a width of $\pi(W_G - 2x_{JN+})$, where x_{JN+} is the depth of the N^+ source diffusion. The channel resistance is then given by

$$R_{CH} = \frac{L_{CH}}{\pi(W_G - 2x_{JN+})\mu_{ni} C_{OX}(V_G - V_{TH})}, \tag{6.107}$$

where L_{CH} is the channel length, μ_{ni} is the inversion layer mobility, C_{OX} is the specific capacitance of the gate oxide, V_G is the gate bias voltage, and V_{TH} is the threshold voltage. The specific on-resistance contributed by the channel in the power VD-MOSFET structure can be obtained from (6.107) by multiplying the cell resistance with the cell area:

$$R_{CH,SP} = \frac{L_{CH} W_{Cell}^2}{\pi(W_G - 2x_{JN+})\mu_{ni} C_{OX}(V_G - V_{TH})}. \tag{6.108}$$

For the conservative power MOSFET design with a gate width of 10 μm and a bar width of 4 μm, the cell pitch is reduced to 14 μm in the ALL topology. For an N^+ source region depth of 1 μm, the specific resistance contributed by the channel at a gate bias of 10 V for the ALL cell topology is 0.715 mΩ cm^2 if the channel length is 1 μm and the gate oxide thickness is 500 Å. This value is slightly smaller than that obtained for the square and hexagonal cell topologies.

For the ALL cell topology, the resistance of the accumulation layer between the edge of the P-base junction and the center of the gate electrode can be derived by taking into account the circular geometry for the current flow. The resistance of a small elemental segment of the accumulation layer located at a distance r from the center of the polysilicon gate pad is given by

$$dR_A = \frac{dr}{2\pi r \mu_{nA} C_{OX}(V_G - V_{TH})}. \tag{6.109}$$

The resistance contributed by the accumulation layer can be obtained by integration of the resistance from the center of the polysilicon pad to the edge of the accumulation layer located at a radius of $(0.5W_G - x_{JP})$. Since the resistance tends to infinity at the center of the polysilicon pad, it is convenient to assume that current flow stops at a small radius δ from the center of the polysilicon pad. In this case, the resistance contributed by the accumulation layer in the ALL cell topology is given by

$$R_A = \frac{1}{2\pi \mu_{nA} C_{OX}(V_G - V_{TH})} \ln\left(\frac{W_G - 2x_{JP}}{\delta}\right). \tag{6.110}$$

If the value for the parameter δ is chosen as

$$\delta = 0.37(W_G - 2x_{JP}), \tag{6.111}$$

then the resistance of the accumulation layer is given by

$$R_A = \frac{1}{2\pi \mu_{nA} C_{OX}(V_G - V_{TH})}. \tag{6.112}$$

The specific on-resistance contributed by the accumulation layer in the ALL cell topology can be obtained from (6.112) by multiplying the above resistance with the cell area:

$$R_{A,SP} = K_A \frac{W_{Cell}^2}{2\pi \mu_{nA} C_{OX}(V_G - V_{TH})}. \tag{6.113}$$

For the conservative power MOSFET design with a cell pitch of 14 μm, the specific resistance contributed by the accumulation layer in the ALL topology at a gate bias of 10 V is 0.685 mΩ cm^2 if the P-base junction depth (x_{JP}) is 2 μm and the gate oxide thickness is 500 Å. This value is slightly larger than that obtained for the square and hexagonal cell topologies.

In the ALL cell topology, the cross-sectional area for the JFET region can be assumed to have a uniform cross-sectional diameter a (with a given by (6.67) as in the linear cell topology). If the JFET region is assumed to extend to the bottom the P-base junction, the resistance of the JFET region can then be obtained by using

$$R_{\text{JFET}} = \frac{4\rho_{\text{JFET}} x_{\text{JP}}}{\pi (W_{\text{G}} - 2x_{\text{P}} - 2W_0)^2}. \tag{6.114}$$

The specific on-resistance contributed by the JFET region in the ALL cell topology can be obtained by multiplying the above resistance with the cell area:

$$R_{\text{JFET,SP}} = \frac{4\rho_{\text{JFET}} x_{\text{JP}} W_{\text{Cell}}^2}{\pi (W_{\text{G}} - 2x_{\text{P}} - 2W_0)^2}. \tag{6.115}$$

For the conservative power MOSFET design with a cell width of 14 μm, the specific resistance contributed by the JFET region in the ALL design is 0.436 mΩ cm^2 if the P-base junction depth (x_{JP}) is 2 μm and resistivity of the JFET region is 0.267 Ω cm based upon a JFET doping concentration of 2×10^{16} cm^{-3}. This value is larger than that obtained for the square and hexagonal cell topologies.

The resistance contributed by the drift region in the ALL cell topology can be analyzed by using model A for the current flow pattern. However, in this case, the current flows in a circular geometry with a cross-sectional area of $(\pi / 4)(a + 2y)^2$ at a depth y below the P-base junction. The resistance contributed by the drift region in the ALL cell topology is then given by

$$R_{\text{D}} = \frac{2\rho_{\text{D}}}{\pi} \left[\frac{2t}{a(a+2t)} \right]. \tag{6.116}$$

The specific on-resistance contributed by the drift region in the ALL topology can be obtained by multiplying the above resistance with the cell area:

$$R_{\text{D,SP}} = \frac{2\rho_{\text{D}}}{\pi} \left[\frac{2t}{a(a+2t)} \right] W_{\text{Cell}}^2. \tag{6.117}$$

For the conservative power MOSFET design with a cell pitch of 14 μm, the specific resistance contributed by the drift region in the ALL topology is 0.554 mΩ cm^2 if the P-base junction depth (x_{JP}) is 2 μm and resistivity of the drift region is 0.50 Ω cm based upon a doping concentration of 1×10^{16} cm^{-3}. This value is larger than that obtained with the linear cell topology.

The specific resistance contributed by the N$^+$ substrate in the ALL topology is the same as that for the linear cell topology:

$$R_{\text{SUB,SP}} = \rho_{\text{SUB}} t_{\text{SUB}}, \tag{6.118}$$

where ρ_{SUB} and t_{SUB} are the resistivity and thickness of the N$^+$ substrate, respectively. The specific resistance contributed by the substrate can be assumed to be 0.06 mΩ cm^2. The total specific on-resistance for the power VD-MOSFET structure with the ALL cell topology is 2.39 mΩ cm^2, which is close to that for the linear cell topology.

6.8 Power U-MOSFET On-Resistance

The power U-MOSFET structure is shown in Fig. 6.53 with its internal resistance components. The same internal resistances encountered in the power VD-MOSFET structure are present in the power U-MOSFET structure with the exception of the JFET region resistance. The JFET region is eliminated within the power U-MOSFET structure because the trench extends beyond the bottom of the P-base region to form a channel connecting the N^+ source region with the N-drift region. The elimination of the JFET region allows a significant reduction of the overall specific on-resistance for the power U-MOSFET structure, not only because its resistance is excluded but also more importantly because the cell pitch can be made much smaller than that of the power VD-MOSFET structure. A smaller cell pitch reduces the specific resistance contributions from the channel, accumulation, and drift regions.

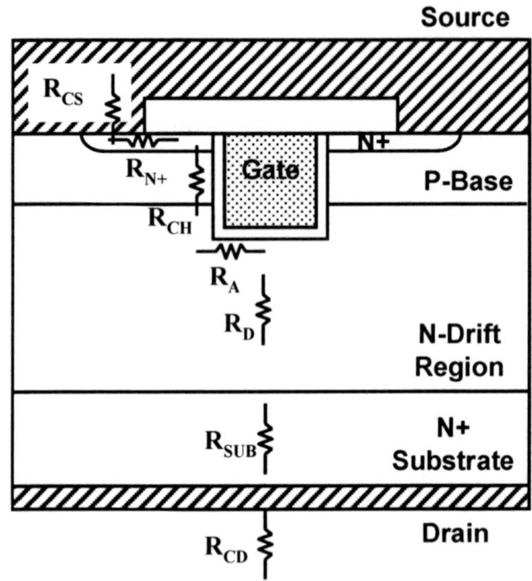

Fig. 6.53 Power U-MOSFET structure with its internal resistances

The total on-resistance for the power U-MOSFET structure is obtained by the addition of all the resistances because they are considered to be in series in the current path between the source and drain electrodes:

$$R_{ON} = R_{CS} + R_{N+} + R_{CH} + R_A + R_D + R_{SUB} + R_{CD}. \tag{6.119}$$

Each of the resistances within the power U-MOSFET structure is analyzed below. It is customary to utilize the linear cell surface topology for the power U-MOSFET structure because the trench surface can be oriented in the preferred direction most favorable for producing high-quality etched surfaces.

A cross section of the power U-MOSFET structure is illustrated in Fig. 6.54 with various dimensions that can be used for the analysis of the on-resistance components. The edges of the masks used to define the boundary for the N^+ source ion implant and the contact window are also shown in the figure by the dimensions W_S and W_C. These edges decide the area available within the structure for making contact to the N^+ source region. The position of the N^+ source definition mask also determines the length (L_{N+}) of the N^+ source region.

In addition, the current flow pattern in the U-MOSFET structure is indicated by the shaded area in the figure. After entering the drift region, the current is assumed to spread from the trench bottom at a 45° angle. Since the mesa width is small, the current paths will usually overlap as illustrated in the figure. Consequently, the area for current transport varies for a portion of the drift region and then becomes uniform for the rest of the drift region.

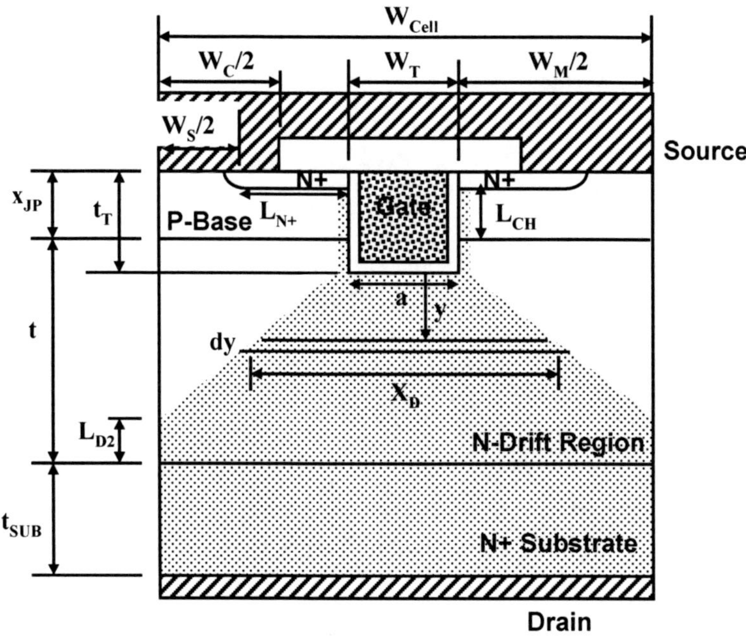

Fig. 6.54 Power U-MOSFET structure with current flow model used for analysis of its internal resistances

6.8.1 Source Contact Resistance

The contact to the N^+ source region is made over a relatively small area in the U-MOSFET structure because the size of the mesa region between the trenches must be minimized to obtain the lowest possible specific on-resistance as demonstrated later in this section. Further, the area for the contact is limited by the need to simultaneously contact the P-base region with the source electrode to suppress the parasitic N–P–N bipolar transistor. During device fabrication, one edge of the N^+ source region is defined by the trench while the other end of the N^+ source region is defined by using a photoresist boundary. The contact between the source metal electrode and the N^+ source region occurs only from this boundary to the edge of the contact window which is defined by another photolithographic masking step. Although the gate electrode is recessed below the semiconductor surface in the U-MOSFET structure because it is imbedded inside the trench, the contact window must still be displaced from the edges of the trenches to prevent gate–source short circuits from occurring.

For computation of the contact resistance between the N^+ source region and its electrode, it is necessary to determine the contact area for the source region. The contact area for the N^+ source region is determined by the difference in width of the contact window (W_C) and the N^+ source ion implant window (W_S). The contact resistance to each of the N^+ source regions within the power U-MOSFET cell structure can be obtained by dividing the specific contact resistance (ρ_C), which is determined by the work function of the contact metal and the doping concentration at the surface of the N^+ region as discussed in Chap. 2, with the contact area:

$$R_{CS} = \frac{2\rho_C}{Z(W_C - W_S)}, \tag{6.120}$$

where Z is the length of the cell in the orthogonal direction to the cross section shown in the figure. The specific contact resistance for the source regions within the power U-MOSFET structure can then be obtained by multiplying the source contact resistance with the cell area ($W_{Cell}Z$), while recognizing that there are two source regions within each cross section shown in the figure:

$$R_{CS,SP} = \rho_C \frac{W_{Cell}}{W_C - W_S}. \tag{6.121}$$

This expression indicates that the contact resistance to the N^+ source regions can be significantly enhanced when the design rules are reduced to shrink the size of the mesa region. In this expression, the cell width is determined by the trench and the mesa widths:

$$W_{Cell} = W_T + W_M. \tag{6.122}$$

To illustrate the amplification of the contact resistance to the source region, consider a power U-MOSFET structural design with a trench width (W_T) of 1 μm, a mesa width (W_M) of 4 μm, a source contact window width (W_C) of 3 μm, and a source boundary width (W_S) of 2 μm. In this design, the width of the contact to each of the N^+ source regions is only 0.5 μm while the cell width (W_{Cell}) is 5 μm. Consequently, the contact resistance to the source gets amplified by a factor of 5 times. It is necessary to ensure that the specific contact resistance (ρ_C) is less than $1 \times 10^{-5}\ \Omega\ \text{cm}^2$, by increasing the surface doping concentration of the N^+ region above $5 \times 10^{19}\ \text{cm}^{-3}$, to avoid a strong impact on the total on-resistance for the power MOSFET structure. It is common practice to also use a low barrier height contact metal such as titanium or titanium silicide to reduce the specific contact resistance.

6.8.2 Source Region Resistance

From the N^+ source region contact, the current must flow along the source region until it reaches the channel. The resistance contributed by the source region is determined by the sheet resistance of the N^+ diffusion (ρ_{SQN+}) and its length (L_{N+}):

$$R_{SN+} = \rho_{SQN+} \frac{L_{N+}}{Z}. \tag{6.123}$$

The length of the N^+ source region can be related to the windows shown in the figure during the fabrication of the cell:

$$L_{N+} = \frac{W_M - W_S}{2}. \tag{6.124}$$

The specific resistance contributed by the current flow through the N^+ source region is obtained by multiplying its resistance with the cell area and taking into account the presence of two source regions within each cell:

$$R_{SN+,SP} = \frac{\rho_{SQN+} L_{N+} W_{Cell}}{2}. \tag{6.125}$$

The specific resistance of the N^+ source region is usually negligible in the power U-MOSFET structure. In the case of the power U-MOSFET structure with cell pitch of 5 μm, the length of the N^+ source region is 1 μm. For a typical sheet resistance of 20 Ω sq^{-1} for the N^+ source region with a junction depth of 1 μm, the specific resistance contributed by current flow through the source region is only 0.0005 mΩ cm^2.

6.8.3 Channel Resistance

The power U-MOSFET structure shown in Fig. 6.54 contains channels formed on both of the vertical sidewalls of the trench-gate structure. The resistance contributed by each of the channels is given by

$$R_{CH} = \frac{L_{CH}}{Z \mu_{ni} C_{OX} (V_G - V_{TH})}, \tag{6.126}$$

where L_{CH} is the channel length, μ_{ni} is the inversion layer mobility, C_{OX} is the specific capacitance of the gate oxide, V_G is the gate bias voltage, and V_{TH} is the threshold voltage. In the case of the power U-MOSFET structure, the channel length is defined by the difference in the depth of the P-base and N^+ source junctions:

$$L_{CH} = x_{JP} - x_{N+}, \tag{6.127}$$

where x_{JP} is the junction depth of the P-base region and x_{N+} is the junction depth of the N^+ source region.

The specific on-resistance contributed by the channel in the power VD-MOSFET structure can be obtained from (6.126) by multiplying the cell resistance with the cell area after accounting for the fact that there are two channels within each cross section of the structure that feed the current from the source into the drift region:

$$R_{CH,SP} = \frac{L_{CH} W_{Cell}}{2\mu_{ni} C_{OX}(V_G - V_{TH})}. \tag{6.128}$$

For the power U-MOSFET structure with a cell width of 5 μm, the specific resistance contributed by the channel at a gate bias of 10 V is 0.229 mΩ cm^2 if the channel length is 1 μm and the gate oxide thickness is 500 Å. This is much smaller than the channel contribution within the power VD-MOSFET structure. The reduction in the specific on-resistance contributed by the channel in the U-MOSFET structure is due to the increase in the channel density.

6.8.4 Accumulation Resistance

In the power U-MOSFET structure, the current flowing through the inversion channel enters the drift region without encountering a JFET region. The current spreads from the edge of the P-base junction (J_1) along the surface of the trench due to the formation of an accumulation layer because of the positive gate bias. It is then distributed into the drift region. For an analytical treatment, it is convenient to first calculate the resistance for current flow via the accumulation layer from the edge of the P-base region to the center of the trench. This distance (L_A) consists of a portion along the vertical sidewalls of the trench and a portion along its bottom surface:

$$L_A = t_T - x_{JP} + \left(\frac{W_T}{2}\right). \tag{6.129}$$

The resistance of the accumulation layer between the edge of the P-base junction and the center of the trench is then given by

$$R_A = \frac{L_A}{Z\mu_{nA} C_{OX}(V_G - V_{TH})}. \tag{6.130}$$

The specific on-resistance contributed by the accumulation layer in the power U-MOSFET structure can be obtained from (6.130) by multiplying the above resistance with the cell area after accounting for the fact that there are two accumulation layer paths within each cross section of the structure that feed the current from the two source regions into the drift region:

$$R_{A,SP} = K_A \frac{L_A W_{Cell}}{2\mu_{nA} C_{OX}(V_G - V_{TH})}. \tag{6.131}$$

In this equation, a coefficient K_A has been introduced to account for the current spreading from the accumulation layer into the drift region as was done for the VD-MOSFET structure. A typical value for this coefficient is 0.6 based upon the current flow observed from numerical simulations of power U-MOSFET structures. The threshold voltage in the expression is for the on-set of formation of the accumulation layer. Although this has a lower value than the threshold voltage previously derived for the formation of the inversion layer, it is convenient to use the same threshold voltage when performing the analytical computations.

For the power U-MOSFET structure with a cell width of 5 μm, the specific resistance contributed by the accumulation layer at a gate bias of 10 V is 0.055 mΩ cm^2 if the P-base junction depth (x_{JP}) is 1.5 μm, the trench depth is 2 μm, and the gate oxide thickness is 500 Å. This value is an order of magnitude smaller than that for the power VD-MOSFET structure because the accumulation layer path (L_A) is only 1 μm in size and the cell pitch is much smaller for the U-MOSFET structure. In some power U-MOSFET structures, a thicker oxide is used at the bottom of the trench to reduce the gate–drain capacitance. In these devices, the accumulation layer resistance will be increased due to the smaller charge within the layer.

6.8.5 Drift Region Resistance

The resistance contributed by the drift region in the power U-MOSFET structure is greater than that for the ideal drift region discussed earlier in this chapter due to current spreading from the trench surface into the drift region. The cross-sectional area for the current flow in the drift region increases from the width *a*, which is the width of the trench for the U-MOSFET structure, as illustrated in Fig. 6.54 by the shaded area.

The resistance of the drift region developed in this section is based upon the current flow pattern shown in Fig. 6.54. In this pattern, it is assumed that the cross-section width (X_D) for current flow increases at a 45° angle from the bottom of the trench. It is also assumed that the width of the mesa region is sufficiently small when compared with the thickness of the drift region to allow the current flow paths to merge when the current reaches the N$^+$ substrate. The resistance of the drift region is now determined by two portions: a first portion with a cross-sectional area that increases with the depth and a second portion with a uniform cross-sectional area for the current flow.

As illustrated in Fig. 6.54, the width of the cross-sectional area for the first portion of the current path at a depth y below the JFET region is given by

$$X_{\mathrm{D}} = a + 2y. \tag{6.132}$$

The resistance of an elemental segment of thickness dy at this depth below the trench bottom surface is then given by

$$\mathrm{d}R_{\mathrm{D}} = \frac{\rho_{\mathrm{D}}\mathrm{d}y}{ZX_{\mathrm{D}}} = \frac{\rho_{\mathrm{D}}\mathrm{d}y}{Z(a+2y)}. \tag{6.133}$$

The resistance contributed by the first portion of the drift region can be obtained by integration of the above elemental resistance over the limits $y = 0$ and $y = (W_{\mathrm{M}}/2)$:

$$R_{\mathrm{D1}} = \frac{\rho_{\mathrm{D}}}{2Z}\ln\left[\frac{a+W_{\mathrm{M}}}{a}\right] = \frac{\rho_{\mathrm{D}}}{2Z}\ln\left[\frac{W_{\mathrm{T}}+W_{\mathrm{M}}}{W_{\mathrm{T}}}\right], \tag{6.134}$$

because $a = W_{\mathrm{T}}$. The resistance contributed by the second portion of the drift region is determined by current flowing through a uniform cross-sectional width (W_{Cell}). The distance (L_{D2}, shown in Fig. 6.54) over which the cross-sectional area is uniform is given by

$$L_{\mathrm{D2}} = t + x_{\mathrm{JP}} - t_{\mathrm{T}} - \frac{W_{\mathrm{M}}}{2}. \tag{6.135}$$

The resistance of this portion of the drift region is then given by

$$R_{\mathrm{D2}} = \frac{\rho_{\mathrm{D}}}{ZW_{\mathrm{Cell}}}\left(t + x_{\mathrm{JP}} - t_{\mathrm{T}} - \frac{W_{\mathrm{M}}}{2}\right). \tag{6.136}$$

The specific on-resistance contributed by the drift region in the power U-MOSFET structure can be obtained by multiplying the above two resistances with the cell area:

$$R_{\mathrm{D,SP}} = \frac{\rho_{\mathrm{D}}W_{\mathrm{Cell}}}{2}\ln\left[\frac{W_{\mathrm{T}}+W_{\mathrm{M}}}{W_{\mathrm{T}}}\right] + \rho_{\mathrm{D}}\left(t + x_{\mathrm{JP}} - t_{\mathrm{T}} - \frac{W_{\mathrm{M}}}{2}\right). \tag{6.137}$$

Consider a power U-MOSFET structure constructed using a drift region with doping concentration of 1×10^{16} cm^{-3} corresponding to a blocking voltage capability of 50 V. For a cell width of 5 μm, the specific resistance contributed by the drift region calculated by using the above model is 0.209 mΩ cm^2. This is 60% smaller than the value for the power VD-MOSFET structure.

6.8.6 N$^+$ Substrate Resistance

When the current reaches the bottom of the N-drift region, it is very quickly distributed throughout the heavily doped N$^+$ substrate. The current flow through the substrate can therefore be assumed to occur with a uniform cross-sectional area. Under this assumption, the specific resistance contributed by the N^+ substrate is given by

$$R_{SUB,SP} = \rho_{SUB} t_{SUB}, \tag{6.138}$$

where ρ_{SUB} and t_{SUB} are the resistivity and thickness of the N^+ substrate, respectively. When processing the wafers used to manufacture power MOSFET structures, it is necessary to use a substrate with a thickness of 500 μm on which the thin (2–50 μm) epitaxial drift region is grown to avoid breakage during handling of the wafers in the processing equipment. The lowest available resistivity for a typical phosphorus-doped silicon wafer is 0.003 Ω cm. The specific resistance contributed by this wafer is 0.15 mΩ cm^2. This value makes a large contribution to the specific on-resistance in a power U-MOSFET structure designed to support low (<50 V) voltages. With the wafer thickness reduced to 200 μm after device fabrication, the specific resistance contributed by the substrate can be reduced to 0.06 mΩ cm^2. It is therefore common practice to reduce the thickness of the wafer just before application of drain electrode metal. In addition, special wafer doping techniques have evolved that allow reduction of the resistivity of the N^+ substrate to 0.001 Ω cm range.

6.8.7 Drain Contact Resistance

Before entering the drain electrode, the current flows through the contact resistance between the drain metal and the N^+ substrate. Due to the uniform current flow at the drain contact, its resistance is not amplified unlike in the case of the source contact. It is therefore satisfactory to achieve a contact resistivity of 1×10^{-5} Ω cm^2. This can be obtained by using a titanium contact layer, to achieve a low barrier height, followed by a coating of nickel and silver. The nickel acts as a barrier between the titanium and the silver. The silver layer is ideal for mounting the chips to the package by using solders.

6.8.8 Total On-Resistance

The total specific on-resistance for the power U-MOSFET structure can be computed by adding all the above seven components. For the case of the power U-MOSFET structure designed with a cell pitch (W_{Cell}) of 5 μm to support 50 V, the total specific on-resistance is 0.613 mΩ cm^2. The contributions from each of the seven components of the on-resistance are summarized in Fig. 6.55 together with their percentage contributions. By comparing the numbers, it can be observed that the channel and drift region resistance are the largest components. The ideal specific on-resistance for a device capable of supporting 50 V obtained by using (6.2) is 0.105 mΩ cm^2. Consequently, the specific on-resistance for this design of the power U-MOSFET structure is six times larger than the ideal case, which is much better than that achieved with the power VD-MOSFET structure. The specific on-resistance for the U-MOSFET structure can be brought closer to the ideal value by using a smaller mesa width as discussed in Sect. 6.9.

Resistance	Value (mOhm-cm^2)	Percentage Contribution
Source Contact ($R_{CS,SP}$)	0.05	8.2
Source ($R_{N+,SP}$)	0.0005	0.0
Channel ($R_{CH,SP}$)	0.229	37.4
Accumulation ($R_{A,SP}$)	0.055	9.0
Drift ($R_{D,SP}$)	0.209	34.1
Substrate ($R_{SUB,SP}$)	0.06	9.8
Drain Contact ($R_{DS,SP}$)	0.01	1.6
Total ($R_{T,SP}$)	0.613	100

Fig. 6.55 On-resistance components within the 50-V power U-MOSFET structure with 5-μm cell pitch

Simulation Example

To validate the above model for the on-resistance of the power U-MOSFET structure, the results of two-dimensional numerical simulations on a structure designed with a 30-V rating are described here. The structure had a drift region thickness of 3 μm with a doping concentration of 1.3×10^{16} cm^{-3} located below the

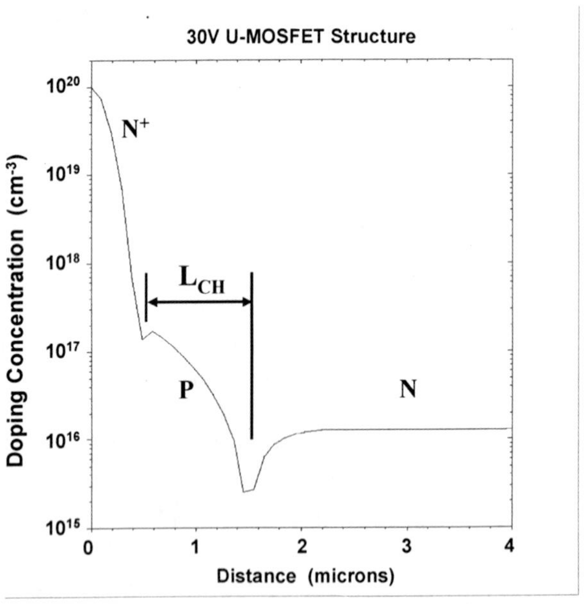

Fig. 6.56 Channel doping profile for the U-MOSFET structure

P-base region. The P-base region and N^+ source regions have depths of 0.5 and 1.5 μm, respectively, as shown in the doping profile in Fig. 6.56, leading to a channel length of 1 μm. The maximum compensated P-type doping concentration in the channel is 1.5×10^{17} cm^{-3}. In conjunction with a gate oxide thickness of 500 Å, a threshold voltage of 3.5 V was obtained for this doping concentration.

To analyze the current flow pattern in the power U-MOSFET structure, the above device structure was turned on by the application of a gate bias of 10 V. The on-state current flow pattern at a small drain bias of 0.1 V is shown in Fig. 6.57. In the figure, the depletion layer boundary is shown by the dotted lines and the junction boundary is delineated by the dashed line. It can be observed that the current flows from the channel and distributes into the drift region via the accumulation layer. The current spreading occurs from both the trench sidewall and the trench bottom. The current flow becomes uniform at a depth of 4 μm. This corresponds to a 45° spreading angle which is in agreement with the model developed for the specific on-resistance of the drift region.

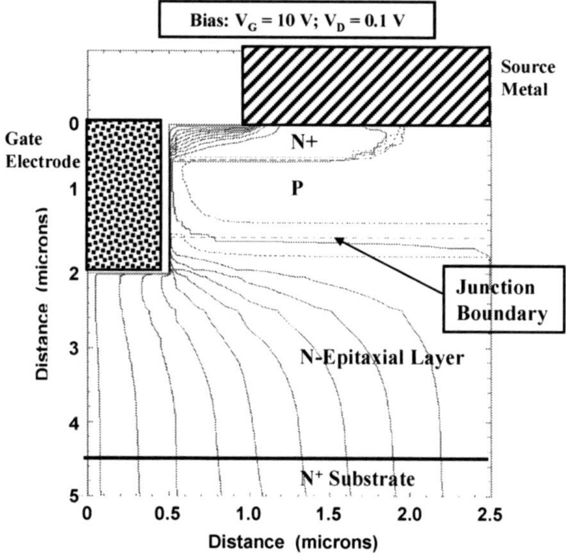

Fig. 6.57 Current flow distribution pattern in the U-MOSFET structure

From the transfer characteristics at a drain bias of 0.1 V, a threshold voltage of 3.5 V was extracted. The total specific on-resistance obtained from the numerical simulations for the above power U-MOSFET structure was 0.35 mΩ cm^2 at a gate bias of 10 V. As discussed previously for the simulations of the power VD-MOSFET structure, to compare this with the values predicted by the analytical model, it is necessary to extract the mobility in the inversion and accumulation layers within the simulated U-MOSFET structure. The channel and accumulation layer mobility were extracted by simulation of a long channel lateral MOSFET structure with the same gate oxide thickness. The inversion layer mobility was found to be 450 cm^2 V^{-1} s^{-1} while that for the accumulation layer was found to be 1,000 cm^2 V^{-1} s^{-1} at a gate bias of 10 V. When these values were used in the analytical model with the dimensions and doping concentrations for this U-MOSFET structure, the calculated value for the specific on-resistance is 0.39 mΩ cm^2 at a gate bias of 10 V. This confirms the validity of the model with a 45° spreading angle for the current flow in the drift region.

6.9 Power U-MOSFET Cell Optimization

In Sect. 6.8, it was demonstrated that the channel and drift region resistance contributions are the dominant components in the power U-MOSFET structure. These components can be reduced by decreasing the width of the mesa region. However, a reduction of the mesa width is inhibited by the placement of the contact to the P-base region at the center of the mesa. This problem can be resolved by locating the contact at selected locations orthogonal to the cross section of the structure.

6.9.1 Orthogonal P-Base Contact Structure

The width of the mesa region in the power U-MOSFET structure can be significantly reduced in size if the contact to the P-base region is located at selective position orthogonal to the cross section. This design approach is illustrated in Fig. 6.58 with a three-dimensional view of a unit cell. Since no space is occupied by the P-base contact in the mesa portion of the cross section, the mesa size can be reduced to the minimum value allowed by the design rules and process

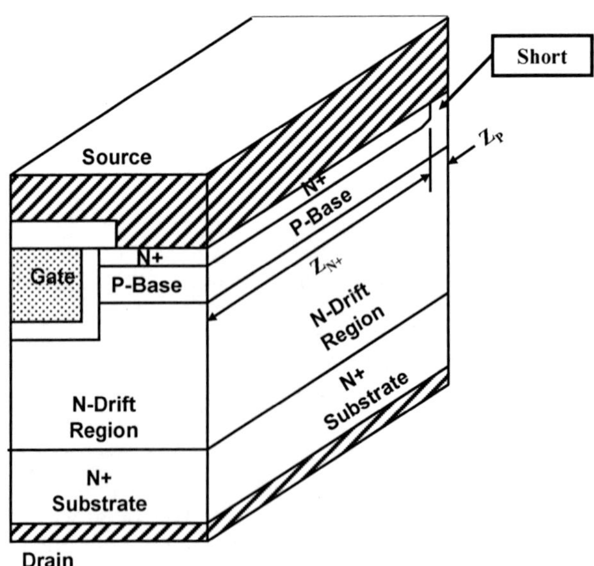

Fig. 6.58 Power U-MOSFET cell structure with P-base contact placed orthogonal to the cross section

technology for etching the trenches. In comparison to the power U-MOSFET structure discussed in the previous section with a cell pitch (W_{Cell}) of 5 μm, the cell structure in Fig. 6.58 can be designed with a cell pitch (W_{Cell}) of just 2 μm. This provides an increase in the channel density by a factor of 2.5 times resulting in a smaller specific on-resistance.

The implementation of the cell structure shown in Fig. 6.58 requires selection of the distance between the contacts to the P-base region along the device fingers. It is necessary to interrupt the N^+ source regions at the contacts to the P-base region. On the one hand, this can reduce the channel length (Z) if the distance (Z_{N+}) between the contacts to the P-base region is made small. On the other hand, if the distance (Z_{N+}) between the contacts to the P-base region is made very large, the parasitic bipolar transistor within the power U-MOSFET structure can get triggered under dynamic switching conditions due to the large P-base resistance. A design with a ratio (Z_{N+}/Z_P) of 10 provides an adequate design compromise.

Resistance	Value (mOhm-cm^2)	Percentage Contribution
Source Contact ($R_{CS,SP}$)	0.02	6.7
Source ($R_{N+,SP}$)	0.0005	0
Channel ($R_{CH,SP}$)	0.092	31.0
Accumulation ($R_{A,SP}$)	0.022	7.3
Drift ($R_{D,SP}$)	0.094	31.5
Substrate ($R_{SUB,SP}$)	0.06	20.1
Drain Contact ($R_{DS,SP}$)	0.01	3.4
Total ($R_{T,SP}$)	0.297	100

Fig. 6.59 On-resistance components within the 50-V power U-MOSFET structure with 2-μm cell pitch

For the case of the power U-MOSFET structure designed with a cell pitch (W_{Cell}) of 2 μm to support 50 V, the total specific on-resistance obtained by adding the seven components is 0.297 mΩ cm^2. The contributions from each of the seven components of the on-resistance are summarized in Fig. 6.59 together with their percentage contributions. By comparing the numbers, it can be observed that the channel and drift region resistance are significantly reduced but are still the largest components. The ideal specific on-resistance for a device capable of supporting 50 V obtained by using (6.2) is 0.105 mΩ cm^2. Consequently, the specific on-resistance for this design of the power U-MOSFET structure is only three times larger than the ideal case. It is worth pointing out that the substrate contribution is significant in this design but can be reduced with wafer thinning technology.

Simulation Example

To further validate the model for the on-resistance of the power U-MOSFET structure, the results of two-dimensional numerical simulations on a structure with cell pitch of 2 μm designed with a 30-V rating are described here. The structure had a drift region thickness of 3 μm with a doping concentration of 1.3×10^{16} cm^{-3} located below the P-base region. The P-base region and N$^+$ source regions have depths of 0.5 and 1.5 μm, respectively, as shown in the doping profile in Fig. 6.56, leading to a channel length of 1 μm. The maximum compensated P-type doping concentration in the channel is 1.5×10^{17} cm^{-3}. In conjunction with a gate oxide thickness of 500 Å, a threshold voltage of 3.5 V was obtained for this doping concentration.

To analyze the current flow pattern in the power U-MOSFET structure with the narrow mesa width, the above device structure was turned on by the application of a gate bias of 10 V. The on-state current flow pattern at a small drain bias of 0.1 V is shown in Fig. 6.60. In the figure, the depletion layer boundary is shown by the dotted lines and the junction boundary is delineated by the dashed line. It can be observed that the current flows from the channel and distributes into the drift region via the accumulation layer. The current spreading occurs from both the trench sidewall and the trench bottom. The current flow becomes uniform at a depth of 3 μm. This corresponds to a 45° spreading angle which is in agreement with the model developed for the specific on-resistance of the drift region.

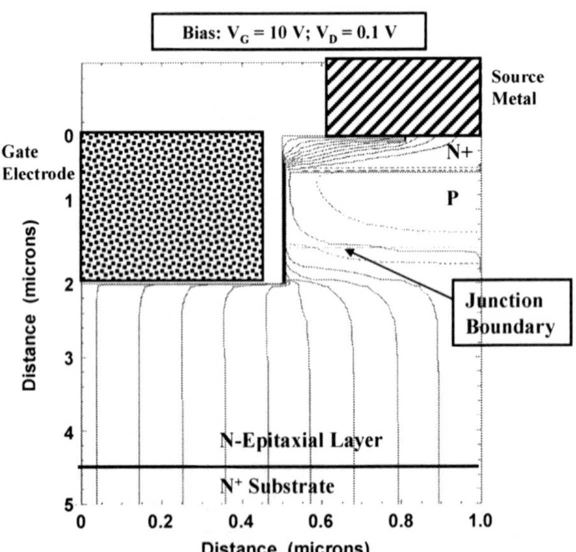

Fig. 6.60 Current flow distribution pattern in the U-MOSFET structure with 2-μm cell pitch

The total specific on-resistance obtained from the numerical simulations for the above power U-MOSFET structure was 0.22 mΩ cm^2 at a gate bias of 10 V. Using an inversion and accumulation layer mobility of 450 cm^2 V^{-1} s^{-1} and

1,000 $cm^2 V^{-1} s^{-1}$, respectively, in the analytical model with the dimensions and doping concentrations for this U-MOSFET structure, the calculated value for the specific on-resistance is 0.21 mΩ cm^2 at a gate bias of 10 V. This provides further confirmation of the validity of the model with a 45° spreading angle for the current flow in the drift region.

6.9.2 Impact of Breakdown Voltage

The breakdown voltage capability for the power U-MOSFET structure can be increased by using a higher resistivity drift region with a larger thickness. Consequently, the resistance contributed by the drift region increases with increasing breakdown voltage. The analytical model developed in the previous section can be utilized to analyze the specific on-resistance for this structure for each case. In the power U-MOSFET structure with a cell pitch (W_{Cell}) of 2 μm, the sum of the contributions from all the components, with the exception of the drift region resistance, is only 0.203 mΩ cm^2. Consequently, the total specific on-resistance for the power U-MOSFET structure approaches the ideal specific on-resistance at high breakdown voltages.

The specific on-resistance for the power U-MOSFET structure calculated by using the above model for the internal resistances is shown in Fig. 6.61 together with the ideal specific on-resistance. The calculated specific on-resistance by using model A for the power VD-MOSFET structure is included in this figure for comparison. For a 30-V power U-MOSFET device, the specific on-resistance is an order of magnitude lower than for the power VD-MOSFET structure. This value is still about an order of magnitude greater than the ideal specific on-resistance because of the substantial contribution from the N^+ substrate.

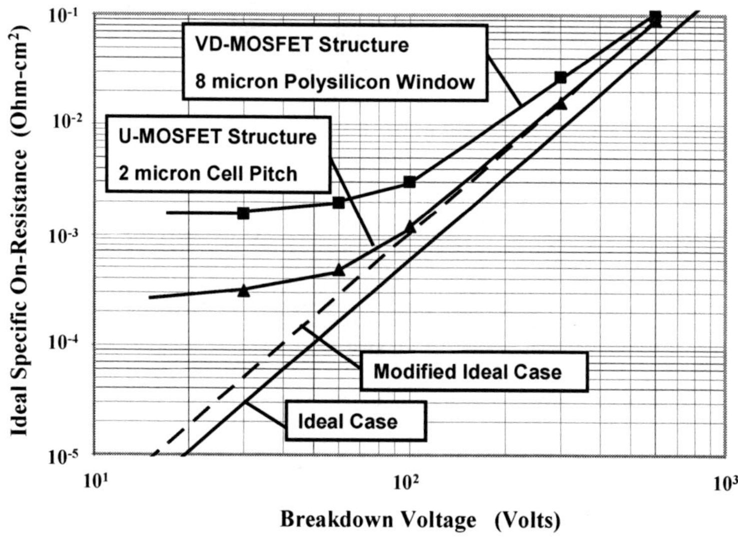

Fig. 6.61 Specific on-resistance of the power U-MOSFET linear cell structure

In the above figure, a modified specific on-resistance plot is provided (see dashed line) after accounting for an edge termination with 80% of the parallel-plane breakdown voltage. It can be observed that the specific on-resistance for the power U-MOSFET structure is very close to the modified ideal specific on-resistance for blocking voltages above 100 V. It can therefore be concluded that further improvements in the performance of the power U-MOSFET structure can be obtained by improving the edge termination design.

It is common practice in the industry to use the planar U-MOSFET architecture for devices with low blocking voltage ratings because of the much lower specific on-resistance when compared with the power VD-MOSFET structure. However, this structure is not as rugged as the power VD-MOSFET structure under adverse operating conditions such as switching inductive loads with the simultaneous presence of high voltage and current.

6.9.3 Ruggedness Improvement

The ruggedness of the power U-MOSFET structure has been found to be inferior to that of the planar VD-MOSFET structure. The primary reason for the failure of power U-MOSFET devices, under inductive load switching operation, has been found to be the enhanced electric field generated at the sharp corners of the trenches. This problem can be overcome by rounding the bottom of the trenches as illustrated in Fig. 6.62.

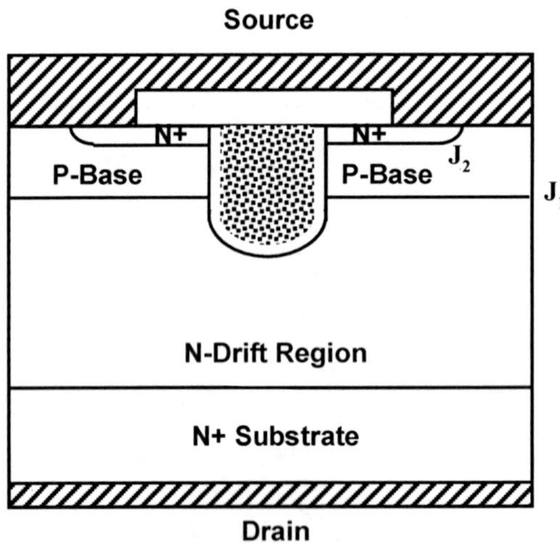

Fig. 6.62 Power U-MOSFET structure with rounded trench bottom surface

A further enhancement in the ruggedness of the power U-MOSFET structure can be obtained by the inclusion of a deep P^+ region as illustrated in Fig. 6.63. The presence of the P^+ regions reduces the electric field at the trench corners. However, it is important to restrict the P^+ regions away from the trench sidewalls to prevent a drastic increase in the threshold voltage. Since these P^+ regions must be formed with additional masking steps within the mesa region and their depth must exceed that of the P-base region to provide the benefit of reducing the electric field at the trench corners, the mesa width for this structure becomes substantially larger. For a 1-μm wide ion implant window for the P^+ regions, the mesa width must be enlarged to 5 μm to accommodate a diffusion depth of 2 μm which matches the trench depth. This makes the cell pitch larger which increases the contribution to the on-resistance from the channel. In addition, it has been reported that these deep regions can produce a JFET effect in the U-MOSFET structure when their depth becomes comparable with the trench depth.[16]

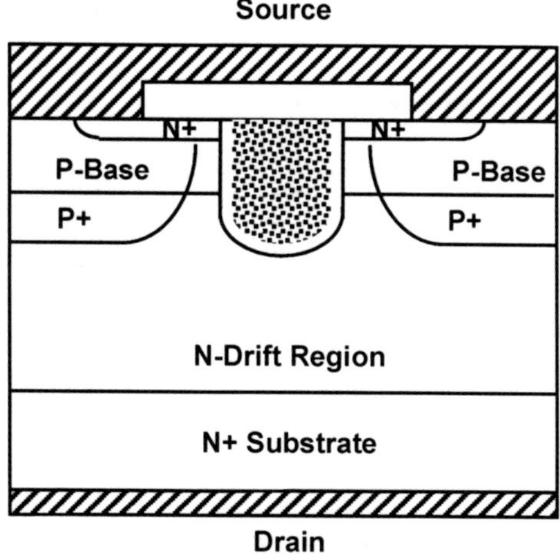

Fig. 6.63 Power U-MOSFET structure with deep P^+ regions

6.10 Square-Law Transfer Characteristics

Current flow between the source and drain electrodes can occur through the power MOSFET structure when the gate bias exceeds the threshold voltage. For a constant drain bias voltage in excess of the pinch-off voltage for the channel, the saturated drain current ($I_{D,sat}$) is related to the gate voltage by (6.49) derived for the lateral MOSFET structure. When applied to the power MOSFET structure, the drain current per unit area is given by

$$J_{D,sat} = \frac{2I_{D,sat}}{W_{Cell}Z} = \frac{\mu_{ni}C_{OX}}{W_{Cell}L_{CH}}(V_G - V_{TH})^2. \quad (6.139)$$

The presence of two channels per unit cell of width W_{Cell} in the power U-MOSFET structure has been accounted for while writing this equation. From this expression, the drain current density increases as the square of the gate bias voltage. This behavior is illustrated in Fig. 6.64 for a case of the power VD-MOSFET structure with cell pitch (W_{Cell}) of 20 μm and the power U-MOSFET structure with cell pitch of 2 μm discussed in the previous sections. These device structures have a channel length of 1 μm, gate oxide thickness of 500 Å, and a threshold voltage of 2 V. The inversion layer mobility was assumed to be 200 cm^2 V^{-1} s^{-1}. A much larger drain current density is obtained for the power U-MOSFET structure because its channel density is an order of magnitude larger than that of the power VD-MOSFET structure.

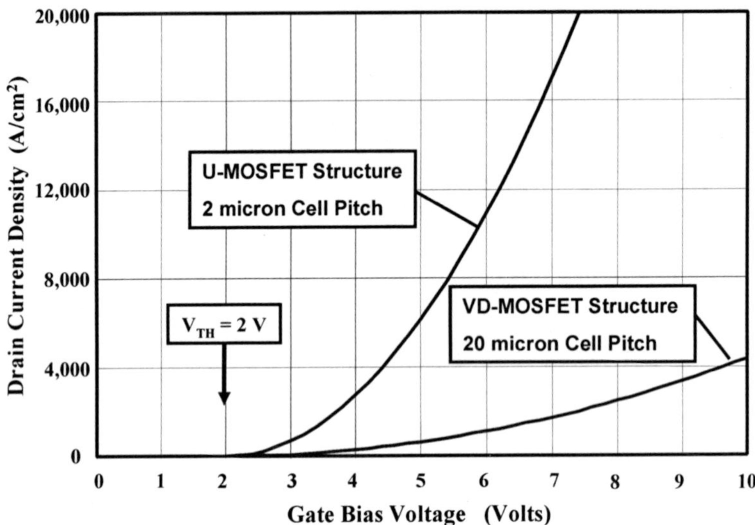

Fig. 6.64 Transfer characteristics for the power VD-MOSFET and U-MOSFET structures

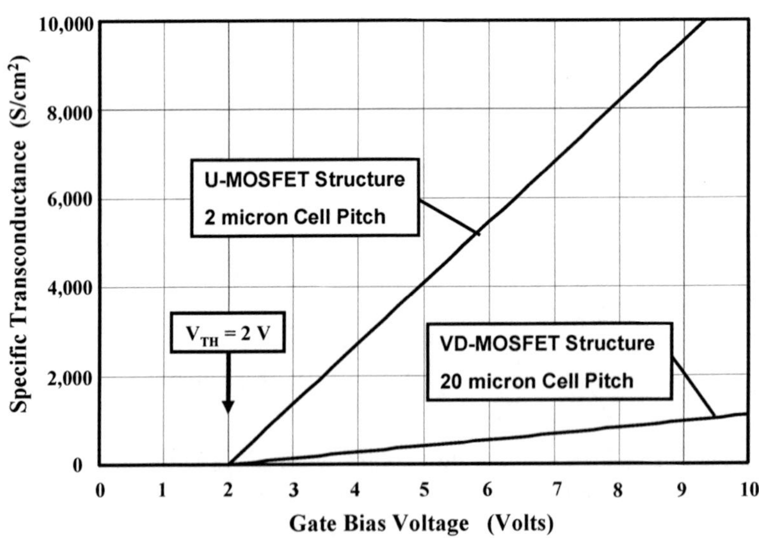

Fig. 6.65 Transconductance for the power VD-MOSFET and U-MOSFET structures

The transconductance for the power MOSFET structure can be derived from (6.139):

$$g_{m,SP} = \frac{dJ_{D,sat}}{dV_G} = \frac{2\mu_{ni}C_{OX}}{W_{Cell}L_{CH}}(V_G - V_{TH}). \tag{6.140}$$

The transconductance increases in proportion to the gate bias voltage once it exceeds the threshold voltage. A much larger transconductance is obtained for the power U-MOSFET structure, as shown in Fig. 6.65, when compared with the power VD-MOSFET structure due to its larger channel density.

The *square-law relationship* between the drain current and gate bias voltage is a consequence of the channel pinch-off physics discussed in Sect. 6.5.5. This behavior introduces a fundamental nonlinearity between the input signal (gate voltage) and the output signal (drain current) if the power MOSFET structure is used for audio or RF amplifiers.[17] The distortion of the signal is schematically illustrated in Fig. 6.66. In an amplifier, the transistor is biased at a quiescent operating point decided by the DC gate bias voltage (V_{GQ}) which produces a quiescent DC drain current (I_{GQ}). The superposition of the AC input signal voltage swings the gate voltage from a minimum value of V_{Gmin} to a maximum value of V_{Gmax}. Due to the nonlinear transfer characteristics exhibited by the power MOSFET structure operating under channel pinch-off conditions, the drain current exhibits a much larger excursion when the gate voltage swings above the DC gate bias voltage than when it swings below the DC gate bias voltage. The distortion in the drain current produced by the nonlinear transfer characteristics creates harmonic signals. In the case of audio applications, these harmonics alter the musical experience of consumers. In the case of RF cellular applications, the harmonics must be suppressed with expensive linearization circuitry to prevent broadcasting into adjacent channels in the cellular network.[13]

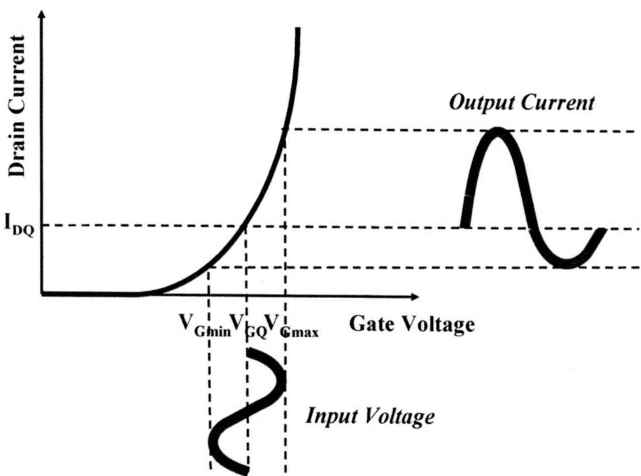

Fig. 6.66 Signal distortion created by the power MOSFET structure operating with channel pinch-off physics

Simulation Example

To confirm the square-law model for the transfer characteristics of the power VD-MOSFET structure, the results of two-dimensional numerical simulations on a structure with cell pitch of 20 μm designed with a 30-V rating are described here. The structure had a drift region thickness of 3 μm with a doping concentration of 1.3×10^{16} cm^{-3} located below the P-base region. The P-base region and N$^+$ source regions have depths of 3 and 1 μm, respectively. The doping concentration in the JFET region was enhanced by using an additional N-type doping concentration of 1×10^{16} cm^{-3} with a depth of 3 μm. The lateral doping profile for this structure taken along the surface under the gate electrode was previously shown in Fig. 6.44. From the profile, it can be observed that the doping concentration of the JFET region has been increased to 2.1×10^{16} cm^{-3} due to the additional doping. This reduced the lateral extension of the P-base region to 2.4 μm when compared with a vertical depth of 3 μm. The N$^+$ source region has a lateral depth of 0.8 μm leading to a channel length of 1.6 μm. The maximum compensated P-type doping concentration in the channel is 1.5×10^{17} cm^{-3}. In conjunction with a gate oxide thickness of 500 Å, a threshold voltage of 3.8 V was obtained for this doping concentration as shown in the transfer characteristics in Fig. 6.67.

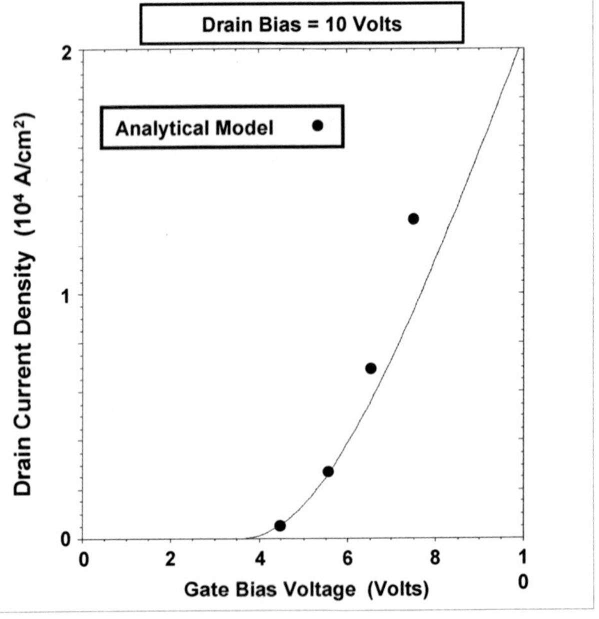

Fig. 6.67 Transfer characteristics for the VD-MOSFET structure with 20-μm cell pitch

The transfer characteristics observed from the numerical simulations display a nonlinear behavior. To determine if this is consistent with a square-law behavior predicted by the analytical model, the figure shows points calculated by using the model with a channel inversion layer mobility of 450 $cm^2\ V^{-1}\ s^{-1}$. It can be seen that the square-law model is an accurate representation of the characteristics obtained from the simulations at gate bias voltages near the threshold voltage. At high gate bias voltages, the drain current obtained by using the analytical square-law model is larger than those observed from the simulations. This difference is related to the decrease in the inversion layer mobility with increasing electric field normal to the surface (see Fig. 2.18) which is neglected in the analytical model.

6.11 Superlinear Transfer Characteristics

As discussed above, the well-accepted theory for current saturation in power MOSFET structures, based upon channel pinch-off, leads to a square-law relationship between the drain current and gate voltage. This is detrimental for audio and RF amplification because the nonlinearity introduces undesired harmonics at the output of power amplifiers. This section describes a new mode of operation that has been recently discovered[18] for MOSFETs that enables obtaining linear transfer characteristics. The physics of superlinear operation is described in this section. The design of silicon RF power MOSFET structures capable of exhibiting this new mode of operation is described elsewhere.[13]

The superlinear mode of operation is based upon maintaining the MOS channel of the MOSFET in the linear regime of operation while achieving current saturation by utilizing the saturated velocity–field curve for semiconductors. It is worth pointing out that this new mode of operation is distinct from the linear transfer characteristics reported for submicron transistors when the electrons in the inversion layer undergo velocity saturation.[19] In the case of these submicron transistors, the velocity saturation for electrons in the channel occurs only if the drain voltage is sufficiently high to push the longitudinal electric field above 5×10^4 V cm^{-1}. At these high electric fields, the on-set of impact ionization creates "hot electrons" at the vicinity of the gate oxide. These hot electrons can add charge in the gate oxide resulting in undesirable shifts in the threshold voltage. Further, a submicron channel length cannot support the high voltages (>30 V) that are necessary for power transistors designed for audio and RF base station applications.

As shown in Sect. 6.5.5, when the MOSFET operates in the linear region, the transfer characteristics become linear (see (6.46)) and the transconductance becomes constant, independent of the gate bias (see (6.48)). This requires maintaining the voltage of the channel at the drain end well below the gate bias voltage (or more accurately well below ($V_G - V_{TH}$)). Of course, the challenge is to simultaneously obtain current saturation under these operating conditions. This can be achieved by utilizing the saturated velocity–field curve for semiconductors (see Fig. 2.13).

The velocity of electrons in silicon increases linearly with increasing electric field at low electric fields. In this regime of operation, it is possible to define an electron mobility that is a function of the doping concentration and the temperature as discussed in Chap. 2. However, at higher electric fields, the mobility reduces due to optical phonon scattering. Eventually, when the electric field exceeds 5×10^4 V cm^{-1}, the electron velocity becomes constant independent of the electric field. This is known as the *saturated electron drift velocity* ($v_{sat,n}$). For silicon at room temperature (300 K), the saturated electron drift velocity is 1×10^7 cm s^{-1}.

Based upon using innovative device structures,[13] it is possible to shield the MOS channel from the drain voltage to maintain it in the linear region while simultaneously increasing the electric field in the drift region, so that the electron velocity undergoes saturation. Under these conditions, the charge in the channel inversion layer is given by

$$Q_{inv} = C_{OX}(V_G - V_T), \tag{6.141}$$

where C_{OX} is the capacitance per unit area of the gate MOS structure, V_G is the applied gate bias, and V_{TH} is the threshold voltage. The gate oxide capacitance per unit area (*specific gate capacitance*) is given by

$$C_{OX} = \left(\frac{\varepsilon_{OX}}{t_{OX}}\right), \tag{6.142}$$

where ε_{OX} is the dielectric constant for silicon dioxide (3.41×10^{-13} F cm^{-1}) and t_{OX} is the gate oxide thickness. This inversion layer charge is injected into the drift region producing the drain current:

$$I_{D,sat} = Q_{inv} v_{sat,n} Z, \tag{6.143}$$

where Z is the width of the channel. Combining (6.141) and (6.143),

$$I_{D,sat} = C_{OX}(V_G - V_T) v_{sat,n} Z = \frac{\varepsilon_{OX}}{t_{OX}}(V_G - V_T) v_{sat,n} Z. \tag{6.144}$$

According to this equation, the transfer characteristic of the MOSFET is linear in spite of the device being operated in the current saturation regime. The corresponding transconductance is given by

$$g_m = C_{OX} v_{sat,n} Z = \frac{\varepsilon_{OX}}{t_{OX}} v_{sat,n} Z, \tag{6.145}$$

indicating that the transconductance is independent of the gate bias. This results in an ideal linear transfer characteristic that is highly desirable for amplification of audio and RF signals.

A more complete description of the drain current–voltage relationship can be derived by taking into account the electron velocity–field curve. As mentioned above, the drain current is determined by the electrons injected from the inversion layer (MOS channel) unto the drift region. If the drain voltage is insufficient to transport them at the saturated drift velocity through the drift region, the drain current is determined by the electron velocity (v_n) corresponding to the electric field in the drift region:

$$I_D = Q_{inv} v_n Z. \tag{6.146}$$

Combining (6.141) and (6.146),

$$I_D = C_{OX}(V_G - V_T)v_n Z = \frac{\varepsilon_{OX}}{t_{OX}}(V_G - V_T)v_n Z, \tag{6.147}$$

with the electron velocity related to the electric field in the drift region[20] by

$$v_n = \frac{10^7 E}{(10^5 + E^{1.3})^{0.77}}, \tag{6.148}$$

where the electric field (E) in the drift region is determined by the drain bias (V_D) and the length (L_D) of the drift region:

$$E = \frac{V_D}{L_D}, \tag{6.149}$$

under the assumption that the electric field is uniform in the drift region. In the case of uniformly doped drift regions in conventional vertical power MOSFET structures, such as the one shown in Fig. 6.3, the electric field is not uniform across the drift region. However, for the devices which are based upon the charge coupling concept,[13,18] the electric field indeed becomes uniform in the drift region. This expression is then valid for analytical calculations of their drain current–voltage relationship.

Combining (6.147) and (6.148),

$$I_D = \frac{\varepsilon_{OX}}{t_{OX}}(V_G - V_T)\left[\frac{10^7 E}{(10^5 + E^{1.3})^{0.77}}\right]Z. \tag{6.150}$$

Using this equation, the output characteristics of the superlinear MOSFET structure can be analytically modeled using appropriate physical parameters. Since the drain resistance is dominant for the superlinear MOSFET structures, this equation models the device characteristics to high operating voltages.

As an example of a superlinear MOSFET structure, consider the case of a gate oxide thickness of 500 Å and a channel length of 1 μm. The output characteristics predicted for gate bias values ranging from 3 to 10 V are shown in Fig. 6.68 for a device with a threshold voltage of 2 V. Note that the drain current has been calculated for a device with a width (Z) of 1 mm. In contrast with the conventional MOSFET characteristics, it can be observed that the curves are equally spaced with increasing gate bias steps. The boundary for the transition from a resistive current–voltage relationship at low drain voltages to a high output resistance domain at large drain voltages is not as well defined as for the channel pinch-off model. This is due to a monotonic variation of the electron drift velocity with increasing drain bias.

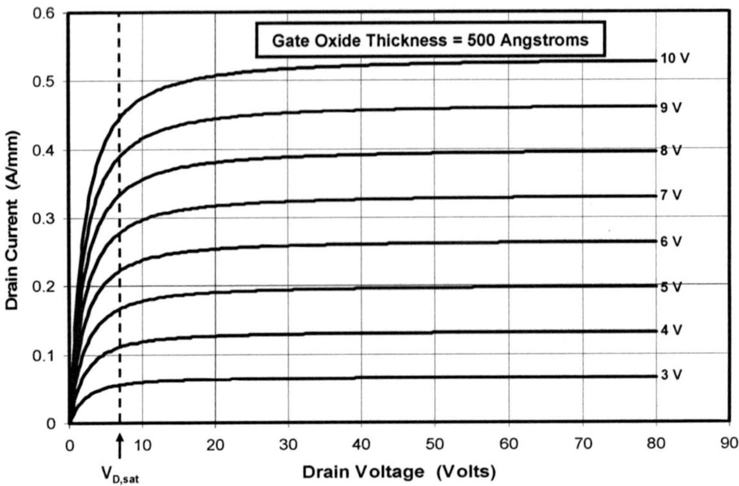

Fig. 6.68 Output characteristics of a superlinear MOSFET structure

An approximate boundary between these zones of operation can be created by defining the transition to occur when the drift velocity becomes 80% of the saturated drift velocity. Using this definition, the boundary has been indicated in Fig. 6.68 with a dashed line marked $V_{D,sat}$. Since the velocity–field curve is independent of the drain current (and hence the gate voltage), the transition voltage ($V_{D,sat}$) is also observed to be independent of the gate bias voltage. This is in contrast to an increase in the transition drain voltage with increasing gate bias for the conventional MOSFET structures as indicated by the dashed line in Fig. 6.37. For a device designed to support 80 V using a drift length of 3.5 μm, the value for $V_{D,sat}$ is about 6 V. This is satisfactory for allowing large drain voltage excursions enabling delivery of high output power in amplifiers.

From Fig. 6.68, it is obvious that the drain current increases linearly with increasing gate voltage. The transconductance of the superlinear MOSFET structure can be derived from (6.150):

$$g_m = \frac{dI_D}{dV_G} = \frac{\varepsilon_{OX}}{t_{OX}}\left[\frac{10^7 E}{(10^5 + E^{1.3})^{0.77}}\right]Z. \qquad (6.151)$$

As indicated by this expression, the transconductance is independent of the gate voltage. This behavior is illustrated for the superlinear MOSFET structure in Fig. 6.69. The impact of reducing the gate oxide from 500 to 250 Å is provided in the figure. The transconductance for the superlinear MOSFET structure is comparable in magnitude with that for the conventional device.

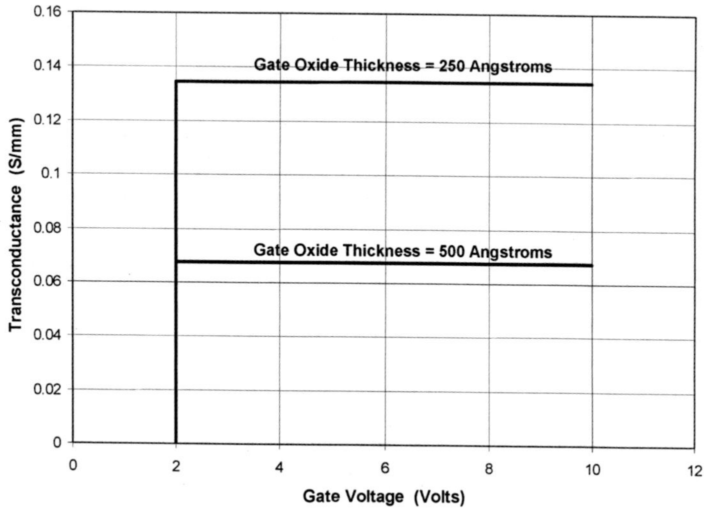

Fig. 6.69 Transconductance of a superlinear MOSFET structure

Power MOSFET structures with superlinear characteristics have been developed for RF cellular base station applications.[13] These devices have been shown to provide an order of magnitude improvement in the distortion of the signals. This improvement is sufficient for the elimination of the feed-forward circuitry required with the power LD-MOSFET structures.

6.12 Output Characteristics

The output characteristics for the lateral n-channel MOSFET structure were derived in Sect. 6.5.5. In the case of the power MOSFET structures, the characteristics are modified at low drain bias voltages by the addition of all the internal resistance to the channel resistance contribution. In addition, the drain current is observed to increase slightly with increasing drain bias voltage even after the on-set of channel pinch-off as illustrated in Fig. 6.70 for a power VD-MOSFET structure. The structure was assumed to have a channel length of 1 μm, a cell pitch of 20 μm, and a gate oxide thickness of 500 Å. A channel mobility of 200 $cm^2\ V^{-1}\ s^{-1}$ was used with a threshold voltage of 2 V. Consequently, practical power MOSFET structures exhibit a finite output resistance after drain current saturation by channel pinch-off.

The slight increase in the drain current with increasing drain voltage beyond channel pinch-off is associated with a reduction of the channel length with increasing drain bias as was illustrated in the figures in Sect. 6.5.5. The reduction of the channel length with increasing drain bias can be modeled as an extension of the depletion region into the channel from the drain side:

$$\Delta L_{CH} = K_O \sqrt{\frac{2\varepsilon_S V_D}{q}\left[\frac{N_D}{N_A(N_A + N_D)}\right]}. \tag{6.152}$$

where N_A and N_D are the doping concentration in the P-base and N-drift regions, respectively. The parameter K_O has been introduced to account for the shielding of the channel potential by the JFET region.

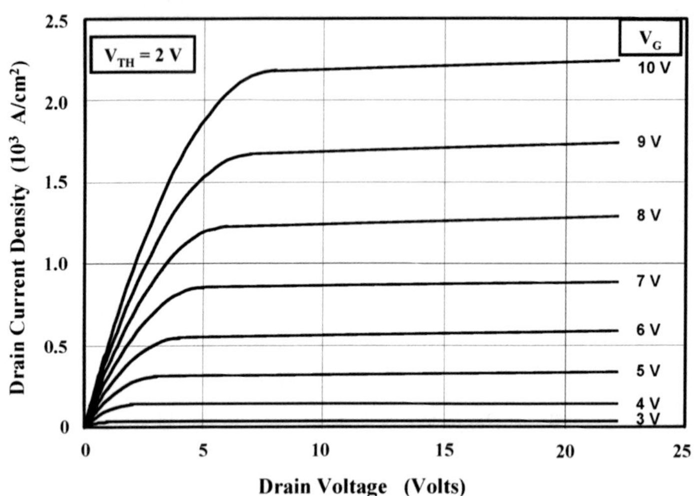

Fig. 6.70 Output characteristics of a power VD-MOSFET structure

After taking into account the depletion of the P-base region, the saturated drain current for the power MOSFET structure is given by

$$I_{D,sat} = \frac{Z\mu_{ni}C_{OX}}{L_{CH} - \Delta L_{CH}}(V_G - V_{TH})^2. \tag{6.153}$$

The presence of two channels in each cross-sectional cell pitch (W_{Cell}) has been accounted for when deriving this expression from (6.49). The saturated current density in the power MOSFET structure can be obtained by dividing the above current by the cell area:

$$J_{D,sat} = \frac{\mu_{ni}C_{OX}}{W_{Cell}(L_{CH} - \Delta L_{CH})}(V_G - V_{TH})^2. \tag{6.154}$$

The specific output resistance can be obtained by taking the first derivative of the drain current density in relation to the drain voltage:

$$\frac{1}{R_{\rm O,SP}} = \frac{{\rm d}J_{\rm D,sat}}{{\rm d}V_{\rm D}} = \frac{\mu_{\rm ni} C_{\rm OX} (V_{\rm G} - V_{\rm TH})^2}{W_{\rm Cell}} \frac{\rm d}{{\rm d}V_{\rm D}} \left(\frac{1}{L_{\rm CH} - b\sqrt{V_{\rm D}}} \right), \tag{6.155}$$

where

$$b = K_{\rm O} \sqrt{\frac{2\varepsilon_{\rm S}}{q} \left[\frac{N_{\rm D}}{N_{\rm A}(N_{\rm A} + N_{\rm D})} \right]}. \tag{6.156}$$

From these equations, an expression for the specific output resistance can be derived:

$$R_{\rm O,SP} = \frac{W_{\rm Cell}}{\mu_{\rm ni} C_{\rm OX} (V_{\rm G} - V_{\rm TH})^2} b\sqrt{V_{\rm D}} \left(\frac{L_{\rm CH}}{b} - V_{\rm D} \right)^2. \tag{6.157}$$

This expression is based upon assuming an abrupt junction with uniform doping concentrations on both sides. In practical devices, a graded doping profile is produced due to the diffusion of the P-base region into the N-drift region as illustrated in Fig. 6.8. The compensation at the junction greatly reduces the doping concentration in the vicinity of the junction, resulting in a depletion region extension into the P-base region at low drain bias voltages below the channel pinch-off voltage. This phenomenon shrinks the effective channel length well below that shown in Fig. 6.44 as $L_{\rm CH}$, namely, the difference in the lateral junction depths of the P-base and N^+ source regions. The smaller effective channel length increases the transconductance of the power MOSFET structures. Fortunately, as the drain bias increases, the depletion of the P-base region is suppressed by its increasing doping concentration away from the P-base/N-drift junction. This phenomenon produces output characteristics with a large output resistance.

Simulation Example

The output characteristics of a power VD-MOSFET structure obtained by using two-dimensional numerical simulations are described here. The structure was designed with a 30-V rating with cell pitch of 20 μm. It had a drift region thickness of 3 μm with a doping concentration of 1.3×10^{16} cm^{-3} located below the P-base region. The P-base region and N^+ source regions had vertical depths of 3 and 1 μm, respectively. The doping concentration in the JFET region was enhanced by using an additional N-type doping concentration of 1×10^{16} cm^{-3} with a depth of 3 μm. The lateral doping profile for this structure taken along the surface under the gate electrode was previously shown in Fig. 6.44. From the profile, it can be observed that the doping concentration of the JFET region has been increased to 2.1×10^{16} cm^{-3} due to the additional doping. This reduced the lateral extension of the P-base region to 2.4 μm when compared with a vertical depth of 3 μm. The N^+ source region has a lateral depth of 0.8 μm leading to a channel length of 1.6 μm. The maximum compensated P-type doping concentration in the channel is

1.5×10^{17} cm^{-3}. In conjunction with a gate oxide thickness of 500 Å, a threshold voltage of 3.8 V was obtained for this doping concentration as shown previously in the transfer characteristics in Fig. 6.67.

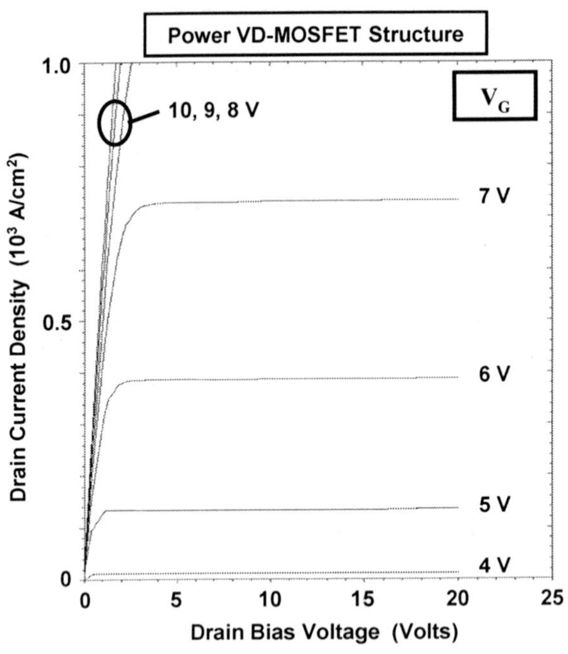

Fig. 6.71 Output characteristics for the power VD-MOSFET structure with 20-μm cell pitch

The output characteristics of the power VD-MOSFET obtained by using the simulations are shown in Fig. 6.71. The structure exhibits excellent current saturation with flat output characteristics. The traces for increasing gate bias voltages are nonuniformly spaced due to the square-law behavior of the transfer characteristics. The saturated drain current at a gate bias of 7 V is 730 A cm^{-2}. The saturated drain current predicted by using the analytical model, based upon channel pinch-off physics with a channel length of 1.6 μm, is 980 A cm^{-2}. Using the data extracted from the numerical simulations, a specific output resistance of 4.2 Ω cm^2 was computed for this structure at a gate bias of 7 V. This value is 2,600 times larger than the specific on-resistance for the structure. The specific output resistance computed by using the analytical model matches the simulations if a channel length of 1.6 μm is used with parameter K_O set at 0.11.

The output characteristics of the power U-MOSFET obtained by using the simulations are shown in Fig. 6.72. This structure also exhibits excellent current saturation with flat output characteristics. The traces for increasing gate bias voltages are nonuniformly spaced due to the square-law behavior of the transfer characteristics. The current density is larger by an order of magnitude when compared with the power VD-MOSFET structure because of the smaller cell pitch and channel length. The saturated drain current at a gate bias of 6 V is 9,400 A cm^{-2}.

The saturated drain current predicted by using the analytical model based upon channel pinch-off physics with a channel length of 1.0 μm is 10,500 A cm^{-2}. Using the data extracted from the numerical simulations, a specific output resistance of 0.255 Ω cm^2 was computed for this structure at a gate bias of 6 V. This value is 1,160 times larger than the specific on-resistance for the structure. The specific output resistance computed by using the analytical model matches the simulations if a channel length of 1.0 μm is used with parameter K_O set at 0.11. It can therefore be concluded that the same value for K_O can be utilized for modeling both power MOSFET structures.

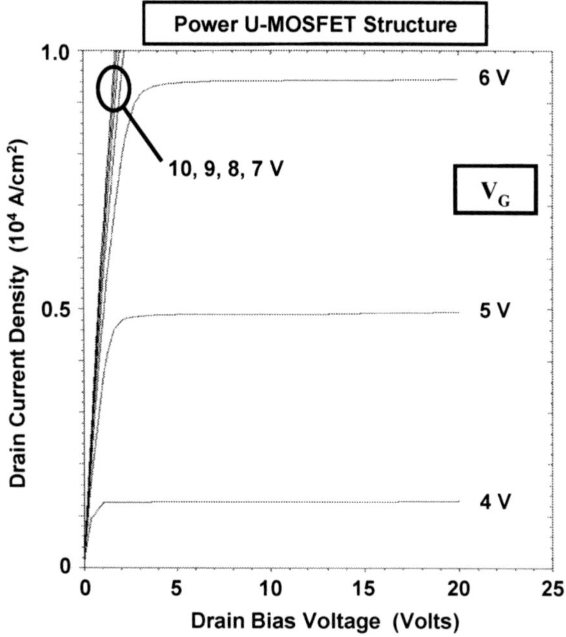

Fig. 6.72 Output characteristics for the power U-MOSFET structure with 2-μm cell pitch

6.13 Device Capacitances

The on-state current flow in a power MOSFET structure occurs by unipolar transport, namely, only electrons are involved for an n-channel device. The absence of minority carrier injection allows interruption of the current flow immediately after reduction of the gate bias below the threshold voltage. The process by which the majority carrier density is returned to equilibrium in a semiconductor[10] is referred to as the *dielectric relaxation time*. The dielectric relaxation time is given by

$$\tau_d = \frac{\varepsilon_S}{\sigma_S}, \tag{6.158}$$

where σ_S is the conductivity of the semiconductor. For typical doping concentrations in the drift regions of silicon power MOSFET structures, the dielectric relaxation time is on the order of picoseconds. Although this implies a very fast switching speed for the power MOSFET structures, in practice the switching speed is limited by the device capacitances.

The input drive signal for a power MOSFET structure is applied to the gate electrode, which is a part of an MOS sandwich. Due to the small thickness of the gate oxide, the MOS sandwich comprises a significant capacitance. Analysis of this capacitance requires taking into account the formation of a depletion layer in the semiconductor under certain bias conditions. The rate at which the power MOSFET structure can be switched between the on- and off-states is determined by the rate at which the input capacitance can be charged or discharged. In addition, the capacitance between the drain and the gate electrodes has been found to play an important role in determining the drain current and voltage transitions during the switching event. These capacitances are analyzed in this section for the power VD-MOSFET and U-MOSFET structures.

6.13.1 Basic MOS Capacitance

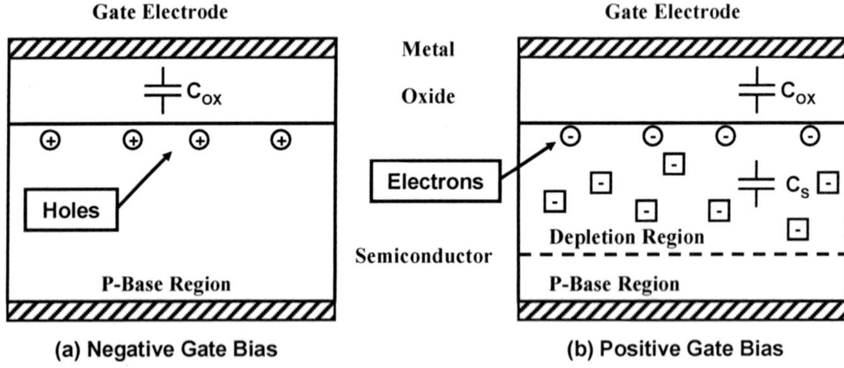

Fig. 6.73 Capacitances within a basic MOS structure

In an n-channel power MOSFET structure, the inversion layer channel is formed over the surface of a P-type base region. An elemental portion of the gate structure in the P-base region is shown in Fig. 6.73 when a negative or positive gate bias is applied to the gate electrode. When a negative bias is applied to the gate electrode, mobile holes are attracted from the bulk toward the oxide–semiconductor interface as illustrated in Fig. 6.73a to form an accumulation layer. As majority carriers in the P-base region, the holes can respond to an AC signal superposed on the negative DC voltage applied to the gate electrode. Consequently, the capacitance for the MOS structure becomes equal to that of the gate oxide under accumulation conditions:

$$C_{\mathrm{ACCUM}} = C_{\mathrm{OX}} = \frac{\varepsilon_{\mathrm{OX}}}{t_{\mathrm{OX}}}, \tag{6.159}$$

where C_{OX} is the specific capacitance of the gate oxide.

When a positive bias is applied to the gate electrode, a depletion region first forms in the P-type semiconductor region as illustrated in Fig. 6.73b. The charge associated with the ionized acceptors in the depletion region is indicated by the square boxes in the figure. The application of an AC signal superposed on the positive DC voltage applied to the gate electrode produces a response only from the P-base region located at the edge of the depletion layer because there are no mobile charges within the depletion region. Consequently, the capacitance for the MOS structure becomes equal to that of the series combination of the gate oxide capacitance and the depletion layer capacitance within the semiconductor. Under depletion-mode operation,

$$\frac{1}{C_{\mathrm{DEPL}}} = \frac{1}{C_{\mathrm{OX}}} + \frac{1}{C_{\mathrm{S}}}, \tag{6.160}$$

where C_{S} is the specific capacitance of the semiconductor depletion region.

When the positive bias applied to the gate electrode is increased, eventually an inversion layer forms at the oxide–semiconductor interface. This is illustrated in Fig. 6.73b by the mobile electrons located just below the oxide. Once the inversion layer forms, any further increase in the DC voltage applied to the gate electrode is supported across the gate oxide as discussed in Sect. 6.5.3. Consequently, the thickness of the depletion region reaches a maximum value. Despite the creation of a layer of mobile electrons at the oxide–semiconductor interface, the application of an AC signal superposed on the positive DC voltage applied to the gate electrode produces a response only from the P-base region located at the edge of the depletion layer. Any variation in the charge in the inversion layer must be produced by the carrier generation process. The space-charge generation lifetime in silicon is usually far larger than the period of the high frequency (typically 1 MHz) AC signals used to measure the gate capacitance. Consequently, the charge in the inversion layer cannot respond to the AC signal applied to the gate electrode and the capacitance for the MOS structure becomes equal to that of the series combination of the gate oxide capacitance and the depletion layer capacitance within the semiconductor. Under inversion-mode operation,

$$\frac{1}{C_{\mathrm{INV}}} = \frac{1}{C_{\mathrm{OX}}} + \frac{1}{C_{\mathrm{SMIN}}}, \tag{6.161}$$

where C_{SMIN} is the minimum specific capacitance of the semiconductor corresponding to the maximum depletion region width.

The capacitance associated with the depletion region formed in the semiconductor can be derived from the space charge in the semiconductor, which was analyzed in Sect. 6.5.2:

$$C_S = \frac{dQ_S}{d\psi_S}, \tag{6.162}$$

where ψ_S is the semiconductor surface potential. Using (6.21) for the semiconductor charge,

$$C_S = \frac{\varepsilon_S\{1 - e^{-(q\psi_S/kT)} + (n_{P0}/p_{P0})[e^{q\psi_S/kT} - 1]\}}{\sqrt{2}L_D F(q\psi_S/kT, n_{P0}/p_{P0})}. \tag{6.163}$$

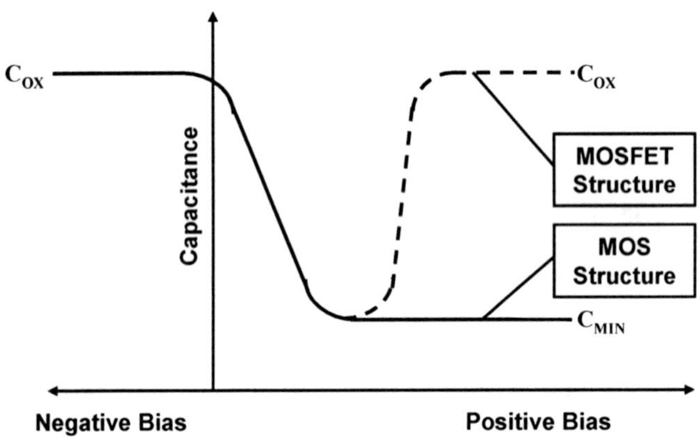

Fig. 6.74 Capacitance variation for a basic MOS structure with P-type substrate

The variation of the capacitance of the MOS structure with gate bias in accordance with the above equations is illustrated in Fig. 6.74 by the solid line. The capacitance asymptotes to that of the gate oxide when a negative gate bias is applied. When a positive gate bias is applied, the capacitance reduces due to the formation of the depletion region in the semiconductor. The capacitance asymptotes to its minimum value once the depletion region width reaches its maximum value.

However, this behavior is not observed in power MOSFET structures. In the power MOSFET structure, the inversion layer charge can be supplied from the N^+ source region circumventing the slow minority carrier generation process. Consequently, at large positive gate bias voltages, the capacitance asymptotes to that of the gate oxide as shown by the dashed lines in Fig. 6.74. Consequently, the input capacitance of the power MOSFET structure can be computed by using the oxide capacitance after accounting for the gate geometry. Furthermore, a deep depletion region forms in the drift region under the gate oxide due to the presence of the adjacent junction that sweeps out the minority carriers. Consequently, the capacitance between the gate and the drain electrodes must be computed by taking this effect into account.

6.13.2 Power VD-MOSFET Structure Capacitances

The capacitances within the power VD-MOSFET structure can be determined by using the analysis in Sect. 6.13.1 in conjunction with the geometry of the device structure. The capacitances that are pertinent for analyzing the switching performance of the power VD-MOSFET structure are indicated in Fig. 6.75 where the portion of the structure in the vicinity of the gate electrode is shown. The input (or gate) capacitance of the structure is primarily determined by the overlap of the gate electrode with the N^+ source and P-base regions. Due to high doping concentrations of the N^+ source and P-base regions, it is satisfactory to assume that these capacitances are determined by the oxide capacitance.

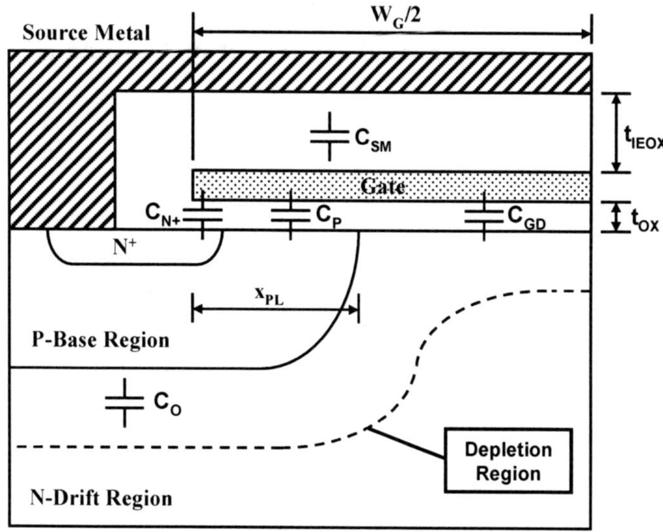

Fig. 6.75 Capacitances within the power VD-MOSFET structure

It is common design practice in power VD-MOSFET structures to overlap the source electrode over the gate electrode. This allows the formation of a large source metal area avoiding the need for defining narrow metal fingers that are traditionally used for lateral devices in integrated circuits. The formation of a large source metal can be accomplished with a very thick metal layer (typically 4–10 μm of aluminum) to reduce its resistance between points at which source bond wires are applied to chip. However, the overlap of the source electrode with the gate electrode produces an additional input capacitance. This capacitance can be reduced by using a relatively thick (typically 6,000 Å) interelectrode dielectric film (t_{IEOX}).

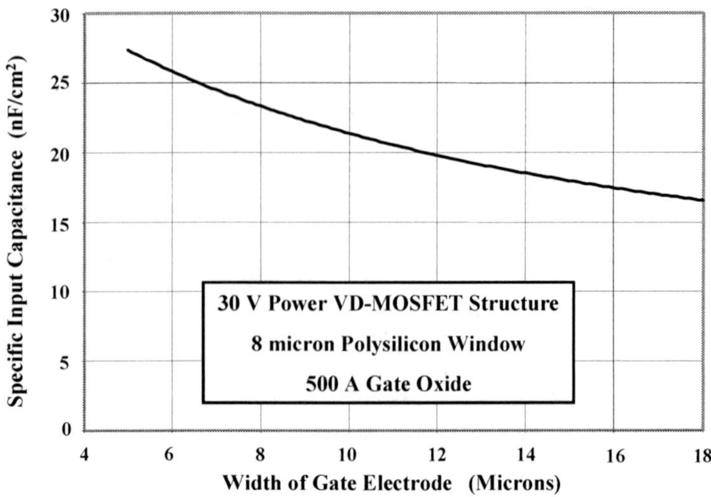

Fig. 6.76 Input capacitance for the power VD-MOSFET structure

The specific input (or gate) capacitance for the power VD-MOSFET structure can be obtained by combining the above capacitance contributions:

$$C_{\mathrm{IN,SP}} = C_{\mathrm{N+}} + C_{\mathrm{P}} + C_{\mathrm{SM}} = \frac{2x_{\mathrm{PL}}}{W_{\mathrm{Cell}}}\left(\frac{\varepsilon_{\mathrm{OX}}}{t_{\mathrm{OX}}}\right) + \frac{W_{\mathrm{G}}}{W_{\mathrm{Cell}}}\left(\frac{\varepsilon_{\mathrm{OX}}}{t_{\mathrm{IEOX}}}\right), \tag{6.164}$$

where t_{OX} and t_{IEOX} are the thicknesses of the gate and interelectrode oxides, respectively. The calculated input capacitance for the power VD-MOSFET structure (assumed to be independent of the gate and drain bias voltages) for the case of a structure with 20-μm cell pitch and polysilicon window of 8 μm is provided in Fig. 6.76. At a polysilicon gate width of 10 μm, the specific input capacitance calculated by using the above expression is 21.35 nF cm^{-2} when a lateral junction depth of 2.4 μm is used for the P-base region.

The capacitance between the gate and drain electrodes is determined by the width of the JFET region where the gate electrode overlaps the N-drift region. The MOS structure in this portion of the power VD-MOSFET structure operates under deep depletion conditions when a positive voltage is applied to the drain. Using (6.160) for the specific capacitance of the MOS structure under depletion conditions, the gate–drain (or reverse transfer) capacitance for the power VD-MOSFET structure is given by

$$C_{\mathrm{GD,SP}} = \frac{W_{\mathrm{G}} - 2x_{\mathrm{PL}}}{W_{\mathrm{Cell}}}\left(\frac{C_{\mathrm{OX}}C_{\mathrm{S,M}}}{C_{\mathrm{OX}} + C_{\mathrm{S,M}}}\right), \tag{6.165}$$

where $C_{S,M}$ is the semiconductor capacitance under the gate oxide, which decreases with increasing drain bias voltage. The specific capacitance of the semiconductor depletion region can be obtained by computation of the depletion layer width.

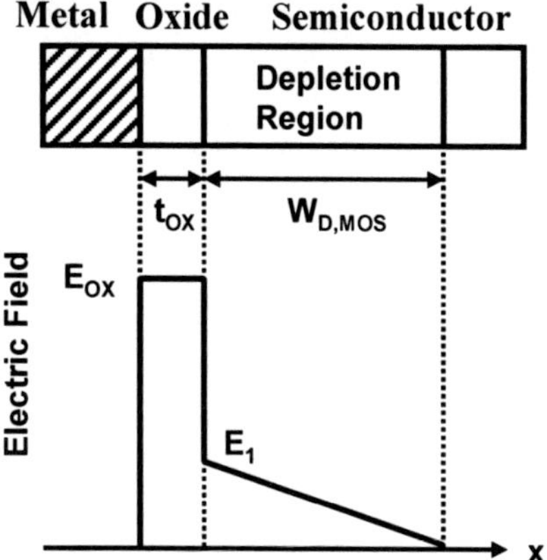

Fig. 6.77 Electric field profile for the MOS structure

The electric field distribution for the MOS structure under deep depletion conditions is illustrated in Fig. 6.77. The applied drain bias is shared between the oxide and the semiconductor:

$$V_D = V_{OX} + V_S = E_{OX} t_{OX} + \frac{1}{2} E_1 W_{D,MOS}, \tag{6.166}$$

where V_{OX} is the voltage supported by the oxide and V_S is the voltage supported within the semiconductor. The electric field (E_1) in the semiconductor at the oxide interface and the electric field (E_{OX}) in the oxide are interrelated via Gauss's law:

$$E_{OX} = \frac{\varepsilon_S}{\varepsilon_{OX}} E_1. \tag{6.167}$$

In addition, the electric field in the semiconductor (E_1) is related to the depletion layer width by

$$E_1 = \frac{qN_D}{\varepsilon_S} W_{D,MOS}. \tag{6.168}$$

Combining these relationships,

$$V_D = \frac{qN_D}{C_{OX}} W_{D,MOS} + \frac{qN_D}{2\varepsilon_S} W_{D,MOS}^2, \tag{6.169}$$

where C_{OX} is the specific capacitance of the gate oxide (ε_{OX}/t_{OX}). The solution for this quadratic equation provides the depletion layer width in the semiconductor under the gate oxide:

$$W_{D,MOS} = \frac{\varepsilon_S}{C_{OX}}\left\{\sqrt{1+\frac{2V_D C_{OX}^2}{q\varepsilon_S N_D}}-1\right\}. \tag{6.170}$$

The specific capacitance for the semiconductor is then obtained by using

$$C_{S,M} = \frac{\varepsilon_S}{W_{D,MOS}}. \tag{6.171}$$

The gate–drain (or reverse transfer) capacitance can be computed by using (6.165) with the above equations to determine the semiconductor capacitance as a function of the drain bias voltage.

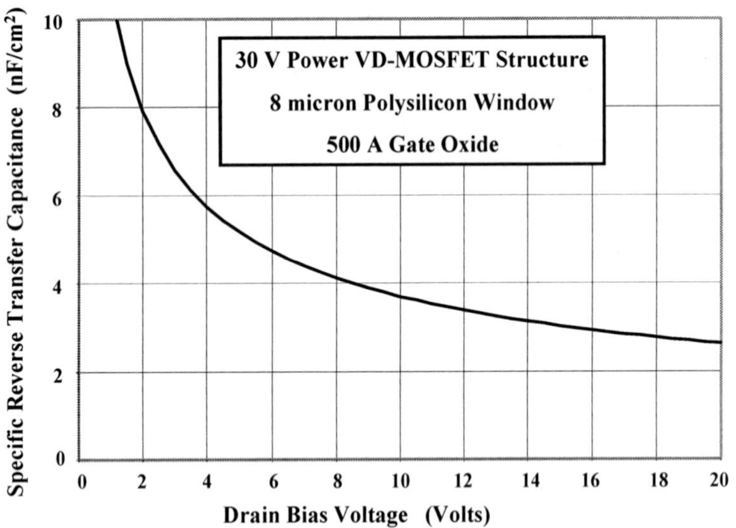

Fig. 6.78 Gate–drain capacitance for the power VD-MOSFET structure

As an example, the specific gate transfer capacitance obtained by using the above analytical formulae is shown in Fig. 6.78 for the case of a power VD-MOSFET structure with 20-μm cell pitch, polysilicon window of 8 μm, and polysilicon gate width of 12 μm. This structure has a gate oxide thickness of 500 Å and a lateral junction depth of 2.4 μm for the P-base region. The gate–drain (reverse transfer) capacitance decreases with increasing drain bias voltage due to the expansion of the depletion region in the semiconductor. At a drain bias of 15 V, the specific reverse transfer capacitance has a magnitude of 3.0 nF cm^{-2} for this design.

The reverse transfer capacitance for the power VD-MOSFET structure is strongly dependent on the gate width because this parameter determines the area for the overlap between the gate and the drain regions. As the gate width is increased, the specific reverse transfer capacitance increases in spite of an increase in the cell pitch. This is illustrated in Fig. 6.79 where the reverse transfer capacitance is plotted as a function of the gate width. The polysilicon window was kept constant at 8 μm in all these power VD-MOSFET structures. All the other structural parameters were also maintained at the same values for comparison of the devices. It can be observed that the specific transfer capacitance can be greatly reduced when the gate width is made small. However, this produces an increase in the specific on-resistance.

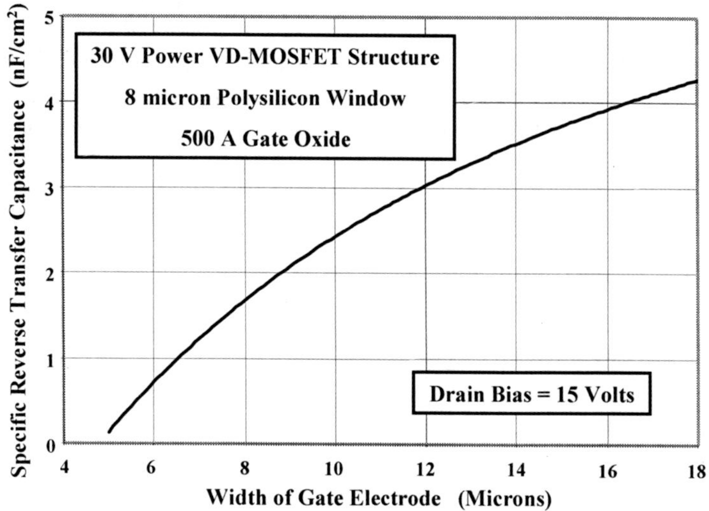

Fig. 6.79 Gate–drain capacitance for power VD-MOSFET structures

An additional capacitance that is present in the power VD-MOSFET structure is associated with the coupling of the drain to the source electrode. This is referred to as the *output capacitance* (C_O). Since the P-base region intervenes between the source and drain electrodes in the power VD-MOSFET structure, the output capacitance is associated with the capacitance of the junction between the P-base region and the N-drift region as indicated in Fig. 6.75. The specific junction capacitance can be derived by using

$$C_{S,J} = \frac{\varepsilon_S}{W_{D,J}}, \tag{6.172}$$

where the depletion region thickness at the junction ($W_{D,J}$) is related to the drain bias voltage:

$$W_{D,J} = \sqrt{\frac{2\varepsilon_S(V_D + V_{bi})}{qN_D}}. \quad (6.173)$$

This depletion layer width is larger than that under the gate oxide because all the applied drain voltage must be supported across the P–N junction. The specific output capacitance for the power VD-MOSFET structure can then be obtained by assuming (model A) that the area of the junction within the cell is $(W_{PW} + 2x_{PL})Z$:

$$C_O = \left(\frac{W_{PW} + 2x_{PL}}{W_{Cell}}\right)C_{S,J}, \quad (6.174)$$

where x_{PL} is the lateral extension of the P-base region under the gate electrode. However, the above expression is found to predict a larger output capacitance than observed with numerical simulations as shown later in this section. Consequently, an alternate model (model B) for the specific output capacitance for the power VD-MOSFET structure is proposed by assuming that the area of the junction within the cell is $(W_{PW} + x_{PL})Z$:

$$C_O = \left(\frac{W_{PW} + x_{PL}}{W_{Cell}}\right)C_{S,J}. \quad (6.175)$$

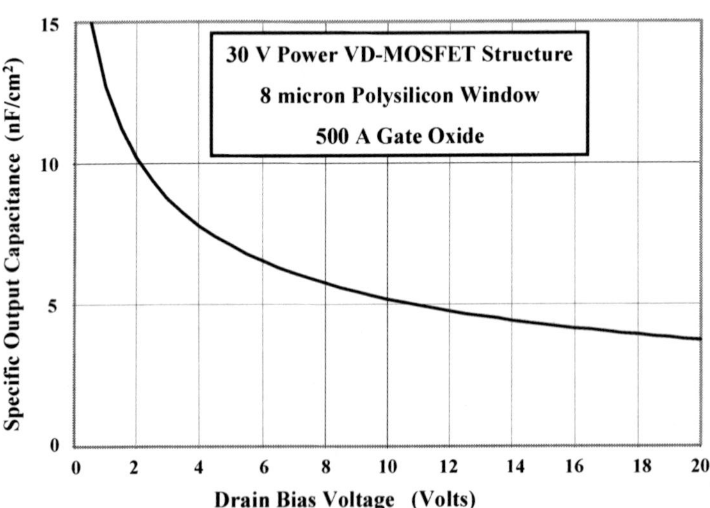

Fig. 6.80 Output capacitance for the power VD-MOSFET structure

As an example, the specific output capacitance obtained by using the above analytical model B is shown in Fig. 6.80 for the case of a power VD-MOSFET structure with 20-μm cell pitch, polysilicon window of 8 μm, and polysilicon gate width of 12 μm. This structure has a gate oxide thickness of 500 Å and a lateral junction depth of 2.4 μm for the P-base region. A built-in potential of 0.8 V was assumed for the P-base/N-drift junction, and the drift region has a doping concentration of 1.3×10^{16} cm^{-3}. The specific output capacitance decreases with increasing drain bias voltage due to the expansion of the depletion region under the P-base region. At a drain bias of 15 V, the specific output capacitance has a magnitude of 4.3 nF cm^{-2} for this design.

Simulation Example

To confirm the model for the capacitances of the power VD-MOSFET structure, the results of two-dimensional numerical simulations on a structure (with a cell pitch of 20 μm, a gate width of 12 μm, and a polysilicon window width of 8 μm) designed with a 30-V rating are described here. The structure had a drift region thickness of 3 μm with a doping concentration of 1.3×10^{16} cm^{-3} located below the P-base region. The P-base region and N$^+$ source regions have vertical depths of 3 and 1 μm, respectively. The doping concentration in the JFET region was enhanced by using an additional N-type doping concentration of 1×10^{16} cm^{-3} with a depth of 3 μm. The lateral doping profile for this structure taken along the surface under the gate electrode was previously shown in Fig. 6.44. From the profile, it can be observed that the doping concentration of the JFET region has been increased to 2.1×10^{16} cm^{-3} due to the additional doping. This reduced the lateral extension of the P-base region to 2.4 μm when compared with a vertical depth of 3 μm. The N$^+$ source region has a lateral depth of 0.8 μm leading to a channel length of 1.6 μm. The maximum compensated P-type doping concentration in the channel is

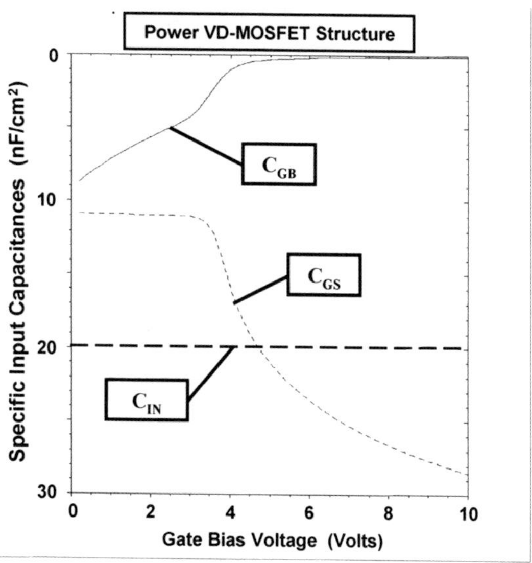

Fig. 6.81 Input capacitances for the VD-MOSFET structure with 20-μm cell pitch

1.5×10^{17} cm^{-3}. In conjunction with a gate oxide thickness of 500 Å, a threshold voltage of 3.8 V was obtained for this doping concentration as shown in the transfer characteristics in Fig. 6.67.

The input capacitances can be extracted by performing the numerical simulations with a small AC signal superposed on the DC gate bias voltage. The values obtained for the above power VD-MOSFET structure with a cell pitch of 20 μm are shown in Fig. 6.81. The input capacitance is comprised of two components – the first is between the gate electrode and the source electrode (C_{GS}) and the second is between the gate electrode and the base electrode (C_{GB}). The total input capacitance can be obtained by the addition of these capacitances because they are in parallel and share a common contact electrode in the actual power VD-MOSFET structure. From the figure, a total specific input capacitance of about 20 nF cm^{-2} is observed which is approximately independent of the gate bias voltage. At gate bias voltages below the threshold voltage, there is significant coupling between the P-base region and the gate electrode leading to a large contribution to the input capacitance from this path. When the gate bias voltage exceeds the threshold voltage, the inversion layer screens the P-base region from the gate electrode and couples it with the N$^+$ source region. Consequently, the contribution from C_{GB} decreases to zero while that from C_{GS} increases as the gate bias voltage is increased. The specific input capacitance extracted from the numerical simulations is in excellent agreement with that calculated with the analytical model (see Fig. 6.76) providing validation for the model.

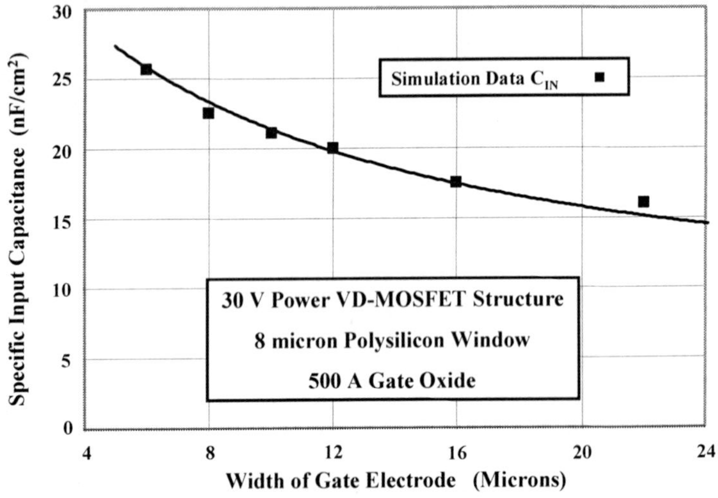

Fig. 6.82 Input capacitances for VD-MOSFET structures

The input capacitance was also extracted by using numerical simulations for the power VD-MOSFET structures with various gate widths. The width of the polysilicon window was kept at 8 μm for all these structures. All other structural parameters were also the same as those provide above. The specific input capacitance can be observed to reduce with increasing gate width as shown in Fig. 6.82. The values extracted from the simulations are in good agreement with those calculated by using the analytical model providing further validation for the utility of the model.

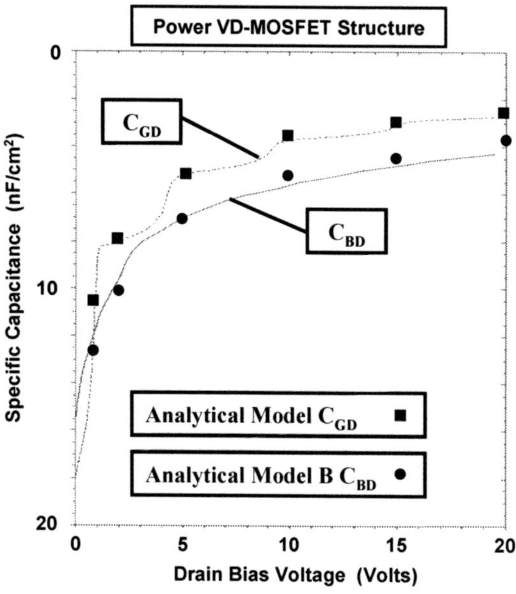

Fig. 6.83 Reverse transfer and output capacitances for the VD-MOSFET structure with 20-μm cell pitch

The drain–gate (reverse transfer) capacitance can be extracted by performing the numerical simulations with a small AC signal superposed on the DC drain bias voltage. The values obtained for the above power VD-MOSFET structure with a cell pitch of 20 μm are shown in Fig. 6.83. The gate-to-drain and base-to-drain capacitances are shown in the figure for comparison. Both of these capacitances decrease with increasing drain bias voltage. For this power VD-MOSFET structure, the reverse transfer (gate–drain) capacitance is comparable with the output (base–drain) capacitance. This implies a strong feedback path between the drain and the gate electrodes, which is detrimental to the switching speed and power loss for the power MOSFET structure. The values for the reverse transfer and output capacitances obtained by using the analytical model at various drain bias voltages are also shown in Fig. 6.83 by the square and circular symbols, respectively. The excellent agreement with the simulation values provides validation for the analytical models for both of these capacitances.

The reverse transfer capacitance was also extracted by using numerical simulations for the power VD-MOSFET structures with various gate widths. The width of the polysilicon window was kept at 8 μm for all these structures. All other structural parameters were also the same as those provide above. The specific reverse transfer capacitance can be observed to reduce drastically at smaller gate widths as shown in Fig. 6.84. The values extracted from the simulations are in good agreement with those calculated by using the analytical model providing validation for the model.

To understand the differences between the two models for the output capacitance, its value was extracted by using numerical simulations for the power VD-MOSFET structures with various gate widths. The width of the polysilicon window was kept at 8 μm for all these structures. All other structural parameters were also the same as those provide above. The specific output capacitance can be observed to decrease as the gate width is made larger as shown in Fig. 6.85. The values extracted from the simulations fall in between those obtained from both the analytical models.

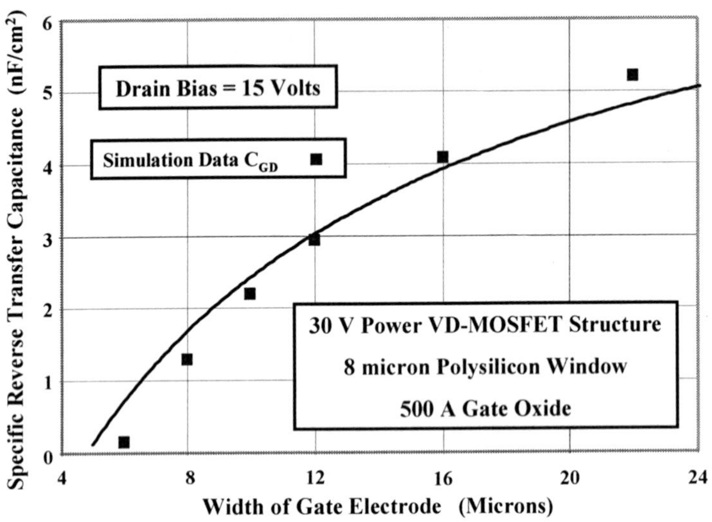

Fig. 6.84 Reverse transfer capacitances for VD-MOSFET structures

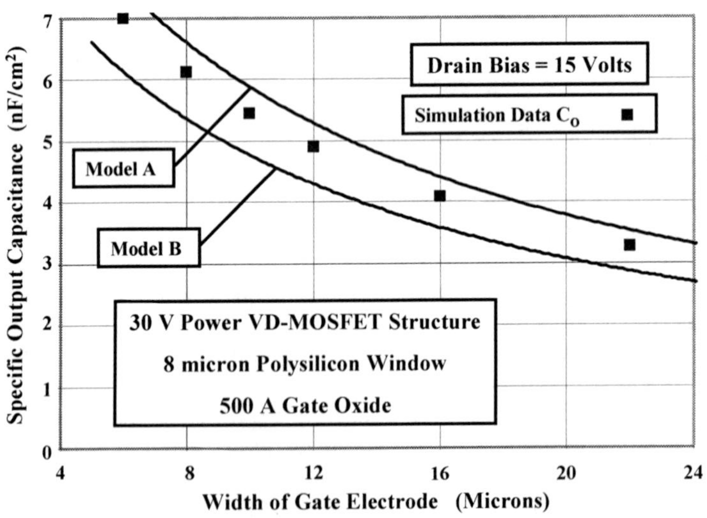

Fig. 6.85 Output capacitances for VD-MOSFET structures at a drain bias of 15 V

According to the analytical models, the specific output capacitance is a strong function of the drain bias voltage. To validate this behavior, the values calculated by using model B are shown in Fig. 6.83 by the circular symbols. The excellent agreement with the results of the numerical simulations provides credence for the model.

6.13.3 Power U-MOSFET Structure Capacitances

The capacitances within the power U-MOSFET structure can be analyzed by using the basic MOS capacitances analyzed in Sect. 6.13.1 in conjunction with the geometry of the device structure. The capacitances that are pertinent for analyzing the switching performance of the power U-MOSFET structure are indicated in Fig. 6.86 where the portion of the structure in the vicinity of the gate electrode is shown. The input (or gate) capacitance of the structure is primarily determined by the overlap of the gate electrode with the N^+ source and P-base regions along the sidewalls of the trench. Due to high doping concentrations of the N^+ source and P-base regions, it is satisfactory to assume that these capacitances are determined by the oxide capacitance.

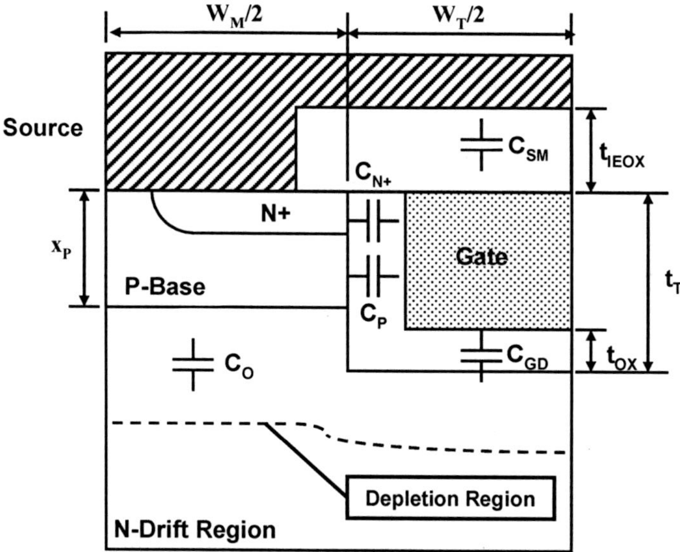

Fig. 6.86 Capacitances within the power U-MOSFET structure

As in the case of the power VD-MOSFET structure, it is common design practice in power U-MOSFET structures to overlap the source electrode over the gate electrode. This allows the formation of a large source metal area avoiding the need for defining narrow metal fingers that are traditionally used for lateral devices in integrated circuits. The formation of a large source metal can be accomplished with a very thick metal layer (typically 4–10 μm of aluminum) to reduce its resistance between points at which source bond wires are applied to chip. However, the overlap of the source electrode with the gate electrode produces an additional input capacitance. This capacitance can be reduced by using a relatively thick (typically 6,000 Å) interelectrode dielectric film (t_{IEOX}).

The specific input (or gate) capacitance for the power U-MOSFET structure can be obtained by combining the above capacitance contributions:

$$C_{IN,SP} = C_{N+} + C_P + C_{SM} = \frac{2x_P}{W_{Cell}}\left(\frac{\varepsilon_{OX}}{t_{OX}}\right) + \frac{W_T}{W_{Cell}}\left(\frac{\varepsilon_{OX}}{t_{IEOX}}\right), \quad (6.176)$$

where t_{OX} and t_{IEOX} are the thicknesses of the gate and interelectrode oxides, respectively. The input capacitance for the power U-MOSFET structure can be assumed to be independent of the gate and drain bias voltages. For the case of a structure with 2-μm cell pitch and 1-μm wide mesa and trench regions, the specific input capacitance calculated by using the above expression is 105 nF cm^{-2} when a vertical junction depth of 1.5 μm is used for the P-base region with a gate oxide thickness of 500 Å. This value is much larger than that for the power VD-MOSFET structure. Thus, the reduction of the specific on-resistance with the power U-MOSFET structure is attended with an increase in the input gate capacitance. The specific input capacitance of the power U-MOSFET structure can be reduced by increasing the mesa width as shown in Fig. 6.87. However, this is accompanied by an increase in the specific on-resistance.

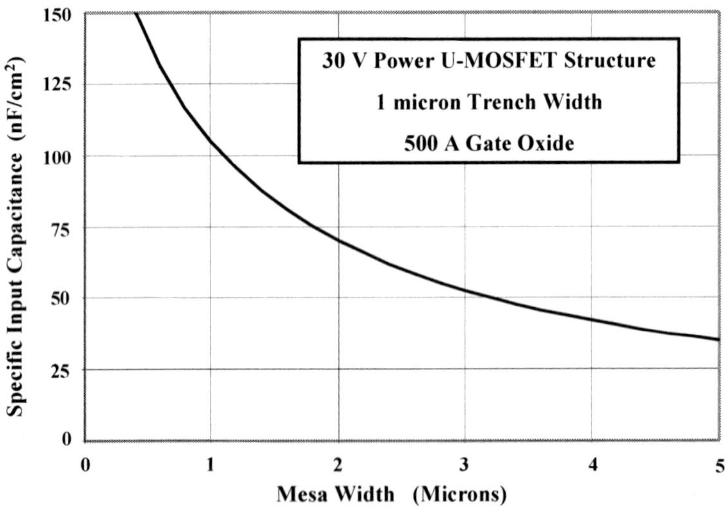

Fig. 6.87 Input capacitance for the power U-MOSFET structure

The capacitance between the gate and drain electrodes is determined by the width of the bottom of the trench. The MOS structure in this portion of the power U-MOSFET structure operates under deep depletion conditions when a positive voltage is applied to the drain. Using (6.160) for the specific capacitance of the MOS structure under depletion conditions, the gate–drain (or reverse transfer) capacitance for the power U-MOSFET structure is given by

$$C_{GD,SP} = \left[\frac{W_T + 2(t_T - x_P)}{W_{Cell}}\right]\left(\frac{C_{OX}C_{S,M}}{C_{OX} + C_{S,M}}\right), \tag{6.177}$$

where $C_{S,M}$ is the semiconductor capacitance under the gate oxide, which decreases with increasing drain bias voltage. The specific capacitance of the semiconductor depletion region ($C_{S,M}$) can be obtained by using (6.171) with the depletion region width under the gate oxide given by (6.170) as previously derived for the power VD-MOSFET structure. The gate–drain (or reverse transfer) capacitance can be computed by using (6.177) with the above equations to determine the semiconductor capacitance as a function of the drain bias voltage.

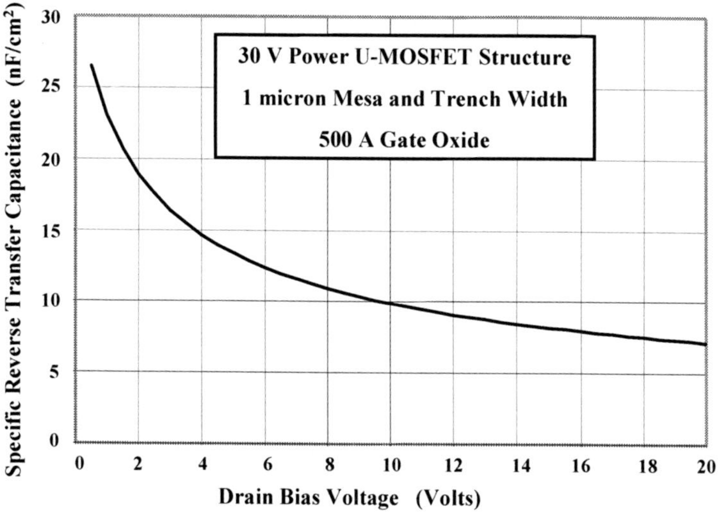

Fig. 6.88 Gate–drain capacitance for the power U-MOSFET structure

As an example, the specific gate transfer capacitance obtained by using the above analytical formulae is shown in Fig. 6.88 for the case of a power U-MOSFET structure with 2-μm cell pitch, mesa width of 1 μm, and trench width of 1 μm. This structure has a gate oxide thickness of 500 Å and a vertical junction depth of 1.5 μm for the P-base region. The trench extends to a depth of 2 μm. The gate–drain (reverse transfer) capacitance decreases with increasing drain bias voltage due to the expansion of the depletion region in the semiconductor. At a drain bias of 15 V, the specific reverse transfer capacitance has a magnitude of 8.2 nF cm^{-2} for this design, which is much larger than that for the power VD-MOSFET structure.

As in the case of the specific input capacitance, the reverse transfer capacitance for the power U-MOSFET structure can be reduced by increasing the mesa width. This is illustrated in Fig. 6.89 where the specific reverse transfer capacitance at a drain bias of 15 V is plotted as a function of the gate width. The trench width was kept constant at 1 μm in all these power U-MOSFET structures. All the other structural parameters were also maintained at the same values for comparison of the devices. It can be observed that the specific transfer capacitance can be greatly reduced when the mesa width is enlarged. However, this produces an increase in the specific on-resistance.

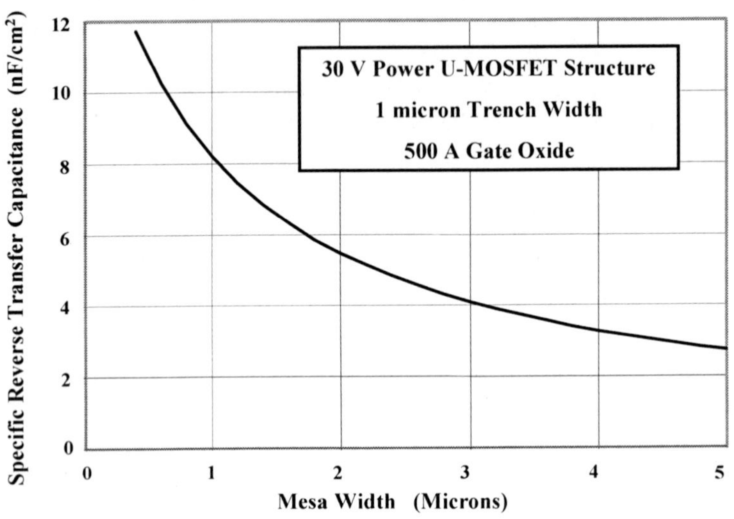

Fig. 6.89 Gate–drain capacitances for power U-MOSFET structures

As in the case of the power VD-MOSFET structure, the output capacitance for the power U-MOSFET structure is associated with the capacitance of the junction between the P-base region and the N-drift region as indicated in Fig. 6.86 because the P-base region intervenes between the source and drain electrodes. The specific junction capacitance can be derived by using (6.172) in conjunction with (6.173). The specific output capacitance for the power U-MOSFET structure can then be obtained by assuming (for model A) that the area of the junction within the cell is ($W_M Z$):

$$C_O = \left(\frac{W_M}{W_{Cell}} \right) C_{S,J}, \tag{6.178}$$

where W_M is the mesa width. This model predicts a higher value for the specific output capacitance than observed with the simulations as shown later in this section. This is due to the screening of the P-base region from the drain potential by the presence of the trench electrode which is located deeper than the junction. The effective junction area for determination of the output capacitance is reduced by the screening to

$$A_{\text{EFF}} = [W_{\text{M}} - 2K_{\text{S}}(t_{\text{T}} - x_{\text{P}} - t_{\text{OX}})]Z, \tag{6.179}$$

where K_S is a screening parameter. Here, the fact that the gate electrode extends above the trench bottom by the gate oxide thickness has been taken into account. The specific output capacitance obtained by using this model B is then given by

$$C_{\text{O}} = \left[\frac{W_{\text{M}} - 2K_{\text{S}}(t_{\text{T}} - x_{\text{P}} - t_{\text{OX}})}{W_{\text{Cell}}}\right] C_{\text{S,J}}. \tag{6.180}$$

As an example, the specific output capacitance obtained by using the above analytical model B is shown in Fig. 6.90 for the case of a power U-MOSFET structure with 5-μm cell pitch, mesa width of 4 μm, and trench width of 1 μm. This structure has a gate oxide thickness of 500 Å and a trench depth of 2 μm. A built-in potential of 0.8 V was assumed for the P-base/N-drift junction, and the drift region has a doping concentration of 1.3×10^{16} cm^{-3}. A screening factor (K_S) of 0.75 was used in this case. The specific output capacitance decreases with increasing drain bias voltage due to the expansion of the depletion region under the P-base region. At a drain bias of 15 V, the specific output capacitance has a magnitude of 5.5 nF cm^{-2} for this design.

The specific output capacitance for the power U-MOSFET structure increases with enlargement of the mesa region. This is illustrated in Fig. 6.91 for the case of a structure with trench width of 1 μm at a drain bias of 15 V by using the equations derived for both models. All the structural parameters for the device are the same as those used for the previous graphs in this section. It can be seen

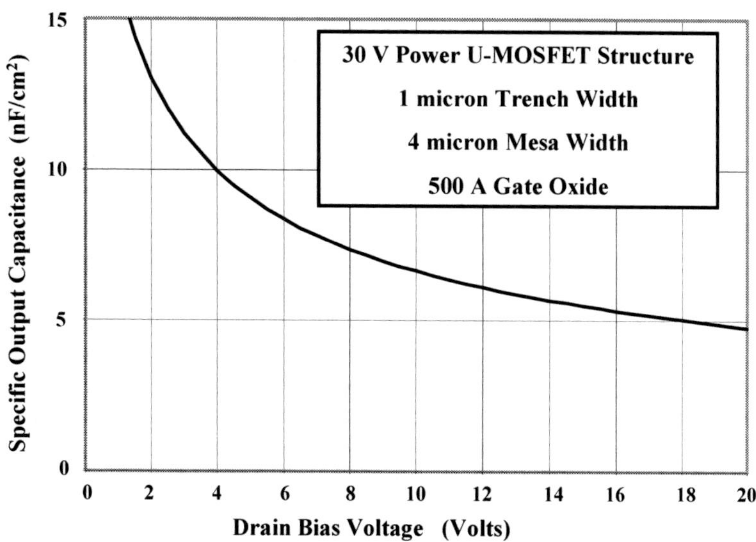

Fig. 6.90 Output capacitance for the power U-MOSFET structure

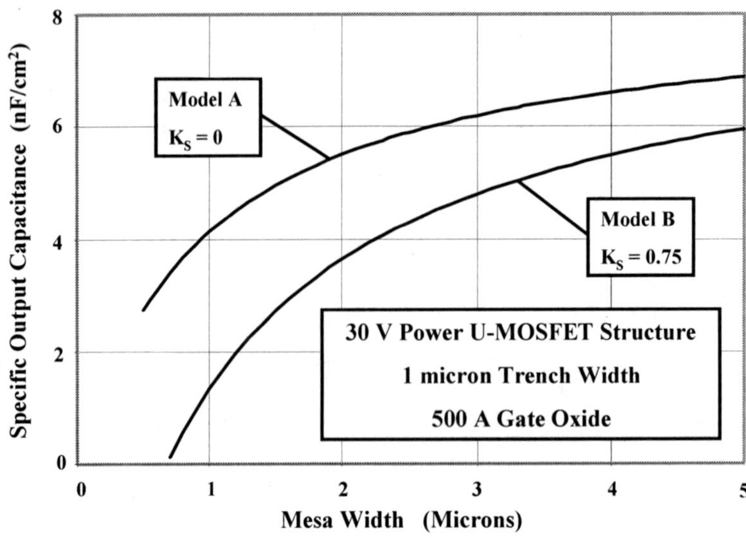

Fig. 6.91 Output capacitances for power U-MOSFET structures

that the model A without screening ($K_S = 0$) predicts a much larger specific output capacitance when compared with model B using a screening factor $K_S = 0.75$ especially at small mesa widths. As discussed in the simulation results below, a screening factor (K_S) of 0.75 is observed to provide agreement with the values extracted from numerical simulations.

Simulation Example

To confirm the model for the capacitances of the power U-MOSFET structure, the results of two-dimensional numerical simulations on a structure designed with a 30-V rating are described here for a cell pitch of 2 μm, a trench width of 1 μm, and a mesa width of 1 μm. The structure had a drift region thickness of 3 μm with a doping concentration of 1.3×10^{16} cm^{-3} located below the P-base region. The P-base region and N$^+$ source regions have vertical depths of 1.5 and 0.5 μm, respectively. The trench had a depth of 2 μm with a gate oxide thickness of 500 Å.

The input capacitances can be extracted by performing the numerical simulations with a small AC signal superposed on the DC gate bias voltage. The values obtained for the above power U-MOSFET structure with a cell pitch of 2 μm are shown in Fig. 6.92. The input capacitance is comprised of two components – the first is between the gate electrode and the source electrode (C_{GS}) and the second is between the gate electrode and the base electrode (C_{GB}). The total input capacitance can be obtained by the addition of these capacitances because they are in parallel and share a common contact electrode in the actual power U-MOSFET structure. From the figure, a total specific input capacitance of about 100 nF cm^{-2} is observed which is independent of the gate bias voltage. At gate bias voltages below the threshold voltage, there is significant coupling between the

P-base region and the gate electrode leading to a large contribution to the input capacitance from this path. When the gate bias voltage exceeds the threshold voltage, the inversion layer screens the P-base region from the gate electrode and couples it with the N^+ source region. Consequently, the contribution from C_{GB} decreases to zero while that from C_{GS} increases as the gate bias voltage is increased. The specific input capacitance extracted from the numerical simulations is in excellent agreement with that calculated with the analytical model (see Fig. 6.87) providing validation for the model.

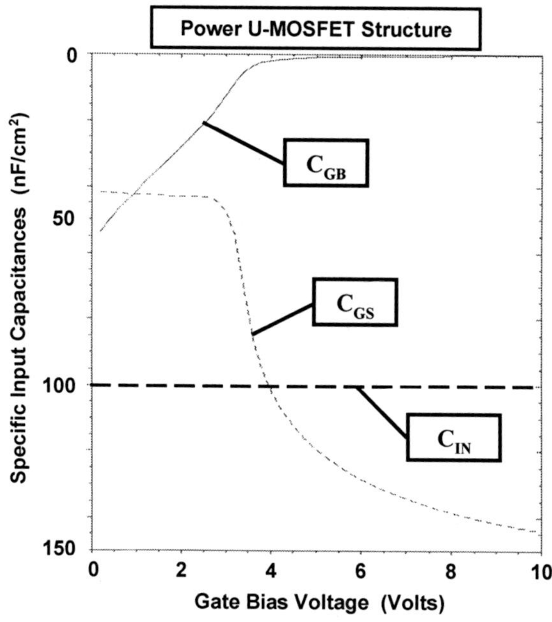

Fig. 6.92 Input capacitances for the U-MOSFET structure with 2-μm cell pitch

The input capacitance was also extracted by using numerical simulations for the power U-MOSFET structures with two mesa widths. The width of the trench was kept at 1 μm for these structures. All other structural parameters were also the same as those provide above. The specific input capacitance can be observed to reduce with increasing mesa width as shown in Fig. 6.93. The values extracted from the simulations are in good agreement with those calculated by using the analytical model providing further validation for the utility of the model.

The drain–gate (reverse transfer) capacitance can be extracted by performing the numerical simulations with a small AC signal superposed on the DC drain bias voltage. The values obtained for the above power U-MOSFET structure with a cell pitch of 2 μm are shown in Fig. 6.94. The gate-to-drain and base-to-drain capacitances are shown in the figure for comparison. Both of these capacitances decrease with increasing the drain bias voltage. The base-to-drain capacitance is determined by the depletion layer width under the P-base/N-drift junction. For this power U-MOSFET structure, the reverse transfer capacitance (gate–drain capacitance) is much greater

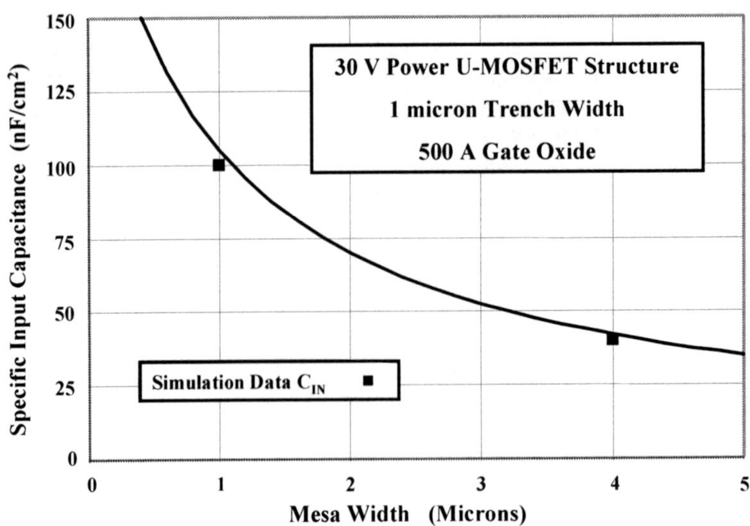

Fig. 6.93 Input capacitances for U-MOSFET structures

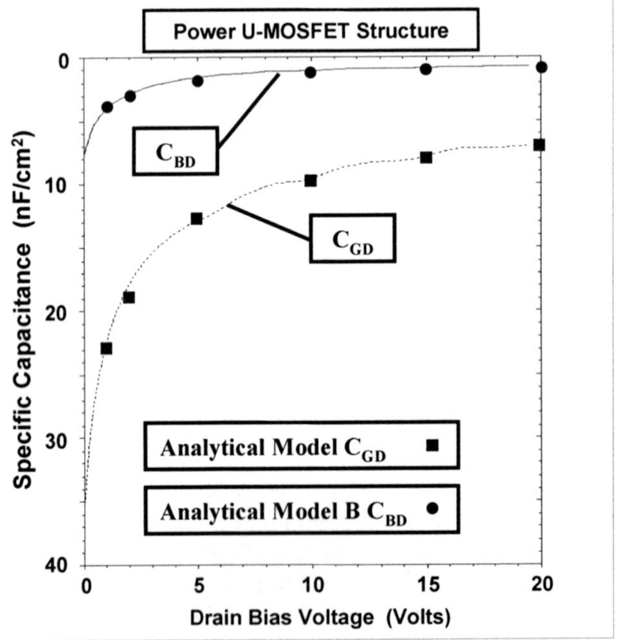

Fig. 6.94 Reverse transfer and output capacitances for the U-MOSFET structures with 2-μm cell pitch

than the base–drain capacitance. This implies a very strong feedback path between the drain and the gate electrodes which is detrimental to the switching speed and power loss for the power MOSFET structure. The values for the reverse transfer capacitance obtained by using the analytical model at various drain bias voltages are also shown in Fig. 6.94 by the square symbols. The excellent agreement with the simulation values provides validation for the analytical model. The specific reverse transfer capacitance for the U-MOSFET structure is much larger than that of the power VD-MOSFET structure.

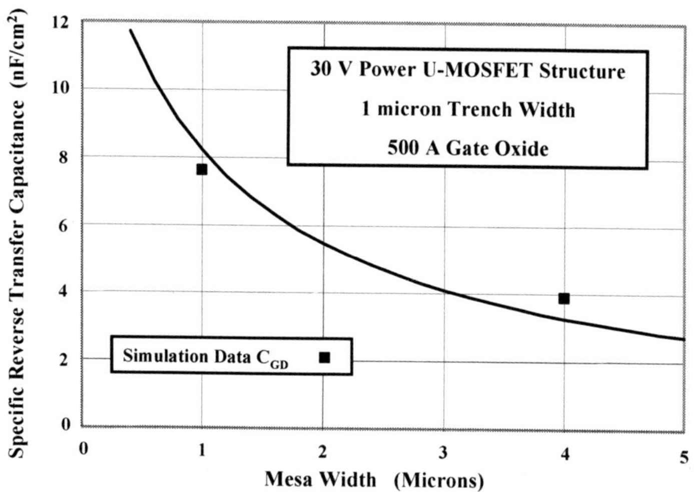

Fig. 6.95 Reverse transfer capacitances for U-MOSFET structures

The reverse transfer capacitance was also extracted by using numerical simulations for the power U-MOSFET structures with two mesa widths. The width of the trench was kept at 1 μm for these structures. All other structural parameters were also the same as those provide above. The specific reverse transfer capacitance can be observed to reduce at larger mesa widths as shown in Fig. 6.95. The values extracted from the simulations are in good agreement with those calculated by using the analytical model providing validation for the model.

According to the analytical models, the specific output capacitance is a strong function of the drain bias voltage. To validate this behavior, the values calculated by using model B with a screening factor (K_S) of 0.75 are shown in Fig. 6.94 by the circular symbols. The excellent agreement with the results of the numerical simulations provides credence for the model when the screening of the mesa region by the trench-gate electrode is taken into account.

To understand the differences between the two models for the output capacitance, its value was extracted by using numerical simulations for the power U-MOSFET structures with two mesa widths. The width of the trench was kept at 1 μm for these structures. All other structural parameters were also the same as those provide above. The specific output capacitance can be observed to increase as the mesa width is enlarged as shown in Fig. 6.96. The values extracted from the simulations are lower than those predicted by the model. This can be attributed to the highly graded doping profile in the vicinity of the junction as shown in Fig. 6.56. The low doping concentration at the junction and doping gradient in its vicinity produce an increase in the depletion layer thickness which is not accounted for by (6.173), which was based upon an abrupt junction approximation.

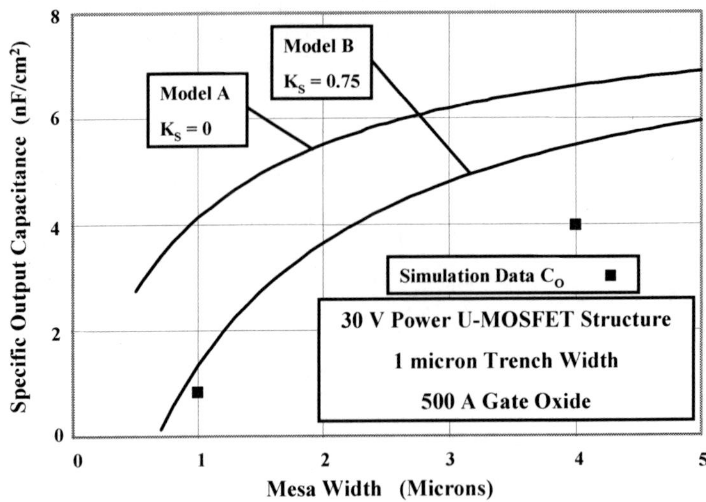

Fig. 6.96 Output capacitances for U-MOSFET structures at a drain bias of 15 V

6.13.4 Equivalent Circuit

The symbol for the n-channel power MOSFET structure is provided in Fig. 6.97. The connection between the source electrode and P-base region is indicated by the arrow in the symbol. The direction of the arrow represents the direction for current flow when the body diode in the structure is forward biased. For the n-channel

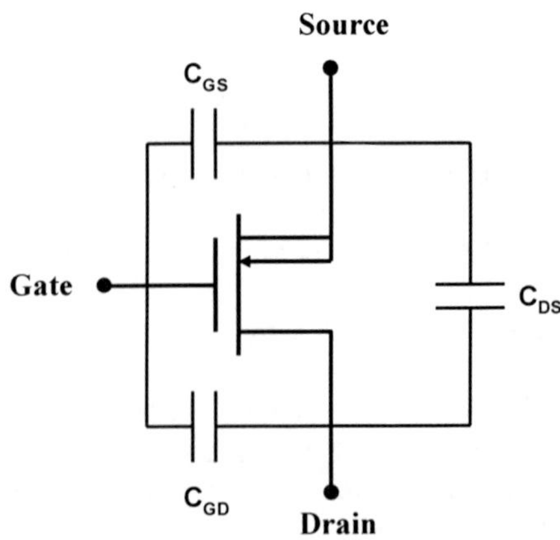

Fig. 6.97 Capacitances for the power MOSFET structure

power MOSFET structure, this would occur when the drain potential is negative with respect to the source potential.

The capacitances between the various terminals of the power MOSFET structure are also shown in Fig. 6.97. The gate-to-source capacitance (C_{GS}) is also referred to as the input capacitance. The capacitance (C_{DS}) between the drain and source terminals is also referred to as the output capacitance. The capacitance (C_{GD}) between the gate and the drain terminals is sometimes referred to as the Miller capacitance. All of these capacitances play a role in determining the switching performance for the power MOSFET structure.

To reduce the on-state power losses, it is necessary to increase the area of the power MOSFET structure to reduce its on-resistance. However, this leads to an increase in the input capacitance. A larger gate drive current is necessary to charge a greater input capacitance within the same time frame. This is undesirable because it increases the cost of the gate drive circuit. A larger chip area also increases the drain–gate capacitance. This prolongs the switching time interval as discussed later in this chapter. A special effort is undertaken during modern power MOSFET design to reduce the drain–gate capacitance in order to reduce power losses in high frequency power converters.

6.14 Gate Charge

As discussed in the previous sections, the gate–drain capacitance is highly nonlinear and varies considerably with drain bias voltage. To aid circuit design and analysis, it is therefore customary to describe power MOSFET structures in terms of the gate charge. For example, the gate–drain charge can be related to the gate–drain capacitance by

$$Q_{GD} = \int_{V_{ON}}^{V_{DS}} C_{GD}(V_D)\,dV_D, \tag{6.181}$$

where V_{DS} is the drain supply voltage and V_{ON} is the on-state voltage. In writing this expression, it has been assumed that the power MOSFET is being turned off with a voltage supply attached to the drain through a load resistance.

6.14.1 Charge Extraction

It is standard practice in the industry to extract the gate charge by the application of a constant current source at the gate terminal while turning on the power MOSFET structure from the blocking state. The linearized current and voltage waveforms observed during the turn-on process are shown in Fig. 6.98. During the turn-on process, the gate current is used to charge the capacitances C_{GS} and C_{GD} shown in Fig. 6.97. The capacitance between the gate and source terminals (C_{GS}) remains approximately constant as discussed in the previous sections. The capacitance between the gate and drain terminals (C_{GD}) also remains constant as long as the drain voltage remains constant at the supply voltage (V_{DS}). This is applicable for the time interval from zero to time t_2. Consequently, the gate voltage increases at a constant rate given by

$$\frac{\mathrm{d}v_{\mathrm{GS}}}{\mathrm{d}t} = \frac{J_{\mathrm{G}}}{C_{\mathrm{GS}} + C_{\mathrm{GD}}(V_{\mathrm{DS}})}, \tag{6.182}$$

where the gate–drain capacitance must be computed at the drain supply voltage (V_{DS}). The gate voltage increases linearly with time during this phase:

$$v_{\mathrm{GS}}(t) = \frac{J_{\mathrm{G}}t}{C_{\mathrm{GS}} + C_{\mathrm{GD}}(V_{\mathrm{DS}})}. \tag{6.183}$$

The gate voltage reaches the threshold voltage at time t_1 given by

$$t_1 = \frac{V_{\mathrm{TH}}[C_{\mathrm{GS}} + C_{\mathrm{GD}}(V_{\mathrm{DS}})]}{J_{\mathrm{G}}}. \tag{6.184}$$

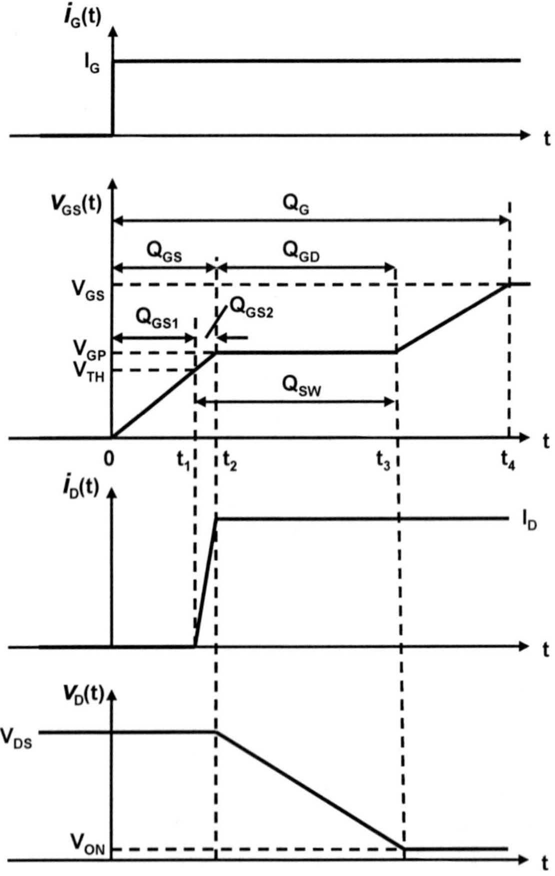

Fig. 6.98 Linearized waveforms for the power MOSFET structure during turn-on with a constant gate current

After this time interval, the drain current begins to increase with increasing gate bias voltage:

$$J_{\mathrm{D}}(t) = g_{\mathrm{m}}[v_{\mathrm{GS}}(t) - V_{\mathrm{TH}}]. \tag{6.185}$$

If the transconductance is assumed to be constant (independent of the gate bias and drain current level), the drain current increases linearly with time as shown in the figure. However, in power MOSFET structures operating under channel pinch-off conditions as discussed in Sect. 6.5.5, the transconductance is given by

$$g_{\mathrm{m}} = \frac{2\mu_{\mathrm{ni}} C_{\mathrm{OX}}}{W_{\mathrm{Cell}} L_{\mathrm{CH}}}[v_{\mathrm{GS}}(t) - V_{\mathrm{TH}}]. \tag{6.186}$$

The drain current density is then given by

$$J_{\mathrm{D}}(t) = \frac{2\mu_{\mathrm{ni}} C_{\mathrm{OX}}}{W_{\mathrm{Cell}} L_{\mathrm{CH}}}[v_{\mathrm{GS}}(t) - V_{\mathrm{TH}}]^2. \tag{6.187}$$

Using (6.183),

$$J_{\mathrm{D}}(t) = \frac{2\mu_{\mathrm{ni}} C_{\mathrm{OX}}}{W_{\mathrm{Cell}} L_{\mathrm{CH}}}\left[\frac{J_{\mathrm{G}} t}{C_{\mathrm{GS}} + C_{\mathrm{GD}}(V_{\mathrm{DS}})} - V_{\mathrm{TH}}\right]^2. \tag{6.188}$$

The drain current increases up to time t_2 when it reaches the on-state current density (J_{ON}). The time t_2 derived by using the above equation is

$$t_2 = \left(\frac{C_{\mathrm{GS}} + C_{\mathrm{GD}}(V_{\mathrm{DS}})}{J_{\mathrm{G}}}\right)\left(V_{\mathrm{TH}} + \sqrt{\frac{J_{\mathrm{ON}} W_{\mathrm{Cell}} L_{\mathrm{CH}}}{2\mu_{\mathrm{ni}} C_{\mathrm{OX}}}}\right). \tag{6.189}$$

After time interval t_2, the drain voltage begins to decrease while the drain current remains constant as shown in Fig. 6.98. Since the drain current is constant, the gate voltage also remains constant at

$$V_{\mathrm{GP}} = V_{\mathrm{TH}} + \sqrt{\frac{J_{\mathrm{ON}} W_{\mathrm{Cell}} L_{\mathrm{CH}}}{2\mu_{\mathrm{ni}} C_{\mathrm{OX}}}}. \tag{6.190}$$

This is referred to as the *gate plateau voltage*. The gate–drain capacitance is discharged by the gate current during this time interval producing a reduction of the drain voltage. None of the gate current flows into the gate–source capacitance during this time because the gate voltage is constant. The rate of change in the drain voltage is given by

$$\frac{\mathrm{d}v_{\mathrm{D}}}{\mathrm{d}t} = -\frac{J_{\mathrm{G}}}{C_{\mathrm{GD}}(v_{\mathrm{D}})}. \tag{6.191}$$

The rate of change in the drain voltage will be constant as shown in the figure if the gate–drain capacitance does not vary with drain voltage. However, as discussed in Sect. 6.13, the gate–drain capacitance increases with decreasing drain voltage producing a reduction in the rate of change of the drain voltage. The gate–drain capacitance is of the form

$$C_{\mathrm{GD}}(v_{\mathrm{D}}) = K_{\mathrm{G}}\left(\frac{C_{\mathrm{OX}}C_{\mathrm{S,M}}}{C_{\mathrm{OX}}+C_{\mathrm{S,M}}}\right), \tag{6.192}$$

where the semiconductor capacitance ($C_{\mathrm{S,M}}$) is determined by the width of the depletion layer under the gate oxide ($W_{\mathrm{D,MOS}}$) given by (6.170). In this expression, K_{G} is a geometry factor to account for the gate–drain overlap area. From (6.165) for the power VD-MOSFET structure,

$$K_{\mathrm{G}}(\text{VD-MOSFET}) = \left(\frac{W_{\mathrm{G}}-2x_{\mathrm{PL}}}{W_{\mathrm{Cell}}}\right), \tag{6.193}$$

while using (6.177) for the power U-MOSFET structure:

$$K_{\mathrm{G}}(\text{U-MOSFET}) = \left(\frac{W_{\mathrm{T}}+2(t_{\mathrm{T}}-x_{\mathrm{P}})}{W_{\mathrm{Cell}}}\right). \tag{6.194}$$

At time t_3, the drain voltage reaches the on-state voltage drop (V_{ON}) of the power MOSFET structure. This time can be obtained by combining the above expressions to get

$$-\frac{C_{\mathrm{OX}}K_{\mathrm{G}}}{J_{\mathrm{G}}}\sqrt{\frac{q\varepsilon_{\mathrm{S}}N_{\mathrm{D}}}{q\varepsilon_{\mathrm{S}}N_{\mathrm{D}}+2v_{\mathrm{D}}(t)C_{\mathrm{OX}}^{2}}}\,\mathrm{d}v_{\mathrm{D}} = \mathrm{d}t. \tag{6.195}$$

Since the drain voltage changes from V_{DS} to V_{ON} during the time interval from t_2 to t_3,

$$\int_{V_{\mathrm{DS}}}^{V_{\mathrm{ON}}} -\frac{C_{\mathrm{OX}}K_{\mathrm{G}}}{J_{\mathrm{G}}}\sqrt{\frac{q\varepsilon_{\mathrm{S}}N_{\mathrm{D}}}{q\varepsilon_{\mathrm{S}}N_{\mathrm{D}}+2v_{\mathrm{D}}(t)C_{\mathrm{OX}}^{2}}}\,\mathrm{d}v_{\mathrm{D}} = \int_{t_2}^{t_3}\mathrm{d}t, \tag{6.196}$$

leading to a solution for the time interval:

$$t_3 - t_2 = \frac{2K_{\mathrm{G}}}{J_{\mathrm{G}}}\frac{q\varepsilon_{\mathrm{S}}N_{\mathrm{D}}}{C_{\mathrm{OX}}}\left[\sqrt{1+\frac{2V_{\mathrm{DS}}C_{\mathrm{OX}}^{2}}{q\varepsilon_{\mathrm{S}}N_{\mathrm{D}}}} - \sqrt{1+\frac{2V_{\mathrm{ON}}C_{\mathrm{OX}}^{2}}{q\varepsilon_{\mathrm{S}}N_{\mathrm{D}}}}\right]. \tag{6.197}$$

During this time interval, the drain voltage decreases as given by

$$v_{\mathrm{D}}(t)=\frac{q\varepsilon_{\mathrm{S}}N_{\mathrm{D}}}{2C_{\mathrm{OX}}^{2}}\left\{\left[\sqrt{1+\frac{2V_{\mathrm{DS}}C_{\mathrm{OX}}^{2}}{q\varepsilon_{\mathrm{S}}N_{\mathrm{D}}}}-\frac{J_{\mathrm{G}}C_{\mathrm{OX}}(t-t_{2})}{2K_{\mathrm{G}}q\varepsilon_{\mathrm{S}}N_{\mathrm{D}}}\right]^{2}-1\right\}. \tag{6.198}$$

Beyond time t_3, the gate–drain capacitance again becomes constant albeit at a large value due to the small on-state drain bias voltage. The gate voltage then increases in accordance with (6.182) with a larger value for the gate–drain capacitance:

$$v_{\mathrm{GS}}(t)=\frac{J_{\mathrm{G}}t}{C_{\mathrm{GS}}+C_{\mathrm{GD}}(V_{\mathrm{ON}})}. \tag{6.199}$$

Thus, the rate of rise of the gate voltage after time t_3 is smaller than the rate of rise of the gate voltage up to time t_2. Eventually, the gate voltage reaches its supply voltage (V_{GS}) at time t_4 given by

$$t_{4}-t_{3}=\frac{V_{\mathrm{GS}}-V_{\mathrm{GP}}}{J_{\mathrm{G}}}[C_{\mathrm{GS}}+C_{\mathrm{GD}}(V_{\mathrm{ON}})]. \tag{6.200}$$

The largest power dissipated by the power MOSFET structure during the switching event occurs during the time span ($t_3 - t_1$) because both the drain current and the voltage are large. Consequently, this is the time interval of most interest when comparing power MOSFET structures. For comparison of power MOSFET structures (and in datasheets for applications), it is common practice to define a gate charge associated with various time intervals in the gate waveform as indicated in Fig. 6.98. These charges are defined as:

Pre-threshold gate charge (Q_{GS1})
Post-threshold gate charge (Q_{GS2})
Gate charge (Q_{GS})
Gate–drain charge (Q_{GD})
Gate switching charge (Q_{SW})
Total gate charge (Q_{G})

Some of these charges are interrelated as follows:

$$Q_{\mathrm{GS}}=Q_{\mathrm{GS1}}+Q_{\mathrm{GS2}}, \tag{6.201}$$

$$Q_{\mathrm{SW}}=Q_{\mathrm{GS2}}+Q_{\mathrm{GD}}. \tag{6.202}$$

The gate switching charge (Q_{SW}) is considered the most significant parameter for power MOSFET performance because it is associated with the time interval ($t_3 - t_1$) when the most power loss occurs during the switching interval. The gate–drain charge (Q_{GD}) is usually a dominant portion of the switching charge and is therefore also given importance during device design and power MOSFET structural comparisons.

Since a constant gate current I_G is applied to the power MOSFET structure during the turn-on event, the various components of the gate charge can be derived by its multiplication with the appropriate time intervals:

$$Q_{GS1} = J_G t_1 = V_{TH}[C_{GS} + C_{GD}(V_{DS})], \tag{6.203}$$

$$Q_{GS2} = J_G(t_2 - t_1) = [C_{GS} + C_{GD}(V_{DS})]\sqrt{\frac{J_{ON}W_{Cell}L_{CH}}{2\mu_{ni}C_{OX}}}, \tag{6.204}$$

$$Q_{GS} = J_G t_2 = [C_{GS} + C_{GD}(V_{DS})]\left(V_{TH} + \sqrt{\frac{J_{ON}W_{Cell}L_{CH}}{2\mu_{ni}C_{OX}}}\right), \tag{6.205}$$

$$Q_{GD} = J_G(t_3 - t_2) = \frac{2K_G q\varepsilon_S N_D}{C_{OX}}\left[\sqrt{1 + \frac{2V_{DS}C_{OX}^2}{q\varepsilon_S N_D}} - \sqrt{1 + \frac{2V_{ON}C_{OX}^2}{q\varepsilon_S N_D}}\right], \tag{6.206}$$

$$\begin{aligned} Q_{SW} &= J_G(t_3 - t_1) = [C_{GS} + C_{GD}(V_{DS})]\sqrt{\frac{J_{ON}W_{Cell}L_{CH}}{2\mu_{ni}C_{OX}}} \\ &\quad + \frac{2K_G q\varepsilon_S N_D}{C_{OX}}\left[\sqrt{1 + \frac{2V_{DS}C_{OX}^2}{q\varepsilon_S N_D}} - \sqrt{1 + \frac{2V_{ON}C_{OX}^2}{q\varepsilon_S N_D}}\right], \end{aligned} \tag{6.207}$$

$$\begin{aligned} Q_G &= J_G t_4 = [C_{GS} + C_{GD}(V_{DS})]V_{GP} \\ &\quad + \frac{2K_G q\varepsilon_S N_D}{C_{OX}}\left[\sqrt{1 + \frac{2V_{DS}C_{OX}^2}{q\varepsilon_S N_D}} - \sqrt{1 + \frac{2V_{ON}C_{OX}^2}{q\varepsilon_S N_D}}\right] \\ &\quad + [C_{GS} + C_{GD}(V_{ON})](V_G - V_{GP}). \end{aligned} \tag{6.208}$$

It is worth pointing out that these gate charge expressions are not dependent on the gate drive current.

As an example, the waveforms obtained by using the above analytical expressions are shown in Fig. 6.99 for the 30-V power VD-MOSFET structure with 20-μm cell pitch, a polysilicon gate width of 12 μm, and a gate oxide thickness of 500 Å. A gate drive current density of 0.1 A cm^{-2} was used to turn on the device from a steady-state blocking voltage of 20 V. To account for the enhanced doping concentration in the JFET region, the doping concentration used to compute the gate–drain capacitance was chosen as the average value of the peak doping at the gate oxide interface and the doping concentration of the drift region. The gate geometry factor (K_G) obtained for this structure by using the structural dimensions is 0.36.

The gate voltage initially increases linearly with time. After reaching the threshold voltage of 3.5 V, the drain current can be observed to increase in a nonlinear manner because the transconductance is a function of the gate bias voltage. In this example, the drain current density increases until it reaches an on-state current density of 200 A cm^{-2}. This transition occurs rapidly when compared with the time taken for the drain voltage to decrease during the next time interval. The on-state current density determines the gate plateau voltage which has a value of 5.3 V in this example. During the gate plateau voltage phase, the drain voltage decreases in a nonlinear manner until it reaches the on-state voltage drop. After this time, the gate voltage again increases but at a slower rate than during the initial turn-on phase. The gate charge values obtained for the power VD-MOSFET structure by using the above equations are Q_{GS1} = 80 nC cm^{-2}, Q_{GS2} = 42 nC cm^{-2}, Q_{GS} = 122 nC cm^{-2}, Q_{GD} = 206 nC cm^{-2}, Q_{SW} = 248 nC cm^{-2}, and Q_G = 503 nC cm^{-2}. It can be concluded that the gate–drain charge (Q_{GD}) is the dominant portion (83%) of the gate switching charge (Q_{SW}).

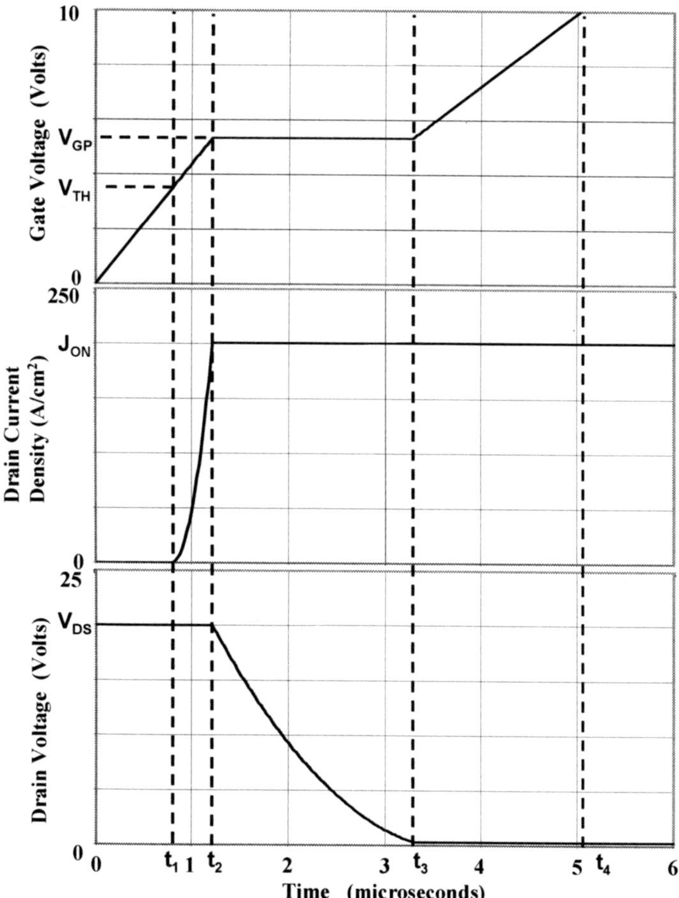

Fig. 6.99 Analytically computed waveforms for a 30-V power MOSFET structure with 20-μm cell pitch

In the above model, the reduction of the specific on-resistance during the time period from t_3 to t_4 due to the increasing gate bias voltage has been ignored. At time t_3, the voltage drop $v_{DS}(t)$ is determined by the on-state current density and the on-resistance for a gate bias equal to the gate plateau voltage (V_{GP}) and not the final gate bias voltage (V_{GS}). As the gate voltage increases during the time period from t_3 to t_4, the on-resistance decreases producing a gradual decline in the on-state voltage drop, which is not shown in Fig. 6.99. Consequently, the gate–drain capacitance also increases during this period and the gate voltage does not increase at a linear rate as shown in the figure.

Simulation Example

To confirm the model for the gate charge of the power VD-MOSFET structure, the results of two-dimensional numerical simulations on a structure (with a cell pitch of 20 μm, a gate width of 12 μm, and a gate oxide thickness of 500 Å) designed with a 30-V rating are described here. The structure had a drift region thickness of 3 μm with a doping concentration of 1.3×10^{16} cm^{-3} located below the P-base region. The P-base region and N$^+$ source regions have vertical depths of 3 and 1 μm, respectively. The doping concentration in the JFET region was enhanced to a surface concentration of 2.3×10^{16} cm^{-3} with a diffusion depth of 3 μm.

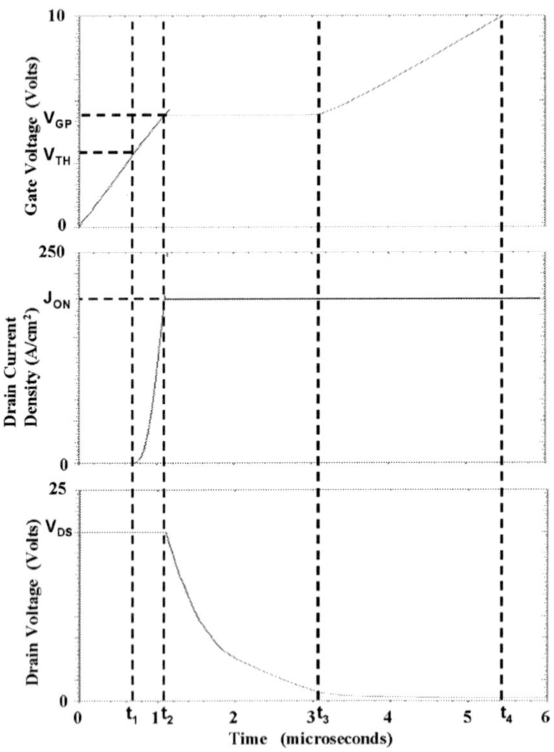

Fig. 6.100 Turn-on waveforms for the power VD-MOSFET structure with 20-μm cell pitch

The gate charge waveforms obtained by using an input gate current density of 0.1 A cm^{-2} when turning on the power VD-MOSFET structure from a blocking state with drain bias of 20 V are shown in Fig. 6.100. The on-state current density is 200 A cm^{-2} at a DC gate bias of 10 V at the end of the turn-on transient. The gate voltage increases at a constant rate at the beginning of the turn-on process as predicted by the analytical model. When the gate voltage reaches the threshold voltage (3.5 V), the drain current begins to increase. The drain current increases as predicted by the analytical model in a quadratic manner until it reaches the on-state current density of 200 A cm^{-2}.

Once the drain current reaches the on-state value, the gate voltage remains constant at the plateau voltage (V_{GP}). The plateau voltage for this structure is 5.3 V for the drain current density of 200 A cm^{-2} as governed by the transconductance of the device. The drain voltage decreases during the plateau phase in a nonlinear manner. After the end of the plateau phase, the gate voltage again increases until it reaches the gate supply voltage. Although the increase in gate voltage is nonlinear at the beginning of this transition, it becomes linear over most of the time after the plateau phase. The waveforms obtained by using the analytical model (see Fig. 6.99) are very similar to those observed in the numerical simulations providing validation for the model.

Specific Gate Charge	Numerical Simulation (nC/cm^2)	Analytical Model (nC/cm^2)
Q_{GS1}	75	80
Q_{GS2}	35	42
Q_{GS}	110	122
Q_{GD}	205	206
Q_{SW}	240	248
Q_G	550	503

Fig. 6.101 Gate charge extracted from numerical simulations for the power VD-MOSFET structure with 20-μm cell pitch

The values for the various components of the gate charge extracted from the numerical simulations are compared with those calculated by using the analytical model in Fig. 6.101. There is very good agreement between these values, indicating that the analytical model is a good representation of the physics of turn-on for power MOSFET structures.

6.14.2 Voltage and Current Dependence

The gate charge is a function of the initial DC blocking voltage (V_{DS}) and the final on-state current density (J_{ON}) which determines the on-state voltage drop (V_{ON}) at the end of the turn-on process. This can be illustrated with the help of an example. Consider the 30-V power VD-MOSFET structure with 20-μm cell pitch, polysilicon gate width of 12 μm, and a gate oxide thickness of 500 Å. The gate charge waveforms obtained by using the above expressions are shown in Fig. 6.102 for three values for the initial DC blocking voltage. A small decrease in the charge (Q_{GS}) required to reach the gate plateau voltage is observed with increasing DC drain bias voltage because of the reduction of the gate–drain capacitance (C_{GD}). The increase in the gate–drain charge (Q_{GD}) with increasing DC drain bias voltage is quite prominent. This increase is associated with the charging of the gate–drain capacitance over a larger voltage range when the initial blocking voltage (V_{DS}) is increased.

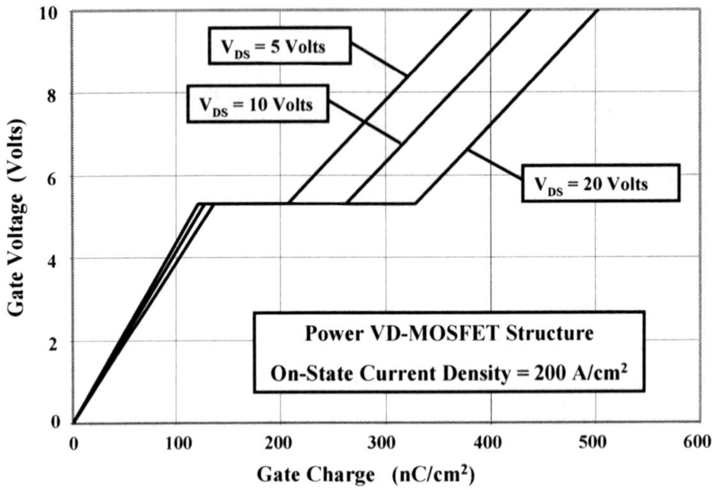

Fig. 6.102 Gate charge waveforms for the power VD-MOSFET structure with 20-μm cell pitch

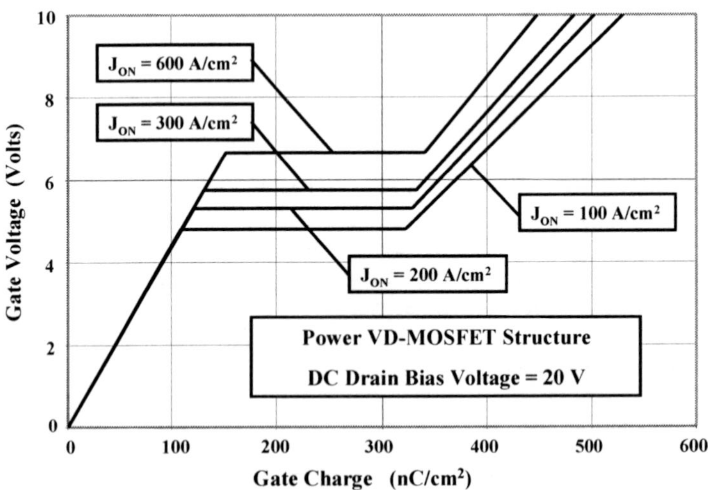

Fig. 6.103 Gate charge waveforms for the power VD-MOSFET structure with 20-μm cell pitch

The impact of changing the on-state current density on the gate charge waveforms obtained by using the above expressions is shown in Fig. 6.103 by using four values for the on-state current density with the initial drain bias kept at 20 V. As expected, the gate plateau voltage increases with increasing on-state current density. Consequently, the gate charge (Q_{GS}) required to reach the plateau voltage also increases. The gate–drain charge (Q_{GD}) is nearly independent of the on-state current density. The increase in the on-state voltage drop, which accompanies an increase in the on-state current density, produces a small decrease in the gate–drain charge (Q_{GD}) not only due to the smaller voltage excursion but also because of the reduction of the final value for the gate–drain capacitance (C_{GD}). The slope of the waveforms after the plateau phase changes with the on-state current density because of the differences between the gate–drain capacitances (C_{GD}) during on-state operation.

Simulation Example

To confirm the model for the gate charge of the power VD-MOSFET structure, the results of two-dimensional numerical simulations on a structure designed with a 30-V rating are described here with a cell pitch of 20 μm, a gate width of 12 μm, and a gate oxide thickness of 500 Å. The structure had a drift region thickness of 3 μm with a doping concentration of 1.3×10^{16} cm^{-3} located below the P-base region. The P-base region and N$^+$ source regions have vertical depths of 3 and 1 μm, respectively. The doping concentration in the JFET region was enhanced to a surface concentration of 2.3×10^{16} cm^{-3} with a diffusion depth of 3 μm.

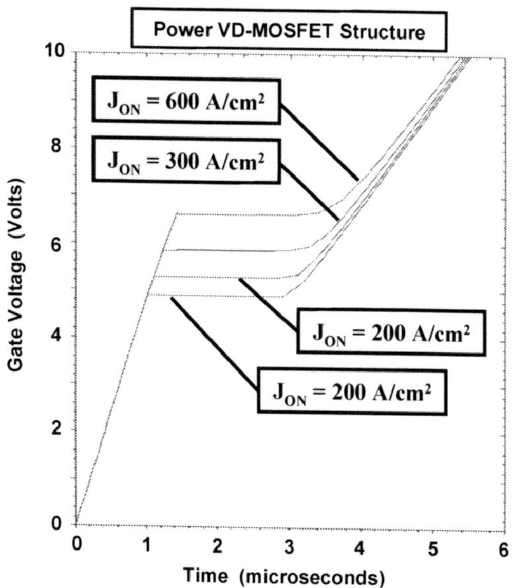

Fig. 6.104 Gate charge waveforms for the power VD-MOSFET structure with 20-μm cell pitch

The gate charge waveforms obtained from the numerical simulations for the case of four on-state current densities are shown in Fig. 6.104 while maintaining the same initial drain bias of 20 V and a gate charging current density of 0.1 A cm^{-2}. The behavior observed in the numerical simulations is very similar to that predicted by the analytical model (see Fig. 6.103). As the on-state current density is increased, the plateau voltage increases as predicted by the model. This prolongs the time taken for the current to reach the on-state value. The duration for the plateau phase is not significantly impacted by the on-state current density which is also consistent with the analytical model. The values for the various components of the gate charge extracted from the simulations are compared with those calculated by using the analytical model in Fig. 6.105. There is very good agreement between these values, indicating that the analytical model is a good representation of the physics of turn-on for power MOSFET structures.

	Numerical Simulation (nC/cm^2)	**Analytical Model (nC/cm^2)**
V_{GP} (J_{ON} = 100 A/cm^2)	4.9	4.8
Q_{GD} (J_{ON} = 100 A/cm^2)	190	212
Q_{SW} (J_{ON} = 100 A/cm^2)	220	242
V_{GP} (J_{ON} = 200 A/cm^2)	5.3	5.3
Q_{GD} (J_{ON} = 200 A/cm^2)	190	206
Q_{SW} (J_{ON} = 200 A/cm^2)	230	242
V_{GP} (J_{ON} = 300 A/cm^2)	5.8	5.7
Q_{GD} (J_{ON} = 300 A/cm^2)	185	201
Q_{SW} (J_{ON} = 300 A/cm^2)	240	253
V_{GP} (J_{ON} = 600 A/cm^2)	6.6	6.7
Q_{GD} (J_{ON} = 600 A/cm^2)	180	188
Q_{SW} (J_{ON} = 600 A/cm^2)	255	261

Fig. 6.105 Gate charge extracted from numerical simulations of the power VD-MOSFET structure with 20-μm cell pitch

	Numerical Simulation (nC/cm^2)	**Analytical Model (nC/cm^2)**
Q_{GD} (V_{DS} = 20 V)	190	206
Q_{SW} (V_{DS} = 20 V)	230	242
Q_{GD} (V_{DS} = 10 V)	130	134
Q_{SW} (V_{DS} = 10 V)	180	179
Q_{GD} (V_{DS} = 5 V)	90	84
Q_{SW} (V_{DS} = 5 V)	150	131

Fig. 6.106 Gate charge extracted from numerical simulations of the power VD-MOSFET structure with 20-μm cell pitch

As a further validation of the analytical model, the gate charge waveforms were obtained by using numerical simulations with three values for the initial drain bias voltage while using the same on-state current density of 200 A cm^{-2} and a gate charging current density of 0.1 A cm^{-2}. The values for the various components of the gate charge extracted from the simulations are compared with those calculated by using the analytical model in Fig. 6.106. There is very good agreement between these values, indicating that the analytical model is a good representation of the physics of turn-on for power MOSFET structures.

6.14.3 VD-MOSFET vs. U-MOSFET Structure

There is a substantial difference between the gate charge waveforms for the power U-MOSFET and power VD-MOSFET structures. The much smaller cell pitch for the power U-MOSFET structure was previously shown to lead to much larger specific input capacitance. This has a detrimental impact on the gate charge waveform for the power U-MOSFET structure. To illustrate this, the gate charge waveform for the 30-V power U-MOSFET structure – with 2-μm cell pitch, a trench width of 1 μm, and a gate oxide thickness of 500 Å – is compared with the power VD-MOSFET structure in Fig. 6.107 when switched on from a drain bias of 20 V to on-state operation at a current density of 300 A cm^{-2} by using a 10-V gate supply voltage. Based upon the structural parameters, the gate geometry factor (K_G) for the power VD-MOSFET structure is 0.36 while that for the power U-MOSFET structure is 1.0. The gate plateau voltage for the power U-MOSFET structure is smaller due to its lower threshold voltage (V_{TH} = 3 V) and larger transconductance.

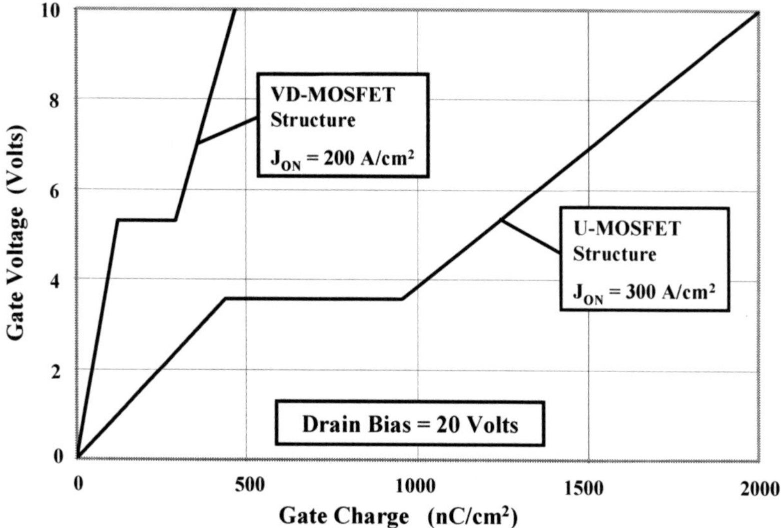

Fig. 6.107 Gate charge waveforms for the power VD-MOSFET and U-MOSFET structures

It can be observed that all the components of the gate charge are much larger for the power U-MOSFET structure. The gate charge values obtained for the power VD-MOSFET structure by using the analytical model are $Q_{GS1} = 80$ nC cm^{-2}, $Q_{GS2} = 42$ nC cm^{-2}, $Q_{GS} = 122$ nC cm^{-2}, $Q_{GD} = 206$ nC cm^{-2}, $Q_{SW} = 248$ nC cm^{-2}, and $Q_G = 503$ nC cm^{-2}. In the case of the power U-MOSFET structure with a cell pitch of 2 μm switched on by using the same circuit conditions, the gate charge values obtained by using the analytical model are $Q_{GS1} = 338$ nC cm^{-2}, $Q_{GS2} = 63$ nC cm^{-2}, $Q_{GS} = 401$ nC cm^{-2}, $Q_{GD} = 516$ nC cm^{-2}, $Q_{SW} = 579$ nC cm^{-2}, and $Q_G = 1{,}965$ nC cm^{-2}. It is apparent that the gate charge for the power U-MOSFET structure is much larger than that for the power VD-MOSFET structure. Consequently, the much lower specific on-resistance achieved with the power U-MOSFET structure, which is beneficial for reducing on-state power loss, is offset by a much large gate charge which leads to high switching losses.

Simulation Example

To confirm the model for the gate charge of the power U-MOSFET structure, the results of two-dimensional numerical simulations on a structure designed with a 30-V rating are described here with a cell pitch of 2 μm, a trench width of 1 μm, and a gate oxide thickness of 500 Å. The structure had a drift region thickness of 3 μm with a doping concentration of 1.3×10^{16} cm^{-3} located below the P-base region. The P-base region and N$^+$ source regions have vertical depths of 1.5 and 0.5 μm, respectively. The trench had a depth of 2 μm.

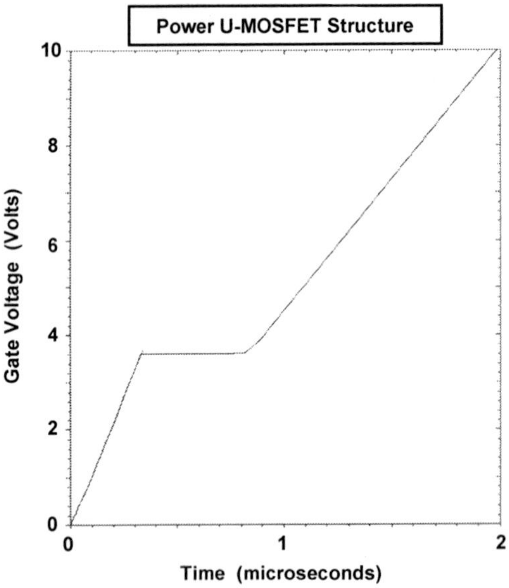

Fig. 6.108 Gate charge waveforms for the power U-MOSFET structure with 2-μm cell pitch

The gate charge waveform obtained by using an input gate current density of 1 A cm^{-2} when turning on the power U-MOSFET structure from a blocking state with drain bias of 20 V is shown in Fig. 6.108. The on-state current density was chosen as 300 A cm^{-2} because of the much lower specific on-resistance for the power U-MOSFET structure. The gate voltage increases at a constant rate at the beginning of the turn-on process as predicted by the analytical model. When the gate voltage reaches the threshold voltage (3.0 V), the drain current begins to increase (not shown in the figure) until it reaches the on-state current density of 300 A cm^{-2}.

Once the drain current reaches the on-state value, the gate voltage remains constant at the plateau voltage (V_{GP}). The plateau voltage for this structure is 3.56 V for the drain current density of 300 A cm^{-2} as governed by the transconductance of the device. After the end of the plateau phase, the gate voltage again increases until it reaches the gate supply voltage. The waveform obtained for the power U-MOSFET structure by using the analytical model (see Fig. 6.107) is very similar to that observed in the numerical simulations providing validation for the model.

The values for the various components of the gate charge extracted from the numerical simulations are compared with those calculated by using the analytical model in Fig. 6.109. There is good agreement between these values, indicating that the analytical model is a good representation of the physics of turn-on for the power U-MOSFET structure as well.

Specific Gate Charge	Numerical Simulation (nC/cm^2)	Analytical Model (nC/cm^2)
Q_{GS1}	275	337
Q_{GS2}	60	63
Q_{GS}	335	400
Q_{GD}	505	515
Q_{SW}	565	578
Q_G	1960	1965

Fig. 6.109 Gate charge extracted from numerical simulations of the power U-MOSFET structure with 2-μm cell pitch

6.14.4 Impact of VD-MOSFET and U-MOSFET Cell Pitch

The gate charge for the power VD-MOSFET structure is altered by changes to the width of the gate electrode because this varies not only the gate overlap area over the N-drift region but also the specific capacitances due to a change in the cell pitch. To illustrate this, consider the case of a 30-V power VD-MOSFET structure with a polysilicon window of 8 μm and a gate oxide thickness of 500 Å. The gate–drain charge for these structures, when switched on from a drain bias of 20 V to on-state operation at a current density of 200 A cm^{-2} by using a 10-V gate supply voltage, is plotted in Fig. 6.110 when the gate width is varied. It can be observed that the gate–drain charge increases monotonically with increasing gate width. It can be concluded that the gate–drain charge can be minimized by making the gate width as small as possible to reduce the area for overlap of the gate with the N-drift region. However, it was previously shown that the specific on-resistance becomes extremely large when the gate width is made small.

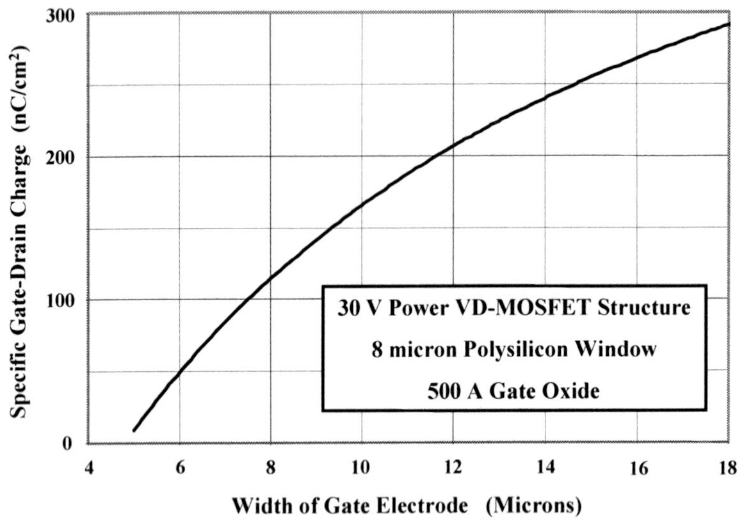

Fig. 6.110 Gate–drain charge for the power VD-MOSFET structures

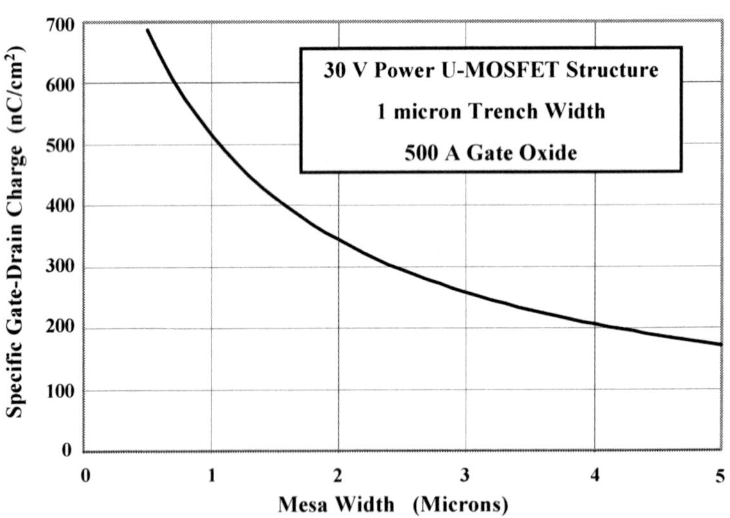

Fig. 6.111 Gate–drain charge for the power U-MOSFET structures

In the case of the power U-MOSFET structure, the gate charge can be altered by changing the width of the mesa region due to its impact on the cell pitch. To illustrate this, consider the case of a 30-V power U-MOSFET structure with a trench width of 1 μm and a gate oxide thickness of 500 Å. The gate charge for these structures, when switched on from a drain bias of 20 V to on-state operation at a current density of 300 A cm^{-2} by using a 10-V gate supply voltage, is plotted in Fig. 6.111 when the mesa width is varied. It can be observed that the gate–drain charge decreases monotonically with increasing mesa width. It can be concluded that the gate–drain charge can be reduced by making the mesa width as large as possible. However, it was previously shown that the specific on-resistance becomes larger when the mesa width is made larger.

Simulation Example

Further validation of the model for computing the gate charge for power MOSFET structures can be derived by analysis of the impact of changes in the cell structure. To confirm the model for the gate charge of the power VD-MOSFET structure, the results of two-dimensional numerical simulations on structures designed with a 30-V rating are described here. The gate width for these structures was varied while maintaining a polysilicon window of 8 μm and a gate oxide thickness of 500 Å. All the structures had a drift region thickness of 3 μm with a doping concentration of 1.3×10^{16} cm^{-3} located below the P-base region. The P-base region and N^+ source regions have vertical depths of 1.5 and 0.5 μm, respectively. The doping concentration in the JFET region was enhanced to a surface concentration of 2.3×10^{16} cm^{-3} with a diffusion depth of 3 μm.

The gate–drain charge acquired from numerical simulations using an input gate current density of 0.1 A cm^{-2} when turning on the power VD-MOSFET structure

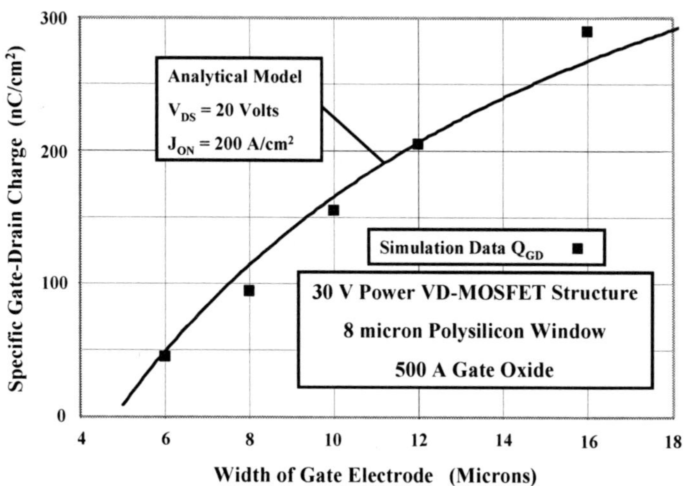

Fig. 6.112 Gate–drain charge for the power VD-MOSFET structures with 8-μm polysilicon window

from a blocking state with drain bias of 20 V is shown in Fig. 6.112. The on-state current density for all devices was maintained at 200 A cm^{-2} at a DC gate bias of 10 V at the end of the turn-on transient. The gate–drain charge was observed to increase with increasing gate width as expected due to the larger gate overlap with the N-drift region. A good agreement is observed between the values obtained by using the analytical model and those extracted from the simulations, indicating that the model can be utilized over a broad range of power VD-MOSFET structural parameters.

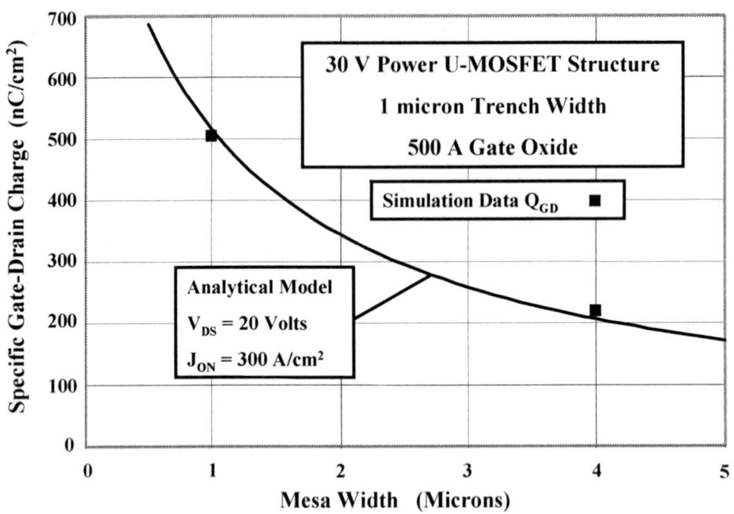

Fig. 6.113 Gate–drain charge for the power U-MOSFET structures with 1-μm trench width

The gate–drain charge acquired from numerical simulations using an input gate current density of 1 A cm^{-2} when turning on the power U-MOSFET structure from a blocking state with drain bias of 20 V is shown in Fig. 6.113. The on-state current density for all devices was maintained at 300 A cm^{-2} at a DC gate bias of 10 V at the end of the turn-on transient. The gate–drain charge was observed to decrease with increasing gate width as expected due to the larger cell pitch. A good agreement is observed between the values obtained by using the analytical model and those extracted from the simulations, indicating that the model can be utilized over a broad range of power U-MOSFET structural parameters.

6.15 Optimization for High Frequency Operation

In Sect. 6.7.1 for the power VD-MOSFET structure, it was shown that the specific on-resistance can be minimized by adjusting the width of gate electrode. Similarly, in Sect. 6.9 for the power U-MOSFET structure, it was shown that the specific on-resistance can be reduced by decreasing the width of mesa region. These procedures are appropriate for the design of power MOSFET structures used in low frequency applications, such as automotive electronics, where the on-state power loss is dominant. In this situation, an arbitrarily large die area can be chosen, within the constraints of increasing cost, to reduce the absolute on-resistance of the power MOSFET device used in the application. In other applications with high operating frequencies, such as switch-mode power supplies, the switching losses become an equally important consideration during system design. In this case, it becomes necessary to choose an optimum chip area to achieve minimum total power dissipation within the power MOSFET device.[21]

6.15.1 Input Switching Power Loss

Significant switching power losses can arise from the charging and discharging of the large input capacitance in power MOSFET devices at high frequencies. The input capacitance (C_{IN}) of the power MOSFET structure must be charged to the gate supply voltage (V_{GS}) when turning on the device and then discharged to zero volts when turning off the device during each period of the operating cycle. The energy loss during the turn-on and turn-off event is given by

$$E_{ON} = E_{OFF} = \frac{1}{2} C_{IN} V_{GS}^2. \tag{6.209}$$

The total power loss can be obtained by summing the on-state power dissipation for a duty cycle $\delta = t_{ON}/T$ and the switching power losses:

$$P_T = P_{ON} + P_{SW} = \delta R_{ON} I_{ON}^2 + C_{IN} V_{GS}^2 f, \tag{6.210}$$

where R_{ON} is the on-resistance of the power MOSFET structure, I_{ON} is the on-state current, and f is the operating frequency. In writing this equation, the switching power losses due to the drain current and voltage transitions have been neglected. These switching power losses are treated in Sect. 6.15.2.

As discussed in the previous sections, each power MOSFET structure and cell design have a unique combination of specific on-resistance and specific input capacitance. When expressed using these parameters, the total power dissipation is given by

$$P_T = \frac{R_{ON,SP}}{A} \delta I_{ON}^2 + C_{IN,SP} A V_{GS}^2 f, \tag{6.211}$$

where A is the active area of the power MOSFET device. From this expression, it can be seen that the on-state power loss will decrease as the device active area is increased while the switching loss will increase in proportion to the active area.

A minimum total power loss occurs for each power MOSFET structure at an optimum active area. This is demonstrated in Fig. 6.114 for the 30-V power VD-MOSFET structure with a cell pitch of 20 μm at an operating frequency of 100 KHz using a 50% duty cycle. This device has a specific on-resistance of 1.6 mΩ cm^2 and a specific input capacitance of 20 nF cm^{-2}. The plot in the figure was computed by using an on-state current of 10 A and a gate bias voltage of 10 V. The smallest total power loss occurs for this particular device structure at an active area of 0.63 cm^2.

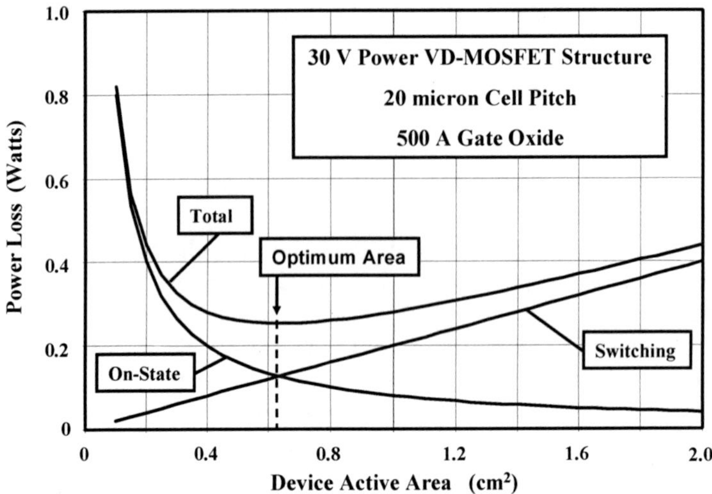

Fig. 6.114 Optimization of the active area for the power VD-MOSFET structures with 20-μm cell pitch

It can be observed from the figure that the on-state and switching power losses are equal at the optimum active area. By equating these terms in (6.211), an expression for the optimum active area is obtained:

$$A_{\mathrm{OPT}} = \sqrt{\frac{R_{\mathrm{ON,SP}}}{C_{\mathrm{IN,SP}}}} \left(\frac{I_{\mathrm{ON}}}{V_{\mathrm{GS}}} \right) \left(\sqrt{\frac{\delta}{f}} \right). \tag{6.212}$$

The second and third terms in this equation are determined by the circuit operating conditions. However, from the first term in this expression, a useful technology figure of merit can be defined:

$$\mathrm{FOM}(A) = \frac{R_{\mathrm{ON,SP}}}{C_{\mathrm{IN,SP}}}. \tag{6.213}$$

This figure of merit has a unique value based upon the power MOSFET structural design. For the 30-V power VD-MOSFET structure with 20-μm cell pitch, this technology figure of merit has a value of 80,000 Ohm^{2}cm^{4}s^{-1}. According to (6.212), a larger optimum area is required with increasing on-state current levels, larger duty cycles, and decreasing operating frequency.

In the power electronic community, there is trend toward increasing the operating frequency for switch-mode power supplies to reduce the size and weight of the magnetic components. The impact of changes in the operating frequency upon the optimum active area for the power VD-MOSFET structure with the 20-μm cell pitch is provided in Fig. 6.115. The location of the optimum area is indicated by the vertical arrows in the figure. It can be observed that the optimum area becomes smaller when the operating frequency is increased. This is beneficial for reducing the cost of the power MOSFET device. However, the minimum total power loss becomes larger with increasing frequency. This is detrimental to obtaining high efficiency for power conversion as well as for reducing the thermal budget and heat sinking capability. From the above equations, an expression for the minimum total power dissipation can be obtained:

$$P_{\mathrm{T}}(\min) = 2I_{\mathrm{ON}}V_{\mathrm{GS}}\sqrt{\delta R_{\mathrm{ON,SP}}C_{\mathrm{IN,SP}}f}. \tag{6.214}$$

The minimum total power dissipated in the power MOSFET device doubles when the frequency is quadrupled. Consequently, the ability to migrate to higher operating frequencies in power conversion circuits is dependent on making enhancements to the power MOSFET technology.

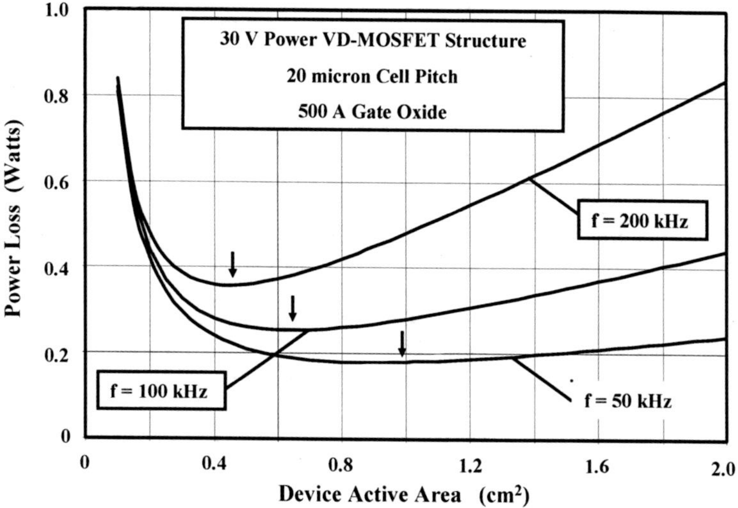

Fig. 6.115 Optimization of the active area for the power VD-MOSFET structures with 20-μm cell pitch

A second technology figure of merit related to the minimum power dissipation can be defined as

$$\mathrm{FOM}(B) = R_{\mathrm{ON,SP}}C_{\mathrm{IN,SP}}. \tag{6.215}$$

For the 30-V power VD-MOSFET structure with 20-μm cell pitch, this technology figure of merit has a value of 32 ps.

In the previous sections, it was shown that the power U-MOSFET structure with 2-μm cell pitch has a much smaller specific on-resistance and larger input capacitance than the above power VD-MOSFET structure. For this power U-MOSFET design, the specific on-resistance is 0.31 mΩ cm^2 with a specific input capacitance of 105 nF cm^{-2}. Consequently, the technology FOM(A) for this power U-MOSFET design is 2,952 Ohm^2 cm^4 s^{-1}, which leads to much smaller optimum active area for the device as shown in Fig. 6.116. The optimum active area is reduced from 0.63 cm^2 for the power VD-MOSFET structure to only 0.12 cm^2 for the power U-MOSFET structure. Even with a more complex process technology for the power U-MOSFET devices, this leads to a smaller device cost. However, the technology FOM(B) for the power U-MOSFET structure has a value of 33 ps, which is nearly identical to that for the power VD-MOSFET structure. This results in the same minimum total power dissipation for both structures, indicating that there is no gain in power conversion efficiency by replacing the power VD-MOSFET structure with the power U-MOSFET structure.

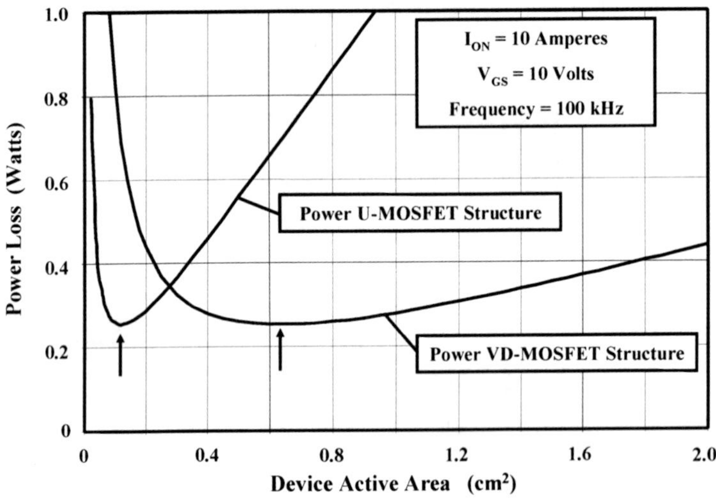

Fig. 6.116 Comparison of the optimum active area for the power VD-MOSFET structure with the power U-MOSFET structure

The power VD-MOSFET structure can be optimized for improving its high frequency performance using the same approach as the optimization of its specific on-resistance by varying the width of the gate electrode. As an example, the figure-of-merit values obtained for a power VD-MOSFET structure with a polysilicon window of 8 μm and a gate oxide of 500 Å are provided in Fig. 6.117. It can be observed that both of the figures of merit go through minima as the gate electrode width is increased. The area at which the minima occur is indicated in the figure by the vertical arrows. For the FOM(A), the optimum gate electrode width is 8.3 μm while for the FOM(B) the optimum gate electrode width is 11.6 μm. In this situation, it is better to select the smaller gate width of 8.3 μm as the optimum design because the FOM(B) increases by only 10% from its minimum value. The optimum gate electrode width of 8.3 μm obtained for high frequency operation is slightly smaller than the optimum gate electrode width of 9.4 μm for minimizing the specific on-resistance of the power VD-MOSFET structure.

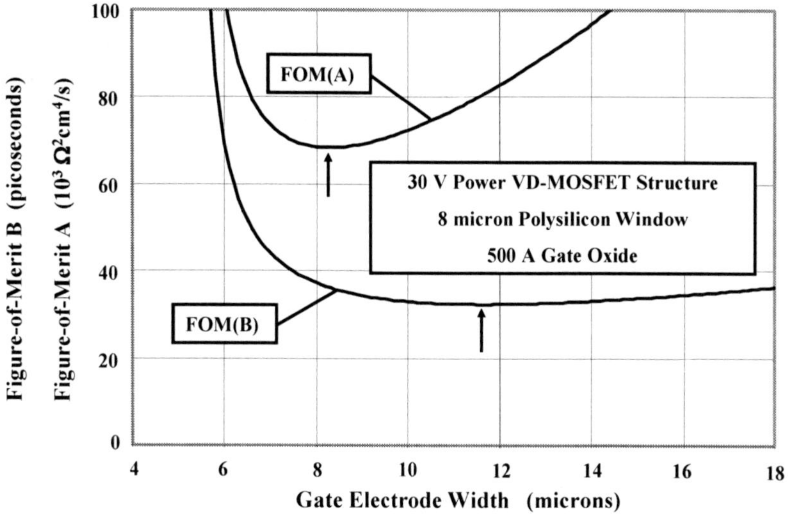

Fig. 6.117 Optimization of the figure of merits for the power VD-MOSFET structure

The power U-MOSFET structure can also be optimized for improving its high frequency performance using the same approach as the optimization of its specific on-resistance by varying the width of the mesa region. As an example, the figure-of-merit values obtained for a power U-MOSFET structure with a trench width of 1 μm and a gate oxide of 500 Å are provided in Fig. 6.118 when the mesa width is varied from 0.5 to 5 μm. It can be observed that neither figure of merit

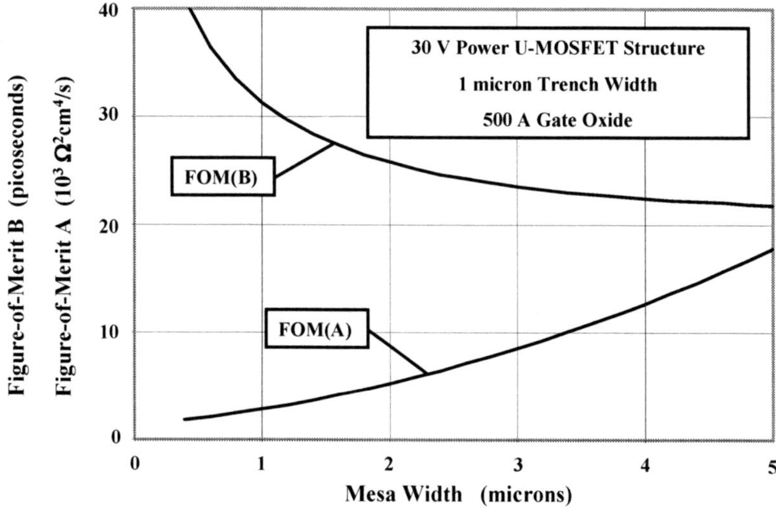

Fig. 6.118 Optimization of the figure of merits for the power U-MOSFET structure

goes through a minimum as the mesa width is increased. The FOM(*A*) increases monotonically with increasing mesa width, indicating that a small mesa width is desirable for reducing the active area and cost for the power U-MOSFET device. However, the FOM(*B*) decreases monotonically with increasing mesa width, indicating that a large mesa width is desirable for reducing the total power dissipation. As a compromise, a power U-MOSFET design with a mesa width of 2 μm would be suitable for this example. The mesa width of 2 μm obtained for high frequency operation is also desirable for reducing the specific input and gate–drain capacitances of the power U-MOSFET structure as shown in Figs. 6.87 and 6.88.

6.15.2 Output Switching Power Loss

In most applications for power MOSFET structures with high operating frequency, the switching losses associated with the drain current and voltage transitions become a dominant portion of the total power loss. These power losses were previously discussed in the context of Fig. 1.4 for both the turn-on and turn-off events during each switching cycle. In Sect. 6.14, it was demonstrated that the time period associated with the increase of the drain current and decrease of the drain voltage is determined by the charging of the device capacitances. It is therefore common practice in the industry to use the following figures of merit to compare the performance of power MOSFET products:

$$\text{FOM}(C) = R_{\text{ON,SP}} Q_{\text{GD,SP}} \tag{6.216}$$

and

$$\text{FOM}(D) = R_{\text{ON,SP}} Q_{\text{SW,SP}}. \tag{6.217}$$

Although FOM(*D*) encompasses both the drain current and voltage transitions, it is customary to use FOM(*C*) because the gate–drain charge tends to dominate in the switching gate charge. One advantage of using these expressions is that the figure of merit becomes independent of the active area of the power MOSFET device. These figures of merit therefore provide a good measure of the quality of the power MOSFET structure, cell design, and process technology.

As an example, the 30-V power VD-MOSFET structure with 20-μm cell pitch can be compared with the power U-MOSFET structure with 2-μm cell pitch. The FOM(*C*) for the power VD-MOSFET structure is 330 mΩ nC vs. 160 mΩ nC for the power U-MOSFET structure. In the case of FOM(*D*), the value for the power VD-MOSFET structure is 387 mΩ nC vs. 191 mΩ nC for the power U-MOSFET structure. Consequently, the power U-MOSFET structure has become the preferred choice for switch-mode power supply applications. Recent enhancements to the trench device structure, including selectively increasing the thickness of the oxide at the bottom of the trench, have enabled improving the FOM(*C*) to below 30 mΩ nC.

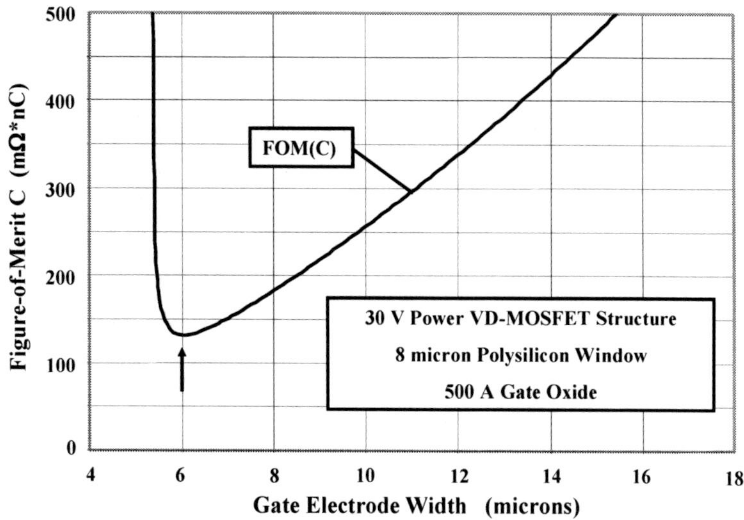

Fig. 6.119 Optimization of the figure of merit (*C*) for the power VD-MOSFET structure

The figure of merit for the power VD-MOSFET structure is governed by the width of the gate electrode. This is demonstrated in Fig. 6.119 for the case of a structure with polysilicon window of 8 μm and gate oxide thickness of 500 Å. A sharp minimum of the FOM(*C*) is observed at a gate electrode width of 6 μm, which is well below the width for achieving the lowest specific on-resistance. The lowest FOM(*C*) for the power VD-MOSFET structure is 130 mΩ nC making it competitive with that for the power U-MOSFET structure. However, its much large specific on-resistance (2.7 mΩ cm^2) is a disadvantage from the cost and packaging viewpoint.

Recently, a planar power MOSFET structure (named the SSCFET structure), shown in Fig. 6.120, has been reported with improved FOM(*C*) by utilizing the screening of the gate with a deep P^+ region.[22] A retrograde doping profile[23] was also incorporated in this structure to achieve a potential barrier below the gate electrode while reducing the resistance of the JFET region. With the power SSCFET architecture, the cell pitch can be greatly reduced to the same size (2 μm) as that used in power U-MOSFET devices and the gate width (W_G) can be reduced to 1 μm. In addition, the screening of the gate region by the deep P^+ regions enables reduction of the electric field at location *A*, which suppresses reach-through breakdown problems. This allows supporting the rated blocking voltage with much shorter channels than required in either the power VD-MOSFET or U-MOSFET structures. Consequently, the specific on-resistance for the 30-V power SSCFET structure becomes comparable with that for the power U-MOSFET structure. At the same time, the screening of the gate electrode by the P^+ regions reduces the gate–drain capacitance and charge. These features have enabled obtaining a FOM(*C*) for the power SSCFET structure of 20 mΩ nC, making it an attractive choice for power supply applications.

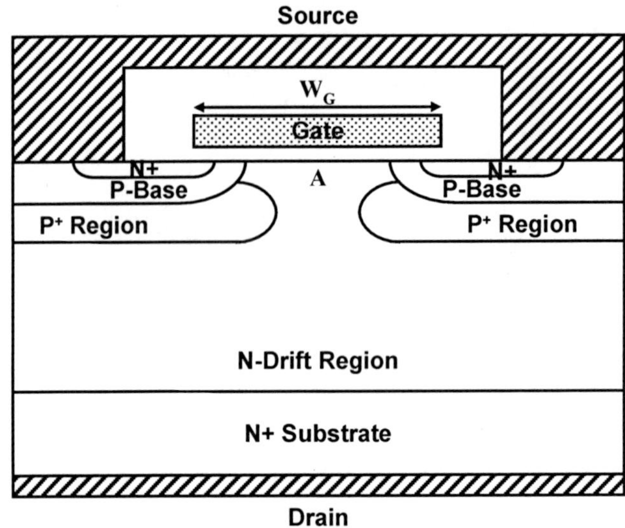

Fig. 6.120 Power SSCFET structure

6.15.3 Gate Propagation Delay

It is necessary to use a refractory gate electrode material during the fabrication of the power MOSFET structures because of the high temperatures required to anneal and diffuse the ion-implanted regions. The most commonly used gate electrode material is polysilicon. For a typical thickness of 0.5 μm, heavily doped N-type polysilicon has a sheet resistance of 30 Ω sq^{-1}. As illustrated in the cross sections for the power MOSFET structures (see Figs. 6.3 and 6.4), the gate electrode is submerged below the source electrode by using an intermetal dielectric to enable the formation of a large source metal bonding area over the active area of the device. The contact to the gate electrode must then be achieved at selected locations orthogonal to the cross section shown in the figures.

A top view and cross section for a typical layout for the gate fingers are shown in Fig. 6.121 for the case of a linear cell geometry. The active area is located below the source metal while a gate contact pad is formed on the left-hand side. When a gate bias is applied to the device, the gate voltage is impressed at the gate contact pad instantaneously if the package resistance is neglected. However, the gate voltage is not immediately available at the end of the fingers because of the finite time for charging the gate capacitance. This propagation delay for the gate voltage impacts the on-resistance of the device during the turn-on transient.

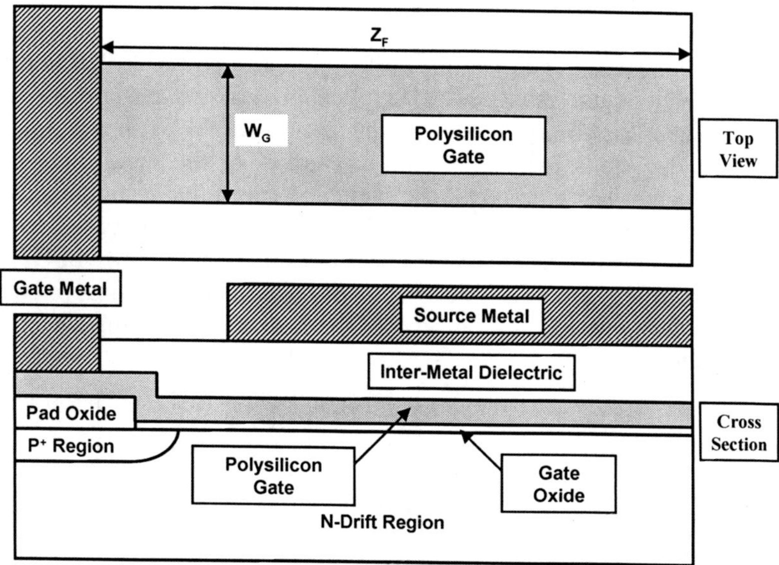

Fig. 6.121 Gate electrode contact configuration

A worst-case analysis of the *R–C* time constant for propagation of the gate voltage can be performed by assuming that all the resistance of the gate polysilicon finger charges all the input gate capacitance for the finger. The resistance for the gate finger is given by

$$R_{\mathrm{GF}} = \rho_{\mathrm{SQ,POLY}} \frac{Z_{\mathrm{F}}}{W_{\mathrm{G}}}, \tag{6.218}$$

where $\rho_{\mathrm{SQ,POLY}}$ is the sheet resistance of the polysilicon gate electrode and Z_{F} is the length of the gate finger. The input capacitance for each gate finger is given by

$$C_{\mathrm{GF}} = C_{\mathrm{OX}} Z_{\mathrm{F}} W_{\mathrm{G}}, \tag{6.219}$$

where C_{OX} is the specific capacitance of the gate oxide. Combining these expressions, the *R–C* time constant for charging the gate is given by

$$\tau_{\mathrm{G}} = R_{\mathrm{GF}} C_{\mathrm{GF}} = \rho_{\mathrm{SQ,POLY}} C_{\mathrm{OX}} Z_{\mathrm{F}}^2. \tag{6.220}$$

From this equation, it can be concluded that a small gate charging time constant can be achieved by using short gate fingers. However, this reduces the proportion of the active area within the device degrading its on-resistance. For a polysilicon gate electrode and a gate oxide thickness of 500 Å, the propagation time constant is 0.1 μs for a gate finger length of 2 mm. This can be reduced by replacing the polysilicon gate electrode with a silicided gate electrode whose sheet resistance is one tenth of that for polysilicon.

6.16 Switching Characteristics

The power MOSFET structure is most often used as a power switch in circuits for energy conversion and management applications. When designed with a low blocking voltage capability (<100 V), the power MOSFET has a low on-state voltage drop due to its low specific on-resistance. At the same time, the power MOSFET structure has a fast inherent switching speed due to unipolar operation. The switching behavior of power MOSFET structures is governed by the gate drive circuit and the nature of the load. Most often, the power MOSFET device is used to control the current in inductive loads such as the windings of motors. In these circuits, a freewheeling diode carries the load current during a portion of the operating cycle.

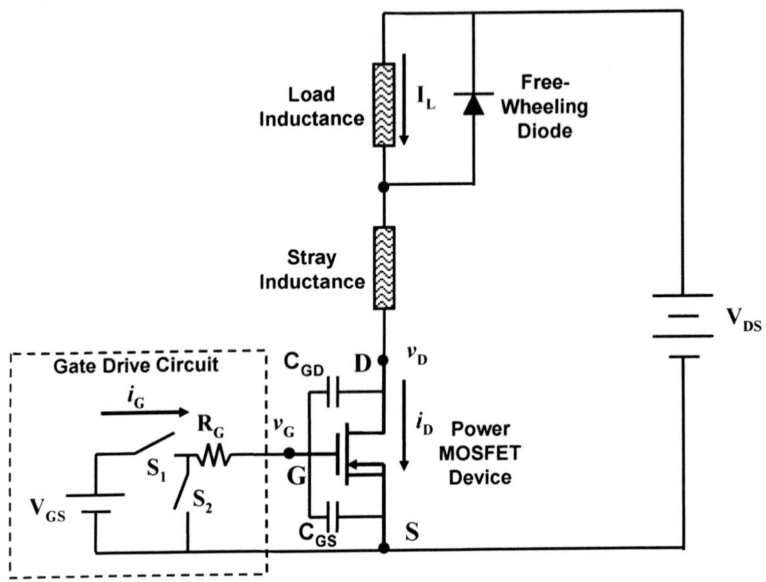

Fig. 6.122 Power MOSFET device operating in an inductive load circuit

The operation of a power MOSFET device in an inductive load circuit is illustrated in Fig. 6.122. The device is switched on and off by a control or gate drive circuit, which can be represented (Thevenin's equivalent) as a DC voltage (V_{GS}) with a series resistance (R_G). The load current I_L transfers between the power MOSFET device and the freewheeling diode during each operating cycle. The inductor is charged (i.e., its current increases) when the power MOSFET device is turned on while it is discharged when the load current flows via the diode. However, the change in the inductor current is small during one cycle, allowing the assumption that the current I_L is constant.[24] The stray inductance is included to account for package and board parasitic elements.

In the above circuit, the load current flows through the freewheeling diode whenever the power MOSFET device is in the off-state. The operation of the power MOSFET in the off-state is ensured by the switch S_2 in the control circuit being in the closed position while switch S_1 is in an open position. These switches are usually comprised of lateral MOSFET structures within the control (integrated) circuit. The power MOSFET device can be switched on by opening switch S_2 and closing switch S_1 in the control circuit. The current then transfers from the diode to the power MOSFET device. During the turn-off transient, this operation is reversed to transfer the current from the power MOSFET device back to the diode.

6.16.1 Turn-On Transient

Prior to the turn-on transient, the power MOSFET device is operating in its off-state because switch S_2 is closed and switch S_1 is open. The load current is therefore flowing through the freewheeling diode. These initial conditions are defined by $v_G = 0$, $i_D = 0$, and $v_D = V_{DS}$. When switch S_2 is opened and switch S_1 is closed shortly thereafter by the control circuit during the turn-on process, the gate bias voltage source (V_{GS}) begins to charge the capacitances of the power MOSFET device. Since no drain current can flow through the power MOSFET device until the gate voltage exceeds its threshold voltage, the drain voltage initially remains at the drain bias voltage. The gate–drain capacitance $C_{GD}(V_{DS})$ remains constant because the drain voltage is constant. Consequently, the time constant for charging the gate of the power MOSFET device is $R_G[C_{GS} + C_{GD}(V_{DS})]$ resulting in a gate voltage given by

$$v_G(t) = V_{GS}\{1 - e^{-t/R_G[C_{GS}+C_{GD}(V_{DS})]}\}. \tag{6.221}$$

The gate voltage reaches the threshold voltage at time:

$$t_1 = R_G[C_{GS} + C_{GD}(V_{DS})]\ln\left(\frac{V_{GS}}{V_{GS} - V_{TH}}\right). \tag{6.222}$$

This time can be considered to be a *delay time* before any drain current begins to flow in the circuit after the turn-on is initiated by the control circuit.

Once the gate voltage exceeds the threshold voltage, drain current begins to flow during the second phase of the turn-on process. For the power MOSFET structure operating with channel pinch-off physics, the drain current is given by

$$i_D(t) = g_m[v_G(t) - V_{TH}] = \frac{\mu_{ni} C_{OX} Z}{2L_{CH}}[v_G(t) - V_{TH}]^2. \tag{6.223}$$

Although the drain current increases during the second phase, the drain voltage remains at the drain supply voltage (V_{DS}) because the diode cannot sustain any voltage until all of the load current is transferred to the power MOSFET device. Since the drain voltage remains constant, the drain–gate capacitance is also invariant during the second phase of the turn-on process. Consequently, the gate voltage continues to increase at an exponential rate as described by (6.221) with the same time constant. The drain current increases as the square of the gate voltage as described by (6.223) with a nonlinear waveform as illustrated in Fig. 6.123.

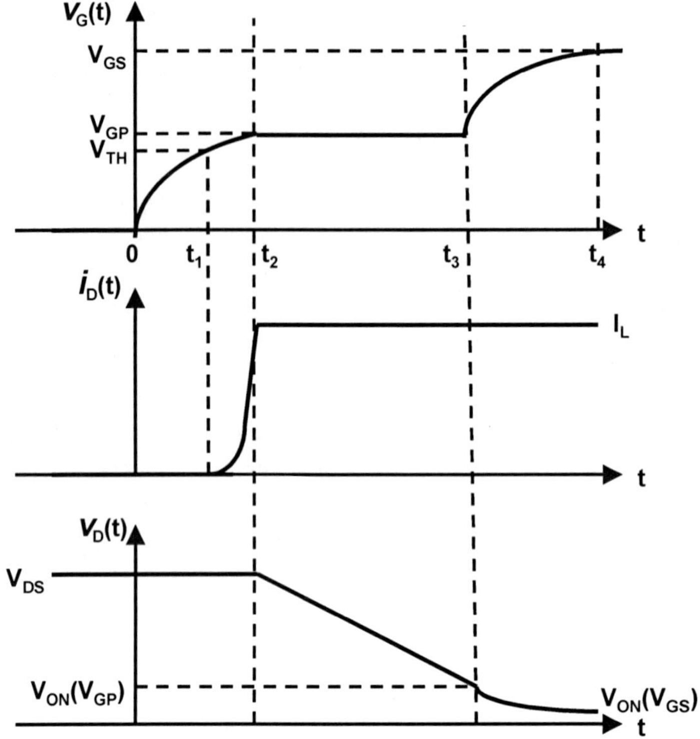

Fig. 6.123 Waveforms for the power MOSFET structure during turn-on with a gate voltage source

The drain current increases until it becomes equal to the load current (I_L) at the end of the second phase (time t_2) given by

$$t_2 = R_G[C_{GS} + C_{GD}(V_{DS})] \ln \left[\frac{V_{GS}\mu_{ni}C_{OX}Z}{V_{GS}\mu_{ni}C_{OX}Z - L_{CH}\sqrt{I_L} - V_{TH}\mu_{ni}C_{OX}Z} \right]. \quad (6.224)$$

All of the load current has transferred from the diode to the power MOSFET device at time t_2 and the diode is now able to support voltage. (The reverse recovery process for the diode will be ignored to simplify the analysis here.) The drain voltage of the power MOSFET device therefore begins to reduce at this time. Since the drain current is constant (equal to the load current), the gate voltage at time t_2 is given by

$$v_{\rm G}(t_2) = V_{\rm GP} = \frac{I_{\rm L}}{g_{\rm m}} + V_{\rm TH}. \tag{6.225}$$

Combining this expression with the dependence of the transconductance on the gate bias voltage, an expression for the plateau voltage is obtained:

$$V_{\rm GP} = V_{\rm TH} + \sqrt{\frac{I_{\rm L} L_{\rm CH}}{\mu_{\rm ni} C_{\rm OX} Z}}. \tag{6.226}$$

The gate voltage remains constant at the plateau voltage until the drain voltage has reduced to the on-state voltage drop corresponding to the product of the load current and the on-resistance of the device *at a gate bias equal to the plateau voltage*.

Since the gate voltage is constant during the plateau phase, all the gate current $i_{\rm G}(t)$ is used to charge the gate–drain or Miller capacitance. The gate current during the plateau phase is given by

$$i_{\rm GP} = \frac{V_{\rm GS} - V_{\rm GP}}{R_{\rm G}} = \frac{1}{R_{\rm G}}\left[V_{\rm GS} - \left(V_{\rm TH} + \sqrt{\frac{I_{\rm L} L_{\rm CH}}{\mu_{\rm ni} C_{\rm OX} Z}}\right)\right]. \tag{6.227}$$

As this current charges the gate–drain capacitance, its voltage decreases at a rate given by

$$\frac{{\rm d}v_{\rm GD}}{{\rm d}t} = -\frac{i_{\rm GP}}{C_{\rm GD}(v_{\rm D})}. \tag{6.228}$$

Since the gate–source voltage is constant at $V_{\rm GP}$ during this time, the drain voltage also decreases at the same rate:

$$\frac{{\rm d}v_{\rm D}}{{\rm d}t} = \frac{{\rm d}v_{\rm GD}}{{\rm d}t} = -\frac{i_{\rm GP}}{C_{\rm GD}(v_{\rm D})} = -\frac{V_{\rm GS} - V_{\rm GP}}{R_{\rm G} C_{\rm GD}(v_{\rm D})}. \tag{6.229}$$

As indicated in this expression, the gate–drain capacitance is a function of the drain voltage. The drain voltage decreases in accordance with (6.198) derived earlier in Sect. 6.14 and the time duration $(t_3 - t_2)$ is given by (6.197). Alternately, the gate–drain capacitance can be assumed to have a constant average $(C_{\rm GD,av})$ value during the transient. In this case, the drain voltage decreases linearly with time as given by

$$v_{\rm D}(t) = V_{\rm DS} - \frac{(V_{\rm GS} - V_{\rm GP})t}{R_{\rm G} C_{\rm GD,av}}. \tag{6.230}$$

This behavior is illustrated in Fig. 6.123.

At the end of the plateau phase (at time t_3), the drain voltage becomes equal to the on-state voltage drop corresponding to the plateau gate bias voltage:

$$v_D(t_3) = I_L R_{ON}(V_{GP}), \tag{6.231}$$

where $R_{ON}(V_{GP})$ is the on-resistance of the power MOSFET device at a gate bias equal to the plateau voltage. Using these relationships, the time interval $(t_3 - t_2)$ is given by

$$t_3 - t_2 = \frac{R_G C_{GD,av}}{V_{GS} - V_{GP}}[V_{DS} - I_L R_{ON}(V_{GP})]. \tag{6.232}$$

Beyond this point in time, the gate voltage increases exponentially as shown in the figure until it reaches the gate supply voltage. The time constant for this exponential rise is different from the initial phase due to the large gate–drain capacitance. The increasing gate voltage produces a reduction of the on-resistance of the power MOSFET device, resulting in a small reduction of the drain voltage during this fourth phase of the turn-on process.

The largest power dissipation during the turn-on transient occurs during the time interval $(t_3 - t_1)$ when either the drain current or the drain voltage has large magnitudes. This time interval can be reduced by decreasing the gate–drain capacitance (C_{GD}), which has been an important objective over the years when creating improved device technologies such as the SSCFET.

6.16.2 Turn-Off Transient

After carrying the load current during its duty cycle, the power MOSFET device is switched off to transfer the current back to the freewheeling diode. Prior to the turn-off transient, the device is operating in its on-state because switch S_1 is closed and switch S_2 is open. These initial conditions are defined by $v_G = V_{GS}$, $i_D = I_L$, and $v_D = V_{ON}(V_{GS})$. To initiate the turn-off process, switch S_1 is opened and switch S_2 is subsequently closed by the control circuit. The gate electrode of the power MOSFET device is then connected to the source via the gate resistance to discharge its capacitances. However, no changes in the drain current or voltage occur until the gate voltage reaches the magnitude required to operate the power MOSFET device at a saturated drain current equal to the load current. (The small increase in the drain voltage, due to the increase in on-resistance resulting from the reduction of the gate bias voltage, has been neglected here.) This gate plateau voltage is given by (6.226). During this time interval, the gate–drain capacitance $C_{GD}(V_{ON})$ remains constant because the drain voltage is constant. Consequently, the time constant for discharging the gate of the power MOSFET device is $R_G[C_{GS} + C_{GD}(V_{ON})]$ and the gate voltage decreases exponentially with time as given by

$$v_G(t) = V_{GS} e^{-t/R_G[C_{GS}+C_{GD}(V_{ON})]}. \tag{6.233}$$

The time t_4 for reaching the gate plateau voltage can be obtained by using this equation with (6.226) for the plateau voltage:

$$t_4 = R_G[C_{GS} + C_{GD}(V_{ON})]\ln\left[\frac{V_{GS}}{V_{GP}}\right]. \tag{6.234}$$

This time can be considered as a *turn-off delay time* before the drain voltage begins to increase after the turn-off is initiated by the control circuit.

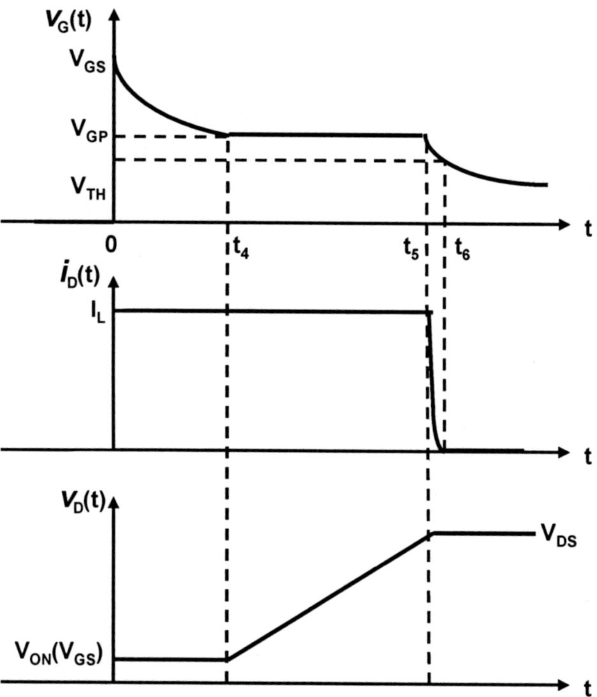

Fig. 6.124 Waveforms for the power MOSFET structure during turn-off with a gate voltage source

The drain voltage begins to increase at time t_4 but the drain current remains constant at the load current I_L, because the current cannot be transferred to the diode until the voltage at the drain of the MOSFET device exceeds the supply voltage V_{DS} by one diode drop to forward bias the diode. Since the drain current is constant, the gate voltage also remains constant at the gate plateau voltage. Consequently,

$$i_{GP} = \frac{V_{GP}}{R_G} = \frac{1}{R_G}\left(V_{TH} + \sqrt{\frac{I_L L_{CH}}{\mu_{ni} C_{OX} Z}}\right). \tag{6.235}$$

Since all of the gate current is used to discharge the gate–drain capacitance during the plateau phase because there is no change in the voltage across the gate–source capacitance,

$$\frac{dv_D}{dt} = \frac{dv_{GD}}{dt} = \frac{i_{GP}}{C_{GD}(v_D)}. \tag{6.236}$$

As indicated in this expression, the gate–drain capacitance is a function of the drain voltage. Alternately, the gate–drain capacitance can be assumed to have a constant average ($C_{GD,av}$) value during the transient. In this case, the drain voltage increases linearly with time as given by

$$v_D(t) = V_{ON} + \frac{1}{R_G C_{GD,av}}\left(V_{TH} + \sqrt{\frac{I_L L_{CH}}{\mu_{ni} C_{OX} Z}}\right)(t - t_4). \tag{6.237}$$

This behavior is illustrated in Fig. 6.124 during the period from t_4 to t_5. This time interval can be determined by assuming that the drain voltage becomes equal to the drain supply voltage (V_{DS}) plus one diode drop (V_{FD}) at time t_5:

$$t_5 - t_4 = R_G C_{GS,av} \frac{V_{DS} + V_{FD} - V_{ON}}{V_{GP}}. \tag{6.238}$$

At the end of the plateau phase (at time t_5), the load current begins to transfer from the power MOSFET device to the freewheeling diode. Since the drain voltage remains constant, the gate–drain capacitance can also be assumed to remain constant during this phase. The current flowing through the gate resistance (R_G) discharges both the gate–drain and gate–source capacitances, leading to an exponential fall in gate voltage from the plateau voltage:

$$v_G(t) = V_{GP} e^{-(t-t_5)/R_G[C_{GS}+C_{GD}(V_{DS})]}, \tag{6.239}$$

as illustrated in the figure. The drain current follows the gate voltage as given by (6.223) until it reaches zero when the gate bias becomes equal to the threshold voltage at time t_6. The time interval for reaching the threshold voltage can be obtained by using (6.239):

$$t_6 - t_5 = R_G[C_{GS} + C_{GD}(V_{DS})]\ln\left(\frac{V_{GP}}{V_{TH}}\right). \tag{6.240}$$

Beyond this point in time, the gate voltage decreases exponentially as shown in the figure until it reaches zero. The time constant for this exponential decay is different from the initial phase due to the smaller gate–drain capacitance.

The largest power dissipation during the turn-off transient occurs during the time interval ($t_6 - t_4$) when either the drain current or the drain voltage has a large magnitude. This time interval can be reduced by decreasing the gate–drain capacitance (C_{GD}). This has been an important objective during improved device structural design.

6.16.3 Switching Power Losses

When used as a switch in power conversion circuits, the power MOSFET device incurs power dissipation during the on-state due to its finite on-resistance and during the two switching intervals for each operating cycle. The total power dissipation is given by

$$P_{\mathrm{T}} = P_{\mathrm{ON}} + P_{\text{TURN-ON}} + P_{\text{TURN-OFF}}. \tag{6.241}$$

The on-state power loss is given by

$$P_{\mathrm{ON}} = \frac{t_{\mathrm{ON}}}{T} R_{\mathrm{ON}} I_{\mathrm{L}}^2, \tag{6.242}$$

where the first term is the duty cycle (on-time t_{ON} divided by the period T) for the power MOSFET device. The switching power losses are

$$P_{\text{TURN-ON}} = \frac{t_3 - t_1}{2T} I_{\mathrm{L}} V_{\mathrm{DS}}, \tag{6.243}$$

$$P_{\text{TURN-OFF}} = \frac{t_6 - t_4}{2T} I_{\mathrm{L}} V_{\mathrm{DS}}, \tag{6.244}$$

under the assumption that the drain current and voltage excursions are approximately linear with time. Since the switching power losses are proportional to the frequency of operation, they become of greater significance in applications such as switch-mode power supplies, where the operating frequency is in the range of 200–500 kHz. For these applications, power MOSFET structures with reduced gate–drain capacitance and charge are required to achieve high efficiencies.[22]

6.16.4 [d*V*/d*t*] Capability

During operation in a circuit with high operating frequency, the drain voltage of the power MOSFET device undergoes a very rapid change, i.e., a very high [d*V*/d*t*] is imposed on the drain. One example of such a circuit is the *voltage regulator module (VRM)* used to supply power to microprocessors and graphics chips. The commonly used circuit configuration for the VRM is the sync-buck topology shown in Fig. 6.125. The circuit is used to convert a backplane input voltage (V_{IN}) in the range of 12–20 V into a well-regulated, low-voltage DC source (V_{OUT}) in the range of 3.3–0.9 V for delivery of power to microprocessors. Two power MOSFET devices are used in this topology: the control-FET as the switch to regulate the voltage and the sync-FET to act as a synchronous rectifier. The control-FET must exhibit excellent high speed switching capability while the sync-FET must have a low on-resistance with low input capacitance as well. The sync-FET is subjected to a high [d*V*/d*t*] during each cycle as the control-FET turns on and off. This stress can cause device failure to occur due to two operating problems as discussed below.

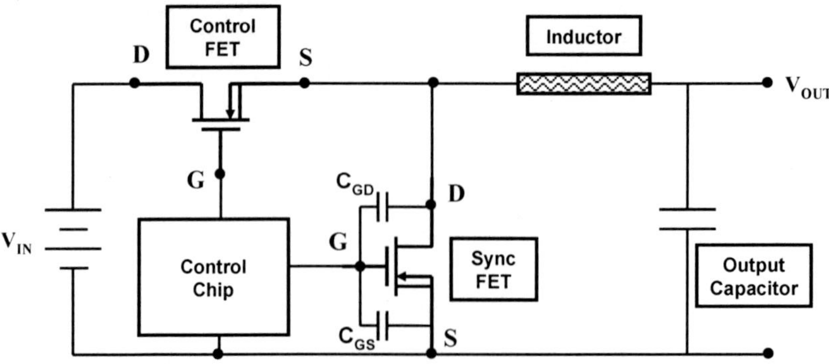

Fig. 6.125 Sync-buck circuit topology for voltage regulator modules

Mode 1: Capacitive Turn-On

When the voltage at the drain of the sync-FET is subjected to a high [dV/dt], the voltage at its drain is shared between the two capacitors shown in the circuit leading to a gate voltage given by

$$v_G(t) = \frac{C_{GD}}{C_{GD} + C_{GS}} \left[\frac{dV_D}{dt} \right] t. \tag{6.245}$$

In writing this expression, it has been assumed that the impedance of the gate–source capacitance is much smaller than the resistance (R_G) in the gate control circuit. If it is assumed that the ramp produces a voltage equal to the input voltage (neglecting any ringing in the circuit due to stray inductances), the maximum voltage induced at the gate is given by

$$v_{G,max} = \frac{C_{GD}}{C_{GD} + C_{GS}} V_{IN}. \tag{6.246}$$

If the maximum gate voltage exceeds the threshold voltage, the sync-FET will be turned on without initiation by the control chip. During this event, the control-FET is also in its on-state. Consequently, the input power supply is short circuited to ground due to the inadvertent turn-on of the sync-FET by the high [dV/dt]. This produces a large *shoot-through current*, which can lead to destructive failure of both the transistors. Based upon the above expression, the sync-FET structure must have a small ratio for the gate–drain to gate–source (C_{GD}/C_{GS}) capacitances to avoid the shoot-through problem. This has been achieved by screening the gate from the drain in the SSCFET structure.[22]

An alternate analysis of the sync-buck circuit topology can be performed under the assumption that the resistance (R_G) of the control circuit is much lower than the impedance of the gate–source capacitance of the sync-FET. In this case, the current induced by the high [dV/dt] at the drain electrode of the sync-FET produces a current through the gate–drain capacitance:

$$i_{\mathrm{GD}} = C_{\mathrm{GD}} \left[\frac{\mathrm{d}V_{\mathrm{D}}}{\mathrm{d}t} \right]. \tag{6.247}$$

When this current flows through the gate resistance (R_{G}) of the control circuit, it produces a voltage at the gate terminal of the sync-FET:

$$v_{\mathrm{G}} = i_{\mathrm{GD}} R_{\mathrm{G}} = R_{\mathrm{G}} C_{\mathrm{GD}} \left[\frac{\mathrm{d}V_{\mathrm{D}}}{\mathrm{d}t} \right]. \tag{6.248}$$

If the gate voltage exceeds the threshold voltage, the sync-FET can be inadvertently turned on leading to destructive failure because of the shoot-through current flow. The largest [dV/dt] that is allowable prior to shoot-through is given by

$$\left[\frac{\mathrm{d}V_{\mathrm{D}}}{\mathrm{d}t} \right]_{\max} = \frac{V_{\mathrm{TH}}}{R_{\mathrm{G}} C_{\mathrm{GD}}}. \tag{6.249}$$

Since the maximum operating frequency for the VRM circuit is limited by the largest allowable [dV/dt], this expression indicates that it is important to reduce the gate–drain capacitance for the sync-FET.

Mode 2: Bipolar Turn-On

The power MOSFET structure contains an inherent parasitic N–P–N bipolar transistor formed between the two internal junctions. The turn-on of the bipolar transistor is suppressed by short circuiting the junction between the N$^+$ source and P-base regions as shown in the cross section in Fig. 6.126. In spite of this, the

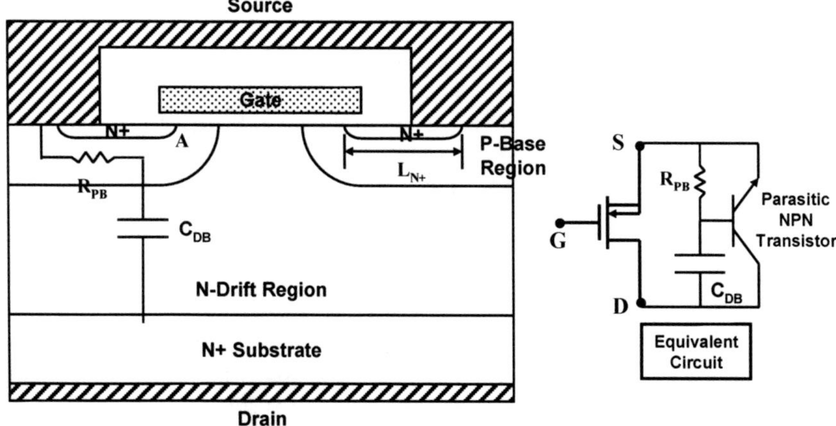

Fig. 6.126 Path for displacement current in the power VD-MOSFET structure

emitter–base junction of the parasitic bipolar transistor can be turned on at location A because of the finite resistance R_{PB} of the P-base region.

Under the presence of a high [dV/dt] at the drain terminal, a displacement current flows through the drain–base capacitance (C_{DB}):

$$i_{D} = C_{DB}\left[\frac{dV_{D}}{dt}\right]. \tag{6.250}$$

When this displacement current flows through the resistance of the P-base region, the voltage drop across the resistance forward biases the emitter–base junction at location A. Injection from the junction begins to occur when the voltage drop is equal to the built-in potential (V_{bi}) for the junction. The largest [dV/dt] that can be applied at the drain terminal before initiating the injection from the junction is given by

$$\left[\frac{dV_{D}}{dt}\right]_{max} = \frac{V_{bi}}{R_{PB}C_{DB}}. \tag{6.251}$$

A larger [dV/dt] capability can be obtained for the power VD-MOSFET structure by reducing the base resistance. The base resistance is given by

$$R_{PB} = \rho_{SQ,PB}\frac{L_{N+}}{Z}, \tag{6.252}$$

where $\rho_{SQ,PB}$ is the sheet resistance of the P-base region including the impact of compensation by the N^{+} source region. This is referred to as the *pinch sheet resistance*. A small value for the base resistance can be achieved by reducing the length L_{N+} for the N^{+} source region by using improved process design rules. Alternately, the sheet resistance for the P-base region can be reduced by the inclusion of a deep P^{+} region as illustrated in Fig. 6.127. It is necessary to use an additional masking step to define the ion implantation window for the deep P^{+} region and its lateral extension must be restricted to avoid encroachment into the channel to avoid an increase in the threshold voltage of the power VD-MOSFET structure.

The P-base resistance in the presence of the deep P^{+} region is given by

$$R_{PB} = \rho_{SQ,PB}\frac{L_{1}}{Z} + \rho_{SQ,P+}\frac{L_{2}}{Z}, \tag{6.253}$$

where $\rho_{SQ,P+}$ is the sheet resistance of the deep P^{+} region. The lengths L_{1} and L_{2} are indicated in the figure. For a typical power VD-MOSFET structure with L_{N+} of 5 μm, the P-base resistance is 1.5 Ω per cm of channel width (Z) based upon a P-base pinch sheet resistance of 3,000 Ω sq^{-1}. The P-base resistance can be reduced to 0.3 Ω per cm of channel width (Z) by including the P^{+} region with a sheet resistance of 50 Ω sq^{-1} if the P^{+} region is diffused until L_{1} is 1 μm. Consequently, it is possible to improve the [dV/dt] capability by a factor of 5 times by the addition of the P^+ region.

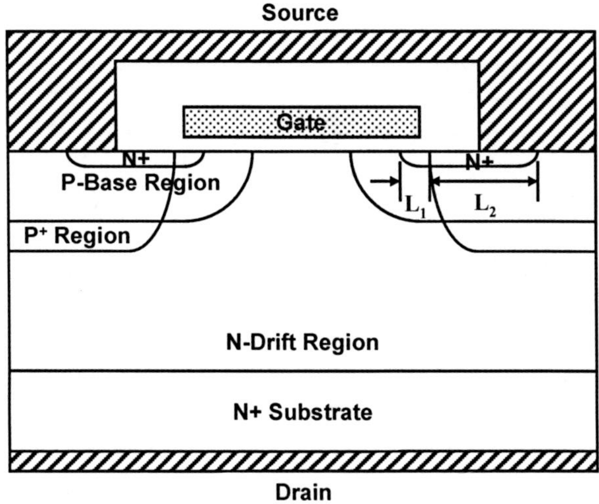

Fig. 6.127 Power VD-MOSFET structure with a deep P^+ region

With effective shorting of the P-base region with the N^+ source region, the breakdown voltage of the power MOSFET structure becomes equal to that of the diode between the P-base region and the N-drift region. From the point of view of the parasitic N–P–N transistor, this corresponds to the open-emitter breakdown voltage (BV_{CB0}). When injection from the emitter begins to occur, the transistor breakdown voltage reduces to the open-base breakdown voltage (BV_{CE0}), which can be half the magnitude of BV_{CB0}. Since the operating voltage at the drain is usually greater than half the rated breakdown voltage capability of the device, the power MOSFET structure will then undergo destructive failure under normal circuit operating conditions. These problems are aggravated as the device heats up during operation because of a reduction in the built-in potential (V_{bi}) and an increase in the sheet resistance of the P-base region due to a reduction of the mobility. Consequently, despite the additional costs, it is common practice to include a deep P^+ region when fabricating power VD-MOSFET structures to ensure adequate ruggedness in applications.

6.17 Safe Operating Area

The safe operating area (SOA) defines the space within the i–v characteristics of a power MOSFET structure where it can be operated without destructive failure as long as the device temperature stays within the maximum junction temperature limit. For the power MOSFET structure, one limit to the SOA is determined by a maximum drain current ($I_{D,max}$) as indicated in Fig. 6.128 by a horizontal line. The maximum drain current is determined by the fusing current for the wire bonds used on the source side of the device. For the example shown in the figure, this is assumed to be at 25 A.

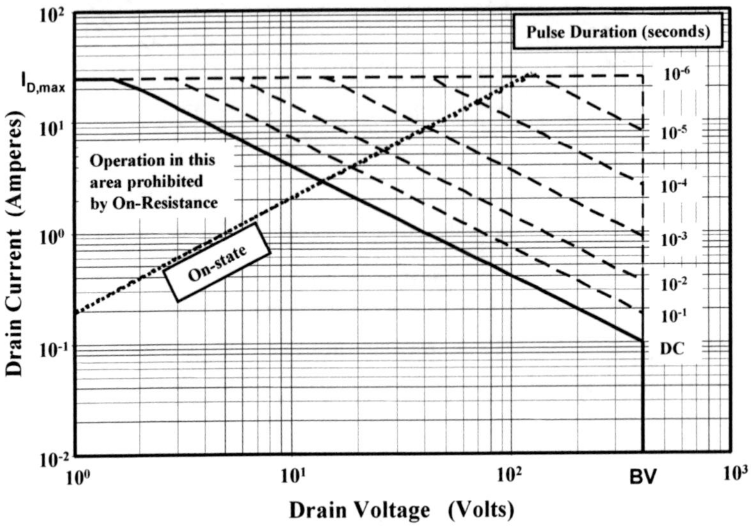

Fig. 6.128 Typical safe operating area for a power MOSFET structure

A second important limit to the SOA is the blocking voltage rating of the power MOSFET structure, which is indicated by the vertical line at a voltage BV assumed to be 400 V for the example in the figure. A third limit to the operating area for the device is based upon its on-state voltage drop, which depends on the on-resistance (R_{ON}) of the device. This limit is given by

$$I_D = \frac{V_D}{R_{ON}}. \tag{6.254}$$

In the figure, a 5-Ω on-resistance was assumed for the device.

The other boundaries for the SOA of the power MOSFET structure are determined by the maximum junction temperature ($T_{J,max}$). When operated under DC bias conditions, the junction temperature is determined by

$$T_J - T_A = P_{Diss} R_\theta = I_D V_D R_\theta, \tag{6.255}$$

where R_θ is the thermal resistance of the package. Based upon this equation, the drain current is limited to

$$I_D = \frac{T_{J,max} - T_A}{V_D R_\theta}. \tag{6.256}$$

The SOA delineated by the above limitations is shown in the figure by the solid line when a maximum junction temperature of 150°C is assumed with an ambient temperature of 50°C.

Under pulsed operation, a much larger drain current can be carried by the power MOSFET structure due to the reduced duty cycle and power dissipation. For the case of single pulses, the thermal impedance is a function of the pulse duration (t_P):

$$Z_\theta = \frac{Z_{DC} Z_K \sqrt{t_P}}{Z_{DC} + Z_K \sqrt{t_P}}, \tag{6.257}$$

where Z_{DC} is the thermal impedance in steady state. When this is factored into (6.256), the drain current becomes limited to

$$I_D = \frac{(T_{J,max} - T_A)[Z_{DC} + Z_K \sqrt{t_P}]}{V_D Z_{DC} Z_K \sqrt{t_P}}. \tag{6.258}$$

The drain current values computed by using this expression are plotted in the figure for the case of various pulse widths. In this example, the transient thermal impedance was computed by using a steady-state value (Z_{DC}) of 2.5°C W^{-1} with a coefficient Z_K of 10°C $W^{-1} s^{-1/2}$. The SOA becomes enlarged as the pulse width is reduced until at a pulse width of 1 μs an ideal square-shaped SOA is obtained. This SOA is determined solely by thermal limitations. However, the power MOSFET structure contains a parasitic bipolar transistor. If this parasitic bipolar transistor becomes activated, the SOA can be greatly reduced leading to destructive failure of the device. In addition, the SOA can be reduced by the body bias effect as discussed below.

6.17.1 Bipolar Second Breakdown

A reduction of the blocking voltage capability can occur in the power bipolar transistors due to a phenomenon termed *second breakdown*. This phenomenon can also occur in power MOSFET devices due to the presence of a parasitic N–P–N bipolar transistor within the n-channel power MOSFET structure. When the drain bias applied to the power MOSFET structure is close to its avalanche breakdown voltage, significant drain current flows due to impact ionization. This current must be removed via the contact to the P-base region as illustrated in Fig. 6.129. The lateral current flow path through the P-base region produces a forward bias across the emitter–base junction of the parasitic bipolar transistor at location *A*. If this voltage drop becomes close to the built-in potential of the junction, the emitter will begin to inject electrons leading to the activation of the parasitic bipolar transistor. The breakdown voltage will then reduce from the open-emitter breakdown voltage (BV_{CB0}) to the open-base breakdown voltage (BV_{CE0}).

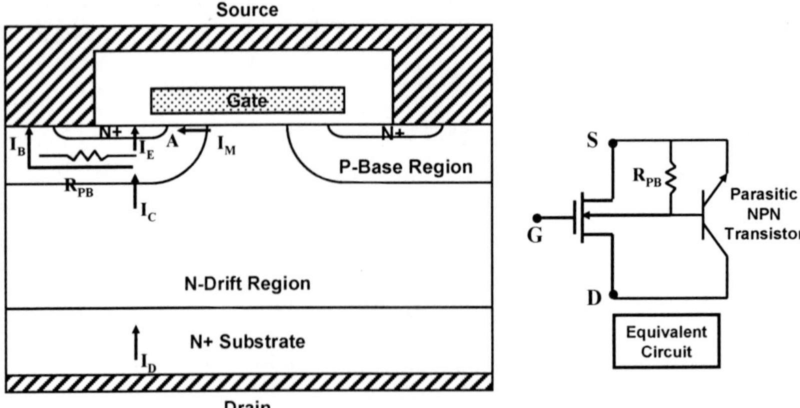

Fig. 6.129 Impact of parasitic bipolar transistor in the power MOSFET structure

The power VD-MOSFET structure is illustrated in Fig. 6.129 together with the equivalent circuit for the analysis of the bipolar second breakdown problem. The normal channel current (I_M) for the MOSFET structure is also indicated in the figure together with the bipolar transistor currents. These currents can be related by using Kirchhoff's laws as follows:

$$I_D = I_C + I_M, \tag{6.259}$$

$$I_S = I_E + I_M + I_B \sim I_E + I_M, \tag{6.260}$$

$$I_B = I_C - I_E, \tag{6.261}$$

and

$$I_C = \alpha I_E = \gamma_E \alpha_T M I_E. \tag{6.262}$$

Due to the narrow width and low doping concentration of the P-base region, the emitter injection efficient (γ_E) and the base-transport factor (α_T) can be assumed to nearly unity for the parasitic bipolar transistor in a power MOSFET structure.

The potential at location A in the P-base region is determined by the voltage drop across the lumped resistance R_{PB} due to the base current I_B:

$$V_A = R_{PB} I_B. \tag{6.263}$$

This forward bias across the emitter–base junction of the parasitic bipolar transistor produces an emitter current given by

$$I_E = I_0 e^{qV_A/kT}. \tag{6.264}$$

Combining the above equations,

$$I_E = I_0 \exp\left[\frac{qR_{PB}}{kT}(M-1)I_E\right]. \tag{6.265}$$

Using a first-order expansion of the exponential term in (6.265),

$$I_E = \frac{I_0}{1-(qR_{PB}/kT)(M-1)I_0}. \tag{6.266}$$

The multiplication coefficient depends upon the applied drain bias:

$$M = \frac{1}{1-(V_D/\mathrm{BV})^6}. \tag{6.267}$$

As the drain bias is increased, the multiplication coefficient (M) increases leading to an increase in the emitter current. The drain bias at which the emitter current becomes very large can be obtained by equating the denominator of (6.266) to zero:

$$V_{D,SB} = \frac{\mathrm{BV}}{[qR_{PB}I_0/kT]^{1/6}}. \tag{6.268}$$

The bipolar second breakdown mode can be suppressed by reduction of the P-base resistance with the addition of a P^+ region as illustrated in Fig. 6.127.

6.17.2 MOS Second Breakdown

The voltage drop across the P-base resistance due to the avalanche breakdown-induced current produces a *body bias effect* that can also lead to a reduction of the voltage that can be supported by the power MOSFET structure. A body bias coefficient can be defined to account for the increase in the drain current due to a change in the P-base potential (V_A) in the vicinity of the channel:

$$\gamma_B = \frac{\delta I_D}{\delta V_A}. \tag{6.269}$$

The source current is then given by

$$I_S = I_M + \gamma_B V_A = I_M + \gamma_B R_{PB} I_B. \tag{6.270}$$

The base current generated by the avalanche multiplication at high drain bias voltages is given by

$$I_B = I_D - I_S = (M-1)I_S. \tag{6.271}$$

Combining these expressions,

$$I_B = (M-1)(I_M + \gamma_B R_B I_B), \tag{6.272}$$

leading to

$$I_{\mathrm{B}} = \frac{(M-1)I_{\mathrm{M}}}{1-\gamma_{\mathrm{B}}R_{\mathrm{PB}}(M-1)} \tag{6.273}$$

and

$$I_{\mathrm{D}} = \frac{I_{\mathrm{M}}}{1-\gamma_{\mathrm{B}}R_{\mathrm{PB}}(M-1)}. \tag{6.274}$$

As the drain voltage increases, this expression indicates that the current will increase catastrophically when its denominator becomes zero. By using (6.267), the drain bias at which the second breakdown occurs is then obtained:

$$V_{\mathrm{D,SB}} = \frac{\mathrm{BV}}{[\gamma_{\mathrm{B}}R_{\mathrm{PB}}]^{1/6}}. \tag{6.275}$$

The MOS second breakdown mode can also be suppressed by reduction of the P-base resistance with the addition of a P^+ region as illustrated in Fig. 6.127. In modern power VD-MOSFET designs, the second breakdown phenomena are well suppressed with the addition of the P^+ regions. Consequently, these devices exhibit the SOA shown in Fig. 6.128.

6.18 Integral Body Diode

In the power n-channel MOSFET structure, high voltage is supported across the junction formed between the P-base region and the N-drift region when a positive drain bias is applied. The same junction becomes forward biased if a negative bias

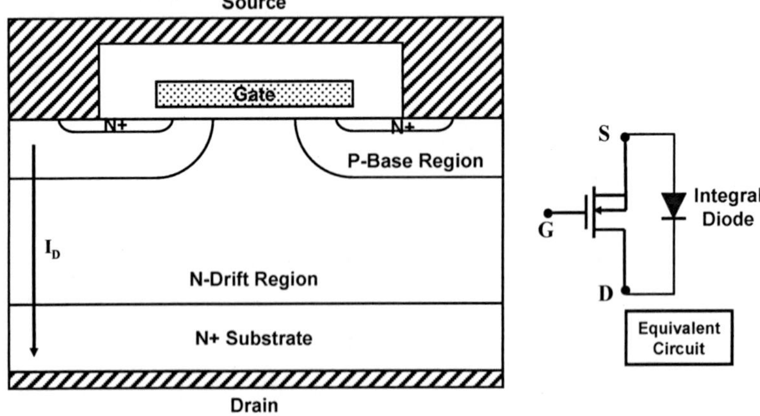

Fig. 6.130 Integral diode within the power MOSFET structure

is applied to the drain terminal. This P–N junction diode can therefore be utilized to conduct current through the power MOSFET structure in the third quadrant of operation. The equivalent circuit for the power MOSFET structure with the integral diode is shown in Fig. 6.130. The current flow path through the integral diode within the structure is shown on the cross section in the figure. In motor control circuits, it is desirable to utilize this current path to avoid the cost of an additional fly-back diode. The primary difficult with utilization of the integral diode in the power MOSFET structure is its poor reverse recovery characteristic due to the high lifetime in the N-drift region. In addition, the presence of the N^+ source region can lead to the turn-on of the parasitic N–P–N bipolar transistor during the reverse recovery transient leading to destructive failure.

6.18.1 Reverse Recovery Enhancement

In Chap. 5, it was demonstrated that the reverse recovery performance of P-i-N rectifiers can be improved by using lifetime control processes. Methods for reducing the lifetime consist of the addition of deep level centers such as gold or platinum or by the use of high energy particle bombardment. The application of these techniques to the power MOSFET structure is hindered by the degradation of the gate oxide and its interface properties with the silicon. The diffusion of gold into an MOS structure has been found to produce an accumulation of the gold near the oxide interface, resulting in a strong shift in the threshold voltage and a reduction of the channel mobility.

Based upon the results of radiation damage studies on power MOSFET structures, it has been established that a positive charge is developed in the oxide due to the particle bombardment. The threshold voltage for n-channel power MOSFET structures is shifted to below zero preventing the device from blocking voltage in the first quadrant. Fortunately, it was discovered that this oxide charge can be removed by annealing the devices at relatively low temperatures (between 150 and 200°C) while maintaining the deep levels created by the particle bombardment within the bulk silicon regions.[25] Using this procedure, it has been demonstrated that the reverse recovery time (t_{RR}) of the integral diode can be reduced from 400 to 100 ns by adjusting the radiation dose. Electron irradiation provides a clean and precise process for improving the reverse recovery behavior of the integral diode without compromising any of the power MOSFET device characteristics.

6.18.2 Impact of Parasitic Bipolar Transistor

During the reverse recovery process, the current in the integral diode is reversed until it begins to support a positive drain bias voltage. The path for the reverse recovery current is illustrated in Fig. 6.131. The entire P-base/N-drift junction collects the current which then flows into the contact to the P-base region. This current path occurs via the finite resistance (R_{PB}) of the P-base region. The voltage drop across the resistance forward biases the junction between the N^+ source region and the P-base region. If the voltage across this junction at point *A* exceeds the built-in potential, the emitter of the parasitic bipolar transistor becomes activated. Under these conditions, the blocking voltage capability of the power MOSFET structures is degraded from the open-emitter breakdown voltage (BV_{CB0}) to the open-base breakdown voltage (BV_{CE0}). The drain blocking voltage during the reverse recovery transient can exceed the open-base breakdown voltage of the parasitic N–P–N transistor resulting in destructive failure. This problem can be overcome by the addition of the P^+ region in the structure as shown in Fig. 6.127.

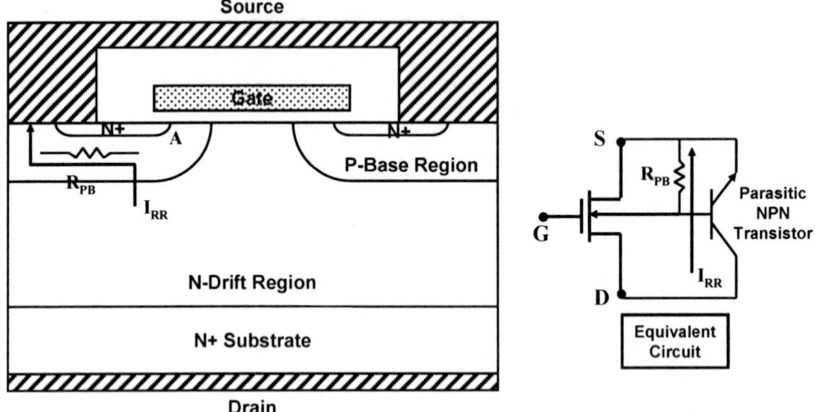

Fig. 6.131 Reverse recovery current path in the integral diode within the power MOSFET structure

6.19 High-Temperature Characteristics

Power MOSFET structures have superior high-temperature operating capability when compared with bipolar devices with the same voltage ratings. Commercial devices are available with a maximum junction temperature capability of up to 200°C. However, changes in the semiconductor properties with temperature have an impact on the electrical characteristics of the power MOSFET structures as discussed in this section.

6.19.1 Threshold Voltage

As discussed in Sect. 6.5.4, the threshold voltage for a power MOSFET structure is given by

$$V_{TH} = \frac{\sqrt{4\varepsilon_S kTN_A \ln(N_A/n_i)}}{C_{OX}} + \frac{2kT}{q}\ln\left(\frac{N_A}{n_i}\right) - \frac{Q_{OX}}{C_{OX}}. \quad (6.276)$$

The threshold voltage for power MOSFET structures decreases with increasing temperature due to the increase in the intrinsic carrier concentration. The change in the threshold voltage is shown in Fig. 6.132 for the case of a P-base doping concentration of 1×10^{17} cm^{-3}. A gate oxide thickness of 500 Å was assumed with a fixed charge of 2×10^{11} cm^{-2} together with N^{+} polysilicon as the gate electrode material. The threshold voltage has been normalized to its value at room temperature. It can be observed from the figure that the threshold voltage changes by about a factor of 2 times over the temperature range of −50 to 200°C. The reduction of the threshold voltage at elevated temperatures must be taken into account during the design of power MOSFET structures, so that its value does not fall below 1 V at the highest anticipated operating junction temperature.

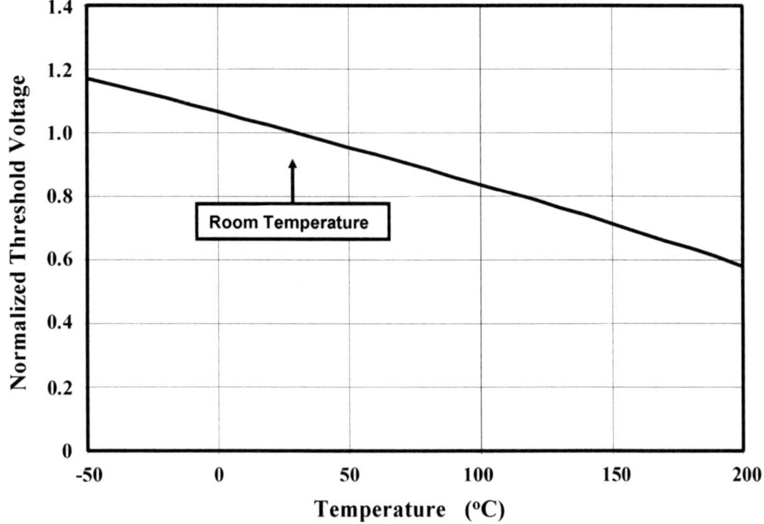

Fig. 6.132 Impact of increasing temperature on the threshold voltage of the power MOSFET structure

6.19.2 On-Resistance

The on-resistance of power MOSFET structures is observed to increase with increasing temperature. This is attributed to the reduction of the mobility in the drift region and the inversion layer. The decrease in the mobility in the drift region is proportional to $T^{-2.42}$ as discussed in Chap. 2. The change in the inversion layer mobility is smaller (T^{-1}) because of the high doping concentration of the P-base region. The reduction in the threshold voltage with temperature that was discussed in Sect. 6.19.1 also contributes to the variation of the on-resistance.

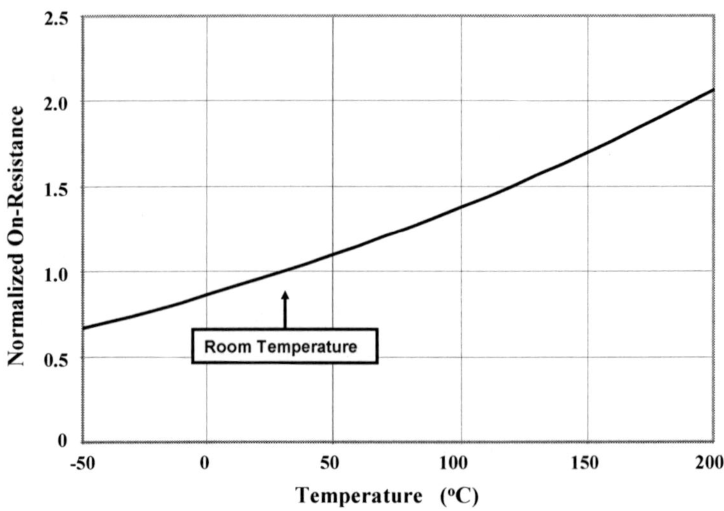

Fig. 6.133 Impact of increasing temperature on the on-resistance of the power VD-MOSFET structure

As an example, the change in the on-resistance with temperature is provided in Fig. 6.133 for the case of the 30-V power VD-MOSFET structure with a gate width of 10 μm and a cell pitch of 18 μm. The device has a channel length of 1.6 μm and a gate oxide thickness of 500 Å. It can be observed that the on-resistance increases by about 70% at 150°C and 100% at 200°C. This increase in on-resistance produces larger on-state power losses in power MOSFET structures when operated at high ambient temperatures, such as in automotive electronics, as well as due to self-heating during operation in power circuits.

Although an increase in the resistance is undesirable due to the enhanced power dissipation, it is an important feature for power MOSFET structures because it prevents current localization within the structure. If the current density increases in a local portion of the device, there is an increase in the local temperature because of the higher power dissipation at this location. The increase in the on-resistance due to the increased temperature favors a redistribution of the current to other locations within the device. This prevents the formation of "hot spots" that can lead to current filamentation and destructive failure. Another benefit of a positive temperature coefficient for the on-resistance is that multiple power MOSFET devices can be connected in parallel for enhancing the power-handling capability with good sharing of the total current.

6.19.3 Saturation Transconductance

In Sect. 6.5.5, it was shown that the transconductance for the MOSFET structure in the saturated current regime of operation is given by

$$g_{ms} = \frac{Z\mu_{ni}C_{OX}}{L_{CH}}(V_G - V_{TH}). \tag{6.277}$$

The temperature-dependent terms in this expression are the channel mobility and the threshold voltage. The reduction of the channel inversion layer mobility is compensated by the reduction of the threshold voltage when the temperature increases. The net result is a reduction of the transconductance with increasing temperature. This is demonstrated in Fig. 6.134 for a power VD-MOSFET structure with the same parameters as given in Sect. 6.19.2. It can be observed that the transconductance is reduced to 70% of its value at room temperature when the device temperature increases to 200°C. This has an impact upon the switching times for the power MOSFET structure.

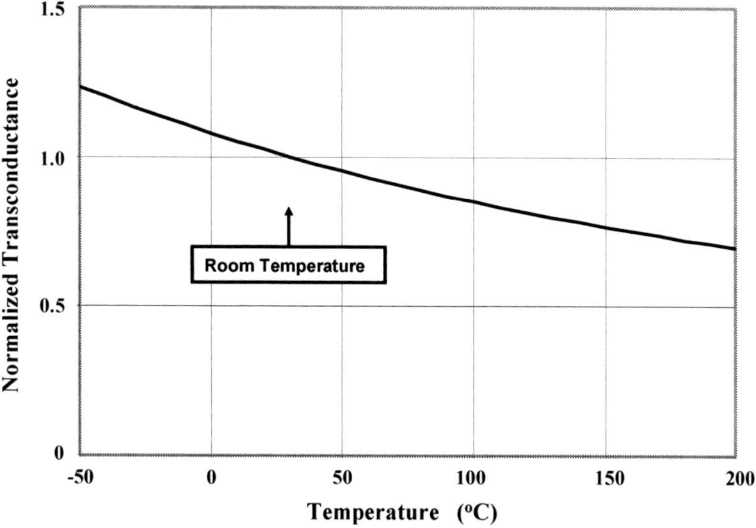

Fig. 6.134 Impact of increasing temperature on the transconductance of the power VD-MOSFET structure

6.20 Complementary Devices

Most of the discussion and analysis in this chapter has been focused on the n-channel power MOSFET structure because of its predominance in applications. The n-channel power MOSFET structure is favored over the p-channel power MOSFET structure because the larger mobility for electrons than holes in silicon results in smaller specific on-resistance for n-channel devices. However, p-channel power MOSFET devices are preferable in many portable power management applications because of the drive simplicity. The p-channel structure can be inserted between the power source and the load to regulate the power with simply pulling down the gate voltage to ground.

6.20.1 The p-Channel Structure

The p-channel power MOSFET structures are shown in Fig. 6.135. When compared with the n-channel counterpart, all the N-type semiconductor regions are replaced by P-type regions. The basic operating principles for the p-channel device are the same as those discussed in the previous sections for the n-channel structure. However, all the bias voltages must be changed from positive to negative values for the p-channel structure. During the 1990s, the planar VD-MOSFET structure shown in Fig. 6.135a was replaced with the trench-gate U-MOSFET structure shown in Fig. 6.135b due to its smaller specific on-resistance.[26,27]

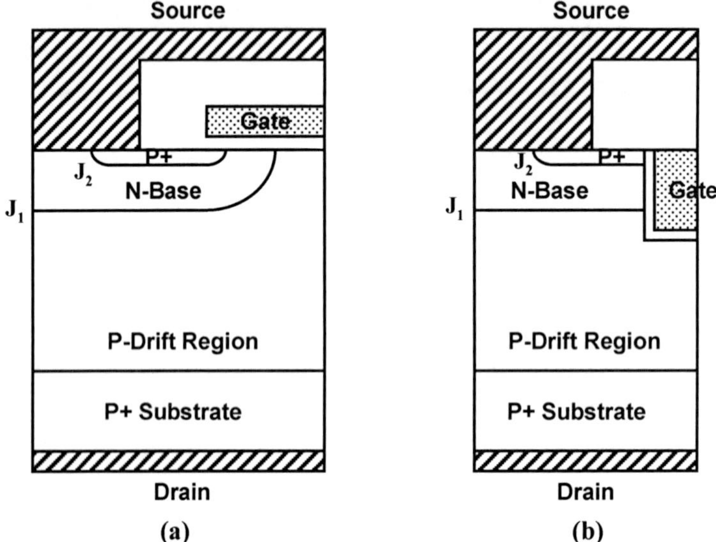

Fig. 6.135 The p-channel power MOSFET structures

6.20.2 On-Resistance

The components of the on-resistance for the p-channel structure are identical to those for the corresponding n-channel structure. However, the magnitude for the specific on-resistance for the various components is larger in the p-channel structure due to the lower mobility for holes than electrons in silicon. In the drift region, the bulk mobility for holes is about three times smaller (see Chap. 3). In the channel region, the inversion layer mobility for holes is also about three times smaller than for electrons (see Chap. 3). Consequently, all the components for the on-resistance for the p-channel device increase by a factor of 3 times when compared with the n-channel structure, resulting in a threefold increase in the total specific on-resistance as well. In addition, the optimization of the VD-MOSFET cell for the p-channel device produces the same optimum value for the width of the gate electrode.

6.20.3 Deep-Trench Structure

Due to their relatively high specific on-resistance, the p-channel power MOSFET structures have been utilized primarily in applications with low blocking voltage requirements. In the case of DC bus voltages of less than 5 V, the breakdown voltage for the device can be reduced to only 12 V. In this case, it is feasible to utilize the deep-trench architecture illustrated in Fig. 6.136 without a drift region. Although first proposed for n-channel devices,[28] this INV-FET structure has also been used for making p-channel devices with low specific on-resistance.[27] In the deep-trench architecture, the trench is extended from the upper surface into the N^+ or P^+ substrate. When the gate bias is applied to turn on the device, an inversion layer is formed along the trench sidewalls within the base region extending all the way from the source to the drain region, allowing a reduction of the specific on-resistance due to elimination of the drift region component in the traditional trench U-MOSFET structure. This concept is useful only for devices with low blocking voltage capability because the drain voltage must be supported across the gate oxide during the off-state. For a typical gate oxide thickness of 500 Å, the blocking voltage is limited to about 20 V from reliability considerations even though the oxide rupture voltage is 50 V. With the INV-FET structure, n-channel devices with a blocking voltage of 26 V and a specific on-resistance of 0.3 mΩ cm^2 have been reported by using a cell pitch of 6 μm.[28] A specific on-resistance of 0.2 mΩ cm^2 has been achieved for the p-channel structure with a blocking voltage of 15 V by reducing the cell pitch to 2.4 μm.[27]

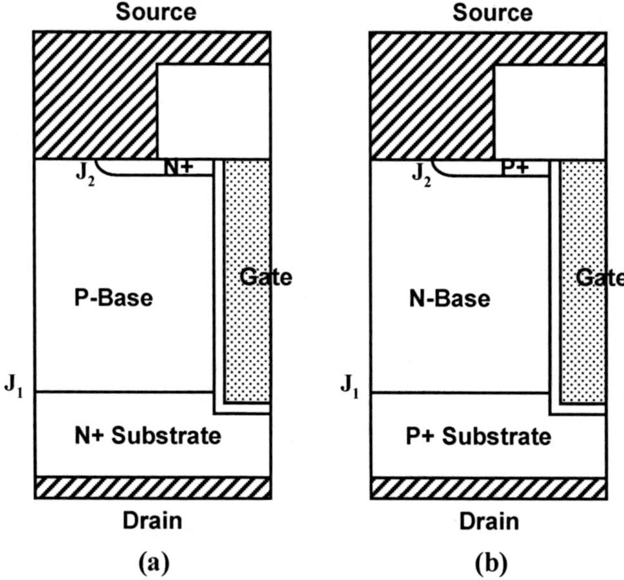

Fig. 6.136 INV-FET power MOSFET structures

6.21 Silicon Power MOSFET Process Technology

The first power MOSFET devices were developed by using the V-groove process. This process fell out of favor due to reliability problems associated with the sharp point of the V-groove gate structure and instabilities associated with the potassium hydroxide-based etching solutions. The planar VD-MOS process gained favor during the 1970s due to its simplicity and compatibility with main stream CMOS technology. The process for planar VD-MOSFET structures was refined during the 1980s with the utilization of improved design rules to reduce the size of the diffusion window leading to a lower specific on-resistance. However, the presence of the JFET region in the power VD-MOSFET structure was found to be a major impedement to reducing the specific on-resistance especially for devices designed for low blocking voltages (<100 V). With the availability of reactive ion etching (RIE)-based trench formation chemistry and production equipment used for manufacturing DRAMs, it became feasible to make power MOSFET devices with the trench-gate structure. The power U-MOSFET devices have become prevalent since the 1990s to achieve a lower specific on-resistance. The processes used for manufacturing the VD-MOSFET and U-MOSFET devices are briefly described in this section.

6.21.1 Planar VD-MOSFET Process

Most modern power VD-MOSFET devices are manufactured with a P^+ region included in the structure to provide adequate ruggedness as described in previous sections. Although some self-aligned processes for fabricating this region have been proposed and demonstrated, most products are manufactured by using an additional masking step for this region. This process is briefly described here with the aid of Fig. 6.137.

The power VD-MOSFET structure is fabricated by starting with an N^+ substrate on which an N-type epitaxial layer has been grown with doping concentration and thickness chosen to achieve the desired blocking voltage rating. A thick field oxide is first grown for the formation of the appropriate edge termination (not shown in Fig. 6.137). Next, a termination mask is used to define an opening for the active area and windows for the floating field rings. The doping concentration in the JFET region can be enhanced by the addition of an N-type ion implant in the active area at this stage in the process. The JFET dopant extends to the dashed line in the figure during the drive-in cycle for the P^+ region. A photoresist mask is then used to define the location of the P^+ implant within the VD-MOSFET cell window. The same implant can be utilized to form the P^+ region for the edge termination. A cross section of the power VD-MOSFET structure is shown in Fig. 6.137a after driving in the P-type dopant (usually boron).

The gate oxide is now grown on the silicon surface immediately followed by the deposition of the polysilicon gate electrode. A cross section of the power VD-MOSFET structure is shown in Fig. 6.137b after patterning the polysilicon. A good alignment between the P^+ region and the polysilicon is required to prevent encroachment of the P-type dopant into the channel. The P-base region is now implanted using the polysilicon as a mask. A cross section of the power VD-MOSFET structure is shown in Fig. 6.137c after driving in the P-type dopant (usually boron). A photoresist mask is now used as illustrated in Fig. 6.137d to define the location of the N^+ source region. During the ion implantation of the N-type dopant (usually phosphorus), the polysilicon acts as one of the boundaries while the photoresist is used to provide a region for short circuiting the N^+ source and P-base regions.

The intermetal dielectric is now deposited to cover the polysilicon gate. This dielectric is comprised of layers of silicon dioxide or oxynitrides that offer good step coverage to passivate the edges of the polysilicon as well as its upper surface. A mask is used to etch windows in the intermetal dielectric film to open contacts to the N^+ source and P-base regions as illustrated in Fig. 6.137e. The same mask is used to form the contact to the gate electrode at the gate pads. The alignment of the contact mask with the edges of the polysilicon is a critical yield-limiting step during the fabrication of power VD-MOSFET devices. To achieve a low total on-resistance, these devices must be fabricated with a very large channel width (Z), typically 1–20 m in size. Any defects in the intermetal dielectric film, especially at the edges of the polysilicon, can lead to short circuits between the gate and source making the device unusable.

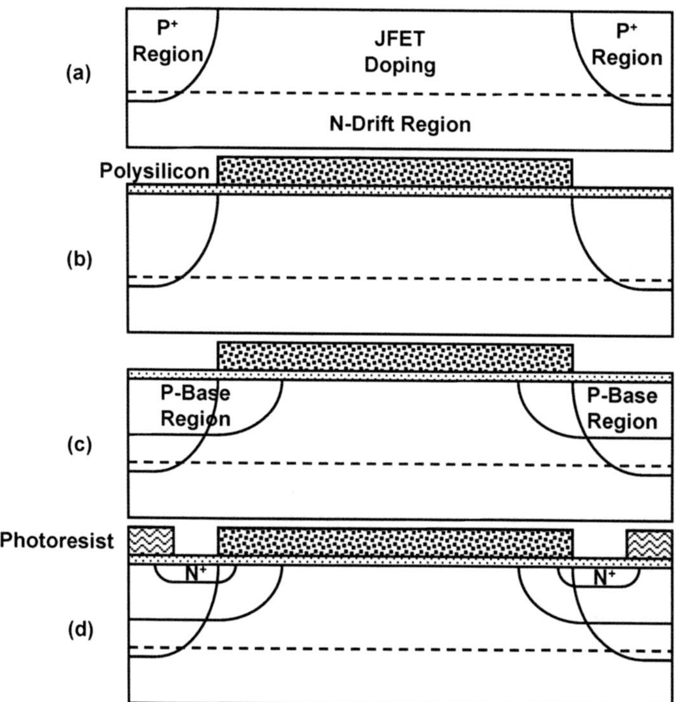

Fig. 6.137 Fabrication process for the power VD-MOSFET structure

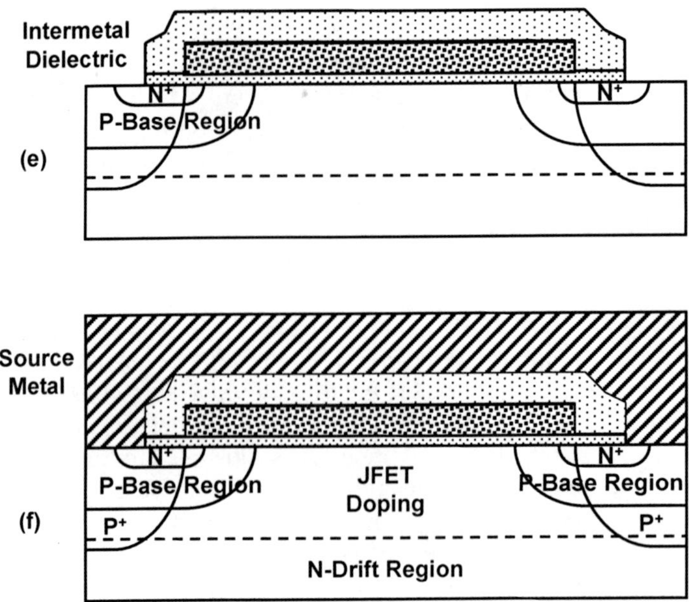

Fig. 6.137 (continued) Fabrication process for the power VD-MOSFET structure

The source metal is now deposited and patterned to define the source electrode as well as gate pads to complete the structure as illustrated in Fig. 6.137f. It is customary to use a titanium-based contact layer followed by a thick (4–8 μm) aluminum layer containing 1% silicon to suppress spiking. The thickness of the N^+ substrate is then reduced by grinding and polishing to reduce its contribution to the on-resistance, followed by deposition of a solderable metal layer on the back surface to complete the drain contact and electrode. A typical metal stack consists of a thin titanium film, a thin nickel barrier layer, and a thick silver layer. The wafers are usually annealed after the formation of the source and drain contacts to improve the contact resistance. The typical device fabrication sequence requires six mask levels with an additional mask for defining windows for wire bonds in a passivation glass deposited on top of the source and gate electrodes.

6.21.2 Trench U-MOSFET Process

The trench-gate structure for power U-MOSFET devices requires its own unique process sequence as illustrated in Fig. 6.138. Although a variety of process options have been proposed, a typical case is briefly described here. As in the case of the power VD-MOSFET structure, an N^+ substrate with an N-type epitaxial layer is used as the starting material. A thick field oxide is first grown for the formation of the appropriate edge termination. Next, a termination mask is used to define an opening for the active area and windows for the floating field rings. A mask is then used to define the location of the P^+ implant within the U-MOSFET cell window. The same implant can be utilized to form the P^+ region for the edge termination.

A cross section of the power U-MOSFET structure is shown in Fig. 6.138a after driving in the P-type dopant (usually boron).

The P-base region is now implanted over the entire active area followed by a drive-in cycle for the P-type dopant (usually boron). A photoresist mask is now used to define the N^+ source region. This establishes the locations of the short circuits between the N^+ source and P-base regions. A cross section of the power U-MOSFET structure is shown in Fig. 6.138b after driving in the N-type dopant (usually phosphorus).

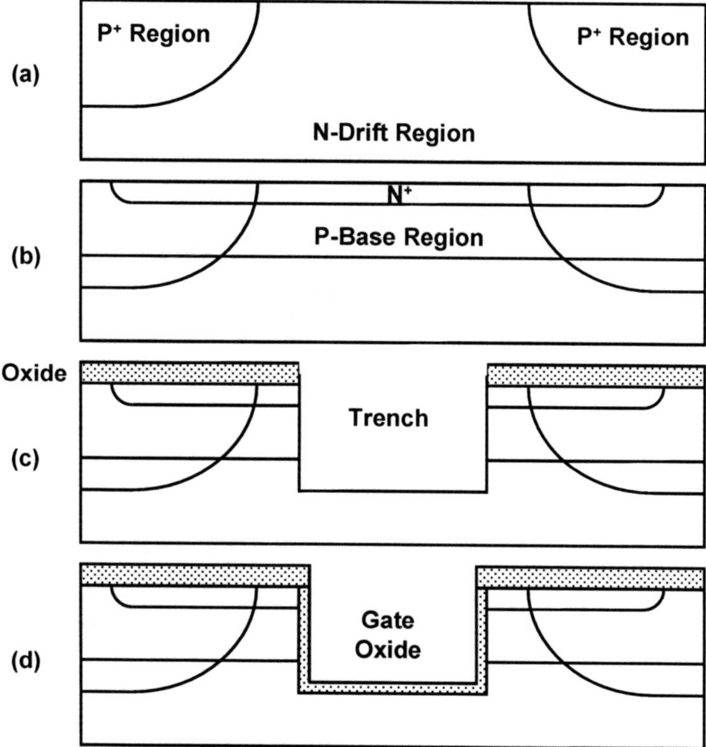

Fig. 6.138 Fabrication process for the power U-MOSFET structure

An oxide, to be utilized as an RIE etch mask, is then either deposited or grown on the silicon surface. This oxide is patterned to open windows where the trenches are to be formed. The RIE process is used to form vertically walled trenches in the silicon. It is important to obtain a smooth trench surface to achieve high inversion layer mobility as discussed in Chap.2. Processes have also been developed for rounding the bottom of the trench to reduce the electric field at the corners.[6] A cross section of the power U-MOSFET structure is shown in Fig. 6.138c after the formation of the trenches.

The gate oxide is now grown of the sidewalls and bottom surface of the trench. A cross section of the power U-MOSFET structure is shown in Fig. 6.138d after this step in the process. The polysilicon gate electrode is now deposited using a conformal process to refill the trenches. A cross section of the power U-MOSFET structure is shown in Fig. 6.138e illustrating a cusp in the polysilicon at the middle of the trench. It is important to use a sufficient polysilicon thickness to make the cusp small and to avoid the formation of voids within the refill. The polysilicon is then planarized to obtain a smoother topology as shown in Fig. 6.138f followed by patterning it. The intermetal dielectric is now deposited to cover the polysilicon gate. As in the case of the VD-MOSFET process, this dielectric is comprised of layers of silicon dioxide or oxynitrides that offer good step coverage to passivate the edges of the polysilicon as well as its upper surface. A mask is used to etch windows in the intermetal dielectric film to open contacts to the N^+ source and P-base regions as illustrated in Fig. 6.138g. The same mask is used to form the contact to the gate electrode at the gate pads.

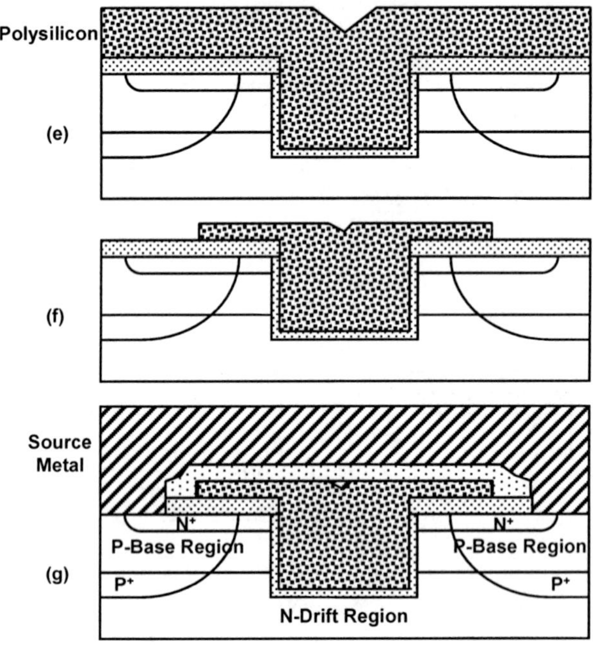

Fig. 6.138 (continued) Fabrication process for the power U-MOSFET structure

The alignment of the contact mask with the edges of the polysilicon is a critical yield-limiting step during the fabrication of power U-MOSFET devices. To achieve a low total on-resistance, these devices must be fabricated with a very large channel width (Z), typically 1–20 m in size. Any defects in the intermetal dielectric film, especially at the edges of the polysilicon, can lead to short circuits between the gate and source making the device unusable. The source metal is now deposited and patterned to define the source electrode as well as gate pads to complete the structure as illustrated in Fig. 6.138g. It is customary to use a titanium-based contact layer followed by a thick (4–8 μm) aluminum layer containing 1% silicon to suppress spiking. The thickness of the N^+ substrate is then reduced by grinding and polishing to reduce its contribution to the on-resistance, followed by deposition of a solderable metal layer on the back surface to complete the drain contact and electrode. A typical metal stack consists of a thin titanium film, a thin nickel barrier layer, and a thick silver layer. The wafers are usually annealed after the formation of the source and drain contacts to improve the contact resistance.

6.22 Silicon Carbide Devices

As previously discussed in the introductory chapter, the specific on-resistance of the drift region can be greatly reduced by replacing silicon with silicon carbide. The Baliga's figure of merit for 4H-SiC is more than 2,000 times superior to that for silicon. This makes silicon carbide a very attractive semiconductor for the development of high-voltage power MOSFET structures.

The fabrication of power MOSFET devices from silicon carbide has been curtailed by two issues. First, the quality of the interface between the thermally grown oxide and the silicon carbide surface has been poor resulting in low inversion layer mobility. Second, the high electric field generated within the silicon carbide produces very high electric field within the gate oxide leading to its rupture during operation. Innovative methods to overcome these challenges have evolved during the 1990s. A detailed discussion of these silicon carbide-based power MOSFET structures is available in a separate book.[29] A brief discussion of two approaches to building silicon carbide-based MOS-gated power switches is provided here.

6.22.1 The Baliga-Pair Configuration

In the Baliga-Pair configuration, a low breakdown voltage, normally-off, silicon MOSFET and a high-voltage, normally-on, silicon carbide JFET/MESFET are connected together as shown in Fig. 6.139. Any trench-gate or planar-gate JFET/ MESFET structures discussed in reference 29 can be used to provide the high blocking voltage capability. It is important that the silicon carbide FET structure be designed for normally-on operation with a low specific on-resistance. It is also necessary for the silicon carbide FET to be able to block the drain bias voltage with a gate potential less than the breakdown voltage of the silicon power MOSFET. The silicon power MOSFET can be either a planar DMOS structure or a trench-gate UMOS structure, discussed earlier in this chapter, with a low specific on-resistance. The source of the silicon carbide FET is connected to the drain of the silicon power MOSFET. Note that the gate of the silicon carbide FET is connected directly to the reference or ground terminal. The path formed between the drain and the gate contact of the silicon carbide FET creates the fly-back diode.

The composite switch is controlled by the signal applied to the gate of the silicon power MOSFET.

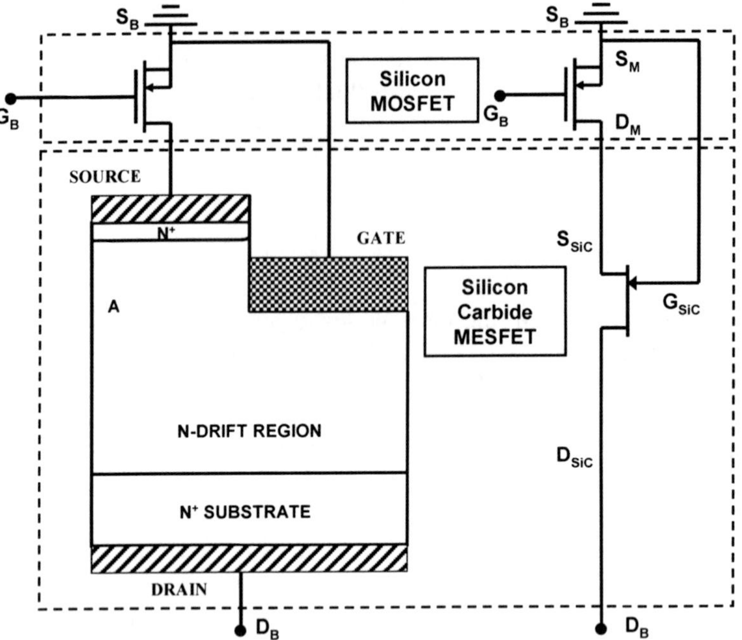

Fig. 6.139 *Baliga-Pair* power switch configuration

Voltage Blocking Mode

The Baliga-Pair switch configuration can block current flow when the gate of the silicon power MOSFET is shorted to ground by the external drive circuit. With zero gate bias, the silicon power MOSFET supports any bias applied to its drain terminal (D_M) unless the voltage exceeds its breakdown voltage. Consequently, at lower voltages applied to the drain terminal (D_B) of the composite switch, the voltage is supported across the silicon power MOSFET because the silicon carbide JFET is operating in its normally-on mode. However, as the voltage at the drain (D_M) of the silicon MOSFET increases, an equal positive voltage develops at the source (S_{SiC}) of the silicon carbide FET. Since the gate (G_{SiC}) of the silicon carbide FET is connected to the ground terminal, this produces a reverse bias across the gate–source junction of the silicon carbide FET. Consequently, a depletion layer extends from the gate contact/junction into the channel of the silicon carbide FET. When the depletion region pinches off the channel at location *A*, further increase in the bias applied to the drain (D_B) of the composite switch is supported across the silicon carbide FET.

After the channel in the silicon carbide JFET/MESFET is pinched off, the potential at the source of the silicon carbide FET is isolated from the drain bias applied to the silicon carbide FET. The voltage across the silicon power MOSFET is consequently also clamped to a value close to the pinch-off voltage of the silicon carbide FET. This feature enables utilization of a silicon power MOSFET structure with a low breakdown voltage. Such silicon power MOSFET devices have very low specific on-resistance with a mature technology available for their production. From this point of view, it is desirable to utilize silicon power MOSFETs with breakdown voltages of below 50 V.

If the Baliga-Pair is designed to support a drain bias of 3,000 V, the ability to utilize a silicon power MOSFET with a breakdown voltage of 30 V requires designing the silicon carbide FET, so that the channel is pinched off at a gate bias of below 20 V. Thus, the blocking gain of the silicon carbide FET should be in excess of 150. This is feasible for both the trench-gate and planar-gate architectures for silicon carbide FET structures discussed in reference 29.

Forward Conduction Mode

The composite switch shown in Fig. 6.139 can be turned on by the application of a positive gate bias to the gate terminal (G_B). If the gate bias is well above the threshold voltage of the silicon power MOSFET, it operates with a low on-resistance. Under these conditions, any voltage applied to the drain terminal (D_B) produces current flow through the normally-on silicon carbide FET and the silicon MOSFET. Due to the low specific on-resistance of both structures, the total on-resistance of the Baliga-Pair configuration is also very small. Depending upon the size of the two devices, an on-resistance of less than 10 mΩ is feasible even when the switch is designed to support 3,000 V. This indicates that the Baliga-Pair configuration will have an on-state voltage drop of about 1 V with a nominal on-state current density of 100 A cm^{-2} flowing through the devices. This is well below typical values of around 4 V for an IGBT designed to support such high voltages resulting in much lower power loss in the applications.

Current Saturation Mode

One of the reasons for the success of the silicon power MOSFET and IGBT in power electronic applications is the gate-controlled current saturation capability of these devices. This feature enables controlling the rate of rise of current in power circuits by tailoring the input gate voltage waveform rather than by utilizing snubbers that are required for devices like gate turn-off thyristors. In addition, current saturation is essential for survival of short-circuit conditions where the device must limit the current.

The current saturation capability is inherent in the Baliga-Pair configuration. If the gate voltage applied to the Baliga-Pair configuration is close to the threshold voltage of the silicon MOSFET, it will enter its current saturation mode when the drain bias increases. This produces a constant current through both the silicon MOSFET and the silicon carbide FET while the drain bias applied to the composite switch increases. At lower drain bias voltages applied to the drain terminal (D_B), the voltage is supported across the silicon power MOSFET. As this voltage increases, the channel in the silicon carbide FET gets pinched off and further voltage is then supported by the silicon carbide FET. Under these bias conditions, both the devices sustain current flow while supporting voltage. The level of the current flowing through the devices is controlled by the applied gate bias. In this sense, the Baliga-Pair behaves like a silicon power MOSFET from the point of view of the external circuit on both the input and output side. This feature makes the configuration attractive for use in power electronic systems because the existing circuit topologies can be used. SOA of the composite switch is mainly determined by the silicon carbide FET because it supports a majority of the applied drain voltage. The excellent breakdown strength, thermal conductivity, and wide band gap of silicon carbide ensure good SOA for the FET structures.

Switching Characteristics

The transition between the on and off modes for the Baliga-Pair configuration is controlled by the applied gate bias. During turn-on and turn-off, the gate bias must charge and discharge the capacitance of the silicon power MOSFET. Since silicon power MOSFETs are extensively used for high frequency power conversion, their input capacitance and gate charge have been optimized by the industry. The switching speed of the Baliga-Pair is consequently very high because of the availability of silicon power MOSFETs designed for high frequency applications. The main limitation to the switching speed of the Baliga-Pair is related to parasitic inductances in the package.

Fly-Back Diode

The Baliga-Pair contains an inherent high-quality fly-back rectifier. When the drain bias is reversed to a negative value, the gate–drain contact/junction of the silicon carbide FET becomes forward biased. Since the gate of the silicon carbide FET is directly connected to the ground terminal, current can flow through this path when the drain voltage is negative in polarity. From this standpoint, it is preferable to use a metal–semiconductor contact for the gate rather than a P–N junction. The Schottky gate contact provides for a lower on-state voltage drop by proper choice of the work function for the gate contact. In addition, the Schottky contact has no significant reverse recovery current. This greatly reduces switching losses in both the rectifier and the FETs. Thus, the Baliga-Pair configuration replaces not only the power switch (such as the IGBT or GTO) in applications but also the power rectifier that is normally used across the switch.

Simulation Example

To demonstrate the operation of the Baliga-Pair configuration, two-dimensional numerical simulations are described here by concatenating a silicon power

MOSFET structure with a silicon carbide MESFET structure, as illustrated in Fig. 6.140. This allows observation of the potential and current distribution within both devices during all modes of operation. The silicon carbide MESFET had a drift region with doping concentration of 1×10^{16} cm^{-3} and thickness of 20 μm designed to support 3,000 V. A trench-gate region with a depth of 1 μm was chosen. The spacing between the gate regions (dimension *a* in the figure) was chosen as 0.6 μm because this provides a good tradeoff between normally-on operation with low specific on-resistance and a good blocking gain. A work function of 4.5 eV was used for the gate contact corresponding to a barrier height of 0.8 eV.

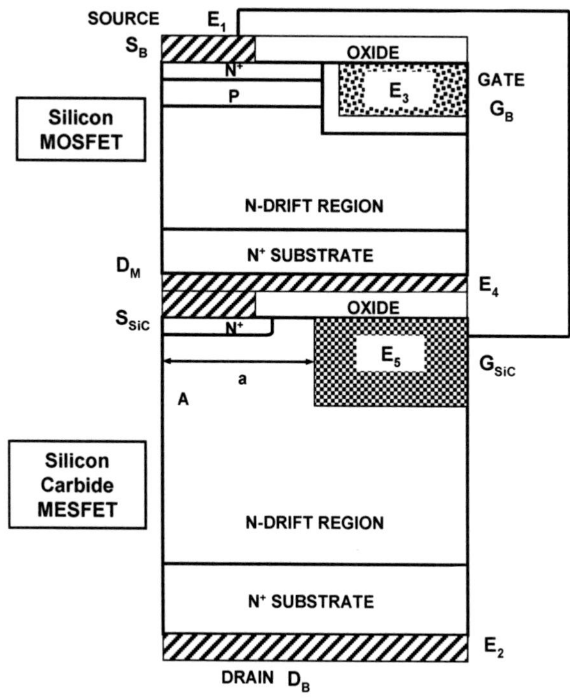

Fig. 6.140 Baliga-Pair simulation model using concatenated silicon MOSFET and 4H-SiC MESFET structures

The silicon power MOSFET had a trench-gate design to reduce its on-resistance and make the structure compatible with the trench-gate silicon carbide FET for simulations. The gate oxide for the MOSFET was chosen as 500 Å with a peak P-base doping concentration of 3×10^{17} cm^{-3}. This resulted in a threshold voltage of 4.5 V for the MOSFET which is acceptable for high-voltage power switches that can be driven using 10-V gate signals. The drift region for the silicon power MOSFET had a doping concentration of 1.5×10^{16} cm^{-3} and thickness of 2.5 μm designed to support 30 V.

The gate electrode (E_5) for the silicon carbide MESFET was connected to the source electrode (E_1) of the silicon power MOSFET during all the simulations. The electrode (E_4) served as both the source contact to the silicon carbide

MESFET and the drain contact for the silicon power MOSFET. This electrode was treated as a floating electrode (zero current boundary conditions) whose potential was monitored to provide insight into the voltage sharing between the two devices. The high-voltage DC bias was applied to the drain electrode (E_2) of the silicon carbide MESFET while the input control signal was applied to the gate electrode (E_3) of the silicon power MOSFET.

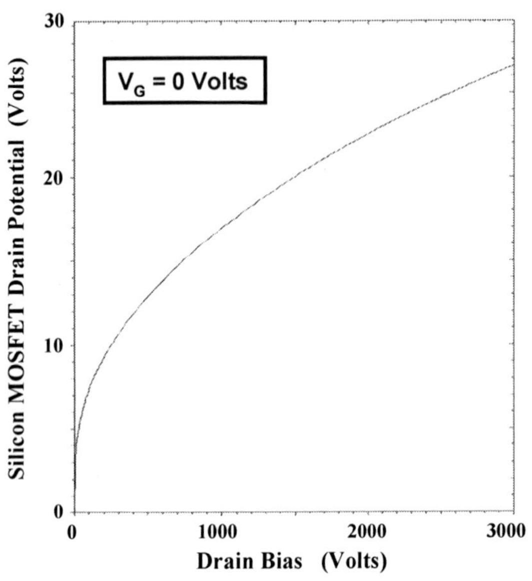

Fig. 6.141 Silicon MOSFET drain voltage in the Baliga-Pair configuration in the blocking mode

To operate the Baliga-Pair configuration in the voltage blocking mode, the gate electrode was held at zero bias and a drain bias was applied to the drain (E_2) of the silicon carbide MESFET structure. The composite switch was able to support 3,000 V with the potential on the drain electrode (E_4) of the silicon MOSFET remaining below 30 V. The voltage developed across the silicon power MOSFET can be monitored by examining the potential at the floating electrode (E_4). As the applied drain voltage increases, the potential at this electrode also increases as shown in Fig. 6.141. The voltage at the floating electrode increases to 26.5 V when the applied voltage to the drain of the composite switch reaches 3,000 V. Consequently, the maximum drain voltage experienced by the silicon power MOSFET is maintained below 30 V for a composite switch capable of supporting 3,000 V. This is an important feature of the Baliga-Pair configuration because it allows the use of silicon power MOSFETs with relatively low breakdown voltages (<50 V). Such silicon power MOSFETs are commercially available with very low on-resistance.

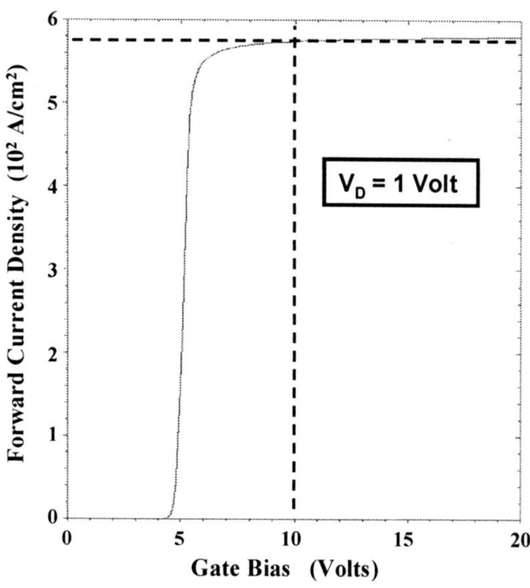

Fig. 6.142 Linear region transfer characteristics for the Baliga-Pair configuration

The Baliga-Pair configuration can be operated in the on-state by the application of a positive gate bias to the silicon power MOSFET. When the gate bias is well above the threshold voltage of the silicon power MOSFET, it can carry a high-current density with very low on-state voltage drop due to its low specific on-resistance. Under these conditions, the voltage difference between the source and gate of the silicon carbide MESFET becomes very small. Consequently, the channel of the silicon carbide MESFET remains undepleted allowing current flow with low on-resistance.

The transfer characteristic for the Baliga-Pair configuration was obtained by increasing the gate bias applied to the silicon power MOSFET while maintaining a bias of 1 V at the drain of the silicon carbide MESFET. From the transfer curve shown in Fig. 6.142, it can be seen that a gate bias of 10 V is sufficient to operate the Baliga-Pair with a low net on-resistance. At this gate bias, the current density flowing through the structures was 570 A cm^{-2} corresponding to a specific on-resistance of 1.9 mΩ cm^2. The silicon carbide MESFET contributes about 90% of this on-resistance with the rest contributed by the silicon power MOSFET.

The ability to limit the current flow by current saturation within a power switch is an attractive attribute from an application's standpoint. The extension of current saturation to high voltages is very desirable because this provides a broad SOA for the switching loci during circuit operation. These features were analyzed for the Baliga-Pair configuration by the application of various gate bias voltages and examining the resulting output characteristics as shown in Fig. 6.143. It can be seen that the composite switch exhibits an excellent current saturation capability to very high drain bias voltages with very high output resistance until the gate bias voltage exceeds 9 V. Beyond this gate bias voltage, current compression is observed – a phenomenon commonly observed in transistors at very high-current densities.[13] Note that the current density has reached a magnitude of 2×10^4 A cm^{-2}

at a gate bias of 9 V. Although such high-current density can be a problem from a power dissipation point of view, these results demonstrate that the Baliga-Pair has a very broad SOA within which inductive load switching can be performed.

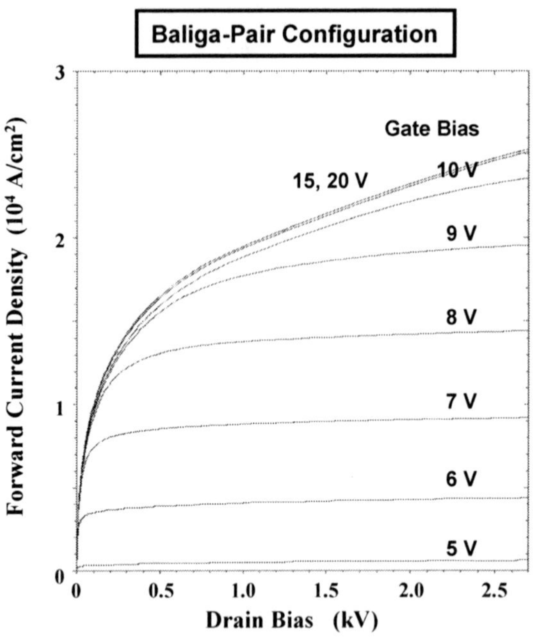

Fig. 6.143 Baliga-Pair configuration output characteristics

The switching performance of the Baliga-Pair configuration was examined under inductive load conditions by initially biasing the composite switch in its blocking state with a drain bias of 2,000 V while keeping the gate bias of the silicon MOSFET structure at zero. The gate was then biased using a constant input current to charge the gate capacitance. This method is commonly used for characterization of power MOSFETs. The gate voltage at the silicon power MOSFET structure increases in response to the stimulus as shown in Fig. 6.144. When the gate voltage exceeds the threshold voltage at time t_1, the drain current begins to rise. The simulations were designed to maintain the drain current density at a constant value once it reached a magnitude of 200 A cm^{-2}. At this time t_2, the drain voltage falls rapidly until time t_3 resulting in a high dV/dt at the output terminals of the FETs. The plateau in the gate voltage until time t_4 is associated with the charging of the Miller capacitance of the silicon MOSFET.

During the switching event for turning on the Baliga-Pair, high power dissipation occurs during the transition time ($t_2 - t_1$) for the drain current and the transition time ($t_3 - t_2$) for the drain voltage. The switching energy associated with these events is given by

$$E_{\text{off}} = 0.5 V_{\text{D}} I_{\text{D}} (t_3 - t_1). \tag{6.278}$$

For the simulated case with $V_D = 2{,}000$ V and $J_D = 200$ A cm^{-2}, the calculated switching energy is 30 mJ cm^{-2}. This would produce a power dissipation of 30 W cm^{-2} at a typical operating frequency of 1 kHz. In comparison, the on-state power dissipation is given by

$$P_{ON} = R_{DSon,SP} J_D^2. \tag{6.279}$$

For the simulated Baliga-Pair configuration with specific on-resistance of 2 mΩ cm^2, the conduction power loss is found to be 80 W cm^{-2}. If the turn-off switching losses are assumed to be comparable with the turn-on power loss computed above, the total power loss is 140 W cm^{-2}, which is acceptable from the point of view of cooling the devices with available heat sink technology. This demonstrates that the Baliga-Pair is capable of providing excellent switching performance for high-voltage power systems.

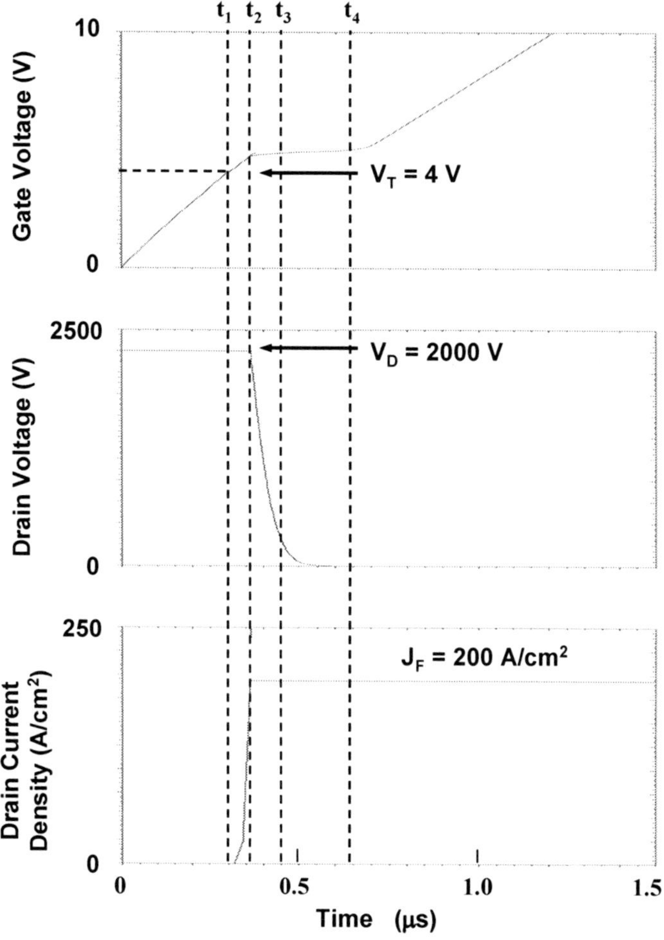

Fig. 6.144 Switching performance of the Baliga-Pair configuration

It is also interesting to observe the behavior of the drain potential of the silicon power MOSFET during the switching transient. As can be seen in Fig. 6.145, the drain voltage remains close to its value during operation in the blocking mode until time t_2 and then reduces to the on-state voltage drop of the silicon power MOSFET. It is worth pointing out that the drain voltage of the silicon power MOSFET ramps down during the entire time interval from t_2 to t_4 while the drain voltage on the silicon carbide MESFET ramps down in a much shorter time interval from t_2 to t_3. Since most of the voltage is supported by the silicon carbide MESFET, this is a favorable outcome that results in smaller turn-off energy and switching losses.

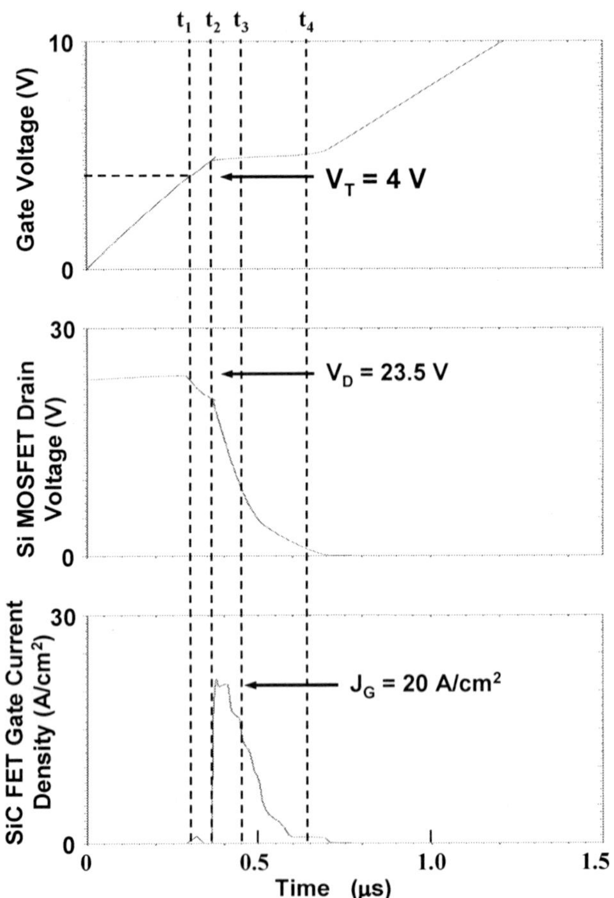

Fig. 6.145 Silicon power MOSFET drain potential during switching of the Baliga-Pair configuration

In addition, the current flow through the gate of MESFET is shown in Fig. 6.145 during the switching transient. This current remains at approximately a constant value during the time interval from t_2 to t_3 when the drain voltage of the MESFET is ramping down. This indicates that the Miller and output capacitances of the MESFET are being charged through the gate electrode of the MESFET. This is responsible for the faster reduction of the drain voltage for the silicon carbide MESFET without being limited by the performance of the silicon MOSFET. These observations indicate that the Baliga-Pair configuration has an excellent switching behavior for power electronic applications.

The Baliga-Pair configuration contains an inherent fly-back rectifier that can be useful in H-bridge and other power electronic circuit applications. The performance of this rectifier can be examined by doing simulations with a negative bias applied to the drain terminal of the silicon carbide MESFET while connecting the gate of the silicon power MOSFET to the source terminal. The resulting characteristic is shown in Fig. 6.146 at room temperature. Excellent characteristics are observed with an on-state voltage drop of only 0.7 V. This indicates that a larger work function could be used for the gate of the silicon carbide MESFET to reduce its leakage current if necessary.

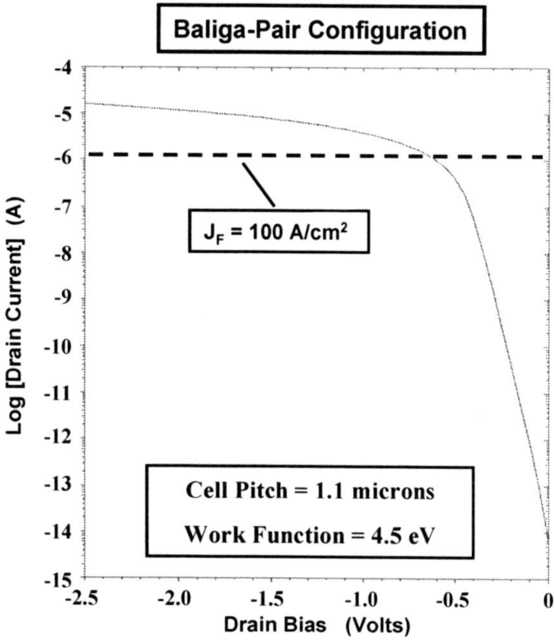

Fig. 6.146 On-state characteristics of the fly-back diode in the Baliga-Pair configuration

The simulations discussed above indicate that the Baliga-Pair configuration provides an excellent power switch for high-voltage power system applications. In terms of packaging, the two transistors within the Baliga-Pair configuration replace the IGBT and the fly-back rectifier, resulting in a similar level of complexity and cost. The attractive features of the Baliga-Pair configuration, also called the *cascode configuration*, have been experimentally confirmed.[30] The turn-off time for the Baliga-Pair configuration was found to be half that for a silicon carbide MOSFET due to the smaller Miller capacitance.

6.22.2 Planar Power MOSFET Structure

The diffusion rate for dopants is extremely small in silicon carbide even at very high temperatures. Consequently, it is not feasible to fabricate the power VD-MOSFET structure. However, a planar power MOSFET structure can be fabricated from silicon carbide by staggering the location of the ion implants for the P-base and N^+ source regions.[31] The basic structure of the planar power MOSFET is shown in Fig. 6.147 together with the location of the ion implantation edges. These devices have been called DIMOSFETs because of the double-implant process used for their fabrication.[32]

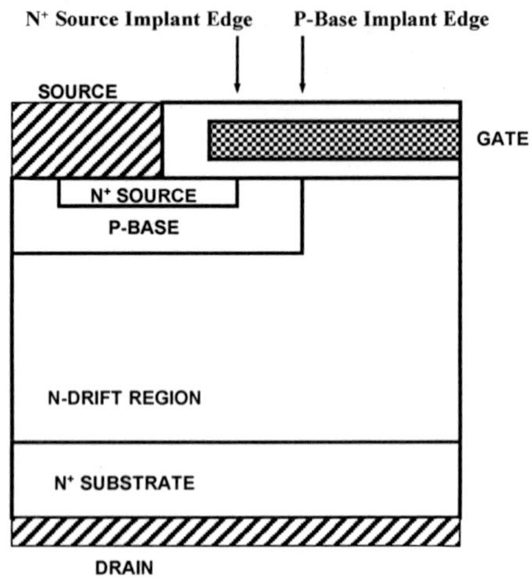

Fig. 6.147 Planar power MOSFET structure

Blocking Characteristics

The voltage is supported by a depletion region formed on both sides of the P-base/N-drift junction during the forward-blocking mode in the planar power MOSFET structure. The maximum blocking voltage is determined by the electric field at this junction becoming equal to the critical electric field for breakdown if the parasitic N^+/P/N bipolar transistor is completely suppressed. This suppression is accomplished by short circuiting the N^+ source and P-base regions using the source metal as shown on the upper left-hand side of the cross section. However, a large leakage current can occur when the depletion region in the P-base region reaches through to the N^+ source region. The doping concentration and thickness of the P-base region must be designed to prevent the reach-through phenomenon from limiting the breakdown voltage. This problem becomes aggravated in silicon carbide devices due to the much larger electric field at the blocking junction.

The applied drain voltage is supported by the N-drift region and the P-base region with a triangular electric field distribution as shown in Fig. 6.148 if the

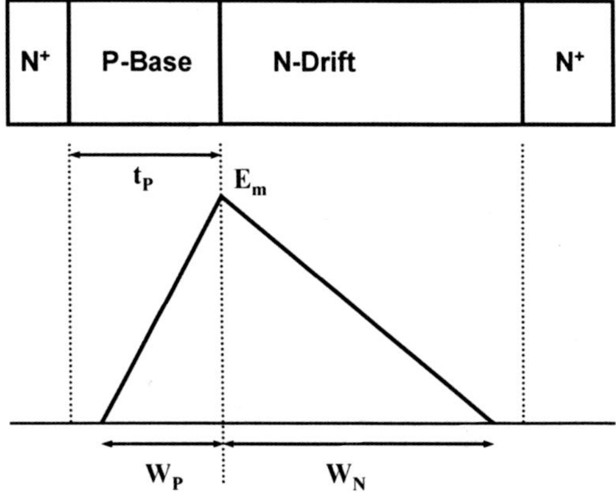

Fig. 6.148 Reach-through in a power MOSFET structure

doping is uniform on both sides. The maximum electric field occurs at the P-base/N-drift junction. The depletion width on the P-base side is related to the maximum electric field by

$$W_{\mathrm{P}} = \frac{\varepsilon_{\mathrm{S}} E_{\mathrm{m}}}{q N_{\mathrm{A}}}, \tag{6.280}$$

where N_A is the doping concentration in the P-base region. The minimum P-base thickness required to prevent reach-through limited breakdown can be obtained by assuming that the maximum electric field at the P-base/N-drift junction reaches the critical electric field for breakdown when the P-base region is completely depleted:

$$t_{\mathrm{P}} = \frac{\varepsilon_{\mathrm{S}} E_{\mathrm{C}}}{q N_{\mathrm{A}}}, \tag{6.281}$$

where E_C is the critical electric field for breakdown in the semiconductor.

The calculated minimum P-base thickness for 4H-SiC power MOSFETs is compared with that for silicon in Fig. 6.149. At any given P-base doping concentration, the thickness for 4H-SiC is about six times larger than for silicon. This implies that the minimum channel length required for silicon carbide devices is much larger than for silicon devices, resulting in a big increase in the on-resistance. The enhancement of the on-resistance is compounded by the lower channel inversion layer mobility observed for silicon carbide.

The minimum thickness of the P-base region required to prevent reach-through breakdown decreases with increasing doping concentration as shown in Fig. 6.149. The typical P-base doping concentration for silicon power MOSFETs is 1×10^{17} cm^{-3} to obtain a threshold voltage between 1 and 3 V for a gate oxide thickness of 500–1,000 Å. At this doping level, the P-base thickness can be reduced to 0.5 μm without reach-through limiting the breakdown voltage. In contrast, for 4H-SiC, it is necessary to increase the P-base doping concentration to above

4×10^{17} cm^{-3} to prevent reach-through with a 0.5-μm P-base thickness. This higher doping concentration increases the threshold voltage as discussed below.

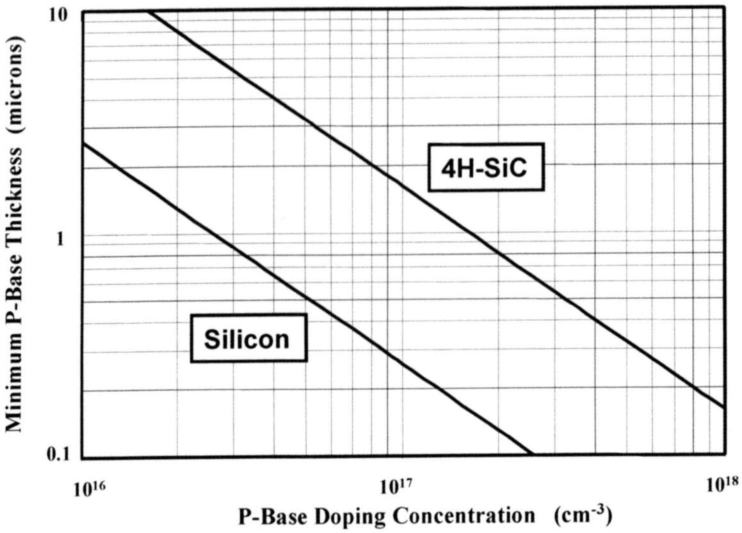

Fig. 6.149 Comparison of minimum P-base thickness to prevent reach-through breakdown in 4H-SiC and silicon

Threshold Voltage

The on-state operation for the silicon carbide planar power MOSFET structure is similar to that for the silicon counterpart. An n-channel region is formed on the surface of the P-base region by the application of a positive gate bias to provide a path from the source to the drain. The on-resistance can be modeled using the same approach as described earlier for the silicon power VD-MOSFET structure.[29] The gate bias must exceed the threshold voltage in order for the formation of the channel. As discussed earlier in this chapter, the threshold voltage can be defined as the gate bias at which on-set of *strong inversion* begins to occur in the channel. This voltage is given by

$$V_{TH} = \frac{\sqrt{4\varepsilon_S kTN_A \ln(N_A / n_i)}}{C_{OX}} + \frac{2kT}{q}\ln\left(\frac{N_A}{n_i}\right), \quad (6.282)$$

where N_A is the doping concentration of the P-base region, k is a Boltzmann's constant, and T is the absolute temperature. The threshold voltage for 4H-SiC planar MOSFETs calculated by using this equation is shown in Fig. 6.150 for the case of a gate oxide thickness of 1,000 Å. The results obtained for a silicon power MOSFET with the same gate oxide thickness are also provided in this figure for comparison.

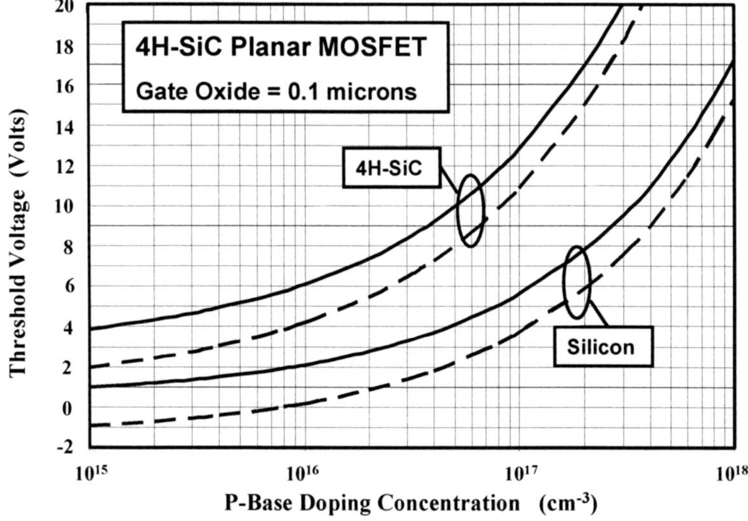

Fig. 6.150 Threshold voltage of 4H-SiC planar MOSFETs. *Dashed lines* include the impact of N^+ polysilicon gate and an oxide fixed charge of 2×10^{11} cm^{-2}

In the case of silicon devices, a threshold voltage of about 4 V is obtained for a P-base doping concentration of 1×10^{17} cm^{-3}. At this doping concentration, the depletion width in the P-base region for silicon devices is less than 0.5 μm, as shown earlier in Fig. 6.149, even when the electric field in the semiconductor approaches the critical electric field for breakdown. This allows the design of silicon power MOSFETs with channel lengths of below 0.5 μm without encountering reach-through breakdown limitations. In contrast, a P-base doping concentration of about 3×10^{17} cm^{-3} is required in 4H-SiC (see Fig. 6.149) to keep the depletion width in the P-base region below 0.5 μm when the electric field in the semiconductor approaches the critical electric field for breakdown. At this doping concentration, the threshold voltage for the 4H-SiC MOSFET approaches 20 V. The much larger threshold voltage for silicon carbide is physically related to its larger band gap (which results in an extremely small value for the intrinsic concentration in (6.282)) as well as the higher P-base doping concentration required to suppress reach-through breakdown. This indicates a fundamental problem for achieving reasonable levels of threshold voltage in silicon carbide power MOSFETs if the conventional silicon structure is utilized.

Oxide Field

The maximum blocking voltage capability of the power MOSFET structure is typically determined by the drift region doping concentration and thickness as already discussed in Chap. 3. However, in the silicon carbide power MOSFET structure, a high electric field develops in the gate oxide under forward-blocking conditions. The electric field developed in the oxide is related to the electric field in the underlying semiconductor by Gauss's law:

$$E_{Oxide} = \left(\frac{\varepsilon_{Semi}}{\varepsilon_{Oxide}} \right) E_{Semi}, \tag{6.283}$$

where ε_{Semi} and ε_{Oxide} are the dielectric constants of the semiconductor and the oxide, respectively, and E_{Semi} is the electric field in the semiconductor. In the case of both silicon and silicon carbide, the electric field in the oxide is about three times larger than in the semiconductor. Since the maximum electric field in the silicon drift region remains below 3×10^5 V cm^{-1}, the electric field in the oxide does not exceed its reliability limit of about 3×10^6 V cm^{-1}. However, for 4H-SiC, the electric field in the oxide reaches a value of 9×10^6 V cm^{-1} when the field in the semiconductor reaches its breakdown strength. This value not only exceeds the reliability limit but also causes rupture of the oxide leading to catastrophic breakdown. It is therefore important to monitor the electric field in the gate oxide when designing and modeling the silicon carbide MOSFET structures.

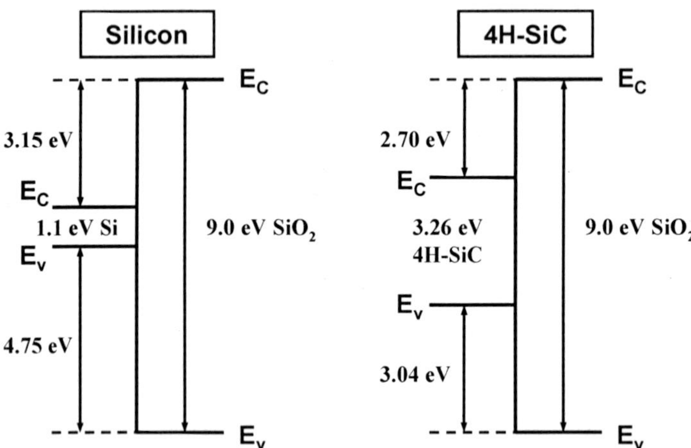

Fig. 6.151 Comparison of the energy band offsets between silicon/oxide and SiC/oxide

For any MOSFET structure, injection of electrons into the gate oxide can occur when the electrons gain sufficient energy in the semiconductor to surmount the potential barrier between the semiconductor and the oxide. This leads to *hot electron injection*-induced instability of the threshold voltage. The energy band offsets between the semiconductor and silicon dioxide are shown in Fig. 6.151 for silicon and 4H-SiC for comparison purposes. Due to the larger band gap of 4H-SiC, the band offset between the conduction band edges for the semiconductor and silicon dioxide is significantly smaller than for the case of silicon.

The band diagram for the 4H-SiC silicon dioxide interface is shown in Fig. 6.152 when a positive bias is applied to the gate. This bias condition is typical for the on-state mode of operation in power MOSFET structures. In this illustration, the gate was assumed to be formed using polysilicon. It can be seen that a narrow barrier is formed at the semiconductor–oxide interface which can be penetrated by the tunneling of electrons from the conduction band into the oxide. This produces a *Fowler–Nordheim tunneling* current that injects electrons into the oxide. This current has been observed in measurements reported on both 6H-SiC[33] and 4H-SiC[34] MOS capacitors. The trapping of these electrons within the gate oxide can cause shifts in the threshold voltage of the MOSFET leading to reliability problems.

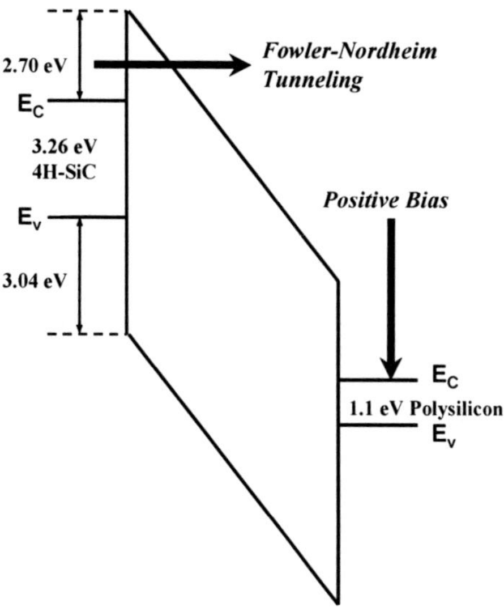

Fig. 6.152 Energy band diagram for the polysilicon/oxide/4H-SiC structure

6.22.3 Shielded Planar Power MOSFET Structures

Two novel silicon carbide power MOSFET structures[35] that provide shielding of the gate oxide from high electric field within the semiconductor are shown in Fig. 6.153. In the case of the structure with the inversion layer channel, the P^+ shielding region extends under both the N^+ source region and the P-base region. It could also extend beyond the edge of the P-base region. In the case of the structure with the accumulation layer channel, the P^+ shielding region extends under the N^+ source region and the N-base region located under the gate. This N-base region can be formed using an uncompensated portion of the N-type drift region or it can be created by adding N-type dopants with ion implantation to control its thickness and doping concentration.

The gap between the P^+ shielding regions is optimized to obtain a low specific on-resistance while simultaneously shielding the gate oxide interface from the high electric field in the drift region. In both structures shown above, a potential barrier is formed at location *A* after the JFET region becomes depleted by the applied drain bias in the blocking mode. This barrier prevents the electric field from becoming large at the gate oxide interface. When a positive bias is applied to the gate electrode, an inversion layer or accumulation layer channel is formed in the structures enabling the conduction of drain current with a low specific on-resistance.

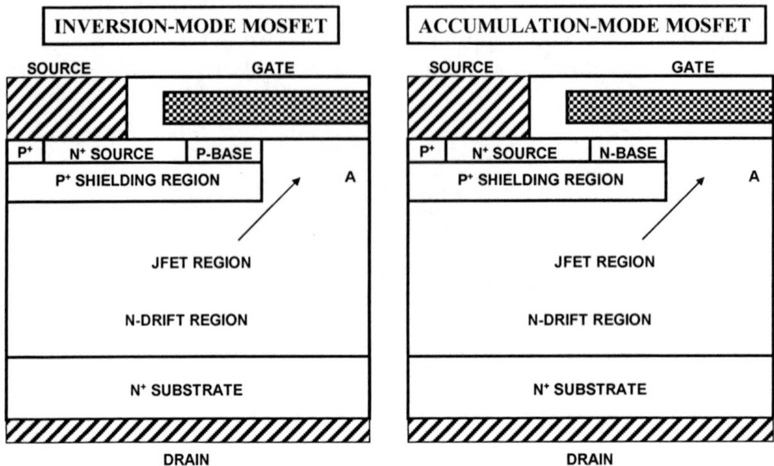

Fig. 6.153 Shielded planar MOSFET structures

Blocking Mode

In the forward-blocking mode of the shielded planar MOSFET structure, the voltage is supported by a depletion region formed on both sides of the P^+ region/N-drift junction. The maximum blocking voltage can be determined by the electric field at this junction becoming equal to the critical electric field for breakdown if the parasitic N^+/P/N bipolar transistor is completely suppressed. This suppression is accomplished by short circuiting the N^+ source and P^+ regions using the source metal as shown on the upper left-hand side of the cross section. This short circuit can be accomplished at a location orthogonal to the cell cross section if desired to reduce the cell pitch while optimizing the specific on-resistance. If the doping concentration of the P^+ region is high, the reach-through breakdown problem discussed earlier in this chapter is completely eliminated. In addition, the high doping concentration in the P^+ region promotes the depletion of the JFET region at lower drain voltages providing enhanced shielding of the channel and gate oxide.

With the shielding provided by the P^+ region, the minimum P-base thickness for 4H-SiC power MOSFETs is no longer constrained by the reach-through limitation. This enables reducing the channel length below the values associated with any particular doping concentration of the P-base region in Fig. 6.149. In addition, the opportunity to reduce the P-base doping concentration enables decreasing the threshold voltage. The smaller channel length and threshold voltage provide the benefits of reducing the channel resistance contribution.

In the case of the accumulation-mode planar MOSFET structure, the presence of the subsurface P^+ shielding region under the N-base region provides the potential required for *completely depleting* the N-base region if its doping concentration and thickness are appropriately chosen. This enables *normally-off operation* of the accumulation-mode planar MOSFET with zero gate bias. It is worth pointing out that this mode of operation is fundamentally different to that of buried channel MOS devices.[36] Buried channel devices contain an undepleted N-type channel region that provides a current path for the drain current at zero gate bias. This region must be depleted by a negative gate bias creating a normally-on device structure.

In the accumulation-mode planar MOSFET structure, the depletion of the N-base region is accompanied by the formation of the potential barrier for the flow of electrons through the channel. The channel potential barrier does not have to have a large magnitude because the depletion of the JFET region screens the channel from the drain bias as well.

The maximum blocking voltage capability of the shielded planar MOSFET structure is determined by the drift region doping concentration and thickness as already discussed in Chap. 3. However, to fully utilize the high breakdown electric field strength available in silicon carbide, it is important to screen the gate oxide from the high field within the semiconductor. In the shielded planar MOSFET structure, this is achieved by the formation of a potential barrier at location *A* by the depletion of the JFET region at a low drain bias voltage.

Threshold Voltage

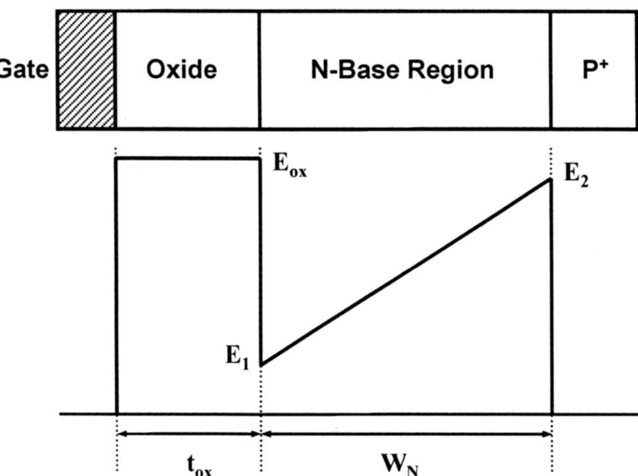

Fig. 6.154 Electric field profile in the gate region for the accumulation-mode MOSFET structure

As discussed previously, the threshold voltage for the inversion-mode silicon carbide power MOSFET structure is large due to the large energy band gap for this semiconductor. This is related to the extremely low values for the intrinsic concentration in (6.282). The band bending required to create a channel in the accumulation-mode planar MOSFET is much smaller than required for the inversion-mode device. This provides the opportunity to reduce the threshold voltage while obtaining the desired normally-off device behavior. A model for the threshold voltage of accumulation-mode MOSFETs has been developed[37] using the electric field profile shown in Fig. 6.154 when the gate is biased at the threshold voltage. In this figure, the electric fields in the semiconductor and oxide are given by

$$E_1 = \frac{V_{bi}}{W_N} - \frac{qN_D W_N}{2\varepsilon_S}, \tag{6.284}$$

$$E_2 = \frac{V_{bi}}{W_N} + \frac{qN_D W_N}{2\varepsilon_S}, \tag{6.285}$$

$$E_{OX} = \frac{\varepsilon_S}{\varepsilon_{OX}} E_1. \tag{6.286}$$

Note that this model is based upon neglecting any voltage supported within the P^+ region under the assumption that it is very heavily doped. Using these electric fields, the threshold voltage is found to be given by

$$V_{TH} = \phi_{MS} + \left(\frac{\varepsilon_S V_{bi}}{\varepsilon_{OX} W_N} - \frac{qN_D W_N}{2\varepsilon_{OX}} \right) t_{OX}. \tag{6.287}$$

The first term in this equation accounts for the work function difference between the gate material and the lightly doped N-base region. The second term represents the effect of the built-in potential of the P^+/N junction that depletes the N-base region.

The analytically calculated threshold voltage for 4H-SiC accumulation-mode MOSFETs is provided in Fig. 6.155, for the case of a gate oxide thickness of 500 Å and N-base thickness of 0.2 μm, as a function of the N-base doping concentration with the inclusion of a metal–semiconductor work function difference of 1 V. For comparison purposes, the threshold voltage for the inversion-mode 4H-SiC MOSFET is also given in this figure for the same gate oxide thickness. A strikingly obvious difference between the structures is a decrease in the threshold voltage for the accumulation-mode structure with increasing doping concentration in the N-base region. This occurs due to the declining influence of the P^+/N junction at the gate oxide interface when the doping concentration of the N-base region is increased. It can also be noted that the temperature dependence of the threshold voltage is smaller for the accumulation-mode structure. Of course, the most important benefit of the accumulation-mode structure is that lower threshold voltages can be achieved than in the inversion-mode structures. At an N-base doping concentration of 1×10^{16} cm^{-3}, the threshold voltage for the accumulation-mode structure is 2.8 V. This will be reduced to just below 2 V in the presence of typical fixed oxide charge in the 4H-SiC/oxide system. Accumulation-mode silicon carbide planar power MOSFET structures have been experimentally demonstrated[38] with low threshold voltages.

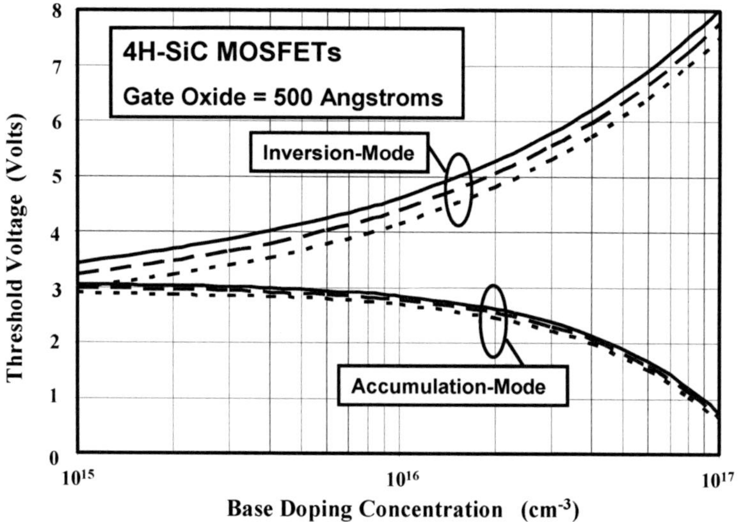

Fig. 6.155 Threshold voltage of 4H-SiC accumulation-mode MOSFETs (*solid line*: 300 K, *dash line*: 400 K, *dotted line*: 500 K)

Simulation Example

To gain insight into the operation of the planar 4H-SiC shielded accumulation-mode MOSFET structure, two-dimensional numerical simulations are described here for a structure with a drift region doping concentration of 1×10^{16} cm^{-3} and a thickness of 20 μm corresponding to a parallel-plane breakdown voltage of 3,000 V. The device had a gate oxide thickness of 400 Å, a P^+ region depth of 1 μm, and a JFET width of 1.5 μm. The N-base region had a thickness of 0.2 μm with a doping concentration of 8.5×10^{15} cm^{-3}. The cell pitch for the structure, corresponding to the cross section shown on the right-hand side of Fig. 6.153, was 3.25 μm.

The blocking capability of the planar 4H-SiC shielded accumulation-mode MOSFET was investigated by maintaining zero gate bias while increasing the drain voltage. It was found that the drain current remains below 1×10^{-13} A up to a drain bias of 3,000 V. This behavior demonstrates that the depletion of the N-base region by the built-in potential of P^+/N junction creates a potential barrier for electron transport through the channel. In addition, the shielding of the N-base region by the underlying P^+ region is very effective for preventing the reach-through breakdown problem, allowing the device to operate up to the full capability of the drift region.

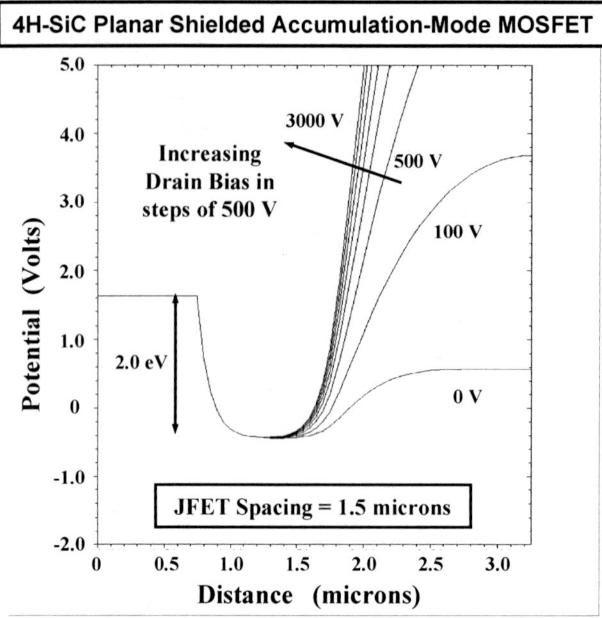

Fig. 6.156 Channel potential barrier in the planar 4H-SiC shielded accumulation-mode MOSFET structure

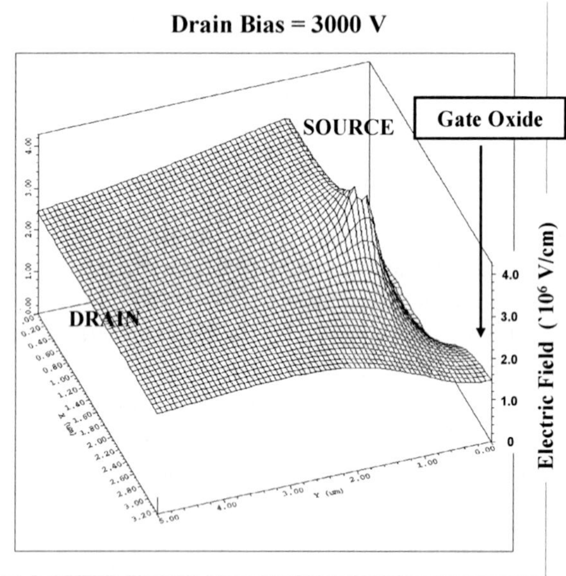

Fig. 6.157 Electric field distribution in the planar 4H-SiC shielded accumulation-mode MOSFET structure

To understand the ability to suppress current flow in the accumulation-mode structure in the absence of the P-base region, it is instructive to examine the potential distribution along the channel. The variation of the potential along the channel is shown in Fig. 6.156 for the planar 4H-SiC shielded accumulation-mode MOSFET described above for various drain bias voltages when the gate bias voltage is held at zero volts. It can be seen that there is a potential barrier in the channel with a magnitude of 2 eV at zero drain bias. This barrier for the transport of electrons from the source to the drain is upheld even when a drain bias of up to 3,000 V is applied. This demonstrates the fundamental operating principle for creating a normally-off device with a depleted N-base region formed by using the built-in potential of the subsurface P^+/N junction.

As in the case of the inversion-mode structure, the P^+ region in the planar 4H-SiC shielded accumulation-mode MOSFET structure can isolate the surface under the gate oxide from the high electric fields developed in the drift region. To demonstrate this, a three-dimensional view of the electric field distribution is shown in Fig. 6.157 for this structure at a drain bias of 3,000 V with zero gate bias. The field distribution shown in this figure indicates that the electric field at the gate oxide is suppressed by the presence of the P^+ region. At the maximum blocking capability of the drift region, the electric field at the gate oxide interface has reached a magnitude of 1.5×10^6 V cm^{-1} while the electric field in the bulk has reached a magnitude of 3×10^6 V cm^{-1}.

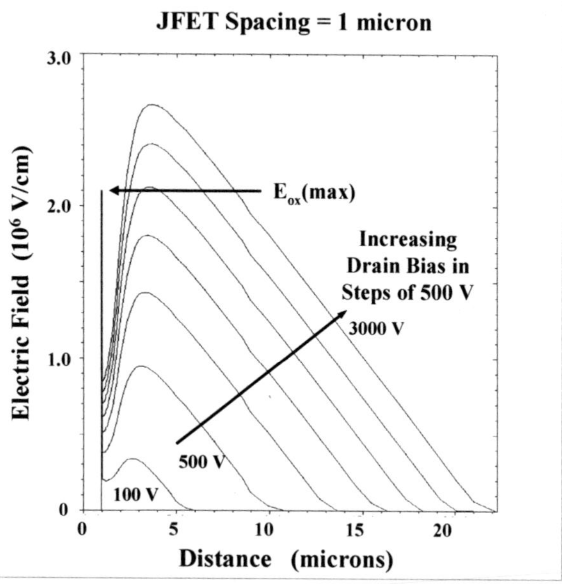

Fig. 6.158 Electric field distribution in the planar 4H-SiC shielded accumulation-mode MOSFET structure

In the planar 4H-SiC shielded accumulation-mode MOSFET structure, the largest electric field at the gate oxide interface occurs at the center of the JFET region. The electric field distribution along the vertical direction at this location is shown in Fig. 6.158 for the case of a planar 4H-SiC shielded accumulation-mode MOSFET structure with a JFET width of 1 μm. It can be seen that the maximum electric field at the gate oxide interface is reduced to 8.5×10^5 V cm^{-1} by the smaller JFET width, resulting in a maximum electric field in the gate oxide of only

2.1×10^6 V cm^{-1} at a drain bias of 3,000 V. This value is sufficiently low for reliable operation of the structure especially due to the planar-gate structure, where there are no localized electric field enhancements under the gate electrode.

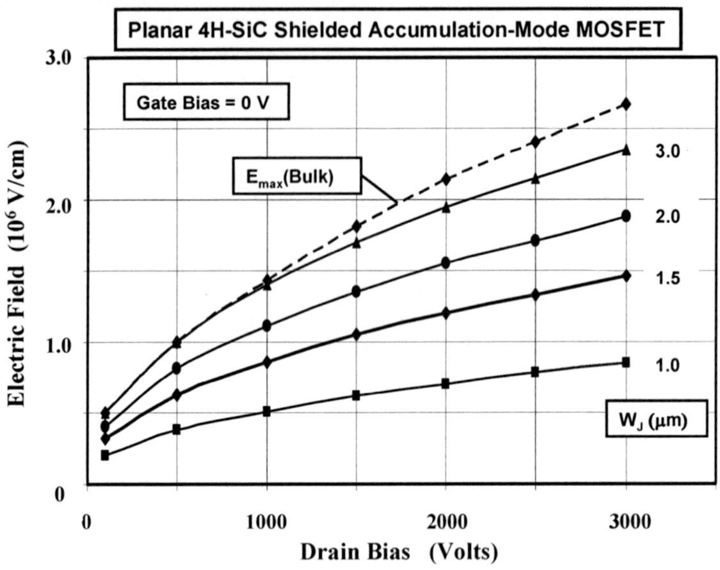

Fig. 6.159 Electric field suppression in the planar 4H-SiC shielded accumulation-mode MOSFET structure

The magnitude of the electric field developed in the vicinity of the gate oxide interface in the planar 4H-SiC shielded accumulation-mode MOSFET structure is dependent upon the width of the JFET region. The electric field at the gate oxide interface becomes smaller as the JFET width is reduced as shown in Fig. 6.159. For a structure with a drift region doping concentration of 1×10^{16} cm^{-3}, the optimum spacing lies between 1 and 2 μm. The JFET width cannot be arbitrarily reduced because the resistance of the JFET region becomes very large when the width approaches the zero-bias depletion width of the P^+/N junction. For a drift region doping concentration of 1×10^{16} cm^{-3}, the zero-bias depletion width is approximately 0.6 μm. Consequently, using a JFET width of 1.5 μm provides adequate space in the JFET region for current flow. A low specific on-resistance of 3 mΩ cm^2 was observed for this structure with an accumulation layer mobility of 90 cm^2 V^{-1} s^{-1}. Accumulation layer mobility values larger than this have been experimentally demonstrated.[38] For a more detailed discussion of these devices, together with more results of numerical simulations, the reader should consult reference 29.

On-Resistance

The on-resistance for the silicon carbide planar MOSFET structure can be modeled by using the same approach previously described for the planar silicon VD-

MOSFET structure.[29] Due to the relatively small thickness for the drift region, it is appropriate to use the current flow pattern shown in Fig. 6.39 for the silicon carbide planar MOSFET structure. The equations developed in Sect. 6.6 can then be utilized for the silicon carbide planar MOSFET structure with appropriate parameters. Due to the relatively high doping concentrations in the drift region, it is usually not necessary to enhance the doping concentration in the JFET region for the silicon carbide planar MOSFET structure. For 4H-SiC structures, it is appropriate to use the following parameters (a) specific contact resistance of $1 \times 10^{-5}\ \Omega\ cm^2$, (b) N^+ source sheet resistance of $50\ \Omega\ sq^{-1}$, (c) inversion layer mobility of $50\ cm^2\ V^{-1}\ s^{-1}$, (d) accumulation layer mobility of $100\ cm^2\ V^{-1}\ s^{-1}$, and (e) N^+ substrate with a resistivity of $0.02\ \Omega$ cm and a thickness of 200 μm.

6.22.4 Shielded Trench-Gate Power MOSFET Structure

The development of silicon carbide power MOSFET devices began with the construction of trench-gate structures with epitaxially grown P-base and N^+ source regions.[39] The breakdown voltage of these structures was found to be limited by the catastrophic rupture of the gate oxide at voltages well below the intrinsic capability of the silicon carbide drift region. This problem is related to the high electric field generated in the gate oxide during the blocking mode as discussed earlier. As in the case of the planar structure, it is possible to overcome this problem by the implementation of a shielding region.[40] In the case of the power U-MOSFET structure, the P^+ shielding region must be located at the bottom of the trench as illustrated in the cross section shown in Fig. 6.160. The shielding region consists of a heavily doped P-type (P^+) region located at the bottom of the trench. The P^+ shielding region is connected to the source electrode at a location

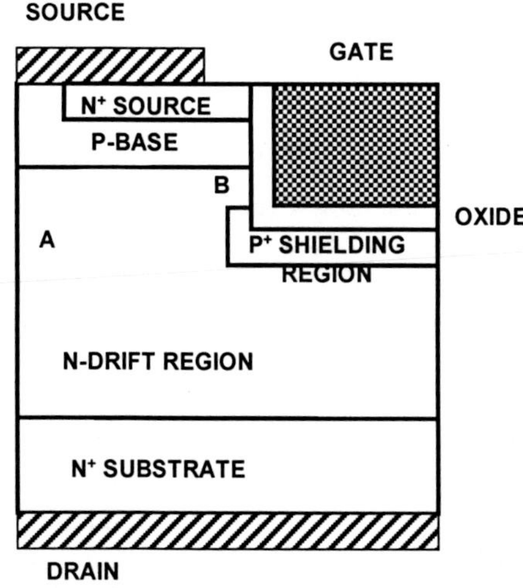

Fig. 6.160 Shielded trench-gate power MOSFET structure

orthogonal to the device cross section shown above. The structure then behaves like a monolithic version of the *Baliga-Pair* configuration discussed in Sect. 6.22.1 with both the JFET and MOSFET formed in the same semiconductor, namely, silicon carbide. From this point of view, the portion of the N-drift region located just below the P-base region serves as both the drain of the MOSFET and the source of the JFET region.

The shielded trench-gate power MOSFET structure can be fabricated by using the same process described earlier for the conventional trench-gate device structure with the addition of an ion implantation step to form the P^+ shielding region. The ion implant used to form the P^+ shielding region must be performed after etching the trenches. It is worth pointing out that this implantation step should not dope the sidewalls of the trenches. This can be accomplished by doing the ion implant orthogonal to the wafer surface. Alternately, a conformal oxide can be deposited on the trench sidewalls and removed from the trench bottom using anisotropic RIE to selectively expose the trench bottom to the P-type dopant during the ion implantation. Another option to reduce the introduction of P-type doping on the trench sidewalls is by using a relatively low P-type doping concentration for the shielding region. The shielding is effective as long as the gate oxide is buffered from the high electric field in the N-drift region.

Blocking Characteristics

The shielded trench-gate power MOSFET operates in the forward-blocking mode when the gate electrode is shorted to the source by the external gate drive circuit. At low drain bias voltages, the voltage is supported by a depletion region formed on both sides of the P-base/N-drift junction. Consequently, the drain potential appears across the MOSFET located at the top of the structure. This produces a positive potential at location *A* in Fig. 6.160, which reverse biases the junction between the P^+ shielding region and the N-drift region because the P^+ shielding region is held at zero volts. The depletion region that extends from the P^+/N junction pinches off the JFET region producing a potential barrier at location *A*. The potential barrier tends to isolate the P-base region from any additional bias applied to the drain electrode. Consequently, a high electric field can develop in the N-drift region below the P^+ shielding region while the electric field at the P-base region remains low. This has the beneficial effects of mitigating the reach-through of the depletion region within the P-base region and in keeping a low electric field in the gate oxide at location *B* where it is exposed to the N-drift region.

The P^+ shielding region must be adequately short circuited to the source terminal in order for the shielding to be fully effective. The location of the P^+ region at the bottom of the trench implies that contact to it must be provided at selected locations orthogonal to the cross section of the device shown in Fig. 6.160. Since the sheet resistance of the ion-implanted P^+ region can be quite high, it is important to provide the contact to the P^+ region frequently in the orthogonal direction during chip design. This must be accomplished without significant loss of channel density if low specific on-resistance is to be realized.

Forward Conduction

In the shielded trench-gate MOSFET structure, current flow between the drain and source can be induced by creating an inversion layer channel on the surface of the P-base region. The current flows from the source region into the drift region through the inversion layer channel formed on the vertical sidewalls of the trench due to the applied gate bias. It must then flow from point *B* in the cross section through the first JFET region which constricts the current into location *A* shown in the cross section. The current then spreads into the N-drift region at a 45° angle and becomes uniform through the rest of the structure. The current path is illustrated in Fig. 6.161 by the shaded area together with the zero-bias depletion boundaries of the junctions indicated by the dashed lines. It can be seen that the addition of the P^+ shielding region introduces *two* JFET regions into the basic trench-gate MOSFET structure. The first one, labeled R_{JFET1} in the figure, is formed between the P-base region and the P^+ shielding region with the current constricted by their zero-bias depletion boundaries. The spacing between these regions (labeled t_B in the figure) must be chosen to prevent it from becoming completely

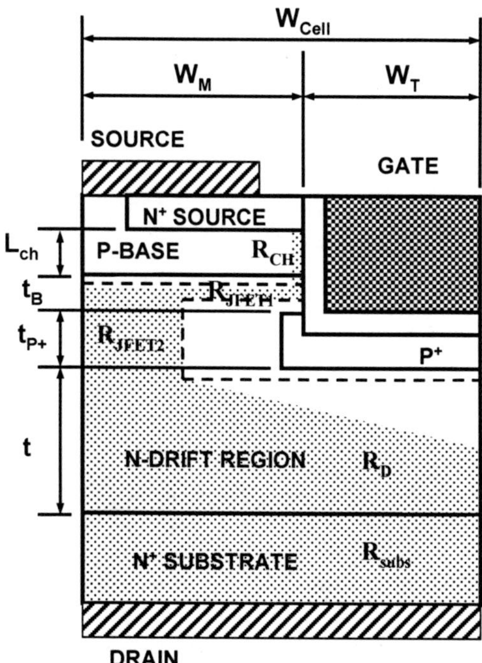

Fig. 6.161 Path for current flow in the shielded trench-gate 4H-SiC power MOSFET structure

depleted. For a typical N-drift region doping concentration of 1×10^{16} cm^{-3}, corresponding to a breakdown voltage of 3,000 V, the zero-bias depletion width is about 0.6 μm. In this case, the spacing (t_B) would have to be about 1.5 μm to ensure the existence of an undepleted path for the transport of electrons. This spacing can be reduced if the doping concentration in the JFET region is selectively increased when compared with the N-drift region.[40]

The second JFET region, labeled R_{JFET2} in the figure, is formed between the P$^+$ shielding regions. Its resistance is determined by the thickness of the P$^+$ shielding region (labeled t_{P+} in the figure), which can be assumed to be twice the junction depth of the P$^+$ shielding region. Since the cross section for current flow through this region is constricted by the zero-bias depletion width of the P$^+$/N junction, it is again advantageous to increase the doping concentration in the JFET region to avoid having to enlarge the mesa width.[40] A smaller mesa width allows maintaining a smaller cell pitch which reduces the specific on-resistance due to a larger channel density.

The total on-resistance for the shielded trench-gate power MOSFET structure is determined by the resistance of all the components in the current path:

$$R_{ON,SP} = R_{CH} + R_{JFET1} + R_{JFET2} + R_D + R_{subs}, \tag{6.288}$$

where R_{CH} is the channel resistance, R_{JFET1} and R_{JFET2} are the resistances of the two JFET regions, R_D is the resistance of the drift region after taking into account current spreading from the channel, and R_{subs} is the resistance of the N$^+$ substrate. These resistances can be analytically modeled by using the current flow pattern indicated by the shaded regions in Fig. 6.161.

The specific channel resistance is given by

$$R_{CH,SP} = \frac{L_{CH} W_{Cell}}{\mu_{inv} C_{OX} (V_G - V_{TH})}, \tag{6.289}$$

where L_{CH} is the channel length determined by the width of the P-base region as shown in Fig. 6.161, μ_{inv} is the mobility for electrons in the inversion layer channel, C_{OX} is the specific capacitance of the gate oxide, V_G is the applied gate bias, and V_{TH} is the threshold voltage.

The specific resistance of the first JFET region can be calculated by using

$$R_{JFET1,SP} = \rho_{JFET} W_{Cell} \left(\frac{x_{P+} + W_{J0}}{t_B - 2W_{J0}} \right), \tag{6.290}$$

where ρ_{JFET} is the resistivity of the JFET region, x_{P+} is the junction depth of the P$^+$ shielding region, and W_{J0} is the zero-bias depletion width *in the JFET region*. The resistivity and zero-bias depletion width used in this equation must be computed by using the enhanced doping concentration of the JFET region. The specific resistance of the second JFET region can be calculated by using

$$R_{\text{JFET2,SP}} = \rho_{\text{JFET}} W_{\text{Cell}} \left(\frac{t_{\text{P+}} + 2W_{\text{J0}}}{W_{\text{M}} - x_{\text{P+}} - W_{\text{J0}}} \right). \tag{6.291}$$

The specific drift region spreading resistance can be obtained by using

$$R_{\text{D,SP}} = \rho_{\text{D}} W_{\text{Cell}} \ln\left(\frac{W_{\text{Cell}}}{W_{\text{M}} - x_{\text{P+}} - W_{\text{J0}}} \right) + \rho_{\text{D}}(t - W_{\text{T}} - x_{\text{P+}} - W_{\text{J0}}), \tag{6.292}$$

where t is the thickness of the drift region below the P^+ shielding region and W_T and W_M are the widths of the trench and mesa regions, respectively, as shown in the figure.

The contribution to the resistance from the N^+ substrate is given by

$$R_{\text{SUB,SP}} = \rho_{\text{subs}} t_{\text{subs}}, \tag{6.293}$$

where ρ_{subs} and t_{subs} are the resistivity and thickness of the substrate, respectively. A typical value for this contribution is $4 \times 10^{-4}\ \Omega\ \text{cm}^2$ for silicon carbide due to the relatively high (0.02 Ω cm) resistivity of available N^+ substrates.

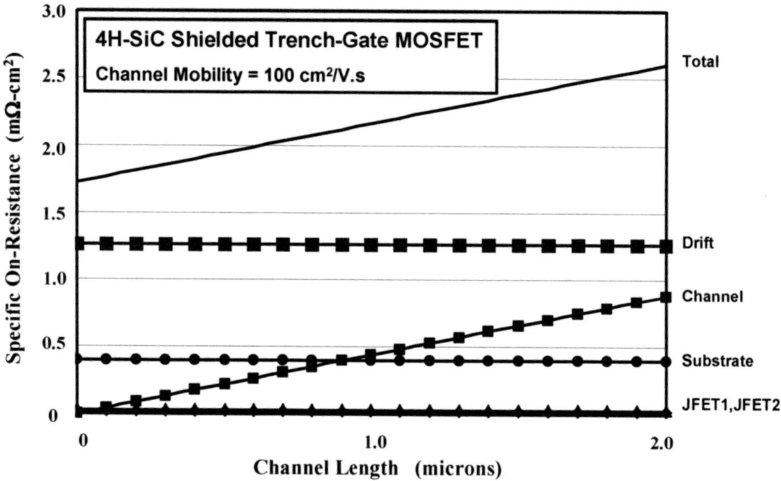

Fig. 6.162 Components of the on-resistance within a 4H-SiC shielded trench-gate MOSFET structure

As an example, the specific on-resistance for the 4H-SiC shielded trench-gate MOSFET structure with a drift region doping concentration of 1×10^{16} cm^{-3} and a thickness of 20 μm (capable of supporting 3,000 V) is shown in Fig. 6.162. A gate oxide thickness of 500 Å was used here with a trench width (W_T) of 0.25 μm and a mesa width (W_M) of 1.25 μm. The effective gate drive voltage ($V_G - V_{TH}$) was assumed to be 5 V based upon a gate voltage of 10 V and a threshold voltage of 5 V. This threshold voltage is consistent with the model for the inversion-mode 4H-SiC MOSFET. The JFET region doping concentration was enhanced to 5×10^{16} cm^{-3} to reduce its resistance contribution. The various components of the on-resistance are shown in Fig. 6.162 when a channel inversion layer mobility of 100 cm^2 V^{-1} s^{-1} was assumed. Inversion layer mobility larger than this has been experimentally observed with silicon dioxide as the gate dielectric.[41] The channel resistance increases with increasing channel length. The other components are unaffected by the increase in channel length because the cell pitch remains unaltered. It can be seen that the drift region resistance is dominant here while the JFET resistances are very small. Due to the shielding of the P-base region, the channel length for the shielded trench-gate MOSFET structure can be reduced when compared with that for the unshielded conventional trench-gate MOSFET structure. Consequently, with a channel length of 0.4 μm, a total specific on-resistance of 1.9 mΩ cm^2 is obtained for the shielded structure, which is within two times the ideal specific on-resistance of the drift region.

Simulation Example

To gain insight into the operation of the shielded trench-gate 4H-SiC power MOSFET structure, two-dimensional numerical simulations are described here for a structure with a drift region doping concentration of 1×10^{16} cm^{-3} and a thickness of 20 μm corresponding to a parallel-plane breakdown voltage of 3,000 V. The device had a gate oxide thickness of 500 Å. To take advantage of the reduced channel length in the shielded structure, the depth of the P-base region was reduced to 0.5 μm. The N$^+$ source region had a depth of 0.1 μm leading to a channel length of 0.4 μm. A trench width (W_T) of 0.55 μm was chosen with a mesa width (W_M) of 1.25 μm, resulting in a cell pitch of 1.8 μm. The P$^+$ region located at the bottom of the 1.5-μm deep trench had a junction depth of 0.2 μm, resulting in a t_{P+} parameter of 0.4 μm. The doping concentration of the JFET region was increased to 5×10^{16} cm^{-3} to reduce the zero-bias depletion widths of the junctions.[40]

The blocking capability of the shielded trench-gate 4H-SiC power MOSFET structure was investigated by maintaining zero gate bias while increasing the drain voltage. It was found that the device could sustain a drain bias of up to 3,000 V without any reach-through current despite the narrow width of the P-base region.

A serious problem for the conventional trench-gate silicon carbide MOSFET structure is associated with the high electric field generated at the gate oxide interface at the bottom of the trench. In order for the channel to span the P-base region, it is necessary for the trench to penetrate through the P-base region into the N-drift region. This exposes the gate oxide located at the bottom of the trench to the high electric field developed in the silicon carbide drift region. This problem is overcome in the shielded trench-gate structure. A three-dimensional view of the electric field distribution for the shielded trench-gate 4H-SiC MOSFET structure is shown in Fig. 6.163 for a drain bias of 3,000 V with zero bias applied to the gate electrode. The maximum electric field is generated at the junction between the P$^+$ shielding region and the N-drift region. The field distribution shown in this figure indicates that the electric field at the junction is close to the breakdown field strength for 4H-SiC (about 3×10^6 V cm^{-1}). Concurrently, the electric field at the P-base region is greatly reduced to about 1×10^6 V cm^{-1} and it is even smaller in the vicinity of the gate oxide. These results demonstrate the ability to fully utilize the breakdown field strength of the semiconductor without problems of rupture or reliability for the gate oxide.

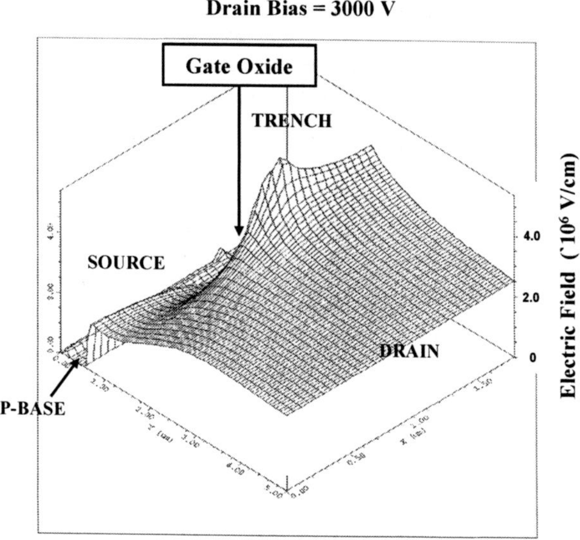

Fig. 6.163 Electric field distribution in the shielded trench-gate 4H-SiC MOSFET structure

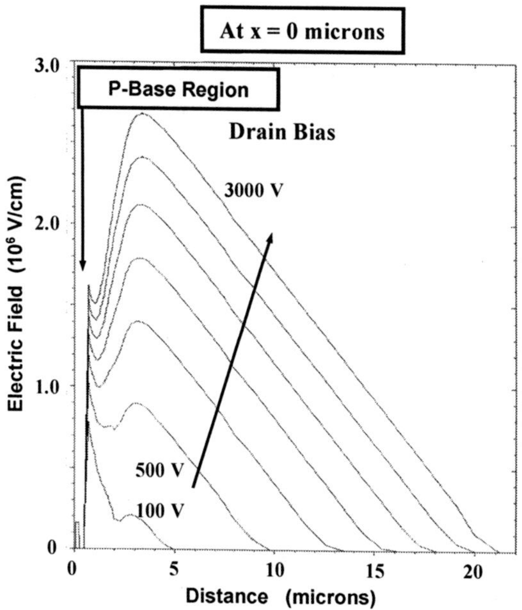

Fig. 6.164 Electric field in the shielded trench-gate 4H-SiC MOSFET structure

From Fig. 6.163, it can be seen that the electric field at the P-base region is significantly reduced when compared with the maximum electric field generated in the semiconductor. The largest electric field at the P-base junction occurs in the middle of the mesa region ($x = 0$ in Fig. 6.160). The behavior of this electric field is shown in Fig. 6.164 for drain bias up to 3,000 V. It can be observed that the formation of a potential barrier due to the presence of the JFET region suppresses the electric field at the P-base region to about one half of the electric field generated in the drift region. This is beneficial for preventing reach-through even with a narrow width for the P-base region.

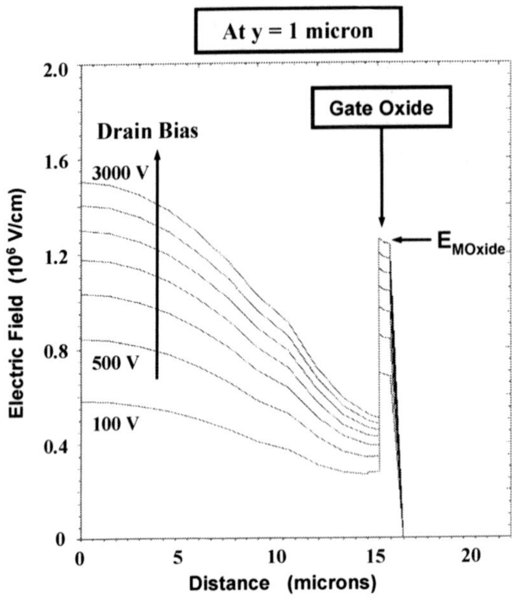

Fig. 6.165 Electric field in the shielded trench-gate 4H-SiC MOSFET structure

The reduction of the electric field in the gate oxide by the shielding provided by the JFET region can be observed in the electric field profile shown in Fig. 6.165 for various drain bias voltages. This profile was taken at a depth of 1 μm from the surface at a location below the P-base region. It can be seen that the electric field in the oxide is only 1.3×10^6 V cm^{-1} – far below its rupture strength – even when the drain bias reaches 3,000 V. The effective screening of the gate oxide in the shielded trench-gate structure from the high electric fields in the drift region provide wide latitude in choice of the gate oxide thickness and fabrication processes.

The on-state voltage drop for the shielded trench-gate 4H-SiC power MOSFET structure is determined by its specific on-resistance. The simulations of the specific on-resistance for the shielded trench-gate 4H-SiC MOSFET structure were performed with different values for the channel inversion layer mobility. The specific on-resistance was obtained by using a drain bias of 1 V and sweeping the gate bias to 20 V to obtain the transfer characteristic. A drain bias of 1 V was used for these simulations because the on-state voltage drop for these structures is typically in this range. The transfer characteristics obtained by using this method are shown in Fig. 6.166 for the device with cell pitch of 1.8 μm. From the transfer characteristics, it can be seen that the threshold voltage that determines the on-resistance is approximately 5 V (allowing operation of this device with a gate bias of 10 V). The impact of changing the channel inversion layer mobility is also shown in this figure. As expected, there is a reduction in the drain current as the channel mobility is reduced.

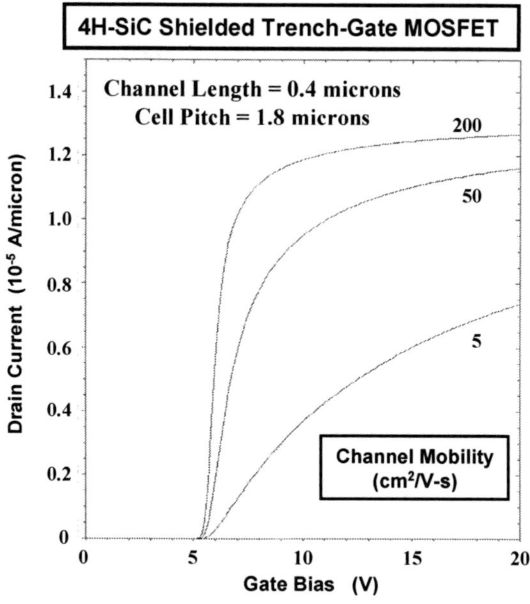

Fig. 6.166 Transfer characteristics of the 4H-SiC shielded trench-gate MOSFET structure

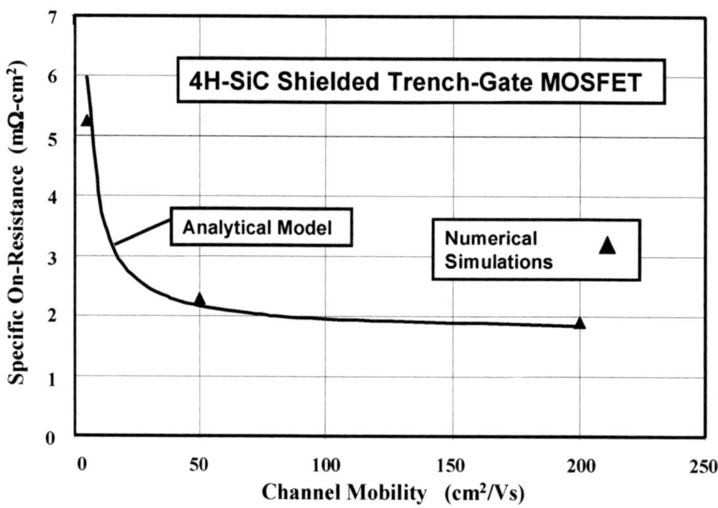

Fig. 6.167 Specific on-resistance for the 4H-SiC shielded trench-gate MOSFETs

The specific on-resistances obtained from the on-state simulations are compared in Fig. 6.167 with the analytically calculated values obtained by using the model developed in the previous section. It can be seen that the predictions of the analytical model are in excellent agreement with the values obtained from the simulations. These results provide confidence in the ability to use the analytical model during the design of the shielded trench-gate 4H-SiC power MOSFET structure. From the figure, it can be concluded that a channel mobility of 50 $cm^2 V^{-1} s^{-1}$ is satisfactory for achieving a low specific on-resistance for a 3-kV shielded trench-gate 4H-SiC power MOSFET structure with cell pitch of 1.8 μm.

Experimental results on an inversion-mode trench-gate 4H-SiC power MOSFET structure with a P^+ shielding region incorporated at the bottom of the trench[42] have been reported by fabricating the devices using 50-μm thick N-type drift regions with doping concentration of 8.5×10^{14} cm^{-3}. An N-type layer with doping concentration of 2×10^{17} cm^{-3} was grown on the drift region to provide the enhanced doping in the JFET region. The P-base region was formed by growth of a 1-μm thick P-type layer with doping concentration of 2×10^{17} cm^{-3} followed by the growth of an N^+ source region with doping concentration of 1×10^{19} cm^{-3}. The P^+ shielding regions were formed (after etching the trenches) with a junction depth of 0.8 μm by using aluminum ion implantation with a dose of 4×10^{13} cm^{-2}. The authors used two sacrificial thermal oxidation steps to remove any surface residue remaining from the implant anneal step. This may have also removed any P-type doping on the trench sidewalls. A relatively thick gate oxide of 0.275 μm was prepared by the thermal oxidation of a deposited layer of polysilicon into the trench to avoid the nonuniformity of thermally grown oxide. The thick gate oxide resulted in a high threshold voltage of 40 V. The devices exhibited a breakdown voltage of 3,000 V by the use of a JTE edge termination. A specific on-resistance of 120 $m\Omega\, cm^2$ was observed at a gate bias of 100 V. The relatively high value is due to the poor channel density in the cell design, the large gate oxide thickness, and the low inversion layer channel mobility (2 $cm^2 V^{-1} s^{-1}$) observed by the authors. However, these results demonstrated the ability to support high voltage in the drift region without encountering gate oxide rupture in a trench-gate device. The authors subsequently reported[43] the fabrication of devices using 115-μm thick N-type drift regions with doping concentration of 7.5×10^{14} cm^{-3}. These devices exhibited a specific on-resistance of 228 $m\Omega\, cm^2$ at a gate bias of 40 V. The epitaxial layer was stated to be capable of supporting 14 kV although the actual measured breakdown voltage was only 5 kV. These results indicate the need for further optimization of the structure.

6.23 Summary

The physics of operation of the power MOSFET structure has been discussed in this chapter. The evolution of the device structural design from the planar-gate VD-MOSFET architecture to the trench-gate power U-MOSFET architecture has allowed significant reduction of the specific on-resistance, especially for devices designed to support low blocking voltages. Since the specific on-resistance now approaches the ideal value, further improvements in the performance of power MOSFET devices have focused on reduction of the input gate capacitance and the gate charge. The planar SSCFET structure provides one such approach for achieving an excellent device figure of merit for high frequency circuit applications. These silicon structures are expected to provide satisfactory performance in low-voltage (<100 V) power electronic applications. Further, improvements to the on-resistance for these silicon devices can be obtained by using charge coupling with a metal electrode located within trenches[44,45] for devices with breakdown voltages below 200 V or by using superjunctions[46] for devices with breakdown voltages above 200 V. A detailed discussion of these structures is beyond the scope of this book.

A significant reduction of the specific on-resistance for power MOSFET devices can be achieved by replacing silicon with wide band-gap semiconductors. The most promising semiconductor for power devices is silicon carbide, especially the 4H-polytype. With 4H-SiC, it is possible to reduce the specific on-resistance by over a factor of 1,000 times in switches designed for operation at above 1,000 V. The commercialization of this technology is expected to result in improvements in the size, weight, and efficiency of high power electronic systems in the future.

Problems

6.1 Determine the ideal specific on-resistances for n-channel silicon power MOSFET structures with breakdown voltages of 30, 60, 100, 500, and 1,000 V. Take into account the variation of the critical electric field and mobility with doping concentration. Compare these values with those obtained by using (6.2).

6.2 Calculate the P-base doping concentration (assuming it is uniformly doped) of an n-channel silicon power MOSFET structure to obtain a threshold voltage of 2 V. The gate oxide thickness is 500 Å. The fixed charge in the gate oxide is 2×10^{11} cm^{-2}. Assume N^{+} polysilicon with a doping concentration of 1×10^{20} cm^{-3} is used as the gate electrode.

6.3 Determine the drift region doping concentration and thickness for an n-channel silicon power MOSFET structure when the structure is designed to support 100 V, assuming all the blocking voltage is supported by the drift region. Assume one-dimensional (parallel-plane) breakdown voltage is achievable in this case.

6.4 Determine the drift region doping concentration and thickness for an n-channel silicon power MOSFET structure when the structure is designed to support 100 V, assuming all the blocking voltage is supported by the drift region. In this case, assume that the breakdown voltage is limited by the edge termination to 80% of the parallel-plane value.

6.5 Calculate the depletion width within the P-base region of the n-channel silicon power MOSFET structure defined in Problem 6.4 when the structure is supporting 100 V.

6.6 Determine the optimum width for the gate electrode for a linear cell topology for the n-channel silicon power MOSFET structure defined in Problem 6.4 using the following parameters (a) inversion mobility of 450 $cm^2 V^{-1} s^{-1}$, (b) accumulation mobility of 1,000 $cm^2 V^{-1} s^{-1}$, (c) JFET region doping concentration of 3×10^{16} cm^{-3}, (d) cell polysilicon window width of 5 μm, (e) N^+ source length of 2.5 μm and sheet resistance of 10 Ω sq^{-1}, (f) contact resistance of 1×10^{-6} Ω cm^2 for both the N^+ source and the drain, (g) N^+ source contact width of 1 μm, (h) N^+ substrate resistivity of 0.002 Ω cm and thickness of 500 μm, (i) gate bias of 5 V, (j) K factor for accumulation spreading of 0.6, (k) N^+ source depth of 1 μm, and (l) P-base depth of 3 μm. What are the absolute and percentage contributions from each of the on-resistance components for the optimum design?

6.7 Calculate the specific on-resistance (in mΩ cm^2) for the ideal drift region for blocking 100 V. Compare the optimum design in Problem 6.6 with the ideal case by taking the ratio of specific on-resistances.

6.8 Repeat Problem 6.6 without the enhanced doping in the JFET region.

6.9 Calculate the specific input capacitance ($C_{IN,SP}$) for the optimum cell design (in nF cm^{-2}) in Problem 6.6.

6.10 Calculate the specific drain overlap capacitance ($C_{GD,SP}$) for the optimum cell design (in nF cm^{-2}) in Problem 6.6 at a drain bias of 10 V.

6.11 Determine the figures of merit [$R_{DSon,SP} \times C_{IN,SP}$] and [$R_{DSon,SP} \times C_{GD,SP}$] (in mΩ nF) for the optimum cell design in Problem 6.6.

6.12 Calculate the specific input capacitance ($C_{IN,SP}$) for the optimum cell design (in nF cm^{-2}) in Problem 6.8.

6.13 Calculate the specific drain overlap capacitance ($C_{GD,SP}$) for the optimum cell design (in nF cm^{-2}) in Problem 6.8 at a drain bias of 10 V.

6.14 Determine the figures of merit [$R_{DSon,SP} \times C_{IN,SP}$] and [$R_{DSon,SP} \times C_{GD,SP}$] (in mΩ nF) for the optimum cell design in Problem 6.8.

6.15 Repeat Problem 6.6 with the ALL cell topology using a bar width of 2 μm.

6.16 Calculate the specific input capacitance ($C_{IN,SP}$) for the optimum cell design (in nF cm^{-2}) in Problem 6.15.

6.17 Calculate the specific drain overlap capacitance ($C_{GD,SP}$) for the optimum cell design (in nF cm^{-2}) in Problem 6.15 at a drain bias of 10 V.

6.18 Determine the figures of merit [$R_{DSon,SP} \times C_{IN,SP}$] and [$R_{DSon,SP} \times C_{GD,SP}$] (in mΩ nF) for the optimum cell design in Problem 6.15.

6.19 Determine the specific on-resistance for a linear cell n-channel silicon U-MOSFET structure with a trench width of 1 μm and a mesa width of 3 μm designed to support 100 V. Assume that the breakdown voltage is limited by the edge termination to 80% of the parallel-plane value. The P-base doping concentration (assuming it is uniformly doped) is chosen to obtain a threshold voltage of 2 V. The gate oxide thickness is 500 Å. The fixed charge in the gate oxide is 2×10^{11} cm^{-2}. Assume N$^+$ polysilicon with a doping concentration of 1×10^{20} cm^{-3} is used as the gate electrode. Use the following parameters (a) inversion mobility of 450 cm^2 V^{-1} s^{-1}, (b) accumulation mobility of 1,000 cm^2 V^{-1} s^{-1}, (c) N$^+$ source length of 1 μm and sheet resistance of 10 Ω sq^{-1}, (d) contact resistance of 1×10^{-6} Ω cm^2 for both the N$^+$ source and the drain, (e) N$^+$ source contact width of 0.5 μm, (f) N$^+$ substrate resistivity of 0.002 Ω cm and thickness of 500 μm, (g) gate bias of 5 V, (h) K factor for accumulation spreading of 0.6, (i) N$^+$ source depth of 0.5 μm, (j) P-base depth of 1.5 μm, (k) trench depth of 2 μm, (l) mesa width is 4 μm, and (m) trench width is 1 μm. What are the absolute and percentage contributions from each of the on-resistance components?

6.20 Calculate the specific input capacitance ($C_{IN,SP}$) for the U-MOSFET structure (in nF cm^{-2}) in Problem 6.19.

6.21 Calculate the specific drain overlap capacitance ($C_{GD,SP}$) for the U-MOSFET structure (in nF cm^{-2}) in Problem 6.19 at a drain bias of 10 V.

6.22 Determine the figures of merit [$R_{DSon,SP} \times C_{IN,SP}$] and [$R_{DSon,SP} \times C_{GD,SP}$] (in mΩ nF) for the U-MOSFET structure in Problem 6.19.

6.23 Obtain the specific gate charge for the optimum design of the VD-MOSFET structure in Problem 6.6. Use an on-state current density of 200 A cm^{-2} and a drain supply voltage of 80 V. Provide values for each of the components of the specific gate charge (in nC cm^{-2}).

6.24 Determine the figures of merit [$R_{DSon,SP} \times Q_{GD,SP}$] and [$R_{DSon,SP} \times Q_{SW,SP}$] (in mΩ nC) for the VD-MOSFET structure in Problem 6.23.

6.25 Obtain the specific gate charge for the U-MOSFET structure in Problem 6.19. Use an on-state current density of 200 A cm^{-2} and a drain supply voltage of 80 V. Provide values for each of the components of the specific gate charge (in nC cm^{-2}).

6.26 Determine the figures of merit [$R_{DSon,SP} \times Q_{GD,SP}$] and [$R_{DSon,SP} \times Q_{SW,SP}$] (in mΩ nC) for the U-MOSFET structure in Problem 6.19.

6.27 Determine the optimum area for the VD-MOSFET structure in Problem 6.6 to minimize power losses at an operating frequency of 200 kHz. Use a gate bias of 5 V, an on-state current of 10 A, and a duty cycle of 50%.

6.28 Determine the optimum area for the U-MOSFET structure in Problem 6.19 to minimize power losses at an operating frequency of 200 kHz. Use a gate bias of 5 V, an on-state current of 10 A, and a duty cycle of 50%.

6.29 Determine the specific on-resistance for a linear cell p-channel silicon U-MOSFET structure with a trench width of 1 μm and a mesa width of 3 μm designed to support 100 V. Assume that the breakdown voltage is limited by the edge termination to 80% of the parallel-plane value. The N-base doping concentration (assuming it is uniformly doped) is chosen to obtain a threshold voltage of 2 V. The gate oxide thickness is 500 Å. The fixed charge in the gate oxide is 2×10^{11} cm^{-2}. Assume P$^+$ polysilicon with a doping concentration of 1×10^{20} cm^{-3} is used as the gate electrode. Use the following parameters (a) inversion mobility of 150 cm^2 V^{-1} s^{-1}, (b) accumulation mobility of 330 cm^2 V^{-1} s^{-1}, (c) P$^+$ source length of 1 μm and sheet resistance of 30 Ω sq^{-1}, (d) contact resistance of 1×10^{-6} Ω cm^2 for both the P$^+$ source and the drain, (e) P$^+$ source contact width of 1 μm, (f) P$^+$ substrate resistivity of 0.006 Ω cm and thickness of 500 μm, (g) gate bias of 5 V, (h) K factor for accumulation spreading of 0.6, (i) P$^+$ source depth of 0.5 μm, (j) N-base depth of 1.5 μm, (k) trench depth of 2 μm, (l) mesa width is 4 μm, and (m) trench width is 1 μm. What are the absolute and percentage contributions from each of the on-resistance components?

6.30 Determine the ideal specific on-resistances for n-channel 4H-SiC power MOSFET structures with breakdown voltages of 300, 600, 1,000, and 5,000 V. Take into account the variation of the critical electric field and mobility with doping concentration.

6.31 Calculate the P-base doping concentration (assuming it is uniformly doped) of an n-channel 4H-SiC planar power MOSFET structure to obtain a threshold voltage of 5 V. The gate oxide thickness is 500 Å. The fixed charge in the gate oxide is 2×10^{11} cm^{-2}. Assume N^+ polysilicon with a doping concentration of 1×10^{20} cm^{-3} is used as the gate electrode.

6.32 Determine the drift region doping concentration and thickness for an n-channel 4H-SiC planar power MOSFET structure when the structure is designed to support 1,000 V, assuming all the blocking voltage is supported by the drift region. Assume one-dimensional breakdown voltage is achievable in this case.

6.33 Calculate the depletion width within the P-base region of the above n-channel 4H-SiC planar power MOSFET structure when the structure is supporting 1,000 V.

6.34 Determine the optimum width for the gate electrode for a linear cell topology for the above n-channel 4H-SiC planar power MOSFET structure using the following parameters (a) inversion mobility of 50 cm^2 V^{-1} s^{-1}, (b) accumulation mobility of 100 cm^2 V^{-1} s^{-1}, (c) cell polysilicon window width of 5 μm, (d) N^+ source length of 2.5 μm and sheet resistance of 50 Ω sq^{-1}, (e) contact resistance of 1×10^{-5} Ω cm^2 for both the N^+ source and the drain, (f) N^+ source contact width of 1 μm, (g) N^+ substrate resistivity of 0.02 Ω cm and thickness of 200 μm, (h) gate bias of 15 V, (i) K factor for accumulation spreading of 0.6, (j) P^+ region depth of 1 μm, and (k) channel length of 1 μm. What are the absolute and percentage contributions from each of the on-resistance components for the optimum design?

6.35 Calculate the specific on-resistance (in mΩ cm^2) for the ideal drift region in 4H-SiC for blocking 1,000 V. Compare the optimum design in Problem 6.34 with the ideal case by taking the ratio of specific on-resistances.

References

[1] D.A. Grant and J. Gowar, "Power MOSFETs: Theory and Applications", Wiley, New York, 1989.

[2] D. Kahng and M.M. Atalla, "Silicon–Silicon Dioxide Field Induced Surface Devices", IRE Solid State Device Research Conference, 1960.

[3] D. Kahng, "A Historical Perspective on the Development of MOS Transistors and Related Devices", IEEE Transactions on Electron Devices, Vol. ED-23, pp. 655–665, 1976.

[4] K.P. Lisiak and J. Berger, "Optimization of Non-Planar Power MOS Transistors", IEEE Transactions on Electron Devices, Vol. ED-25, pp. 1229–1234, 1978.

[5] M.N. Darwish and K. Board, "Optimization of Breakdown Voltage and On-Resistance of VDMOS Transistors", IEEE Transactions on Electron Devices, Vol. ED-31, pp. 1769–1773, 1984.
[6] J. Zeng et al., "An Ultra Dense Trench-Gated Power MOSFET Technology Using a Self-Aligned Process", IEEE International Symposium on Power Semiconductor Devices and ICs, pp. 147–150, 2001.
[7] C. Hu, M.-H. Chi, and V.M. Patel, "Optimum Design of Power MOSFETs", IEEE Transactions on Electron Devices, Vol. ED-31, pp. 1693–1700, 1984.
[8] H.R. Chang and B.J. Baliga, "Numerical and Experimental Analysis of 500 V Power DMOSFET with an Atomic-Lattice-Layout", IEEE Device Research Conference, Abstract VB-5, 1989.
[9] S.M. Sze, "Physics of Semiconductor Devices", Wiley, New York, 1981.
[10] B.G. Streetman and S.K. Banerjee, "Solid State Electronic Devices", 6th Edition, Prentice Hall, Englewood Cliffs, NJ, 2006.
[11] C.G.B. Garrett and W.H. Brattain, "Physical Theory of Semiconductor Surfaces", Physical Review, Vol. 99, pp. 376–386, 1955.
[12] E.H. Nicollian and J.R. Brews, "MOS Physics and Technology", Wiley, New York, 1982.
[13] B.J. Baliga, "Silicon RF Power MOSFETs", World Scientific, Singapore, 2005.
[14] S.K. Ghandhi, "VLSI Fabrication Principles", pp. 155–157, Wiley, New York, 1983.
[15] S.C. Sun and J.D. Plummer, "Modeling of the On-Resistance of LDMOS, VDMOS, and VMOS Power Transistors", IEEE Transactions on Electron Devices, Vol. ED-27, pp. 356–367, 1980.
[16] R.K. Williams et al., "A 30 V P-Channel Trench Gated DMOSFET with 900 mW cm^2 Specific On-Resistance at 2.7 V", IEEE International Symposium on Power Semiconductor Devices and ICs, pp. 53–56, 1996.
[17] A.R. Hambley, "Electrical Engineering", 3rd Edition, pp. 553–556, Prentice Hall, Englewood Cliffs, NJ, 2005.
[18] B.J. Baliga, "MOSFET Devices Having Linear Transfer Characteristics When Operating in Velocity Saturation Mode and Methods of Forming and Operating Same", U.S. Patent Number 6,545,316, April 8, 2003.
[19] Y. Taur and T.K. Ning, "Fundamentals of Modern VLSI Devices", Cambridge University Press, Cambridge, UK.
[20] C. Canali, G. Majni, R. Minder, and G. Ottaviani, "Electron and Hole Drift Velocity Measurements in Silicon", IEEE Transactions on Electron Devices, Vol. ED-22, pp. 1045–1047, 1975.
[21] B.J. Baliga, "Power Semiconductor Device Figure of Merit for High Frequency Applications", IEEE Electron Device Letters, Vol. EDL-10, pp. 455–457, 1989.
[22] B.J. Baliga and D. Alok, "Paradigm Shift in Planar Power MOSFET Technology", Power Electronics Technology Magazine, pp. 24–32, November 2003.
[23] B.J. Baliga, "Vertical Power Devices Having Retrograde Doped Transition Regions Therein", U.S. Patent 6,784,486, August 31, 2004.

[24] B.K. Bose, "Power Electronics and Variable Frequency Drives", IEEE Press, New York, 1997.
[25] B.J. Baliga and J.P. Walden, "Improving the Reverse Recovery of Power MOSFET Integral Diodes by Electron Irradiation", Solid-State Electronics, Vol. 26, pp. 1133–1141, 1983.
[26] R.K. Williams et al., "A 20 V p-Channel with 650 $\mu\Omega$ cm^2 at V_{GS} = 2.7 V", IEEE International Symposium on Power Semiconductor Devices and ICs, pp. 411–414, 1998.
[27] D. Kinzer, D. Asselanis, and R. Carta, "Ultra-Low R_{dson} 12 V p-Channel Trench MOSFET", IEEE International Symposium on Power Semiconductor Devices and ICs, pp. 303–306, 1999.
[28] T. Syau, P. Venkatraman, and B.J. Baliga, "Comparison of Ultra-Low Specific On-Resistance U-MOSFET Structures: The ACCUFET, EXTFET, INVFET, and Conventional U-MOSFETs", IEEE Transactions on Electron Devices, Vol. ED-41, pp. 800–808, 1994.
[29] B.J. Baliga, "Silicon Carbide Power Devices", World Scientific, Singapore, 2006.
[30] A. Mihaila et al., "Static and Dynamic Behavior of SiC JFET/Si MOSFET Cascode Configuration for High-Performance Power Switches", Silicon Carbide and Related Materials – 2001, Material Science Forum, Vol. 389–393, pp. 1239–1242, 2002.
[31] B.J. Baliga and M. Bhatnagar, "Methods of Fabricating Silicon Carbide Field Effect Transistors", U.S. Patent 5,322,802, June 21, 1994.
[32] J.N. Shenoy, J.A. Cooper, and M.R. Melloch, "High Voltage Double-Implanted Power MOSFETs in 6H-SiC", IEEE Electron Device Letters, Vol. 18, pp. 93–95, 1997.
[33] D. Alok, P. McLarty, and B.J. Baliga, "Electrical Properties of Thermal Oxide Grown on N-type 6H-Silicon Carbide", Applied Physics Letters, Vol. 64, pp. 2845–2846, 1994.
[34] J.B. Casady et al., "4H-SiC Power Devices: Comparative Overview of UMOS, DMOS, and GTO Device Structures", Materials Research Society Symposium Proceedings, Vol. 483, pp. 27–38, 1998.
[35] B.J. Baliga, "Silicon Carbide Semiconductor Devices Having Buried Silicon Carbide Conduction Barrier Layers Therein", U.S. Patent 5,543,637, August 6, 1996.
[36] S.T. Sheppard, M.R. Melloch, and J.A. Cooper, "Characteristics of Inversion-Channel and Buried-Channel MOS Devices in 6H-SiC", IEEE Transactions on Electron Devices, Vol. 41, pp. 1257–1264, 1994.
[37] N. Thapar and B.J. Baliga, "Analytical Model for the Threshold Voltage of Accumulation Channel MOS-Gated Devices", Solid-State Electronics, Vol. 42, pp. 1975–1979, 1998.
[38] P.M. Shenoy and B.J. Baliga, "The Planar 6H-SiC ACCUFET", IEEE Electron Device Letters, Vol. 18, pp. 589–591, 1997.

[39] J.W. Palmour et al., "4H-Silicon Carbide Power Switching Devices", Silicon Carbide and Related Materials 1995, Institute of Physics Conference Series, Vol. 142, pp. 813–816, 1996.
[40] B.J. Baliga, "Silicon Carbide Switching Device with Rectifying Gate", U.S. Patent 5,396,085, March 7, 1995.
[41] S. Sridevan and B.J. Baliga, "Lateral n-Channel Inversion-Mode 4H-SiC MOSFETs", IEEE Electron Device Letters, Vol. 19, pp. 228–230, 1998.
[42] Y. Li, J.A. Cooper, and M.A. Capano, "High Voltage (3 kV) UMOSFETs in 4H-SiC", IEEE Transactions on Electron Devices, Vol. 49, pp. 972–975, 2002.
[43] Y. Sui, T. Tsuji, and J.A. Cooper, "On-State Characteristics of SiC Power UMOSFETs on 115 μm Drift Layers", IEEE Electron Device Letters, Vol. 26, pp. 255–257, 2005.
[44] B.J. Baliga, "Power Semiconductor Devices Having Improved High Frequency Switching and Breakdown Characteristics", U.S. Patent 5,998,833, December 7, 1999.
[45] B.J. Baliga, "The Future of Power Semiconductor Technology", Proceedings of the IEEE, Vol. 89, pp. 822–832, 2001.
[46] L. Lorenz et al., "COOLMOS – A New Milestone in High Voltage Power MOS", IEEE International Symposium on Power Semiconductor Devices and ICs, pp. 3–10, 1999.

Chapter 7

Bipolar Junction Transistors

The bipolar junction transistor was invented in 1947 by Shockley, Brattain, and Bardeen.[1,2] Using insights gained from point-contact transistors,[3] the group at Bell Telephone Laboratories found that a much more stable device could be created by using junctions.[4] The multifold advantages of replacing vacuum tubes with a solid-state device rapidly promulgated its use in applications. The availability of single crystal silicon technology with greater purity levels together with the capability to form diffused junctions enabled the fabrication of transistors with high blocking voltage capability. Several decades of effort were undertaken to improve the photolithography techniques and epitaxial deposition capability culminating in the availability of 500-V Darlington power transistors in the late 1970s.[5] During this effort, it was recognized that one of major shortcomings of the power bipolar transistor is its poor current gain. Consequently, these devices were soon eclipsed by the invention[6,7] and rapid commercialization of the insulated gate bipolar transistor (IGBT). The high input impedance of the IGBT simplified the gate control circuit enabling its integration. Together with the larger on-state current density and improved ruggedness of the IGBT when compared with the power bipolar transistor, this enabled major improvements in the size, cost, and performance of power electronic systems for numerous applications. The physics of operation of the IGBT and its applications is discussed in Chap. 9.

In this chapter, the basic structure and operation of the bipolar junction transistor are discussed. The bipolar transistor operates by the injection of minority carriers across one junction, with these carriers collected across a second junction located in close proximity. The physics of current transport by the injection process in the bipolar transistor is reviewed in this chapter to provide a basic understanding of the factors that determine the current gain. From the perspective of power devices, the influence of high-level injection created by operation at elevated current density and the limitations produced by the need to support high

B.J. Baliga, *Fundamentals of Power Semiconductor Devices*, doi: 10.1007/978-0-387-47314-7_7,

voltages are analyzed in detail. Based upon analysis of the minority and majority carrier distributions within the structure, the output characteristics are then generated. The impact of emitter current crowding in the on-state and during turn-off is also included in this chapter.

The bipolar power transistor is primarily used as a switch. The switching time intervals when turning on and turning off the structure are therefore of interest from an application's viewpoint. The presence of minority carrier stored charge within the device is demonstrated to limit the operating frequency of these devices. In addition, the safe operating area (SOA) of the bipolar transistor is shown to be constrained by the on-set of second breakdown phenomena.

Although the power bipolar transistor has been supplanted by the IGBT in all power applications, a good understanding of the physics of current transport in the bipolar transistor is essential for power device specialists. In addition, the IGBT contains a bipolar transistor within its structure. Consequently, the analysis of the IGBT structure requires the application of the physics and concepts introduced in this chapter.

7.1 Power Bipolar Junction Transistor Structure

The basic structure for an N–P–N bipolar power transistor is illustrated in Fig. 7.1. In addition to the N^+ emitter and P-base region for the conventional bipolar transistor,[8] the power transistor contains a lightly doped collector (N-drift) region to allow supporting high blocking voltages. When a positive bias is applied to the collector terminal of the device, the collector–base junction (J_1) becomes reverse biased and supports the voltage. The blocking voltage capability for the device is determined by the doping concentration and thickness of the N-drift region.

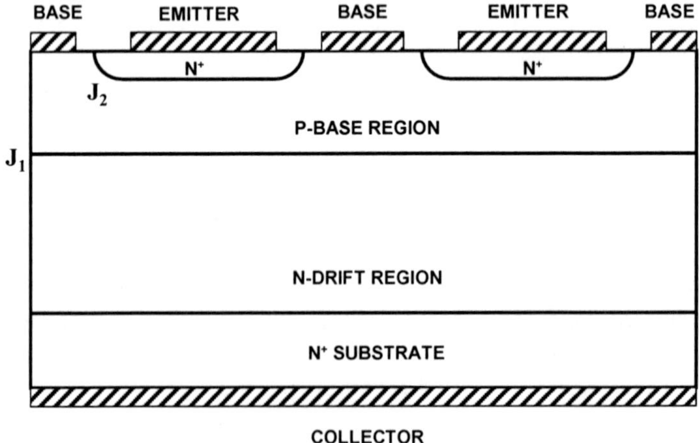

Fig. 7.1 Power bipolar N–P–N transistor structure

Current flow through the N–P–N bipolar transistor is induced by forward biasing the emitter–base junction (J_2) to initiate the injection of electrons. The injected electrons diffuse through the P-base region and arrive at the collector–base junction. When this junction is reverse biased, the electrons collected by the junction are swept through its depletion region producing a collector current. It can be shown that a small base current can produce a large collector current resulting in a substantial current gain. In conjunction with the much larger collector voltage when compared with the base voltage, this produces a large power gain as well.

The bipolar transistor can be switching from its on-state to the blocking state by reversing the bias applied to the base region. The reverse bias not only stops the injection of minority carriers from the emitter–base junction but also removes some of the charge stored in the base region. In the case of the power bipolar transistor, a substantial amount of charge is stored within the thick N-drift region as well. This prolongs the time taken for the transistor to begin supporting voltage, limiting its maximum frequency of operation.

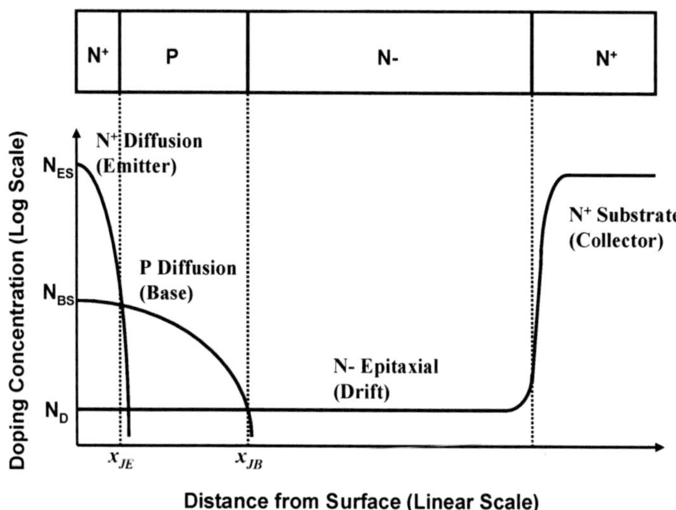

Fig. 7.2 Doping profile for the power bipolar N–P–N transistor structure

The vertical doping profile, extending through one of the N^+ emitter regions, for the power bipolar transistor is illustrated in Fig. 7.2. This profile is achieved by using starting material consisting of an N-type epitaxial layer grown on a heavily doped N-type substrate. The doping concentration and thickness of the epitaxial layer are chosen to obtain the desired voltage-blocking capability for the power bipolar transistor. The P-type base region is then ion implanted and diffused across the entire active area as shown in Fig. 7.1. However, the P-base region must be patterned at the edges of the device to form floating field rings to enhance the breakdown voltage. A mask is now used to define the locations of the N^+ emitter regions. The doping concentration and depth for the emitter must be carefully chosen to obtain a high-current gain as discussed later in this chapter. The doping and thickness of the P-base region, determined by the combination of the P-type and N^+ ion implants, are critical to both the blocking capability and the current gain for the power bipolar transistor. It is customary to interdigitate the emitter and base contacts as illustrated in Fig. 7.1 because the emitter current tends to concentrate at the periphery of the emitter regions due to a current crowding phenomenon.

7.2 Basic Operating Principles

The power bipolar transistor is most often used in the common-emitter circuit configuration as illustrated in Fig. 7.3. In this case, the emitter terminal of the transistor is a common element between the input and output side of the circuit. The input side of the circuit is controlled by the drive circuit, which contains two power supplies that can be used to turn on and turn off the transistor. The voltage source V_{BS1} is used to drive the transistor when it is operating in the current conduction mode while the voltage source V_{BS2} is used to turn off the transistor and maintain it in its voltage-blocking mode. The output side of the circuit contains a high-voltage source (V_{CS}) that delivers power to the load. The transfer of power from the voltage source to the load is controlled by the drive circuit using switches S_1 and S_2.

The bipolar transistor is operated in its current conduction mode by opening switch S_2 and closing switch S_1. This connects the input voltage source V_{BS1} across the base–emitter terminals of the bipolar power transistor through resistor R_{B1}. If the input source voltage exceeds the built-in potential (V_{bi} ~ 0.7 V) of the base–emitter junction, it becomes sufficiently forward biased to produce a base current given by

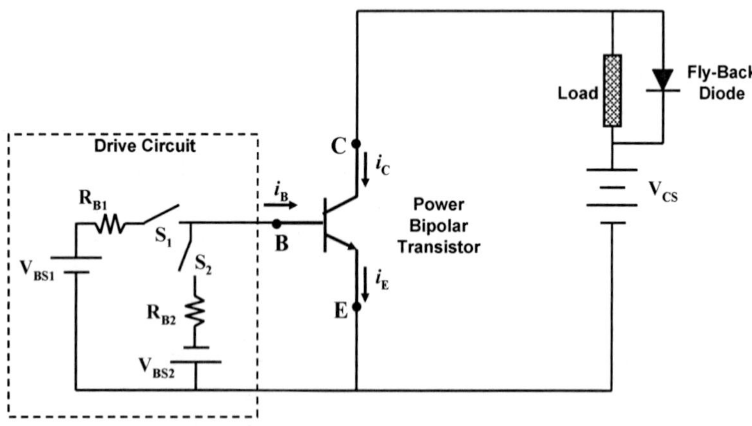

Fig. 7.3 Common-emitter configuration for an N–P–N bipolar transistor

$$i_{\mathrm{B}} = \frac{V_{\mathrm{BS1}} - V_{\mathrm{bi}}}{R_{\mathrm{B1}}}. \tag{7.1}$$

The base current flow is accomplished by the injection of minority carriers (electrons for an N–P–N transistor) from the N^+ emitter region into the P-base region. These minority carriers diffuse from the emitter–base junction (J_2) through the P-base region and are collected by the reverse-biased base–collector junction (J_1). The electrons captured by the base–collector junction are swept through its depletion region producing a collector current (i_C).

When the bipolar transistor is operating in its forward active mode with a reverse-biased base–collector junction, the collector and base currents are related by the *common-emitter current gain* called beta (β):

$$\beta = \frac{i_{\mathrm{C}}}{i_{\mathrm{B}}}. \tag{7.2}$$

Based upon the application of Kirchhoff's current law for the bipolar transistor as a node,

$$i_{\mathrm{E}} = i_{\mathrm{B}} + i_{\mathrm{C}}. \tag{7.3}$$

Combining this relationship with that for the common-emitter current gain,

$$i_{\mathrm{E}} = \left(1 + \frac{i_{\mathrm{C}}}{i_{\mathrm{B}}}\right) i_{\mathrm{B}} = (1+\beta) i_{\mathrm{B}}. \tag{7.4}$$

When the bipolar power transistor is operated as a switch, the power delivered to the load is proportional to ($i_C V_{CS}$), while that utilized from the input control circuit is proportional to ($i_B V_{BS1}$). The common-emitter power gain is then given by

$$G_{\mathrm{P,CE}} = \left(\frac{i_{\mathrm{C}} V_{\mathrm{CS}}}{i_{\mathrm{B}} V_{\mathrm{BS1}}}\right) = \beta \left(\frac{V_{\mathrm{CS}}}{V_{\mathrm{BS1}}}\right). \tag{7.5}$$

It is desirable to control a large collector current, which flows through the load, with a small base current to achieve a large power gain. This requires optimization of the bipolar transistor structure to obtain a large common-emitter current gain. The physical parameters within the bipolar transistor that determine the current gain are discussed in Sect. 7.3.

The bipolar transistor is sometimes used in the common-base circuit configuration as illustrated in Fig. 7.4. In this case, the base terminal of the transistor is used as a common element between the input and output side of the circuit. The input side of the circuit is controlled by the drive circuit, which contains two power supplies that can be used to turn on and turn off the transistor. The voltage source V_{BS1} is used to drive the transistor when it is operating on the current conduction mode while the voltage source V_{BS2} is used to turn off the transistor and maintain it in its voltage-blocking mode. The output side of the circuit contains a high-voltage source (V_{CS}) that delivers power to the load. The transfer of power from the voltage source to the load is controlled by the drive circuit using switches S_1 and S_2.

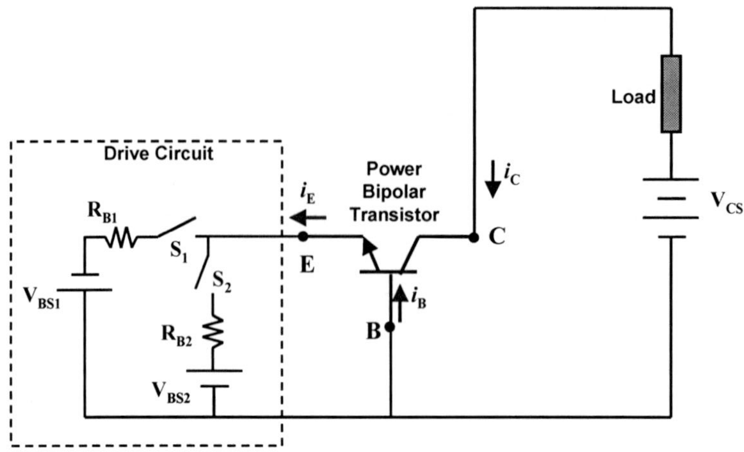

Fig. 7.4 Common-base configuration for an N–P–N bipolar transistor

When the bipolar transistor is operating in its forward active mode with a reverse-biased base–collector junction, the collector and emitter currents are related by the *common-base current gain* called alpha (α):

$$\alpha = \frac{i_C}{i_E}. \tag{7.6}$$

Since only a fraction of the emitter current is delivered to the collector, the common-base current gain is always less than unity. When the bipolar power transistor is operated as a switch in the common-base configuration, the power delivered to the load is proportional to ($i_C V_{CS}$), while that utilized from the input control circuit is proportional to ($i_E V_{BS1}$). The common-base power gain is then given by

$$G_{P,CB} = \left(\frac{i_C V_{CS}}{i_E V_{BS1}}\right) = \alpha\left(\frac{V_{CS}}{V_{BS1}}\right). \tag{7.7}$$

Although the common-base current gain is less than unity, power gain can still be achieved because the collector source voltage can be much larger than the base drive voltage.

The common-base current gain can be related to the common-emitter current gain by using (7.3):

$$\alpha = \frac{i_C}{i_B + i_C} = \frac{\beta i_B}{i_B + \beta i_B} = \frac{\beta}{1+\beta}. \tag{7.8}$$

In a similar manner, the common-emitter current gain can be related to the common-base current gain:

$$\beta = \frac{i_C}{i_E - i_C} = \frac{\alpha i_E}{i_E - \alpha i_E} = \frac{\alpha}{1-\alpha}. \tag{7.9}$$

7.3 Static Blocking Characteristics

In principle, the bipolar power transistor structure is capable of supporting voltage in the first and third quadrants of operation. In the first quadrant of operation with a positive bias applied to the collector electrode, the junction J_1 between the base and collector becomes reverse biased. Due to the incorporation of the N-drift region, this junction can be designed to support a high voltage. Devices with blocking voltages of over 1,200 V in the first quadrant have been developed. In the third quadrant of operation with a negative bias applied to the collector electrode, the junction J_2 between the emitter and base becomes reverse biased. Since this junction is formed between two relatively highly doped regions, the breakdown voltage is usually less than 50 V. For this reason, the power bipolar transistor is usually used as a power switch in a DC circuit with a positive collector power source.

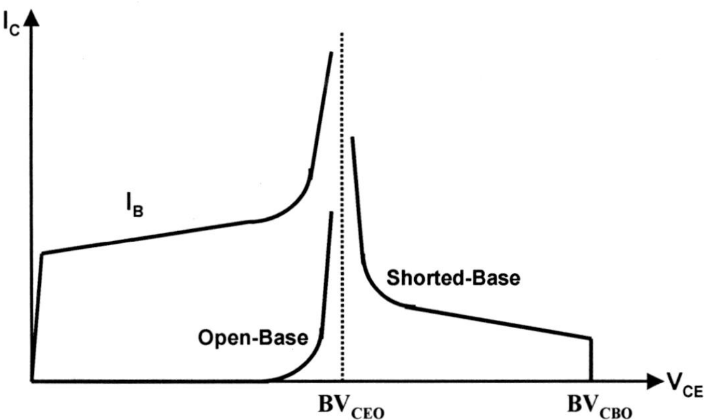

Fig. 7.5 Blocking characteristics for an N–P–N bipolar power transistor

The voltage that can be supported by the power bipolar transistor depends on the bias applied to the base and emitter terminals. A typical set of blocking characteristics are illustrated in Fig. 7.5. It can be observed that the blocking voltage depends upon how the base terminal is connected. The open-base breakdown voltage is much smaller than the blocking voltage capability when the base terminal is shorted to the emitter terminal by the base drive circuit. However, even for the shorted base case, the blocking voltage collapses to that for the open-base case if the collector current increases. The lower blocking voltage capability is also observed when a finite base current is supplied by the drive circuit.

Consequently, the actual blocking voltage rating for the power bipolar transistor is limited to the breakdown voltage of the open-base case.

7.3.1 Open-Emitter Breakdown Voltage

If the emitter terminal is open circuited, the device operates like a diode between the base and collector terminals. In this case, the maximum blocking voltage is determined by the breakdown voltage of the base–collector junction (J_1). The open-emitter breakdown voltage (BV_{CBO}) is governed by the doping concentration and thickness of the lightly doped portion of the collector (N-drift region) and the edge termination for the base–collector junction (J_1). As shown in Sect. 7.3.2, the actual maximum blocking voltage capability for the power bipolar transistor is a fraction of the open-base breakdown voltage. It is necessary to reduce the doping concentration and increase the thickness of the N-drift region to account for this difference. This has an adverse impact on the resistance of the N-drift region, which degrades the output characteristics as discussed later in this chapter. The increase in the thickness of the drift region also increases the turn-off time due to the larger stored charge.

7.3.2 Open-Base Breakdown Voltage

When the base terminal is open circuited and a positive bias is applied to the collector electrode, the base–emitter junction becomes forward biased and the base–collector junction becomes reverse biased. Most of the voltage is supported by the base–collector junction (J_1). However, the leakage current flowing through this junction must also flow through the base–emitter junction (J_2). Consequently,

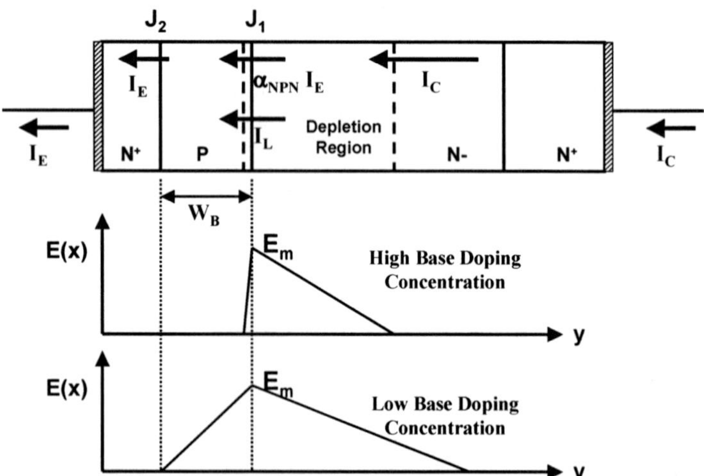

Fig. 7.6 Current transport and electric field profiles within an open-base N–P–N bipolar power transistor

the leakage current is amplified by the gain of the transistor. This reduces the maximum blocking voltage capability.

To analyze the open-base breakdown voltage, consider the transistor structure shown in Fig. 7.6 with a positive bias applied to the collector electrode. A depletion region forms across the reverse-biased base–collector junction (J_1) as indicated in the figure by the dashed lines. The minority carrier generation process produces a finite leakage current at this junction as discussed in Chaps. 2 and 3. Since the base terminal is open circuited, the leakage current must flow through the base–emitter junction.

The injection of minority carriers across the forward-biased base–emitter junction produces a current ($\alpha_{NPN}I_E$) at the collector junction as indicated in the figure. The currents flowing through the open-base transistor structure are interrelated:

$$I_C = \alpha_{NPN} I_E + I_L = I_E. \tag{7.10}$$

Consequently,

$$I_C = I_E = \frac{I_L}{1 - \alpha_{NPN}}. \tag{7.11}$$

From this expression, it can be concluded that the collector (and emitter) current will become very large as the alpha of the transistor approached unity. The criterion for breakdown for the open-base transistor can therefore be defined by:

$$\alpha_{NPN} = \gamma_E \alpha_T M = 1, \tag{7.12}$$

where γ_E is the emitter injection efficiency and α_T is the base transport factor. These terms will be discussed in more detail later in this chapter. It will be assumed here that they are not a strong function of the collector bias voltage. On the other hand, the multiplication coefficient (M) is a strong function of the collector bias voltage, as discussed in Chap. 3:

$$M = \frac{1}{1 - (V_C / \mathrm{BV_{CBO}})^n}. \tag{7.13}$$

At low collector bias voltages, the multiplication factor is equal to unity. In this case, a common-base current gain for low collector voltages is given by

$$\alpha_{NPN}(0) = \gamma_E \alpha_T. \tag{7.14}$$

Using these relationships in (7.12),

$$\alpha_{NPN}(0) \frac{1}{1 - (\mathrm{BV_{CEO}} / \mathrm{BV_{CBO}})^n} = 1, \tag{7.15}$$

where BV_{CEO} is the open-base breakdown voltage. This expression can be rewritten as

$$\frac{BV_{CEO}}{BV_{CBO}} = [1 - \alpha_{NPN}(0)]^{1/n}. \tag{7.16}$$

In terms of the common-emitter current gain, the above expression can be written as

$$\frac{BV_{CEO}}{BV_{CBO}} = \frac{1}{[1 + \beta_{NPN}(0)]^{1/n}}. \tag{7.17}$$

As discussed in Chap. 3, the parameter n has a value of 6 for a P^+/N diode, which is appropriate for the base–collector junction. For a typical current gain (β) of between 50 and 100 at low-current levels in a power bipolar N–P–N transistor, the open-base breakdown voltage (BV_{CEO}) is reduced to half of the open-emitter breakdown voltage (BV_{CBO}). This reduction must be factored into the design of the N-drift region. The smaller doping concentration and larger width of the N-drift region required to obtain the larger open-emitter breakdown voltage produce a larger on-state voltage drop and slower switching speed for the bipolar power transistor.

The above discussion is pertinent when the doping concentration of the P-base region and its width (W_P) are relatively large. This suppresses the extension of the depletion region across the base–collector junction (J_1) into the P-base region as illustrated in Fig. 7.6. However, it will be shown later in this chapter that the current gain can be increased by reducing the doping concentration and width of the P-base region. As illustrated in the lower portion of Fig. 7.6, a reduction of the doping concentration of the P-base region promotes the extension of the depletion region across the base–collector junction (J_1) into the P-base region. If the base width is small, this can lead to the complete depletion of the P-base region before the initiation of impact ionization-induced breakdown at the base–collector junction. In this case, the reach-through limited breakdown voltage will be much smaller than the open-base breakdown voltage given by (7.17). The reach-through breakdown voltage can be analyzed by using the procedure described in Chap. 3 after taking into account the sharing of the collector bias voltage with the lightly doped drift region.

7.3.3 Shorted Base–Emitter Operation

The power N–P–N bipolar transistor is illustrated in Fig. 7.7 with the base and emitter terminals short circuited. When a positive bias is applied to the collector, the base–collector junction is reverse biased allowing the device to support a high voltage. The leakage current at the base–collector junction flows to the base contact as indicated by the dashed line. This current must flow via the resistance (R_B) of the P-base region before it reaches the base contact in this simplified lumped model. The current flow through the base resistance forward biases the base–emitter junction at location *A*. As long as the voltage drop produced across the base resistance is well below the built-in potential of the base–emitter junction, there is no injection initiated from this junction. The device then supports voltage as in the case of operation with an open emitter and the blocking voltage is the same as the open-emitter breakdown voltage (BV_{CBO}). This is indicated by the vertical line on the right-hand side of Fig. 7.5.

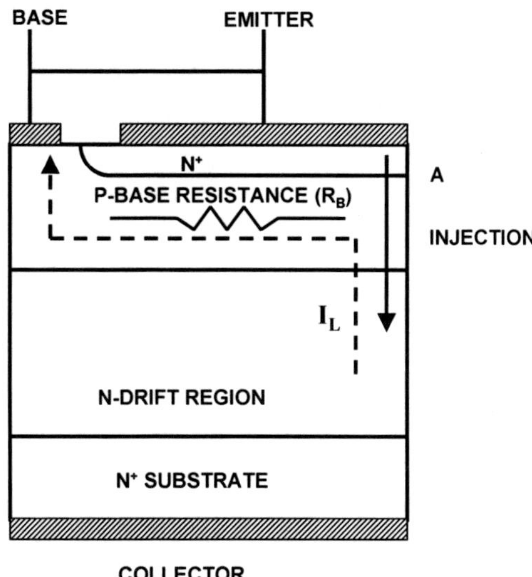

Fig. 7.7 Shorted base–emitter operation of the N–P–N bipolar power transistor

However, as the current flowing through the base–collector junction increases, the voltage drop across the base resistance becomes sufficient to promote the injection of minority carriers across the base–emitter junction at location *A*. This is indicated by the vertical arrow on the right-hand side in Fig. 7.7. Once the base–emitter junction begins to inject minority carriers, the device operates as a bipolar transistor with current gain that increases with increasing collector current. This positive feedback mechanism produces a collapse in the voltage supported by the transistor as shown in Fig. 7.5 until the collector voltage becomes equal to the open-base breakdown voltage (BV_{CEO}). Consequently, it is prudent to assume that the maximum blocking voltage capability for a power bipolar transistor operated with the base and emitter terminals shorted together (BV_{CES}) is equal to the open-base breakdown voltage (BV_{CEO}). In practical application circuits, it is customary to short circuit the base and emitter terminals with the control circuit during the blocking mode. The maximum collector voltage that is applied to the power bipolar transistor must then be less than the open-base breakdown voltage (BV_{CEO}) to avoid destructive failure.

Simulation Example

To gain further insight into the physics of operation for the bipolar power transistor under voltage-blocking conditions, the results of two-dimensional numerical simulations for a typical structure are described here. The total width of the structure, as shown by the cross section in Fig. 7.7, was 200 μm (area = 2×10^{-6} cm^{-2}) with an emitter finger half-width of 190 μm. The structure had a collector drift region doping concentration of 5×10^{13} cm^{-3}. The P-base region had a Gaussian doping profile with a surface concentration of 2×10^{17} cm^{-3} and a depth of 10 μm. The N^+ emitter region had a Gaussian doping profile with a surface concentration of 1×10^{20} cm^{-3} and a depth of 1 μm. The resulting doping profile is shown in Fig. 7.8.

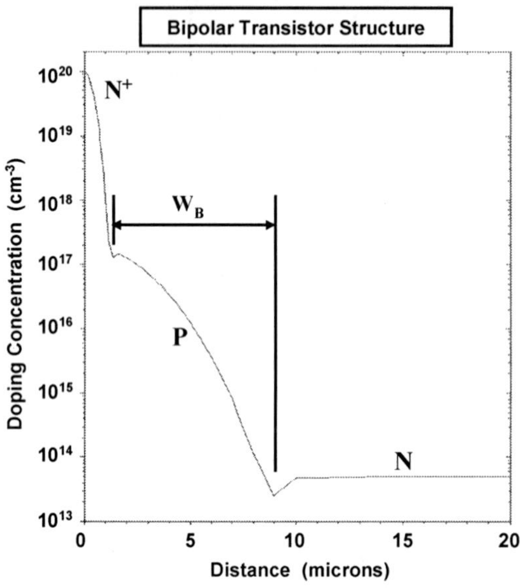

Fig. 7.8 Doping profile for the simulated N–P–N bipolar power transistor structure

The blocking characteristics for the bipolar transistor are shown in Fig. 7.9 for a variety of bias conditions at an ambient temperature of 400 K. The voltage-blocking characteristics of the bipolar transistor operated under the open-emitter configuration are shown by the dashed line. The breakdown voltage (BV_{CBO}) is observed to be about 1,700 V. The voltage-blocking characteristics of the bipolar transistor operated under the open-base configuration are shown by the dotted line. The breakdown voltage (BV_{CEO}) is observed to be 900 V. When the bipolar transistor is operated with the base short circuited to the emitter terminal, the characteristic shown with the solid line is observed with a snapback behavior. At lower collector current levels, the transistor is able to support the open-emitter breakdown voltage but when the collector current density exceeds 1 A cm^{-2}, the blocking voltage drops to the open-base breakdown voltage. The characteristic obtained with a base drive current of 4×10^{-7} A μm^{-1} (current density of 0.2 A cm^{-2}) is also shown in the figure. Using this characteristic, a common-emitter current gain (β) of 25 is obtained. Based upon (7.17), the open-base breakdown voltage is calculated to be 890 V using this gain and the observed open-emitter breakdown voltage providing validation for the simple analytical formulae that govern the breakdown voltage for the power bipolar transistor.

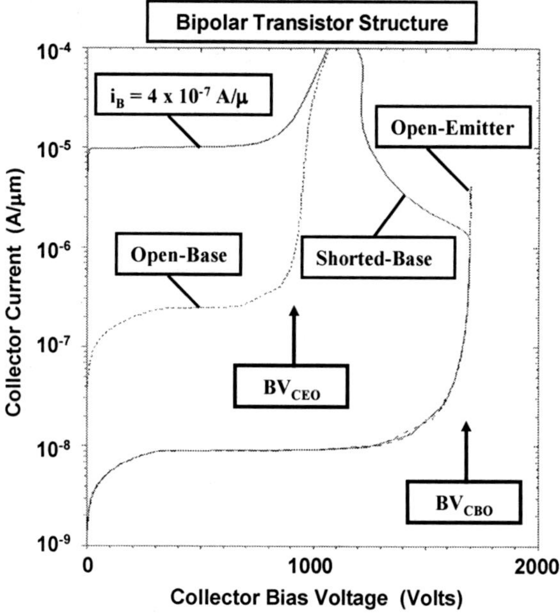

Fig. 7.9 Typical blocking characteristics for an N–P–N bipolar power transistor structure

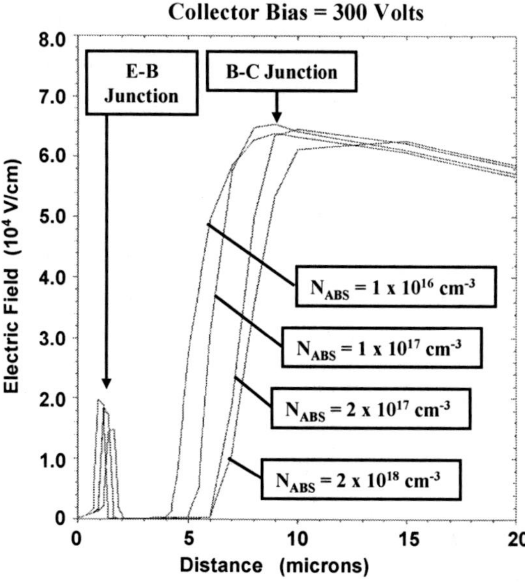

Fig. 7.10 Electric field profiles within N–P–N bipolar power transistors

In Sect. 7.3.2, it was indicated that the breakdown voltage for the bipolar power transistor can be degraded by reach-through of the electric field within the P-base region. This problem becomes aggravated when the doping concentration of the P-base region is reduced to increase the current gain of the transistor. The extension of the electric field in the P-base region depends upon the doping concentration in the N-drift region and the doping profile for the P-base region. As an example, the electric field distributions in the vicinity of the base–collector junction are shown in Fig. 7.10 for four doping profiles for the P-base region. The doping profiles were altered by using four values for the surface concentration (N_{ABS}) for the acceptors used to form the P-base region. It can be observed that the electric field penetrates further into the P-base region toward the emitter–base junction when the surface concentration is reduced below 2×10^{17} cm^{-3}. This limits the ability to obtain a high-current gain in bipolar power transistors.

7.4 Current Gain

Current transport through the bipolar transistor between its collector and emitter terminals can be induced by the application of a base drive current created with the input control circuit. One of the most important characteristics for a bipolar transistor is its current gain (α and β) because this determines the power gain. The current gain for a bipolar transistor can be related its structural parameters. The treatment of the basic bipolar transistor structure is available in many previously published textbooks.[8,9] This section begins with a review of this treatment followed by a discussion of the physics that uniquely impacts the power bipolar transistor structure.

The current flow within the power bipolar transistor is illustrated in Fig. 7.11. The base–emitter junction is assumed to be under forward bias producing a base current (I_B) while the base–collector junction is assumed to be reverse biased by the collector bias supply voltage (see Fig. 7.4). The forward bias across the base–emitter junction (J_2) results in the simultaneous injection of electrons into the base region and the injection of holes into the emitter region. The current I_{nE} is associated with the injection of electrons into the P-base region. The electrons injected into the P-base region diffuse through it and reach the base–collector junction (J_1). This produces a current at the base–collector junction indicated as I_{nC} in the figure. If the voltage across the reverse-biased base–collector junction is

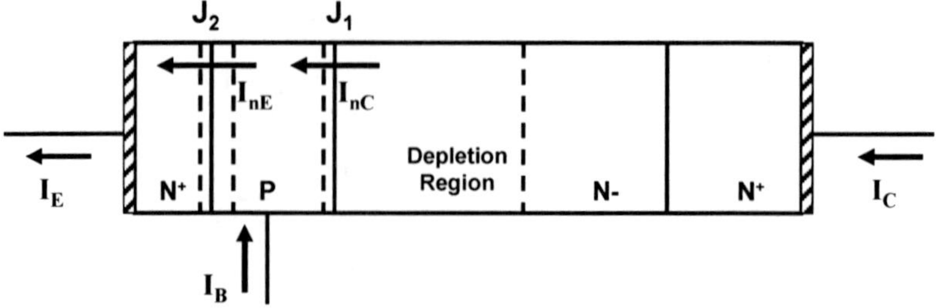

Fig. 7.11 Current flow within the N–P–N bipolar power transistor

large, the electric field at the junction can be sufficiently high to induce impact ionization. Thus, the collector current I_C is not the same as that at the junction (I_{nC}).

The common-base current gain for the power bipolar transistor can be written in terms of the current transport components:

$$\alpha_{NPN} = \frac{\delta I_C}{\delta I_E} = \left(\frac{\delta I_{nE}}{\delta I_E}\right)\left(\frac{\delta I_{nC}}{\delta I_{nE}}\right)\left(\frac{\delta I_C}{\delta I_{nC}}\right). \tag{7.18}$$

It can be concluded that the current gain is determined by three factors indicated with the brackets in this equation. The first term is referred to as the *emitter injection efficiency*:

$$\gamma_E = \left(\frac{\delta I_{nE}}{\delta I_E}\right). \tag{7.19}$$

It is the proportion of the emitter current due to the injection of electrons into the P-base region. This fraction of the emitter current is responsible for producing the collector current. The component of the emitter current due to the injection of holes into the emitter region is wasted from the point of view of deriving collector current flow in the bipolar transistor. The emitter injection efficiency term is always less than unity. In addition, the emitter injection efficiency is reduced by recombination within the depletion region of the base–emitter junction.

The second term in brackets in (7.18) is referred to as the *base transport factor*:

$$\alpha_T = \left(\frac{\delta I_{nC}}{\delta I_{nE}}\right). \tag{7.20}$$

It is a measure of the ability for the electrons injected at the base–emitter junction to reach the base–collector junction. All of the electrons injected into the P-base region are unable to reach the base–collector junction due to recombination within the base region. The base transport factor is always less than unity for a bipolar transistor.

The third term in brackets in (7.18) is referred to as the *collector efficiency*:

$$\gamma_C = \left(\frac{\delta I_C}{\delta I_{nC}}\right). \tag{7.21}$$

This parameter is a measure of the ability of electrons to transport through the collector region. When the base–collector junction is reverse biased as shown in the figure, a high electric field can be developed within its depletion region. The electrons that enter the depletion region from the base region can undergo multiplication due to the impact ionization process if the electric field is sufficiently large. Consequently, the collector efficiency is equal to the multiplication coefficient (*M*) that was discussed in Chap. 3. At low collector bias voltages, the collector efficiency term can be assumed to be equal to unity. At large collector bias voltages, this term can become greater than unity.

The above parameters that determine the common-base current gain can be related to the physical properties of the bipolar transistor. In the case of the power bipolar transistor, it becomes necessary to account for high-level injection effects in the P-base region because of operation at relatively large current densities. At very high collector current densities, a base-widening effect occurs that also degrades the current gain. The analysis of the emitter injection efficiency and base transport factor for a power bipolar transistor is provided below.

7.4.1 Emitter Injection Efficiency

As discussed above, the emitter injection efficiency is a measure of the emitter current due to the injection of electrons into the P-base region. When the base–emitter junction is forward biased, current flow across the junction occurs by the injection of electrons into the base region as well as the injection of holes into the emitter region. The carrier distribution profiles in the base and emitter regions are illustrated in Fig. 7.12 using a linear scale for the concentrations.

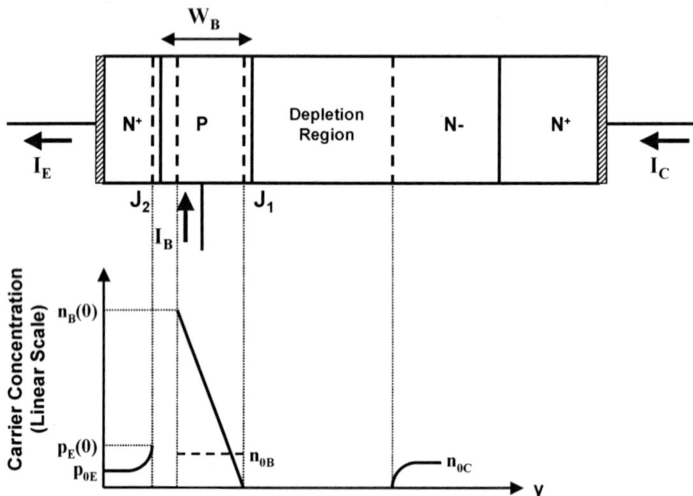

Fig. 7.12 Minority carrier distribution within the N–P–N bipolar power transistor

The injected carrier concentrations on both sides of the base–emitter junction are related to the corresponding minority carrier concentrations in equilibrium by the "Law of the Junction"[8]:

$$n_B(0) = n_{0B} e^{qV_{BE}/kT}, \tag{7.22}$$

$$p_{\mathrm{E}}(0) = p_{0\mathrm{E}} \mathrm{e}^{qV_{\mathrm{BE}}/kT}, \tag{7.23}$$

where V_{BE} is the forward bias across the base–emitter junction. In these equations, $n_{0\mathrm{B}}$ and $p_{0\mathrm{E}}$ are the minority carrier concentrations in equilibrium within the P-base and N^+ emitter regions, respectively. Note that the minority carrier concentration in equilibrium is larger within the P-base region than within the N^+ emitter region due to the larger doping concentration in the emitter for a bipolar transistor.

The minority carriers (holes) injected into the N^+ emitter region diffuse away from the junction, producing a hole current component for the total emitter current. The holes within the emitter region obey the continuity equation:

$$\frac{\mathrm{d}^2 p}{\mathrm{d}y^2} - \frac{p}{L_{\mathrm{pE}}^2} = 0, \tag{7.24}$$

where L_{pE} is the diffusion length for holes in the emitter. The diffusion length for holes is small (submicron) due to the high doping concentration in the emitter (see high doping effects discussed in Chap. 2). If the emitter thickness is much greater than the diffusion length for holes in the emitter, the minority carrier profile exhibits an exponential decay away from the junction as illustrated in the figure (see Chap. 5 for a more detailed discussion of the injected minority carrier concentration for a P–N junction diode). This profile is given by

$$p(y) = p_{\mathrm{E}}(0)\mathrm{e}^{-(y/L_{\mathrm{pE}})}, \tag{7.25}$$

where y is the distance measured moving to the left away from the junction. The hole current density flowing at the base–emitter junction is given by

$$J_{\mathrm{p}}(0) = -qD_{\mathrm{pE}}\left(\frac{\mathrm{d}p}{\mathrm{d}y}\right)_{y=0}, \tag{7.26}$$

where D_{pE} is the diffusion coefficient for holes in the emitter. Note that the diffusion coefficient for holes in the emitter must be computed after taking into account the reduction of the mobility (μ_{pE}) due to heavy doping effects. In Chap. 2, a reduction of the majority carrier mobility with increasing doping concentration was attributed to enhanced Coulombic scattering. Empirical studies[10,11] on the minority carrier mobility within heavily doped regions indicate a similar reduction of the minority carrier mobility. In this case, the mobility for holes in the emitter region should be computed by using the relationship

$$\mu_{\mathrm{pE}}(N_{\mathrm{DE}}) = \frac{2.9\times 10^{15} + 47.7 N_{\mathrm{DE}}^{0.76}}{5.86\times 10^{12} + N_{\mathrm{DE}}^{0.76}}, \tag{7.27}$$

where N_{DE} is the donor concentration in the emitter. Using (7.25) in (7.26) provides the hole current component of the total emitter current:

$$J_p(0) = \frac{qD_{pE}}{L_{pE}} p_E(0) = \frac{qD_{pE}}{L_{pE}} p_{0E} e^{qV_{BE}/kT} \tag{7.28}$$

after incorporating (7.23).

To determine the electron component of the total emitter current, consider the continuity equation for electrons in the P-base region:

$$\frac{d^2n}{dy^2} - \frac{n}{L_{nB}^2} = 0, \tag{7.29}$$

where L_{nB} is the diffusion length for electrons in the base and y is the distance measured moving to the right away from the base–emitter junction. In this section, it will be assumed that the width of the P-base region (W_B in the figure) is much smaller than the diffusion length for electrons in the base region. In this case, recombination in the base region is negligible allowing analysis of the impact of the emitter injection efficiency on the gain of the bipolar transistor. The continuity equation for electrons in the P-base region then becomes

$$\frac{d^2n}{dy^2} = 0, \tag{7.30}$$

indicating that the slope (dn/dy) of the minority carrier concentration is constant in the P-base region. This linear electron concentration profile is illustrated in the figure. Due to the reverse bias at the base–collector junction, the minority carrier concentration at this junction is forced to zero as shown in the figure. Consequently, the electron concentration decreases linearly from an injected concentration [$n_B(0)$] at the base–emitter junction to zero at the base–collector junction:

$$n(y) = n_B(0)\left(1 - \frac{y}{W_B}\right). \tag{7.31}$$

In writing this expression, it has been assumed that the widths of the depletion regions within the P-base region can be neglected. This holds true at small collector bias voltages. The impact of large collector bias voltages on the base width (the Kirk effect) is taken into account later in this chapter.

The electron current component of the total emitter current can be derived from the above electron carrier distribution:

$$J_n(0) = -qD_{nB}\left(\frac{dn}{dy}\right)_{y=0}, \tag{7.32}$$

where D_{nB} is the diffusion coefficient for electrons in the base region. Note that the diffusion coefficient for electrons in the base region must be computed after taking into account the reduction of the mobility (μ_{nB}) due to heavy doping effects in the base region even though its doping is not as high as in the emitter. In this case, the mobility for electrons in the base region should be computed by using the relationship

$$\mu_{nB}(N_{AB}) = \frac{5.1\times10^{18} + 92N_{AB}^{0.91}}{3.75\times10^{15} + N_{AB}^{0.91}}, \tag{7.33}$$

where N_{AB} is the acceptor concentration in the base region. Using (7.31) in (7.32) provides the electron current component of the total emitter current:

$$J_n(0) = \frac{qD_{nB}}{W_B} n_B(0) = \frac{qD_{nB}}{W_B} n_{0B} e^{qV_{BE}/kT} \tag{7.34}$$

after incorporating (7.22). Since the recombination in the base region has been assumed to be negligible, the electron current component at the base–emitter junction is also equal to the collector current density.

The emitter injection efficiency can be obtained by using the electron and hole current components:

$$\gamma_E = \frac{J_n(0)}{J_n(0) + J_p(0)}. \tag{7.35}$$

Substituting (7.28) and (7.34) for the current densities,

$$\gamma_E = \frac{D_{nB}L_{pE}n_{0B}}{D_{nB}L_{pE}n_{0B} + D_{pE}W_B p_{0E}}. \tag{7.36}$$

Using this relationship, the emitter injection efficiency can be computed from the physical parameters for the power bipolar transistor. To facilitate this computation, it is beneficial to relate the minority carrier concentrations in equilibrium to the doping concentrations in the regions:

$$n_{0B} = \frac{n_{iB}^2}{N_{AB}}, \tag{7.37}$$

$$p_{0E} = \frac{n_{iE}^2}{N_{DE}}. \tag{7.38}$$

Note that the intrinsic carrier concentrations in the base and emitter regions are not equal due to the difference in doping concentrations, which impacts the band-gap narrowing for the regions. Using these relationships in (7.36),

$$\gamma_E = \frac{D_{nB}L_{pE}n_{iB}^2 N_{DE}}{D_{nB}L_{pE}n_{iB}^2 N_{DE} + D_{pE}W_B n_{iE}^2 N_{AB}} = \alpha_E. \tag{7.39}$$

The common-base current gain (α_E), as determined purely by the emitter injection efficiency, can be computed by using the above expression.

The common-emitter current gain (β_E), as determined purely by the emitter injection efficiency, can also be obtained by using the electron and hole current components of the total emitter current:

$$\beta_E = \frac{J_n(0)}{J_p(0)}. \tag{7.40}$$

Substituting (7.28) and (7.34) for the current densities,

$$\beta_E = \frac{D_{nB} L_{pE} n_{0B}}{D_{pE} W_B p_{0E}}. \tag{7.41}$$

Using (7.37) and (7.38) for the minority carrier concentrations in equilibrium,

$$\beta_E = \frac{D_{nB} L_{pE} n_{iB}^2 N_{DE}}{D_{pE} W_B n_{iE}^2 N_{AB}}. \tag{7.42}$$

For a power bipolar transistor, it is desirable to obtain a high-current gain to control a large load (collector) current with a small input drive (base) current. Based upon the expressions for the emitter injection efficiency, a high gain can be obtained by using a large doping concentration for the emitter region and a low doping concentration for the base region. In practice, it is not possible to use an arbitrarily high doping concentration for the N^+ emitter region due to heavy doping effects, which produce (a) a reduction of the diffusion length for holes in the emitter and (b) a large increase in the intrinsic carrier concentration in the emitter due to band-gap narrowing. An optimum doping concentration for the emitter has been reported to be about 1×10^{19} cm^{-3}. In this case, typical values for the common-base and common-emitter current gains are 0.96 and 25, respectively, if a base doping concentration of 1×10^{17} cm^{-3} is assumed. In addition, from the above equations for the current gain as determined by the emitter injection efficiency, it is desirable to have a lightly doped base region with a narrow base width. However, this can compromise the blocking voltage capability of the power bipolar transistor due to the reach-through phenomenon as discussed previously in this chapter. A low P-base doping concentration also results in high-level injection effects occurring in the base region at lower current densities. This reduces the emitter injection efficiency as discussed in Sect. 7.4.3.

7.4.2 Emitter Injection Efficiency with Recombination in the Depletion Region

In the previous analysis, the current transport across the base–emitter junction was assumed to occur purely by the diffusion process. However, at low-current densities, it is necessary to account for the recombination current at the base–emitter junction. The recombination current for a P–N junction was discussed in Sect. 5.1.1. The current at the junction produced by recombination within the depletion region is given by

$$J_r = \frac{qn_i W_D}{\tau_{SC}} e^{qV_{BE}/2kT}, \tag{7.43}$$

where W_D is the depletion layer width and τ_{SC} is the space-charge generation lifetime. The emitter injection efficiency at low-current levels can be obtained by including this current component:

$$\gamma_E = \frac{J_n(0)}{J_n(0) + J_p(0) + J_r}. \tag{7.44}$$

Using the previously derived equations (7.28) and (7.34) for the diffusion currents with the above equation for the recombination current,

$$\gamma_E = \frac{D_{nB} L_{pE} n_{0B}}{D_{nB} L_{pE} n_{0B} + D_{pE} W_B p_{0E} + [n_i W_D W_B L_{pE} / \tau_{SC} e^{qV_{BE}/2kT}]}. \tag{7.45}$$

At low forward bias voltages across the base–emitter junction, the last term in square brackets in the denominator of this equation becomes dominant producing a reduction of the emitter injection efficiency and the current gain. As the forward bias across the base–emitter junction is increased, the diffusion currents become much larger than the recombination current and the emitter injection efficiency becomes equal to that discussed in Sect. 7.4.1.

As an example, consider a power bipolar transistor with an N^+ emitter doping concentration of 2×10^{19} cm^{-3} with a diffusion length of 1 μm for the holes, and a P-base doping concentration of 1×10^{17} cm^{-3} with a width of 5 μm. The current gains (β_E) computed for this case with a depletion region width of 0.01 μm and various values for the space-charge generation lifetime are shown in Fig. 7.13. For the case of a space-charge generation lifetime of 1×10^{-7} s, the recombination current begins to dominate leading to a falloff in the current gain when the current density becomes less than 0.1 A cm^{-2}. When the current density becomes more than 1 A cm^{-2}, the current gain approaches 25 as limited by the injection of holes into the emitter region. As the space-charge generation lifetime is increased, a high gain is retained to lower collector current levels. The rate of falloff in the current gain with decreasing collector current density is therefore a strong function of the space-charge generation lifetime.

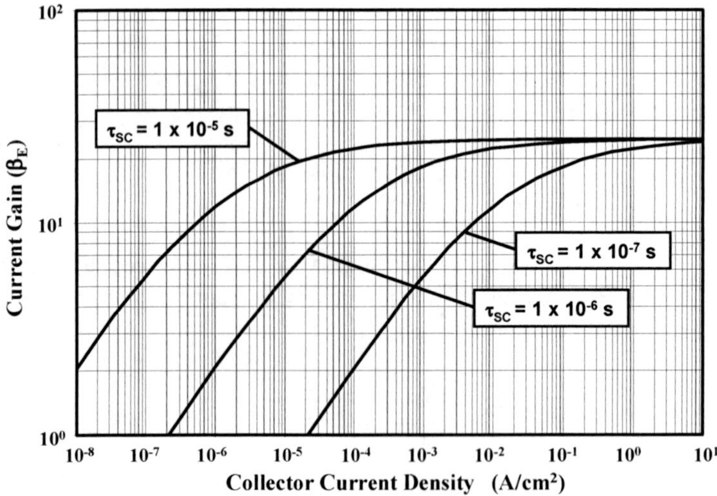

Fig. 7.13 Reduction of current gain at low collector current levels for the N–P–N bipolar power transistor

7.4.3 Emitter Injection Efficiency with High-Level Injection in the Base

From the analysis in Sect. 7.4.1, it may be inferred (see (7.42)) that the doping concentration in the base region for the power bipolar transistor should be reduced to achieve a high-current gain. At low base doping levels, the injected minority (electron) carrier concentration in the P-base region can exceed its doping

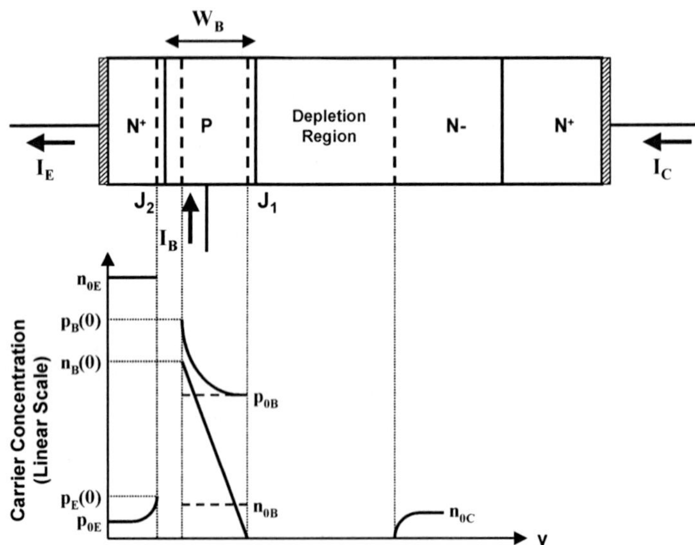

Fig. 7.14 Minority and majority carrier profiles within the N–P–N bipolar power transistor under high-level injection conditions in the base

concentration when the bipolar power transistor is operated at high-current densities. This is referred to as the on-set of *high-level injection in the base region.* To satisfy charge neutrality, the majority (hole) carrier concentration in the base region increases under high-level injection conditions. This enhances the injection of holes from the P-base region into the N^+ emitter region, which results in a reduction of the injection efficiency and current gain of the power bipolar transistor. The reduction of the current gain of a bipolar transistor with increasing current density due to high-level injection in the base region is referred to as the *Webster effect.*[12]

The minority carrier distribution profiles in the base and emitter regions are shown in Fig. 7.14 under high-level injection conditions in the base region. The majority carrier profiles are also shown in the figure. The majority carrier concentration in the emitter remains equal to the emitter doping concentration because of low injection levels in the emitter region. However, the majority carrier concentration in the base is enhanced because the minority carrier density exceeds the doping concentration of the base region in the vicinity of the base–emitter junction. Charge neutrality in the base region requires

$$p_B(0) - p_{0B} = n_B(0) - n_{0B}, \tag{7.46}$$

where $p_B(0)$ and $n_B(0)$ are the majority and minority carrier densities in the P-base region at the base–emitter junction as indicated in the figure. Since the concentration of the minority carriers (n_{0B}) in equilibrium for the base region is much smaller than the other concentrations in the above equation,

$$p_B(0) = p_{0B} + n_B(0). \tag{7.47}$$

The Boltzmann's quasiequilibrium boundary condition for a P–N junction requires

$$\frac{p_E(0)}{p_B(0)} = \frac{n_B(0)}{n_E(0)}\left(\frac{n_{iE}}{n_{iB}}\right)^2. \tag{7.48}$$

Since the doping concentration in the emitter (N_{DE}) is very high,

$$n_E(0) = n_{0E} = N_{DE}. \tag{7.49}$$

Combining the above relationships,

$$p_E(0) = \frac{p_B(0)n_B(0)}{n_E(0)}\left(\frac{n_{iE}}{n_{iB}}\right)^2 = \frac{n_B(0)}{N_{DE}}\left(\frac{n_{iE}}{n_{iB}}\right)^2[p_{0B} + n_B(0)]. \tag{7.50}$$

According to the "Law of the Junction,"

$$n_B(0) = n_{0B}e^{qV_{BE}/kT}, \tag{7.51}$$

where V_{BE} is the forward bias voltage across the base–emitter junction. Substituting this in (7.50),

$$p_E(0) = \frac{p_{0B} n_{0B}}{N_{DE}} \left(\frac{n_{iE}}{n_{iB}} \right)^2 \left[1 + \frac{n_B(0)}{p_{0B}} \right] e^{qV_{BE}/kT}. \tag{7.52}$$

The majority and minority carriers in equilibrium are interrelated through the intrinsic carrier concentrations in the emitter and base regions:

$$p_{0B} n_{0B} = n_{iB}^2, \tag{7.53}$$

$$p_{0E} n_{0E} = p_{0E} N_{DE} = n_{iE}^2. \tag{7.54}$$

Combining these with (7.52) yields

$$p_E(0) = p_{0E} \left[1 + \frac{n_B(0)}{p_{0B}} \right] e^{qV_{BE}/kT}. \tag{7.55}$$

The term within the square brackets in this equation is the increase in the injected hole (minority carrier) concentration in the emitter due to high-level injection in the base region. When the concentration of minority carriers (electrons) injected into the P-base region [$n_B(0)$] is small compared with the majority carrier concentration [p_{0B}] in the base region, this term becomes equal to unity as expected for low-level injection conditions.

The emitter injection efficiency under high-level injection conditions in the base region can be derived by taking into account the enhanced injection of holes into the emitter. Based upon an exponential decrease in the minority carrier (hole) concentration within the emitter,

$$p(y) = p_E(0) e^{-(y/L_{pE})}, \tag{7.56}$$

where y is the distance measured moving to the left away from the junction with $p_E(0)$ given by (7.55). The hole current density flowing at the base–emitter junction is then given by

$$J_p(0) = -qD_{pE} \left(\frac{dp}{dy} \right)_{y=0} = \frac{qD_{pE}}{L_{pE}} p_E(0). \tag{7.57}$$

This is also the base current density (J_B) because it has been assumed in this section that all the base current is used to supply the injection into the emitter. If the diffusion length for electrons in the base region is much larger than the base width (i.e., when recombination in the base is neglected), the minority carrier concentration in the base region has a linear distribution as shown in the figure. The collector current is then equal to the electron current at the base–emitter junction:

$$J_{\mathrm{C}} = J_{\mathrm{n}}(0) = \frac{qD_{\mathrm{nB}}}{W_{\mathrm{B}}} n_{\mathrm{B}}(0) = \frac{qD_{\mathrm{nB}}}{W_{\mathrm{B}}} n_{0\mathrm{B}} \mathrm{e}^{qV_{\mathrm{BE}}/kT}. \tag{7.58}$$

From this equation,

$$n_{\mathrm{B}}(0) = \frac{J_{\mathrm{C}} W_{\mathrm{B}}}{qD_{\mathrm{nB}}}. \tag{7.59}$$

Using this expression in (7.55),

$$p_{\mathrm{E}}(0) = p_{0\mathrm{E}} \left[1 + \frac{J_{\mathrm{C}} W_{\mathrm{B}}}{qD_{\mathrm{nB}} p_{0\mathrm{B}}} \right] \mathrm{e}^{qV_{\mathrm{BE}}/kT}. \tag{7.60}$$

Substituting this expression in (7.57),

$$J_{\mathrm{B}} = J_{\mathrm{p}}(0) = \left(\frac{qD_{\mathrm{pE}} p_{0\mathrm{E}}}{L_{\mathrm{pE}}} \right) \left[1 + \frac{J_{\mathrm{C}} W_{\mathrm{B}}}{qD_{\mathrm{nB}} p_{0\mathrm{B}}} \right] \mathrm{e}^{qV_{\mathrm{BE}}/kT}. \tag{7.61}$$

The common-emitter current gain as determined by emitter injection efficiency under high-level injection conditions in the base region is then obtained by using (7.58) and (7.61):

$$\beta_{\mathrm{E}} = \frac{J_{\mathrm{C}}}{J_{\mathrm{B}}} = \frac{D_{\mathrm{nB}} L_{\mathrm{pE}} n_{0\mathrm{B}}}{D_{\mathrm{pE}} W_{\mathrm{B}} p_{0\mathrm{E}}} \frac{1}{[1 + (J_{\mathrm{C}} W_{\mathrm{B}} / qD_{\mathrm{nB}} p_{0\mathrm{B}})]}. \tag{7.62}$$

Substituting for the minority carrier densities using (7.53) and (7.54),

$$\beta_{\mathrm{E}} = \frac{D_{\mathrm{nB}} L_{\mathrm{pE}} N_{\mathrm{DE}}}{D_{\mathrm{pE}} W_{\mathrm{B}} N_{\mathrm{AB}}} \left(\frac{n_{\mathrm{iB}}}{n_{\mathrm{iE}}} \right)^2 \frac{1}{[1 + (J_{\mathrm{C}} W_{\mathrm{B}} / qD_{\mathrm{nB}} N_{\mathrm{AB}})]}. \tag{7.63}$$

In writing this expression, the majority carrier concentration in the base region ($p_{0\mathrm{B}}$) has been replaced with the doping concentration (N_{AB}). At high collector current densities, the second term in the square brackets becomes dominant and the current gain decreases inversely with increasing current density. The falloff in the current gain occurs when the collector current density approaches and exceeds a Webster current density:

$$J_{\mathrm{W}} = \frac{qD_{\mathrm{nB}} N_{\mathrm{AB}}}{W_{\mathrm{B}}}. \tag{7.64}$$

At current densities well below the Webster current density, the current gain remains constant at a value β_{LL} as given by (7.42) derived earlier under low-level injection conditions for the base region. Thus, the current gain can be expressed as

$$\beta_E = \frac{\beta_{LL}}{1 + (J_C / J_W)}. \quad (7.65)$$

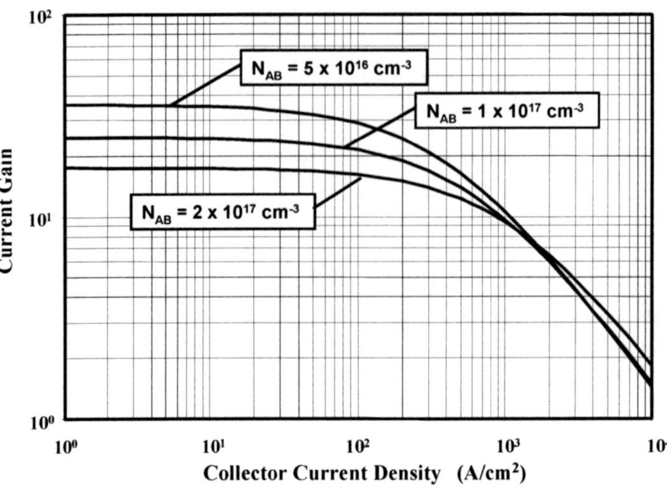

Fig. 7.15 Falloff in the current gain for an N–P–N bipolar power transistor due to high-level injection conditions in the base

As an example, consider the case of a power N–P–N bipolar transistor with an emitter doping concentration of 2×10^{19} cm^{-3} and a base doping concentration of 1×10^{17} cm^{-3}. The common-emitter current gain at low injection levels for this transistor is approximately 25. The Webster current density using these doping concentrations for the base and emitter regions is 640 A cm^{-2}. Consequently, the current gain falls off when the collector current density approaches and exceeds this value as shown in Fig. 7.15. When the doping concentration of the P-base region is reduced by a factor of 2 times, the common-emitter current gain at low injection levels increases to approximately 40. However, the Webster current density using these doping concentrations for the base and emitter regions is also reduced to 410 A cm^{-2}. Consequently, the common-emitter current gain falls off at lower current densities resulting in approximately the same gain at high-current densities (1,000 A cm^{-2}). Similarly, when the doping concentration of the P-base region is increased by a factor of 2 times, the common-emitter current gain at low injection levels decreases to approximately 17.5 and the Webster current density is increased to 1,160 A cm^{-2}. Consequently, the current gain falls off at higher current densities resulting in approximately the same gain at high-current densities. Such high emitter current densities are encountered in power bipolar transistors due to the emitter current crowding phenomenon discussed later in this chapter.

According to (7.65), the current gain becomes inversely proportional to the collector current density when it becomes much larger than the Webster current density:

$$\beta_{\text{EH}} = \frac{\beta_{\text{LL}} J_{\text{W}}}{J_{\text{C}}}. \tag{7.66}$$

Consequently, the base drive current density required to sustain the collector current is given by

$$J_{\text{B}} = \frac{J_{\text{C}}}{\beta_{\text{EH}}} = \frac{J_{\text{C}}^2}{\beta_{\text{LL}} J_{\text{W}}}. \tag{7.67}$$

Thus, the base drive current increases as the square of the collector current for a power bipolar transistor upon the on-set of high-level injection in the base region. This degrades the power gain of the device making the gate control circuit bulky and expensive.

7.4.4 Base Transport Factor

As discussed earlier, the base transport factor is a measure of the ability for the minority carriers injected from the base–emitter junction to reach the base–collector junction. For an N–P–N transistor, it is expressed in terms of the ratio of the electron current at the base–emitter junction to the electron current at the base–collector junction (see (7.20)). In Sects. 7.4.2 and 7.4.3, it was assumed that the diffusion length for electrons (L_{nB}) in the P-base region is much larger than its width (W_{B}). In this case, the base transport factor is equal to unity. However, in a power bipolar transistor, the base width can be relatively large to prevent reach-through breakdown at high collector bias voltages.

The base–emitter junction of the bipolar transistor is forward biased to induce current flow. This produces the injection of electrons into the P-base region. These electrons diffuse through the P-base region and arrive at the base–collector junction producing the collector current. The diffusion equation for electrons in the P-base region under steady-state conditions is

$$\frac{\mathrm{d}^2 n}{\mathrm{d}y^2} - \frac{n}{L_{\text{nB}}^2} = 0, \tag{7.68}$$

where L_{nB} is the diffusion length for electrons in the base and y is the distance measured moving to the right away from the base–emitter junction. The solution for this equation has the form

$$n(y) = A\,\mathrm{e}^{-y/L_{\text{nB}}} + B\,\mathrm{e}^{+y/L_{\text{nB}}}, \tag{7.69}$$

where A and B are constants determined by the following boundary conditions. At the base–emitter junction ($y = 0$), the electron concentration is given by

$$n(0) = n_{\text{B}}(0) = n_{0\text{B}}\mathrm{e}^{qV_{\text{BE}}/kT}, \tag{7.70}$$

due to the forward bias (V_{BE}) across the base–emitter junction. At the base–collector junction, the electron concentration is zero due to the reverse bias:

$$n(W_B) = 0. \tag{7.71}$$

Using these boundary conditions to solve for the constants A and B provides the electron concentration profile:

$$n(y) = n_{0B}\left\{\frac{\sinh[(W_B - y)/L_{nB}]}{\sinh(W_B/L_{nB})}\right\}e^{qV_{BE}/kT}. \tag{7.72}$$

This profile for the injected electrons in the P-base region is illustrated in Fig. 7.16 together with the minority carrier concentration in equilibrium.

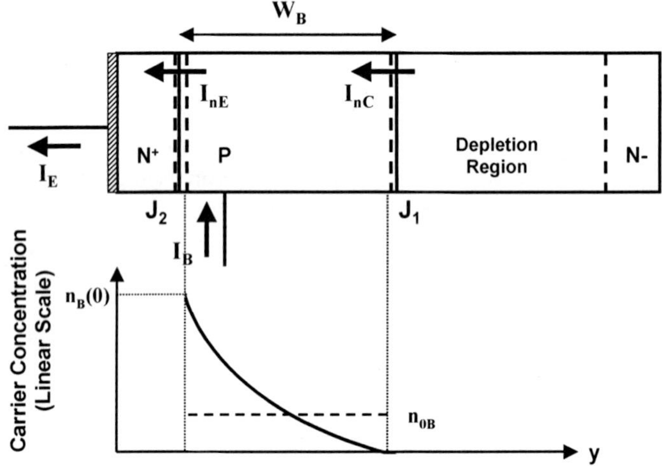

Fig. 7.16 Minority carrier distribution within the N–P–N bipolar power transistor under finite recombination conditions in the base

The electron current at the base–emitter junction (J_{nE}) and the base–collector junction (J_{nC}) can be derived from this electron carrier distribution profile:

$$J_{nE} = -qD_{nB}\left(\frac{dn}{dy}\right)_{y=0} = \frac{qD_{nB}}{L_{nB}}\left(\frac{\cosh(W_B/L_{nB})}{\sinh(W_B/L_{nB})}\right)e^{qV_{BE}/kT}, \tag{7.73}$$

$$J_{nC} = -qD_{nB}\left(\frac{dn}{dy}\right)_{y=W_B} = \frac{qD_{nB}}{L_{nB}}\left(\frac{1}{\sinh(W_B/L_{nB})}\right)e^{qV_{BE}/kT}. \tag{7.74}$$

Using these equations, the base transport factor is obtained:

$$\alpha_T = \frac{J_{nC}}{J_{nE}} = \frac{1}{\cosh(W_B / L_{nB})}. \tag{7.75}$$

The base transport factor is determined by the width of the P-base region relative to the diffusion length for electrons in the base region. When the diffusion length is much larger than the base width, the base transport factor becomes equal to unity. As the base width is increased to suppress reach-through breakdown, the base transport factor becomes less than unity. The variation of the common-base current gain (α_T) is shown in Fig. 7.17 as a function of the base width for various values for the electron diffusion length in the base region. From this graph, it can be concluded that a diffusion length of at least 50 μm is required to achieve a common-base current gain close to unity for a base width of 10 μm. For a typical base doping concentration of 1×10^{17} cm^{-3}, the electron mobility (μ_{nB}) is 750 cm^2 V^{-1} s^{-1} leading to a diffusion coefficient (D_{nB}) of 19.4 cm^2 s^{-1}. Based upon this value, the minority carrier lifetime in the P-base region must exceed 1.3 μs to achieve a high-current gain.

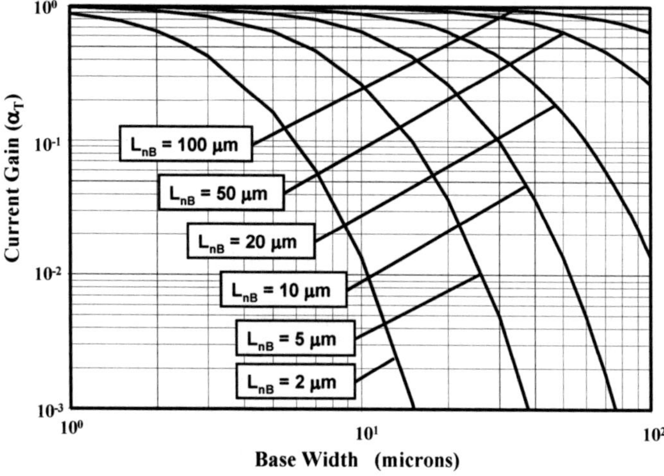

Fig. 7.17 Base transport factor limited common-base current gain for the N–P–N bipolar power transistor

The common-emitter current gain, as determined exclusively by the base transport factor, is given by

$$\beta_T = \frac{\alpha_T}{1 - \alpha_T} = \frac{1}{\cosh(W_B / L_{nB}) - 1}. \tag{7.76}$$

The variation of the common-emitter current gain (β_T) is shown in Fig. 7.18 as a function of the base width for various values for the electron diffusion length in the base region. From this graph, it can be concluded that a diffusion length of 50 μm will produce a common-emitter current gain of 50 for a base width of 10 μm. The current gain drops to only 8 if the diffusion length for electrons in the base is reduced to 20 μm. Consequently, the degradation of the current gain due to finite recombination in the base region must be accounted for during the analysis of power bipolar transistors.

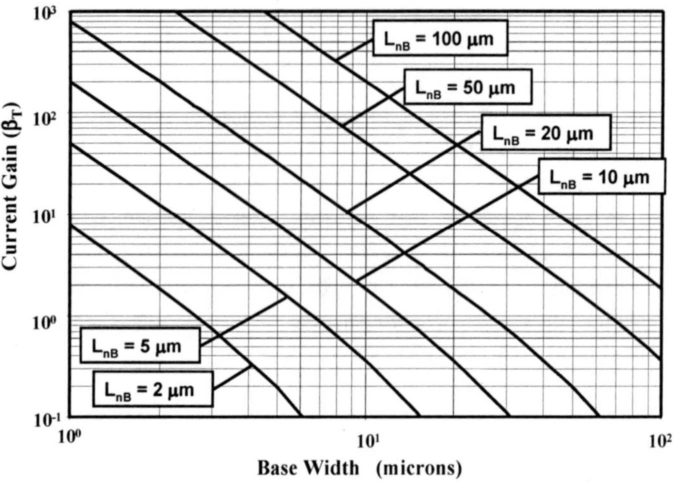

Fig. 7.18 Common-emitter current gain for the N–P–N bipolar power transistor as limited by the base transport factor

7.4.5 Base Widening at High Collector Current Density

When the collector current density is large, another phenomenon that reduces the current gain is an increase in the *effective base width*, which is referred to as the *Kirk effect*.[13] This phenomenon occurs when the bipolar transistor is biased in its forward active regime of operation with a large collector bias voltage. The collector bias is supported across the base–collector junction with a triangular profile at low collector current densities as shown in Fig. 7.19 by the line labeled *a*. This profile is governed by the solution for Poisson's equation with the doping concentration of the N-drift region determining the charge in the depletion region, as previously discussed in Chap. 3:

$$\frac{\mathrm{d}E}{\mathrm{d}y} = -\frac{qN_{\mathrm{D}}}{\varepsilon_{\mathrm{S}}}. \tag{7.77}$$

With a reverse-biased base–collector junction, the current in the bipolar transistor is transported through the collector drift region by electrons drifting under the influence of the electric field in the depletion region. At large collector bias voltages, the electric field in the depletion region is sufficient to accelerate the electrons to their saturated drift velocity ($v_{\mathrm{sat,n}}$), as discussed in Chap. 2. The concentration of the electrons in the depletion region is then related to the collector current density (J_{C}) by

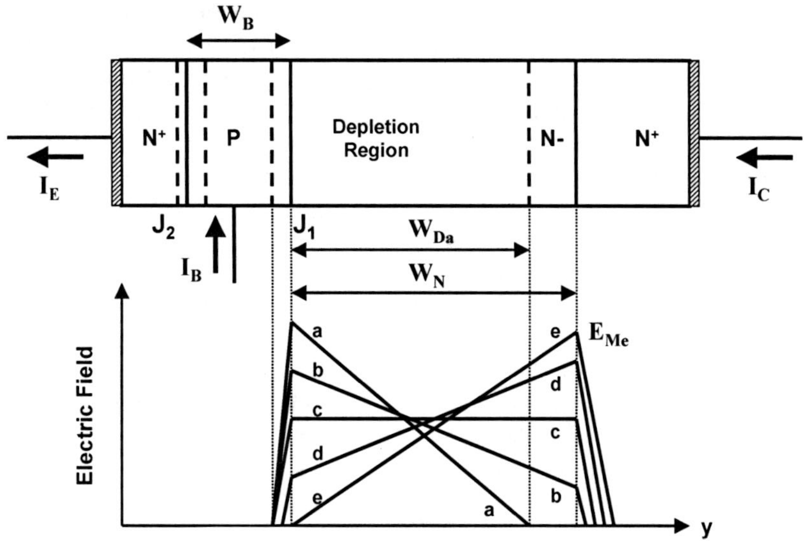

Fig. 7.19 Electric field profiles in the N–P–N bipolar power transistor at various collector current levels

$$n = \frac{J_C}{qv_{\text{sat,n}}}. \tag{7.78}$$

This equation indicates that the concentration of electrons in the depletion region increases in proportion to the collector current density. As the collector current density increases, the electron concentration in the drift region becomes comparable in magnitude to the charge for the donor atoms in the drift region. The compensation of the positive charge due to the donors by the negative charge due to the electrons must be accounted for in determining the electric field profile.

The Poisson's equation that governs the electric field distribution at high collector current densities is given by

$$\frac{dE}{dy} = -\frac{q}{\varepsilon_S}[N_D - n]. \tag{7.79}$$

The electric field profile is then given by

$$E(y) = E(0) - \frac{q}{\varepsilon_S}[N_D - n]y. \tag{7.80}$$

Using (7.78) for the electron concentration,

$$E(y) = E(0) - \frac{q}{\varepsilon_S}\left[N_D - \left(\frac{J_C}{qv_{\text{sat,n}}}\right)\right]y, \tag{7.81}$$

where $E(0)$ is the electric field at the base–collector junction. This expression for the electric field indicates that the profile is linear in shape and that its slope becomes smaller as the collector current density is increased. Thus, for the same applied collector bias, the electric field profile changes to the one labeled *b* with a large collector current density J_{Cb}. Note that the reduction of the slope for the electric field in the drift region promotes its punch-through to the N^+ substrate as illustrated in the figure with the electric field truncated in the N^+ substrate due to its relatively high doping.

As the collector current density is increased, the electric field profile eventually becomes completely flat as shown in the figure by the profile labeled *c*. A flat electric field profile occurs when the charge in the depletion region is equal to zero. In this case, the number of electrons per unit volume transported through the depletion region is exactly equal to the doping concentration for the drift region. This corresponds to the second term in (7.81) becoming equal to zero, which occurs at a collector current density:

$$J_{\rm Cc} = q v_{\rm sat,n} N_{\rm D}. \tag{7.82}$$

When the collector current density is increased even further, the electron concentration in the drift region exceeds the doping concentration of the donors in the drift region. In this case, the net charge in the depletion region becomes negative. The reversal of the net charge governing Poisson's equation for the depletion region produces a reversal of the slope for the electric field profile as shown by the case labeled *d* in the figure. This electric field profile can be described by

$$E(y) = E(0) + \frac{q}{\varepsilon_{\rm S}}\left[\left(\frac{J_{\rm C}}{q v_{\rm sat,n}}\right) - N_{\rm D}\right] y, \tag{7.83}$$

where the term in square brackets is positive. Under the same applied collector bias voltage, the peak in the electric field profile shifts from the base–collector junction to the interface between the N-drift region and the N^+ substrate as shown in the figure.

Eventually, at an even larger collector current density, the electric field becomes equal to zero at the base–collector junction as shown by the profile labeled *e* in the figure. The electric field profile in this case is given by

$$E(y) = \frac{q}{\varepsilon_{\rm S}}\left[\left(\frac{J_{\rm C}}{q v_{\rm sat,n}}\right) - N_{\rm D}\right] y. \tag{7.84}$$

The maximum electric field occurs at the interface between the N-drift region and the N^+ substrate at a distance $y = W_{\rm N}$:

$$E_{\mathrm{Me}} = \frac{q}{\varepsilon_{\mathrm{S}}}\left[\left(\frac{J_{\mathrm{Ce}}}{qv_{\mathrm{sat,n}}}\right) - N_{\mathrm{D}}\right]W_{\mathrm{N}}. \tag{7.85}$$

The collector voltage supported by the electric field profile is given by

$$V_{\mathrm{C}} = \frac{1}{2}E_{\mathrm{Me}}W_{\mathrm{N}} = \frac{q}{2\varepsilon_{\mathrm{S}}}\left[\left(\frac{J_{\mathrm{Ce}}}{qv_{\mathrm{sat,n}}}\right) - N_{\mathrm{D}}\right]W_{\mathrm{N}}^{2}. \tag{7.86}$$

The collector current density can be expressed in terms of the device physical parameters and the applied collector bias voltage from this equation:

$$J_{\mathrm{Ce}} = qv_{\mathrm{sat,n}}\left(\frac{2\varepsilon_{\mathrm{S}}V_{\mathrm{C}}}{qW_{\mathrm{N}}^{2}} + N_{\mathrm{D}}\right) = J_{\mathrm{K}}. \tag{7.87}$$

The collector current density at which the electric field becomes equal to zero at the base–collector junction is referred to as the *Kirk current density* (J_{K}) as indicated in the above equation.

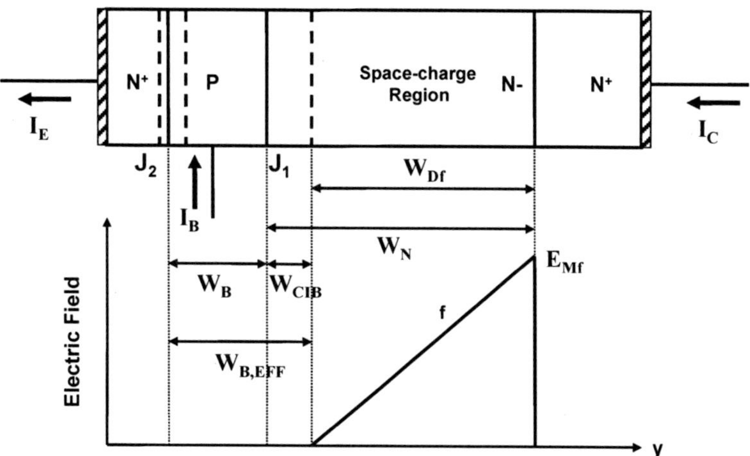

Fig. 7.20 Electric field profile in the N–P–N bipolar power transistor with a current-induced base region

The Kirk current density is of significance for a power bipolar transistor because a *current-induced base region* develops within the collector drift region when the collector current density exceeds its magnitude. The electric field profile *f* for this case is illustrated in Fig. 7.20. Note that a neutral region develops in the drift region adjacent to the base–collector junction. The electrons injected into the P-base region must now diffuse not only through the physical base width (W_{B}) but also through an extra distance called the *current-induced base width* (W_{CIB}). The effective base width for the bipolar transistor then becomes

$$W_{\mathrm{B,EFF}} = W_{\mathrm{B}} + W_{\mathrm{CIB}}. \tag{7.88}$$

This enlargement of the base width reduces the base transport factor and the current gain for the bipolar transistor.

When the electric field profile has taken the form shown in Fig. 7.20, the electron concentration in the drift region has become much larger than the doping concentration of the drift region. This electric field profile can therefore be expressed as

$$E(y) = \left(\frac{J_{\mathrm{Cf}}}{\varepsilon_{\mathrm{S}} v_{\mathrm{sat,n}}} - \frac{qN_{\mathrm{D}}}{\varepsilon_{\mathrm{S}}} \right) y. \tag{7.89}$$

In this expression, J_{Cf} is the collector current density corresponding to the electric field profile shown in Fig. 7.20. If the electric field supported in the N^+ substrate is neglected due to its high doping concentration, the width of the space-charge region is given by

$$W_{\mathrm{Df}} = \sqrt{\frac{2\varepsilon_{\mathrm{S}} V_{\mathrm{C}}}{(J_{\mathrm{Cf}} / v_{\mathrm{sat,n}}) - qN_{\mathrm{D}}}}. \tag{7.90}$$

The width of the current-induced base region is then given by

$$W_{\mathrm{CIB}} = W_{\mathrm{N}} - W_{\mathrm{Df}} = W_{\mathrm{N}} - \sqrt{\frac{2\varepsilon_{\mathrm{S}} V_{\mathrm{C}}}{(J_{\mathrm{Cf}} / v_{\mathrm{sat,n}}) - qN_{\mathrm{D}}}} \tag{7.91}$$

and the effective base width for the bipolar transistor becomes

$$W_{\mathrm{B,EFF}} = W_{\mathrm{B}} + W_{\mathrm{CIB}} = W_{\mathrm{B}} + W_{\mathrm{N}} - \sqrt{\frac{2\varepsilon_{\mathrm{S}} V_{\mathrm{C}}}{(J_{\mathrm{Cf}} / v_{\mathrm{sat,n}}) - qN_{\mathrm{D}}}}. \tag{7.92}$$

The common-base and common-emitter current gains, as limited by recombination in the base region, are then given by (7.75) and (7.76) with the effective base width ($W_{\mathrm{B,EFF}}$) substituted in place of the physical base width (W_{B}). As an example, consider the case of a power N–P–N bipolar transistor with a physical base width of 5 μm (W_{B}) and an N-drift region thickness (W_{N}) of 30 μm with a doping concentration of 1×10^{14} cm^{-3}. The Kirk current density for the on-set of the current-induced base in this case is 736 A cm^{-2} at a collector bias (V_{C}) of 250 V. The width of the current-induced base region is shown in Fig. 7.21 for this case as a function of the collector current density. Note that the current-induced base width becomes equal in magnitude to the physical base width at a collector current density of 1,000 A cm^{-2} in this example.

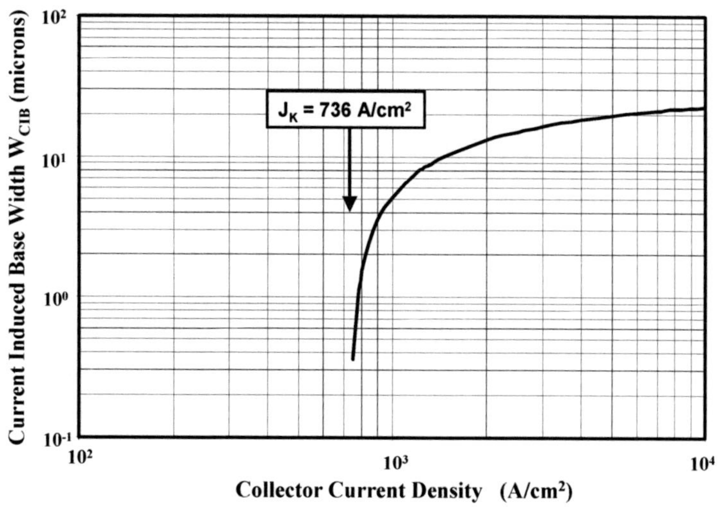

Fig. 7.21 Current-induced base width for the N–P–N bipolar power transistor

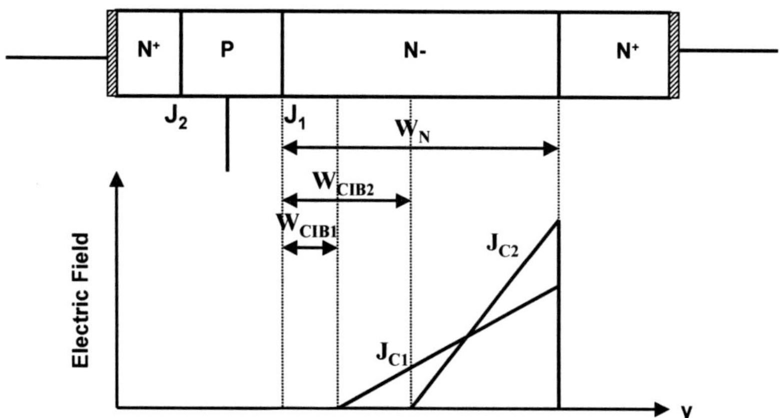

Fig. 7.22 Impact of the collector current density on the electric field profile for the N–P–N bipolar power transistor

The physical basis for the increase in the width of the current-induced base with increasing collector current density is related to the change in the electric field profile as illustrated in Fig. 7.22. At a lower collector current density (J_{C1}), the slope for the electric field profile is more gradual due to the smaller electron density in the drift region. When the collector current density is increased (J_{C2}), the slope for the electric field profile becomes steeper due to the larger electron density in the drift region. The same collector bias voltage is then supported across a smaller depletion width, with a larger maximum electric field, producing an enlargement of the current-induced base width.

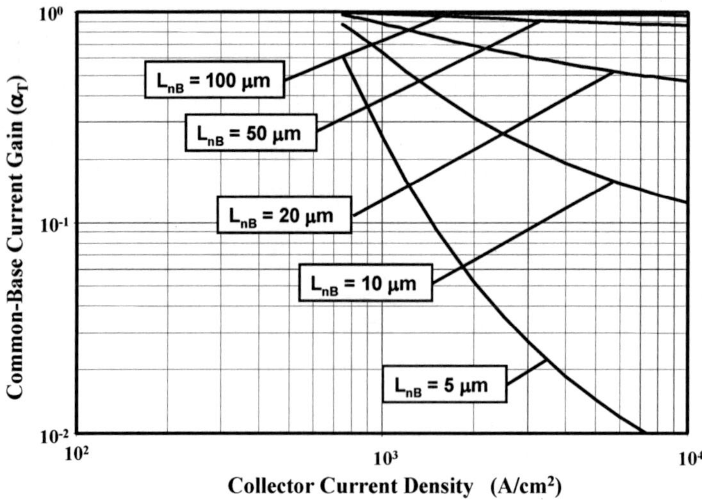

Fig. 7.23 Impact of base widening on the common-base current gain for the N–P–N bipolar power transistor

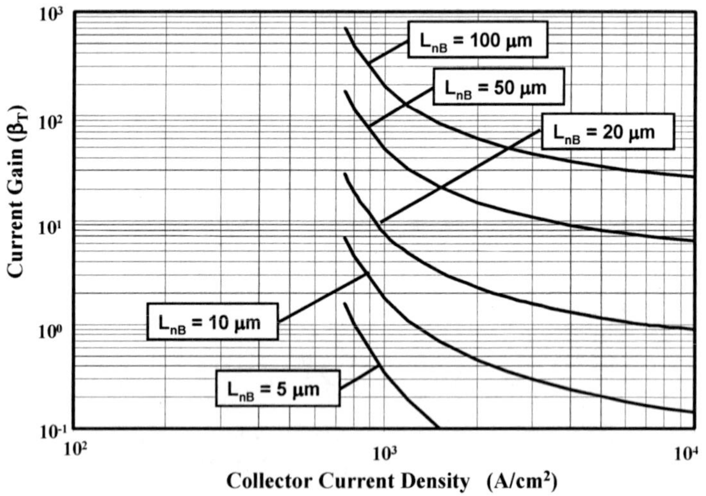

Fig. 7.24 Impact of base widening on the common-emitter current gain for the N–P–N bipolar power transistor

The impact of the current-induced base width on the common-base and common-emitter current gains is shown, respectively, in Figs. 7.23 and 7.24 for the power bipolar transistor with the parameters given above as an example. It can be observed that the current gain begins to decrease when the width of the current-induced base region becomes comparable with the physical base width at a collector current density of about 1,000 A cm^{-2}. This has a strong impact on the current gain when the diffusion length for the electrons is small. It is therefore necessary to account for this phenomenon when designing power bipolar transistors because high collector current densities are created by emitter current crowding under on-state operating conditions.

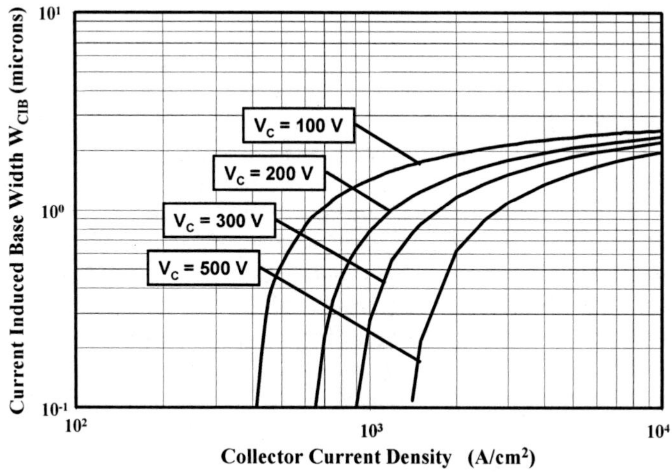

Fig. 7.25 Impact of the collector bias voltage on the current-induced base width for the N–P–N bipolar power transistor

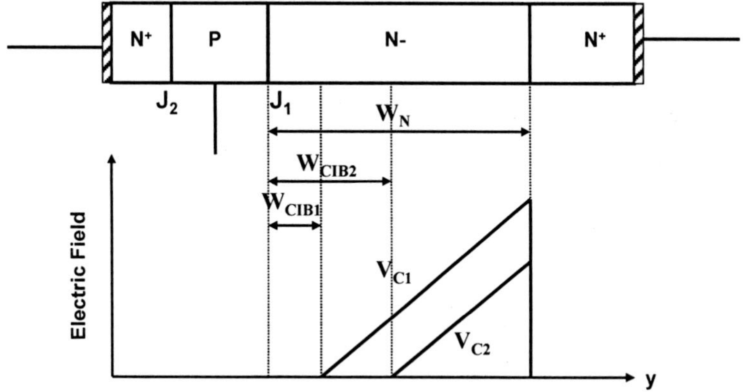

Fig. 7.26 Impact of the collector bias voltage on the electric field profile for the N–P–N bipolar power transistor

The Kirk current density and the width of the current-induced base region are dependent on the collector bias voltage as indicated by (7.87) and (7.91). This is illustrated, for the case of the power bipolar transistor with the parameters previously used in the example, in Fig. 7.25. As the collector bias voltage is reduced, the Kirk current density becomes smaller and the width of the current-induced base becomes larger. The physical basis for this change is related to the change in the electric field profile as illustrated in Fig. 7.26. It can be observed that a smaller collector bias voltage (V_{C2}) is supported over a smaller space-charge region width producing an enlargement of the current-induced base width.

Simulation Example

To provide further insight into the physics of operation of the power bipolar transistor, the results of two-dimensional numerical simulations are provided in this section for a variety of device structures. The doping profile for all the structures is similar to that shown in Fig. 7.8. The emitter and base doping concentrations were varied to elucidate the impact on the current gain. In addition, the impact of band-gap narrowing and Auger recombination in the emitter is provided here.

The baseline N–P–N bipolar power transistor structure (bjt8) had a Gaussian emitter diffusion with a surface concentration of 1×10^{20} cm^{-3} and a Gaussian P-base diffusion with a surface concentration of 2×10^{17} cm^{-3}. The junction depth for the P-base region was 10 μm and that for the emitter region was 1 μm leading to a base width of 9 μm. A lifetime (τ_{n0}, τ_{p0}) of 1 μs was used for all the structures. The output characteristics for the device were derived by using a base drive current density ranging from 1 to 7 A cm^{-2}. The resulting output characteristics are shown in Fig. 7.27. The current gain is observed to be relatively independent of the collector bias voltage once it exceeds 200 V due to the high output resistance.

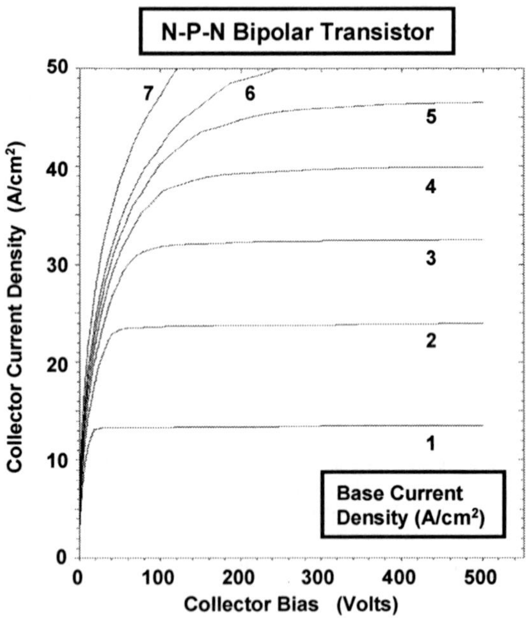

Fig. 7.27 Output characteristics for an N–P–N bipolar power transistor

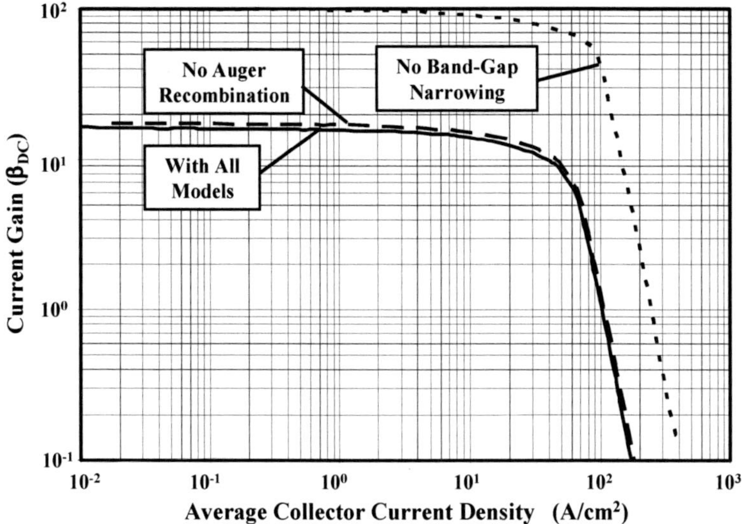

Fig. 7.28 Current gain (β) for an N–P–N bipolar power transistor

The current gain extracted from the simulations for the baseline bipolar power transistor structure (bjt8) is shown in Fig. 7.28 as a function of the collector current density. Note that the average collector current density (total collector current divided by the cell area) has been used for this plot. It can be observed that the current gain (β) is relatively constant with a value of about 15 for collector current densities below 10 A cm^{-2}. As the collector current density is increased beyond this point, the current gain falls off. The rate of falloff in the current gain is much steeper than predicted by either the Webster effect or the Kirk effect. This is due to strong impact of emitter current crowding as the base drive current is increased.

The current crowding at the emitter edge located in proximity to the base contact is shown in Fig. 7.29 for the baseline bipolar power transistor structure. It can be observed that the collector current density is uniform at the lowest base drive current of 0.05 A cm^{-2}. When the base drive current density is raised to 0.5 A cm^{-2}, the collector current density becomes nonuniform with a much larger value near the base contact. The physics responsible for the current crowding phenomenon is discussed in a later section of this chapter. As the base drive current density is increased, the nonuniformity of the collector current density becomes worse resulting in extremely high-current densities at the edge of the emitter. This enhancement of the collector current density exacerbates the Webster and Kirk effects, producing a faster falloff in the current gain with average collector current density than described by the models which are based upon a uniform (or local) current density.

The impact of the semiconductor models on the current gain for the baseline bipolar power transistor can be extracted from the numerical simulations by turning them on and off during the analysis. When the Auger recombination phenomenon is turned off, the current gain increases only slightly as shown in

Fig. 7.28 by the dashed line. In contrast, a much stronger effect is observed when the band-gap narrowing effect is turned off as shown by the dotted line in this figure. The current gain increases from about 14 to 75, which is consistent with the increase in the gain obtained by using (7.42) with equal values for the intrinsic carrier concentration in the emitter and base regions.

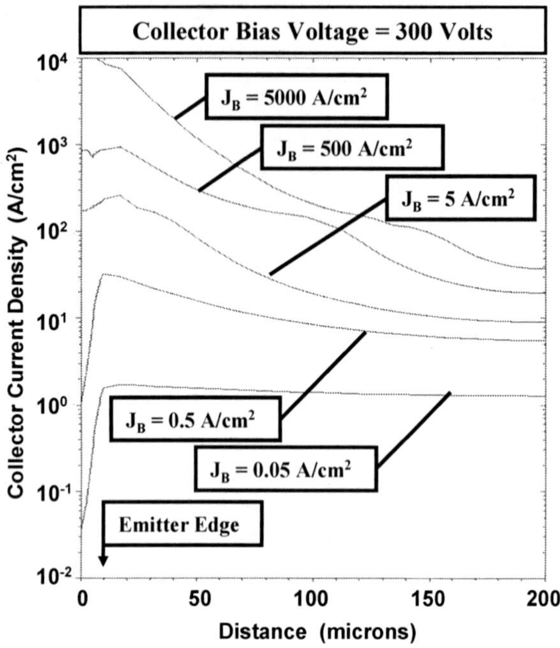

Fig. 7.29 Emitter current crowding in an N–P–N bipolar power transistor

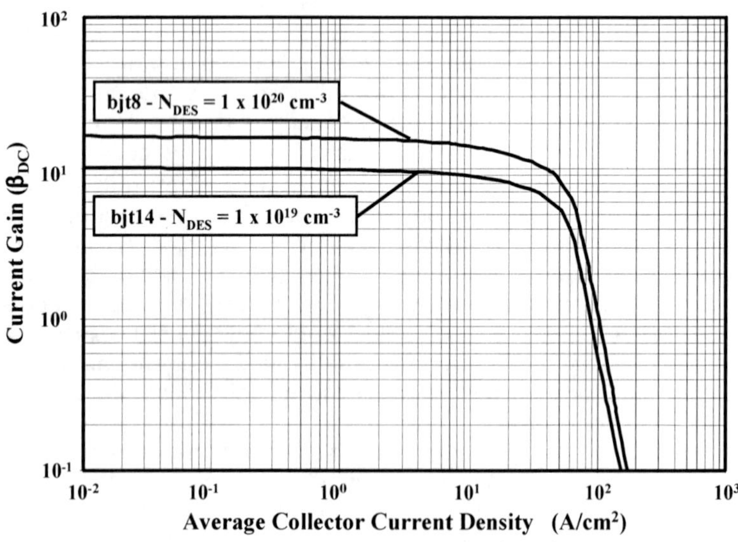

Fig. 7.30 Current gain (β) for N–P–N bipolar power transistors

The impact of the doping concentration of the emitter region on the gain of the bipolar power transistor is illustrated in Fig. 7.30. Here, the surface concentration of the N^+ emitter diffusion was changed from 1×10^{20} cm^{-3} for the baseline bipolar power transistor structure to 1×10^{19} cm^{-3}. It can be observed that the current gain at collector current densities below 10 A cm^{-2} decreases from about 15 to 10 due to the smaller emitter doping concentration. In these structures, the P-base doping profile was created using a surface concentration of 2×10^{17} cm^{-3}. From the doping profile shown in Fig. 7.8 for the baseline structure, it is obvious that the doping concentration in the emitter and base regions varies over many orders of magnitude. To relate the simulation results to the analytical model, it is therefore necessary to define an effective doping concentration for these regions representative of uniformly doped regions with the same thickness. From the doping profile provided in Fig. 7.8, an effective doping concentration of 7×10^{16} cm^{-3} can be estimated for the P-base region. At this doping concentration, the diffusion coefficient for electrons in the base is found to be 23.3 cm^2 s^{-1} and the intrinsic carrier concentration is found to be 1.57×10^{10} cm^{-3} based upon the models and information provided in Chap. 2. The corresponding parameters for the N^+ emitter region for the two cases are provided in Fig. 7.31 together with the value for β obtained by using these values in the analytical model (7.42). The analytical model predicts the appropriate value for the gain with judicious choice of the effective doping concentrations for the emitter and base regions.

Structure	Effective N_{DE} (cm^{-3})	D_{pE} (cm^2/s)	n_{iE} (cm^{-3})	Beta
bjt8	2×10^{19}	1.30	9.7×10^{10}	15.0
bjt14	5×10^{18}	3.52	3.69×10^{10}	9.5

Fig. 7.31 Analytically calculated parameters for N–P–N bipolar power transistors: impact of emitter doping concentration

The impact of changing the P-base doping profile can be observed in Fig. 7.32 where the current gain is plotted for various values for the surface concentration used to produce the P-base region. In these structures, the N^+ emitter doping profile was created using a surface concentration of 1×10^{20} cm^{-3}. As the doping concentration in the P-base region increases, the current gain decreases as expected. The P-base surface concentration must be reduced to 1×10^{17} cm^{-3} to obtain a gain of 20. This makes the structure prone to reach-through limited breakdown. To relate the current gain for these structures with the analytical model, it is necessary to extract an effective doping concentration for the P-base region. This is provided in Fig. 7.33 together with values for the diffusion coefficient for electrons and the intrinsic carrier concentration in the base region. The diffusion coefficient for holes and the intrinsic carrier concentration of the emitter region are given in Fig. 7.31 (structure bjt8) for this case. The current gain computed by using the analytical model (7.42) is provided in Fig. 7.33. Comparing these values with those in Fig. 7.32 obtained with the simulations at collector current densities below 1 A cm^{-2}, it can be concluded that the analytical model predicts the appropriate value for the gain with judicious choice of the effective doping concentrations for the emitter and base regions.

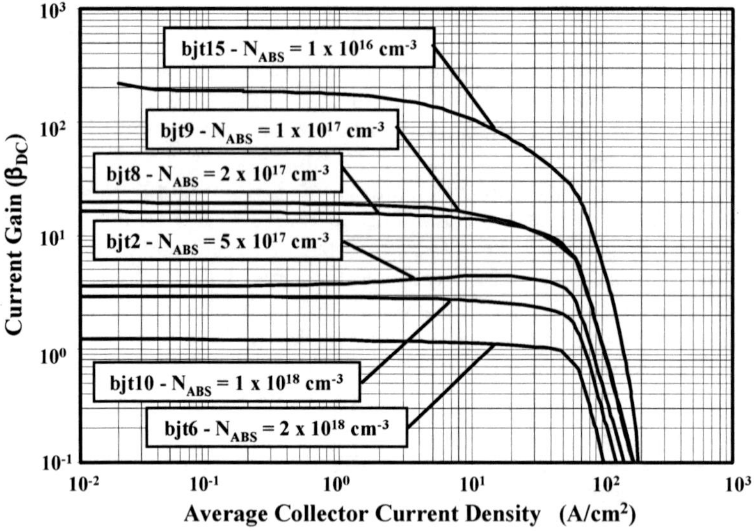

Fig. 7.32 Current gain (β) for N–P–N bipolar power transistors: impact of P-base doping concentration

Structure	Effective N_{AB} (cm^{-3})	D_{nB} (cm^2/s)	n_{iB} (cm^{-3})	Beta
bjt6	6×10^{17}	8.94	1.95×10^{10}	1.03
bjt10	3×10^{17}	12.85	1.77×10^{10}	2.45
bjt2	2×10^{17}	15.6	1.70×10^{10}	4.12
bjt8	7×10^{16}	23.3	1.57×10^{10}	15.0
bjt9	5×10^{16}	25.5	1.54×10^{10}	22.1
bjt15	6×10^{15}	33.3	1.50×10^{10}	228

Fig. 7.33 Analytically calculated parameters for N–P–N bipolar power transistors: impact of P-base doping concentration

As described earlier in the section, the reduction of the current gain at very high-current levels (the Kirk effect) is associated with a widening of the effective base width due to the formation of a current-induced neutral region within the N-drift region adjacent to the base region. The formation of the current-induced base region can be observed by examination of the electric field profile in the bipolar power transistor. As an example, the electric field profiles are shown in Fig. 7.34 for the baseline bipolar power transistor structure (bjt8) under increasing base drive current conditions at an *x* location of 30 μm from the left-hand side in the cross section shown in Fig. 7.7. At the lowest base drive current density of

0.05 A cm^{-2}, the peak of the electric field occurs at the junction between the P-base region and the N-drift region similar to the case of normal reverse-blocking conditions for a P–N junction. When the base drive current density is increased to 0.5 A cm^{-2}, the slope of the electric field profile becomes slightly smaller due to the additional (negative) charge associated with the electrons transported through the drift region. When the base drive current density is increased to 5 A cm^{-2}, the slope of the electric field profile becomes flat, indicating that the additional (negative) charge associated with the electrons transported through the drift region is now equal to the background charge (positive) due to donors in the drift region (5×10^{13} cm^{-3}). At an even larger base drive current density of 50 A cm^{-2}, the slope of the electric field profile reverses with its peak now located at the interface between the N-drift region and the N$^+$ substrate. Under these conditions, a current-induced base region is observed below the P-base region. When the base drive current density is increased to 500 A cm^{-2}, there is an increase in the width of the current-induced base region. These electric field profiles are similar to those shown in Figs. 7.19 and 7.22 for the analysis of the Kirk effect, confirming the validity of the underlying physical mechanism responsible for the reduction of the gain at very high collector current levels.

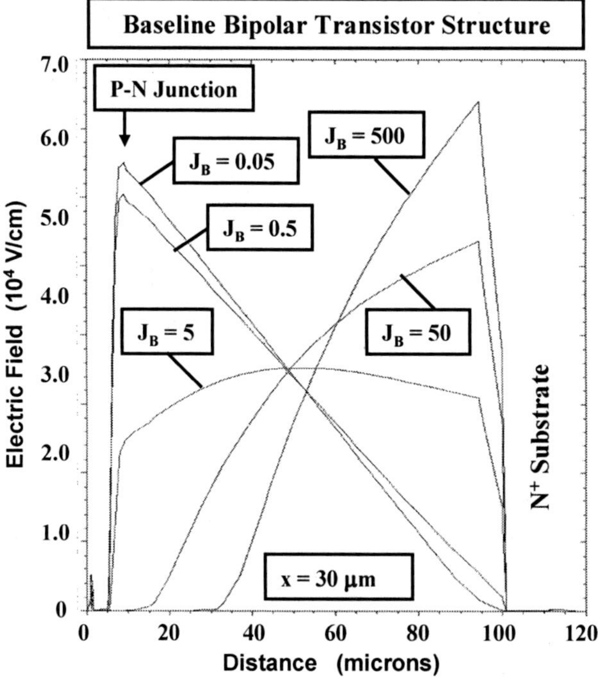

Fig. 7.34 Electric field profiles for an N–P–N bipolar power transistor when operating at high collector current densities

The impact of changes to the collector bias voltage on the width of the current-induced base region can be observed in Fig. 7.35. These profiles were obtained by using a base drive current density of 50 A cm^{-2}. The width of the current-induced base region becomes larger when the collector bias is reduced leading to a reduction of the current gain. These results are consistent with plots in Fig. 7.26 providing further validation for the model.

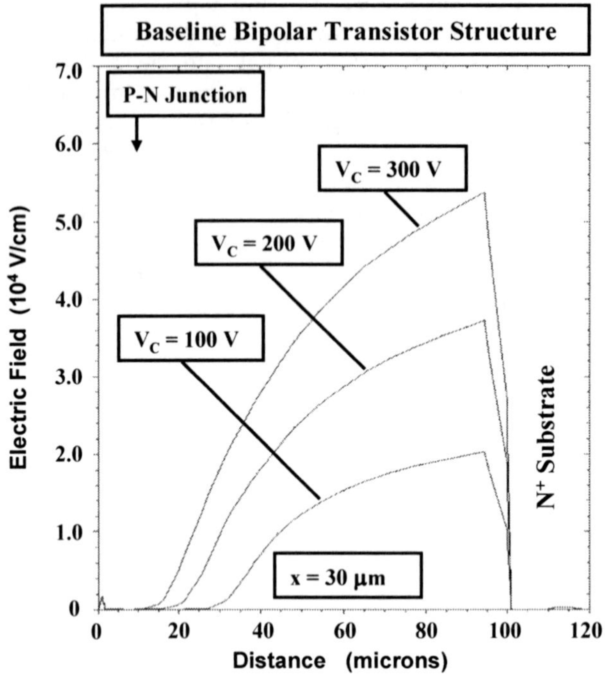

Fig. 7.35 Electric field profiles for an N–P–N bipolar power transistor when operating at high collector current densities

7.5 Emitter Current Crowding

The analytical treatment for the current transport within the bipolar power transistor was performed in the previous sections under one-dimensional conditions. This analysis is pertinent for a local region under the emitter within the transistor. The current distribution within the emitter in a bipolar power transistor is however not uniform. The drive current applied at the base terminal of the bipolar power transistor must flow through the base region to reach the center of the emitter finger as illustrated in Fig. 7.36. A voltage drop is produced by this current flow in the P-base region due to its finite resistance (R_B). The largest forward bias across the emitter–base junction then occurs in the vicinity of the base contact with smaller base–emitter voltage available at the center of the emitter finger (location A) to induce injection of electrons into the base region. Consequently, the emitter (and collector) current density are nonuniform with their highest values under the emitter region located in proximity with the base contact. Analytical treatment of the emitter current crowding phenomenon can be performed under low-level and high-level injection conditions in the P-base region.

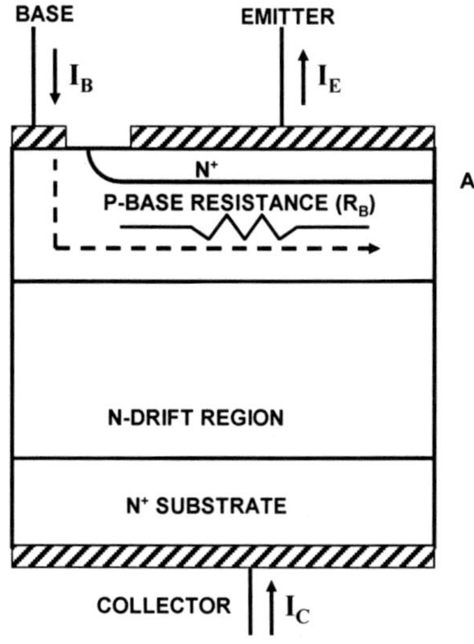

Fig. 7.36 Base drive current flow within the N–P–N bipolar power transistor

7.5.1 Low-Level Injection in the Base

Analytical treatment for the emitter current crowding under low-level injection conditions can be performed under certain simplifying assumptions.[14,15] A cross section of the bipolar power transistor is illustrated in Fig. 7.37 with half of the emitter finger with a width of W_E. A linear geometry is assumed for the transistor with an emitter finger length L_E orthogonal to the cross section in the figure. Consider a segment of the emitter finger of width dx located at a distance of x from the edge of the emitter located closest to the base contact. The emitter current flowing through the segment is given by

$$\mathrm{d}I_{\mathrm{E}}(x) = J_{\mathrm{E}}(x)L_{\mathrm{E}}\mathrm{d}x, \tag{7.93}$$

where $J_E(x)$ is the local emitter current density at a distance x from the edge of the emitter located closest to the base contact. The base current required to support this emitter current is given by

$$\mathrm{d}I_{\mathrm{B}}(x) = (1-\alpha)\mathrm{d}I_{\mathrm{E}}(x) = (1-\alpha)J_{\mathrm{E}}(x)L_{\mathrm{E}}\mathrm{d}x. \tag{7.94}$$

In writing this equation, it has been assumed that the current gain is independent of the emitter current density to simplify the analysis.

The voltage drop produced in the base region across the segment due to the base current flow is given by

$$\mathrm{d}V_{\mathrm{BE}}(x) = \mathrm{d}I_{\mathrm{B}}(x)\mathrm{d}R_{\mathrm{B}} = \mathrm{d}I_{\mathrm{B}}(x)\left(\frac{\rho_{\mathrm{B0}}\mathrm{d}x}{W_{\mathrm{B}}L_{\mathrm{E}}}\right), \tag{7.95}$$

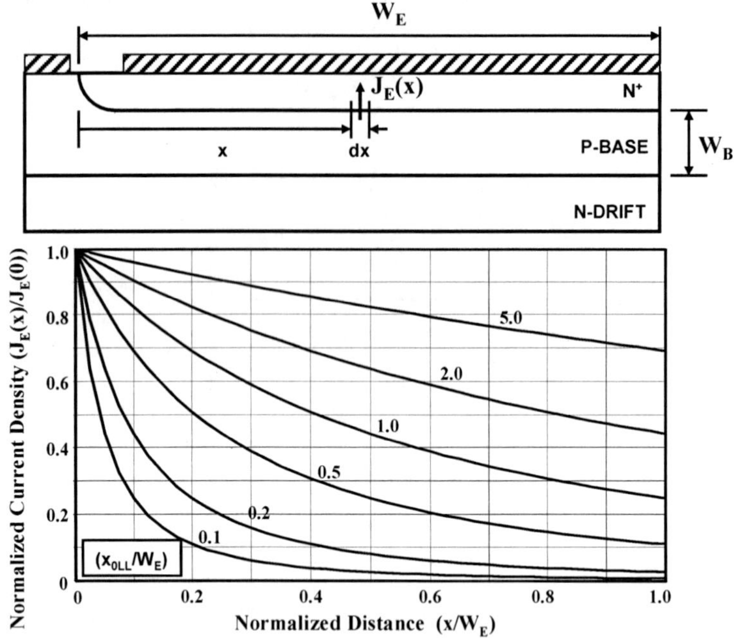

Fig. 7.37 Emitter current distribution within the bipolar power transistor

where ρ_{B0} is the resistivity of the P-base region:

$$\rho_{B0} = \frac{1}{q\mu_p N_{AB}}, \tag{7.96}$$

where N_{AB} is the acceptor doping concentration for the P-base region. In writing this equation, any depletion of the P-base region due to the applied collector bias has been neglected. Transposing dx and then taking the derivative of (7.95),

$$\frac{d^2V_{BE}(x)}{dx^2} = \frac{\rho_{B0}}{W_B L_E}\frac{dI_B(x)}{dx}. \tag{7.97}$$

Using (7.94),

$$\frac{d^2V_{BE}(x)}{dx^2} = \frac{\rho_{B0}}{W_B}(1-\alpha)J_E(x). \tag{7.98}$$

The emitter current density at location x is determined by the local forward voltage drop across the base–emitter junction:

$$J_E(x) = J_{S,LL}e^{qV_{BE}/kT}, \tag{7.99}$$

where $J_{S,LL}$ is the diode saturation current density under low-level injection conditions (see (5.14) derived for a diode operating under low-level injection conditions). Using this expression in (7.98) yields

$$\frac{d^2V_{BE}(x)}{dx^2} = \frac{\rho_{B0}}{W_B}(1-\alpha)J_{S,LL}e^{qV_{BE}(x)/kT}. \tag{7.100}$$

A general solution for this equation is of the form

$$V_{BE}(x) = A\ln(Bx + C), \tag{7.101}$$

where A, B, and C are constants. Substitution into (7.100) allows extraction of these constants:

$$A = -\frac{2kT}{q}, \tag{7.102}$$

$$B = -\sqrt{\frac{q\rho_{B0}}{2kTW_B}(1-\alpha)J_{S,LL}}, \tag{7.103}$$

$$C = e^{-qV_{BE}(0)/2kT}, \tag{7.104}$$

where $V_{BE}(0)$ is the base–emitter bias at the edge of the emitter closest to the base contact. Using this voltage distribution in (7.99) yields

$$J_E(x) = \frac{J_E(0)}{(1 + x/x_{0LL})^2}, \tag{7.105}$$

where x_{0LL} is a *current crowding parameter under low-level injection conditions in the P-base region* given by

$$x_{0LL} = \sqrt{\frac{2kTW_B}{q(1-\alpha)\rho_{B0}J_E(0)}}. \tag{7.106}$$

The emitter current density $J_E(0)$ at the edge of the emitter closest to the base contact can be obtained by using (7.99) with the applied base–emitter bias $V_{BE}(0)$ at the base contact.

The distribution of the emitter current within the bipolar power transistor is shown in the lower part of Fig. 7.37 for various cases of the ratio (x_{0LL}/W_E). It can be observed from the plots that the emitter current density is larger at the edge of the emitter closest to the base contact. This is referred to as the *emitter current crowding*. The emitter current crowding is observed to worsen as the ratio (x_{0LL}/W_E) becomes smaller. Using a typical base width of 10 μm, a current gain (α) of 0.95, and a P-base doping concentration of 1×10^{17} cm^{-3} yields a value for x_{0LL} of 62 μm at an emitter current density of 100 A cm^{-2} at the edge of the emitter closest to the base contact. This corresponds to a ratio (x_{0LL}/W_E) of 0.31 for an emitter width of 200 μm, which would result in significant current crowding as shown in Fig. 7.37. According to (7.106), the value for x_{0LL} becomes larger when the current gain increases. This is due to the smaller base current required to sustain current flow within the transistor, which produces a smaller voltage drop along the base resistance. The value for x_{0LL} becomes larger when the base resistivity is reduced by increasing its doping level. However, the resulting improvement in current distribution is obtained with a lower current gain, partially canceling the benefit derived from the lower base resistivity.

The current crowding parameter is a function of the doping concentration of the P-base region because this determines not only the resistivity of the P-base region but also the current gain (α) of the transistor. As the doping concentration of the P-base region is increased, the resistivity decreases while the current gain becomes smaller due to a reduction of the emitter injection efficiency. The net result is a relatively weak dependence of the current crowding parameter on the doping concentration in the P-base region with a minimum value at a doping concentration of about 5×10^{17} cm^{-3} as illustrated in Fig. 7.38.

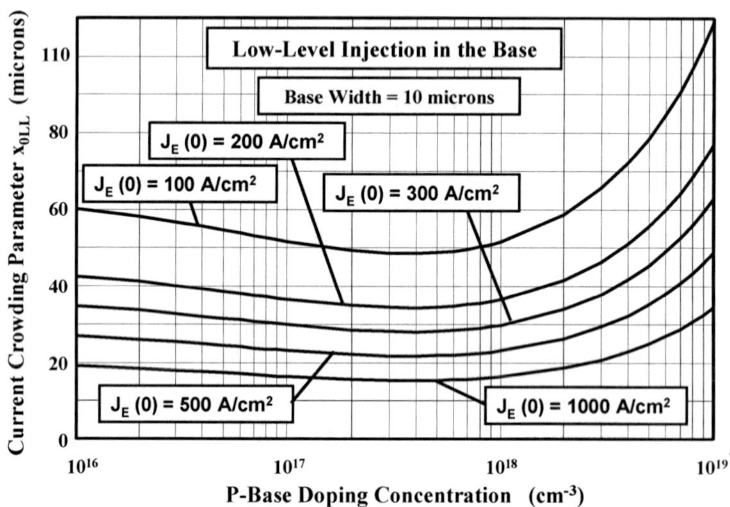

Fig. 7.38 Current crowding parameter for the N–P–N bipolar power transistor for low-level injection conditions in the P-base region

However, the current crowding parameter decreases significantly with increasing emitter current density at the edge of the emitter closest to the base contact. This can also be observed in Fig. 7.38 where the emitter current density at the edge of the emitter has been used as a parametric variable. This implies that the current crowding will become worse as the transistor is driven harder to try to obtain a larger collector current. The enhanced current crowding aggravates the current gain reduction mechanisms producing a more rapid drop-off in the current gain (see simulation results in Sect. 7.4). Note that the impact of high emitter current density on the current gain has been ignored when making the plots in Fig. 7.38 to simplify the analysis. Inclusion of the impact of the reduction of the current gain due to high-level injection in the P-base region will make the current crowding parameter smaller, enhancing the current crowding of the emitter current density.

7.5.2 High-Level Injection in the Base

Analytical treatment for the emitter current crowding under high-level injection conditions in the P-base region can be performed under certain simplifying assumptions.[16,17] A cross section of the bipolar power transistor is illustrated in Fig. 7.39 with half of the emitter finger with a width of W_E. A linear geometry is assumed for the transistor with an emitter finger length L_E orthogonal to the cross section in the figure. Consider a segment of the emitter finger of width dx located at a distance of x from the edge of the emitter located closest to the base contact. The emitter current flowing through the segment is given by

$$\mathrm{d}I_{\mathrm{E}}(x) = J_{\mathrm{E}}(x) L_{\mathrm{E}} \mathrm{d}x, \tag{7.107}$$

where $J_E(x)$ is the local emitter current density at a distance x from the edge of the emitter located closest to the base contact. The base current required to support this emitter current is given by

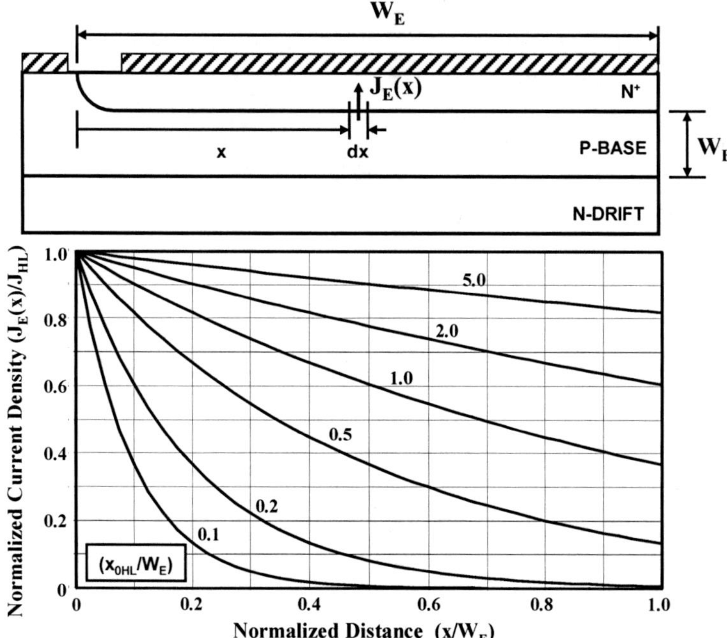

Fig. 7.39 Emitter current distribution within the bipolar power transistor

$$\mathrm{d}I_{\mathrm{B}}(x) = (1-\alpha)\mathrm{d}I_{\mathrm{E}}(x) = (1-\alpha)J_{\mathrm{E}}(x)L_{\mathrm{E}}\mathrm{d}x. \tag{7.108}$$

In writing this equation, it has been assumed that the current gain is independent of the emitter current density to simplify the analysis.

The voltage drop produced in the base region across the segment due to the base current flow is given by

$$\mathrm{d}V_{\mathrm{BE}}(x) = \mathrm{d}I_{\mathrm{B}}(x)\mathrm{d}R_{\mathrm{B}} = \mathrm{d}I_{\mathrm{B}}(x)\left(\frac{\rho_{\mathrm{B}}\mathrm{d}x}{W_{\mathrm{B}}L_{\mathrm{E}}}\right), \tag{7.109}$$

where ρ_{B} is the resistivity of the P-base region. Unlike the case for low-level injection where the resistivity is determined by the doping concentration of the P-base region, the resistivity for the P-base region under high-level injection conditions is given by

$$\rho_{\mathrm{B}} = \rho_{\mathrm{B0}}\frac{J_{\mathrm{HL}}}{J_{\mathrm{E}}(x)}, \tag{7.110}$$

where J_{HL} is the emitter current density for the on-set of high-level injection in the P-base region. The emitter current density at location x is determined by the local forward voltage drop across the base–emitter junction:

$$J_{\mathrm{E}}(x) = J_{\mathrm{S,HL}}\mathrm{e}^{qV_{\mathrm{BE}}/2kT}, \tag{7.111}$$

where $J_{\mathrm{S,HL}}$ is the diode saturation current density under high-level injection conditions (see (5.57) derived for a diode operating under high-level injection conditions). Taking the first derivative with respect to x yields

$$\frac{\mathrm{d}J_{\mathrm{E}}(x)}{\mathrm{d}x} = \frac{qJ_{\mathrm{E}}(x)}{2kT}\frac{\mathrm{d}V_{\mathrm{BE}}(x)}{\mathrm{d}x}. \tag{7.112}$$

Using (7.109) in the above expression,

$$\frac{\mathrm{d}J_{\mathrm{E}}(x)}{\mathrm{d}x} = \frac{qJ_{\mathrm{E}}(x)}{2kT}\frac{\rho_{\mathrm{B}}\mathrm{d}I_{\mathrm{B}}(x)}{W_{\mathrm{B}}L_{\mathrm{E}}}. \tag{7.113}$$

Using (7.110) for the conductivity-modulated resistivity gives

$$\frac{\mathrm{d}J_{\mathrm{E}}(x)}{\mathrm{d}x} = \frac{qJ_{\mathrm{HL}}}{2kT}\frac{\rho_{\mathrm{B0}}\mathrm{d}I_{\mathrm{B}}(x)}{W_{\mathrm{B}}L_{\mathrm{E}}}. \tag{7.114}$$

Taking the derivative of this equation yields

$$\frac{\mathrm{d}^2J_{\mathrm{E}}(x)}{\mathrm{d}x^2} = \frac{qJ_{\mathrm{HL}}}{2kT}\frac{\rho_{\mathrm{B0}}}{W_{\mathrm{B}}L_{\mathrm{E}}}\frac{\mathrm{d}I_{\mathrm{B}}(x)}{\mathrm{d}x}. \tag{7.115}$$

Utilizing (7.108), a second-order differential equation for the emitter current distribution is obtained:

$$\frac{d^2 J_E(x)}{dx^2} - \frac{qJ_{HL}}{2kT}\frac{\rho_{B0}}{W_B}(1-\alpha)J_E(x) = 0. \tag{7.116}$$

The solution for this differential equation is

$$J_E(x) = J_E(0)e^{-x/x_{0HL}}, \tag{7.117}$$

where x_{0HL} is a *current crowding parameter under high-level injection conditions in the P-base region* given by

$$x_{0HL} = \sqrt{\frac{2kTW_B}{q(1-\alpha)\rho_{B0}J_{HL}}}. \tag{7.118}$$

The distribution of the emitter current within the bipolar power transistor is shown in the lower part of Fig. 7.39 for various cases of the ratio (x_{0HL}/W_E). It can be observed from the plots that the emitter current density is larger at the edge of the emitter closest to the base contact. The emitter current crowding is observed to worsen as the ratio (x_{0HL}/W_E) becomes smaller. Using a typical base width of 10 μm, a current gain (α) of 0.95, and a P-base doping concentration of 1×10^{17} cm^{-3} yields a value for x_{0HL} of 62 μm at an emitter high-level injection current density of 100 A cm^{-2}. This corresponds to a ratio (x_{0HL}/W_E) of 0.31 for an emitter width of 200 μm, which would result in significant current crowding as shown in Fig. 7.39. According to (7.118), the value for x_{0HL} becomes larger when the current gain increases. This is due to the smaller base current required to sustain current flow within the transistor, which produces a smaller voltage drop along the base resistance. The value for x_{0HL} becomes larger when the base resistivity is reduced by increasing its doping level. However, the resulting improvement in current distribution is obtained with a lower current gain, partially canceling the benefit derived from the lower base resistivity.

The current crowding parameter under high-level injection conditions is a function of the doping concentration of the P-base region because this determines not only the resistivity of the P-base region but also the current gain (α) of the transistor. As the doping concentration of the P-base region is increased, the resistivity decreases while the current gain becomes smaller due to a reduction of the emitter injection efficiency. The net result is a relatively weak dependence of the current crowding parameter on the doping concentration in the P-base region with a minimum value at a doping concentration of about 5×10^{17} cm^{-3} as illustrated in Fig. 7.38 by the uppermost curve. However, the current crowding parameter under high-level injection conditions is not dependent on the emitter current density because of a proportionate decrease in the resistivity of the P-base region.

Simulation Example

To validate the models for the emitter current crowding, the results of two-dimensional numerical simulations for the power bipolar transistor are described here. In all cases, the emitter current distribution was obtained at a collector bias of 300 V and a base drive current density of 5 A cm^{-2}. The collector current density is shown in Fig. 7.40 for three device structures designed with P-base surface concentrations of 2×10^{17}, 5×10^{17}, and 1×10^{18} cm^{-3}, respectively. The width of the P-base region for all the structures is close to 9 μm as indicated by the doping profile in Fig. 7.8. In all the cases, a much greater collector current density is observed at the edge of the emitter closest to the base contact. When the doping concentration of the P-base region is increased, the current gain becomes smaller due to the reduced injection efficiency. This leads to a reduction of the collector current density for these structures as observed in Fig. 7.40.

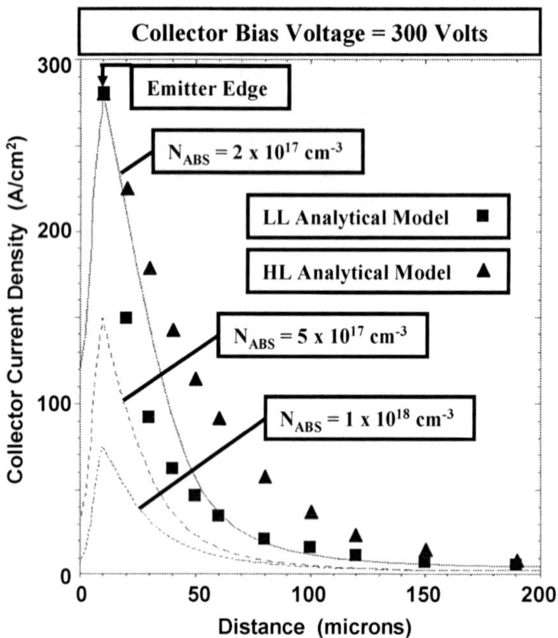

Fig. 7.40 Emitter current crowding in bipolar power transistors

To relate the results of the numerical simulations to the two models described in this section, the collector current density computed by using these models is shown by the square and triangular symbols in Fig. 7.40 for the device (baseline structure bjt8) with a P-base surface concentration of 2×10^{17} cm^{-3}. For the models, an effective P-base doping concentration of 7×10^{16} cm^{-3} was used for consistency with the results discussed in Sect. 7.4 on current gain. It can be observed, on the one hand, that the values computed by using the model based upon low-level injection conditions in the P-base region are in agreement with the simulation results when the collector current density falls below 100 A cm^{-2}. On the other hand, the values computed by using the model based upon high-level injection conditions in the P-base region are in agreement with the simulation results when the collector current density is greater than 150 A cm^{-2}. Using (7.59), the electron concentration $n_B(0)$ is 2.4×10^{16} cm^{-3} at a collector current density of 100 A cm^{-2}. Consequently, a significant portion of the P-base region operates under high-level injection conditions when the collector current density approaches and exceeds this value. The observed emitter current distribution is therefore determined by a portion of the emitter near the base contact operating under high-level injection conditions while the rest of the emitter operates under low-level injection conditions. Consequently, both models for the emitter current crowding within the bipolar power transistor are required to describe the current distribution within the emitter finger.

7.5.3 Emitter Geometry

From the previous sections, it is apparent that most of the collector current flows at the edge of the emitter finger located closest to the base contact electrode. Consequently, a large amount of surface area can be wasted in the bipolar power transistor if the emitter finger is made wider than about twice the parameter x_0 described in the previous sections. Based upon the information provided in Fig. 7.38, this corresponds to a typical emitter width of about 100 μm. In addition, it is necessary to surround the emitter finger with the base contact fingers to ensure maximum utility of the emitter fingers. A typical device layout that accomplishes these objectives is illustrated in Fig. 7.41. This layout is referred to as the *interdigitated finger geometry*. Note the presence of base contact fingers on the outside edges of the emitter fingers. The base contact fingers can be made narrower than the emitter contact fingers because of the smaller currents flowing through them. The edge termination surrounds the P-base region to protect the base–collector junction.

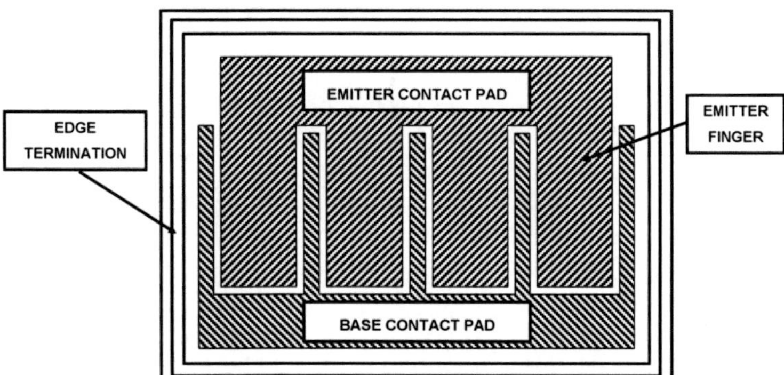

Fig. 7.41 Die layout for a typical bipolar power transistor

7.6 Output Characteristics

The typical output characteristics for the power bipolar transistor structure are shown in Fig. 7.42. The slope of the output characteristics has been exaggerated to discuss the output resistance of the device. The finite output resistance for the bipolar transistor is related to a change in the current gain with increasing collector bias voltage. As the collector bias is increased, the depletion region across the base–collector junction expands within the P-base region even though most of the voltage is supported across the collector drift region. The electrons injected from the base–emitter junction have to diffuse only through the undepleted portion of the P-base region. Consequently, the current gain increases due to an increase in the injection efficiency and base transport factor because the base width that determines the current gain becomes smaller than the physical base width. For a constant base drive current, the collector current increases with increasing collector bias voltage because of the increase in the current gain.

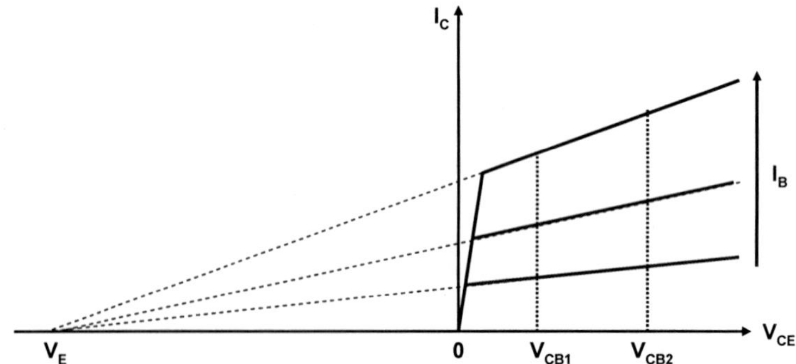

Fig. 7.42 Typical output characteristics for the bipolar power transistor

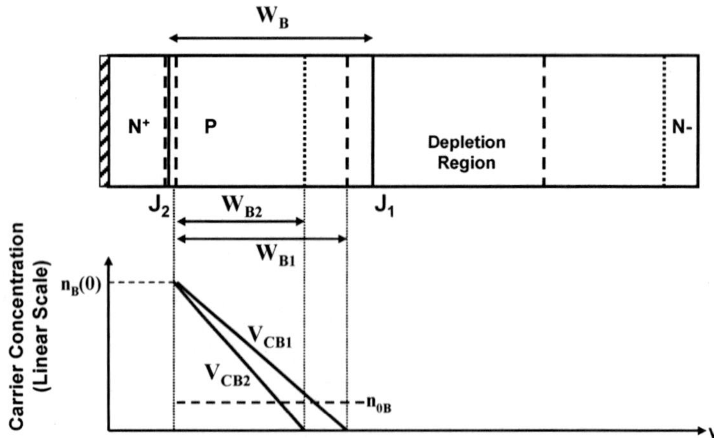

Fig. 7.43 Effect of collector bias on the minority carrier profile in the base region for the bipolar power transistor

The impact of the depletion of the P-base region by the applied collector bias voltage on the minority carrier profile is illustrated in Fig. 7.43. The depletion layer boundaries are shown by the dashed line at a bias V_{CB1} while those for a larger bias V_{CB2} are indicated by the dotted lines. If the forward bias across the base–emitter junction (and the base current) is held constant, the injected electron concentration at the base–emitter junction $n_B(0)$ is equal for the two cases as shown in the lower portion of the figure. The minority carrier density is forced to zero at the edge of the depletion region of the reverse-biased base–collector junction. Consequently, the slope of the minority carrier profile in the P-base region becomes greater for the case of a larger collector bias voltage. Since the collector current density is proportional to the slope of the minority carrier profile at the base–collector junction, this implies an increase in the collector current with increasing collector bias voltage as illustrated in Fig. 7.42.

The shape of the output characteristics can be analyzed by taking into account the width of the depletion region within the P-base region. The depletion layer width in the P-base region is given by[8]

$$W_{BD} = \sqrt{\frac{2\varepsilon_S V_{CB}}{qN_{AB}}\left(\frac{N_D}{N_{AB}+N_D}\right)}, \tag{7.119}$$

where V_{CB} is the base–collector reverse bias voltage, N_{AB} is the doping concentration in the P-base region, and N_D is the doping concentration in the N-drift region. This expression accounts for a finite doping concentration on both sides of the P–N junction. The width of the undepleted portion of the P-base region is then given by

$$W_{B,UD} = W_B - W_{BD} = W_B - \sqrt{\frac{2\varepsilon_S V_{CB}}{qN_{AB}}\left(\frac{N_D}{N_{AB}+N_D}\right)} \tag{7.120}$$

corresponding to a base–collector bias V_{CB}. (The collector–emitter voltage is approximately equal to the base–collector voltage here.) The collector current density for this case is given by

$$J_C = qD_{nB}\frac{dn}{dx} = qD_{nB}\frac{n_B(0)}{W_{B,UD}}. \tag{7.121}$$

Using (7.120),

$$J_C = qD_{nB}\frac{n_B(0)}{W_B}\frac{1}{1-\sqrt{2\varepsilon_S V_{CB} N_D/[qN_{AB}W_B^2(N_{AB}+N_D)]}}. \tag{7.122}$$

The collector current density at small collector bias voltages is given by

$$J_{C0} = qD_{nB}\frac{n_B(0)}{W_B} \tag{7.123}$$

because the depletion of the P-base region can be neglected. Consequently,

$$J_C = J_{C0} \frac{1}{1 - \sqrt{2\varepsilon_S V_{CB} N_D / [q N_{AB} W_B^2 (N_{AB} + N_D)]}}. \tag{7.124}$$

When the base doping concentration (N_{AB}) is much greater than the drift region doping concentration (N_D), the term within the square root in the denominator becomes much smaller than unity. In this case, the application of Maclaurin series gives

$$J_C \approx J_{C0} \left[1 + \frac{\varepsilon_S N_D V_{CB}}{q N_{AB} W_B^2 (N_{AB} + N_D)} \right]. \tag{7.125}$$

This expression implies a linear output characteristic as illustrated in Fig. 7.42. Extrapolation of the collector current density to zero, as indicated by the dashed lines in the figure, provides a voltage intercept on the negative *x*-axis which is referred[18] to as the *Early voltage* (V_E). From (7.125), the Early voltage is given by

$$V_E = \frac{q W_B^2}{\varepsilon_S} \left(\frac{N_{AB}}{N_D} \right) (N_{AB} + N_D). \tag{7.126}$$

Note that, according to this equation, the Early voltage is independent of the collector current density resulting in all the extrapolated output characteristics intersecting the *x*-axis at the same voltage.[8]

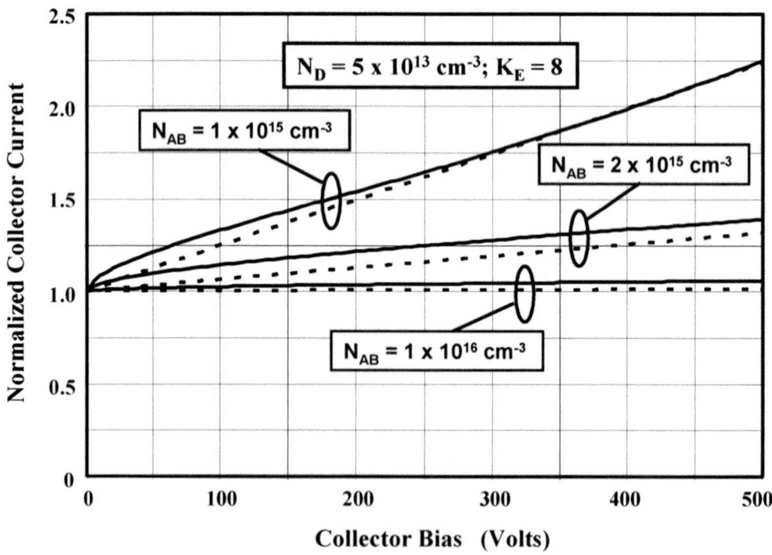

Fig. 7.44 Normalized output characteristics for the bipolar power transistor

The output characteristics computed by using the model in (7.124) are shown in Fig. 7.44 for three values for the doping concentration of the P-base region by the solid lines. In all the cases, a doping concentration of 5×10^{13} cm^{-3} was used for the N-drift region together with a P-base width of 10 μm. As expected, the slope of the output characteristics decreases when the doping concentration of the P-base region is increased. It is worth pointing out that the output characteristics are not perfectly straight lines.

When the linear output characteristic model described by (7.125) is used, the increase in the collector current is found to be much smaller than that predicted by (7.124). A match to the nonlinear model can be obtained by including a constant (K_E) in (7.125):

$$J_C \approx J_{C0}\left[1+\frac{K_E \varepsilon_S N_D V_{CB}}{qN_{AB}W_B^2(N_{AB}+N_D)}\right]. \tag{7.127}$$

The results obtained for the three values for the P-base doping concentration are shown in Fig. 7.44 by the dashed lines for $K_E = 8$.

The specific output impedance for the bipolar power transistor can be derived from the output characteristics described by (7.124):

$$\frac{dJ_C}{dV_{CB}} = \frac{1}{R_{O,sp}}. \tag{7.128}$$

Taking the derivative of (7.124) yields

$$R_{O,sp} = \sqrt{\frac{2qN_{AB}W_B^2(N_{AB}+N_D)V_{CB}}{\varepsilon_S N_D J_{C0}}}\left(1-\sqrt{\frac{2\varepsilon_S N_D V_{CB}}{qN_{AB}W_B^2(N_{AB}+N_D)}}\right). \tag{7.129}$$

According to this relationship, the output impedance is a function of the collector bias voltage and the collector current density. A large output resistance is desirable for power transistors. This can be achieved for the bipolar power transistor by using a larger doping concentration for the P-base region. However, this will reduce the current gain making the control circuit more expensive and bulky.

Simulation Example

To validate the model for the output characteristics of the bipolar power transistor, the results of two-dimensional numerical simulations are described for several devices with different P-base doping concentrations. In all cases, a doping concentration of 5×10^{13} cm^{-3} was used for the N-drift region. The physical base width (W_B) for all the devices was 9 μm. The output characteristics for the baseline structure (bjt8) were previously shown in Fig. 7.27. Due to the relatively large surface doping concentration (2×10^{17} cm^{-3}) for the base region for this structure, the output characteristics are flat with a high output resistance.

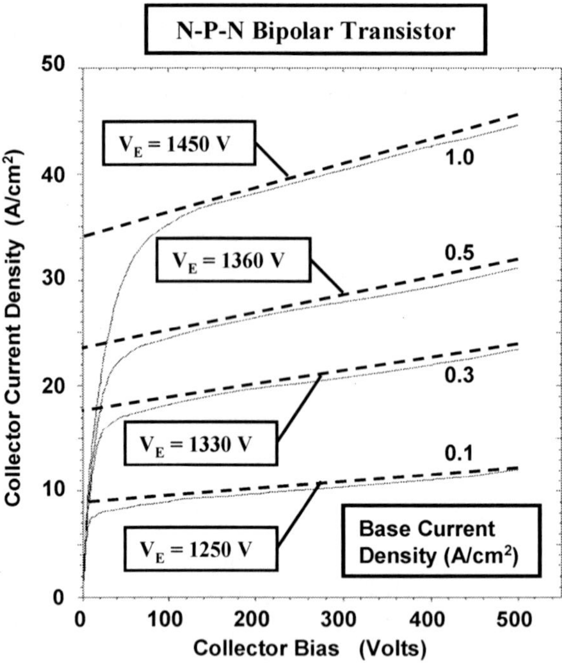

Fig. 7.45 Output characteristics for a bipolar power transistor

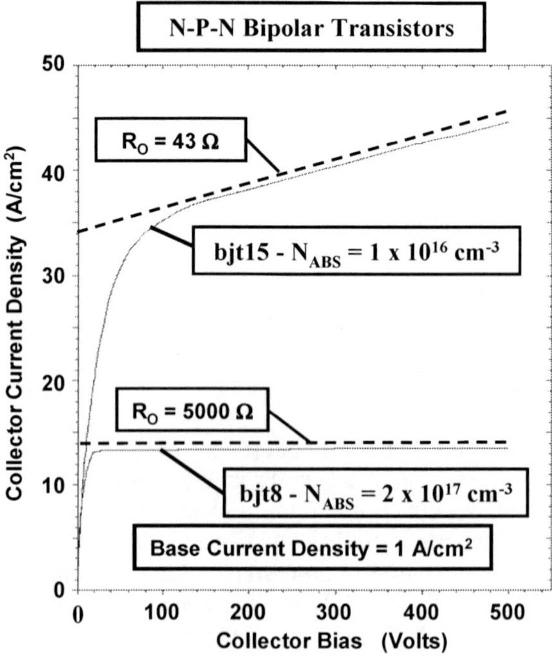

Fig. 7.46 Comparison of the output characteristics for bipolar power transistors

The output characteristics for the bipolar power transistor structure (bjt15) with a P-base surface concentration reduced to 1×10^{16} cm^{-3} are shown in Fig. 7.45. Due to the much larger current gain for this structure, the base drive current has been reduced by an order of magnitude. This structure (bjt15) exhibits output characteristics with a distinct slope indicating a low output resistance. The dashed lines in the figure have been drawn (slightly above the simulation data for clarity) to provide a linear fit to the output characteristics. The Early voltage extracted by using these lines is also given in the figure. The values are approximately equal, indicating that all the lines intersect the *x*-axis at the same voltage.

The impact of reducing the surface concentration of the P-base diffusion on the output characteristics of the bipolar power transistor can be observed in Fig. 7.46. In this figure, the characteristics for the structure with P-base surface concentration of 1×10^{16} and 2×10^{17} cm^{-3} are displayed at the same base drive current density. The dashed lines in the figure have been drawn (slightly above the simulation data for clarity) to provide a linear fit to the output characteristics. The specific output resistance extracted by using these lines is 43 Ω cm^2 for the structure with P-base surface concentration of 1×10^{16} cm^{-3} and 5,000 Ω cm^2 for the structure with P-base surface concentration of 2×10^{17} cm^{-3}. From these results, it is apparent that the P-base doping concentration must be typically above 1×10^{17} cm^{-3} to obtain a high output resistance for the bipolar power transistor.

7.7 On-State Characteristics

The output characteristics for a bipolar power transistor are shown in Fig. 7.47 at low collector bias voltages for the common-emitter configuration. In addition to the *forward active region* where the collector current is determined by the base drive current via the current gain of the transistor, two new regions of operation are evident. At very low collector bias voltages, the transistor operates in its *saturation*

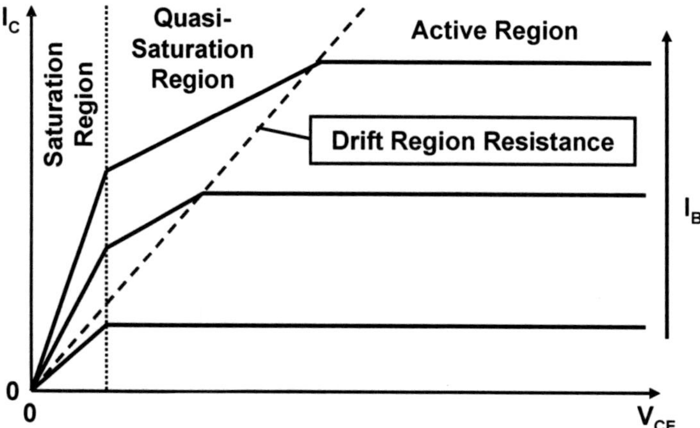

Fig. 7.47 Output characteristics for the bipolar power transistor at low collector bias voltages

region. In addition, there is a gradual transition between the saturation region and the active region called the *quasisaturation region*.

In the saturation region of operation, the collector bias voltage becomes sufficiently low, so that the base–collector junction becomes forward biased because the collector bias voltage falls below the base drive voltage. This produces an injection of minority carriers into the very lightly doped N-drift region. The injected carrier concentration in the N-drift region exceeds its doping level at typical operating collector current densities leading to conductivity modulation of the N-drift region. This is responsible for the low on-state voltage drop when the bipolar power transistor is operated in the saturation region. However, as the collector bias voltage increases, conductivity modulation does not extend over the entire N-drift region, which produces an increase in the on-state voltage drop as shown in the quasisaturation region. These regions of operation are discussed in more detail below.

7.7.1 Saturation Region

The voltages across the terminals of the bipolar power transistor are illustrated in Fig. 7.48 when it is biased in the saturation mode. Due to the forward bias across both the base–emitter and base–collector junctions, the voltage drop between the base and emitter terminals, as well as between the base and collector terminals, becomes approximately 0.7 V. Consequently, the voltage drop between the collector and emitter terminals becomes equal to nearly zero. In practical devices, the two junctions are not symmetrical because of differing doping profiles and there is a finite voltage drop across the N-drift region in spite of its conductivity modulation. These factors result in a finite on-state voltage drop between the collector and emitter terminals, which increases with increasing collector current flow as shown in Fig. 7.47. However, the resistance exhibited by the device is much smaller than the unmodulated drift region resistance as given by

$$R_{\mathrm{D,sp}} = \rho_{\mathrm{D}} W_{\mathrm{D}} = \frac{W_{\mathrm{D}}}{q\mu_{\mathrm{n}} N_{\mathrm{D}}}. \tag{7.130}$$

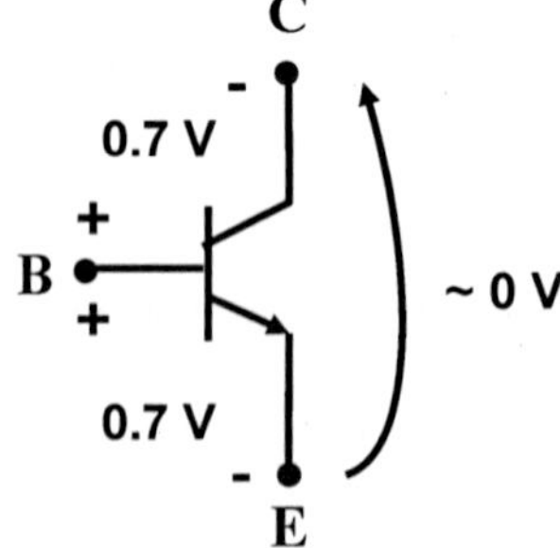

Fig. 7.48 Voltages across the bipolar power transistor in the saturation mode of operation

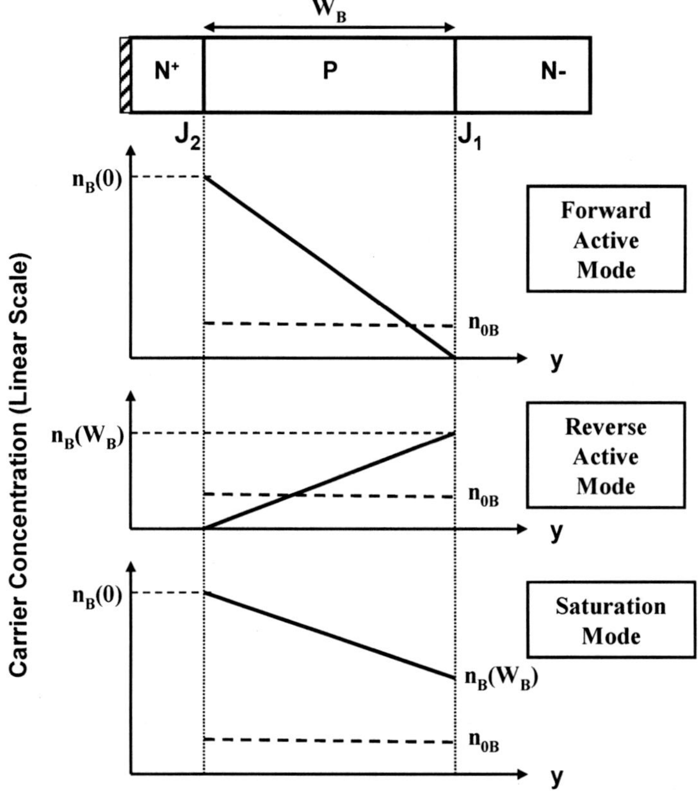

Fig. 7.49 Carrier distribution in the base region of the bipolar power transistor in the saturation mode of operation

The carrier distribution profile in the P-base region of the bipolar power transistor can be obtained by superposition of the profiles for the forward active and reverse active modes of operation. The carrier profiles for these two modes are illustrated in the upper part of Fig. 7.49. In the forward active mode, the base–emitter junction is forward biased leading to an injected carrier density of $n_B(0)$, while the base–collector junction is reverse biased leading to zero carrier density at W_B. Similarly, in the reverse active mode, the base–collector junction is forward biased leading to an injected carrier density of $n_B(W_B)$, while the base–emitter junction is reverse biased leading to zero carrier density at 0. The injected carrier profile for the saturation mode is obtained by the addition of these two profiles as illustrated in the lower part of the figure. The collector current density is then given by

$$J_{C,sat} = \frac{n_B(0) - n_B(W_B)}{W_B}. \tag{7.131}$$

This collector current density is smaller than that obtained in the forward active mode of operation, namely,

$$J_{C,fam} = \frac{n_B(0)}{W_B}. \tag{7.132}$$

Consequently, the bipolar power transistor is operating with a smaller current gain in the saturation mode than in the active region as is apparent from the characteristics shown in Fig. 7.47.

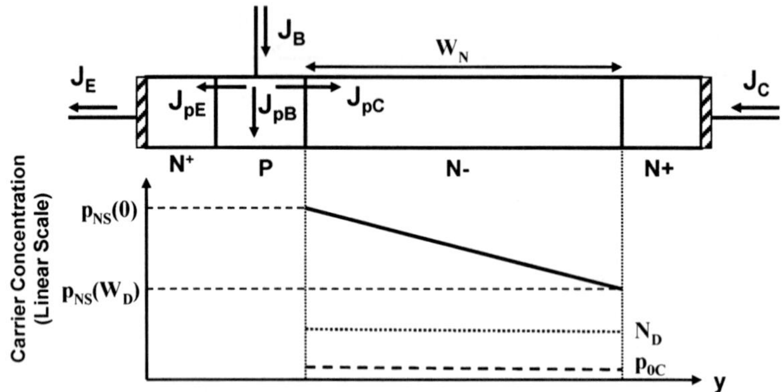

Fig. 7.50 Carrier distribution in the N-drift region of the bipolar power transistor in the saturation mode of operation

The voltage drop across the power bipolar transistor in the saturation mode can be analyzed by computing voltage developed in the N-drift region under conductivity modulation by the high-level injection of minority carriers.[19,20] The bipolar power transistor is illustrated in Fig. 7.50 with a high concentration of minority carriers, which exceeds the doping level (N_D) in the drift region, through the entire drift region. Since charge neutrality must be satisfied in the drift region,

$$n(y) = p(y) + N_D. \tag{7.133}$$

Due to high-level injection conditions in the N-drift region,

$$\frac{dn}{dy} = \frac{dp}{dy}. \tag{7.134}$$

The conduction equation for holes in the N-drift region is

$$J_{pC} = q\mu_p pE - qD_p \frac{dp}{dy} \tag{7.135}$$

while that for electrons in the N-drift region is

$$J_{nC} = q\mu_n nE + qD_n \frac{dn}{dy}. \tag{7.136}$$

As indicated in Fig. 7.50, the base current supplies recombination in the emitter, base, and collector regions. The hole current in the collector drift region (J_{pC}) represents this recombination current. If the current gain is assumed to be large, this current density can be neglected for computation of the electric field profile from (7.135), resulting in

$$E = \frac{kT}{q}\frac{1}{p}\frac{dp}{dy} \tag{7.137}$$

after utilizing the Einstein relationship between the mobility and the diffusion constant. Combining (7.134) and (7.136) with (7.137),

$$J_{nC} = q\mu_n (p + N_D)\left(\frac{kT}{qp}\frac{dp}{dx}\right) + qD_n \frac{dp}{dy}. \tag{7.138}$$

Applying the Einstein relationship between the mobility and the diffusion constant and assuming that the total collector current density (J_C) is approximately equal to the electron current density (J_{nC}) due to the high-current gain,

$$J_C = 2qD_n\left(1 + \frac{N_D}{2p}\right)\frac{dp}{dy}. \tag{7.139}$$

The minority carrier distribution in the N-drift region obtained from solving this differential equation is

$$p(y) = p_{NS}(0) - \frac{J_C y}{2qD_n} + \frac{N_D}{2}\ln\left[\frac{p_{NS}(0)}{p(y)}\right]. \tag{7.140}$$

Since the N-drift region of high-voltage bipolar power transistors is very lightly doped, the last term in the above equation can be neglected giving

$$p(y) = p_{NS}(0) - \frac{J_C y}{2qD_n}. \tag{7.141}$$

This expression indicates a linear carrier distribution as illustrated in Fig. 7.50.

The carrier concentration at the base–collector junction can be obtained by using Boltzmann's relationship:

$$p_{NS}(0) = p_{0C} e^{qV_{BCJ}/kT}, \tag{7.142}$$

where V_{BCJ} is the voltage across the base–collector junction. The voltage across the base–collector junction is not equal to the base–collector bias because of the internal voltage drop across the drift region (V_D) within the bipolar power transistor:

$$V_{BCJ} = V_{BC} - V_D. \tag{7.143}$$

The voltage drop in the drift region can be computed from the electric field given by (7.137):

$$V_D = -\int_0^{W_N} E\,dy = -\frac{kT}{q}\int_{p_{NS}(0)}^{p_{NS}(W_N)} \frac{dp(y)}{p(y)} = \frac{kT}{q}\ln\left[\frac{p_{NS}(0)}{p_{NS}(W_N)}\right]. \tag{7.144}$$

Combining these equations yields a quadratic expression for the carrier concentration [$p_{NS}(0)$] in terms of the base–collector voltage and the collector current density:

$$p_{NS}^2(0) - p_{NS}(0)p_{0C}e^{qV_{BC}/kT} + \frac{J_C W_N}{2qD_n}p_{0C}e^{qV_{BC}/kT} = 0, \tag{7.145}$$

from which the carrier distribution can be obtained.

The voltage drop across the N-drift region in the saturation mode can then be obtained by using (7.144). Using (7.141) to obtain the carrier concentration at the interface between the collector drift region and the N$^+$ substrate,

$$V_D = \frac{kT}{q}\ln\left[\frac{p_{NS}(0)}{p_{NS}(0) - (J_C W_N / 2qD_n)}\right]. \tag{7.146}$$

The on-state voltage drop in the saturation mode is approximately equal to this voltage. Typical values for the on-state voltage drop in the saturation mode are in the range of 100–300 mV.

It is worth pointing out that there is a substantial amount of charge stored within the N-drift region when the bipolar power transistor is operating in the saturation mode. This stored charge is given by

$$Q_{Sat} = \frac{q[p_{NS}(0) + p_{NS}(W_D)]W_N}{2}. \tag{7.147}$$

The charge in the drift region must be sustained by supplying a base current. The component of the base current flowing into the collector drift region responsible for supporting the recombination of these carriers is given by

$$J_{pC} = \frac{Q_{Sat}}{\tau_{HL}} = \frac{q[p_{NS}(0) + p_{NS}(W_N)]W_N}{2\tau_{HL}}. \tag{7.148}$$

7.7.2 Quasisaturation Region

The carrier distribution in the base region for the bipolar power transistor, when operating in the quasisaturation region, is similar to that shown in Fig. 7.49. Consequently, the bipolar power transistor operates with a smaller current gain in the quasisaturation mode than in the active region as is apparent from the characteristics shown in Fig. 7.47. However, its current gain is larger than when operating in the saturation region.

The voltage drop across the power bipolar transistor in the quasisaturation mode can be analyzed by computing the voltage developed in the N-drift region using the same approach as for the saturation region. However, high-level injection does not extend throughout the N-drift region in the quasisaturation mode as illustrated in Fig. 7.51. Equation (7.141) derived in Sect. 7.7.1 for the carrier distribution is also applicable for the quasisaturation mode within the portion of the N-drift region with width W_{NM} that operates under high-level injection conditions. At this location, the minority carrier concentration becomes equal to the majority carrier concentration in equilibrium (N_{D}). The width of the conductivity-modulated region is obtained by using this information in (7.141):

$$W_{\mathrm{NM}} = \frac{2qD_{\mathrm{n}}}{J_{\mathrm{C}}}[p_{\mathrm{NS}}(0) - N_{\mathrm{D}}]. \tag{7.149}$$

Since the injected carrier concentration is far greater than the majority carrier concentration in the drift region,

$$W_{\mathrm{NM}} = \frac{2qD_{\mathrm{n}}}{J_{\mathrm{C}}} p_{\mathrm{NS}}(0). \tag{7.150}$$

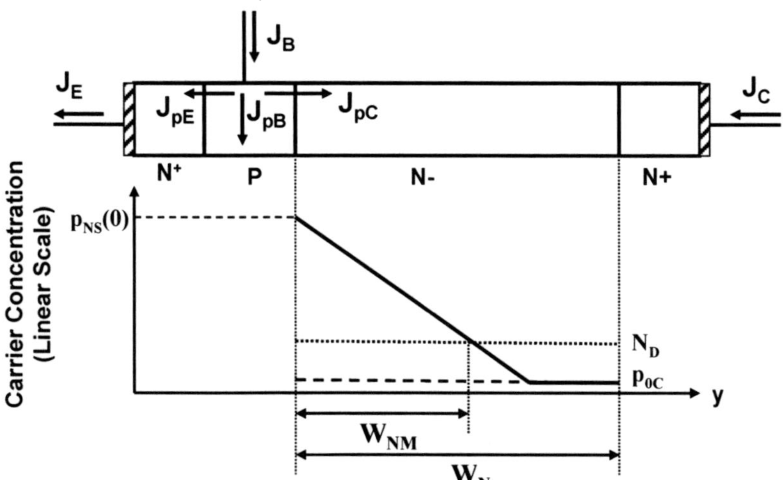

Fig. 7.51 Carrier distribution in the N-drift region of the bipolar power transistor in the quasisaturation mode of operation

The voltage drop across the drift region in the quasisaturation mode becomes dominated by the resistive voltage drop produced by the collector current flow through the unmodulated portion of the drift region:

$$V_{\mathrm{D}} = \frac{J_{\mathrm{C}}(W_{\mathrm{N}} - W_{\mathrm{NM}})}{q\mu_{\mathrm{n}}N_{\mathrm{D}}}. \tag{7.151}$$

Consequently, a substantial voltage drop can occur across the drift region in the quasisaturation mode. As the collector bias is increased, the unmodulated portion of the drift region becomes larger. Eventually, the entire drift region becomes unmodulated at the boundary between the quasisaturation region and the active region. This boundary is then defined by the specific resistance of the drift region (see (7.130)) as illustrated in Fig. 7.47. The characteristics illustrated in this figure are based upon assuming a uniformly doped P-base region with much larger doping concentration than the N-drift region. In practical devices fabricated by using a diffused P-base region with a highly graded doping profile, the collector current continues to increase beyond the boundary defined by the drift region resistance because of an increase in the current gain due to the reduction of the effective base width by the extension of the depletion layer in the P-base region with increasing collector bias voltage. In addition, the current crowding at the edge of the emitter closest to the base contact makes the local collector current density much larger than the average collector current density. This increases the voltage drop across the unmodulated drift region making the quasisaturation boundary move to larger collector bias voltages.

As in the case of the saturation mode, there is a substantial amount of charge stored within the N-drift region when the bipolar power transistor is operating in the quasisaturation mode. This stored charge is given by

$$Q_{\text{Q-Sat}} = \frac{qp_{\mathrm{NS}}(0)W_{\mathrm{NM}}}{2}. \tag{7.152}$$

The component of the base current flowing into the collector drift region is responsible for supporting the recombination of these carriers:

$$J_{\mathrm{pC}} = \frac{Q_{\text{Q-Sat}}}{\tau_{\mathrm{HL}}} = \frac{qp_{\mathrm{NS}}(0)W_{\mathrm{NM}}}{2\tau_{\mathrm{HL}}}. \tag{7.153}$$

The stored charge in the quasisaturation mode is smaller than that in the saturation mode. When the bipolar power transistor is switched from the on-state to the blocking state, the stored charge in the drift region must be removed by a reverse base drive current before the transistor can support voltage across the base–collector junction. It is therefore preferable to operate the bipolar power transistor in the quasisaturation mode during the on-state to achieve a faster switching speed. However, this produces larger on-state power dissipation due to the larger on-state voltage drop in the quasisaturation region.

Simulation Example

To validate the model for the quasisaturation characteristics of the bipolar power transistor, the results of two-dimensional numerical simulations are described here for the baseline device structure (bjt8). For this structure, a doping concentration of 5×10^{13} cm^{-3} was used for the N-drift region with a thickness of 80 μm. The specific resistance for the drift region obtained by using these parameters is 0.735 Ω cm^2.

The output characteristics for the baseline bipolar power transistor structure (bjt8) are shown in Fig. 7.52 at lower collector bias voltages to display the saturation and quasisaturation regions. The saturation region occurs at collector bias voltages below 0.5 V. To operate in the saturation region to obtain a low on-state voltage drop with reasonable (20 A cm^{-2}) collector current density, it is necessary to increase the base drive current density to above 10 A cm^{-2} resulting in a very low-current gain. The quasisaturation region extends to much larger collector bias voltages in the range of 15–20 V even for a relatively modest collector current density of 20 A cm^{-2}.

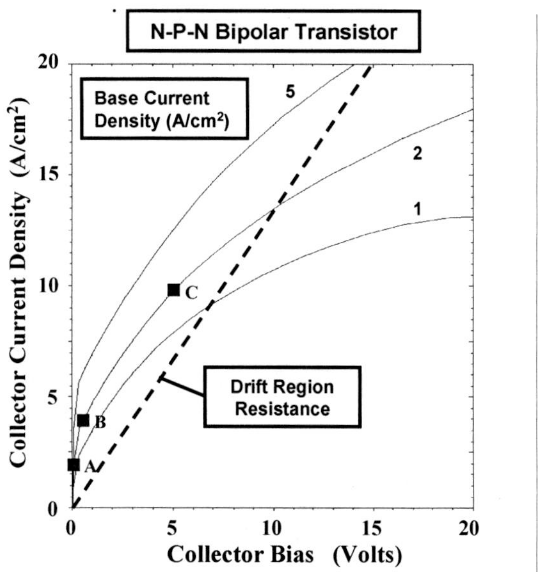

Fig. 7.52 Quasisaturation characteristics for a bipolar power transistor

The minority carrier distribution profiles in the N-drift region for the baseline bipolar power transistor structure (bjt8) are shown in Fig. 7.53. Note that the distance scale begins at the edge of the base–collector junction in this plot. When the device is operated in the saturation region (at point *A* in Fig. 7.52), the injected minority carrier density is far greater than the background doping concentration throughout the drift region, confirming the high-level injection assumption used for the analysis of the saturation region (see profile *A* in Fig. 7.53).

When the device is operated at the boundary between the saturation and quasisaturation regions (at point *B* in Fig. 7.52), the injected minority carrier density becomes equal to the majority carrier concentration at the boundary between the N-drift region and the N$^+$ substrate (see profile *B* in Fig. 7.53). Since the conductivity modulation still extends throughout the drift region, the on-state voltage drop is low at this operating point. When the device is operated at a larger collector bias voltage (at point *C* in Fig. 7.52), the high-level injection extends only through a portion of the drift region (see profile *C* in Fig. 7.53). Since a portion of the drift region is not conductivity modulated, the on-state voltage drop is larger at this operating point. The carrier profile is approximately linear in shape for all the cases as predicted by the analytical model.

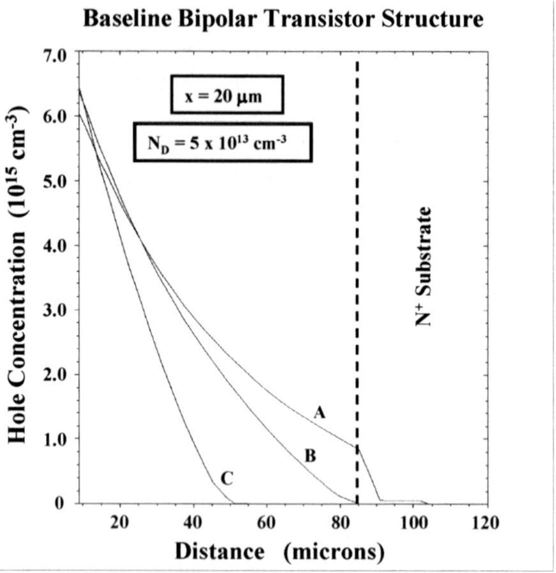

Fig. 7.53 Minority carrier profiles in the drift region for a bipolar power transistor operating in the saturation and quasisaturation regions

7.8 Switching Characteristics

Bipolar power transistors were developed as switches in power circuits used for applications such as motor control. In these circuits, the device is switched between the on-state and the off-state to regulate the current being delivered to the load. The bipolar power transistor is usually used in the common-emitter configuration as shown in Fig. 7.3. The loads can be either resistive or inductive in nature. For a resistive load, the locus for the collector current–voltage trajectory is determined by the load resistance as indicated in Fig. 7.54 by the solid line. For an inductive load, the locus for the collector current–voltage trajectory is determined by the dashed line indicated in Fig. 7.54. The device switches from the blocking state, while supporting the collector supply voltage (V_{CS}) at point *A*, to the on-state with a voltage drop V_{ON} at point *B* during the turn-on transient. The device switches from the on-state to the collector supply voltage during the turn-off transient. These transitions are discussed in this section because they determine the power losses incurred when the bipolar power transistor is operated at high frequencies.

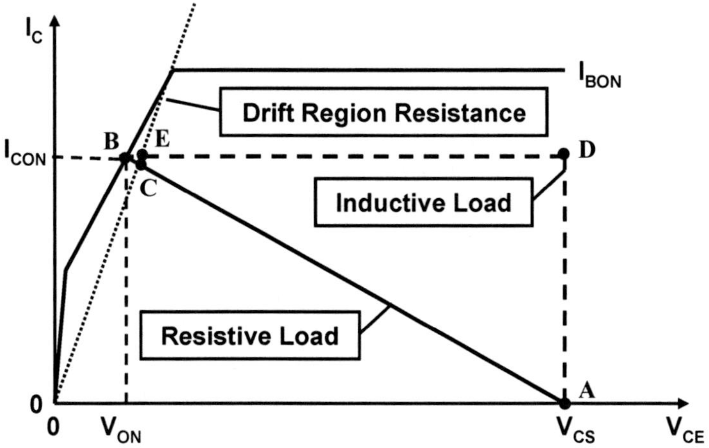

Fig. 7.54 Switching loci for the bipolar power transistor

7.8.1 Turn-On Transition

At the beginning of the turn-on transition process, the bipolar power transistor is biased with the base–emitter junction under reverse bias due to the external drive circuit shown in Fig. 7.3. Although it is sufficient to short circuit the base and emitter terminals to maintain the power bipolar transistor in the blocking state, the reverse bias voltage (V_{BS2}) is necessary to produce a reverse base current to accelerate the turn-off process. The base current can be assumed to be equal to zero in the blocking state with the base–collector junction supporting the applied collector bias voltage with no collector current flow.

When the device is turned on by opening switch S_2 and closing switch S_1 in Fig. 7.3, a base current begins to flow limited by the base drive resistance R_{B1}. If the switching time for the switches S_1 and S_2 is small, the base current is initially given by

$$I_{B0} = \frac{V_{BS1} + V_{BS2}}{R_{B1}}, \tag{7.154}$$

because the potential at the base electrode is initially equal to the reverse bias voltage (V_{BS2}). Once the base–emitter junction becomes forward biased, the base current is given by

$$I_B = \frac{V_{BS1} - V_{bi}}{R_{B1}}, \tag{7.155}$$

where V_{bi} is the built-in potential (~0.8 V) for the base–emitter junction.

During the turn-on process, collector current flow does not begin to occur until the electrons injected from the base–emitter junction arrive at the base–collector junction. This time interval is given by the *base transit time*[8]:

$$t_{\mathrm{T}} = \frac{W_{\mathrm{B}}^2}{2D_{\mathrm{n}}}, \tag{7.156}$$

where W_B is the base width. For a base region with doping concentration of 1×10^{17} cm^{-3} and width of 10 μm, the transit time is found to be 24 ns. The collector current begins to increase after this time interval.

The stored electron charge in the P-base region of the N–P–N bipolar power transistor begins to develop after the on-set of the injection of electrons from the emitter region. The charge has a linear profile in the P-base region as illustrated in Fig. 7.55 with a maximum value of $n_B(0, t)$ at the base–emitter junction and zero at the base–collector junction. The charge control equation that governs the buildup of the electron charge in the base region is

$$\frac{\mathrm{d}Q_{\mathrm{nB}}(t)}{\mathrm{d}t} = \beta J_{\mathrm{B}} - \frac{Q_{\mathrm{nB}}(t)}{\tau_{\mathrm{n}}}, \tag{7.157}$$

where $Q_{nB}(t)$ is the electron charge per unit area in the base region and τ_n is the recombination lifetime for electrons. If the recombination in the base region is assumed to be negligible,

$$\frac{\mathrm{d}Q_{\mathrm{nB}}(t)}{\mathrm{d}t} = \beta J_{\mathrm{B}}. \tag{7.158}$$

Consequently,

$$Q_{\mathrm{nB}}(t) = \beta J_{\mathrm{B}} t. \tag{7.159}$$

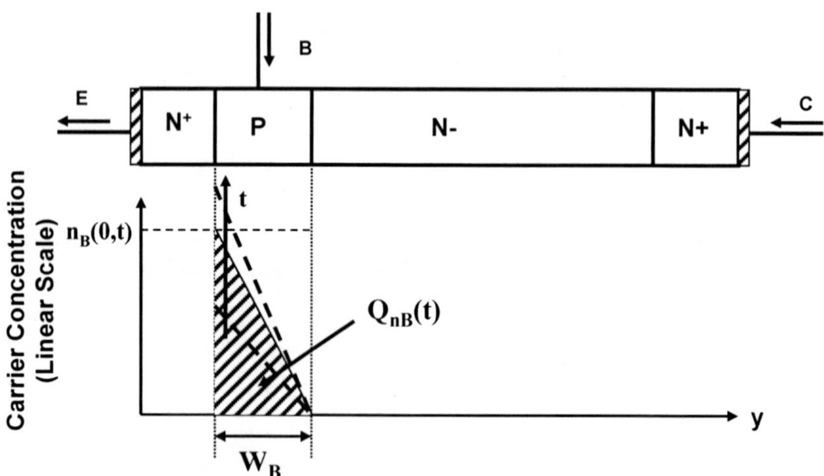

Fig. 7.55 Buildup of the stored charge in the base region of the bipolar power transistor during the turn-on transient

The electron stored charge in the P-base region is given by the triangular shaded area in Fig. 7.55:

$$Q_{\mathrm{nB}}(t) = \frac{1}{2} q n_{\mathrm{B}}(0,t) W_{\mathrm{B}}. \tag{7.160}$$

Using this relationship with (7.159) yields

$$n_{\mathrm{B}}(0,t) = \frac{2\beta J_{\mathrm{B}}}{q W_{\mathrm{B}}} t, \tag{7.161}$$

which describes the growth of the injected carrier concentration at the base–emitter junction. The dashed lines in Fig. 7.55 indicate the evolution of the injected carrier concentration with time. The collector current corresponding to this carrier concentration is

$$J_{\mathrm{C}}(t) = q D_{\mathrm{n}} \frac{\mathrm{d}n}{\mathrm{d}x} = q D_{\mathrm{n}} \frac{n_{\mathrm{B}}(0,t)}{W_{\mathrm{B}}} = \frac{2 D_{\mathrm{n}} \beta J_{\mathrm{B}}}{W_{\mathrm{B}}^2} t. \tag{7.162}$$

Based upon this expression, it can be concluded that the collector current will increase linearly with time during the turn-on process.

The transition time for the collector current to increase from zero to the on-state current density ($J_{\mathrm{C,ON}}$) is then given by

$$t_{\mathrm{I,ON}} = \frac{W_{\mathrm{B}}^2 J_{\mathrm{C,ON}}}{2 D_{\mathrm{n}} \beta J_{\mathrm{B}}} = \frac{W_{\mathrm{B}}^2}{2 D_{\mathrm{n}}} \tag{7.163}$$

because $J_{\mathrm{C,ON}} = \beta J_{\mathrm{B}}$. It is advantageous to reduce the transition time for the collector current to decrease the switching losses. This can be achieved by decreasing the width of the P-base region, which also increases the current gain (β). However, this can lead to a reduction of the breakdown voltage due to the reach-through phenomenon.

Resistive Load

The above expressions were derived under the assumption that the base–collector junction is reverse biased during the transient. This is applicable for the resistive load from operating point *A* in the blocking state (see Fig. 7.54) until point *C* where the load line intersects the boundary of the quasisaturation region as defined by the resistance of the N-drift region. Once the bipolar power transistor enters its quasisaturation region, this assumption is no longer valid because the base–collector junction becomes forward biased with the injection of holes into the N-drift region. The rate of reduction of the collector voltage (and therefore the increase of the collector current) now becomes limited by the development of the stored charge in the N-drift region. As the stored charge in the N-drift region increases, the unmodulated portion of the drift region becomes smaller resulting in a smaller voltage drop.

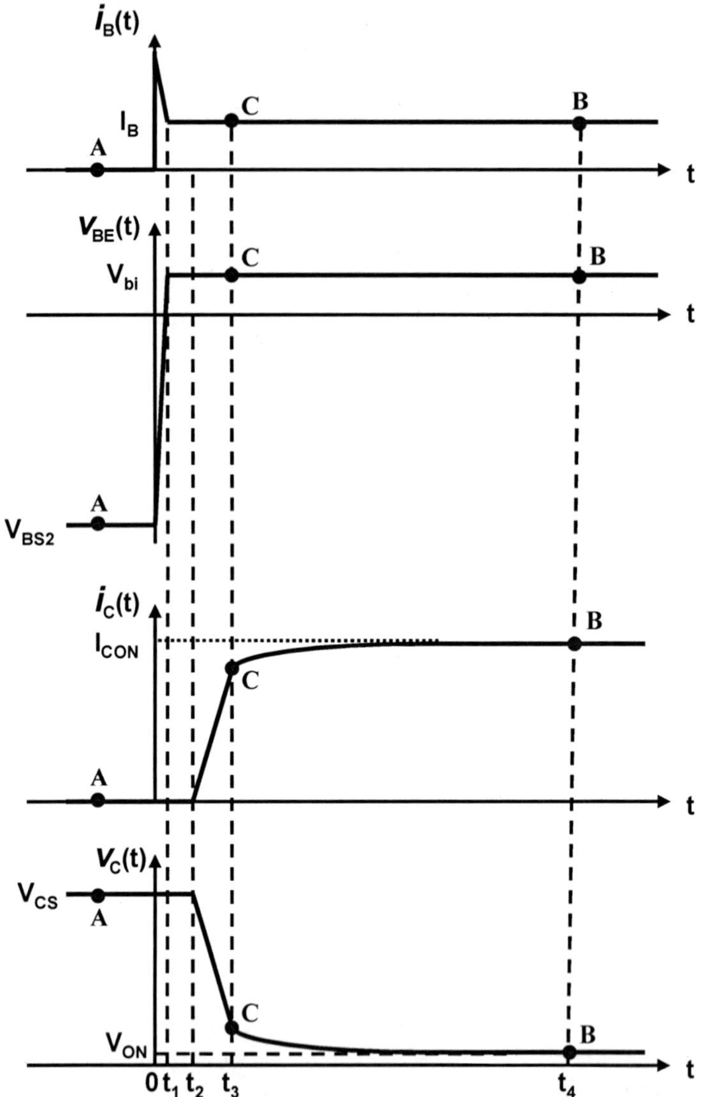

Fig. 7.56 Switching waveforms for the bipolar power transistor during the turn-on transient with a resistive load

The waveforms for the terminal currents and voltages are illustrated in Fig. 7.56 during turn-on with a resistive load with points labeled corresponding to points in Fig. 7.54. Initially, there is a surge in the base current because the potential across the base resistor is ($V_{BS1} + V_{BS2}$). During the time interval 0 to t_1, the base–emitter voltage changes from the reverse bias potential V_{BS2} to the forward-biased junction potential V_{bi}. At this time, the injection of electrons from the base–emitter junction begins to occur. The collector current does not begin to increase until after the transit time for electrons, i.e., the time interval $(t_2 - t_1) = t_T$. The collector current then increases at a linear rate until time $t_3 = (t_2 + t_{I,ON})$. The collector voltage simultaneously decreases at a linear rate due to the resistive load. At time t_3, the bipolar power transistor enters the quasisaturation region of operation at point C on the characteristics. The collector voltage gradually decreases to the steady-state on-voltage drop at time t_4. This slow reduction of the collector voltage produces significant power dissipation during the turn-on event.

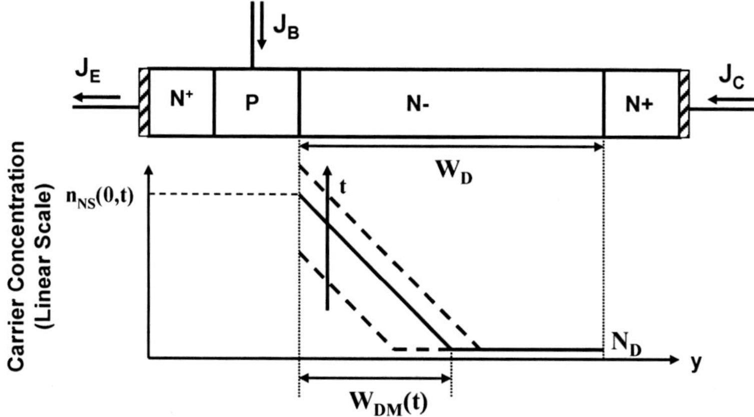

Fig. 7.57 Buildup of the stored charge in the drift region of the bipolar power transistor during the turn-on transient

Analytical modeling of the voltage drop in the quasisaturation region during the turn-on process can be performed by using the assumption that the minority (and hence the majority) carrier concentration at the base–collector junction increases exponentially with time. The carrier distribution within the N-drift region is linear as determined by the physics described earlier for the quasisaturation region:

$$p_{NS}(y,t) = p_{NS}(0,t) - \frac{J_C(t)}{2qD_n} y \approx p_{NS}(0,t) - \frac{J_{C,ON}}{2qD_n} y, \qquad (7.164)$$

because the collector current density remains close to the on-state current density during this portion of the turn-on transient. The buildup of the stored charge in the N-drift region is illustrated in Fig. 7.57 with the dashed lines indicating the evolution of the charge. Note that the slope of the carrier distribution profile remains constant according to the above equation.

The width of the modulated portion of the drift region can then be determined by using (7.150):

$$W_{\mathrm{DM}}(t) = \frac{2qD_{\mathrm{n}}}{J_{\mathrm{C}}(t)} p_{\mathrm{NS}}(0,t) = \frac{2qD_{\mathrm{n}}}{J_{\mathrm{C,ON}}} p_{\mathrm{NS}}(0,F)(1 - e^{-t/\tau}), \tag{7.165}$$

where $p_{\mathrm{NS}}(0, F)$ is the final steady-state hole concentration at the base–collector junction at the end of the turn-on process and τ is the minority carrier lifetime. Although the collector current density increases slightly during the transition through the quasisaturation region, its value can be assumed to be approximately equal to the final steady-state collector current density ($J_{\mathrm{C,ON}}$). The collector voltage of the bipolar power transistor is then given by

$$V_{\mathrm{QS}}(t) = \rho_{\mathrm{D}}[W_{\mathrm{D}} - W_{\mathrm{DM}}(t)] J_{\mathrm{C,ON}} K_{\mathrm{CC}}, \tag{7.166}$$

where K_{CC} is the current crowding parameter to account for the larger local current density near the edge of the emitter closest to the base contact.

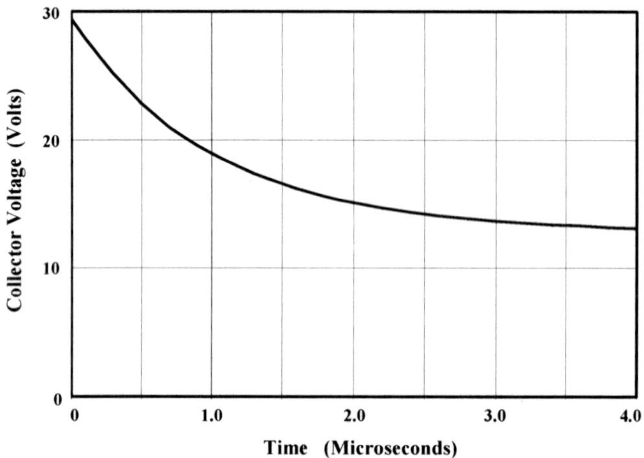

Fig. 7.58 Reduction of the on-state voltage drop for the bipolar power transistor during the turn-on transient

As an example, consider a bipolar power transistor with a drift region doping concentration of 5×10^{13} cm^{-3} and thickness of 80 μm. The slow reduction of the collector voltage is shown in Fig. 7.58 after the transistor enters the quasisaturation region during the turn-on process. The current crowding parameter was assumed to have a value of 2. It can be observed that the on-state voltage drop decreases from about 29 V to about 12 V in 4 μs. The power dissipation within the bipolar transistor is significantly enhanced due to the large collector current density during this time interval.

Simulation Example

To validate the model for the turn-on process for the bipolar power transistor with a resistive load, the results of two-dimensional numerical simulations are described here for the baseline device structure (bjt8). For this structure, a doping concentration of 5×10^{13} cm^{-3} was used for the N-drift region with a thickness of 80 μm. The device was initially biased with a negative base drive voltage of 5 V with a base resistance to limit the current to obtain a steady-state blocking condition with 800 V applied to the collector terminal. The base drive voltage was rapidly switched to positive 5 V to initiate the turn-on process. A load resistance was attached between the collector terminal and its bias supply to emulate the resistive load conditions.

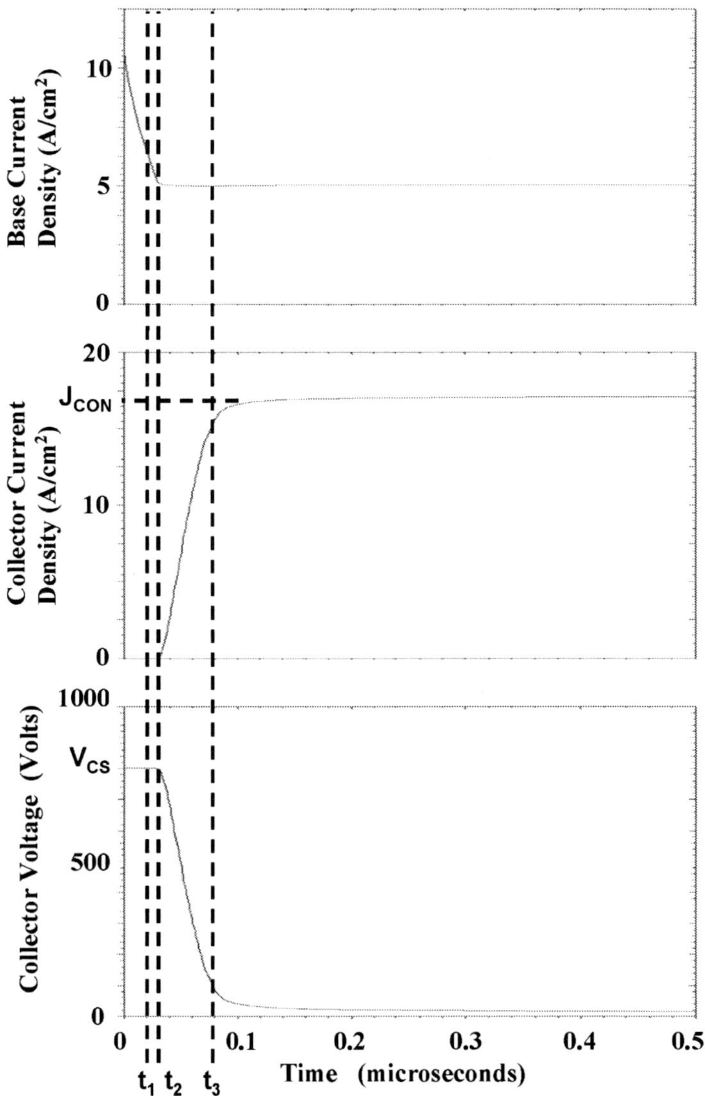

Fig. 7.59 Typical turn-on waveforms for a bipolar power transistor with a resistive load

The terminal current–voltage waveforms obtained from the numerical simulations are shown in Fig. 7.59. There is an initial surge in the base drive current after which it settles down to the steady-state base drive current. The base electrode voltage crosses zero at the time t_1 indicated in the figure while the collector current begins to rise at the time t_2 indicated in the figure. The transition time for the collector current is about 50 ns. This delay is consistent with the transit time for electrons in the P-base region. The collector current increases linearly with time as described by the analytical model until the device enters the quasisaturation region at time t_3. The collector voltage then reduces slowly with time from an initial value of about 25 V. The reduction of the collector voltage occurs over a time frame of about 3 μs. This behavior is consistent with the behavior predicted by the analytical model.

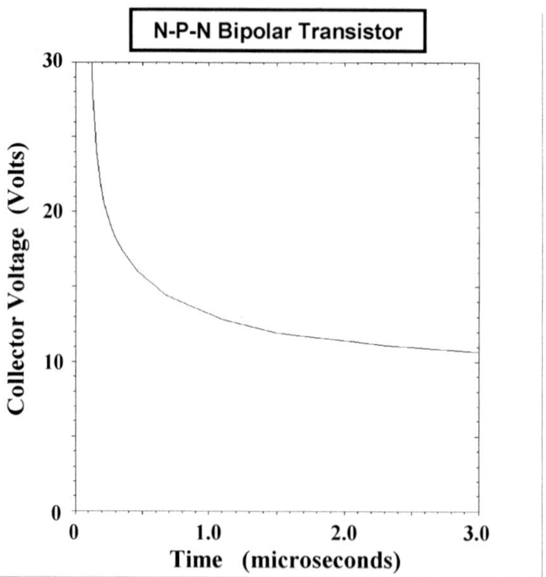

Fig. 7.60 Reduction of the on-state voltage drop for the bipolar power transistor during the turn-on transient

The slow reduction of the collector voltage during the transition through the quasisaturation region is shown in Fig. 7.60. It can be observed that the collector voltage is about 22 V at the beginning of this time interval and then reduces to the steady-state value of about 10 V in 3 μs. The time taken for this slow decay of the collector voltage is consistent with a lifetime of 1 μs in the N-drift region and in agreement with the assumptions of the analytical model.

The current distribution within the bipolar power transistor is shown in Fig. 7.61 during the turn-on process. During the initial stages of the turn-on process, the current density increases as shown by the solid lines up to a time of 75 ns. A very high-current density (ten times the steady-state average current density) develops at the edge of the emitter finger closest to the base contact. The current then spreads over the emitter finger as shown by the dashed lines until the distribution reaches the steady-state condition with a current density at the edge of the emitter of twice the average current density. The larger current density at the edge of the emitter during the time interval from 190 to 461 ns enhances the voltage drop in the collector drift region making the power dissipation larger.

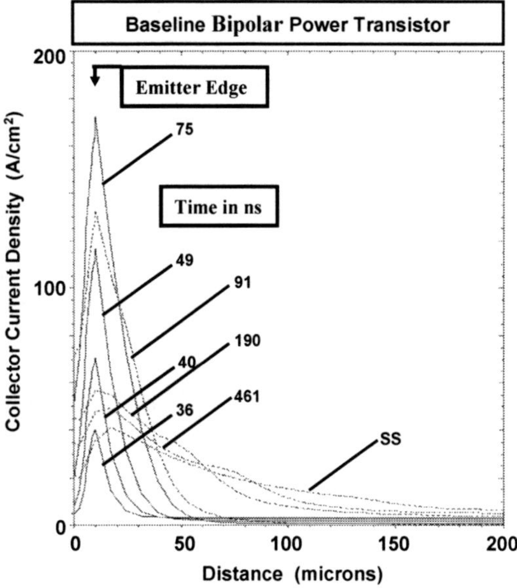

Fig. 7.61 Current distribution during turn-on of the bipolar power transistor

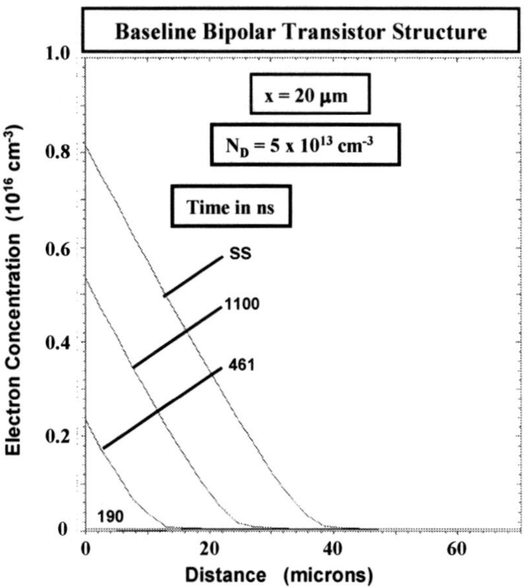

Fig. 7.62 Carrier distribution in the drift region during turn-on of the bipolar power transistor

The modulation of the conductivity of the N-drift region during the turn-on transient can be observed in Fig. 7.62 where the electron concentration is plotted at various time instances. The stored charge in the N-drift region begins to build up after 190 ns at which time the collector voltage is at 22 V. The profile for the electrons is linear in the N-drift region as described by the analytical model. The slope for the carrier distribution remains the same for all the time instances as predicted by the analytical model (see (7.164)).

Inductive Load

In the case of an inductive load, the locus for the current–voltage trajectory during the turn-on process is determined as shown by the dashed lines in Fig. 7.54. The collector current increases until it reaches the on-state value at point D while the collector voltage remains constant at V_{CS}. All the current flowing through the fly-back diode (see Fig. 7.3) has been transferred to the transistor at this point in time. The collector voltage then reduces while the collector current remains constant. The previously derived expressions for the delay time (7.156) and the transition time (7.163) are valid for the inductive load situation as well. These expressions were derived under the assumption that the base–collector junction is reverse biased during the transient. This is applicable for the inductive load from operating point A in the blocking state until point E in Fig. 7.54. Once the bipolar power transistor enters its quasisaturation region, the base–collector junction becomes forward biased with the injection of holes into the N-drift region. The rate of reduction of the collector voltage (and therefore the increase of the collector current) now becomes limited by the development of the stored charge in the N-drift region. As the stored charge in the N-drift region increases, the unmodulated portion of the drift region becomes smaller resulting in a smaller voltage drop.

The waveforms for the terminal currents and voltages are illustrated in Fig. 7.63 during turn-on with an inductive load. Initially, there is a surge in the base current because the potential across the base resistor is (V_{BS1} + V_{BS2}). During the time interval 0 to t_1, the base–emitter voltage changes from the reverse bias potential V_{BS2} to the forward-biased junction potential V_{bi}. At this time, the injection of electrons from the base–emitter junction begins to occur. The collector current does not begin to increase until after the transit time for electrons, i.e., the time interval $(t_2 - t_1) = t_T$. The collector current then increases at a linear rate until time $t_3 = (t_2 + t_{I,ON})$ when it becomes equal to the steady-state on-current. The collector voltage remains constant during this time.

The collector voltage reduces with time, while the collector current remains constant, after time t_3. The time taken for the collector voltage to decrease to the voltage at point E in Fig. 7.54 can be estimated by assuming that the space-charge region at the base–collector junction is removed by the collector current flow. Thus,

$$t_{V,ON} = t_4 - t_3 = \frac{Q_{SC}}{J_{C,ON}} = \frac{qN_D W_{DN} + qN_{AB} W_{DB}}{J_{C,ON}}, \tag{7.167}$$

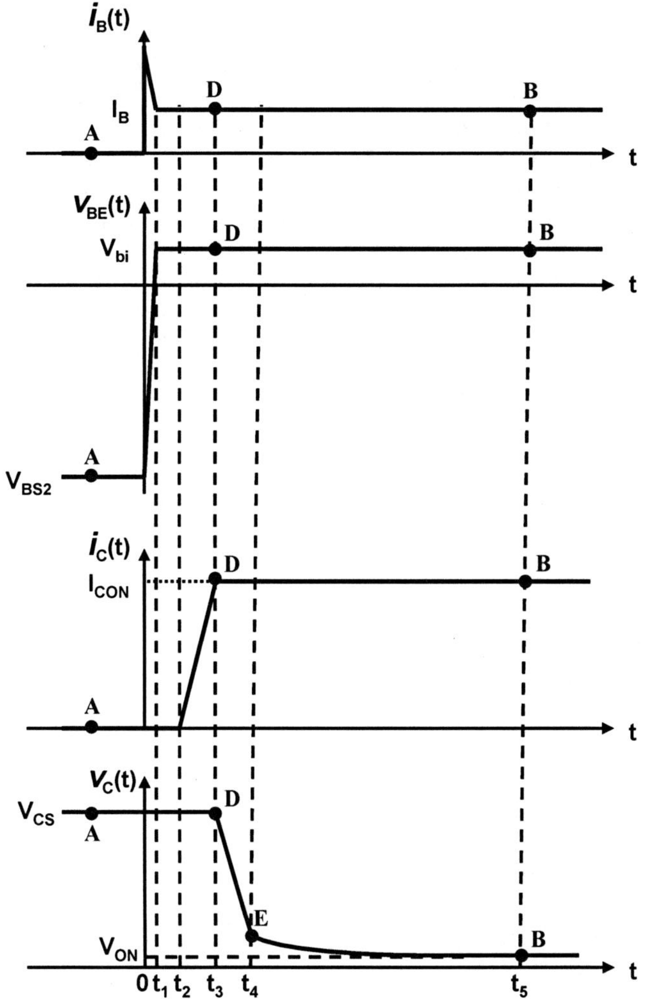

Fig. 7.63 Switching waveforms for the bipolar power transistor during the turn-on transient with an inductive load

which accounts for the space charge on both sides of the base–collector junction. A typical value for this voltage transition time is between 10 and 50 ns.

At time t_4, the bipolar power transistor enters the quasisaturation region of operation at point E on the characteristics. The collector voltage gradually decreases to the steady-state on-voltage drop at time t_5. This slow reduction of the collector voltage produces significant power dissipation during the turn-on event. The same analytical model for the slow reduction of the collector voltage in the quasisaturation region developed for the resistive load case can be used for the inductive load case as well.

Simulation Example

To validate the model for the turn-on process for the bipolar power transistor with an inductive load, the results of two-dimensional numerical simulations are described here for the baseline device structure (bjt8). For this structure, a doping concentration of 5×10^{13} cm^{-3} was used for the N-drift region with a thickness of 80 μm. The device was initially biased with a negative base drive voltage of 5 V with a base resistance to limit the current to obtain a steady-state blocking condition with 800 V applied to the collector terminal. The base drive voltage was rapidly switched to positive 5 V to initiate the turn-on process. The base electrode voltage crosses zero at the time t_1 indicated in Fig. 7.64 while the collector current begins to rise at the time t_2 indicated in the figure. This delay is consistent with the transit time for electrons in the P-base region.

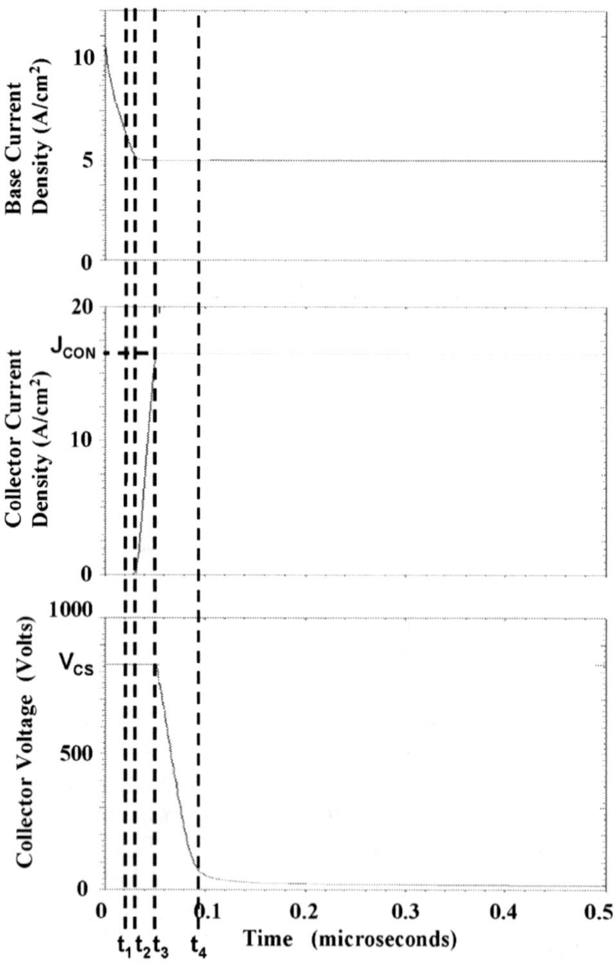

Fig. 7.64 Typical turn-on waveforms for a bipolar power transistor with an inductive load

There is an initial surge in the base drive current after which it settles down to the steady-state base drive current. The transition time for the collector current is about 50 ns. Then, the collector current increases linearly with time as described by the analytical model until the device enters the quasisaturation region at time t_3. After this, the collector voltage decreases rapidly over a time interval of about 40 ns and then reduces slowly with time from an initial value of about 25 V. The reduction of the collector voltage occurs over a time frame of about 3 μs. This behavior is consistent with the behavior predicted by the analytical model.

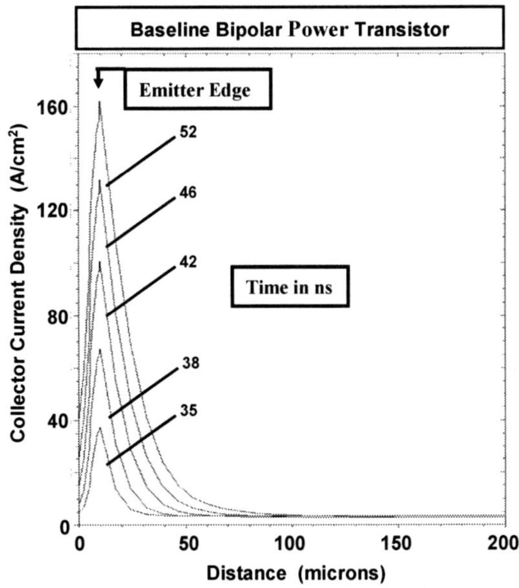

Fig. 7.65 Current distribution during the first phase of the turn-on of the bipolar power transistor with an inductive load

The current distribution within the bipolar power transistor is shown in Fig. 7.65 during the first phase of the turn-on process for the inductive load case when the collector current is increasing while the collector voltage is maintained at the initial blocking voltage. During the initial stages of the turn-on process, the current density increases up to a time of 52 ns. A very high-current density (ten times the steady-state average current density) develops at the edge of the emitter finger closest to the base contact.

During the second phase of the turn-on process with an inductive load, the collector current remains constant while the collector voltage decreases rapidly. The current distribution during the second phase is shown in Fig. 7.66 with the distribution at the beginning of this phase indicated by the dashed line. The current density first increases at the edge of the emitter closest to the base contact. The current then spreads along the emitter finger until the distribution reaches the steady-state condition with a current density at the edge of the emitter of twice the average current density. The larger current density at the edge of the emitter during the time interval from 63 to 405 ns enhances the voltage drop in the collector drift region making the power dissipation larger.

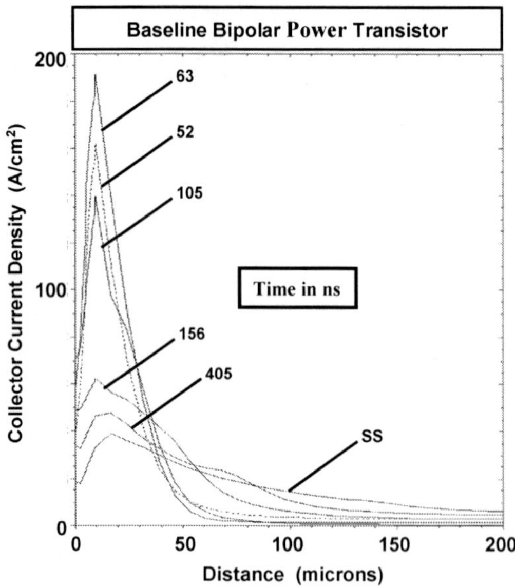

Fig. 7.66 Current distribution during the second phase of the turn-on of the bipolar power transistor

7.8.2 Turn-Off Transition

The bipolar power transistor is operated with the base–emitter junction under forward bias by the external drive circuit shown in Fig. 7.3 at beginning of the turn-off transition process with switch S_1 closed and switch S_2 open. The turn-off transition is initiated by opening switch S_1 and closing switch S_2. Initially, a reverse base current begins to flow limited by the base drive resistance R_{B2}. If the switching time for the switches S_1 and S_2 is small, the reverse base current is

$$I_{BR} = \frac{V_{BS2} + V_{bi}}{R_{B2}}, \tag{7.168}$$

because the potential at the base electrode is initially equal to the forward bias junction potential (V_{bi}). As the base–emitter junction becomes reverse biased, the base current reduces to

$$I_{BR} = \frac{V_{BS2}}{R_{B2}}. \tag{7.169}$$

During the turn-off process, collector current is squeezed toward the center of the emitter finger because the reverse base drive current first extracts the stored charge located closest to the base contact terminal. Consequently, the current density at the center of the emitter increases with time under inductive load conditions, resulting in high local power dissipation within the transistor. The turn-off process occurs by the removal of the stored charge in the P-base region as well as a portion of the stored charge located in the collector drift region. The base–collector junction is unable to support a high voltage until the charge at the base–collector junction is removed allowing a space-charge layer to be established. The delay between the rise of the bipolar transistor collector voltage and the application of the reverse base drive current is referred to as the *storage time*. Bipolar power transistors exhibit large values for the storage time which inhibits their ability to operate at higher frequencies.

Storage Time Model

Emitter current crowding in the on-state confines most of the stored charge near the edge of the emitter closest to the base contact at the beginning of the turn-off transient. Consequently, a simple one-dimensional analysis for the storage time can be performed by using charge control analysis. In the initial on-state condition, excess electron charge is present in the P-base region due to the injection of electrons from the emitter. In addition, excess electron and hole charge is present in the N-drift region due to the injection of holes into the drift region across the forward-biased base–collector junction. The electron concentration profile in the P-base region is shown in Fig. 7.67 as a straight line because the recombination in the base has been neglected. The hole concentration profile in the N-drift region is also shown in the figure. As discussed previously, under quasisaturation conditions, the hole concentration profile is linear and extends partially through the drift region to a depth W_{NM}. Due to high-level injection conditions in the N-drift region,

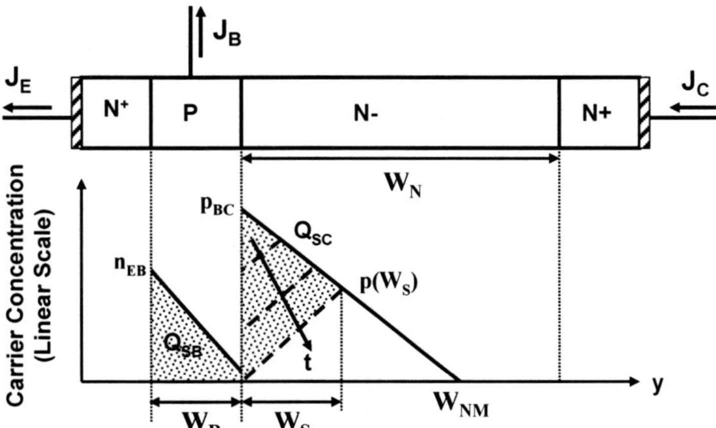

Fig. 7.67 Stored charge in the base and collector regions of the bipolar power transistor during the turn-off transient

the electron concentration profile matches that for the holes in the conductivity-modulated portion.

Due to the high doping concentration in the P-base region when compared with the N-drift region for bipolar power transistors, the injected electron concentration in the P-base region at the base–collector junction in the on-state is much smaller than that at the base–emitter junction as illustrated in Fig. 7.67. The stored electron charge in the P-base region (shaded area in the P-base region in the figure) is therefore given by

$$Q_{\mathrm{SB}} = \frac{1}{2} q n_{\mathrm{EB}} W_{\mathrm{B}}, \tag{7.170}$$

where n_{EB} is the injected electron concentration at the emitter–base junction. If the P-base region is operating under low-level injection conditions, the initial on-state collector current density is given by

$$J_{\mathrm{C,ON}} = qD_{\mathrm{n}} \frac{\mathrm{d}n}{\mathrm{d}y} = qD_{\mathrm{n}} \frac{n_{\mathrm{EB}}}{W_{\mathrm{B}}}. \tag{7.171}$$

Consequently,

$$n_{\mathrm{EB}} = \frac{J_{\mathrm{C,ON}} W_{\mathrm{B}}}{qD_{\mathrm{n}}}. \tag{7.172}$$

Using this in (7.170) yields

$$Q_{\mathrm{SB}} = \frac{J_{\mathrm{C,ON}} W_{\mathrm{B}}^2}{2D_{\mathrm{n}}}. \tag{7.173}$$

The stored charge in the collector drift region is also partially removed by the reverse base current. The progressive removal of the holes from the N-drift region by the reverse base drive current is indicated by the dashed lines in Fig. 7.67. The slope of the carrier profiles indicated by the dashed lines remains constant during this process because of the constant reverse base drive current. The slope of the carrier concentration indicated by the solid line also remains constant because the collector current is constant for an inductive load operating condition. Eventually, the hole concentration at the base–collector junction is reduced to zero enabling the junction to begin supporting a high collector voltage. The stored charge in collector drift region (shaded area in the N-drift region in the figure) removed by the reverse base drive current is therefore given by

$$Q_{\mathrm{SC}} = \frac{1}{2} q p_{\mathrm{BC}} W_{\mathrm{S}}, \tag{7.174}$$

where p_{BC} is the injected hole concentration at the base–collector junction.

The reverse base drive current is related to the concentration profiles for the holes defined by the dashed lines in the figure:

$$J_{\mathrm{BR}} = 2qD_{\mathrm{p}}\frac{\mathrm{d}p}{\mathrm{d}y} = 2qD_{\mathrm{p}}\frac{p(W_{\mathrm{S}})}{W_{\mathrm{S}}}, \tag{7.175}$$

where $p(W_S)$ is the hole concentration at the distance W_S from the base–collector junction. Note that this equation takes into account high-level injection conditions within the N-drift region. This concentration is also determined by the initial hole concentration profile:

$$p(W_{\mathrm{S}}) = p_{\mathrm{BC}}\frac{W_{\mathrm{NM}} - W_{\mathrm{S}}}{W_{\mathrm{NM}}}. \tag{7.176}$$

Combining (7.175) and (7.176),

$$W_{\mathrm{S}} = \frac{2qD_{\mathrm{p}}W_{\mathrm{NM}}p_{\mathrm{BC}}}{J_{\mathrm{BR}}W_{\mathrm{NM}} + 2qD_{\mathrm{p}}p_{\mathrm{BC}}}. \tag{7.177}$$

The initial on-state collector current density is also related to the initial hole concentration profile in the N-drift region:

$$J_{\mathrm{C,ON}} = 2qD_{\mathrm{n}}\frac{\mathrm{d}n}{\mathrm{d}y} = 2qD_{\mathrm{n}}\frac{\mathrm{d}p}{\mathrm{d}y} = 2qD_{\mathrm{n}}\frac{p_{\mathrm{BC}}}{W_{\mathrm{NM}}}. \tag{7.178}$$

Consequently,

$$p_{\mathrm{BC}} = \frac{J_{\mathrm{C,ON}}W_{\mathrm{NM}}}{2qD_{\mathrm{n}}}. \tag{7.179}$$

Using this expression in (7.177),

$$W_{\mathrm{S}} = \frac{D_{\mathrm{p}}W_{\mathrm{NM}}J_{\mathrm{C,ON}}}{D_{\mathrm{n}}J_{\mathrm{BR}} + D_{\mathrm{p}}J_{\mathrm{C,ON}}}. \tag{7.180}$$

Substituting (7.179) and (7.180) in (7.174) yields

$$Q_{\mathrm{SC}} = \frac{D_{\mathrm{p}}W_{\mathrm{NM}}^{2}J_{\mathrm{C,ON}}^{2}}{4D_{\mathrm{n}}(D_{\mathrm{n}}J_{\mathrm{BR}} + D_{\mathrm{p}}J_{\mathrm{C,ON}})}. \tag{7.181}$$

The stored charge given by the sum of Q_{SB} and Q_{SC} is removed by the reverse base current flow (J_{BR}) during the storage time interval (t_S):

$$J_{\mathrm{BR}}t_{\mathrm{S}} = Q_{\mathrm{SB}} + Q_{\mathrm{SC}}. \tag{7.182}$$

Using (7.173) and (7.181) provides a solution for the storage time:

$$t_S = \left(\frac{J_{C,ON}}{J_{BR}} \right) \left[\frac{W_B^2}{2D_n} + \frac{W_{NM}^2}{4D_n} \frac{D_p J_{C,ON}}{D_n J_{BR} + D_p J_{C,ON}} \right]. \tag{7.183}$$

From this expression, it can be concluded that the storage time can be reduced by increasing the reverse base drive current at the disadvantage of a larger, more expensive, drive circuit.

For the case of a bipolar power transistor with a base width of 10 μm and a modulated drift region with width of 40 μm in the initial on-state, the storage time obtained by using this equation is 230 ns if the on-state current density is 17.5 A cm^{-2} and the reverse base drive current is 5 A cm^{-2}. These operating conditions correspond to an injected electron concentration n_{EB} of 3×10^{15} cm^{-3} at the emitter–base junction, an injected hole concentration p_{BC} of 6×10^{15} cm^{-3} at the base–collector junction, a stored charge Q_{SB} of 2.5×10^{-7} C cm^{-2} in the P-base region, and a stored charge Q_{SC} of 1.12×10^{-6} C cm^{-2} in the N-drift region. These values are consistent with the assumption of low-level injection conditions in the P-base region and high-level injection conditions in the N-drift region used for the analysis.

Voltage Rise Time Model

Once the carrier concentration becomes zero at the base–collector junction, it is able to support a large collector voltage by the formation of a depletion region. For a one-dimensional structure, the extension of the depletion region is constrained by the remaining stored charge in the N-drift region. However, in a two-dimensional structure, the depletion layer expands at the edge of the emitter closest to the base contact by pushing the stored charge into the remaining segment of the emitter finger. A simple model for the growth of the depletion layer can be derived by using charge control analysis for a one-dimensional structure representing the edge of the emitter finger near the base contact. At an elapsed time t_e after the storage phase, the charge removed to form the depletion region is given by

$$Q_{SD}(t_e) = qN_D W_D(t_e), \tag{7.184}$$

where $W_D(t_e)$ is the width of the depletion region at this time. This charge is also equal to the product of the reverse base drive current and the elapsed time:

$$J_{BR} t_e = Q_{SD}(t_e) = qN_D W_D(t_e). \tag{7.185}$$

The depletion region width at time t_e is then obtained:

$$W_D(t_e) = \frac{J_{BR} t_e}{qN_D}. \tag{7.186}$$

The voltage supported across this depletion region is given by

$$V_C(t_e) = \frac{qN_D}{2\varepsilon_S} W_D^2(t_e). \tag{7.187}$$

Using (7.186) in (7.187) yields

$$V_C(t_e) = \frac{J_{BR}^2 t_e^2}{2q\varepsilon_S N_D}, \tag{7.188}$$

indicating that the collector voltage will increase as the square of the time interval after the storage time. The time taken for the collector voltage to increase to the collector supply voltage, called the *voltage rise time* (t_V), can be derived from (7.188):

$$t_V = \frac{\sqrt{2q\varepsilon_S N_D V_{CS}}}{J_{BR}}. \tag{7.189}$$

The voltage rise time can be reduced by increasing the reverse base drive current at the disadvantage of a larger, more expensive, drive circuit. In the case of a bipolar power transistor with a drift region doping concentration of 5×10^{13} cm^{-3} switching to a collector supply voltage of 800 V using a reverse base drive current of 5 A cm^{-2}, the voltage rise time predicted by this equation is 23 ns.

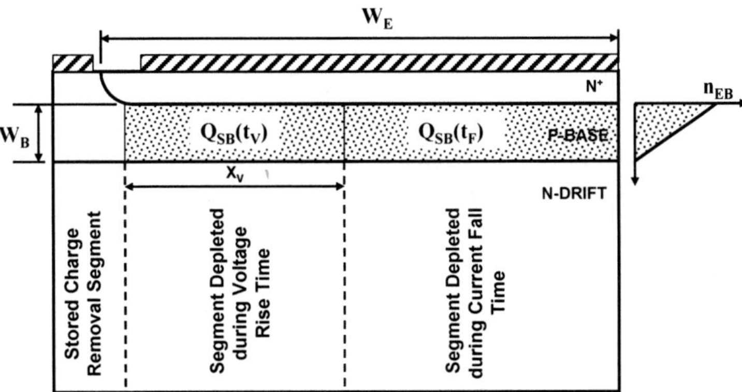

Fig. 7.68 Current constriction in the bipolar power transistor during the turn-off transient

Once the collector voltage reaches the supply voltage, the load current is transferred to the fly-back diode shown in Fig. 7.3. This process results in a reduction of the collector current of the bipolar power transistor while the collector voltage remains at the collector supply potential. At the beginning of this transient, a portion of the emitter finger of width X_V has been turned off as illustrated in Fig. 7.68. Within the N-drift region of this portion, the collector current flows by the transport of electrons at the saturated drift velocity. Within the P-base region of this segment, the collector current is sustained by the injected electron concentration whose profile is shown on the right-hand side. The width of this segment can be obtained by charge control analysis.

The total stored charge in the base region removed during the voltage rise time is given by

$$Q_{SB}(t_V) = \frac{1}{2} q n_{EB} W_B X_V L_E, \tag{7.190}$$

where L_E is the length of the emitter finger orthogonal to the cross section. Using (7.172) to relate the electron concentration at the emitter–base junction to the on-state collector current density,

$$Q_{SB}(t_V) = \frac{J_{C,ON} W_B^2 X_V L_E}{2D_n}. \tag{7.191}$$

The above charge is also equal to the charge removed by the reverse base drive current during the voltage rise time:

$$\frac{J_{C,ON} W_B^2 X_V L_E}{2D_n} = J_{BR} W_E L_E t_V. \tag{7.192}$$

From this expression, the width of the finger that has been turned off during the voltage rise time is obtained:

$$X_V = \frac{2D_n W_E J_{BR} t_V}{J_{C,ON} W_B^2}. \tag{7.193}$$

For a typical power bipolar transistor with an emitter width of 200 μm and P-base width of 10 μm, the width (X_V) that has been turned off at the end of the voltage rise time is found to be 110 μm when the on-state collector current density is 15 A cm^{-2} and reverse base current density is 5 A cm^{-2}. Thus, approximately half of the emitter finger is turned off at this time while the collector current remains constant, leading to an increase in the current density in the remaining portion of the emitter finger.

Current Fall Time Model

The collector current decreases after the collector voltage reaches the collector supply voltage because the current transfers to the fly-back diode. The time taken for the collector current to reduce to zero is determined by the removal of the stored charge remaining in the P-base region after the voltage transient. This stored charge [$Q_{SB}(t_F)$] is indicated by the shaded region on the right-hand side of the P-base region in Fig. 7.68.

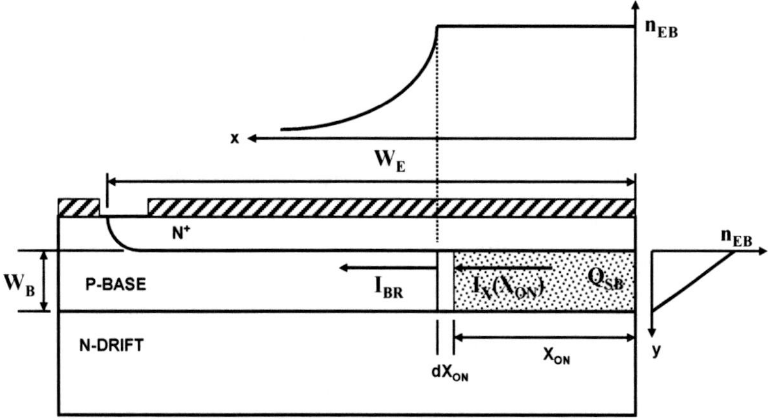

Fig. 7.69 Charge distributions for the current fall time analysis in the bipolar power transistor during the turn-off transient

The fall time for the collector current can be derived by analysis of the extraction of the stored charge in the P-base region by the reverse base drive current. The stored charge (electrons) has a linearly decreasing concentration in the *y*-direction as illustrated on the right-hand side of Fig. 7.69 within the on-portion of the emitter finger. Its concentration along the *x*-direction is uniform within the on-portion of the emitter finger and decays exponentially, with a characteristic diffusion length for electrons, away from the on-portion within the off-portion of the emitter finger as illustrated at the top of Fig. 7.69 at the emitter–base junction. These concentration profiles are based upon the assumption that the collector current density is uniform within the on-portion of the emitter finger and remains constant during the collector fall time. Under these conditions, the electron distribution is given by

$$n(x,y) = n_{\mathrm{EB}}\left(1-\frac{y}{W_{\mathrm{B}}}\right)\mathrm{e}^{-(x-X_{\mathrm{ON}})/L_{\mathrm{n}}}, \tag{7.194}$$

where L_n is the diffusion length for electrons in the P-base region. The stored charge within the segment $\mathrm{d}X_{\mathrm{ON}}$ is given by

$$\mathrm{d}Q_{\mathrm{SB}} = \frac{1}{2}qn_{\mathrm{EB}}W_{\mathrm{B}}L_{\mathrm{E}}\mathrm{d}X_{\mathrm{ON}}. \tag{7.195}$$

Using (7.172) for the electron concentration at the emitter–base junction,

$$\mathrm{d}Q_{\mathrm{SB}} = \frac{J_{\mathrm{C,ON}}W_{\mathrm{B}}^{2}L_{\mathrm{E}}}{2D_{\mathrm{n}}}\mathrm{d}X_{\mathrm{ON}}. \tag{7.196}$$

Consequently,

$$\frac{dQ_{SB}}{dt} = \frac{J_{C,ON} W_B^2 L_E}{2D_n} \frac{dX_{ON}}{dt}. \tag{7.197}$$

The stored charge within the segment dX_{ON} is also determined by the difference between the current entering the segment from the on-portion in the x-direction $[I_x(X_{ON})]$ and the current leaving the segment (I_{BR}):

$$\frac{dQ_{SB}}{dt} = I_x(X_{ON}) - I_{BR}. \tag{7.198}$$

The current entering the segment in the x-direction from the on-portion can be obtained by using

$$I_x(X_{ON}) = \int_0^{W_B} J_x L_E dy \tag{7.199}$$

with the current density in the x-direction given by

$$J_x = qD_n \left.\frac{\delta n}{\delta x}\right|_{x=X_{ON}}. \tag{7.200}$$

Using (7.194) for the electron distribution yields

$$J_x = q\frac{D_n}{L_n} n_{EB}\left(1 - \frac{y}{W_B}\right). \tag{7.201}$$

Substituting this expression in (7.199),

$$I_x(X_{ON}) = \frac{qD_n n_{EB} W_B L_E}{2L_n}. \tag{7.202}$$

Using (7.172) for the electron concentration at the emitter–base junction,

$$I_x(X_{ON}) = \frac{J_{C,ON} W_B^2 L_E}{2L_n}. \tag{7.203}$$

Substituting (7.197) and (7.203) into (7.198) yields

$$\frac{W_B^2 J_{C,ON} L_E}{2D_n} \frac{dX_{ON}}{dt} = \frac{J_{C,ON} W_B^2 L_E}{2L_n} - I_{BR}. \tag{7.204}$$

Expressing the reverse base current in terms of a reverse base drive current density,

$$I_{BR} = J_{BR} W_E L_E \tag{7.205}$$

provides the expression

$$\frac{dX_{ON}}{dt} = \frac{D_n}{L_n} - \frac{2D_n W_E J_{BR}}{W_B^2 J_{C,ON}}. \tag{7.206}$$

The first term on the right-hand side of this expression is smaller in magnitude than the second term, indicating that the width of the on-portion (X_{ON}) is shrinking with time. Defining a velocity for the movement of the on-portion as

$$v_{ON} = \frac{2D_n W_E J_{BR}}{W_B^2 J_{C,ON}} - \frac{D_n}{L_n} \tag{7.207}$$

and substituting into (7.206),

$$\frac{dX_{ON}}{dt} = -v_{ON}. \tag{7.208}$$

The evolution of the width of the on-portion with time can be obtained by integration of this expression from an initial on-portion width of ($W_E - X_V$):

$$X_{ON}(t) = (W_E - X_V) - v_{ON} t_e, \tag{7.209}$$

where t_e is the elapsed time after the end of the voltage rise transient.

Since the current density in the on-portion has been assumed to remain constant during this transient, the collector current is given by

$$I_C = J_{C,ON} X_{ON}(t) L_E = J_{C,ON}[(W_E - X_V) - v_{ON} t_e] L_E. \tag{7.210}$$

The average collector current density, defined as the ratio of the collector current to the device area ($W_E L_E$), is then given by

$$J_C(t) = \frac{J_{C,ON}}{W_E}[(W_E - X_V) - v_{ON} t_e]. \tag{7.211}$$

The collector current decreases linearly with time according to this analysis. The time taken for the collector current to decrease to zero, defined as the *current fall time*, can be obtained from the above expression:

$$t_F = \frac{W_E - X_V}{v_{ON}}. \tag{7.212}$$

For a typical power bipolar transistor with an emitter width of 200 μm and P-base width of 10 μm, the velocity for the movement of the on-portion is found to be 4.5×10^5 cm s^{-1} (or 4.5 μm ns^{-1}) when the on-state collector current density is 15 A cm^{-2} and reverse base current density is 5 A cm^{-2}. Using a width of 110 μm for the portion turned off during the voltage transient, the collector current fall time is then found to be 20 ns.

Turn-Off Waveforms for an Inductive Load

The waveforms for the current and voltage for the bipolar power transistor can be constructed based upon the above analysis for the storage time, the voltage rise time, and the current fall time. For the typical device structure with an emitter width of 200 μm and a base width of 10 μm, the waveforms obtained by using the analytical models are shown in Fig. 7.70 for the case of a collector on-state current density of 15 A cm^{-2} and reverse base drive current density of 5 A cm^{-2}.

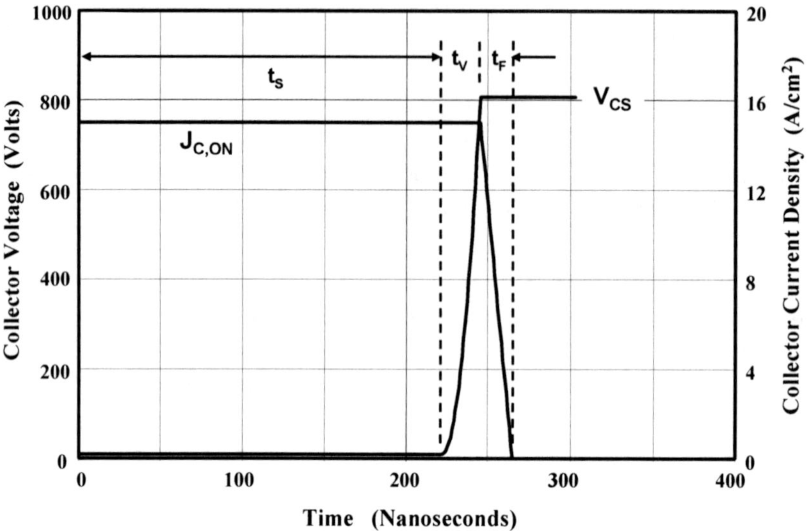

Fig. 7.70 Analytically calculated waveforms for the bipolar power transistor during the turn-off transient

The power dissipated during the turn-off transient can be obtained from the above analysis by using the time intervals for the voltage and current transients. The energy loss per turn-off event is given by

$$E_{OFF} = \int_0^{t_e} J_C(t)V_C(t)\mathrm{d}t = \frac{J_{C,ON}V_{CS}}{3}t_V + \frac{J_{C,ON}V_{CS}}{2}t_F. \tag{7.213}$$

The power dissipation (per unit area) can then be obtained by taking the product of the switching energy and the operating frequency. Due to the short transition times for the voltage and current for the bipolar power transistor, the switching power loss is low making it an attractive power switch in high frequency applications such as lamp ballasts. However, the on-state voltage drop for the structure is relatively large producing high power dissipation in the on-state. In addition, its maximum operating frequency is limited by the long storage time interval.

Simulation Example

To validate the model for the turn-off process for the bipolar power transistor with an inductive load, the results of two-dimensional numerical simulations are described here for the baseline device structure (bjt8). For this structure, a doping concentration of 5×10^{13} cm^{-3} was used for the N-drift region with a thickness of 80 μm. The device was initially biased with a positive base drive voltage of 5 V with a base resistance to limit the base current to 5 A cm^{-2}. The base drive voltage was rapidly switched to negative 5 V to initiate the turn-off process while maintaining a constant collector current to emulate the inductive load conditions. The switching waveforms obtained from the numerical simulations are shown in Fig. 7.71.

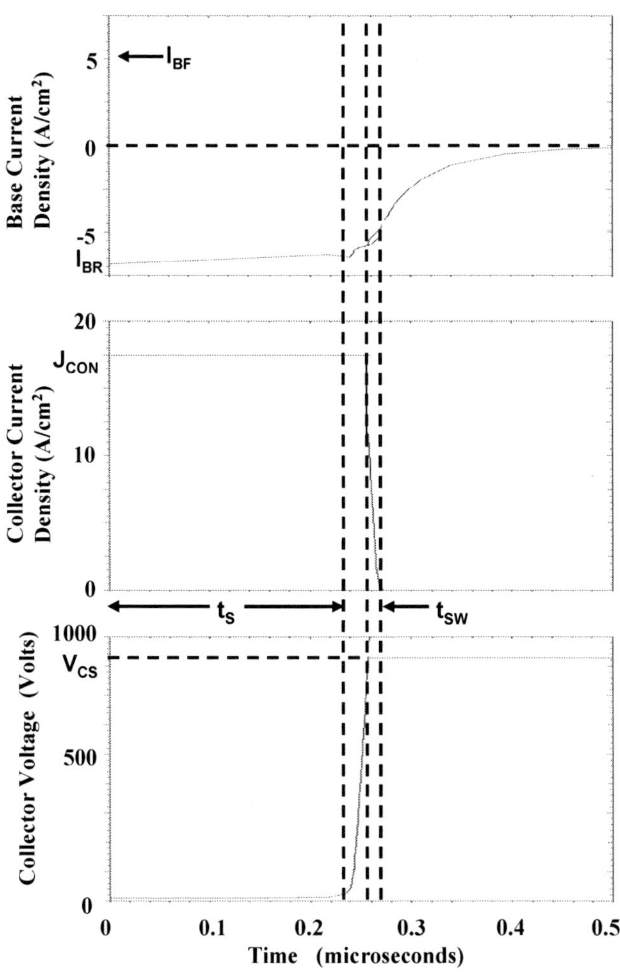

Fig. 7.71 Typical turn-off waveforms for a bipolar power transistor with an inductive load

The base drive current immediately switches to about $-5\ \text{A cm}^{-2}$ upon initiation of the turn-off process. The collector current and voltage remain constant during a storage time interval of 235 ns. The analytical model provides an accurate prediction of this storage time. The collector voltage then begins to increase as the square of the elapsed time as predicted by the analytical model. The collector voltage reaches 800 V after 22 ns (voltage rise time). The analytical model is able to accurately predict this voltage transition time. After this, the collector current decreases linearly with time as predicted by the analytical model. The transition time for the collector current (fall time) is 15 ns. This is also consistent with the behavior predicted by the analytical model.

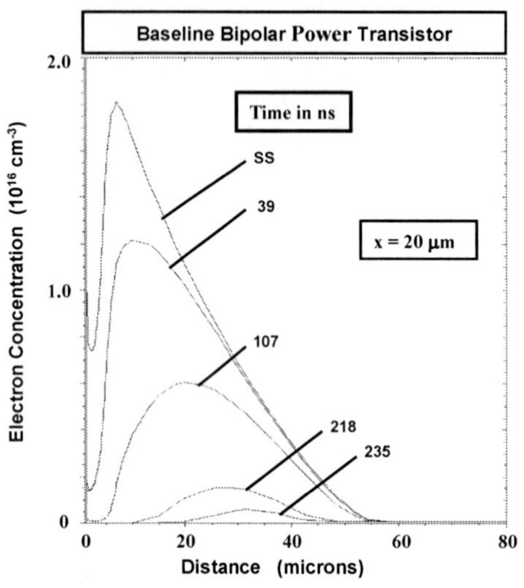

Fig. 7.72 Vertical electron concentration profile within the bipolar power transistor

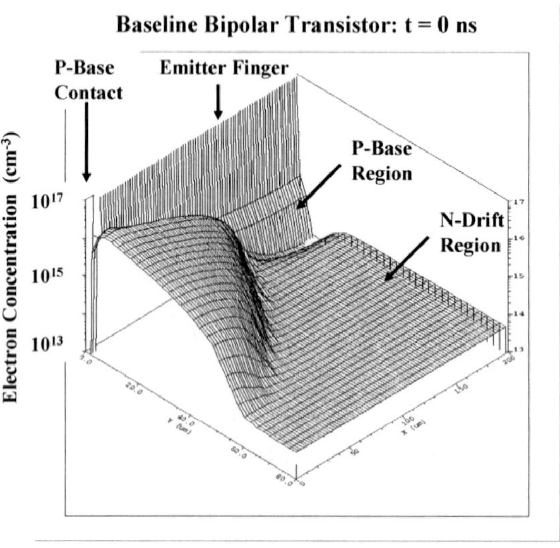

Fig. 7.73 Initial electron charge distribution within the bipolar power transistor

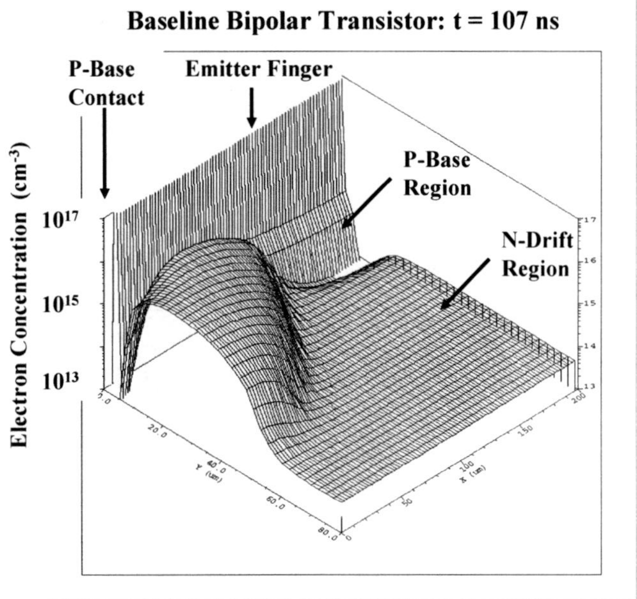

Fig. 7.74 Three-dimensional view of the electron charge distribution within the bipolar power transistor during the turn-off transient

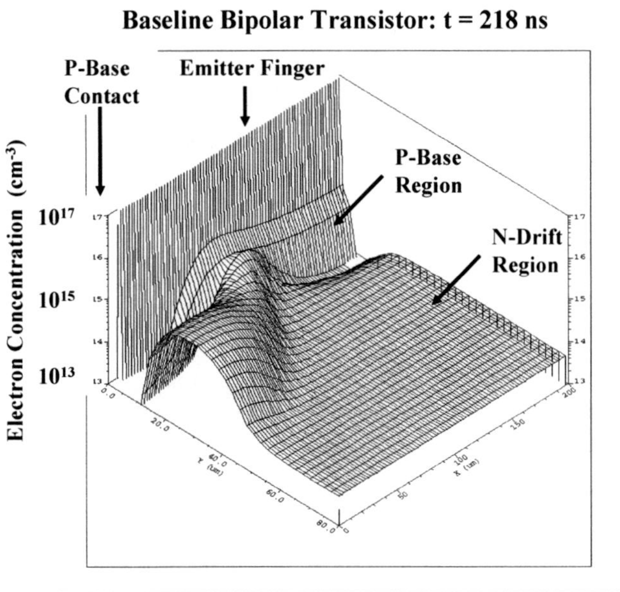

Fig. 7.75 Three-dimensional view of the electron charge distribution within the bipolar power transistor during the turn-off transient

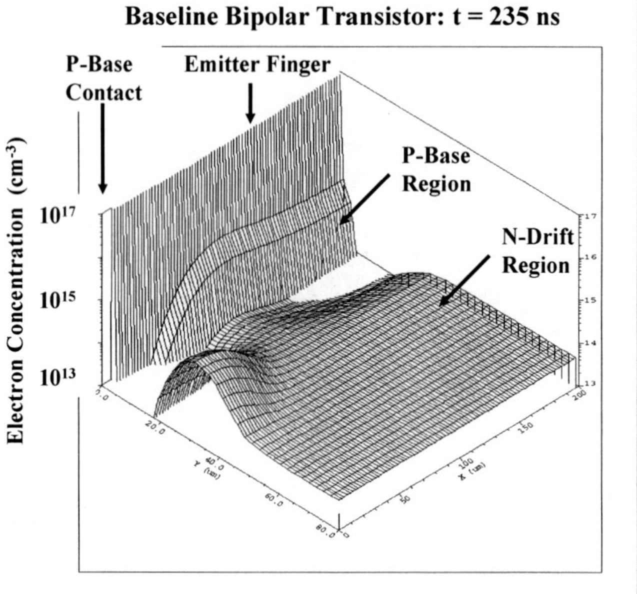

Fig. 7.76 Three-dimensional view of the electron charge distribution within the bipolar power transistor during the turn-off transient

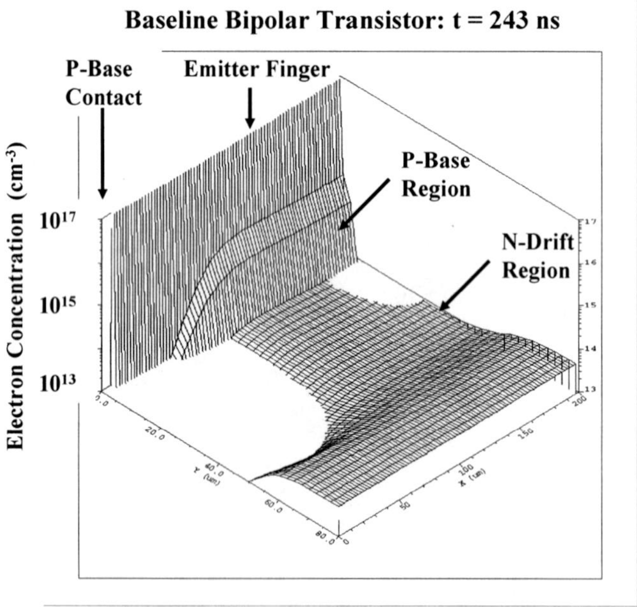

Fig. 7.77 Three-dimensional view of the electron charge distribution within the bipolar power transistor during the turn-off transient

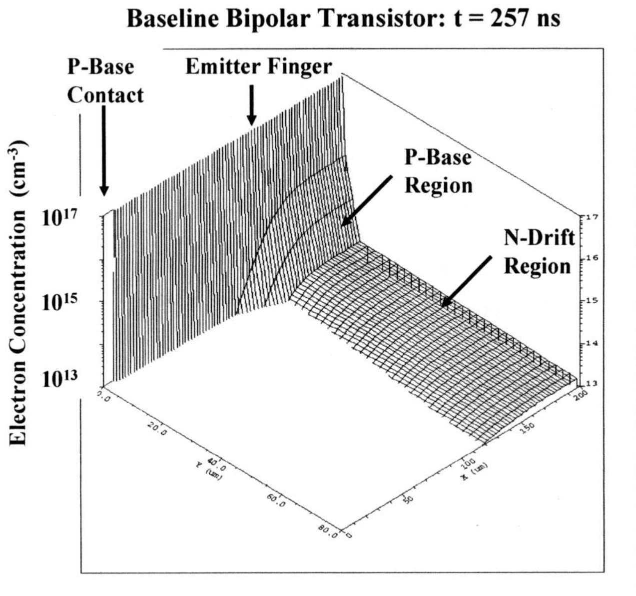

Fig. 7.78 Three-dimensional view of the electron charge distribution within the bipolar power transistor during the turn-off transient

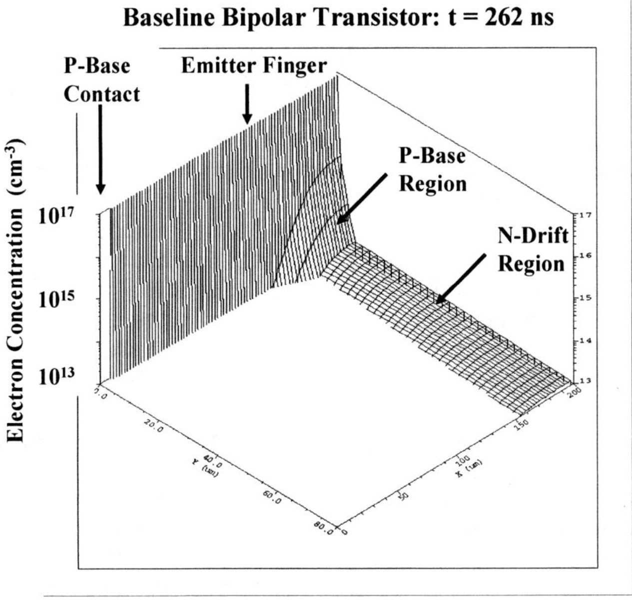

Fig. 7.79 Three-dimensional view of the electron charge distribution within the bipolar power transistor during the turn-off transient

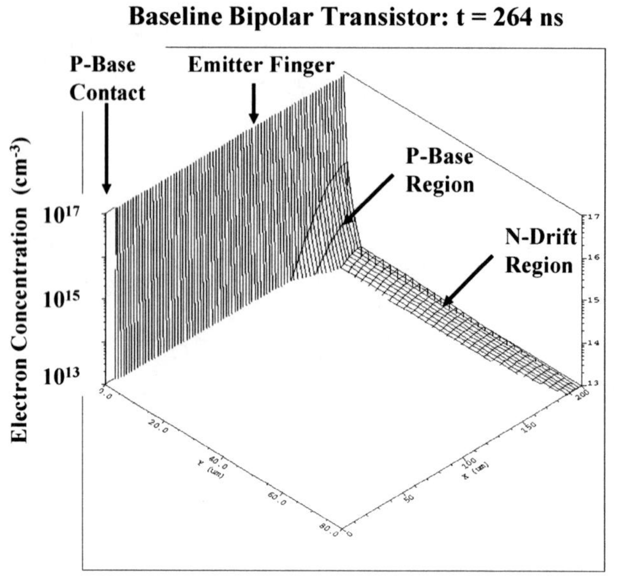

Fig. 7.80 Three-dimensional view of the electron charge distribution within the bipolar power transistor during the turn-off transient

To gain further insight into the turn-off physics, it is instructive to examine the electron distribution within the bipolar power transistor structure at various stages of the turn-off process. The electron concentration profile near the edge of the emitter closest to the base contact is shown in Fig. 7.72. It can be observed that the profile at t = 39 and 107 ns has the shape illustrated in Fig. 7.67 allowing analysis of the storage time with a one-dimensional model.

The physics determining the turn-off process for the bipolar power transistor can be understood by examination of the electron distribution within the two-dimensional structure. Three-dimensional views of the electron concentration are provided in Figs. 7.73–7.80 at various stages during the turn-off process. The initial (t = 0 ns) on-state electron distribution during operation in the quasisaturation region is shown in Fig. 7.73. It can be observed that the highest concentration of electrons is resident at the edge of emitter closest to the base contact because of the emitter current crowding phenomenon. The concentration of electrons in the N-drift region is observed to be much greater than that within the P-base region.

The electron distribution during the storage time is provided in Figs. 7.74, 7.75, and 7.76 as the time progresses to 107, 218, and 235 ns, respectively. It can be observed from these figures that the stored charge is being extracted mostly from the edge of the structure in proximity with the base contact. This allows a one-dimensional analysis for the storage time. At t = 235 ns, the electron concentration in this region becomes equal to zero at the base–collector junction located at a depth of about 10 μm, enabling the bipolar power transistor to begin supporting higher collector voltages.

The electron distribution during the voltage rise time is provided in Figs. 7.77 and 7.78 as the time progresses to 243 and 257 ns, respectively. It can be observed from these figures that a depletion region is forming at the edge of the structure in proximity with the base contact. This allows analysis of the voltage rise time with a one-dimensional analysis. The depletion layer width increases with time in the *y*-direction due to the larger voltage being supported across the bipolar power transistor [V_{CE}(243 ns) = 105 V; V_{CE}(257 ns) = 828 V]. In addition, the depletion region expands in the *x*-direction as described in the analytical model (see Fig. 7.68). At the end of the voltage rise time, a portion of the emitter finger with a width (X_{ON}) of 80 μm is still carrying the initial on-state collector current while a segment with a width (X_V) of 110 μm has been turned off. The analytical model provides an accurate assessment of these widths.

It can be observed from these figures that the collector current is sustained in the on-segment by the injection of electrons in the P-base region and the transport of electrons through the collector region at the saturated drift velocity. The concentration of electrons in the N-drift region is therefore given by

$$n_{\rm D} = \frac{J_{\rm C}}{qv_{\rm sat,n}}. \qquad (7.214)$$

For an initial on-state collector current density of 17.5 A cm^{-2}, the electron concentration obtained by using the above expression with a saturated drift velocity of 1×10^7 cm s^{-1} is between 1 and 2×10^{13} cm^{-3} after taking into account the reduced conduction area within the structure at this stage of the turn-off process. This is in agreement with the observed electron concentration in the N-drift region within the on-segment.

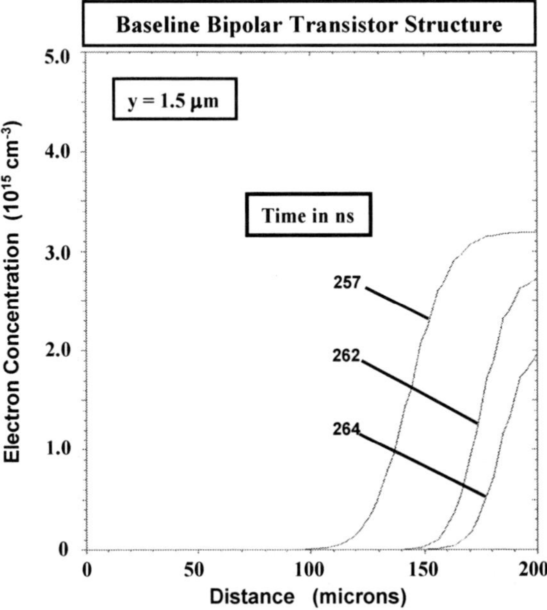

Fig. 7.81 Electron concentration profile within the P-base region of the bipolar power transistor during the turn-off transient

The electron distribution during the current fall time is provided in Figs. 7.79 and 7.80 as the time progresses to 262 and 264 ns, respectively. It can be observed that the on-segment shrinks in size as described by the two-dimensional analytical model for this part of the turn-off process. To provide further insight into the turn-off process, the electron concentration profile along the *x*-direction is provided in Fig. 7.81 during various points of the collector fall time interval. It can be observed that the electrons are diffusing from the on-segment into the off-segment as assumed in the two-dimensional analytical model (see Fig. 7.69). The electron concentration remains approximately constant during a portion of the transient as assumed in the analytical model and then decreases toward the end of the current fall interval. This results a slightly smaller collector current fall time than predicted by the model.

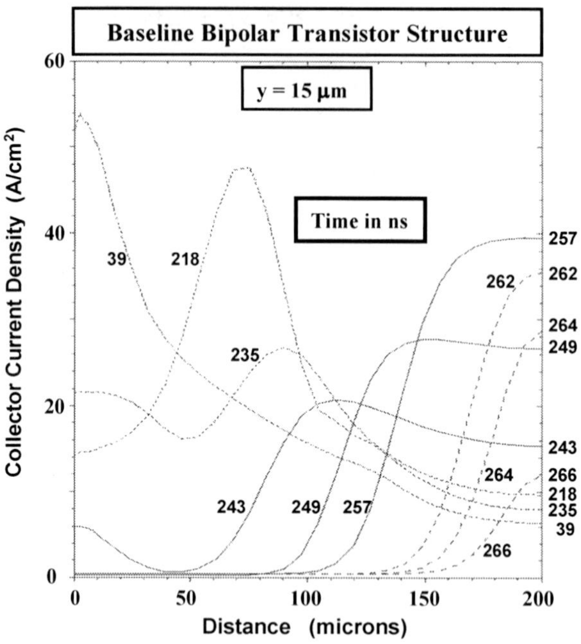

Fig. 7.82 Collector current density within the P-base region of the bipolar power transistor during the turn-off transient

To complete the description of the turn-off process, the evolution of the collector current density within the emitter finger is provided in Fig. 7.82. The collector current density profiles during the storage phase are shown by the dotted lines at time t = 39, 218, and 235 ns. In these cases, a high-current density is observed at the edge of the emitter finger near the base contact. The collector current density profiles during the voltage rise phase are shown by the solid lines at time t = 243, 249, and 257 ns. In these cases, the current density increases at the center of the emitter finger because the area for the on-segment is decreasing while the total collector current remains constant. The collector current density profiles during the current fall phase are shown by the dashed lines at time t = 262, 264, and 266 ns. In these cases, the current density is observed to initially remain approximately constant and then decreases rapidly as the collector current decreases.

The high-current densities observed within the emitter finger of the bipolar power transistor produce local hot spots due to the enhanced power dissipation. These hot spots also promote current filamentation which can limit the SOA for the structure as discussed in Sect. 7.9.

7.9 Safe Operating Area

The trajectory for the collector current and voltage during the turn-off and turn-on events is illustrated in Fig. 7.83 for a bipolar power transistor. These switching loci are applicable for the circuit shown in Fig. 7.3 with an inductive load and a P-i-N rectifier used as the fly-back diode. During the turn-off process shown in Fig. 7.83a, the voltage increases with a constant current flowing through the transistor until the voltage reaches the collector supply voltage (V_{CS}). The current then transfers from the transistor to the fly-back rectifier. An overshoot in the collector voltage is observed due to the presence of finite stray inductances in the collector circuit.

During the turn-on process shown in Fig. 7.83b, the collector current in the bipolar power transistor increases until it becomes equal to the load current. During this time, the load current is being transferred from the fly-back diode to the transistor. The collector current then continues to increase because of the reverse recovery current of the P-i-N rectifier until it settles down to the on-state collector current ($I_{C,ON}$). The bipolar power transistor must be capable of handling the large overshoot in current because the reverse recovery current for the P-i-N rectifier is typically as large as the load current.

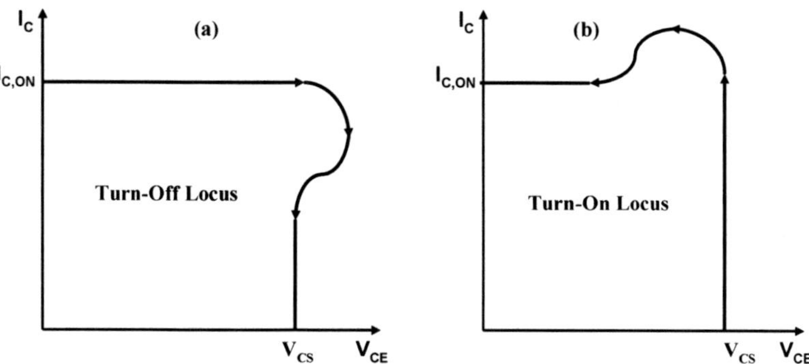

Fig. 7.83 Typical switching loci for the bipolar power transistor

The bipolar power transistor must be capable of operating in all portions of the turn-off and turn-on loci without destructive failure. Due to the simultaneous large collector current and voltage during the transients, considerable power dissipation occurs within the bipolar power transistor. The steady-state thermally defined limit to the current–voltage locus is determined by the maximum permissible junction temperature [$T_{J(max)}$] and the thermal resistance (R_{TH}):

$$P_{\mathrm{D(max)}} = I_{\mathrm{C}} V_{\mathrm{CE}} = \frac{T_{\mathrm{J(max)}} - T_{\mathrm{A}}}{R_{\mathrm{TH}}}, \tag{7.215}$$

where T_A is the ambient temperature.[21] For a constant thermal resistance, this expression indicates that the maximum collector current will reduce inversely with increasing collector voltage.

A larger instantaneous power can be tolerated during the short switching intervals. However, the SOA for the bipolar power transistor can become limited by the formation of hot spots accompanied by current filaments. The localization of the collector current can occur during operation with the emitter–base junction under forward bias as in the case of turn-on or with the emitter–base junction under reverse bias as in the case of turn-off. These limitations, commonly referred to as the *second breakdown*, are discussed below.

7.9.1 Forward-Biased Second Breakdown

The bipolar power transistor can undergo destructive failure due to thermal runaway when biased in the forward active region of operation. The emitter current becomes localized due to a positive thermal feedback phenomenon leading to the formation of a current filament with very high-current density. The high power dissipation within the current filament raises the local temperature, which favors even greater current flow in this area. This in turn produces a further rise in the local temperature. Eventually, the local temperature exceeds the eutectic temperature between silicon and the aluminum metallization leading to the melting of the emitter metal at the hot spot. The metal then penetrates through the junctions producing a short circuit, which destroys the device. The failure mechanism responsible for limiting the performance of the bipolar power transistor, when the emitter–base junction is forward biased, is referred to as the *forward-biased second breakdown* phenomenon. The current–voltage locus within which the bipolar power transistor can be operated without destructive failure while the emitter–base junction is forward biased is referred to as the *forward-biased safe operating area* (FBSOA).

Consider a bipolar power transistor driven by a voltage source (V_{BE}). The collector current density is then given by

$$J_{\mathrm{C}} = \frac{q D_{\mathrm{nB}} n_{\mathrm{0B}}}{W_{\mathrm{B}}} \mathrm{e}^{qV_{\mathrm{BE}}/kT}. \tag{7.216}$$

The minority carrier concentration in the base under equilibrium is given by

$$n_{\mathrm{0B}} = \frac{n_{\mathrm{iB}}^2}{N_{\mathrm{AB}}}, \tag{7.217}$$

where n_{iB} is the intrinsic carrier concentration and N_{AB} is the doping concentration in the P-base region. Using this expression in (7.216),

$$J_{\mathrm{C}} = \frac{qD_{\mathrm{nB}}n_{\mathrm{i}}^{2}}{W_{\mathrm{B}}N_{\mathrm{AB}}}\mathrm{e}^{qV_{\mathrm{BE}}/kT}. \tag{7.218}$$

For a collector bias voltage V_{CE}, the steady-state power dissipation is

$$P_{\mathrm{D}} = J_{\mathrm{C}}V_{\mathrm{CE}}. \tag{7.219}$$

If the collector current density increases in a local region, the power dissipation in this region also becomes larger than in the rest of the transistor. The local temperature then increases in accordance with

$$T_{\mathrm{J}} = P_{\mathrm{D}}R_{\mathrm{TH}} + T_{\mathrm{A}} = J_{\mathrm{C}}V_{\mathrm{CE}}R_{\mathrm{TH}} + T_{\mathrm{A}}. \tag{7.220}$$

An increase in the local temperature produces an increase in the collector current density in this region because the intrinsic carrier concentration increases and is a strong function of temperature (see Chap. 2). This provides a positive feedback mechanism for the local collector current density and temperature to increase until destructive failure occurs at the hot spot because of the melting of the aluminum metallization at about 600°C.

The localization of the collector current can be prevented by driving the bipolar power transistor with a current source instead of a voltage source. This can be achieved by using a base–emitter voltage source (V_{BS1}) with a series resistance as illustrated in Fig. 7.3. The base drive current during on-state operation is then given by

$$I_{\mathrm{B}} = \frac{V_{\mathrm{BS1}} - V_{\mathrm{bi}}}{R_{\mathrm{B1}}} \approx \frac{V_{\mathrm{BS1}}}{R_{\mathrm{B1}}} \tag{7.221}$$

if the base drive supply voltage is much greater than the built-in potential (0.8 V) for the P–N junction. The base current is then controlled by the external drive circuit, with the base–emitter voltage of the bipolar power transistor related to the base current density by (see (7.28))

$$\mathrm{e}^{qV_{\mathrm{BE}}/kT} = \frac{L_{\mathrm{pE}}J_{\mathrm{B}}}{qD_{\mathrm{pE}}p_{0\mathrm{E}}} = \frac{L_{\mathrm{pE}}N_{\mathrm{DE}}J_{\mathrm{B}}}{qD_{\mathrm{pE}}n_{\mathrm{iE}}^{2}}. \tag{7.222}$$

Using this expression in (7.218),

$$J_{\mathrm{C}} = \frac{D_{\mathrm{nB}}L_{\mathrm{pE}}N_{\mathrm{DE}}n_{\mathrm{iB}}^{2}}{D_{\mathrm{pE}}W_{\mathrm{B}}N_{\mathrm{AB}}n_{\mathrm{iE}}^{2}}J_{\mathrm{B}}. \tag{7.223}$$

Based upon this expression, it can be concluded that the collector current density is relatively insensitive to local temperature fluctuations because the strong temperature dependence of the intrinsic concentrations in the base and emitter regions is cancelled out. This prevents the formation of current filaments and hot spots within the emitter fingers of bipolar power transistors. However, the base drive circuit requires a relatively large supply voltage which increases the power dissipated in the control electronics.

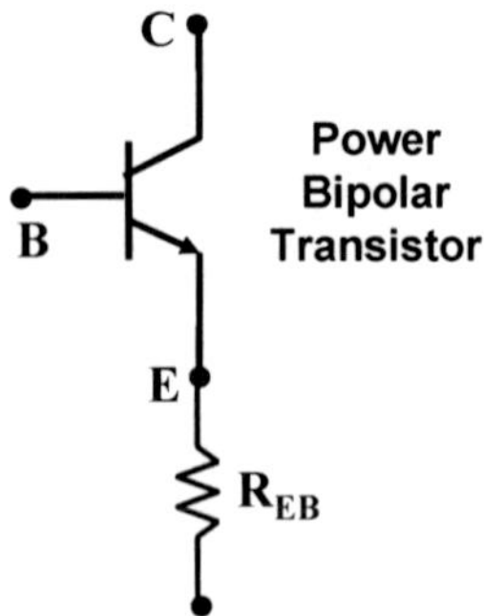

Fig. 7.84 Emitter ballasting resistance for the bipolar power transistor

A commonly used approach to increasing the uniformity of the collector current distribution within a bipolar power transistor is by the addition of an *emitter ballasting resistance* as illustrated in Fig. 7.84. When the emitter current increases at a local region within the emitter finger, the voltage drop across the emitter ballast resistance also increases. This reduces the forward bias for the base–emitter junction in this location suppressing the injection of electrons into the P-base region. This in turn reduces the collector current density at this location providing a stabilizing influence of the current distribution. It is worth pointing out that this approach cannot be implemented with an emitter ballasting resistance external to the bipolar transistor structure. The emitter ballasting resistance must be distributed within the all portions of the emitter fingers to provide the stabilizing influence. In addition, since the entire emitter current flows through the emitter ballasting resistance, its magnitude must be small to avoid increasing the on-state voltage drop and power dissipation.

An elegant implementation of the emitter ballasting resistance in the bipolar power transistor can be achieved as illustrated in Fig. 7.85 by restricting the emitter metallization to the center of the emitter finger. The emitter ballast resistance is derived from the sheet resistance of the N^+ emitter diffusion. Due to the high doping concentration of the N^+ emitter region required for obtaining a good injection efficiency, its sheet resistance is low (~10 Ω sq^{-1}). This enables the formation of the emitter resistance with low values that are compatible with obtaining only a small increase in the on-state voltage drop.

During operation in the on-state, it was previously shown that the emitter current crowds at the edge of the emitter closest to the base contact as indicated by the arrow in the figure. Most of the emitter current is consequently forced to flow through the emitter ballast resistance. This produces the additional voltage drop required to promote uniform emitter current distribution. A significant advantage of this approach, to introducing an emitter ballast resistance that is uniformly distributed within the bipolar power transistor, is the absence of any additional process steps during device fabrication. The emitter ballast resistance is created during the layout of the emitter metallization relative to the edges of the emitter fingers.

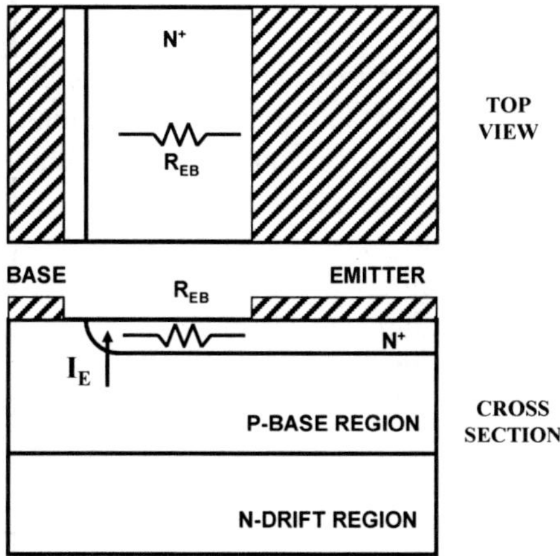

Fig. 7.85 Distributed emitter ballasting resistance within the bipolar power transistor structure

7.9.2 Reverse-Biased Second Breakdown

The SOA for the bipolar power transistor is limited by different physics when switching from the on-state to the blocking state. During the turn-off transient, the base drive current is reversed to remove the stored charge within the P-base and the N-drift regions. The failure mechanism encountered during the turn-off process is consequently referred to as the *reverse-biased second breakdown* phenomenon. The current–voltage locus within which the bipolar power transistor can be operated without destructive failure while the emitter–base junction is reverse biased is referred to as the *reverse-biased safe operating area* (RBSOA).

As previously demonstrated in the section on turn-off physics for the bipolar power transistor, the collector current constricts to the middle of the emitter finger due to removal of charge from the proximity of the base contact by the reverse base drive current. In the case of an inductive load, the collector current remains constant during this stage of the turn-off process producing an increase in the collector current density in the middle of the emitter finger. As the collector voltage increases, the electric field in the N-drift region becomes large resulting in the transport of electrons at the saturated drift velocity. The electron concentration in the drift region at the middle of the emitter finger is then given by

$$n_{\mathrm{D}} = \frac{J_{\mathrm{C}}}{q v_{\mathrm{sat,n}}}, \tag{7.224}$$

where J_{C} is the local collector current density.

The electron concentration exceeds the background doping level when the local current density becomes large. As an example, this would occur at a current density of 80 A cm^{-2} for a drift region with doping concentration of 5×10^{13} cm^{-3}. The electric field then reverses in slope as shown previously in Fig. 7.19. When the collector current density is further increased and exceeds the Kirk current density (J_K), a current-induced base region forms at the base–collector junction with a peak electric field located at the interface between the N-drift region and the N$^+$ substrate. The electric field profiles under these conditions are illustrated in Fig. 7.86 with and without taking into account the current crowding at the middle of the emitter finger. It can be observed that a much larger maximum electric field occurs at the interface between the N-drift region and the N$^+$ substrate in the presence of current crowding. This precipitates avalanche breakdown at a lower voltage than expected either for the background doping concentration of the N-drift region or with a uniform current distribution within the structure.

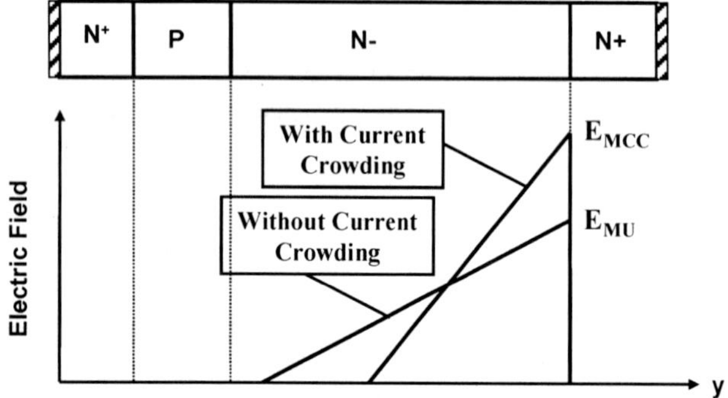

Fig. 7.86 Electric field profiles within the bipolar power transistor structure with and without current crowding during turn-off

When the electron concentration becomes far greater than the background doping concentration in the N-drift region, the Poisson's equation in the drift region is given by

$$\frac{d^2V(y)}{dy^2} = -\frac{dE}{dy} = -\frac{Q}{\varepsilon_S} = \frac{qn_D}{\varepsilon_S} = \frac{J_C}{\varepsilon_S v_{sat,n}} \tag{7.225}$$

by using (7.224) for the electron concentration in the N-drift region. The avalanche breakdown voltage in the presence of a high collector current density can be solved by using this equation with the Fulop's formula for the impact ionization coefficient:

$$\mathrm{BV}(J_C) = 5.34 \times 10^{13} \left(\frac{J_C}{qv_{sat,n}} \right)^{-3/4}. \tag{7.226}$$

If the bipolar power transistor is turned off with a reverse base drive current that is substantially smaller than the local collector current density at the middle of the collector finger, the current gain of the transistor must be accounted for leading to

$$\mathrm{BV}_{\mathrm{RBSOA}} = \mathrm{BV}(J_{\mathrm{C}})(1-\alpha)^{1/n}. \tag{7.227}$$

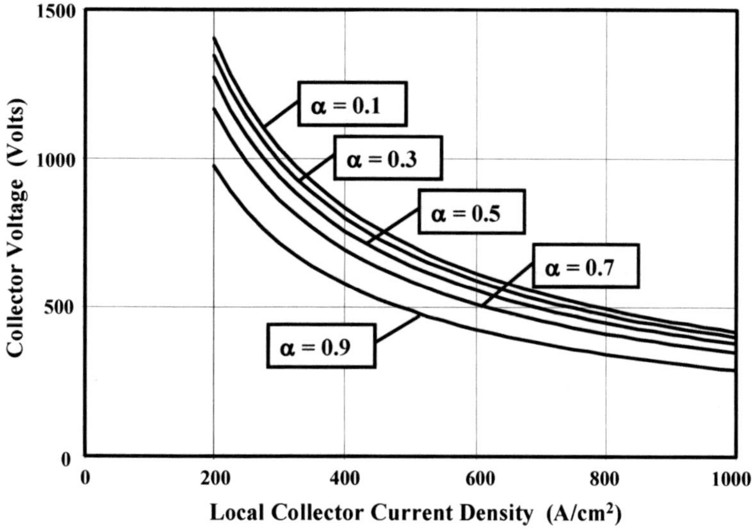

Fig. 7.87 Safe operating area for the bipolar power transistor structure during turn-off

The RBSOA computed by using the above equation is shown in Fig. 7.87 for various values for the common-base current gain. It can be observed from this figure that the maximum collector voltage that can be sustained by the bipolar power transistor is substantially reduced when the local collector current density becomes large. The RBSOA also degrades when the current gain of the transistor is increased. It is therefore necessary to ensure that the local current density at the middle of the emitter finger does not become extremely large during turn-off not only from a thermal standpoint but also to prevent local avalanche breakdown that can destroy the device.

One approach to improving the RBSOA of the bipolar power transistor without degrading the current gain during on-state operation is illustrated in Fig. 7.88. In this structure, the depth of the N^+ emitter region is made shallower in the middle of the emitter finger when compared with its depth at the edges of the emitter finger adjacent to the P-base contact. The deeper N^+ diffusion at the emitter edges produces a higher gain in this portion because of the narrower P-base width when compared with the middle of the emitter finger. Since most of the emitter current flows at the emitter edges during on-state operation due to the emitter current crowding phenomenon, this ensures a high gain in the conduction mode. At the same time, the current gain becomes smaller during turn-off due to the larger width of the P-base region at the middle of the emitter finger.

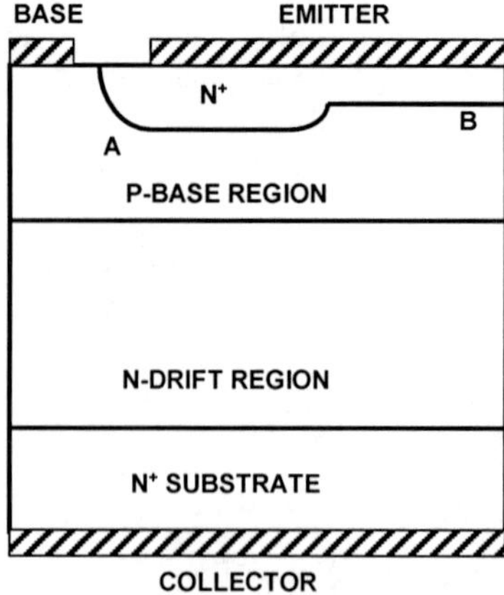

Fig. 7.88 Bipolar power transistor structure with reduced current gain at the middle of the emitter finger

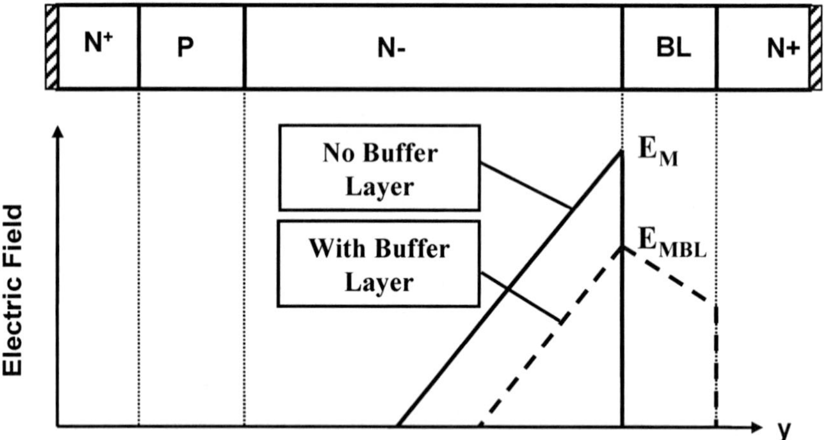

Fig. 7.89 Bipolar power transistor structure with buffer layer in the collector drift region

Another approach to improving the RBSOA of the bipolar power transistor is by the incorporation of a buffer layer in the N-drift region. The buffer layer is placed at the interface between the N-drift region and the N^+ substrate as illustrated in Fig. 7.89. The presence of the buffer layer reduces the peak electric field at the interface, as shown by the dashed line, suppressing premature avalanche breakdown. To be effective in reducing the electric field, the doping concentration of the buffer layer must be comparable with the electron concentration flowing through the drift region under the enhanced collector current density at the middle of the emitter finger. Typical doping concentration for the buffer layer is 5×10^{14} cm^{-3}, corresponding to a collector current density of 800 A cm^{-2}.

7.9.3 Boundary for Safe Operating Area

The SOA for the bipolar power transistor defines the collector current–voltage space within which the device can be operated without destructive failure. One of the limits to the SOA is set by the maximum allowable collector current at low collector bias voltages. This limit is determined by the current-handling capability of the emitter bond wires as shown by the horizontal line in Fig. 7.90. A second limit to the SOA is set by the maximum collector bias voltage at lower collector current levels. This boundary is determined by the breakdown voltage of the structure, including the impact of the edge termination, as shown by the vertical line in Fig. 7.90.

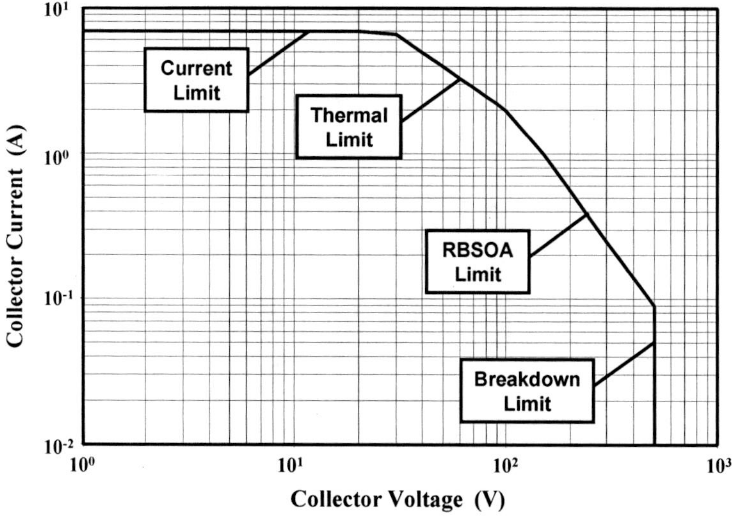

Fig. 7.90 Safe operating area boundaries for a typical bipolar power transistor

The third limit to the SOA is set by thermal considerations. The thermal limit is determined by the power dissipation within the device producing an increase in the junction temperature. The maximum allowable power dissipation is given by (7.215). When $T_{J,max}$ (the maximum junction temperature for reliable operation), T_A (the ambient temperature), and R_{TH} (the thermal resistance of the package) are fixed, this expression indicates that the maximum allowable collector current will decrease inversely with increasing collector voltage. This boundary is shown in Fig. 7.90 as the line with a slope of −1 on the log–log plot.

The fourth limit to the SOA is determined by the on-set of the second breakdown phenomenon during the turn-off process. According to (7.227) in conjunction with (7.226), the maximum allowable collector voltage should decrease inversely as the 3/4 power of the collector current. However, this does not account for the impact of current crowding in the emitter during the turn-off process. In practice, the RBSOA limit is found to follow a slope of −2 on the log–log plot as shown in the figure.[21]

7.10 Darlington Configuration

In the previous sections, it has been demonstrated that the current gain of the bipolar power transistor is quite low (~5) due to various other design constraints. Consequently, a very large external base drive current is required to maintain the transistor in its on-state as well as during turn-off. This makes the base drive circuit bulky and expensive from an application's standpoint. The Darlington configuration[22] provides an approach to obtaining a high-current gain within a monolithic bipolar transistor device structure. In the Darlington configuration, another bipolar power transistor is integrated together with the output bipolar power transistor, as illustrated in Fig. 7.91, to produce a larger current gain.

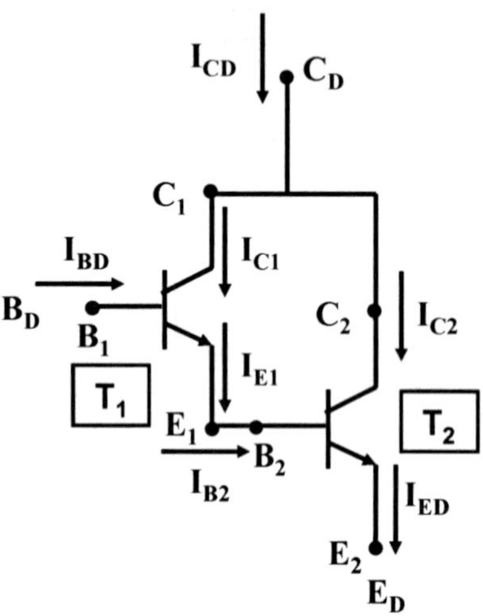

Fig. 7.91 Darlington configuration for the bipolar power transistor

The base drive current for the composite structure is supplied to the input bipolar transistor T_1. The emitter current of this bipolar transistor is then fed to the base terminal of the second (output) bipolar transistor T_2. Note that the collectors for both the bipolar transistors are connected together creating the collector terminal for the composite device structure.

Since the base drive current I_{B2} for the output bipolar transistor is equal to the emitter current I_{E1} for the input bipolar transistor,

$$I_{B2} = I_{E1} = (1+\beta_1)I_{B1} = (1+\beta_1)I_{BD}, \tag{7.228}$$

where β_1 is the common-emitter current gain of the input transistor. The collector current for the output transistor T_2 is then given by

$$I_{C2} = \beta_2 I_{B2} = \beta_2(1+\beta_1)I_{BD}, \tag{7.229}$$

where β_2 is the common-emitter current gain of the output transistor. The collector current for the input transistor T_1 is given by

$$I_{C1} = \beta_1 I_{B1} = \beta_1 I_{BD}. \tag{7.230}$$

Consequently, the collector current for the composite Darlington transistor structure is given by

$$I_{CD} = I_{C1} + I_{C2} = \beta_1 I_{BD} + \beta_2 (1 + \beta_1) I_{BD}. \tag{7.231}$$

From this expression, the common-emitter current gain for the Darlington configuration is obtained:

$$\beta_{BD} = \frac{I_{CD}}{I_{BD}} = \beta_1 + \beta_2 (1 + \beta_1) = \beta_1 + \beta_2 + \beta_1 \beta_2. \tag{7.232}$$

Thus, a significant increase in the current gain can be derived by utilizing the Darlington configuration.

The improved current gain obtained by using the Darlington configuration is accompanied by a substantial increase in the on-state voltage drop. Since the base drive current for the output transistor is derived via the emitter current from the input transistor, the collector potential for the Darlington configuration consists of the voltage drop across the input transistor plus the base–emitter voltage drop for the output transistor:

$$V_{CE,D} = V_{CE1} + V_{BE2}. \tag{7.233}$$

Even if the input transistor is driven into saturation to reduce its on-state voltage drop, the voltage drop for the Darlington configuration becomes in excess of the voltage drop across a forward-biased diode.

Several other considerations influence the design of the Darlington bipolar transistor configuration. Firstly, a high leakage current can arise from the amplification of the leakage current of the input transistor by the current gain of the output transistor. Secondly, a path for removal of the stored charge in the output transistor by the reverse base drive current must be provided to avoid a very long storage phase. This is accomplished by the addition of shunting resistances across the emitter–base junctions of both the internal transistors as illustrated in Fig. 7.92. The shunting resistances reduce the current gains for the internal transistors at low-current levels preventing the amplification of the leakage current. The shunt resistance R_1 also provides a path for extraction of the stored charge within the output transistor with the reverse base drive current during turn-off. In addition, it is convenient to integrate an antiparallel diode with the transistors for use as a fly-back diode in H-bridge circuits commonly used for motor control applications.

A three-dimensional view of the layout of the monolithic Darlington bipolar transistor configuration is illustrated in Fig. 7.93. The input bipolar transistor is located on the left-hand side of the chip with its shunting resistance created by overlapping the emitter metal over the emitter–base junction of transistor T_1 on the edge away from its base contact. The emitter metal E_1 of the input transistor is not connected to the external terminals and is usually covered by a polyimide insulating layer. This metal layer also serves as the base terminal for the output transistor whose emitter is interdigitated with it in the central portion of the chip. The second resistance R_2 is created on the right-hand side of the chip. The antiparallel diode D is formed by overlapping the emitter terminal E_2 with the P-base region.

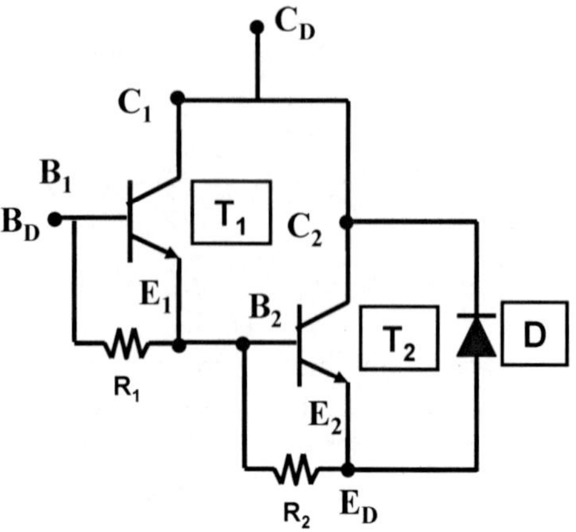

Fig. 7.92 Darlington configuration for the bipolar power transistor with internal shunting resistances and antiparallel diode

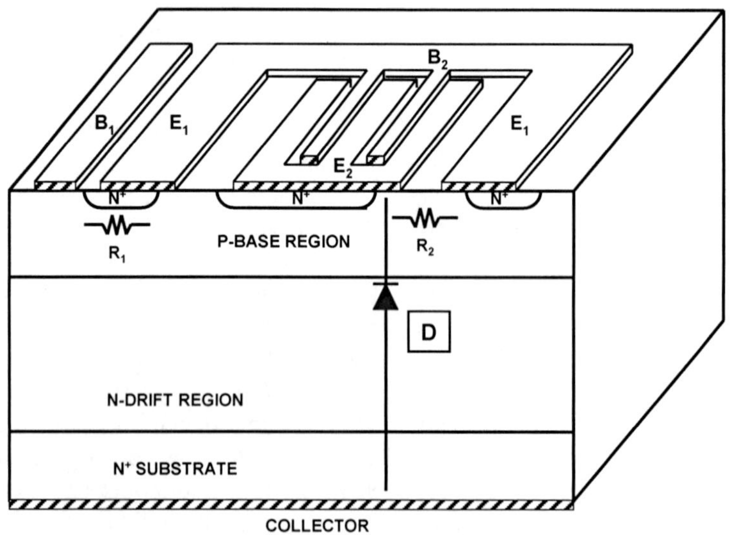

Fig. 7.93 Implementation of the monolithic Darlington bipolar power transistor configuration with internal shunting resistances and antiparallel diode

The relative size of the input and output transistors must be optimized to achieve the best gain and total area for the Darlington transistor configuration.[24] A good balance between the operating current density of the input transistor and the output transistor is required to match the roll-off in the current gain with total collector current density. It is also possible to extend the Darlington concept to multiple stages of bipolar transistors. Although this enables increasing the current gain, the on-state voltage drop also grows because of additional diode voltage drops for the stages.

7.11 Summary

The physics of operation of the bipolar power transistor has been described in this chapter. The bipolar transistor is a current-controlled structure whose current gain depends upon the doping concentrations of the emitter and base regions as well as the width of the base region. In the case of the bipolar power transistor, a falloff in the current gain occurs due to high-level injection in the base region, which degrades its power gain. When optimized to support large voltages, the common-emitter current gain of the power bipolar transistor becomes smaller than 10 at typical operating current levels. The large stored charge within the P-base and N-drift regions of the bipolar power transistor in the on-state reduces the maximum operating frequency. The large resistance of the N-drift region produces a quasisaturation region which increases the on-state voltage drop. These characteristics produce significant power losses within the device and the control circuit.

The bipolar power transistor is also prone to failure modes associated with emitter current crowding during the on-state and turn-off modes. All of these complexities have motivated the development of alternative power switches for applications. The most successful among these alternatives is the IGBT.[25] Its development in the early 1980s has resulted in the complete displacement of the bipolar power transistor leading to its extinction from the power semiconductor landscape. However, the physics of operation for the bipolar power transistor continues to have relevance because it is incorporated within the IGBT structure.

Problems

7.1 A bipolar transistor has a common-emitter current gain of 50. Determine its common-base current gain.

7.2 The open-emitter breakdown voltage of the above bipolar transistor is 1,000 V. Determine its open-base breakdown voltage.

7.3 Consider an $N^+PN–N^+$ bipolar power transistor structure with uniformly doped emitter, base, and collector drift regions. The N^+ emitter region has a doping concentration of 2×10^{19} cm^{-3} and thickness of 10 μm. The P-base region has a doping concentration of 2×10^{17} cm^{-3} and thickness of 10 μm. The N-collector drift region has a doping concentration of 2×10^{14} cm^{-3} and thickness of 40 μm. The Shockley–Read–Hall (low-level, high-level, and space-charge generation) lifetimes are 10 ns in the emitter region, 1 μs in the base region, and 10 μs in the collector drift region. Calculate the emitter injection efficiency at low injection levels in the P-base region ignoring band-gap narrowing and Auger recombination.

7.4 Determine the base transport factor for the power bipolar transistor described in Problem 7.3.

7.5 Determine the common-emitter and common-base current gains for the power bipolar transistor described in Problem 7.3 at low injection levels in the P-base region.

7.6 Determine the open-emitter breakdown voltage (BV_{CB0}) for the power bipolar transistor described in Problem 7.3 using the critical electric field for the doping concentration in the collector drift region. Make sure to take punch-through into account for the drift region. Confirm that the depletion region has not penetrated the entire base region.

7.7 Calculate the open-base breakdown voltage (BV_{CE0}) for the power bipolar transistor described in Problem 7.3 using $n = 6$ in the equation for the multiplication factor.

7.8 Determine the leakage current density at 300 K at a collector bias of 200 V due to space-charge generation in the collector drift region. What is the open-base leakage current density under these conditions?

7.9 Calculate the quasisaturation specific resistance for the power bipolar transistor described in Problem 7.3.

7.10 Determine the Webster current density for the power bipolar transistor described in Problem 7.3.

7.11 What is the common-emitter current gain for the power bipolar transistor described in Problem 7.3 at a current density of 1,000 A cm^{-2}?

7.12 Determine the Kirk current density for the power bipolar transistor described in Problem 7.3 at a collector bias of 300 V.

7.13 What is the Early voltage for the power bipolar transistor described in Problem 7.3?

7.14 Determine the on-state voltage drop ($V_{CE,sat}$) for the power bipolar transistor described in Problem 7.3 in the saturation region of operation. Assume that the injected hole concentration at the B–C junction [$p_{NS}(0)$] is 2×10^{16} cm^{-3} and the injected hole concentration is equal to the doping concentration at the N-drift/N$^+$ substrate interface. What is the collector current density under these operating conditions? Calculate the stored charge in the collector drift region under these conditions.

7.15 The power bipolar transistor described in Problem 7.3 is turned off from the initial on-state conditions described in Problem 7.14 by using a reverse base drive current density of 5 A cm^{-2}. Determine the storage time during the turn-off process.

7.16 What is the voltage rise time under the turn-off conditions described in Problem 7.15 if the collector bias voltage is 300 V? Determine the width of the emitter finger that has been turned off at this point in time if the emitter width (W_E) is 200 μm.

7.17 Determine the velocity for shrinking the on-portion of the emitter finger during the turn-off process. What is the collector current fall time for the conditions described in Problems 7.15 and 7.16?

7.18 Calculate the energy loss incurred during the turn-off process described in Problems 7.15–7.17.

7.19 A Darlington power transistor is designed with a common-emitter current gain of 10 for the input transistor and 5 for the output transistor. What is the current gain for the Darlington transistor?

7.20 What is the typical on-state voltage drop for the Darlington transistor?

References

[1] "The Solid-State Century", Scientific American (Special issue), January 1998.

[2] W. Shockley, "The Path to the Conception of the Junction Transistor", IEEE Transactions on Electron Devices, Vol. ED-31, pp. 1523–1546, 1984.

[3] J. Bardeen and W.H. Brattain, "The Transistor, A Semiconductor Triode", Physical Review, Vol. 74, pp. 230–231, 1948.

[4] W. Shockley, "The Theory of p–n Junctions in Semiconductors and p–n Junction Transistors", Bell System Technical Journal, Vol. 28, pp. 435–489, 1949.

[5] M.S. Adler et al., "The Evolution of Power Device Technology", IEEE Transactions on Electron Devices, Vol. ED-31, pp. 1570–1591, 1984.
[6] B.J. Baliga, "Enhancement and Depletion Mode Vertical Channel MOS Gated Thyristors", Electronics Letters, Vol. 15, pp. 645–647, 1979.
[7] B.J. Baliga et al., "The Insulated Gate Rectifier (IGR): A New Power Switching Device", IEEE International Electron Devices Meeting, Abstract 10.6, pp. 264–267, 1982.
[8] B.G. Streetman and S.K. Banerjee, "Solid State Electronic Devices", 6th Edition, Prentice Hall, Englewood Cliffs, NJ, 2006.
[9] S.M. Sze, "Physics of Semiconductor Devices", 2nd Edition, pp. 265–270, Wiley, New York, 1981.
[10] R. Mertens and R.J. Van Overstraeten, "Measurement of the Minority Carrier Parameters in Heavily Doped Silicon", Solid-State Electronics, Vol. 19, pp. 857–862, 1976.
[11] G.E. Possin, M.S. Adler, and B.J. Baliga, "Measurement of the p–n Product in Heavily Doped Epitaxial Emitters", IEEE Transactions on Electron Devices, Vol. ED-31, pp. 3–17, 1984.
[12] W.M. Webster, "On the Variation of Junction-Transistor Current Amplification Factor with Emitter Current", Proceedings of the IRE, Vol. 42, pp. 914–916, 1954.
[13] C.T. Kirk, "A Theory of Transistor Cut-Off Frequency Fall-Off at High Current Densities", IEEE Transactions on Electron Devices, Vol. ED-9, pp. 164–174, 1966.
[14] A. Blicher, "Field-Effect and Bipolar Power Transistor Physics", pp. 140–143, Academic, San Diego, CA, 1981.
[15] R.J. Hauser, "The Effects of Distributed Base Potential on Emitter-Current Injection Density and Effective Base Resistance for Stripe Transistor Geometries", IEEE Transactions on Electron Devices, Vol. ED-11, pp. 238–242, 1964.
[16] S.K. Ghandhi, "Semiconductor Power Devices", pp. 157–161, Wiley, New York, 1977.
[17] N.H. Fletcher, "Some Aspects of the Design of Power Transistors", Proceedings of the IRE, pp. 551–559, 1955.
[18] J.M. Early, "Effects of Space-Charge Layer Widening in Junction Transistors", Proceedings of the IRE, Vol. 40, p. 1401, 1952.
[19] S.K. Ghandhi, "Semiconductor Power Devices", pp. 146–150, Wiley, New York, 1977.
[20] W.J. Chudobiak, "The Saturation Characteristics of *npvn* Power Transistors", IEEE Transactions on Electron Devices, Vol. ED-17, pp. 843–852, 1970.
[21] F.F. Oettinger, D.L. Blackburn, and S. Rubin, "Thermal Characteristics of Power Transistors", IEEE Transactions on Electron Devices, Vol. ED-23, pp. 831–838, 1976.
[22] S. Darlington, "Semiconductor Signal Translating Device", U.S. Patent 2,663,806, December 22, 1953.

[24] C.F. Wheatley and W.G. Einthoven, "On the Proportioning of Chip Area for Multi-Stage Darlington Power Transistors", IEEE Transactions on Electron Devices, Vol. ED-23, pp. 870–878, 1976.

[25] B.J. Baliga et al., "The Insulated Gate Transistor", IEEE Transactions on Electron Devices, Vol. ED-31, pp. 821–828, 1984.

Chapter 8

Thyristors

The power thyristor was developed as a replacement for the thyratron, a vacuum tube used for power applications prior to the advent of solid-state devices. The thyristor exhibits bistable characteristics allowing operation in either a blocking mode with low off-state current or a current conduction state with low on-state voltage drop. The simple construction of these structures using P–N junctions enabled commercialization of devices in the 1950s. These devices were found to be attractive from an application's viewpoint because they eliminated the need for the cumbersome filaments required in vacuum tubes and were much more rugged and smaller in size.

The power thyristor provides both forward and reverse voltage-blocking capability making it well suited for AC power circuit applications. The device can be triggered from the forward-blocking off-state to the on-state by using a relatively small gate control current. Once triggered into the on-state, the thyristor remains stable in the on-state even without the gate drive current. In addition, the device automatically switches to the reverse-blocking off-state upon reversal of the voltage in an AC circuit. These features greatly simplify the gate control circuit, relative to that required for the power transistor, reducing its cost and size. Furthermore, a structure, called the *triac*, has been developed that enables triggering the device into the on-state even during operation in the third quadrant.

Due to significant interest in the development of solid-state devices for the control of motors operating from DC power sources, a structure, called the *gate turn-off thyristor* (GTO), was also developed in the 1960s. In this device, the structure is modified to enable the switching of the device from the on-state to the off-state while operating in the first quadrant. This is performed by the application of a large reverse gate drive current, akin to that used for turning off the bipolar power transistor. In spite of the bulky and expensive gate control circuits required for the GTO, it was widely adopted for the control of motors in traction (electric

B.J. Baliga, *Fundamentals of Power Semiconductor Devices*, doi: 10.1007/978-0-387-47314-7_8,

streetcars and electric locomotives) applications until recently. The scaling of the power-handling capability of the insulated gate bipolar transistor (IGBT) to handle very high (megawatt) power levels in the twenty-first century has resulted in the displacement of these devices by the IGBT in traction applications.

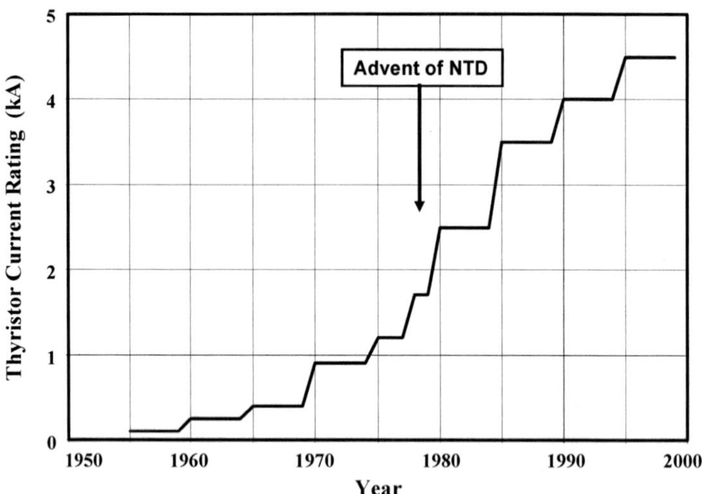

Fig. 8.1 Growth in current ratings for the thyristor

The ability to control operation between blocking states and on-states for a thyristor by using a third terminal was first reported in the 1950s.[1,2] The extensive application potential for these devices to home appliances and power distribution systems generated strong interest in making improvements in the power ratings for the thyristors. The growth of the current-handling capability for the power thyristors is charted in Fig. 8.1. Starting with a modest current of 100 A in the 1950s, the current rating has been scaled to approach 5,000 A for a single device. These high-current levels are required for power distribution systems such as high-voltage DC (HVDC) transmission networks. From the figure, it can be observed that the most rapid increase in the current-handling capability took place at the end of the 1970s. This outcome can be traced to the development of the neutron transmutation doping (NTD) process (see Chap. 2) in the mid-1970 time frame. Using the NTD process, it became possible to obtain larger diameter silicon wafers with uniform properties enabling the observed scaling of the current-handling capability of thyristors.

There was a concomitant increase in the voltage-blocking capability for the thyristor as illustrated in the chart in Fig. 8.2. Beginning with devices capable of operating up to a few hundred volts in the 1950s, the voltage rating for thyristors has been escalated to 8,000 V. The increase in the voltage rating had to be accomplished by the availability of higher resistivity silicon wafers. This was initially achieved by the development of the float-zone process. However, the resistivity variation produced by this process was inadequate for utilization in the large diameter wafers desired to increase the current ratings. The NTD process was instrumental in providing the breakthrough required to create large diameter silicon wafers with low N-type doping concentration and high uniformity in the resistivity across the wafers. Consequently, a substantial gain in the voltage rating occurred in the late 1970s after the commercial availability of NTD silicon as indicated in the chart.

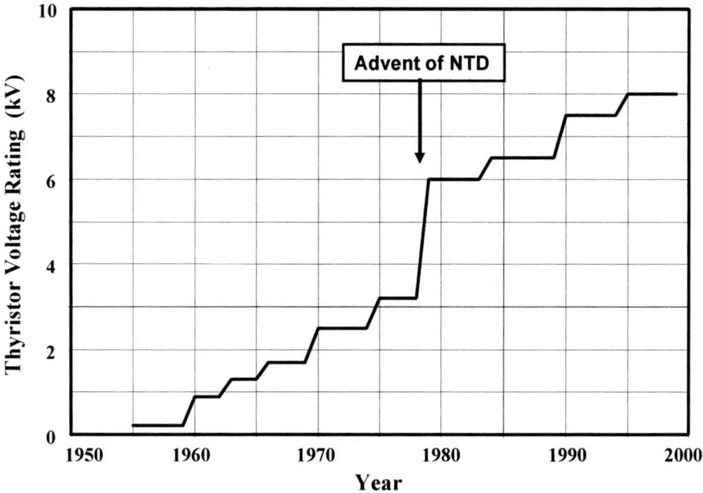

Fig. 8.2 Growth in voltage ratings for the thyristor

Today, single thyristors are available with the capability to block over 8,000 V and conduct 5,000 A in the on-state. Consequently, a single thyristor device can control 40 MW of power. Such devices are attractive for power distribution networks to reduce the total number of devices required in a HVDC station. The reduction of the number of devices connected in series and parallel provides the added benefits of a smaller number of other components that are needed in the system to ensure proper voltage and current distribution between the multiple thyristors. Light-triggered thyristors have also been developed to enable stacking them in a series string to hold off the very high voltages (in excess of 100 kV) that are commonplace for power distribution.

The basic structure and operation of the thyristor are discussed in this chapter. The thyristor contains two coupled bipolar transistors that provide an internal positive feedback mechanism that allows the device to sustain itself in the on-state. This internal feedback mechanism makes it difficult to turn off the structure by external means. To enable operation at elevated temperatures, it is necessary to short circuit the emitter and base regions of the thyristor. The impact of this on the gate control current and switching behavior is analyzed in this chapter. Analytical models are developed for all the operating modes, including the switching transient for the GTO.

8.1 Power Thyristor Structure and Operation

The basic structure for an N^+–P–N–P^+ power thyristor is illustrated in Fig. 8.3. The structure is usually constructed by starting with a lightly doped N-type silicon wafer whose resistivity is chosen based upon the blocking voltage rating for the device. The anode P^+ region is formed by the diffusion of dopants from the backside of the wafer to a junction depth x_{JA}. The P-base and N^+ cathode regions are formed by the diffusion of dopants from the front of the wafer to a depth of x_{JB} and x_{JK}, respectively. Electrodes are formed on the front side of the wafer to contact the cathode and P-base regions, and on the backside of the wafer to contact the anode region. No contact electrode is usually attached to the N-drift (N-base) region.

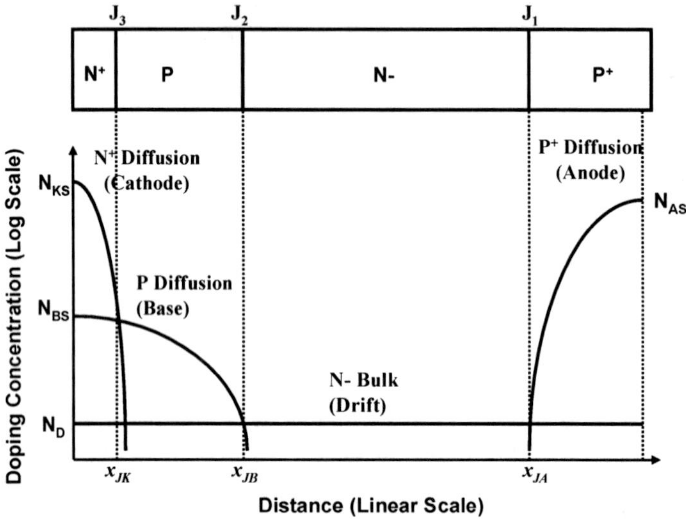

Fig. 8.3 Power thyristor structure and its doping profile

Thyristors are used in applications at very high power levels which require a large voltage-blocking capability, typically in excess of 3,000 V. To achieve such high-voltage blocking capability in both forward and reverse operating quadrants, the P^+ anode/N-drift junction and the P-base/N-drift junction must have a highly graded doping profile. This increases the blocking voltage capability due to voltage supported on the more highly doped side of the junction and allows the utilization of the positive and negative bevels at the edges to suppress premature breakdown at the junction termination. The highly graded junctions can be produced by using gallium and aluminum as the P-type dopants instead of boron, the more commonly used dopant for integrated circuits as well as other power devices discussed in the book. These dopants have much larger diffusion coefficients than boron (see Fig. 2.21), allowing the production of large junction depths (20–100 μm) in a reasonable processing time. Gallium is used as the dopant for the P^+ anode region because of its high solid solubility. Aluminum is used as the dopant for the P-base region due to its lower solid solubility. The doping concentration for the P-base region of the thyristor must be in the range of 1×10^{16}–1×10^{17} cm^{-3} to obtain a reasonable gain for the internal N–P–N bipolar transistor and for enabling edge termination with a negative bevel for the P-base/N-drift junction.

The output characteristics for the thyristor structure are illustrated in Fig. 8.4. The thyristor structure contains three P–N junctions that are in series as indicated in Fig. 8.3. When a negative bias is applied to the anode terminal of the device, the P$^+$ anode/N-drift junction (J_1) and the N$^+$ cathode/P-base junction (J_3) become reverse biased while the P-base/N-drift junction (J_2) becomes forward biased. Due to high doping concentrations on both sides of the N$^+$ cathode/P-base junction (J_3), it is capable of supporting less than 50 V. Consequently, most of negative bias applied to the anode terminal is supported by the P$^+$ anode/N-drift junction (J_1). The reverse-blocking voltage capability for the device is determined by the doping concentration and thickness of the N-drift region. Note that an open-base bipolar transistor is formed within the thyristor structure between junction J_1 and junction J_2. Consequently, the breakdown voltage is not determined by the avalanche breakdown voltage but by the open-base transistor breakdown voltage (see Sect. 3.7). The width of the N-drift region between these two junctions must be carefully optimized to maximize the blocking voltage capability and minimize the on-state voltage drop.

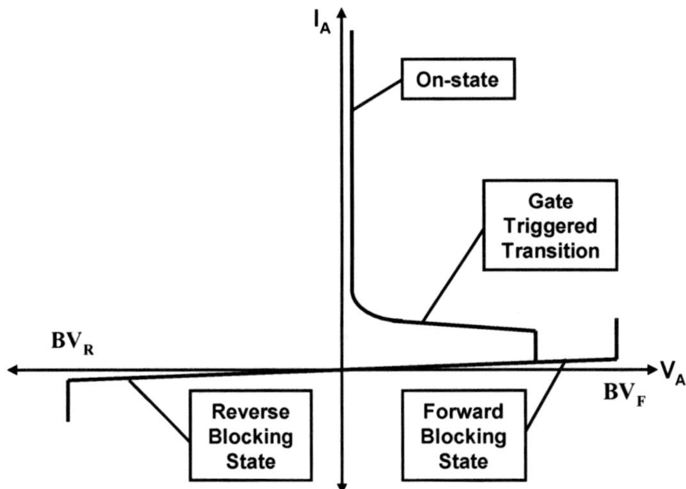

Fig. 8.4 Output characteristics of the power thyristor structure

When a positive bias is applied to the anode terminal of the thyristor, the P$^+$ anode/N-drift junction (J_1) and the N$^+$ cathode/P-base junction (J_3) become forward biased while the junction (J_2) between the P-base region and the N-drift region becomes reverse biased. The applied positive bias is mostly supported across the N-drift region. As in the case of reverse-blocking operation, the blocking voltage capability is determined by open-base transistor breakdown rather than avalanche breakdown. Although not perfectly symmetrical as discussed in a subsequent section of this chapter, the reverse- and forward-blocking capability for the thyristor structure are approximately equal making it suitable for use in AC power circuits.

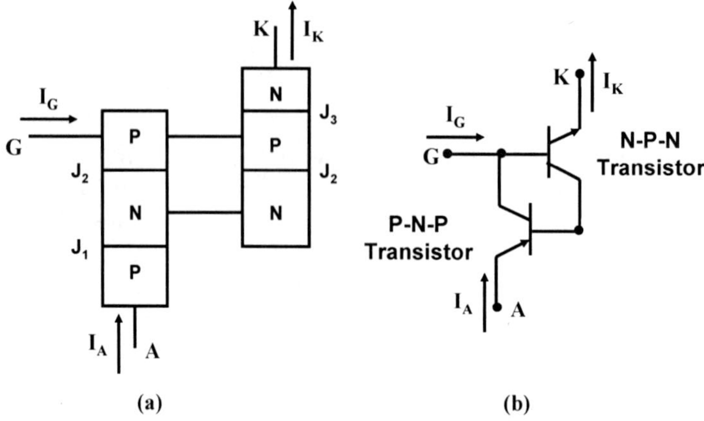

Fig. 8.5 Two coupled transistor equivalent circuits for the power thyristor structure

Current flow through the thyristor can be induced in the first quadrant of operation by using a current supplied through the gate terminal to trigger the device into its on-state. The gate current forward biases the N^+ cathode/P-base junction (J_3) to initiate the injection of electrons. The injected electrons trigger a positive feedback mechanism produced by the two coupled bipolar transistors within the thyristor structure. The first bipolar transistor is an N–P–N transistor formed between the N^+ cathode/P-base/N-drift regions while the second bipolar is a P–N–P transistor formed between the P^+ anode/N-drift/P-base regions as illustrated in Fig. 8.5a. The equivalent circuit for the thyristor, shown in Fig. 8.5b, is based upon these coupled bipolar transistors. Once current flow is initiated through the transistors, they are able to provide the base drive current for each other by a process referred to as the *regenerative action*. In this process, the collector current of the N–P–N transistor provides the base drive current for the P–N–P transistor and the collector current of the P–N–P transistor provides the base drive current for the N–P–N transistor. The regenerative action inherent within the thyristor structure allows stable operation of the device in its on-state without any external gate drive current. This is one of its advantages when compared with the bipolar transistors.

Once the thyristor is operating in its on-state, the *i–v* characteristics can be shown to become similar to that for a P-i-N rectifier, resulting in the anode current increasing exponentially with the on-state voltage drop (or anode–cathode voltage) as discussed in Chap. 5. Consequently, thyristors can be designed with very high-voltage blocking capability with low on-state voltage drops making them excellent power devices for circuits used in power distribution systems.

The power thyristor can be switched from its on-state to the blocking state by reversing the bias applied to the anode electrode. The reverse bias applied to the anode electrode forces the thyristor to undergo a reverse recovery process similar to that observed in a P-i-N rectifier. Once the thyristor has entered the reverse-blocking mode, a positive voltage can once again be applied to the anode without turning on the device until a gate control signal is applied.

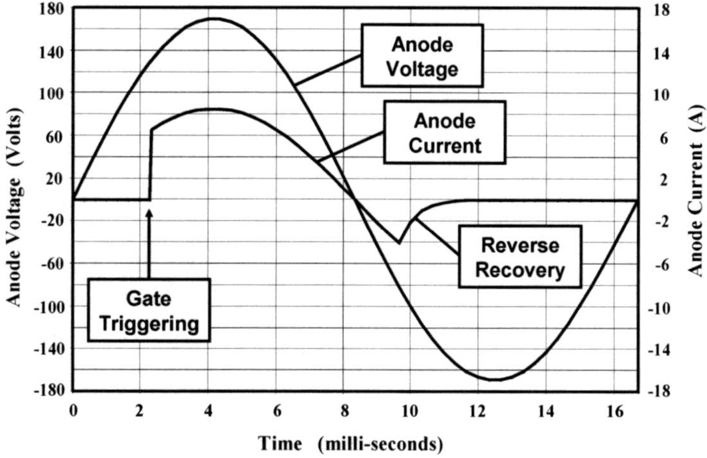

Fig. 8.6 Phase control operation of the power thyristor structure

A typical set of anode current and voltage waveforms for a thyristor operating from a standard 120-V rms AC power source are illustrated in Fig. 8.6. The device remains in its blocking state with no anode current flow until it is triggered by the gate signal at about 2.6 ms after the voltage crosses zero. The current then follows the anode voltage, for a resistive load as depicted in the figure, until the voltage crosses zero and becomes negative. The thyristor then undergoes a reverse recovery transient as shown in the figure with negative current flow for a short duration. Subsequently, it is able to support the reverse anode bias for the rest of the cycle with no current flow. In this *phase control* scheme of operation, the amount of power delivered to a load, such as heating element, can be regulated by adjusting the gate-triggering point to control the duration for anode current flow in the positive half-cycle. A device, called the triac, has also been developed based upon the thyristor structure, which enables triggering current flow in both the first and third quadrant of operation.

8.2 Blocking Characteristics

One of the important attributes of the power thyristor is the capability to support large voltages. As discussed in Sect. 8.1, the thyristor structure provides both forward-blocking capability in the first quadrant of operation and reverse-blocking capability in the third quadrant of operation. The voltage-blocking capability in both quadrants is approximately equal making the thyristor well suited for AC power circuits. However, the physics for blocking voltages in the first and third quadrants is not identical as discussed in this section.

8.2.1 Reverse-Blocking Capability

When a negative bias is applied to the anode terminal of the thyristor, the N^+ cathode/P-base junction (J_3) and the P^+ anode/N-drift junction (J_1) become reverse biased while the junction between the P-base region and the N-drift region becomes forward biased. The P-base region of the thyristor is formed by the diffusion of P-type dopants with a surface concentration ranging between 1×10^{17} and 1×10^{16} cm^{-3}, while the N^+ cathode region is very heavily doped as shown in Fig. 8.3. Due to the high doping concentration on both sides of junction J_3, it can only support relatively small voltages (<50 V). Consequently, almost all of the applied negative bias to the anode terminal is supported across the P^+ anode/N-drift junction (J_1).

The reverse-blocking voltage is supported across the P^+ anode/N-drift junction with a depletion layer extending mostly within the N-drift region as illustrated in Fig. 8.7. The maximum electric field occurs at the P^+ anode/N-drift junction (J_1). On the one hand, if the width of the N-drift region is very large when compared with the depletion layer width, the blocking voltage capability is limited by avalanche breakdown when the maximum electric field (E_m) becomes equal to

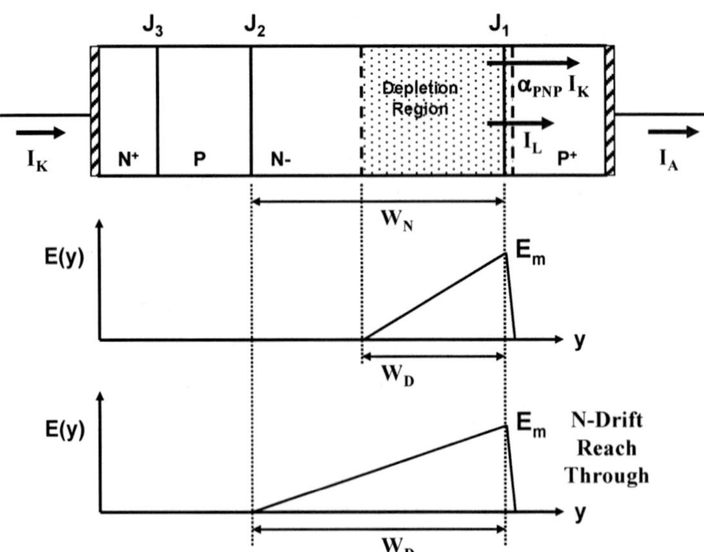

Fig. 8.7 Electric field distribution during the reverse-blocking mode in the power thyristor structure

the critical electric field. This corresponds to the multiplication coefficient (M) becoming equal to infinity. The avalanche breakdown voltage is given by

$$\mathrm{BV_{PP}} = 5.34 \times 10^{13} N_\mathrm{D}^{-3/4}. \tag{8.1}$$

However, a very large width for the N-drift region is unacceptable because it increases the on-state voltage drop of the thyristor.

On the other hand, if the width of the N-drift region is made small when compared with the depletion layer width, the depletion region can extend through the entire N-drift region as illustrated at the bottom of Fig. 8.7 while the maximum electric field at the junction is well below the critical electric field for breakdown. This phenomenon is referred to as the *reach-through* condition. When the depletion region from the reverse-biased junction J_1 reaches the forward-biased junction J_2, holes are injected from junction J_2 producing an abrupt increase in the anode current. This phenomenon limits the maximum blocking capability for the thyristor at the reach-through breakdown voltage given by

$$\mathrm{BV_{RT}} = \frac{qN_\mathrm{D}}{2\varepsilon_\mathrm{S}} W_\mathrm{N}^2, \tag{8.2}$$

where N_D is the doping concentration of the N-drift region and W_N is its width.

The actual breakdown voltage for the thyristor in the reverse-blocking mode falls between the above limits, and is governed by the open-base transistor breakdown phenomenon.[3] To analyze this phenomenon for the thyristor, consider the currents flowing at the boundary of the depletion region as illustrated in Fig. 8.7. The current consists of the leakage current due to the generation process within the depletion region and the cathode current amplified by the current gain of the P–N–P transistor. Based upon the application of Kirchhoff's current law to the thyristor structure in the absence of a gate current,

$$I_\mathrm{A} = \alpha_\mathrm{PNP} I_\mathrm{K} + I_\mathrm{L} = I_\mathrm{K} \tag{8.3}$$

leading to the relationship

$$I_\mathrm{A} = \frac{I_\mathrm{L}}{1 - \alpha_\mathrm{PNP}}. \tag{8.4}$$

From this expression, it can be concluded that the anode current will increase very rapidly when the common-base current gain of the P–N–P bipolar transistor within the thyristor structure approaches unity. As the negative anode bias is increased, the width of the undepleted portion of the N-drift region becomes smaller producing an increase in the base transport factor (α_T). Concurrently, the maximum electric field at junction J_1 becomes larger leading to an increase in the multiplication coefficient. Both phenomena produce an increase in the common-base current gain with increasing anode bias until it becomes equal to unity resulting in open-base transistor breakdown.

Based upon the above analysis, the open-base transistor breakdown condition is given by

$$\alpha_{\text{PNP}} = (\gamma_{\text{E}}\alpha_{\text{T}})_{\text{PNP}} M = 1. \tag{8.5}$$

The injection efficiency of the P-base/N-drift junction is close to unity because of the relatively high doping concentration in the P-base region and the low doping concentration of the N-drift region. The magnitude of the other two terms in the above equation is a function of the anode bias.

The base transport factor is determined by the width (l) of the undepleted portion of the N-drift region:

$$\alpha_{\text{T}} = \frac{1}{\cosh(l / L_{\text{P}})} \tag{8.6}$$

with

$$l = W_{\text{N}} - \sqrt{\frac{2\varepsilon_{\text{S}} V_{\text{A}}}{qN_{\text{D}}}}, \tag{8.7}$$

where V_A is the applied reverse bias to the anode electrode. As the reverse bias increases, the width of the undepleted portion of the N-drift region shrinks resulting in an increase in the base transport factor.

The multiplication factor is determined by the anode bias relative to the avalanche breakdown voltage of the P^+ anode/N-drift junction ($BV_{PP,A}$):

$$M = \frac{1}{1 - (V_{\text{A}} / \text{BV}_{\text{PP,A}})^n}, \tag{8.8}$$

where $n = 6$ for the case of a P^+/N diode. The multiplication coefficient also increases with increasing anode bias. The open-base transistor breakdown voltage (and the reverse-blocking capability of the thyristor) is determined by the anode voltage, at which the multiplication factor becomes equal to the reciprocal of the base transport factor.

As an example of the design procedure for optimization of the drift region doping concentration and width, consider the case of a power thyristor that must have a reverse-blocking voltage of 2,000 V. In the case of avalanche breakdown, there is a unique value of 8×10^{13} cm^{-3} for the drift region doping concentration to obtain this blocking voltage. However, in the case of open-base transistor breakdown, many combinations of the drift region doping concentration and width can be used to obtain this blocking voltage capability. This is illustrated in Fig. 8.8 where the open-base breakdown voltage is plotted as a function of the drift region width for three cases of the drift region doping concentration. A lifetime of 10 μs was used in the N-drift region for this analysis.

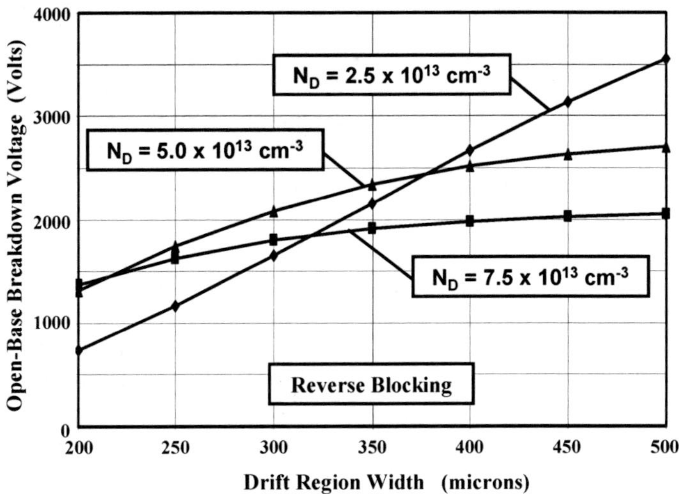

Fig. 8.8 Open-base breakdown voltage for the power thyristor structure in the reverse-blocking mode

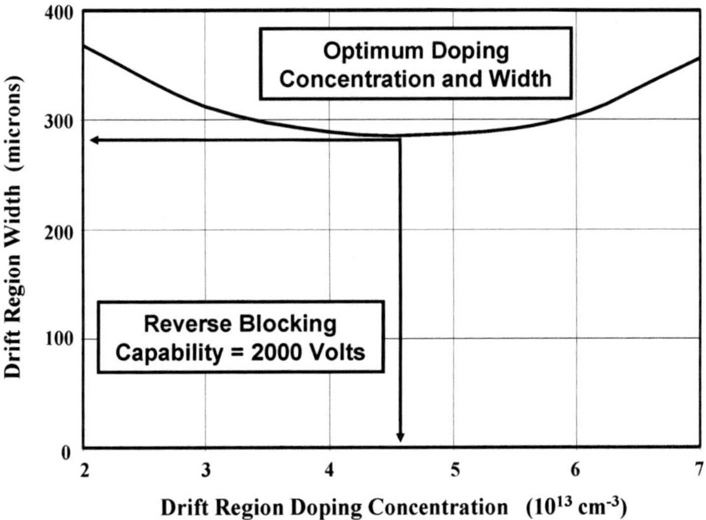

Fig. 8.9 Optimum width and doping concentration of the drift region for the power thyristor structure to obtain a reverse-blocking capability of 2,000 V

It can be observed from Fig. 8.8 that the open-base breakdown voltage becomes equal to 2,000 V at a drift region width of 334 μm for a drift region doping concentration of 2.5×10^{13} cm^{-3}. In this case, the base transport factor becomes close to unity under breakdown conditions. When the doping concentration of drift region is increased to 5×10^{13} cm^{-3}, the drift region thickness is reduced to 287 μm to achieve the same open-base breakdown voltage of 2,000 V. The drift region thickness increases to 422 μm, if the doping concentration of drift region is increased to 7.5×10^{13} cm^{-3}, to achieve the same open-base breakdown voltage of 2,000 V. In this case, the multiplication coefficient becomes large under open-base breakdown conditions. These examples demonstrate that there is an optimum drift region doping concentration to obtain a minimum drift region width to achieve an open-base breakdown voltage of 2,000 V. The location of the optimum design with a width of 287 μm and a doping concentration of 4.6×10^{13} cm^{-3} is illustrated for this case in Fig. 8.9.

The reverse-blocking voltage capability for the thyristor is also dependent upon the ambient temperature. As temperature increases, the impact ionization coefficients are reduced leading to a decrease in the multiplication coefficient. Concurrently, the base transport factor increases due to an increase in the minority carrier lifetime. In general, the reduction of the impact ionization coefficient is dominant as temperature goes up. Consequently, the reverse-blocking voltage capability for the thyristor increases with increasing temperature. This is accompanied by a significant increase in the leakage current with temperature due to the enhanced generation rate for minority carriers in the depletion and neutral regions. Note that the current produced by the space-charge generation current and the diffusion current is amplified by the gain of the open-base P–N–P transistor (see (8.4)). Consequently, if the reverse-blocking voltage is measured at a constant anode current level, the breakdown voltage for the thyristor is observed to first increase and then decrease as temperature increases.

8.2.2 Forward-Blocking Capability

When a positive bias is applied to the anode terminal of the thyristor, the N$^+$ cathode/P-base junction (J_3) and the P$^+$ anode/N-drift junction (J_1) become forward biased while the junction between the P-base region and the N-drift region becomes reverse biased. The P-base region of the thyristor is more heavily doped than the N-drift region as shown in Fig. 8.3. Consequently, most of the applied positive bias to the anode terminal is supported across the N-drift region. However, due to the graded doping profile at junction J_2, the depletion layer extends into P-base region as well. If the doping concentration of the P-base region is too low, the electric field can extend through its entire width as illustrated at the bottom of Fig. 8.10. Once the P-base region is depleted, injection of electrons occurs from the N$^+$ cathode/P-base junction (J_3) resulting in current flow, which limits the blocking voltage capability.

The P-base region reach-through limited breakdown voltage is given by

$$\mathrm{BV}_{\mathrm{RT,P}} = \frac{q}{2\varepsilon_{\mathrm{S}}} \frac{N_{\mathrm{AP}}(N_{\mathrm{AP}} + N_{\mathrm{AD}})}{N_{\mathrm{D}}} W_{\mathrm{P}}^{2}, \tag{8.9}$$

where N_{D} is the doping concentration of the N-drift region, N_{AP} is the doping concentration of the P-base region, and W_{P} is its width. The doping concentration in the P-base region of thyristors must be sufficiently larger than that in the N-drift region to suppress this breakdown mode.

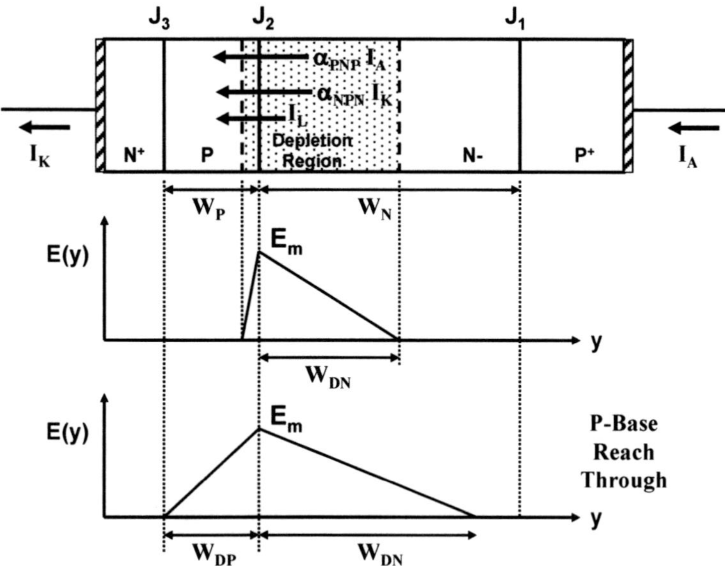

Fig. 8.10 Electric field distribution during the forward-blocking mode in the power thyristor structure

The forward-blocking voltage is then supported across the P-base/N-drift junction (J_2) with a depletion layer extending mostly within the N-drift region as illustrated in the middle of Fig. 8.10. The maximum electric field occurs at the P-base/N-drift junction (J_2). On the one hand, if the width of the N-drift region is made very large, the blocking voltage capability becomes limited by avalanche breakdown when the maximum electric field (E_m) becomes equal to the critical electric field. This corresponds to the multiplication coefficient (M) becoming equal to infinity. However, a very large width for the N-drift region is unacceptable because it increases the on-state voltage drop of the thyristor. On the other hand, if the width of the N-drift region is made small, the depletion region can extend through the entire N-drift region while the maximum electric field at the junction is well below the critical electric field for breakdown. This results in an N-drift region reach-through limited breakdown as discussed earlier for the reverse-blocking mode. The maximum blocking capability for the thyristor at the N-drift region reach-through limit is given by

$$BV_{RT,N} = \frac{qN_D}{2\varepsilon_S} W_N^2, \tag{8.10}$$

where N_D is the doping concentration of the N-drift region and W_N is its width.

The actual breakdown voltage for the thyristor in the forward-blocking mode falls between the above limits because it is governed by the open-base transistor breakdown phenomenon. To analyze this phenomenon for the thyristor, consider the currents flowing at the boundary of the depletion region as illustrated in

Fig. 8.10. The current consists of the leakage current due to the generation process within the depletion region, the anode current amplified by the current gain of the P–N–P transistor, and the cathode current amplified by the current gain of the N–P–N transistor. Based upon the application of Kirchhoff's current law to the thyristor structure with no gate current,

$$I_A = \alpha_{NPN} I_K + \alpha_{PNP} I_A + I_L = I_K \tag{8.11}$$

leading to the relationship

$$I_A = \frac{I_L}{1 - \alpha_{NPN} - \alpha_{PNP}}. \tag{8.12}$$

From this expression, it can be concluded that the anode current will increase very rapidly when the sum of the common-base current gains of the P–N–P and N–P–N bipolar transistors within the thyristor structure approaches unity. As the positive anode bias is increased, the width of the undepleted portion of the N-drift region becomes smaller producing an increase in the base transport factor ($\alpha_{T,PNP}$). Concurrently, the maximum electric field at junction J_2 becomes larger leading to an increase in the multiplication coefficient. Both phenomena produce an increase in the common-base current gain of the P–N–P transistor (α_{PNP}) with increasing anode bias until it becomes equal to ($1 - \alpha_{NPN}$) resulting in open-base transistor breakdown. The common-base current gain of the N–P–N transistor (α_{NPN}) is close to unity in the thyristor structure due to the high injection efficiency and base transport factor. Consequently, a smaller increase in the common-base current gain of the P–N–P transistor (α_{PNP}) is sufficient to produce open-base transistor breakdown in the forward-blocking mode when compared with the reverse-blocking mode.

Based upon the above analysis, the open-base transistor breakdown condition in the forward-blocking mode is given by

$$\alpha_{NPN} + \alpha_{PNP} = [(\gamma_E \alpha_T)_{NPN} + (\gamma_E \alpha_T)_{PNP}]M = 1. \tag{8.13}$$

Note that, although the injection efficiency and base transport factor for the two transistors are different, they share the same multiplication factor due to the common reverse-biased collector junction for both transistors.

The multiplication factor is determined by the anode bias relative to the avalanche breakdown voltage of the P-base/N-drift junction ($BV_{PP,B}$):

$$M = \frac{1}{1 - (V_A / BV_{PP,B})^n}, \tag{8.14}$$

where $n = 6$ for the case of a P^+/N diode. The open-base transistor breakdown condition (and the forward-blocking capability of the thyristor) is then given by

$$M = \frac{1}{(\gamma_E \alpha_T)_{NPN} + (\gamma_E \alpha_T)_{PNP}}. \tag{8.15}$$

In comparison, the open-base transistor breakdown condition (and the reverse-blocking capability of the thyristor) is given by

$$M = \frac{1}{(\gamma_E \alpha_T)_{PNP}}. \tag{8.16}$$

Based upon these expressions, it can be concluded that the forward-blocking voltage will always be lower than the reverse-blocking capability for a thyristor.

In the forward-blocking mode, the injection efficiency and base transport factor for the N–P–N transistor remain approximately independent of the anode bias. The injection efficiency of the P^+ anode/N-drift junction is close to unity because of the high doping concentration in the anode region and the low doping concentration of the drift region. However, the base transport factor for the P–N–P transistor is a strong function of the anode bias. The base transport factor is determined by the width of the undepleted portion of the N-drift region:

$$\alpha_T = \frac{1}{\cosh(l / L_P)}, \tag{8.17}$$

where l is the width of the undepleted portion of the N-drift region given by

$$l = W_N - \sqrt{\frac{2\varepsilon_S V_A}{qN_D}}, \tag{8.18}$$

where V_A is the applied forward bias to the anode electrode. As the forward bias increases, the width of the undepleted portion of the N-drift region shrinks resulting in an increase in the base transport factor.

As an example of the design procedure for optimization of the drift region doping concentration and width, consider the case of a power thyristor that must have a forward-blocking voltage of 2,000 V. In the case of avalanche breakdown, there is a unique value of 8×10^{13} cm^{-3} for the drift region doping concentration to obtain this blocking voltage. However, in the case of open-base transistor breakdown, many combinations of the doping concentration and width can be used to obtain this blocking voltage capability. This is illustrated in Fig. 8.11 where the open-base breakdown voltage is plotted as a function of the drift region width for three cases of the drift region doping concentration. A lifetime of 10 μs was used in the N-drift region for this analysis with a common-base current gain of 0.65 for the N–P–N transistor based upon a diffusion length for electrons equal to the width of the P-base region. It can be observed from the figure that the open-base breakdown voltage becomes equal to 2,000 V at a drift region width of 515 μm for a drift region doping concentration of 2.5×10^{13} cm^{-3}. When the doping concentration of drift region is increased to 5×10^{13} cm^{-3}, the drift region thickness is reduced to 436 μm to achieve the same open-base breakdown voltage of 2,000 V. If the doping concentration of drift region is increased to 7.5×10^{13} cm^{-3}, the drift region thickness increases to 542 μm to achieve the same open-base breakdown voltage of 2,000 V.

This demonstrates that there is an optimum drift region doping concentration to obtain a minimum drift region width to achieve the open-base breakdown voltage of 2,000 V.

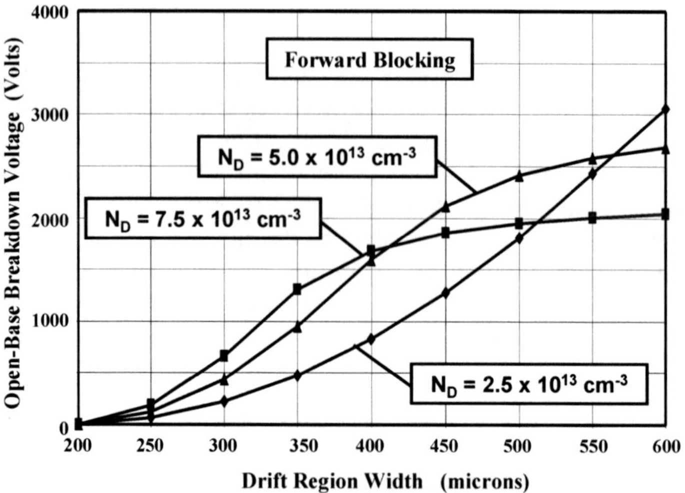

Fig. 8.11 Open-base breakdown voltage for the power thyristor structure in the forward-blocking mode

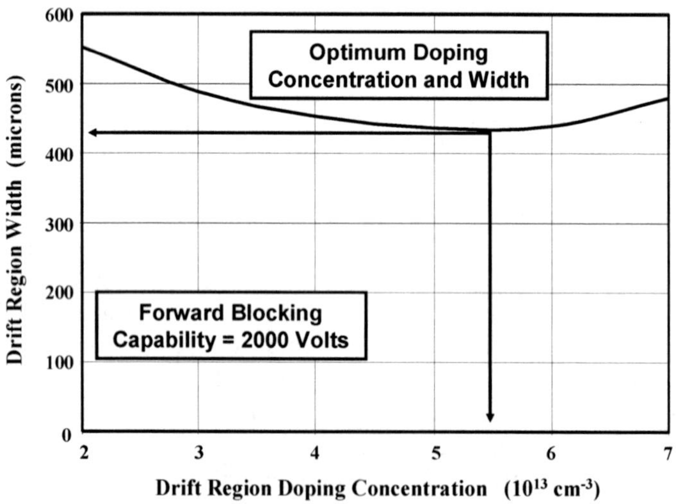

Fig. 8.12 Optimum width and doping concentration of the drift region for the power thyristor structure in the forward-blocking mode

The location of the optimum design is provided for the case of a forward-blocking voltage of 2,000 V in Fig. 8.12. In comparison with the reverse-blocking case, the optimum drift region doping concentration is slightly larger. More importantly, the width of drift region required to achieve the forward-blocking voltage has increased from 287 μm for the reverse-blocking case to 435 μm. This has an adverse impact on the on-state voltage drop.

The forward-blocking voltage capability for the thyristor is also dependent upon the ambient temperature. As temperature increases, the impact ionization coefficients are reduced leading to a decrease in the multiplication coefficient. Concurrently, the base transport factor increases due to an increase in the minority carrier lifetime. In general, the reduction of the impact ionization is dominant as temperature goes up. Consequently, the forward-blocking voltage capability for the thyristor increases with increasing temperature until the leakage current becomes sufficient to turn on the structure due to the regenerative action. The forward-blocking capability then degrades rapidly with further increase in the temperature.

8.2.3 Cathode Shorting

A substantial improvement in the forward-blocking capability, especially at elevated temperatures, can be obtained by suppressing the current gain of the N–P–N transistor within the thyristor structure. A high-current gain for the N–P–N transistor is necessary at high-current levels to create the regenerative action that is required to sustain the thyristor in its on-state mode with a low on-state voltage drop. It is therefore desirable to suppress the current gain of the N–P–N transistor only at low (leakage) current levels. This is possible by utilizing *emitter or cathode shorts* within the thyristor structure as illustrated in Fig. 8.13. The cathode shorts are produced by interrupting the N^+ cathode region, so that the cathode electrode overlaps the N^+ cathode/P-base junction at periodic intervals within the thyristor. In the presence of the cathode shorts, the leakage current collected by the P-base region from the reverse-biased junction J_2 is diverted to the cathode shorts as

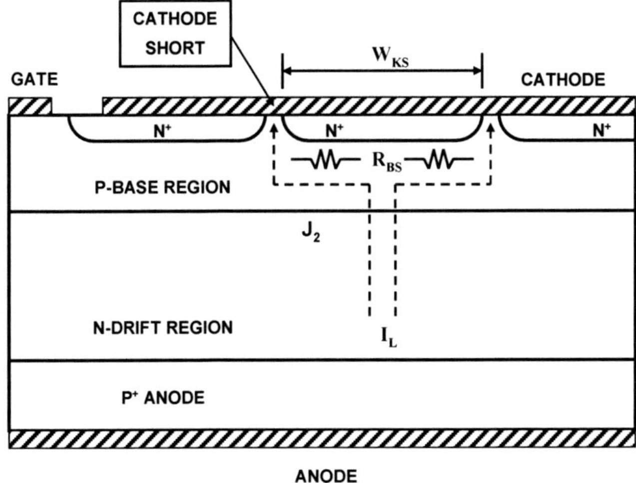

Fig. 8.13 Cathode-shorted power thyristor structure

shown in the figure without traversing the N^+ cathode/P-base junction. This prevents the injection of electrons from the N^+ cathode in response to the holes being collected at junction J_2, resulting in suppressing the current gain of the N–P–N transistor.

The current gain for the N–P–N bipolar transistor can be analyzed in the presence of the emitter short by using the current flow shown in Fig. 8.14. The upper part of the figure illustrates the current flow to the cathode short in the thyristor structure while the lower part of the figure shows the current flow in a lumped equivalent circuit. In practice, the leakage current is collected across the entire junction J_2 resulting in a larger current level near the emitter–base short. The current (I_S) flowing into the emitter–base short produces a voltage drop across the resistance (R_{BS}) within the P-base region. This voltage drop forward biases the emitter–base junction at the center of the N^+ emitter at location A in the figure.

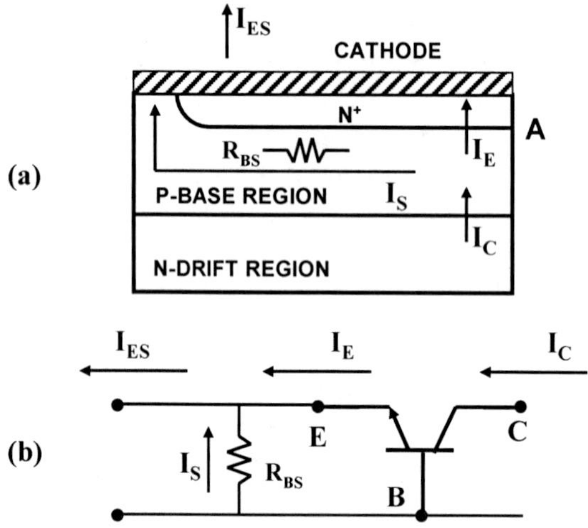

Fig. 8.14 Analysis of current gain for an N–P–N bipolar transistor with emitter shorting

The emitter current (I_E) of the N–P–N bipolar transistor depends upon the forward bias across the emitter–base junction:

$$I_E = I_O(e^{qV_{BE}/kT} - 1), \tag{8.19}$$

where V_{BE} is the forward bias across the emitter–base junction at location A. The current I_O is the saturation current for the emitter–base junction:

$$I_O = \frac{qD_n n_{P0}}{L_n} A, \tag{8.20}$$

where A is the junction area. The total current flowing at the emitter terminal of the shorted N–P–N bipolar transistor consists of its emitter current plus the current flowing via the shunting resistance (R_{BS}):

$$I_{\mathrm{ES}} = I_{\mathrm{E}} + I_{\mathrm{S}} = I_{\mathrm{O}}(\mathrm{e}^{qV_{\mathrm{BE}}/kT} - 1) + \frac{V_{\mathrm{BE}}}{R_{\mathrm{BS}}}. \tag{8.21}$$

The common-base current gain in the presence of the emitter short is given by

$$\alpha_{\mathrm{NPN,S}} = \frac{I_{\mathrm{C}}}{I_{\mathrm{ES}}} = \alpha_{\mathrm{NPN}}\left[\frac{\mathrm{e}^{qV_{\mathrm{BE}}kT} - 1}{(\mathrm{e}^{qV_{\mathrm{BE}}/kT} - 1) + (V_{\mathrm{BE}}/I_{\mathrm{O}}R_{\mathrm{BS}})}\right], \tag{8.22}$$

where

$$\alpha_{\mathrm{NPN}} = \frac{I_{\mathrm{C}}}{I_{\mathrm{E}}} \tag{8.23}$$

is the common-base current gain of the N–P–N transistor without the emitter short.

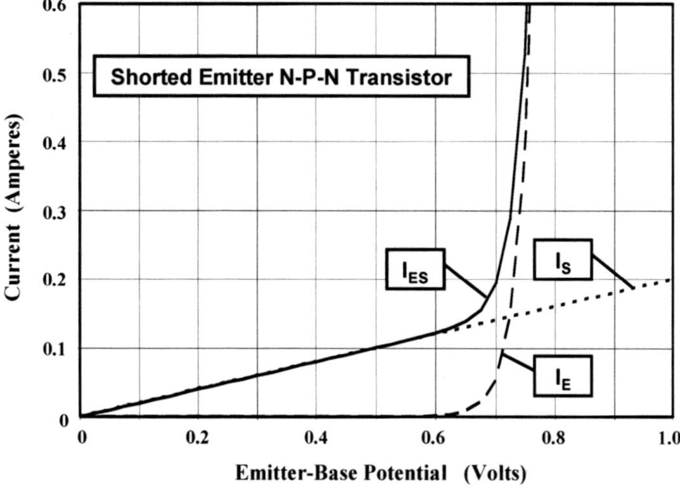

Fig. 8.15 Current distribution within an N–P–N bipolar transistor with emitter shorting

The relative magnitude of the current flowing through the emitter–base junction and the shunting resistance is illustrated in Fig. 8.15 with increasing forward bias voltage across the emitter–base junction. The emitter–base junction does not begin to inject a substantial amount of electrons until the voltage drop across the junction exceeds 0.7 V at room temperature. Consequently, all the collector current flows through the shunting resistance when the emitter–base voltage drop is less than 0.6 V. However, when the voltage drop across the emitter–base junction exceeds 0.7 V, the emitter current increases very rapidly and becomes dominant at larger voltages across the junction.

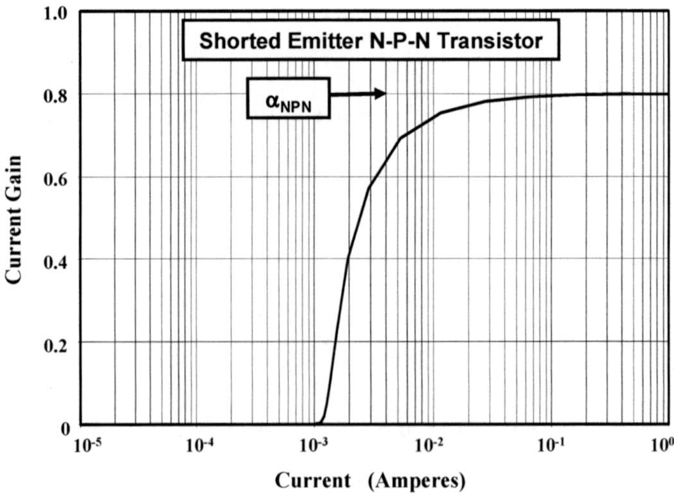

Fig. 8.16 Current gain for an N–P–N bipolar transistor with emitter shorting

The change in the common-base current gain for the N–P–N bipolar transistor with emitter shorting is illustrated in Fig. 8.16. It can be observed from this figure that the current gain is very low at small current levels and becomes equal to the current gain without the emitter short at high-current levels. This is the desired behavior to improve the forward-blocking capability while retaining strong regenerative action in the on-state. With the inclusion of the emitter shorts in a thyristor structure, the forward-blocking capability can be made very close to the reverse-blocking capability.

8.2.4 Cathode Shorting Geometry

In Sect. 8.2.3, it was demonstrated that the current gain begins to increase when the current increases beyond a certain level. As the current gain of the N–P–N transistor increases, the thyristor can be switched from the blocking state to the on-state. An increase in the current occurs in the thyristor structure with increasing temperature, while it is operating in the forward-blocking mode, because of the strong dependence of the space-charge generation and diffusion components of the leakage current upon the temperature. Consequently, as the temperature increases, the forward-blocking capability of the thyristor begins to degrade.

The leakage current at which the thyristor will switch from the forward-blocking mode to the on-state can be determined for a linear cathode short geometry by using two-dimensional analysis. A cross section of thyristor structure is shown in Fig. 8.17 for a region from the center of the cathode to the cathode short. When the thyristor is operating in the forward-blocking mode, a uniform leakage current density (J_L) can be assumed to be arriving at the reverse-biased junction J_2. This leakage current flows through the P-base region to the cathode electrode at the cathode short (at location B in the figure).

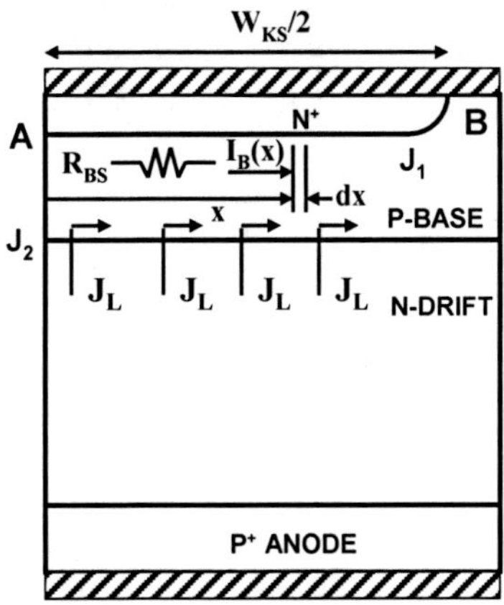

Fig. 8.17 Linear cathode shorting geometry for the thyristor

The current collected within a small segment (dx) of the P-base region along the cathode is given by

$$\mathrm{d}I_{\mathrm{B}} = J_{\mathrm{L}} Z\,\mathrm{d}x, \tag{8.24}$$

where Z is the width of the cathode orthogonal to the cross section shown in the figure. The current flowing in the P-base region at a distance x from the center of the cathode finger (at location A) is then given by

$$I_{\mathrm{B}}(x) = \int_0^x J_{\mathrm{L}} Z\,\mathrm{d}x = J_{\mathrm{L}} Zx. \tag{8.25}$$

The resistance of the P-base region of thickness dx is given by

$$\mathrm{d}R_{\mathrm{BS}} = \frac{\rho_{\mathrm{SB}}}{Z}\mathrm{d}x, \tag{8.26}$$

where ρ_{SB} is the sheet resistance of the P-base region under the N$^+$ cathode. Combining the above expressions, the voltage drop produced across the segment due to the current flow in the P-base region is

$$\mathrm{d}V_{\mathrm{B}}(x) = I_{\mathrm{B}}(x)\mathrm{d}R_{\mathrm{BS}} = J_{\mathrm{L}}\rho_{\mathrm{SB}} x\,\mathrm{d}x. \tag{8.27}$$

Consequently, the total voltage drop produced in the P-base region between points A and B is

$$V_{\mathrm{B}}(A) = \int_0^{W_{\mathrm{KS}}/2} \mathrm{d}V_{\mathrm{B}}\mathrm{d}x = J_{\mathrm{L}}\rho_{\mathrm{SB}}\frac{W_{\mathrm{KS}}^2}{8}. \tag{8.28}$$

The center of the cathode finger is forward biased by this voltage due to the leakage current flowing through the P-base region. The maximum allowable leakage current can be determined by assuming that the thyristor will switch to the on-state when the forward bias for the N^+ cathode/P-base junction becomes equal to V_{bi} (~0.7 V) at the center of the cathode finger. Using this criterion in (8.28),

$$J_{L,max} = \frac{8V_{bi}}{\rho_{SB} W_{KS}^2}, \tag{8.29}$$

with a corresponding leakage current given by

$$I_{L,max} = J_{L,max} Z \frac{W_{KS}}{2} = \frac{4V_{bi} Z}{\rho_{SB} W_{KS}}. \tag{8.30}$$

Based upon this equation, it can be concluded that the forward-blocking capability at elevated temperatures can be enhanced for a thyristor by reducing the sheet resistance of the P-base region and the width of the cathode finger.

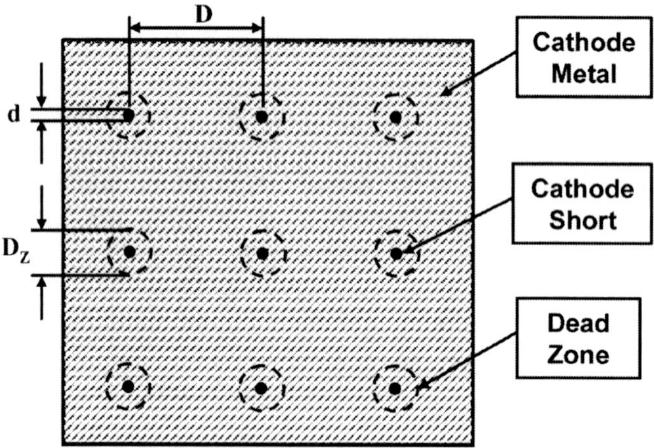

Fig. 8.18 Surface topology of the cathode shorting geometry for the thyristor

In power thyristors with larger active areas, typically fabricated using entire silicon wafers with diameters ranging from 100 to 150 mm, the shorting of the cathode is accomplished by using a two-dimensional array of shorts that are uniformly distributed on the top surface of the device. This can be achieved by appropriate masking of the N^+ cathode diffusion as illustrated in Fig. 8.18 for a square array of cathode shorts of diameter d spaced apart at a distance D. For this geometry, the largest forward bias produced in the P-base region by the leakage current is given by

$$V_{B,max} = J_L \rho_{SB} A_S, \tag{8.31}$$

where the shorting geometry area factor is given by[4]

$$A_S = \frac{1}{16}\left\{d^2 + D^2\left[2\ln\left(\frac{D}{d}\right) - 1\right]\right\}. \tag{8.32}$$

Other geometries for the cathode shorts, such as triangular and hexagonal arrays, have also been proposed and implemented in practical devices.[5]

The presence of the cathode shorts creates dead zones within the thyristor where no regenerative action can occur. The size of the dead zone is larger than the actual diameter of the short as illustrated in the figure by the circles with dashed lines. The fractional area lost due to the presence of the cathode short is given by

$$F_S = \frac{\pi}{4}\left(\frac{D_Z}{D}\right)^2, \tag{8.33}$$

where D_Z is the diameter of the dead zone. Placing the cathode shorts at close proximity can degrade the available active area for the conduction of current producing an increase in the on-state voltage drop of the thyristor.

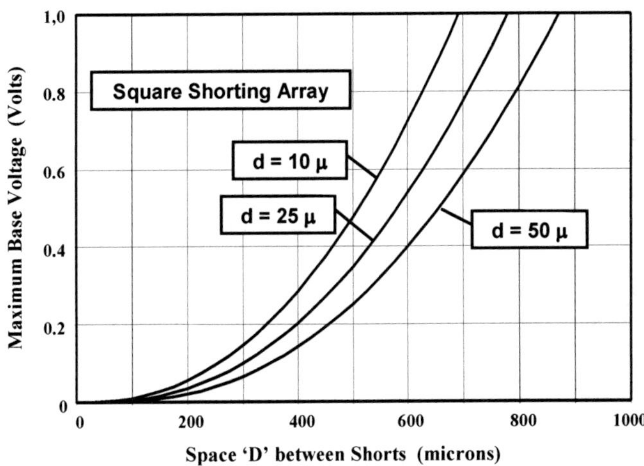

Fig. 8.19 Maximum voltage developed in the P-base region of a thyristor with emitter shorts in a square array

As an example, the maximum voltage developed within the P-base region calculated by using the area factor given by (8.32) is shown in Fig. 8.19 for the square array of cathode shorts. A leakage current density of 1 A cm^{-2} was used for these calculations with a P-base sheet resistance of 450 Ω sq^{-1}. Due to the large active area for typical thyristors designed with high-current handling capability, it is customary to utilize large geometries for the lithography during device fabrication. For the case of a short diameter (*d*) of 1 mil (25 μm), the spacing between the shorts must be made less than 20 mil (500 μm) to keep the maximum voltage generated in the P-base region to less than 0.5 V. Even for this case, the loss of current conduction area is only 3% if a dead zone diameter of 100 μm is assumed.

Simulation Example

To gain further insight into the physics of operation for the power thyristor under voltage-blocking conditions, the results of two-dimensional numerical simulations for a typical structure are described here. The total width of the structure, as shown by the cross section in Fig. 8.17, was 1,000 μm (area = 1×10^{-5} cm^{-2}) with a cathode finger width of 980 μm. The structure had an N-drift region doping concentration of 5×10^{13} cm^{-3} and width of 360 μm. The P-base region had a Gaussian doping profile with a surface concentration of 5×10^{17} cm^{-3} and a depth of 25 μm. The N^+ cathode region had a Gaussian doping profile with a surface concentration of 1×10^{20} cm^{-3} and a depth of 10 μm. The P^+ anode region had a Gaussian doping profile with a surface concentration of 1×10^{19} cm^{-3} and a depth of 10 μm. The resulting doping profile is shown in Fig. 8.20. A lifetime (τ_{p0}, τ_{n0}) of 10 μs was used for the N-drift region.

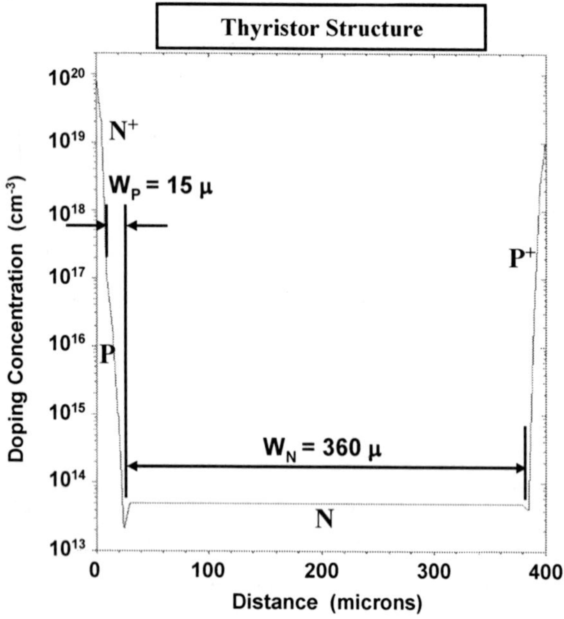

Fig. 8.20 Doping profile for the simulated N–P–N–P power thyristor structure

The reverse-blocking characteristics for the thyristor structure are shown in Fig. 8.21 at various ambient temperatures ranging from 300 to 500 K. The breakdown voltage is indicated by an abrupt increase in the anode current. At room temperature (300 K), the reverse-blocking voltage obtained with the simulations is 2,350 V. The value predicted by the analytical model (see plot with doping concentration of 5×10^{13} cm^{-3} in Fig. 8.8 at an N-drift region width of 360 μm) is in excellent agreement with this value providing validity for the model. The reverse-blocking capability increases with increasing temperature because of a reduction in the impact ionization coefficients.

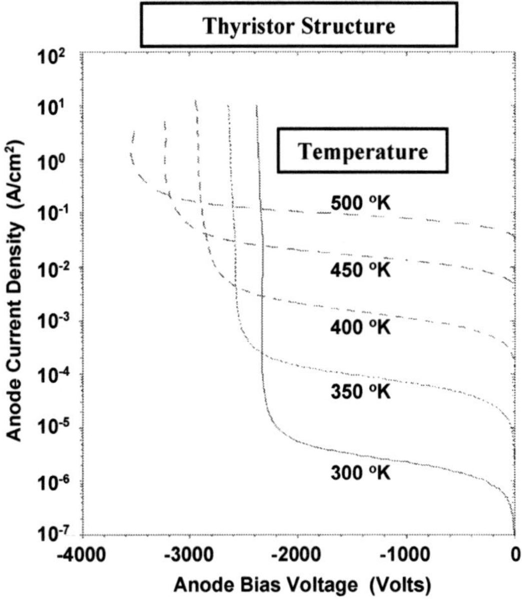

Fig. 8.21 Typical reverse-blocking characteristics for the N–P–N–P power thyristor structure

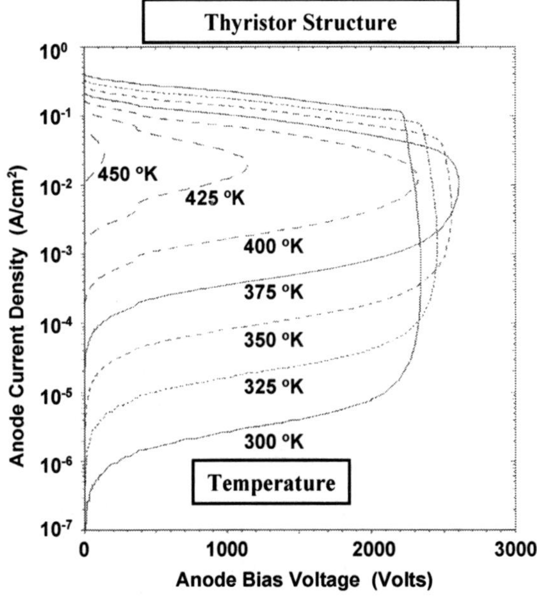

Fig. 8.22 Typical forward-blocking characteristics for the N–P–N–P power thyristor structure

The forward-blocking characteristics for the thyristor structure are shown in Fig. 8.22 at various ambient temperatures ranging from 300 to 450 K. The forward-blocking capability is limited by the snapback of the characteristics when the anode current exceeds a critical value that is dependent upon the temperature. The thyristor is being triggered from the blocking state to the on-state when the leakage current (amplified by the gain of the internal transistors) exceeds this critical value. At room temperature (300 K), the forward-blocking voltage obtained with the simulations is 2,350 V – the same as that in the reverse-blocking mode due to the utilization of the cathode short. Consequently, the analytical model allows an accurate prediction of the forward-blocking capability.

Initially, the forward-blocking capability increases with increasing temperature due to a reduction in the impact ionization coefficients. However, beyond 375 K, there is a rapid degradation of the blocking voltage capability in spite of the cathode shorts. This is due to the growth in the leakage current density which is sufficient to produce an increase in the current gain of the N–P–N transistor as shown in Fig. 8.16. The leakage current density at which the thyristor switches from the blocking state to the on-state at room temperature is 0.11 A cm^{-2}. This value is also predicted by the analytical model using (8.29) if an effective doping concentration of 6×10^{15} cm^{-3} is assumed for the P-base region, which is consistent with the doping profile shown in Fig. 8.20.

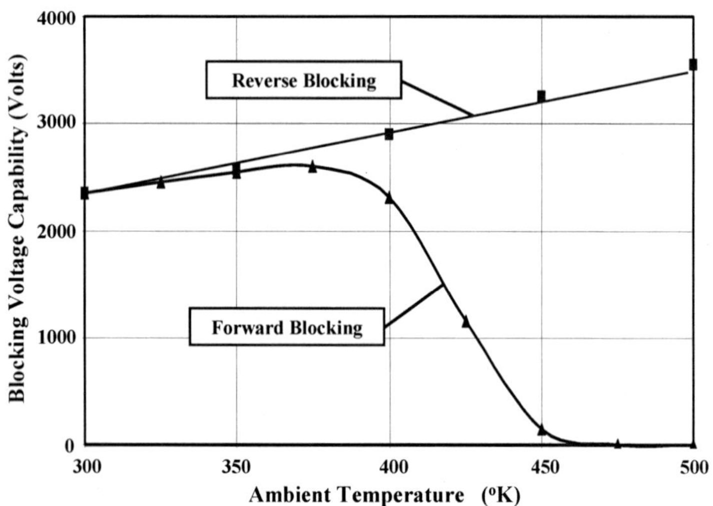

Fig. 8.23 Forward- and reverse-blocking voltage for the N–P–N–P power thyristor structure

The forward- and reverse-blocking capability obtained from the simulations are plotted in Fig. 8.23 as a function of the ambient temperature. It can be observed that the forward-blocking capability is the same as the reverse-blocking capability at room temperature (300 K). As the temperature increases, the reverse-blocking capability increases monotonically while the forward-blocking capability reaches a maximum value at 375 K and then degrades with increasing temperature. This thyristor design is capable of supporting 2,000 V in both the first and third quadrants of operation up to a temperature of 400 K (125°C), which is typical for power thyristors.

The voltage is primarily supported in the thyristor within the N-drift region. This is illustrated in Fig. 8.24 where the electric field profiles are shown during operation in the reverse- and forward-blocking modes at two voltages. It can be observed that the P^+ anode/N-drift junction (J_1) becomes reverse biased during the reverse-blocking mode with the depletion region extending toward the left-hand side with increasing (negative) anode bias. In contrast, the P-base/N-drift junction (J_2) becomes reverse biased during the forward-blocking mode with the depletion region extending toward the right-hand side with increasing (positive) anode bias. The electric field profiles used in Figs. 8.7 and 8.10 to develop the analytical models for the blocking voltage are consistent with those observed with the simulations.

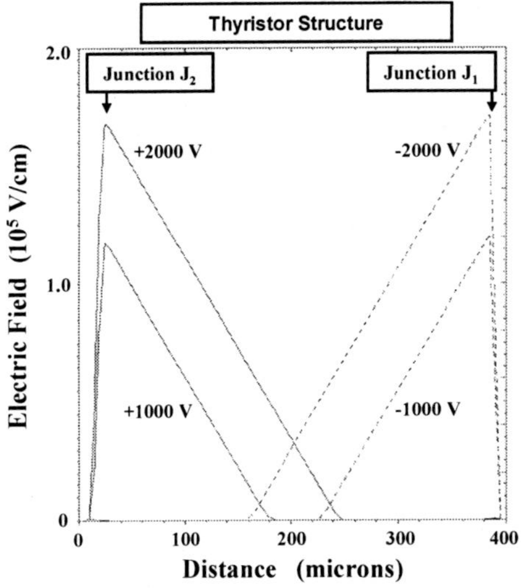

Fig. 8.24 Electric field profiles in the forward- and reverse-blocking mode for the N–P–N–P power thyristor structure

8.3 On-State Characteristics

One of the attributes of the thyristor structure is its excellent forward conduction characteristic even when designed to support large voltage levels. The thyristor structure can be triggered from the forward-blocking mode at the anode supply voltage (V_{AS}) into the forward conduction mode by the application of a small gate current to initiate the turn-on process as illustrated in Fig. 8.25 by the dashed line. This gate current serves to increase the current gain of the N–P–N transistor until the combined gain of the integral N–P–N and P–N–P transistors within the thyristor structure is sufficient to sustain its regenerative action. The thyristor then operates in its on-state with a forward conduction characteristic similar to that observed for a P-i-N rectifier. In the on-state, strong conductivity modulation of the N-drift region occurs due to high-level injection of holes, allowing the thyristor to carry high-current levels with a low on-state voltage drop. The range of anode currents, over which the thyristor can maintain on-state operation, is limited at the lower end by its holding current (I_H) as shown in the figure. When the anode current is reduced below the holding current level, the regenerative action can no longer be sustained and the thyristor reverts to the forward-blocking mode.

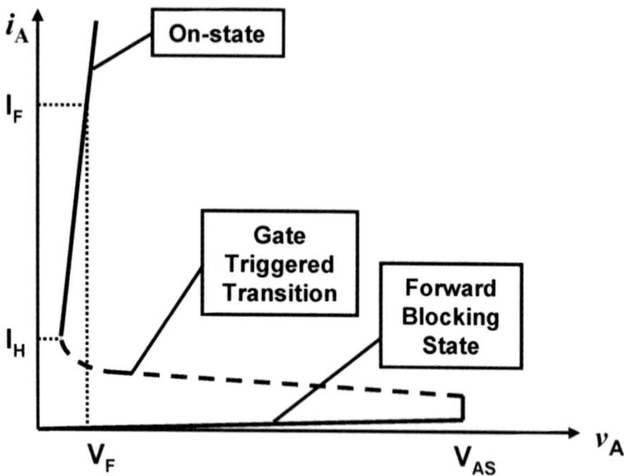

Fig. 8.25 Forward conduction characteristics for the power thyristor structure

8.3.1 On-State Operation

When a positive bias is applied to the anode terminal of the thyristor, the N^+ cathode/P-base junction (J_3) and the P^+ anode/N-base junction (J_1) become forward biased while the junction (J_2) between the P-base region and the N-base region becomes reverse biased during the forward-blocking mode of operation. However, once the thyristor has been triggered into its on-state mode of operation, injection of minority carriers occurs across the forward-biased junctions J_1 and J_3. The holes injected into the N-base region from the P^+ anode region diffuse through the N-base region and are collected by junction J_2. When these holes enter the P-base region, they create a base drive current for the N–P–N transistor because of the addition of majority carriers to the P-base region. This base drive current promotes the injection of electrons from the N^+ cathode region into the P-base region. These electrons diffuse through the P-base region and are collected by junction J_2. When these electrons enter the N-base region, they create a base drive current for the P–N–P transistor because of the addition of majority carriers to the N-base region. This base drive current promotes the injection of more holes from the P^+ anode region into the N-base region. This creates a positive internal feedback mechanism within the thyristor structure which is referred to as the *regenerative action*. The regenerative action within the thyristor structure can sustain current flow without the need for an external gate drive current once the device has been triggered into its on-state.

In the on-state mode, the base current due to electron flow in the N-base region can be related to the anode current (I_A) injected from the P^+ emitter of the transistor by $[(1 - \alpha_{PNP})I_A)]$ if the P–N–P transistor is operating in its active region. However, the electron current collected by junction J_2 due to injection from the N^+ cathode region is related to the cathode current by ($\alpha_{NPN}I_K$) if the N–P–N transistor is operating in its active region. Since the anode and cathode currents are equal during on-state operation and the current gains are close to unity, this implies that a much larger concentration of electrons is supplied to the base region of the P–N–P transistor than required for operation in the active region. The transistor is therefore forced into its saturation region of operation with junction J_2 forward biased. Consequently, all three junctions within the thyristor structure operate in forward bias during on-state operation. The net voltage drop across the thyristor becomes the voltage drop across junction J_1 plus junction J_3 minus the voltage drop across junction J_2. This results in the on-state voltage drop for a thyristor becoming equal to that across a P-i-N rectifier in the on-state.

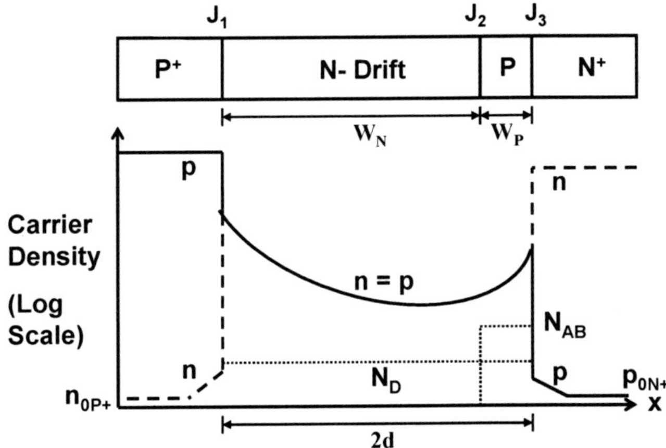

Fig. 8.26 Carrier distribution within the power thyristor structure in the on-state

The power thyristor can be treated as a P-i-N rectifier for analysis of its forward conduction characteristics. In this case, it is assumed that the junction J_2 is strongly forward biased resulting in high-level injection in not only the N-base region but also the P-base region as illustrated in Fig. 8.26. The thyristor can then be regarded as a P-i-N rectifier between the P^+ anode and N^+ cathode regions. In this context, the N^+ cathode region is referred to as a *remote emitter* because it provides electrons to the N-base region through the intervening P-base region. The electron and hole concentrations within the N-base and P-base regions take a catenary distribution in accordance with the analysis for the P-i-N rectifier in Sect. 5.1.3:

$$n(x) = p(x) = \frac{\tau_{\mathrm{HL}} J_{\mathrm{A}}}{2qL_{\mathrm{a}}} \left[\frac{\cosh(x/L_{\mathrm{a}})}{\sinh(d/L_{\mathrm{a}})} - \frac{\sinh(x/L_{\mathrm{a}})}{2\cosh(d/L_{\mathrm{a}})} \right]. \tag{8.34}$$

The distance d for the thyristor structure is given by

$$d = \frac{W_{\mathrm{N}} + W_{\mathrm{P}}}{2} \tag{8.35}$$

as indicated in the figure. A minimum on-state voltage drop occurs for the thyristor structure when the ambipolar diffusion length (L_a) is equal to the distance d (see analysis for the P-i-N rectifier in Chap. 5). The typical on-state voltage drop for the thyristor calculated by using the P-i-N rectifier model under high-level injection conditions is approximately 1 V at an operating anode current density of 100 A cm^{-2}.

Due to the very high-voltage ratings (>3 kV) for typical thyristors used in HVDC systems, they must be constructed using very high resistivity drift regions with large thickness. The minority carrier lifetime in the drift region must be made as large as possible to obtain a sufficiently large ambipolar diffusion length. The aluminum and gallium diffusions used to obtain the relatively deep diffusions also provide a gettering mechanism that reduces the concentration of recombination centers within the N-drift region, resulting in a high lifetime and diffusion length. With these processes, it is possible to achieve an on-state voltage drop ranging from 1 to 2 V.

8.3.2 Gate-Triggering Current

The thyristor structure can be triggered from the forward-blocking state to the on-state by the application of a gate drive current. The current flow within the thyristor structure during the forward-blocking mode in the presence of the gate current is illustrated in Fig. 8.27. The currents flowing at the junction J_2 due to the anode and cathode currents are derived from the common-base current gains of the two internal transistors. Together with the leakage current (I_{L}), these currents constitute the anode current:

$$I_{\mathrm{A}} = \alpha_{\mathrm{PNP}} I_{\mathrm{A}} + \alpha_{\mathrm{NPN}} I_{\mathrm{K}} + I_{\mathrm{L}}. \tag{8.36}$$

Applying Kirchhoff's current law to the thyristor structure,

$$I_{\mathrm{K}} = I_{\mathrm{A}} + I_{\mathrm{G}}. \tag{8.37}$$

Combining these equations yields

$$I_{\mathrm{A}} = \frac{\alpha_{\mathrm{NPN}} I_{\mathrm{G}} + I_{\mathrm{L}}}{1 - \alpha_{\mathrm{PNP}} - \alpha_{\mathrm{NPN}}}. \tag{8.38}$$

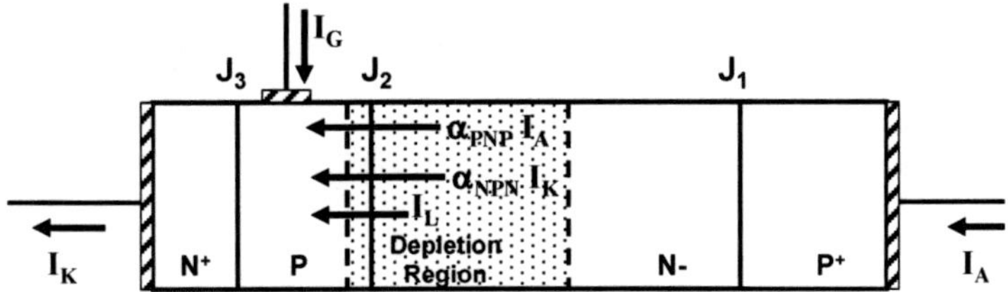

Fig. 8.27 Current flow in the power thyristor structure after application of a gate drive current

For the thyristor structure with cathode shorts, the gate drive current serves to raise the gain of the N–P–N transistor until the denominator of (8.38) becomes equal to zero. This implies a very large anode current flow corresponding to operation in the on-state. A simple analysis for the gate current required to trigger the thyristor into the on-state can be performed by assuming that the triggering occurs when the forward bias across the N^+ cathode/P-base junction becomes equal to the built-in potential of a P–N junction. Consider the cross section of the thyristor shown in Fig. 8.28. When the gate current is applied via the electrode, it flows within the P-base region to the cathode short located closest to the gate terminal as indicated by the dashed line. The gate current produces a voltage drop across the resistance of the P-base region (R_{BG}). This voltage forward biases the N^+ cathode/P-base junction with a maximum value at the edge of the cathode located closest to the gate electrode, producing electron injection as indicated by the arrow in the figure.

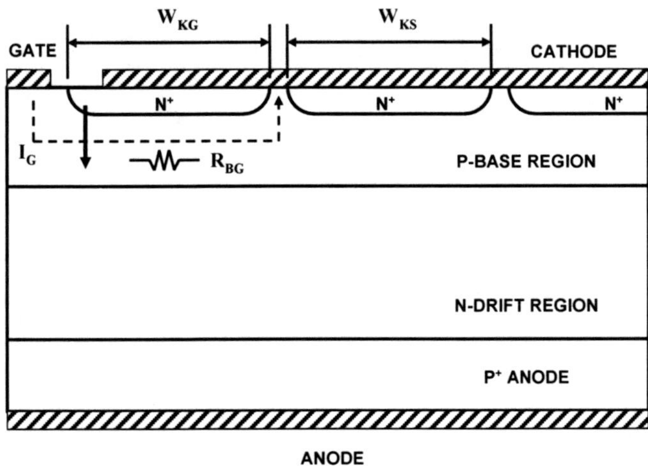

Fig. 8.28 Gate current flow path in the power thyristor structure

For the linear thyristor geometry with a length Z orthogonal to the cross section shown in Fig. 8.28, the voltage drop produced by the gate current flow prior to the turn-on of the thyristor is given by

$$I_G R_{BG} = I_G \rho_{SB} \frac{W_{KG}}{Z}, \tag{8.39}$$

where ρ_{SB} is the sheet resistance of the P-base region and W_{KG} is the total width of the cathode region between the gate electrode and the first cathode short. The gate-triggering current is obtained by equating this voltage drop to the built-in potential for a P–N junction:

$$I_{GT} = \frac{V_{bi} Z}{\rho_{SB} W_{KG}}. \tag{8.40}$$

According to this expression, the gate-triggering current can be reduced by increasing the width of the cathode between the gate electrode and the first cathode short. However, care should be taken to make sure that this does not compromise the forward-blocking capability at elevated temperatures.

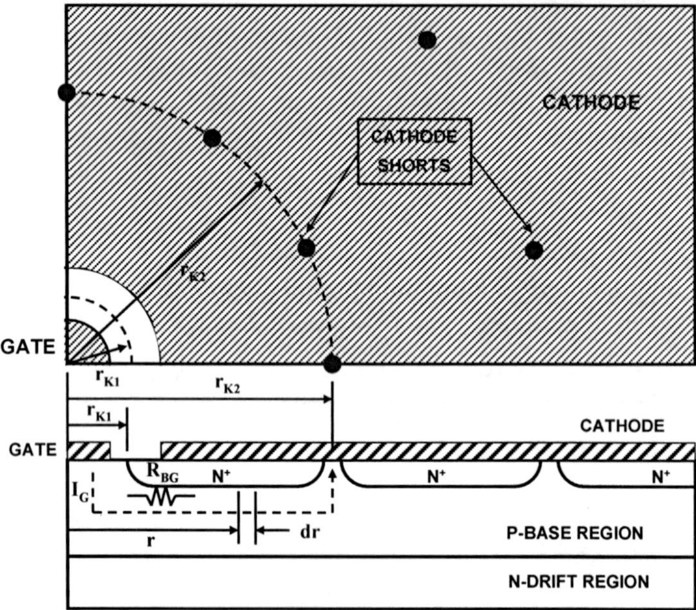

Fig. 8.29 Power thyristor structure with central gate electrode and concentric cathode shorts

For a typical high-voltage power thyristor, the sheet resistance of the P-base region is about 500 Ω sq^{-1}. Using (8.40), the gate-triggering current is found to be 15 mA per cm of cathode finger length (Z). Thus, a relatively small gate current is sufficient to turn on the thyristor from its forward-blocking mode. In practical devices, the high-current thyristors are fabricated by using an entire circular silicon wafer. The wafer is enclosed within a "hockey-puck" package designed to remove the heat developed in the active area of the thyristor. Since it is most convenient to introduce the gate wire at the center of the package, the gate electrode must be located at the center of the wafer. An illustration of this thyristor design is shown in Fig. 8.29.

The circular gate geometry can be analyzed by using cylindrical coordinates as indicated in the figure. In this case, the first row of shorts is regarded as a continuous contact between the cathode electrode and the P-base region along the circular dashed line with a radius of r_{K2}. The resistance (R_{BG}) of the P-base region between the edge of the N^+ cathode located closest to the gate contact and the first row of cathode shorts can be obtained by considering a segment of width dr located at a radius r. The resistance of the segment is given by

$$\mathrm{d}R_{BG} = \frac{\rho_{PB}}{W_P}\frac{\mathrm{d}r}{2\pi r} = \frac{\rho_{SB}}{2\pi}\frac{\mathrm{d}r}{r}, \tag{8.41}$$

where ρ_{PB} is the resistivity of the P-base region. The resistance of the P-base region is then given by

$$R_{BG} = \int_{r_{K1}}^{r_{K2}} \frac{\rho_{SB}}{2\pi}\frac{\mathrm{d}r}{r} = \frac{\rho_{SB}}{2\pi}\ln\left(\frac{r_{K2}}{r_{K1}}\right). \tag{8.42}$$

The thyristor is triggered from the forward-blocking mode to the on-state when the gate current is sufficient to produce a voltage drop V_{bi} across this resistance. Using this criterion,

$$I_{GT} = \frac{2\pi V_{bi}}{\rho_{SB}\ln(r_{K2}/r_{K1})}. \tag{8.43}$$

A gate-triggering current of 30 mA can be achieved for a thyristor with a P-base sheet resistance of 500 Ω sq^{-1} by placing the first row of cathode shorts at a radius (r_{K2}) of 0.42 cm if the inner radius (r_{K1}) of the N^+ cathode is 0.3 cm. This is typical for a large area thyristor whose gate electrode has a radius of 0.28 cm.

8.3.3 Holding Current

The thyristor structure can maintain itself in the on-state without an external gate drive current due to the internal regenerative action. However, when the anode current is reduced, the thyristor will revert to the forward-blocking mode at a current level referred to as the *holding current*. This occurs because the current gain of the N–P–N transistor rapidly falls off below a certain current level as previously shown in Fig. 8.16. Once the sum of the common-base current gains of the internal transistors becomes smaller than unity, the regenerative action is unable to sustain the thyristor in its on-state. The holding current (I_H) is indicated on the device characteristics in Fig. 8.25.

Analysis of the holding current can be performed with the aid of Fig. 8.30 for the case of a linear cathode shorting geometry. In this case, it is assumed that the thyristor is operating in the on-state with a uniform cathode current density (J_K) flowing across the N^+ cathode/P-base junction. Although most of the cathode current flows into the P-base/N-drift junction due to the high gain of the N–P–N transistor, a small fraction flows into the P-base region due to recombination of minority carriers. This base current for the N–P–N transistor flows into the cathode short at location *B*.

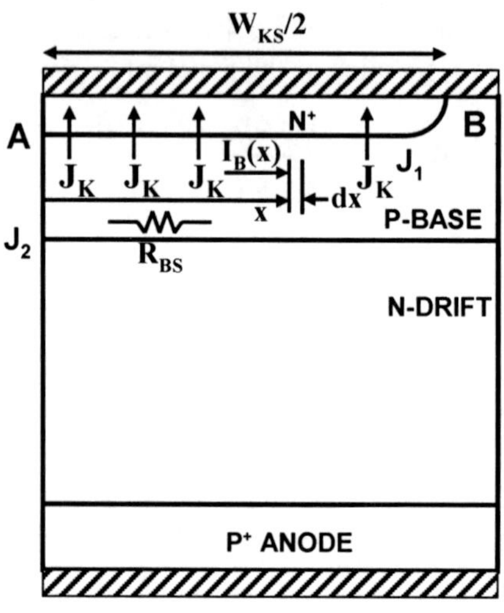

Fig. 8.30 Holding current analysis for the linear emitter shorting geometry for the thyristor

The base current $I_B(x)$ at a distance x from the center of the cathode (location *A*) is given by

$$I_B = (1 - \alpha_{NPN}) J_K Z x, \tag{8.44}$$

where Z is the length of the structure orthogonal to the cross section. The voltage drop produced across a segment dx located at a distance x from the center of the cathode is then given by

$$\mathrm{d}V_B(x) = I_B(x)\frac{\rho_{SB}}{Z}\mathrm{d}x = (1 - \alpha_{NPN}) J_K \rho_{SB} x \, \mathrm{d}x. \tag{8.45}$$

The voltage developed at the middle of the cathode (location *A*) is obtained by integration of the above voltage:

$$V_B(A) = \int_0^{W_{KS}/2} (1-\alpha_{NPN}) J_K \rho_{SB} x \, dx = (1-\alpha_{NPN}) J_K \rho_{SB} \frac{W_{KS}^2}{8}. \quad (8.46)$$

As the cathode (and anode) current density is reduced, the voltage drop at the middle of the cathode becomes smaller until it falls below that for a forward-biased P–N diode (V_{bi}). At this current level, the injection of electrons from the N^+ cathode/P-base junction ceases and the gain of the N–P–N transistor becomes very low, resulting in the thyristor no longer being able to sustain the regenerative action. Using this criterion to define the holding current density in the above equation yields

$$J_H = \frac{8V_{bi}}{(1-\alpha_{NPN})\rho_{SB} W_{KS}^2}. \quad (8.47)$$

A small holding current density is desirable for a thyristor structure because it allows operation over a broad range of on-state current densities. Based upon the above equation, it can be concluded that the holding current density will decrease with increasing temperature due to a reduction of the built-in potential and an increase in the sheet resistance of the P-base region. From a design viewpoint, the holding current density can be reduced by increasing the sheet resistance for the P-base region and placing the cathode shorts further apart. However, this will reduce the leakage current at which the thyristor will be turned on, degrading its high temperature forward-blocking capability.

For a typical thyristor structure with cathode short spacing (W_{KS}) of 1,000 μm and a P-base sheet resistance of 500 Ω sq^{-1}, the holding current density calculated by using the above equation is 5 A cm^{-2} for a current gain of 0.65 for the N–P–N transistor. This is acceptable because typical on-state current densities for thyristors range from 50 to 300 A cm^{-2}.

Simulation Example

To gain further insight into the physics of operation for the power thyristor in the on-state, the results of two-dimensional numerical simulations for the typical structure (described in Sect. 8.2) are discussed here. The total width of the structure, as shown by the cross section in Fig. 8.17, is 1,000 μm (area = 1×10^{-5} cm^{-2}) with a cathode finger width of 980 μm, and its doping profile is shown in Fig. 8.20. To operate the thyristor in its on-state, a gate drive current of 2×10^{-6} A μm^{-1} (0.2 A cm^{-2}) was applied while sweeping the anode voltage. The forward conduction characteristics obtained at various temperatures are shown in Fig. 8.31 for the case of a lifetime (τ_{p0}, τ_{n0}) of 10 μs. The constant anode current at very low anode bias voltages is due to the gate drive current. The anode current increases exponentially with increasing anode voltage as predicted by the analytical model based upon using the P-i-N diode physics when the anode bias is increased beyond 0.5 V at 300 K. The on-state voltage drop is 1.2 V at an anode current density of 100 A cm^{-2}. It

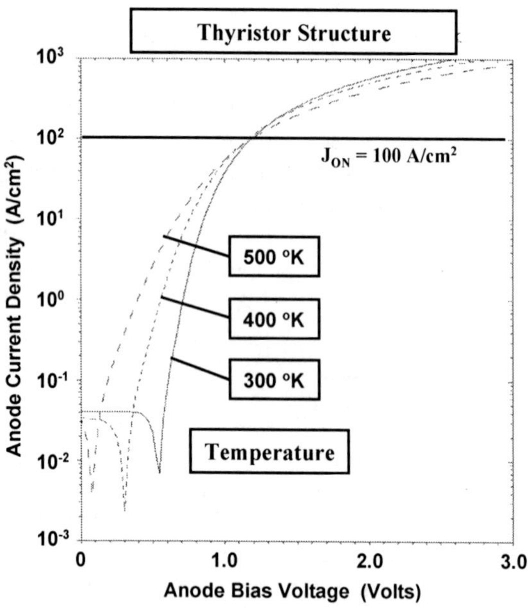

Fig. 8.31 Typical forward conduction characteristics for the N–P–N–P power thyristor structure

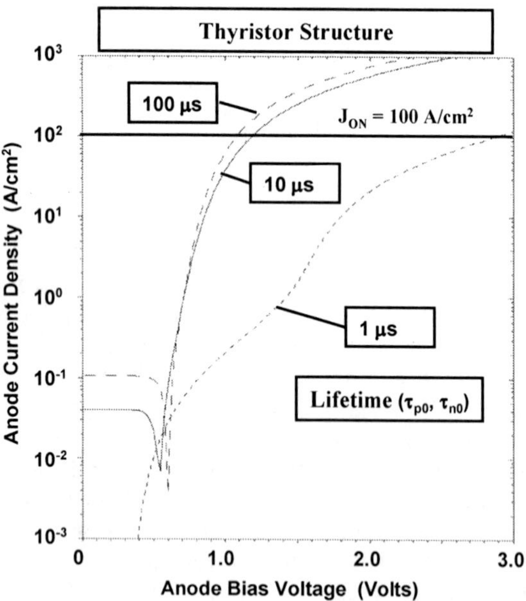

Fig. 8.32 Impact of lifetime changes on the forward conduction characteristics for the N–P–N–P power thyristor structure

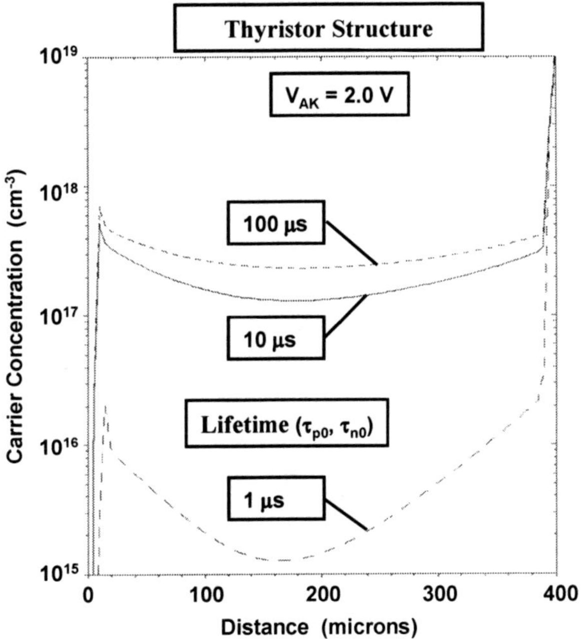

Fig. 8.33 Carrier distribution within the N–P–N–P power thyristor structure during the on-state

can be observed from the figure that the on-state voltage drop increases with increasing temperature for anode current density above 100 A cm^{-2}. This is a favorable outcome for preventing the development of current localization and hot spots within the thyristor.

The forward conduction characteristics for the thyristor structure are shown in Fig. 8.32 when the lifetime is varied. When the lifetime (τ_{p0}, τ_{n0}) is increased from 10 to 100 μs, the on-state voltage drop reduces slightly because the (d/L_a) ratio increases from unity. In contrast, when the lifetime (τ_{p0}, τ_{n0}) is reduced from 10 to 1 μs, the on-state voltage drop increases by a significant amount to 2.9 V because the (d/L_a) ratio decreases below unity. This is consistent with the P-i-N diode model for the on-state characteristics for a thyristor.

The reason for the increase in the on-state voltage drop for the lowest lifetime case can be understood by examination of the free carrier distribution profile within the thyristor. The hole carrier distribution is provided in Fig. 8.33 for the three lifetime cases at an anode forward bias of 2 V. The hole concentration exceeds the doping concentration of the P-base region when the lifetime is 10 and 100 μs, producing a catenary profile through the combination of the P-base region and the N-drift region. This is consistent with the P-i-N diode-based analytical model for the thyristor. However, when the lifetime is reduced to 1 μs, the hole concentration becomes much smaller and does not exceed the P-base doping concentration. The low hole concentration and its steep gradients enhance the voltage drop across the drift region resulting in a large on-state voltage drop.

The holding current for the thyristor was extracted with the numerical simulations by initially biasing the device in its on-state current and then reducing the anode current until the device switched to the blocking state. A typical set of *i*–*v* characteristics are shown in Fig. 8.34 at various temperatures. At 300 K, the holding current density is observed to be 0.4 A cm^{-2}. The analytical model predicts a slightly larger value due to the assumptions made for the sheet resistance of the P-base region and the current gain of the N–P–N transistor. It can be observed that the holding current decreases with increasing temperature as predicted by the analytical model due to a reduction of the built-in potential and an increase in the sheet resistance of the P-base region.

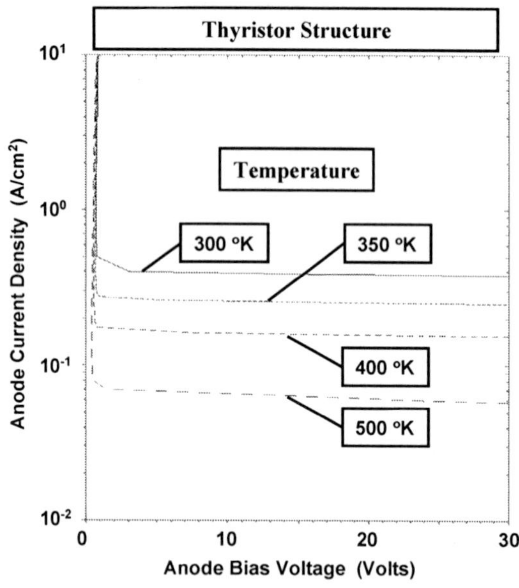

Fig. 8.34 Holding current for the N–P–N–P power thyristor structure

8.4 Switching Characteristics

Thyristors are typically used to regulate power flow in AC power circuits by controlling their on-state duration as illustrated in Fig. 8.6. During circuit operation, the thyristor is switched from the forward-blocking state to the on-state by the application of a gate drive current. It then switches from the on-state to the reverse-blocking mode when the anode voltage is reversed from positive to negative at the end of the first half-cycle of the AC input voltage waveform. These switching transitions are not instantaneous. During the turn-on process, the rise in the current is delayed by the need to establish the internal regenerative action. This turns on the portion of the thyristor structure in proximity with the gate terminal. Due to the large area for high-current thyristors, a further time delay is associated with the spreading of the current from the near the gate terminal to the rest of the active area of the device.

The switching of the thyristor from the on-state to the reverse-blocking state produces substantial power dissipation. As discussed in Sect. 8.3, a high concentration of minority carriers is stored within the N-drift region due to high-level injection conditions in the on-state. These carriers must be removed before the thyristor is capable of supporting voltage. When the anode voltage reverses at the end of the first half-cycle, a negative anode current flow is observed as shown in Fig. 8.6 to remove most of the stored charge. The thyristor undergoes a reverse recovery process akin to that previously discussed for the P-i-N rectifier.

It can be observed from Fig. 8.6 that the largest positive rate of change of voltage [dV/dt] is impressed upon the anode of the thyristor at the beginning of each half-cycle. This [dV/dt] increases with increasing frequency of operation. It has been found that the thyristor can be triggered into the on-state without the application of a gate drive current when the [dV/dt] is sufficiently large. This phenomenon determines the maximum frequency of operation for thyristors. These processes are analyzed in this section. For each case, analytical expressions are derived to allow an improved understanding of the underlying physics.

8.4.1 Turn-On Time

The thyristor can be triggered from the forward-blocking mode to the on-state by the application of a gate drive current. The gate drive current flows from the gate terminal to the first row of cathode shorts as illustrated in Fig. 8.28. The voltage drop in the P-base region due to the gate current flow forward biases the N^+ emitter/P-base junction (J_3) at the edge of the cathode closest to the gate terminal, producing the injection of electrons as indicated by the arrow in Fig. 8.28. This does not immediately produce anode current flow. The injected electrons diffuse through the P-base region in a finite time interval referred to as the *base transit time*. For the N–P–N transistor, the base transit time is given by

$$t_{\mathrm{t,NPN}} = \frac{W_{\mathrm{P}}^2}{2D_{\mathrm{n}}}, \tag{8.48}$$

where W_P is the thickness of the P-base region. The base transit time for the P-base region is typically 50 ns. Once the electrons cross the P-base/N-base junction (J_2), they immediately promote the injection of holes from the P^+ anode/N-base junction (J_1) to preserve charge neutrality in the N-base region. The injection of carriers at the P^+ anode/N-base junction initiates current flow through the device. Consequently, the anode current begins to flow after a *delay time* interval which is equal to the transit time for the N–P–N transistor. The delay time is typically 50 ns in duration.

After injection from the P^+ anode/N-base junction, the holes diffuse through the N-base region until they are collected at the P-base/N-base junction (J_2). The time taken for this process is the base transit time for the P–N–P transistor as given by

$$t_{t,PNP} = \frac{(W_N - W_{DN})^2}{2D_p}, \tag{8.49}$$

where W_N is the thickness of the N-base region and W_{DN} is the thickness of the depletion region as shown in Fig. 8.10. The base transit time for the N-base region is a strong function of the initial anode bias voltage before turning on the thyristor. The above transit time is the value as limited by the diffusion of holes in the neutral region under the assumption that the applied anode bias is supported across only the depletion region. The typical transit time determined by the diffusion process is 50 μs for a thyristor with an N-base region width of 350 μm.

During the turn-on process, a high concentration of holes is injected into the N-drift region from the P^+ anode region and a high concentration of electrons is injected into the N-drift region from the N^+ cathode region (remote emitter). The depletion region cannot be sustained under these conditions and the anode voltage is distributed throughout the N-base region. The electric field in the conductivity-modulated N-base region is then determined by the applied anode bias and the thickness of the N-base region:

$$E_{NB} = \frac{V_A}{W_N}. \tag{8.50}$$

Due to the presence of this electric field, the holes travel through the N-base region at a drift velocity given by

$$v_P = \mu_P E_{NB} = \frac{\mu_P V_A}{W_N}. \tag{8.51}$$

The transit time for the holes is therefore given by

$$t_{t,PNP} = \frac{W_N}{v_P} = \frac{W_N^2}{\mu_P V_A}, \tag{8.52}$$

which is much smaller than the transit time determined by the diffusion process. The typical transit time determined by the drift process is only 5 ns for a thyristor with an N-base region width of 350 μm at an anode bias of 500 V.

An analytical model for the increase in the anode current during the turn-on transient for a one-dimensional thyristor structure can be derived based upon charge control principles.[6] This analysis takes into consideration the internal feedback mechanism between the N–P–N and P–N–P transistors within the thyristor structure to determine the growth of the stored charge within the N-base region and the P-base region. Due to relatively short time for the turn-on transient when compared with the lifetime, the recombination within the N-base region and the P-base region can be assumed to be negligible during this analysis.

The increase in the stored electrons within the N-base region (Q_{SN}) during the turn-on process occurs due to the electrons collected at the P-base/N-base junction (J_2). Since these electrons are supplied by the injection from the N$^+$ cathode (emitter of the N–P–N transistor),

$$\frac{dQ_{SN}}{dt} = \alpha_{NPN} J_K(t), \tag{8.53}$$

where J_K is the cathode current density.

The increase in the stored holes within the P-base region (Q_{SP}) during the turn-on process occurs due to the holes collected at the P-base/N-base junction (J_2) and the holes supplied by the gate drive current. Since the holes arriving at the P-base/N-base junction (J_2) are supplied by injection from the P$^+$ cathode (emitter of the P–N–P transistor),

$$\frac{dQ_{SP}}{dt} = \alpha_{PNP} J_A(t) + J_G, \tag{8.54}$$

where J_A is the anode current density.

The collector current for the N–P–N transistor is related to the stored charge in its base region via the transit time:

$$J_{C,NPN} = \alpha_{NPN} J_K(t) = \frac{Q_{SP}}{t_{t,NPN}}, \tag{8.55}$$

where $t_{t,NPN}$ is the transit time for the N–P–N transistor. Similarly, the collector current for the P–N–P transistor is related to the stored charge in its base region via the transit time:

$$J_{C,PNP} = \alpha_{PNP} J_A(t) = \frac{Q_{SN}}{t_{t,PNP}}, \tag{8.56}$$

where $t_{t,PNP}$ is the transit time for the P–N–P transistor.

Combining (8.53) and (8.55),

$$\frac{dQ_{SN}}{dt} = \frac{Q_{SP}}{t_{t,NPN}}. \tag{8.57}$$

Taking the first derivative of this equation yields

$$\frac{d^2 Q_{SN}}{dt^2} = \frac{1}{t_{t,NPN}} \frac{dQ_{SP}}{dt}. \tag{8.58}$$

Combining this equation with (8.54) and (8.56),

$$\frac{d^2 Q_{SN}}{dt^2} - \frac{Q_{SN}}{t_{t,NPN} t_{t,PNP}} = \frac{J_G}{t_{t,NPN}}. \tag{8.59}$$

The solution for this second-order differential equation is

$$Q_{SN}(t) = J_G t_{t,PNP} \left(e^{t/\sqrt{t_{t,NPN} t_{t,PNP}}} - 1 \right). \tag{8.60}$$

Substituting this expression into (8.56) yields an equation that describes the increase in the anode current during the turn-on process:

$$J_A(t) = \frac{J_G}{\alpha_{PNP}} \left(e^{t/\sqrt{t_{t,NPN} t_{t,PNP}}} - 1 \right). \tag{8.61}$$

Based upon this equation, it can be concluded that the anode current will grow exponentially with time after the delay phase. The time constant for the rise in the anode current is observed to be the geometric mean of the transit times for the N–P–N and P–N–P transistors.

The *rise time* is defined as the time taken by the anode current density to increase to the on-state value. Using (8.61), the rise time can be obtained:

$$t_R = \sqrt{t_{t,NPN} t_{t,PNP}} \ln \left(\frac{\alpha_{PNP} J_{A,SS}}{J_G} + 1 \right), \tag{8.62}$$

where $J_{A,SS}$ is the steady-state (on-state) anode current density. The rise time is determined by the transit times for the internal N–P–N and P–N–P transistors within the thyristor structure.

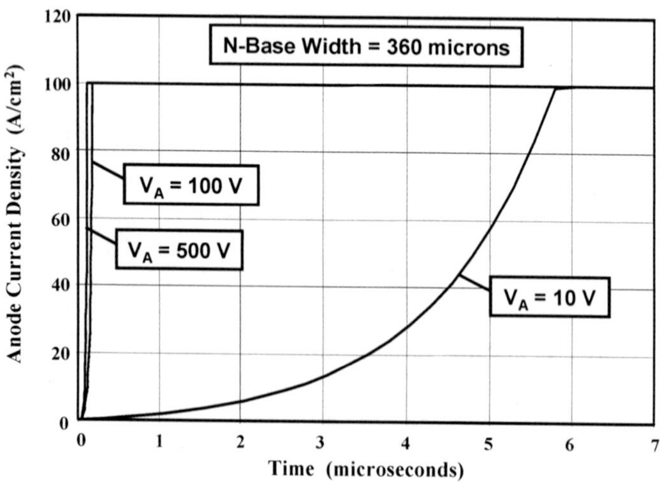

Fig. 8.35 Turn-on transient for a one-dimensional thyristor

As an example, the rise in the anode current with time is shown in Fig. 8.35 for the case of a one-dimensional thyristor structure with a P-base region width of 15 μm and an N-base region width of 360 μm. This thyristor structure is capable of blocking 2,000 V in both directions when a lifetime (τ_{p0}, τ_{n0}) of 10 μs is chosen. The transit time for the N–P–N transistor is calculated to be 43 ns and that for the P–N–P transistor is calculated to be 50 μs when limited by the diffusion process. This holds true for the case of an anode bias of 10 V. In this case, the current reaches its steady-state value with a rise time of 5.8 μs. However, when the anode voltage is increased to 100 V, the transit time for the P–N–P transistor is greatly reduced to 26 ns due to the enhanced drift current. Consequently, the rise time for the anode current also decreases to only 0.14 μs. A further increase in the anode voltage to 500 V produces a reduction of the rise time to only 0.07 μs.

Power thyristors are designed with a large active area to achieve a high-current handling capability. Once the anode current flow is produced at the edge of the cathode located near the gate electrode, the current spreads over the rest of the active area. This is a relatively slow process due to the long distance over which the current must be distributed. The time taken for the distribution of the current has been characterized by a *spreading velocity*. The spreading velocity has been empirically derived to be about 5,000 cm s^{-1} for typical thyristors.[7–10] As the current spreads over the active area of the thyristor, the current density decreases producing a corresponding reduction in the on-state voltage drop. For the case of a resistive load, the current gradually increases during this time due to the smaller on-state voltage drop.

Analytical treatment of the current spreading can be performed by assuming that the process is driven by the diffusion of free carriers from the on-portion of the thyristor into the off-portion of the structure. If recombination is neglected because the lifetime is much greater than the turn-on transient time, the time-dependent diffusion equation for electrons within the N-base region is given by

$$\frac{\partial \delta n(x,t)}{\partial t} = \frac{\partial^2 \delta n(x,t)}{\partial x^2}, \tag{8.63}$$

where δn is the excess electron concentration. The solution for this equation is a Gaussian distribution:

$$\partial n(x,t) = \frac{n(0,t)}{2\sqrt{\pi D_n t}} e^{-x^2/4D_n t}, \tag{8.64}$$

with $n(0, t)$, the electron concentration at $x = 0$, being related to the anode current density previously derived for the one-dimensional thyristor structure (see (8.61)).

The current flow through the N-base region (in the y-direction) can be assumed to be controlled by the drift component as previously discussed for the transit time analysis. After accounting for high-level injection conditions in the on-portion of the structure,

$$J_A(0,t) = [q\mu_p p(0,t) + q\mu_n n(0,t)]E_{NB}, \quad (8.65)$$

where E_{NB} is the electric field in the conductivity-modulated N-base region. Using (8.50) for this electric field and recognizing that the electron and hole concentrations are equal due to high-level injection conditions,

$$n(0,t) = \frac{J_A(t)W_N}{q(\mu_p + \mu_n)V_A}. \quad (8.66)$$

Substituting this expression into (8.64) with the anode current density given by (8.61) yields

$$\partial n(x,t) = \frac{J_G W_N \left(e^{t/\sqrt{t_{t,NPN}t_{t,PNP}}} - 1\right)}{2q(\mu_p + \mu_n)V_A \alpha_{PNP}\sqrt{\pi D_n t}} e^{-x^2/4D_n t}. \quad (8.67)$$

The total electron concentration $n(x, t)$ in the N-base region is then obtained by adding the doping concentration N_D to the above excess electron concentration.

During the turn-on transient, the electron concentration at the edge of the cathode located in proximity with the gate electrode $n(0, t)$ increases with time due to the increasing anode current. Concurrently, the electrons diffuse from the on-portion toward the center of the cathode. An example of the electron distribution in the N-base region is shown in Fig. 8.36 at various points in time for the case of an anode bias of 10 V. The on-portion can be observed to expand from the gate electrode toward the right-hand side if the on-portion is defined by the electron concentration exceeding a certain value (e.g., 5×10^{15} cm^{-3}).

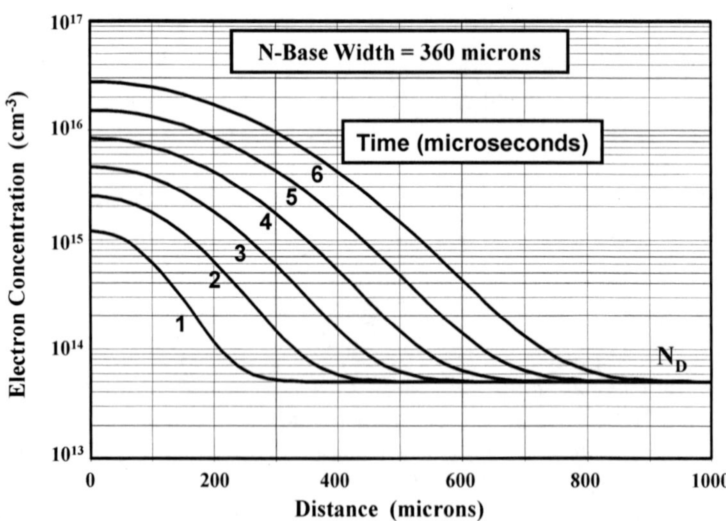

Fig. 8.36 Electron distribution during the turn-on transient for a two-dimensional thyristor

The width of the on-portion can be derived from (8.67) by setting the excess electron concentration to a fixed value N_{ON} at a distance X_{ON}, the boundary for the on-region:

$$N_{\mathrm{ON}}=\frac{J_{\mathrm{G}}W_{\mathrm{N}}\left(\mathrm{e}^{t/\sqrt{t_{\mathrm{t,NPN}}t_{\mathrm{t,PNP}}}}-1\right)}{2q(\mu_{\mathrm{p}}+\mu_{\mathrm{n}})V_{\mathrm{A}}\alpha_{\mathrm{PNP}}\sqrt{\pi D_{\mathrm{n}}t}}\mathrm{e}^{-X_{\mathrm{ON}}^{2}/4D_{\mathrm{n}}t}. \tag{8.68}$$

From this equation,

$$X_{\mathrm{ON}}=\sqrt{(4D_{\mathrm{n}}t)\ln\left[\frac{J_{\mathrm{G}}W_{\mathrm{N}}\left(\mathrm{e}^{t/\sqrt{t_{\mathrm{t,NPN}}t_{\mathrm{t,PNP}}}}-1\right)}{2qN_{\mathrm{ON}}(\mu_{\mathrm{n}}+\mu_{\mathrm{p}})V_{\mathrm{A}}\alpha_{\mathrm{PNP}}\sqrt{\pi D_{\mathrm{n}}t}}\right]}. \tag{8.69}$$

From Fig. 8.36, it can be observed that the rate at which the on-region expands is relatively constant at a value of 100 μm μs^{-1} (10^4 cm s^{-1}). This value is comparable with the range of empirical values (4×10^3 to 10^4 cm s^{-1}) reported in the literature.[10]

Simulation Example

To gain further insight into the physics of turn-on for the power thyristor, the results of two-dimensional numerical simulations for the typical structure (described in Sect. 8.3) are discussed here. The total width of the structure, as shown by the cross section in Fig. 8.17, is 1,000 μm (area = 1×10^{-5} cm^{-2}) with a cathode finger width of 980 μm. To turn on the thyristor, the gate drive current was abruptly increased from 0 to 2×10^{-6} A μm^{-1} (0.2 A cm^{-2}) while the anode voltage was maintained fixed with a load resistance in series with the thyristor structure.

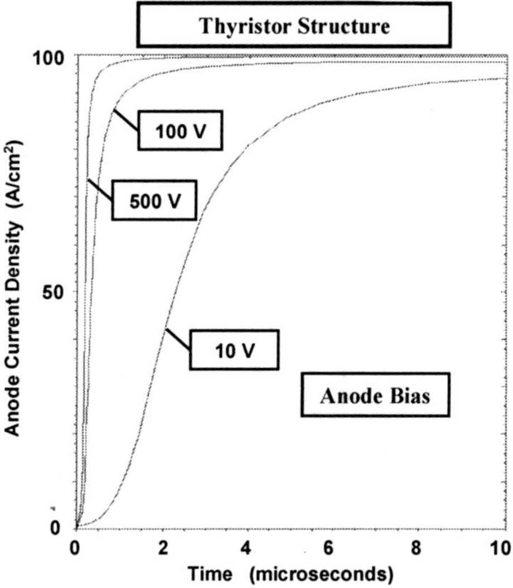

Fig. 8.37 Turn-on characteristics for the N–P–N–P power thyristor structure

The turn-on characteristics for the thyristor structure are shown in Fig. 8.37 for three values of the anode bias. In each case, the load resistance was altered to obtain the same anode current density (100 A cm^{-2}) after the thyristor is in the on-state. In all cases, there is a short delay time of about 50 ns, which is consistent with the delay time computed by using (8.48) for the base transit time of the N–P–N transistor. The anode current then increases exponentially with time as described by the analytical equation (8.61). For the case of an anode bias of 10 V, the anode current approaches 90% of its steady-state value after about 5 μs. This behavior is consistent with the analytical model for turn-on based upon the diffusion of carriers through the N-base region. When the anode bias is increased to 100 and 500 V, the anode current density increases much more rapidly. This behavior is consistent with the analytical model for turn-on based upon the drift of carriers through the N-base region. In all cases, a gradual increase in the anode current density is also observed at the end of the turn-on transient due to the current spreading phenomenon. The time span for this transient is relatively independent of the anode bias voltage.

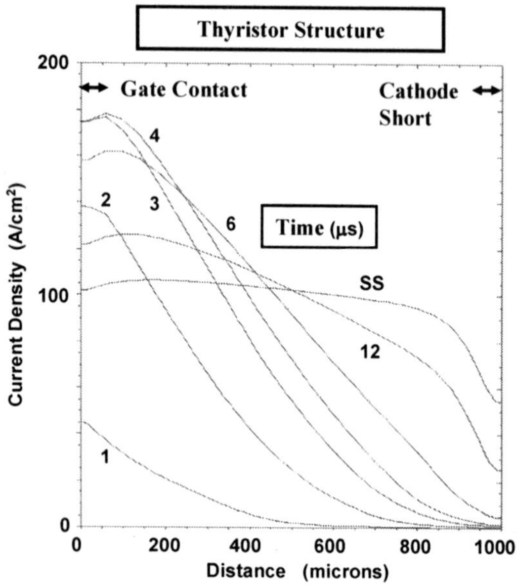

Fig. 8.38 Current distribution within the power thyristor structure during turn-on

The current spreading phenomenon can be examined by observing the current distribution within the thyristor structure during various time intervals of the turn-on process as shown in Fig. 8.38. From Fig. 8.38, it can be observed that the current initially increases at the edge of the cathode near the gate contact. The current density peaks at 3 μs. After this time, the current density decreases as the on-region expands over the cathode. In the steady-state (SS) case, the current density becomes relatively uniform at 100 A cm^{-2} with the exception of the region close to the cathode short. A "dead zone" of about 100 μm occurs around the cathode short as discussed in Sect. 8.2.4.

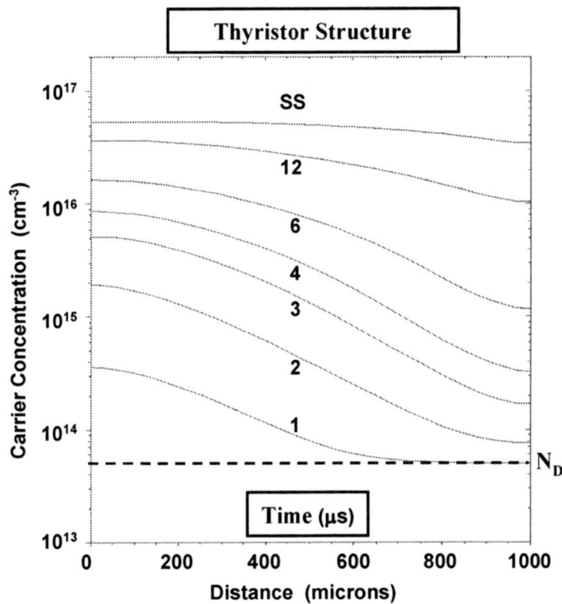

Fig. 8.39 Carrier distribution within the power thyristor structure during turn-on

The electron carrier distribution in the N-base region is shown in Fig. 8.39 during various instances of the turn-on transient. It can be observed that the electron concentration increases in the vicinity of the gate electrode and also spreads toward the center of the cathode as time progresses. This behavior is qualitatively similar to that shown in Fig. 8.36 based upon the analytical model for the diffusion of electrons. Based upon the increasing width of the conductivity-modulated region, a spreading velocity ranging between 1.0 and 1.5×10^4 cm s^{-1} can be extracted from the plot. Based upon this, it can be concluded that the analytical model provides a good estimate for the current spreading within the thyristor during the turn-on process.

8.4.2 Gate Design

As discussed in Sect. 8.4.1, the turn-on of thyristors first occurs at the edges of the gate electrode and then spreads throughout the cathode area. When the anode current increases rapidly (i.e., with a high dI_A/dt), the current density at the edges of the gate becomes large. In conjunction with the high anode voltage during the initial stages of the turn-on process, this produces high local power dissipation which can produce destructive failure. This problem can be solved by increasing the gate periphery to distribute the area over which the turn-on occurs. A larger gate current is required to turn on the thyristor as the gate periphery is made larger.

Two examples for the gate design for power thyristors are illustrated in Fig. 8.40. The "spoke" gate design shown in Fig. 8.40a extends the gate electrode metal from the central gate wire bonding pad toward the edges of the wafer. This allows simultaneous turn-on of parts of the thyristor that are located at the center of the wafer and at the circumference of the wafer. An even better design, called the *involute structure*, is shown in Fig. 8.40b. An involute pattern is generated when a string is tightly unwound from the edges of a cylinder. If two involutes are unwound from a cylinder of radius r with an angular displacement θ, it can be shown that all points between them will be equidistant with a spacing of $(r\theta)$. Consequently, all portions of the cathode within the thyristor will operate in a similar manner. The involute design greatly increases the gate periphery enabling improvement of the $[dI/dt]$ capability of the thyristor. However, the gate drive current must be proportionally increased when turning on the structure. The involute design is utilized for thyristors used in inverter circuits operated at relatively high frequencies.

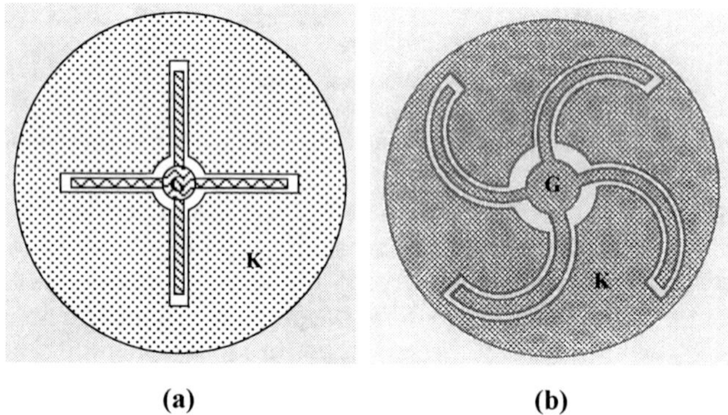

(a) **(b)**

Fig. 8.40 Typical gate designs for thyristors

8.4.3 Amplifying Gate Design

Thyristors with extended gate designs require substantial gate drive current due to the large periphery for the gate electrode. The gate drive current can be reduced by first turning on a small portion of the thyristor located around the gate electrode (called the *amplifying gate thyristor*) and using its cathode current to serve as the gate drive current for the rest of the (main) thyristor.[11] A top view and cross section of the amplifying gate thyristor structure are illustrated in Fig. 8.41. The amplifying gate region surrounds the gate contact pad located at the center of the device. The gate extensions discussed in Sect. 8.4.2 can be attached to the amplifying gate thyristor cathode metal to improve the $[dI/dt]$ capability. Note that the metal electrode for the amplifying gate region overlaps the N^+ cathode/P-base junction at its edge located away from the gate electrode, and the electrode for the amplifying gate region is not connected to either the gate or cathode terminals of the thyristor structure.

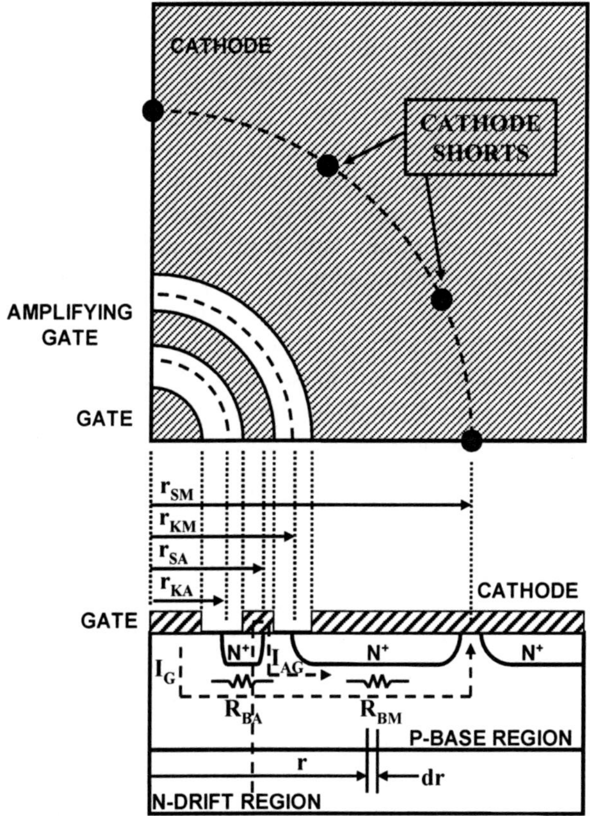

Fig. 8.41 Amplifying gate thyristor structure

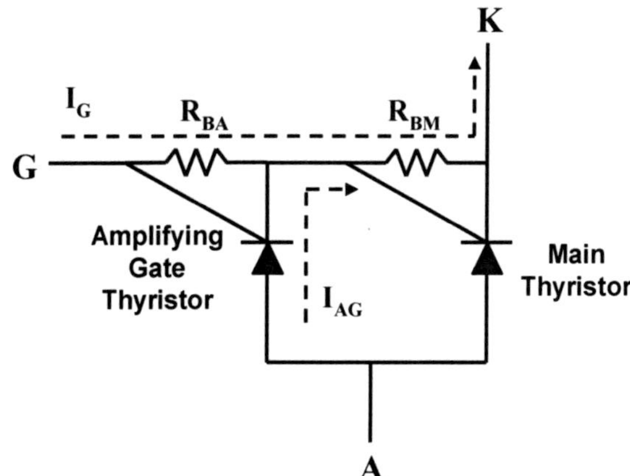

Fig. 8.42 Circuit representation for the amplifying gate thyristor structure

An equivalent circuit for the amplifying gate thyristor structure is provided in Fig. 8.42. It consists of the main thyristor on the right-hand side and the amplifying gate thyristor on the left-hand side. The resistances within the P-base region under the cathodes for the two thyristors are also shown in the figures. These resistances shunt the N^+ cathode/P-base junctions for both these thyristors. The gate current (I_G), applied to turn on the thyristor, flows from the gate contact to the first row of emitter shorts within the main thyristor. This is illustrated by the dashed line labeled I_G in Fig. 8.42 and the cross section of the structure in the lower portion of Fig. 8.41. When the gate current flows through the resistances indicated in the figures, a positive voltage is induced within the P-base region toward the gate side in relation to the cathode potential. This produces a forward bias across the N^+ cathode/P-base junctions for both these thyristors. With appropriate design of the device geometry, as discussed below, the forward bias across the N^+ cathode/P-base junction of the amplifying gate thyristor can be made larger than that for the main thyristor. Consequently, the amplifying gate thyristor can be triggered into its on-state prior to the turn-on of the main thyristor. Since the cathode electrode for the amplifying gate thyristor is connected to the P-base region, its cathode current is forced to flow into the cathode shorts for the main thyristor as shown by the dashed lines labeled I_{AG} in both figures. This current then serves as the gate drive current for the main thyristor. Most of the gate drive current for the main thyristor is therefore derived from the anode circuit via the amplifying gate thyristor.

The analysis of the geometrical design for the amplifying gate thyristor structure can be performed for the circular geometry shown in Fig. 8.41 by using cylindrical coordinates as indicated in the lower portion of the figure. The radii that define the boundaries for the N^+ cathode regions are provided in the figure. The resistance of the P-base region under the N^+ cathode regions can be analyzed by considering a small segment at a radius r. The resistance of this segment is given by

$$\mathrm{d}R_B = \frac{\rho_{PB}\mathrm{d}r}{2\pi r W_P} = \frac{\rho_{S,PB}}{2\pi}\frac{\mathrm{d}r}{r}, \tag{8.70}$$

where $\rho_{S,PB}$ is the sheet resistance of the P-base region obtained by taking the ratio of the resistivity (ρ_{PB}) of the P-base region and its thickness (W_P) under the N^+ cathode region.

The resistance of the P-base region under the N^+ cathode region of the amplifying gate thyristor is obtained by integration of the above resistance between the radii that define its edges:

$$R_{BA} = \int_{r_{KA}}^{r_{SA}} \mathrm{d}R_B = \frac{\rho_{S,PB}}{2\pi}\ln\left(\frac{r_{SA}}{r_{KA}}\right). \tag{8.71}$$

In the same manner, the resistance of the P-base region under the N^+ cathode region of the main thyristor is obtained by integration of the incremental resistance between the radii that define its edges:

$$R_{\rm BM} = \int_{r_{\rm KM}}^{r_{\rm SM}} {\rm d}R_{\rm B} = \frac{\rho_{\rm S,PB}}{2\pi} \ln\left(\frac{r_{\rm SM}}{r_{\rm KM}}\right). \tag{8.72}$$

The design criterion for the amplifying gate thyristor requires making the resistance R_{BA} greater than the resistance R_{BM}, so that the amplifying gate thyristor will be triggered by the applied gate current prior to the turn-on of the main thyristor. Based upon the above equations for these resistances, this can be achieved by making

$$\frac{r_{\rm SA}}{r_{\rm KA}} > \frac{r_{\rm SM}}{r_{\rm KM}}. \tag{8.73}$$

It was previously shown (see Sect. 8.3.2) that the gate-triggering current for the thyristor structure is determined by the cathode shorting geometry. This is not applicable for the amplifying gate thyristor structure. Instead, the gate-triggering current for the amplifying gate thyristor structure is determined by the dimensions of the amplifying gate region. Using the criterion that the amplifying gate thyristor turns on when its N^+ cathode/P-base junction is forward biased by the junction built-in potential when the gate current flows through the resistance R_{BA}, the gate-triggering current is given by

$$I_{\rm GT} = \frac{2\pi V_{\rm bi}}{\rho_{\rm S,PB} \ln(r_{\rm SA} / r_{\rm KA})}. \tag{8.74}$$

For a typical thyristor structure, the sheet resistance of the P-base region is about 500 Ω sq^{-1}. A gate drive current of 10 mA can be achieved for this thyristor by utilizing a ratio (r_{SA}/r_{KA}) of 1.11 based upon a built-in potential of 0.8 V. The gate electrode pad in the center of the thyristor is typically designed with a radius of 100 mil (0.25 cm) to allow connection of the gate wire. An inner radius (r_{KA}) of 0.5 cm with an outer radius (r_{SA}) of 0.55 cm for the N^+ cathode region of the amplifying gate thyristor satisfies the design requirements in this case.

8.4.4 [d*V*/d*t*] Capability

Due to the high operating voltages in thyristor applications, a high rate of change of the anode voltage commonly occurs during circuit operation. The rapid change in the anode voltage produces a displacement current at the reverse-biased P-base/N-base junction during operation in the first quadrant. The transport of the displacement current within the P-base region acts like a gate drive current. As the rate of change in anode voltage is increased, the displacement current can be sufficient to trigger the thyristor into the on-state without an external gate drive current. This disruption of circuit operation can lead to destructive failure of the thyristor due to enhanced duty cycles.

The maximum rate of change of the anode voltage that can be tolerated by the thyristor before turning on is determined by the design of the cathode shorting geometry. The analysis of the maximum [dV/dt] capability is performed in this section by taking into account a linear cathode shorting arrangement as illustrated in Fig. 8.43. A uniform displacement current density (J_D) is produced across the reverse-blocking junction J_2 by the increasing anode voltage. The displacement current density changes with increasing anode voltage due to a reduction of the junction capacitance.

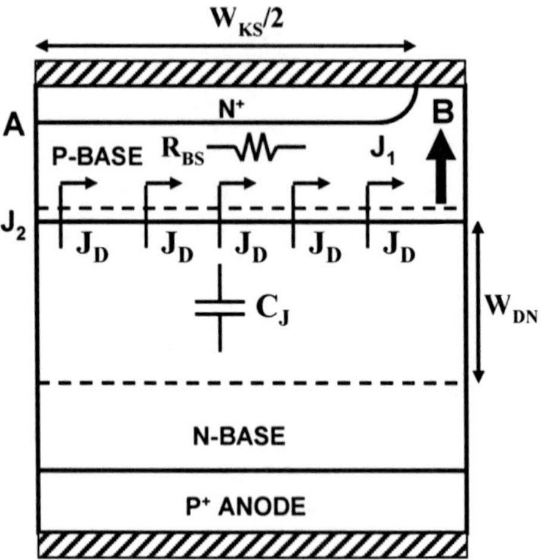

Fig. 8.43 Displacement current flow within the thyristor structure

The current flow pattern illustrated in Fig. 8.43 is identical to that analyzed previously for the leakage current flow in the thyristor (see Fig. 8.17). As long as the thyristor remains in its forward-blocking mode, the displacement current collected across junction J_2 flows into the cathode short (location B in the figure). When the displacement current flows through the resistance (R_{BS}) of the P-base region, it forward biases the junction J_1 at the center of the N^+ cathode (location A in the figure). When the forward bias is equal to the built-in potential for junction J_1, injection of electrons is initiated at location A, which triggers the thyristor into its on-state.

The displacement current that will trigger the thyristor into its on-state can be derived by using the procedure provided in Sect. 8.2.4:

$$J_{D,max} = \frac{8V_{bi}}{\rho_{SB} W_{KS}^2}. \tag{8.75}$$

The displacement current density is related to the junction capacitance:

$$J_{\mathrm{D}} = C_{\mathrm{J}}\left[\frac{\mathrm{d}V}{\mathrm{d}t}\right]. \tag{8.76}$$

Combining the above equations, a relationship for the maximum [dV/dt] capability is obtained:

$$\left[\frac{\mathrm{d}V}{\mathrm{d}t}\right]_{\max} = \frac{8V_{\mathrm{bi}}}{C_{\mathrm{J}}\rho_{\mathrm{SB}}W_{\mathrm{KS}}^{2}}. \tag{8.77}$$

The junction capacitance is dependent upon the anode voltage:

$$C_{\mathrm{J}} = \frac{\varepsilon_{\mathrm{S}}}{W_{\mathrm{DN}}} = \sqrt{\frac{q\varepsilon_{\mathrm{S}}N_{\mathrm{D}}}{2(V_{\mathrm{A}}+V_{\mathrm{bi}})}} \tag{8.78}$$

under the assumption that most of the applied anode voltage (V_{A}) is supported within a depletion region of width W_{DN} in the N-base region. Using this expression in (8.77),

$$\left[\frac{\mathrm{d}V}{\mathrm{d}t}\right]_{\max} = \frac{8V_{\mathrm{bi}}}{\rho_{\mathrm{SB}}W_{\mathrm{KS}}^{2}}\sqrt{\frac{2(V_{\mathrm{A}}+V_{\mathrm{bi}})}{q\varepsilon_{\mathrm{S}}N_{\mathrm{D}}}}. \tag{8.79}$$

Based upon this equation, it can be concluded that the [dV/dt] capability can be improved by reducing the sheet resistance of the P-base region and making the distance between the cathode shorts smaller. This is in conflict with obtaining low holding and gate drive currents. Note that the [dV/dt] capability degrades as the anode voltage is reduced, as well as when the device temperature increases (due to a reduction of the built-in potential and an increase in the sheet resistance of the P-base region).

In the case of phase control operation as illustrated in Fig. 8.6, the anode voltage is given by

$$v(t) = V_{\mathrm{M}}\sin(2\pi f t), \tag{8.80}$$

where V_{M} is the maximum value for the sinusoidal voltage applied to the anode. The maximum rate of change in the sinusoidal voltage, obtained by taking the first derivative of the voltage waveform, occurs at $t = 0$:

$$\left[\frac{\mathrm{d}V}{\mathrm{d}t}\right]_{\max} = 2\pi f V_{\mathrm{M}}. \tag{8.81}$$

The maximum frequency of operation before the loss of gate control is then given by

$$f_{\max} = \frac{[\mathrm{d}V/\mathrm{d}t]_{\max}}{2\pi V_{\mathrm{M}}}. \tag{8.82}$$

Combining this relationship with (8.79), while recognizing that the maximum rate of change in the anode voltage occurs when the anode voltage crosses zero, yields

$$f_{\max} = \frac{1}{\pi V_{\mathrm{M}} \rho_{\mathrm{SB}} W_{\mathrm{KS}}^2} \sqrt{\frac{32 V_{\mathrm{bi}}^3}{q \varepsilon_{\mathrm{S}} N_{\mathrm{D}}}}. \tag{8.83}$$

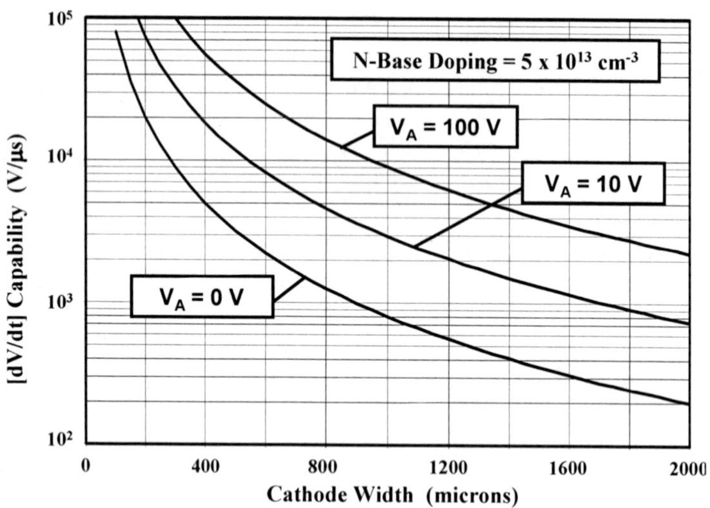

Fig. 8.44 [dV/dt] capability for a typical thyristor structure

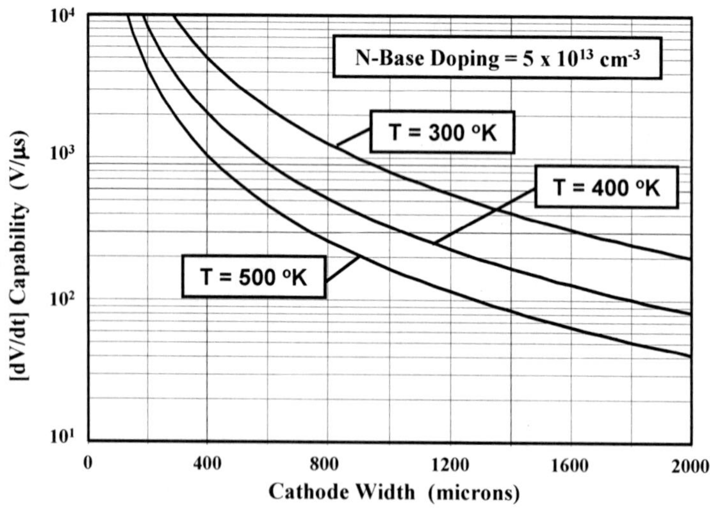

Fig. 8.45 [dV/dt] capability for a typical thyristor structure

The [dV/dt] capability computed by using (8.79), for a thyristor with a P-base sheet resistance of 350 Ω sq^{-1} and an N-base doping concentration of 5 × 10^{13} cm^{-3}, is plotted in Fig. 8.44 for three cases of the anode voltage. As expected, the lowest [dV/dt] capability occurs when the anode voltage is assumed to be zero. For the case of a cathode width of 1,000 μm, the [dV/dt] capability predicted by the analytical solution is about 1,000 V μs^{-1}. This is a typical value for high-voltage thyristor designs.

The [dV/dt] capability for thyristors degrades with increasing temperature as discussed earlier. As an example, the [dV/dt] capability computed by using (8.79), for a thyristor with a P-base sheet resistance of 350 Ω sq^{-1} at 300 K and an N-base doping concentration of 5 × 10^{13} cm^{-3}, is plotted in Fig. 8.45 at three temperatures. For the case of a cathode width of 1,000 μm, the [dV/dt] capability predicted by the analytical solution is reduced from about 1,000 V μs^{-1} at room temperature to 150 V μs^{-1} at 500 K. It is therefore important to ensure adequate [dV/dt] capability for thyristors at elevated temperatures after taking into account self-heating during operation.

The maximum operating frequency, limited by the [dV/dt] capability of the thyristor, computed by using (8.83) is shown in Fig. 8.46 for the case of three doping concentrations in the N-base region. For the case of an N$^+$ cathode width of 1,000 μm, the maximum operating frequency is indicated to be in excess of 20 kHz. However, thyristors are not capable of operating at such high frequencies because of the relatively slow reverse recovery process. The reverse recovery process was illustrated in Fig. 8.6 for a thyristor operating at 60 Hz. In this case, the reverse recovery transient is completed during the second half-cycle well before the anode voltage reverts to a positive value. The stored charge produced during on-state operation can then be assumed to be completely eliminated before the application of the high [dV/dt] at the end of the period.

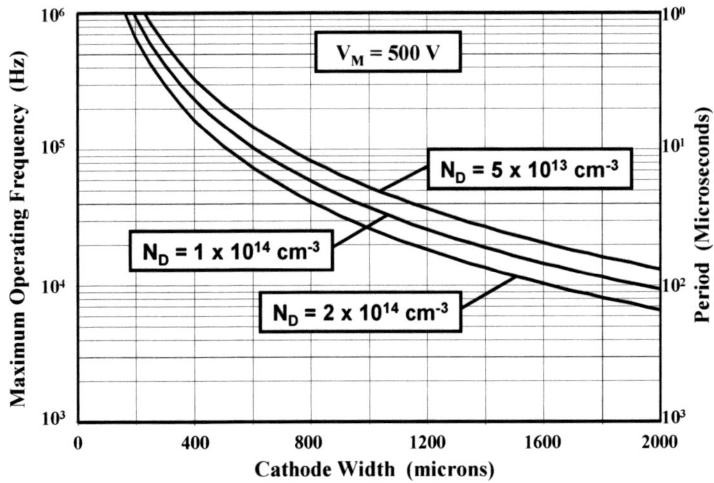

Fig. 8.46 Maximum operating frequency for a typical thyristor structure

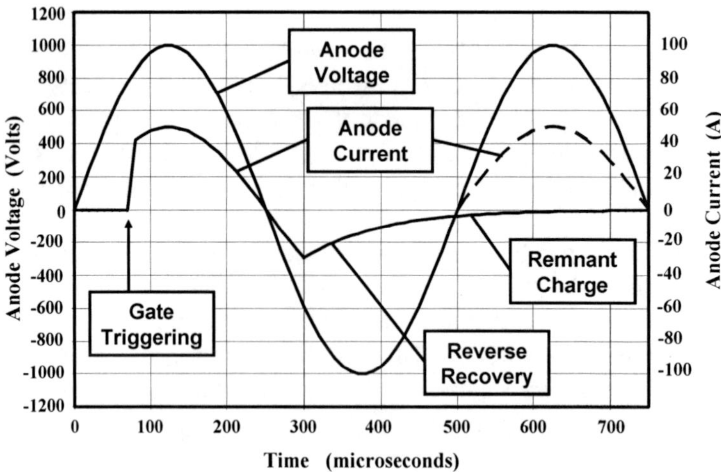

Fig. 8.47 Reverse recovery during high frequency operation of a thyristor structure

The impact of the reverse recovery process on the [dV/dt] capability is illustrated in Fig. 8.47 for the case of operation at 2 kHz. The minority carrier lifetime used to generate the current waveform is 100 μs, which is typical for a high-voltage thyristor designed with low on-state voltage drop. It can be observed that the reverse recovery transient is not completed at the end of the first period (t = 500 μs). Consequently, a substantial amount of stored charge is still remnant in the N-base region when the anode voltage becomes positive with a high [dV/dt] at the start of the second period. A large current is generated in the P-base region as a result of the extraction of the stored charge via junction J_2. This current is sufficient to trigger the thyristor into its on-state as illustrated in the figure by the dashed line.

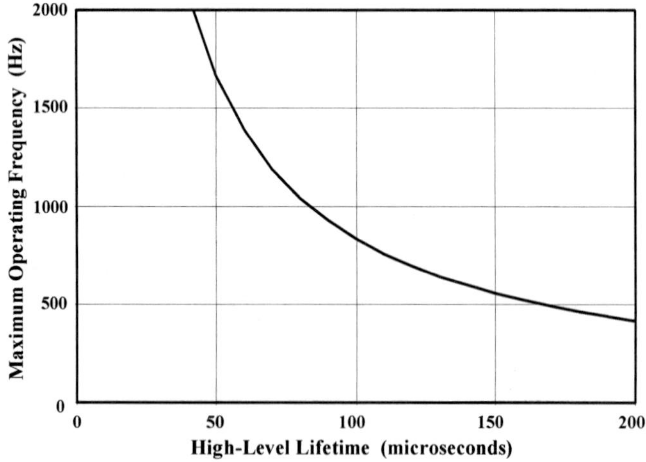

Fig. 8.48 Maximum operating frequency for a typical thyristor structure limited by the reverse recovery process

The reverse recovery process typically occurs over a time interval of three times the high-level recombination lifetime (τ_{HL}). To prevent [dV/dt]-induced turn-on of the thyristor, it is desirable for the reverse recovery transient to be completed within one quarter of the time period. Using these criteria, the maximum operating frequency is given by

$$f_{max} = \frac{1}{12\tau_{HL}}. \tag{8.84}$$

The maximum operating frequency limited by the reverse recovery process is plotted in Fig. 8.48. Thyristors designed for operation at above 2,000 V require a high-level lifetime of 100 μs. This limits their operating frequency to below 1 kHz.

Simulation Example

To gain further insight into the physics of turn-on for the power thyristor due to a rapid rate of change in the anode voltage, the results of two-dimensional numerical simulations for the typical structure (described in the previous sections) are discussed here. The total width of the structure used for the numerical simulations, which is twice the width of the cross section in Fig. 8.43, was 1,000 μm (area = 1 × 10^{-5} cm^{-2}) with a cathode finger width (W_{KS}) of 980 μm between the cathode shorts.

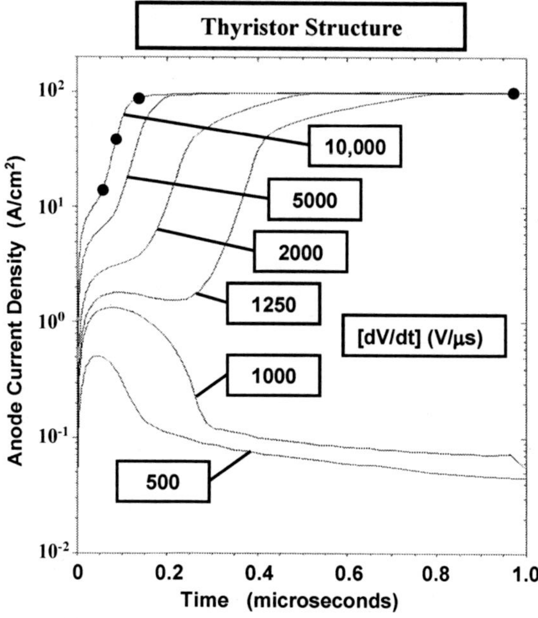

Fig. 8.49 Turn-on characteristics for the thyristor structure with various [dV/dt] applied to the anode electrode

The anode voltage was changed from 0 to 1,000 V at various time intervals with a load resistance in series with the thyristor structure. The turn-on characteristics for the thyristor structure are shown in Fig. 8.49. It can be observed that the thyristor does not turn on for the cases of [d*V*/d*t*] up to 1,000 V μs^{-1}. An initial anode current flow is observed due to the displacement current but this is insufficient to trigger the thyristor into the on-state. However, when the [d*V*/d*t*] exceeds 1,000 V μs^{-1}, the thyristor turns on and operates in the on-state with a current density of 100 A cm^{-2}. The prediction of the analytical model (see Fig. 8.44 for a cathode width of 980 μm) is consistent with the observed maximum [d*V*/d*t*] capability of 1,000 V μs^{-1} obtained in the numerical simulations.

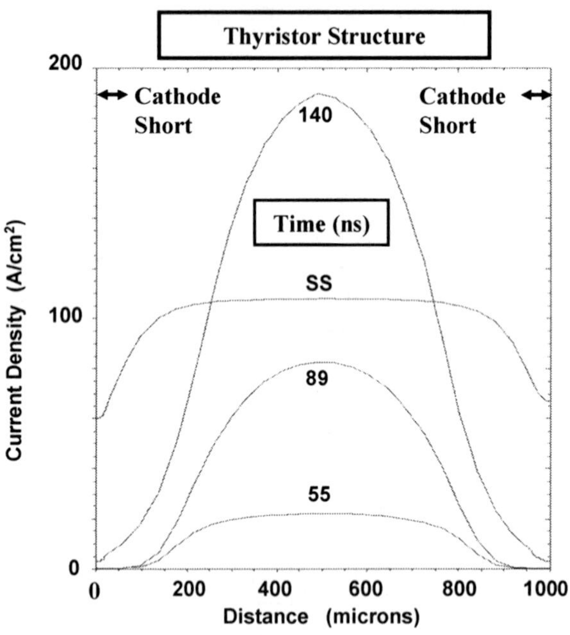

Fig. 8.50 Current distribution within the thyristor structure during turn-on due to the applied [d*V*/d*t*]

The current distribution within the thyristor structure is shown in Fig. 8.50 when it is turned on by the applied [d*V*/d*t*] to the anode electrode. The time instances corresponding to the plots are indicated in Fig. 8.49 by the dots. At $t = 55$ ns, the current flow is due to the displacement current. Subsequently, the thyristor turns on at the middle of the cathode as assumed in the analytical model. This produces an increase in the current density at the middle of the cathode as shown by the cases of $t = 89$ ns and $t = 140$ ns. Eventually, the entire cathode carries the current with a reduced current density at the cathode shorts. The results observed from the numerical simulations provide confirmation of the approach used to develop the analytical solution for the [d*V*/d*t*] capability for thyristors.

8.4.5 Turn-Off Process

Thyristors are most commonly used in phase control circuits to regulate the power delivered from a sinusoidal AC voltage source to the load. The typical waveforms for the voltage and current in a phase control circuit were previously shown in Fig. 8.6. In this figure, it can be observed that the anode current switches off during the second half of the cycle when the anode voltage changes from positive to negative polarity. When the anode voltage crosses zero, the anode current continues to flow in the reverse direction with an approximately constant rate of change or constant [dI/dt] until the P^+ anode/N-base junction is able to support voltage. The anode current decreases to zero in an exponential manner once this junction becomes reverse biased. This behavior is similar to the reverse recovery process discussed in Chap. 5 for the P-i-N power rectifier. Analysis of the turn-off time for the thyristor can therefore be performed by using the analysis described for the P-i-N rectifier.

Simulation Example

To gain further insight into the physics of turn-off for the power thyristor under a constant [dI/dt], the results of two-dimensional numerical simulations for the typical structure (described in the previous sections) are discussed here. The total width of the structure used for the numerical simulations was 1,000 μm (area = 1×10^{-5} cm^{-2}) with a cathode finger width of 980 μm between the gate contact and cathode short. The thyristor was switched from on-state operation at a current density of 100 A cm^{-2} with a constant [dI/dt] (rate of 150 A cm^{-2} in 10 μs). This reverse ramp rate was applied until the anode voltage reached −1,000 V. The anode voltage was then held constant allowing the anode current to decay to zero.

The anode current and voltage waveforms for the thyristor are shown in Fig. 8.51. These waveforms are very similar to those exhibited by the P-i-N rectifier (see Chap. 5 for the nonpunch-through case). The anode current flows in the reverse direction until the anode voltage becomes equal to the reverse bias voltage and then decreases exponentially to zero. These waveforms confirm that the thyristor structure can be analyzed by using the reverse recovery analysis previously discussed in Chap. 5 for the P-i-N rectifier. In the case of the baseline simulation structure with a lifetime (τ_{p0}, τ_{n0}) of 10 μs, the reverse recovery process requires a total time interval of about 20 μs.

The free carrier distribution within the thyristor structure provides further insight into the turn-off process. This information is shown in Fig. 8.52 at various points in time during the turn-off transient, which are indicated by the dots in Fig. 8.51. The initial on-state carrier distribution has the expected catenary shape. At a time t = 6.6 μs, the carrier distribution is relatively flat because the anode current is crossing zero at this time (time t_1 in Fig. 8.51). The carrier concentration at the P^+ anode/N-base junction becomes equal to zero at t = 9 μs. After this point in time (t_2), the thyristor begins to support a reverse voltage as observed in Fig. 8.51. It can be observed that a space-charge region forms at the P^+ anode/N-base

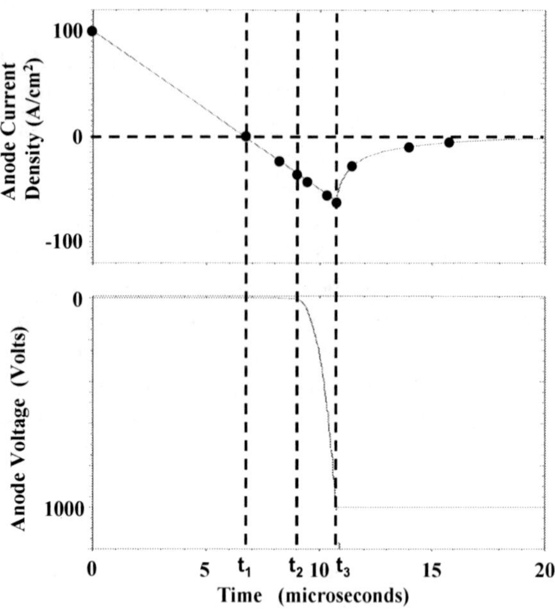

Fig. 8.51 Turn-off characteristics for the thyristor structure with constant [d*I*/d*t*] applied to the anode electrode

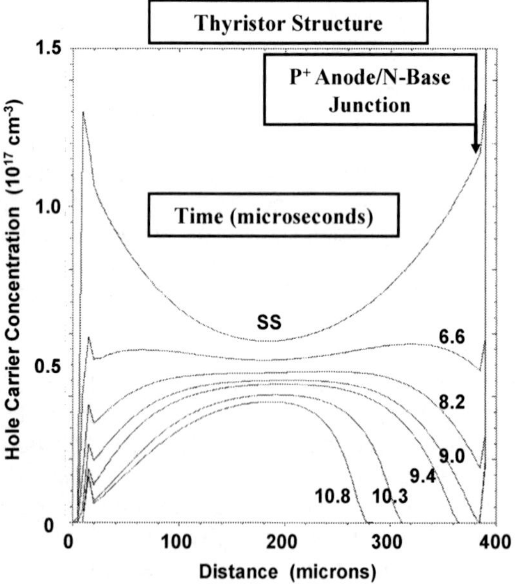

Fig. 8.52 Free carrier distribution within the thyristor structure with constant [d*I*/d*t*] applied to the anode electrode

junction allowing the thyristor to support an increasing voltage with time. At time $t = 10.8$ μs, the thyristor voltage reaches the supply voltage of 1,000 V as observed in Fig. 8.51.

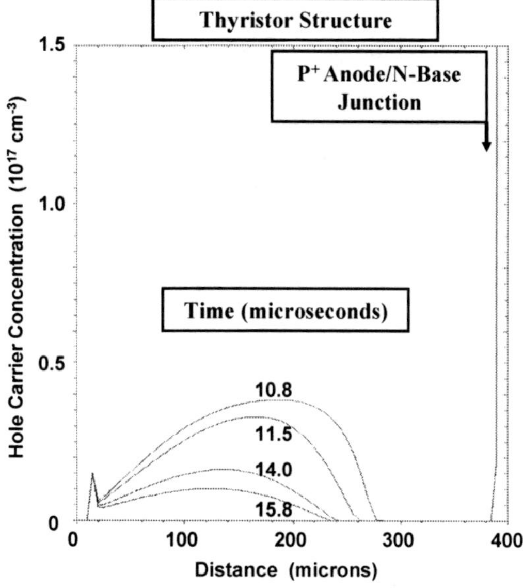

Fig. 8.53 Free carrier distribution within the thyristor structure with constant [dI/dt] applied to the anode electrode

After the anode voltage reaches the supply voltage, the remaining free carriers are removed by a combination of the recombination process and the reverse current flow. This is shown in Fig. 8.53. Note that the depletion region expands slightly during this time interval. The space-charge region width is determined by the sum of the donor charge and the holes transported through the drift region. The hole density in the space-charge region decreases as the anode current decays to zero. The resulting reduction of the charge in the drift region produces a small expansion of its width with time.

8.5 Light-Activated Thyristors

In high power transmission and distribution systems, the system operating voltages are extremely large (typically 100,000 V). The maximum blocking voltage capability for silicon thyristors falls far below these voltage levels. Consequently, it is necessary to connect a large number of thyristors in series to hold off the operating voltage. A technical difficulty encountered with this arrangement is the large potential difference between the gate electrodes for the devices. Until the 1970s, the only available approach was to use transformers to deliver the gate control signal to each thyristor with a local power source to supply the gate turn-on current. More elegant approaches began with the implementation of indirect optical triggering of thyristors by using optical fibers to deliver the signal to local power sources for turning on the device.[12] Subsequently, light-activated thyristor structures were developed that enabled their turn-on directly by the optical signal delivered from the fiber optical cable.[11,13]

The basic concept for the light-activated thyristor structure is based upon the generation of the gate-triggering current within the structure by creating hole–electron pairs with an optical source. For a silicon thyristor, the available GaAs/GaAlAs heterojunction lasers are appropriate because their emission wavelengths (820–900 nm) allow the radiation to penetrate to depths of 10–50 μm into the silicon structure. The hole–electron pairs can then be generated within the depletion region of the reverse-biased P-base/N-base junction during the forward-blocking mode of operation. The lasers also provide sufficiently high power levels for efficient coupling to optical fibers and have short optical delay times.

8.5.1 [d*I*/d*t*] Capability

The main difficulty encountered with the development of light-activated thyristors has been to achieve the triggering of the thyristor with the small power levels generated by the optical signal while retaining a good [d*I*/d*t*] and [d*V*/d*t*] capability. Using (8.61) for the increase in the anode current during the turn-on transient, the rate of change in the anode current is given by

$$\frac{\mathrm{d}J_\mathrm{A}}{\mathrm{d}t} = \frac{J_\mathrm{G}}{\alpha_\mathrm{PNP}\sqrt{t_\mathrm{t,NPN}t_\mathrm{t,PNP}}}\,\mathrm{e}^{t/\sqrt{t_\mathrm{t,NPN}t_\mathrm{t,PNP}}}. \tag{8.85}$$

Based upon this expression, it can be concluded that the [d*I*/d*t*] capability will be compromised if the gate current density is reduced. Fortunately, the direct optical triggering of thyristors can be integrated with the amplifying gate concept to achieve an acceptable [d*I*/d*t*] capability.

A top view and cross section of a light-activated structure containing an amplifying gate region are illustrated in Fig. 8.54. Unlike the conventional thyristor structure with amplifying gate region, the light-activated structure has a window in the metallization for the central gate region. The gate metal is used to form a cathode short across its N^+ cathode/P-base junction on the outer periphery. An optical fiber is brought through the package and located above the window in the gate metallization in the center of the device. The optical gate signal creates a photon flux (ϕ_0), indicated by the downward vertical arrow in the figure, within the gate region. The photon flux generates hole–electron pairs within the semiconductor which creates a current density (J_O) within the gate region. This current flows to the cathode short formed in the main cathode region as indicated by the dotted line. The optically generated current produces a forward bias across the N^+ cathode/P-base junction of the gate region at its center (point *A* in the figure). With appropriate gate design, the forward bias voltage can be made sufficiently high to initiate the injection of electrons from the N^+ cathode into the P-base region at point *A* causing latch-up of the thyristor formed in the gate region. The current (I_G) in this thyristor flows, as shown by the dashed line in the figure, to the cathode short formed in the main cathode region. This current turns on the amplifying gate thyristor which then turns on the main thyristor. With appropriate design of the amplifying gate structure, the light-activated thyristor can be turned on with a high rate of increase [d*I*/d*t*] of the anode current.

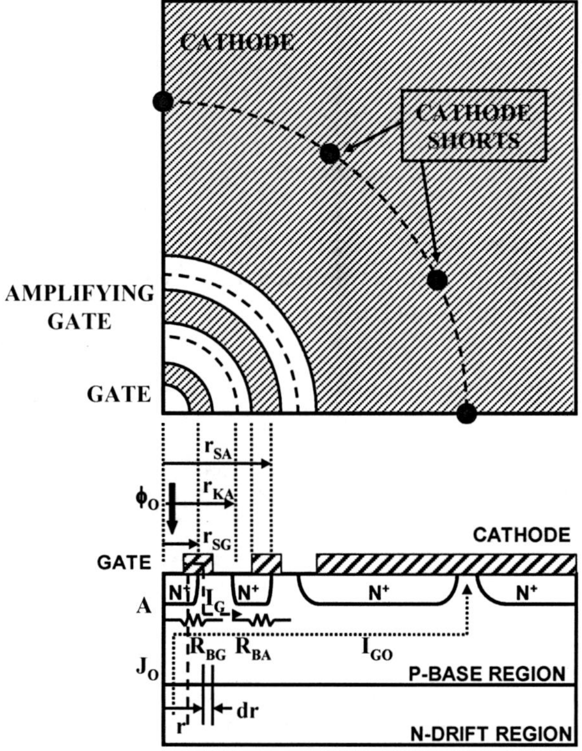

Fig. 8.54 Light-activated thyristor structure

8.5.2 Gate Region Design

The design of the light-activated thyristor requires that the photon-induced current is sufficient to turn on the thyristor in the gate region but not the thyristor within the amplifying gate structure.[11] The optically induced gate current flows through the P-base region after it is collected across the P-base/N-base junction. The lateral current flowing through the P-base region at a radius *r* from the center is given by

$$I_{PB}(r) = \pi r^2 J_O, \tag{8.86}$$

where J_O is the optically induced current density being collected by the P-base/N-base junction. The voltage drop produced across a narrow segment dr by this current is given by

$$\mathrm{d}V_{PB}(r) = I_{PB}(r)\mathrm{d}R(r). \tag{8.87}$$

The resistance of the segment d$R(r)$ is given by

$$\mathrm{d}R(r) = \frac{\rho_{S,PB}}{2\pi r}\mathrm{d}r, \tag{8.88}$$

where $\rho_{S,PB}$ is the sheet resistance of the P-base region under the N$^+$ cathode. Combining the above equations,

$$\mathrm{d}V_{PB}(r) = \frac{J_O\rho_{S,PB}}{2} r\,\mathrm{d}r. \tag{8.89}$$

The forward bias across the N$^+$ cathode/P-base junction at point A due to the optically induced gate current can be obtained by integration of the above voltage to the location (r_{SG}) of the cathode short in the gate region:

$$V_A = \int_0^{r_{SG}} \frac{J_O\rho_{S,PB}}{2} r\,\mathrm{d}r = \frac{J_O\rho_{S,PB}r_{SG}^2}{4}. \tag{8.90}$$

The thyristor within the gate region gets triggered into its on-state when the voltage drop at point A becomes equal to the built-in junction potential. Using this criterion in the above equation,

$$r_{SG} = \sqrt{\frac{4V_{bi}}{J_O\rho_{S,PB}}}. \tag{8.91}$$

For a typical high-voltage thyristor with a sheet resistance of 500 Ω sq^{-1} for the P-base region under the cathode, a design with radius (r_{SG}) of 0.25 cm will allow triggering the thyristor in the gate region if the optically induced current density is assumed to be 0.1 A cm^{-2}. This level of optically induced current can be derived from a light pulse provided by a laser and delivered using an optical fiber as discussed below.

8.5.3 Optically Generated Current Density

The optically induced current density can be obtained based upon the analysis of the gate region with a uniform incident photon flux per unit area (ϕ_0) as shown in Fig. 8.55. In the analysis, it will be assumed that the surface of the silicon has a thin antireflection coating such that the reflection coefficient is reduced to zero. If the incident photons have energy ($h\nu$) that exceeds the band gap for silicon (1.1 eV), they will create hole–electron pairs in the semiconductor. The hole–electron pair generation rate is given by[14]

$$G(y) = \phi_0 \alpha \mathrm{e}^{-\alpha y}, \tag{8.92}$$

where α is the absorption coefficient at the wavelength of the optical signal. An exponential decay in the generation rate occurs with the depth in the silicon due to the absorption of photons as shown in the lower portion of the figure. The absorption coefficient (α) for silicon at a typical wavelength of 0.85 μm for the GaAs/GaAlAs heterojunction lasers is 800 cm^{-1}. This corresponds to a characteristic decay length ($1/\alpha$) of 12.5 μm for the generation rate.

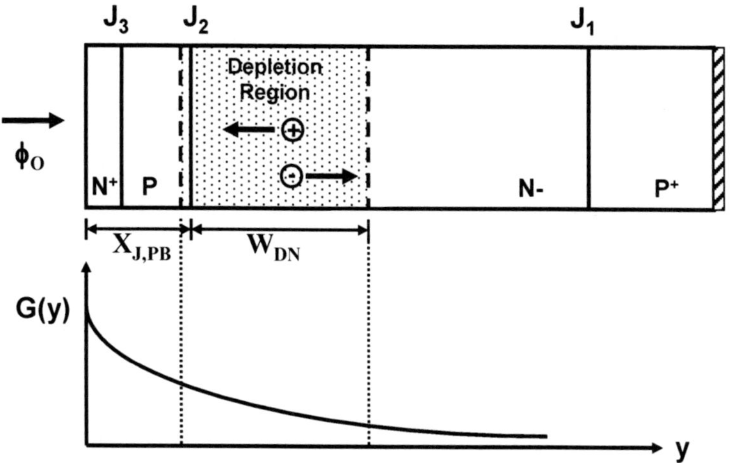

Fig. 8.55 Optically induced current in the gate region of the light-activated thyristor structure

The electron–hole pairs that are generated by the optical stimulus are collected by the reverse-biased P-base/N-base junction (J_2) due to both the drift process within the depletion region and the diffusion process in the neutral regions adjacent to the depletion region. The depletion region width (W_{DN}) for light-activated thyristors is large because they are designed to operate at high voltages. Consequently, it can be assumed that most of the optically induced current is created by the electron–hole pairs generated within the depletion region. The current produced by the optical stimulus is then given by

$$J_{\mathrm{O}} = q \int_{X_{\mathrm{J,PB}}}^{W_{\mathrm{DN}}+X_{\mathrm{J,PB}}} G(y)\mathrm{d}y. \tag{8.93}$$

The depletion region width within the P-base region has been neglected when writing this equation. Performing the integration after using (8.92) yields

$$J_{\rm O} = q\phi_0 e^{-\alpha X_{\rm J,PB}}(1 - e^{-\alpha W_{\rm DN}}). \tag{8.94}$$

The depletion layer width ($W_{\rm DN}$) is usually much larger than the reciprocal of the absorption coefficient for high-voltage thyristors. Under this assumption,

$$J_{\rm O} = q\phi_0 e^{-\alpha X_{\rm J,PB}}. \tag{8.95}$$

The incident photon flux per unit area can also be related to the incident optical power per unit area (P_0) delivered via the optical fiber:

$$\phi_0 = \frac{P_0 \lambda(\mu\text{m})}{1.24}, \tag{8.96}$$

where the wavelength (λ) of the optical signal is in microns. In the case of a typical wavelength of 0.85 μm for the GaAs/GaAlAs heterojunction lasers, the incident photon flux density is equal to 0.685 times the incident optical power density. For a typical thyristor structure with a P-base junction depth of 25 μm operated with a GaAs/GaAlAs heterojunction laser, the optically generated current density ($J_{\rm O}$) is calculated to be about 0.1 A cm^{-2} by using these equations. This value was therefore utilized when designing the gate region of the light-activated thyristor.

8.5.4 Amplifying Gate Design

The above design of the light-activated thyristor is based upon the assumption that the optically induced current is sufficiently large to turn on the thyristor incorporated within the gate region but not the thyristor located within the amplifying gate region. The entire optically induced current generated in the gate region flows through the P-base region of the amplifying gate structure as shown by the dotted line in Fig. 8.54. This current flow produces a forward bias ($V_{\rm AG}$) across the N$^+$ cathode/P-base junction of the thyristor within the amplifying gate structure. It is necessary to ensure that this current flow will not directly trigger the amplifying gate structure because the available optically induced gate current is small resulting in a low [dI/dt] capability. It is preferable to trigger the amplifying gate structure with the current derived from triggering the thyristor located within the gate region.

The optically induced current flowing through the P-base region of the amplifying gate structure is given by

$$I_{\rm GO} = \pi r_{\rm SG}^2 J_{\rm O}. \tag{8.97}$$

When this current flows through the resistance ($R_{\rm BA}$) of the P-base region (see (8.71)), it produces a forward bias across the N$^+$ cathode/P-base junction of the thyristor within the amplifying gate structure given by

$$V_{AG} = I_{GO} R_{BA} = \frac{J_O \rho_{S,PB}}{2} r_{SG}^2 \ln\left(\frac{r_{SA}}{r_{KA}}\right). \tag{8.98}$$

For the typical design for the amplifying gate structure in Sect. 8.4.3 with a ratio (r_{SA}/r_{KA}) of 1.11 and using an optically induced current density of 0.1 A cm^{-2} with a P-base sheet resistance of 500 Ω sq^{-1}, a forward bias of less than 0.2 V is produced for the case of a gate region with a radius (r_{SG}) of 0.25 cm. This amplifying gate design is therefore compatible with the gate structure for optical triggering of the thyristor.

8.6 Self-Protected Thyristors

Thyristors are used in power circuits with large available energy that is usually delivered to the load. Under adverse circuit operating conditions, the thyristor may be subjected to destructive failure modes. One such adverse operating condition occurs when the anode voltage for the thyristor becomes much larger than the nominal supply voltage due to switching high anode current through stray inductances in series with the device. Even though this event may be short-lived, the voltage can exceed the breakdown voltage of the thyristor leading to a high local current at its edge termination. This can produce sufficient local power dissipation to produce destructive failure of the thyristor. It is therefore preferable to build in a structure within the thyristor during its design such that the device turns on in the gate area under these transient high-voltage conditions.[15] The thyristor recovers to its normal mode of operation under gate control after one AC cycle, allowing operation without destructive failure.

Another failure mode identified for thyristors is turn-on by a rapid change in the anode voltage above its rated [dV/dt] capability. Under these conditions, it is again preferable that the thyristor gets turned on in the gate area rather than at a random location under the cathode electrode. Turn-on in the gate area allows utilization of the amplifying gate structure to spread the turn-on area to reduce local power dissipation and prevent destructive failure.

8.6.1 Forward Breakdown Protection

The protection of a thyristor structure against failure when the anode voltage exceeds its rated blocking voltage capability is sometimes referred to as the *forward breakover protection*. The term breakover refers to the reduction of the anode voltage with increasing anode current when the thyristor latches up. The basic approach used to achieve this function is to ensure that the thyristor will turn on in the gate region when the anode voltage exceeds its rated forward-blocking capability instead of at a random location at its edge termination.

A thyristor structure that provides forward breakover protection is shown in Fig. 8.56. This structure is similar to the thyristor structure with an amplifying gate region with the gate region altered to recess the gate contact and incorporate a P^+ region under the contact. The depth of the trench used to recess the gate contact and the depth of the P^+ diffusion are chosen such that the doping concentration of the P-base region is enhanced under the gate contact in the vicinity of the P-base/N-base junction as illustrated in the figure. The larger doping concentration enhances the electric field locally under the gate region when the thyristor is in its forward-blocking mode as illustrated in the figure on the right-hand side. The enhanced electric field ($E_{M,A}$) at location A produces impact ionization at a higher level than for the rest of the blocking junction. The breakdown-induced local current (I_{BV}) flows to the cathode short under the main cathode electrode as shown by the dotted line in the figure under the assumption that the gate electrode is open circuited by the external gate drive circuit. Since this current flows through the resistance R_{BA}, it acts as the gate drive current for turning on the amplifying gate thyristor. When the amplifying gate thyristor turns on, its current (I_{AG}) flows through the resistance R_{BM} turning on the main thyristor. Consequently, the thyristor design shown in the figure ensures turning on the device, for anode voltages above its rated forward-blocking voltage capability, at the gate electrode in the same manner as turning on the thyristor with the application of a gate drive signal. This ensures turn-on without destructive failure.

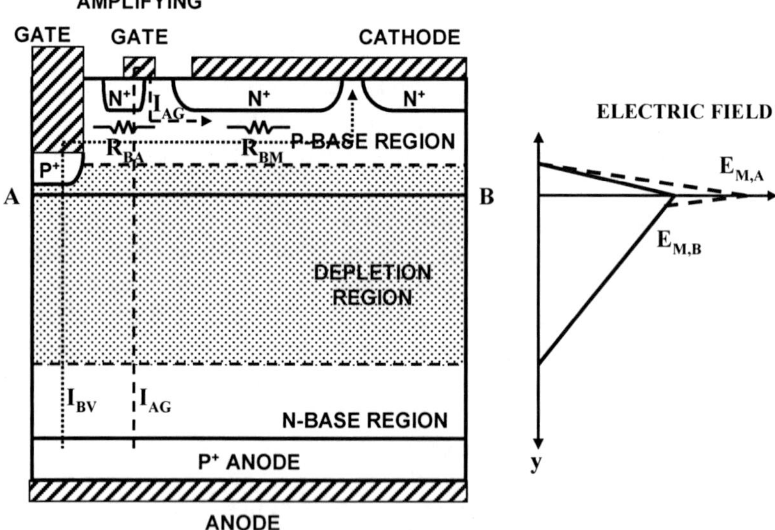

Fig. 8.56 Protection of the thyristor structure against anode voltages above the blocking voltage rating

In order for the above self-protection scheme to work, it is necessary to make sure that the breakdown will occur in the gate region. This can be achieved through the proper choice of the depth of the trench etched in the gate region and the depth of the P^+ region under the gate contact. The relatively large dimensions for the junction depth of the P-base region and the depletion region extension in the P-base region provide latitude while processing the trench in the gate region.

Simulation Example

To gain insight into the design of the self-protected thyristor structure, the results of two-dimensional numerical simulations for a typical structure are discussed here. The device structure used was the same as the baseline thyristor structure discussed in the previous sections with the addition of a trench and P^+ region under the gate contact. The depth of the P^+ diffusion was maintained at 5 μm and the depth of the trench in the gate region was varied. The total width of the structure used for the numerical simulations was 1,000 μm (area = 1×10^{-5} cm^{-2}) with a cathode finger width of 980 μm between the gate contact and cathode short.

The breakover phenomenon was initiated by increasing the anode voltage while maintaining an open circuit at the gate terminal. The resulting *i–v* characteristics are shown in Fig. 8.57 for various cases of the depth of the trench under the gate electrode. The breakdown behavior for the baseline thyristor structure without the self-protection feature is included in the figure for comparison. It can be observed that the breakdown voltage is reduced in the presence of the self-protection region. With a trench depth of 20 μm, the breakdown voltage is reduced to 2,000 V in comparison with 2,300 V for the unprotected structure. The breakdown-induced current can be used to trigger the thyristor by incorporation of the amplifying gate structure. When the trench depth is increased to 25 μm, the breakdown voltage is

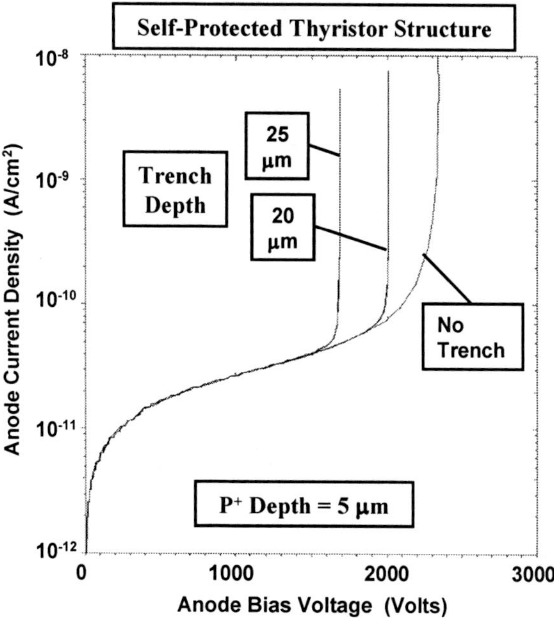

Fig. 8.57 Breakdown voltages for thyristor structures with self-protection in gate region

further reduced to 1,650 V. Consequently, the depth of the trench under the gate contact can be adjusted to obtain the desired breakdown voltage with the breakdown current being located under the gate contact.

The enhancement in the electric field under the gate contact of the self-protected thyristor structure can be observed in Fig. 8.58. The three-dimensional view of the electric field distribution in the vicinity of the gate contact is taken at an anode bias of 1,500 V. It can be observed that the electric field under the gate contact is enhanced by a factor of about 1.5 times when compared with the electric field at the P-base/N-base junction under the cathode region of the main thyristor. The electric field enhancement occurs due to the introduction of a high doping concentration on the P-side of the junction. This localized increase in the electric field is sufficient to reduce the breakdown voltage under the gate contact. The current flow induced by the breakdown under the gate contact flows through the P-base region to the cathode short located at x = 1,000 μm. This current can be utilized to turn on the thyristor by incorporation of an amplifying gate structure. A larger enhancement of the electric field can be obtained by increasing the depth of the trench formed under the gate contact because this adds more P-type doping concentration at the junction formed under the gate contact.

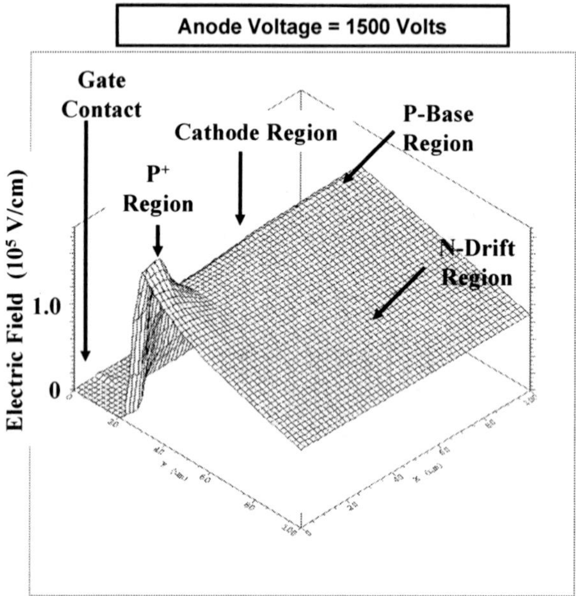

Fig. 8.58 Electric field distribution for a thyristor structure with self-protection in gate region

8.6.2 [dV/dt] Turn-On Protection

The turn-on of the thyristor under a rapid change in the anode voltage was previously discussed in Sect. 8.4.4. The design criterion for the cathode shorting geometry was developed based upon the displacement current produced by the applied [d*V*/d*t*] across the P-base/N-base junction. The [d*V*/d*t*] rating for the thyristor is usually made smaller than the design value for the [d*V*/d*t*] capability as determined by the cathode short design. However, the thyristor may still be subjected to spurious high-voltage spikes that occasionally exceed the [d*V*/d*t*] rating during operation in circuits. Under these conditions, the thyristor may turn on at a random location under the cathode electrode. The high inrush current produced at this location can result in destructive failure of the thyristor. It is preferable that the thyristor structure turns on in the gate region under the spurious voltage spike because the turn-on can be controlled by the amplifying gate region. Such a thyristor design is considered to be self-protected against [d*V*/d*t*]-induced turn-on.

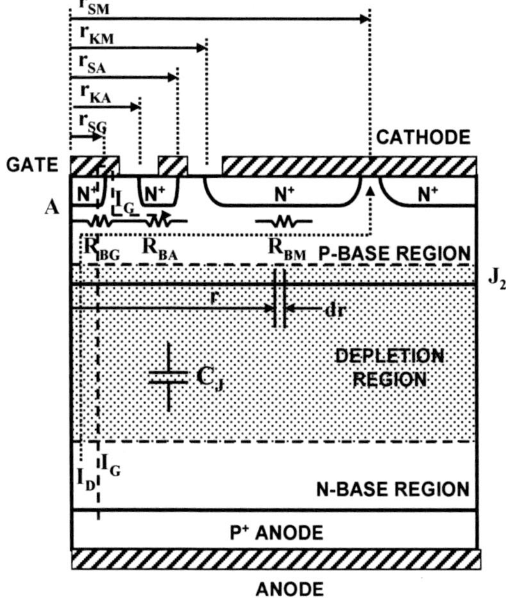

Fig. 8.59 Protection of the thyristor structure against [d*V*/d*t*]-induced turn-on

A thyristor design which is self-protected against [d*V*/d*t*]-induced turn-on is illustrated in Fig. 8.59. This design is similar to that for the amplifying gate thyristor with the inclusion of an N^+ cathode region under the gate electrode. The gate electrode overlaps the junction between this N^+ cathode region and the P-base region. This provides a contact between the gate electrode and the P-base region to allow introduction of the gate current into the P-base region during normal triggering of the thyristor.

A displacement current flows across the entire reverse-biased P-base/N-base junction (J_2) due to the junction capacitance (C_J) when a rapid change in the anode voltage occurs in the positive direction during circuit operation. This displacement current density (J_D) is shown in Fig. 8.43. The displacement current collected by the P-base/N-base junction (J_2) flows to the first row of cathode shorts in the main thyristor as indicated by the dotted line in Fig. 8.59. Since this current flows below the N$^+$ cathode region under the gate electrode, the junction becomes forward biased at the center of the gate region (location A in the figure). As the [dV/dt] increases, the forward bias across the junction under the gate electrode can exceed the built-in potential leading to the injection of electrons at location A, which initiates the turn-on of the thyristor formed under the gate electrode. The current flow in the thyristor formed under the gate contact, indicated by the dashed line (I_G) in the figure, serves as a gate current to trigger the thyristor in the amplifying gate structure.

The design criteria for achieving self-protection of the thyristor can be derived by analysis of the displacement current flowing in the gate region and the rest of the thyristor structure. Consider the case of a thyristor illustrated in Fig. 8.59 with the circular gate geometry. The lateral current flowing in the P-base region due to the collection of the displacement current (J_D) across the P-base/N-base junction at a radius r is given by

$$I_{\mathrm{D,PB}} = \pi r^2 J_{\mathrm{D}}. \tag{8.99}$$

This current produces a voltage drop across the segment dr given by

$$\mathrm{d}V(r) = I_{\mathrm{D,PB}} \mathrm{d}R(r), \tag{8.100}$$

where d$R(r)$ is the resistance of the segment. This resistance can be related to the sheet resistance ($\rho_{\mathrm{S,PB}}$) of the P-base region under the N$^+$ cathode:

$$\mathrm{d}R(r) = \frac{\rho_{\mathrm{S,PB}} \mathrm{d}r}{2\pi r}. \tag{8.101}$$

Combining these relationships,

$$\mathrm{d}V(r) = \frac{1}{2} J_{\mathrm{D}} \rho_{\mathrm{S,PB}} r \, \mathrm{d}r. \tag{8.102}$$

The forward bias across the N$^+$ cathode/P-base junction under the gate electrode is then given by

$$V_{\mathrm{J,G}} = \int_0^{r_{\mathrm{SG}}} \mathrm{d}V(r) = \int_0^{r_{\mathrm{SG}}} \frac{1}{2} J_{\mathrm{D}} \rho_{\mathrm{S,PB}} r \, \mathrm{d}r = \frac{J_{\mathrm{D}} \rho_{\mathrm{S,PB}}}{4} r_{\mathrm{SG}}^2, \tag{8.103}$$

where r_{SG} is the radius of the N$^+$ cathode region formed under the gate contact. The thyristor under the gate region will turn on when the forward bias across the N$^+$ cathode/P-base junction under the gate electrode becomes equal to the built-in potential. The resulting criterion for the design of the gate region is

$$r_{SG} = \sqrt{\frac{4V_{bi}}{J_D \rho_{S,PB}}}. \tag{8.104}$$

Since the displacement current density is given by (8.76),

$$r_{SG} = \sqrt{\frac{4V_{bi}}{\rho_{S,PB} C_J [dV/dt]}}, \tag{8.105}$$

where the junction capacitance (C_J) can be obtained by using (8.78).

For proper operation of the above self-protection scheme, it is necessary to ensure that the displacement current arising from the applied [dV/dt] does not trigger the amplifying gate or the main thyristor structures prior to triggering the thyristor in the gate region. The design criteria for achieving this can be derived by examination of the forward bias voltage developed across the N^+ cathode/P-base junctions for the thyristors in the amplifying gate region and the main thyristor.

Using the above approach for the gate region, the forward bias across the N^+ cathode/P-base junction for the thyristor in the amplifying gate region is given by

$$V_{J,AG} = \int_{r_{KA}}^{r_{SA}} dV(r) = \int_{r_{KA}}^{r_{SA}} \frac{1}{2} J_D \rho_{S,PB} r \, dr = \frac{J_D \rho_{S,PB}}{4} (r_{SA}^2 - r_{KA}^2), \tag{8.106}$$

where r_{KA} and r_{SA} are the inner and outer radii of the N^+ cathode region formed in the amplifying gate structure. Similarly, the forward bias across the N^+ cathode/ P-base junction for the main thyristor is given by

$$V_{J,M} = \int_{r_{KM}}^{r_{SM}} dV(r) = \int_{r_{KM}}^{r_{SM}} \frac{1}{2} J_D \rho_{S,PB} r \, dr = \frac{J_D \rho_{S,PB}}{4} (r_{SM}^2 - r_{KM}^2), \tag{8.107}$$

where r_{KM} and r_{SM} are the inner and outer radii of the N^+ cathode region formed in the main thyristor structure. The triggering of the thyristor under an applied [dV/dt] at the gate region can be ensured by making

$$V_{J,G} > V_{J,AG} > V_{J,M}, \tag{8.108}$$

which can be achieved by making

$$r_{SG}^2 > (r_{SA}^2 - r_{KA}^2) > (r_{SM}^2 - r_{KM}^2). \tag{8.109}$$

The desired [dV/dt] at which the thyristor is triggered into the on-state with self-protection can be achieved by using the above equations for selecting the radius of N^+ cathode region under the gate electrode. From Sect. 8.4.3 for the amplifying gate structure design, the inner and outer radii for the N^+ cathode region of the amplifying gate thyristor were found to be 0.5 and 0.55 cm, respectively, with a gate electrode radius of 0.25 cm. By using a radius of 0.24 cm for the N^+ cathode region under the gate contact, a self-protected [dV/dt] capability of about 100 V μs^{-1}

can be obtained while ensuring that the criterion defined by (8.109) is met by the design. Inner and outer radii for the N^+ cathode region for the main thyristor of 0.6 and 0.7 cm, respectively, will also satisfy the criterion defined by (8.109).

8.7 The Gate Turn-Off Thyristor Structure

As discussed in the previous sections, the thyristor structure contains a set of coupled transistors that provide a regenerative action during the conduction of current in the on-state. These devices are designed for operation in AC circuits where the anode voltage cycles between positive and negative values. The regenerative action is disrupted whenever the anode voltage reverses from positive to negative. The turn-off of the device then occurs with a reverse recovery process to establish blocking voltage capability. Such device structures are not suitable for applications in DC circuits unless expensive commutation circuits[16] are added to reverse the anode voltage polarity.

Until the development of the IGBT, power bipolar transistors were the device of choice for DC circuits. As discussed in Chap. 7, the current gain and on-state characteristics of bipolar power transistors degrade rapidly with increasing voltage rating. This has precluded the development of devices with voltage ratings above 2,000 V that are suitable for traction (electric locomotive) applications. The development of a thyristor structure that can be designed to turn on and turn off the current flow under control by a gate signal in a DC circuit was motivated by this need. Such thyristors have been named gate turn-off (GTO) thyristors. The GTO is turned on in the same manner as the thyristor structures described in the previous sections. The turn-off for the GTO is accomplished by the application of a large reverse gate current. The gate current must be sufficient to remove the stored charge from the P-base region and disrupt the regenerative action of the internal coupled transistors.

In this section, the basic operation for the GTO is first discussed. A simple one-dimensional analysis for turning off the GTO is then developed. A two-dimensional analysis for the removal of the stored charge is included to provide a more accurate computation of the storage time. This analysis allows analytical description of the turn-off waveforms for the anode current and voltage. The maximum controllable current for the GTO is then shown to be limited by the breakdown voltage of the N^+ cathode/P-base junction. It is demonstrated that the GTO structure exhibits a long tail in the anode current that can lead to high power losses during turn-off. An anode-shorted structure is described that can reduce this switching power loss.

8.7.1 Basic Structure and Operation

The basic structure for the GTO thyristor is illustrated in Fig. 8.60. Although similar to the conventional thyristor structure, it is important to note that the GTO structure does not contain cathode shorts. In addition, the width (W_{KS}) of the cathode region for the GTO structure is made much shorter than for the conventional thyristor to facilitate turning off the anode current. The electric field profile for this structure is shown on the right-hand side of the figure during the forward-blocking mode of operation. The voltage is supported across the P-base/N-base junction (J_2). The forward-blocking capability is determined by the open-base breakdown voltage of the N–P–N transistor as described earlier for the conventional thyristor. The above structure has approximately the same reverse-blocking voltage capability when the voltage is supported by the P$^+$ anode/N-base junction (J_1).

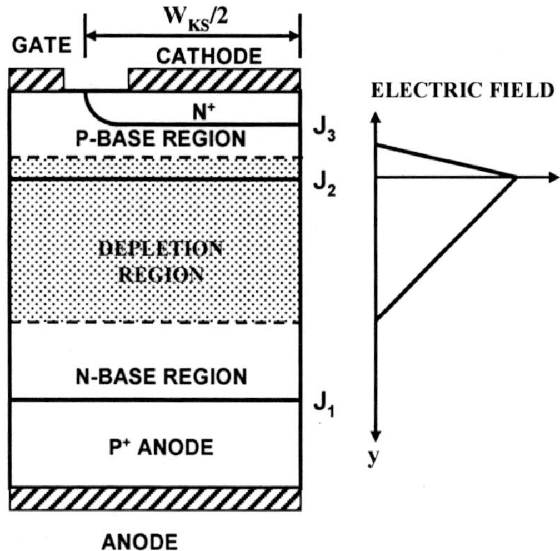

Fig. 8.60 Symmetric gate turn-off thyristor structure

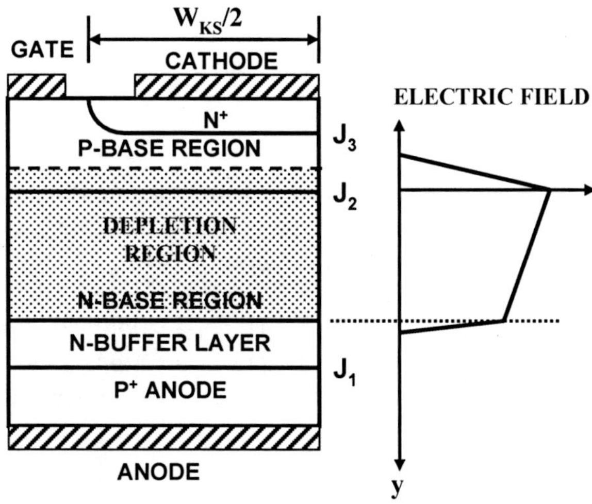

Fig. 8.61 Asymmetric gate turn-off thyristor structure

Since the GTO structure is intended for use in DC circuits, its reverse-blocking capability does not have to match its forward-blocking capability. An asymmetric GTO structure that takes advantage of this is illustrated in Fig. 8.61. Here, an N-buffer layer is added in the N-base region adjacent to the P^+ anode region. The N-buffer layer has a much larger doping concentration than the lightly doped portion of the N-base region. These changes result in a trapezoidal shape for the electric field profile as illustrated on the right-hand side of the figure. The same forward-blocking capability can be achieved for the asymmetric GTO structure with a smaller net thickness for the N-base region than necessary for the symmetric structure. This enables reduction of the on-state voltage drop. The presence of the N-buffer layer also reduces the current gain of the P–N–P transistor, which will be shown to improve the turn-off gain of the GTO. The N-buffer layer is often used in conjunction with an anode short to reduce the turn-off time as discussed later in this section.

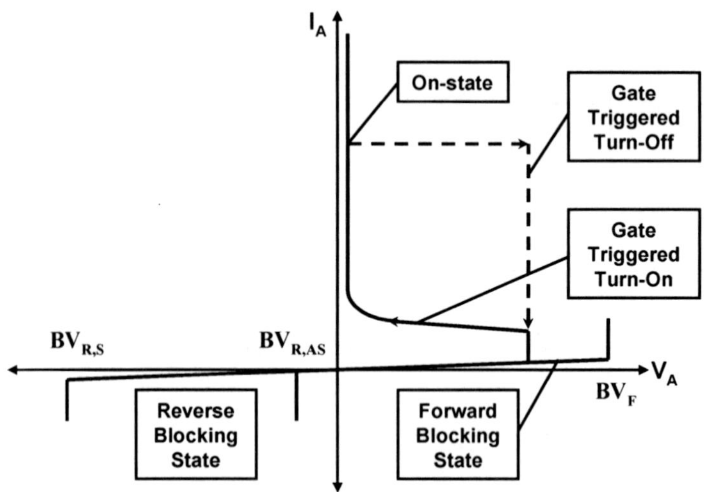

Fig. 8.62 Output characteristics of the GTO structure

The output characteristics for the GTO structure are illustrated in Fig. 8.62. The symmetric structure has a high reverse-blocking capability ($BV_{R,S}$) while the asymmetric structure can support only a relatively small voltage ($BV_{R,AS}$) in the reverse-blocking mode. The GTO can be triggered into the on-state while operating in the forward-blocking mode by the application of a small gate current. Once the device enters its regenerative mode of operation, it can sustain the on-state current flow without any gate drive signal. The device can be turned off without reversing the anode voltage by the application of a large reverse gate current. The *i–v* trajectory during the turn-off transient is shown by the dashed lines for an inductive load.

The GTO structure can be turned off by the application of either a constant reverse gate current (step drive) or a gate current that increases with time in the reverse direction (ramp drive) as shown in the upper part of Fig. 8.63. In both cases, the regenerative action within the thyristor does not cease until a time interval called the *storage time* (t_S) has elapsed. The storage time can be reduced by increasing the reverse gate current for the step drive or the ramp rate for the ramp drive. After the storage time interval, the anode voltage increases to the DC supply voltage. The anode current then exhibits an abrupt decrease followed by a long slow decay of the current. This slow decay of the anode current is referred to as the *current tail*. Due to high anode voltage during the time duration of the current tail, a large switching power loss occurs that limits the frequency of operation of the GTO structure. In addition, considerable power loss is associated with the voltage rise time interval (t_V) during which the anode voltage increases to the supply voltage while the anode current remains at the on-state value for an inductive load.

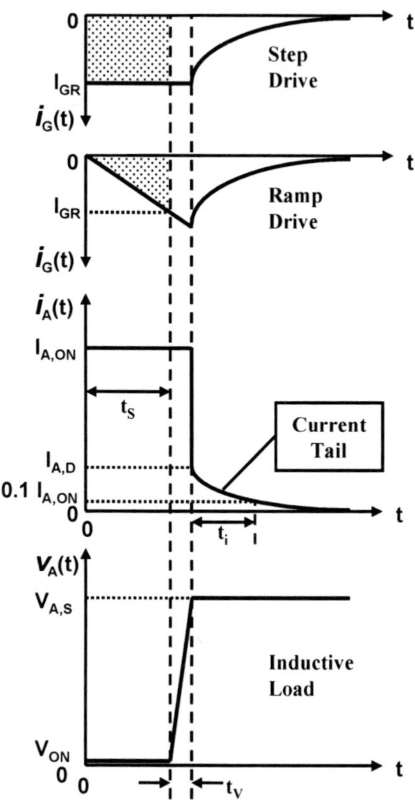

Fig. 8.63 Switching characteristics of the GTO structure

8.7.2 One-Dimensional Turn-Off Criterion

A one-dimensional analysis of the gate current required to turn off the thyristor structure can be developed by taking into consideration the currents flowing at the

P-base/N-base junction (J_2). During on-state operation, the regenerative coupling between the inherent N–P–N and P–N–P transistors within the thyristor structure produces the currents indicated in Fig. 8.64 at the P-base/N-base junction (J_2). The collector current for the N–P–N transistor provides the base drive current for the P–N–P transistor while the collector current for the P–N–P transistor provides the base drive current for the N–P–N transistor. The regenerative action can be disrupted by the application of a reverse gate current that deprives the N–P–N transistor of its base drive current. If sufficient gate drive current is applied, the regenerative action will be interrupted leading to the turn-off of the thyristor structure.

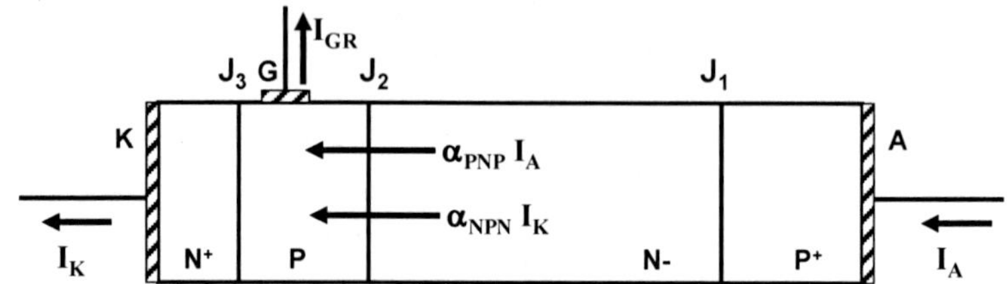

Fig. 8.64 Current components during turn-off for the GTO structure

The base drive current required to maintain current flow in the N–P–N transistor is $[(1 - \alpha_{NPN})I_K]$. The base drive current for the N–P–N transistor supplied by the P–N–P transistor is ($\alpha_{PNP}I_A$) while the base drive current for the N–P–N transistor removed during turn-off is the reverse gate drive current (I_{GR}). To make the net base drive current supplied to the N–P–N transistor less than the base drive current required to maintain current flow, it is necessary to satisfy the condition

$$[(\alpha_{PNP}I_A) - I_{GR}] < [(1 - \alpha_{NPN})I_K]. \tag{8.110}$$

Applying Kirchhoff's current law to the GTO structure,

$$I_A = I_K + I_{GR}. \tag{8.111}$$

The criterion developed for turning off the GTO by combining these relationships is

$$I_{GR} > \left(\frac{\alpha_{NPN} + \alpha_{PNP} - 1}{\alpha_{NPN}} \right) I_A. \tag{8.112}$$

Using a typical value of 0.9 for the common-base current gain for the N–P–N transistor, the gate current must exceed 0.2 times the anode current according to this equation to achieve interruption of the regenerative action within the thyristor, if the common-base current gain for the P–N–P transistor is reduced to 0.28.

The above criterion for turn-off of the GTO structure under gate control can also be expressed by defining a turn-off gain (β):

$$\beta = \frac{I_{\mathrm{A}}}{I_{\mathrm{GR}}}. \tag{8.113}$$

Based upon (8.112), the maximum turn-off gain is given by

$$\beta_{\mathrm{M}} = \left(\frac{\alpha_{\mathrm{NPN}}}{\alpha_{\mathrm{NPN}} + \alpha_{\mathrm{PNP}} - 1} \right). \tag{8.114}$$

A maximum turn-off current gain of 5 can be obtained according to this equation by using a typical value of 0.9 for the common-base current gain for the N–P–N transistor and reducing the current gain of the P–N–P transistor to 0.28. Nonetheless, a large reverse gate current is required to turn off the GTO structure. This increases the complexity and cost of the gate drive circuit for the GTO.

From the above relationships, it can be concluded that a larger turn-off current gain can be achieved by making the current gain of the N–P–N transistor close to unity while reducing the current gain for the P–N–P transistor. Based upon the need to maintain a high-current gain for the N–P–N transistor, cathode shorts are not utilized during the design of the GTO structure. Instead, the width of the cathode between the gate electrodes is made much smaller than for conventional thyristors, and the gate electrode is distributed across the entire wafer. A high [d*V*/d*t*] capability and good high temperature forward-blocking capability are ensured by shorting the gate electrode to the cathode electrode by the external gate drive circuit except when triggering the device between the on- and off-states.

The turn-off current gain can be increased by reducing the current gain for the P–N–P transistor by several methods. In the case of the asymmetric GTO structure, the addition of the highly doped N-type buffer layer reduces the emitter injection efficiency of the P–N–P transistor. A further reduction of the emitter injection efficiency of the P–N–P transistor can be obtained by adding anode shorts as discussed later in this section. Another approach to reducing the current gain for the P–N–P transistor is by reducing the lifetime in the N-base region to decrease the base transport factor. A reduction of the lifetime in the N-base region also facilitates reduction of the switching losses because of a faster decay of the tail current. However, a reduction of the lifetime in the N-base region produces an increase in the on-state voltage drop. It is customary to make a tradeoff between the on-state and switching power losses for the GTO structure as in the case of all bipolar devices.

8.7.3 One-Dimensional Storage Time Analysis

The storage time for the GTO structure can limit its operating frequency. During the storage time interval, the applied reverse gate current extracts the stored charge within the P-base region. Prior to the application of the reverse gate current, the GTO structure is operating in its on-state like a P-i-N rectifier. In the P-i-N rectifier model for the thyristor in its on-state, a large density of free carriers is stored within the P-base and N-base regions (see Sect. 8.3.1). The stored charge can be represented by an average free carrier density (see Sect. 5.1.3) given by

$$n_a = \frac{J_{A,ON}\tau_{HL}}{q(W_P + W_N)}. \tag{8.115}$$

In the simple one-dimensional analysis, the storage time is determined by the reverse gate current removing all the stored charge ($Q_{S,PB}$) within the P-base region. Consequently, for the step drive case,

$$J_{GR}t_S = Q_{S,PB} = qn_aW_P. \tag{8.116}$$

Combining the above relationships yields

$$t_S = \left(\frac{W_P\tau_{HL}}{W_P + W_N}\right)\left(\frac{J_{A,ON}}{J_{GR}}\right) = \left(\frac{W_P\tau_{HL}}{W_P + W_N}\right)\beta_T, \tag{8.117}$$

where β_T is the turn-off gain.

From the above equation developed with the one-dimensional analysis, it can be concluded that the storage time can be reduced by decreasing the high-level time and increasing the gate drive current, i.e., by operating with a lower turn-off current gain. A reduction of the high-level lifetime increases the on-state voltage drop producing greater power dissipation while a larger gate current increases the complexity and cost of the gate drive circuit as well as the gate drive power losses. For a typical GTO structure with P-base and N-base widths of 40 and 300 μm, respectively, with a high-level lifetime of 20 μs, the storage time calculated by using the above equation is 4.7 μs when the device is turned off at a current gain of 2. In the case of the ramp drive case, the storage time becomes twice as large for the same value of the peak reverse gate current (compare shaded areas in Fig. 8.63).

8.7.4 Two-Dimensional Storage Time Model

A simple two-dimensional analysis for the storage time during the turn-off of the GTO structure can be performed by assuming that the reverse gate current is used to remove the charge stored within the P-base region. The stored charge in the P-base region due to the on-state current flow is indicated by the shaded region in Fig. 8.65. If a linear GTO structure topology is assumed with a depth Z orthogonal to the cross section, the total stored charge in the P-base region is given by

$$Q_{S,PB} = qW_P\frac{W_K}{2}Zn_a, \tag{8.118}$$

where the average carrier concentration within the P-base region due to the on-state current flow is given by

$$n_a = \frac{\tau_{HL}}{q(W_P + W_N)} \left(\frac{2I_{A,ON}}{W_K Z} \right). \tag{8.119}$$

Combining these relationships,

$$Q_{S,PB} = \left(\frac{W_P}{W_P + W_N} \right) I_{A,ON} \tau_{HL}. \tag{8.120}$$

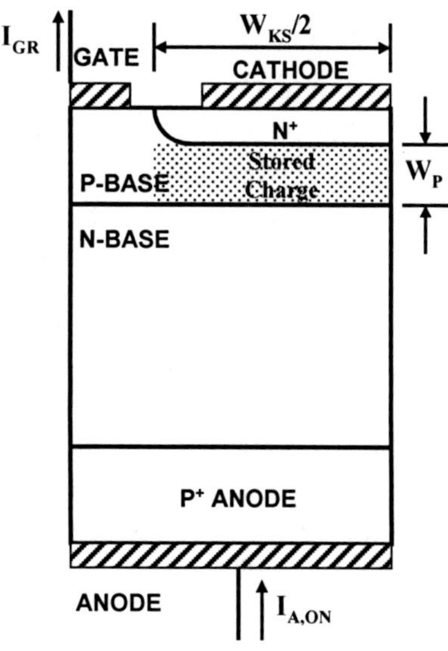

Fig. 8.65 Stored charge in the P-base region of the two-dimensional GTO structure

In the case of a step drive with constant reverse gate current during the turn-off transient, the charge removed by the gate current (as illustrated by the shaded area in Fig. 8.63) is given by

$$Q_R = I_{GR} t_{S,STEP}. \tag{8.121}$$

Equating this to the stored charge in the P-base region provides the storage time:

$$t_{S,STEP} = \left(\frac{W_P}{W_P + W_N} \right) \left(\frac{I_{A,ON}}{I_{GR}} \right) \tau_{HL} = \left(\frac{W_P}{W_P + W_{NB}} \right) \beta_T \tau_{HL}. \tag{8.122}$$

In the case of a ramp drive with a ramp rate a during the turn-off transient, the charge removed by the gate current (as illustrated by the shaded area in Fig. 8.63) is given by

$$Q_R = \frac{1}{2} I_{GR} t_{S,RAMP} = \frac{1}{2} a t_{S,RAMP}^2. \tag{8.123}$$

Equating this to the stored charge in the P-base region provides the storage time:

$$t_{S,RAMP} = \sqrt{2\left(\frac{W_P}{W_P + W_N}\right)\left(\frac{I_{A,ON}}{a}\right)\tau_{HL}}. \tag{8.124}$$

For a GTO structure with a P-base width of 40 μm and an N-base width of 300 μm with a high-level lifetime of 20 μs, the storage time obtained by using the above equation is 9.7 μs if the ramp rate for the gate is 50 A cm^{-2} μs^{-1} and the on-state current density is 100 A cm^{-2}.

8.7.5 One-Dimensional Voltage Rise Time Model

Once the stored charge in the P-base region has been removed by the reverse gate drive current, the P-base/N-base junction (J_2) begins to support voltage across a space-charge region as illustrated in Fig. 8.66. Unlike the depletion region formed across this junction under the steady-state forward-blocking mode, the space-charge region formed during the turn-off of the GTO contains a large concentration of holes due to the transport of these carriers from the stored charge remaining within the N-base region. Since the electric field in the space-charge region is large, it can be assumed that the holes are transported at their saturated drift velocity ($v_{sat,p}$) in this region. The hole concentration (p_{SC}) within the space-charge

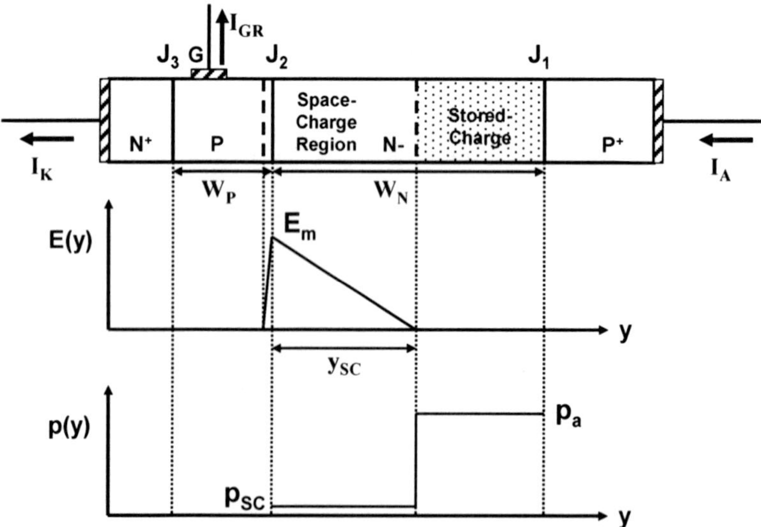

Fig. 8.66 Electric field and free carrier distributions during turn-off for the GTO structure

region is therefore related to the on-state anode current density that continues to flow during the voltage rise time interval:

$$p_{SC} = \frac{J_{A,ON}}{qv_{sat,p}}. \tag{8.125}$$

As an example, the hole concentration within the space-charge region is 6.25×10^{13} cm^{-3} at an anode current density of 100 A cm^{-2}. Since this is comparable with the doping concentration in the N-base region (typically 5×10^{13} cm^{-3}), the presence of the holes must be accounted for when deriving the width of the space-charge region.

The width of the space-charge region (y_{SC}) at any point in time is determined by the solution of Poisson's equation with a net charge given by the sum of the positive charge from the ionized donors and the positive charge from the holes being transported through the region. The anode voltage at any time during the voltage rise time interval is then related to the space-charge region width by

$$y_{SC}(t) = \sqrt{\frac{2\varepsilon_S V_A(t)}{q(N_D + p_{SC})}}. \tag{8.126}$$

Taking the time derivative of this expression yields

$$\frac{dy_{SC}(t)}{dt} = \sqrt{\frac{\varepsilon_S}{2q(N_D + p_{SC})V_A(t)}}\frac{dV_A}{dt}. \tag{8.127}$$

As the space-charge region expands, it extracts some of the remaining stored charge at its boundary. The anode current density is also related to this extraction of the stored charge:

$$J_{A,ON} = qp_a \frac{dy_{SC}(t)}{dt}, \tag{8.128}$$

where p_a is the concentration of the holes in the stored charge region as shown in the figure. Using (8.127),

$$J_{A,ON} = qp_a \sqrt{\frac{\varepsilon_S}{2q(N_D + p_{SC})V_A(t)}}\frac{dV_A}{dt}. \tag{8.129}$$

This equation can be rewritten in the form

$$\sqrt{\frac{2(N_D + p_{SC})J_{A,ON}^2}{q\varepsilon_S p_a^2}}dt = \frac{dV_A}{\sqrt{V_A(t)}}. \tag{8.130}$$

Integration of this expression yields

$$\sqrt{\frac{2(N_{\mathrm{D}}+p_{\mathrm{SC}})J_{\mathrm{A,ON}}^{2}}{q\varepsilon_{\mathrm{S}}p_{\mathrm{a}}^{2}}}t=2\sqrt{V_{\mathrm{A}}(t)}. \tag{8.131}$$

The voltage transient during the rise time interval is then obtained:

$$V_{\mathrm{A}}(t)=\frac{(N_{\mathrm{D}}+p_{\mathrm{SC}})J_{\mathrm{A,ON}}^{2}}{2q\varepsilon_{\mathrm{S}}p_{\mathrm{a}}^{2}}t^{2}. \tag{8.132}$$

The average hole concentration (p_a) in the stored charge region is determined by the free carrier distribution within the GTO during the on-state:

$$p_{\mathrm{a}}=\frac{J_{\mathrm{A,ON}}\tau_{\mathrm{HL}}}{q(W_{\mathrm{P}}+W_{\mathrm{N}})}. \tag{8.133}$$

Substituting this into (8.132) yields

$$V_{\mathrm{A}}(t)=\frac{q(N_{\mathrm{D}}+p_{\mathrm{SC}})(W_{\mathrm{P}}+W_{\mathrm{N}})^{2}}{2\varepsilon_{\mathrm{S}}\tau_{\mathrm{HL}}^{2}}t^{2}. \tag{8.134}$$

According to this solution for the voltage transient, the anode voltage will rise as the square of time.

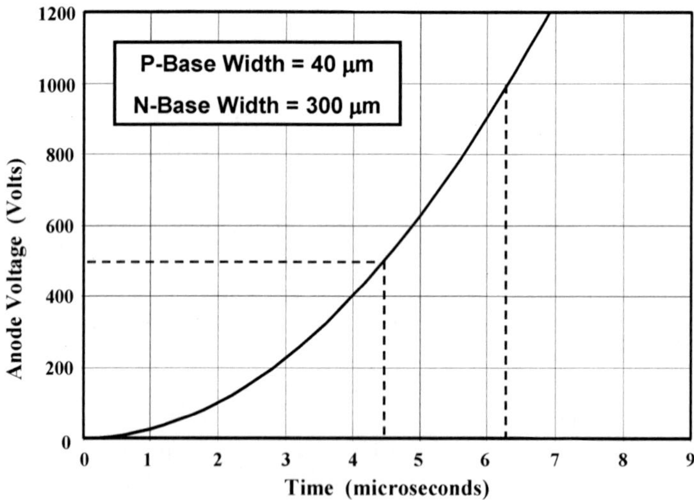

Fig. 8.67 Anode voltage transient during turn-off for the GTO structure

The voltage transient obtained by using the above one-dimensional analysis is illustrated in Fig. 8.67 for the case of a GTO structure with a P-base width of 40 μm and an N-base width of 300 μm. The doping concentration of the N-base region was assumed to be 5×10^{13} cm^{-3} with a high-level lifetime of 20 μs.

The device was assumed to be turned off from an initial on-state current density of 100 A cm^{-2} leading to a hole concentration within the space-charge region of 6.25×10^{13} cm^{-3}. It can be observed that the anode voltage reaches 500 and 1,000 V in 4.5 and 6.3 μs, respectively. This waveform is applicable after the end of the storage time interval.

The time taken for the anode voltage to reach a supply voltage (V_S), defined as the *voltage rise time* (t_V), can be derived from (8.134):

$$t_V = \sqrt{\frac{2\varepsilon_S V_S}{q(N_D + p_{SC})(W_P + W_N)^2}}\tau_{HL}. \tag{8.135}$$

The values for the voltage rise times given in the previous paragraph can be computed by using this expression. Based upon this expression, it can be concluded that the voltage rise time will be larger when the high-level lifetime is increased, which in turn will lead to larger power dissipation during this phase of operation.

8.7.6 One-Dimensional Current Fall Time Model

In the case of switching an inductive load with a clamping diode, the anode voltage becomes constant at the supply voltage after the voltage rise time interval. The anode current then decreases as illustrated in Fig. 8.63 with a sharp drop in value followed by a current tail. The sudden drop in the anode current is associated with the cessation of cathode current flow at the end of the voltage rise time interval. Just prior to the turn-off of the thyristor, the base drive current for the P–N–P transistor ($I_{B,PNP}$) due to the electrons supplied by the N–P–N transistor is given by ($\alpha_{NPN} I_K$) as indicated in Fig. 8.64. This base drive current for the P–N–P transistor produces an anode current given by

$$I_{AT} = \beta_{PNP} I_{B,PNP} = \left(\frac{\alpha_{PNP}}{1-\alpha_{PNP}}\right)\alpha_{NPN} I_K. \tag{8.136}$$

When the cathode current is interrupted by the cessation of regenerative action during the turn-off of the GTO structure, the anode current is abruptly reduced by the above amount leading to the sudden fall in the anode current. The anode current at the start of the current tail is therefore given by

$$I_{A,D} = I_{A,ON} - I_{AT} = I_{A,ON} - \left(\frac{\alpha_{PNP}}{1-\alpha_{PNP}}\right)\alpha_{NPN} I_K, \tag{8.137}$$

with the cathode current related to the anode current by (8.111). For a typical common-base current gain of 0.9 for the N–P–N transistor and 0.65 for the P–N–P transistor, an 80% fall in the anode current is predicted by this equation, resulting in an initial tail current of 20% of the initial on-state anode current.

The decay of the anode current after the initial abrupt reduction is governed by the recombination of the excess holes and electrons that are trapped within the N-base region. These stored charges are shown by the shaded area in Fig. 8.66. In the absence of diffusion, the continuity equation for holes in the N-base region is given by

$$\frac{\mathrm{d}\delta p_{\mathrm{N}}}{\mathrm{d}t} = -\frac{\delta p_{\mathrm{N}}}{\tau_{\mathrm{HL}}}, \tag{8.138}$$

where δp_{N} is the excess hole concentration in the N-base region. The solution for this equation is

$$\delta p_{\mathrm{N}}(t) \approx p_{\mathrm{N}}(t) = p_{\mathrm{a}}\mathrm{e}^{-t/\tau_{\mathrm{HL}}}, \tag{8.139}$$

because high-level injection conditions prevail in the N-base region with the initial concentration of holes in the stored charge region being equal to the average hole concentration due to the injection of carriers in the on-state. Since this concentration (p_{a}) is given by (8.133) because of high-level injection conditions in the N-base region,

$$p_{\mathrm{N}}(t) = \frac{J_{\mathrm{A,ON}}\tau_{\mathrm{HL}}}{q(W_{\mathrm{P}} + W_{\mathrm{N}})}\mathrm{e}^{-t/\tau_{\mathrm{HL}}}. \tag{8.140}$$

The anode current flow that supports the recombination of carriers within the stored charge region can be analyzed by examination of the carrier distribution on both sides of the P^+ anode/N-base junction (J_1). The free carrier concentration in the N-base region in the vicinity of junction J_1 can be assumed to be independent of distance but a function of time as given by (8.140). The high concentration of electrons in the N-base region produces the injection of electrons into the P^+ anode region as indicated in Fig. 8.68. These injected electrons diffuse away from the junction producing an exponential decay in concentration as shown in the figure. The free carrier concentrations on the two sides of the junction are related by

$$\frac{p_{\mathrm{A}}}{p_{\mathrm{N}}(t)} = \frac{n_{\mathrm{N}}(t)}{n_{\mathrm{A}}(0,t)} = \mathrm{e}^{qV_{\mathrm{A}}/kT}, \tag{8.141}$$

where $p_{\mathrm{N}}(t)$ and $n_{\mathrm{N}}(t)$ are the hole and electron concentrations in the N-base region at junction J_1 given by (8.140), p_{A} is the hole concentration in the P^+ anode region, and $n_{\mathrm{A}}(0, t)$ is the electron concentration at junction J_1 in the P^+ anode region. The hole concentration in the P^+ anode region can be assumed to be equal to its acceptor doping concentration (N_{AA}). Using this in (8.141) yields

$$n_{\mathrm{A}}(0,t) = \frac{n_{\mathrm{N}}(t)p_{\mathrm{N}}(t)}{N_{\mathrm{AA}}} = \frac{p_{\mathrm{N}}^2(t)}{N_{\mathrm{AA}}}. \tag{8.142}$$

The electrons diffuse into the P$^+$ anode region with an exponential distribution due to low-level injection conditions:

$$n_A(y,t) = n_A(0,t)e^{-y/L_{nP+}}, \tag{8.143}$$

where L_{nP+} is the diffusion length for electrons in the P$^+$ anode region.

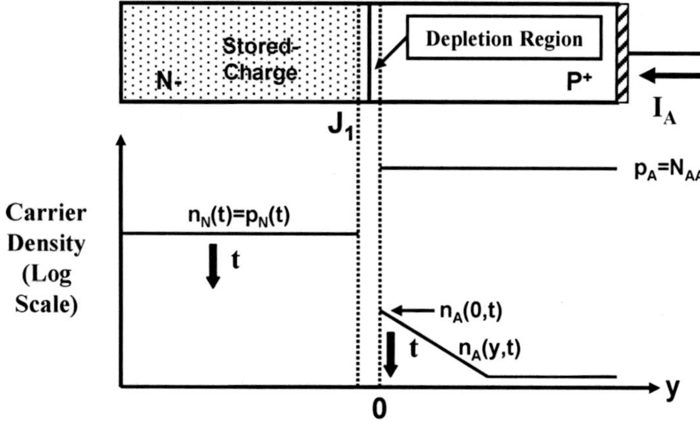

Fig. 8.68 Free carrier distribution at the P+ anode/N-base junction during turn-off

The anode current is produced by the diffusion of the injected electrons in the P$^+$ anode side of the junction:

$$J_A(t) = -qD_n \left.\frac{\partial n_A(y,t)}{\partial y}\right|_{y=0} = \frac{qD_n n_A(0,t)}{L_{nP+}} \tag{8.144}$$

by using (8.143). Making use of (8.140) and (8.142) yields

$$J_A(t) = \frac{qD_n p_N^2(t)}{L_{nP+}N_{AA}} = \frac{D_n J_{A,ON}^2 \tau_{HL}^2}{qL_{nP+}N_{AA}(W_P + W_N)^2} e^{-2t/\tau_{HL}} = J_{A,D}\, e^{-2t/\tau_{HL}}. \tag{8.145}$$

This equation indicates that the anode current varies as the square of the carrier density in the stored charge region during the anode current tail. Consequently, the anode current decreases exponentially with time with a time constant of one half of the high-level lifetime even though the free carrier density in the stored charge region is decreasing exponentially with time with a time constant equal to the high-level lifetime.

Consider the case of a GTO structure with P-base and N-base widths of 40 and 300 μm, respectively, a high-level lifetime of 20 μs in the N-base region, and an anode doping concentration of 1×10^{19} cm^{-3}. The diffusion coefficient for electrons (D_n) in the anode region for this doping concentration is 3 cm^2 V^{-1} s^{-1}. The diffusion length for electrons (L_n) in the anode region is then 2.74 μm if the low-level recombination lifetime in this region is 25 ns. Using these values in

(8.145) with an on-state anode current density of 100 A cm^{-2} yields an anode current density ($J_{A,D}$) of 24 A cm^{-2} at the beginning of the anode current tail.

The anode current waveform predicted by the above equation is shown in Fig. 8.69. In this figure, a small portion with constant anode current is included prior to the anode current transient for clarity. It is common practice to define a current turn-off time (t_i) as the duration for the anode current to decrease to one tenth of its on-state value. This time can be derived from (8.145):

$$t_i = \frac{\tau_{HL}}{2}\ln\left[\frac{10J_{A,D}}{J_{A,ON}}\right] = \frac{\tau_{HL}}{2}\ln\left[\frac{10D_n J_{A,ON}\tau_{HL}^2}{qL_{nP+}N_{AA}(W_P+W_N)^2}\right]. \quad (8.146)$$

Using the values given in the previous paragraph, the current turn-off time is found to be 8.6 μs. This is a relatively long time interval during which there is a substantial anode current density flowing through the GTO structure while its anode voltage is high. Consequently, the high power dissipation associated with the current tail limits the frequency of operation for the GTO structure.

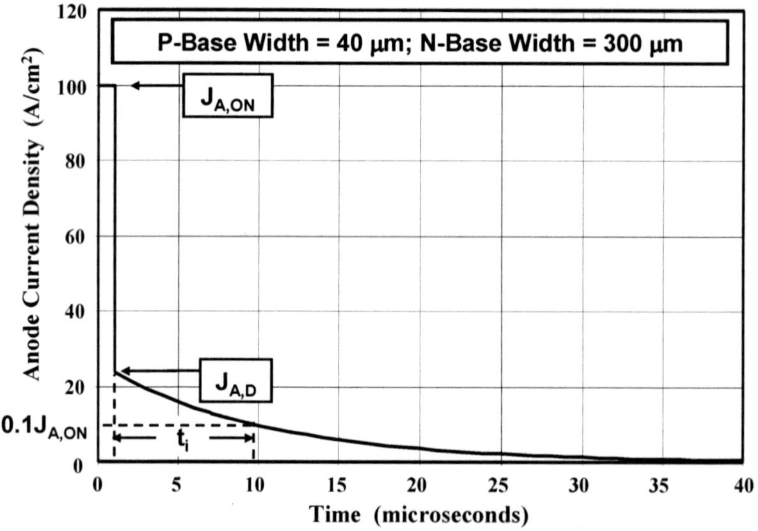

Fig. 8.69 Anode current transient during turn-off for the GTO structure

Simulation Example

To gain insight into the operation of the GTO structure during its turn-off, the results of two-dimensional numerical simulations for a typical structure are discussed here for the case of a ramp gate drive. The device structure used has the cross section shown in Fig. 8.60 with a width of 200 μm. No cathode short was used in the structure. The width of the cathode in the structure was 190 μm with a gate contact electrode of 5 μm in width.

The doping profile for the GTO structure used in the simulations is provided in Fig. 8.70. The widths of the P-base and N-base regions are 40 and 300 μm, respectively, as indicated in the figure. The N^+ cathode and P^+ anode regions have a thickness of 10 and 50 μm, respectively.

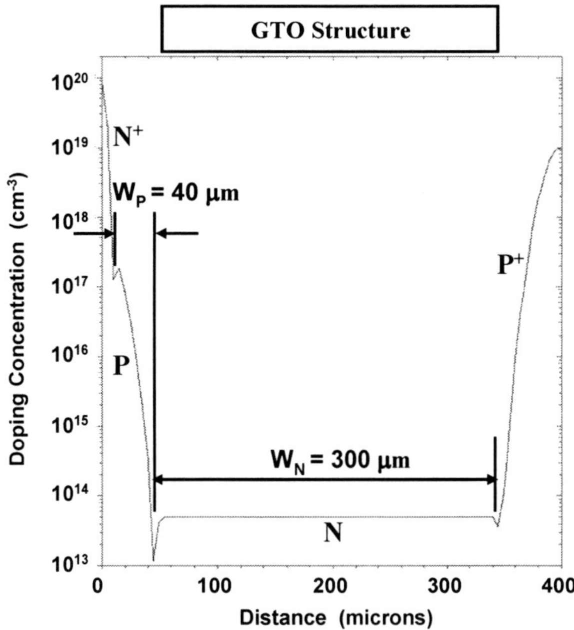

Fig. 8.70 Doping profile for the GTO structure

The waveforms for the anode voltage and current, as well as the gate current, are shown in Fig. 8.71. The simulations were performed with a gate ramp rate of $-5\ \mathrm{A\ cm^{-2}\ \mu s^{-1}}$. Two cases for the anode supply voltage were used to examine the impact on the anode voltage rise time and anode current fall time. The storage time extracted from the simulations is 7.5 μs. The simple two-dimensional analysis for the storage time provides a good estimate for this time interval. After the storage time, the anode current increases as the square of the time as predicted by the analytical model. The anode voltage rise time for the case of supply voltages of 500 and 1,000 V obtained in the simulations is 4 and 5.1 μs, respectively. The one-dimensional analysis for the voltage rise time provides a good prediction for this time interval. The anode current waveform exhibits a sharp drop at the end of the voltage rise time interval and then decays slowly over a period of about 30 μs. After the abrupt drop in the anode current, the cathode current ceases making the anode and gate currents equal in magnitude. The one-dimensional model for the anode current waveform (see Fig. 8.69) is also consistent with these results.

To gain further insight into the turn-off process within the GTO structure, it is instructive to look at the free carrier distribution at various points of time. The electron concentration is shown in Fig. 8.72 when the GTO structure is operating in its initial on-state with an anode current density of $100\ \mathrm{A\ cm^{-2}}$. The electron concentration is almost uniform throughout the P-base and N-base regions in both the *x*- and *y*-directions. This provides justification for using the average carrier concentration (n_a) as given by (8.115) during the turn-off analysis.

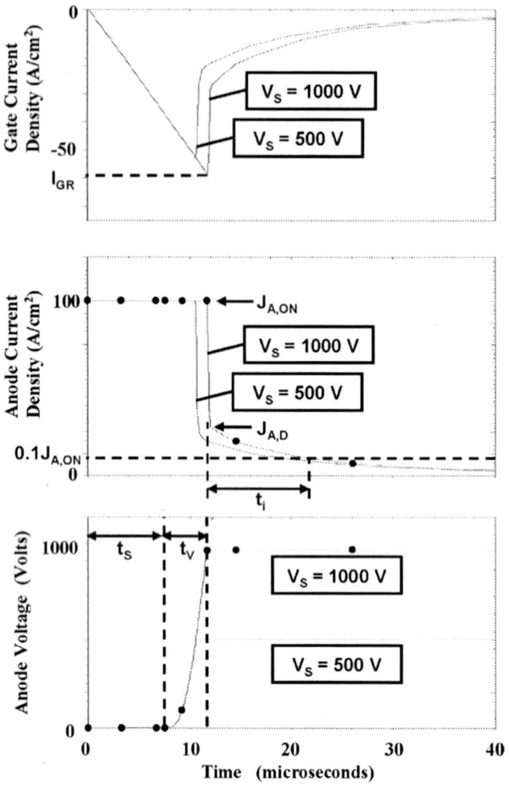

Fig. 8.71 GTO turn-off waveforms

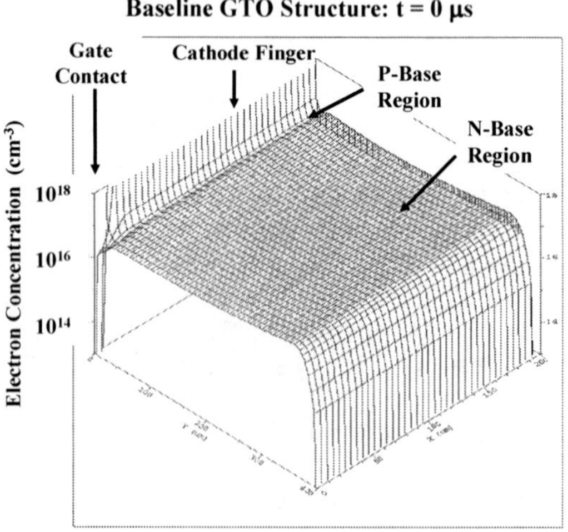

Fig. 8.72 Initial electron charge distribution within the GTO structure

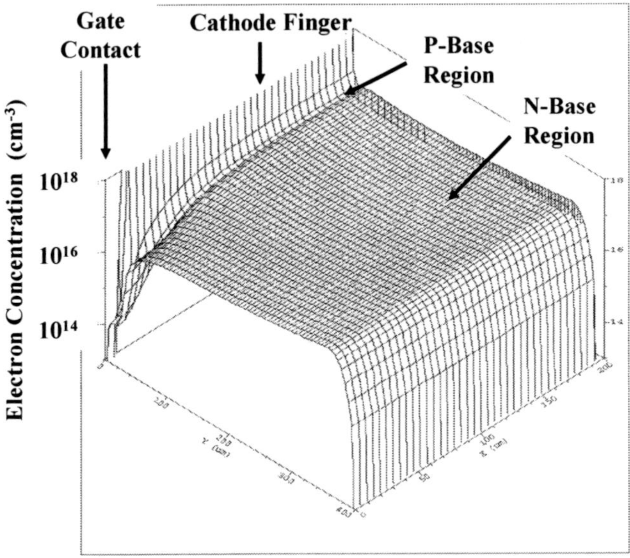

Fig. 8.73 Three-dimensional view of the electron charge distribution within the GTO structure during the turn-off transient

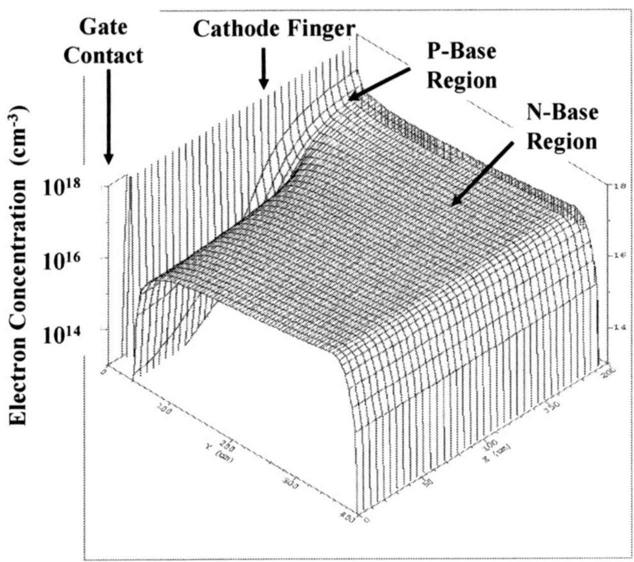

Fig. 8.74 Three-dimensional view of the electron charge distribution within the GTO structure during the turn-off transient

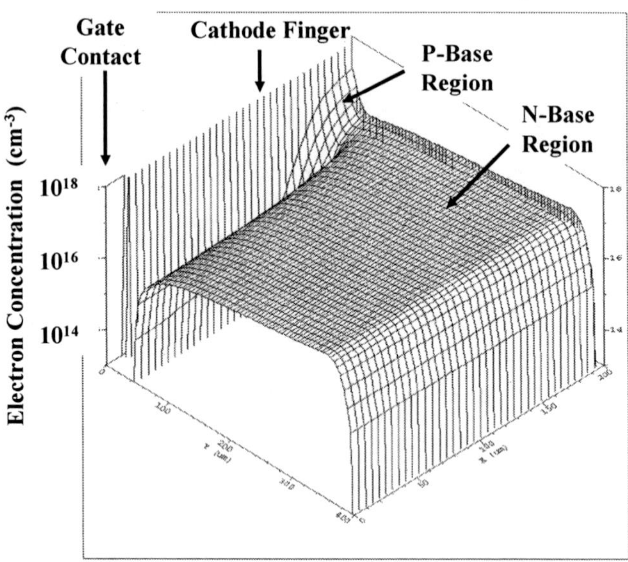

Fig. 8.75 Three-dimensional view of the electron charge distribution within the GTO structure during the turn-off transient

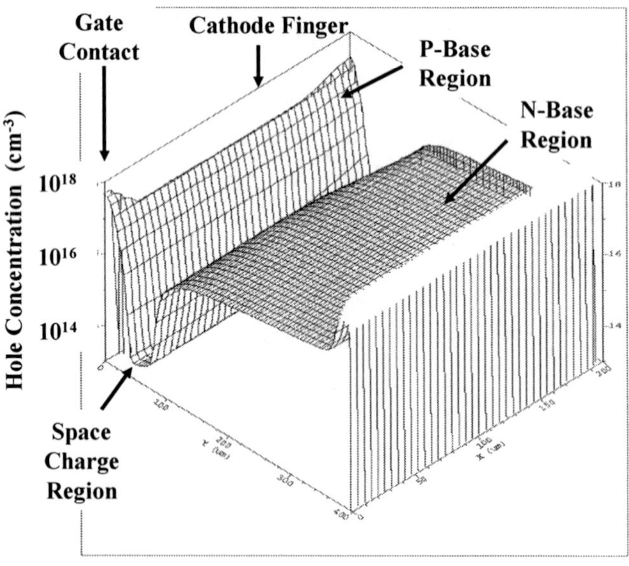

Fig. 8.76 Three-dimensional view of the hole charge distribution within the GTO structure during the turn-off transient

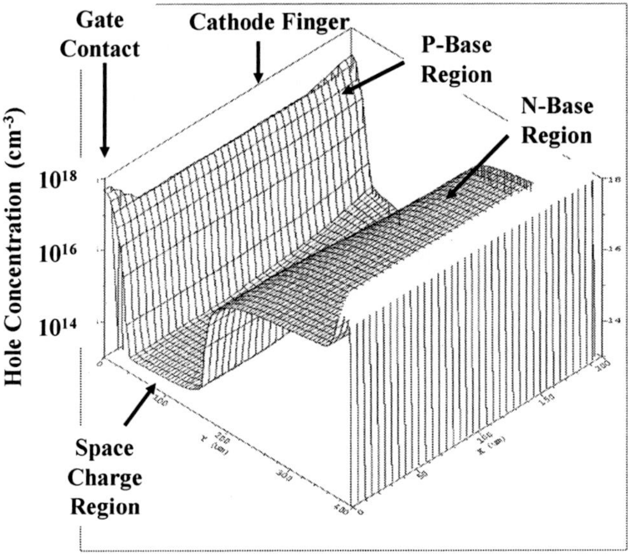

Fig. 8.77 Three-dimensional view of the hole charge distribution within the GTO structure during the turn-off transient

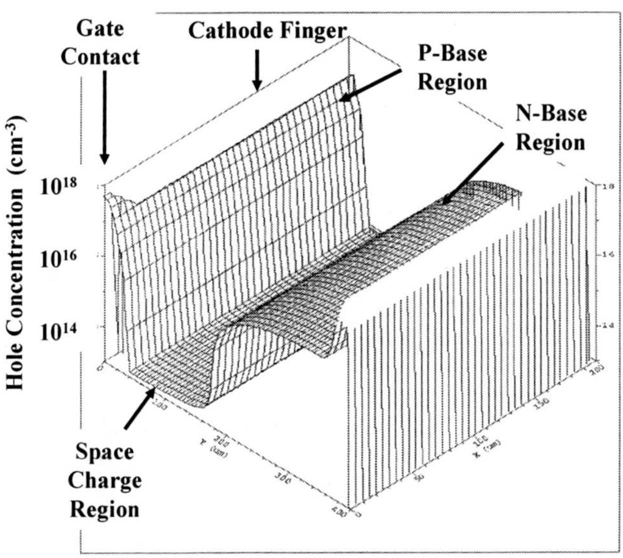

Fig. 8.78 Three-dimensional view of the hole charge distribution within the GTO structure during the turn-off transient

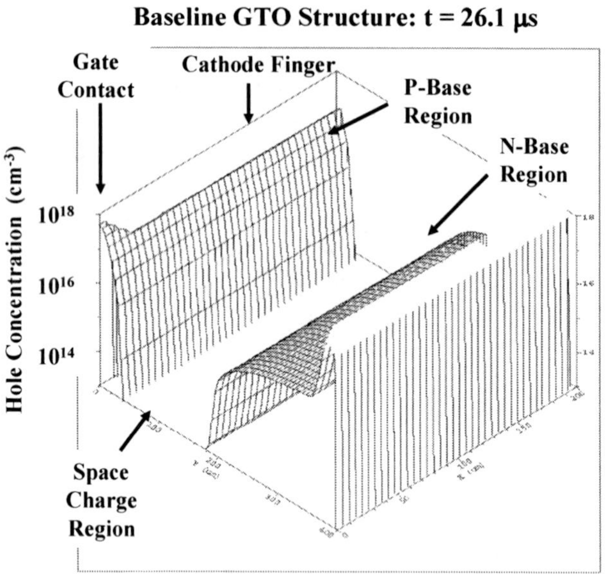

Baseline GTO Structure: t = 26.1 μs

Gate Contact

Cathode Finger

P-Base Region

N-Base Region

Hole Concentration (cm^{-3})

10^{18}

10^{16}

10^{14}

Space Charge Region

Fig. 8.79 Three-dimensional view of the hole charge distribution within the GTO structure during the turn-off transient

Three-dimensional views of the carrier distribution within the GTO structure reveal the physics of operation during the turn-off process. These distributions are shown at various points in time during the storage phase, the anode voltage rise time phase, and the anode current reduction phase as indicated by the dots on the waveforms in Fig. 8.71. It is instructive to examine the minority carrier concentration in the P-base region during the storage phase. The electron carrier distribution is therefore provided in Fig. 8.73 during the storage time interval at 3.3 μs. It can be observed that the reverse gate current has removed some stored charge from the P-base region in the vicinity of the gate contact. The electron concentration in the N-base region remains close to the initial on-state distribution. The electron carrier distribution is shown in Fig. 8.74 during the storage time interval at 6.6 μs. It can be observed that the reverse gate current has now removed more charge from the P-base region while the electron concentration in the N-base region still remains close to the initial on-state distribution. The electron carrier distribution is shown in Fig. 8.75 at the end of the storage time interval (at 7.7 μs). It can be observed that the reverse gate current has removed almost all the stored charge from the P-base region while the electron concentration in the N-base region still remains close to the initial on-state distribution, justifying the model used for the storage time. At the end of the storage time, the device begins to support voltage – anode voltage was found to increase to 2 V from the on-state voltage drop of 1.2 V.

It is instructive to examine the minority carrier distribution within the N-base region after the storage phase. The hole carrier distribution is therefore provided in Fig. 8.76 during the voltage rise time interval at 9.2 μs. It can be observed that some of the stored charge has been removed from the N-base region in the vicinity of the P-base/N-base junction. A space-charge region is visible at this junction with a small hole concentration due to the anode current flow. The presence of these holes was accounted for by the term p_{SC} in (8.126). The hole distribution within the structure can be observed to be still remarkably one-dimensional in nature. This provides justification for using a one-dimensional analysis for the voltage rise time. The anode voltage has risen to 100 V at this point in time. The hole carrier distribution is provided in Fig. 8.77 at the end of the voltage rise time interval at 11.7 μs when the anode voltage has reached a supply voltage of 1,000 V. It can be observed that more of the stored charge has been removed from the N-base region in the vicinity of the P-base/N-base junction. The space-charge region at this junction has widened to support the larger anode voltage.

After the voltage rise time, the anode current decreases abruptly followed by a long current tail. The hole carrier distribution is provided in Fig. 8.78 during the anode current tail interval at 14.6 μs when the anode current density has decayed to 19 A cm^{-2}. It can be observed that the hole concentration in the space-charge region is smaller because of this reduced current density. The space-charge region at the junction has widened slightly to support the same anode voltage with a smaller space-charge density. Finally, the hole carrier distribution is provided in Fig. 8.79 during the anode current tail interval at 26.1 μs when the anode current density has decayed to 7 A cm^{-2}. It can be observed that the hole concentration in the space-charge region has fallen below 10^{13} cm^{-3} because of the reduced current density. The space-charge region at the junction has widened slightly to support the same anode voltage with the smaller space-charge density. Eventually, all the stored charge in the N-base region is removed by the recombination process.

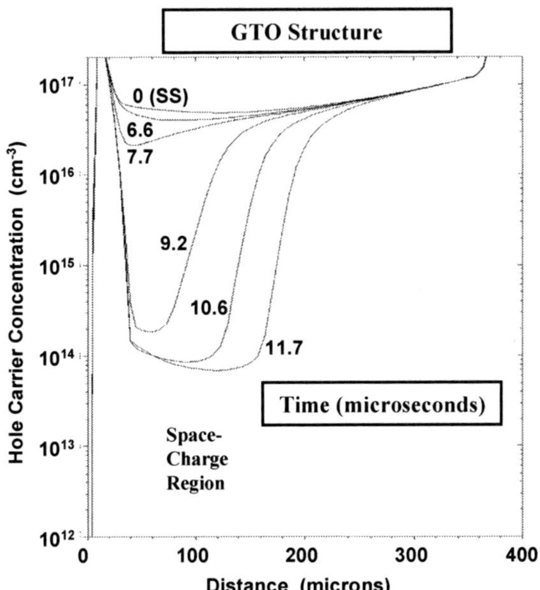

Fig. 8.80 Hole carrier distribution at the center of the cathode finger during the turn-off transient

The three-dimensional views of the carrier distributions shown in the above figures demonstrate that the hole carrier distribution is remarkably one dimensional in nature even though the turn-off process is two dimensional in nature. This observation justifies creating one-dimensional analytical models for the turn-off waveforms that provide a good description of the GTO switching behavior. Further insight into the turn-off process, especially during the anode current tail, can be obtained by examination of the carrier density at the center of the N^+ cathode. A one-dimensional view of the minority carrier distribution in the GTO structure at the center of the cathode is shown in Fig. 8.80 from the initial steady-state operating point ($t = 0$ μs) to the end of the voltage rise time ($t = 11.7$ μs). The initial carrier distribution has a catenary form within the P-base and N-base regions because the GTO operates like a P-i-N rectifier in the on-state. The carrier concentration is almost constant through these regions. This carrier distribution does not change significantly until the end of the storage phase ($t = 7.7$ μs). After this time, a space-charge region begins to form as indicated in the figure. The hole concentration in the space-charge region is about 7×10^{13} cm^{-3}, which is consistent with the value for p_{SC} obtained by using the analytical model with the carriers moving at the saturated drift velocity. The width of the space-charge region can be observed to be about 110 μm, which is consistent with the predictions of the analytical model (see (8.126)).

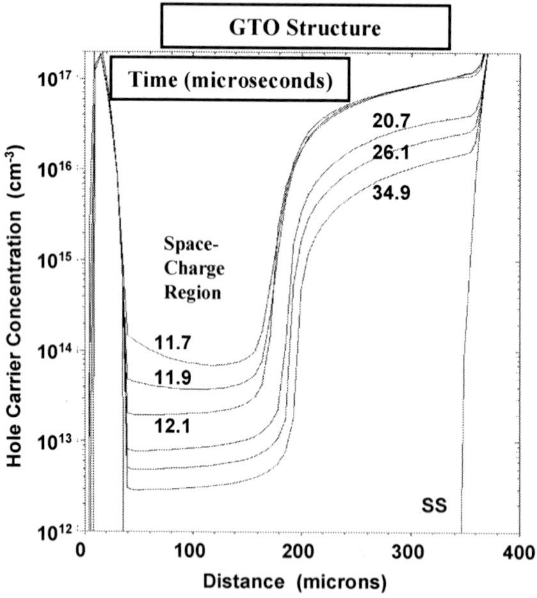

Fig. 8.81 Hole carrier distribution at the center of the cathode finger during the turn-off transient

A one-dimensional view of the minority carrier distribution in the GTO structure at the center of the cathode is shown in Fig. 8.81 from the end of the storage phase ($t = 11.7$ μs) to the completion of the turn-off process. When the anode current reduces abruptly between $t = 11.7$ μs and $t = 12.1$ μs, the hole concentration decreases in the space-charge region but is not altered within the rest of the N-base region. This is consistent with the assumption used in the analytical model that the hole concentration within the stored charge in the N-base region at the beginning of the anode current decay phase is equal to the initial on-state carrier concentration (see (8.139)). Subsequently, the hole concentration in the stored charge region decreases due to the recombination process. This results in a smaller anode current and a smaller hole concentration within the space-charge region as well. The slight expansion of the space-charge region during the anode current tail can also be observed in Fig. 8.81.

8.7.7 Switching Energy Loss

The power loss incurred during the switching transient limits the maximum operating frequency for the GTO structure. During the storage time interval, the voltage across the device is equal to its on-state voltage drop. Consequently, although the storage time for the GTO structure is very long, it can be accounted for as a part of the on-time interval. From the point of view of limiting the maximum frequency of operation, consider the case of a device operating at a 50% duty cycle with the on-time being determined by the storage time. In this case, since half the period (T) is equal to the storage time (t_S), the maximum operating frequency is given by

$$f_{max} = \frac{1}{T} = \frac{1}{2t_S}. \tag{8.147}$$

A storage time of 10 μs would correspond to a maximum operating frequency of 50 kHz.

In practice, the maximum operating frequency for the GTO structure is limited by the turn-off losses. The turn-off losses are associated with the voltage rise time interval and the current fall time interval. The energy loss for each event can be computed by integration of the power loss, as given by the product of the instantaneous current and voltage. During the voltage rise time interval, the anode current is constant while the voltage increases as the square of the time. The energy loss during the voltage rise time interval is therefore given by

$$E_{OFF,V} = \int_0^{t_V} J_{A,ON} V_A(t) \mathrm{d}t. \tag{8.148}$$

Using (8.134) for the anode voltage waveform,

$$\begin{aligned} E_{OFF,V} &= \frac{qJ_{A,ON}(N_D + p_{SC})(W_P + W_N)^2}{2\varepsilon_S \tau_{HL}^2} \int_0^{t_V} t^2 \mathrm{d}t \\ &= \frac{qJ_{A,ON}(N_D + p_{SC})(W_P + W_N)^2}{6\varepsilon_S \tau_{HL}^2} t_V^3. \end{aligned} \tag{8.149}$$

For the example shown in Fig. 8.67 with an anode supply voltage of 1,000 V, the energy loss per unit area during the voltage rise time is found to be 0.22 J cm^{-2} if the on-state current density is 100 A cm^{-2}.

During the anode current tail, the anode voltage is constant while the current decreases exponentially with time. The energy loss during the current tail is therefore given by

$$E_{\mathrm{OFF,I}} = \int_0^\infty J_{\mathrm{A}}(t)V_{\mathrm{S}}\mathrm{d}t. \tag{8.150}$$

Using (8.145) for the anode current waveform,

$$\begin{aligned} E_{\mathrm{OFF,I}} &= \int_0^\infty V_{\mathrm{S}}J_{\mathrm{A,D}}\mathrm{e}^{-2t/\tau_{\mathrm{HL}}}\mathrm{d}t \\ &= \frac{J_{\mathrm{A,D}}V_{\mathrm{S}}\tau_{\mathrm{HL}}}{2}. \end{aligned} \tag{8.151}$$

For the example shown in Fig. 8.69 with an anode supply voltage of 1,000 V and an on-state current density of 100 A cm^{-2}, the energy loss per unit area during the current tail is found to be 0.24 J cm^{-2}.

Thus, the total energy loss per unit area ($E_{\mathrm{OFF,V}} + E_{\mathrm{OFF,I}}$) during the turn-off process for the GTO structure is found to be 0.46 J cm^{-2}. For operation of the GTO structure at a 50% duty cycle with an on-state current density of 100 A cm^{-2}, the power dissipation per unit area is 60 W cm^{-2} due to an on-state voltage drop of 1.2 V. When operated at a frequency of 150 Hz, the turn-off power loss per unit area is found to be 69 W cm^{-2} by using the energy loss per unit area given above. Thus, the total power loss per unit area for the GTO structure is 129 W cm^{-2} when operated at 150 Hz, which is a reasonable power density for typical power device packaging technology. This analysis indicates that the maximum operating frequency for the GTO structure is limited to relatively low frequencies due to the high turn-off energy losses. The turn-off energy loss can be reduced by decreasing the high-level lifetime but this is accompanied by an increase in the on-state voltage drop, which increases the on-state power dissipation and negates the overall improvement in performance.

8.7.8 Maximum Turn-Off Current

The largest anode current that can be turned off under gate control is referred to as the *maximum turn-off current or the maximum controllable current*. The maximum turn-off current has been found to be limited by the on-set of breakdown of the N$^+$ cathode/P-base junction during the turn-off process. This limitation can be understood by examination of the gate current flow at the end of the storage phase as illustrated in Fig. 8.82. At this point in time, the stored charge in the P-base region has been confined to the middle of the cathode finger as shown in the figure by the shaded area. The reverse gate drive current must traverse the entire P-base region under the N^+ cathode to remove the remaining stored charge. The reverse gate current flows through the P-base region producing a voltage drop at point A due to the resistance R_{BM}. If this voltage drop becomes equal to the breakdown voltage of the N^+ cathode/P-base junction, impact ionization-induced current begins to flow at point A which sustains the gate current preventing further removal of the stored charge. Consequently, the GTO structure will be unable to turn off.

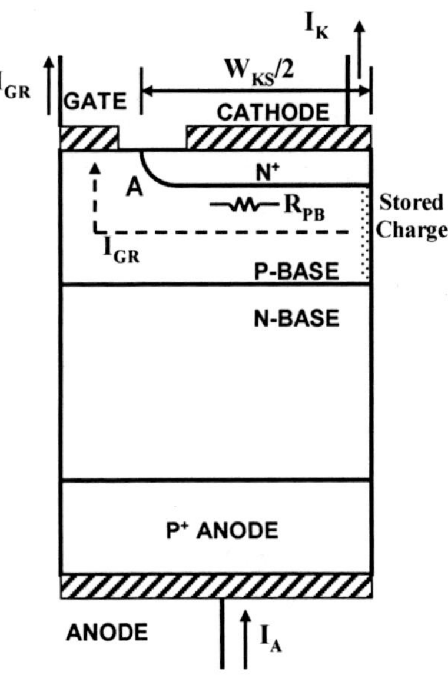

Fig. 8.82 Gate current flow at the end of the storage phase during the turn-off for the GTO structure

The maximum reverse gate current that can flow before the on-set of breakdown of the N^+ cathode/P-base junction is given by

$$I_{GR,M} = \frac{BV_{GK}}{R_{PB}} = \frac{2BV_{GK}Z}{\rho_{SB}W_{KS}}, \tag{8.152}$$

where BV_{GK} is the breakdown voltage of the N^+ cathode/P-base junction and ρ_{SB} is the sheet resistance of the P-base region under the N^+ cathode. The maximum anode current that can be turned off by this reverse gate current is given by

$$I_{A,M} = \beta_M I_{GR,M} = \frac{2\beta_M BV_{GK}Z}{\rho_{SB}W_{KS}}. \tag{8.153}$$

Using the maximum turn-off gain obtained from the one-dimensional analysis (8.114),

$$I_{A,M} = \frac{2BV_{GK}Z}{\rho_{SB}W_{KS}}\left(\frac{\alpha_{NPN}}{\alpha_{NPN}+\alpha_{PNP}-1}\right). \quad (8.154)$$

The maximum turn-off current density can then be obtained by dividing the above current by the device area:

$$J_{A,M} = \frac{4BV_{GK}}{\rho_{SB}W_{KS}^{2}}\left(\frac{\alpha_{NPN}}{\alpha_{NPN}+\alpha_{PNP}-1}\right). \quad (8.155)$$

The typical GTO structure has a maximum turn-off gain of about 5, a typical P-base region sheet resistance of 350 Ω sq^{-1} and a breakdown voltage for the N^+ cathode/P-base junction of 30 V. For the case of an N^+ cathode width of 400 μm, the maximum turn-off current density for this GTO structure by using the above equation is 1,000 A cm^{-2}.

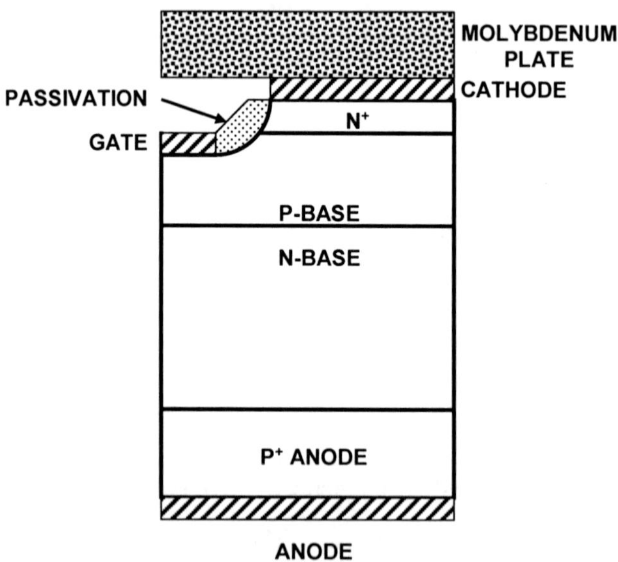

Fig. 8.83 Recessed gate architecture for the GTO structure

The maximum turn-off current density can be improved by increasing the breakdown voltage of the N^+ cathode/P-base junction. An improvement in the breakdown voltage of the junction at point *A* can be achieved by eliminating the cylindrical junction curvature of the planar junction. The etching of a mesa at the gate region allows removal of the junction curvature and simultaneously recesses the gate contact, so that a large molybdenum plate can be placed on top of the wafer to contact the cathode electrode. This commonly used GTO architecture is illustrated in Fig. 8.83.

8.7.9 Cell Design and Layout

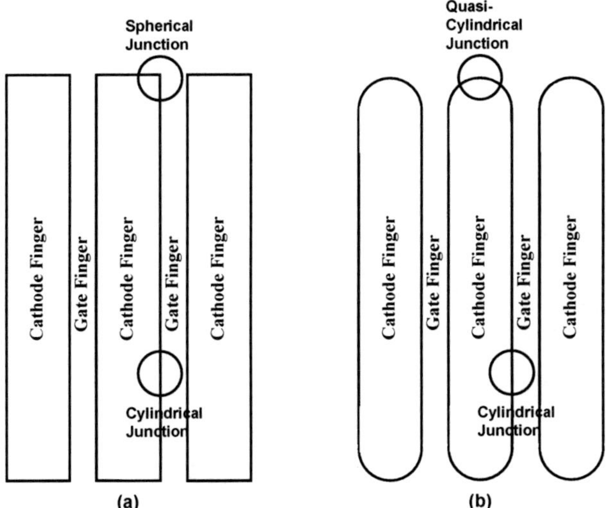

Fig. 8.84 Cathode cell designs for the GTO structure

The simplest cathode finger design from a layout standpoint utilizes rectangular cathode fingers as illustrated in Fig. 8.84a. In this case, a cylindrical junction is formed at the edges of the fingers and a spherical junction is formed at the corners of the fingers as indicated in the figure. The breakdown voltage of the N^+ cathode/P-base junction is then limited by spherical junction curvature as discussed in Chap. 3. An increase in the breakdown voltage for the N^+ cathode/P-base junction can be achieved by using the modified layout with rounded fingers as illustrated in Fig. 8.84b. A typical width for the cathode finger in the GTO structure of 400 μm results in a radius of 200 μm for the rounding of the fingers. Since this radius is far greater than the typical depth of 10 μm for the N^+ cathode/P-base junction, a quasicylindrical junction is formed at the ends of the fingers which increase the breakdown voltage to that for cylindrical curvature. Consequently, the commonly used design for the cathode finger contains rounded ends as shown in Fig. 8.84b.

A typical layout for the GTO structure across a circular wafer is illustrated in Fig. 8.85. The gate contact electrode is shown in the center of the wafer as in the case of the conventional thyristor structure. However, due to the high-current required to turn off the GTO structure, it is common to also place a gate contact to the GTO structure on the periphery of the wafer. All the cathode fingers are usually made with the same width and length to achieve uniform and simultaneous turn-off over the entire device area. Unless this is accomplished, the cathode current can constrict to only a few fingers during the turn-off process, producing a very high local power dissipation that leads to destructive failure of the GTO structure.[17]

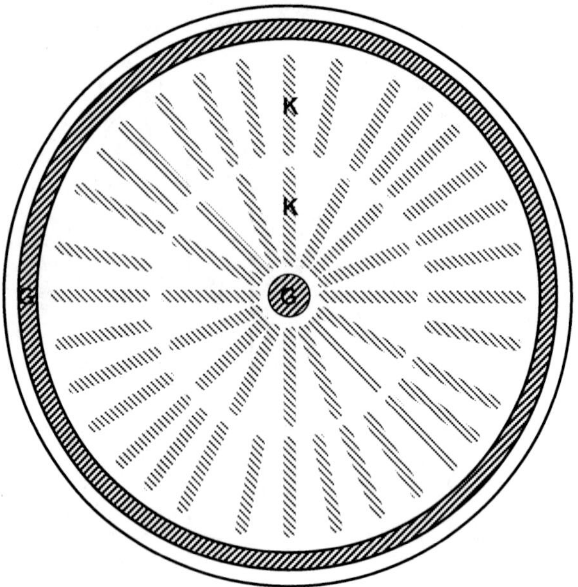

Fig. 8.85 Typical layout for the GTO structure

8.8 The Triac Structure

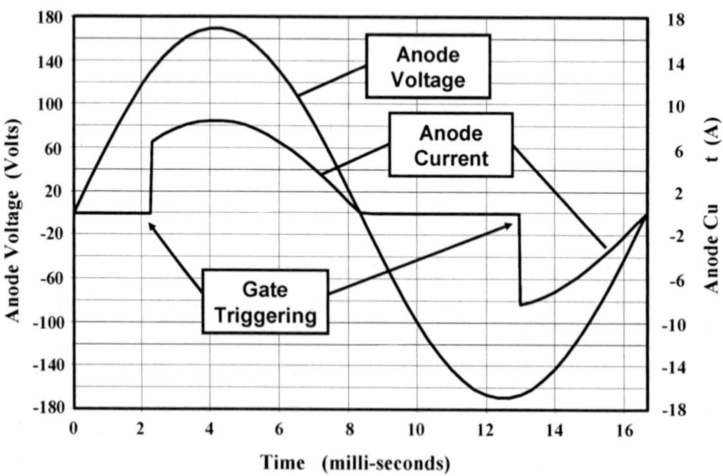

Fig. 8.86 Phase control power delivery during both half-cycles of an AC voltage source

Power is most commonly delivered to applications using AC sinusoidal sources as in the case of residences. The operation of household appliances such as mixers and food processors are some typical examples of these applications. The power delivery in these applications is done by using phase control circuits. This was previously discussed with the aid of Fig. 8.6 for the case of a thyristor as the power switch. The thyristor can conduct current during the positive half-cycle of the AC voltage source after it is triggered at an appropriate time during the half-cycle. It supports the AC voltage during the negative half-cycle preventing power delivery to the load. More efficient operation is achieved by delivering power to loads during both half-cycles of the AC voltage source, with the current conduction time during both half-cycles controlled by the power circuit as illustrated in Fig. 8.86.

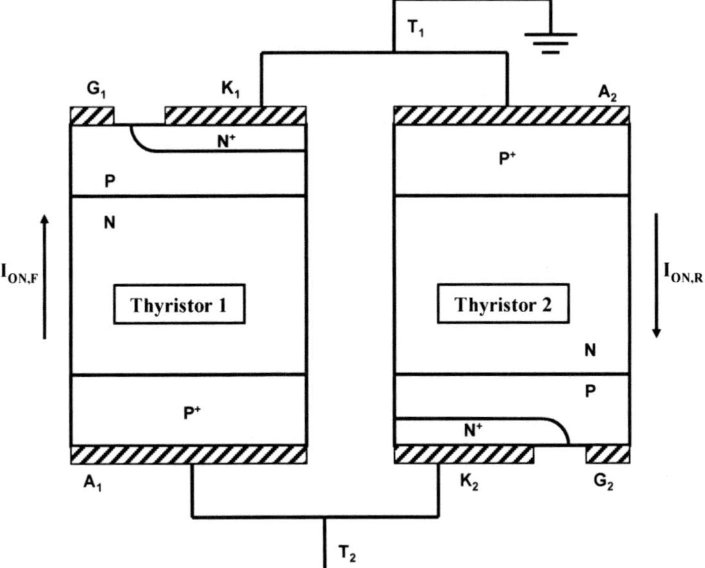

Fig. 8.87 Back-to-back thyristor configuration

This capability can be achieved by connecting two thyristors in a back-to-back configuration as shown in Fig. 8.87. In this configuration, one of the terminals is used as the reference or ground terminal as indicated for terminal T_1 in the figure. The anode and gate voltages must be applied in reference to this terminal. During the first half-cycle, when the voltage applied to terminal T_2 is positive, thyristor 1 operates in the forward-blocking mode while thyristor 2 operates in the reverse-blocking mode. Thyristor 1 can be triggered into its on-state during any time in the first half-cycle of operation allowing current to flow in the load. Similarly, during the second half-cycle, when the voltage applied to terminal T_2 is negative, thyristor 2 operates in the forward-blocking mode while thyristor 1 operates in the reverse-blocking mode. Consequently, thyristor 2 can be triggered into its on-state during any time in the second half-cycle of operation allowing current to flow in the load. Thus, phase control operation can be achieved with the back-to-back thyristor configuration with power flow to the load during both half-cycles. However, this solution has several disadvantages. Firstly, two thyristors are required to achieve the function. Secondly, the voltage at gate terminal (G_2) for thyristor 2 is close to its cathode voltage, which is operating at a high (and variable) voltage because it is connected to terminal T_2. This complicates the gate drive circuitry.

For obtaining phase control operation in both half-cycles of the AC voltage supply, it is preferable to use a device with the output characteristics shown in Fig. 8.88. This device has symmetrical forward- and reverse-blocking capability, and can be triggered into the on-state in both the first and third quadrants. In addition, it is desirable that the gate triggering is achievable by using a single gate terminal. A thyristor-based device structure that has been developed[18] with these features is referred to as the *triode AC switch or triac*.

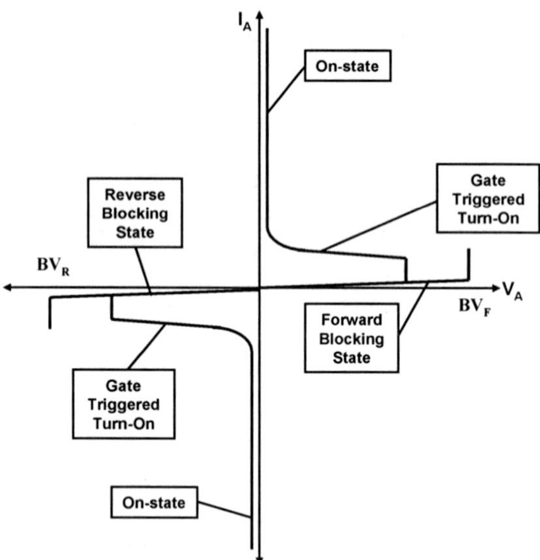

Fig. 8.88 Output characteristics for the triac

8.8.1 Basic Structure and Operation

The triac structure illustrated in Fig. 8.89 can be considered to be the monolithic combination of two thyristors in a back-to-back configuration. However, to create this structure, it is necessary to form two P-base regions on opposite surfaces of the structure. This introduces a lightly doped portion to the anode region for both the thyristors. The gate region is formed on only one side of the device. The gate electrode makes contact to the P-base region as in the case of conventional thyristor structures. Cathode shorts are incorporated for both thyristors within the triac to obtain a good blocking voltage and [dV/dt] capability.

When a positive bias is applied to terminal T_2 of the triac structure, the voltage is supported across junction J_2 with the blocking voltage capability determined by the open-base P–N–P transistor. The device can be triggered into its on-state with thyristor 1 carrying the current in the first quadrant. When a negative bias is applied to terminal T_2 of the triac structure, the voltage is supported across junction J_1 with the blocking voltage capability again determined by the open-base P–N–P transistor. The device can be triggered into its on-state with thyristor 2 carrying the current in the third quadrant. In both cases, most of the voltage is supported within the N-base region resulting in a symmetric blocking voltage capability. Since different portions of the triac structure carry current in the first and third quadrants of operation, its area is twice that for the thyristor structure. The main advantage of the triac structure is that it can be triggered into the on-state using a single gate terminal referenced to one of the high-current carrying terminals. The gate-triggering mechanisms are discussed in the subsequent sections.

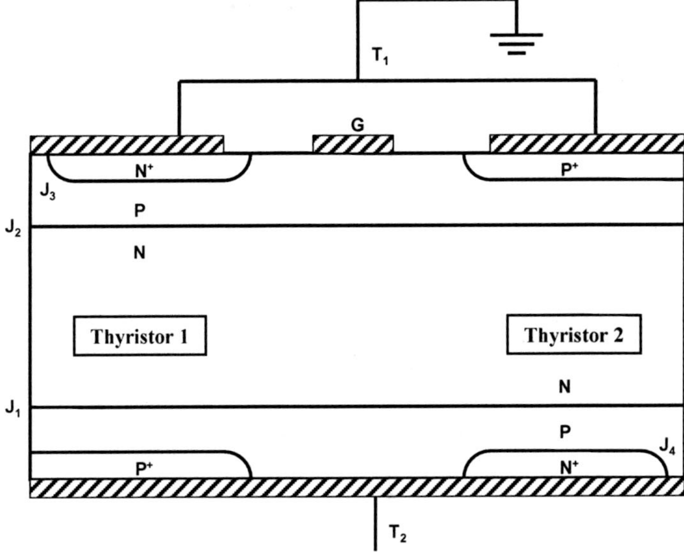

Fig. 8.89 Monolithic triac structure

8.8.2 Gate-Triggering Mode 1

When the triac is operated in the first quadrant with a positive bias applied to terminal T_2, thyristor 1 is operating in its forward-blocking mode. A positive bias applied to the gate terminal will induce a gate current flow to the cathode short as shown by the dashed line in Fig. 8.90. This produces a voltage drop in the P-base resistance under the N^+ cathode region which forward biases junction J_3. The largest forward bias occurs at location A near the gate contact. If the voltage drop at point A becomes close to the built-in potential, injection of electrons is initiated from junction J_3, which promotes the injection of holes from junction J_1, ultimately triggering thyristor 1 into its on-state. This triggering mode is similar to that for conventional thyristor structures.

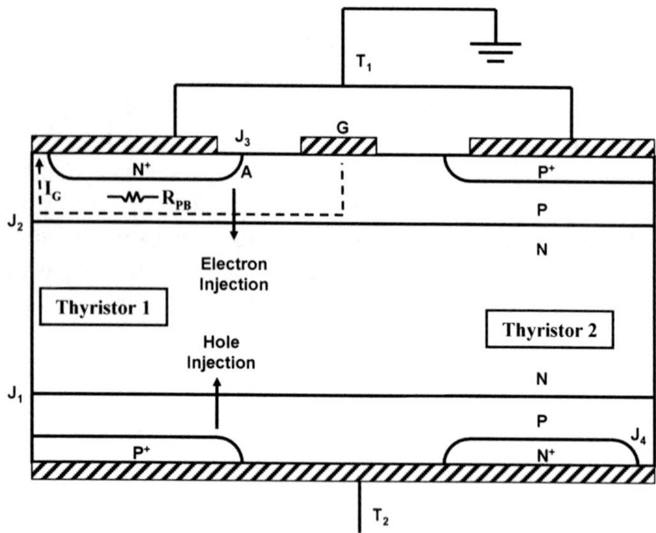

Fig. 8.90 Gate-triggering mode 1 for the triac structure

8.8.3 Gate-Triggering Mode 2

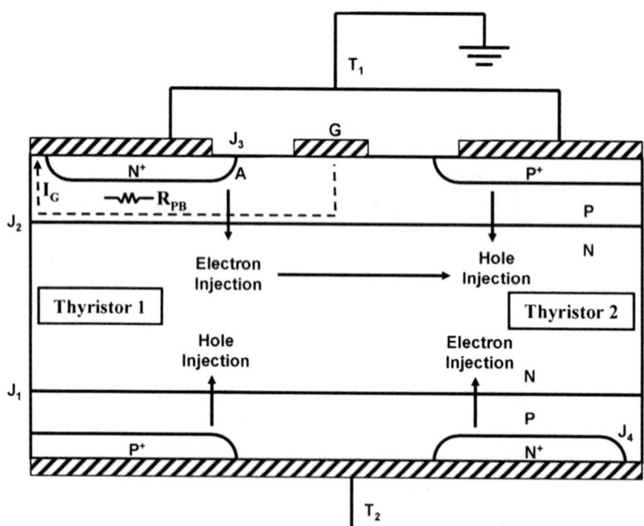

Fig. 8.91 Gate-triggering mode 2 for the triac structure

When the triac is operated in the third quadrant with a negative bias applied to terminal T_2, thyristor 2 is operating in its forward-blocking mode. A positive bias applied to the gate terminal will induce a gate current flow to the cathode short as shown by the dashed line in Fig. 8.91. This produces a voltage drop in the P-base resistance under the N^+ cathode region which forward biases junction J_3. The largest forward bias occurs at location A near the gate contact. If the voltage drop at point A becomes close to the built-in potential, injection of electrons is initiated from junction J_3. Although junction J_2 is forward biased, it collects the injected electrons and transfers them to the N-base region as indicated in the figure. These electrons act as the base drive current for the P–N–P transistor within thyristor 2, promoting the injection of holes from junction J_2. These holes diffuse through the N-base region and are collected by junction J_1. They then provide the base drive current for the N–P–N transistor for thyristor 2. This initiates the injection of electrons from junction J_4. These electrons diffuse through the P-base region and are collected at junction J_1. They then serve as the base drive current for the P–N–P transistor. This creates the regenerative action within thyristor 2 triggering it into its on-state. Since the electrons that are responsible for turning on thyristor 2 are derived from N^+ cathode region located within thyristor 1, this mechanism is referred to as the *remote gate triggering*. The triac structure can therefore be turned on using the same gate terminal when the voltage at terminal T_2 is either positive or negative.

8.8.4 [dV/dt] Capability

One of the problems observed with the triac structure is inadvertent turn-on under the application of a rapid change in the voltage applied to terminal T_2. This problem occurs because the device is usually conducting current just before the voltage across it is reversed at the end of each half-cycle as can be observed in Fig. 8.86. For an AC sinusoidal voltage waveform, the highest [dV/dt] occurs when the voltage crosses zero. The combination of the high [dV/dt] and the presence of the stored charge can lead to the undesirable turn-on of the triac without application of the gate control signal.

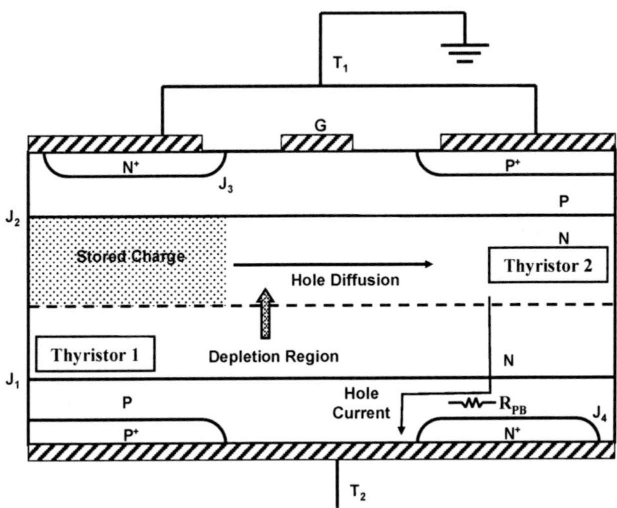

Fig. 8.92 Impact of stored charge on the triac structure

Consider the case of thyristor 1 operating in its on-state during the first half-cycle of the voltage waveform. A large amount of stored charge is then produced within the N-base region as indicated by the shaded region in Fig. 8.92. When the voltage at terminal T_2 changes from positive to negative, the junction J_1 becomes reverse biased. The depletion region at this junction expands as the anode voltage increases removing the stored holes. Due to the diffusion of the stored charge from the thyristor 1 region toward the thyristor 2 region, some of the holes produce a current within thyristor 2 as shown in the figure. This current produces a voltage drop across resistor R_{PB}, which forward biases junction J_4 at the middle of the N^+ cathode of thyristor 2. It has been found that the stored charge within the triac structure is sufficient to forward bias the N^+ cathode of thyristor 2 to produce the injection of electrons leading to its turn-on. Thyristor 2 will then conduct current for the entire second half-cycle without a gate-triggering signal. This disrupts the phase control operation.[19] The problems with [dV/dt] turn-on during commutation have restricted the triac structure to operation at below 1,200 V and 100 A.[20] However, they have been extensively used for home appliance controls. In these applications, the electromagnetic interference (EMI) created by the uncontrolled rate of rise of the device current in the triac during their turn-on has been found to produce flicker in television reception. These devices have consequently been displaced by IGBTs in modern household applications.

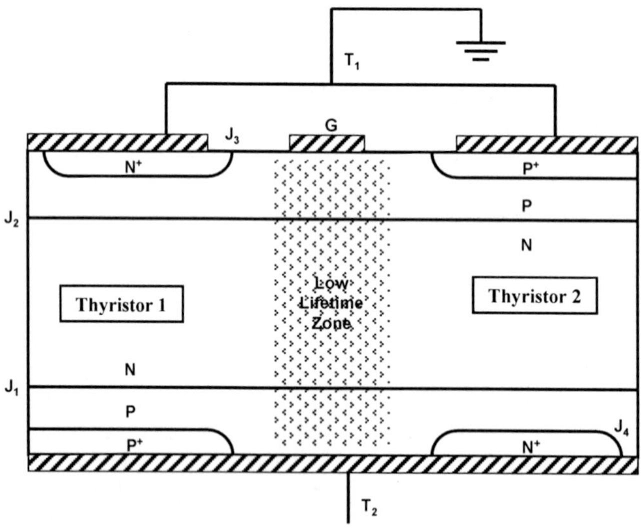

Fig. 8.93 Improving the [dV/dt] performance of the triac structure

One option for improving the [dV/dt] capability is to isolate the two thyristors and suppress the migration of the stored charge between the two segments. This can be accomplished by creating a low lifetime zone between the two thyristors as illustrated in Fig. 8.93. At the same time, a high lifetime is maintained in the thyristor regions to produce devices with low on-state voltage drop. The low lifetime zone can be produced by using electron irradiation with a tungsten mask to protect the thyristor regions within the triac structure.

8.9 Summary

The physics of operation of the thyristor structure has been described in this chapter. The thyristor structure contains two coupled transistors that create a regenerative action within the device that can sustain on-state current flow with no external stimulus. Very high-voltage devices can be synthesized with low on-state voltage drop and high-current handling capability. Thyristors are extensively used in HVDC transmission systems with special gate structures that allow optically induced triggering. They are also employed for driving very high power motors.

A thyristor structure (GTO) capable of turning off current in DC power circuits has been developed for motor drives used in traction (electric locomotives) applications. These devices found extensive use in high speed rail transport systems until recently. The development of the IGBT[21] in the early 1980s followed by scaling its voltage- and current-handling capability in 1990s has resulted in the displacement of the GTO in traction applications. However, the physics of operation for the thyristor continues to have relevance because a four-layer (parasitic thyristor) device is formed within the IGBT structure. The design of the IGBT requires a good understanding of thyristor physics to prevent its destructive failure under certain circuit operating conditions.

Problems

8.1 Consider an $N^+PN–P^+$ power thyristor structure with uniformly doped N^+ cathode, P-base, N-drift, and P^+ anode regions. The N^+ cathode region has a doping concentration of 2×10^{19} cm^{-3} and thickness of 10 μm. The P-base region has a doping concentration of 2×10^{17} cm^{-3} and thickness of 20 μm. The N-drift region has a doping concentration of 5×10^{13} cm^{-3} and thickness of 300 μm. The P^+ anode region has a doping concentration of 2×10^{19} cm^{-3} and thickness of 50 μm. The Shockley–Read–Hall (low-level, high-level, and space-charge generation) lifetime is 10 ns in the N^+ cathode and P^+ anode regions, 10 μs in the P-base and N-drift regions. Ignore band-gap narrowing and Auger recombination. Use an ambipolar diffusion constant D_a of 15 cm^2 s^{-1} for the on-state calculations. The structure has a linear cell geometry with an emitter width of 0.5 cm and length of 1 cm. What is the blocking voltage capability for the device?

8.2 Determine the cathode short distance (D) for a square shorting array to prevent turn-on of the above thyristor structure. Use a maximum forward bias for the cathode/base junction of 0.5 V at a leakage current density of 1 A cm^{-2}. Assume a cathode short size (d) of 25 μm.

8.3 Determine the on-state current for the thyristor structure described in Problem 8.1 at a forward bias of 1.0 V.

8.4 Calculate the gate-triggering current for the thyristor structure described in Problem 8.1. Use the cathode shorting distance from Problem 8.2.

8.5 What is the holding current for the thyristor structure described in Problem 8.1? Use the cathode shorting distance from Problem 8.2 and a current gain of 0.5 for the N–P–N transistor in the on-state.

8.6 Determine the [dV/dt] capability for the thyristor structure described in Problem 8.1 when the device is blocking 100 V. What is the maximum operating frequency for the thyristor if the sinusoidal anode voltage has a maximum value of 1,000 V?

8.7 Calculate the current rise time when turning on the thyristor structure described in Problem 8.1 with an anode supply voltage of 10 V. What is the current rise time if the anode supply voltage is increased to 100 V? Use a gate current density of 1 A cm^{-2}.

8.8 Using the typical current spreading velocity in a thyristor, determine the current spreading time for the thyristor structure described in Problem 8.1.

8.9 Design an amplifying gate thyristor structure using a gate electrode pad with radius of 0.25 cm to achieve a gate-triggering current of 200 mA. Use the parameters given in Problem 8.1 for the semiconductor regions.

8.10 Design a light-triggered thyristor using an optically generated current density of 0.1 A cm^{-2} in the semiconductor. The central gate region has a radius of 0.25 cm.

8.11 A gate turn-off thyristor structure has uniformly doped N$^+$ cathode, P-base, N-drift, and P$^+$ anode regions with the following parameters. The N$^+$ cathode region has a doping concentration of 2×10^{19} cm^{-3} and thickness of 10 μm. The P-base region has a doping concentration of 2×10^{17} cm^{-3} and thickness of 40 μm. The N-drift region has a doping concentration of 5×10^{13} cm^{-3} and thickness of 300 μm. The P$^+$ anode region has a doping concentration of 2×10^{19} cm^{-3} and thickness of 50 μm. The Shockley–Read–Hall (low-level, high-level, and space-charge generation) lifetime is 10 ns in the N$^+$ cathode and P$^+$ anode regions, 10 μs in the P-base and N-drift regions. Ignore band-gap narrowing and Auger recombination.

Use an ambipolar diffusion constant D_a of 15 $cm^2 s^{-1}$ for the on-state calculations. The structure has a linear cell geometry with an emitter width of 200 μm and length of 1 cm. Determine the on-state current at an on-state voltage drop of 1.0 V.

8.12 Determine the storage time if the gate turn-off thyristor structure described in Problem 8.11 is turned off at a current gain of 2 with a constant gate drive current.

8.13 The gate turn-off thyristor structure described in Problem 8.11 is turned off using a DC supply voltage of 1,000 V. Determine the voltage rise time.

8.14 Determine the current waveform during turn-off for the gate turn-off thyristor structure described in Problem 8.13. Provide the magnitude of the sudden drop in the anode current. What is the time constant for the exponential decay in the anode current? Determine the current turn-off time.

8.15 Determine the energy loss during the turn-off of the gate turn-off thyristor structure described in Problems 8.13 and 8.14. Provide the energy loss values associated with the voltage rise time and the current fall time.

8.16 What is the maximum controllable anode current for the gate turn-off thyristor structure described in Problem 8.11. Assume one-dimensional parallel-plane breakdown voltage for the cathode/base junction.

References

[1] I.M. Mackintosh, "The Electrical Characteristics of Silicon p–n–p–n Triodes", Proceeding of the IRE, Vol. 46, p. 1229, 1958.

[2] R.W. Aldrich and N. Holonyak, "Multiterminal p–n–p–n Switches", Proceeding of the IRE, Vol. 46, p. 1236, 1958.

[3] A. Herlet, "The Maximum Blocking Capability of Silicon Thyristors", Solid-State Electronics, Vol. 8, pp. 655–671, 1965.

[4] P.S. Raderecht, "A Review of the Shorted Emitter Principle as Applied to p–n–p–n Silicon Controlled Rectifiers", International Journal of Electronics, Vol. 31, pp. 541–564, 1971.

[5] A. Munoz-Yague and P. Leturcq, "Optimum Design of Thyristor Gate–Emitter Geometry", IEEE Transactions on Electron Devices, Vol. ED-23, pp. 917–924, 1976.

[6] S.K. Ghandhi, "Power Semiconductor Devices", pp. 210–214, Wiley, New York, 1977.

[7] R.L. Longini and J. Melngailis, "Gated Turn-On of Four Layer Switch", IEEE Transactions on Electron Devices, Vol. ED-10, pp. 178–185, 1963.
[8] W.H. Dodson and R.L. Longini, "Probed Determination of Turn-On Spread of Large Area Thyristors", IEEE Transactions on Electron Devices, Vol. ED-13, pp. 478–484, 1966.
[9] H.J. Ruhl, "Spreading Velocity of the Active Area Boundary in a Thyristor", IEEE Transactions on Electron Devices, Vol. ED-17, pp. 672–680, 1970.
[10] Y. Yamasaki, "Experimental Observation of the Lateral Plasma Propagation in a Thyristor", IEEE Transactions on Electron Devices, Vol. ED-22, pp. 65–68, 1975.
[11] V.A.K. Temple and A.P. Ferro, "High-Power Dual Amplifying Gate Light Triggered Thyristors", IEEE Transactions on Electron Devices, Vol. ED-23, pp. 893–898, 1976.
[12] T. Horigome et al., "A 100 kV Thyristor Converter for High Voltage DC Transmission", IEEE Transactions on Electron Devices, Vol. ED-17, pp. 809–815, 1976.
[13] D. Silber, W. Winter, and M. Fullmann, "Progress in Light Activated Power Thyristors", IEEE Transactions on Electron Devices, Vol. ED-23, pp. 899–904, 1984.
[14] S.M. Sze, "Physics of Semiconductor Devices", 2nd Edition, pp. 754–756, Wiley, New York, 1981.
[15] F.J. Niedernostheide et al., "Light Triggered Thyristors with Integrated Protection Functions", IEEE International Symposium on Power Semiconductor Devices and ICs, pp. 267–270, 2000.
[16] F.F. Mazda, "Power Electronics Handbook", Chapter 11, pp. 227–245, Butterworths, Boston, 1993.
[17] A.A. Jaecklin, "Performance Limitations of a GTO with Near-Perfect Technology", IEEE Transactions on Electron Devices, Vol. ED-39, pp. 1507–1513, 1992.
[18] F.E. Gentry, R.I. Scace, and J.K. Flowers, "Bidirectional Triode P–N–P–N Switches", Proceedings of the IEEE, Vol. 53, pp. 355–369, 1965.
[19] G.D. Bergman, "Gate Isolation and Commutation in Bi-Directional Thyristors", International Journal of Electronics, Vol. 21, pp. 17–35, 1966.
[20] P.D. Taylor, "Thyristor Design and Realization", pp. 147–151, Wiley, New York, 1987.
[21] B.J. Baliga et al., "The Insulated Gate Transistor", IEEE Transactions on Electron Devices, Vol. ED-31, pp. 821–828, 1984.

Chapter 9

Insulated Gate Bipolar Transistors

The insulated gate bipolar transistor (IGBT) concept was discovered[1] and developed in the early 1980s to provide an improved alternate power device to bipolar power transistors.[2] As discussed in Chap. 7, one of the major shortcomings of the bipolar power transistor is its low-current gain when designed to operate at high voltages. This increases the size, weight, and cost of its discrete gate drive circuit. In addition, the bipolar power transistor needs snubber circuits to compensate for its poor safe operating area (SOA). The resulting system solution obtained with bipolar power transistors is cumbersome in design, difficult to manufacture, and expensive for implementation in consumer and industrial applications.

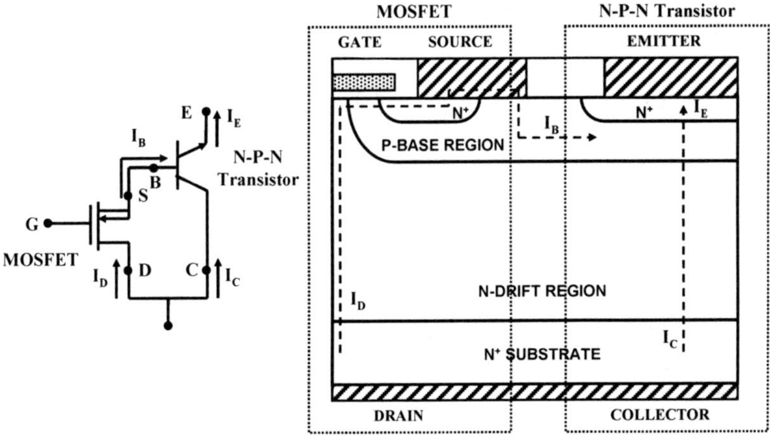

Fig. 9.1 Hybrid MOS–bipolar power devices

B.J. Baliga, *Fundamentals of Power Semiconductor Devices*, doi: 10.1007/978-0-387-47314-7_9,

One of the approaches taken to create an improved high-voltage switch is to use a hybrid solution with a power MOSFET structure providing the gate drive current to a bipolar power transistor.[3] The equivalent circuit for this hybrid MOS–bipolar configuration is shown in Fig. 9.1 together with its monolithic implementation. Note that the MOSFET and bipolar transistor are located in different portions of the silicon chip in this concept. The gate drive signal for this hybrid configuration is applied to the power MOSFET, providing the advantages of compact, low cost, integrated gate drive circuits made possible by the high input impedance, voltage-controlled operation. The source current for the power MOSFET is used to provide the base drive current for the bipolar power transistor. The bipolar power transistor is designed to carry most of the device current because of its superior on-state characteristics. Both the transistors can be simultaneously fabricated by forming a single P-region as the base of the bipolar power transistor and the base of the power MOSFET, and a single N^+ diffusion to create the emitter of the bipolar power transistor and the source of the power MOSFET.

Unfortunately, the hybrid MOS–bipolar configuration has many disadvantages. Firstly, the power MOSFET must have the same high-voltage blocking capability as the bipolar power transistor. The high specific on-resistance for high-voltage power MOSFET structures, together with the low-current gain for the bipolar power transistor, results in a large area occupied by the power MOSFET in spite of typically carrying a tenth of the total device current. This reduces the average on-state current density making the die size and cost large. Secondly, a reverse base drive current cannot be applied in this circuit to the bipolar power transistor resulting in a very long turn-off time. Attempts made to improve upon this circuit by using an additional MOSFET in series with the emitter of the bipolar transistor have not found widespread applications.[4]

The IGBT represents the true MOS–bipolar integration in the sense of intermingling the physics of operation of the MOSFET structure with the physics of operation of bipolar transistors.[5] In this concept, the MOSFET structure is used to provide the base drive current for the bipolar transistor with the bipolar transistor used to modulate the conductivity of the drift region for the MOSFET structure. The reduced resistance of the power MOSFET structure allows providing much greater base drive current to the bipolar transistor in this configuration. Consequently, the bipolar power transistor structure can be altered from the traditional narrow-base N–P–N transistor structure to the wide-base P–N–P transistor structure. Although this change reduces the current gain of the bipolar transistor, it greatly enhances the SOA for the composite structure. Moreover, the IGBT structure has inherent forward- and reverse-blocking capability enabling its utilization in both DC and AC circuit applications.

The first observation of the IGBT mode of operation in a four-layer vertical device structure was reported in 1979.[1] In this work, the ability to increase and reduce the current flow in a four-layer structure under the control of a gate voltage without latch-up of the thyristor was experimentally demonstrated over a limited current range for the first time. The first IGBT devices with substantial current ratings were reported in the literature in 1982 with a symmetric (equal forward and reverse) blocking voltage capability of 600 V in recognition of their prospective applications in DC and AC circuits.[6] It was soon apparent that the asymmetrical IGBT structure, with only forward-blocking capability, could be optimized with superior on-state and switching speed capability.[7,8] The ability to scale the voltage rating of the IGBT structure with modest increase in the on-state voltage drop was also recognized immediately after its invention.[5] This provided a strong impetus to scale up the voltage rating of IGBT structures as shown in Fig. 9.2. This chart has been created using data published on IGBT devices at the IEEE International Electron Devices Meeting until 1990 and the IEEE International Symposium on Power Semiconductor Devices and ICs beyond this time frame. Within a short span of 10 years, devices capable of supporting over 3,000 V were developed. Continued interest in using the IGBT for traction (electric locomotive drives) applications led to the scaling up of the voltage ratings to 6,500 V for these devices by work done in Europe and Japan.

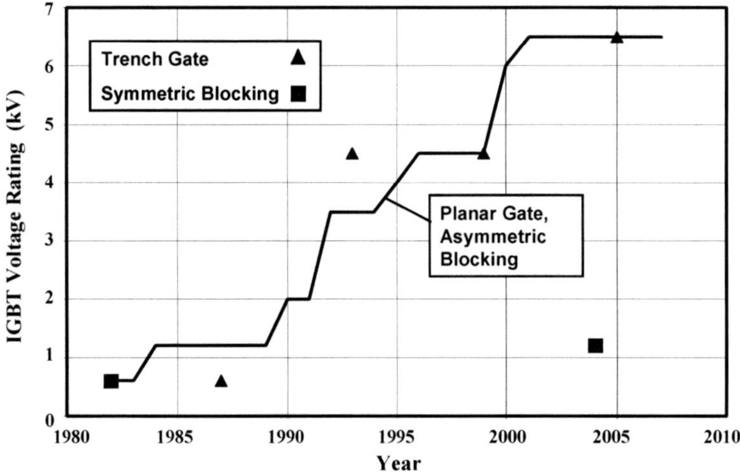

Fig. 9.2 Growth in voltage ratings for IGBTs

Interest in reducing the on-state voltage drop for IGBTs motivated the utilization of a trench-gate architecture in place of the planar-gate configuration in 1987.[9] Subsequently, the observation of an enhanced carrier concentration in the vicinity of the trench-gate region, named the injection enhancement effect,[10] has stimulated the development of devices that have the same high-voltage and current-handling capability as planar devices. In addition, although IGBT structures with reverse-blocking capability were first reported in 1982,[6] these structures lay dormant for many years due to higher degree of interest in DC circuit applications as well as the difficulties for formation of passivated reverse-blocking junctions. There has been a recent resurgence of interest in these structures for applications in matrix (AC-to-AC) converters.[11–14]

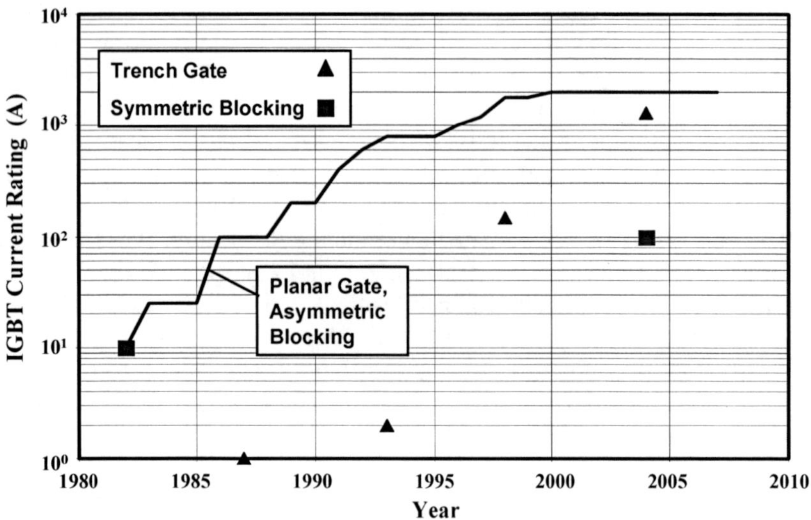

Fig. 9.3 Growth in current-handling capability for IGBTs

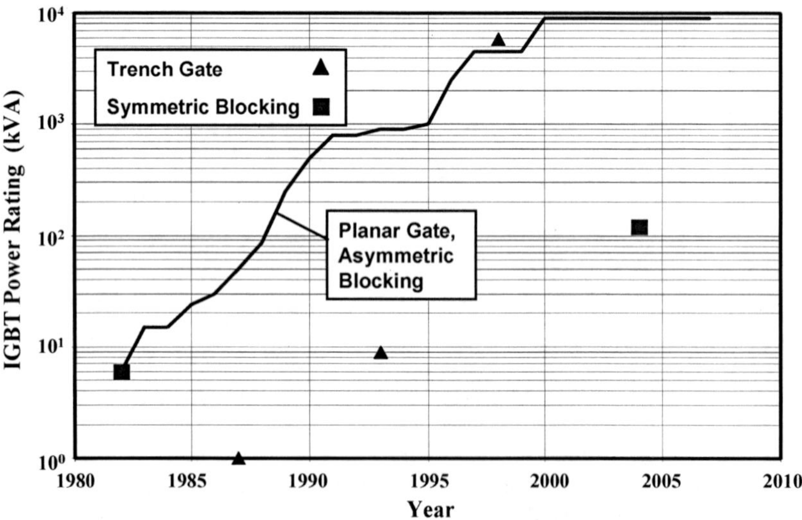

Fig. 9.4 Growth in power-handling capability for IGBTs

The growth in the current-handling capability for the IGBT devices is charted in Fig. 9.3 using the same reference sources used for the voltage ratings. The first IGBTs reported in 1982 had a modest current rating of 10 A, which quickly scaled to 25 A for use in adjustable speed motor drives for air-conditioning applications. Since then, interest in Europe and Japan in using the IGBT for traction applications has motivated a steady growth in the current-handling capability. This has been accomplished by scaling the die size as well as improving the packaging. Press pack IGBTs have now become available with the capability of switching 2,000 A. These devices have utilized the planar-gate structure due its larger die yield and ruggedness. The current-handling capability of trench-gate IGBT devices has generally lagged behind in time when compared with planar-gate devices. However, trench-gate IGBTs have now been developed capable of handling 1,300 A.

The extraordinarily rapid growth in the power-handling capability of IGBT devices is charted in Fig. 9.4. The chart has been developed by using the previously cited reference sources augmented by review papers that have reported these trends.[5,15] The first IGBTs reported in 1982 were capable of delivering 6 kW of power to a load. Since then, a rapid growth in power-handling capability has been achieved by scaling both the voltage and the current ratings. A single IGBT module is now available with the ability to control 9 MW of power. Consequently, the IGBT structure has not only completely displaced the bipolar power transistor in medium power applications but is now displacing the gate turn-off thyristor (GTO) in high power applications.

In this chapter, the basic structure and operating characteristics of the IGBT are first described. The analysis of forward- and reverse-blocking capability is then performed followed by discussion of the on-state characteristics of IGBTs with various collector structures. The IGBT structure is shown to behave like a P–i–N rectifier in the on-state, producing a low on-state voltage drop even for structures designed to support large voltages. The switching behavior of the IGBT is then analyzed including discussion of lifetime control processes for optimization of the tradeoff between on-state and switching losses. Various methods to suppress the parasitic thyristor within the IGBT structure are discussed in detail because this is essential to the design of functional devices. The analysis and design of trench-gate IGBT structures are also provided in this chapter.

9.1 Basic Device Structures

The basic structure for an n-channel symmetric blocking IGBT structure is illustrated in Fig. 9.5. The current-carrying terminals for the IGBT structure have been labeled the emitter and collector terminals because this made the device structure pin compatible with existing bipolar transistor circuits when the IGBT was initially developed. Unfortunately, this is contradictory to the physics of operation of the internal bipolar transistor within the IGBT structure. The control terminal for the IGBT is labeled as a gate terminal because it has an MOS structure similar to that in a power MOSFET device.

The IGBT structure is formed by using four (N–P–N–P) alternating semiconductor layers. This creates a basic thyristor structure. However, the thyristor structure is made inoperative by including a deep P^+ diffusion and short circuiting the P-base region to the N^+ emitter region using the emitter electrode as shown in the figure. When thyristor operation is suppressed, the IGBT structure operates as a transistor as discussed in more detail below. The symmetric blocking IGBT structure is often referred to as the *nonpunch-through (NPT) IGBT structure* because the electric field does not extend through the entire width of the lightly doped portion of the N-drift region.

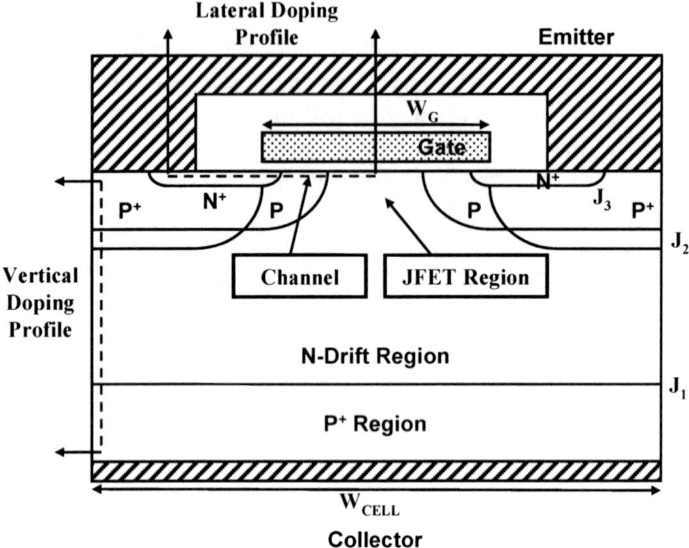

Fig. 9.5 Symmetric IGBT device structure

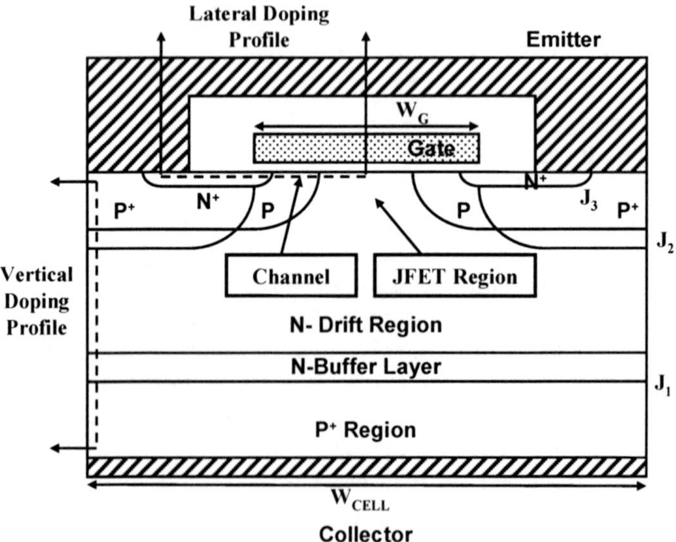

Fig. 9.6 Asymmetric IGBT device structure

The cross section for the asymmetric IGBT structure is shown in Fig. 9.6. This IGBT structure is optimized for DC circuit applications where no reverse bias is applied to the device because it operates exclusively in the first quadrant of the *i–v* characteristics. The main distinguishing feature of the asymmetric IGBT structure is the introduction of an N-buffer layer within the N-drift region. This layer is sometimes referred to as the *field-stop layer*. In the asymmetric IGBT structure, the forward-blocking voltage can be supported by a thinner N-drift region resulting in reducing the on-state voltage drop. This structure is often referred to as the *punch-through (PT) IGBT structure* because the electric field "punches through" the entire lightly doped portion of the N-drift region.

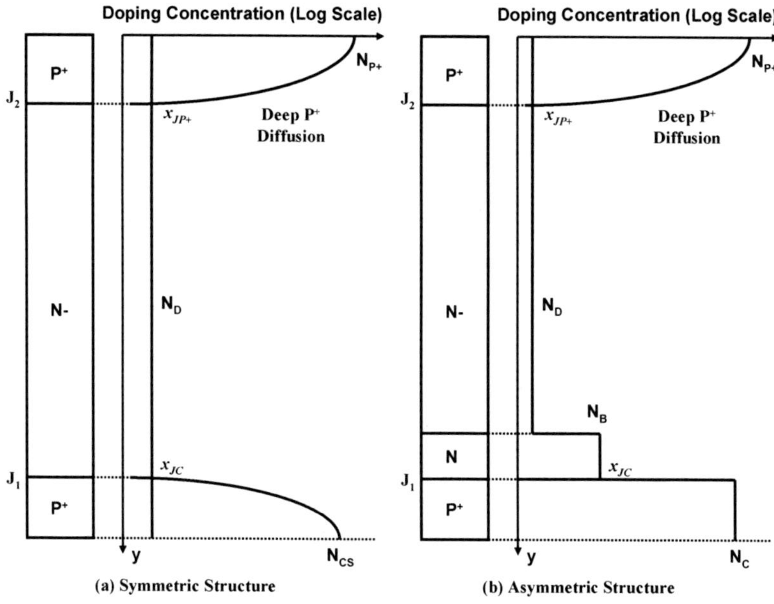

Fig. 9.7 Vertical doping profiles for the IGBT device structures

The symmetric IGBT structure is usually fabricated by starting with a lightly doped N-type silicon wafer whose resistivity is chosen based upon the blocking voltage rating for the device. The collector P^+ region is formed by the diffusion of dopants from the backside of the wafer to a junction depth x_{JC}. The asymmetric IGBT structure is usually fabricated by the growth of the N-type drift region on top of a P^+ substrate. The initial growth of the N-type drift region is performed with a larger doping concentration to create the N-buffer layer. The rest of the N-type drift region is then grown with a low doping concentration to the thickness necessary to support the forward-blocking voltage. In both IGBT structures, a deep P^+ region is formed at the center of the window in the polysilicon gate electrode to suppress the latch-up of the parasitic thyristor. Its depth (x_{JP+}) is chosen such that the P-type dopant does not penetrate into the semiconductor below the gate electrode due to lateral diffusion. The doping profiles taken in the vertical direction along the dashed lines shown in the device cross sections are illustrated in Fig. 9.7.

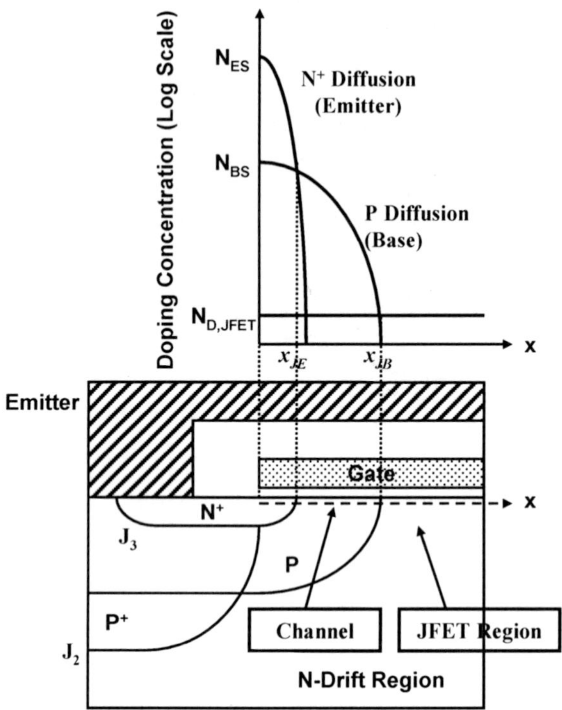

Fig. 9.8 Lateral doping profile for the IGBT device structure

The gate oxide is grown after the formation of the deep P^+ region, immediately followed by the deposition of the polysilicon gate electrode. The P-base is then created in both IGBT structures by the ion implantation of boron as the P-type dopant using the polysilicon as the mask. The P-type dopant is diffused to a depth of x_{JB} to form the P-base region. Next, a mask is used in the center of the polysilicon window during the ion implantation of the N^+ emitter to allow the short circuiting of the junction J_3 between the N^+ emitter and the P-base regions. The N-type dopant is diffused to a depth of x_{JE} to form the N^+ emitter region. The N^+ emitter and P-base regions are consequently self-aligned to the edge of the polysilicon gate electrode, enabling control of the channel length by their relative diffusion depths. Electrodes are formed on the front side of the wafer to contact the emitter and gate regions, and on the backside of the wafer to contact the collector region. No contact electrode is usually attached to the N-drift (N-base) region.

The doping profile taken in the lateral direction, just under the gate oxide along the dashed lines shown in the device cross sections, is illustrated in Fig. 9.8. The N^+ emitter and P-base regions have Gaussian doping profiles with surface concentrations of N_{ES} and N_{BS}, respectively. The doping concentration of the JFET region is sometimes enhanced to $N_{D,JFET}$ by an additional diffusion of N-type dopants prior to the formation of the P^+ region on the top surface of the wafer. The larger doping concentration in the JFET region reduces the lateral diffusion depth x_{JB} of the P-base region. This must be accounted for when designing the process to obtain an adequate channel length. The maximum P-type doping concentration of the P-base region, after accounting for compensation by the N-type emitter dopant, determines the threshold voltage for the IGBT structure. This concentration is usually controlled by adjusting the dose of the P-type ion implantation, which determines the surface concentration N_{BS} of the P-base region.

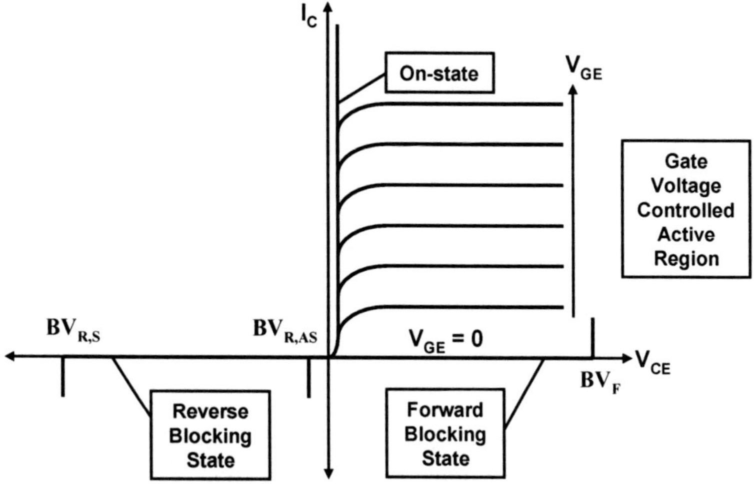

Fig. 9.9 Output characteristics of the IGBT device structures

9.2 Device Operation and Output Characteristics

The IGBT structure is capable of supporting high voltages when the gate electrode is short circuited to the emitter electrode by the external gate control circuit. When a positive bias is applied to the collector of either the symmetric or asymmetric IGBT structure, junction J_1 is forward biased while junction J_2 becomes reverse biased. The applied voltage is supported across the reverse-biased deep P^+/N-base (or N-drift) region junction J_2 with the depletion region extending in the N-base region toward junction J_1. The maximum forward-blocking voltage capability of the symmetric IGBT structure is determined by the open-base breakdown voltage of the P–N–P transistor. The thickness of the N-base region and the minority carrier lifetime in this region determine the breakdown voltage. In the case of the asymmetric IGBT structure, the lightly doped portion of the N-base (N-drift) region becomes completely depleted at a relatively low voltage. The electric field stops at the interface between the N-drift region and the N-buffer layer because of the relatively high doping concentration (10^{16}–10^{17} cm^{-3}) of the N-buffer layer. The N-buffer layer thickness and doping concentration must be sufficient to suppress reach-through and keep the gain of the P–N–P transistor as low as possible. In this case, the forward-blocking capability of the asymmetric IGBT structure will be determined by the thickness of the lightly doped portion of the N-base region.

When a negative bias is applied to the collector of the IGBT structure, junction J_1 is reverse biased while junction J_2 becomes forward biased. The applied voltage is supported across the reverse-biased junction J_1 with the depletion region extending in the N-base region toward junction J_2. The maximum reverse-blocking voltage capability of the symmetric IGBT structure is determined by the open-base breakdown voltage of the P–N–P transistor. The thickness of the N-base region and the minority carrier lifetime in this region determine the breakdown voltage. The reverse-blocking voltage capability is consequently equal to the forward-blocking voltage capability. In the case of the asymmetric IGBT structure, the junction J_1 cannot support a high voltage because of the high doping concentration (10^{16}–10^{17} cm^{-3}) of the N-buffer layer and the P$^+$ collector region (10^{19} cm^{-3}). This is acceptable for utilization of the asymmetric IGBT structure in DC circuits, especially for motor control where an antiparallel diode is connected across the IGBT device.

The application of a positive bias to the gate of the IGBT structures creates an inversion layer channel under the gate electrode, which connects the N$^+$ emitter region to the N-base region. This allows the transport of electron current from the N$^+$ emitter region to the N-base region when a positive voltage is applied to the collector terminal. The electron current serves as the base drive current for the P–N–P transistor, promoting the injection of holes from the P$^+$ collector/N-base junction J_1. The injected holes produce the emitter current of the P–N–P bipolar transistor. Consequently, current flow occurs from the collector terminal to the emitter terminal with a bipolar component associated with the wide-base P–N–P transistor and a unipolar component via the channel of the MOSFET region. The N-base (N-drift) region of the IGBT structures operates with high-level injection conditions during current flow. This reduces the resistance of the N-base region allowing high-current flow with low on-state voltage drop. With sufficiently large gate bias to ensure operation of the MOSFET region in its linear (triode) mode, the on-state characteristics of the IGBT structure look like those of a P–i–N rectifier as illustrated in Fig. 9.9.

When the gate bias voltage is made only slightly above the threshold voltage of the MOSFET region, the channel enters pinch-off conditions resulting in limiting the electron current delivered to the N-base region. Since this limits the base drive current for the P–N–P bipolar transistor, its current also becomes saturated. The IGBT structure is therefore capable of exhibiting an active region with the collector current saturated at a value determined by the applied gate bias as illustrated in Fig. 9.9. These characteristics are valuable for providing short-circuit protection features in applications. The SOA for the IGBT has been found to be excellent eliminating the need for snubbers that were essential when bipolar power transistors were used in applications.

The switching of the IGBT device in power circuits can be performed by toggling the gate bias between zero and a positive bias V_{GS} above its threshold voltage. When the gate bias is abruptly increased from zero to V_{GS}, the collector current of the IGBT increases rapidly with an on-state voltage drop that is determined by the modulation of the conductivity of the N-base region. The rate of rise in collector current can be controlled by tailoring the rate of rise of the gate bias voltage. When the gate bias is rapidly decreased from V_{GS} to zero, the collector current is turned off with a current tail associated with the removal of the stored charge in the N-base region. The turn-off switching losses can be reduced by decreasing the lifetime in the N-base region but this is accompanied by an increase in the on-state voltage drop. It is therefore customary to perform a tradeoff between the on-state voltage drop and the switching losses when designing IGBT devices.

As mentioned earlier, the IGBT structure consists of four alternating semiconductor layers that also constitute a thyristor structure as indicated in Fig. 9.10. It is necessary to completely suppress the operation of this thyristor structure to preserve the IGBT mode of operation over a broad range of currents and voltages. The suppression of the triggering and latch-up of the parasitic thyristor can be accomplished by numerous techniques that are discussed in detail later in this chapter. The successful implementation of these methods for suppressing the parasitic thyristor has been crucial to the success of the IGBT device as a product for power electronic applications.

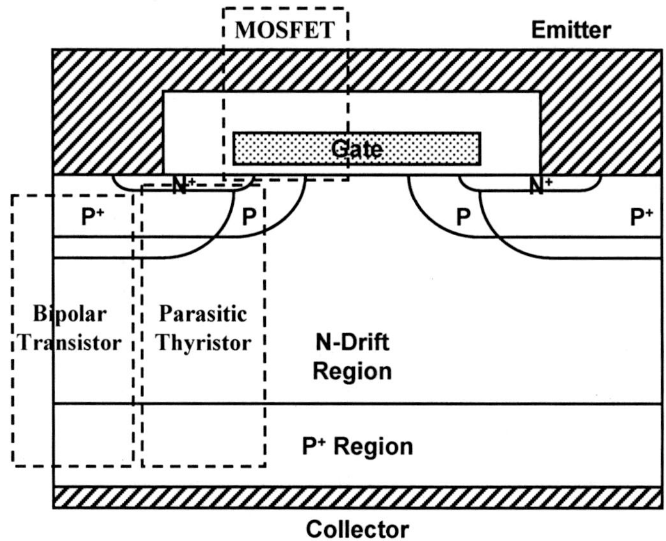

Fig. 9.10 Parasitic thyristor within the IGBT device structure

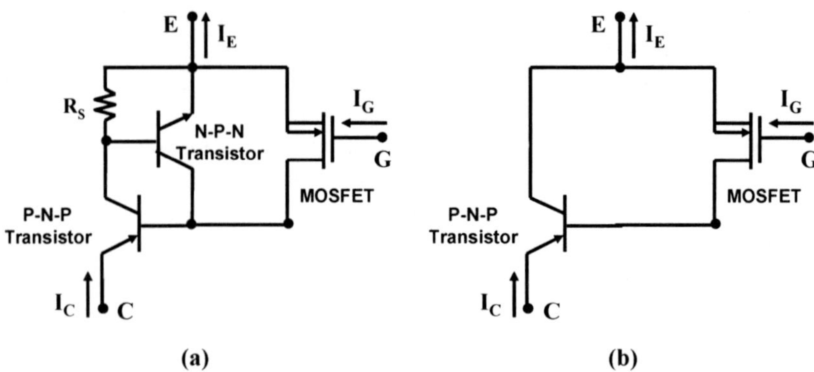

Fig. 9.11 Equivalent circuits for the IGBT device structure

9.3 Device Equivalent Circuit

The equivalent circuit representation of the IGBT structure is provided in Fig. 9.11a. The circuit contains the coupled P–N–P and N–P–N transistors that form the parasitic thyristor as well as the MOSFET which provides the base drive current for the P–N–P transistor. Although the N^+ emitter and P-base regions in the IGBT structure are short circuited by the emitter electrode, there is a finite resistance for bipolar current flow to the emitter electrode indicated by the resistor R_S. If this resistance is made extremely small, the N–P–N transistor within the IGBT structure cannot be activated. In this case, the equivalent circuit for the IGBT becomes a bipolar transistor driven by a MOSFET transistor as illustrated in Fig. 9.11b.

9.4 Blocking Characteristics

As discussed in the previous section, the symmetric IGBT structure provides both forward-blocking capability in the first quadrant of operation and reverse-blocking capability in the third quadrant of operation. The voltage-blocking capability in both quadrants is approximately equal making the IGBT well suited for AC power circuits. In the case of the asymmetric IGBT structure, the operating characteristics for the device are optimized for operation in the first quadrant while sacrificing the reverse-blocking capability.

9.4.1 Symmetric Structure Forward-Blocking Capability

When a positive bias is applied to the collector terminal of the IGBT structure, the deep P^+ region/N-base junction (J_2) becomes reverse biased while the junction (J_1) between the P^+ collector region and the N-base region becomes forward biased. The collector voltage is supported across the P^+ region/N-base junction (J_2) with a depletion layer extending mostly within the N-base region as illustrated in Fig. 9.12. Here, a one-dimensional view of only the portion under the deep P^+ region is illustrated because this governs the blocking voltage capability for the IGBT structure. The maximum electric field occurs at the P^+ region/N-base junction (J_2). On the one hand, if the width of the N-base region is very large when compared with the depletion layer width, the blocking voltage capability is limited by avalanche breakdown when the maximum electric field (E_m) becomes equal to the critical electric field. This corresponds to the multiplication coefficient (M) becoming equal to infinity. The avalanche breakdown voltage is given by

$$\mathrm{BV_{PP}} = 5.34 \times 10^{13} N_D^{-3/4}. \tag{9.1}$$

However, a very large width for the N-base region is unacceptable because it increases the on-state voltage drop of the IGBT structure.

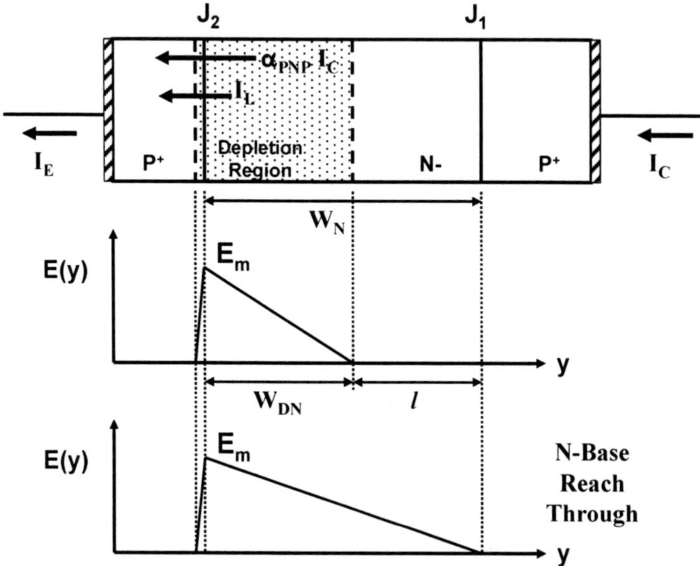

Fig. 9.12 Electric field distribution during the forward-blocking mode in the symmetric IGBT structure

On the other hand, if the width of the N-base region is made small when compared with the depletion layer width, the depletion region can extend through the entire N-base region as illustrated at the bottom of Fig. 9.12 while the maximum electric field at the junction is well below the critical electric field for breakdown. This phenomenon is referred to as the *reach-through* condition. When the depletion region from the reverse-biased junction J_2 reaches the forward-biased junction J_1, holes are injected from junction J_1 producing an abrupt increase in the anode current. This phenomenon limits the maximum blocking capability of the IGBT structure at the reach-through breakdown voltage given by

$$BV_{RT} = \frac{qN_D}{2\varepsilon_S} W_N^2, \tag{9.2}$$

where N_D is the doping concentration of the N-base region and W_N is its width.

The actual breakdown voltage for the IGBT structure in the forward-blocking mode falls between the above limits, and is governed by the open-base transistor breakdown phenomenon. To analyze this phenomenon for the IGBT structure, consider the currents flowing at the boundary of the depletion region as illustrated in Fig. 9.12. The current consists of the leakage current due to the generation process within the depletion region and the collector current amplified by the current gain of the P–N–P transistor. Based upon the application of Kirchhoff's current law to the IGBT structure in the absence of a gate current,

$$I_C = \alpha_{PNP} I_C + I_L = I_E, \tag{9.3}$$

leading to the relationship

$$I_C = \frac{I_L}{1 - \alpha_{PNP}}. \tag{9.4}$$

From this expression, it can be concluded that the collector current will increase very rapidly when the common-base current gain of the P–N–P bipolar transistor within the IGBT structure approaches unity. As the positive collector bias is increased, the width of the undepleted portion of the N-base region becomes smaller producing an increase in the base transport factor (α_T). Concurrently, the maximum electric field at junction J_2 becomes larger leading to an increase in the multiplication coefficient. Both phenomena produce an increase in the common-base current gain with increasing collector bias until it becomes equal to unity resulting in open-base transistor breakdown.

Based upon the above analysis, the open-base transistor breakdown condition for the IGBT structure is given by

$$\alpha_{PNP} = (\gamma_E \alpha_T)_{PNP} M = 1. \tag{9.5}$$

The injection efficiency (γ_E) of the P^+ collector/N-base junction (J_1) is close to unity because of the relatively high doping concentration in the P^+ collector region and the low doping concentration of the N-base region. The magnitude of the other two terms in the above equation is a function of the collector bias.

The base transport factor is determined by the width (l) of the undepleted portion of the N-drift region (see Fig. 9.12):

$$\alpha_T = \frac{1}{\cosh(l / L_p)} \tag{9.6}$$

with

$$l = W_{\mathrm{N}} - \sqrt{\frac{2\varepsilon_{\mathrm{S}} V_{\mathrm{C}}}{qN_{\mathrm{D}}}}, \tag{9.7}$$

where V_{C} is the applied positive bias to the collector electrode and L_{p} is the diffusion length for holes in the N-base region. As the collector bias increases, the width of the undepleted portion of the N-base region shrinks resulting in an increase in the base transport factor.

The multiplication factor is determined by the collector bias relative to the avalanche breakdown voltage of the deep P$^+$ region/N-base junction ($\mathrm{BV_{PP}}$):

$$M = \frac{1}{1 - (V_{\mathrm{C}} / \mathrm{BV_{PP}})^n}, \tag{9.8}$$

where $n = 6$ for the case of a P$^+$/N diode. The multiplication coefficient also increases with increasing collector bias. The open-base transistor breakdown voltage (and the forward-blocking capability of the IGBT structure) is determined by the collector voltage at which the multiplication factor becomes equal to the reciprocal of the base transport factor.

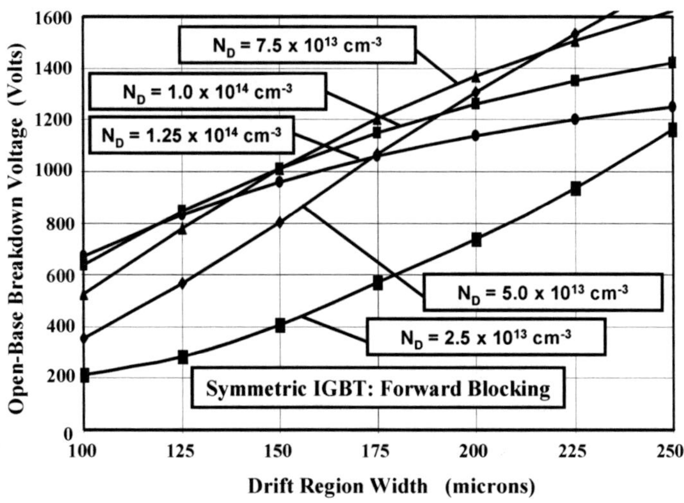

Fig. 9.13 Open-base breakdown voltage for the symmetric IGBT structure in the forward-blocking mode

As an example of the design procedure for optimization of the drift region doping concentration and width, consider the case of an IGBT structure that must have a forward-blocking voltage of 1,200 V. In the case of avalanche breakdown, there is a unique value of 1.6×10^{14} cm^{-3} for the doping concentration of the drift region to obtain this blocking voltage. However, in the case of open-base transistor breakdown, many combinations of the drift region doping concentration and width can be used to obtain this blocking voltage capability. This is illustrated in Fig. 9.13 where the open-base breakdown voltage is plotted as a function of the drift region width for five cases of the drift region doping concentration. A lifetime of 10 μs was used in the N-drift region for this analysis.

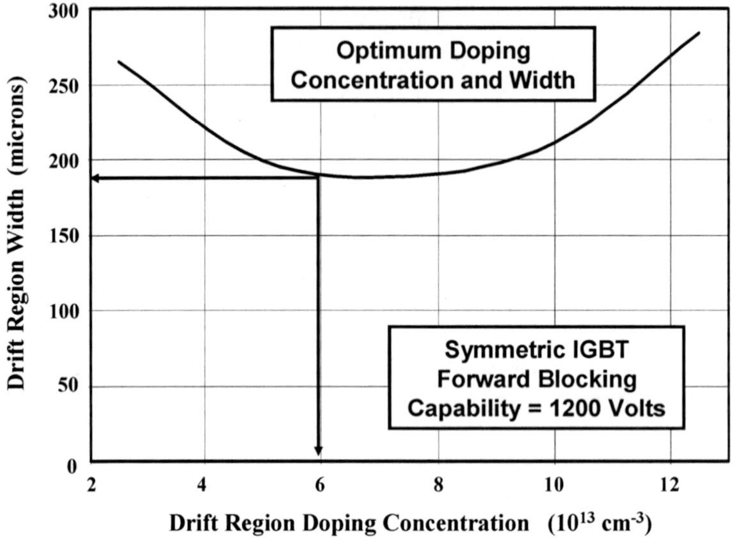

Fig. 9.14 Optimum width and doping concentration of the drift region for the symmetric IGBT structure to obtain a forward-blocking capability of 1,200 V

It can be observed from Fig. 9.13 that the open-base breakdown voltage becomes equal to 1,300 V (10% greater than a design target of 1,200 V) at a drift region width greater than 250 μm for a drift region doping concentration of 2.5×10^{13} cm^{-3}. In this case, the base transport factor becomes close to unity under breakdown conditions. When the doping concentration of drift region is increased to 5×10^{13} cm^{-3}, the drift region thickness is reduced to 200 μm to achieve the same open-base breakdown voltage of 1,300 V. If the doping concentration of drift region is increased to 1.25×10^{14} cm^{-3}, the drift region thickness increases to 300 μm to achieve the same open-base breakdown voltage of 1,300 V. In this case, the multiplication coefficient becomes large under open-base breakdown conditions. These examples demonstrate that there is an optimum drift region doping concentration to obtain a minimum drift region width to achieve an open-base breakdown voltage of 1,300 V. The location of the optimum design with a width of 190 μm and a doping concentration of 6×10^{13} cm^{-3} is illustrated for this case in Fig. 9.14.

The forward-blocking voltage capability for the symmetric IGBT structure is also dependent upon the ambient temperature. As temperature increases, the impact ionization coefficients are reduced leading to a decrease in the multiplication coefficient. Concurrently, the base transport factor increases due to an increase in the minority carrier lifetime. In general, the reduction of the impact ionization is dominant as temperature goes up. Consequently, the forward-blocking voltage capability for the symmetric IGBT structure increases with increasing temperature.

9.4.2 Symmetric Structure Reverse-Blocking Capability

When a negative bias is applied to the collector terminal of the IGBT structure, the P^+ collector/N-base junction (J_1) becomes reverse biased while the junction (J_2) between the deep P^+ region and the N-base region becomes forward biased. The reverse-blocking voltage is supported across the P^+ region/N-base junction (J_1) with a depletion layer extending mostly within the N-base region as illustrated in Fig. 9.15. Here, a one-dimensional view of only the portion under the deep P^+ region is illustrated because this governs the blocking voltage capability for the IGBT structure. The maximum electric field occurs at the P^+ collector/N-base junction (J_1). On the one hand, if the width of the N-base region is very large when compared with the depletion layer width, the blocking voltage capability is limited by avalanche breakdown when the maximum electric field (E_m) becomes equal to the critical electric field. This corresponds to the multiplication coefficient (M) becoming equal to infinity. The avalanche breakdown voltage is given by (9.1). However, a very large width for the N-base region is unacceptable because it increases the on-state voltage drop of the IGBT structure.

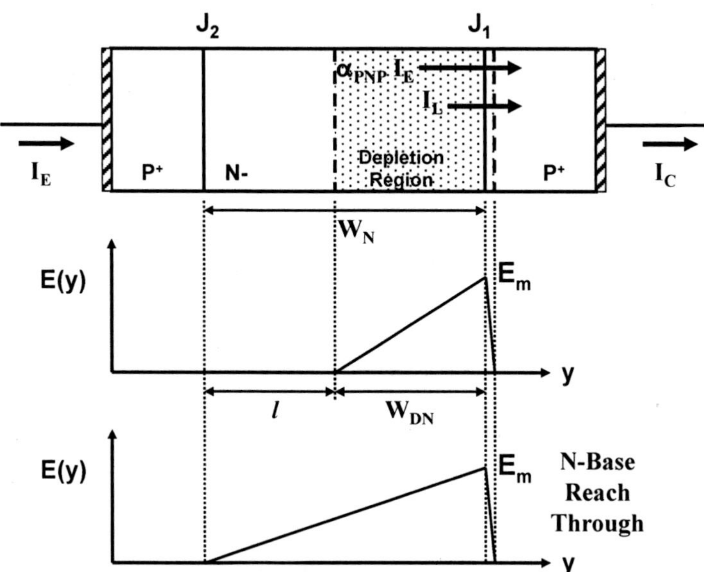

Fig. 9.15 Electric field distribution during the reverse-blocking mode in the symmetric IGBT structure

On the other hand, if the width of the N-base region is made small when compared with the depletion layer width, the depletion region can extend through the entire N-base region as illustrated at the bottom of Fig. 9.15 while the maximum electric field at the junction is well below the critical electric field for breakdown. This phenomenon is referred to as the reach-through condition. When the depletion region from the reverse-biased junction J_1 reaches the forward-biased junction J_2, holes are injected from junction J_2 producing an abrupt increase in the collector current. This phenomenon limits the maximum blocking capability for the symmetric IGBT structure at the reach-through breakdown voltage given by (9.2).

The actual breakdown voltage for the symmetric IGBT structure in the reverse-blocking mode falls between the above limits, and is governed by the open-base transistor breakdown phenomenon. Since the depletion layer extends in the symmetric IGBT structure in the same N-base region as during the forward-blocking mode, the open-base breakdown voltage in the reverse-blocking mode is the same as for the forward-blocking mode. The forward-blocking analysis is therefore pertinent to determination of the reverse-blocking voltage as well. The equations derived in Sect. 9.4.1 can therefore be used for determination of the reverse-blocking capability.

9.4.3 Symmetric Structure Leakage Current

The leakage current in reverse-blocking junctions is produced by space-charge generation within the depletion region as discussed in Chap. 5. In the case of the symmetric IGBT structure in the forward-blocking mode, the space-charge generation current at the reverse-biased deep P$^+$/N-base junction J_2 is amplified by the gain of the internal P–N–P transistor:

$$J_{\mathrm{L}} = \frac{J_{\mathrm{SCG}}}{1-\alpha_{\mathrm{PNP}}}, \tag{9.9}$$

where the space-charge generation current density is given by

$$J_{\mathrm{SCG}} = \frac{qW_{\mathrm{D}}n_{\mathrm{i}}}{\tau_{\mathrm{SC}}} = \frac{n_{\mathrm{i}}}{\tau_{\mathrm{SC}}}\sqrt{\frac{2q\varepsilon_{\mathrm{S}}V_{\mathrm{C}}}{N_{\mathrm{D}}}}. \tag{9.10}$$

The space-charge generation current increases with increasing collector bias. Concurrently, the current gain (α_{PNP}) of the P–N–P transistor is also a function of the collector bias voltage. The base transport factor and the multiplication factor increase when the collector bias increases. At collector bias voltages less than 80% of the open-base breakdown voltage, the multiplication factor remains close to unity. It is therefore sufficient to account for the increase in the base transport factor with collector bias as given by (9.6) and (9.7).

As an example, consider the case of a symmetric IGBT structure that is designed with a forward-blocking capability of 1,200 V. This would be satisfied by using an N-base region with a doping concentration of 5×10^{13} cm^{-3} and a width of 200 μm. The leakage current computed by using the above analysis is shown in Fig. 9.16 for the case of a space-charge generation lifetime (τ_{SC}) of 20 μs. The leakage current density (J_{SC}) at 300 K due to just space-charge generation, without amplification by the gain of the P–N–P transistor, is also shown in the figure. It can be seen that this current increases with collector bias voltage due to the expansion of the width of the depletion region. However, an increase in the leakage current by an order of magnitude occurs at a collector bias of 800 V when the gain of the P–N–P transistor is included in the analysis. In this figure, the leakage current at elevated temperatures is also plotted as a function of the collector bias. Due to the significant increase in the intrinsic carrier concentration with temperature, the leakage current increases by a factor of 500 times when the temperature increases from 300 to 400 K.

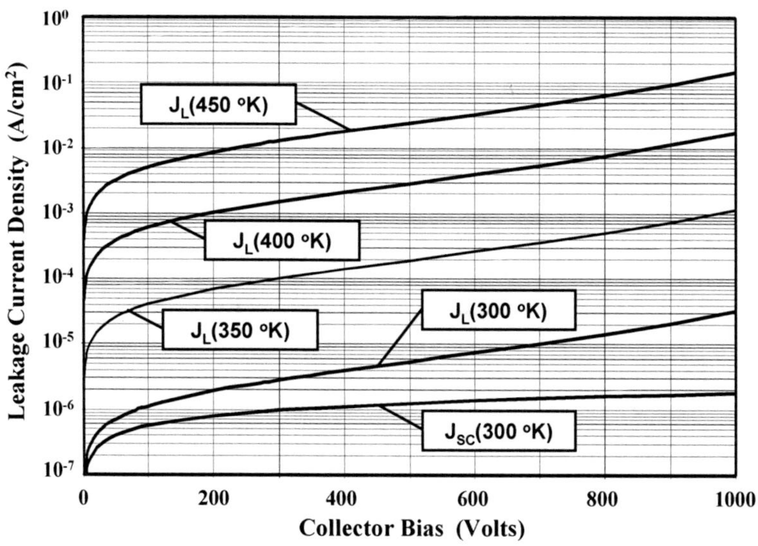

Fig. 9.16 Leakage current for the symmetric IGBT structure

The changes in the leakage current due to variations of the lifetime can also be modeled by using the equations derived in this section. When the space-charge generation lifetime is reduced, the space-charge generation current increases as described by (9.10). At the same time, the current gain of the P–N–P transistor is reduced due to the concomitant decrease in the minority carrier low-level lifetime and diffusion length. This behavior can be observed in Fig. 9.17 where the leakage current density is plotted for the above structure for three values of the low-level lifetime. The space-charge generation lifetime was assumed to be equal to twice the low-level lifetime. When the low-level lifetime is increased from 10 to 100 μs, the leakage current density at low collector bias voltages becomes smaller. However, the leakage current increases more rapidly with collector voltage due to the larger minority carrier diffusion length leading to a crossover of the current at 900 V. When the low-level lifetime is reduced from 10 to 1 μs, the leakage current density at low collector bias voltages increases but the current does not increase as rapidly with increasing collector bias due to the smaller minority carrier diffusion length.

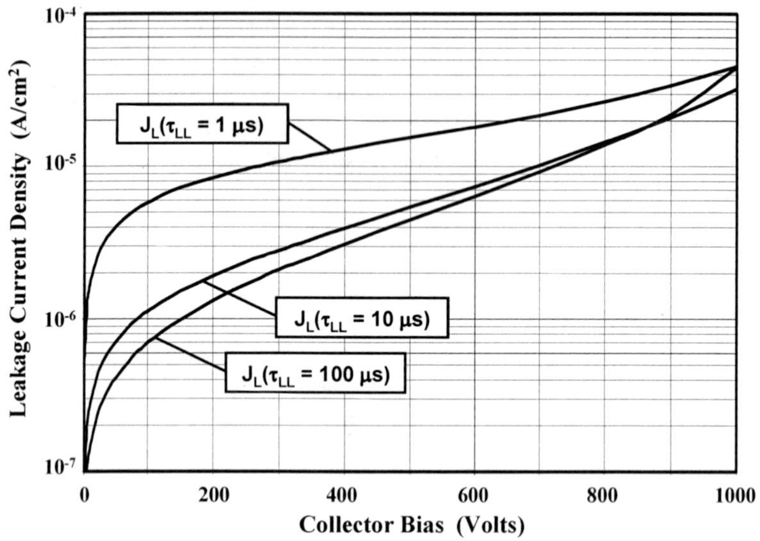

Fig. 9.17 Leakage current for the symmetric IGBT structure

Simulation Example

To gain further insight into the physics of operation for the symmetric IGBT structure under voltage-blocking conditions, the results of two-dimensional numerical simulations for a typical structure are described here. The simulations were performed by using a cell with half the width of the structure shown in Fig. 9.5, i.e., extending from the center of the diffusion window to the center of the polysilicon gate electrode. This cell has a width of 15 μm (area = 1.5×10^{-7} cm^{-2}) with an opening of 7 μm in the polysilicon gate electrode. The symmetric IGBT structure used for the simulations was formed by starting with a uniformly doped N-drift region with concentration of 5×10^{13} cm^{-3} and thickness of 220 μm. The P-base region has a Gaussian doping profile with a surface concentration of 4×10^{17} cm^{-3} and a vertical depth of 3 μm. The N^+ emitter region has a Gaussian doping profile with a surface concentration of 1×10^{20} cm^{-3} and a vertical depth of 1 μm. The deep P^+ region has a Gaussian doping profile with a surface concentration of 1×10^{19} cm^{-3} and a vertical depth of 5 μm. The doping concentration in the JFET region was enhanced by using a Gaussian doping profile with a surface concentration of 1×10^{16} cm^{-3} and a vertical depth of 5 μm. The P^+ collector region was formed at the bottom of the device using a Gaussian doping profile with a surface concentration of 1×10^{19} cm^{-3} and a depth of 20 μm. The doping profile in the vertical direction through the deep P^+ region is shown in Fig. 9.18, indicating that the net width of the N-base region is 195 μm after accounting for the diffusions.

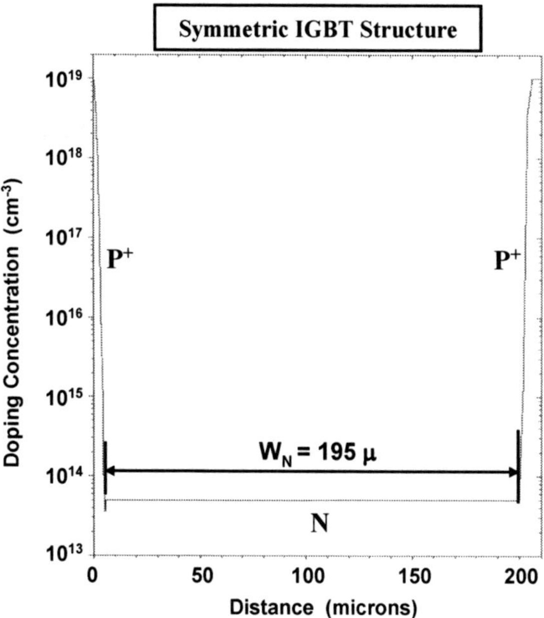

Fig. 9.18 Vertical doping profile for the simulated symmetric IGBT structure

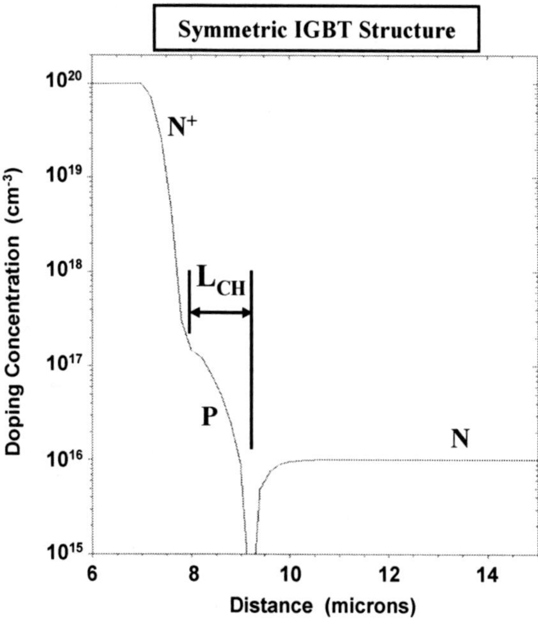

Fig. 9.19 Lateral doping profile for the simulated symmetric IGBT structure

The lateral doping profile for the simulated symmetric IGBT structure is provided in Fig. 9.19. The channel for the MOSFET region in the IGBT structure is formed by the relative diffusion of the P-base and N$^+$ emitter regions with a common polysilicon boundary located at x = 7 μm. Note that the JFET doping enhances the concentration of the N-base region to 1×10^{16} cm^{-3}. The peak concentration in the P-base region is 1.5×10^{17} cm^{-3} which results in a reasonable threshold voltage while preventing reach-through problems. The channel length can be observed to be 1.5 μm.

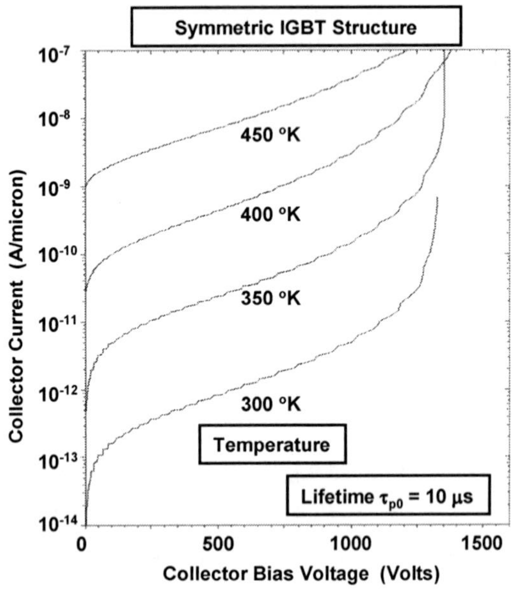

Fig. 9.20 Forward-blocking characteristics for the symmetric IGBT structure

The forward-blocking capability was obtained by increasing the collector bias while maintaining the gate electrode at 0 V. The characteristics obtained at various temperatures are provided in Fig. 9.20 for the case of a minority carrier lifetime (τ_{p0}) of 10 μs. In all cases, the leakage current increases rapidly with increasing collector voltage. This occurs due to the increase in the space-charge generation volume and the increase in the current gain (α_{PNP}) of the open-base P–N–P transistor. The simple analytical theory (see Fig. 9.16) provides a very good qualitative and quantitative description of the leakage current behavior as a function of both the collector voltage and the temperature.

The leakage currents obtained at 300 K using the numerical simulations are provided in Fig. 9.21 as a function of the lifetime in the N-base region. It can be observed that the leakage current decreases at low collector voltages when the lifetime is increased from 10 to 100 μs. However, the blocking voltage capability is compromised and falls below 1,200 V due to the increase in the current gain of the P–N–P transistor. When the lifetime is reduced to 1 μs, the leakage current at low collector voltages increases but the rate of increase with voltage becomes smaller.

The behavior obtained by using the simple analytical model for the leakage current (see Fig. 9.17) is in remarkably good qualitative and quantitative agreement with these simulation results.

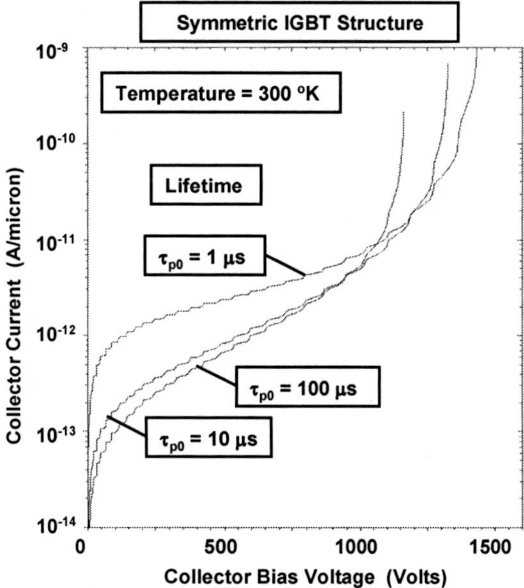

Fig. 9.21 Forward-blocking characteristics for the symmetric IGBT structure

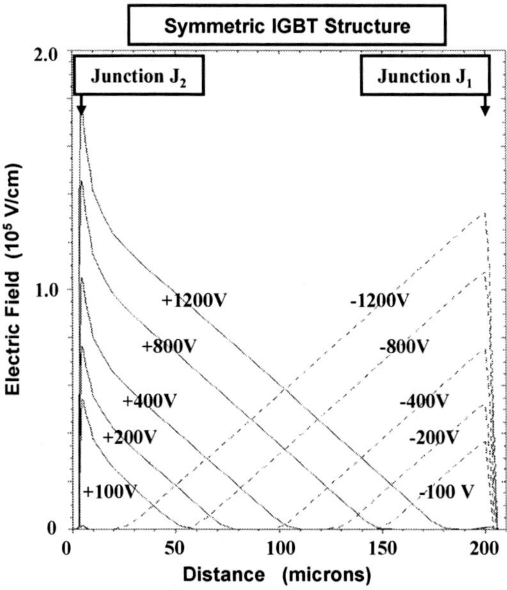

Fig. 9.22 Electric field profiles in the forward- and reverse-blocking mode for the symmetric IGBT structure

The voltage is primarily supported within the N-base region in the symmetric IGBT structure. This is illustrated in Fig. 9.22 where the electric field profiles are shown during operation in the reverse- and forward-blocking modes at five voltages. It can be observed that the P^+ collector/N-base junction (J_1) becomes reverse biased during the reverse-blocking mode with the depletion region extending toward the left-hand side with increasing (negative) collector bias. In contrast, the deep P^+/N-base junction (J_2) becomes reverse biased during the forward-blocking mode with the depletion region extending toward the right-hand side with increasing (positive) collector bias. The electric field profiles used in Figs. 9.12 and 9.15 to develop the analytical models for the blocking voltage are consistent with those observed with the simulations.

9.4.4 Asymmetric Structure Forward-Blocking Capability

When a positive bias is applied to the collector terminal of the asymmetric IGBT structure, the deep P^+ region/N-base junction (J_2) becomes reverse biased while the junction (J_1) between the P^+ collector region and the N-base region becomes forward biased. The forward-blocking voltage is supported across the deep P^+ region/N-base junction (J_2) with a depletion layer extending mostly within the N-base region

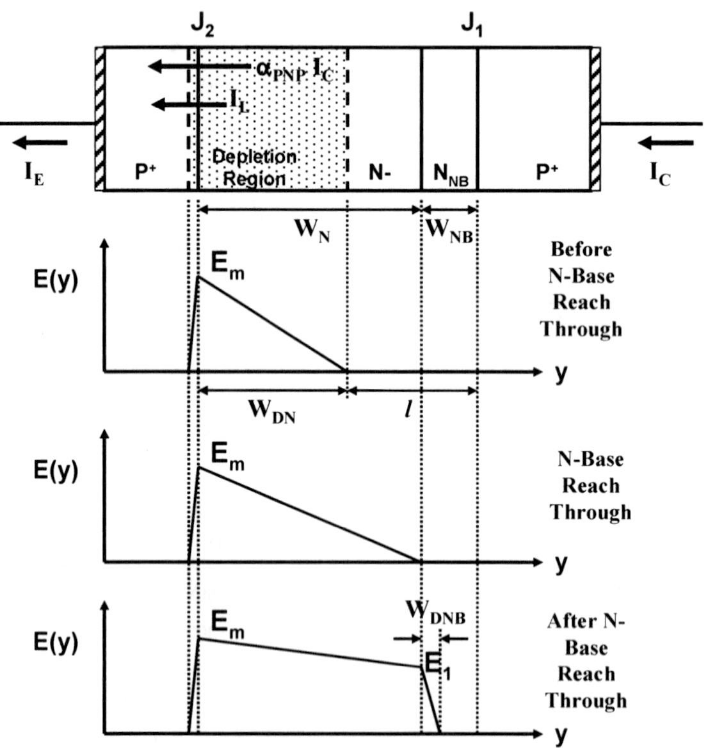

Fig. 9.23 Electric field distribution during the forward-blocking mode in the asymmetric IGBT structure

as illustrated in Fig. 9.23. Here, a one-dimensional view of only the portion under the deep P^+ region is illustrated because this governs the blocking voltage capability for the IGBT structure. The maximum electric field occurs at the P^+ region/N-base junction (J_2). On the one hand, if the width of the N-base region is very large when compared with the depletion layer width, the blocking voltage capability is limited by avalanche breakdown when the maximum electric field (E_m) becomes equal to the critical electric field. This corresponds to the multiplication coefficient (M) becoming equal to infinity. The avalanche breakdown voltage is given by

$$\mathrm{BV_{PP}} = 5.34 \times 10^{13} N_D^{-3/4}, \tag{9.11}$$

where N_D is the doping concentration in the lightly doped portion of the N-base region. However, a very large width for the N-base region is unacceptable because it increases the on-state voltage drop of the thyristor.

On the other hand, if the width of the N-base region is made small when compared with the depletion layer width, the depletion region can extend through the entire lightly doped portion of the N-base region as illustrated in the middle of Fig. 9.23 while the maximum electric field at the junction is well below the critical electric field for breakdown. The depletion region reaches through the lightly doped portion of the N-base region at a collector bias given by

$$V_{RT} = \frac{qN_D}{2\varepsilon_S} W_N^2, \tag{9.12}$$

where W_N is the width of the lightly doped portion of the N-base region.

Due to the presence of the N-buffer layer in the asymmetric IGBT structure, the device can continue to support voltages after the depletion region reaches through the lightly doped portion of the N-base region. The electric field distribution for this case is illustrated at the bottom of Fig. 9.23. (The vertical axes for the three electric field plots in the figure have different scales to allow displaying the field distributions. The slope of the electric field with distance within the lightly doped portion of the N-base region is the same for all three cases as determined by the doping concentration in the region.) The electric field has a trapezoidal shape typical of punch-through structures. As discussed in Chap. 3, the avalanche breakdown voltage for the punch-through structure is given by

$$\mathrm{BV_{PT}} = E_C W_N - \frac{qN_D}{2\varepsilon_S} W_N^2, \tag{9.13}$$

where E_C is the critical electric field for breakdown corresponding to the doping concentration N_D in the lightly doped portion of the N-base region.

The actual breakdown voltage for the asymmetric IGBT structure in the forward-blocking mode falls below the punch-through breakdown voltage because it is governed by the open-base transistor breakdown phenomenon. To analyze this phenomenon for the IGBT structure, consider the currents flowing at the boundary of the depletion region as illustrated in Fig. 9.23. The current consists of the leakage current due to the generation process within the depletion region and the collector current amplified by the current gain of the P–N–P transistor. Based upon the application of Kirchhoff's current law to the IGBT structure in the absence of a gate current,

$$I_{\mathrm{C}} = \alpha_{\mathrm{PNP}} I_{\mathrm{C}} + I_{\mathrm{L}} = I_{\mathrm{E}}, \tag{9.14}$$

leading to the relationship

$$I_{\mathrm{C}} = \frac{I_{\mathrm{L}}}{1 - \alpha_{\mathrm{PNP}}}. \tag{9.15}$$

From this expression, it can be concluded that the collector current will increase very rapidly when the common-base current gain of the P–N–P bipolar transistor within the asymmetric IGBT structure approaches unity. As discussed earlier for the symmetric IGBT structure, when the positive collector bias is increased, the width of the undepleted portion of the N-base region becomes smaller producing an increase in the base transport factor (α_{T}). Concurrently, the maximum electric field at junction J_2 becomes larger leading to an increase in the multiplication coefficient. Both phenomena produce an increase in the common-base current gain with increasing collector bias.

However, for the asymmetric IGBT structure, the emitter injection efficiency becomes smaller than unity due to the high doping concentration of the N-buffer layer. The emitter injection efficiency for the P$^+$ collector/N-buffer junction (J_1) can be obtained by using an analysis similar to that described for the bipolar power transistor in Chap. 7 (see (7.39)):

$$\gamma_{\mathrm{E}} = \frac{D_{\mathrm{p,NB}} L_{\mathrm{nE}} N_{\mathrm{AE}}}{D_{\mathrm{p,NB}} L_{\mathrm{nE}} N_{\mathrm{AE}} + D_{\mathrm{nE}} W_{\mathrm{NB}} N_{\mathrm{DNB}}}, \tag{9.16}$$

where $D_{\mathrm{p,NB}}$ and D_{nE} are the diffusion coefficients for minority carriers in the N-buffer and P$^+$ collector regions, N_{AE} and L_{nE} are the doping concentration and diffusion length for minority carriers in the P$^+$ collector region, and N_{DNB} and W_{NB} are the doping concentration and width of the N-buffer layer. In determining the diffusion coefficients and the diffusion length, it is necessary to account for the high doping concentrations in the P$^+$ collector region and N-buffer layer. In addition, the lifetime within the highly doped P$^+$ collector region is reduced due to heavy doping effects, which shortens the diffusion length.

Based upon the above analysis, the open-base transistor breakdown condition for the asymmetric IGBT structure is given by

$$\alpha_{\mathrm{PNP}} = (\gamma_{\mathrm{E}} \alpha_{\mathrm{T}})_{\mathrm{PNP}} M = 1. \tag{9.17}$$

Before the advent of reach-through of the lightly doped portion of the N-base region, the base transport factor (α_{T}) is determined by the width (l) of the undepleted portion of the N-drift region (see Fig. 9.23):

$$\alpha_{\mathrm{T}} = \frac{1}{\cosh(l / L_{\mathrm{p}})} \tag{9.18}$$

with

$$l = W_{\mathrm{N}} - \sqrt{\frac{2\varepsilon_{\mathrm{S}} V_{\mathrm{C}}}{q N_{\mathrm{D}}}}, \tag{9.19}$$

where V_{C} is the applied forward bias to the collector electrode. As the forward bias increases, the width of the undepleted portion of the N-base region shrinks resulting in an increase in the base transport factor. However, when the collector bias exceeds the reach-through voltage (V_{RT}), the electric field is truncated by the high doping concentration of the N-buffer layer making the undepleted width equal to the width of the N-buffer layer. The base transport factor is then given by

$$\alpha_{\mathrm{T}} = \frac{1}{\cosh(W_{\mathrm{NB}} / L_{\mathrm{p,NB}})}, \tag{9.20}$$

which is independent of the collector bias. Here, $L_{\mathrm{p,NB}}$ is the diffusion length for holes in the N-buffer layer. This analysis neglects the depletion region extension (shown as W_{DNB} in Fig. 9.23) within the N-buffer layer.

The diffusion length for holes ($L_{\mathrm{p,NB}}$) in the N-buffer layer depends upon the diffusion coefficient and the minority carrier lifetime in the N-buffer layer. The diffusion coefficient varies with the doping concentration in the N-buffer layer based upon the concentration dependence of the mobility. In addition, the minority carrier lifetime has been found to be dependent upon the doping concentration.[16] The variation of the minority carrier lifetime with doping concentration is inherent in the lifetime analysis provided in Chap. 2 (see (2.76)) because the Fermi level position varies with doping concentration:

$$n_0 = N_{\mathrm{D}} = n_{\mathrm{i}} \mathrm{e}^{(E_{\mathrm{F}} - E_{\mathrm{i}})/kT}. \tag{9.21}$$

Combining this relationship with (2.76) yields

$$\frac{\tau_{\mathrm{LL}}}{\tau_{\mathrm{p0}}} = \left[1 + \frac{n_{\mathrm{i}}}{N_{\mathrm{D}}} \mathrm{e}^{(E_{\mathrm{r}} - E_{\mathrm{i}})/kT}\right] + \zeta \frac{n_{\mathrm{i}}}{N_{\mathrm{D}}} \mathrm{e}^{-(E_{\mathrm{r}} - E_{\mathrm{i}})/kT}, \tag{9.22}$$

where E_{r} is the recombination center position within the band gap.

The variation of the low-level lifetime computed by using the above expression is plotted in Fig. 9.24 for two cases of the recombination level position and the ζ parameter. A reduction of the low-level lifetime with increasing doping concentration is observed when the recombination center is located well below midgap and has a large ζ parameter. However, recombination centers located close to midgap have been found to dominate in the recombination process. In this case, there is no significant variation of the low-level lifetime with doping concentration when the concentration of the recombination center is kept constant as shown in the figure.

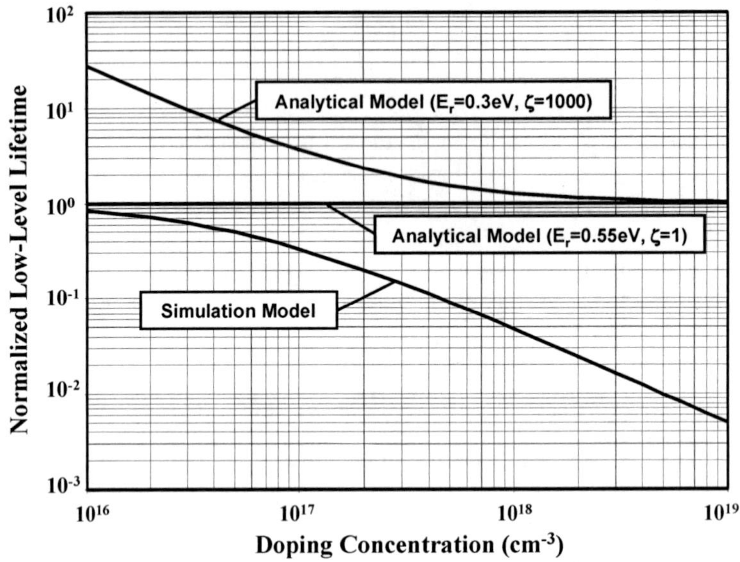

Fig. 9.24 Low-level lifetime variation with doping concentration

However, it has been empirically observed that the low-level lifetime decreases with increasing doping concentration.[16,17] This is usually modeled by using the relationship

$$\frac{\tau_{\mathrm{LL}}}{\tau_{\mathrm{p0}}} = \frac{1}{1 + (N_{\mathrm{D}} / N_{\mathrm{REF}})}, \tag{9.23}$$

where N_{REF} is a reference doping concentration[18] whose value will be assumed to be 5×10^{16} cm^{-3}. This behavior is also plotted in Fig. 9.24 as the simulation model. The reduction of the low-level lifetime indicated by this model must be taken into account for computing the diffusion length of minority carriers when the doping concentration in the N-buffer layer is increased.

The multiplication factor given by (9.8) is determined by the collector bias relative to the avalanche breakdown voltage of the deep P$^+$ region/N-base junction (BV_{PP}) *without the punch-through phenomenon.* To apply this formulation to the punch-through case relevant to the asymmetric IGBT structure, it is necessary to relate the maximum electric field at the junction for the two cases. The electric field at the interface between the lightly doped portion of the N-base region and the N-buffer layer is given by

$$E_1 = E_m - \frac{qN_D W_N}{\varepsilon_S}. \tag{9.24}$$

The voltage supported by the device is given by

$$V_C = \left(\frac{E_m + E_1}{2}\right) W_N = E_m W_N - \frac{qN_D}{2\varepsilon_S} W_N^2. \tag{9.25}$$

From this expression, the maximum electric field is given by

$$E_m = \frac{V_C}{W_N} + \frac{qN_D W_N}{2\varepsilon_S}. \tag{9.26}$$

The corresponding equation for the nonpunch-through case is

$$E_m = \sqrt{\frac{2qN_D V_{NPT}}{\varepsilon_S}}. \tag{9.27}$$

Consequently, the nonpunch-through voltage that determines the multiplication coefficient M corresponding to the applied collector bias V_C for the punch-through case is given by

$$V_{NPT} = \frac{\varepsilon_S E_m^2}{2qN_D} = \frac{\varepsilon_S}{2qN_D}\left(\frac{V_C}{W_N} + \frac{qN_D W_N}{2\varepsilon_S}\right)^2. \tag{9.28}$$

The multiplication coefficient can be computed by using this nonpunch-through voltage:

$$M = \frac{1}{1 - (V_{NPT} / \mathrm{BV_{PP}})^n}, \tag{9.29}$$

where $n = 6$ for the case of a P^+/N diode. The multiplication coefficient increases with increasing collector bias. The open-base transistor breakdown voltage (and the forward-blocking capability of the asymmetric IGBT structure) is determined by the collector voltage at which the multiplication factor becomes equal to the reciprocal of the product of the base transport factor and the emitter injection efficiency.

As an example of the design procedure for optimization of the drift region doping concentration and width, consider the case of an IGBT structure that must have a forward-blocking voltage of 1,200 V. In the case of avalanche breakdown, there is a unique value of 1.6×10^{14} cm^{-3} for the doping concentration of the drift region to obtain this blocking voltage. However, in the case of the asymmetric IGBT structure, it is advantageous to use a much lower doping concentration for the lightly doped portion of the N-base region to reduce its width. The strong conductivity modulation of the N-base region during on-state operation favors a smaller thickness for the N-base region independent of its original doping concentration. The doping concentration of the N-buffer layer must be sufficiently large to prevent reach-through of the electric field to the P^+ collector region. Doping concentrations above 1×10^{16} cm^{-3} are sufficient to accomplish this goal. A larger doping concentration in the N-buffer layer produces a reduction of the emitter injection efficiency which results in an increase in the blocking voltage. This is illustrated in Fig. 9.25 where the open-base breakdown voltage is plotted as a function of the N-buffer region doping concentration. A lifetime of 1 μs was used in the N-drift region for this analysis with the lifetime in the N-buffer layer scaled in accordance with (9.23). It can be observed that the base transport factor decreases with increasing doping concentration in the N-buffer layer due to a reduction of the lifetime in accordance with (9.20). The emitter injection efficiency is also reduced with increasing doping concentration in the N-buffer layer in accordance with (9.16). The reduction of the product of the base transport factor and the emitter injection efficiency produces the observed increase in the open-base breakdown voltage.

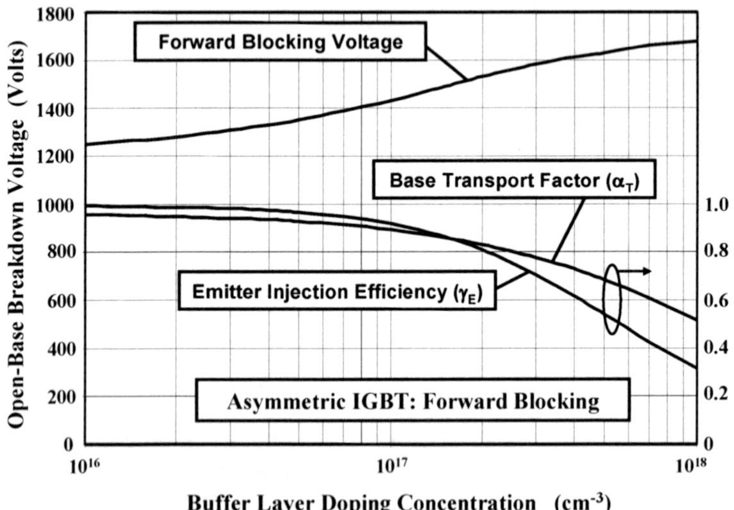

Fig. 9.25 Open-base breakdown voltage for the asymmetric IGBT structure in the forward-blocking mode

The forward-blocking voltage capability for the asymmetric IGBT structure is also dependent upon the ambient temperature. As temperature increases, the impact ionization coefficients are reduced leading to a decrease in the multiplication coefficient. Concurrently, the base transport factor increases due to an increase in the minority carrier lifetime. In general, the reduction of the impact ionization is dominant as temperature goes up. Consequently, the forward-blocking voltage capability for the asymmetric IGBT structure increases with increasing temperature.

9.4.5 Asymmetric Structure Reverse-Blocking Capability

When a negative bias is applied to the collector terminal of the IGBT structure, the P^+ collector/N-base junction (J_1) becomes reverse biased while the junction (J_2) between the deep P^+ region and the N-base region becomes forward biased. The reverse-blocking voltage is supported across the P^+ collector/N-buffer layer junction (J_1) with a depletion layer extending mostly within the N-buffer layer as illustrated in Fig. 9.26. The maximum electric field occurs at the P^+ collector/N-buffer layer junction (J_1). The base transport factor for the P–N–P transistor is small under these conditions because of the large undepleted width of the lightly doped portion of the N-base region. Consequently, the blocking voltage capability is limited by avalanche breakdown when the maximum electric field (E_m) becomes equal to the critical electric field at the P^+ collector/N-buffer layer junction (J_1). Based upon the analysis of avalanche breakdown voltage in Chap. 3, the reverse-blocking voltage can be obtained from the doping concentration of the N-buffer layer:

$$\mathrm{BV}_{\mathrm{RB}} = 5.34 \times 10^{13} N_{\mathrm{NB}}^{-3/4} \tag{9.30}$$

and the depletion layer extension in the N-buffer layer at breakdown can be obtained by using

$$W_{\mathrm{D,NB}} = 2.67 \times 10^{10} N_{\mathrm{NB}}^{-7/8}. \tag{9.31}$$

The magnitudes of the breakdown voltage and depletion layer width at breakdown are provided in Fig. 9.27 for the typical range of doping concentrations in the N-buffer layer. The breakdown voltage ranges from 53 to 3.4 V as the doping in the N-buffer layer is increased from 1×10^{16} to 4×10^{17} cm^{-3}. In this doping range, the maximum depletion region extension in the N-buffer layer is only 2.6 μm. It is therefore sufficient to use an N-buffer layer thickness of 10 μm

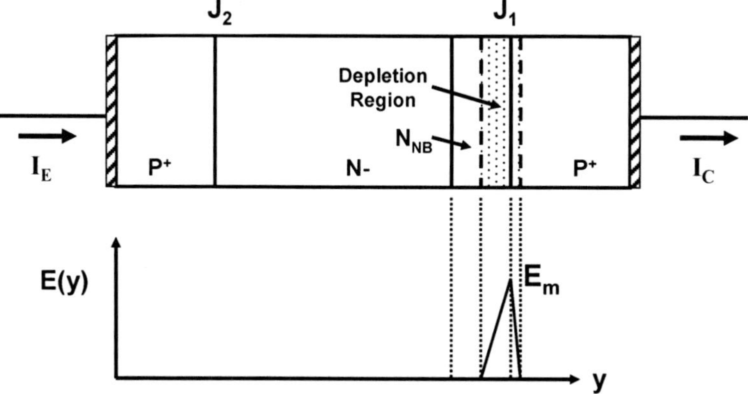

Fig. 9.26 Electric field distribution during the reverse-blocking mode in the asymmetric IGBT structure

to ensure absence of reach-through of the N-base region during forward- and reverse-blocking operation.

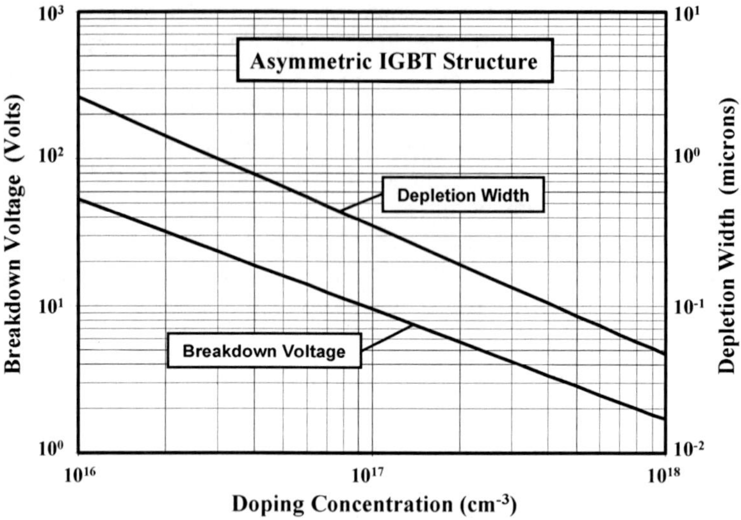

Fig. 9.27 Reverse breakdown voltage for the asymmetric IGBT structure and the corresponding depletion layer width

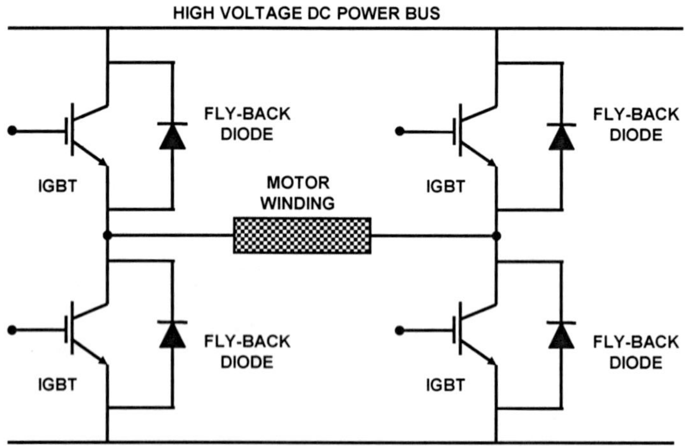

Fig. 9.28 Typical totem pole configuration used in PWM motor drives

The asymmetric IGBT structure is commonly used in H-bridge circuits for motor control applications, such as variable speed drives. In this topology, a reverse conducting or fly-back diode is connected across the IGBT, as shown in Fig. 9.28, to carry current during some circuit operating conditions.[19] When a power MOSFET is used in this totem pole configuration, its internal body diode can complete for the reverse current flow with the fly-back diode. This problem does not occur for the asymmetric IGBT structure if its reverse-blocking capability exceeds 4 V. Even without special termination of the P^+ collector/N-buffer layer junction (J_1) at the edges of the chip, the asymmetric IGBT structure can support this voltage if the N-buffer layer doping does not exceed 4×10^{17} cm^{-3}.

9.4.6 Asymmetric Structure Leakage Current

The leakage current in reverse-blocking junctions is produced by space-charge generation within the depletion region as discussed in Chap. 5. In the case of the asymmetric IGBT structure in the forward-blocking mode, the space-charge generation current at the reverse-biased deep P^+/N-base junction J_2 is amplified by the gain of the internal P–N–P transistor:

$$J_{\mathrm{L}} = \frac{J_{\mathrm{SCG}}}{1-\alpha_{\mathrm{PNP}}}. \tag{9.32}$$

The space-charge generation current density is given by

$$J_{\mathrm{SCG}} = \frac{qW_{\mathrm{D}}n_{\mathrm{i}}}{\tau_{\mathrm{SC}}} = \frac{n_{\mathrm{i}}}{\tau_{\mathrm{SC}}}\sqrt{\frac{2q\varepsilon_{\mathrm{S}}V_{\mathrm{C}}}{N_{\mathrm{D}}}} \tag{9.33}$$

at low collector bias voltages *before the depletion region in the lightly doped portion of the N-base region reaches through to the interface between the lightly doped portion of the N-base region and the N-buffer layer*. The space-charge generation current increases with increasing collector bias in this regime of operation for the asymmetric IGBT structure.

Concurrently, the current gain (α_{PNP}) of the P–N–P transistor is also a function of the collector bias voltage because the base transport factor increases when the collector bias increases. Prior to the complete depletion of the lightly doped portion of the N-base region, the multiplication factor remains close to unity. It is therefore sufficient to account for the increase in the base transport factor with collector bias as given by (9.18) and (9.19).

Once the lightly doped portion of the N-base region becomes completely depleted, the electric field becomes truncated at the interface between the lightly doped portion of the N-base region and the N-buffer layer as illustrated at the bottom of Fig. 9.23. The space-charge generation width then becomes independent of the collector bias because the depletion width in the N-buffer layer is small. Under these bias conditions, the base transport factor also becomes independent of the collector bias as given by (9.20). Consequently, the leakage current becomes independent of the collector bias until the on-set of avalanche multiplication.

The transport of minority carriers through the N-base region of the P–N–P transistor occurs through the N-buffer layer and the lightly doped portion of the N-base region that has not been depleted by the applied collector bias. The base transport factor is therefore given by

$$\alpha_{\mathrm{T}} = \alpha_{\mathrm{T,N\text{-}Buffer}}\alpha_{\mathrm{T,N\text{-}Base}}. \tag{9.34}$$

The base transport factor associated with the N-buffer layer can be obtained from the decay of the hole current within the N-buffer layer as given by low-level injection theory:

$$\alpha_{\mathrm{T,N\text{-}Buffer}} = \frac{J_{\mathrm{p}}(W_{\mathrm{NB}}-)}{J_{\mathrm{p}}(y_{\mathrm{N}})} = \frac{\gamma_{\mathrm{E}} J_{\mathrm{C}} \mathrm{e}^{-W_{\mathrm{NB}}/L_{\mathrm{p,NB}}}}{\gamma_{\mathrm{E}} J_{\mathrm{C}}} = \mathrm{e}^{-W_{\mathrm{NB}}/L_{\mathrm{p,NB}}}, \tag{9.35}$$

where $J_{\mathrm{p}}(y_{\mathrm{N}})$ is the hole current density at the P^{+} collector/N-buffer layer junction (J_1), $J_{\mathrm{p}}(W_{\mathrm{NB}}-)$ is the hole current density at the interface between the lightly doped portion of the N-base region and the N-buffer layer, W_{NB} is the width of the N-buffer layer, and $L_{\mathrm{p,NB}}$ is the minority carrier diffusion length in the N-buffer layer. The base transport factor for the lightly doped portion of the N-base region is similar to that previously derived for the bipolar power transistor in Sect. 7.4.4 under low-level injection conditions appropriate for computation of the leakage current:

$$\alpha_{\mathrm{T,N\text{-}Base}} = \frac{1}{\cosh[(W_{\mathrm{N}} - W_{\mathrm{DN}})/L_{\mathrm{p,N}}]}, \tag{9.36}$$

where W_{N} is the width of the lightly doped portion of the N-base region and $L_{\mathrm{p,N}}$ is the minority carrier diffusion length in the lightly doped portion of the N-base region. In this expression, the depletion width W_{DN} prior to punch-through is given by

$$W_{\mathrm{DN}} = \sqrt{\frac{2\varepsilon_{\mathrm{S}} V_{\mathrm{C}}}{q N_{\mathrm{D}}}}. \tag{9.37}$$

As the collector bias voltage increases, the base transport factor for the P–N–P transistor increases until it becomes equal to that for the N-buffer layer.

As an example, consider the case of an asymmetric IGBT structure that is designed with a forward-blocking capability of 1,200 V. This would be satisfied by using an N-base region with a lightly doped portion having a doping concentration of $5 \times 10^{13}\ \mathrm{cm}^{-3}$ and a width of 100 μm, and an N-buffer layer with a thickness of 10 μm. The leakage current computed by using the above analysis is shown in Fig. 9.29 for the case of a space-charge generation lifetime (τ_{SC}) of 1 μs in the lightly doped portion of the N-base region for two cases of doping concentration in the N-buffer layer. In performing this analysis, the reduction of the lifetime with increasing doping concentration in the N-buffer layer (see (9.23)) was taken into account. The space-charge generation current is also included in the figure for comparison purposes. It can be seen that the space-charge generation leakage current increases with collector bias voltage (to 2.2×10^{-5} A cm^{-2}) due to the expansion of the width of the depletion region until about 400 V. This corresponds to a reach-through voltage of 386 V obtained by using (9.12). The space-charge generation leakage current then becomes independent of the collector bias. The leakage current for the asymmetric IGBT structure is larger than the space-charge generation current due to the current gain of the P–N–P transistor. For the case of an N-buffer layer doping of 1×10^{17} cm^{-3}, the current gain of the P–N–P transistor is 0.44 after the lightly doped portion of the N-base region becomes completely depleted. The leakage current density for this case is 3.9×10^{-5} A cm^{-2}. When the doping concentration of the N-buffer layer is reduced to 1×10^{16} cm^{-3}, the current gain of the P–N–P transistor increases to 0.70 after the lightly doped portion of the N-base region becomes completely depleted. Consequently, the leakage current density for this case increases to 7×10^{-5} A cm^{-2}.

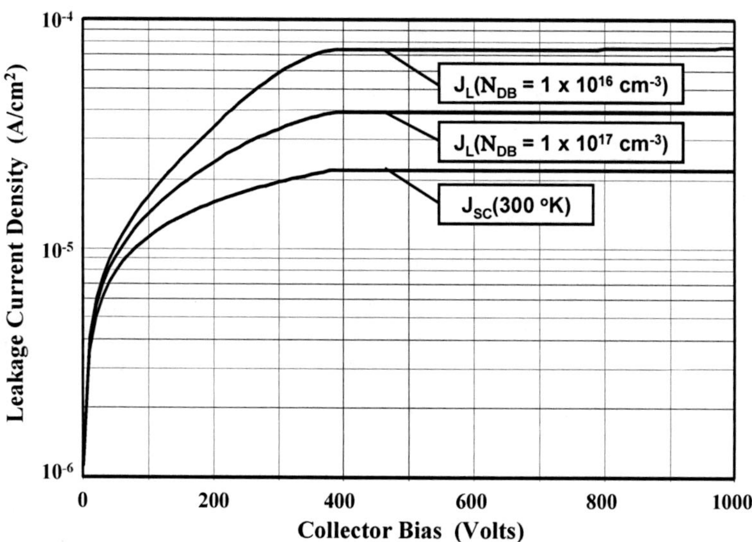

Fig. 9.29 Leakage current for the asymmetric IGBT structure

The leakage current for the asymmetrical IGBT structure at elevated temperatures can also be obtained by using the above analytical model. As an example, consider the case of an N-buffer layer doping concentration of 1×10^{16} cm^{-3}. The leakage current density at various temperatures calculated by using the analytical model is provided in Fig. 9.30 together with the space-charge generation current at 300 K. When the temperature increases from 300 to 400 K, the leakage current increases by a factor of 500 times because of the rapid increase in the intrinsic carrier concentration.

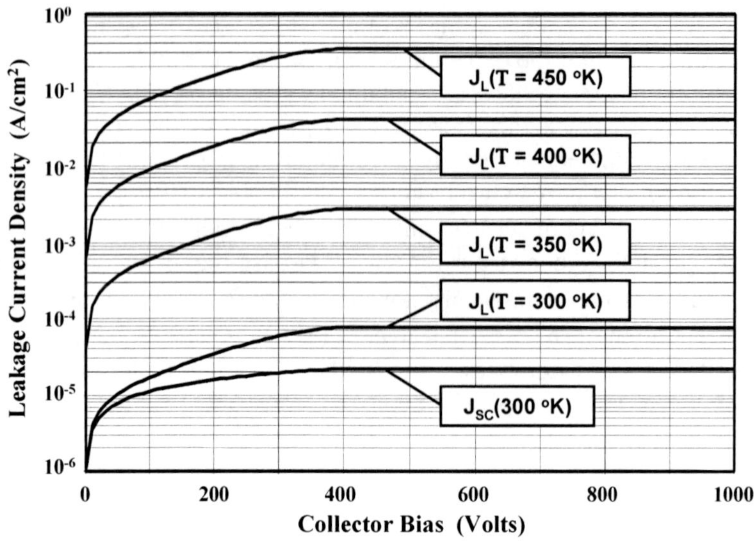

Fig. 9.30 Leakage current for the asymmetric IGBT structure

Simulation Example

To gain further insight into the physics of operation for the asymmetric IGBT structure under voltage-blocking conditions, the results of two-dimensional numerical simulations for a typical structure are described here. The simulations were performed by using a cell with half the width of the structure shown in Fig. 9.6, i.e., extending from the center of the diffusion window to the center of the polysilicon gate electrode. This cell has a width of 15 μm (area = 1.5×10^{-7} cm^{-2}) with an opening of 7 μm in the polysilicon gate electrode. The asymmetric IGBT structure used for the simulations was formed by using a uniformly doped P$^+$ substrate with a doping concentration of 1×10^{19} cm^{-3} with an epitaxial N-base region consisting of a lightly doped portion with concentration of 5×10^{13} cm^{-3} and thickness of 100 μm and an N-buffer layer with thickness of 10 μm. The doping concentration in the N-buffer layer was varied to examine its impact on the forward-blocking capability and leakage current.

The P-base region has a Gaussian doping profile with a surface concentration of 4×10^{17} cm^{-3} and a vertical depth of 3 μm. The N$^+$ emitter region has a Gaussian doping profile with a surface concentration of 1×10^{20} cm^{-3} and a vertical depth of 1 μm. The deep P$^+$ region has a Gaussian doping profile with a surface concentration of 1×10^{19} cm^{-3} and a vertical depth of 5 μm. The doping concentration in the JFET region was enhanced by using a Gaussian doping profile with a surface concentration of 1×10^{16} cm^{-3} and a depth of 5 μm. The doping profile in the vertical direction through the deep P$^+$ region is shown in Fig. 9.31, indicating that the net width of the N-base region is 95 μm after accounting for the deep P$^+$ region. The lateral doping profile for the simulated asymmetric IGBT structure is the same as that used for the symmetric IGBT structure (see Fig. 9.19).

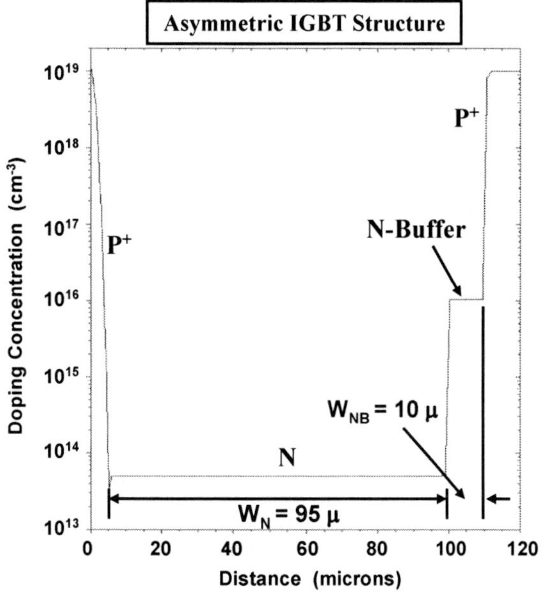

Fig. 9.31 Doping profile for the simulated asymmetric IGBT structure

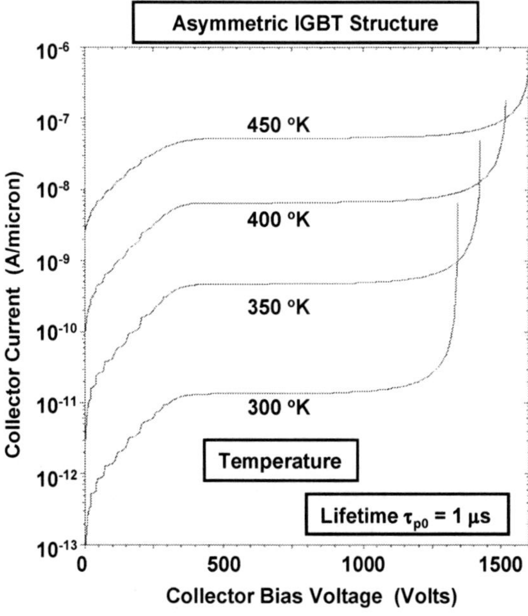

Fig. 9.32 Forward-blocking characteristics for the asymmetric IGBT structure

The forward-blocking capability was obtained by increasing the collector bias while maintaining the gate electrode at 0 V. The characteristics obtained at various temperatures are provided in Fig. 9.32 for the case of a minority carrier lifetime (τ_{p0}) of 1 μs. In all cases, the leakage current increases rapidly with increasing collector voltage until about 400 V. This occurs due to the increase in the space-charge generation volume and the increase in the current gain (α_{PNP}) of the open-base P–N–P transistor until the collector bias becomes equal to the reach-through voltage obtained by using the analytical solution given by (9.12). The leakage current then becomes independent of the collector voltage until close to the breakdown voltage. This behavior is well described by the analytical model (see Fig. 9.30). The leakage current density obtained by using the analytical model is within a factor of 2 of the values derived from the numerical simulations at all the temperatures. Thus, the simple analytical theory provides a very good qualitative and quantitative description of the leakage current behavior as a function of both the collector voltage and the temperature.

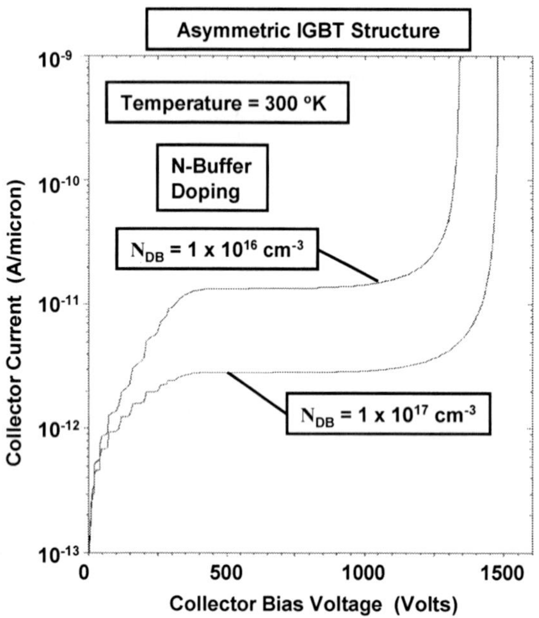

Fig. 9.33 Forward-blocking characteristics for the asymmetric IGBT structure

The leakage currents obtained by using the numerical simulations are provided in Fig. 9.33 for two doping concentrations in the N-buffer layer. A lifetime of 1 μs (τ_{p0}) was used in the lightly doped portion of the N-base region for both cases. It can be observed that the leakage current decreases when the N-buffer layer doping concentration is increased due to the reduction of the emitter injection efficiency and the base transport factor of the P–N–P transistor. The reduced current gain of the P–N–P transistor also results in an increase in the open-base breakdown voltage. The behavior obtained by using the simple analytical model for the leakage current (see Fig. 9.29) is in good qualitative and quantitative agreement with these simulation results.

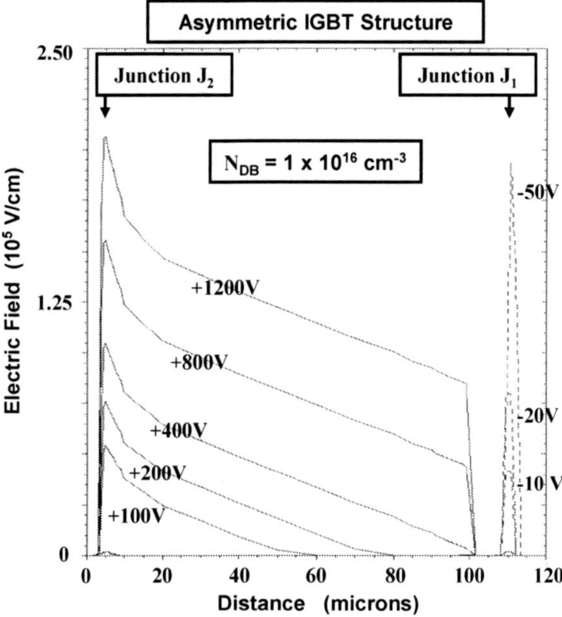

Fig. 9.34 Electric field profiles in the forward- and reverse-blocking mode for the asymmetric IGBT structure

The voltage is primarily supported within the lightly doped portion of the N-base region in the asymmetric IGBT structure during operation in the forward-blocking mode. This is illustrated in Fig. 9.34 where the electric field profiles are shown during operation in the reverse- and forward-blocking modes at several collector voltages. It can be observed that the deep P^+/N-base junction (J_2) becomes reverse biased during the forward-blocking mode with the depletion region extending toward the right-hand side with increasing (positive) collector bias. The electric field has a triangular shape until the entire lightly doped portion of the N-base region becomes completely depleted. This occurs at a collector bias of 400 V in precise agreement with the value obtained by using the analytical solution (see (9.12)). The electric field profile then takes a trapezoidal shape due to the high doping concentration in the N-buffer layer. The electric field profiles used in Fig. 9.23 to develop the analytical models for the blocking voltage are consistent with those observed with the simulations.

The P^+ collector/N-base junction (J_1) becomes reverse biased during the reverse-blocking mode with the depletion region confined to the N-buffer layer for the asymmetric IGBT structure. This can be observed in Fig. 9.34 for the electric field profiles shown with the dashed lines. A very high electric field occurs when the collector voltage reaches –50 V. The reverse breakdown voltage obtained with the simulations for this case was found to be 54 V. The analytical model for the reverse breakdown voltage (see (9.30) and Fig. 9.27) is able to predict this accurately.

9.5 On-State Characteristics

As already discussed in Sect. 9.2, the IGBT structure is designed to support a large voltage when the gate bias is below the threshold voltage. However, if a gate bias is applied with a value much larger than the threshold voltage, the MOSFET within the IGBT structure operates in its linear region. Electrons supplied into the N-base region from the channel of the MOSFET promote the injection of holes from the P^+ collector/N-base junction. Due to the low doping concentration used in the N-base region to achieve the high blocking voltage capability, the concentration of the injected holes exceeds the doping concentration in the N-base region even at moderate collector current densities. Consequently, the N-base region operates under high-level injection conditions with strong modulation of its conductivity. This reduces the resistance within the IGBT structure during on-state operation allowing it to carry high-current densities with a low on-state voltage drop. The output characteristics of the IGBT structure then resemble those for a P–i–N rectifier when the IGBT is operated with a large gate bias as illustrated in the Fig. 9.35. However, if the gate bias is made only slightly above the threshold voltage, the MOSFET region will enter its saturation mode and the IGBT collector current will also saturate as shown in the figure. In this mode, it is appropriate to treat the IGBT structure as a MOSFET providing the base drive current for a P–N–P transistor.

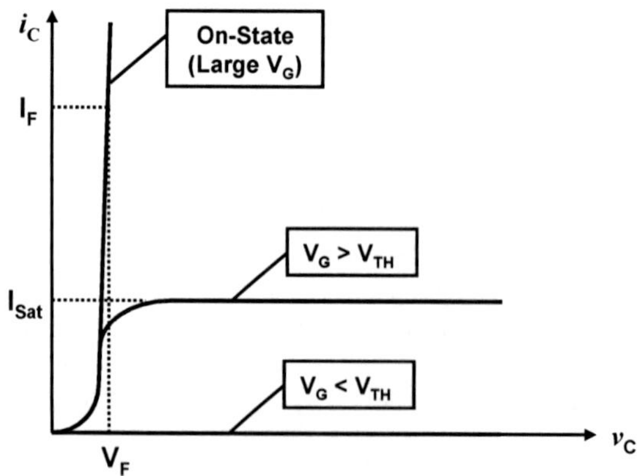

Fig. 9.35 Forward conduction characteristics for the IGBT structure

9.5.1 On-State Model

A simple model that can be used to analyze the on-state characteristics of the IGBT structure consists of a P–i–N rectifier connected in series with a MOSFET operating in its linear region. This equivalent circuit for the IGBT structure is shown in Fig. 9.36 together with the cross section of the device, indicating the corresponding regions in the device that constitute the MOSFET and P–i–N rectifier. The current flow path pertinent to the MOSFET/P–i–N rectifier model is indicated by the shaded area in the figure. For most of the current flow in the N-base region, the current density remains uniform allowing simplification of the model for the P–i–N rectifier to a one-dimensional structure. The collector current then transfers to the MOSFET at location *A* in the figure. The MOSFET current flow is governed by operation of the MOSFET in its linear region.

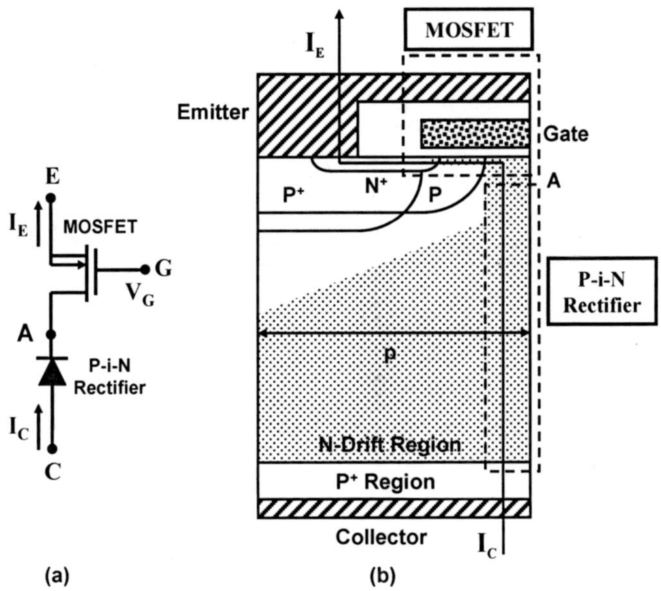

Fig. 9.36 On-state MOSFET/P–i–N equivalent circuit for the IGBT structure

The application of a large positive gate bias to induce operation of the IGBT structure in the on-state leads to the formation of an accumulation layer under the gate electrode where it overlaps the N-base region. During on-state current flow, the electrons arriving in the accumulation layer via the channel of the MOSFET can be viewed as being injected into the N-base region. The corresponding injection of holes from the P^+ collector region produces high-level injection conditions within the N-base region. The electron and hole concentrations within the N-base region take a catenary distribution under the gate electrode as illustrated in Fig. 9.37. In accordance with the analysis for the P–i–N rectifier in Sect. 5.1.3,

$$n(y) = p(y) = \frac{\tau_{\mathrm{HL}} J_{\mathrm{C}}}{2qL_{\mathrm{a}}}\left[\frac{\cosh(y/L_{\mathrm{a}})}{\sinh(d/L_{\mathrm{a}})} - \frac{\sinh(y/L_{\mathrm{a}})}{2\cosh(d/L_{\mathrm{a}})}\right], \tag{9.38}$$

where the distance *d* is half the width (W_{N}) of the N-base region as indicated in the figure.

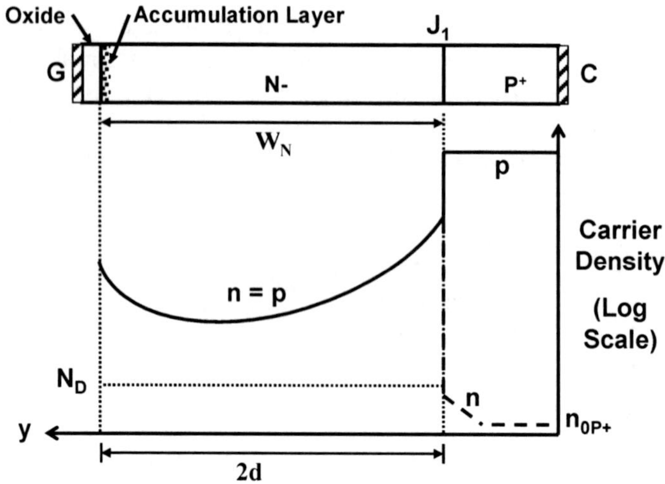

Fig. 9.37 Carrier distribution within the P–i–N rectifier portion of the IGBT structure in the on-state

The voltage drop across the P–i–N rectifier portion in the IGBT structure can be obtained (see Chap. 5) by using

$$V_{\mathrm{F,PiN}} = \frac{2kT}{q} \ln\left[\frac{J_{\mathrm{C}} W_{\mathrm{N}}}{4qD_{\mathrm{a}} n_{\mathrm{i}} F(W_{\mathrm{N}}/2L_{\mathrm{a}})}\right]. \tag{9.39}$$

As in the case of the P–i–N rectifier, a minimum on-state voltage drop occurs when the ambipolar diffusion length (L_a) is equal to the distance d or one half the width (W_N) of the N-base region. The typical contribution to the on-state voltage drop for the IGBT structure from the P–i–N rectifier portion is approximately 1 V at a collector current density of 100 A cm^{-2}. The voltage drop in the P–i–N portion produces a "knee" in the on-state characteristics when it is plotted on a linear scale because the current increases exponentially with the on-state voltage drop as given by

$$J_{\mathrm{C}} = \frac{4qD_{\mathrm{a}} n_{\mathrm{i}}}{W_{\mathrm{N}}} F\left(\frac{W_{\mathrm{N}}}{2L_{\mathrm{a}}}\right) \mathrm{e}^{qV_{\mathrm{PiN}}/2kT}. \tag{9.40}$$

In the MOSFET/P–i–N rectifier model, all of the collector current in the IGBT structure flows through the MOSFET before leaving through the emitter terminal. The voltage drop across the MOSFET portion must be added to the voltage drop across the P–i–N portion to determine the total on-state voltage drop for the IGBT structure. The voltage drop across the channel of the MOSFET when operated in its linear regime (see Sect. 6.6.3 for the channel resistance R_{CH}) is given by

$$V_{\text{F,MOSFET}} = I_{\text{C}} R_{\text{CH}} = \frac{I_{\text{C}} L_{\text{CH}}}{Z \mu_{\text{ni}} C_{\text{OX}} (V_{\text{G}} - V_{\text{TH}})}, \tag{9.41}$$

where Z is the length of the IGBT structure orthogonal to the cross section shown in Fig. 9.36. The collector current is related to the collector current density by

$$I_{\text{C}} = J_{\text{C}} p Z, \tag{9.42}$$

where p is the cell pitch indicated in Fig. 9.36. Applying this relationship,

$$V_{\text{F,MOSFET}} = \frac{p L_{\text{CH}} J_{\text{C}}}{\mu_{\text{ni}} C_{\text{OX}} (V_{\text{G}} - V_{\text{TH}})}. \tag{9.43}$$

The on-state voltage drop for the IGBT structure is then given by the addition of the voltage drops across the P–i–N and MOSFET portions of the device:

$$V_{\text{F,IGBT}} = \frac{2kT}{q} \ln\left[\frac{J_{\text{C}} W_{\text{N}}}{4qD_{\text{a}} n_{\text{i}} F(W_{\text{N}}/2L_{\text{a}})}\right] + \frac{p L_{\text{CH}} J_{\text{CH}}}{\mu_{\text{ni}} C_{\text{OX}} (V_{\text{G}} - V_{\text{TH}})}. \tag{9.44}$$

At low on-state collector current densities with a large gate bias voltage, the first term in the above equation becomes dominant. In this regime of operation, the collector current in the IGBT structure increases exponentially with increasing on-state voltage drop similar to that observed for the P–i–N rectifier. At larger collector current densities, the second term in the above equation becomes significant, adding a resistance in series with the diode voltage drop.

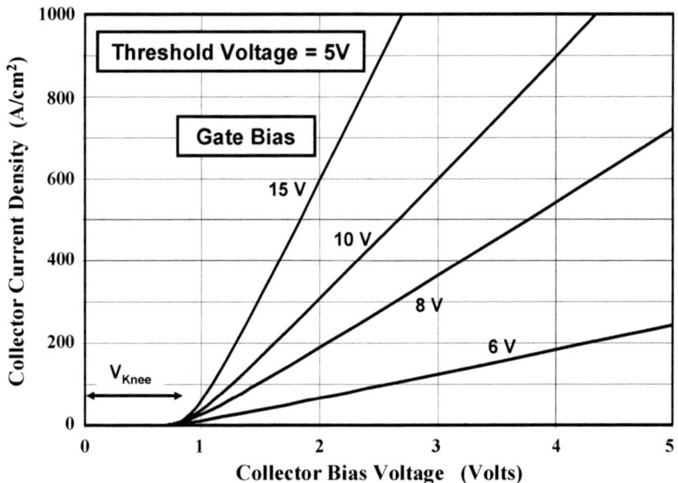

Fig. 9.38 On-state characteristics for the symmetric IGBT structure using linear region model for MOSFET portion

As an example, the on-state *i–v* characteristics are provided in Fig. 9.38 for the symmetric IGBT structure with an N-base width of 200 μm. The lifetime (τ_{p0}, τ_{n0}) was assumed to be 10 μs in the N-base region. The structure has a cell pitch of 15 μm and a channel length of 1.5 μm. A channel mobility of 200 cm^2 V^{-1} s^{-1} was used with a gate oxide thickness of 500 Å. It can be seen that very little current flow occurs until the collector bias exceeds a *knee* voltage of about 0.8 V as in the case of a P–i–N rectifier. The current then increases with a differential resistance which becomes smaller with increasing gate bias.

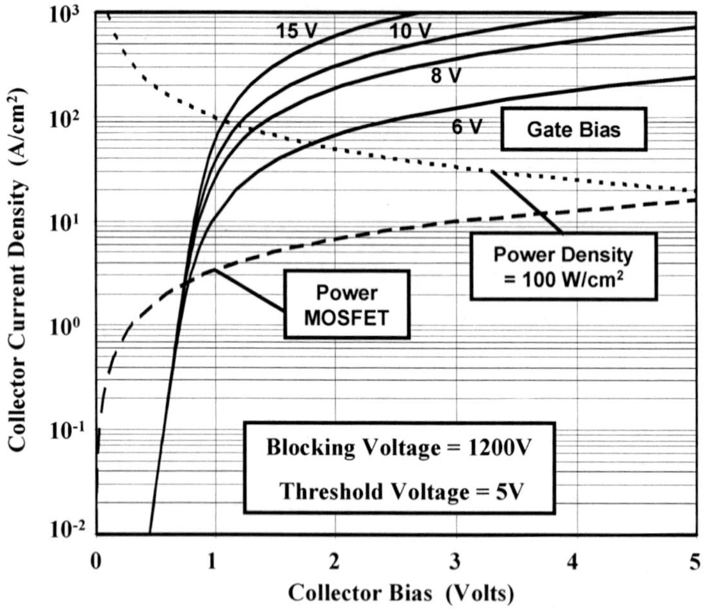

Fig. 9.39 On-state characteristics for the symmetric IGBT structure

The exponential increase in the collector current with collector bias for the above symmetric IGBT structure can be observed in Fig. 9.39. From this figure, it can also be seen that the characteristics are similar to those of the P–i–N rectifier at large gate bias voltages. At a gate bias of 5 V above the threshold voltage (V_G of 10 V in the figure), the on-state voltage drop is found to be 1.26 V at an on-state current density of 100 A cm^{-2}. Since the specific on-resistance for a power MOSFET capable of supporting 1,200 V is 0.3 Ω cm^2, the forward voltage drop for the power MOSFET at this current density would be 30 V. In practice, the on-state operating point is dictated by maintaining the power dissipation at a value required to keep the junction temperature below 125°C from reliability considerations. Based upon typical thermal impedances, the power dissipation for power devices is about 100 W cm^{-2}. The locus of the power dissipation is indicated in Fig. 9.39 by the dotted line. The intersection of this line with the on-state characteristics defines the operating point for the device. In the case of the 1,200-V symmetric IGBT structure, the operating point is at a current density of 85 A cm^{-2} with an on-state voltage drop of 1.18 V. In contrast, for a silicon power MOSFET that is designed to support 1,200 V, the operating point is at a current density of 18 A cm^{-2} with an on-state voltage drop of 5.44 V. Consequently, the chip area for the IGBT structure is nearly five times smaller than that of the power MOSFET structure. This not only reduces the cost of the power device but also reduces the gate drive current because of the smaller input capacitance.

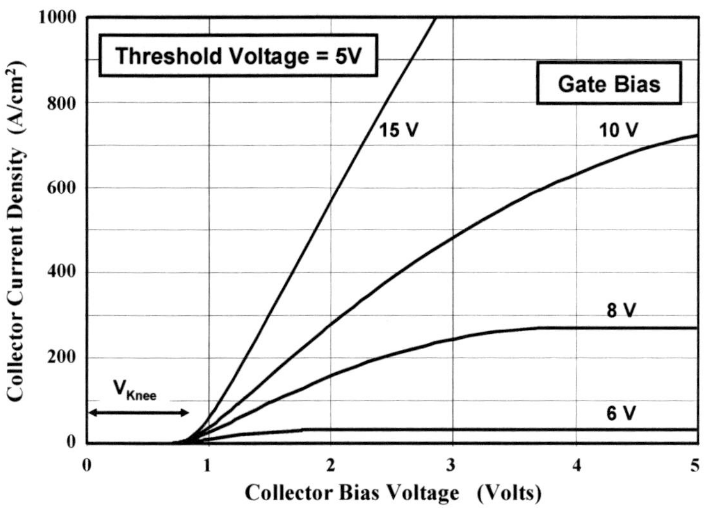

Fig. 9.40 On-state characteristics for the symmetric IGBT structure including channel pinch-off model for MOSFET portion

The on-state characteristics for the IGBT structure obtained by using the MOSFET/P–i–N model can be analyzed to include the saturation of the collector current at large collector current levels when the gate bias is close to the threshold voltage. In this case, the voltage drop across the MOSFET channel is no longer assumed to be small when compared with the gate bias voltage. When the channel pinch-off phenomenon is included in the analysis of the channel resistance, the voltage drop across the MOSFET channel is related to the collector current by (see Sect. 6.5.5)

$$I_{\mathrm{C}} = J_{\mathrm{C}} pZ = \frac{\mu_{\mathrm{ni}} C_{\mathrm{OX}} Z}{2L_{\mathrm{CH}}}[2(V_{\mathrm{G}} - V_{\mathrm{TH}})V_{\mathrm{F,MOSFET}} - V_{\mathrm{F,MOSFET}}^2]. \tag{9.45}$$

Solving for the voltage drop across the MOSFET channel yields

$$V_{\mathrm{F,MOSFET}} = (V_{\mathrm{G}} - V_{\mathrm{TH}})\left[1 - \sqrt{1 - \frac{2pL_{\mathrm{CH}}J_{\mathrm{C}}}{\mu_{\mathrm{ni}} C_{\mathrm{OX}} (V_{\mathrm{G}} - V_{\mathrm{TH}})^2}}\right]. \tag{9.46}$$

The voltage drop across the IGBT structure can be obtained by adding the voltage drop across the P–i–N portion (given by (9.39)) to this voltage drop across the MOSFET portion. The results are shown in Fig. 9.40 for the same structure that was used to obtain Fig. 9.38. The saturation of the collector current is now evident when the gate bias becomes close to the threshold voltage.

Simulation Example

To gain further insight into the physics of operation for the IGBT structure in the on-state, the results of two-dimensional numerical simulations for the typical symmetric structure (described in the previous section) are discussed here. The simulations were performed by using a cell with half the width of the structure shown in Fig. 9.6, i.e., extending from the center of the diffusion window to the center of the polysilicon gate electrode. This cell has a width of 15 μm (area = 1.5×10^{-7} cm^{-2}) with an opening of 7 μm in the polysilicon gate electrode. To operate the IGBT structure in its on-state, various gate voltages were applied while sweeping the collector voltage.

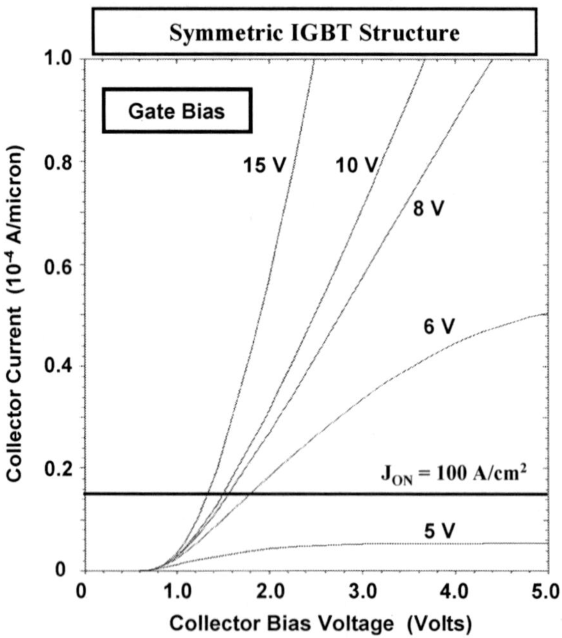

Fig. 9.41 Typical forward conduction characteristics for the IGBT structure

The forward conduction characteristics are shown in Fig. 9.41 for the case of a lifetime (τ_{p0}, τ_{n0}) of 10 μs. It can be observed that the collector current increases only after the collector bias exceeds a "knee voltage" of about 0.8 V. The on-state voltage drop decreases when the gate bias is increased. At a gate bias of 5 and 6 V (close to the threshold voltage of 4.7 V), the current becomes saturated. The MOSFET/P–i–N model provides an accurate depiction of these phenomena (see

Fig. 9.40). The on-state voltage drop is 1.26 V at a collector current density of 100 A cm^{-2} for a gate bias of 10 V. The MOSFET/P–i–N model provides an accurate quantitative assessment of this value.

The forward conduction characteristics for the symmetric IGBT structure are shown in Fig. 9.42 using a logarithmic scale for the collector current. The collector current can be observed to increase exponentially with collector bias voltage for current densities below 10 A cm^{-2} when a gate bias of 15 V is applied to maintain the MOSFET channel in its linear mode of operation. This behavior is accurately modeled by the P–i–N portion of the equivalent circuit for the IGBT structure. When the gate bias is reduced, the resistance contributed by the channel increases producing a nonexponential rise in the collector current with the applied collector bias. At the smallest gate bias of 5 V, a strong saturation of the collector current becomes evident. Thus, the on-state *i–v* characteristics predicted by the MOSFET/P–i–N model (see Fig. 9.39) provide an accurate description of the IGBT structure.

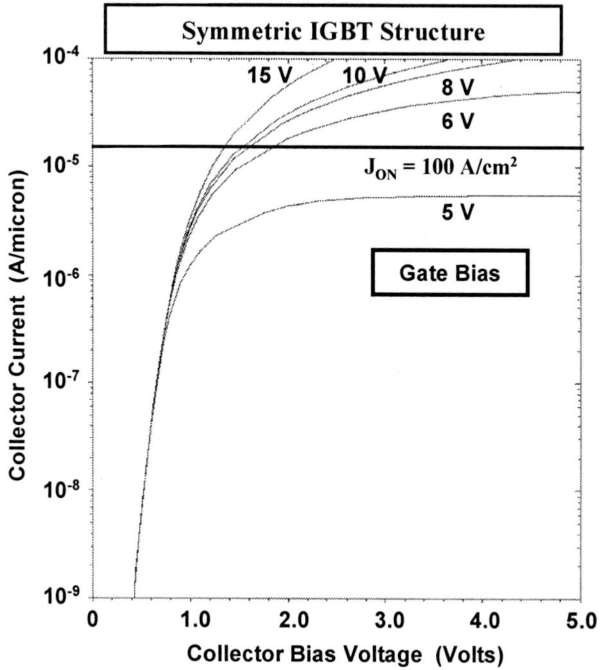

Fig. 9.42 Typical forward conduction characteristics for the symmetric IGBT structure

9.5.2 On-State Carrier Distribution: Symmetric Structure

According to the MOSFET/P–i–N model for the on-state characteristics of the symmetric IGBT structure, the free carrier distribution within the N-base region exhibits a catenary distribution in accordance with (9.38). Although this is a reasonable approximation at the middle of the gate electrode, a low free carrier concentration occurs under the deep P$^+$ region because this junction is reverse biased when the IGBT structure is operating in its on-state. Consequently, the boundary conditions for the IGBT structure are different from those for the P–i–N rectifier.[20] The reduced free carrier concentration in the portion of the device away from the middle of the gate electrode tends to increase the on-state voltage drop.

A one-dimensional analysis of the free carrier distribution can be performed for the portion of the IGBT structure under the deep P$^+$ region by solving the continuity equation for minority carriers (holes in the N-base region) under steady-state high-level injection conditions (see (5.23)):

$$\frac{d^2p}{dy^2} - \frac{p}{L_a^2} = 0, \tag{9.47}$$

where L_a is the ambipolar diffusion length in the N-base region given by

$$L_a = \sqrt{D_a \tau_{HL}}. \tag{9.48}$$

The general solution for the carrier distribution is given by

$$p(y) = A\cosh\left(\frac{y}{L_a}\right) + B\sinh\left(\frac{y}{L_a}\right), \tag{9.49}$$

where A and B are parameters determined by the boundary conditions. If the distance y is defined as starting from the P$^+$ collector/N-base junction (J_1) and ending at the deep P$^+$/N-base junction (J_2) (i.e., at $y = W_N$), these boundary conditions are

$$p(0) = p_0 \tag{9.50}$$

due to the injection of holes from the P$^+$ collector/N-base junction (J_1) and

$$p(W_N) = 0 \tag{9.51}$$

due to the reverse-biased deep P$^+$/N-base junction (J_2). Using these boundary conditions yields

$$A = p_0 \tag{9.52}$$

and

$$B = -\frac{p_0}{\tanh(W_N / L_a)}. \tag{9.53}$$

Substituting into (9.49) and performing simplification of the terms yields

$$p(y) = p_0 \frac{\sinh[(W_N - y)/L_a]}{\sinh(W_N / L_a)}. \tag{9.54}$$

The hole concentration (p_0) at the P$^+$ collector/N-base junction (J_1) can be related to collector current (J_C) by using

$$2qD_p\left(\frac{dp}{dy}\right)_{y=0} = J_p(0) = \gamma_{E,ON}J_C, \tag{9.55}$$

where $\gamma_{E,ON}$ is the injection efficiency of the P$^+$ collector/N-base junction (J_1) in the on-state and $J_p(0)$ is the portion of the total collector current due to holes at the junction. Under low-level injection conditions, the injection efficiency for the P$^+$ collector/N-base junction (J_1) is close to unity due to the low doping concentration within the N-base region for the symmetric IGBT structure. However, during on-state operation with high collector current density, the electron concentration within the N-base region at the P$^+$ collector/N-base junction (J_1) becomes large due to high-level injection conditions. This promotes increased injection of electrons into the P$^+$ collector region reducing the injection efficiency.

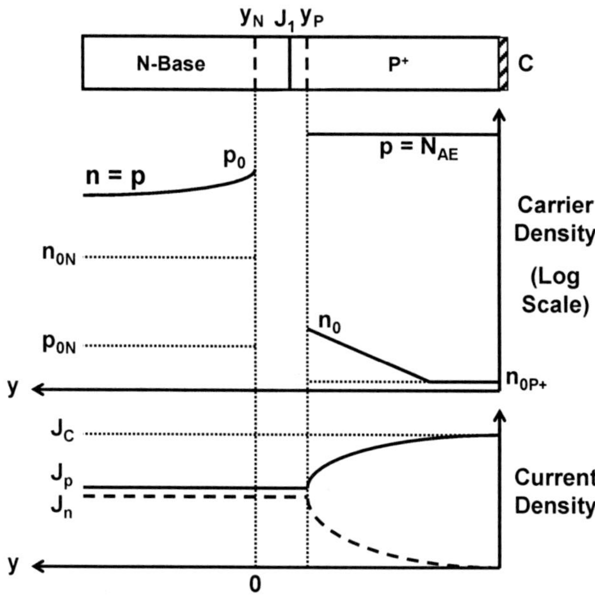

Fig. 9.43 Impact of high-level injection at the P$^+$ collector/N-base junction in the symmetric IGBT structure

The free carrier concentrations are illustrated in the vicinity of the P$^+$ collector/N-base junction (J_1) in Fig. 9.43 for the IGBT structure in the on-state. The electron and hole concentrations are much larger than the minority and majority carrier concentrations in equillibrium on the N-base side. The electron and hole current densities in the N-base region do not vary rapidly with distance. The injected electron concentration in the P$^+$ collector region (which is behaving as an emitter within the IGBT structure) is enhanced by the high concentration of electrons on the N-side of the junction. These electrons decay at the characteristic diffusion length (L_{nE}) for the P$^+$ collector region. The electron and hole components of the collector current are illustrated at the bottom of the figure. The electron current in the P$^+$ collector region decreases within the P$^+$ collector region. The electron and hole currents are considered to be continuous through the depletion region.[21]

Using the Boltzmann quasiequilibrium boundary condition for the P–N junction,

$$\frac{p(y_N)}{p(y_P)} = \frac{n(y_P)}{n(y_N)} = e^{-(q\Delta\psi/kT)}, \tag{9.56}$$

where $p(y_N)$ and $p(y_P)$ are the hole concentrations on the two sides of the junction, $n(y_N)$ and $n(y_P)$ are the electron concentrations on the two sides of the junction, and $\Delta\psi$ is the potential barrier across the junction under forward bias operation. Due to low-level injection conditions in the P$^+$ collector region, the hole concentration is equal to the acceptor doping concentration (N_{AE}):

$$p(y_P) = p_{0E} = N_{AE}. \tag{9.57}$$

The electron and hole concentrations at the junction on the N-base side are equal due to high-level injection conditions:

$$n(y_N) = p(y_N) = p_0. \tag{9.58}$$

Using these relationships in (9.56) yields

$$n(y_P) = n_0 = \frac{p_0^2}{N_{AE}}. \tag{9.59}$$

The diffusion current due to electrons in the P$^+$ collector region under low-level injection conditions is given by

$$J_n(y_P) = \frac{qD_{nE}n_0}{L_{nE}} = \frac{qD_{nE}p_0^2}{L_{nE}N_{AE}}. \tag{9.60}$$

Since the electron current is continuous through the depletion region,

$$J_n(y_N) = J_n(y_P) = \frac{qD_{nE}p_0^2}{L_{nE}N_{AE}}. \tag{9.61}$$

The injection efficiency for holes at the P$^+$ collector/N-base junction (J_1) is then determined by

$$\gamma_{E,ON} = \frac{J_p(y_N)}{J_C} = 1 - \frac{J_n(y_N)}{J_C}. \tag{9.62}$$

A diffused P^+ collector region with a surface concentration of 1×10^{19} cm^{-3} has an effective doping concentration of 1×10^{18} cm^{-3} due to the graded doping profile. Using an injected hole carrier concentration (p_0) of 1×10^{17} cm^{-3} in the N-base region at a collector current density of 100 A cm^{-2}, the hole and electron currents at the junction are found to be 25 and 75 A cm^{-2}, respectively. Thus, the injection efficiency is typically reduced to about 0.25 due to the high-level injection conditions in the N-base region.

The electron current density in the N-base region at the P^+ collector/N-base junction ($y = 0$) is given by

$$J_{\mathrm{n}}(0) = q\mu_{\mathrm{n}} n(0) E(0) + qD_{\mathrm{n}} \left(\frac{\mathrm{d}n}{\mathrm{d}y} \right)_{y=0} = (1 - \gamma_{\mathrm{E,ON}}) J_{\mathrm{C}}. \tag{9.63}$$

From this equation, an expression for the electric field in the N-base region at the junction ($y = 0$) is obtained:

$$E(0) = \frac{(1 - \gamma_{\mathrm{E,ON}}) J_{\mathrm{C}}}{q\mu_{\mathrm{n}} p_0} - \frac{kT}{q p_0} \left(\frac{\mathrm{d}p}{\mathrm{d}y} \right)_{y=0} \tag{9.64}$$

based upon making the electron concentration $n(0)$ at the junction equal to the hole concentration (p_0) due to high-level injection conditions in the N-base region. The hole current density at the P^+ collector/N-base region ($y = 0$) is given by

$$J_{\mathrm{p}}(0) = q\mu_{\mathrm{p}} p(0) E(0) - qD_{\mathrm{p}} \left(\frac{\mathrm{d}p}{\mathrm{d}y} \right)_{y=0}. \tag{9.65}$$

Using (9.64) in (9.65) yields

$$J_{\mathrm{p}}(0) = \left(\frac{\mu_{\mathrm{p}}}{\mu_{\mathrm{n}}} \right) (1 - \gamma_{\mathrm{E,ON}}) J_{\mathrm{C}} - 2qD_{\mathrm{p}} \left(\frac{\mathrm{d}p}{\mathrm{d}y} \right)_{y=0}. \tag{9.66}$$

However, the hole current density is also given by

$$J_{\mathrm{p}}(0) = \gamma_{\mathrm{E,ON}} J_{\mathrm{C}}. \tag{9.67}$$

Combining these relationships,

$$\left(\frac{\mathrm{d}p}{\mathrm{d}y} \right)_{y=0} = \left(\frac{\mu_{\mathrm{p}}}{\mu_{\mathrm{n}}} \right) \left(\frac{J_{\mathrm{C}}}{2qD_{\mathrm{p}}} \right) \left[1 - \gamma_{\mathrm{E,ON}} - \left(\frac{\mu_{\mathrm{n}}}{\mu_{\mathrm{p}}} \right) \gamma_{\mathrm{E,ON}} \right]. \tag{9.68}$$

Based upon (9.54),

$$\left(\frac{\mathrm{d}p}{\mathrm{d}y} \right)_{y=0} = \frac{p_0}{L_{\mathrm{a}} \tanh(W_{\mathrm{n}} / L_{\mathrm{a}})}. \tag{9.69}$$

Using (9.69) in (9.66) together with (9.67) provides a solution for the hole concentration at the P^+ collector/N-base junction:

$$p_0 = \frac{J_C L_a}{2qD_p}\left(\frac{\mu_p}{\mu_n}\right)\tanh(W_N / L_a)\left[1-\gamma_{E,ON}-\left(\frac{\mu_n}{\mu_p}\right)\gamma_{E,ON}\right]. \tag{9.70}$$

Combining this expression with (9.61) and (9.62) yields a quadratic solution for the hole concentration (p_0) at the P^+ collector/N-base junction:

$$ap_0^2 + bp_0 + c = 0 \tag{9.71}$$

with coefficients

$$a = \frac{qD_{nE}}{L_{nE}N_{AE}J_C}\left(1-\frac{\mu_n}{\mu_p}\right), \tag{9.72}$$

$$b = -\frac{2qD_p}{L_a J_C \tanh(W_N / L_a)}\left(\frac{\mu_p}{\mu_n}\right), \tag{9.73}$$

$$c = -\left(\frac{\mu_n}{\mu_p}\right). \tag{9.74}$$

The value of the hole concentration (p_0) at the P^+ collector/N-base junction can then be computed by using

$$p_0 = -\frac{b}{2a}\left[1+\sqrt{1-\left(\frac{4ac}{b^2}\right)}\right]. \tag{9.75}$$

As an example, the hole carrier densities calculated by using (9.54) with the above analytical solution for the hole concentration (p_0) at the P^+ collector/N-base junction are provided in Fig. 9.44 for the symmetric IGBT structure with an N-base width of 200 μm at an on-state collector current density of 100 A cm^{-2}. The high-level lifetime was varied from 200 to 0.2 μs in the N-base region. In this analysis, a diffusion length for electrons (L_{nE}) of 0.5 μm was obtained in the P^+ collector region for a high-level lifetime of 20 μs in the N-base region after scaling the lifetime in the P^+ collector region in proportion to the lifetime in the N-base region. A doping concentration of 1×10^{18} cm^{-3} was used for the P^+ collector region as being representative of a diffused region with surface concentration of 1×10^{19} cm^{-3}. It can be seen that there is a high concentration of holes (indicated by p_0 in the figure) at the P^+ collector/N-base junction (J_1) which reduces to zero at the deep P^+/N-base junction (J_2) at $y = 200$ μm. The injected hole concentration (p_0) at the P^+ collector/N-base junction (J_1) becomes smaller when the lifetime is reduced. In the case of the smallest lifetime of 0.2 μs, high-level injection conditions do not prevail throughout the N-base region because the hole concentration falls below the doping concentration indicated by the dashed line. The analytical model derived above is not applicable for the upper half of the N-base region in this case.

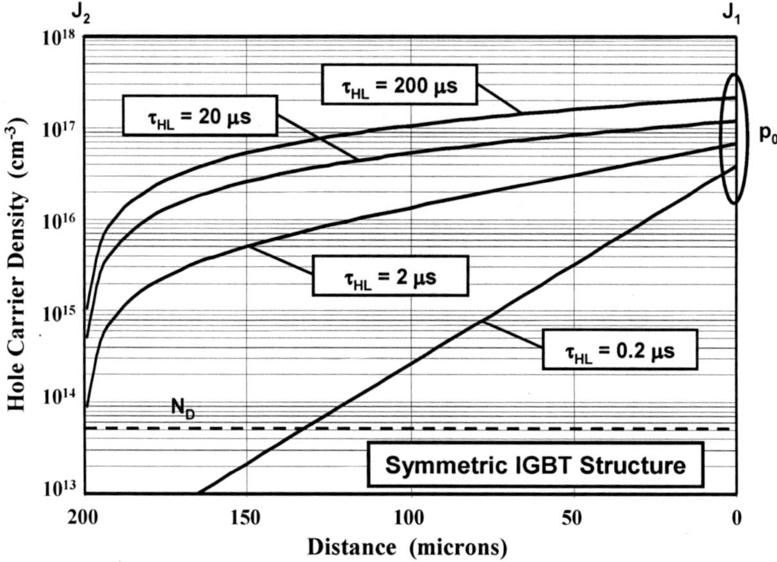

Fig. 9.44 On-state hole distribution within the symmetric IGBT structure under the deep P^+ region

Simulation Example

To gain further insight into the physics of operation for the IGBT structure in the on-state, the results of two-dimensional numerical simulations for the typical symmetric structure (described in the previous section) are discussed here. The hole concentration was obtained in the on-state at a current density of 100 A cm^{-2} for various values for the high-level lifetime. A three-dimensional view of the free carrier distribution is shown in Fig. 9.45 for the case of a high-level lifetime of 20 μs. It can be observed that the hole concentration reduces to zero at the deep P^+/N-base junction. There is a slight increase in the hole concentration under the gate electrode but even in this portion, the carrier profile is closer to that described by (9.54) rather than a catenary distribution applicable to P–i–N rectifiers.

A one-dimensional view of the hole distribution under the deep P^+ region obtained from simulations performed with the various high-level lifetime values is provided in Fig. 9.46. It can be observed that the hole concentration at the P^+ collector/N-base junction (J_1) is much larger than the background doping concentration and goes to zero at the deep P^+/N-base junction (J_2). The hole concentration at the P^+ collector/N-base junction (J_1) becomes smaller when the lifetime is reduced. The predictions of the analytical model (see Fig. 9.44) are in very good qualitative and quantitative agreement with the results of the simulations for practical lifetime values (>1 μs) for the symmetric IGBT structure, providing validation for the utility of the analytical model.

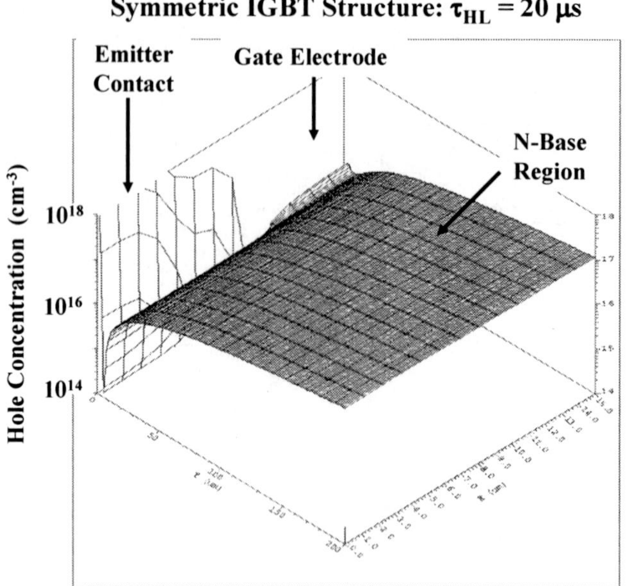

Fig. 9.45 Hole distribution in the on-state for the symmetric IGBT structure

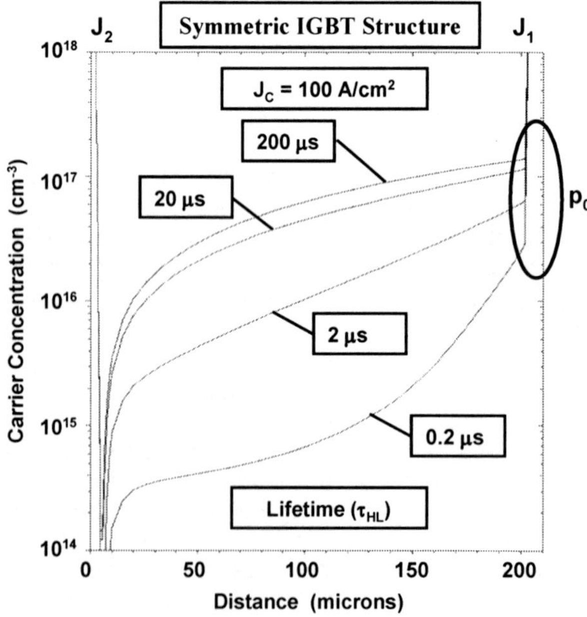

Fig. 9.46 Hole distribution in the on-state for the IGBT structure

9.5.3 On-State Voltage Drop: Symmetric Structure

In Sect. 9.5.2, it was established that the free carrier distribution within the N-base region of the IGBT structure deviates from that for the P–i–N rectifier. The lower carrier concentration in the upper portion of the IGBT structure tends to increase the voltage drop in the N-base region when compared with the P–i–N rectifier. It is therefore necessary to derive the voltage drop in the N-base region for the IGBT structure using the carrier distribution given by (9.54). In addition, a more accurate prediction for the on-state voltage drop for the IGBT structure can be obtained by addition of the accumulation and JFET region contributions to the channel contribution from the MOSFET portion.

The on-state voltage drop for the IGBT structure can be obtained by using

$$V_{\mathrm{ON}} = V_{\mathrm{P+N}} + V_{\mathrm{NB}} + V_{\mathrm{MOSFET}}, \tag{9.76}$$

where $V_{\mathrm{P+N}}$ is the voltage drop across the P^+ collector/N-base junction (J_1), V_{NB} is the voltage drop across the N-base region after accounting for conductivity modulation due to high-level injection conditions, and V_{MOSFET} is the voltage drop across the MOSFET portion. The voltage drop across the P^+ collector/N-base junction (J_1) can be obtained from the increase in the minority carrier concentration at the junction boundary:

$$V_{\mathrm{P+N}} = \frac{kT}{q}\ln\left(\frac{p_0}{p_{0\mathrm{N}}}\right) = \frac{kT}{q}\ln\left(\frac{p_0 N_{\mathrm{D}}}{n_{\mathrm{i}}^2}\right), \tag{9.77}$$

where the minority carrier density in equilibrium ($p_{0\mathrm{N}}$) has been related to the doping concentration in the N-base region. The increase in the hole concentration at the junction (p_0) can be obtained from the analysis provided in Sect. 9.5.2.

The voltage drop across the N-base region (V_{NB}) can be obtained by integration of the electric field within the N-base region. The electric field can be derived by analysis of the hole and electron current density in the N-base region:

$$J_{\mathrm{p}}(y) = q\mu_{\mathrm{p}}\left[p(y)E(y) - \frac{kT}{q}\frac{\mathrm{d}p}{\mathrm{d}y}\right], \tag{9.78}$$

$$J_{\mathrm{n}}(y) = q\mu_{\mathrm{n}}\left[n(y)E(y) + \frac{kT}{q}\frac{\mathrm{d}n}{\mathrm{d}y}\right]. \tag{9.79}$$

The increase in y along the left direction (see Figs. 9.37 and 9.43) has been accounted for when writing these expressions. Continuity of collector current flow requires

$$J_{\mathrm{n}}(y) + J_{\mathrm{p}}(y) = J_{\mathrm{C}}. \tag{9.80}$$

Due to high-level injection conditions in the N-base region, the electron and hole concentrations, as well as their derivatives, can be assumed to be equal. Combining these expressions with this knowledge and solving for the electric field yields

$$E(y) = \frac{J_{\mathrm{C}}}{qp(y)(\mu_{\mathrm{n}} + \mu_{\mathrm{p}})} - \frac{kT}{q}\left(\frac{\mu_{\mathrm{n}} - \mu_{\mathrm{p}}}{\mu_{\mathrm{n}} + \mu_{\mathrm{p}}}\right)\frac{1}{p(y)}\frac{\mathrm{d}p}{\mathrm{d}y}. \tag{9.81}$$

Using the free carrier distribution given by (9.54),

$$E(y) = \frac{J_{\mathrm{C}}\sinh(W_{\mathrm{N}}/L_{\mathrm{a}})}{qp_0(\mu_{\mathrm{n}} + \mu_{\mathrm{p}})\sinh[(W_{\mathrm{N}} - y)/L_{\mathrm{a}}]} - \frac{kT}{q}\left(\frac{\mu_{\mathrm{n}} - \mu_{\mathrm{p}}}{\mu_{\mathrm{n}} + \mu_{\mathrm{p}}}\right)\frac{1}{L_{\mathrm{a}}\tanh[(W_{\mathrm{N}} - y)/L_{\mathrm{a}}]}. \tag{9.82}$$

The voltage drop across the N-base region can be obtained by integrating the above electric field from $y = 0$ at the P^+ collector/N-base junction (J_1) to the edge of depletion region of the deep P^+/N-base junction (J_2), namely at $y = (W_{\mathrm{N}} - W_{\mathrm{ON}})$, where W_{ON} is the depletion width at the deep P^+/N-base junction (J_2) in the on-state. The voltage drop associated with the first part of the above electric field expression is

$$V_{\mathrm{NB1}} = \frac{2L_{\mathrm{a}}J_{\mathrm{C}}\sinh(W_{\mathrm{N}}/L_{\mathrm{a}})}{qp_0(\mu_{\mathrm{n}} + \mu_{\mathrm{p}})}\{\tanh^{-1}[\mathrm{e}^{-(W_{\mathrm{ON}}/L_{\mathrm{a}})}] - \tanh^{-1}[\mathrm{e}^{-(W_{\mathrm{N}}/L_{\mathrm{a}})}]\}. \tag{9.83}$$

The depletion width (W_{ON}) across the deep P^+/N-base junction (J_2) in the on-state depends on the on-state voltage drop. The voltage drop associated with the second part of the above electric field expression is

$$V_{\mathrm{NB2}} = \frac{kT}{q}\left(\frac{\mu_{\mathrm{n}} - \mu_{\mathrm{p}}}{\mu_{\mathrm{n}} + \mu_{\mathrm{p}}}\right)\ln\left[\frac{\tanh(W_{\mathrm{ON}}/L_{\mathrm{a}})\cosh(W_{\mathrm{ON}}/L_{\mathrm{a}})}{\tanh(W_{\mathrm{N}}/L_{\mathrm{a}})\cosh(W_{\mathrm{N}}/L_{\mathrm{a}})}\right]. \tag{9.84}$$

The voltage drop across the MOSFET portion includes contributions from the JFET region, the accumulation layer, and the channel. As discussed in Sect. 9.5.2, the free carrier concentration becomes small in the vicinity of the deep P^+/N-base junction (J_2). Consequently, it is appropriate to compute the resistance of the JFET region based upon its doping concentration. Since IGBT structures have low doping concentrations in the N-base region to obtain high blocking voltage ratings, the JFET region can become completely depleted by the built-in potential of the deep P^+/N-base junction (J_2). For instance, the zero bias depletion width is 5 μm for a typical doping concentration of $5 \times 10^{13}\ \mathrm{cm}^{-3}$ in the N-base region. For a typical gate electrode width (W_{G}) of 16 μm and a P-base depth of 3 μm, the available space for current flow in the JFET region is then reduced to zero. This can not only lead to a high on-state voltage drop for the IGBT structure but also produce a negative resistance (snapback) in the forward *i–v* characteristics that is detrimental to circuit operation. This problem can be overcome by enhancing the doping concentration in the JFET region by the ion implantation and diffusion of phosphorus. For the purposes of formulating an analytical model, the JFET region will be assumed to have an effective doping concentration of 5×10^{15} cm^{-3} based upon a Gaussian profile with a surface concentration of 1×10^{16} cm^{-3}. The resistivity (ρ_{JFET}) for the JFET region is then about 1.25 Ω cm.

Based upon the analysis for the power MOSFET structure (see (6.72)),

$$V_{JFET} = J_C R_{JFET,SP} = \frac{J_C \rho_{JFET}(x_{JP} + W_0)W_{CELL}}{W_G - 2x_{JP} - 2W_0}, \tag{9.85}$$

where x_{JP} is the junction depth of the deep P$^+$ region. The voltage drop across the accumulation layer in the IGBT structure can also be derived based upon the analysis for the power MOSFET structure (see (6.66)):

$$V_{ACC} = J_C R_{A,SP} = \frac{J_C K_A (W_G - 2x_{JP})W_{CELL}}{4\mu_{nA} C_{OX}(V_G - V_{TH})}. \tag{9.86}$$

The accumulation layer coefficient (K_A) for the IGBT structure can be assumed to have the same value (0.6) as for power MOSFET structures. The voltage drop across the channel of the MOSFET portion can also be obtained by using the analysis for the power MOSFET structure (see (6.63)):

$$V_{CH} = J_C R_{CH,SP} = \frac{J_C L_{CH} W_{CELL}}{2\mu_{ni} C_{OX}(V_G - V_{TH})}. \tag{9.87}$$

As an example, the components of the on-state voltage drop determined by using the above analytical model are provided in Fig. 9.47 for the case of the symmetric IGBT structure with blocking voltage capability of 1,200 V. The device has an N-base width of 200 μm with a depth of 5 μm for the deep P$^+$ region. All the parameters used for the structure are the same as those used to compute the free carrier concentration in Fig. 9.44. The MOSFET portion has an effective JFET region doping concentration of 5×10^{15} cm^{-3} corresponding to a diffused region with surface concentration of 1×10^{16} cm^{-3} and depth of 5 μm, a gate oxide thickness of 500 Å, and a channel length of 1.5 μm. The mobility for electrons in the inversion and accumulation layers was assumed to be 450 and 1,000 cm^2 V^{-1} s^{-1}, respectively (the same as those used for analysis of power MOSFET structures in Chap. 7). The IGBT structure had a cell pitch (W_{CELL}) of 30 μm with a gate electrode width (W_G) of 16 μm. The gate bias and threshold voltages used were 15 and 5 V, respectively. From the figure, it can be concluded that the voltage drop across the P$^+$ collector/N-base junction and the MOSFET portion is dominant when the high-level lifetime is large (>20 μs). When the high-level lifetime is reduced, the voltage drop across the N-base region increases and becomes dominant for lifetime value below 2 μs. A very rapid increase in the on-state voltage drop is observed when the high-level lifetime is reduced below 1 μs. This limits the maximum switching speed for the symmetric IGBT structure.

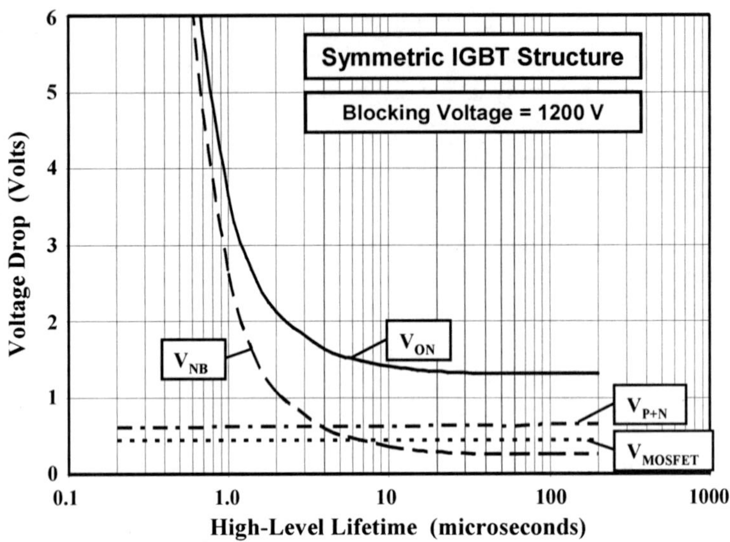

Fig. 9.47 On-state voltage drop for the symmetric IGBT structure

Simulation Example

To corroborate the analytical model for the on-state voltage drop, the results of two-dimensional numerical simulations for the typical symmetric structure (described in the previous section) are discussed here. A three-dimensional view of the doping distribution at the upper portion of the device structure is provided in Fig. 9.48. The N^+ emitter and deep P^+ diffusions are located on the upper left-hand side while the MOSFET portion is located on the upper right-hand side. The enhanced doping of the JFET region can be observed at the upper right-hand side with a surface concentration of 1×10^{16} cm^{-3} and a depth of 5 μm. From the profile, an effective doping concentration of 5×10^{15} cm^{-3} is appropriate in the analytical model.

The on-state voltage drop was obtained at a current density of 100 A cm^{-2} at a gate bias of 15 V for various values for the high-level lifetime. It was found that the on-state voltage drop increases rapidly when the high-level lifetime is reduced below 2 μs. The on-state voltage drop obtained from the numerical simulations is compared with those obtained by using the analytical model in Fig. 9.49. The analytical model provides an accurate prediction of the on-state voltage drop until the high-level lifetime becomes smaller than 1 μs. When the lifetime becomes smaller than 1 μs, the assumptions of high-level injection conditions within the N-base region used for the analytical model are not satisfied in the upper portion of the structure.

In spite of this, the analytical model is adequate for the analysis of the on-state voltage drop for the symmetric IGBT structure over a broad range of parameters.

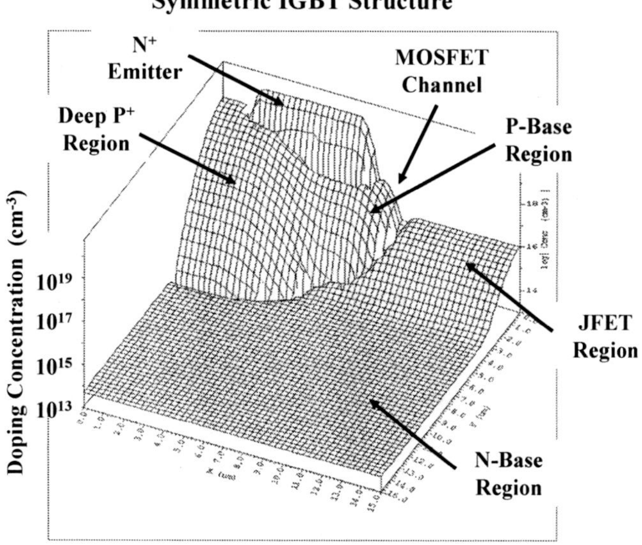

Fig. 9.48 Doping distribution within the IGBT structure at the upper portion

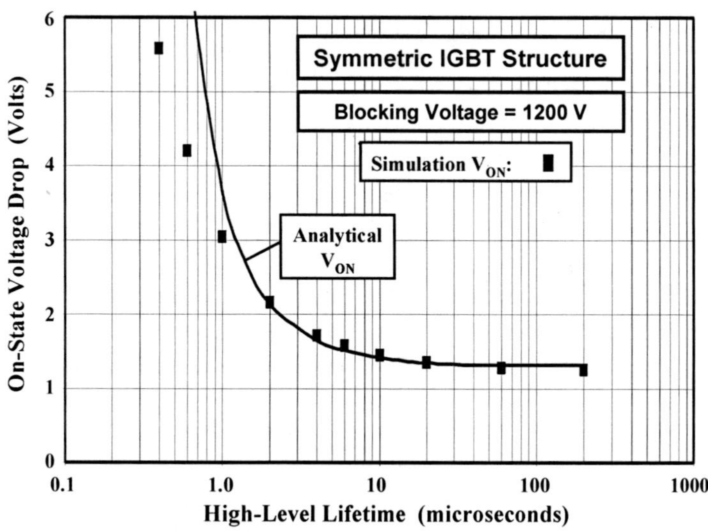

Fig. 9.49 On-state voltage drop for the symmetric IGBT structure

9.5.4 On-State Carrier Distribution: Asymmetric Structure

Based upon the analysis in Sect. 9.5.3, the on-state voltage drop for the symmetric IGBT structure was observed to increase rapidly when the recombination lifetime in the N-base region is reduced. This limits the maximum switching frequency for the symmetric IGBT structure as discussed later in this chapter. An improvement in the switching speed, while maintaining a low on-state voltage drop, can be derived by using the asymmetric IGBT structure due to its smaller N-base width for the same forward-blocking voltage capability. As an example, it was demonstrated in Sect. 9.4 that an N-base width of 110 μm can be utilized for the asymmetric IGBT structure when compared with 200 μm for the symmetric IGBT structure to achieve the same forward-blocking capability of 1,200 V.

If the buffer layer concentration in the asymmetric IGBT is relatively low ($<5 \times 10^{16}$ cm^{-3}), the injected hole concentration in the N-buffer layer during on-state operation can exceed its doping concentration. Consequently, both the N-buffer layer and the lightly doped portion of the N-base region operate under high-level injection conditions. The analytical model for the free carrier distribution developed in Sect. 9.5.2 for the symmetric IGBT structure is then also applicable to the asymmetric IGBT structure. Modifying (9.54) to the relevant width of the N-base region for the asymmetric IGBT structure yields the hole distribution profile:

$$p(y) = p_0 \frac{\sinh[(W_N + W_{NB} - y)/L_a]}{\sinh[(W_N + W_{NB})/L_a]}. \tag{9.88}$$

The hole concentration (p_0) at the P$^+$ collector/N-buffer layer junction (J_1) can also be obtained by using the equations developed in Sect. 9.5.2 for the symmetric IGBT structure (see (9.70)).

As an example, the hole carrier densities calculated by using (9.88) are provided in Fig. 9.50 for the asymmetric IGBT structure with an N-base region with lightly doped portion of width 100 μm and an N-buffer layer width of 10 μm at an on-state collector current density of 100 A cm^{-2}. The high-level lifetime was varied from 200 to 0.2 μs in the N-base region. In this analysis, a diffusion length for electrons (L_{nE}) of 0.06 μm was obtained in the P$^+$ collector region for the case of a high-level lifetime of 20 μs in the N-base region after the lifetime in the P$^+$ collector region was scaled in proportion to the lifetime in the N-base region. The smaller diffusion length obtained in this case is due to the abrupt junction doping profile at junction J_1, which makes the effective doping concentration in the P$^+$ collector region equal to its uniform doping concentration of 1×10^{19} cm^{-3}. It can be seen that there is a high concentration of holes (indicated by p_0 in the figure) at the P$^+$ collector/N-base junction (J_1) which reduces to zero at the deep P$^+$/N-base junction (J_2) at $y = 110$ μm. The injected hole concentration (p_0) at the P$^+$ collector/N-base junction (J_1) becomes smaller when the lifetime is reduced.

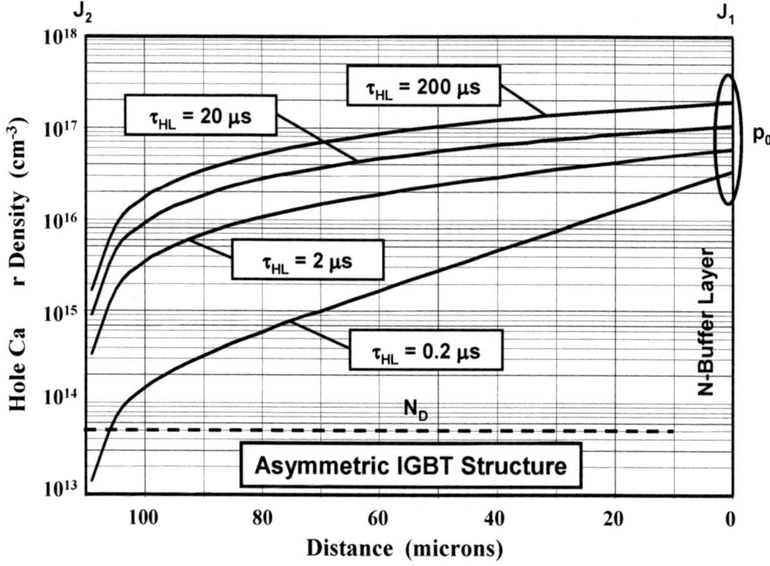

Fig. 9.50 On-state hole distribution within the asymmetric IGBT structure under the deep P^+ region for low buffer layer doping concentrations

Simulation Example

To gain further insight into the physics of operation for the asymmetric IGBT structure in the on-state, the results of two-dimensional numerical simulations for the typical asymmetric structure with a relatively low (1×10^{16} cm^{-3}) doping concentration in the N-buffer layer are discussed here. This structure had an N-base region consisting of a lightly doped (5×10^{13} cm^{-3}) portion with a width of 100 μm and an N-buffer layer with a width of 10 μm. The hole concentration was obtained in the on-state at a current density of 100 A cm^{-2} for various values for the high-level lifetime.

A one-dimensional view of the hole distribution under the deep P^+ region obtained from simulations performed with the various high-level lifetime values is provided in Fig. 9.51. It can be observed that the hole concentration at the P^+ collector/N-base junction (J_1) is larger than the background doping concentration even within the N-buffer layer and goes to zero at the deep P^+/N-base junction (J_2). The hole concentration at the P^+ collector/N-base junction (J_1) becomes smaller when the lifetime is reduced. The predictions of the analytical model for low N-buffer layer doping concentrations (see Fig. 9.50) are in very good qualitative and quantitative agreement with the results of the simulations for practical lifetime values (>0.2 μs) for the asymmetric IGBT structure, providing validation for the model for the case of relatively low buffer layer doping levels.

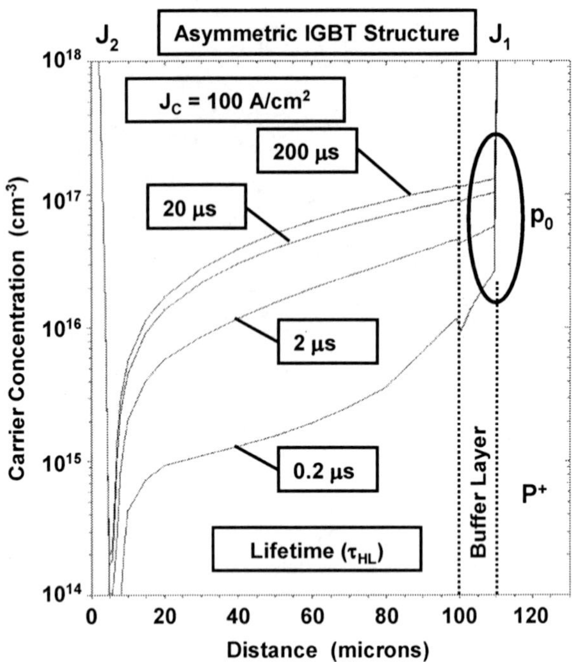

Fig. 9.51 Hole distribution in the on-state for the asymmetric IGBT structure with N-buffer layer doping concentration of 1×10^{16} cm^{-3}

An alternate approach (to changing the lifetime in the N-base region) for controlling the on-state and switching characteristics for the asymmetric IGBT structure is by adjusting the doping concentration in the N-buffer layer.[6,22] When the doping level in the N-buffer layer is increased, the injection efficiency at the P$^+$ collector/N-buffer junction (J_1) is reduced producing a smaller injected hole concentration in the N-base region. This increases the on-state voltage drop for the asymmetric IGBT structure and simultaneously reduces the turn-off time during switching. This provides an approach to trading off the on-state and switching power losses in a manner similar to that obtained with lifetime control.

When the doping concentration in the N-buffer layer is made greater than 1×10^{17} cm^{-3}, high-level injection conditions do not prevail within the N-buffer layer and are confined to the lightly doped portion of the N-base region. It is therefore necessary to use low-level injection analysis at junction J_1 and high-level injection analysis within the lightly doped portion of the N-base region. The carrier distributions and currents are illustrated in Fig. 9.52 under these conditions on both sides of the P–N junction (J_1). The component of the current due to the injection of holes into the N-side of the junction is given by[21]

$$J_{\text{p}}(y_{\text{N}}) = \frac{D_{\text{p,NB}}p_{0,\text{NB}}L_{\text{n,P+}}}{D_{\text{p,NB}}p_{0,\text{NB}}L_{\text{n,P+}} + D_{\text{n,P+}}n_{0,\text{P+}}L_{\text{p,NB}}}J_{\text{C}}, \tag{9.89}$$

where $D_{\text{p,NB}}$ and $D_{\text{n,P+}}$ are the diffusion coefficients for holes in the N-buffer layer and electrons in the P^+ collector region, $p_{0,\text{NB}}$ and $n_{0,\text{P+}}$ are the minority carrier concentrations in the N-buffer layer and the P^+ collector region in equilibrium, and $L_{\text{p,NB}}$ and $L_{\text{n,P+}}$ are the minority carrier diffusion coefficients in the N-buffer layer and the P^+ collector region.

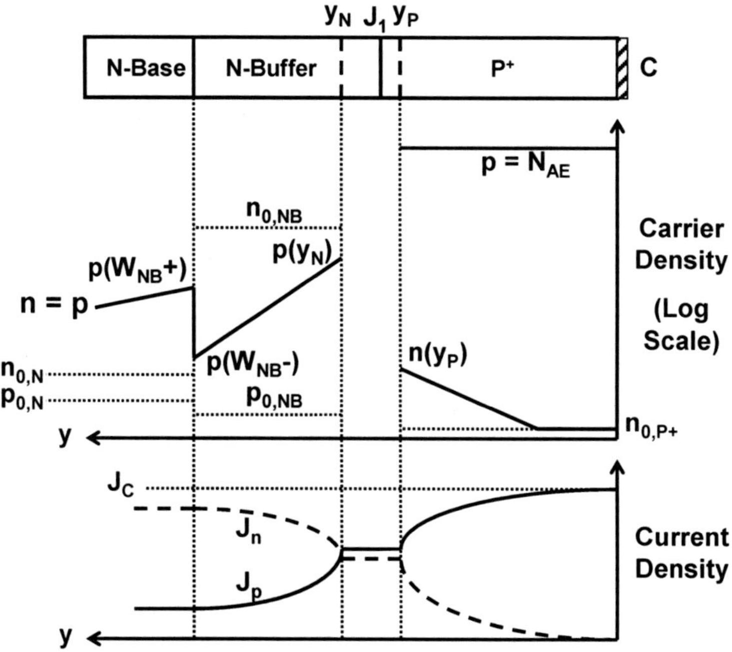

Fig. 9.52 On-state carrier and current distributions at the P^+ collector/N-buffer junction within the asymmetric IGBT structure under low-level injection in the N-buffer layer

The concentration of the injected minority carriers on both sides of the junction is given by

$$p(y_{\text{N}}) = \frac{p_{0,\text{NB}}L_{\text{p,NB}}L_{\text{n,P+}}}{qD_{\text{p,NB}}p_{0,\text{NB}}L_{\text{n,P+}} + qD_{\text{n,P+}}n_{0,\text{P+}}L_{\text{p,NB}}}J_{\text{C}} + p_{0,\text{NB}}, \tag{9.90}$$

$$n(y_{\text{P}}) = \frac{n_{0,\text{P+}}L_{\text{p,NB}}L_{\text{n,P+}}}{qD_{\text{p,NB}}p_{0,\text{NB}}L_{\text{n,P+}} + qD_{\text{n,P+}}n_{0,\text{P+}}L_{\text{p,NB}}}J_{\text{C}} + n_{0,\text{P+}}. \tag{9.91}$$

The minority carriers diffuse away from the junction at their respective characteristic diffusion lengths leading to an exponential decay in their concentrations.

This is shown in Fig. 9.52 as a straight line due to the logarithmic scale for the free carrier densities. Within the P^+ collector region, the minority carrier density decreases to its equilibrium concentration. Within the N-buffer layer, the minority carrier density decreases to

$$p(W_{NB}-) = \frac{p_{0,NB} L_{p,NB} L_{n,P+}}{qD_{p,NB} p_{0,NB} L_{n,P+} + qD_{n,P+} n_{0,P+} L_{p,NB}} J_C \, e^{-(W_{NB}/L_{p,NB})} \tag{9.92}$$

if the minority carrier concentration in equilibrium is neglected. This equation provides the hole concentration on the right-hand side of the boundary between the lightly doped portion of the N-base region and the N-buffer layer as indicated in Fig. 9.52.

In accordance with low-level injection theory for a P–N junction, the hole current density on the N-side decreases exponentially away from the junction. Consequently, the hole current density at the interface between the lightly doped portion of the N-base region and the N-buffer layer is given by

$$J_p(W_{NB}-) = \frac{D_{p,NB} p_{0,NB} L_{n,P+}}{D_{p,NB} p_{0,NB} L_{n,P+} + D_{n,P+} n_{0,P+} L_{p,NB}} J_C \, e^{-(W_{NB}/L_{p,NB})}. \tag{9.93}$$

The free carrier distribution in the lightly doped portion of the N-base region can be obtained by solving the continuity equation for minority carriers (holes in the N-base region) under steady-state high-level injection conditions. This solution has the same form as previously derived for the asymmetric structure, namely (see (9.88))

$$p(y) = p(W_{NB}+) \frac{\sinh[(W_N + W_{NB} - y)/L_a]}{\sinh[(W_N + W_{NB})/L_a]} \tag{9.94}$$

which is valid for $y > W_{NB}$. The hole concentration ($p(W_{NB}+)$) at the left-hand side of the interface between the lightly doped portion of the N-base region and the N-buffer layer can be obtained by recognizing that the hole current density at the interface must be continuous:

$$J_p(W_{NB}+) = J_p(W_{NB}-). \tag{9.95}$$

The hole current density at the left-hand side of the interface between the lightly doped portion of the N-base region and the N-buffer layer is given by

$$J_p(W_{NB}+) = 2qD_p \left(\frac{dp}{dy} \right)_{y=W_{NB+}} = \frac{2qD_p p(W_{NB}+)}{L_a \tanh[(W_N + W_{NB})/L_a]} \tag{9.96}$$

by using (9.94). Combining (9.95) with (9.96) yields

$$p(W_{\mathrm{NB}}+) = \frac{L_{\mathrm{a}} \tanh[(W_{\mathrm{N}} + W_{\mathrm{NB}}) / L_{\mathrm{a}}]}{2qD_{\mathrm{p}}} J_{\mathrm{p}}(W_{\mathrm{NB}}-). \tag{9.97}$$

In conjunction with (9.93), this expression provides a solution for the hole concentration at the left-hand side of the interface between the lightly doped portion of the N-base region and the N-buffer layer. The hole and electron concentration profiles in the lightly doped portion of the N-base region can then be calculated by using (9.94). The discontinuity in the hole concentration at the interface between the lightly doped portion of the N-base region and the N-buffer layer is illustrated in Fig. 9.52.

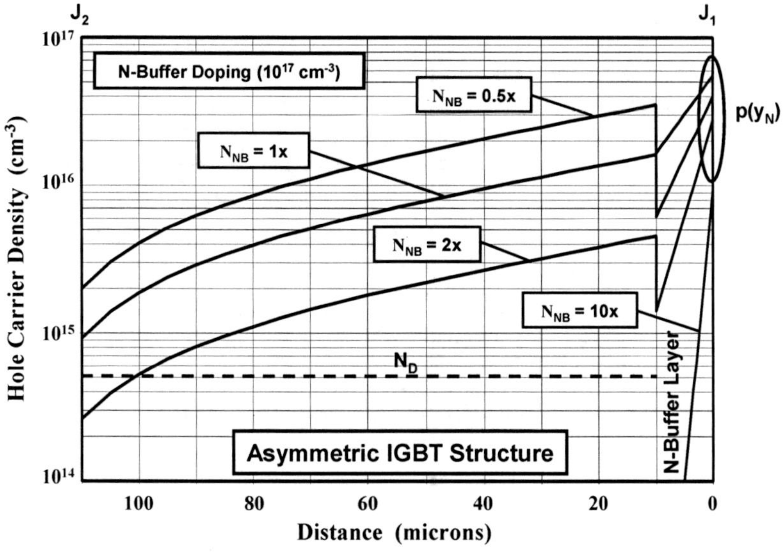

Fig. 9.53 On-state hole distribution within the asymmetric IGBT structure under the deep P^+ region for high buffer layer doping concentrations

As an example, the hole carrier densities calculated by using the above analytical solutions for the hole concentrations are provided in Fig. 9.53 for the asymmetric IGBT structure at an on-state collector current density of 100 A cm^{-2}. The lightly doped portion of the N-base region has a width of 100 μm and the N-buffer layer has a thickness of 10 μm. The high-level lifetime in the N-base region is 1 μs while the lifetime in the N-buffer layer is scaled with its doping concentration. In this analysis, a doping concentration of 1×10^{19} cm^{-3} was used for the P^+ collector region as being representative of a heavily doped substrate region on which the N-base region was formed using epitaxial growth. The N-buffer layer doping concentration was varied from 5×10^{16} to 1×10^{18} cm^{-3}. It can be seen that the injected concentration of holes (indicated by $p(y_{\mathrm{N}})$ in the figure) at the P^+ collector/N-buffer junction (J_1) is below the doping concentration in the

N-buffer layer. The injected hole concentration decays exponentially from the P^+ collector/N-buffer junction (J_1) within the N-buffer layer. The concentration of holes on the left-hand side of the interface between the lightly doped portion of the N-base region and the N-buffer layer is larger than on its right-hand side. The hole profile within the lightly doped portion of the N-base region decreases with increasing distance in a manner similar to that for the symmetric IGBT structure. The hole concentration becomes smaller when the doping concentration in the N-buffer layer is increased. For the largest doping concentration of 1×10^{18} cm^{-3} in the N-buffer layer, no conductivity modulation of the lightly doped portion of the N-base region is observed.

Simulation Example

To gain further insight into the physics of operation for the asymmetric IGBT structure in the on-state, the results of two-dimensional numerical simulations for the asymmetric structure with high doping concentrations in the N-buffer layer are discussed here. This structure had an N-base region consisting of a lightly doped (5×10^{13} cm^{-3}) portion with a width of 100 μm and an N-buffer layer with a width of 10 μm. The hole concentration was obtained in the on-state at a current density of 100 A cm^{-2} for various values for the doping concentration in the N-buffer layer.

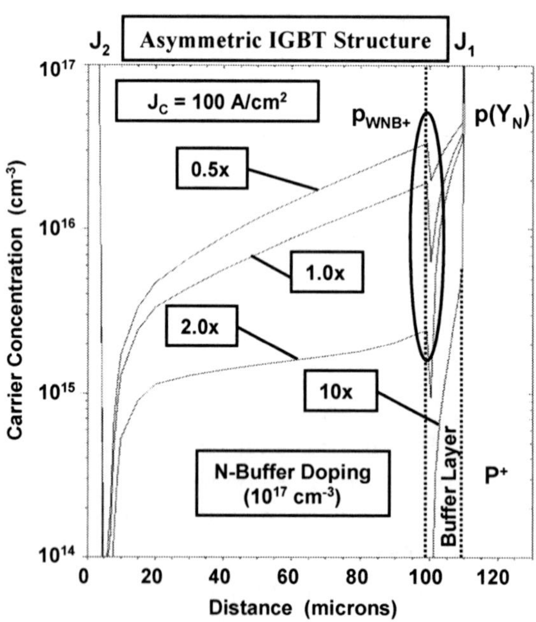

Fig. 9.54 Hole distribution in the on-state for the asymmetric IGBT structure with high N-buffer doping concentrations

A one-dimensional view of the hole distribution under the deep P^+ region obtained from simulations performed with the various N-buffer layer doping values is provided in Fig. 9.54. It can be observed that the hole concentration at the P^+ collector/N-base junction (J_1) is smaller than the background doping concentration within the N-buffer layer and goes to zero at the deep P^+/N-base junction (J_2). The hole concentration at the P^+ collector/N-base junction (J_1) becomes smaller when the N-buffer layer doping concentration is increased. The predictions of the analytical model for high N-buffer layer doping concentrations (see Fig. 9.53) are in very good qualitative and quantitative agreement, including the discontinuity at the interface between the lightly doped portion of the N-base region and the N-buffer layer, with the results of the simulations for the asymmetric IGBT structure, providing validation for the analytical model for the case of relatively high buffer layer doping levels.

9.5.5 On-State Voltage Drop: Asymmetric Structure

In Sect. 9.5.4, it was established that the free carrier distribution within the N-base region of the asymmetric IGBT structure is identical to that for the symmetric IGBT structure when the doping concentration in the N-buffer layer is low ($<5 \times 10^{16}$ cm^{-3}). Since the width of the N-base region (sum of the width of the lightly doped part and the N-buffer layer) for the asymmetric IGBT structure is much smaller than that for the symmetric IGBT structure, the on-state voltage drop becomes smaller especially when the lifetime is reduced.

The on-state voltage drop for the asymmetric IGBT structure can be obtained by using

$$V_{\mathrm{ON}} = V_{\mathrm{P+N}} + V_{\mathrm{NB}} + V_{\mathrm{MOSFET}}, \tag{9.98}$$

where $V_{\mathrm{P+N}}$ is the voltage drop across the P^+ collector/N-base junction (J_1), V_{NB} is the voltage drop across the N-base region (both portions) after accounting for conductivity modulation due to high-level injection conditions, and V_{MOSFET} is the voltage drop across the MOSFET portion. In the asymmetric IGBT structure, the junction (J_1) between the P^+ collector region and the N-base region is formed with the more heavily doped N-buffer layer. Consequently, the voltage drop across the junction (J_1) must be modified to

$$V_{\mathrm{P+N}} = \frac{kT}{q}\ln\left(\frac{p_0}{p_{0,\mathrm{NB}}}\right) = \frac{kT}{q}\ln\left(\frac{p_0 N_{\mathrm{D,NB}}}{n_{\mathrm{i}}^2}\right), \tag{9.99}$$

where $N_{\mathrm{D,NB}}$ is the doping concentration in the N-buffer layer.

As an example, the components of the on-state voltage drop determined by using the analytical model are provided in Fig. 9.55 for the case of the asymmetric IGBT structure with blocking voltage capability of 1,200 V. The device has an N-base width consisting of a lightly doped portion with a thickness of 100 μm and an N-buffer layer with a thickness of 10 μm. All the parameters used for the structure are the same as those used to compute the free carrier concentration in Fig. 9.53. The MOSFET portion has an effective JFET region doping concentration of

5×10^{15} cm^{-3} corresponding to a diffused region with surface concentration of 1×10^{16} cm^{-3} and depth of 5 μm, a gate oxide thickness of 500 Å, and a channel length of 1.5 μm. The mobility for electrons in the inversion and accumulation layers was assumed to be 450 and 1,000 cm^2 V^{-1} s^{-1}, respectively (the same as those used for analysis of power MOSFET structures in Chap. 7). The IGBT structure had a cell pitch (W_{CELL}) of 30 μm with a gate electrode width (W_G) of 16 μm. The gate bias and threshold voltages used were 15 and 5 V, respectively.

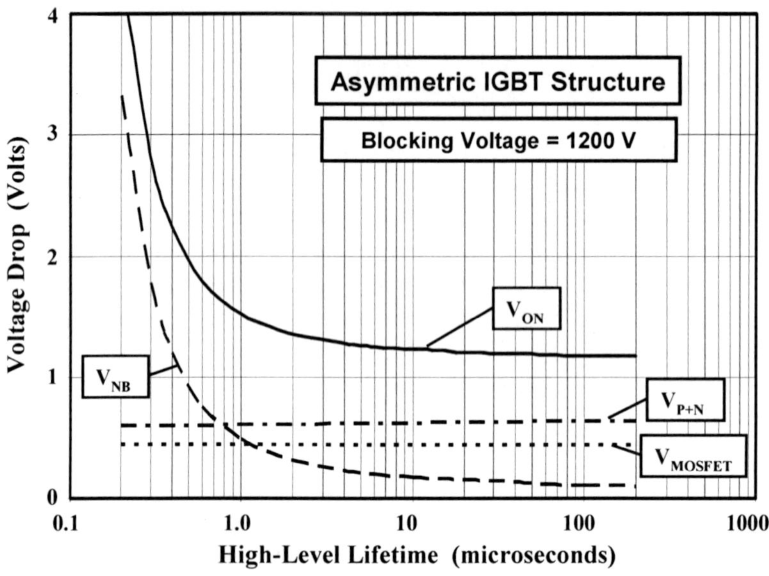

Fig. 9.55 On-state voltage drop for the asymmetric IGBT structure with N-buffer layer doping below 1×10^{17} cm^{-3}

From the above figure, it can be concluded that the voltage drop across the P$^+$ collector/N-base junction and the MOSFET portion is dominant when the high-level lifetime is large (>2 μs). When the high-level lifetime is reduced, the voltage drop across the N-base region increases and becomes dominant for lifetime value below 0.5 μs. A very rapid increase in the on-state voltage drop is observed when the high-level lifetime is reduced below 0.3 μs. In comparison with the symmetric IGBT structure with the same voltage-blocking capability (discussed in Sect. 9.5.3), a much lower on-state voltage drop is obtained with the asymmetric IGBT structure for small values for the high-level lifetime. For example, at a high-level lifetime of 0.8 μs, the on-state voltage drop for the symmetric IGBT structure is 5 V while that for the asymmetric IGBT structure is only 1.6 V. The asymmetric IGBT structure is therefore preferred for power circuits operated from a DC power bus (as in the case of variable speed motor control).

Simulation Example

To corroborate the analytical model for the on-state voltage drop for the asymmetric IGBT structure with N-buffer layer doping concentration below 1×10^{17} cm^{-3}, the results of two-dimensional numerical simulations for the typical asymmetric structure (described in the previous section) are discussed here. The on-state voltage drop was obtained at a current density of 100 A cm^{-2} at a gate bias of 15 V for various values for the high-level lifetime. It was found that the on-state voltage drop increases rapidly when the high-level lifetime is reduced below 0.3 μs. The on-state voltage drop obtained from the numerical simulations is compared with those obtained by using the analytical model in Fig. 9.56. It can be concluded that the analytical model provides an accurate prediction of the on-state voltage drop over a broad range of high-level lifetime values.

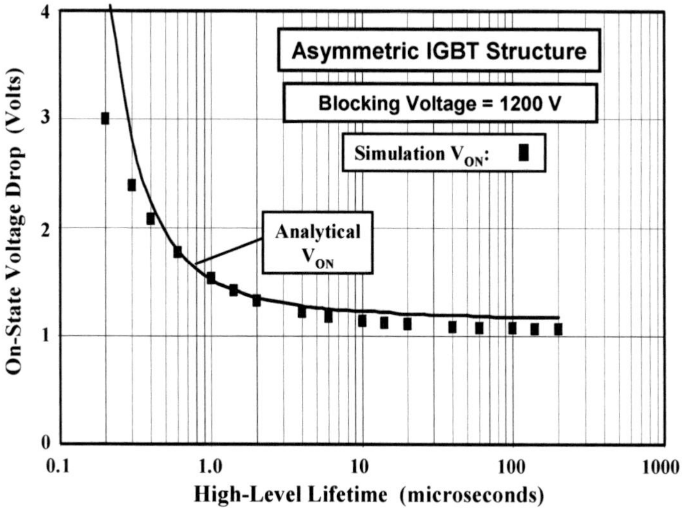

Fig. 9.56 On-state voltage drop for the asymmetric IGBT structure with N-buffer layer doping concentration of 1×10^{16} cm^{-3}

When the doping concentration in the N-buffer layer is increased beyond 5×10^{16} cm^{-3}, the model for the voltage drop contributed by the N-base region must be modified to account for low-level injection conditions within the N-buffer layer. The total voltage drop across the asymmetric IGBT structure is still given by

$$V_{ON} = V_{P+N} + V_{NB} + V_{MOSFET}, \tag{9.100}$$

where V_{P+N} is the voltage drop across the P$^+$ collector/N-base junction (J_1), V_{NB} is the voltage drop across the N-base region (both portions) after accounting for conductivity modulation due to high-level injection conditions, and V_{MOSFET} is the voltage drop across the MOSFET portion. In the asymmetric IGBT structure with large N-buffer layer doping concentrations, the junction (J_1) between the

P^+ collector region and the N-base region operates at low injection levels. Consequently, the voltage drop across the junction (J_1) must be modified to

$$V_{\mathrm{P+N}} = \frac{kT}{q}\ln\left(\frac{p(y_{\mathrm{N}})}{p_{0,\mathrm{NB}}}\right) = \frac{kT}{q}\ln\left(\frac{p(y_{\mathrm{N}})N_{\mathrm{D,NB}}}{n_{\mathrm{i}}^2}\right), \tag{9.101}$$

where $p(y_N)$ is the injected hole concentration at the junction in the N-buffer layer (see Fig. 9.52) and $N_{D,NB}$ is the doping concentration in the N-buffer layer.

The voltage drop across the N-base region consists of the voltage drop across the N-buffer layer and the lightly doped portion of the N-base region. Since the N-buffer layer operates under low injection levels, there is only a small ohmic voltage drop (<10 mV) due to majority carrier transport[21] within this region which can be neglected. The voltage drop across the lightly doped portion of the N-base region can be obtained by using the solutions (see (9.83) and (9.84)) derived previously for the symmetric IGBT structure with p_0 replaced by $p(W_{NB}+)$. The voltage drop across the MOSFET portion in the asymmetric IGBT structure is identical to that for the symmetric IGBT structure.

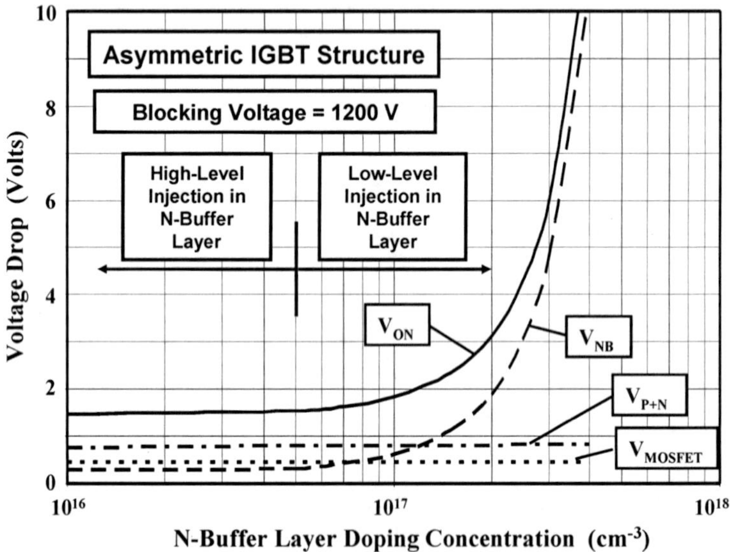

Fig. 9.57 On-state voltage drop for the asymmetric IGBT structure

As an example, the components of the on-state voltage drop determined by using the analytical model are provided in Fig. 9.57 for the case of the asymmetric IGBT structure with blocking voltage capability of 1,200 V. The device has an N-base width consisting of a lightly doped portion with a thickness of 100 μm and an N-buffer layer with a thickness of 10 μm. All the parameters used for the structure are the same as those used to compute the free carrier concentration in Fig. 9.53. The high-level lifetime in the lightly doped portion of the N-base region is 2 μs.

The MOSFET portion has an effective JFET region doping concentration of 5×10^{15} cm^{-3} corresponding to a diffused region with surface concentration of 1×10^{16} cm^{-3} and depth of 5 μm, a gate oxide thickness of 500 Å, and a channel length of 1.5 μm. The mobility for electrons in the inversion and accumulation layers was assumed to be 450 and 1,000 cm^2 V^{-1} s^{-1}, respectively (the same as those used for analysis of power MOSFET structures in Chap. 7). The IGBT structure had a cell pitch (W_{CELL}) of 30 μm with a gate electrode width (W_{G}) of 16 μm. The gate bias and threshold voltages used were 15 and 5 V, respectively. Note that the model with high injection level in the N-buffer layer was used for N-buffer layer doping levels below 5×10^{16} cm^{-3} and the model with low injection level in the N-buffer layer was used for N-buffer layer doping levels above 5×10^{16} cm^{-3}.

From the above figure, it can be concluded that the voltage drop across the P$^+$ collector/N-base junction and the MOSFET portion is dominant when the doping concentration in the N-buffer layer is less than 1×10^{17} cm^{-3}. When the doping concentration in the N-buffer layer is increased beyond 1×10^{17} cm^{-3}, the voltage drop across the N-base region increases and becomes dominant. A very rapid increase in the on-state voltage drop is observed when the doping concentration in the N-buffer layer is increased beyond 2.5×10^{17} cm^{-3}.

Simulation Example

To corroborate the analytical model for the on-state voltage drop for the asymmetric IGBT structure with large doping concentrations in the N-buffer layer, the results of two-dimensional numerical simulations for the typical asymmetric structure

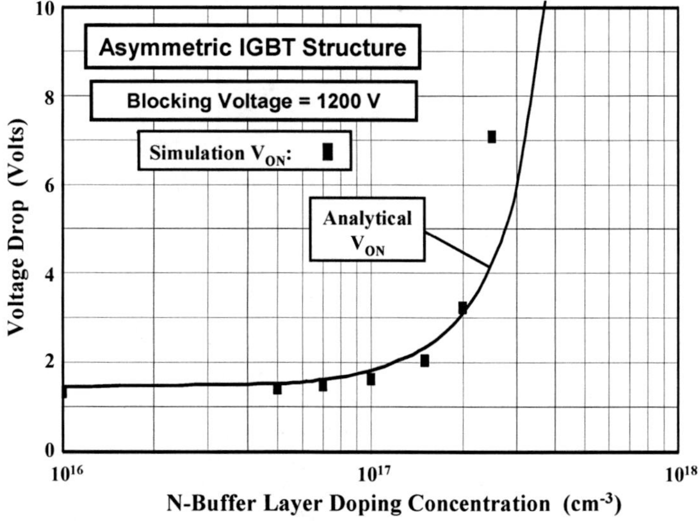

Fig. 9.58 On-state voltage drop for the asymmetric IGBT structure

(described in the previous section) are discussed here. The on-state voltage drop was obtained at a current density of 100 A cm^{-2} at a gate bias of 15 V for various values for the high-level lifetime. It was found that the on-state voltage drop increases rapidly when the doping concentration in the N-buffer layer is increased beyond 2×10^{17} cm^{-3}. The on-state voltage drop obtained from the numerical simulations is compared with those obtained by using the analytical model in Fig. 9.58. It can be concluded that the analytical model provides an accurate prediction of the on-state voltage drop over a broad range of doping concentrations in the N-buffer layer.

9.5.6 On-State Carrier Distribution: Transparent Emitter Structure

The traditional method for improving the switching speed of bipolar power devices is based upon reducing the lifetime in the drift region to accelerate the recombination of the stored charge. Although this method is widely practiced for the IGBT structure, an alternate technique for improving the switching speed has evolved based upon the use of a P^+ collector region with a relatively low doping concentration and a very small thickness.[23–25] Since the P^+ collector region in the IGBT structure behaves as an emitter for minority carriers into the N-base region during on-state operation, this technique is referred to as the *transparent emitter structure*. The basic concept for the transparent emitter structure is to reduce the stored charge in the N-base region during on-state operation by decreasing the injection efficiency of the P^+ collector region. A large value for the high-level lifetime is maintained in the N-base region in these devices to obtain a reasonably low on-state voltage drop.

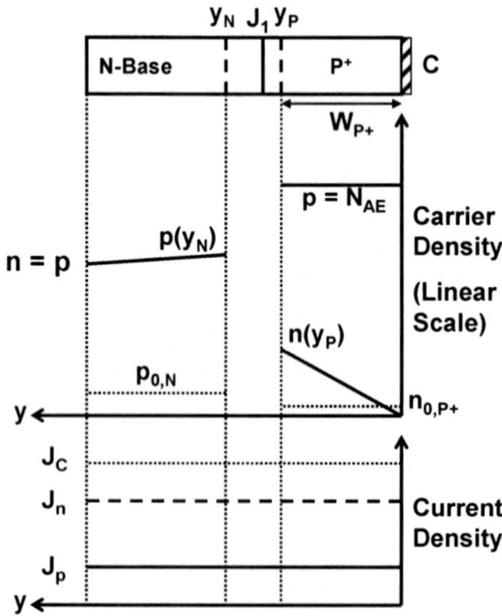

Fig. 9.59 On-state carrier and current distributions at the P^+ collector/N-base junction within the IGBT structure with transparent emitter

Recombination in the P$^+$ collector region becomes dominant during on-state operation when the doping concentration and thickness of the P$^+$ collector region in the symmetric IGBT are reduced while a large lifetime is maintained in the N-base region. Under forward bias conditions for the P$^+$ collector/N-base junction, the electrons injected into the P$^+$ collector region diffuse to the collector contact without significant recombination within the region if its thickness is much smaller than the diffusion length. The electron concentration at the contact can be assumed to be zero because the recombination rate is extremely large at an ohmic contact. This produces a linear minority carrier concentration profile, as well as a constant electron current density, in the P$^+$ collector region as illustrated in Fig. 9.59.

The electron current density in the P$^+$ collector region is then given by

$$J_{\mathrm{n}} = qD_{\mathrm{nE}}\left(\frac{\mathrm{d}n}{\mathrm{d}y}\right) = \frac{qD_{\mathrm{nE}}n(y_{\mathrm{P}})}{W_{\mathrm{P+}}}. \tag{9.102}$$

Substituting for the electron concentration ($n(y_{\mathrm{P}})$) at the P$^+$ collector/N-base region junction (J_1) using (9.59),

$$J_{\mathrm{n}} = \frac{qD_{\mathrm{nE}}}{W_{\mathrm{P+}}}\frac{p_0^2}{N_{\mathrm{AE}}}, \tag{9.103}$$

where N_{AE} is the effective doping concentration of the P$^+$ collector region. The hole current density in the P$^+$ collector region is then given by

$$J_{\mathrm{p}} = J_{\mathrm{C}} - J_{\mathrm{n}}. \tag{9.104}$$

These currents are continuous across the junction as shown in the figure.

The free carrier distribution in the N-base region is governed by high-level injection conditions with no recombination when the lifetime is large:

$$\frac{\mathrm{d}^2 p}{\mathrm{d}y^2} = 0. \tag{9.105}$$

The general solution for this case is a linear carrier distribution given by

$$p(y) = Ay + B. \tag{9.106}$$

The constants A and B are determined by the following boundary conditions:

$$p(0) = p_0, \tag{9.107}$$

$$p(W_{\mathrm{N}}) = 0, \tag{9.108}$$

leading to the solution

$$p(y) = \frac{p_0}{W_N}(W_N - y). \tag{9.109}$$

The hole and electron current densities in the N-base region are given by

$$J_p(y) = q\mu_p\left[p(y)E(y) - \frac{kT}{q}\frac{dp}{dy}\right], \tag{9.110}$$

$$J_n(y) = q\mu_n\left[n(y)E(y) + \frac{kT}{q}\frac{dn}{dy}\right]. \tag{9.111}$$

In the above expressions, the electron and hole concentrations, as well as their derivatives, can be assumed to be equal due to high-level injection conditions in the N-base region. In addition, continuity of collector current flow requires

$$J_n(y) + J_p(y) = J_C. \tag{9.112}$$

Combining these expressions and solving for the electric field yields

$$E(y) = \frac{J_C}{qp(y)(\mu_n + \mu_p)} - \frac{kT}{q}\left(\frac{\mu_n - \mu_p}{\mu_n + \mu_p}\right)\frac{1}{p(y)}\frac{dp}{dy}. \tag{9.113}$$

Using the free carrier distribution given by (9.109),

$$E(y) = \left[\frac{J_C W_N}{qp_0(\mu_n + \mu_p)} + \frac{kT}{q}\left(\frac{\mu_n - \mu_p}{\mu_n + \mu_p}\right)\right]\frac{1}{W_N - y}. \tag{9.114}$$

The hole current density within the N-base region at the P^+ collector/N-buffer layer junction (J_1) can then be obtained by using (9.110) with the hole carrier distribution given by (9.109) and the above electric field:

$$J_p(y_N) = \frac{\mu_p J_C}{\mu_n + \mu_p} + \left(\frac{2\mu_n\mu_p}{\mu_n + \mu_p}\right)\frac{kT}{W_N}p_0. \tag{9.115}$$

Using (9.103) and (9.104), the hole current density in the P^+ collector region at the P^+ collector/N-buffer layer junction (J_1) is given by

$$J_p(y_P) = J_C - \frac{qD_{nE}}{W_{P+}}\frac{p_0^2}{N_{AE}}. \tag{9.116}$$

The hole current densities ($J_p(y_P)$ and $J_p(y_N)$) on the two sides of the junction can be equated because the current is continuous at the junction. This yields a quadratic solution for the hole concentration (p_0) at the P^+ collector/N-base junction:

$$ap_0^2 + bp_0 + c = 0, \tag{9.117}$$

with coefficients

$$a = \frac{qD_{\text{nE}}}{W_{\text{P+}} N_{\text{AE}}}, \tag{9.118}$$

$$b = -\left(\frac{2\mu_{\text{n}}\mu_{\text{p}}}{\mu_{\text{n}} + \mu_{\text{p}}}\right)\frac{kT}{W_{\text{N}}}, \tag{9.119}$$

$$c = -\left(\frac{\mu_{\text{n}}}{\mu_{\text{n}} + \mu_{\text{p}}}\right) J_{\text{C}}. \tag{9.120}$$

The value of the hole concentration (p_0) at the P^+ collector/N-base junction can then be computed by using

$$p_0 = -\frac{b}{2a}\left[1 - \sqrt{1 - \left(\frac{4ac}{b^2}\right)}\right]. \tag{9.121}$$

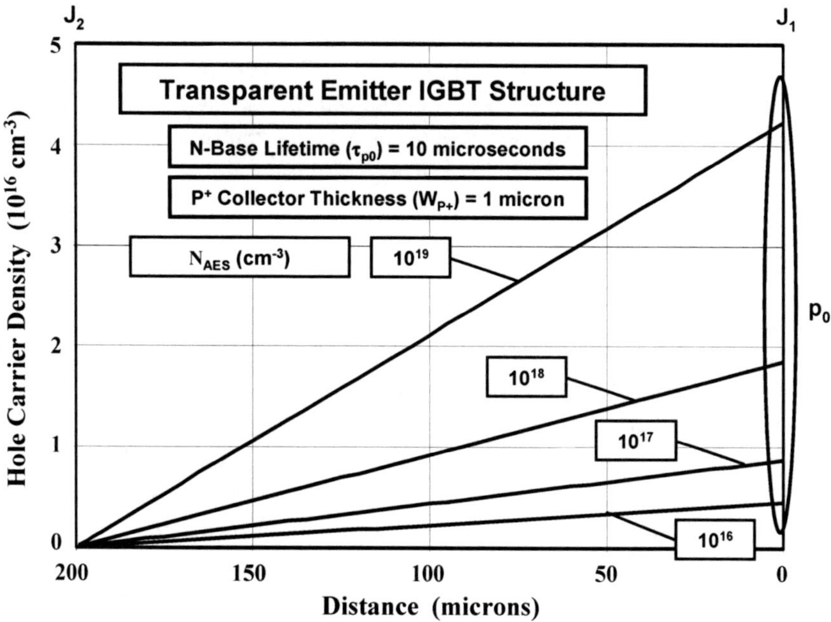

Fig. 9.60 On-state hole distribution within the transparent emitter IGBT structure under the deep P^+ region

As an example, the hole carrier densities calculated by using (9.109) with the above analytical solution for the hole concentration (p_0) at the P$^+$ collector/N-base junction are provided in Fig. 9.60 for the transparent emitter IGBT structure with an N-base width of 200 μm at an on-state collector current density of 100 A cm^{-2}. A high-level lifetime of 20 μs (τ_{p0} of 10 μs) was used in the N-base region. Since it is customary to form the P$^+$ collector region using a diffused layer with a typical thickness (W_{P+}) of 1 μm, an effective doping concentration must be used in the analytical solution. Due to the large variation of the doping concentration within the P$^+$ collector region, it is appropriate to use an effective doping concentration related to the geometric mean of the surface concentration of the P$^+$ diffusion and the concentration at the junction (which is equal to the doping concentration of the N-base region):

$$N_{AE}(\text{Effective}) = K_{TE1}\sqrt{N_{AES}N_D}. \tag{9.122}$$

where N_{AES} is the surface concentration of the P$^+$ diffusion and N_D is the doping concentration of the N-base region. A value of 20 for the constant K_{TE1} was found to predict the free carrier concentrations within the N-base region over a broad range of values for the surface concentration of the P$^+$ diffusion. In this analysis, the diffusion coefficient for electrons (D_{nE}) must be scaled with the effective doping concentration of P$^+$ collector region. It can be seen that the hole concentration varies linearly from junction J_1 to junction J_2. The concentration of holes (indicated by p_0 in the figure) at the P$^+$ collector/N-base junction (J_1) is much smaller than for the case of the symmetric IGBT structure discussed in Sect. 9.5.2 (see Fig. 9.44). The injected hole concentration (p_0) at the P$^+$ collector/N-base junction (J_1) becomes smaller when surface concentration of the P$^+$ collector region is reduced.

Simulation Example

To gain further insight into the physics of operation for the transparent emitter IGBT structure in the on-state, the results of two-dimensional numerical simulations for the typical structure with 1-μm thick P$^+$ collector region are discussed here. This structure had an N-base region with a doping concentration of 5×10^{13} cm^{-3}, a width of 200 μm, and a lifetime (τ_{p0}) of 10 μs. The hole concentration was obtained in the on-state at a current density of 100 A cm^{-2} for various values for the doping concentration of the P$^+$ collector region. The P$^+$ collector region was formed by using a Gaussian profile with a junction depth of 1 μm using various surface doping concentrations.

A one-dimensional view of the hole distributions under the deep P$^+$ region obtained from simulations performed with the various surface concentrations for the P$^+$ collector region is provided in Fig. 9.61. It can be observed that the hole concentration in the N-base region varies linearly with distance as predicted by the the analytical model and goes to zero at the deep P$^+$/N-base junction (J_2). The hole concentration at the P$^+$ collector/N-base junction (J_1) becomes smaller when the surface concentration for the P$^+$ collector region is reduced. The predictions of the analytical model for various surface doping concentrations for the P$^+$ collector region (see Fig. 9.60) are in very good qualitative and quantitative agreement with the results of the simulations when a value for K_{TE1} of 20 is used in (9.122).

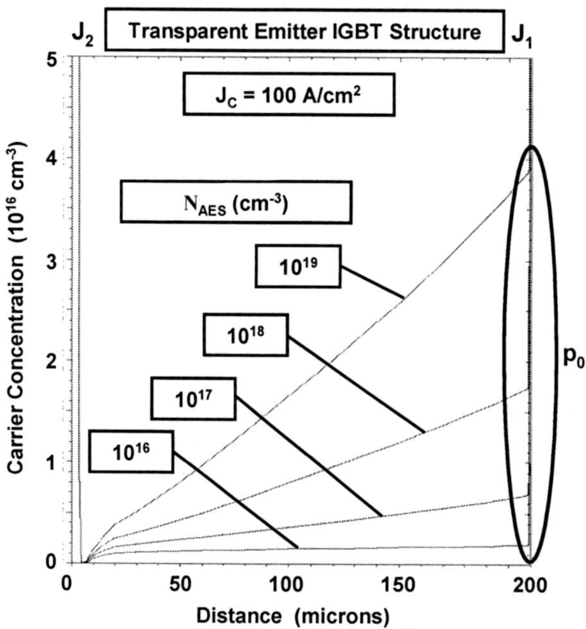

Fig. 9.61 Hole distribution in the on-state for the transparent emitter IGBT structure

9.5.7 On-State Voltage Drop: Transparent Emitter Structure

The on-state voltage drop for the transparent emitter IGBT structure can be obtained by using the same approach used earlier for the symmetric IGBT structure by the addition of V_{P+N}, the voltage drop across the P$^+$ collector/N-base junction (J_1); V_{NB}, the voltage drop across the N-base region after accounting for conductivity modulation due to high-level injection conditions; and V_{MOSFET}, the voltage drop across the MOSFET portion:

$$V_{ON} = V_{P+N} + V_{NB} + V_{MOSFET}. \tag{9.123}$$

The voltage drop across the P$^+$ collector/N-base junction (J_1) can be obtained from the increase in the minority carrier concentration at the junction boundary:

$$V_{P+N} = \frac{kT}{q}\ln\left(\frac{p_0}{p_{0N}}\right) = \frac{kT}{q}\ln\left(\frac{p_0 N_D}{n_i^2}\right), \tag{9.124}$$

where the minority carrier density in equilibrium (p_{0N}) has been related to the doping concentration in the N-base region. The increase in the hole concentration at the junction (p_0) can be obtained from the analysis provided in Sect. 9.5.6.

The voltage drop across the N-base region (V_{NB}) can be obtained by integration of the electric field within the N-base region. An expression for the electric field distribution in the N-base region for the transparent emitter IGBT structure was derived in Sect. 9.5.6 (see (9.114)). The voltage drop across the N-base region can be obtained by integrating the electric field from $y = 0$ at the P$^+$ collector/N-base junction (J_1) to the edge of depletion region of the deep P$^+$/N-base junction (J_2), namely at $y = (W_N - W_{ON})$, where W_{ON} is the depletion width at the deep P$^+$/N-base junction (J_2) in the on-state. The voltage drop in the N-base region is given by

$$V_{NB} = \left[\frac{J_C W_N}{q(\mu_n + \mu_p)p_0} + \frac{kT}{q}\left(\frac{\mu_n - \mu_p}{\mu_n + \mu_p}\right)\right]\ln\left(\frac{W_N}{W_{ON}}\right). \tag{9.125}$$

The depletion width across the deep P$^+$/N-base junction (J_2) in the on-state depends on the on-state voltage drop. The voltage drop across the MOSFET portion includes contributions from the JFET region, the accumulation layer, and the channel. The same expressions derived in Sect. 9.5.3 can be used for the transparent emitter IGBT structure as well.

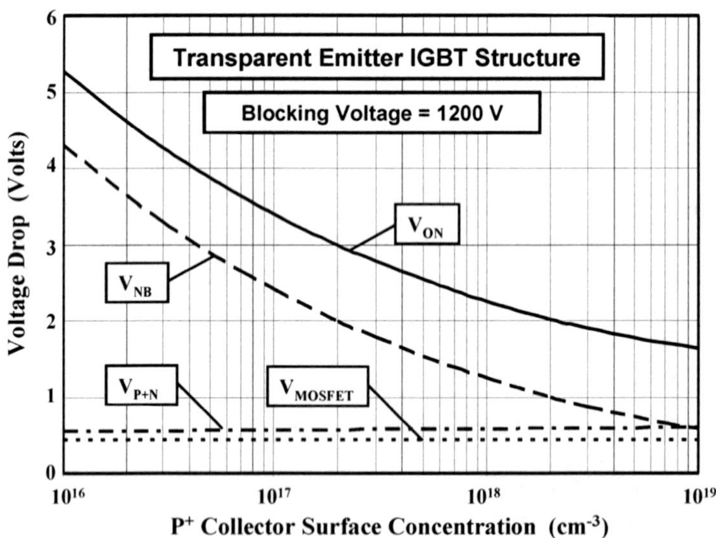

Fig. 9.62 On-state voltage drop for the transparent emitter IGBT structure

As an example, the components of the on-state voltage drop determined by using the above analytical model are provided in Fig. 9.62 for the case of the transparent emitter IGBT structure with blocking voltage capability of 1,200 V.

The device has an N-base width of 200 μm with a lifetime (τ_{p0}) of 10 μs. The depth of the deep P$^+$ collector region was 1 μm for all the cases. All the other parameters used for the structure are the same as those used to compute the free carrier concentration in Fig. 9.60. The MOSFET portion has an effective JFET region doping concentration of 5×10^{15} cm^{-3} corresponding to a diffused region with surface concentration of 1×10^{16} cm^{-3} and depth of 5 μm, a gate oxide thickness of 500 Å, and a channel length of 1.5 μm. The mobility for electrons in the inversion and accumulation layers was assumed to be 450 and 1,000 cm^2 V^{-1} s^{-1}, respectively (the same as those used for analysis of power MOSFET structures in Chap. 7). The IGBT structure had a cell pitch (W_{CELL}) of 30 μm with a gate electrode width (W_G) of 16 μm. The gate bias and threshold voltages used were 15 and 5 V, respectively. Note that the voltage drop across the junction (V_{P+N}) is less than for the symmetric IGBT structure due to the reduced hole injection level in the N-base region. From the figure, it can be concluded that the voltage drop across the N-base region is comparable with that across the P$^+$ collector/N-base junction and the MOSFET portion when the surface concentration of the P$^+$ collector region is 1×10^{19} cm^{-3}. When the surface concentration of the P$^+$ collector region is reduced, the voltage drop across the N-base region increases and becomes dominant. The on-state voltage drop is observed to increase gradually with reduction of the surface concentration of the P$^+$ collector region. The analytical solution indicates that the injection efficiency of the transparent emitter structure is typically reduced to 30%.

Simulation Example

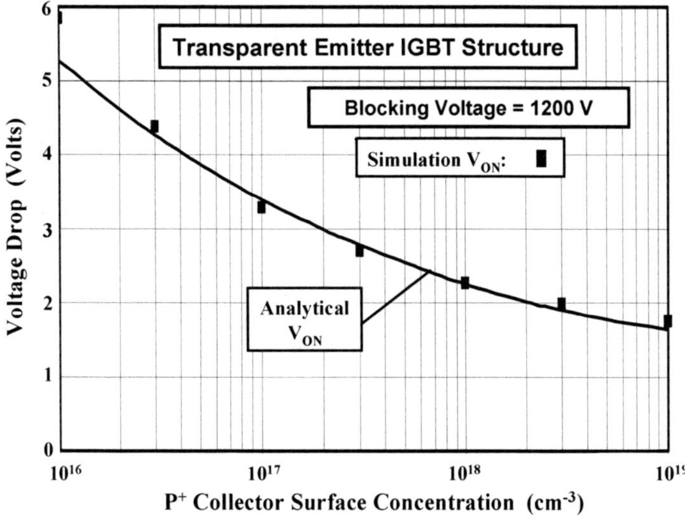

Fig. 9.63 On-state voltage drop for the transparent emitter IGBT structures

To corroborate the analytical model for the on-state voltage drop for the transparent emitter IGBT structure, the results of two-dimensional numerical simulations for the typical case with 1-μm thick P^+ collector region are discussed here. The structure is the same as the one described in Sect. 9.5.6. The on-state voltage drop was obtained at a current density of 100 A cm^{-2} at a gate bias of 15 V for various values for the surface concentration of the thin P^+ collector region. It was found that the on-state voltage drop increases gradually as the surface concentration of the P^+ collector region is reduced. The injection efficiency of the transparent emitter was found to be 30% as predicted by the analytical model. The on-state voltage drops obtained from the numerical simulations are compared with those obtained by using the analytical model in Fig. 9.63. The analytical model provides an accurate prediction of the on-state voltage drop for the entire broad range of surface concentrations of the P^+ collector region.

9.6 Current Saturation Model

The P–i–N rectifier/MOSFET model for the IGBT structure provides a good description of its on-state mode of operation. However, when the IGBT structure operates in the current saturation mode, it is necessary to account for the separate hole and electron current paths within the structure. This can be accomplished by using the P–N–P transistor/MOSFET model for the IGBT structure.[26] The equivalent circuit for an n-channel IGBT structure based upon the P–N–P transistor/MOSFET model consists of an n-channel MOSFET, providing the base drive current to a P–N–P transistor as shown in Fig. 9.64a. The P–N–P transistor and MOSFET portions are identified by the dashed boxes in Fig. 9.64b.

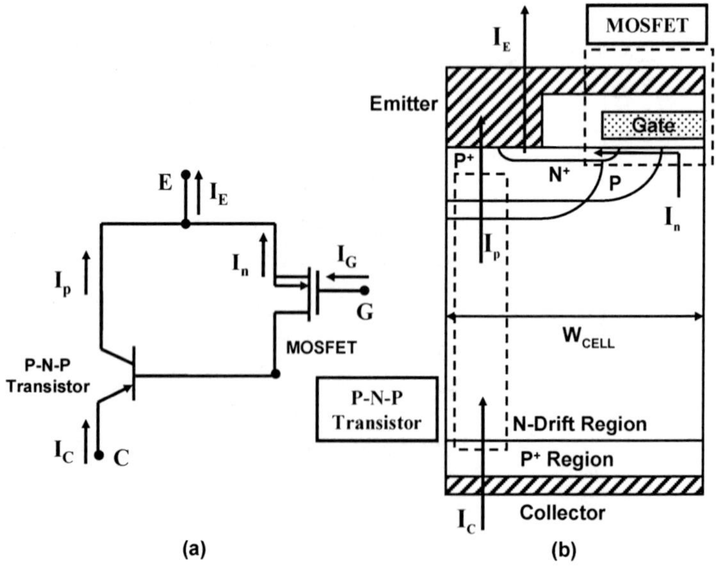

Fig. 9.64 Equivalent circuit for the IGBT structure in the current saturation mode

The emitter current for the IGBT structure includes both the hole current flow via the P–N–P transistor and the electron current via the MOSFET portion:

$$I_E = I_p + I_n. \tag{9.126}$$

In the IGBT structure, the electron current serves as the base drive current for the P–N–P transistor. Consequently, these currents are interrelated by the common-base current gain of the P–N–P transistor:

$$I_p = \left(\frac{\alpha_{PNP}}{1 - \alpha_{PNP}} \right) I_n. \tag{9.127}$$

Combining the above relationships,

$$I_E = \frac{I_n}{1 - \alpha_{PNP}}. \tag{9.128}$$

Under steady-state operating conditions, the gate current for the IGBT structure is zero due to the high impedance of the MOS gate structure. Consequently, the collector current is also given by the above expression.

The current gain of the P–N–P transistor is determined by the product of the emitter injection efficiency and the base transport factor because the multiplication coefficient is unity at low bias voltages:

$$\alpha_{PNP} = \gamma_E \alpha_{T,PNP}. \tag{9.129}$$

As discussed in the previous section, the injection efficiency for the IGBT structures is less than unity due to high-level injection conditions in the N-base region. The base transport factor is given by

$$\alpha_{T,PNP} = \frac{1}{\cosh(l / L_a)}, \tag{9.130}$$

where L_a is the ambipolar diffusion length and l is the width of the undepleted portion of the N-base region, which can be obtained by using

$$l = W_N - W_{DN} = W_N - \sqrt{\frac{2\varepsilon_S V_C}{qN_D}}. \tag{9.131}$$

In the current saturation mode, the depletion region extends from the junction (J_2) between the deep P^+ region and the N-base region toward junction J_1. In this simplified model, the expansion of the depletion region is assumed to be governed by the doping concentration (N_D) in the N-base region as shown by (9.131).

Current saturation in the IGBT structure is induced by making the gate bias voltage close to the threshold voltage, so that the channel of the MOSFET portion reaches the pinch-off condition when the collector bias is increased. When pinch-off occurs in the channel of the MOSFET portion, its current becomes saturated at a value given by (6.49). In the IGBT structure, this is the electron current flowing through the MOSFET portion:

$$I_{\mathrm{n}} = \frac{\mu_{\mathrm{ni}} C_{\mathrm{OX}} Z}{2L_{\mathrm{CH}}} (V_{\mathrm{G}} - V_{\mathrm{TH}})^2, \tag{9.132}$$

where Z is the width of the IGBT structure orthogonal to the cross section shown in Fig. 9.64. Substituting this expression into (9.128),

$$I_{\mathrm{E,SAT}} = I_{\mathrm{C,SAT}} = \frac{\mu_{\mathrm{ni}} C_{\mathrm{OX}} Z}{2L_{\mathrm{CH}}(1-\alpha_{\mathrm{PNP}})} (V_{\mathrm{G}} - V_{\mathrm{TH}})^2. \tag{9.133}$$

From this expression, it can be concluded that the saturated current for the IGBT structure is enhanced (typically by a factor of 2) when compared with a MOSFET structure with the same channel design due to the gain of the P–N–P transistor within the IGBT structure.

The transconductance for the IGBT structure in the current saturation mode can be derived from (9.133):

$$g_{\mathrm{m,SAT}} = \frac{\mathrm{d}I_{\mathrm{C,SAT}}}{\mathrm{d}V_{\mathrm{G}}} = \frac{\mu_{\mathrm{ni}} C_{\mathrm{OX}} Z}{L_{\mathrm{CH}}(1-\alpha_{\mathrm{PNP}})} (V_{\mathrm{G}} - V_{\mathrm{TH}}). \tag{9.134}$$

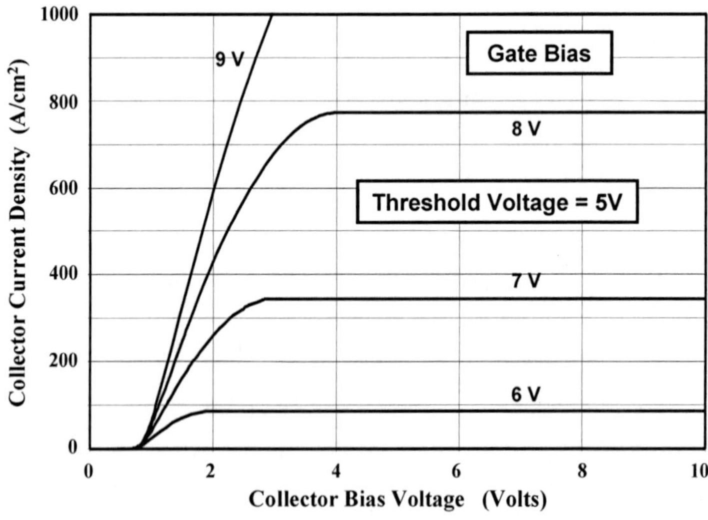

Fig. 9.65 Current saturation characteristics for the IGBT structure at low collector bias voltages

The transconductance for the IGBT structure is also enhanced (typically by a factor of 2) when compared with a MOSFET structure with the same channel design due to the gain of the P–N–P transistor within the IGBT structure.

The *i–v* characteristics for the IGBT structure at low collector bias voltage must include the voltage drop across the forward-biased junction J_1. This can be modeled as a P–i–N rectifier due to high-level injection within the N-base region with a uniform current density J_C:

$$V_{F,PiN} = \frac{2kT}{q} \ln\left[\frac{J_C W_N}{4qD_a n_i F(W_N / 2L_a)}\right]. \tag{9.135}$$

For the P–N–P transistor/MOSFET model, the channel current is given by

$$I_n = pZJ_C(1 - \alpha_{PNP}). \tag{9.136}$$

The voltage drop across the MOSFET portion obtained by using this channel current is

$$V_{F,MOSFET} = (V_G - V_{TH})\left[1 - \sqrt{1 - \frac{2pL_{CH}J_C(1 - \alpha_{PNP})}{\mu_{ni} C_{OX} (V_G - V_{TH})^2}}\right]. \tag{9.137}$$

The voltage drop across the IGBT structure can be obtained by adding the voltage drop across the P–i–N portion to the voltage drop across the MOSFET portion. The results are shown in Fig. 9.65 for the same structure that was used to obtain Fig. 9.40. The saturation of the collector current is evident when the gate bias becomes close to the threshold voltage and the saturated current has a larger magnitude at each gate bias voltage than that obtained with the P–i–N/MOSFET model (Fig. 9.40).

Simulation Example

The results of two-dimensional numerical simulations for the typical symmetric IGBT structure under current saturation are discussed here. The structure is the same as the one described in the previous section on the symmetric IGBT structure. The threshold voltage for this structure is about 4.5 V. At a gate bias of 5 V, the device exhibits current saturation at a collector current density of 20 A cm^{-2}. The saturated current density increases with gate bias voltage. Unlike the power MOSFET structure, the IGBT also exhibits a knee in the characteristics leading to low-current levels at collector bias voltages below 0.8 V. It can be concluded that the characteristics obtained by using the analytical model (see Fig. 9.65) provide a good description of the behavior of the IGBT structure.

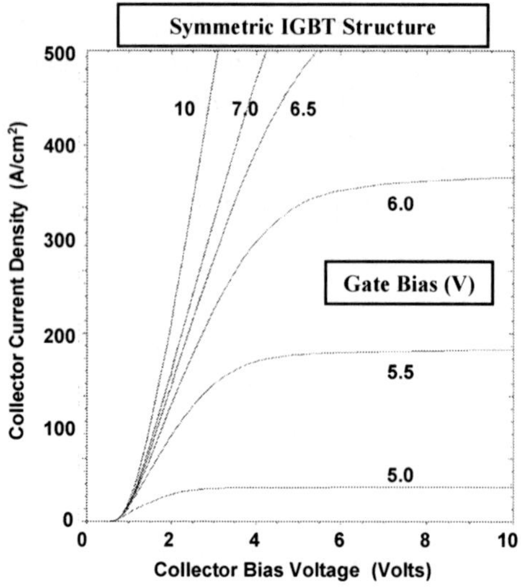

Fig. 9.66 Current saturation in the symmetric IGBT structure

9.6.1 Carrier Distribution: Symmetric Structure

A more accurate model for the output characteristics of the symmetric IGBT structure can be derived by analysis of the free carrier distribution when the device is operating with both a high collector current density and a high collector bias voltage as is relevant for the current saturation mode. The case where the re-combination in the N-base region can be neglected is considered here, namely,

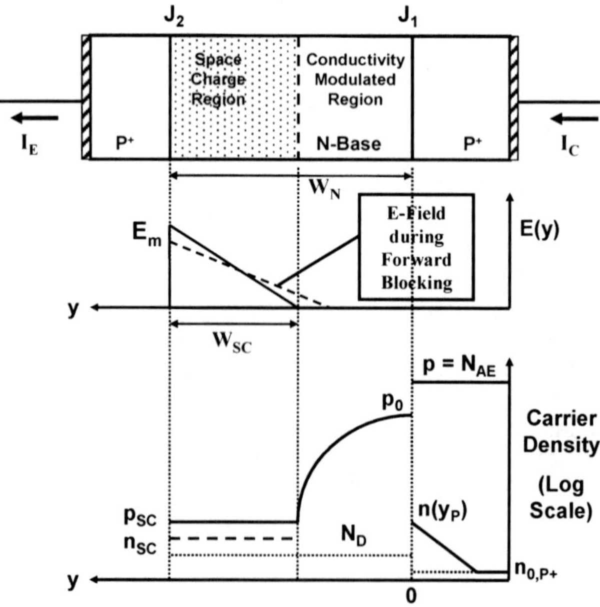

Fig. 9.67 Carrier and electric field profiles for the symmetric IGBT structure in the current saturation mode

when the lifetime in the N-base region is large. Due to the high collector bias voltage in the current saturation mode, a space-charge region forms at the deep P^+/N-base junction (J_2). Unlike the depletion region formed in the forward-blocking mode, the space-charge region contains electrons and holes produced by the large collector current flow. This is illustrated in Fig. 9.67 by the concentrations p_{SC} and n_{SC}. The electric field within the space-charge region is sufficient to increase the velocity of these carriers to the saturated drift velocity over most of the space-charge region. Consequently, these carrier densities can be related to the hole and electron current densities within the space-charge region:

$$n_{SC} = \frac{J_{n,PNP}}{qv_{sat,n}}, \tag{9.138}$$

$$p_{SC} = \frac{J_C}{qv_{sat,p}}. \tag{9.139}$$

The electron current via the channel of the MOSFET portion in the IGBT structure is the base drive current for the P–N–P transistor. Under current saturation conditions for the IGBT structure, the MOSFET channel current is saturated at a value dictated by the gate bias voltage relative to the threshold voltage as given by (9.132). The current can be expressed as an electron current density:

$$J_{B,PNP} = J_{n,PNP} = \frac{I_{D,SAT}}{(W_{CELL}/2)Z} = \frac{\mu_{ni} C_{OX}}{W_{CELL} L_{CH}} (V_G - V_{TH})^2. \tag{9.140}$$

The electron charge in the space-charge region is therefore given by

$$n_{SC} = \frac{\mu_{ni} C_{OX}}{qv_{sat,n} W_{CELL} L_{CH}} (V_G - V_{TH})^2. \tag{9.141}$$

The free carrier distribution in the N-base region is governed by high-level injection conditions with no recombination:

$$\frac{d^2 p}{dy^2} = 0. \tag{9.142}$$

The general solution for this case is a linear carrier distribution given by

$$p(y) = Ay + B. \tag{9.143}$$

The constants A and B are determined by the following boundary conditions

$$p(0) = p_0, \tag{9.144}$$

$$p(W_N - W_{SC}) = p_{SC}, \tag{9.145}$$

leading to the solution

$$p(y) = p_0 - \left(\frac{p_0 - p_{SC}}{W_N - W_{SC}} \right) y. \tag{9.146}$$

This *linear* carrier distribution in the N-base region is illustrated at the bottom of Fig. 9.67 on a logarithmic scale. The hole concentration (p_0) in the N-base region at the P$^+$ collector/N-base junction (J_1) can be obtained by relating it to the hole current flow at the junction inside the P$^+$ collector region.

The hole and electron current densities in the N-base region are given by the continuity equations:

$$J_p(y) = q\mu_p \left[p(y)E(y) - \frac{kT}{q} \frac{dp}{dy} \right], \tag{9.147}$$

$$J_n(y) = q\mu_n \left[n(y)E(y) + \frac{kT}{q} \frac{dn}{dy} \right]. \tag{9.148}$$

In the above expressions, the electron and hole concentrations, as well as their derivatives, can be assumed to be equal due to high-level injection conditions in the N-base region. In addition, continuity of collector current flow requires

$$J_n(y) + J_p(y) = J_C. \tag{9.149}$$

Combining these expressions and solving for the electric field yields

$$E(y) = \frac{J_C}{qp(y)(\mu_n + \mu_p)} - \frac{kT}{q} \left(\frac{\mu_n - \mu_p}{\mu_n + \mu_p} \right) \frac{1}{p(y)} \frac{dp}{dy}. \tag{9.150}$$

The electric field in the N-base region at junction J_1 (at $y = 0$) obtained by using the free carrier distribution given by (9.146) is

$$E(0) = \frac{J_C}{q(\mu_n + \mu_p)p_0} + \frac{kT}{qp_0} \left(\frac{p_0 - p_{SC}}{W_N - W_{SC}} \right). \tag{9.151}$$

The hole current density within the N-base region at the P$^+$ collector/N-base region junction (J_1) can then be obtained by using (9.147) with the hole carrier distribution given by (9.146) and the above electric field:

$$J_p(0) = \frac{\mu_p J_C}{\mu_n + \mu_p} + \left(\frac{2\mu_n \mu_p}{\mu_n + \mu_p} \right) \frac{kT}{W_N - W_{SC}} p_0. \tag{9.152}$$

This is also referred to as the hole current density $J_p(y_N)$.

The electron current density $J_n(y_P)$ in the P$^+$ collector region at the P$^+$ collector/N-base junction (J_1) is given by (9.61). Using this equation, the hole current density in the P$^+$ collector region at the P$^+$ collector/N-base junction (J_1) is given by

$$J_{\rm p}(y_{\rm P}) = J_{\rm C} - \frac{qD_{\rm nE}}{L_{\rm PnE}} \frac{p_0^2}{N_{\rm AE}}. \tag{9.153}$$

The hole current densities ($J_{\rm p}(y_{\rm P})$ and $J_{\rm p}(y_{\rm N})$) on the two sides of the junction can be equated because the current is continuous at the junction. This yields a quadratic solution for the hole concentration (p_0) at the P$^+$ collector/N-base junction:

$$ap_0^2 + bp_0 + c = 0 \tag{9.154}$$

with coefficients

$$a = \frac{qD_{\rm nE}}{L_{\rm nE} N_{\rm AE}}, \tag{9.155}$$

$$b = -\left(\frac{2\mu_{\rm n}\mu_{\rm p}}{\mu_{\rm n} + \mu_{\rm p}}\right) \frac{kT}{W_{\rm N} - W_{\rm SC}}, \tag{9.156}$$

$$c = -\left(\frac{\mu_{\rm n}}{\mu_{\rm n} + \mu_{\rm p}}\right) J_{\rm C}. \tag{9.157}$$

The value of the hole concentration (p_0) at the P$^+$ collector/N-base junction can be computed by using

$$p_0 = -\frac{b}{2a}\left[1 - \sqrt{1 - \left(\frac{4ac}{b^2}\right)}\right]. \tag{9.158}$$

Since the collector current density depends upon the current gain of the P–N–P transistor, an iterative procedure is required to solve for the hole concentration p_0.

An elegant solution for the hole concentration p_0 can be derived under the assumption that the collector current density does not vary significantly with collector bias voltage due to current saturation. The hole concentration p_0 can then be obtained by analysis with a low collector bias voltage. The injection efficiency of the P$^+$ collector/N-base junction at low collector bias voltage can be obtained by combining (9.61) and (9.62):

$$\gamma_{\rm E,S} = 1 - \frac{qD_{\rm nE}}{J_{\rm C0} L_{\rm nE}} \left(\frac{p_0^2}{N_{\rm AE}}\right), \tag{9.159}$$

where $J_{\rm C0}$ is the saturated collector current at low collector bias voltages. Since the base transport factor for the P–N–P transistor has been assumed to be unity because of no recombination in the N-base region and the multiplication factor is unity due to the low collector bias voltage,

$$J_{C0} = \frac{J_{B,PNP}}{1-\gamma_{E,S}}. \tag{9.160}$$

Combining (9.159) and (9.160) provides a solution for the hole concentration p_0:

$$p_0 = \sqrt{\frac{J_{B,PNP} L_{nE} N_{AE}}{q D_{nE}}}. \tag{9.161}$$

The collector current density corresponding to the above hole carrier concentration can be derived by combining (9.152) and (9.153):

$$J_{C0} = 2\frac{kT}{W_N}\mu_p p_0 + \left(\frac{\mu_n + \mu_p}{\mu_n}\right)\frac{q D_{nE}}{L_{nE} N_{AE}} p_0^2. \tag{9.162}$$

In this model, the hole concentration in the space-charge region is assumed to remain independent of the collector bias voltage at a value

$$p_{SC} = \frac{J_{C0}}{q v_{sat,p}}. \tag{9.163}$$

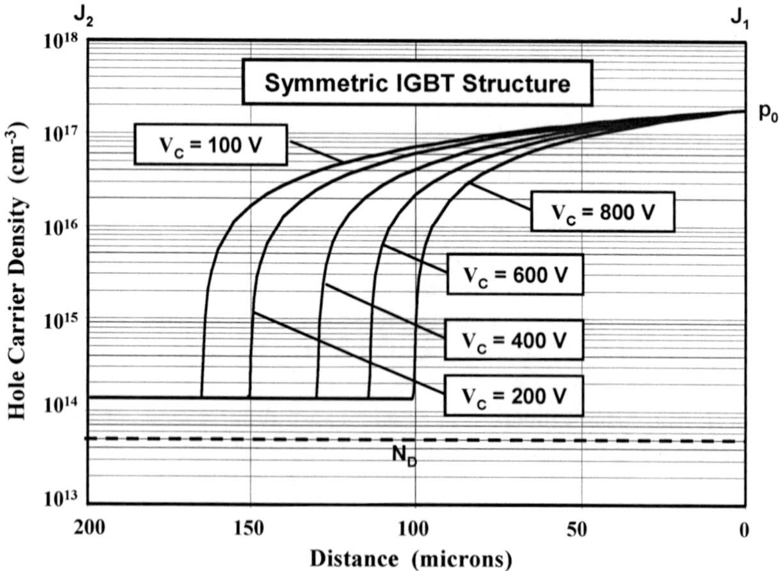

Fig. 9.68 Hole distribution within the symmetric IGBT structure during the current saturation mode at various collector bias voltages

As an example, the hole carrier profiles calculated by using the above analytical solution for the hole concentration (p_0) at the P^+ collector/N-base junction are provided in Fig. 9.68 for the symmetric IGBT structure with an N-base width of 200 μm at gate bias of 6 V. The device was assumed to have a threshold voltage of 5 V. For a gate oxide thickness of 500 Å and a channel length of 1.5 μm, the base drive current density ($J_{B,PNP}$) for the P–N–P transistor (channel current for the MOSFET portion) is found to be 136 A cm^{-2}. For a doping concentration of 1×10^{19} cm^{-3} for the P^+ collector region, an effective doping concentration of 2×10^{18} cm^{-3} was assumed leading to an injection efficiency of 0.39. The hole concentration (p_0) at the P^+ collector/N-base region is then found to be 1.8×10^{17} cm^{-3} as can be observed in Fig. 9.68.

Using the above values for the base drive current density and current gain of the P–N–P transistor, the collector current density at low collector bias voltages is found to be 224 A cm^{-2} at a gate bias of 6 V. Using this collector current density, the hole concentration in the space-charge region is found to be 1.4×10^{14} cm^{-3} as observed in Fig. 9.68. The hole concentration decreases linearly from p_0 at the P^+ collector/N-base junction to p_{SC} at the edge of the space-charge layer. The width of the space-charge layer can be computed at each collector bias voltage by using

$$W_{SC} = \sqrt{\frac{2\varepsilon_S V_C}{q(N_D + p_{SC} - n_{SC})}}. \qquad (9.164)$$

It can be observed from Fig. 9.68 that the space-charge region expands with increasing collector bias voltage. Since the hole concentration exceeds the electron concentration in this analytical model, the width of the space-charge layer is smaller than the width of the depletion region at each collector bias voltage.

Simulation Example

To gain further insight into the physics of operation for the symmetric IGBT structure in the current saturation mode, the results of two-dimensional numerical simulations for the typical structure are discussed here. This structure had an N-base region with a doping concentration of 5×10^{13} cm^{-3}, a width of 200 μm, and a lifetime (τ_{p0}) of 10 μs. The threshold voltage for this structure was found to be about 4.5 V. The hole concentration was obtained at a gate bias of 5.5 V for various values of the collector bias voltage. The saturated collector current density at this gate bias at low collector bias voltages was found to be 360 A cm^{-2} while the hole current density was found to be 150 A cm^{-2}. This corresponds to an injection efficiency of 0.42, which is well predicted by the analytical model.

A one-dimensional view of the hole distribution under the deep P^+ region obtained from simulations performed with the various collector bias voltages is provided in Fig. 9.69. The hole concentration in the N-base region varies linearly with distance as predicted by the analytical model and goes to a constant value

(p_{SC}) at the edge of the space-charge region. The hole concentration (p_0) at the P^+ collector/N-base junction remains independent of the collector bias voltage as assumed in the analytical model. The predicted values for p_0 and p_{SC} by the analytical model are in very good quantitative agreement with the results of the simulations.

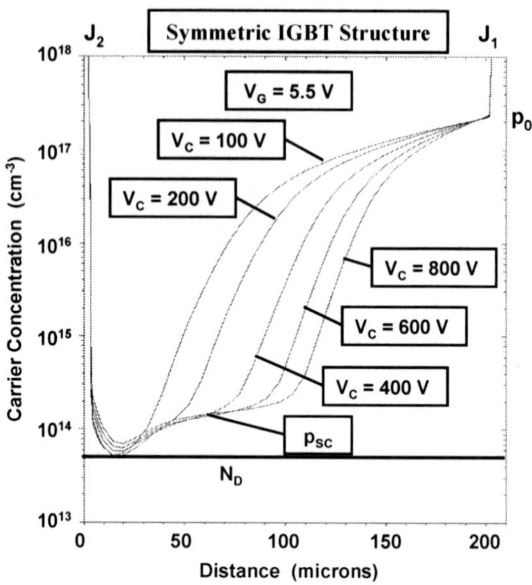

Fig. 9.69 Hole distribution in the current saturation mode for the symmetric IGBT structure

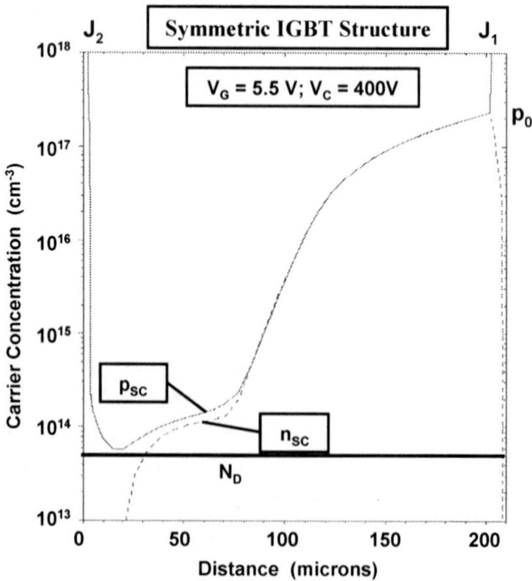

Fig. 9.70 Hole and electron distributions in the current saturation mode for the symmetric IGBT structure

During operation in the current saturation mode, the space-charge region contains both holes and electrons. This can be observed in Fig. 9.70 where these concentrations are shown at a collector bias of 400 V at a gate bias of 5.5 V. The electron and hole concentrations predicted by the analytical solutions (see (9.138) and (9.139)) are in good agreement with those observed with the simulations. The electron concentration (n_{SC}) in the space-charge region is smaller than the hole concentration (p_{SC}) because the electron current density ($J_{B,PNP}$) is smaller than the hole current density ($J_C(0)$). The electron and hole concentrations are equal within the rest of the N-base region because it is operating under high-level injection conditions.

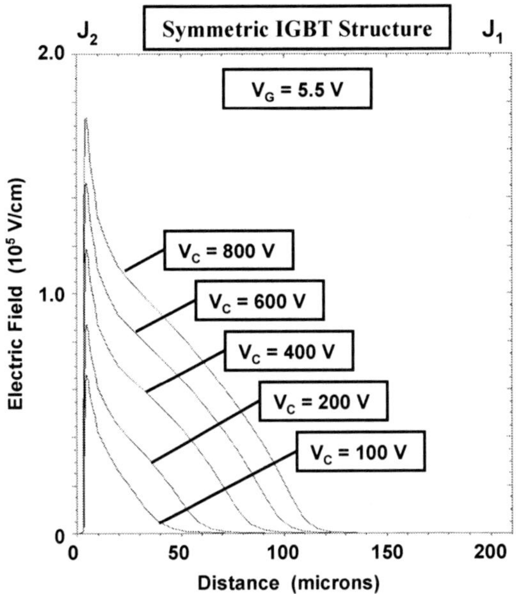

Fig. 9.71 Electric field distribution in the current saturation mode for the symmetric IGBT structure

The electric field profiles in the symmetric IGBT structure during operation in the current saturation mode are shown in Fig. 9.71 at various collector bias voltages for the case of a gate bias voltage of 5.5 V. It can be observed that the electric field has approximately a triangular shape which is consistent with the profile assumed in the analytical model (see Fig. 9.67). Using the hole and electron concentrations in the space-charge layer (as given by (9.138) and (9.139)) with a doping concentration of 5×10^{13} cm^{-3} in the N-base region, the space-charge layer widths (W_{SC}) computed by using the analytical solution (see (9.164)) are 35, 50, 71, 86, and 100 μm at collector bias voltages of 100, 200, 400, 600, and 800 V, respectively. From Fig. 9.71, it can be concluded that the width of the space-charge layer obtained with the analytical solution is a good representation of the operation of the symmetric IGBT structure in the current saturation mode. In contrast, the depletion layer widths (W_D) computed by using the background doping concentration of 5×10^{13} cm^{-3} in the N-base region are much larger: 51, 72, 102, 125, and 144 μm at collector bias voltages of 100, 200, 400, 600, and 800 V, respectively. It is therefore important to take into account the hole and electron concentrations within the high field region when doing the analysis of the symmetric IGBT structure in the current saturation mode.

9.6.2 Output Characteristics: Symmetric Structure

The output characteristics for the symmetric blocking IGBT structure can be derived by analysis of the increase in the collector current with increasing collector voltage. The collector current density ($J_C(0)$) at low collector bias voltages can be obtained for any gate bias voltage by first computing the base drive current density ($J_{B,PNP}(V_G)$) for the P–N–P transistor by using (9.140). Using this current density, the corresponding value for the hole concentration ($p_0(V_G)$) at the P$^+$ collector/N-base junction can be obtained for this gate bias voltage by using (9.161). The collector current density (J_{C0}) at low collector bias voltages can then be obtained for each gate bias voltage by using (9.162).

The collector current density increases with increasing collector bias voltage in accordance with

$$J_C(V_G,V_C) = \frac{J_{B,PNP}(V_G)}{1-\alpha_{PNP}} = \frac{J_{B,PNP}(V_G)}{1-(\gamma_{E,S}\alpha_T M)}, \tag{9.165}$$

where the injection efficiency can be obtained by using (9.153):

$$\gamma_{E,S} = \frac{J_p(x_P)}{J_C(0)} = 1 - \frac{qD_{nE}}{J_C(0)L_{PnE}}\frac{p_0^2}{N_{AE}}. \tag{9.166}$$

The injection efficiency decreases with increasing gate bias voltage due to the larger concentration of holes injected into the N-base region.

In (9.165), the multiplication factor is given by

$$M = \frac{1}{1-(V_C/\mathrm{BV}_{SC})^n}, \tag{9.167}$$

where BV_{SC} is the parallel-plane breakdown voltage after accounting for the additional charge due to the free carriers in the space-charge region:

$$\mathrm{BV}_{SC} = 5.34\times10^{13}(N_D + p_{SC} - n_{SC})^{-3/4}. \tag{9.168}$$

Due to the presence of both electrons and holes in the space-charge region during the impact ionization process, it is appropriate to use a value of 5 for the coefficient n in (9.167).

Although the carrier concentration (p_0) at the P$^+$ collector/N-base junction was analyzed under the assumption that the base transport factor is unity, this approximation cannot be used for the output characteristics because the collector current increases gradually due to an increase in the base transport factor with increasing collector bias voltage. To obtain an accurate analysis for the base transport factor as a function of the collector bias voltage, it is necessary to perform analysis of the carrier concentration with inclusion of recombination in the N-base region. The continuity equation for minority carriers (holes in the N-base region) under high-level steady-state conditions then becomes

$$\frac{\mathrm{d}^2 p}{\mathrm{d}y^2} - \frac{p}{L_\mathrm{a}^2} = 0, \tag{9.169}$$

where L_a is the ambipolar diffusion length in the N-base region given by

$$L_\mathrm{a} = \sqrt{D_\mathrm{a}\tau_\mathrm{HL}}. \tag{9.170}$$

The general solution for the carrier distribution is given by

$$p(y) = A\cosh\left(\frac{y}{L_\mathrm{a}}\right) + B\sinh\left(\frac{y}{L_\mathrm{a}}\right), \tag{9.171}$$

where A and B are parameters determined by the boundary conditions. If the distance y is defined as starting from the P^+ collector/N-base interface as indicated in Fig. 9.67 and because the above expression is valid to the edge of the space-charge region (i.e., at $y = (W_\mathrm{N} - W_\mathrm{SC})$), these boundary conditions are

$$p(0) = p_0 \tag{9.172}$$

and

$$p(W_\mathrm{N} - W_\mathrm{SC}) = p_\mathrm{SC}. \tag{9.173}$$

Using these boundary conditions in (9.171) yields

$$A = p_0 \tag{9.174}$$

and

$$B = \frac{p_\mathrm{SC} - p_0\cosh[(W_\mathrm{N} - W_\mathrm{SC})/L_\mathrm{a}]}{\sinh[(W_\mathrm{N} - W_\mathrm{SC})/L_\mathrm{a}]}. \tag{9.175}$$

Substituting into (9.171) yields

$$p(y) = p_0\cosh\left(\frac{y}{L_\mathrm{a}}\right) + \frac{p_\mathrm{SC} - p_0\cosh[(W_\mathrm{N} - W_\mathrm{SC})/L_\mathrm{a}]}{\sinh[(W_\mathrm{N} - W_\mathrm{SC})/L_\mathrm{a}]}\sinh\left(\frac{y}{L_\mathrm{a}}\right). \tag{9.176}$$

The continuity equations for the hole and electron current densities in the conductivity-modulated portion of the N-base region are

$$J_{\mathrm{p}}(y) = q\mu_{\mathrm{p}} p(y) E(y) - qD_{\mathrm{p}} \frac{\mathrm{d}p}{\mathrm{d}y} \tag{9.177}$$

and

$$J_{\mathrm{n}}(y) = q\mu_{\mathrm{n}} n(y) E(y) - qD_{\mathrm{n}} \frac{\mathrm{d}n}{\mathrm{d}y}. \tag{9.178}$$

Due to high-level injection conditions in the conductivity-modulated portion of the N-base region, the hole and electron concentrations are equal to satisfy charge neutrality. Since the electron current density is equal to the total collector current density minus the hole current density, (9.178) becomes

$$J_{\mathrm{C}} - J_{\mathrm{p}}(y) = q\mu_{\mathrm{n}} p(y) E(y) - qD_{\mathrm{n}} \frac{\mathrm{d}p}{\mathrm{d}y}. \tag{9.179}$$

Solving for the electric field using this equation and substituting this field into (9.177),

$$J_{\mathrm{p}}(y) = \left(\frac{\mu_{\mathrm{p}}}{\mu_{\mathrm{p}} + \mu_{\mathrm{n}}} \right) J_{\mathrm{C}} - 2qD_{\mathrm{p}} \left(\frac{\mu_{\mathrm{n}}}{\mu_{\mathrm{p}} + \mu_{\mathrm{n}}} \right) \frac{\mathrm{d}p}{\mathrm{d}y}. \tag{9.180}$$

The base transport factor for the N-base region at low collector bias voltages can be obtained by using (9.180) for the hole current density together with (9.176) for the hole carrier distribution in the conductivity-modulated region:

$$\alpha_{\mathrm{T}} = \frac{J_{\mathrm{p}}(W_{\mathrm{N}} - W_{\mathrm{SC}})}{J_{\mathrm{p}}(0+)}. \tag{9.181}$$

The current density at the edge of the space-charge region ($y = (W_{\mathrm{N}} - W_{\mathrm{SC}})$) obtained by using the above equation is

$$J_{\mathrm{p}}(W_{\mathrm{N}} - W_{\mathrm{SC}}) = \left\{ \begin{array}{l} \left(\dfrac{\mu_{\mathrm{p}}}{\mu_{\mathrm{p}} + \mu_{\mathrm{n}}} \right) - \left(\dfrac{\mu_{\mathrm{n}} K_{\mathrm{S}}}{\mu_{\mathrm{p}} + \mu_{\mathrm{n}}} \right) \\ \left[\begin{array}{l} \sinh\left(\dfrac{W_{\mathrm{N}} - W_{\mathrm{SC}}}{L_{\mathrm{a}}} \right) \tanh\left(\dfrac{W_{\mathrm{N}} - W_{\mathrm{SC}}}{L_{\mathrm{a}}} \right) \\ -\cosh\left(\dfrac{W_{\mathrm{N}} - W_{\mathrm{SC}}}{L_{\mathrm{a}}} \right) \end{array} \right] \end{array} \right\} J_{\mathrm{C}}, \tag{9.182}$$

where

$$K_{\rm S} = \left(\frac{\mu_{\rm p} + \mu_{\rm n}}{\mu_{\rm n}} \right) \gamma_{\rm E} - \left(\frac{\mu_{\rm p}}{\mu_{\rm n}} \right). \tag{9.183}$$

In this expression, the width of the space-charge region can be obtained by using (9.164). The hole current density at the left-hand side of junction (J_1) was also previously derived in Sect. 9.5.2 (see (9.60)–(9.62)):

$$J_{\rm p}(0+) = \gamma_{\rm E} J_{\rm C} \tag{9.184}$$

with the injection efficiency given by

$$\gamma_{\rm E,S} = 1 - \frac{qD_{\rm nE} p_0^2}{J_{\rm C0} L_{\rm nE} N_{\rm AE}}. \tag{9.185}$$

The base transport factor in the conductivity modulated lightly doped N-base region is enhanced by the combination of drift and diffusion due to the high-level injection conditions. With the current gain obtained by using the above base transport factor, the collector current density can be obtained for any applied gate bias voltage from (9.165).

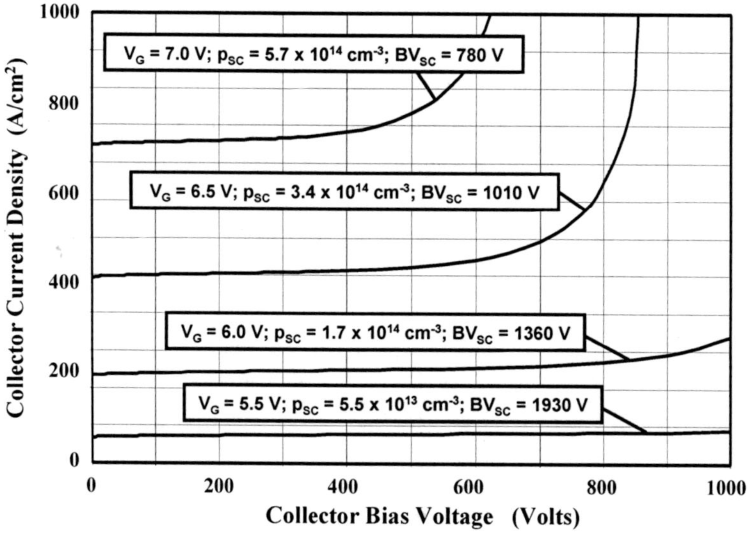

Fig. 9.72 Output characteristics for the symmetric IGBT structure

The output characteristics of the symmetric IGBT structure calculated by using the above analytical model with the device parameters given in Sect. 9.5.3 are provided in Fig. 9.72. The emitter injection efficiency ranges from 0.6 to 0.4 as the collector current density increases with increasing gate bias. The base transport factor at low collector bias voltages is approximately 0.85. This confirms that the hole concentration (p_0) at junction (J_1) can be computed under the assumption of a base transport factor equal to unity. The base transport factor increases from 0.85

to unity with increasing collector bias voltage, producing a gradual increase in the collector current. When the collector bias voltage approaches the breakdown voltage (BV_{SC}) *including the impact of the mobile free carriers in the space-charge region*, the collector current increases rapidly. The on-set of this increase in collector current occurs at lower collector bias voltages when the gate bias voltage is increased because of a reduction of the breakdown voltage (BV_{SC}) due to the larger concentration of holes (p_{SC}) at larger gate bias voltages. The values for these parameters are provided in Fig. 9.72 at each gate bias voltage for reference.

Simulation Example

The output characteristics of the typical symmetric IGBT structure discussed in the previous section were obtained by using two-dimensional numerical simulations at various gate bias voltages. This device structure has a threshold voltage of about 4.5 V. The output characteristics are shown in Fig. 9.73. At a gate bias voltage of 5.5 V, the output characteristic is flat with high output impedance up to a collector bias voltage of 800 V. The output resistance then decreases due to the on-set of significant avalanche multiplication. At larger gate bias voltages, the output impedance begins to degrade due to the on-set of impact ionization at lower collector bias voltages. It can be concluded that the output characteristics obtained by using the analytical model (see Fig. 9.72) provide a good description of the behavior of the symmetric IGBT structure.

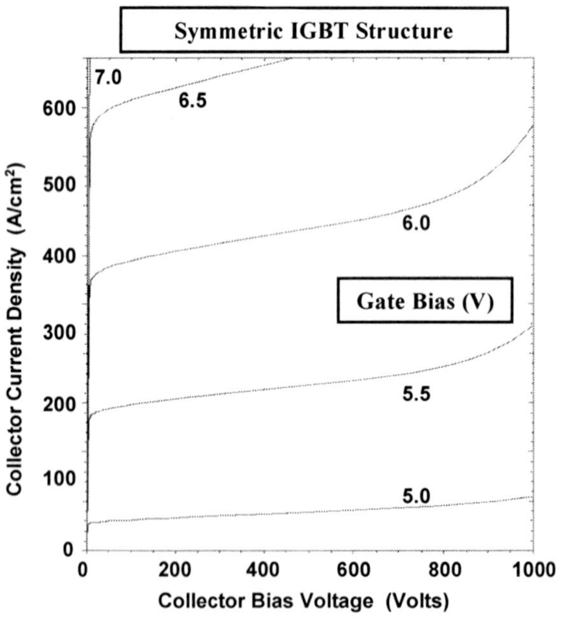

Fig. 9.73 Output characteristics of the symmetric IGBT structure

9.6.3 Output Resistance: Symmetric Structure

The output resistance for the IGBT structure is important in determining its ability to withstand short-circuit conditions and limiting the load current. It is desirable to have a large output resistance for optimum circuit performance. From the output characteristic shown in Fig. 9.72, it is apparent that the symmetric IGBT structure exhibits a high output resistance at low collector bias voltages. However, the output resistance degrades as the collector voltage is increased. This behavior can be quantified by using the relationship between the collector current density and the collector voltage derived in the previous section:

$$J_{\mathrm{C}}(V_{\mathrm{G}},V_{\mathrm{C}})=\frac{J_{\mathrm{B,PNP}}(V_{\mathrm{G}})}{1-\alpha_{\mathrm{PNP}}}=\frac{J_{\mathrm{B,PNP}}(V_{\mathrm{G}})}{1-(\gamma_{\mathrm{E,S}}\alpha_{\mathrm{T}}M)}. \tag{9.186}$$

The injection efficiency decreases with increasing gate bias voltage due to the larger concentration of holes injected into the N-base region but is independent of the collector bias voltage. At lower collector bias voltages where the multiplication coefficient is close to unity, the output resistance is governed by the increase in the base transport factor.

At large collector bias voltages, the base transport factor (α_{T}) becomes close to unity and thus independent of the collector bias voltage. However, the multiplication factor is dependent on the collector bias voltage as given by

$$M=\frac{1}{1-(V_{\mathrm{C}}/\mathrm{BV}_{\mathrm{SC}})^{n}}, \tag{9.187}$$

where $\mathrm{BV}_{\mathrm{SC}}$ is the parallel-plane breakdown voltage after accounting for the additional charge due to the free carriers in the space-charge region. Due to the presence of both electrons and holes in the space-charge region during the impact ionization process, it is appropriate to use a value of 5 for the coefficient n in (9.187). Substituting (9.187) into (9.186) yields

$$J_{\mathrm{C}}(V_{\mathrm{G}},V_{\mathrm{C}})=\frac{J_{\mathrm{B,PNP}}(V_{\mathrm{G}})[(\mathrm{BV}_{\mathrm{SC}})^{n}-(V_{\mathrm{C}})^{n}]}{(1-\gamma_{\mathrm{E,S}})(\mathrm{BV}_{\mathrm{SC}})^{n}-(V_{\mathrm{C}})^{n}}. \tag{9.188}$$

Taking the first derivative of the collector current density with respect to the collector bias voltage and then inverting the expression provides an analytical expression for the specific output resistance for the symmetric IGBT structure:

$$R_{\mathrm{O,sp}}=\frac{[(1-\gamma_{\mathrm{E,S}})(\mathrm{BV}_{\mathrm{SC}})^{n}-(V_{\mathrm{C}})^{n}]^{2}}{n\gamma_{\mathrm{E,S}}J_{\mathrm{B,PNP}}(\mathrm{BV}_{\mathrm{SC}})^{n}(V_{\mathrm{C}})^{n-1}}. \tag{9.189}$$

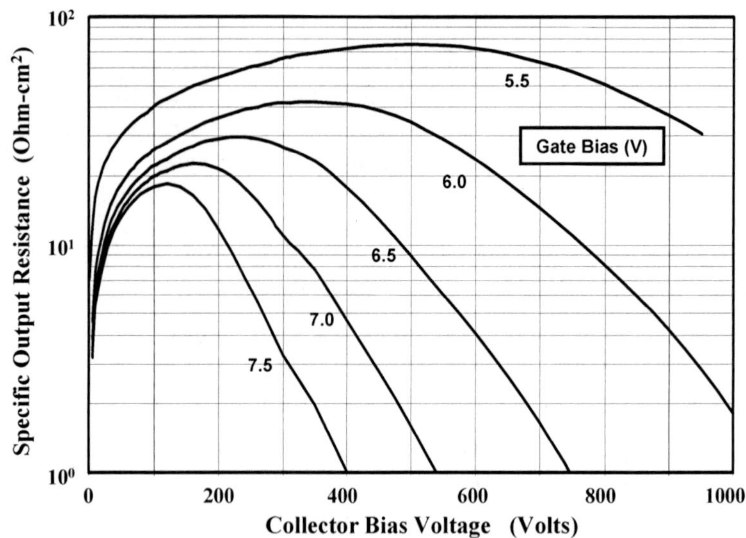

Fig. 9.74 Specific output resistance for the symmetric IGBT structure

The specific output resistances computed by using the above analysis are provided in Fig. 9.74 for the typical symmetric IGBT structure. This structure had the same device parameters as used to produce the output characteristics shown in Fig. 9.72. The output resistance initially increases with increasing collector bias voltage due to a reduced rate of change in the base transport factor. However, at larger collector bias voltages, the output resistance degrades because of the rapid increase in the multiplication coefficient.

9.6.4 Carrier Distribution: Asymmetric Structure

As in the case of the symmetric IGBT structure, a model for the output characteristics of the asymmetric IGBT structure can be derived by analysis of the free carrier distribution when the device is operating with both a high collector current density and a high collector bias voltage as is relevant for the current saturation mode. In the case of the asymmetric IGBT structure, the injection efficiency is reduced by the high doping concentration in the N-buffer layer. In addition, the transport of holes through the N-buffer layer is inhibited by recombination due to the smaller lifetime in this highly doped portion of the N-base region, producing a reduction of the base transport factor. The base transport factor of the lightly doped portion of the N-base region is determined by high-level injection conditions. As the collector bias voltage is increased, the space-charge layer width increases producing an increase in the base transport factor for the lightly doped portion of the N-base region until it becomes equal to unity when the space-charge layer punches through to the N-buffer layer. Consequently, the current gain of the P–N–P transistor becomes much less than unity and a function of the collector bias voltage for the asymmetric IGBT structure until the space-charge layer punches through to the N-buffer layer.

As discussed previously in Sect. 9.6.1 for the symmetric IGBT structure, a space-charge region forms at the deep P$^+$/N-base junction (J_2) for the asymmetric IGBT structure with holes and electrons transported within this region at their saturated drift velocity. Consequently, these carrier densities can be related to the hole and electron current densities within the space-charge region:

$$n_{\mathrm{SC}} = \frac{J_{\mathrm{n}}}{qv_{\mathrm{sat,n}}}, \tag{9.190}$$

$$p_{\mathrm{SC}} = \frac{J_{\mathrm{p}}}{qv_{\mathrm{sat,p}}}. \tag{9.191}$$

The electron current via the channel of the MOSFET portion in the IGBT structure is the base drive current for the P–N–P transistor. Under current saturation conditions for the IGBT structure, the MOSFET channel current is saturated at a value dictated by the gate bias voltage relative to the threshold voltage as given by (9.132). The current can be expressed as an electron current density:

$$J_{\mathrm{n}} = J_{\mathrm{B,PNP}} = \frac{I_{\mathrm{D,SAT}}}{(W_{\mathrm{CELL}}/2)Z} = \frac{\mu_{\mathrm{ni}}C_{\mathrm{OX}}}{W_{\mathrm{CELL}}L_{\mathrm{CH}}}(V_{\mathrm{G}} - V_{\mathrm{TH}})^2. \tag{9.192}$$

The electron charge in the space-charge region is therefore given by

$$n_{\mathrm{SC}} = \frac{\mu_{\mathrm{ni}}C_{\mathrm{OX}}}{qv_{\mathrm{sat,n}}W_{\mathrm{CELL}}L_{\mathrm{CH}}}(V_{\mathrm{G}} - V_{\mathrm{TH}})^2. \tag{9.193}$$

A one-dimensional analysis of the free carrier distribution can be performed for the portion of the IGBT structure under the deep P$^+$ region by solving the continuity equation for minority carriers (holes in the N-base region) under high-level steady-state conditions:

$$\frac{\mathrm{d}^2 p}{\mathrm{d}y^2} - \frac{p}{L_{\mathrm{a}}^2} = 0, \tag{9.194}$$

where L_{a} is the ambipolar diffusion length in the N-base region given by

$$L_{\mathrm{a}} = \sqrt{D_{\mathrm{a}}\tau_{\mathrm{HL}}}. \tag{9.195}$$

The general solution for the carrier distribution is given by

$$p(y) = A\cosh\left(\frac{y}{L_{\mathrm{a}}}\right) + B\sinh\left(\frac{y}{L_{\mathrm{a}}}\right), \tag{9.196}$$

where A and B are parameters determined by the boundary conditions.

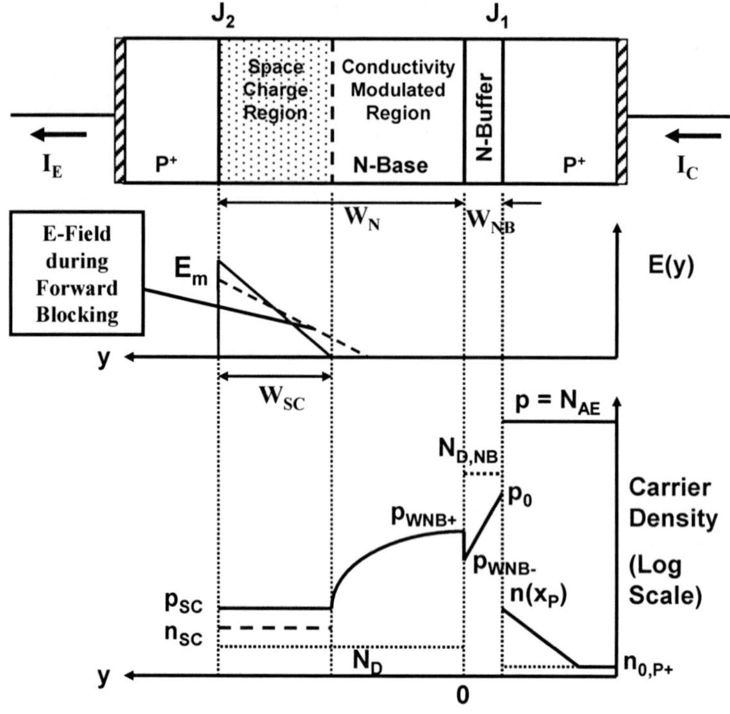

Fig. 9.75 Electric field and hole distributions for the asymmetric IGBT structure before space-charge layer punch-through

If the distance y is defined as starting from the N-buffer/N-base interface as indicated in Fig. 9.75 and because the above expression is valid to the edge of the space-charge region (i.e., at $y = (W_N - W_{SC})$), these boundary conditions are

$$p(W_{NB}+) = p_{WNB+} \tag{9.197}$$

and

$$p(W_N - W_{SC}) = p_{SC}. \tag{9.198}$$

Using these boundary conditions in (9.196) yields

$$A = p_{WNB+} \tag{9.199}$$

and

$$B = \frac{p_{SC} - p_{WNB+}\cosh[(W_N - W_{SC})/L_a]}{\sinh[(W_N - W_{SC})/L_a]}. \tag{9.200}$$

Substituting into (9.196) yields

$$p(y) = p_{\text{WNB+}} \cosh\left(\frac{y}{L_{\text{a}}}\right) + \frac{p_{\text{SC}} - p_{\text{WNB+}} \cosh[(W_{\text{N}} - W_{\text{SC}})/L_{\text{a}}]}{\sinh[(W_{\text{N}} - W_{\text{SC}})/L_{\text{a}}]} \sinh\left(\frac{y}{L_{\text{a}}}\right). \tag{9.201}$$

The continuity equations for the hole and electron current densities in the conductivity-modulated portion of the N-base region are

$$J_{\text{p}}(y) = q\mu_{\text{p}} p(y) E(y) - qD_{\text{p}} \frac{\text{d}p}{\text{d}y} \tag{9.202}$$

and

$$J_{\text{n}}(y) = q\mu_{\text{n}} n(y) E(y) - qD_{\text{n}} \frac{\text{d}n}{\text{d}y}. \tag{9.203}$$

Due to high-level injection conditions in the conductivity-modulated portion of the N-base region, the hole and electron concentrations are equal to satisfy charge neutrality. Since the electron current density is equal to the total collector current density minus the hole current density, (9.203) becomes

$$J_{\text{C}} - J_{\text{p}}(y) = q\mu_{\text{n}} p(y) E(y) - qD_{\text{n}} \frac{\text{d}p}{\text{d}y}. \tag{9.204}$$

Solving for the electric field using this equation and substituting this field into (9.202),

$$J_{\text{p}}(y) = \left(\frac{\mu_{\text{p}}}{\mu_{\text{p}} + \mu_{\text{n}}}\right) J_{\text{C}} - 2qD_{\text{p}} \left(\frac{\mu_{\text{n}}}{\mu_{\text{p}} + \mu_{\text{n}}}\right) \frac{\text{d}p}{\text{d}y}. \tag{9.205}$$

The hole concentration ($p(W_{\text{NB}}+) = p_{\text{WNB+}}$) at the left-hand side of the interface between the lightly doped portion of the N-base region and the N-buffer layer can be obtained by recognizing that the hole current density at the interface must be continuous:

$$J_{\text{p}}(W_{\text{NB}}+) = J_{\text{p}}(W_{\text{NB}}-). \tag{9.206}$$

The hole current density at the left-hand side of the interface between the lightly doped portion of the N-base region and the N-buffer layer can be obtained by using (9.202) with $y = 0$ and the carrier concentration given by (9.201). The hole current density at the right-hand side of the interface between the lightly doped portion of the N-base region and the N-buffer layer was previously derived in Sect. 9.5.4 (see (9.93)):

$$J_{\mathrm{p}}(W_{\mathrm{NB}}-) = \gamma_{\mathrm{E}} J_{\mathrm{C}} \mathrm{e}^{-W_{\mathrm{NB}}/L_{\mathrm{p,NB}}} \tag{9.207}$$

because the N-buffer layer is assumed to be operating under low-level injection levels. In this equation, the injection efficiency is given by

$$\gamma_{\mathrm{E}} = \frac{D_{\mathrm{p,NB}} p_{0,\mathrm{NB}} L_{\mathrm{n,P+}}}{D_{\mathrm{p,NB}} p_{0,\mathrm{NB}} L_{\mathrm{n,P+}} + D_{\mathrm{n,P+}} n_{0,\mathrm{P+}} L_{\mathrm{p,NB}}}. \tag{9.208}$$

Combining these equations yields a solution for the hole concentration on the left-hand side of the interface between the lightly doped portion of the N-base region and the N-buffer layer:

$$p_{\mathrm{WNB+}} = \frac{K_{\mathrm{AS}} J_{\mathrm{C}} L_{\mathrm{a}}}{2qD_{\mathrm{p}}} \tanh[(W_{\mathrm{N}} - W_{\mathrm{SC}})/L_{\mathrm{a}}], \tag{9.209}$$

where

$$K_{\mathrm{AS}} = \left(\frac{\mu_{\mathrm{p}} + \mu_{\mathrm{n}}}{\mu_{\mathrm{n}}}\right) \gamma_{\mathrm{E}} \mathrm{e}^{-W_{\mathrm{NB}}/L_{\mathrm{p,NB}}} - \left(\frac{\mu_{\mathrm{p}}}{\mu_{\mathrm{n}}}\right). \tag{9.210}$$

As the collector bias voltage increases in the current saturation mode, the collector current remains approximately constant while the space-charge layer width becomes larger. According to the above equation, this should produce a reduction of the hole concentration ($p_{\mathrm{WNB+}}$) in the lightly doped portion of the N-base region at the N-buffer layer interface with increasing collector bias voltage. These results are applicable until the space-charge layer punches through to the N-buffer layer. Since the collector current density is approximately constant in the saturation mode, the hole concentration ($p_{\mathrm{WNB+}}$) in the lightly doped portion of the N-base region at the N-buffer layer interface can be computed without an iterative process by using the collector current density (J_{C0}) at low collector bias voltages and assuming that the space-charge layer width is zero.

Within the N-buffer layer, the hole concentration decays exponentially away from the P^+ collector/N-buffer layer junction:

$$p(y) = p_0 \mathrm{e}^{-y/L_{\mathrm{p,NB}}} \tag{9.211}$$

with the concentration at the junction provided by low-level junction theory:

$$p_0 = \frac{p_{0,\mathrm{NB}} J_{\mathrm{C0}}}{(qD_{\mathrm{p}} p_{0,\mathrm{NB}}/L_{\mathrm{p,NB}}) + (qD_{\mathrm{n}} n_{0,\mathrm{P+}}/L_{\mathrm{nE}})}. \tag{9.212}$$

The collector current density J_{C0} at low collector bias voltages can be obtained from the base drive current which is a function of the gate bias voltage:

$$J_{\mathrm{C0}} = \frac{J_{\mathrm{B,PNP}}}{1 - \alpha_{\mathrm{PNP,0}}}. \tag{9.213}$$

The current gain of the P–N–P transistor at low collector bias voltages is given by

$$\alpha_{\mathrm{PNP,0}} = \gamma_{\mathrm{E}} \alpha_{\mathrm{T,0}} = \gamma_{\mathrm{E}} \alpha_{\mathrm{T,N\text{-}Buffer}} \alpha_{\mathrm{T,N\text{-}Base,0}}. \tag{9.214}$$

In this expression, the injection efficiency can be obtained by using (9.208). The base transport factor for the P–N–P transistor cannot be obtained by using the low-level injection theory for a bipolar transistor (i.e., (7.75)). For the asymmetric IGBT structure, the base transport factor is associated with current transport through the N-buffer layer at low injection levels and the current transport through the lightly doped portion of the N-base region at high-level injection conditions as indicated by the two terms in the above equation. The base transport factor associated with the N-buffer layer can be obtained from the decay of the hole current within the N-buffer layer as given by low-level injection theory:

$$\alpha_{\mathrm{T,N\text{-}Buffer}} = \frac{J_{\mathrm{p}}(W_{\mathrm{NB}}-)}{J_{\mathrm{p}}(x_{\mathrm{N}})} = \frac{\gamma_{\mathrm{E}} J_{\mathrm{C}} \mathrm{e}^{-W_{\mathrm{NB}}/L_{\mathrm{p,NB}}}}{\gamma_{\mathrm{E}} J_{\mathrm{C}}} = \mathrm{e}^{-W_{\mathrm{NB}}/L_{\mathrm{p,NB}}}. \tag{9.215}$$

The base transport factor for the N-base region at low collector bias voltages can be obtained by using (9.205) for the hole current density together with (9.201) for the hole carrier distribution in the conductivity-modulated region:

$$\alpha_{\mathrm{T,N\text{-}Base,0}} = \frac{J_{\mathrm{p}}(W_{\mathrm{N}})}{J_{\mathrm{p}}(W_{\mathrm{NB}}+)}. \tag{9.216}$$

Using (9.201) and (9.205),

$$J_{\mathrm{p}}(W_{\mathrm{NB}}+) = \left[\left(\frac{\mu_{\mathrm{p}}}{\mu_{\mathrm{p}} + \mu_{\mathrm{n}}}\right) + \left(\frac{\mu_{\mathrm{n}}}{\mu_{\mathrm{p}} + \mu_{\mathrm{n}}}\right) K_{\mathrm{AS}}\right] J_{\mathrm{C}} \tag{9.217}$$

and

$$J_{\mathrm{p}}(W_{\mathrm{N}}) = \left\{\frac{\left(\dfrac{\mu_{\mathrm{p}}}{\mu_{\mathrm{p}} + \mu_{\mathrm{n}}}\right) - \left(\dfrac{\mu_{\mathrm{n}} K_{\mathrm{AS}}}{\mu_{\mathrm{p}} + \mu_{\mathrm{n}}}\right)}{\left[\sinh\left(\dfrac{W_{\mathrm{N}}}{L_{\mathrm{a}}}\right) \tanh\left(\dfrac{W_{\mathrm{N}}}{L_{\mathrm{a}}}\right) - \cosh\left(\dfrac{W_{\mathrm{N}}}{L_{\mathrm{a}}}\right)\right]}\right\} J_{\mathrm{C}}. \tag{9.218}$$

The base transport factor in the conductivity-modulated lightly doped portion of the N-base region is enhanced by the combination of drift and diffusion due to the high-level injection conditions.

Computing the current gain with the above base transport factors, the collector current density (J_{C0}) at low collector bias voltages can be obtained for any applied gate bias voltage from (9.213). This collector current density can then be used to obtain the carrier concentrations within the entire N-base region with the distribution in the N-buffer layer given by (9.211) and in the conductivity-modulated portion of the N-base region by (9.201). The space-charge layer width in (9.201) is given by

$$W_{SC} = \sqrt{\frac{2\varepsilon_S V_C}{q(N_D + p_{SC} - n_{SC})}} \tag{9.219}$$

with the carrier concentrations provided by (9.190) and (9.191). The hole carrier distribution is illustrated in Fig. 9.75 for the case of a typical collector bias voltage prior to punch-through of the space-charge layer to the N-buffer layer.

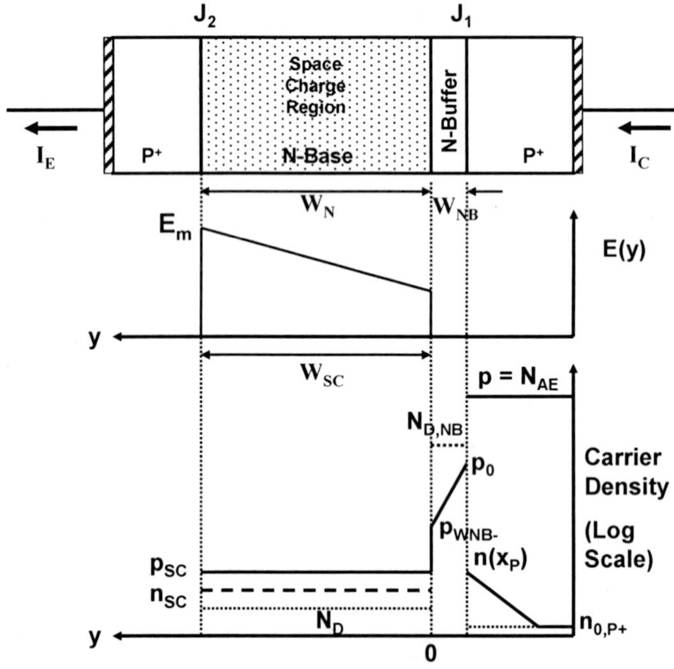

Fig. 9.76 Electric field and hole distributions for the asymmetric IGBT structure after space-charge layer punch-through

At large collector bias voltages, the space-charge layer extends through the entire width of the lightly doped portion of the N-base region. A high electric field is prevalent throughout the lightly doped portion of the N-base region resulting in the electron and hole concentrations given by (9.190) and (9.191), respectively. The distribution of holes within the N-buffer layer is governed by low-level injection physics if the doping concentration in the N-buffer layer is at least

1×10^{17} cm^{-3}. As previously discussed in Sect. 9.5.4, the hole concentration varies exponentially with distance within the N-buffer layer with concentrations p_0 and p_{WNB-} at its edges. There is a discontinuity in the hole concentration at the interface between the lightly doped portion of the N-base region and the N-buffer layer. The electric field and hole distributions for this case are illustrated in Fig. 9.76.

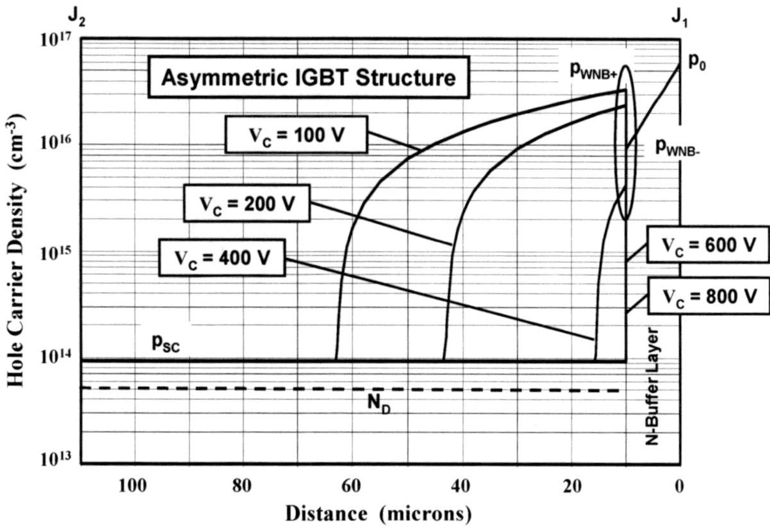

Fig. 9.77 Hole distribution within the asymmetric IGBT structure during the current saturation mode at various collector bias voltages

The hole distribution within the asymmetric IGBT structure obtained by using the above analytical model is illustrated in Fig. 9.77 at various collector bias voltages for the device structure previously described in Sect. 9.5.4. A high-level lifetime (τ_{HL}) of 2 μs was used in the lightly doped portion of the N-base region. In this example, the gate bias voltage is 6 V. Using the parameters of the asymmetric structure provided in Sect. 9.5.4, this produces an electron (base drive) current density of 136 A cm^{-2}. The emitter injection efficiency obtained by using an N-buffer layer doping concentration of 1×10^{17} cm^{-3} is 0.72. The base transport factor for the N-buffer layer using a diffusion length based on scaling the lifetime with doping concentration is 0.155. The base transport factor for the N-base region at low collector bias voltages is 0.78 leading to a common-base current gain of 0.12 at low collector bias voltages. Using this current gain, the collector current density at low collector bias voltages is found to be 150 A cm^{-2}. The hole concentration (p_0) at the P$^+$ collector/N-buffer layer junction (J_1) obtained at this collector current density by using (9.90) is 6.0×10^{16} cm^{-3}. This concentration does not vary with collector bias voltage. The hole and electron concentrations within the space-charge region are 9.3×10^{13} and 8.5×10^{13} cm^{-3}, respectively. These values are assumed to be independent of the collector bias voltage in the analytical model. The space-charge layer widths obtained by using the analytical model are 47, 67, and 94 μm at a collector bias of 100, 200, and 400 V, respectively. The space-charge layer punches through to the N-buffer layer at a collector bias of 450 V at this collector current density.

Simulation Example

To gain further insight into the physics of operation for the asymmetric IGBT structure in the current saturation mode, the results of two-dimensional numerical simulations for the typical structure are discussed here. This structure had an N-base region consisting of a lightly doped (5×10^{13} cm^{-3}) portion with a width of 100 μm and an N-buffer layer with a width of 10 μm. A lifetime (τ_{p0}) of 1 μs was used in the lightly doped portion of the N-base region. The device structural parameters are provided in Sect. 9.5.4. The threshold voltage for this structure was found to be about 4.5V. The saturated collector current density for a gate bias of 5.5 V at low collector bias voltages was found to be 170 A cm^{-2}. The hole concentration profile was obtained at a gate bias of 5.5 V for various values of the collector bias voltage.

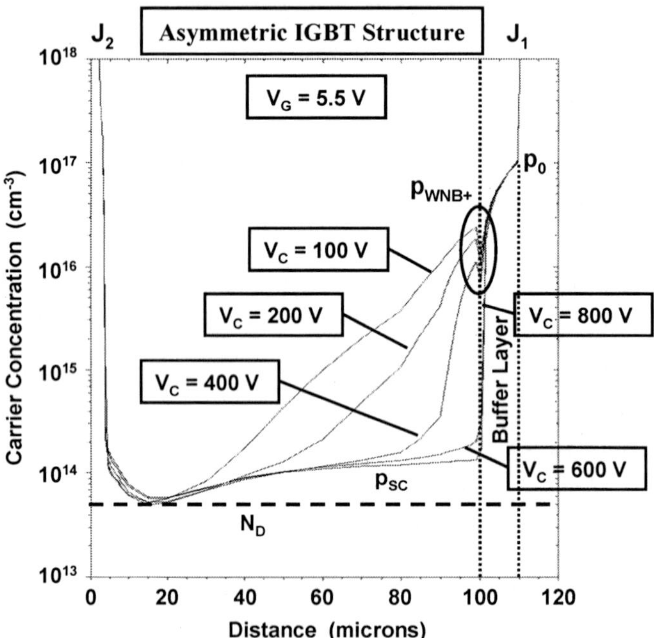

Fig. 9.78 Hole distribution in the current saturation mode for the asymmetric IGBT structure

A one-dimensional view of the hole distribution under the deep P$^+$ region obtained from simulations performed with the various collector bias voltages is provided in Fig. 9.78. The hole concentration (p_0) at the P$^+$ collector/N-buffer layer junction (J_1) remains independent of the collector bias voltage as assumed in the analytical model. The hole concentration within the N-buffer layer decreases approximately exponentially with distance as described by the analytical model. The hole concentration at junction (J_1) is equal to the doping concentration in the N-buffer layer. Consequently, the assumption of low-level injection conditions used for the analytical model is not valid throughout the N-buffer layer. In spite of this, the analytical model provides a good description of the hole carrier distribution in the asymmetric IGBT structure. The hole concentration (p_{WNB+}) in the lightly doped portion of the N-base region at its interface with the N-buffer region deceases with increasing collector bias voltage as predicted by the analytical model. Within the lightly doped portion of the N-base region, the hole concentration decreases with distance and goes to a constant value (p_{SC}) at the edge of the space-charge region. The predicted values for p_{WNB+} and p_{SC} by the analytical model are in good quantitative agreement with the results of the simulations.

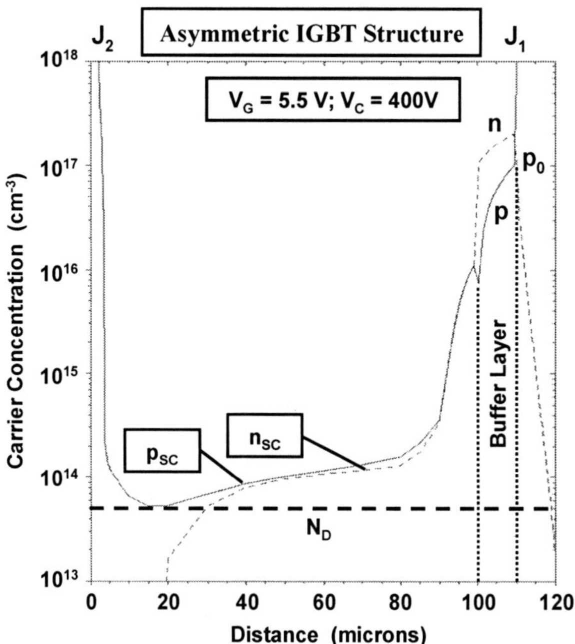

Fig. 9.79 Hole and electron distributions in the current saturation mode for the asymmetric IGBT structure

During operation in the current saturation mode, the space-charge region contains both holes and electrons. This can be observed in Fig. 9.79 where these concentrations are shown at a collector bias of 400 V for a gate bias of 5.5 V. The electron and hole concentrations predicted by the analytical solutions (see (9.190) and (9.191)) are in good agreement with those observed with the simulations. The electron concentration (n_{SC}) in the space-charge region is smaller than the hole concentration (p_{SC}) because the electron current density ($J_{B,PNP}$) is smaller than the hole current density (J_{C0}). The electron and hole concentrations are equal within the rest of the lightly doped portion of the N-base region because it is operating under high-level injection conditions. However, in the N-buffer layer, the electron concentration is enhanced above the doping concentration of 1×10^{17} cm^{-3} because the hole concentration near the P$^+$ collector/N-base junction is comparable with the doping concentration. In spite of this, the assumption of low-level injection conditions in the N-buffer layer used to derive the analytical model provides a sufficiently accurate description of the operating physics for the asymmetrical IGBT structure.

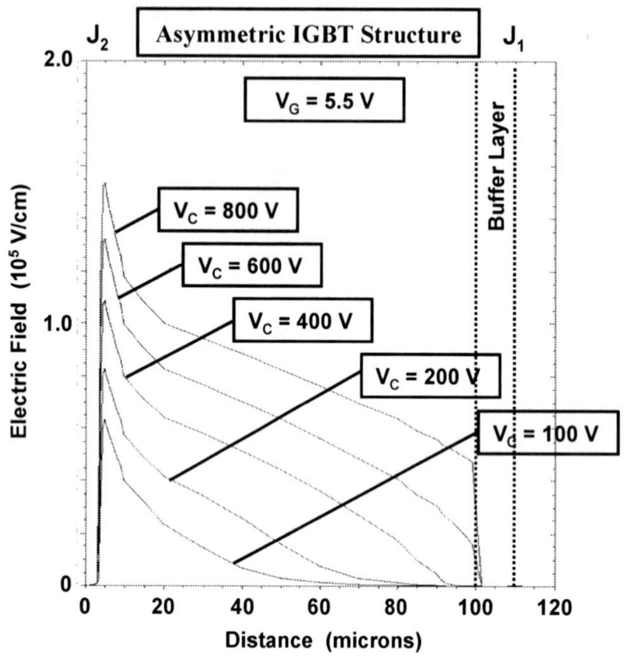

Fig. 9.80 Electric field distribution in the current saturation mode for the asymmetric IGBT structure

The electric field profiles in the asymmetric IGBT structure during operation in the current saturation mode are shown in Fig. 9.80 at various collector bias voltages for the case of a gate bias voltage of 5.5 V. It can be observed that the electric field has approximately a triangular shape until the space-charge layer punches through to the N-buffer layer. This behavior is consistent with the profile assumed in the analytical model. The space-charge layer widths (W_{SC}) observed with the simulations are 45, 65, and 90 μm at collector bias voltages of 100, 200, and 400 V, respectively. The analytical model predicts these values accurately.

9.6.5 Output Characteristics: Asymmetric Structure

The output characteristics for the asymmetric IGBT structure can be derived by analysis of the increase in the collector current with increasing collector voltage. The collector current density (J_{C0}) at low collector bias voltages can be obtained for any gate bias voltage by first computing the base drive (electron) current density ($J_{B,PNP}(V_G)$) for the P–N–P transistor by using (9.192). Using this current density, the corresponding value for the collector current density (J_{C0}) at low collector bias voltages can be obtained for this gate bias voltage by using (9.213) and the procedure described in Sect. 9.6.4.

The collector current density increases with increasing collector bias voltage in accordance with

$$J_C(V_G, V_C) = \frac{J_{B,PNP}(V_G)}{1-\alpha_{PNP}} = \frac{J_{B,PNP}(V_G)}{1-[\gamma_{E,S}\alpha_T(V_C)M(V_C)]}, \tag{9.220}$$

where the injection efficiency can be obtained by using (9.208). The injection efficiency for the asymmetric IGBT structure is independent of the collector and gate bias voltages. In (9.220), the base transport factor (α_T) consists of the combination of the base transport factor for the N-buffer layer and the base transport factor for the lightly doped portion of the N-base region. The base transport factor ($\alpha_{T,N\text{-Buffer}}$) for the N-buffer layer, as given by (9.215), is also independent of the collector and gate bias voltages. The base transport factor for the lightly doped portion of the N-base region is given by

$$\alpha_{T,N\text{-Base}} = \frac{J_p(W_N - W_{SC})}{J_p(W_{NB}+)}, \tag{9.221}$$

where the denominator is provided in (9.217). The numerator in (9.221) can be derived by using the same procedure used to derive (9.218) in Sect. 9.6.4:

$$J_p(W_N - W_{SC}) = \left\{ \begin{array}{l} \left(\dfrac{\mu_p}{\mu_p+\mu_n}\right) - \left(\dfrac{\mu_n K_{AS}}{\mu_p+\mu_n}\right) \\ \left[\begin{array}{l} \sinh\left(\dfrac{W_N - W_{SC}}{L_a}\right)\tanh\left(\dfrac{W_N - W_{SC}}{L_a}\right) \\ -\cosh\left(\dfrac{W_N - W_{SC}}{L_a}\right) \end{array} \right] \end{array} \right\} J_C. \tag{9.222}$$

The base transport factor ($\alpha_{T,N\text{-Base}}$) in the conductivity-modulated lightly doped portion of the N-base region is enhanced by the combination of drift and diffusion due to the high-level injection conditions. It increases with increasing collector bias voltage due to the widening of the space-charge layer and becomes equal to unity when the space-charge layer punches through to the N-buffer region. This produces a small increase in the collector current density with increasing collector bias voltage until punch-through of the space-charge layer occurs.

After punch-through of the space-charge layer, the base transport factor becomes independent of the collector voltage while the multiplication factor begins to increase. The multiplication factor in (9.220) is given by

$$M = \frac{1}{1-(V_{\rm NPT}/{\rm BV}_{\rm SC})^n}, \tag{9.223}$$

where BV_{SC} is the parallel-plane breakdown voltage after accounting for the additional charge due to the free carriers in the space-charge region:

$$\text{BV}_{\text{SC}} = 5.34\times10^{13}(N_{\text{D}} + p_{\text{SC}} - n_{\text{SC}})^{-3/4}. \tag{9.224}$$

In (9.223), the nonpunch-through voltage (V_{NPT}) corresponding to the applied collector bias voltage must be used to account for the trapezoidal electric field distribution as previously discussed in Sect. 9.4.4:

$$V_{\rm NPT} = \frac{\varepsilon_{\rm S}}{2qN_{\rm D}}\left[\frac{V_{\rm C}}{W_{\rm N}} + \frac{q(N_{\rm D} + p_{\rm SC} - n_{\rm SC})W_{\rm N}}{2\varepsilon_{\rm S}}\right]^2. \tag{9.225}$$

Due to the presence of both electrons and holes in the space-charge region during the impact ionization process, it is appropriate to use a value of 5 for the coefficient n in (9.223).

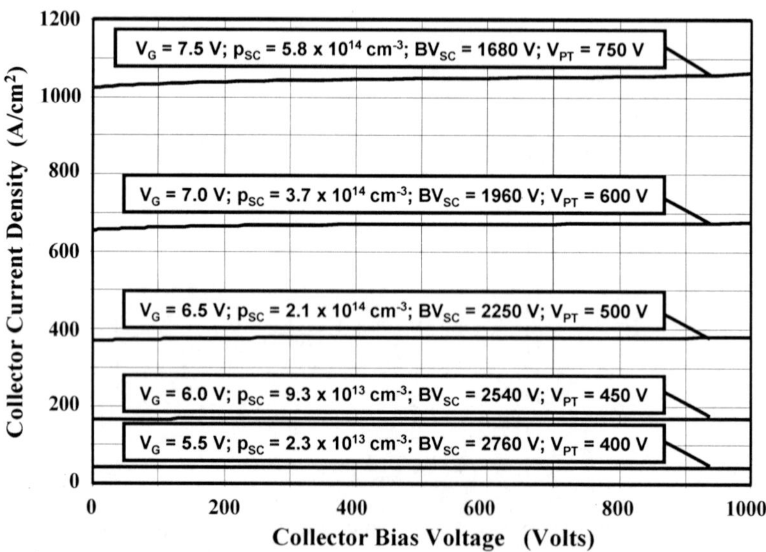

Fig. 9.81 Output characteristics for the asymmetric IGBT structure

The output characteristics of the asymmetric IGBT structure with the device parameters provided in the preceding sections are shown in Fig. 9.81. It can be observed that the saturated collector current increases slightly up to the punch-through bias voltage (V_{PT}) for each gate bias. The collector current remains relatively constant beyond this collector bias because the space-charge layer has extended throughout the lightly doped portion of the N-base region. The values for the parameters p_{SC}, V_{PT}, and BV_{SC} are included in Fig. 9.81 at each gate bias voltage for reference. The punch-through voltage increases with increasing gate bias voltage due to operation at a larger collector current density with more carriers within the space-charge layer. The multiplication coefficient remains close to unity for all the gate bias voltage cases because the breakdown voltage (BV_{SC}) *including the impact of the mobile free carriers in the space-charge region* is much larger than the collector bias voltage. In this asymmetric IGBT structure, the hole and electron concentrations in the space-charge layer are nearly equal for all values of the gate bias voltage. Consequently, the breakdown voltage (BV_{SC}) remains much larger than the collector bias voltages in Fig. 9.81.

Simulation Example

The output characteristics of the asymmetric IGBT structure discussed in the previous section were obtained by using two-dimensional numerical simulations at various gate bias voltages. This device structure has a threshold voltage of about 4.5 V. The output characteristics are shown in Fig. 9.82. At all gate bias voltages, the collector current density increases until a collector bias of about 500 V. The collector current density then becomes independent of the collector bias voltage because the space-charge layer punches through to the N-buffer layer. At collector bias voltages above 800 V, the output resistance decreases due to the on-set of some avalanche multiplication. It can be concluded that the output characteristics obtained by using the analytical model (see Fig. 9.81) provide a good description of the behavior of the asymmetric IGBT structure.

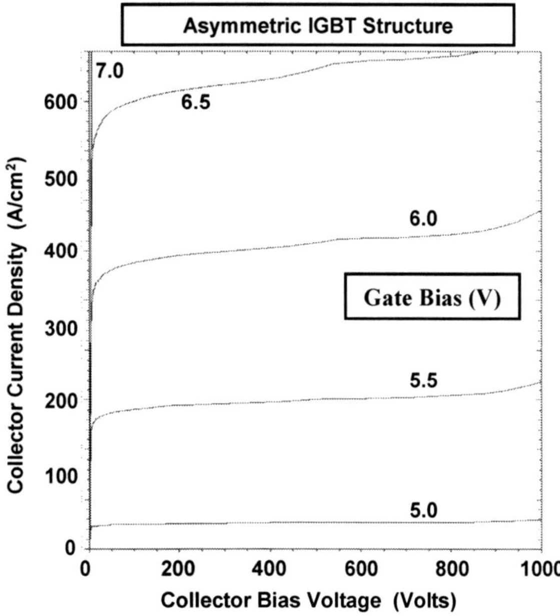

Fig. 9.82 Output characteristics of the asymmetric IGBT structure

9.6.6 Output Resistance: Asymmetric Structure

From the output characteristic shown in Fig. 9.81, it is apparent that the asymmetric IGBT structure exhibits a high output resistance over a broad range of collector bias voltages. This behavior can be quantified by using the relationship between the collector current density and the collector voltage derived in Sect. 9.6.5:

$$J_C(V_G, V_C) = \frac{J_{B,PNP}(V_G)}{1 - \alpha_{PNP}} = \frac{J_{B,PNP}(V_G)}{1 - (\gamma_{E,S}\alpha_T M)}, \tag{9.226}$$

where the base transport factor consists of the combination of the base transport factor for the N-buffer layer and the base transport factor for the lightly doped portion of the N-base region. The base transport factor for the N-buffer layer is independent of the collector bias voltage. At lower collector bias voltages, the base transport factor for the lightly doped portion of the N-base region increases with increasing collector bias voltage. In this range of collector bias voltages, the multiplication coefficient remains close to unity.

After the space-charge layer punches through to the N-buffer layer, the base transport factor for the P–N–P transistor becomes independent of the collector bias voltage. However, the multiplication factor increases at large collector bias voltage as given by

$$M = \frac{1}{1 - (V_{NPT}/\mathrm{BV}_{SC})^n}, \tag{9.227}$$

where BV_{SC} is the parallel-plane breakdown voltage after accounting for the additional charge due to the free carriers in the space-charge region and V_{NPT} is the nonpunch-through voltage given by (9.225). Due to the presence of both electrons and holes in the space-charge region during the impact ionization process, it is appropriate to use a value of 5 for the coefficient n in (9.227). Substituting (9.227) into (9.226) yields

$$J_C(V_G, V_C) = \frac{J_{B,PNP}(V_G)[(\mathrm{BV}_{SC})^n - (V_{NPT})^n]}{(1 - \gamma_E\alpha_{T,N\text{-Buffer}})(\mathrm{BV}_{SC})^n - (V_{NPT})^2}. \tag{9.228}$$

Taking the first derivative of the collector current density with respect to the collector bias voltage and then inverting the expression provides the specific output resistance for the asymmetric IGBT structure:

$$R_{O,sp} = \frac{[(1 - \gamma_E\alpha_{T,N\text{-Buffer}})(\mathrm{BV}_{SC})^n - (V_{NPT})^n]^2}{n\gamma_E\alpha_{T,N\text{-Buffer}}J_{B,PNP}(\mathrm{BV}_{SC})^n(V_{NPT})^{n-1}}. \tag{9.229}$$

The specific output resistances computed by using the above analysis for the typical asymmetric IGBT structure are provided in Fig. 9.83 for various gate bias voltages. This structure had the same device parameters as used to produce the output characteristics shown in Fig. 9.81. Initially, the output resistance increases with increasing collector bias voltage because the base transport factor for the lightly doped portion of the N-base region increases more gradually with increasing collector bias voltage. The output resistance reaches a peak value when the space-charge layer punches through to the N-buffer layer. Beyond this point, the output resistance decreases due to an increase in collector current produced by the increasing multiplication factor. The punch-through voltage changes with gate bias due to a higher hole and electron concentration in the space-charge region with increasing collector current density.

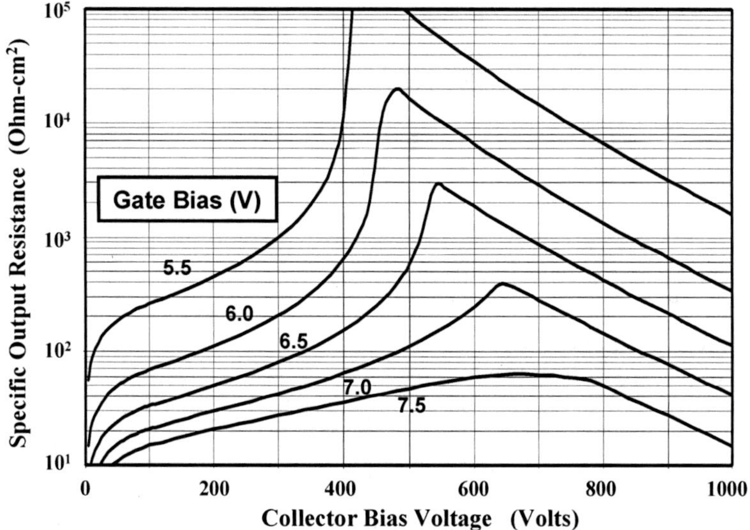

Fig. 9.83 Specific output resistance for the asymmetric IGBT structure

9.6.7 Carrier Distribution: Transparent Emitter Structure

The transparent emitter IGBT structure can be analyzed by using the same approach used for the symmetric IGBT structure with no recombination in the N-base region. The only difference between the structures is the injection efficiency at the P^+ collector/N-base junction. The electron current is provided as the base drive current for the P–N–P transistor via the channel of the MOSFET portion in the IGBT structure. Under current saturation conditions for the IGBT structure, the MOSFET channel current is saturated at a value dictated by the gate bias voltage relative to the threshold voltage as given by (9.140). The current can be expressed as an electron current density:

$$J_{\text{B,PNP}} = J_{\text{n,PNP}} = \frac{I_{\text{D,SAT}}}{(W_{\text{CELL}}/2)Z} = \frac{\mu_{\text{ni}} C_{\text{OX}} Z}{W_{\text{CELL}} L_{\text{CH}}} (V_{\text{G}} - V_{\text{TH}})^2. \tag{9.230}$$

The electron charge in the space-charge region is therefore given by

$$n_{SC} = \frac{J_{n,PNP}}{qv_{sat,n}} = \frac{\mu_{ni}C_{OX}}{qv_{sat,n}W_{CELL}L_{CH}}(V_G - V_{TH})^2. \tag{9.231}$$

In the transparent emitter structure, the width of the P^+ collector region is made very small allowing analysis of current transport within this region without recombination. Consequently, the current component due to the injection of electrons into the P^+ collector region is given by (see Sect. 9.5.6)

$$J_n = \frac{qD_{nE}}{W_{P+}}\frac{p_0^2}{N_{AE}}, \tag{9.232}$$

where N_{AE} is the effective doping concentration of the P^+ collector region. The injection efficiency of the P^+ collector/N-base junction at low collector bias voltage is then given by

$$\gamma_{E,S} = 1 - \frac{J_n}{J_{C0}} = 1 - \frac{qD_{nE}}{J_{C0}W_{P+}}\left(\frac{p_0^2}{N_{AE}}\right), \tag{9.233}$$

where J_{C0} is the saturated collector current at low collector bias voltages. Since the base transport factor for the P–N–P transistor has been assumed to be unity because of no recombination in the N-base region and the multiplication factor is unity due to the low collector bias voltage,

$$J_{C0} = \frac{J_{B,PNP}}{1 - \gamma_{E,S}}. \tag{9.234}$$

Combining (9.233) and (9.234) provides an elegant solution for the hole concentration p_0:

$$p_0 = \sqrt{\frac{J_{B,PNP}W_{P+}N_{AE}}{qD_{nE}}}. \tag{9.235}$$

The collector current density corresponding to the above hole carrier concentration can be obtained by using (see corresponding equation (9.162) for the symmetric IGBT structure)

$$J_{C0} = 2\frac{kT}{W_N}\mu_p p_0 + \left(\frac{\mu_n + \mu_p}{\mu_n}\right)\frac{qD_{nE}}{W_{P+}N_{AE}}p_0^2. \tag{9.236}$$

In this model, the hole concentration in the space-charge region is assumed to remain independent of the collector bias voltage at a value

$$p_{SC} = \frac{J_{C0}}{qv_{sat,p}}. \tag{9.237}$$

Applying the approach previously used for the symmetric IGBT structure with no recombination in the N-base region,

$$p(y) = p_0 - \left(\frac{p_0 - p_{SC}}{W_N - W_{SC}} \right) y. \tag{9.238}$$

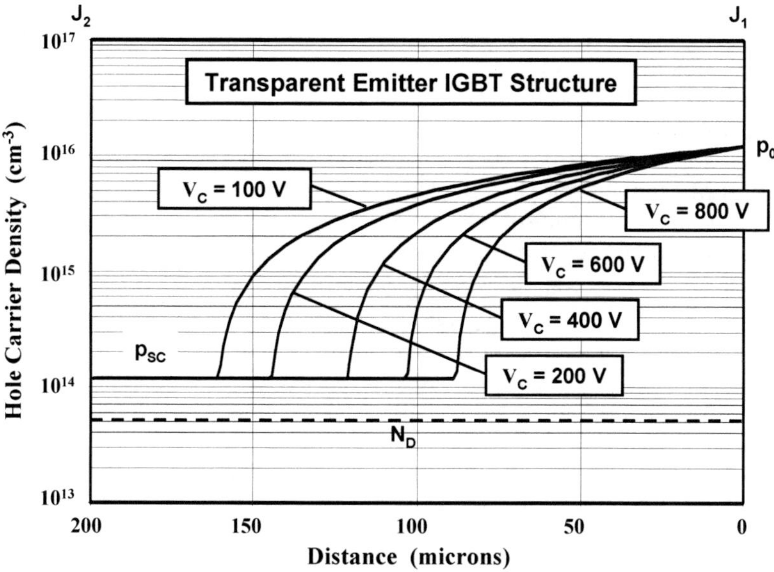

Fig. 9.84 Hole distribution within the transparent emitter IGBT structure during the current saturation mode at various collector bias voltages

As an example, the hole carrier profiles calculated by using the above analytical solution for the hole concentration (p_0) at the P$^+$ collector/N-base junction are provided in Fig. 9.84 for the transparent emitter IGBT structure with an N-base width of 200 μm at gate bias of 6 V. The device was assumed to have a threshold voltage of 5 V. For a gate oxide thickness of 500 Å and a channel length of 1.5 μm, the base drive current density ($J_{B,PNP}$) for the P–N–P transistor (channel current for the MOSFET portion) is found to be 136 A cm^{-2}. For a surface doping concentration of 1×10^{17} cm^{-3} for the P$^+$ collector region, an effective doping concentration of 4.5×10^{16} cm^{-3} was obtained by using (9.122) leading to an injection efficiency of 0.28. The hole concentration (p_0) at the P$^+$ collector/N-base region is then found to be 1.2×10^{16} cm^{-3} as can be observed in Fig. 9.84.

Using the above values for the base drive current density and current gain of the P–N–P transistor, the collector current density at low collector bias voltages is found to be 189 A cm^{-2} at a gate bias of 6 V. Using this collector current density, the hole concentration in the space-charge region is found to be 1.2×10^{14} cm^{-3} as observed in Fig. 9.84. The hole concentration decreases linearly from p_0 at the P^+ collector/N-base junction to p_{SC} at the edge of the space-charge layer. The width of the space-charge layer can be computed at each collector bias voltage by using

$$W_{SC} = \sqrt{\frac{2\varepsilon_S V_C}{q(N_D + p_{SC} - n_{SC})}}. \tag{9.239}$$

It can be observed from Fig. 9.84 that the space-charge region expands with increasing collector bias voltage.

Simulation Example

To gain further insight into the physics of operation for the transparent emitter IGBT structure in the current saturation mode, the results of two-dimensional numerical simulations for the typical structure are discussed here. This structure had an N-base region with a doping concentration of 5×10^{13} cm^{-3}, a width of 200 μm, and a lifetime (τ_{p0}) of 10 μs. The threshold voltage for this structure was found to be about 4.5 V. The hole concentration was obtained at a gate bias of 5.5 V for various values of the collector bias voltage. The saturated collector current density at this gate bias at low collector bias voltages was found to be 170 A cm^{-2}.

A one-dimensional view of the hole distribution under the deep P^+ region obtained from simulations performed with the various collector bias voltages is provided in Fig. 9.85. The hole concentration in the N-base region varies linearly

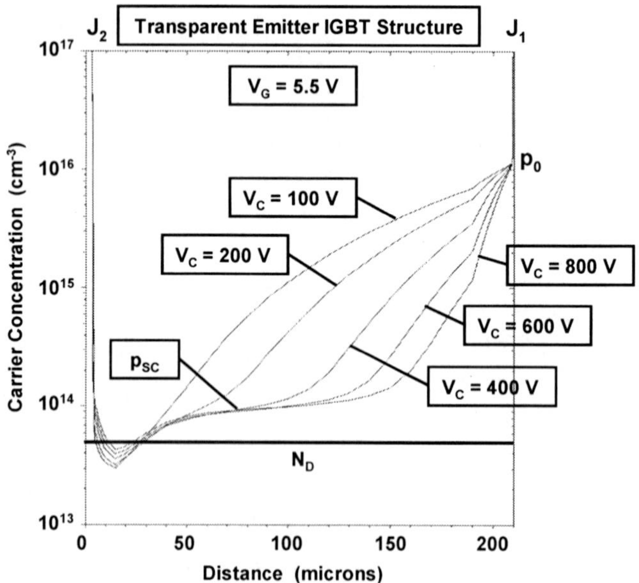

Fig. 9.85 Hole distribution in the current saturation mode for the transparent emitter IGBT structure

with distance as predicted by the analytical model and goes to a constant value (p_{SC}) at the edge of the space-charge region. The hole concentration (p_0) at the P$^+$ collector/N-base junction remains independent of the collector bias voltage as assumed in the analytical model. The predicted values for p_0 and p_{SC} by the analytical model are in very good quantitative agreement with the results of the simulations. The electric field profiles for the transparent emitter IGBT structure are similar to those previously shown for the symmetric IGBT structure.

9.6.8 Output Characteristics: Transparent Emitter Structure

The output characteristics for the transparent emitter IGBT structure can be derived by analysis of the increase in the collector current with increasing collector voltage. The collector current density (J_{C0}) at low collector bias voltages can be obtained for any gate bias voltage by first computing the base drive current density ($J_{B,PNP}(V_G)$) for the P–N–P transistor by using (9.230). Using this current density, the corresponding value for the hole concentration ($p_0(V_G)$) at the P$^+$ collector/N-base junction can be obtained for this gate bias voltage by using (9.235). The collector current density (J_{C0}) at low collector bias voltages can then be obtained for each gate bias voltage by using (9.236).

The collector current density increases with increasing collector bias voltage in accordance with

$$J_C(V_G, V_C) = \frac{J_{B,PNP}(V_G)}{1-\alpha_{PNP}} = \frac{J_{B,PNP}(V_G)}{1-(\gamma_{E,S}\alpha_T M)}, \tag{9.240}$$

where the injection efficiency can be obtained by using (9.233):

$$\gamma_{E,S} = \frac{J_p(y_P)}{J_{C0}} = 1 - \frac{qD_{nE}}{J_{C0}W_{P+}}\frac{p_0^2}{N_{AE}}. \tag{9.241}$$

The injection efficiency decreases with increasing gate bias voltage due to the larger concentration of holes injected into the N-base region.

In (9.240), the multiplication factor is given by

$$M = \frac{1}{1-(V_C/\mathrm{BV}_{SC})^n}, \tag{9.242}$$

where BV_{SC} is the parallel-plane breakdown voltage after accounting for the additional charge due to the free carriers in the space-charge region:

$$\mathrm{BV}_{SC} = 5.34\times 10^{13}(N_D + p_{SC} - n_{SC})^{-3/4}. \tag{9.243}$$

Due to the presence of both electrons and holes in the space-charge region during the impact ionization process, it is appropriate to use a value of 5 for the coefficient n in (9.242).

Although the carrier concentration (p_0) at the P$^+$ collector/N-base junction was analyzed under the assumption that the base transport factor is unity, this approximation cannot be used for the output characteristics because the collector current increases gradually due to an increase in the base transport factor with increasing collector bias voltage. To obtain an accurate analysis for the base transport factor as a function of the collector bias voltage, it is necessary to perform analysis of the carrier concentration with inclusion of recombination in the N-base region. This analysis is identical to that previously described in Sect. 9.6.2 for the symmetric IGBT structure.

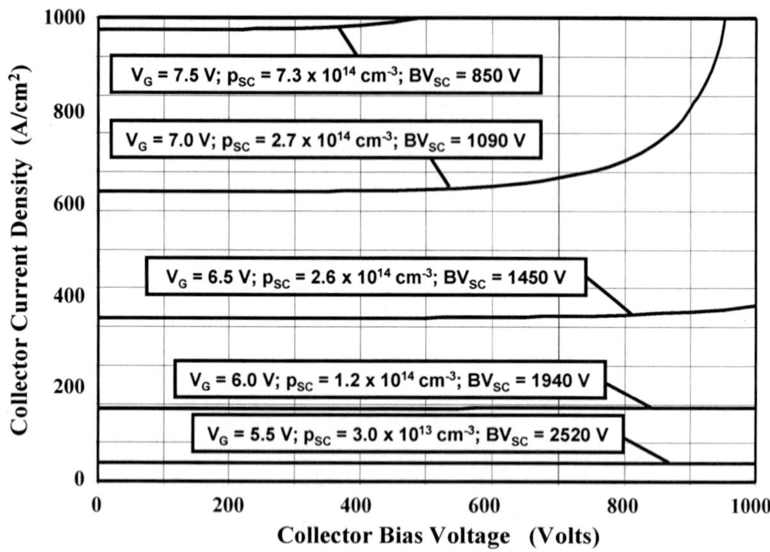

Fig. 9.86 Output characteristics for the transparent emitter IGBT structure

The output characteristics of the transparent emitter IGBT structure computed by using the analytical model with the device parameters given in Sect. 9.5.6 are provided in Fig. 9.86. The emitter injection efficiency ranges from 0.27 to 0.29 as the collector current density increases with increasing gate bias. The base transport factor at low collector bias voltages is approximately 0.98. This confirms that the hole concentration (p_0) at junction (J_1) can be computed under the assumption of a base transport factor equal to unity. The base transport factor increases from 0.98 to unity with increasing collector bias voltage producing a small increase in the collector current. When the collector bias voltage approaches the breakdown voltage (BV_{SC}) *including the impact of the mobile free carriers in the space-charge region*, the collector current increases rapidly. The on-set of this increase in collector current occurs at lower collector bias voltages when the gate bias voltage is increased because of a reduction of the breakdown voltage (BV_{SC}) due to the larger concentration of holes (p_{SC}) at larger gate bias voltages. The values for these parameters are provided in Fig. 9.86 at each gate bias voltage for reference.

Simulation Example

The output characteristics of the transparent emitter IGBT structure discussed in the previous section were obtained by using two-dimensional numerical simulations at various gate bias voltages. This device structure has a threshold voltage of about 4.5 V. The output characteristics are shown in Fig. 9.87. At each gate bias voltage, the saturated collector current density is reduced when compared with that for the symmetric IGBT structure due to the smaller emitter injection efficiency. At a gate bias voltage of 5.5 V, the output characteristics are flat with high output impedance up to a collector bias voltage of 1,000 V. At larger gate bias voltages, the output impedance begins to degrade due to the larger concentration of holes and electrons in the space-charge layer. It can be concluded that the output characteristics obtained by using the analytical model (see Fig. 9.86) provide a good description of the behavior of the transparent emitter IGBT structure at gate bias voltages close to the threshold voltage.

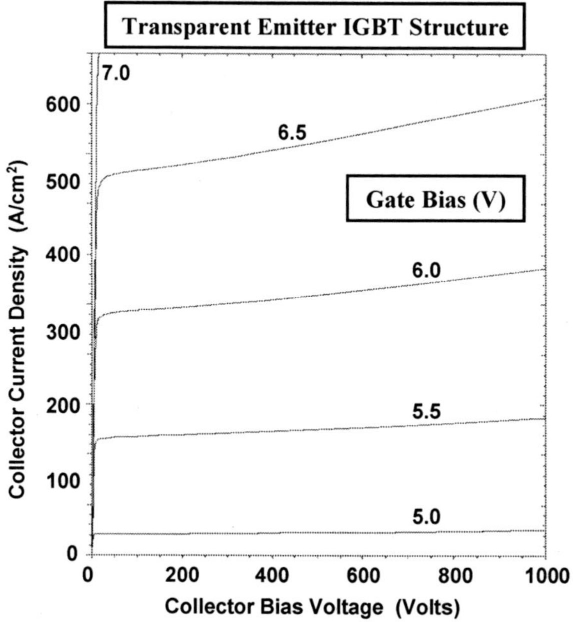

Fig. 9.87 Output characteristics of the transparent emitter IGBT structure

9.6.9 Output Resistance: Transparent Emitter Structure

From the output characteristic shown in Fig. 9.86, it is apparent that the transparent emitter IGBT structure exhibits a high output resistance at low collector and gate bias voltages. However, the output resistance degrades as the collector voltage is increased. This behavior is similar to that previously described for the symmetric blocking IGBT structure. The output resistance for the transparent emitter structure with P^+ collector doping concentration of 1×10^{17} cm^{-3} described in Sect. 9.6.7 is provided in Fig. 9.88. This structure exhibits larger values for the output resistance when compared with the symmetric IGBT structure. The specific output resistances obtained from the output characteristics derived by using numerical simulations range from 150 to 10 Ω cm^2 as the gate bias is increased from 0.5 to 2.5 V above the threshold voltage. The values obtained with the analytical model are therefore of the proper order of magnitude.

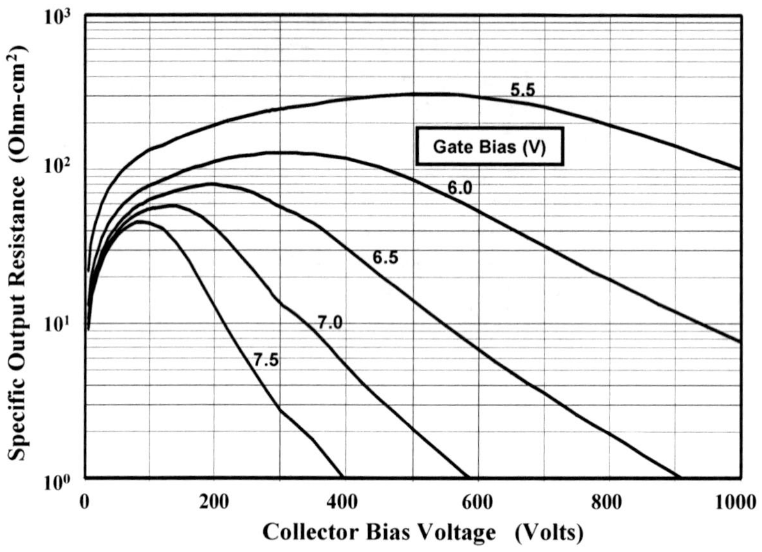

Fig. 9.88 Specific output resistance for the transparent emitter IGBT structure

9.7 Switching Characteristics

A popular application for IGBT devices is for variable frequency motor control in heating, ventilating, and air-conditioning applications. The U.S. Department of Energy estimates that two thirds of the electricity in the country is used for powering motors. The evolution of IGBT-based variable speed drives has been determined to produce an efficiency enhancement of 25% leading to the saving of an astounding 2 quads (2×10^{15} btu) of electrical energy per year.[27] The totem pole circuit used for variable speed motor control was previously shown in Fig. 9.28. The typical waveforms for the current and voltage for the IGBT and power rectifier in this circuit configuration are illustrated in Fig. 9.89. During the time interval from t_1 to t_2, the IGBT current ramps up at a constant rate when it is turned on. After the current reaches a peak value (I_{PT}) determined by the sum of the motor current (I_M) and the peak reverse recovery current (I_{PR}) of the rectifier, the IGBT current reduces to the motor current. The power loss during turn-on is governed by the time interval from t_1 to t_3, which is determined by the reverse recovery behavior of the power rectifier.

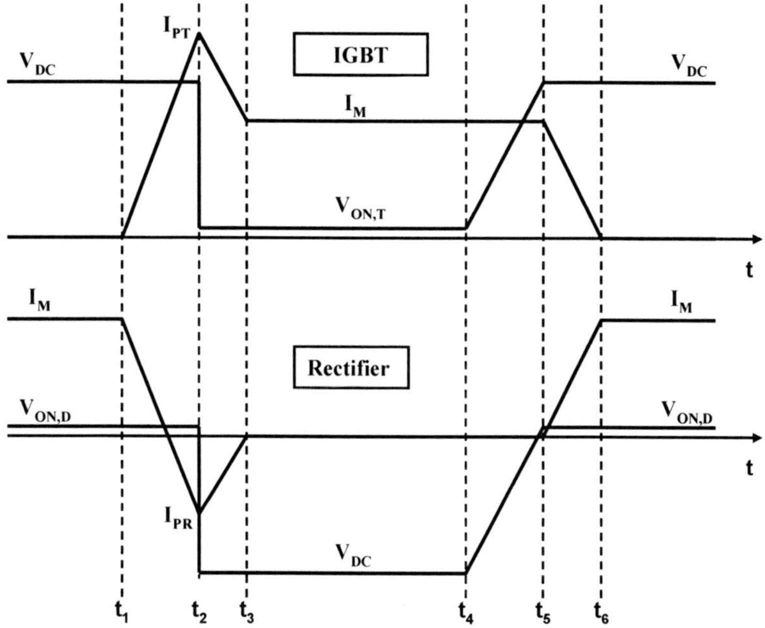

Fig. 9.89 Linearized waveforms for the IGBT and rectifier in a typical motor control application

During the time interval from t_3 to t_4, the IGBT is maintained in its on-state. As discussed in Sect. 9.5, a large amount of stored charge exists within the N-base region of the IGBT structure during the on-state. During the next time interval from t_4 to t_6, the IGBT is turned off. In order for the IGBT structure to support a high voltage, it is necessary to remove the stored charge within the N-base region. Consequently, the turn-off time duration and the turn-off power loss are determined by the physics of carrier removal within the IGBT structure.

The IGBT structure is also utilized in circuits, such as automotive ignition control and for triggering discharge lamps, where the current in the IGBT must increase rapidly when it is turned on. During this process, a high voltage can be developed across the IGBT structure because of the time taken to produce the conductivity modulation of the N-base region. This process is discussed first in this section followed by analysis of the turn-off physics for the IGBT structure.

9.7.1 Turn-On Physics: Forward Recovery

When the IGBT structure is turned on with a high rate of increase in the collector current, the voltage drop across the device can be substantially larger than the on-state voltage drop under steady-state operation because of the time taken to modulate the conductivity of the N-base region. The voltage drop across the IGBT structure becomes larger than the on-state voltage drop while the collector current is increasing and then settles down to the steady-state on-state voltage drop. This phenomenon is referred to as the *forward recovery* of the IGBT structure.

During operation under steady-state conditions previously analyzed in Sect. 9.5, it was found that the voltage drop across the drift region in the IGBT structure becomes independent of the collector current density because the free carrier concentration in the drift region increases in proportion to the collector current density (see (9.83) and (9.84) in conjunction with (9.70)). However, the drift region in the IGBT structure does not get modulated in proportion to the current density in the transient case due to the finite rate for the diffusion of minority carriers. A portion of the drift region remains without conductivity modulation when the current increases at a rapid rate. Since this portion of the drift region has a high resistance due to its low doping concentration, the voltage drop across the IGBT is much greater than under steady-state operation.

The increase in the forward voltage drop when the IGBT structure is turned on with a high ramp rate for the collector current can be analyzed by solving for the injected carrier distribution as a function of both time and position in the drift region. The excess majority carrier density is of interest here to determine the conductivity modulation of the drift region. Since the conductivity modulation advances from the P^+ collector/N-base junction (J_1), it is convenient to define the distance (y) increasing from zero at this junction (as defined in Fig. 9.43). The excess majority carrier concentration (δn) injected into the drift region is governed by the continuity equation:

$$\frac{\partial \delta n}{\partial t} = \frac{1}{q}\frac{\partial J_{\mathrm{n}}}{\partial y} - \frac{\partial n}{\tau_{\mathrm{n}}}. \tag{9.244}$$

If the current flow is dominated by diffusion, the electron current density is given by

$$J_{\mathrm{n}} = qD_{\mathrm{n}}\frac{\partial \delta n}{\partial y} \tag{9.245}$$

leading to the diffusion equation[28] for electrons:

$$\frac{\partial \delta n}{\partial t} = D_{\mathrm{n}}\frac{\partial^2 \delta n}{\partial y^2} - \frac{\partial n}{\tau_{\mathrm{n}}}. \tag{9.246}$$

Since the forward voltage overshoot occurs only under high ramp rates for the current within a time duration which is small when compared with the recombination lifetime, the recombination process can be neglected leading to

$$\frac{\partial \delta n(y,t)}{\partial t} = D_{\mathrm{n}}\frac{\partial^2 \delta n(y,t)}{\partial y^2}. \tag{9.247}$$

This equation governs the distribution of the excess electrons in the drift region as a function of both space and time. The solution for the excess electron concentration is of the form

$$\delta n(y,t) = A(t)\mathrm{e}^{-\left(y/\sqrt{4D_{\mathrm{n}}t}\right)}, \tag{9.248}$$

where the term $A(t)$ is determined by the current density, which is increasing as a function of time. The current density at the edge of the P^+ collector/N-base junction (at $y = 0$) under high-level injection conditions is given by (see derivation for (5.34))

$$J = J_{\mathrm{n}}(0) = 2qD_{\mathrm{n}}\left(\frac{\mathrm{d}\delta n}{\mathrm{d}y}\right)_{y=0}. \tag{9.249}$$

Using the carrier distribution in (9.248),

$$J = 2qD_{\mathrm{n}}\frac{A(t)}{\sqrt{4D_{\mathrm{n}}t}}. \tag{9.250}$$

During the forward recovery transient, the current density increases at a constant rate a until the collector current density reaches its on-state value. Thus,

$$J = at. \tag{9.251}$$

From these equations, the coefficient $A(t)$ is obtained:

$$A(t) = \frac{at^{3/2}}{q\sqrt{D_{\mathrm{n}}}}. \tag{9.252}$$

This equation determines the excess electron concentration at the edge of the P–N junction, which is increasing as a function of time due to the increase in the current density. The excess electron concentration then decays away from the P–N junction with a diffusion length given by

$$L(t) = \sqrt{4D_{\mathrm{n}}t}. \tag{9.253}$$

The evolution of the excess electron concentration within the drift region is then given by

$$\delta n(y,t) = \frac{at^{3/2}}{q\sqrt{D_{\mathrm{n}}}}\mathrm{e}^{-\left(y/\sqrt{4D_{\mathrm{n}}t}\right)}. \tag{9.254}$$

The total electron concentration within the drift region can then be obtained by adding the electron concentration due to the donor dopant atoms to the excess electron concentration:

$$n(y,t) = \delta n(y,t) + N_{\mathrm{D}} = \frac{at^{3/2}}{q\sqrt{D_{\mathrm{n}}}} \mathrm{e}^{-\left(y/\sqrt{4D_{\mathrm{n}}t}\right)} + N_{\mathrm{D}}. \tag{9.255}$$

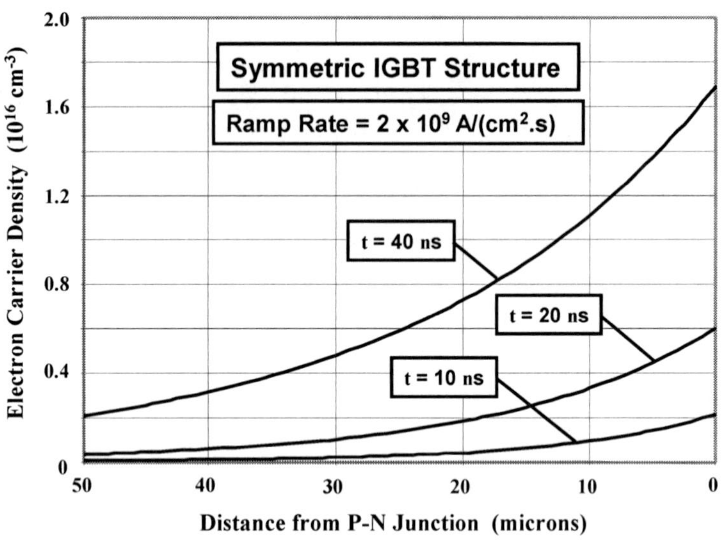

Fig. 9.90 Electron concentration in the symmetric IGBT structure during forward recovery

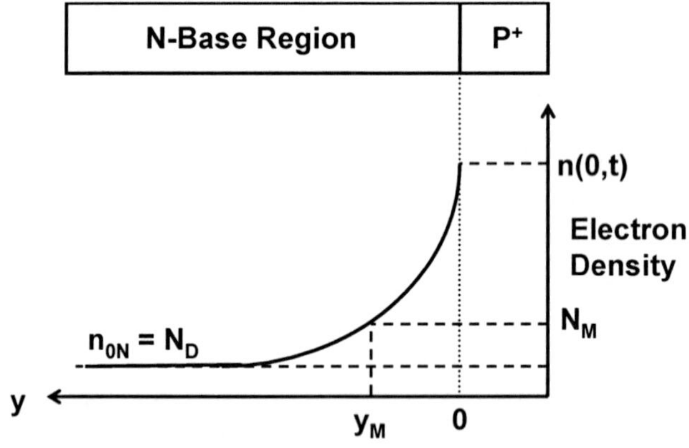

Fig. 9.91 Electron concentration in the symmetric IGBT structure during the forward recovery process

The total electron concentration is plotted in Fig. 9.90 for three instances of time in the case of a ramp rate of 2×10^9 A cm^{-2} s^{-1} for the typical symmetric IGBT structure with a drift region thickness of 200 μm and doping concentration of 5×10^{13} cm^{-3}. As time progresses from 10 to 20 to 40 ns, the electron concentration at the P–N junction ($y = 0$) increases due to the increase in the current density from 20 to 40 to 80 A cm^{-2}. Concurrently, the electrons are distributed further into the drift region by the diffusion process.

The voltage drop across the drift region can be calculated by taking the product of the current density at any time and the resistance of the drift region. Based upon the above electron concentration distribution, it can be concluded that the drift region resistance reduces with time because a larger proportion of the drift region becomes conductivity modulated as the current ramps up. To analyze the reduction of the resistance of the drift region, consider a segment at distance y from the P–N junction with a small thickness dy. The resistivity at this location is given by

$$\mathrm{d}R = \rho(y)\mathrm{d}y = \frac{\mathrm{d}y}{q\mu_\mathrm{n}(y,t)n(y,t)}, \tag{9.256}$$

where the electron mobility is a function of position and time because the electron concentration depends on these parameters. (It is necessary to account for carrier–carrier scattering for the determination of the mobility due to the relatively large concentration of both minority and majority carriers.) The resistance of the drift region during the forward recovery transient is then given by

$$R_\mathrm{N\text{-}Base}(t) = \int_0^{W_\mathrm{N}} \mathrm{d}R = \int_0^{W_\mathrm{N}} \frac{\mathrm{d}y}{q\mu_\mathrm{n}(y,t)n(y,t)}. \tag{9.257}$$

A simple analytical solution for this resistance cannot be derived due to the dependence of the mobility on the electron carrier concentration.

An alternative approach for the analysis of the drift region resistance is based upon defining a "modulation concentration" (N_M) as indicated in Fig. 9.91 and assuming that the resistance of the drift region with electron concentration above this value is negligible when compared with rest of the drift region, where no modulation is assumed to occur. The conductivity-modulated portion of the drift region then has a distance y_M as shown in the figure. This distance can be obtained by using (9.255):

$$y_\mathrm{M}(t) = \sqrt{4D_\mathrm{n}t}\,\ln\left[\frac{A(t)}{N_\mathrm{M} - N_\mathrm{D}}\right]. \tag{9.258}$$

The expansion of this conductivity-modulated region with time during the ramp-up of the current is shown in Fig. 9.92 for three cases of the ramp rates when a value of 1×10^{14} cm^{-3} was used for the "modulation concentration" (N_M). In each case, the conductivity-modulated region is formed after a time delay during which the current increases to the level required to achieve high-level injection in the drift region. This time delay is therefore shorter for the faster ramp rates. The conductivity-modulated region then grows as shown in the figure. A faster rate of growth occurs with increasing ramp rates. In the case of the ramp rate of 5×10^9 A cm^{-2} s^{-1}, a slight change in the rate of growth of the conductivity-modulated region occurs at 20 ns when the collector current density becomes constant at the on-state value (100 A cm^{-2}).

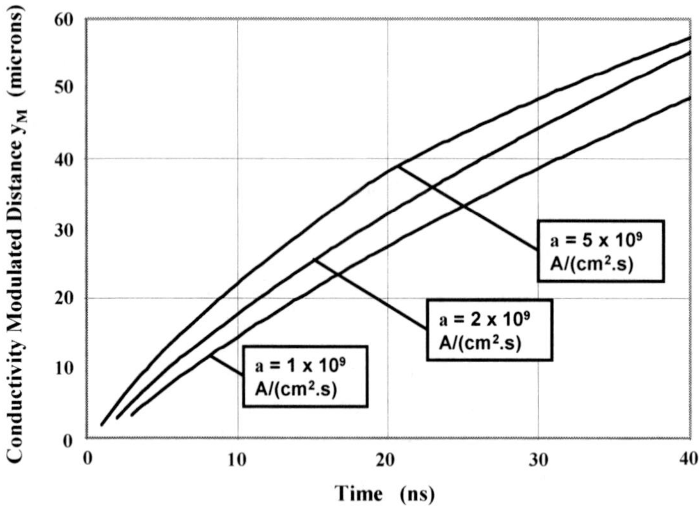

Fig. 9.92 Growth of the conductivity-modulated region during the forward recovery process

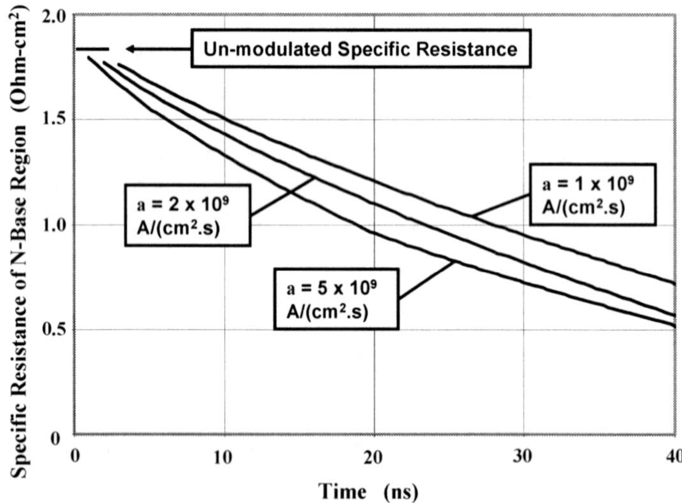

Fig. 9.93 Decrease of the N-base region resistance during the forward recovery process

The resistance of the N-base region decreases as the conductivity-modulated portion expands with time as shown in Fig. 9.93 for the three cases of the ramp rate. In this IGBT structure, the drift region had a width of 200 μm with a doping concentration of 5×10^{13} cm^{-3} resulting in an unmodulated specific resistance of 1.84 Ω cm^2. The voltage drop across the drift region ($v_D(t)$) can be obtained by multiplying the specific resistance of the unmodulated region and the current density pertaining to each time instant:

$$v_D(t) = R_{\text{N-Base}}(t) J_F(t). \tag{9.259}$$

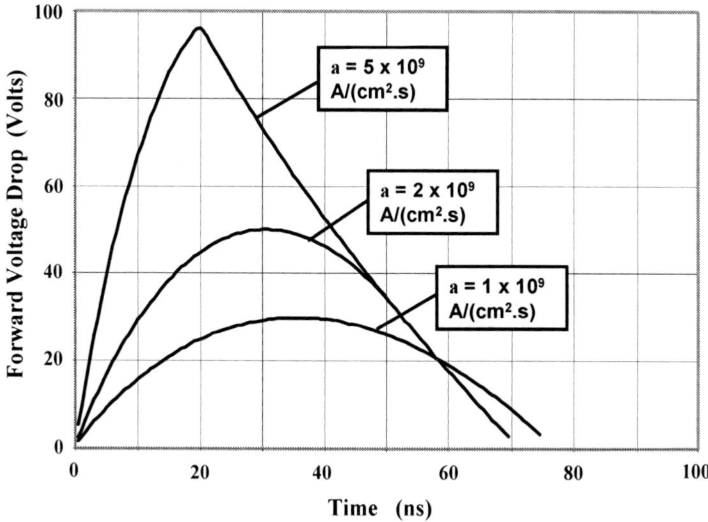

Fig. 9.94 Voltage drop across the symmetric IGBT structure during the forward recovery process

The total voltage drop across the symmetric IGBT structure during the forward recovery transient consists of the junction voltage drop plus the voltage drop across the unmodulated portion of the drift region. The evolution of the forward voltage drop across the symmetric IGBT structure with time during the transient as described by the analytical model is shown in Fig. 9.94. The portion of the transient below 1 V is not shown because the model is based upon high-level injection in the drift region. The maximum forward voltage drop (or overshoot voltage) can be observed to become larger with increasing rate of rise of the current density. For this symmetric IGBT structure, the maximum forward voltage drop increases from 30 to 50 to 95 V as the ramp rate increases from 1×10^9 to 2×10^9 to 5×10^9 A cm^{-2} s^{-1}. The time at which the maximum forward voltage drop is observed for each of these cases decreases from 35 to 30 to 20 ns. According to the model, the voltage overshoot is not a function of the minority carrier lifetime.

Simulation Example

To validate the above model for the forward recovery transient in the symmetric IGBT structure, the results of numerical simulations for the typical 1,200-V structure described in the previous sections are provided here. The structure had a drift region thickness of 200 μm with a doping concentration of 5×10^{13} cm^{-3}. The lifetime (τ_{p0} and τ_{n0}) in the drift region was 10 μs. The cathode current was ramped from zero to a steady-state value of 100 A cm^{-2} using ramp rates of 1×10^9, 2×10^9, and 5×10^9 A cm^{-2} s^{-1}. The voltage drop across the rectifier during the transient is shown in Fig. 9.95 for the three cases. The maximum forward voltage drops observed for the ramp rates of 1×10^9, 2×10^9, and 1×10^{10} A cm^{-2} s^{-1} are 30, 46, and 82 V, respectively. The peak in the voltage overshoot occurs at 30, 25, and 15 ns for the three cases. The predictions of the analytical model are in good agreement with these values providing validation of the model.

The electron carrier concentration in the drift region can be extracted from the transient simulations at various points in time. As an example, the electron concentration profiles are shown in Fig. 9.96 for the case of a ramp rate of 2×10^9 A cm^{-2} s^{-1}. The electron concentration exhibits an exponential distribution when proceeding away from the P–N junction with the carrier concentration increasing with time. As time progresses during the turn-on transient, the electron density modulates a greater portion of the drift region. The electron distribution predicted by the analytical model (see Fig. 9.90) is consistent with that observed with the numerical simulations.

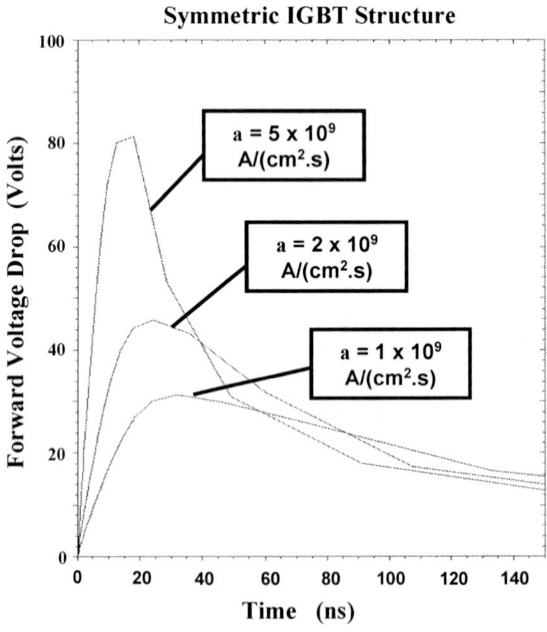

Fig. 9.95 Forward voltage overshoot in the symmetric IGBT structure

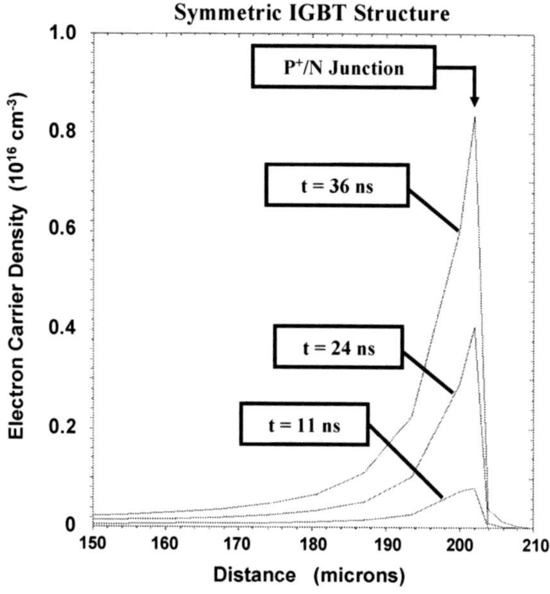

Fig. 9.96 Carrier distributions in the symmetric IGBT structure during the forward recovery process

9.7.2 Turn-Off Physics: No-Load Conditions

Consider an IGBT structure operating in its on-state with no load connected in series with a DC voltage source. When the gate bias voltage for the IGBT structure is abruptly reduced to zero, the collector current will initially exhibit an abrupt reduction followed by a current tail. The initial abrupt reduction of the collector current is associated with the termination of the channel current of the MOSFET portion of the IGBT structure.[29] This is followed by a bipolar current associated with the removal of the stored charge (holes) within the N-base region. Typical waveforms for the collector voltage and current under no-load conditions are illustrated in Fig. 9.97. The collector voltage remains at the on-state voltage drop throughout the no-load turn-off duration.

The collector current waveform observed during the no-load turn-off conditions can be analyzed by examination of the components of the current flowing within the IGBT structure. The electron current component (I_{CH}) flowing through the channel of the MOSFET portion in the IGBT structure serves to provide the base drive current ($I_{B,PNP}$) for the internal P–N–P transistor. Since the collector current of the IGBT structure is the emitter current for the P–N–P transistor, the base current of the P–N–P transistor can be related to the collector current for the IGBT structure by

$$I_C = \frac{I_{B,PNP}}{1 - \alpha_{PNP,0}}. \tag{9.260}$$

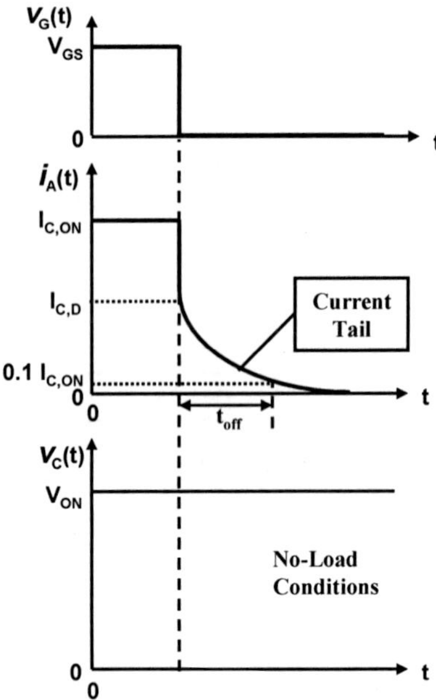

Fig. 9.97 Typical waveforms for no-load turn-off for an IGBT structure

The current gain ($\alpha_{\mathrm{PNP,0}}$) of the P–N–P transistor at low collector bias voltages is given by

$$\alpha_{\mathrm{PNP,0}} = \gamma_{\mathrm{E}} \alpha_{\mathrm{T,0}}. \tag{9.261}$$

In this expression, the injection efficiency under high-level injection conditions in the N-base region can be obtained by using (9.159). The base transport factor for the P–N–P transistor cannot be obtained by using the low-level injection theory for a bipolar transistor (i.e., (7.75)). For the symmetric IGBT structure and the asymmetric IGBT structure with relatively low N-buffer layer doping levels, the base transport factor is associated with current transport through the N-base region at high injection levels. The base transport factor for the N-base region at low collector bias voltages can be obtained by using (9.181).

For the asymmetric IGBT structure discussed in the previous sections, the emitter injection efficiency and base transport factor at low collector bias voltages are 0.61 and 0.81 at an on-state current density of 100 A cm^{-2}. This leads to a current gain of 0.5 for the P–N–P transistor. Using (9.260), the channel current in the IGBT is found to be about 50 A cm^{-2} for this case. During turn-off with no-load conditions, the collector current will then abruptly reduce by about half followed by the current tail. The time constant for the current tail is determined by the lifetime in the N-buffer layer because it is controlled by the rate of recombination of the stored charge.

Simulation Example

To validate the above analytical model for the turn-off transient for the IGBT structure, the results of numerical simulations for the typical 1,200-V asymmetric structure described in the previous sections are provided here. The structure had an N-base region thickness of 100 μm with a doping concentration of 5×10^{13} cm^{-3} for the lightly doped portion, while the N-buffer layer had a thickness of 10 μm with a doping concentration of 1×10^{17} cm^{-3}. The lifetime (τ_{p0} and τ_{n0}) in the drift region was 1 μs. The device was turned off from a steady-state collector current of 100 A cm^{-2} under no-load conditions by ramping the gate voltage to zero in 10 ns. The waveforms for the gate voltage and the collector current are shown in Fig. 9.98. The collector current begins to reduce after the gate bias falls below the threshold voltage. There is sharp drop in collector current as the gate bias falls to zero followed by a long current tail. The analytical model provided in this section provides an accurate description of this turn-off process.

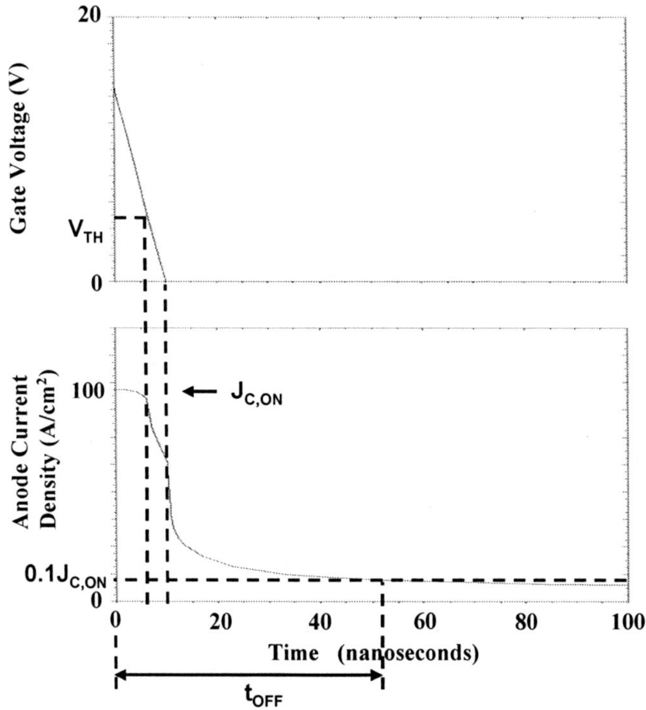

Fig. 9.98 Turn-off waveforms under no-load conditions for the asymmetric IGBT structure

9.7.3 Turn-Off Physics: Resistive Load

The IGBT device is sometimes used to control power delivered to resistive loads, such as burners on a stove or the heating element in a space heater. The basic power circuit consists of the IGBT and the resistive load connected in series with the power source. Based upon the application of Kirchhoff's voltage law for the circuit and Ohms law for the resistor, it can be concluded that the IGBT current and voltage waveforms will be mirror images of each other as shown in Fig. 9.99.

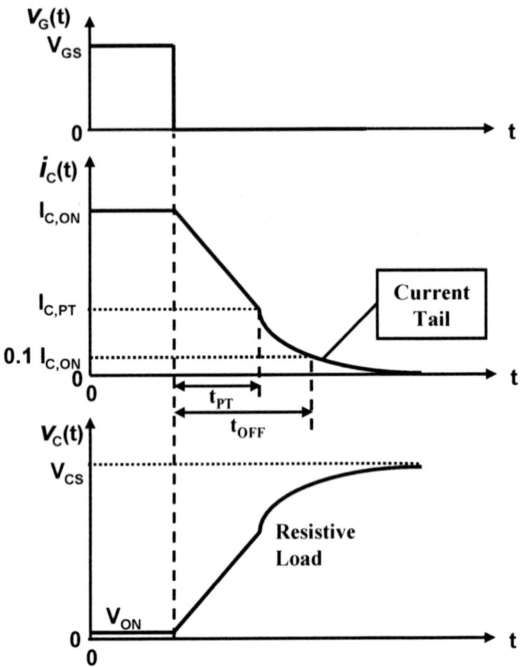

Fig. 9.99 Typical waveforms for resistive load turn-off for an asymmetric IGBT structure

During the turn-off process, the gate bias is abruptly reduced to zero as shown in Fig. 9.99. This stops the electron component of the current that is being delivered via the channel of the MOSFET portion in the IGBT structure. However, the collector current cannot change abruptly with a resistive load because the IGBT structure cannot abruptly support a high voltage due to the presence of the large density of stored charge in the N-base region. Consequently, the entire initial on-state collector current is carried by holes immediately after the gate bias is reduced to zero. The voltage is then supported within the IGBT structure by the formation of a space-charge region at the P-base/N-base junction (J_2). Due to the substantial collector current flowing through the device during the initial stages of the turn-off process, a large concentration of holes is present in the space-charge layer. This must be accounted for in modeling the width of the space-charge layer. The space-charge layer expands with time allowing the device to support larger collector voltages. This produces a corresponding reduction of the collector current. In the case of the resistive load, the collector current decreases approximately linearly with time until the space-charge region punches through to the N-buffer layer for an asymmetric IGBT structure. Once punch-through occurs, the remaining stored charge in the N-buffer layer is removed by recombination at the relatively small minority carrier lifetime in the buffer layer due to its high doping concentration.

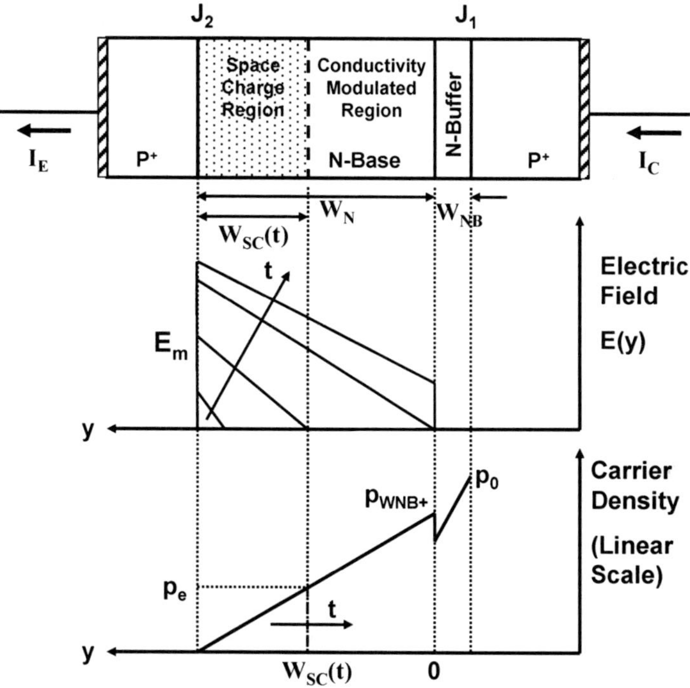

Fig. 9.100 Stored charge and electric field distributions for resistive load turn-off conditions in an asymmetric IGBT structure

The analysis of the turn-off waveforms for the asymmetric IGBT structure can be performed by using the charge control principle. In this analysis, it will be assumed that recombination in the N-base region can be neglected in the on-state. This results in a linear free carrier (hole) distribution within the lightly doped portion of the N-base region during on-state operation as illustrated in Fig. 9.100 (The more complex hole distribution obtained for the asymmetric IGBT structure in Sect. 9.5.4 with finite recombination in the lightly doped portion of the N-base region is nearly linear in shape.):

$$p(y) = p_{\mathrm{WNB+}}\left(1 - \frac{y}{W_{\mathrm{N}}}\right). \tag{9.262}$$

In writing this expression, the hole concentration is assumed to be approximately zero at the edge of the space-charge region during the on-state and the space-charge layer width is assumed to be zero due to the small on-state voltage drop of the IGBT structure. The concentration p_{WNB+} in the on-state at the interface between the lightly doped portion of the N-base region and the N-buffer layer was previously derived in Sect. 9.5.4 (see (9.97)).

It will be assumed that the above hole distribution does not change during the turn-off process in the conductivity-modulated portion of the N-base region. Consequently, the concentration of holes at the edge of the space-charge region (p_e in Fig. 9.100) increases during the turn-off process as the space-charge width increases:

$$p_e(t) = p_{WNB+}\left[\frac{W_{SC}(t)}{W_N}\right]. \tag{9.263}$$

According to the charge control principle, the charge removed by the expansion of the space-charge layer must equal the charge removed due to collector current flow:

$$J_C(t) = qp_e(t)\frac{dW_{SC}(t)}{dt} = qp_{WNB+}\left[\frac{W_{SC}(t)}{W_N}\right]\frac{dW_{SC}(t)}{dt} \tag{9.264}$$

by using (9.263).

The width of the space-charge layer is dependent upon the collector voltage and the space-charge density:

$$W_{SC}(t) = \sqrt{\frac{2\varepsilon_S V_C(t)}{q[N_D + p_{SC}(t)]}}. \tag{9.265}$$

Although the hole concentration in the space-charge layer was neglected when compared with the concentration p_{WNB+} when writing (9.262), it cannot be neglected when determining the space-charge layer width because it is comparable with the doping concentration in the lightly doped portion of the N-base region. The hole concentration in the space-charge layer can be related to the collector current density under the assumption that the carriers are moving at the saturated drift velocity in the space-charge layer:

$$p_{SC}(t) = \frac{J_C(t)}{qv_{sat,p}}. \tag{9.266}$$

Since the hole concentration in the space-charge region decreases with time due to a reduction of the collector current density, the slope of the electric field profile in the space-charge region also becomes smaller with time as illustrated in Fig. 9.100.

Based upon the application of Kirchhoff's voltage law for the circuit and Ohms law for the load resistor, the collector voltage for the IGBT is given by

$$V_C(t) = V_{CS} - J_C(t)R_L, \tag{9.267}$$

where R_L is the specific load resistance given by

$$R_L = \frac{V_{CS}}{J_{C,ON}} \tag{9.268}$$

if the collector supply voltage is much larger than the on-state voltage drop of the IGBT structure.

Using these relationships in (9.265),

$$W_{SC}(t) = \sqrt{\frac{2\varepsilon_S v_{sat,p}[V_{CS} - J_C(t)R_L]}{qN_D v_{sat,p} + J_C(t)}}. \tag{9.269}$$

Solving for the collector current density,

$$J_C(t) = \frac{2\varepsilon_S v_{sat,p} V_{CS} - qN_D v_{sat,p} W_{SC}^2}{W_{SC}^2 + 2\varepsilon_S v_{sat,p} R_L}. \tag{9.270}$$

Substituting this expression into (9.264) yields a differential equation for the space-charge layer width:

$$qp_{WNB+}\left[\frac{W_{SC}(t)}{W_N}\right]\frac{dW_{SC}(t)}{dt} = \frac{2\varepsilon_S v_{sat,p} V_{CS} - qN_D v_{sat,p} W_{SC}^2}{W_{SC}^2 + 2\varepsilon_S v_{sat,p} R_L}. \tag{9.271}$$

The solution for this equation has the form

$$W_{SC}(t) = K_1 + K_2 e^{-t/\tau_R}. \tag{9.272}$$

Applying the boundary condition that the space-charge layer width is zero at the start of the turn-off process yields

$$W_{SC}(t) = K_1(1 - e^{-t/\tau_R}). \tag{9.273}$$

The coefficients K_1 and τ_R can be obtained by substituting the above expression into the differential equation:

$$K_1 = \sqrt{\frac{2\varepsilon_S V_{CS}}{qN_D}} = W_D(V_{CS}) \tag{9.274}$$

and

$$\tau_R = \frac{p_{WNB+}\varepsilon_S(V_{CS} + qv_{sat,p}R_L N_D)}{2qv_{sat,p}W_N N_D^2}. \tag{9.275}$$

As indicated in (9.274), the coefficient K_1 is found to be equal to the depletion layer width corresponding to the collector bias supply voltage for a nonpunch-through diode structure. Based upon the solution in (9.273), the rate of increase in the space-charge layer width becomes smaller with time. This phenomenon is associated with the larger hole concentration (p_e) at the edge of the space-charge layer as time progresses as well as the reduction of the collector current density with time.

The variation of the collector current density as a function of time can be derived by substituting (9.273) into (9.270):

$$J_C(t) = \frac{qv_{sat,p}N_D V_{CS}[1-(1-e^{-t/\tau_R})^2]}{V_{CS}(1-e^{-t/\tau_R})^2 + qv_{sat,p}N_D R_L}. \tag{9.276}$$

This expression is valid until the time at which the space-charge layer punches through to the N-buffer layer. According to the analytical model, the collector current density waveform has a dependence on the collector bias supply voltage even if the initial value ($J_{C,ON}$) is kept the same by adjusting the load resistance.

The punch-through time (t_{PT}) can be obtained by using (9.273) with the space-charge layer width equal to the width (W_N) of the lightly doped portion of the N-base region:

$$t_{PT} = \tau_R \ln\left[\frac{W_D(V_{CS})}{W_D(V_{CS}) - W_N}\right]. \tag{9.277}$$

According to the analytical model, the punch-through time is dependent upon the collector bias supply voltage.

The collector current density at the on-set of the punch-through condition can be obtained by making the space-charge layer width equal to the width of the lightly doped portion of the N-base region in (9.270):

$$J_{C,PT} = \frac{2\varepsilon_S v_{sat,p} V_{CS} - qN_D v_{sat,p} W_N^2}{W_N^2 + 2\varepsilon_S v_{sat,p} R_L}. \tag{9.278}$$

According to the analytical model, the punch-through collector current density is dependent on the collector bias supply voltage even if the initial value ($J_{C,ON}$) is kept the same by adjusting the load resistance.

After the space-charge layer punches through to the N-buffer layer, the remaining stored charge in the N-buffer layer is removed by the recombination process because the electric field cannot penetrate into the N-buffer layer. The recombination of holes in the N-buffer layer occurs under low-level injection conditions resulting in an exponential decay in the collector current:

$$J_C(t) = J_{C,PT} e^{-t/\tau_{p0,NB}}, \tag{9.279}$$

where $\tau_{p0,NB}$ is the minority carrier lifetime in the N-buffer layer. This lifetime is much smaller than in the lightly doped portion of the N-base region due to its higher doping concentration ($N_{D,NB}$):

$$\tau_{p0,NB} = \frac{\tau_{p0,N}}{1 + (N_{D,NB} / N_S)}, \tag{9.280}$$

where $\tau_{p0,N}$ is the minority carrier lifetime in the N-base region and N_S is a lifetime scaling factor.

The turn-off time is defined as the time taken for the collector current density to reduce to one tenth of its initial on-state value. Under resistive load switching conditions, the turn-off time can be obtained by using (9.279):

$$\tau_{off,R} = \tau_{p0,NB} \ln\left(\frac{J_{C,PT}}{0.1 J_{C,ON}}\right). \tag{9.281}$$

According to the analytical model, the turn-off time is dependent upon the collector bias supply voltage because τ_R and $J_{C,PT}$ are a function of the collector bias supply voltage.

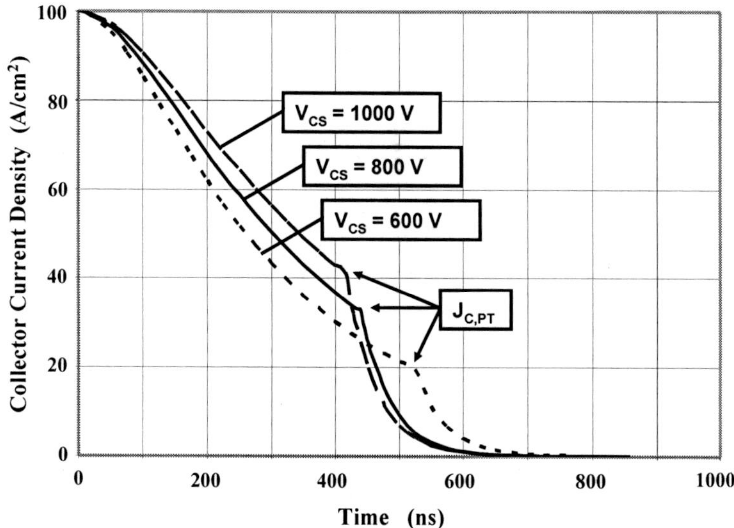

Fig. 9.101 Collector current turn-off waveforms for resistive load turn-off at various collector supply voltages for an asymmetric IGBT structure

As an example, the collector current turn-off waveforms computed by using the analytical model for the typical asymmetric IGBT structure with an N-buffer layer doping concentration of 1×10^{17} cm^{-3} are provided in Fig. 9.101. The width of the lightly doped portion of the N-base region for this structure is 100 μm and the minority carrier lifetime in this portion of the N-base region is 1 μs. The turn-off was performed from an on-state current density of 100 A cm^{-2} with the specific load resistance adjusted in accordance with (9.268) for each collector bias supply voltage. It can be seen from the figure that the collector current decreases approximately linearly with time until the advent of punch-through of the space-charge layer to the N-buffer layer. The current then decreases rapidly in an exponential manner. The punch-through time and the punch-through collector current density are dependent upon the collector bias supply voltage.

Simulation Example

To validate the above model for the turn-off transient for the asymmetric IGBT structure under resistive load conditions, the results of numerical simulations for the typical 1,200-V asymmetric structure described in the previous sections are provided here. The structure had an N-base region thickness of 100 μm with a doping concentration of 5×10^{13} cm^{-3} for the lightly doped portion, while the N-buffer layer had a thickness of 10 μm with a doping concentration of 1×10^{17} cm^{-3}. The lifetime (τ_{p0} and τ_{n0}) in the N-base region was 1 μs. The device was turned off from a steady-state collector current of 100 A cm^{-2} with various collector bias supply voltages (and corresponding resistive loads) by ramping the gate voltage to zero in 10 ns.

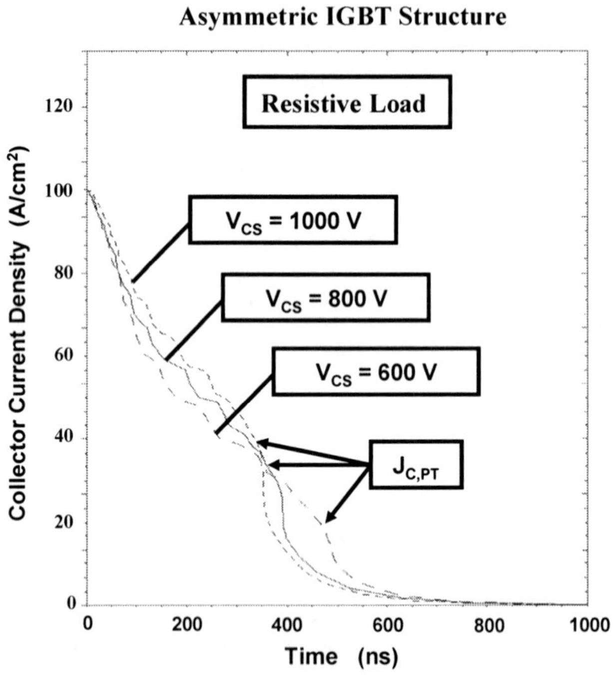

Fig. 9.102 Collector current turn-off waveforms under resistive load conditions for the asymmetric IGBT structure

The waveforms for the collector current are shown in Fig. 9.102 for three values for the collector bias supply voltage. The collector current decreases approximately linearly with time until the punch-through time. In this segment, there is a weak dependence on the collector bias supply voltage. The collector current then decreases exponentially at a much more rapid rate. The collector current waveforms predicted by the analytical model (see Fig. 9.101) are in remarkably good agreement with the shape and dependence on the collector bias supply voltage in spite of the various approximations that were used for developing the model.

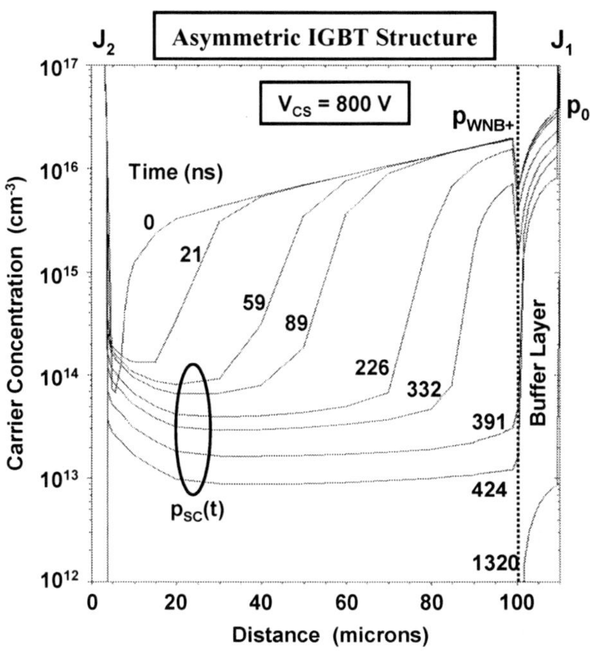

Fig. 9.103 Hole distribution during resistive load turn-off conditions for the asymmetric IGBT structure

The changes in the distribution of holes within the asymmetric IGBT structure are shown in Fig. 9.103 at various time instances during turn-off with a collector bias supply voltage of 800 V. During the initial 220 ns of the turn-off process, the hole concentration decreases within the space-charge region but remains the same as the initial on-state concentration in the conductivity-modulated portion. This validates the assumption used in the analytical model (see Fig. 9.100) for the removal of charge from the N-base region. It can also be observed that the hole concentration (p_{SC}) in the space-charge region decreases with time as predicted by the analytical model. The space-charge layer punches through to the N-buffer layer at 400 ns. This is in good agreement with the predictions of the analytical model as well (see Fig. 9.101). After the space-charge layer punches through to the N-buffer layer, the holes remaining in the N-buffer layer decrease rapidly due to recombination.

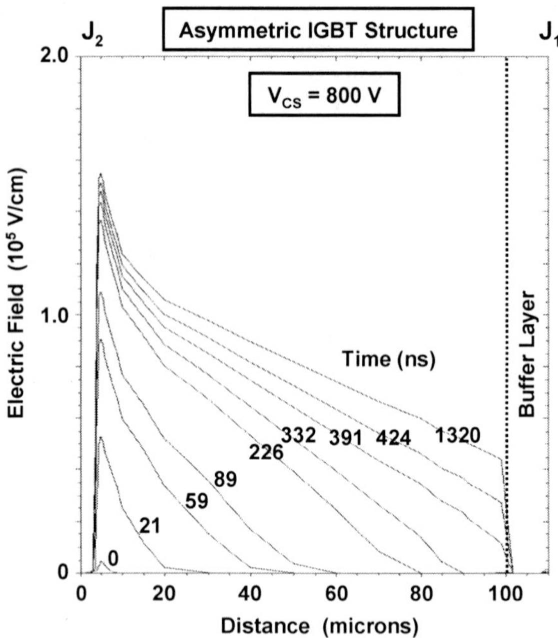

Fig. 9.104 Electric field distribution during resistive load turn-off conditions for the asymmetric IGBT structure

The changes in the distribution of the electric field within the asymmetric IGBT structure are shown in Fig. 9.104 at various time instances during turn-off with a collector bias supply voltage of 800 V. It can be observed that the slope of the electric field profile decreases with time because of the smaller hole concentration in the space-charge layer. The analytical model (see Fig. 9.100) takes this into account when analyzing the rate of spreading of the space-charge layer. The electric field punches through to the N-buffer layer at about 400 ns. These simulation results validate the assumptions used in deriving the analytical model.

9.7.4 Turn-Off Physics: Inductive Load

The IGBT device is most often used to control power delivered to inductive loads, such as the winding of motors used in a wide variety of consumer and industrial applications. Variable speed motor control using IGBT devices has become increasing commonplace due to improvements in efficiency by as much as 25% over constant speed motor control with dampers for regulating loads.[27] The basic power circuit consists of the IGBT and the inductive load connected in series with the power source, with a clamping diode across the load to transfer the current when the IGBT is turned off. For an inductive load, the voltage across the IGBT structure must first increase to the collector bias supply voltage (plus a small diode on-state voltage drop) before the collector current can begin to reduce and get transferred to the diode.

Due to the widespread application of IGBT structures for this type of application, many papers have been published on the analysis and optimization of the three basic types of IGBT structures for inductive load applications. Some examples are papers on the symmetric (or nonpunch-through) structures,[30] the asymmetric (or punch-through) structure,[31] and the transparent emitter structure.[23–25] The papers provide an empirical description of the basic turn-off waveforms for each of the structures and the resulting power losses. Due to the significant differences between the performance of the three structures and the utilization of all of them for inductive load applications, the analysis of all three structures is included in this section.

Symmetric IGBT Structure

The IGBT collector current and voltage turn-off waveforms for an inductive load are illustrated in Fig. 9.105 for all of the IGBT structures. When the gate voltage is abruptly switched off, the collector current continues to flow at the on-state value due to the inductive load. This current must be sustained by a bipolar current since the channel current has ceased. Initially, the collector voltage increases in a linear manner until it reaches the collector supply voltage (plus a diode drop). A large amount of power dissipation occurs during the voltage rise time because of the

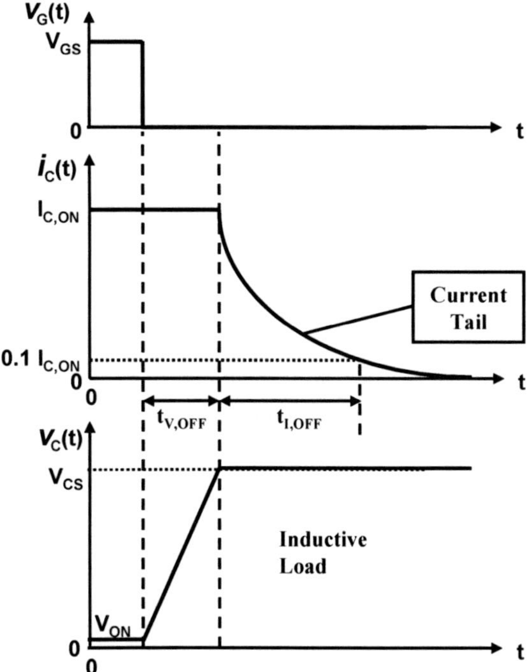

Fig. 9.105 Typical waveforms for inductive load turn-off for an IGBT structure

high collector current. When the collector voltage reaches the collector bias supply voltage, the current begins to transfer from the IGBT structure to the diode. The current then decreases exponentially with time limited by the recombination of the large amount of stored charge in the N-base region near the P^+ collector/N-base junction. The long tail in the collector current produces large power dissipation because of the high collector voltage during this phase of the turn-off process.

During the first phase of the turn-off process, the entire on-state collector current is carried by holes immediately after the gate bias is reduced to zero. The voltage is then supported within the IGBT structure by the formation of a space-charge region at the P-base/N-base junction (J_2). Due to the substantial collector current flowing through the device during the initial stages of the turn-off process, a large concentration of holes is present in the space-charge layer. This must be accounted for in modeling the width of the space-charge layer. The space-charge layer expands with time allowing the device to support larger collector voltages. For the symmetric IGBT structure, the space-charge layer cannot extend through the entire N-base region when the collector bias voltage reaches the collector supply voltage. Consequently, a large concentration of holes is still present in the conductivity-modulated portion of the N-base region. These holes must be removed by recombination under high-level injection conditions producing a long current tail as shown in the figure.

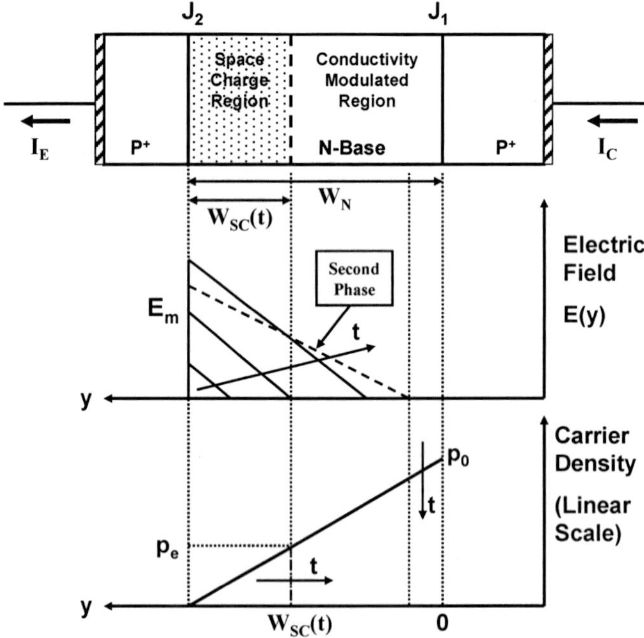

Fig. 9.106 Stored charge and electric field distributions for inductive load turn-off conditions in a symmetric IGBT structure

The analysis of the turn-off waveforms for the symmetric IGBT structure can be performed by using the charge control principle. In this analysis, it will be assumed that recombination in the N-base region can be neglected in the on-state. This results in a linear free carrier (hole) distribution within the lightly doped portion of the N-base region during on-state operation as illustrated in Fig. 9.106 (The more complex hole distribution obtained for the symmetric IGBT structure in Sect. 9.5.2 with finite recombination in the lightly doped portion of the N-base region is nearly linear in shape.):

$$p(y) = p_0\left(1 - \frac{y}{W_N}\right). \tag{9.282}$$

In writing this expression, the hole concentration is assumed to be approximately zero at the edge of the space-charge region during the on-state and the space-charge layer width is assumed to be zero due to the small on-state voltage drop of the IGBT structure. The concentration p_0 in the on-state at the P^+ collector/N-base junction was previously derived in Sect. 9.5.2 (see (9.70)).

It will be assumed that the above hole distribution does not change during the first phase of the turn-off process in the conductivity-modulated portion of the N-base region. Consequently, the concentration of holes at the edge of the space-charge region (p_e) increases during the turn-off process as the space-charge width increases:

$$p_e(t) = p_0\left[\frac{W_{SC}(t)}{W_N}\right]. \tag{9.283}$$

According to the charge control principle, the charge removed by the expansion of the space-charge layer must equal the charge removed due to collector current flow:

$$J_{C,ON} = qp_e(t)\frac{dW_{SC}(t)}{dt} = qp_0\left[\frac{W_{SC}(t)}{W_N}\right]\frac{dW_{SC}(t)}{dt} \tag{9.284}$$

by using (9.283). Unlike the resistive load turn-off case, the collector current density is constant at its on-state value during the voltage rise time. Integrating this equation and applying the boundary condition of zero width for the space-charge layer at time zero provides the solution for the evolution of the space-charge region width with time:

$$W_{SC}(t) = \sqrt{\frac{2W_N J_{C,ON} t}{qp_0}}. \tag{9.285}$$

The space-charge layer expands toward the right-hand side as indicated by the horizontal time arrow in Fig. 9.106 while the hole concentration in the conductivity-modulated portion remains unchanged.

The collector voltage supported by the symmetric IGBT structure is related to the space-charge layer width by

$$V_C(t) = \frac{q(N_D + p_{SC})W_{SC}^2(t)}{2\varepsilon_S}. \tag{9.286}$$

Although the hole concentration in the space-charge layer was neglected when compared with the concentration p_0 when writing (9.282), it cannot be neglected when determining the space-charge layer width because it is comparable with the doping concentration in the lightly doped portion of the N-base region. The hole concentration in the space-charge layer can be related to the collector current density under the assumption that the carriers are moving at the saturated drift velocity in the space-charge layer:

$$p_{SC} = \frac{J_{C,ON}}{qv_{sat,p}}. \tag{9.287}$$

Unlike the resistive load case, the hole concentration in the space-charge region remains constant during the voltage rise time because the collector current density is constant. Consequently, the slope of the electric field profile in the space-charge region also becomes independent of time as illustrated in Fig. 9.106.

Applying the solution for the evolution of the space-charge layer from (9.285) in (9.286),

$$V_C(t) = \frac{W_N(N_D + p_{SC})J_{C,ON}}{\varepsilon_S p_0}t. \tag{9.288}$$

The analytical model for turn-off of the symmetric IGBT structure under inductive load conditions predicts a linear increase in the collector voltage with time.

The end of the first phase of the turn-off process, where the collector voltage increases while the collector current remains constant, occurs when the collector voltage reaches the collector supply voltage (V_{CS}). This time interval ($t_{V,OFF}$) can be obtained by making the collector voltage equal to the collector supply voltage in (9.288):

$$t_{V,OFF} = \frac{\varepsilon_S p_0 V_{CS}}{W_N(N_D + p_{SC})J_{C,ON}}. \tag{9.289}$$

According to the analytical model, the voltage rise time is proportional to the collector bias supply voltage. However, it is only weakly dependent on the on-state current density (through p_{SC}) because the hole concentration p_0 is proportional to the on-state current density.

The width of the space-charge layer at the end of the voltage transient can be obtained by using the collector supply voltage:

$$W_{SC}(t_{V,OFF}) = \sqrt{\frac{2\varepsilon_S V_{CS}}{q(N_D + p_{SC})}}. \tag{9.290}$$

The width of the space-charge layer at the end of the first phase depends upon the collector supply voltage and the initial on-state current density (via p_{SC}). To avoid reach-through breakdown, this width must be less than the width (W_N) of the N-base region.

During the second phase of the turn-off process, the decay of the collector current after the voltage rise time is governed by the recombination of the excess holes and electrons that are trapped within the N-base region in the vicinity of the P^+ collector/N-base junction because the space-charge layer cannot extend through the entire N-base region. This is indicated by the vertical time arrow in Fig. 9.106. In the absence of diffusion, the continuity equation for holes in the N-base region is given by

$$\frac{d\delta p_N}{dt} = -\frac{\delta p_N}{\tau_{HL}}, \tag{9.291}$$

where δp_N is the excess hole concentration in the N-base region. The solution for this equation is

$$\delta p_N(t) \approx p_N(t) = p_0 e^{-t/\tau_{HL}} \tag{9.292}$$

because high-level injection conditions prevail in the N-base region. The concentration of holes in the stored charge region at the beginning of the second phase has been assumed to be equal to the hole concentration (p_0) at the P^+ collector/N-base junction due to the injection of carriers in the on-state.

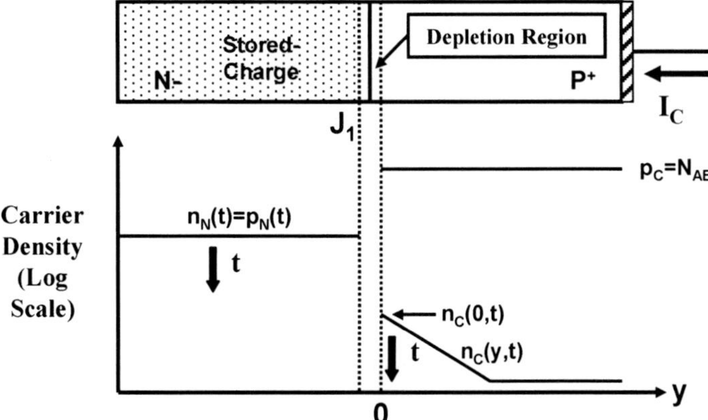

Fig. 9.107 Free carrier distribution at the P^+ collector/N-base junction during turn-off for the symmetric IGBT structure

The collector current flow that supports the recombination of carriers within the stored charge region can be analyzed by examination of the carrier distribution on both sides of the P^+ collector/N-base junction (J_1). The free carrier concentration in the N-base region in the vicinity of junction J_1 can be assumed to be independent of distance but a function of time as given by (9.292). The high concentration of electrons in the N-base region produces the injection of electrons into the P^+ collector region as indicated in Fig. 9.107. These injected electrons diffuse away from the junction producing an exponential decay in concentration as shown in the figure.

The free carrier concentrations on the two sides of the junction are related by

$$\frac{p_C}{p_N(t)} = \frac{n_N(t)}{n_C(0,t)} = e^{qV_C/kT}, \tag{9.293}$$

where $p_N(t)$ and $n_N(t)$ are the hole and electron concentrations in the N-base region at junction J_1 given by (9.292), p_C is the hole concentration in the P^+ collector region, and $n_C(0, t)$ is the electron concentration at junction J_1 in the P^+ collector region. The hole concentration in the P^+ collector region can be assumed to be equal to its acceptor doping concentration (N_{AE}). Using this in (9.293) yields

$$n_C(0,t) = \frac{n_N(t)p_N(t)}{N_{AE}} = \frac{p_N^2(t)}{N_{AE}}. \tag{9.294}$$

The electrons diffuse into the P^+ collector region with an exponential distribution due to low-level injection conditions:

$$n_C(y,t) = n_C(0,t)e^{-y/L_{nE}}, \tag{9.295}$$

where L_{nE} is the diffusion length for electrons in the P^+ collector region.

The collector current is produced by the diffusion of the injected electrons in the P^+ collector side of the junction:

$$J_C(t) = -qD_{nE}\left.\frac{\partial n_C(y,t)}{\partial y}\right|_{y=0} = \frac{qD_{nE}n_C(0,t)}{L_{nE}} \tag{9.296}$$

by using (9.295). Making use of (9.292) and (9.294) yields

$$J_C(t) = \frac{qD_{nE}p_N^2(t)}{L_{nE}N_{AE}} = \frac{qD_{nE}p_0^2}{L_{nE}N_{AE}}e^{-2t/\tau_{HL}}. \tag{9.297}$$

This equation indicates that the collector current varies as the square of the carrier density in the stored charge region during the collector current tail. Since the collector current density at the beginning of the second phase is equal to the on-state current density, the equation can also be written as

$$J_C(t) = J_{C,ON} e^{-2t/\tau_{HL}}. \tag{9.298}$$

Consequently, the collector current tail decreases exponentially with time with a time constant of one half of the high-level lifetime even though the free carrier density in the stored charge region is decreasing exponentially with time with a time constant equal to the high-level lifetime.

During the second phase of the turn-off process, the hole concentration (p_{SC}) in the space-charge layer decreases because the collector current density is becoming smaller. This reduces the net charge in the space-charge layer resulting in a change in the slope of the electric field profile as illustrated in Fig. 9.106 by the dashed line. This results in a small expansion of the space-charge layer during the second phase of the turn-off transient.

The collector current turn-off time ($t_{I,OFF}$) is defined as the time taken for the current to decay to one tenth of the on-state value as illustrated in Fig. 9.105. Based upon (9.298), the collector current turn-off time is given by

$$\tau_{I,OFF} = \frac{\tau_{HL}}{2} \ln(10) = 1.15\tau_{HL}. \tag{9.299}$$

According to the analytical model, the turn-off time is independent of the collector bias supply voltage and the on-state current density. It can be shortened by reducing the lifetime in the N-base region using the lifetime control techniques discussed in Chap. 2.

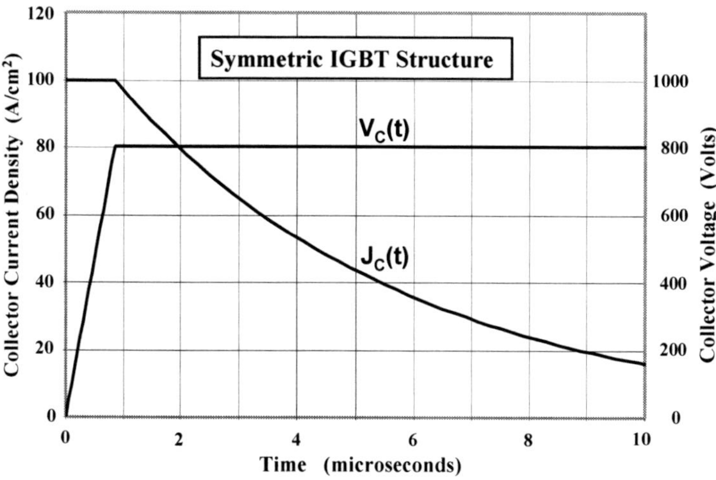

Fig. 9.108 Collector current and voltage transients during turn-off for the symmetric IGBT structure with an inductive load

Consider the case of a symmetric IGBT structure with N-base region with a width of 200 μm, a doping concentration of 5×10^{13} cm^{-3}, and a high-level lifetime of 10 μs. During the first phase of the turn-off process with an inductive load from an initial on-state collector current density of 100 A cm^{-2}, the hole concentration in the space-charge layer is 6.25×10^{13} cm^{-3}, which is comparable with the doping concentration. The presence of the hole charge must therefore be included when computing the width of the space-charge layer during the voltage transient. For this structure, the hole concentration (p_0) at the P$^+$ collector/N-base junction in the on-state is found to be 1.1×10^{17} cm^{-3}. The collector voltage transient obtained by using these values is provided in Fig. 9.108. The collector voltage rise time ($t_{V,OFF}$) is found to be 0.85 μs. After the voltage transient is completed, the collector current decays exponentially as shown in Fig. 9.108. The collector current turn-off time ($t_{I,OFF}$) is found to be 11.5 μs.

Simulation Example

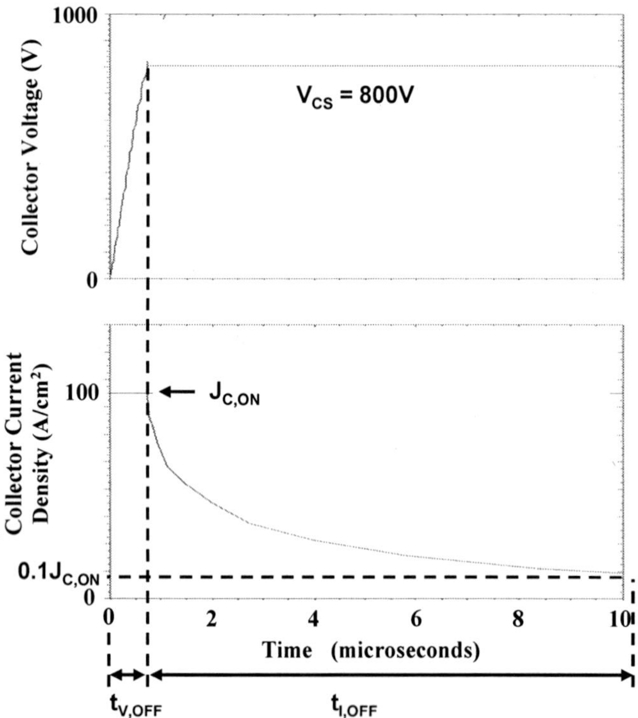

Fig. 9.109 Collector current and voltage turn-off waveforms under inductive load conditions for the symmetric IGBT structure

To validate the above model for the turn-off transient for the symmetric IGBT structure under inductive load conditions, the results of numerical simulations for the typical 1,200-V symmetric structure described in the previous sections are provided here. The structure had an N-base region thickness of 200 μm with a doping concentration of 5×10^{13} cm^{-3}. The lifetime (τ_{p0} and τ_{n0}) in the N-base region was 10 μs leading to a high-level lifetime (τ_{HL}) of 20 μs. The device was turned off from an on-state collector current of 100 A cm^{-2} with a collector bias supply voltage of 800 V by ramping the gate voltage to zero in 10 ns. During the first phase of the turn-off process, the collector current was held constant allowing the collector voltage to increase as a function of time. After the collector voltage reached 800 V, the collector voltage was held constant and the collector current was allowed to decay with time.

The waveforms for the collector current and voltage obtained from the numerical simulations are shown in Fig. 9.109. During the first phase of the turn-off process, the collector voltage increases linearly with time as predicted by the analytical model (see (9.288)). The collector voltage rise time obtained from the simulations is 0.74 μs. The value computed by using (9.289) for the collector voltage rise time is in good agreement with this observation. During the second phase of the turn-off process, the collector current decreases exponentially with time as predicted by the analytical model. The collector current fall time obtained from the simulations is 11.25 μs. The value computed by using (9.299) for the collector current fall time is in excellent agreement with this observation. This provides validation for the assumption in the model that the current turn-off process requires analysis of the injection of carriers into the P^+ collector region.

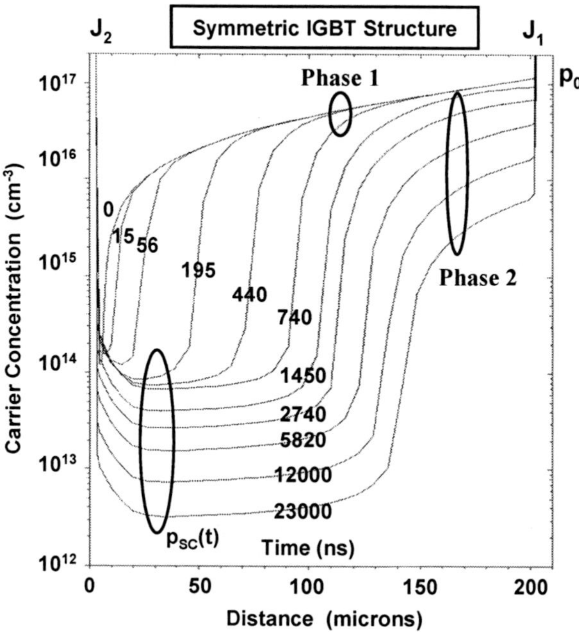

Fig. 9.110 Hole distribution during inductive load turn-off conditions for the symmetric IGBT structure

The changes in the distribution of holes within the symmetric IGBT structure are shown in Fig. 9.110 at various time instances during inductive load turn-off with a collector bias supply voltage of 800 V. During the first phase extending to 740 ns of the turn-off process, the hole concentration within the conductivity-modulated portion of the N-base region remains the same as determined by the initial on-state carrier distribution. This validates the assumption used in the analytical model (see Fig. 9.106) for the removal of charge from the N-base region. During this time period, the space-charge region expands with a constant hole concentration (p_{SC}) of about 6.5×10^{13} cm^{-3} as determined by the on-state collector current density. This provides validation for the analytical formulations developed for the space-charge concentration and space-charge layer width (see (9.287) and (9.290)). During the second phase, it can also be observed that the hole concentration (p_{SC}) in the space-charge region decreases with time as predicted by the analytical model. At the same time, the hole concentration in the conductivity-modulated portion also reduces with time. There is a relatively prolonged time interval for the removal of the stored charge in the conductivity-modulated portion of the N-base region as limited by the recombination lifetime.

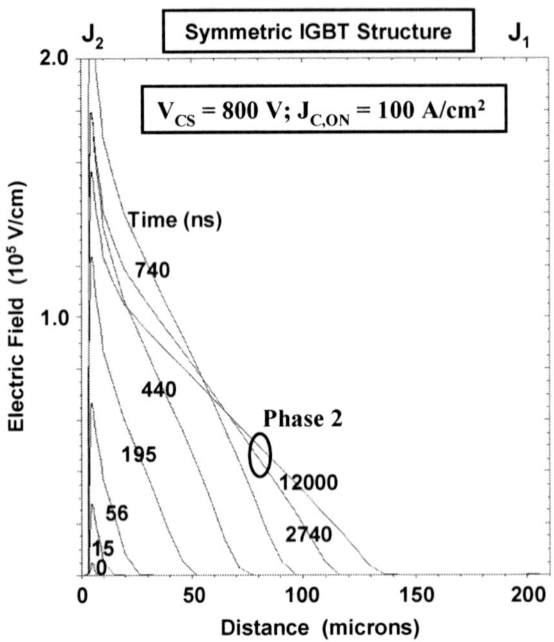

Fig. 9.111 Electric field distribution during inductive load turn-off conditions for the symmetric IGBT structure

The changes in the distribution of the electric field within the symmetric IGBT structure are shown in Fig. 9.111 at various time instances during turn-off with a collector bias supply voltage of 800 V. It can be observed that the slope of the electric field profile remains constant during the first phase as assumed in the analytical model. The width of the space-charge layer at the end of the first phase is 95 μm. The analytical model (see (9.290)) provides an accurate determination of this value. During the second phase, the slope of the electric field decreases as predicted by the analytical model. This produces a small increase in the width of the space-charge layer during the second phase of the turn-off process.

Consequently, the simulation results validate the assumptions used in deriving the analytical model.

Asymmetric IGBT Structure

The asymmetric IGBT structure is usually designed to support a blocking voltage that is substantially larger than the collector supply voltage in its applications. Consequently, although the depletion region punches through to the N-buffer layer at the breakdown voltage in the forward-blocking mode, the space-charge layer does not typically punch through to the N-buffer layer during inductive load switching because of the smaller collector supply voltage and because of the additional positive charge due to the hole concentration (p_{SC}) within the space-charge layer. Consequently, the collector current and voltage waveforms for the asymmetric IGBT structure under inductive load switching conditions are similar in appearance to those shown previously in Fig. 9.105 for the case of the symmetric IGBT structure. However, the time duration for the collector voltage rise time and the collector current fall time is substantially smaller. This is because the smaller stored charge within the N-base region in the on-state for the asymmetric IGBT structure makes the voltage transient much faster. In addition, the low lifetime in the N-buffer layer due to its higher doping level makes the

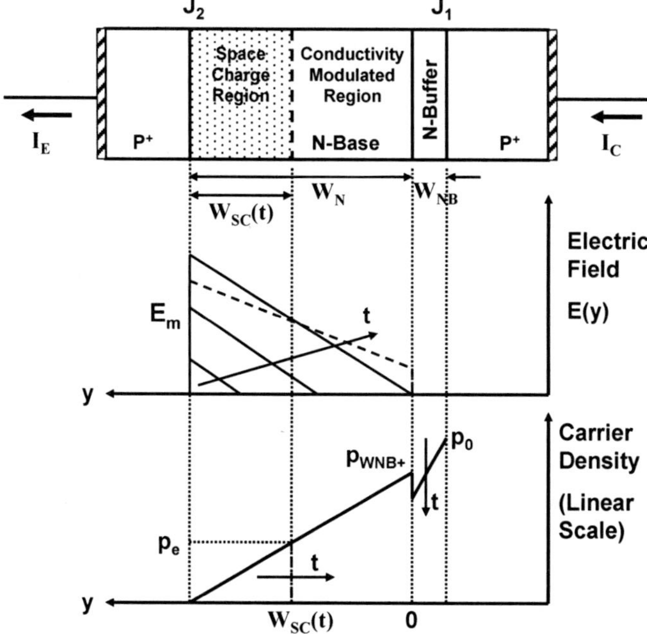

Fig. 9.112 Stored charge and electric field distributions for inductive load turn-off conditions in an asymmetric IGBT structure

decay of the collector current in the asymmetric IGBT structure much faster than for the symmetric IGBT structure.

The analysis of the turn-off waveforms for the asymmetric IGBT structure can be performed by using the charge control principle. In this analysis, it will be assumed that recombination in the N-base region can be neglected in the on-state. This results in a linear free carrier (hole) distribution within the lightly doped portion of the N-base region during on-state operation as illustrated in Fig. 9.112 (The more complex hole distribution obtained for the asymmetric IGBT structure in Sect. 9.5.4 with finite recombination in the lightly doped portion of the N-base region is nearly linear in shape.):

$$p(y) = p_{\text{WNB+}}\left(1 - \frac{y}{W_{\text{N}}}\right). \tag{9.300}$$

In writing this expression, the hole concentration is assumed to be approximately zero at the edge of the space-charge region during the on-state and the space-charge layer width is assumed to be zero due to the small on-state voltage drop of the IGBT structure. The concentration $p_{\text{WNB+}}$ in the on-state at the interface between the lightly doped portion of the N-base region and the N-buffer layer was previously derived in Sect. 9.5.4.

It will be assumed that the above hole distribution does not change during the first phase of the turn-off process in the conductivity-modulated portion of the N-base region. Consequently, the concentration of holes at the edge of the space-charge region (p_{e}) increases during the turn-off process as the space-charge width increases:

$$p_{\text{e}}(t) = p_{\text{WNB+}}\left[\frac{W_{\text{SC}}(t)}{W_{\text{N}}}\right]. \tag{9.301}$$

The solution for the evolution of the space-charge layer with time is then obtained by using the same approach as for the symmetric structure:

$$W_{\text{SC}}(t) = \sqrt{\frac{2W_{\text{N}}J_{\text{C,ON}}t}{qp_{\text{WNB+}}}}. \tag{9.302}$$

The space-charge layer expands toward the right-hand side as indicated by the horizontal time arrow in Fig. 9.112 with the hole concentration in the conductivity-modulated portion remaining unchanged.

The collector voltage supported by the asymmetric IGBT structure is related to the space-charge layer width by

$$V_{\text{C}}(t) = \frac{q(N_{\text{D}} + p_{\text{SC}})W_{\text{SC}}^2(t)}{2\varepsilon_{\text{S}}}. \tag{9.303}$$

Although the hole concentration in the space-charge layer was neglected when compared with the concentration p_{WNB+} when writing (9.300), it cannot be neglected when determining the space-charge layer width because it is comparable with the doping concentration in the lightly doped portion of the N-base region. The hole concentration in the space-charge layer can be related to the collector current density under the assumption that the carriers are moving at the saturated drift velocity in the space-charge layer:

$$p_{SC} = \frac{J_{C,ON}}{q v_{sat,p}}. \tag{9.304}$$

The hole concentration in the space-charge region remains constant during the voltage rise time because the collector current density is constant. Consequently, the slope of the electric field profile in the space-charge region also becomes independent of time as illustrated in Fig. 9.112.

Applying the solution for the evolution of the space-charge layer from (9.302) in (9.303),

$$V_C(t) = \frac{W_N(N_D + p_{SC})J_{C,ON}}{\varepsilon_S p_{WNB+}} t. \tag{9.305}$$

The analytical model for turn-off of the asymmetric IGBT structure under inductive load conditions predicts a linear increase in the collector voltage with time. The rate of increase in the collector voltage for the asymmetric IGBT structure is much faster than for the symmetric IGBT structure because the concentration p_{WNB+} is an order of magnitude smaller than the concentration p_0 for the symmetric structure.

The end of the first phase of the turn-off process, where the collector voltage increases while the collector current remains constant, occurs when the collector voltage reaches the collector supply voltage (V_{CS}). This time interval ($t_{V,OFF}$) can be obtained by making the collector voltage equal to the collector supply voltage in (9.305):

$$t_{V,OFF} = \frac{\varepsilon_S p_{WNB+} V_{CS}}{W_N(N_D + p_{SC})J_{C,ON}}. \tag{9.306}$$

According to the analytical model, the voltage rise time is proportional to the collector bias supply voltage. However, it is only weakly dependent on the on-state current density (through p_{SC}) because the hole concentration p_{WNB+} is proportional to the on-state current density. The voltage rise time for the asymmetric IGBT structure is much smaller than for the symmetric IGBT structure because the concentration p_{WNB+} is an order of magnitude smaller than the concentration p_0 for the symmetric structure.

The width of the space-charge layer at the end of the voltage transient can be obtained by using the collector supply voltage:

$$W_{\mathrm{SC}}(t_{\mathrm{V,OFF}}) = \sqrt{\frac{2\varepsilon_{\mathrm{S}}V_{\mathrm{CS}}}{q(N_{\mathrm{D}} + p_{\mathrm{SC}})}}. \tag{9.307}$$

The width of the space-charge layer at the end of the first phase depends upon the collector supply voltage and the initial on-state current density (via p_{SC}). This expression is valid only if the above space-charge layer width is smaller than the width (W_{N}) of the lightly doped portion of the N-base region.

During the second phase of the turn-off process, the decay of the collector current after the voltage rise time is governed by the recombination of the excess holes that are trapped within the N-base and N-buffer region. In a typical case, the space-charge layer extends through most of the width (W_{N}) of the lightly doped portion of the N-base region. Consequently, the decay of the collector current is governed by recombination of holes in the N-buffer layer as indicated by the vertical time arrow in Fig. 9.112. In the absence of diffusion, the continuity equation for holes in the N-buffer layer is given by

$$\frac{\mathrm{d}\delta p_{\mathrm{NB}}}{\mathrm{d}t} = -\frac{\delta p_{\mathrm{NB}}}{\tau_{\mathrm{p0,NB}}}, \tag{9.308}$$

where δp_{NB} is the excess hole concentration in the N-buffer layer. The solution for this equation is

$$\delta p_{\mathrm{NB}}(y_{\mathrm{N}}, t) \approx p_{\mathrm{NB}}(y_{\mathrm{N}}, t) = p(y_{\mathrm{N}})\,\mathrm{e}^{-t/\tau_{\mathrm{p0,NB}}} \tag{9.309}$$

because low-level injection conditions prevail in the N-buffer layer. The concentration of holes in the stored charge region at the beginning of the second phase has been assumed to be equal to the hole concentration (p_0) at the P^+ collector/N-base junction due to the injection of carriers in the on-state.

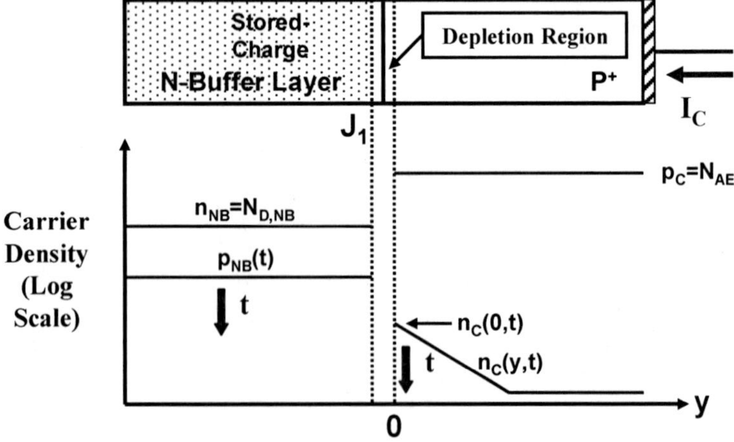

Fig. 9.113 Free carrier distribution at the P^+ collector/N-buffer layer junction during turn-off for the asymmetric IGBT structure

The collector current flow that supports the recombination of carriers within the stored charge region can be analyzed by examination of the carrier distribution on both sides of the P^+ collector/N-base junction (J_1). The free carrier concentration in the N-buffer layer in the vicinity of junction J_1 can be assumed to be independent of distance but a function of time as given by (9.309). The presence of excess holes in the N-buffer layer produces the injection of electrons into the P^+ collector region as indicated in Fig. 9.113. These injected electrons diffuse away from the junction producing an exponential decay in concentration as shown in the figure. The free carrier concentrations on the two sides of the junction are related by

$$\frac{p_C}{p_{NB}(y_N,t)} = \frac{n_{NB}}{n_C(0,t)} = e^{qV_C/kT}, \tag{9.310}$$

where $p_{NB}(y_N, t)$ is the hole concentration in the N-buffer layer at junction J_1 given by (9.309), n_{NB} is the electron concentration in the N-buffer layer, p_C is the hole concentration in the P^+ collector region, and $n_C(0, t)$ is the electron concentration at junction J_1 in the P^+ collector region. The hole concentration in the P^+ collector region can be assumed to be equal to its acceptor doping concentration (N_{AE}) and the electron concentration in the N-buffer layer can be assumed to be equal to the donor doping concentration ($N_{D,NB}$) due to low-level injection conditions. Using this in (9.310) yields

$$n_C(0,t) = \frac{N_{D,NB}}{N_{AE}} p_{NB}(y_N,t). \tag{9.311}$$

The electrons diffuse into the P^+ collector region with an exponential distribution due to low-level injection conditions:

$$n_C(y,t) = n_C(0,t)e^{-y/L_{nE}}, \tag{9.312}$$

where L_{nE} is the diffusion length for electrons in the P^+ collector region.

The collector current is produced by the diffusion of the injected electrons in the P^+ collector side of the junction:

$$J_C(t) = -qD_{nE}\left.\frac{\partial n_C(y,t)}{\partial y}\right|_{y=0} = \frac{qD_{nE}n_C(0,t)}{L_{nE}} \tag{9.313}$$

by using (9.312). Making use of (9.309) and (9.311) yields

$$J_C(t) = \frac{qD_{nE}N_{D,NB}p_{NB}(y_N,t)}{L_{nE}N_{AE}} = \frac{qD_{nE}N_{D,NB}p(y_N)}{L_{nE}N_{AE}}e^{-t/\tau_{p0,NB}}. \tag{9.314}$$

This equation indicates that the collector current varies in proportion to the stored charge in the N-buffer layer during the collector current tail. Since the collector current density at the beginning of the second phase is equal to the on-state current density, the equation can also be written as

$$J_C(t) = J_{C,ON} e^{-t/\tau_{p0,NB}}. \tag{9.315}$$

Consequently, the collector current decreases exponentially with time with a time constant equal to the low-level lifetime in the N-buffer layer.

During the second phase of the turn-off process, the hole concentration (p_{SC}) in the space-charge layer decreases because the collector current density is becoming smaller. This reduces the net charge in the space-charge layer resulting in a change in the slope of the electric field profile as illustrated in Fig. 9.112 by the dashed line. This results in a small expansion of the space-charge layer during the second phase of the turn-off transient.

The collector current turn-off time ($t_{I,OFF}$) is defined as the time taken for the current to decay to one tenth of the on-state value as illustrated in Fig. 9.105. Based upon (9.315), the collector current turn-off time is given by

$$\tau_{I,OFF} = \tau_{p0,NB} \ln(10) = 2.3\tau_{p0,NB}. \tag{9.316}$$

According to the analytical model, the turn-off time is independent of the collector bias supply voltage and the on-state current density. It can be reduced by reducing the lifetime in the N-buffer layer by selective lifetime control techniques, such as deep proton ion implantation.

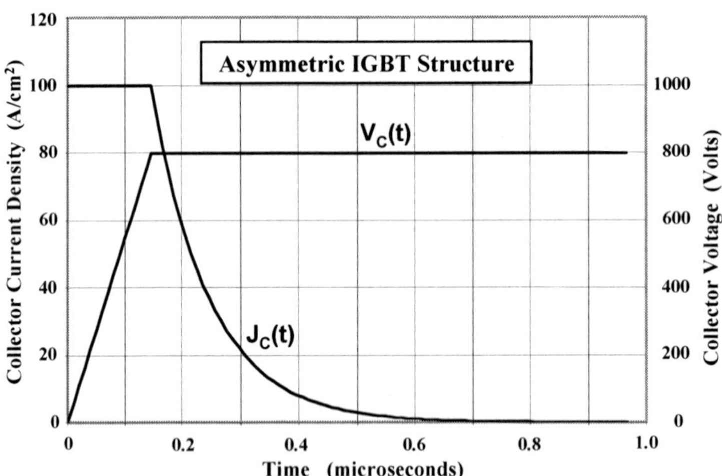

Fig. 9.114 Collector current and voltage transients during turn-off for the asymmetric IGBT structure with an inductive load

Consider the case of an asymmetric IGBT structure with N-base region with a lightly doped portion with a width of 100 μm, a doping concentration of 5×10^{13} cm^{-3}, and a lifetime of 1 μs; and an N-buffer layer with a width of 10 μm, a doping concentration of 1×10^{17} cm^{-3}, and a lifetime of 0.1 μs. During the first phase of the turn-off process with an inductive load from an initial on-state collector current density of 100 A cm^{-2}, the hole concentration in the space-charge layer is 6.25×10^{13} cm^{-3}, which is comparable with the doping concentration. The presence of the hole charge must therefore be included when computing the width of the space-charge layer during the voltage transient. For this structure, the hole concentration (p_{WNB+}) at the interface between the lightly doped portion of the N-base region and the N-buffer layer in the on-state is found to be 2×10^{16} cm^{-3}. The collector voltage transient obtained by using these values is provided in Fig. 9.114. The collector voltage rise time ($t_{V,OFF}$) is found to be 0.15 μs. After the voltage transient is completed, the collector current decays exponentially as shown in Fig. 9.114. The collector current turn-off time ($t_{I,OFF}$) is found to be 0.23 μs.

Simulation Example

To validate the above model for the turn-off transient for the asymmetric IGBT structure under inductive load conditions, the results of numerical simulations for the typical 1,200-V asymmetric structure described in the previous sections are provided here. The structure had an N-base region with a lightly doped portion with

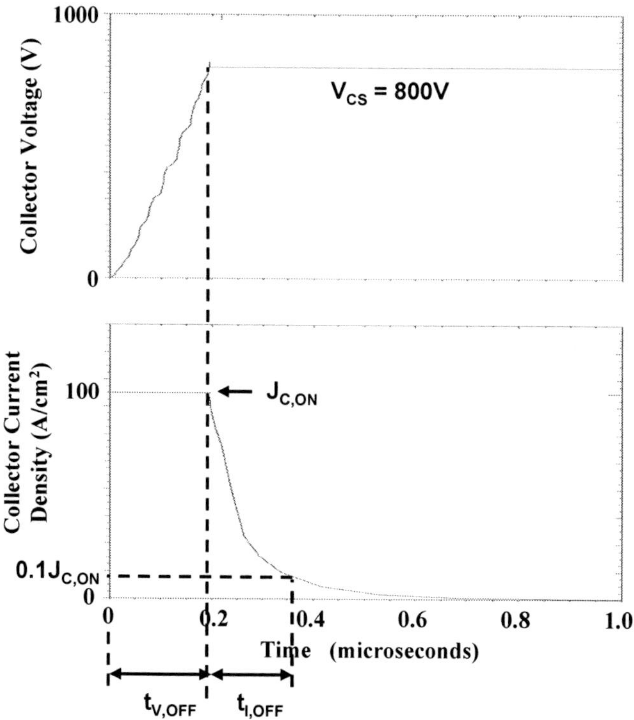

Fig. 9.115 Collector current and voltage turn-off waveforms under inductive load conditions for the asymmetric IGBT structure

thickness of 100 μm and doping concentration of 5×10^{13} cm^{-3}, and an N-buffer layer with thickness of 10 μm and doping concentration of 1×10^{17} cm^{-3}. The lifetime (τ_{p0} and τ_{n0}) in the N-base region was 1 μs. The device was turned off from a steady-state collector current of 100 A cm^{-2} with a collector bias supply voltage of 800 V by ramping the gate voltage to zero in 10 ns. During the first phase of the turn-off process, the collector current was held constant allowing the collector voltage to increase as a function of time. After the collector voltage reached 800 V, the collector voltage was held constant and the collector current was allowed to decay with time.

The waveforms for the collector current and voltage obtained from the numerical simulations are shown in Fig. 9.115. During the first phase of the turn-off process, the collector voltage increases linearly with time as predicted by the analytical model (see (9.305)). The collector voltage rise time obtained from the simulations is 0.19 μs. The value computed by using (9.306) for the collector voltage rise time is in good agreement with this observation. During the second phase of the turn-off process, the collector current decreases exponentially with time as predicted by the analytical model. The collector current fall time obtained from the simulations is 0.18 μs. The value computed by using (9.316) for the collector current fall time by using the scaled lifetime in the N-buffer layer is in good agreement with this observation. This provides validation for the assumption in the model that the current turn-off process is determined by the recombination of minority carriers in the N-buffer layer under low-level injection conditions.

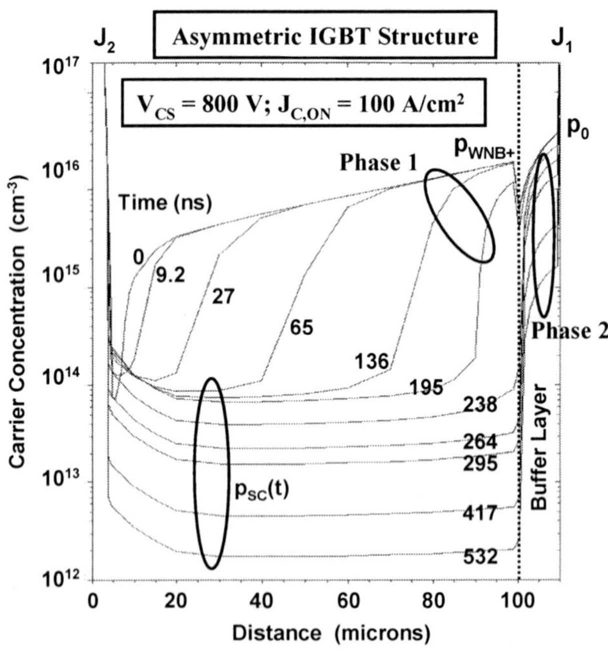

Fig. 9.116 Hole distribution during inductive load turn-off conditions for the asymmetric IGBT structure

The changes in the distribution of holes within the asymmetric IGBT structure are shown in Fig. 9.116 at various time instances during inductive load turn-off with a collector bias supply voltage of 800 V. During the first phase extending to 195 ns of the turn-off process, the hole concentration within the conductivity-modulated portion of the N-base region and in the N-buffer layer remains the same at the initial on-state carrier distribution. This validates the assumption used in the analytical model (see Fig. 9.112) for the removal of charge from the N-base region. During this time period, the space-charge region expands with a constant hole concentration (p_{SC}) of ~ 8×10^{13} cm^{-3} as determined by the on-state collector current density, providing validation for the analytical formulations developed for the space-charge concentration and space-charge layer width (see (9.304) and (9.307)). During the second phase, it can also be observed that the hole concentration in the N-buffer layer decreases with time. At the same time, the hole concentration (p_{SC}) in the space-charge region decreases with time due to a reduction of the collector current density as predicted by the analytical model. The stored charge in the N-buffer layer is removed rapidly due to the low minority carrier lifetime in this region.

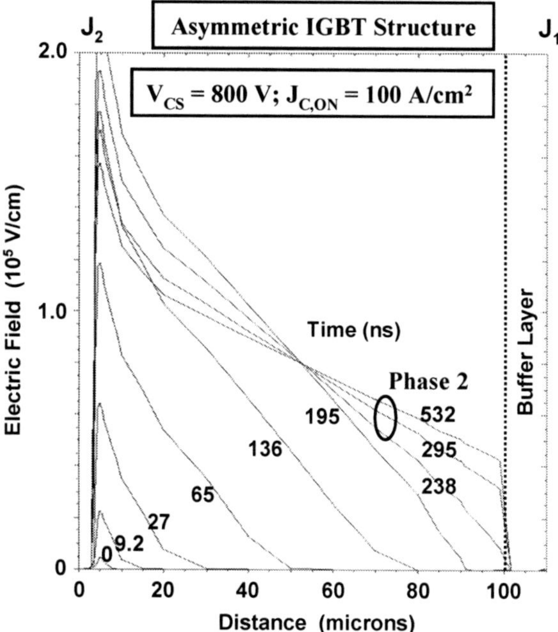

Fig. 9.117 Electric field distribution during inductive load turn-off conditions for the asymmetric IGBT structure

The changes in the distribution of the electric field within the asymmetric IGBT structure are shown in Fig. 9.117 at various time instances during turn-off with a collector bias supply voltage of 800 V. It can be observed that the slope of the electric field profile remains constant during the first phase as assumed in the analytical model. The width of the space-charge layer at the end of the first phase is 95 μm. The analytical model (see (9.307)) provides an accurate determination of this value. During the second phase, the slope of the electric field decreases as predicted by the analytical model. This produces extension of the space-charge layer through the entire lightly doped portion of the N-base region during the second phase of the turn-off process as assumed in the analytical model. Consequently, the simulation results validate the assumptions used in deriving the analytical model.

Transparent Emitter IGBT Structure

The transparent emitter IGBT structure is usually designed with a low P^+ collector doping concentration and thickness to suppress the injection efficiency while maintaining a high lifetime in the N-base region.[24] This results in a smaller injected hole concentration within the N-base region which accelerates the turn-off process when compared with the symmetric IGBT structure. Although the collector current and voltage waveforms for the transparent emitter IGBT structure under inductive load switching conditions are similar in appearance to those shown previously in Fig. 9.105 for the case of the symmetric IGBT structure, the time duration for the collector voltage rise time and the collector current fall time is substantially smaller. In addition, the injection of carriers into the P^+ collector region makes the decay of the collector current in the transparent emitter IGBT structure much faster than for the symmetric IGBT structure.

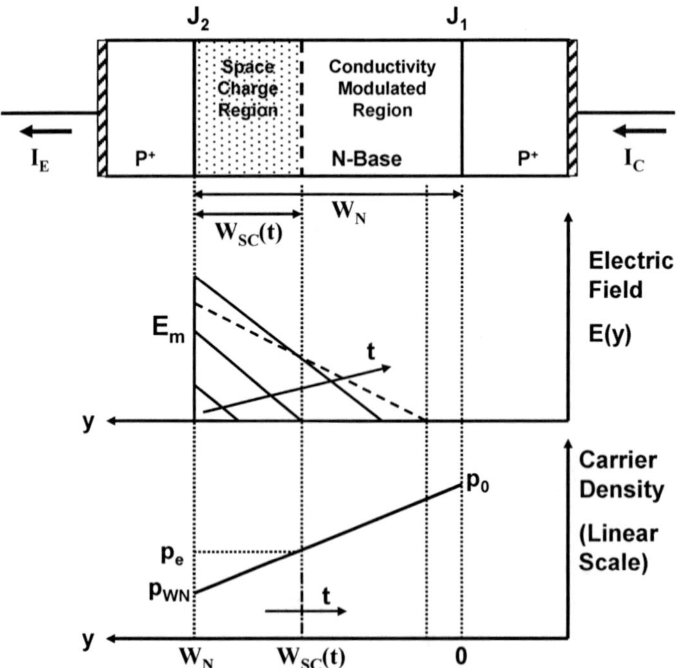

Fig. 9.118 Stored charge and electric field distributions for inductive load turn-off conditions in the transparent emitter IGBT structure

The analysis of the turn-off waveforms for the transparent emitter IGBT structure can be performed by using the charge control principle. In this analysis, it will be assumed that recombination in the N-base region can be neglected because the lifetime in the N-base region is usually large for this structure. This produces a linear free carrier (hole) distribution within the lightly doped portion of the N-base region during on-state operation as previously illustrated for the symmetric IGBT structure in Fig. 9.106. However, in the transparent emitter IGBT structure, the concentration for holes injected into the N-base region becomes much smaller than for the symmetrical IGBT structure. Consequently, the assumption of a negligible hole concentration at $y = W_N$ produces faster expansion of the space-charge layer than warranted for this structure. It is instead necessary to use the linear hole distribution shown in Fig. 9.118 with a finite hole concentration (p_{WN}) at $y = W_N$:

$$p(y) = p_0 - \left(\frac{p_0 - p_{WN}}{W_N}\right) y. \tag{9.317}$$

The concentration p_0 in the on-state at the P$^+$ collector/N-base junction was previously derived in Sect. 9.5.6.

It will be assumed that the above hole distribution does not change during the first phase of the turn-off process in the conductivity-modulated portion of the N-base region. Consequently, the concentration of holes at the edge of the space-charge region (p_e) increases during the turn-off process as the space-charge width increases:

$$p_e(t) = p_{WN} + (p_0 - p_{WN})\left[\frac{W_{SC}(t)}{W_N}\right]. \tag{9.318}$$

The solution for the evolution of the space-charge layer with time is then obtained by using the same approach as for the symmetric structure:

$$W_{SC}(t) = \frac{W_N}{p_0 - p_{WN}}\left[\sqrt{\frac{2(p_0 - p_{WN})J_{C,ON}t}{qW_N} + p_{WN}^2} - p_{WN}\right]. \tag{9.319}$$

The collector voltage supported by the transparent emitter IGBT structure is related to the space-charge layer width by

$$V_C(t) = \frac{q(N_D + p_{SC})W_{SC}^2(t)}{2\varepsilon_S}. \tag{9.320}$$

Although the hole concentration in the space-charge layer was neglected when compared with the concentration p_0 when writing (9.317), it cannot be neglected when determining the space-charge layer width because it is comparable with the doping concentration in the lightly doped portion of the N-base region. The hole concentration in the space-charge layer can be related to the collector current density under the assumption that the carriers are moving at the saturated drift velocity in the space-charge layer:

$$p_{SC} = \frac{J_{C,ON}}{qv_{sat,p}}. \tag{9.321}$$

The hole concentration in the space-charge region remains constant during the voltage rise time because the collector current density is constant. Consequently, the slope of the electric field profile in the space-charge region also becomes independent of time as illustrated in Fig. 9.118.

The analytical model for turn-off of the transparent emitter IGBT structure under inductive load conditions predicts a nonlinear increase in the collector voltage with time. The rate of increase in the collector voltage for the transparent emitter IGBT structure is much faster than for the symmetric IGBT structure because the concentration p_0 is an order of magnitude smaller than the concentration p_0 for the symmetric structure.

The end of the first phase of the turn-off process, where the collector voltage increases while the collector current remains constant, occurs when the collector voltage reaches the collector supply voltage (V_{CS}). This time interval ($t_{V,OFF}$) can be obtained by making the collector voltage equal to the collector supply voltage in (9.320) and using (9.319):

$$t_{V,OFF} = \frac{\varepsilon_S V_{CS}(p_0 + p_{SC})}{J_{C,ON} W_N (N_D + p_{SC})} + \frac{p_{WN}}{J_{C,ON}} \sqrt{\frac{2q\varepsilon_S V_{CS}}{N_D + p_{SC}}}. \tag{9.322}$$

According to the analytical model, the voltage rise time is strongly dependent on the collector bias supply voltage. However, it is only weakly dependent on the on-state current density because the hole concentrations p_0 and p_{WN} are proportional to the on-state current density. The voltage rise time for the transparent emitter IGBT structure is much smaller than for the symmetric IGBT structure because the concentrations p_0 and p_{WN} are an order of magnitude smaller than the concentration p_0 for the symmetric structure.

The width of the space-charge layer at the end of the voltage transient can be obtained by using the collector supply voltage:

$$W_{SC}(t_{V,OFF}) = \sqrt{\frac{2\varepsilon_S V_{CS}}{q(N_D + p_{SC})}}. \tag{9.323}$$

The width of the space-charge layer at the end of the first phase depends upon the collector supply voltage and the initial on-state current density (via p_{SC}).

During the second phase of the turn-off process, the decay of the collector current after the voltage rise time is governed by the removal of the excess holes within the N-base region. In a typical case, the space-charge layer extends through only a fraction of the width of the N-base region leaving a large number of holes and electrons in the vicinity of the P^+ collector/N-base junction. In the transparent emitter IGBT structure, the lifetime in the N-base region is large making the recombination process very slow in this region. Consequently, the collector current reduction is governed by recombination within the transparent emitter, i.e., the P^+ collector region for the IGBT structure. As discussed previously in Sect. 9.5.6, the electron concentration within the P^+ collector region becomes linear from the junction to zero at the contact as illustrated in Fig. 9.119 under the assumptions of an infinite surface recombination velocity at the ohmic contact and no recombination inside the P^+ collector region.

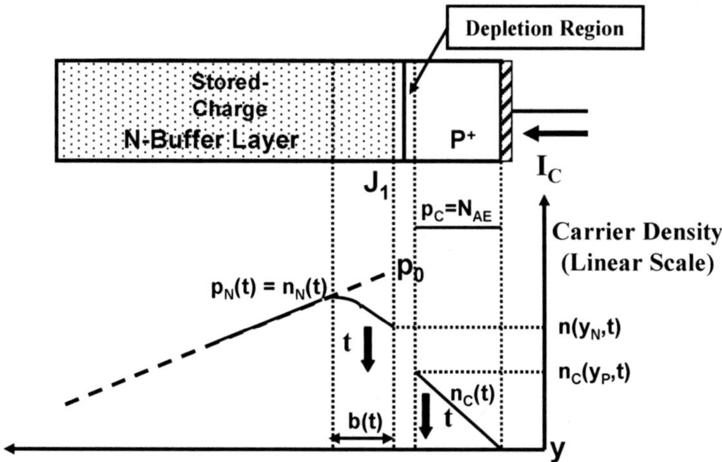

Fig. 9.119 Free carrier distribution at the P^+ collector/N-base junction during the second phase of turn-off for the transparent emitter IGBT structure

The electron concentration ($n(y_N, t)$) in the N-base region at the junction is related to the electron concentration ($n_C(0, t)$) at the junction in the P^+ collector region (see derivation for (9.59)) by

$$n(y_N, t) = \sqrt{N_{AE} n_C(y_P, t)}. \tag{9.324}$$

The reduced electron concentration in the N-base region at the junction produces a diffusion of the electrons toward the junction as indicated by the reversed slope for the free carrier distribution in Fig. 9.119. Analysis of the removal of the electrons from the N-base region by the process of their diffusion toward the junction followed by recombination within the P^+ collector region can be performed by taking the approach used to obtain the carrier distribution in the presence of surface recombination[32]:

$$n(y, t) = p_0 e^{-\alpha t} \cos \varsigma [b(t) - y]. \tag{9.325}$$

In the case of the turn-off process for the transparent emitter IGBT structure, the distance b through which the electrons diffuse increases with time. The rate of increase in this parameter will be assumed to be as the square root of time.

The effective lifetime[32] for the removal of electrons is then given by

$$\tau_{\text{TE}} = \frac{b(t)}{S_{\text{r}}}, \tag{9.326}$$

where S_r is the surface recombination velocity at the edge of the N-base region, i.e., at the junction. Based upon the above equations, the surface recombination velocity at the edge of the N-base region is found to be inversely proportional to the effective doping concentration in the P^+ collector region. Since the collector current density at the beginning of the second phase is equal to the on-state current density, the collector current density during the second phase is given by

$$J_{\text{C}}(t) = J_{\text{C,ON}} e^{-t/\tau_{\text{TE}}} \tag{9.327}$$

with the decay constant given by

$$\tau_{\text{TE}} = \sqrt{K_{\text{TE2}} t \sqrt{N_{\text{AE}}}}. \tag{9.328}$$

This expression indicates that the rate of change of the collector current is faster at the beginning of the second phase. This is because the electrons being removed from the N-base region are located closer to the junction. As time progresses, electrons must diffuse further resulting in a longer effective time constant. Substituting (9.328) into (9.327), the collector current density during the second phase is given by

$$J_{\text{C}}(t) = J_{\text{C,ON}} e^{-\sqrt{t}/\left(K_{\text{TE2}} \sqrt{N_{\text{AE}}}\right)^{1/2}}. \tag{9.329}$$

During the second phase of the turn-off process, the hole concentration (p_{SC}) in the space-charge layer decreases because the collector current density is becoming smaller. This reduces the net charge in the space-charge layer resulting in a change in the slope of the electric field profile as illustrated in Fig. 9.118 by the dashed line. This results in a small expansion of the space-charge layer during the second phase of the turn-off transient.

The collector current turn-off time ($t_{\text{I,OFF}}$) is defined as the time taken for the current to decay to one tenth of the on-state value as illustrated in Fig. 9.105. Based upon (9.329), the collector current turn-off time is given by

$$\tau_{\text{I,OFF}} = K_{\text{TE2}} \sqrt{N_{\text{AE}}} \ln^2(10) = 5.3 K_{\text{TE2}} \sqrt{N_{\text{AE}}}. \tag{9.330}$$

Consider the case of a transparent emitter IGBT structure with N-base region with a width of 200 μm, a doping concentration of 5×10^{13} cm^{-3}, and a lifetime of 10 μs. During the first phase of the turn-off process with an inductive load from an initial on-state collector current density of 100 A cm^{-2}, the hole concentration in the space-charge layer is 6.25×10^{13} cm^{-3}, which is comparable with the doping concentration. The presence of the hole charge must therefore be included when computing the width of the space-charge layer during the voltage transient. For this structure, the hole concentration (p_0) in the N-base region at the P^+ collector junction in the on-state is found to be 7×10^{15} cm^{-3} for the case of an effective P^+ collector concentration of 5×10^{16} cm^{-3}. The slightly nonlinear collector voltage transient obtained by using these values is provided in Fig. 9.120. The collector voltage rise time ($t_{V,OFF}$) is found to be 58 ns. After the voltage transient is completed, the collector current decays "super" exponentially as shown in Fig. 9.120. The collector current turn-off time ($t_{I,OFF}$) is found to be 265 ns based upon a constant (K_{TE2}) of 2.2×10^{-16} s cm$^{3/2}$.

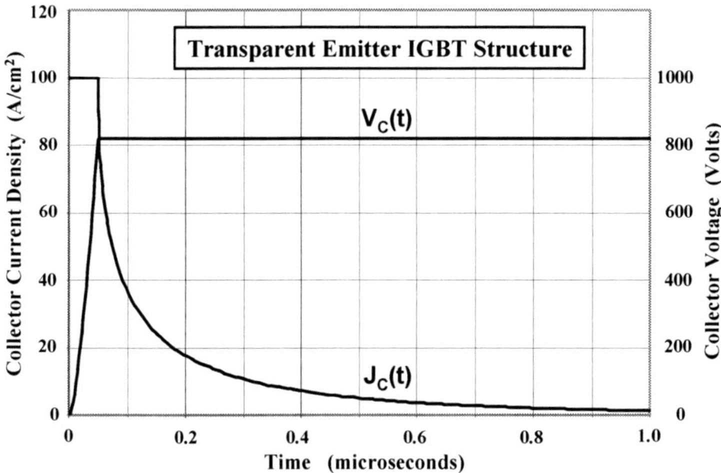

Fig. 9.120 Collector current and voltage transients during turn-off for the transparent emitter IGBT structure with an inductive load

Simulation Example

To validate the above model for the turn-off transient for the transparent emitter IGBT structure under inductive load conditions, the results of numerical simulations for the typical 1,200-V structure described in the previous sections are provided here. The structure had an N-base region with thickness of 200 μm and doping concentration of 5×10^{13} cm^{-3}, and a P^+ collector region with thickness of 1 μm and surface doping concentration of 1×10^{17} cm^{-3}. The lifetime (τ_{p0} and τ_{n0}) in the N-base region was 10 μs. The device was turned off from a steady-state collector current of 100 A cm^{-2} with a collector bias supply voltage of 800 V by ramping the gate voltage to zero in 10 ns. During the first phase of the turn-off process, the collector current was held constant allowing the collector voltage to increase as a function of time. After the collector voltage reached 800 V, the collector voltage was held constant and the collector current was allowed to decay with time.

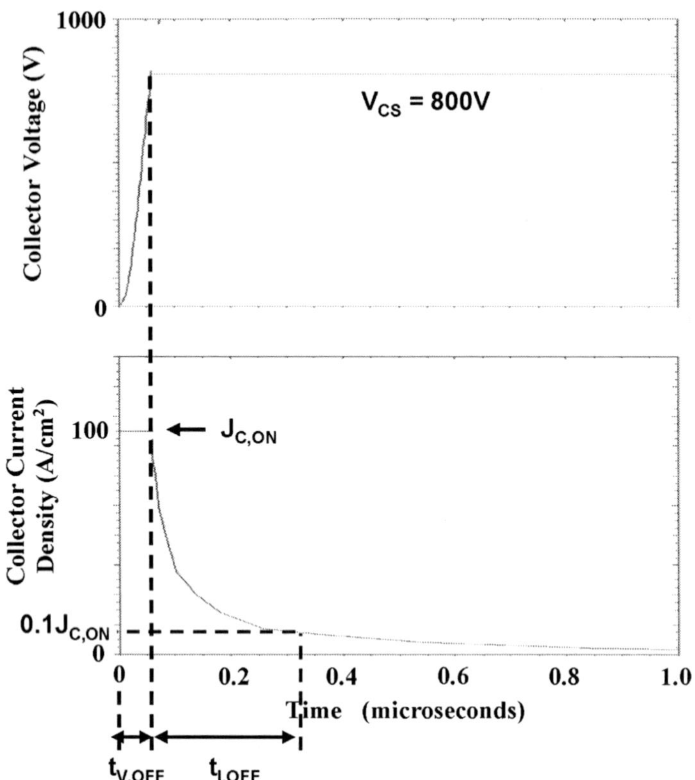

Fig. 9.121 Collector current and voltage turn-off waveforms under inductive load conditions for the transparent emitter IGBT structure

The waveforms for the collector current and voltage obtained from the numerical simulations are shown in Fig. 9.121. During the first phase of the turn-off process, the collector voltage increases nonlinearly with time as predicted by the analytical model. The collector voltage rise time obtained from the simulations is 60 ns. The value computed by using (9.322) for the collector voltage rise time is in excellent agreement with this observation. During the second phase of the turn-off process, the collector current decreases "super" exponentially with time as described by the analytical model. The collector current fall time obtained from the simulations is 270 ns. The value computed by using (9.330) for the collector current fall time, by using the surface recombination-based model, is in good agreement with this observation.

The changes in the distribution of holes within the asymmetric IGBT structure are shown in Fig. 9.122 at various time instances during inductive load turn-off with a collector bias supply voltage of 800 V. During the first phase extending to 60 ns of the turn-off process, the hole concentration within the conductivity-modulated portion of the N-base region remains the same as determined by the initial on-state carrier distribution. This validates the assumption used in the analytical model (see Fig. 9.118) for the removal of charge from the N-base region. During this time period, the space-charge region expands with a constant hole concentration (p_{SC}) of $\sim 8 \times 10^{13}$ cm^{-3} as determined by the on-state collector current density. This provides validation for the analytical formulations developed for the space-charge concentration and space-charge layer width (see (9.321) and (9.323)). During the second phase, it can be observed that the hole concentration has a negative gradient in the conductivity-modulated portion of the N-base region, indicating the diffusion of holes into the transparent emitter region. At the same time, the hole concentration (p_{SC}) in the space-charge region decreases with time due to a reduction of the collector current density as predicted by the analytical model.

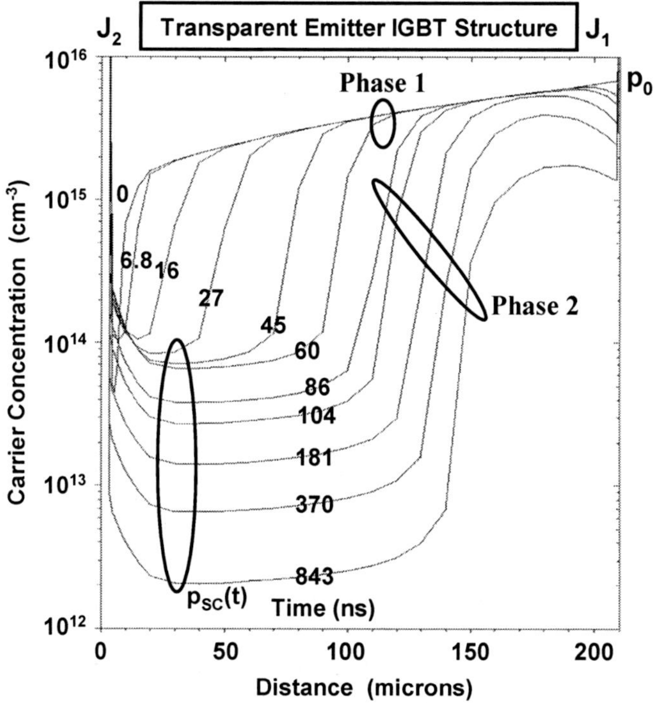

Fig. 9.122 Hole distribution during inductive load turn-off conditions for the transparent emitter IGBT structure

The changes in the distribution of the electric field within the transparent emitter IGBT structure are shown in Fig. 9.123 at various time instances during turn-off with a collector bias supply voltage of 800 V. It can be observed that the slope of the electric field profile remains constant during the first phase as assumed in the analytical model. The width of the space-charge layer at the end of the first phase is 97 μm. The analytical model (see (9.323)) provides an accurate determination of this value. During the second phase, the slope of the electric field decreases as predicted by the analytical model. This produces an increase in the width of the space-charge layer during the second phase of the turn-off process as assumed in the analytical model. Consequently, the simulation results validate the assumptions used in deriving the analytical model.

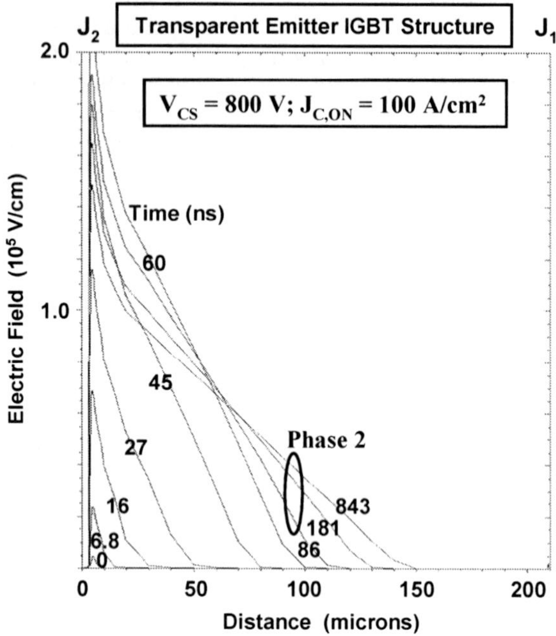

Fig. 9.123 Electric field distribution during inductive load turn-off conditions for the transparent emitter IGBT structure

9.7.5 Energy Loss per Cycle

The device current and voltage waveforms for the IGBT structure were previously shown in Fig. 9.89 for the typical motor control application. The power loss during the turn-on of the IGBT structure is mainly determined by the reverse recovery behavior of the fly-back diode in the totem pole circuit. In contrast, the power loss during turn-off is governed by the IGBT structure as discussed in the previous sections. It is therefore customary to characterize the IGBT structure in terms of the energy loss per cycle during the turn-off process. The turn-off power dissipation for the IGBT structure at any operating frequency can then be derived by multiplying the energy loss per cycle and the operating frequency.

The energy loss during turn-off can be obtained by analysis of the power loss during the two phases of the turn-off process as illustrated in Fig. 9.124. During the first phase of the turn-off process, the voltage increases linearly with time (even for the transparent emitter structure, the voltage waveform is nearly linear in nature) while the collector current density remains constant for an inductive load. The energy loss during the first phase of the turn-off process can therefore be calculated for all three IGBT structures by using

$$E_{\mathrm{V,OFF}} = \frac{1}{2} J_{\mathrm{C,ON}} V_{\mathrm{CS}} t_{\mathrm{V,OFF}}. \tag{9.331}$$

During the second phase of the turn-off process, the collector current decreases exponentially with time while the collector voltage remains at the collector supply voltage. The energy loss during this transient can therefore be obtained by integration of the current–voltage product:

$$E_{\mathrm{I,OFF}} = \int_0^{\infty} V_{\mathrm{CS}} J_{\mathrm{C}}(t)\mathrm{d}t. \tag{9.332}$$

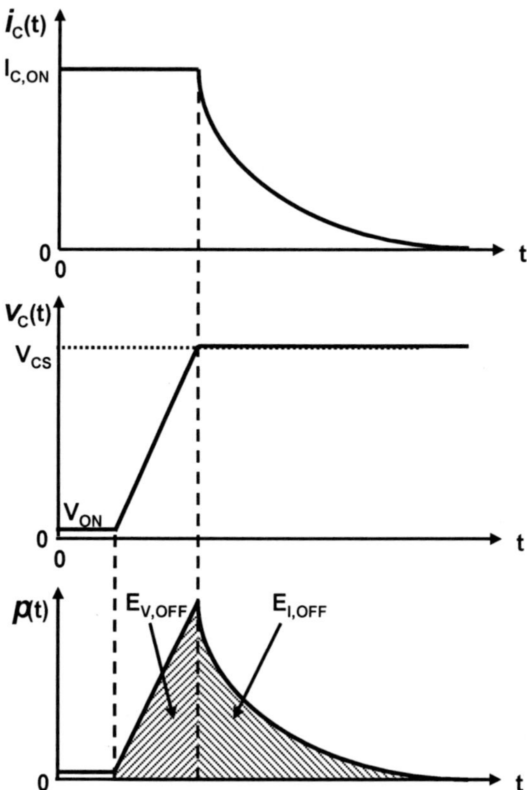

Fig. 9.124 Energy loss during inductive load turn-off for the IGBT structure

Symmetric IGBT Structure

For the symmetric IGBT structure, during the second phase, the collector current density decreases exponentially with time:

$$J_{\mathrm{C}}(t) = J_{\mathrm{C,ON}} \mathrm{e}^{-t/\tau_{\mathrm{OFF}}} \tag{9.333}$$

with the time constant τ_{OFF} equal to half the high-level lifetime in the N-base region. Substituting this expression into (9.332) and performing the integration yields

$$E_{\mathrm{I,OFF}} = J_{\mathrm{C,ON}} V_{\mathrm{CS}} \tau_{\mathrm{OFF}} = J_{\mathrm{C,ON}} V_{\mathrm{CS}} \left(\frac{\tau_{\mathrm{HL,N\text{-}Base}}}{2} \right). \tag{9.334}$$

Asymmetric IGBT Structure

For the asymmetric IGBT structure, during the second phase, the collector current density decreases exponentially with time:

$$J_{\mathrm{C}}(t) = J_{\mathrm{C,ON}} \mathrm{e}^{-t/\tau_{\mathrm{OFF}}} \tag{9.335}$$

with the time constant τ_{OFF} equal to the low-level lifetime in the N-buffer layer. Substituting this expression into (9.332) and performing the integration yields

$$E_{\mathrm{I,OFF}} = J_{\mathrm{C,ON}} V_{\mathrm{CS}} \tau_{\mathrm{OFF}} = J_{\mathrm{C,ON}} V_{\mathrm{CS}} \tau_{\mathrm{p0,N\text{-}Buffer}}. \tag{9.336}$$

Transparent Emitter IGBT Structure

For the transparent emitter IGBT structure, during the second phase, the collector current density decreases "super" exponentially with time:

$$J_{\mathrm{C}}(t) = J_{\mathrm{C,ON}} \mathrm{e}^{-\sqrt{t}/\left(K_{\mathrm{TE2}}\sqrt{N_{\mathrm{AE}}}\right)^{1/2}}. \tag{9.337}$$

Substituting this expression into (9.332) and performing the integration yields

$$E_{\mathrm{I,OFF}} = J_{\mathrm{C,ON}} V_{\mathrm{CS}} \left(2K_{\mathrm{TE2}} \sqrt{N_{\mathrm{AE}}} \right). \tag{9.338}$$

Structure	$t_{\mathrm{V,OFF}}$ (μs)	$E_{\mathrm{V,OFF}}$ (mJ/cm^2)	$t_{\mathrm{I,OFF}}$ (μs)	$E_{\mathrm{I,OFF}}$ (mJ/cm^2)	E_{OFF} (mJ/cm^2)
Symmetric IGBT Structure	**0.85**	**34**	**11.5**	**800**	**834**
Asymmetric IGBT Structure	**0.15**	**6.0**	**0.23**	**8.0**	**14.0**
Transparent Emitter IGBT Structure	**0.058**	**2.3**	**0.27**	**8.0**	**10.3**

Fig. 9.125 Comparison of turn-off energy loss per cycle for IGBT structures

The total energy loss per cycle is obtained by adding the energy loss incurred during the two phases. The energy loss incurred by the three IGBT structures (for the typical design cases described in the previous sections) is compared in Fig. 9.125. The values for the voltage rise time and the current fall time are also provided in the figure. Due to much shorter switching times for the asymmetric and transparent emitter IGBT structures, their energy loss per cycle is nearly 100 times smaller than for the symmetric IGBT structure. These structures are therefore preferred for motor control applications. The asymmetric IGBT structure is usually favored for devices with blocking voltage ratings below 1,000 V because it can be manufactured using epitaxial silicon wafers while the transparent emitter structure is favored for higher voltage devices because it can be manufactured using bulk silicon wafers.

9.8 Power Loss Optimization

In a typical application, such as variable speed motor control, power losses occur within the IGBT structure during turn-on, on-state operation, during turn-off, and in the off-state. The off-state losses are usually negligible due to the low leakage currents at moderate temperatures of operation. The turn-on losses are strongly dependent on the reverse recovery behavior of the fly-back diode. The power losses incurred within the IGBT structure are therefore optimized by minimizing the power loss during the on-state and during the turn-off event. Unfortunately, techniques used to reduce the turn-off loss produce an increase in the on-state power loss. It is therefore customary to perform a tradeoff between these power loss components. For circuits operating at a low switching frequency, it is preferable to minimize the on-state power loss. In contrast, for circuits operating at a high switching frequency, it is preferable to minimize the turn-off power loss. The optimization of the IGBT structures requires the development of a tradeoff curve between the on-state voltage drop and the turn-off energy loss per cycle. The turn-off power loss at any operating frequency can then be obtained by multiplying the energy loss per cycle and the operating frequency.

9.8.1 Symmetric Structure

For the symmetric IGBT structure, the tradeoff curve between on-state voltage drop and turn-off energy loss per cycle is generated by varying the lifetime in the N-base region. This tradeoff curve can be created by using the analytical models developed in earlier sections for the on-state voltage drop and the energy loss per cycle. As an example, the tradeoff curve obtained for the symmetric IGBT structure with 1,200-V blocking voltage capability is shown in Fig. 9.126. This structure has an N-base region width of 200 μm. It can be observed that the on-state voltage drop increases very rapidly when the lifetime is reduced to make the energy loss less than 50 mJ cm^{-2}. The values computed by using the analytical models are in good agreement with those reported in the literature.[24] An improvement in the tradeoff curve can be obtained by operating the IGBT structure at a lower on-state current density. However, this is accomplished at the expense of a larger die size (and cost) for achieving any desired current rating for the device.

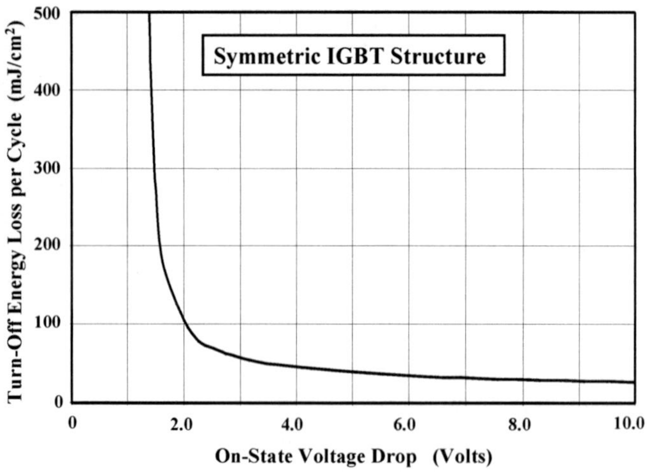

Fig. 9.126 Tradeoff curve for the symmetric IGBT structure obtained by varying the lifetime in the N-base region

Simulation Example

To validate the above model for the energy loss during turn-off for the symmetric IGBT structure under inductive load conditions, the results of numerical simulations for the typical 1,200-V structure described in the previous sections are provided

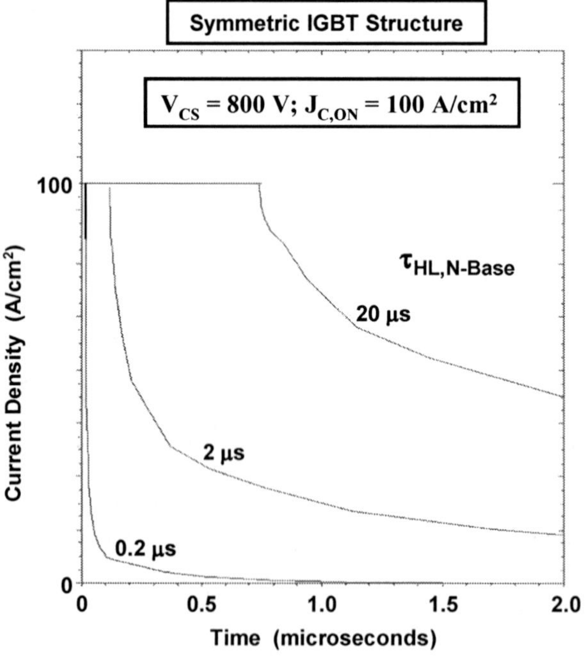

Fig. 9.127 Collector current turn-off waveforms under inductive load conditions for the symmetric IGBT structure with various high-level lifetime values in the N-base region

here for three cases of the lifetime in the N-base region. The structure had an N-base region with a lightly doped portion with thickness of 200 μm and doping concentration of 5×10^{13} cm^{-3}. The device was turned off from a steady-state collector current of 100 A cm^{-2} with a collector bias supply voltage of 800 V by ramping the gate voltage to zero in 10 ns. During the first phase of the turn-off process, the collector current was held constant allowing the collector voltage to increase as a function of time. After the collector voltage reached 800 V, the collector voltage was held constant and the collector current was allowed to decay with time. The waveforms for the collector current obtained from the numerical simulations are shown in Fig. 9.127. It can be seen that the turn-off process becomes faster when the lifetime in the N-base region is reduced resulting in a lower energy loss per cycle.

9.8.2 Asymmetric Structure

For the asymmetric IGBT structure, the tradeoff curve between the on-state voltage drop and the energy loss per cycle during turn-off is usually obtained by varying the doping concentration in the N-buffer layer. This tradeoff curve can be generated by using the analytical models developed in earlier sections for the on-state voltage drop and the energy loss per cycle. As an example, the tradeoff curve obtained for the asymmetric IGBT structure with 1,200-V blocking voltage capability is shown in Fig. 9.128. This structure has a width of 100 μm for the lightly doped portion of the N-base region and 10 μm for the N-buffer layer. It can

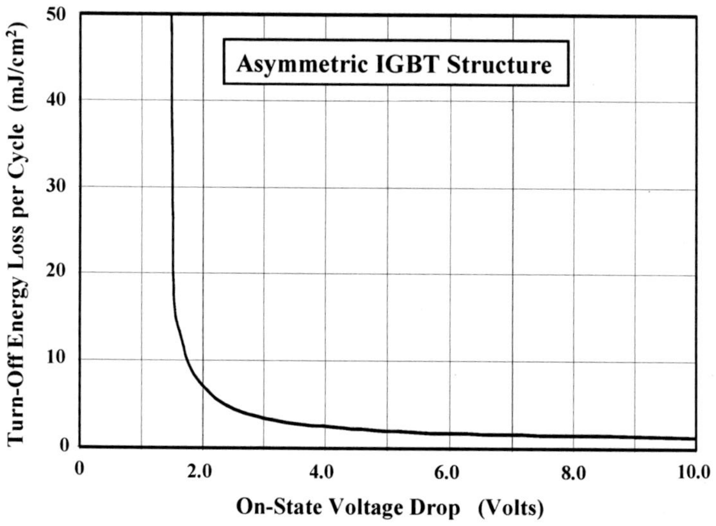

Fig. 9.128 Tradeoff curve for the asymmetric IGBT structure obtained by varying the doping concentration in the N-buffer layer

be observed that the on-state voltage drop increases very rapidly when the buffer layer doping concentration is increased to make the energy loss less than 5 mJ cm^{-2}. The values computed by using the analytical models are in good agreement with those reported in the literature.[23] An improvement in the tradeoff curve can be obtained by operating the IGBT structure at a lower on-state current density. However, this is accomplished at the expense of a larger die size for achieving any desired current rating for the device.

Simulation Example

To validate the above model for the energy loss during turn-off for the asymmetric IGBT structure under inductive load conditions, the results of numerical simulations for the typical 1,200-V structure described in the previous sections are provided here for several cases of the doping concentration in the N-buffer layer. The structure had an N-base region with a lightly doped portion with thickness of 100 μm and doping concentration of 5 × 10^{13} cm^{-3}, and an N-buffer layer with thickness of 10 μm. The device was turned off from a steady-state collector current of 100 A cm^{-2} with a collector bias supply voltage of 800 V by ramping the gate voltage to zero in 10 ns. During the first phase of the turn-off process, the collector current was held constant allowing the collector voltage to increase as a function of time. After the collector voltage reached 800 V, the collector voltage was held constant and the collector current was allowed to decay with time. The waveforms for the collector

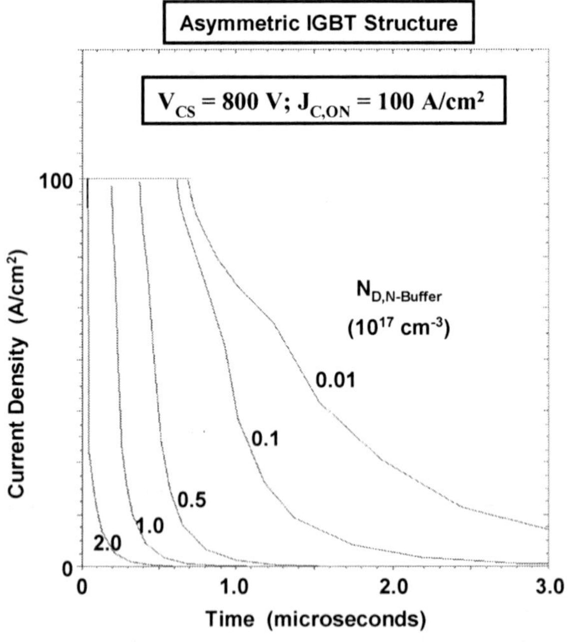

Fig. 9.129 Collector current turn-off waveforms under inductive load conditions for the asymmetric IGBT structure with various doping concentrations in the N-buffer layer

current obtained from the numerical simulations are shown in Fig. 9.129. It can be seen that the turn-off process becomes faster when the doping concentration in the N-buffer layer is increased resulting in a lower energy loss per cycle.

9.8.3 Transparent Emitter Structure

For the transparent emitter IGBT structure, the tradeoff curve between the on-state voltage drop and the energy loss per cycle during turn-off is generated by varying the surface doping concentration of the P^+ collector region. This tradeoff curve can be generated by using the analytical models developed in earlier sections for the on-state voltage drop and the energy loss per cycle. As an example, the tradeoff curve obtained for the transparent emitter IGBT structure with 1,200-V blocking voltage capability is shown in Fig. 9.130. This structure has a width of 200 μm for the N-base region and a thickness of 1 μm for the P^+ collector region. It can be observed that the on-state voltage drop increases rapidly when the doping concentration of the P^+ collector region is reduced to make the energy loss less than 15 mJ cm^{-2}. The values computed by using the analytical models are in good agreement with those reported in the literature.[23–25] An improvement in the tradeoff curve can be obtained by operating the IGBT structure at a lower on-state current density. However, this is accomplished at the expense of a larger die size for achieving any desired current rating for the device.

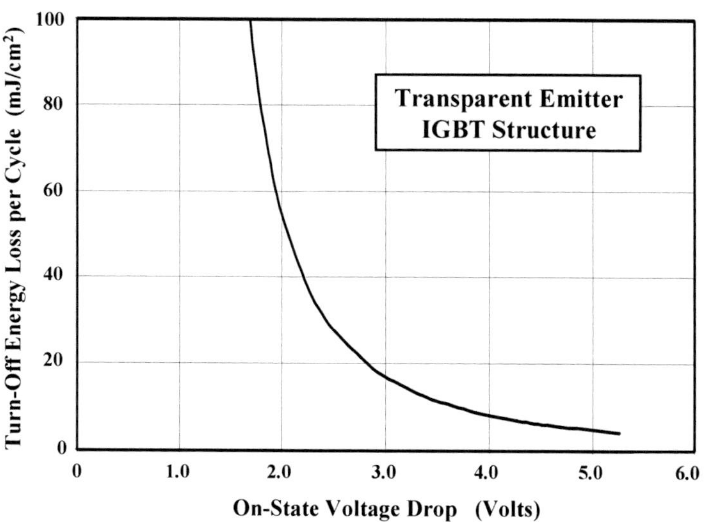

Fig. 9.130 Tradeoff curve for the transparent emitter IGBT structure obtained by varying the doping concentration in the P^+ collector region

Simulation Example

To validate the above model for the energy loss during turn-off for the transparent emitter IGBT structure under inductive load conditions, the results of numerical simulations for the typical 1,200-V structure described in the previous sections are provided here for several cases of the surface doping concentration of the P^+ collector region. The structure had an N-base region with a thickness of 200 μm and doping concentration of 5×10^{13} cm^{-3}. The lifetime (τ_{p0} and τ_{n0}) in the N-base region was 10 μs. The P^+ collector region had a thickness of 1 μm. The device was turned off from a steady-state collector current of 100 A cm^{-2} with a collector bias supply voltage of 800 V by ramping the gate voltage to zero in 10 ns. During the first phase of the turn-off process, the collector current was held constant allowing the collector voltage to increase as a function of time. After the collector voltage reached 800 V, the collector voltage was held constant and the collector current was then allowed to decay with time. The waveforms for the collector current obtained from the numerical simulations are shown in Fig. 9.131. It can be seen that the turn-off process becomes faster when the doping concentration in the P^+ collector region is reduced resulting in a lower energy loss per cycle.

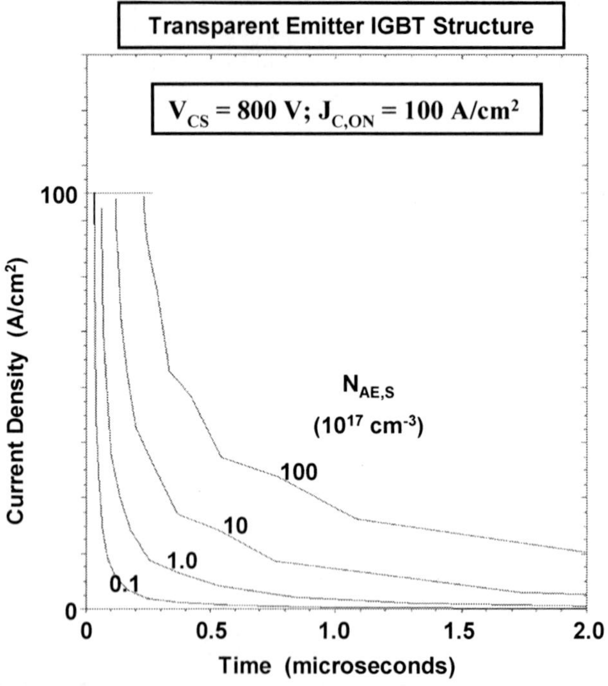

Fig. 9.131 Collector current turn-off waveforms under inductive load conditions for the transparent emitter IGBT structure with various doping levels in the P^+ collector region

9.8.4 Comparison of Tradeoff Curves

A comparison of the tradeoff curves for the three basic IGBT structures can be performed by using Fig. 9.132. All the structures are designed to support 1,200 V in the off-state. It is apparent that the symmetric blocking structure has the worst tradeoff curve while the asymmetric structure has the best tradeoff curve. The most favored IGBT structure used for power electronic applications depends on whether the device must support voltage in both the first and third quadrants. The first IGBT structure reported in the literature was a symmetric blocking design for use in AC and DC circuits.[6] In recognition of the potential improvements in performance for DC circuits, the asymmetric IGBT structure was developed soon thereafter.[8] When the blocking voltage is less than 1,000 V, the asymmetric IGBT structure is usually favored because it can be fabricated by epitaxial growth of the N-buffer layer followed by the growth of the lightly doped portion of the N-base region to a thickness of less than 100 μm. When the blocking voltage exceeds 2,000 V, the transparent emitter IGBT structure is favored because it can be fabricated from bulk silicon wafers with a lightly doped P^+ collector region formed on the back of the wafer by ion implantation. For applications in appliance controls and matrix converters, the symmetric IGBT is necessary to produce the desired reverse-blocking capability.

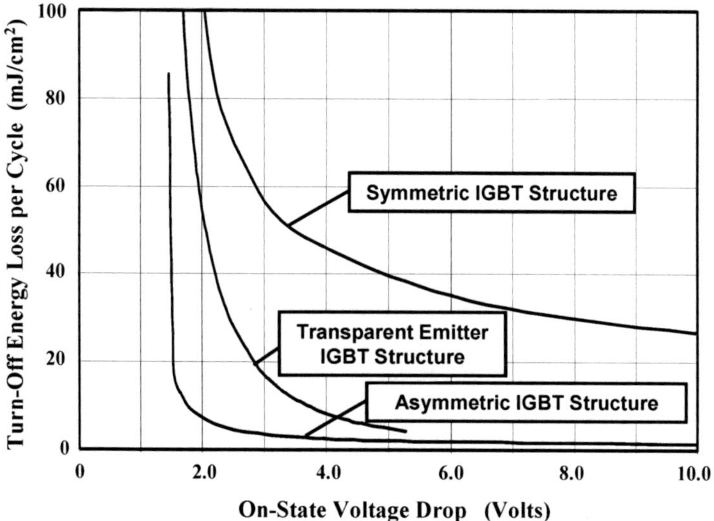

Fig. 9.132 Comparison of the tradeoff curves for the three IGBT structures

9.9 Complementary (P-Channel) Structure

Power switches are often used in AC power circuits for the operation of appliances and numerical controls. Although the triac has been extensively used for such applications in the past as discussed in Chap. 7, it is desirable to have an MOS-gated power device whose rate of turn-on can be controlled by the input gate signal to reduce electromagnetic interference (EMI). An AC power switch configuration can be created by using two n-channel symmetric IGBT structures as illustrated in Fig. 9.133a. In this switch topology, the emitter of one of the devices must be connected to the collector of the other device. If the terminal T_1 is regarded as the reference terminal, high voltage is applied to the other terminal T_2. When the voltage at terminal T_2 is positive, current flow can be produced through the composite switch by turning on IGBT-1 with the application of a positive gate bias relative to the reference terminal T_1. When the voltage at terminal T_2 is negative, current flow can be produced through the composite switch by turning on IGBT-2 with the application of a positive gate bias relative to terminal T_2. However, this requires shifting the gate bias signal for IGBT-2 to a high potential because the voltage at terminal T_2 is large. This complicates the design of the gate circuit and increases its cost.

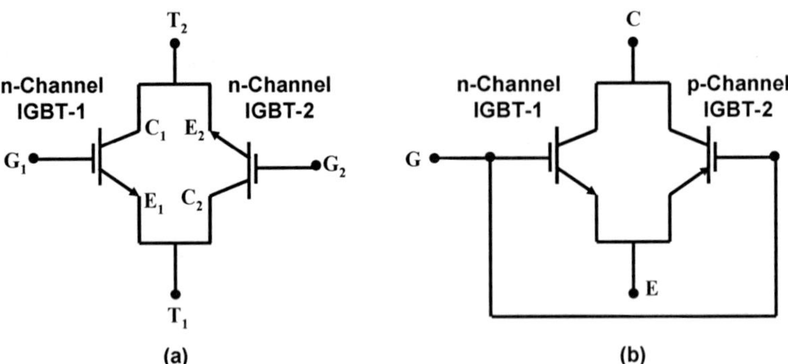

Fig. 9.133 Options for creating an AC switch using two IGBT structures

An alternate AC switch configuration is shown in Fig. 9.133b where one n-channel and one p-channel IGBT structure are utilized. In this switch topology, the emitter terminals for both devices are connected together creating the reference terminal. When the voltage at the collector terminal is positive, current flow can be produced through the composite switch by turning on n-channel IGBT-1 with the application of a positive gate bias relative to the emitter terminal. When the voltage at the collector terminal is negative, current flow can be produced through the composite switch by turning on p-channel IGBT-2 with the application of a negative gate bias relative to the emitter terminal. No level shifting of the gate signal is required making the gate drive circuit less complex and expensive.

As discussed in Chap. 6, the area for the silicon p-channel MOSFET structure is approximately three times larger than that required for the n-channel MOSFET structure to achieve the same on-state resistance because the mobility for holes is smaller than for electrons. This severe penalty in size is not observed for the IGBT structure because of conductivity modulation of the drift region. The resistances of the thick drift region in the n-channel and p-channel IGBT structures are equal during on-state operation. However, the channel resistance for the p-channel IGBT is larger than that for the n-channel IGBT structure.

The p-channel symmetric IGBT structure is illustrated in Fig. 9.134 together with its equivalent circuit. It is constructed by replacing all the layers in the n-channel IGBT structure with their complementary counterparts. Consequently, the equivalent circuit for the p-channel IGBT consists of a p-channel MOSFET driving a wide-base N–P–N transistor. For the same blocking voltage capability, the open-base N–P–N transistor has approximately the same width for the P-base region as width of the N-base region in the P–N–P transistor of the n-channel IGBT structure. For the same lifetime in the wide-base regions for both devices, the gain of the N–P–N transistor in the p-channel IGBT structure can be larger than the gain for the P–N–P transistor in the n-channel IGBT structure because of the larger mobility for electrons. For high lifetime values in the drift region, it has been observed that the on-state voltage drop for the p-channel IGBT structure is very close to that for the n-channel IGBT structure making it an excellent complementary device.[33] As the lifetime is reduced, a greater portion of the collector current flows through the channel of the MOSFET portion in the IGBT structure. The on-state voltage drop for the p-channel IGBT structure becomes larger than that for the n-channel IGBT structure at smaller lifetime values due to the larger channel resistance.

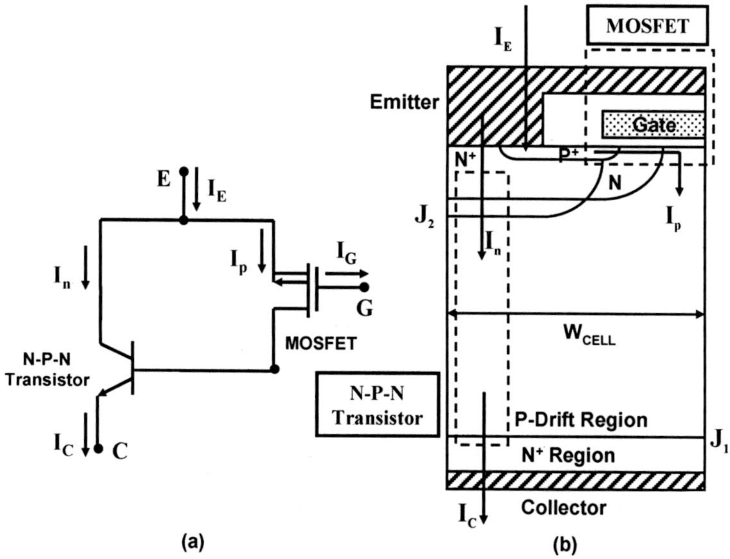

Fig. 9.134 p-channel IGBT structure and its equivalent circuit

9.9.1 On-State Characteristics

The on-state characteristics for the p-channel IGBT structure can be modeled by using the approach previously taken for the n-channel IGBT structure in Sect. 9.5. The on-state voltage drop for the p-channel IGBT structure can be obtained by using

$$V_{ON} = V_{N+P} + V_{PB} + V_{MOSFET}, \tag{9.339}$$

where V_{N+P} is the voltage drop across the N^+ collector/P-base junction (J_1), V_{PB} is the voltage drop across the P-base region after accounting for conductivity modulation due to high-level injection conditions, and V_{MOSFET} is the voltage drop across the MOSFET portion. The voltage drop across the N^+ collector/P-base junction (J_1) can be obtained from the increase in the minority carrier concentration at the junction boundary:

$$V_{N+P} = \frac{kT}{q}\ln\left(\frac{n_0}{n_{0P}}\right) = \frac{kT}{q}\ln\left(\frac{n_0 N_A}{n_i^2}\right), \tag{9.340}$$

where the minority carrier density in equilibrium (n_{0P}) has been related to the doping concentration (N_A) in the P-base region. The increase in the electron concentration at the junction (n_0) can be obtained from the analysis provided in Sect. 9.5.2.

The voltage drop across the P-base region (V_{PB}) can be obtained by integration of the electric field within the P-base region (see Sect. 9.5.3). The first part of the voltage drop is

$$V_{PB1} = \frac{2L_a J_C \sinh(W_P / L_a)}{q n_0 (\mu_n + \mu_p)}\{\tanh^{-1}[e^{-(W_{ON}/L_a)}] - \tanh^{-1}[e^{-(W_P/L_a)}]\}, \tag{9.341}$$

where W_P is the width of the P-base region. The depletion width (W_{ON}) across the deep N^+/P-base junction (J_2) in the on-state depends on the on-state voltage drop. The second part of the voltage drop is

$$V_{PB2} = \frac{kT}{q}\left(\frac{\mu_n - \mu_p}{\mu_n + \mu_p}\right)\ln\left[\frac{\tanh(W_{ON}/L_a)\cosh(W_{ON}/L_a)}{\tanh(W_P/L_a)\cosh(W_P/L_a)}\right]. \tag{9.342}$$

The voltage drop across the MOSFET portion includes contributions from the JFET region, the accumulation layer, and the channel. As discussed in Sect. 9.5.2, the free carrier concentration becomes small in the vicinity of the deep N^+/P-base junction (J_2). Consequently, it is appropriate to compute the resistance of the JFET region based upon its doping concentration. Since p-channel IGBT structures have low doping concentrations in the P-base region to obtain high blocking voltage ratings, the JFET region can become completely depleted by the built-in potential of the deep N^+/P-base junction (J_2). This problem can be overcome by enhancing the doping concentration in the JFET region by the ion implantation and diffusion of boron. For the purposes of formulating an analytical model, the JFET region will be assumed to have an effective doping concentration of 5×10^{15} cm^{-3} based upon a Gaussian profile with a surface concentration of 1×10^{16} cm^{-3}. The resistivity (ρ_{JFET}) for the JFET region is then about 3 Ω cm.

Based upon the analysis for the power MOSFET structure (see (6.72)),

$$V_{\mathrm{JFET}} = J_{\mathrm{C}} R_{\mathrm{JFET,SP}} = \frac{J_{\mathrm{C}} \rho_{\mathrm{JFET}} (x_{\mathrm{JN}} + W_0) W_{\mathrm{CELL}}}{W_{\mathrm{G}} - 2x_{\mathrm{JN}} - 2W_0}, \tag{9.343}$$

where x_{JN} is the junction depth of the deep N^+ region. The voltage drop across the accumulation layer in the IGBT structure can also be derived based upon the analysis for the power MOSFET structure (see (6.66)):

$$V_{\mathrm{ACC}} = J_{\mathrm{C}} R_{\mathrm{A,SP}} = \frac{J_{\mathrm{C}} K_{\mathrm{A}} (W_{\mathrm{G}} - 2x_{\mathrm{JN}}) W_{\mathrm{CELL}}}{4\mu_{\mathrm{pA}} C_{\mathrm{OX}} (V_{\mathrm{G}} - V_{\mathrm{TH}})}. \tag{9.344}$$

The accumulation layer coefficient (K_{A}) for the IGBT structure can be assumed to have the same value (0.6) as for power MOSFET structures. The voltage drop across the channel of the MOSFET portion can also be obtained by using the analysis for the power MOSFET structure (see (6.63)):

$$V_{\mathrm{CH}} = J_{\mathrm{C}} R_{\mathrm{CH,SP}} = \frac{J_{\mathrm{C}} L_{\mathrm{CH}} W_{\mathrm{CELL}}}{2\mu_{\mathrm{pi}} C_{\mathrm{OX}} (V_{\mathrm{G}} - V_{\mathrm{TH}})}. \tag{9.345}$$

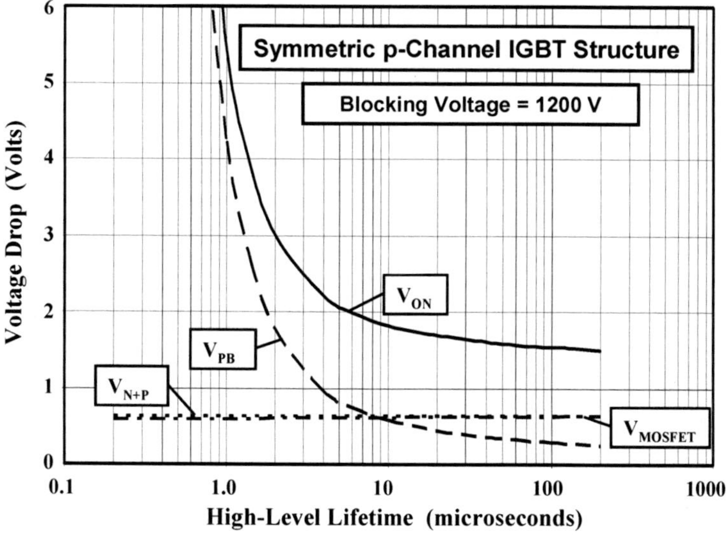

Fig. 9.135 On-state voltage drop for the symmetric P-channel IGBT structure

As an example, the components of the on-state voltage drop determined by using the above analytical model are provided in Fig. 9.135 for the case of the symmetric p-channel IGBT structure with blocking voltage capability of 1,200 V. The device has a p-base width of 200 μm with a depth of 5 μm for the deep N^+ region. All the structural parameters used are the same as those used for the n-channel symmetric IGBT structure in Sects. 9.5.2 and 9.5.3. The MOSFET portion has an effective JFET region doping concentration of 5×10^{15} cm^{-3}, a gate oxide thickness of 500 Å, and a channel length of 1.5 μm. The mobility for holes in the inversion and accumulation layers was assumed to be 200 and 400 cm^2 V^{-1} s^{-1}, respectively. The IGBT structure has a cell pitch (W_{CELL}) of 30 μm with a gate electrode width (W_G) of 16 μm. The gate bias and threshold voltages used are 15 and 5 V, respectively. From the figure, it can be concluded that the voltage drop across the N$^+$ collector/P-base junction and the MOSFET portion is dominant when the high-level lifetime is large (>20 μs). When the high-level lifetime is reduced, the voltage drop across the N-base region increases and becomes dominant for lifetime value below 5 μs. A very rapid increase in the on-state voltage drop is observed when the high-level lifetime is reduced below 2 μs.

Simulation Example

To corroborate the analytical model for the on-state voltage drop for the p-channel IGBT structure, the results of two-dimensional numerical simulations for the typical symmetric structure are discussed here. The doping profiles for this structure are similar to those for the n-channel device with all the N-type layers replaced by P-type layers, and vice versa. The on-state voltage drop was obtained at a current density of 100 A cm^{-2} at a gate bias of 15 V for various values for the high-level lifetime. It was found that the on-state voltage drop increases rapidly when the high-level lifetime is reduced below 2 μs. The on-state voltage drop obtained from the numerical simulations is compared with those obtained by using the analytical model in Fig. 9.136. The analytical model provides an accurate prediction of the on-state voltage drop.

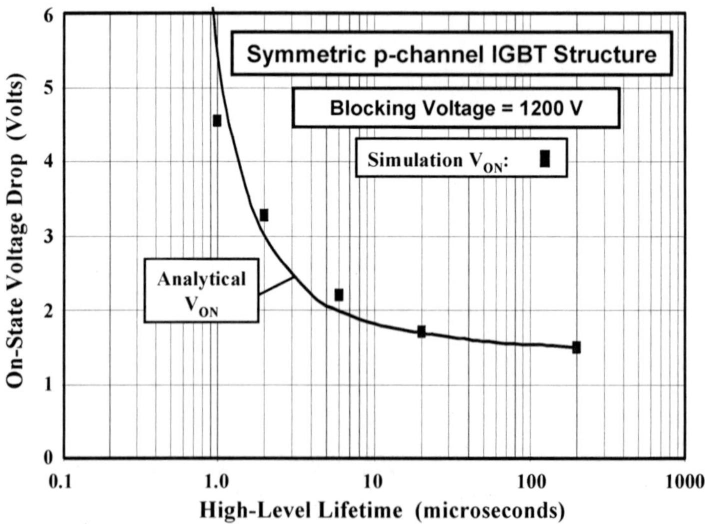

Fig. 9.136 On-state voltage drop for the symmetric p-channel IGBT structure

The electron distribution within the P-base region of the p-channel IGBT structure is shown in Fig. 9.137. It can be observed that the electron concentration is much larger than the background doping concentration in the P-base region confirming its conductivity modulation. This strong conductivity modulation is responsible for the excellent on-state characteristics of the p-channel IGBT structure. The carrier profiles are very similar to those in Fig. 9.46 for the symmetrical n-channel IGBT structure. This justifies using the same model for determination on the on-state characteristics for both structures.

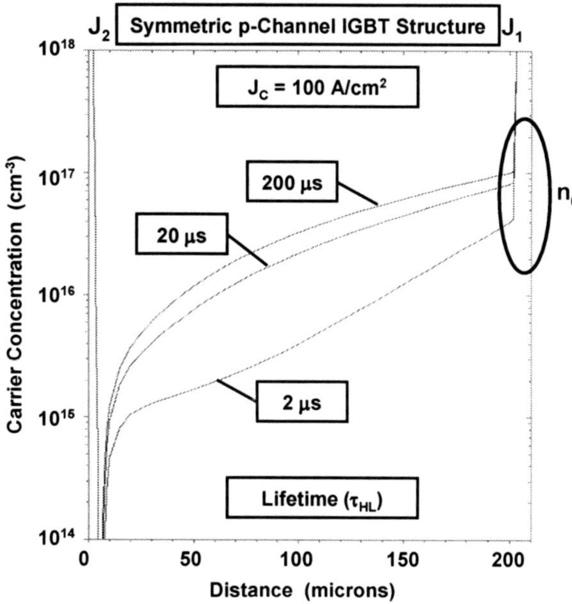

Fig. 9.137 On-state carrier distribution in the symmetric p-channel IGBT structure

9.9.2 Switching Characteristics

The p-channel IGBT structure can be turned off by removal of the gate bias. As in the case of the n-channel structure, this eliminates the channel in the MOSFET portion cutting off the base drive current for the wide-base N–P–N transistor. The turn-off of the symmetric p-channel IGBT occurs by the removal of the stored electrons in the P-base region by recombination at high injection levels. Since this process is identical to that discussed previously for the symmetric n-channel IGBT structure in Sect. 9.7.4, the turn-off energy loss per cycle for the symmetric p-channel IGBT structure is identical to the n-channel structure for the same lifetime in the wide-base region. The equations developed in Sect. 9.7.4 can therefore be used for the symmetric P-channel IGBT structure as well.

9.9.3 Power Loss Optimization

As in the case of the symmetric n-channel IGBT structure, the power loss incurred in the symmetric p-channel IGBT structure can be optimized from an application's perspective by performing a tradeoff between the on-state voltage drop and the turn-off energy loss per cycle. The tradeoff curve obtained by using the analytical models for the on-state voltage drop and the turn-off energy loss per cycle for the symmetric p-channel IGBT structure is provided in Fig. 9.138. From this figure, it can be concluded that the on-state voltage drop increases rapidly if the energy loss per cycle is reduced below 70 mJ cm^{-2}.

For comparison, the tradeoff curve for the symmetric n-channel IGBT structure is also shown in Fig. 9.138. It can be observed that the tradeoff curve for the symmetric p-channel IGBT is worse than that for the n-channel structure.[33] For this reason, most of the IGBT products have been developed by using the n-channel structure. The p-channel IGBT structure is utilized in applications only when there is a need for complementary devices in the power circuit.

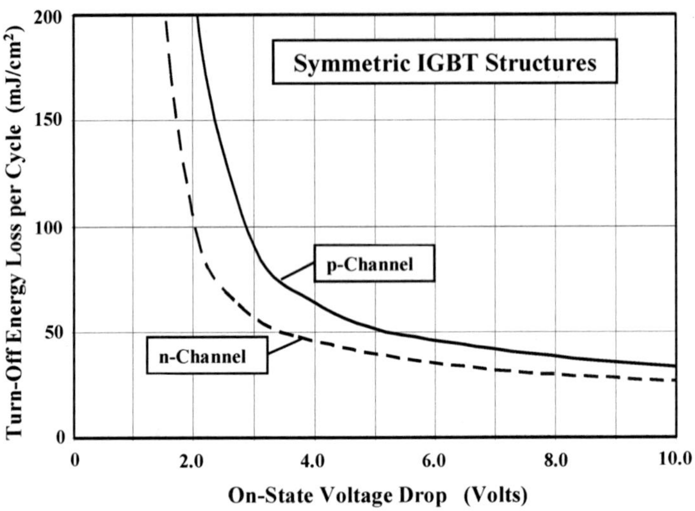

Fig. 9.138 Tradeoff curve for the symmetric p-channel IGBT structure

9.10 Latch-Up Suppression

The IGBT structure normally operates as a wide-base bipolar transistor being driven by a MOSFET structure to provide its base current. However, the construction of the IGBT structure produces four alternating layers of N-type and P-type regions, which constitute a parasitic thyristor. The parasitic P^+–N–P–N^+ thyristor formed within an n-channel IGBT structure is indicated in Fig. 9.10. If the parasitic thyristor is triggered during operation in the first quadrant, it can latch up resulting in current flow that bypasses the MOSFET channel in the IGBT structure. Consequently, once the parasitic thyristor latches up, it is no longer possible to control the device operation with the gate signal. The IGBT structure can undergo destructive failure when the latch-up occurs because of a sudden upsurge in the current flow. When the IGBT structure was first proposed, skeptics considered the sufficient suppression of the parasitic thyristor to be impossible making the device not viable from a commercial standpoint. Fortunately, the latch-up phenomenon was successfully suppressed by a concerted research effort undertaken in the 1980s.

When invented, n-channel IGBT structures were proposed without the N^+ region to eliminate the parasitic thyristor.[34] This approach has not yet been found to be practical. At that time, it was also recognized that the parasitic thyristor can be suppressed by short circuiting the N^+ region to the P-base region as illustrated in Fig. 9.10. Unfortunately, this short circuit occurs only at one end of the N^+ region. During on-state operation, the bipolar current must flow from the N-base region into the P-base region, which acts as the collector of the wide-base N–P–N transistor. The bipolar current collected by the P-base region flows out of the device at the center of the window in the polysilicon gate electrode where the emitter electrode is in contact with the P-base region. The bipolar current collected at the edge of the P-base region near the center of the gate electrode must flow below the N^+ emitter region before being removed by the emitter electrode. The resistance of the P-base region under the N^+ emitter region produces a voltage drop that forward biases the junction J_3 between these regions. When the forward bias becomes sufficient to promote the injection of electrons from the N^+ emitter region, the parasitic thyristor gets triggered resulting in latch-up.

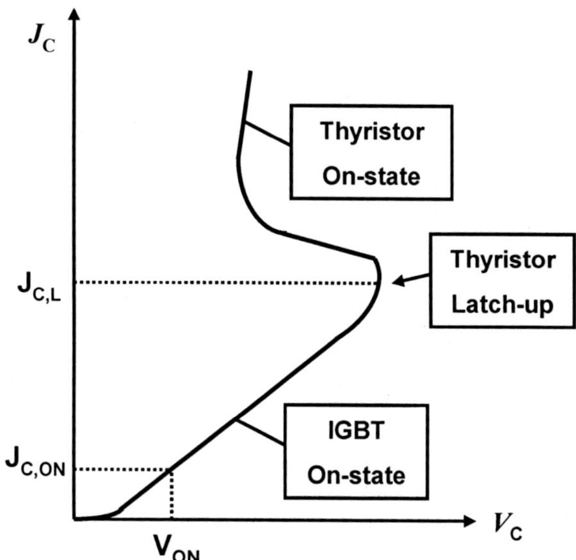

Fig. 9.139 Latch-up characteristics for the IGBT structure

The *i–v* characteristic of the IGBT with the portion after latch-up is shown in Fig. 9.139. It can be observed that there is distinct snapback of the characteristic when latch-up occurs. This is a signature used to define the on-set of latch-up in these devices. To obtain a broad range of operating current levels, the latch-up current density ($J_{C,L}$) is usually made at least ten times larger than the on-state current density ($J_{C,ON}$). The abrupt reduction of the voltage drop across the IGBT structure at the on-set of latch-up can produce a surge in the current flow during circuit operation which can result in destructive failure of the device.

A thorough effort was undertaken during the early years of development of the IGBT structure to find methods to suppress the activation of the parasitic thyristor. The widespread commercial success of the device has resulted from the application of these ideas. In this section, the various approaches that have been explored to suppress the activation of the parasitic thyristor are analyzed. The approaches that have been shown to have a strong impact on latch-up suppression are highlighted. Using these methods, the IGBT structure can be operated up to a high-current density sufficient for various applications. In the ideal case, the device can be designed to undergo current saturation prior to the on-set of latch-up of the parasitic thyristor eliminating this problem.

In order for triggering the parasitic thyristor into its regenerative mode, the current gain of the two coupled bipolar transistors within the IGBT structure must exceed unity as discussed in Chap. 8. At the fundamental level, the suppression of latch-up of the parasitic thyristor can be achieved by reducing the current gain of either the N–P–N transistor or the P–N–P transistor or both. Since the wide-base P–N–P transistor within the n-channel IGBT structure participates in conducting the on-state current, suppressing the gain of the P–N–P transistor has an adverse impact on the on-state voltage drop. It is therefore preferable to reduce the current gain of the N–P–N transistor within the n-channel IGBT structure. Consequently, most of the effort on latch-up suppression has been directed toward reducing the gain of the N–P–N transistor within the n-channel IGBT structure.

9.10.1 Deep P^+ Diffusion

The most effective technique developed to suppress the latch-up of the parasitic thyristor in the IGBT structure is by reducing the resistance for the hole current path. This cannot be accomplished by increasing the doping level of the P-base region because this would increase the channel doping concentration and hence the threshold voltage. The inclusion of a deep P^+ diffusion in the middle of the polysilicon window can be used to reduce the sheet resistance of the P-base region under the N^+ region without changing the doping in the channel of the MOSFET portion.[35] The IGBT structure with the deep P^+ region is compared with the IGBT structure without the P^+ region in Fig. 9.140. The location and depth of the P^+ region must be carefully chosen to ensure that the lateral extension of its diffused profile does not approach the peak doping concentration (approximately 1×10^{17} cm^{-3}) of the channel.

A simple analysis of the latch-up current density for the IGBT structure can be performed by analysis of the forward bias across the junction (J_3) between the N^+ emitter region and the P-base region. In the n-channel IGBT structure, the hole current component is responsible for producing the forward bias across this junction as it flows through the P-base region. The hole current component is given by

$$J_{\mathrm{p}} = \alpha_{\mathrm{PNP,ON}} J_{\mathrm{C}} = \gamma_{\mathrm{E}} \alpha_{\mathrm{T,ON}} J_{\mathrm{C}}. \tag{9.346}$$

During on-state operation, the injection efficiency is reduced due to high-level injection in the N-base region as discussed in Sect. 9.5.2. The base transport factor in the on-state is given by

$$\alpha_{\mathrm{T,ON}} = \frac{1}{\cosh(W_{\mathrm{N}} / L_{\mathrm{a}})}. \tag{9.347}$$

The common-base current gain (α_{PNP}) for the P–N–P transistor in the IGBT structure is typically 0.4 under on-state operating current levels.

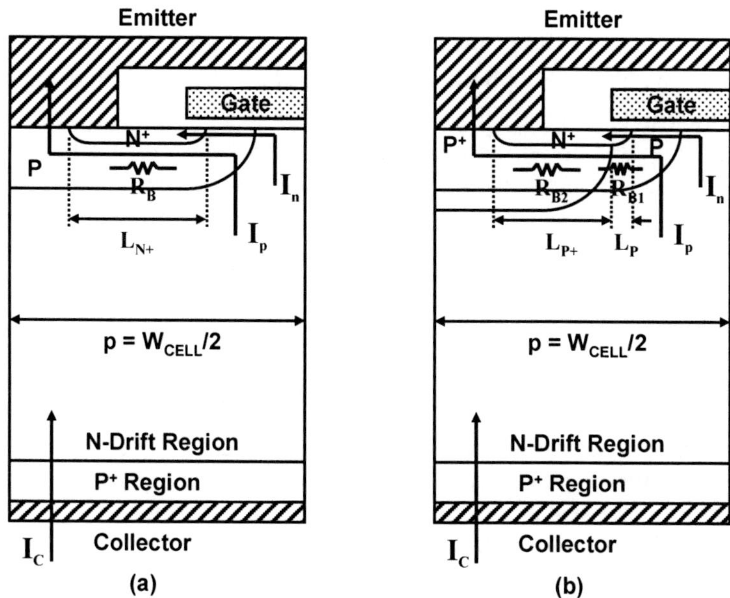

Fig. 9.140 IGBT structures with and without the deep P^+ region

For the IGBT structure shown in Fig. 9.140a without the deep P^+ region, latch-up of the parasitic thyristor will occur when the voltage drop in the P-base region produced by the hole current becomes equal to the built-in potential of the P–N junction:

$$I_{\mathrm{p}} R_{\mathrm{B}} = V_{\mathrm{bi}}. \tag{9.348}$$

The hole current is equal to the product of the hole current density and the cell area ($W_{\mathrm{CELL}} Z$), where Z is the length of the IGBT cell orthogonal to the cross section in the figure. The resistance of the P-base region is given by

$$R_B = \frac{\rho_{SP} L_{N+}}{Z} = \frac{\rho_P L_{N+}}{(x_{J,P} - x_{J,N+})Z} = \frac{L_{N+}}{q\mu_{PB} N_{AP}(x_{J,P} - x_{J,N+})Z}, \tag{9.349}$$

where ρ_{SP} is the pinch sheet resistance of the P-base region below the N^+ emitter region. The resistivity (ρ_P) and the doping concentration (N_{AP}) of the P-base region that are appropriate for this expression are determined by only the portion of the P-type diffusion located below the N^+ region. Combining these relationships provides an expression for the latch-up current density:

$$J_{C,L}(\text{NoP}^+) = \frac{V_{bi}}{\alpha_{PNP,ON}\rho_{SP} L_{N+} p}. \tag{9.350}$$

For a typical P-base surface concentration of about 2×10^{17} cm^{-3}, required to obtain a peak channel doping concentration of 1×10^{17} cm^{-3}, the average doping concentration (N_{AP}) in the P-base region is 5×10^{16} cm^{-3}. Typical values for the vertical junction depths of the P-base ($x_{J,P}$) and N^+ ($x_{J,N+}$) regions are 3 and 1 μm, respectively. For these parameters, the pinch sheet resistance for the P-base region is found to be 2,150 Ω sq^{-1}. For a 16-μm cell pitch (p) with a polysilicon window of 8 μm and a diffusion of the N^+ emitter region up to a 2-μm wide window in the middle of the polysilicon window, the length of the N^+ region (L_{N+}) including its lateral extensions is 8 μm. Using these values, the latch-up current density obtained by using the analytical model is 725 A cm^{-2} for a built-in potential of 0.8 V and a current gain of 0.4. Although this may seem to be adequately above the typical on-state current density of 100 A cm^{-2}, the degradation of the latch-up current density with increasing temperature must be taken into account.[36] When the temperature of the IGBT structure increases due to power dissipation during circuit operation, the latch-up current density is reduced by (1) a reduction of the mobility for holes in the P-base region, (2) a reduction of the built-in potential for the injection at the P–N junction, and (3) an increase in the current gain due to an increase in the lifetime. As an example, when the temperature is increased by 100°C, the mobility for holes is reduced by a factor of 2 times and the built-in potential is reduced to 84%. The net impact of these changes is a reduction of the latch-up current density to only 300 A cm^{-2} at 100°C. This is inadequate because the IGBT structure must be able to handle the large reverse recovery current of the fly-back diode during circuit operation as illustrated in Fig. 9.89.

The improvement in the latch-up current density for the IGBT structure by the addition of the deep P^+ region can be analyzed by using the structure shown in Fig. 9.140b. In this case, the hole current component flows through a portion of the P-base region with a high doping concentration associated with the deep P^+ region and a portion of the P-base region with smaller doping concentration near the edge of the gate electrode. The resistance of the P-base region is then given by

$$R_B = R_{B1} + R_{B2}, \tag{9.351}$$

where the resistance of the portion of the P-base region with smaller doping concentration near the edge of the gate electrode is given by

$$R_{B1} = \frac{\rho_P L_P}{(x_{J,P} - x_{J,N+})Z} = \frac{\rho_{SP} L_P}{Z} \tag{9.352}$$

and

$$R_{B2} = \frac{\rho_{SP+} L_{P+}}{Z} = \frac{\rho_{P+} L_{P+}}{(x_{J,P+} - x_{J,N+})Z} = \frac{L_{P+}}{q\mu_{pP+} N_{AP+}(x_{J,P+} - x_{J,N+})Z}. \tag{9.353}$$

The resistivity (ρ_{P+}) and the doping concentration (N_{AP+}) of the P^+ region that are appropriate for this expression are determined by only the portion of the P-type diffusion located below the N^+ region. Combining these relationships provides an expression for the latch-up current density:

$$J_{C,L}(\text{with } P^+) = \frac{V_{bi}}{\alpha_{PNP,ON}(\rho_{SP} L_P + \rho_{SP+} L_{P+})p}. \tag{9.354}$$

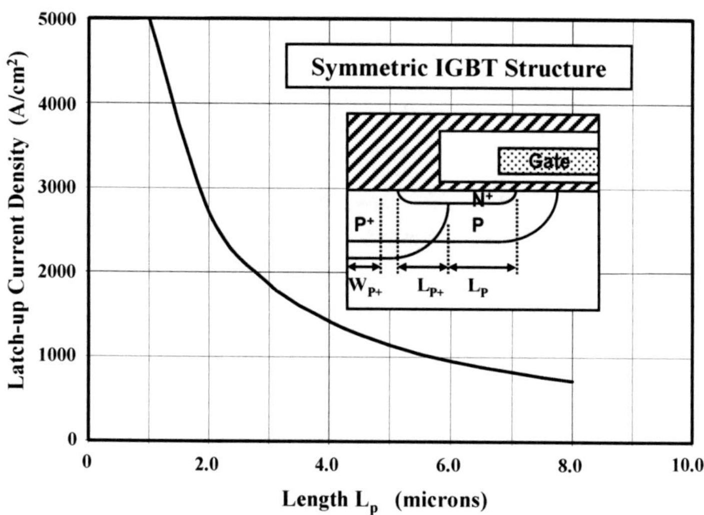

Fig. 9.141 Latch-up current density for the n-channel IGBT structure with the deep P^+ region

For a typical deep P^+ region with surface concentration of 1×10^{19} cm^{-3}, the average doping concentration (N_{AP+}) in the P^+ region is 5×10^{18} cm^{-3}. A typical value for the junction depth of the P^+ region ($x_{J,P+}$) is 5 μm. For these parameters, the pinch sheet resistance for the P^+ region is found to be 50 Ω sq^{-1}. The latch-up current density depends on width of the window (W_{P+}) through which the deep P^+ diffusion is performed as shown in Fig. 9.141. For the case of a length of the N^+ region (L_P) extending beyond the deep P^+ region of 1 μm, the latch-up current density obtained by using the analytical model is 5,000 A cm^{-2} for a built-in potential of 0.8 V. This value provides a sufficiently large margin that takes into account high temperature operation. The values obtained by using the analytical model are in good agreement with those reported for experimental devices in the literature.[35]

Simulation Example

To corroborate the analytical model for the latch-up current density for the IGBT structure, the results of two-dimensional numerical simulations for the typical symmetric n-channel structure are discussed here. A three-dimensional view of the doping profile for a structure with a deep P^+ region with width (W_{P+}) of 2 μm is shown in Fig. 9.142. The length of the N^+ region beyond the deep P^+ region is approximately 3 μm for this case.

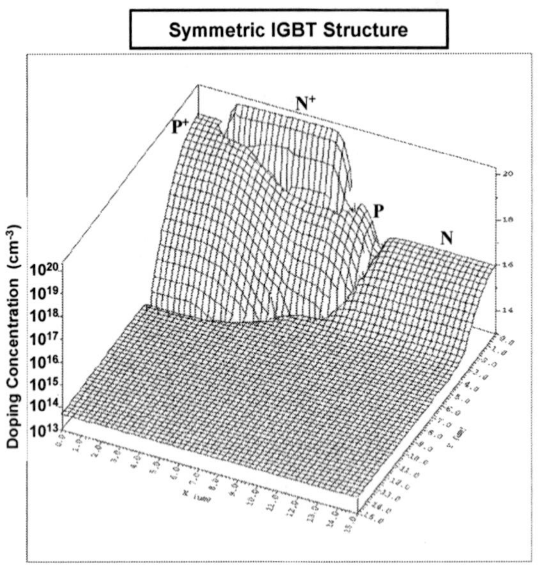

Fig. 9.142 Three-dimensional view of the doping distribution in the symmetric N-channel IGBT structure

The latch-up current density was obtained by using a gate bias of 15 V by sweeping the collector voltage until the collector current exhibited the snapback signature, indicating latch-up of the parasitic thyristor. To examine the impact of the position of the P^+ diffusion window, structures with various values for the diffusion window (W_{P+}) for the deep P^+ region, as well as the case without the deep P^+ region, were analyzed. The lifetime in the N-base region was 10 μs for all the structures. The on-state *i–v* characteristic for the IGBT structure with a deep P^+ window of 2 μm is shown in Fig. 9.143 with the electron and hole components. It can be observed from this figure that the common-base current gain is decreasing with increasing collector current density. This phenomenon is due to the reduction of the injection efficiency as the device enters into high-level injection conditions in the N-base region. At the on-state current density of 100 A cm^{-2}, the common-base current gain ($\alpha_{PNP,ON}$) is found to be 0.4.

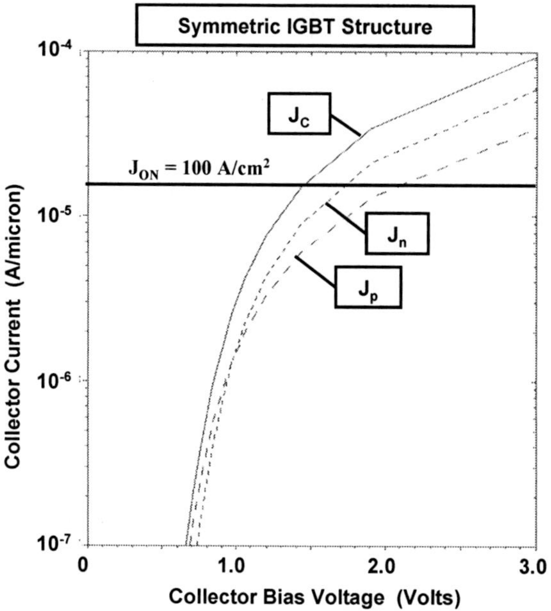

Fig. 9.143 On-state current components for the symmetric IGBT structure

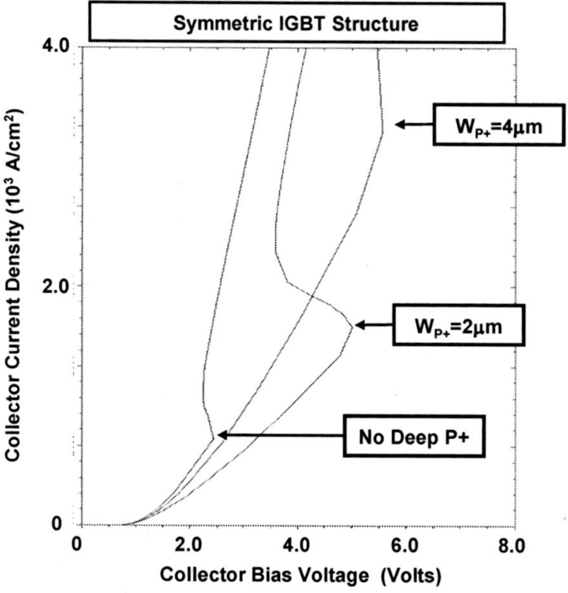

Fig. 9.144 Latch-up characteristics of symmetric IGBT structures

The latch-up characteristics for the symmetric n-channel IGBT structures at 300 K with two values of the deep P^+ diffusion window are compared with the device structure without a deep P^+ region in Fig. 9.144. The latch-up current density for the structure without the deep P^+ region is 700 A cm^{-2} as predicted by the analytical model. The latch-up current densities for the structure with the deep P^+ regions are 1,670 and 3,330 A cm^{-2} corresponding to L_P values of 1.5 and 3.0 μm extracted from the doping distributions. The analytical model provides an accurate estimation of the latch-up current density in spite of the complex nonuniform dopant distributions in the actual structures.

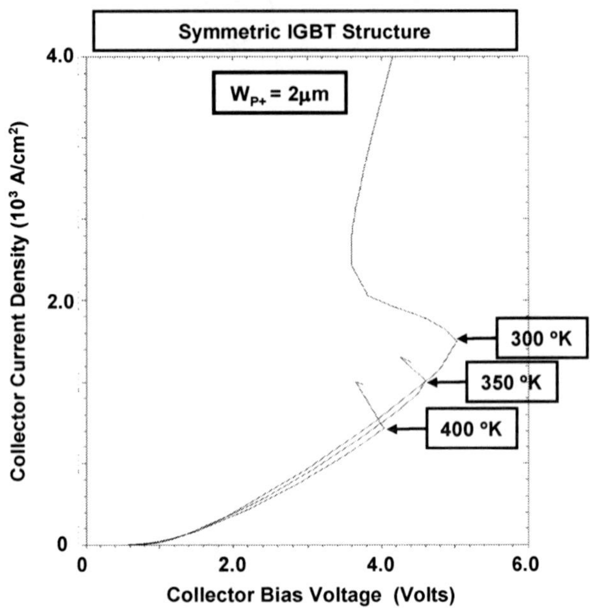

Fig. 9.145 Latch-up characteristics of a symmetric IGBT structure at various temperatures

The latch-up characteristics for the symmetric n-channel IGBT structure with a deep P^+ region formed by using a diffusion window of 2 μm are shown in Fig. 9.145 at various ambient temperatures. The latch-up current density is 1,670, 1,330, and 930 A cm^{-2} at 300, 350, and 400 K, respectively. The latch-up current density decreases by about a factor of 2 from 300 to 400 K as predicted by the analytical model.

9.10.2 Shallow P^+ Layer

An alternate method for reducing the resistance in the P-base region to suppress latch-up of the parasitic thyristor in the IGBT structure utilizes a shallow P^+ region instead of the deep P^+ region described in Sect. 9.10.1. A cross section of the structure is shown in Fig. 9.146 together with the doping profile. For the most effective improvement in the latch-up current level, it is preferable to align the P^+

ion implantation to the polysilicon gate electrode, so that the sheet resistance of the P-base region is reduced along its entire length. However, this limits the maximum depth of the P$^+$ region below the semiconductor surface because the ion implant energy cannot be increased beyond the point of penetration through the gate electrode. By ion implantation of boron (chosen due to its low mass and higher range in silicon) with an energy of 120 keV, the shallow P$^+$ region can be placed about 1 μm below the surface. The dose of the shallow P$^+$ region is constrained to 1×10^{14} cm^{-2} (corresponding to a peak concentration of about 1×10^{18} cm^{-3}) by the compensation of the N$^+$ emitter region.

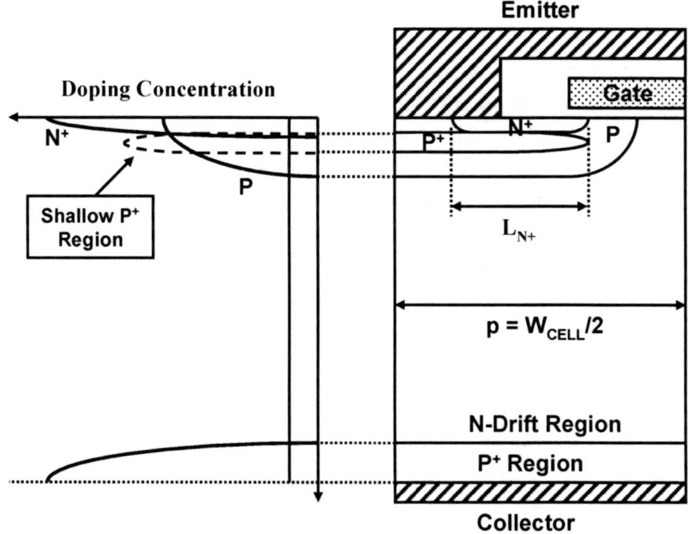

Fig. 9.146 IGBT structure with the shallow P$^+$ region and its doping profile

The latch-up current density for the IGBT structure with the shallow P$^+$ layer can be derived by using the approach used in Sect. 9.10.1. Since the shallow P$^+$ layer extends under the entire N$^+$ region, the resistance of the P-base region is given by

$$R_{\mathrm{B}} = \frac{\rho_{\mathrm{SPB}} L_{\mathrm{N+}}}{Z}, \tag{9.355}$$

where ρ_{SPB} is the pinch sheet resistance of the P-base region below the N$^+$ region, including the shallow P$^+$ layer. This sheet resistance can be obtained as the parallel combination of the sheet resistance of the original P-base diffusion and the sheet resistance of the shallow P$^+$ layer:

$$\rho_{\mathrm{SPB}} = \frac{\rho_{\mathrm{SP}} \rho_{\mathrm{SSP}}}{\rho_{\mathrm{SP}} + \rho_{\mathrm{SSP}}}. \tag{9.356}$$

The sheet resistance (ρ_{SP}) of the original P-base region is given by (9.349). The sheet resistance (ρ_{SSP}) of the shallow P$^+$ layer is given by

$$\rho_{\mathrm{SSP}} = \frac{1}{q\mu_{\mathrm{pP+}} N_{\mathrm{AE,P+}} t_{\mathrm{SP}}}, \tag{9.357}$$

where $N_{\mathrm{AE,P+}}$ is the effective doping concentration of the shallow P^+ region and t_{SP} is its thickness. The latch-up current density can then be computed by using

$$J_{\mathrm{C,L}}(\text{shallow P}^+) = \frac{V_{\mathrm{bi}}}{\alpha_{\mathrm{PNP,ON}} \rho_{\mathrm{SPB}} L_{\mathrm{N+}} p}. \tag{9.358}$$

For the shallow P^+ layer described previously, the effective doping concentration is about 5×10^{17} cm^{-3} with a thickness of about 1 μm. For these parameters, the sheet resistance (ρ_{SSP}) for the shallow P^+ layer is found to be 680 Ω sq^{-1} when compared with 2,150 Ω sq^{-1} for the original P-base region. Consequently, the latch-up current density is enhanced by about three times from 725 to 2,290 A cm^{-2} for the device structure with dimensions provided in the previous section. This improved performance has been experimentally confirmed by using high energy boron ion implantation self-aligned to the gate electrode.[37] One of the problems encountered with this approach is the compensation of the N^+ region by the dopants from the shallow P^+ layer. This can increase the resistance of the N^+ region, which can lead to a larger on-state voltage drop.

Simulation Example

To corroborate the analytical model for the latch-up current density for the IGBT structure with the shallow P^+ layer, the results of two-dimensional numerical simulations for the typical n-channel structure are discussed here. The vertical doping profile for the structure with the shallow P^+ region is provided in Fig. 9.147.

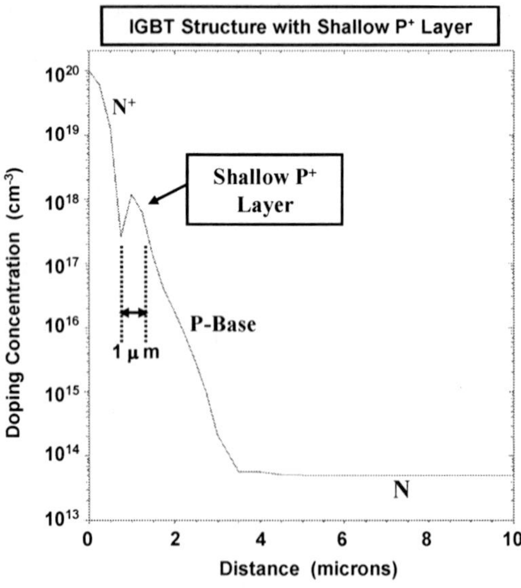

Fig. 9.147 Vertical doping profile in the IGBT structure with shallow P^+ layer

The latch-up current density for the IGBT structure with the shallow P^+ layer was obtained by using a gate bias of 15 V by sweeping the collector voltage until the collector current exhibited the snapback signature, indicating latch-up of the parasitic thyristor. The latch-up characteristic for this structure is compared with the symmetric n-channel IGBT structure with just the P-base region in Fig. 9.148. The latch-up current density for the structure with the shallow P^+ region is 2,000 A cm^{-2}, an improvement by a factor of about 3 times as predicted by the analytical model. No change in the on-state voltage drop is observed.

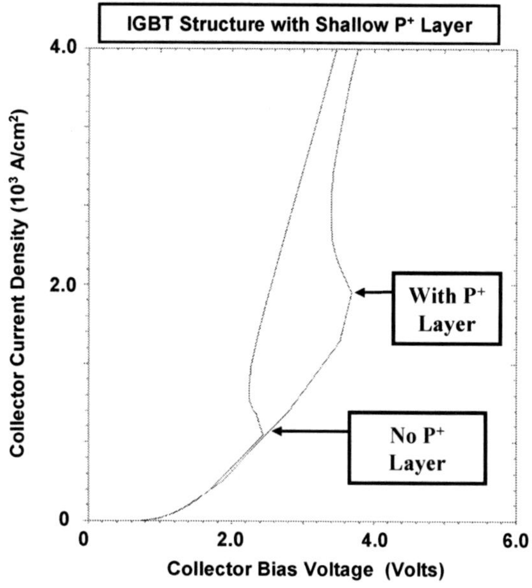

Fig. 9.148 Improvement in the latch-up characteristics of symmetric IGBT structure by addition of the shallow P^+ layer

9.10.3 Reduced Gate Oxide Thickness

To increase the latching current density of the IGBT structure, it is desirable to reduce the sheet resistance of the P-base region by increasing its doping concentration. This can be accomplished without an undesirable increase in the threshold voltage by simultaneously reducing the gate oxide thickness. This has the added benefit of reducing the channel resistance which improves the on-state voltage drop of the IGBT structure.

The threshold voltage for the IGBT structure can be derived by using the same methodology used for the power MOSFET structure (see Sect. 6.5.4):

$$V_{\mathrm{TH}} = \frac{t_{\mathrm{OX}}\sqrt{4\varepsilon_{\mathrm{S}} kTN_{\mathrm{A}} \ln(N_{\mathrm{A}}/n_{\mathrm{i}})}}{\varepsilon_{\mathrm{OX}}} + \frac{2kT}{q}\ln\left(\frac{N_{\mathrm{A}}}{n_{\mathrm{i}}}\right). \tag{9.359}$$

In writing this equation, the impact of the work function of the gate electrode and oxide charge is not included for simplicity.

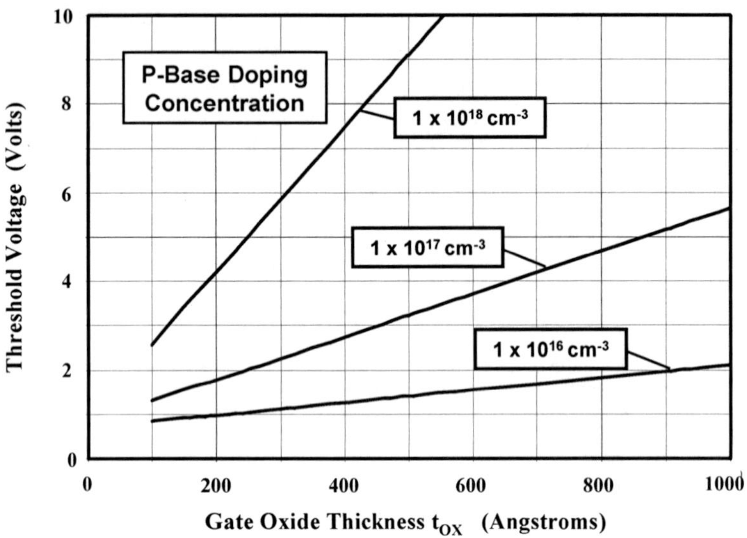

Fig. 9.149 Dependence of threshold voltage on gate oxide thickness

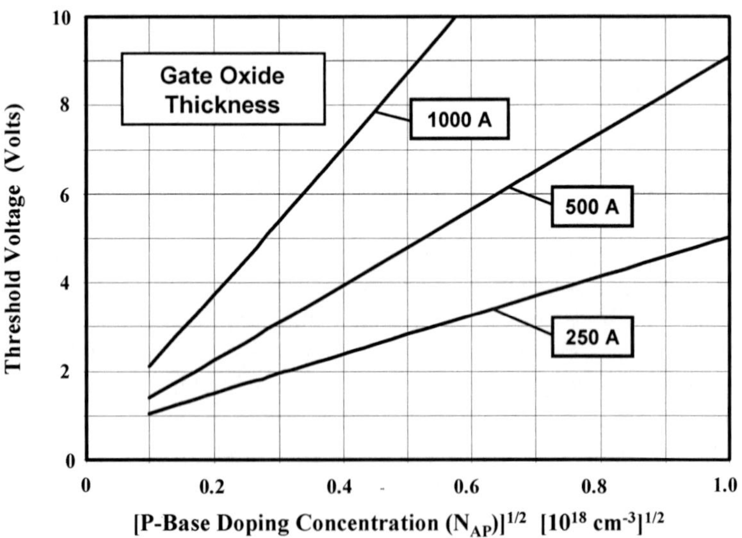

Fig. 9.150 Dependence of threshold voltage on P-base doping concentration

The threshold voltage computed by using (9.359) is plotted in Fig. 9.149 as a function of the gate oxide thickness. It is apparent that the threshold voltage increases linearly with gate oxide thickness in accordance with the first term in the equation. The threshold voltage obtained by using this equation is plotted in Fig. 9.150 as a function of the *square root* of the P-base doping concentration. It is apparent that there is a linear variation of the threshold voltage in this plot in accordance with the first term in the equation. From these plots, it can be concluded that the first term in the equation is dominant allowing computation of the threshold voltage by using

$$V_{\mathrm{TH}} = \frac{t_{\mathrm{OX}}}{\varepsilon_{\mathrm{OX}}}\sqrt{4\varepsilon_{\mathrm{S}} kTN_{\mathrm{A}} \ln\left(\frac{N_{\mathrm{A}}}{n_{\mathrm{i}}}\right)}. \tag{9.360}$$

Based upon this equation, the threshold voltage increases linearly with increasing oxide thickness and approximately as the square root of the doping concentration in the semiconductor. Consequently, a reduction of the gate oxide thickness by a factor 2 times allows increasing the doping concentration of the P-base region by a factor of 4 times while maintaining the same threshold voltage. This should produce an increase in the latch-up current density by a factor of 4 times as well. In practice, the improvement in latch-up current density is smaller due to a reduction of the mobility with increasing doping level.

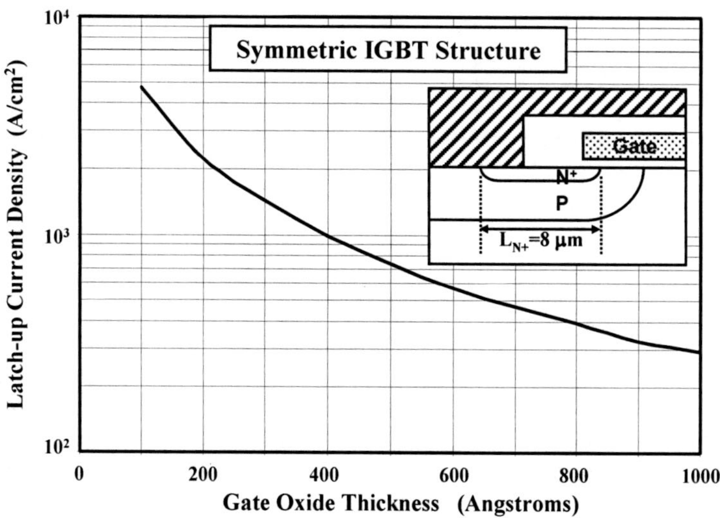

Fig. 9.151 Latch-up current density for IGBT structures with different gate oxide thickness

The improvement in the latch-up current density that can be achieved by scaling down the gate oxide thickness while maintaining a fixed threshold voltage is demonstrated in Fig. 9.151 for the case of the symmetric IGBT structure. This structure has no deep P^+ region as illustrated in the inset in the figure. This structure has an N-base region with width of 200 μm and high-level lifetime of 20 μs resulting in a common-base current gain of 0.4. The P-base doping concentration used for the computation is based upon using (9.359) to obtain the maximum P-base doping concentration in the P-base region and dividing this by a factor of 2 to obtain the effective doping concentration in the P-base region. It can be observed from this figure that the latch-up current density can be increased from about 300 to 1,500 A cm^{-2} by scaling the gate oxide down from 1,000 to 250 Å. The improvements projected by the analytical model were experimentally confirmed during the early stages of IGBT development, providing strong impetus to scale the gate oxide down in products.

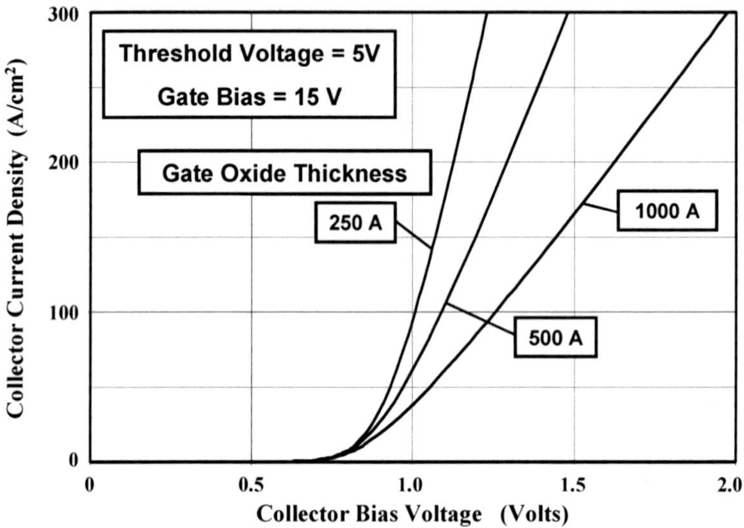

Fig. 9.152 On-state characteristics of IGBT structures with different gate oxide thickness

A reduction of the gate oxide thickness produces a reduction of the channel resistance in the MOSFET portion of the IGBT structure. The resulting improvement in the on-state characteristics can be predicted by using the P–i–N/MOSFET model. As an example, the on-state characteristics for the symmetric IGBT structure are shown in Fig. 9.152 for three cases of the gate oxide thickness. This structure has an N-base region with width of 200 μm and high-level lifetime of 20 μs. The channel length is 1.5 μm. From this figure, it can be concluded that the on-state voltage drop improves from 1.26 to 1.00 V when the gate oxide thickness is reduced from 1,000 to 250 Å. In actual devices, the improvement becomes greater when the lifetime in the N-base region is reduced because a larger proportion of the collector current flows through the channel in these cases.

Simulation Example

To corroborate the analytical model for the improvement in the latch-up current density for the IGBT structure with scaled gate oxide thickness, the results of two-dimensional numerical simulations for the typical symmetric n-channel structure without a deep P^+ region are discussed here. The surface doping concentration of the P-base diffusion was adjusted as the gate oxide thickness was changed to maintain a constant threshold voltage.

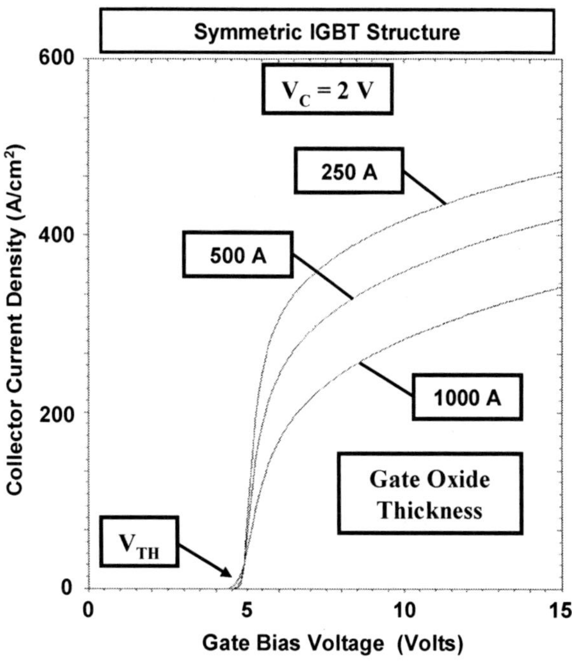

Fig. 9.153 Transfer characteristics for the IGBT structures with different gate oxide thicknesses

The transfer characteristics for the IGBT structures obtained at a collector bias of 2 V are shown in Fig. 9.153 for the case of three gate oxide thicknesses. The threshold voltages for all three cases are identical. The collector current density at a gate bias of 15 V increases when the gate oxide thickness is reduced as expected from the improvement in the on-state characteristics predicted by the analytical model.

The latch-up current densities for the IGBT structures with the different gate oxide thicknesses were obtained by using a gate bias of 15 V by sweeping the collector voltage until the collector current exhibited the snapback signature, indicating latch-up of the parasitic thyristor. The latch-up characteristics for the structures with three gate oxide thicknesses can be compared in Fig. 9.154. The latch-up current density improves by a factor of about 3.5 times, which is a little less than the 5 times improvement predicted by the analytical model (see Fig. 9.151), when the gate oxide thickness is reduced from 1,000 to 250 Å.

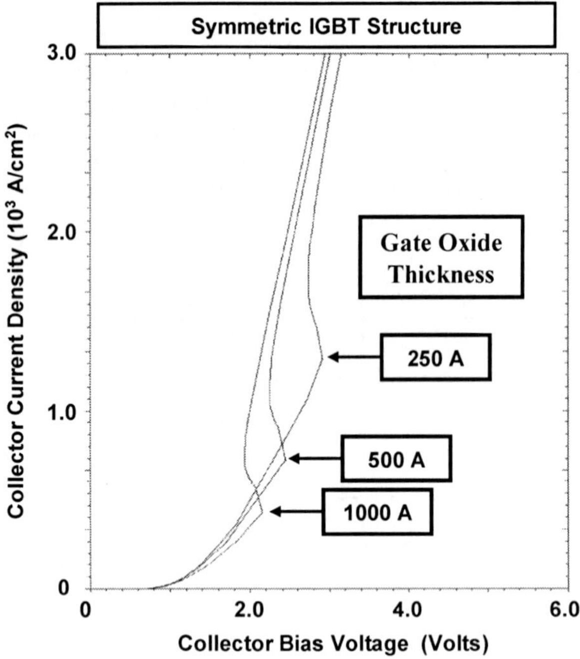

Fig. 9.154 Latch-up characteristics of symmetric IGBT structure with different gate oxide thicknesses

9.10.4 Bipolar Current Bypass

The latch-up of the IGBT structure is produced by the forward biasing of the N$^+$ emitter/P-base junction by the flow of the bipolar current in the P-base region under the N$^+$ emitter region. The forward biasing of the junction can be mitigated by providing an alternate path (called the *bypass*) for the bipolar current. A cross section of the structure with the bipolar current bypass is shown in Fig. 9.155. This structure is created by eliminating the N$^+$ emitter region on one side of the device cell structure.[38]

The latch-up current density for the IGBT structure with the bipolar bypass path can be derived by using the approach used in the previous sections. A simple view of the structure is that half of the bipolar current (I_{P2}) flows via the bypass path while the rest of the bipolar current (I_{P1}) flows under the N$^+$ emitter region. The latch-up current density is then given by

$$J_{C,L}(\text{bypass}) = \frac{2V_{bi}}{\alpha_{PNP,ON}\rho_{SP}L_{N+}W_{CELL}}. \tag{9.361}$$

Unfortunately, this much improvement is not observed in the IGBT structure because the bipolar current does not flow uniformly as assumed in the above analysis. To preserve charge neutrality, the holes in the N-base region tend to flow

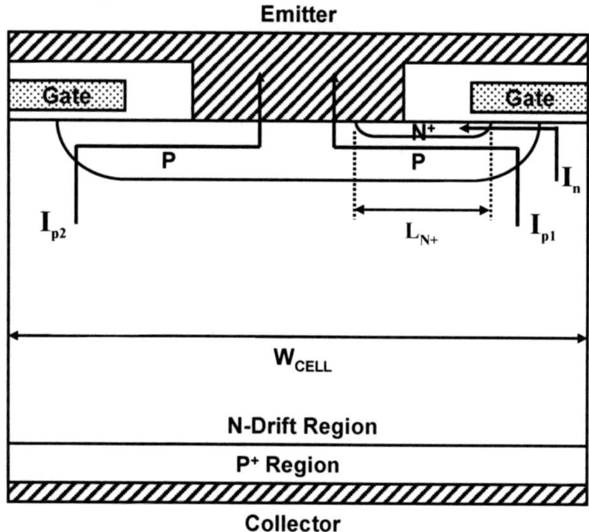

Fig. 9.155 IGBT structure with the bipolar current bypass path

toward the supply of electrons via the channel of the MOSFET portion in the IGBT structure. Consequently, the bipolar current component (I_{P1}) that flows under the N$^+$ emitter region is almost equal to the total bipolar current.

The IGBT structure with the bipolar current bypass path has a larger on-state voltage drop because of the reduction in the channel density. The reduced channel density also produces a reduction in the saturated current level. This can be taken advantage of to produce current saturation prior to the on-set of latch-up.[38]

Simulation Example

To clarify the operation of the IGBT structure with the bipolar bypass path, the results of two-dimensional numerical simulations for the typical symmetric N-channel structure are discussed here. This structure had an N-base region with thickness of 200 μm. The gate oxide thickness for the MOSFET portion was 500 Å with a channel length of 1.5 μm. The P-base region was formed by using a surface concentration of 4×10^{17} cm^{-3} and a vertical depth of 3 μm. A doping concentration in the JFET region was enhanced by using N-type doping with a surface concentration of 1×10^{16} cm^{-3} and depth of 5 μm. No deep P$^+$ region was utilized to highlight the impact of the bipolar bypass path.

The latch-up current density for the IGBT structure was obtained by using a gate bias of 15 V by sweeping the collector voltage until the collector current exhibited the snapback signature, indicating latch-up of the parasitic thyristor. The latch-up characteristic for this structure is compared with the symmetric n-channel IGBT structure with just the P-base region in Fig. 9.156. The latch-up current density for the structure with the bipolar bypass is 470 A cm^{-2} when compared with 730 A cm^{-2} for the structure without the bipolar bypass path.

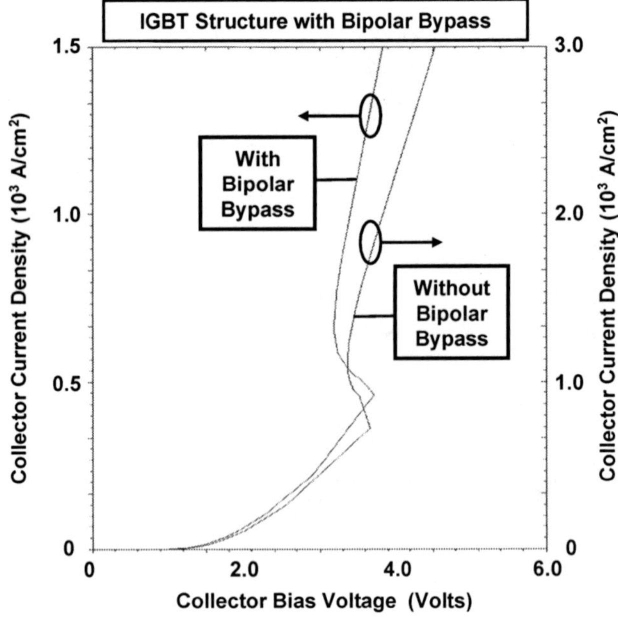

Fig. 9.156 Latch-up characteristics of symmetric IGBT structure with and without the bipolar bypass path

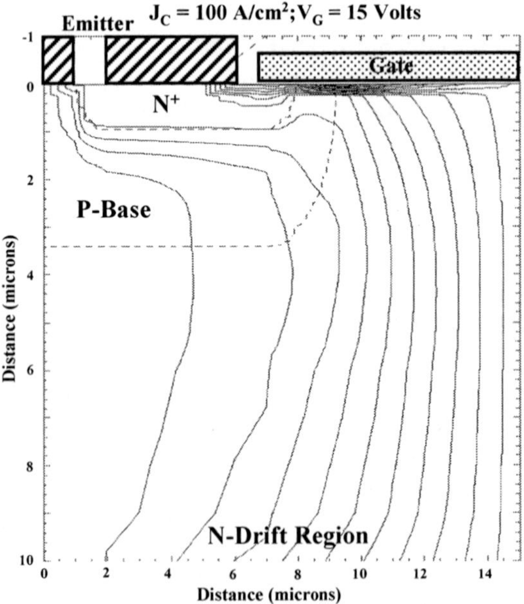

Fig. 9.157 Current distribution (*flow lines*) in the symmetric IGBT structure without the bipolar bypass path

To understand the reason for the lack of improvement in latch-up current density with the bipolar bypass path, it is necessary to examine the current distribution within the IGBT structures during on-state operation. The current distributions for the structures with and without the bipolar bypass path are provided in Figs. 9.157 and 9.158. From these figures, it can be observed that the current tends to congregate in the vicinity of the MOSFET channel. This produces a symmetric current distribution around the polysilicon window for the structure without the bipolar current bypass path. In contrast, the current distribution for the IGBT structure with the bipolar bypass path is asymmetric around the polysilicon window. Since most of the bipolar current flows under the N^+ emitter region in the structure with the bipolar current bypass path, latch-up occurs at nearly the same absolute collector current level as the structure without the bipolar current bypass path. Since the area of the structure with the bipolar current bypass path is twice as large as for the structure without the bipolar current bypass path, the latch-up current density for the structure with the bipolar current bypass path is smaller than for the structure without the bipolar current bypass path.

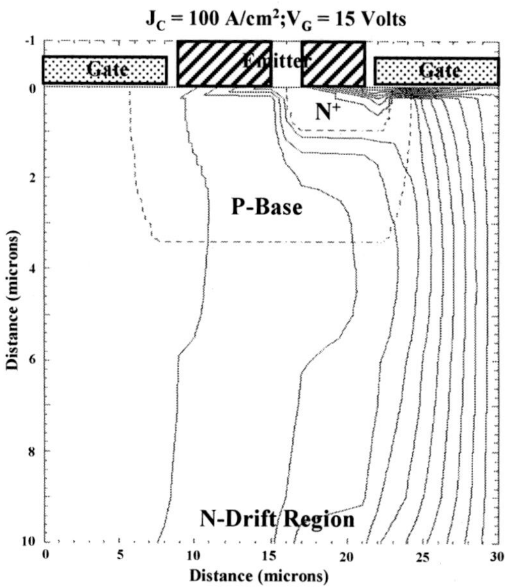

Fig. 9.158 Current distribution (*flow lines*) in the symmetric IGBT structure with the bipolar bypass path

9.10.5 Diverter Structure

The latch-up of the IGBT structure is produced by the forward biasing of the N^+ emitter/P-base junction (J_3) by the flow of the bipolar current in the P-base region under the N^+ emitter region. In the previous section, it was demonstrated that the bipolar current tends to congregate close to the channel of the MOSFET portion in the IGBT structure. The forward biasing of the N^+ emitter/P-base junction (J_3) can be mitigated by diverting the bipolar current flowing in the vicinity of the channel with the incorporation of a diverter region[39] near the channel. A cross section of the structure with the P^{+} diverter region is shown in Fig. 9.159.

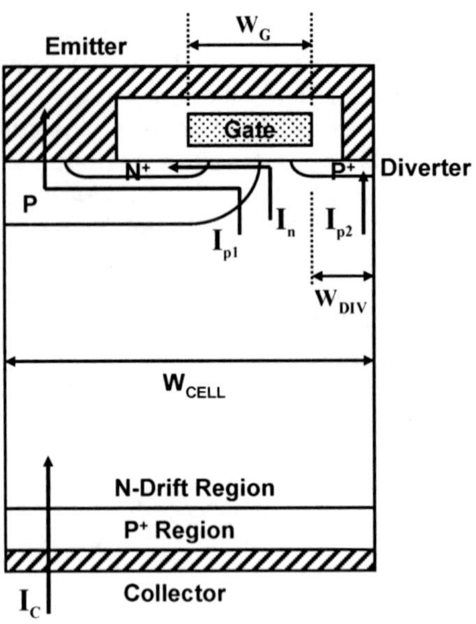

Fig. 9.159 IGBT structure with the diverter region

The latch-up current density for the IGBT structure with the diverter can be derived by analysis of the fraction of the bipolar current that flows into the diverter region. A simple approach for this analysis is to assume that the fraction of the bipolar current (I_{P2}) that flows into the diverter is proportional to the area in the cell on the right-hand side of the middle of the gate electrode. The width of this region is (W_{DIV} + (W_G/2)). The bipolar current flowing into the P-base region is then given by

$$I_{P1} = \frac{W_{CELL} - W_{DIV} - (W_G / 2)}{W_{CELL}} I_C \alpha_{PNP,ON}. \tag{9.362}$$

Using this current as the basis for creating the forward bias across the N^{+} emitter/P-base junction yields the following latch-up current density:

$$J_{C,L}(\text{div}) = \left[\frac{W_{CELL}}{W_{CELL} - W_{DIV} - (W_G / 2)}\right] \frac{V_{bi}}{\alpha_{PNP,ON} \rho_{SP} L_{N+} W_{CELL}}. \tag{9.363}$$

As an example, if a 1-μm deep P^{+} diverter region is formed by using a 2-μm wide diverter window (W_{DIV}) in the IGBT structure with a cell width (W_{CELL}) of 16 μm, the latch-up current density increases by about 50%.

The introduction of the diverter into the IGBT structure produces an increase in the on-state voltage drop because of the constriction of the channel current flow. The presence of the P^+ diverter region reduces the area for spreading the current from the channel. For this reason, it is important to maintain a shallow junction depth for the P^+ diverter region. Since the diverter is connected to the emitter electrode, the same metal layer can be used to simultaneously make contact to the N^+ emitter, P-base, and P^+ diverter regions. This allows using a single large area metal electrode at the top of the device (as in the case of the basic IGBT structure) eliminating any fine line patterning of the thick contact metal for the emitter.

Simulation Example

To clarify the operation of the IGBT structure with the diverter region, the results of two-dimensional numerical simulations for the typical symmetric n-channel structure are discussed here. The latch-up current density for the IGBT structure was obtained by using a gate bias of 15 V by sweeping the collector voltage until the collector current exhibited the snapback signature, indicating latch-up of the parasitic thyristor. The on-state current components for this structure are provided in Fig. 9.160. It can be observed that the bipolar current extracted via the diverter (I_{DIV}) is about one third of the current (I_{PB}) flowing through the contact to the P-base region.

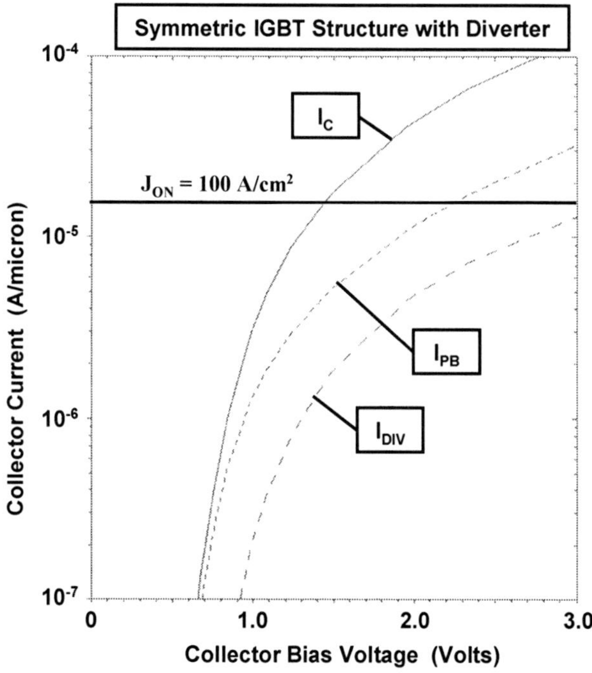

Fig. 9.160 Current components in the symmetric IGBT structure with the diverter region

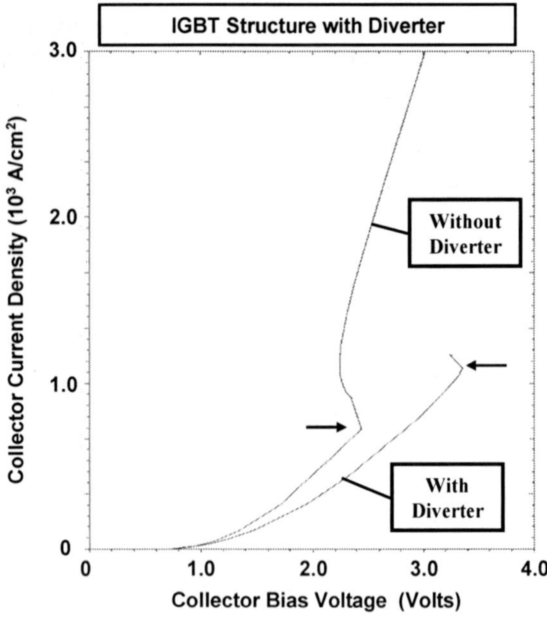

Fig. 9.161 Latch-up characteristics of symmetric IGBT structure with and without the diverter region

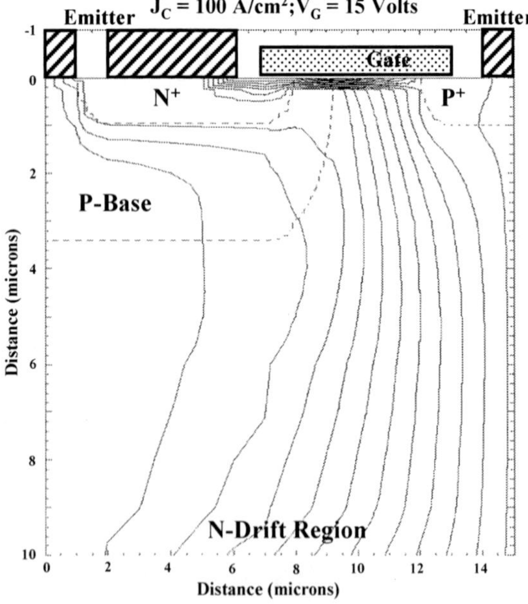

Fig. 9.162 Current distribution (*flow lines*) in the symmetric IGBT structure with the diverter region

The latch-up characteristic for the symmetric IGBT structure with the diverter is compared with the symmetric n-channel IGBT structure with just the P-base region in Fig. 9.161. The point of latch-up is indicated by the horizontal arrows. The latch-up current density for the structure with the diverter is 1,130 A cm^{-2} when compared with 730 A cm^{-2} for the structure without the diverter. The improvement in latch-up current density is consistent with the simple analytical model based upon partitioning of the current. To understand the reason for the improvement in latch-up current density obtained with the diverter, it is necessary to examine the current distribution within the IGBT structures during on-state operation. The current distribution for this structure is provided in Fig. 9.162. It can be observed that some of the current in the vicinity of the MOSFET channel flows out of the P^+ diverter. The current distribution in the structure with the diverter remains symmetric around the polysilicon window. Since the bipolar current flow under the N^+ emitter region in the structure with the diverter is reduced, latch-up occurs at a larger collector current density than for the structure without the diverter.

9.10.6 Cell Topology

The layout of the polysilicon gate structure on the surface of the IGBT structure has an impact on the latch-up current density. Many cell topologies can be utilized for the IGBT structure as illustrated in Fig. 9.163. The topology shown in Fig. 9.163a with a linear window in a linear array was used for the analyses described in all the previous sections. The other topologies have been described in the literature for the optimization of the performance of the IGBT structures. The impact of these alternate topologies on the latch-up current density is analyzed in this section.

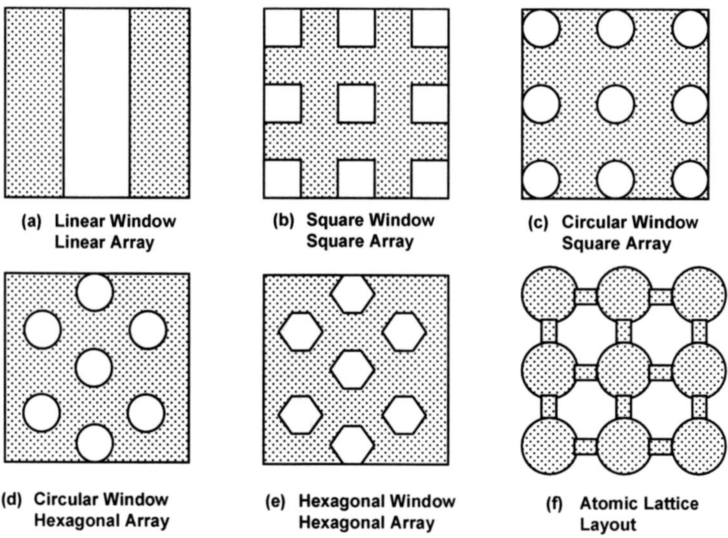

Fig. 9.163 Cell topologies for the IGBT structure

Square Window in a Square Array

The use of a square-shaped polysilicon window in a square array provides a simple design of the IGBT structure from the point of view of the layout tools. However, this layout can significantly degrade the latch-up current density. This occurs due to two reasons. Firstly, the bipolar current that is responsible for promoting latch-up is collected from a greater area at the corners of the polysilicon window as illustrated in Fig. 9.164. To make matters worse, the length (L_P) of the P-base region under the N^+ emitter (see inset cross section in Fig. 9.141) beyond the edge of the deep P^+ region is larger at the corners because the deep P^+ region is also formed by using a smaller square window inside the polysilicon window as illustrated in Fig. 9.164. In this figure, the dashed lines represent the edges of the diffusion windows (polysilicon for the N^+ emitter and P-base region, and a photoresist mask for the deep P^+ diffusion).

The degradation of the latch-up current density in comparison with the linear cell in a linear array can be analyzed by computing the impact of the larger current collected at the corners of the square cell and the larger size for the length (L_{PC}) of the P-base region (under the N^+ emitter beyond the edge of the deep P^+ region) at the corner vs. the length (L_{PE}) of the P-base region (under the N^+ emitter beyond the edge of the deep P^+ region) at the straight edges:

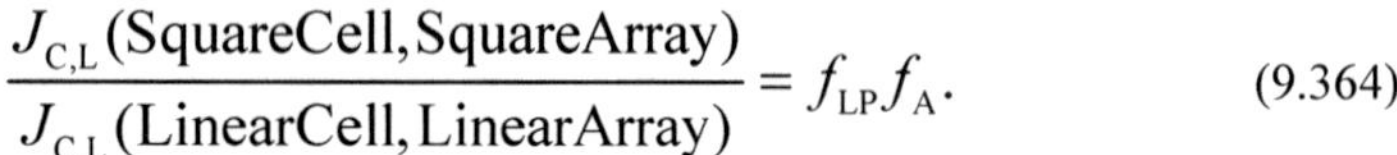

$$\frac{J_{C,L}(\text{SquareCell, SquareArray})}{J_{C,L}(\text{LinearCell, LinearArray})} = f_{LP} f_A. \tag{9.364}$$

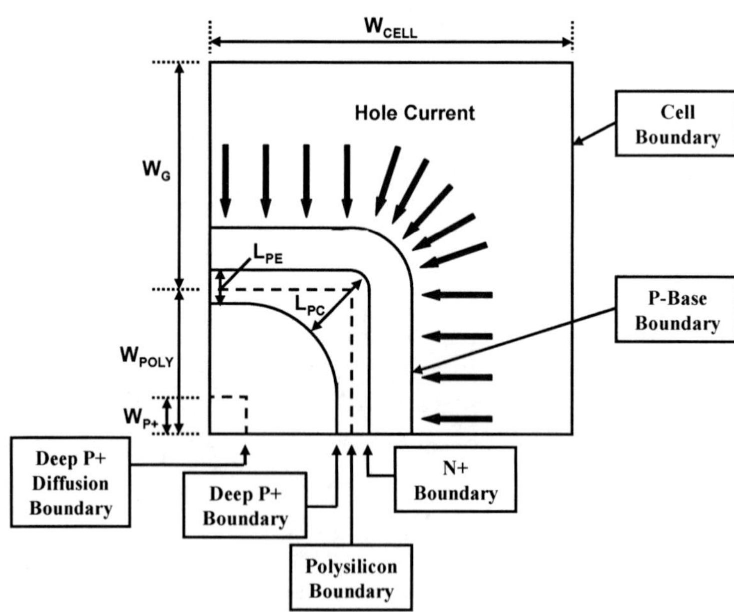

Fig. 9.164 Bipolar current distribution in the IGBT structure with square polysilicon window in a square array

In the above equation, the length factor obtained by using geometrical analysis is given by

$$f_{\mathrm{LP}} = \frac{L_{\mathrm{PE}}}{L_{\mathrm{PC}}} = \frac{(W_{\mathrm{POLY}} - W_{\mathrm{P+}}) - (x_{\mathrm{JP+}} - x_{\mathrm{JN+}})}{\sqrt{2}(W_{\mathrm{POLY}} - W_{\mathrm{P+}}) - (x_{\mathrm{JP+}} - x_{\mathrm{JN+}})}. \tag{9.365}$$

The area factor (f_{A}) can be obtained by assuming that the hole current from the corner of the cell flows into a sector with radius equal to the average of the junction depths of the N^+ and deep P^+ regions:

$$f_{\mathrm{A}} = \frac{\pi(x_{\mathrm{JP+}} + x_{\mathrm{JN+}})W_{\mathrm{POLY}}}{4W_{\mathrm{G}}^2}. \tag{9.366}$$

For typical cell dimensions of 8 μm for the polysilicon window (W_{POLY}), 8 μm for the polysilicon width (W_{G}), and junction depths of 1 and 5 μm for the N^+ and deep P^+ diffusions, the length factor is 0.45 while the area factor is 0.59. This leads to net reduction of the latch-up current density by a factor of 3.8 times. Based upon this analysis, it can be concluded that the topology with a square cell in a square array is highly detrimental to the performance of the IGBT structure. This has been experimentally confirmed in the literature.[40]

Circular Window in a Hexagonal Array

The use of a circular-shaped polysilicon window in a circular array (see Fig. 9.163d) is difficult to replicate from the point of view of the layout tools because the mask edges must be defined by straight boundaries. However, this layout has been approximated by using the hexagonal-shaped polysilicon window in a hexagonal array (see Fig. 9.163e). Both of these topologies can be analyzed by

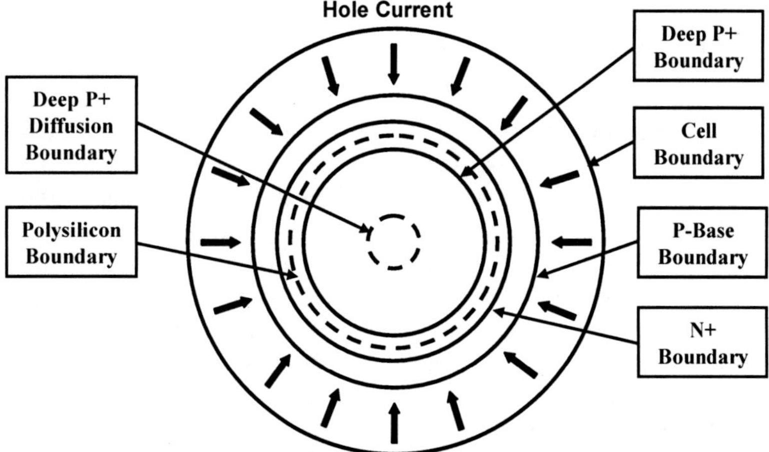

Fig. 9.165 Bipolar current distribution in the IGBT structure with circular polysilicon window in a circular array

using the hole current distribution shown in Fig. 9.165. In this figure, the dashed lines represent the edges of the diffusion windows (polysilicon for the N^+ emitter and P-base region, and a photoresist mask for the deep P^+ diffusion).

As illustrated in the figure, the current distribution is uniform for this topology. In addition, the length (L_P) of the P-base region under the N^+ emitter (see inset cross section in Fig. 9.141) beyond the edge of the deep P^+ region is also uniform around the polysilicon window. In spite of this symmetry, the latch-up current density for this topology is worse than that for the linear cell because the resistance of the bipolar current path is greater due to the circular geometry.

For the IGBT structure, latch-up of the parasitic thyristor will occur when the voltage drop in the P-base region produced by the hole current becomes equal to the built-in potential of the P–N junction:

$$I_p R_B = V_{bi}. \tag{9.367}$$

The hole current for the circular cell arrangement is given by

$$I_p = \pi W_{CELL}^2 J_C \alpha_{PNP,ON}. \tag{9.368}$$

The resistance of the P-base region is dominated by the portion of the P-base region that extends beyond the edge of the deep P^+ diffusion. Using cylindrical coordinates appropriate for the circular cell geometry,

$$R_B = \int_{W_{P+}+x_{JP+}}^{W_{POLY}+x_{JN+}} \frac{\rho_{SB}}{2\pi r} dr = \frac{\rho_{SB}}{2\pi} \ln\left(\frac{W_{POLY}+x_{JN+}}{W_{P+}+x_{JP+}}\right). \tag{9.369}$$

In this expression, W_{POLY} and W_{P+} are the *radii* of the polysilicon window and the deep P^+ diffusion window, respectively. Combining these relationships provides an expression for the latch-up current density:

$$J_{C,L}(CC,CA) = \frac{2V_{bi}}{\alpha_{PNP,ON}\rho_{SP}W_{CELL}^2 \ln[(W_{POLY}+x_{JN+})/(W_{P+}+x_{JP+})]}. \tag{9.370}$$

For comparison with the linear cell topology, it is appropriate to use the same device physical parameters: junction depths of the deep P^+ region ($x_{J,P+}$) and N^+ region ($x_{J,N+}$) of 5 and 1 μm, respectively; a pinch sheet resistance for the P-base region of 2,150 Ω sq^{-1}; a 16-μm cell radius (W_{CELL}) with a polysilicon window radius of 8 μm; and a diffusion of the deep P^+ region from a 2-μm radius window (W_{P+}) in the middle of the polysilicon window. Using these values, the latch-up current density obtained by using the above analytical model is 2,890 A cm^{-2} for a built-in potential of 0.8 V. This value is a factor of 1.8 times smaller than that computed for the linear cell topology. Consequently, the circular and hexagonal topologies degrade the latch-up performance of the IGBT structure in comparison with the linear cell geometry.

Atomic Lattice Layout

In the previous sections, it was demonstrated that the current crowding in the square cell and the enhanced P-base resistance in the circular cell degrade the latch-up current density for the IGBT structure. Both of these problems can be overcome by utilizing the atomic lattice layout (ALL) topology illustrated in Fig. 9.163f. The basic concept behind the ALL topology is to use circular polysilicon islands with the diffusions being performed from outside its boundaries. This is the inverse of the circular cell topology. The polysilicon islands must be interconnected using the thin polysilicon bars to allow application of the gate bias. The moniker "atomic lattice layout" is derived from the resemblance of this topology to atoms in a crystal lattice with bonds between them.[41]

The ALL topology can be analyzed by using the hole current distribution shown in Fig. 9.166. In this figure, the dashed lines represent the edges of the diffusion windows (polysilicon for the N^+ emitter and P-base region, and a photoresist mask for the deep P^+ diffusion). As illustrated in the figure, the current distribution is uniform around the polysilicon window for this topology. In addition, the length (L_P) of the P-base region under the N^+ emitter (see inset cross section in Fig. 9.141) beyond the edge of the deep P^+ region is also uniform around the polysilicon window. However, the latch-up current density for this topology is superior to that for the linear cell because the resistance of the bipolar current path is reduced due to the current radiating outward.

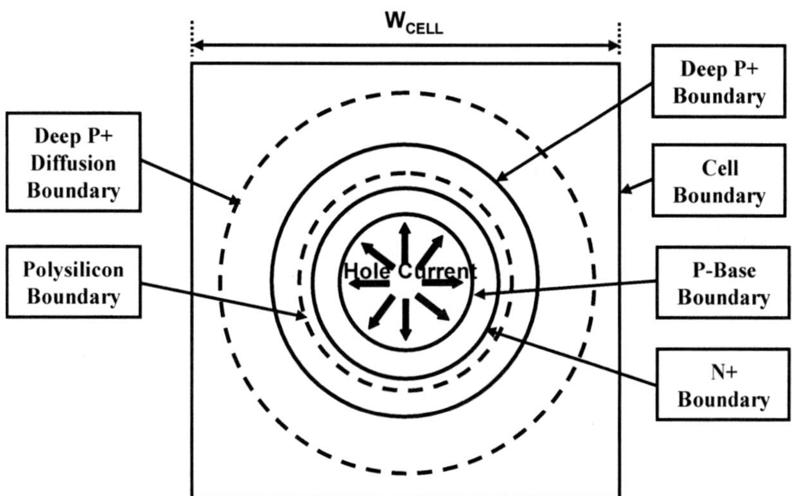

Fig. 9.166 Bipolar current distribution in the IGBT structure with atomic lattice layout

For the IGBT structure, latch-up of the parasitic thyristor will occur when the voltage drop in the P-base region produced by the hole current becomes equal to the built-in potential of the P–N junction:

$$I_p R_B = V_{bi}. \tag{9.371}$$

The hole current for the ALL cell arrangement is given by

$$I_p = \pi W_G^2 J_C \alpha_{PNP,ON}. \tag{9.372}$$

The resistance of the P-base region is dominated by the portion of the P-base region that extends beyond the edge of the deep P^+ diffusion. Using cylindrical coordinates appropriate for the ALL cell geometry,

$$R_B = \int_{W_{POLY}-x_{JN+}}^{W_{P+}-x_{JP+}} \frac{\rho_{SB}}{2\pi r} dr = \frac{\rho_{SB}}{2\pi} \ln\left(\frac{W_{P+}-x_{JP+}}{W_{POLY}-x_{JN+}}\right). \tag{9.373}$$

In this expression, W_{POLY} and W_{P+} are the *radii* of the polysilicon window and the deep P^+ diffusion window, respectively. Combining these relationships provides an expression for the latch-up current density:

$$J_{C,L}(\text{ALL}) = \frac{2V_{bi}}{\alpha_{PNP,ON}\rho_{SP}W_G^2 \ln[(W_{P+}-x_{JP+})/(W_{POLY}-x_{JN+})]}. \tag{9.374}$$

For comparison with the linear cell topology, it is appropriate to use the same device physical parameters: junction depths of the deep P^+ region ($x_{J,P+}$) and N^+ region ($x_{J,N+}$) of 5 and 1 μm, respectively; a pinch sheet resistance for the P-base region of 2,150 Ω sq^{-1}; a polysilicon window radius of 8 μm; and a diffusion of the deep P^+ region from a 14-μm diameter window (W_{P+}). Using these values, the latch-up current density obtained by using the above analytical model is 11,570 A cm^{-2} for a built-in potential of 0.8 V. This value is a factor of 2.2 times larger than that computed for the linear cell topology. Consequently, the ALL topology provides an enhancement of the latch-up performance of the IGBT structure in comparison with the linear cell geometry. This improvement was first experimentally confirmed for p-channel IGBT structures.[41] Subsequently, the ALL cell topology has been shown to produce n-channel IGBT structures that exhibit current saturation without latch-up even at 200°C.[42]

9.10.7 Latch-Up Proof Structure

In the previous sections, improvements in the latch-up current density for the IGBT structure that can be achieved by using a variety of techniques were individually evaluated. In practice, it is prudent to apply a combination of several of these methods to suppress the latch-up phenomenon. When this design approach is taken, an IGBT structure can be created that undergoes current saturation at the gate bias used for on-state operation prior to on-set of latch-up. Such a highly desired device design is referred to as the *latch-up proof* IGBT structure.

A criterion for creating the latch-up proof IGBT structure can be defined by comparison of the saturation current density with the latch-up current density.

The saturated current density for the IGBT structure at a gate bias (V_G) is given by (see Sect. 9.6)

$$J_{C,SAT} = \frac{\mu_{ni} C_{OX}}{2L_{CH} W_{CELL} (1-\alpha_{PNP})} (V_G - V_{TH})^2. \tag{9.375}$$

The latch-up current density for the linear structure with a deep P$^+$ region is given by

$$J_{C,L} = \frac{V_{bi}}{\alpha_{PNP} \rho_{SP} L_P W_{CELL}} \tag{9.376}$$

if the resistance contributed from the deep P$^+$ region is neglected in comparison with the much larger resistance of the P-base region. The latch-up criterion requires

$$J_{C,L} > J_{C,SAT}. \tag{9.377}$$

This leads to the latch-up proof operating condition:

$$(V_G - V_{TH}) < \sqrt{\frac{2V_{bi} L_{CH}}{\rho_{SP} L_P \mu_{ns} C_{OX}} \left(\frac{1-\alpha_{PNP}}{\alpha_{PNP}} \right)}. \tag{9.378}$$

In the previous section, it was demonstrated that the addition of a diverter region to the IGBT structure reduces the bipolar current component in the P-base region, leading to a 50% increase in the latch-up current density. Consequently, the criterion for the latch-up proof IGBT design with the inclusion of a diverter region becomes

$$(V_G - V_{TH}) < \sqrt{\frac{3V_{bi} L_{CH}}{\rho_{SP} L_P \mu_{ns} C_{OX}} \left(\frac{1-\alpha_{PNP}}{\alpha_{PNP}} \right)}. \tag{9.379}$$

As an example of a latch-up proof IGBT design, consider the case of a structure with a deep P$^+$ region that reduces L_P to 1 μm. The maximum gate bias voltage for latch-up proof operation for this structure computed by using (9.378) is plotted in Fig. 9.167. For a gate oxide thickness of 500 Å, latch-up proof operation is not possible at an on-state gate bias of 15 V as indicated by the vertical arrow. The gate bias must be reduced to 12 V for latch-up free operation leading to a significant increase in the on-state voltage drop. The maximum gate bias voltage for latch-up proof operation for the structure when the diverter region is included (as computed by using (9.379)) is also plotted in Fig. 9.167. For a gate oxide thickness of 250 Å, this structure can be operated at an on-state gate bias of 8 V due to the larger transconductance. It is therefore possible to obtain latch-up proof operation with a significant margin as indicated by the vertical arrow.

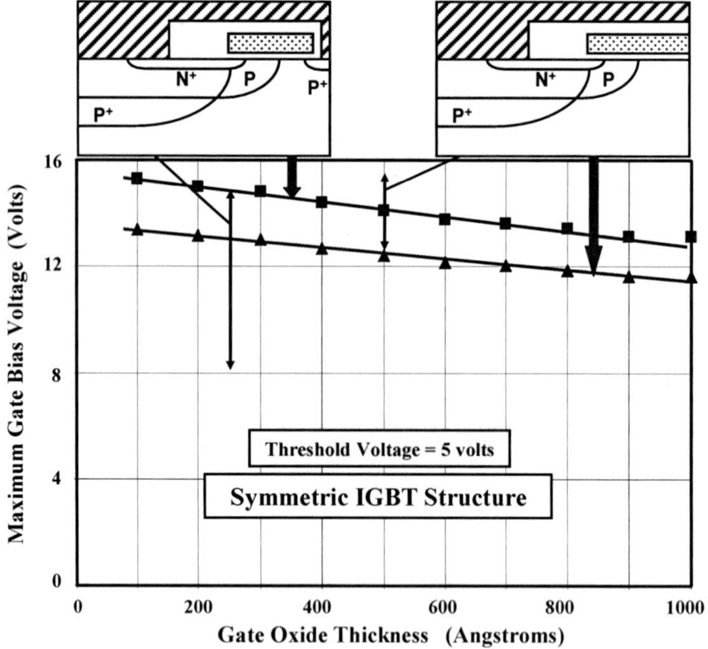

Fig. 9.167 Criterion for latch-up proof operation of the IGBT structure

The pinch sheet resistance for the P-base layer for a gate oxide thickness of 250 Å is 1,000 Ω sq^{-1} for a threshold voltage of 5 V. Using the same structural parameters as in the previous sections, the latch-up current density is found to be 10,000 A cm^{-2} for the IGBT structure with the diverter region. The saturated collector current density computed by using (9.375) for this structure is 3,400 A cm^{-2} at an on-state gate bias of 8 V. This indicates that the IGBT design is latch-up proof because the latch-up current density is larger than the saturated collector current density at the reduced on-state gate bias voltage. Alternately, latch-up proof operation can be obtained at the expense of a larger on-state voltage drop by reducing the channel density.[37,39]

Simulation Example

To demonstrate the latch-up proof operation of the IGBT structure, the results of two-dimensional numerical simulations for the symmetric n-channel structure with a deep P^+ region and a diverter are discussed here. The edge of the diffusion for the deep P^+ region was moved to 4 μm to reduce the length L_P, while not altering the channel doping profile. The output characteristic for the IGBT structure was obtained by using a gate bias of 8 V by sweeping the collector voltage to determine whether the collector current exhibited the snapback signature, indicating latch-up of the parasitic thyristor. It can be observed from Fig. 9.168 that this IGBT structure enters the current saturation mode without the latch-up of the parasitic thyristor. The on-state voltage drop for the latch-up proof structure was 1.49 V at a collector current density of 100 A cm^{-2} at the gate bias of 8 V. The latch-up free operation of the IGBT structure is therefore obtained at the penalty of higher on-state voltage drop.

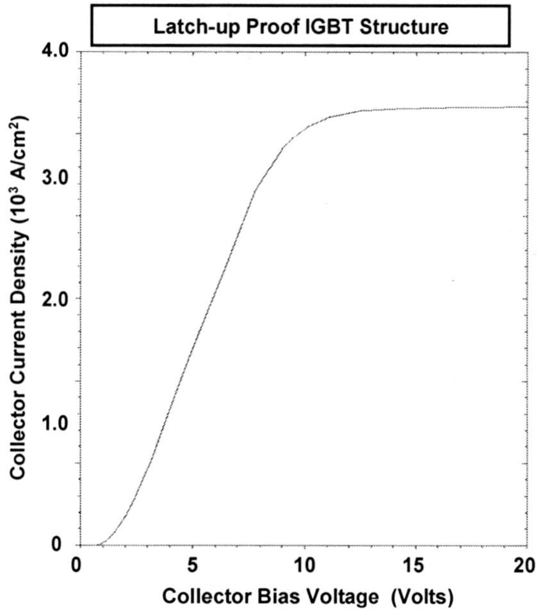

Fig. 9.168 Latch-up proof characteristics of the symmetric IGBT structure with deep P^+ region and diverter

9.11 Safe Operating Area

The area of the current–voltage boundary within which a power device can be operated without destructive failure is referred to as its *safe operating area*. For most power devices, the maximum operating current density at low bias voltages is limited by the temperature rise associated with the power dissipation. In the case of the IGBT structure, the maximum collector current density at low collector bias voltage is limited by the on-set of the latch-up of the parasitic thyristor. At high collector bias voltages and low-current levels, the boundary of operation becomes constrained by breakdown either within the device cell structure or at the edge termination. For the IGBT structure, the open-base transistor breakdown voltage at the edge termination usually limits the maximum operating voltage.

In reference to Fig. 9.89, it was pointed out that the current and voltage imposed on the IGBT structure become simultaneously large during the turn-on and turn-off switching events (at times t_2 and t_5 in the figure). The loci for these transients is mapped in Fig. 9.169. The IGBT structure must be capable of operating without destructive failure over the entire area covered by these loci.

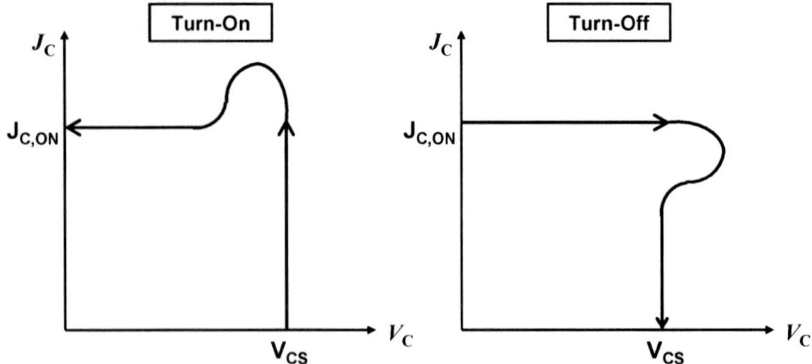

Fig. 9.169 Loci of operation of the IGBT structure during power switching

During the time when the IGBT structure is subjected to high current and voltage simultaneously, a phenomenon called the *avalanche-induced second breakdown* can occur that leads to destructive failure. This phenomenon can be triggered during both the turn-on transient and the turn-off transient. During the turn-on transient, it is said to limit the *forward-biased safe operating area* (FBSOA) because the gate bias is turned on. During the turn-off transient, it is said to limit the *reverse-biased safe operating area* (RBSOA) because the gate bias is turned off. These SOA boundaries are different as discussed below.

9.11.1 Forward-Biased Safe Operating Area

The FBSOA for the IGBT structure limits the loci for the current–voltage trajectories when the device is being turned on in a circuit with an inductive load. Since the gate bias is applied during this mode of operation, electrons are supplied into the N-base region by the channel of the MOSFET portion and holes are supplied to the N-base region by injection from the P^+ collector region for an N-channel IGBT structure. The high operating voltage is supported across a space-charge region formed at the junction (J_2) between the deep P^+ region and the N-base region. The electrons and holes move through the space-charge region at their saturated drift velocities due to the high electric field. This physics of operation was previously described under the current saturation mode in Sect. 9.6.

The concentrations of the holes and electrons in the space-charge region are given by

$$n_{SC} = \frac{J_n}{qv_{sat,n}}, \tag{9.380}$$

$$p_{SC} = \frac{J_p}{qv_{sat,p}}, \tag{9.381}$$

where J_n and J_p are the electron and hole current components in the N-base region. The net positive charge that determines the electric field profile in the space-charge region is then given by

$$N^+ = N_D + p - n = N_D + \frac{J_p}{qv_{sat,p}} - \frac{J_n}{qv_{sat,n}}. \tag{9.382}$$

In the case of the IGBT structure, the hole current component is usually larger than the electron current component. Since the saturated velocities for holes and electrons are nearly equal in silicon, the hole concentration (p_{SC}) in the space-charge region exceeds the electron concentration (n_{SC}). Consequently, the net positive charge in the current saturation mode becomes larger than the background donor concentration. This also enhances the electric field at the junction (J_2) between the deep P^+ region and the N-base region as previously illustrated in Fig. 9.67.

Solving for the avalanche breakdown voltage in the N-base region using the net charge in the space-charge region with Fulop's formula for impact ionization yields

$$BV_{FBSOA} = \frac{5.34 \times 10^{13}}{(N^+)^{3/4}}. \tag{9.383}$$

This breakdown voltage is smaller than the avalanche breakdown voltage determined by the background donor concentration. Since the net positive charge increases with increasing collector current density, the breakdown voltage (BV_{FBSOA}) under current saturation decreases with increasing collector current density. Since the avalanche current is amplified by the gain of the P–N–P transistor, the FBSOA boundary is limited by the product of the current gain and multiplication factor becoming equal to unity.

The maximum voltage that can be sustained by the IGBT structure in the current saturation mode defines the FBSOA. The output characteristics were analytically derived in Sect. 9.6 for the three basic IGBT structures. From the output characteristics shown in Fig. 9.72 for the symmetric IGBT structure obtained by using the analytical model, it can be observed that the maximum sustaining voltages are becoming smaller with increasing collector current levels corresponding to larger gate bias voltages. The boundary of these curves defines the SOA.

The SOA in the current saturation mode as defined by the three limits is illustrated in Fig. 9.170 for the n-channel (solid lines) and p-channel (dashed lines) IGBT structures. The latch-up limit for the p-channel IGBT structure is twice the current density for the n-channel IGBT structure because the sheet resistance of its n-base region under the emitter is smaller due to the larger mobility for electrons. However, its SOA limit is much smaller because the electron current dominates over the hole current within the wide P-base region of the p-channel IGBT structure and the impact ionization coefficient for electrons is an order of magnitude greater than that for holes (see Fig. 2.10). Since the breakdown in the SOA mode shifts from the edge terminations to the cell structure where current transport is occurring, special effort must be used during design to suppress impact ionization within the p-channel IGBT cell structure as discussed in a subsequent section.

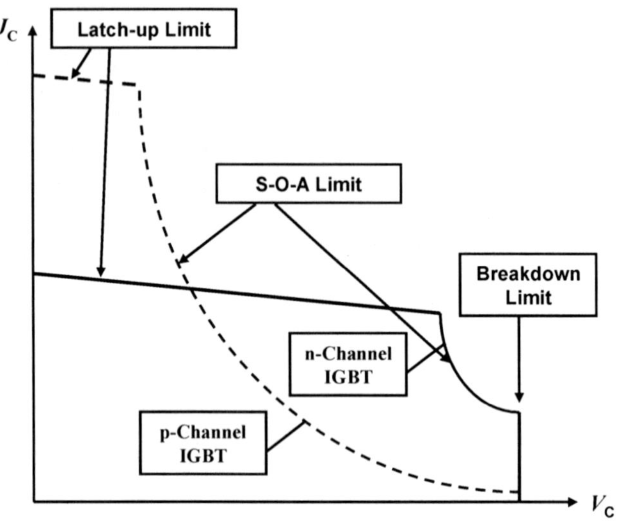

Fig. 9.170 Forward-biased safe operating area for the n- and p-channel IGBT structures

Simulation Example

To illustrate the FBSOA of the IGBT structure, the results of two-dimensional numerical simulations for the symmetric n-channel and p-channel structures are discussed here. The FBSOA limits for the IGBT structures were obtained by sweeping the collector voltage using various gate bias voltages. The output characteristics for the n-channel IGBT structure are displayed in Fig. 9.171. It can be observed in the figure that this IGBT structure enters the latch-up mode at large gate bias voltages and the breakdown mode at low gate bias voltages. (The boundary for the latch-up current density gradually reduces with increasing collector bias voltage due to the increasing gain for the P–N–P transistor.) In between these modes, there is a small region where the SOA is limited by the presence of mobile carriers within the space-charge region. The FBSOA limits are indicated in the figure by using dashed lines.

The output characteristics for the p-channel IGBT structure are displayed in Fig. 9.172. It can be observed in the figure that this IGBT structure enters the latch-up mode at very large gate bias voltages. For most of the gate bias voltages, the SOA is limited by the presence of mobile carriers within the space-charge region. The FBSOA limits are indicated in the figure by using dashed lines.

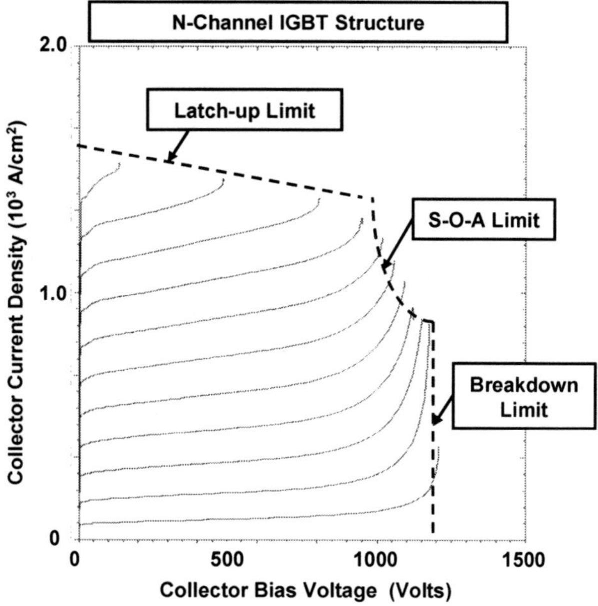

Fig. 9.171 Forward-biased safe operating area for the symmetric N-channel IGBT structure with deep P^+ region

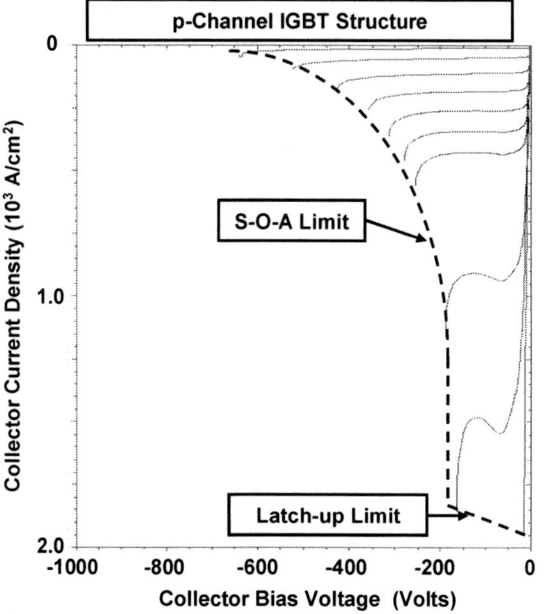

Fig. 9.172 Forward-biased safe operating area for the symmetric P-channel IGBT structure with deep N^+ region

9.11.2 Reverse-Biased Safe Operating Area

The RBSOA for the IGBT structure limits the loci for the current–voltage trajectories when the device is being turned off in a circuit with an inductive load. Since the gate bias is switched off during this mode of operation, all the current is maintained during the voltage rise time by hole current flow through the N-base region in the n-channel IGBT structure. The high operating voltage is supported across a space-charge region formed at the junction (J_2) between the deep P^+ region and the N-base region. The holes move through the space-charge region at their saturated drift velocities due to the high electric field. The physics of operation for the turn-off mode was previously described in Sect. 9.7.4.

During turn-off with an inductive load, the concentration of the holes in the space-charge region is given by

$$p_{SC} = \frac{J_{C,ON}}{q v_{sat,p}}. \tag{9.384}$$

The net positive charge that determines the electric field profile in the space-charge region is then given by

$$N^+ = N_D + p = N_D + \frac{J_{C,ON}}{q v_{sat,p}}. \tag{9.385}$$

Since the net positive charge in the turn-off mode becomes larger than the background donor concentration, the electric field at the junction (J_2) between the deep P^+ region and the N-base region is enhanced. The enhancement of the electric field in the turn-off mode is larger than during the current saturation mode, resulting in a smaller RBSOA when compared with the FBSOA.

Solving for the avalanche breakdown voltage in the N-base region using the net charge in the space-charge region with Fulop's formula for impact ionization yields

$$BV_{RBSOA} = \frac{5.34 \times 10^{13}}{(N^+)^{3/4}}. \tag{9.386}$$

This breakdown voltage is smaller than the avalanche breakdown voltage determined by the background donor concentration. The breakdown voltage (BV_{RBSOA}) during turn-off decreases with increasing collector current density because of the increasing net positive charge (N^+). Since the avalanche current is amplified by the gain of the P–N–P transistor, the RBSOA boundary is limited by the product of the current gain and multiplication factor becoming equal to unity. The breakdown in the RBSOA mode shifts from the edge terminations to the cell structure where current transport is occurring. Consequently, special effort must be used during device design to suppress impact ionization within the p-channel IGBT cell structure as discussed at the end of this section.

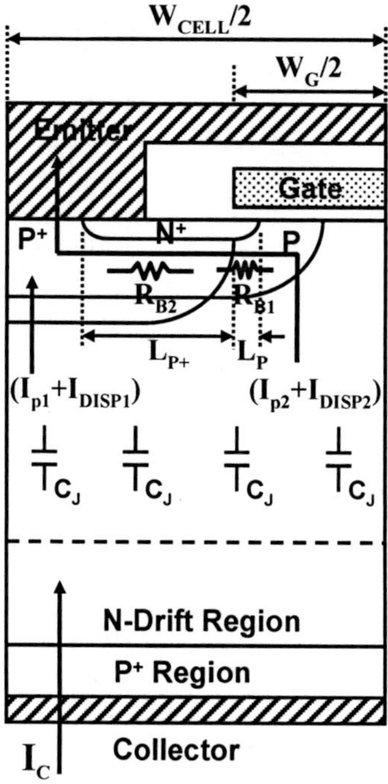

Fig. 9.173 Impact of displacement current on the RBSOA of the IGBT structure

Even at collector bias voltages well below the breakdown voltage given by (9.386), it has been found that the IGBT undergoes destructive failure if the rate of change in the collector voltage during turn-off becomes large. This can be attributed to the on-set of latch-up due to the additional displacement current that occurs during the turn-off process. The impact of the displacement current can be performed by using the capacitance of the junction (J_2) between the deep P^+ region and the N-base region. Since there is no electron current component during the turn-off process, the hole current due to injection from the P^+ collector region becomes uniform across the device cross section. The fraction of the hole current and the displacement current that is flowing under the gate electrode is responsible for forward biasing the N^+ emitter/P-base junction, leading to the latch-up of the parasitic thyristor.

The latch-up criterion which determines the RBSOA limit is then given by

$$(I_{\mathrm{P2}} + I_{\mathrm{DISP2}})R_{\mathrm{B}} = V_{\mathrm{bi}}. \tag{9.387}$$

In this expression, the hole current component is related to the initial on-state current density

$$I_{\mathrm{P2}} = \frac{J_{\mathrm{C,ON}} W_{\mathrm{G}} Z}{2}, \tag{9.388}$$

where Z is the device dimension orthogonal to the cross section. Since the displacement current acts as the base drive current for the P–N–P transistor, the displacement current component in (9.387) is given by

$$I_{\mathrm{DISP2}} = \frac{\alpha_{\mathrm{PNP,0}} J_{\mathrm{DISP}}}{1-\alpha_{\mathrm{PNP,0}}} \frac{W_{\mathrm{G}} Z}{2}. \tag{9.389}$$

The resistance of the P-base region, neglecting the portion with the deep P$^+$ region, is given by

$$R_{\mathrm{B}} \approx R_{\mathrm{B1}} = \frac{\rho_{\mathrm{SP}} L_{\mathrm{P}}}{Z}. \tag{9.390}$$

The displacement current density is the largest during the beginning of the turn-off transient due to the largest junction capacitance. Using this for a worst case analysis,

$$J_{\mathrm{DISP}} = \frac{\varepsilon_{\mathrm{S}}}{W_{\mathrm{ON}}} \left[\frac{\mathrm{d}V_{\mathrm{C}}}{\mathrm{d}t} \right], \tag{9.391}$$

where W_{ON} is the initial depletion layer width in the on-state. This depletion width is even smaller than the zero bias depletion width for the N-base region due to the large concentration of holes in the space-charge region in the on-state.

Combining these relationships yields an expression for the RBSOA limited collector current density:

$$J_{\mathrm{RBSOA}} = \frac{2V_{\mathrm{bi}}}{\rho_{\mathrm{SP}} L_{\mathrm{P}} W_{\mathrm{G}}} - \frac{\alpha_{\mathrm{PNP,0}} \varepsilon_{\mathrm{S}}}{(1-\alpha_{\mathrm{PNP,0}}) W_{\mathrm{ON}}} \left[\frac{\mathrm{d}V_{\mathrm{C}}}{\mathrm{d}t} \right]. \tag{9.392}$$

This expression indicates that the RBSOA limited current density decreases if the [$\mathrm{d}V_{\mathrm{C}}/\mathrm{d}t$] during the turn-off process increases. This phenomenon has been reported in the literature[43–45] by varying the gate drive resistance to tailor the rate of change of the collector voltage.

The RBSOA limit for the n-channel IGBT structure computed by using (9.392) is plotted in Fig. 9.174 as a function of the [$\mathrm{d}V_{\mathrm{C}}/\mathrm{d}t$] during the turn-off process. The parameters used for the device structure are a sheet resistance of 2,150 Ω sq^{-1} for the P-base region, a length L_{P} of 2 μm, and a gate width (W_{G}) of 16 μm. The space-charge layer width in the on-state was assumed to be 1 μm. From this figure, it can be concluded that the RBSOA current density is degraded when the current gain of the P–N–P transistor is larger.

At large collector bias voltages, avalanche multiplication can become significant during the turn-off process. Since the IGBT structure is operating at a high-current density, the breakdown voltage in the cell area becomes reduced due to the presence of a high concentration of holes as indicated by (9.386). The cell breakdown voltage then becomes smaller than the breakdown voltage at the edge termination. Under these conditions, the breakdown phenomenon shifts from the edges of the IGBT chip to the active area and the breakdown voltage becomes influenced by electric field crowding at the corners of the cell structure.

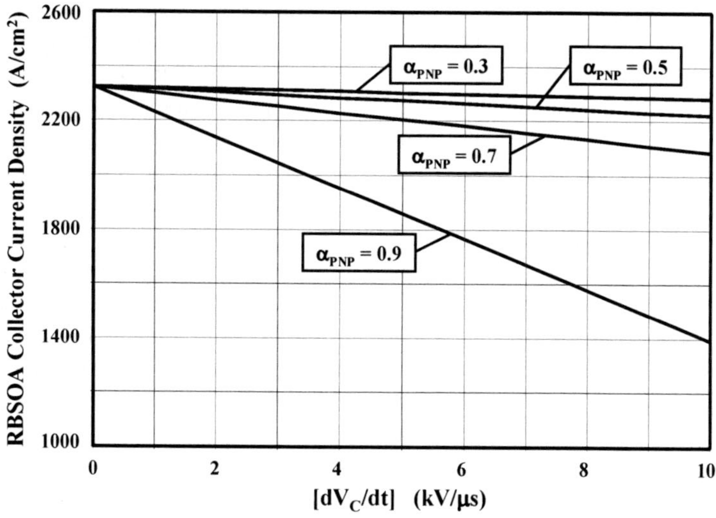

Fig. 9.174 RBSOA limit for the n-channel IGBT structure

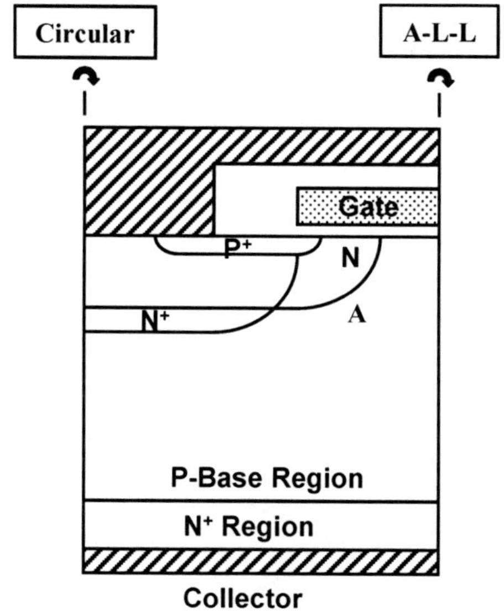

Fig. 9.175 Impact of cell topology on RBSOA of the p-channel IGBT structure

A cross section of a p-channel IGBT structure is illustrated in Fig. 9.175. It can be seen that a cylindrical junction forms at the corner of the N-base region (at location *A* in the figure). This junction curvature produces an enhancement of the electric field as discussed in Chap. 3. Fortunately, the gate electrode behaves as a field plate mitigating some of the electric field enhancement. Despite this, the RBSOA of the p-channel IGBT structure with a linear cell topology becomes degraded by the junction curvature. This effect can be particularly severe if the ends of the linear fingers are terminated with sharp corners when a rectangular-shaped window is opened in the polysilicon window. In this case, a spherical junction is formed at the corners of the polysilicon window degrading the breakdown voltage in the cell even further.

If a circular cell design is used for the p-channel IGBT, it is equivalent to rotation of the cell around an axis located at the left-hand edge of the cross section in Fig. 9.175. This creates additional curvature at the junction which enhances the electric field at location *A* beyond that for the linear cell topology. In contrast, if the ALL cell design is used for the p-channel IGBT, it is equivalent to rotation of the cell around an axis located at the right-hand edge of the cross section in Fig. 9.175. This produces a saddle junction which reduces the electric field at location *A* below that for the linear cell topology. The ALL cell topology has been demonstrated to provide a RBSOA improvement by a factor of 2 times for the p-channel IGBT structure when compared with the linear cell topology.[41] This occurs in spite of a larger $[dV_C/dt]$ for the IGBT structure with the ALL cell topology associated with its smaller input capacitance.

9.11.3 Short-Circuit Safe Operating Area

The IGBT structure is used in power circuits to regulate the energy delivered to various types of loads, such as the windings in a motor, from a DC power source. Occasionally, a short circuit can form across the load as illustrated in Fig. 9.176, resulting in connecting the DC power source directly to the collector of the IGBT structure while its gate bias is still turned on. It would be desirable to detect the

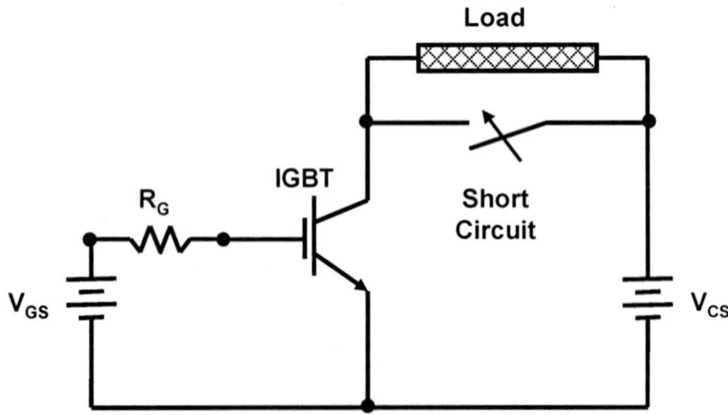

Fig. 9.176 Circuit illustrating short-circuit operation of the IGBT structure

short-circuit condition and turn off the IGBT using feedback to the IGBT control circuit before the device undergoes destructive failure. The time frame of the feedback loop is about 10 μs. It is therefore desirable for the IGBT structure to withstand the high-current flow under short-circuit conditions while supporting the collector power supply voltage. The capability of the IGBT structure to withstand the simultaneous application of a high current and voltage for short time durations is referred to as its *short-circuit safe operating area* (SCSOA).

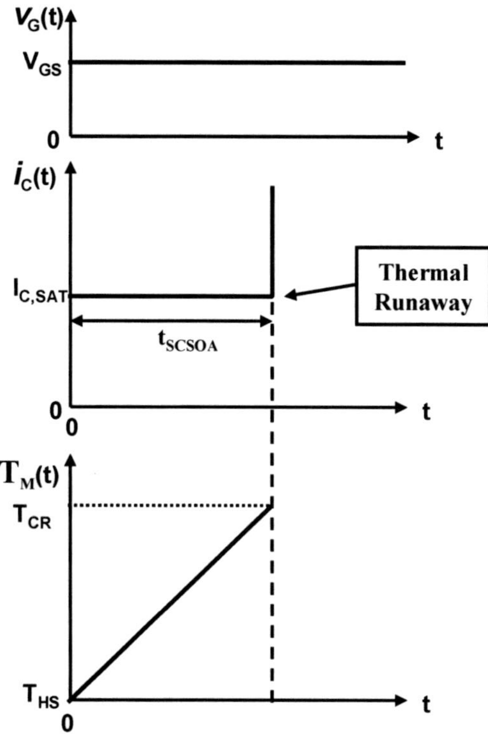

Fig. 9.177 Collector current waveform and temperature rise during the short-circuited turn-off process

Due to the absence of the load impedance during short-circuit operation, the current flow in the IGBT structure becomes limited only by its saturation current as illustrated in Fig. 9.177. The power being dissipated by the IGBT structure per unit area (power density) during the short-circuit condition is therefore given by

$$p_{\mathrm{D}} = J_{\mathrm{C,SAT}} V_{\mathrm{CS}}, \tag{9.393}$$

where V_{CS} is the collector DC supply voltage and $J_{\mathrm{C,SAT}}$ is the saturation current density determined by the gate bias voltage. If this power is assumed to heat the entire volume of silicon (equal to the product of the thickness of the wafer (W_{Si}) and 1-cm^2 area) with no heat removal, the temperature rise is given by

$$\Delta T = \frac{p_D t}{W_{Si} C_V} = \frac{J_{C,SAT} V_{CS} t}{W_{Si} C_V}, \quad (9.394)$$

where C_V is the volumetric specific heat (1.66 J cm^{-3} K for silicon). This expression indicates that the temperature will rise linearly as a function of time, as illustrated in Fig. 9.177, due to the constant power density during short-circuited operation. It is believed that the IGBT structure can withstand this increase in temperature until it reaches a critical value (T_{CR}) when the built-in potential of the junction becomes zero.[46] The device then undergoes a thermal runaway process leading to destructive failure. For silicon, the critical temperature is approximately 700 K. The IGBT structure does not undergo latch-up prior to reaching the critical temperature because the time duration is insufficient to establish the feedback between the internal transistors to produce the regenerative action of the parasitic thyristor.

The time duration over which the IGBT structure can withstand the short-circuit condition can be derived from (9.394):

$$t_{SCSOA} = \frac{(T_{CR} - T_{HS}) W_{Si} C_V}{J_{C,SAT} V_{CS}}, \quad (9.395)$$

where T_{HS} is the initial (heat-sink) temperature. From this expression, it can be concluded that the SCSOA time becomes smaller as the collector supply voltage is increased and as the saturated collector current is increased. As an example, if the symmetric IGBT structure with a thickness of 210 μm is operating at a saturated collector current density of 1,000 A cm^{-2} with a collector supply voltage of 200 V, the short-circuit withstand time is found to be 70 μs if the heat-sink temperature is 300 K. In practice, the SCSOA time is much smaller because the entire volume of silicon is not heated to the critical temperature due to the bottom of the wafer remaining at the heat-sink temperature.

In the presence of a heat sink at the bottom (collector side for the IGBT structure) of the device, the temperature varies from a maximum value at the upper surface to the heat-sink temperature at the bottom. This temperature distribution can be analyzed by the thermal diffusion equation:

$$\frac{\partial T}{\partial t} = D_{TH} \frac{\partial^2 T}{\partial y^2} + \frac{Q_{TH}}{S_v}, \quad (9.396)$$

where D_{TH} is the thermal diffusivity (0.9 cm^2 s^{-1} for silicon) and Q_{TH} is the power generated per unit volume. A solution for the temporal temperature distribution that satisfies this equation is

$$T(y,t) - T_{HS} = (R_T t) e^{-(y/L_T)}, \quad (9.397)$$

where L_T is the thermal diffusion length (40 μm for silicon). The rate of increase in temperature with time (R_T) can be obtained from (9.394):

$$R_{\rm T} = \frac{{\rm d}T}{{\rm d}t} = \frac{K_{\rm T} J_{\rm C,SAT} V_{\rm CS}}{W_{\rm Si} C_{\rm V}}, \tag{9.398}$$

where the constant K_T is introduced to account for the nonuniform temperature distribution through the wafer thickness.

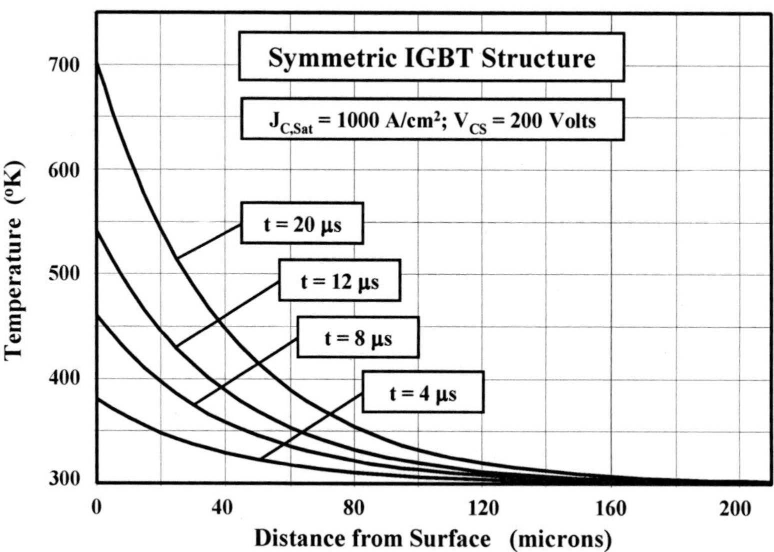

Fig. 9.178 Temperature distribution within the symmetric IGBT structure during the short-circuited turn-off process

As an example of the temperature rise within the device, the temperature distribution for a symmetric IGBT structure with wafer thickness of 210 μm obtained by using (9.397) is provided in Fig. 9.178 for the case of short-circuit operation with a saturated collector current density of 1,000 A cm^{-2} and a collector supply voltage of 200 V. A value of 3.5 was used for the constant K_T. It can be observed from this figure that the maximum temperature occurs at the upper surface while the bottom of the wafer is held at 300 K. Consequently, a much smaller volume of silicon is raised to the higher temperatures making the SCSOA time smaller than that given by (9.395).

The maximum temperature within the device occurs at the upper surface ($y = 0$) during the short-circuited turn-off process. The IGBT structure undergoes thermal runaway when the maximum temperature reaches the critical temperature. Using the above relationships, the time duration over which the IGBT structure can withstand the short-circuit condition is given by

$$t_{\mathrm{SCSOA}} = \frac{(T_{\mathrm{CR}} - T_{\mathrm{HS}})W_{\mathrm{Si}}C_{\mathrm{V}}}{K_{\mathrm{T}}J_{\mathrm{C,SAT}}V_{\mathrm{CS}}}. \tag{9.399}$$

As an example, if the symmetric IGBT structure with a thickness of 210 μm is operating at a saturated collector current density of 1,000 A cm^{-2} with a collector supply voltage of 200 V, the SCSOA time is found to be 20 μs using this equation. If the collector bias voltage is increased to 400 V, the SCSOA time will be reduced to 10 μs. Alternately, if the saturated collector current density is increased to 2,000 A cm^{-2} while maintaining a collector supply voltage of 200 V, the SCSOA time will again be reduced to 10 μs.

The SCSOA for the p-channel IGBT structure has been found to be much worse than that for the n-channel IGBT structure.[47] In the p-channel IGBT structure, avalanche multiplication can be significant due to the larger impact ionization coefficient for electrons moving through the space-charge layer. This reduces the SOA for the p-channel IGBT structure.

Simulation Example

To illustrate the SCSOA of the IGBT structure, the results of two-dimensional *nonisothermal* numerical simulations for the symmetric n-channel structure are discussed here. To suppress latch-up, the device structure with a diverter region

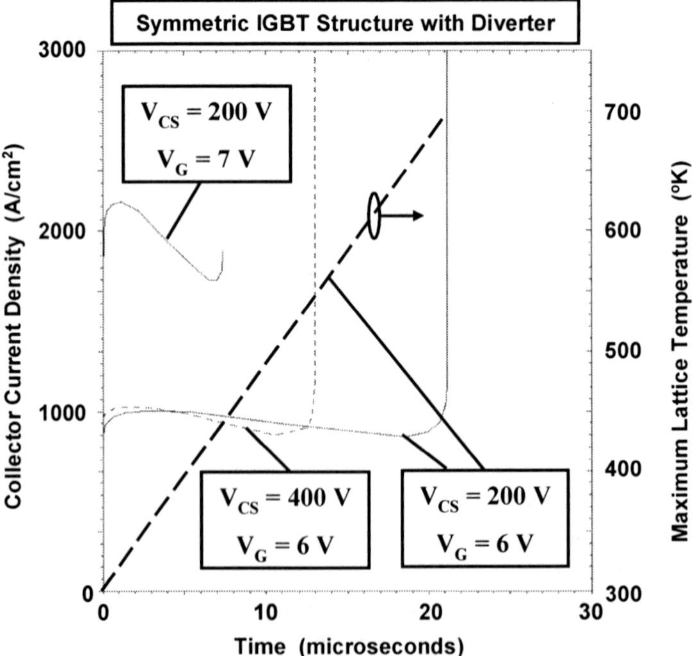

Fig. 9.179 Short-circuit safe operating area for the symmetric n-channel IGBT structure with diverter region

and a gate oxide of 250 Å was chosen. The SCSOA limit for the IGBT structure was obtained by abruptly turning on the gate for the device while maintaining a high collector voltage without any load. The temporal behavior of the collector current was monitored to detect the on-set of thermal runaway. The result obtained with a saturated collector current density of 1,000 A cm^{-2} (obtained at a gate bias of 6 V) and a collector supply voltage of 200 V is provided in Fig. 9.179. It can be observed that the collector current remains approximately constant until 21 μs and then increases abruptly. The predictions of the analytical model are consistent with this value for a value of 3.5 for the factor K_T. The maximum lattice temperature extracted from the simulations is shown in the figure by the bold dashed line for this case. The maximum lattice temperature increases linearly with time as predicted by the simple analytical model and reaches 700 K at the on-set of the thermal runaway process.

When the collector supply voltage was increased from 200 to 400 V, while keeping the same gate bias voltage (6 V), the collector current shown by the dashed lines is observed. The SCSOA time is reduced to about 12 μs, which is approximately by a factor of 2 as predicted by the analytical model. When the gate bias voltage was increased from 6 to 7 V, while maintaining a collector supply voltage of 200 V, the SCSOA time is reduced to about 7 μs, which is inversely proportional to the increase in collector current density by a factor of 2.2 as predicted by the analytical model. The reduction of the saturated collector current during the transient is associated with the reduction of the channel mobility due to the increase in temperature at the top of the device structure.

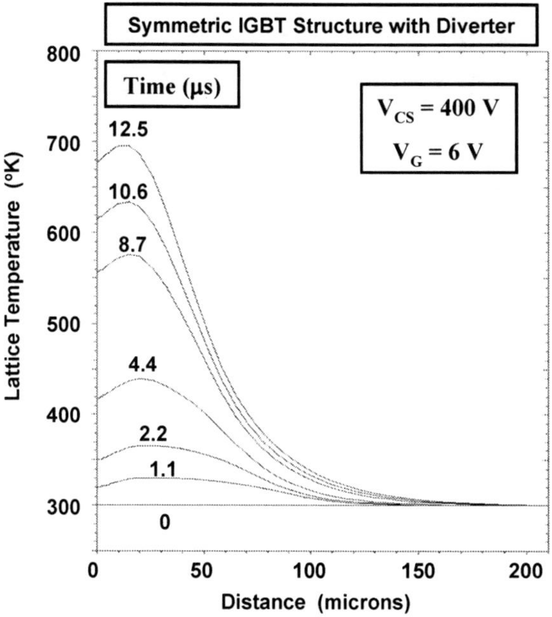

Fig. 9.180 Temperature distribution within the symmetric IGBT structure with diverter region during the SCSOA transient

The temperature distribution within the IGBT structure during the SCSOA transient is provided in Fig. 9.180 for the case of a collector bias voltage of 400 V and a gate bias voltage of 6 V (corresponds to the transient shown by the dashed line in Fig. 9.179). It can be observed that the temperature is largest near the upper surface and decreases exponentially with depth as described by the analytical model (see Fig. 9.178). The maximum temperature increases linearly with time and reaches 700 K at 12.5 μs, at which point in time the device enters thermal runaway.

9.12 Trench-Gate Structure

In the previous sections, the IGBT structures had planar-gate architectures. As in the case of the power MOSFET structure, the performance of the IGBT device can be improved by using a trench-gate structure.[9,48] Trench gates provide higher channel density for the MOSFET portion in the IGBT structure while eliminating the JFET resistance component. In addition, the free carrier concentration in the N-base region near the emitter is enhanced. These phenomena produce a reduction of the on-state voltage drop when compared with a planar device with the same blocking voltage capability, especially for devices with high switching speed.

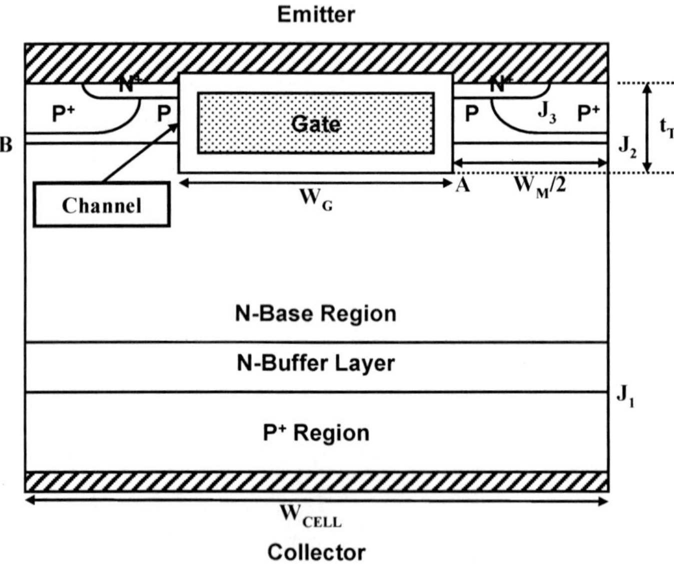

Fig. 9.181 Trench-gate asymmetric n-channel IGBT structure

The trench-gate asymmetric n-channel IGBT structure is illustrated in Fig. 9.181. The gate region is formed after the diffusion of the P-base and N^+ emitter regions by etching a trench using reactive ion etching (RIE). The trench must have a depth greater than the junction depth of the P-base region, so that the channel formed on the trench sidewalls can extend into the N-base region. The gate oxide is formed on the surface of the trench followed by the deposition of polysilicon as the gate electrode. The polysilicon is planarized to recess the gate electrode slightly below the silicon surface. After deposition of an intermetal dielectric film, contact windows are formed to the N^+ emitter and P-base regions. A higher concentration P^+ region is usually included in the middle of the mesa region during device fabrication to improve the contact to the P-base region and suppress the latch-up of the parasitic thyristor. The depth of the P^+ region is usually less than that of the P-base region and especially the depth of the trench to avoid creating a JFET effect. The vertical doping profiles at the middle of the mesa region for the trench-gate device are similar to those shown earlier in Fig. 9.7. The channel doping profile along the trench sidewalls is also similar to that shown in Fig. 9.8 for the planar device but the channel is oriented along the vertical (y-direction) for the trench-gate structure.

9.12.1 Blocking Mode

The trench-gate IGBT structure contains the same set of junctions that are present in the planar IGBT structure with a wide N-base region used to support high voltages. When a positive bias is applied to the collector electrode of the trench-gate asymmetric IGBT structure with zero bias at the gate electrode, the voltage is supported across the reverse-biased junction (J_2) between the P-base and N-base regions. The device can support a high voltage across this junction as limited by the breakdown voltage of the wide-base P–N–P transistor as discussed previously in Sect. 9.4.4. The design methodology described in Sect. 9.4 for the planar IGBT structure is also applicable to the trench-gate structure.

One of the differences between the planar-gate IGBT structure and the trench-gate IGBT structure is the formation of a sharp corner (at location A in Fig. 9.181) within the N-base region due to the rectangular shape of the trenches. This produces a localized enhancement of the electric field at location A, when compared with the electric field in the middle of the mesa at location B, when the device operates in the forward-blocking mode.[48] The high impact ionization at the sharp corner can degrade the breakdown voltage. A commonly used solution to this problem is to use an RIE process that can round out the trench corners.[49] Another approach suitable for wide trenches is by the local oxidation of silicon (LOCOS) process.[50] In addition, the electric field in the gate oxide is larger for the trench-gate structure when compared with the planar-gate structure. In spite of the enhanced electric field, trench-gate IGBT products have been developed with good performance and reliability.[51]

Simulation Example

To illustrate the operation of the trench-gate IGBT structure in the forward-blocking mode, the results of two-dimensional numerical simulations for the asymmetric N-channel structure with 1×10^{16} cm^{-3} doped N-buffer layer are discussed here. This structure had a mesa width (W_M) of 6 μm and a trench width (W_T) of 1 μm. The gate oxide thickness used was 500 Å. The N^+ emitter region was formed by diffusion from a boundary at 2 μm leaving 2 μm in the middle of the mesa for the contact to the P-base region. The channel doping profile is similar to that previously shown in Fig. 9.19 for the planar structure.

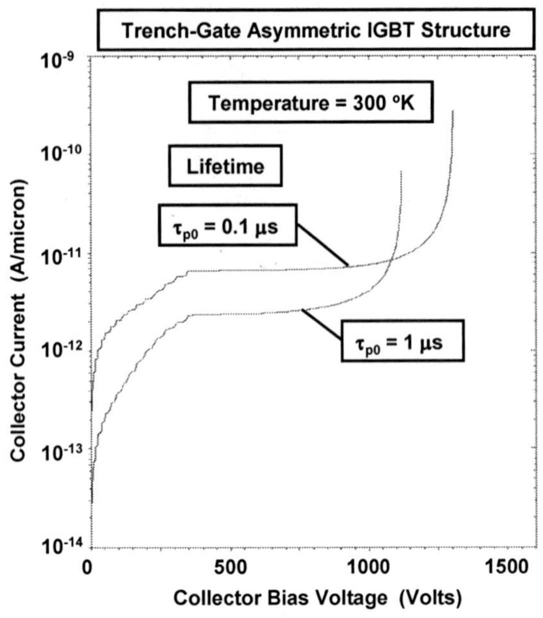

Fig. 9.182 Forward-blocking characteristics for the trench-gate asymmetric n-channel IGBT structure

The blocking voltage capability was analyzed by using various values for the lifetime (τ_{p0}, τ_{n0}) in the N-base region. It can be observed from the characteristics displayed in Fig. 9.182 that the blocking voltage capability increases when the lifetime is reduced as predicted by the analytical model. However, the forward-blocking voltage for the trench-gate structure at a lifetime of 1 μs is reduced to 1,120 V when compared with 1,340 V for the planar-gate structure (see Fig. 9.32) due to the electric field enhancement at the corners of the trench. This illustrates the importance of rounding the trench corners in practical devices. When the lifetime is reduced to 0.1 μs, the forward-blocking voltage for the trench-gate structure improves to 1,305 V.

The impact of the sharp trench corner, located in close proximity to the high-voltage blocking junction in the trench-gate IGBT structure, can be observed by using the three-dimensional plot provided in Fig. 9.183. The largest electric field within the device occurs at the trench corner. The evolution of the electric field near the trench corner is shown in Fig. 9.184 where the profile is taken in the silicon along the *y*-direction at 0.1 μm away from the trench sidewall. At a collector bias of 1,200 V, the electric field at the trench corner is twice as large as that at the P–N junction. This significant enhancement of the electric field at the trench corner degrades the forward-blocking voltage by 220 V when compared with the planar-gate structure.

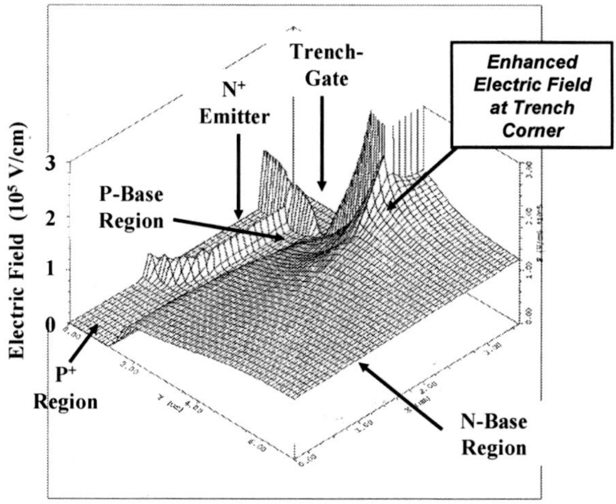

Trench-Gate Asymmetric IGBT Structure

Trench-Gate
N+ Emitter
Enhanced Electric Field at Trench Corner
P-Base Region
P+ Region
N-Base Region
Electric Field (10^5 V/cm)
3
2
1
0

Fig. 9.183 Three-dimensional view of the electric field near the trench corner for the asymmetric IGBT structure at a collector bias of 800 V

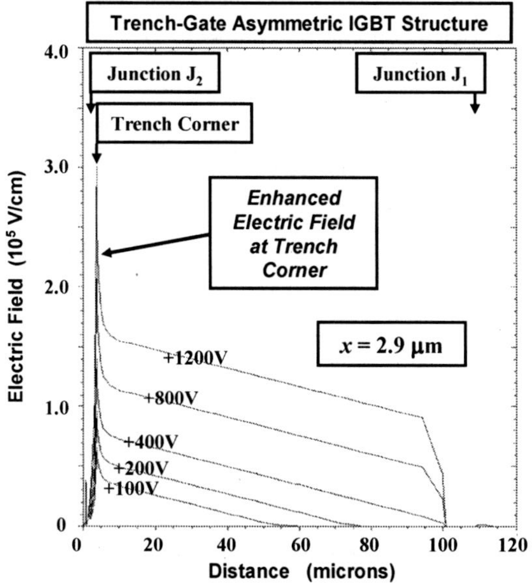

Fig. 9.184 Evolution of the electric field near the trench corner for the symmetric IGBT structure

9.12.2 On-State Carrier Distribution

The on-state carrier distribution in the trench-gate asymmetric IGBT structure is governed by the same high-level injection phenomena previously discussed in Sect. 9.5.2 for the symmetric planar-gate structure, allowing the use of the analytical model for the free carrier distribution developed in Sect. 9.5.4 for the asymmetric structure. According to this analysis, the hole concentration has a maximum value at the P^+ collector/N-base junction (J_1) and reduces to zero at the P-base/N-base junction (J_2). However, in the trench-gate structure, the hole concentration under the gate electrode is enhanced due to the formation of the accumulation layer as illustrated in Fig. 9.37, producing a catenary carrier distribution like that observed in a P–i–N rectifier. This enhanced carrier distribution under the gate electrode produces superior conductivity modulation of the N-base region in the trench-gate IGBT structure because the gate region extends into the N-base region below the P-base region. This phenomenon has been called the *injection enhancement effect* and IGBT structures with deep trench-gate regions have been relabeled as injection-enhanced gate transistors (IEGTs).[10] This obfuscation of the device nomenclature creates unnecessary confusion because the basic phenomenon of carrier enhancement under the gate electrode occurs in all IGBT structures, including the planar-gate structures.[52]

Simulation Example

To illustrate the free carrier distribution in the trench-gate IGBT structure in the on-state, the results of two-dimensional numerical simulations for the asymmetric N-channel structure are discussed here. This structure had the same structural parameters provided in the previous section. The on-state bias conditions are a

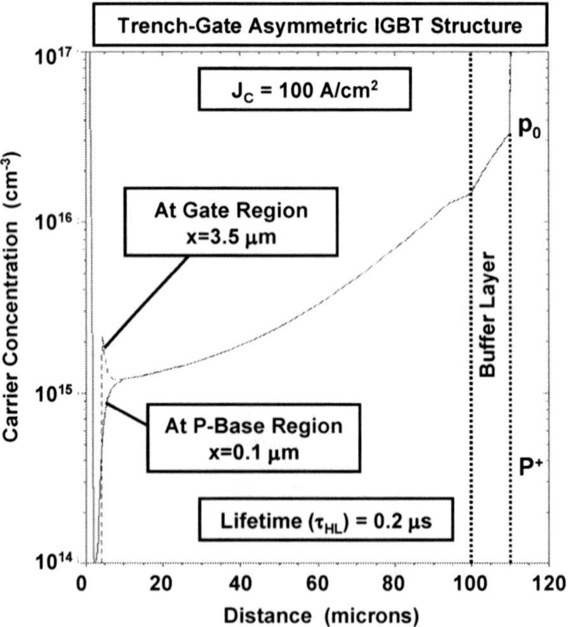

Fig. 9.185 Hole concentration profiles for the trench-gate asymmetric IGBT structure at the P-base and gate regions

collector current density of 100 A cm^{-2} and a gate bias of 15 V. The hole distribution along the vertical direction obtained by using the numerical simulations is provided in Fig. 9.185 at two locations along the *y*-direction. It can be observed from this figure that there is a small enhancement of the hole concentration at the gate electrode.

9.12.3 On-State Voltage Drop

The on-state voltage drop for the trench-gate IGBT structure can be obtained by using the methodology described earlier for the planar-gate structures:

$$V_{\mathrm{ON}} = V_{\mathrm{P+N}} + V_{\mathrm{NB}} + V_{\mathrm{MOSFET}}, \tag{9.400}$$

where $V_{\mathrm{P+N}}$ is the voltage drop across the P$^+$ collector/N-base junction (J_1), V_{NB} is the voltage drop across the N-base region after accounting for conductivity modulation due to high-level injection conditions, and V_{MOSFET} is the voltage drop across the MOSFET portion. Since the free carrier distribution is similar to that for the planar-gate structure, the voltage drop associated with the first two terms in the above equation can be obtained by using the solutions provided in Sect. 9.5.5.

In the trench-gate structure, the JFET region is eliminated. Consequently, the voltage drop across the MOSFET portion includes contributions only from the accumulation layer and the channel. These contributions are also reduced when compared with the planar-gate structure due to the much smaller cell pitch of the trench-gate structure. The voltage drop across the accumulation layer in the trench-gate IGBT structure can be derived based upon the analysis for the power MOSFET structure (see (6.66)). However, it is necessary to account for the accumulation layer formed along the vertical sidewalls and the bottom of the trench:

$$V_{\mathrm{ACC}} = J_{\mathrm{C}} R_{\mathrm{A,SP}} = \frac{J_{\mathrm{C}} K_{\mathrm{A}} [t_{\mathrm{T}} - x_{\mathrm{JP}} + (W_{\mathrm{G}}/2)] W_{\mathrm{CELL}}}{2\mu_{\mathrm{nA}} C_{\mathrm{OX}} (V_{\mathrm{G}} - V_{\mathrm{TH}})}. \tag{9.401}$$

The accumulation layer coefficient (K_{A}) for the IGBT structure can be assumed to have the same value (0.6) as for power MOSFET structures. The voltage drop across the channel of the MOSFET portion can also be obtained by using the analysis for the power MOSFET structure (see (6.63)):

$$V_{\mathrm{CH}} = J_{\mathrm{C}} R_{\mathrm{CH,SP}} = \frac{J_{\mathrm{C}} L_{\mathrm{CH}} W_{\mathrm{CELL}}}{2\mu_{\mathrm{ni}} C_{\mathrm{OX}} (V_{\mathrm{G}} - V_{\mathrm{TH}})}. \tag{9.402}$$

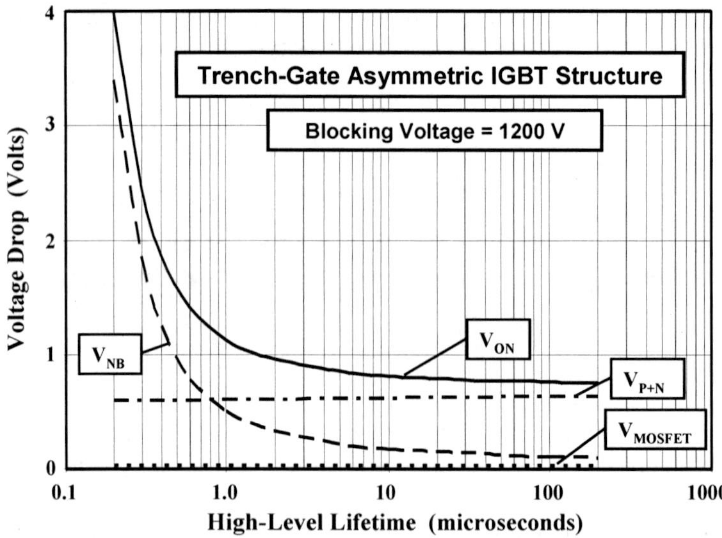

Fig. 9.186 On-state voltage drop for the shallow trench-gate asymmetric IGBT structure

As an example, the components of the on-state voltage drop determined by using the above analytical model are provided in Fig. 9.186 for the case of the asymmetric trench-gate IGBT structure with blocking voltage capability of 1,200 V. The device has an N-base width of 100 μm and an N-buffer layer with thickness of 10 μm; a gate oxide thickness of 500 Å; a trench depth of 4 μm; and a channel length of 1.5 μm. The mobility for electrons in the inversion and accumulation layers was assumed to be 450 and 1,000 cm^2 V^{-1} s^{-1}, respectively (the same as those used for analysis of power MOSFET structures in Chap. 7). The IGBT structure had a cell pitch (W_{CELL}) of 3.5 μm with a gate electrode width (W_G) of 1 μm. The gate bias and threshold voltages used were 15 and 5 V, respectively. From the figure, it can be concluded that the voltage drop across the P$^+$ collector/N-base junction is dominant when the high-level lifetime is large (>2 μs). When the high-level lifetime is reduced, the voltage drop across the N-base region increases and becomes dominant for lifetime value below 0.8 μs. A very rapid increase in the on-state voltage drop is observed when the high-level lifetime is reduced below 0.5 μs. The on-state voltage drop for the trench-gate IGBT structure is about 0.3 V smaller than for the planar-gate structure due to elimination of the JFET resistance.[48]

Simulation Example

To corroborate the analytical model for the on-state voltage drop, the results of two-dimensional numerical simulations for the trench-gate asymmetric structure (described in the previous section) are discussed here. The on-state voltage drop was obtained at a current density of 100 A cm^{-2} at a gate bias of 15 V for various values for the high-level lifetime. It was found that the on-state voltage drop increases rapidly when the high-level lifetime is reduced below 0.5 μs. The on-state voltage drops obtained from the numerical simulations are compared with those obtained by using the analytical model in Fig. 9.187. It can be observed that the analytical model provides an accurate prediction of the on-state voltage drop.

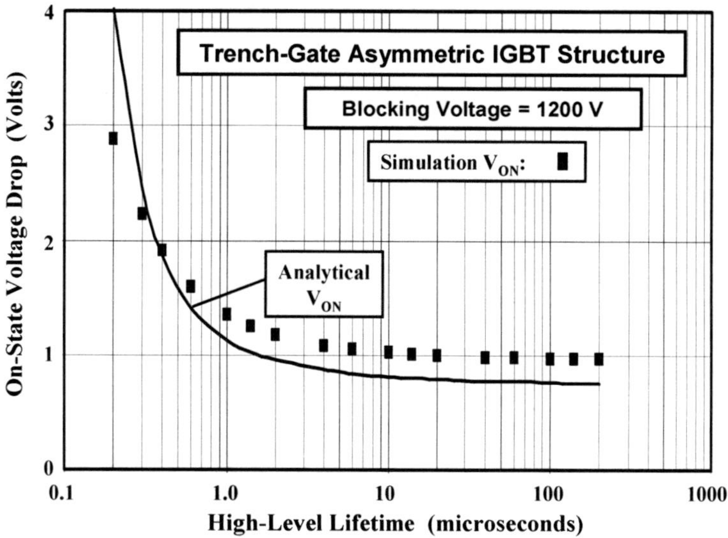

Fig. 9.187 On-state voltage drop for the shallow trench-gate asymmetric IGBT structure

9.12.4 Switching Characteristics

The switching characteristics for the trench-gate IGBT structure can be expected to be very similar to the planar-gate structure because the carrier concentration profiles in the on-state are similar for both cases (compare Fig. 9.51 ($\tau_{HL} = 0.2$ μs case) and Fig. 9.185). Since the physics of the turn-off process is the same for both of the structures, the turn-off time and turn-off energy loss analysis provided in Sect. 9.7.4 can also be applied to the trench-gate structure. The on-state voltage drop for the shallow trench-gate IGBT structure is smaller than that of the planar-gate IGBT structure while its turn-off behavior is similar. Consequently, it can be concluded that the tradeoff curve for the trench-gate IGBT structure must be superior to that for the planar-gate structure. A comparison of the tradeoff curves, for the two types of gate structures for the asymmetric IGBT with 1,200-V forward-blocking capability, is provided in Fig. 9.188. It is apparent that the trench-gate structure has lower overall power loss especially for applications operating at slower switching speeds. This difference has been experimentally confirmed for IGBT structures with various blocking voltage capabilities.[48,52]

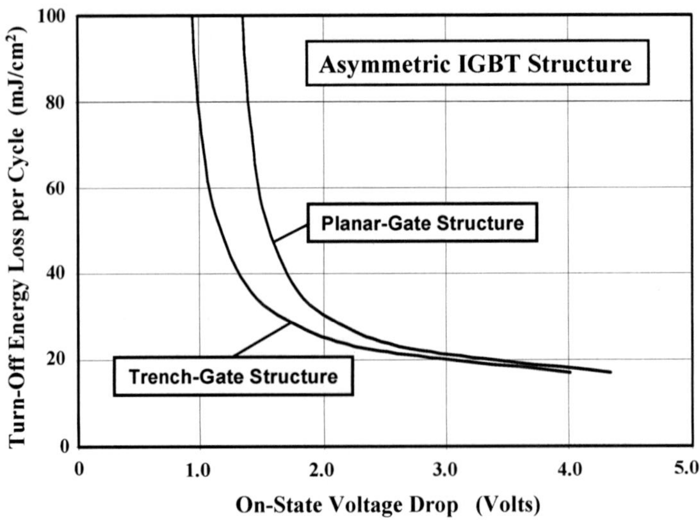

Fig. 9.188 Comparison of the tradeoff curve for the shallow trench-gate and planar-gate asymmetric IGBT structures

9.12.5 Safe Operating Area

The fabrication process for the trench-gate IGBT structure allows reduction of the length of the N^+ emitter finger as well as an improvement in the path for the bipolar current flow. This combination enables the design of devices that can be made immune to the latch-up of the parasitic thyristor. As with all n-channel IGBT structures, latch-up of the parasitic thyristor in the trench-gate IGBT structure occurs when the voltage drop in the P-base region produced by the hole current flow becomes equal to the built-in potential of the P–N junction. As previously discussed for the planar structure, the hole current tends to flow near the electron supply to satisfy charge neutrality. The path for the removal of the hole current is illustrated in Fig. 9.189 for the case of the trench-gate structure.

The latch-up criterion for the trench-gate IGBT structure is

$$I_{\mathrm{p}} R_{\mathrm{B}} = V_{\mathrm{bi}}. \tag{9.403}$$

The hole current is equal to the product of the hole current density and the cell area ($W_{\mathrm{CELL}}Z$), where Z is the length of the IGBT cell orthogonal to the cross section in the figure. The resistance of the P-base region is given by

$$R_{\mathrm{B}} = \frac{\rho_{\mathrm{P}} L_{\mathrm{P}}}{(x_{\mathrm{J,P}} - x_{\mathrm{J,N+}})Z} = \frac{\rho_{\mathrm{SP}} L_{\mathrm{P}}}{Z}, \tag{9.404}$$

where ρ_{SP} is the pinch sheet resistance of the P-base region below the N^+ emitter region. In writing this expression, the resistance of the portion of the P-base region with the P^+ diffusion has been assumed to be negligible. The resistivity (ρ_{P}) and

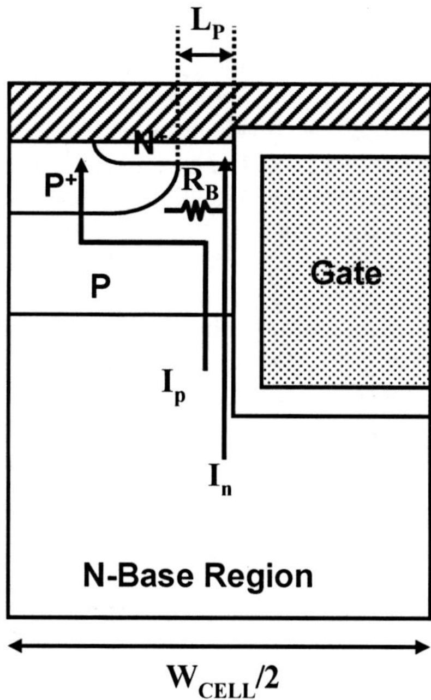

Fig. 9.189 Current flow within the trench-gate IGBT structure

the doping concentration (N_{AP}) of the P-base region that are appropriate for this expression are determined by only the portion of the P-type diffusion located below the N^+ region. Combining these relationships provides an expression for the latch-up current density:

$$J_{C,L}(\text{trench-gate}) = \frac{2V_{bi}}{\alpha_{PNP,ON}\rho_{SP}L_P W_{CELL}}. \tag{9.405}$$

For a typical P-base surface concentration of about 2×10^{17} cm^{-3}, required to obtain a peak channel doping concentration of 1×10^{17} cm^{-3}, the average doping concentration (N_{AP}) in the P-base region is 5×10^{16} cm^{-3}. Typical values for the junction depths of the P-base ($x_{J,P}$) and N^+ emitter region ($x_{J,N+}$) are 3 and 1 μm, respectively. For these parameters, the pinch sheet resistance for the P-base region is found to be 2,150 Ω sq^{-1}. For a 7-μm cell (W_{CELL}) with a trench width of 1 μm and a diffusion of the P^+ region from a 1-μm wide window in the middle of the mesa, the length (L_P) is 2 μm. Using these values, the latch-up current density obtained by using the analytical model is 13,000 A cm^{-2} for a built-in potential of 0.8 V. This is an order to magnitude greater than that of the planar-gate IGBT structure. This improvement in latch-up current density for the trench-gate IGBT structure has been experimentally confirmed.[48]

Simulation Example

To corroborate the analytical model for the latch-up current density for the trench-gate IGBT structure, the results of two-dimensional numerical simulations for the trench-gate asymmetric structure (described in the previous sections) are discussed here. The latch-up was detected by sweeping the collector voltage while maintaining various values for the gate bias voltage and looking for a snapback in the *i–v* characteristics. From the results provided in Fig. 9.190, it can be observed that latch-up occurs at a current density of about 13,000 A cm^{-2} when the gate bias is either 15, 10, or 8 V. Based upon this value, it can be concluded that the analytical model provides an accurate prediction of the latch-up current density. For the gate bias of 7 and 6 V, the device exhibits current saturation up to a collector bias of 20 V without latch-up. Due to the high channel density, the trench-gate IGBT structure can be operated at a gate bias of only 6 V with a low on-state voltage drop. It can therefore be considered a latch-up resistant design.

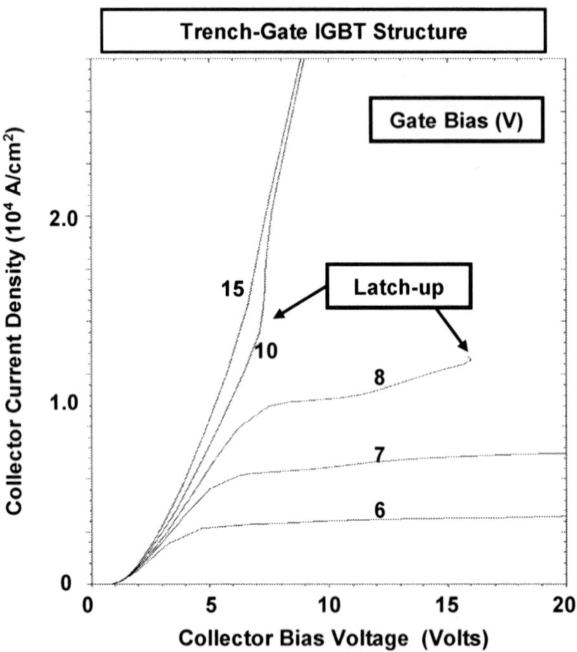

Fig. 9.190 Output characteristics for the trench-gate asymmetric IGBT structure demonstrating latch-up at larger gate bias voltages

The analytical model for the latch-up current density is based upon the assumption that the hole current is not uniformly distributed across the trench-gate IGBT structure and flows close to the MOSFET channel. This phenomenon can be corroborated by examining the current flow pattern in the trench-gate IGBT structure prior to the on-set of latch-up. The current flow lines in the device are shown in Fig. 9.191 at a gate bias of 15 V and a collector bias of 5 V. From the boundaries for the P–N junctions shown in the cross section in the figure, it can be seen that the length of the N^+ emitter region is approximately 2 μm. It can be observed that most of the current flows toward the channel formed on the vertical sidewalls of the trench-gate structure. This justifies the worst case approximation used in the analytical model, namely that all the hole current within the cross section flows under the N^+ emitter region.

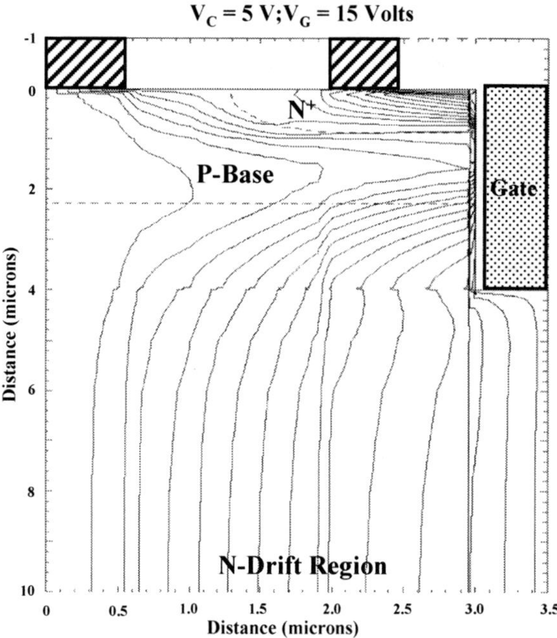

Fig. 9.191 Current distribution in the trench-gate asymmetric IGBT structure

Due to the small cell pitch (W_{CELL}) for the trench-gate IGBT structure, its channel density is much larger than that for the planar IGBT structure. Although this is beneficial for reducing the on-state voltage drop, the large channel density produces very high saturated current levels in the trench-gate IGBT structure which degrades the short-circuit withstand capability. The short-circuit withstand time for the trench-gate IGBT can be computed by using (9.399). For the asymmetric trench-gate IGBT structure, the saturated collector current density at a gate bias of 6 V is about 6,000 A cm^{-2}. This can create problems during short-circuit operation due to the high power dissipation produced by the larger saturated current density.[53] The problem is greater for the asymmetric structure, when compared with the symmetric blocking structure, due to the smaller thickness of silicon used to support the voltage. The short-circuit withstand time predicted by using this saturated current density in (9.399) is about 1 μs at a collector bias of 400 V. By reducing the gate bias to 1 V above the threshold voltage, the saturated current density can be reduced extending the short-circuit withstand time to over 20 μs.

Simulation Example

To illustrate the SCSOA of the trench-gate IGBT structure, the results of two-dimensional *nonisothermal* numerical simulations for the asymmetric n-channel structure are discussed here. To suppress latch-up, a gate bias of 4 V was chosen for these simulations. The SCSOA limit for the IGBT structure was obtained by abruptly turning on the gate voltage for the device while maintaining a high collector voltage without any load. The temporal behavior of the collector current was monitored to detect the on-set of thermal runaway. The results obtained with a saturated collector current density of 200 A cm^{-2} (obtained at a gate bias of 4 V) and collector supply voltages of 200 and 400 V are provided in Fig. 9.192. It can be observed that the collector current first increases gradually with time and then increases abruptly when the lattice temperature (shown by the dashed line) reaches 600 K. The short-circuit withstand time obtained by using the analytical model is much larger than that observed with the simulations because the analytical model does not account for the increasing saturated collector current density during the transient.

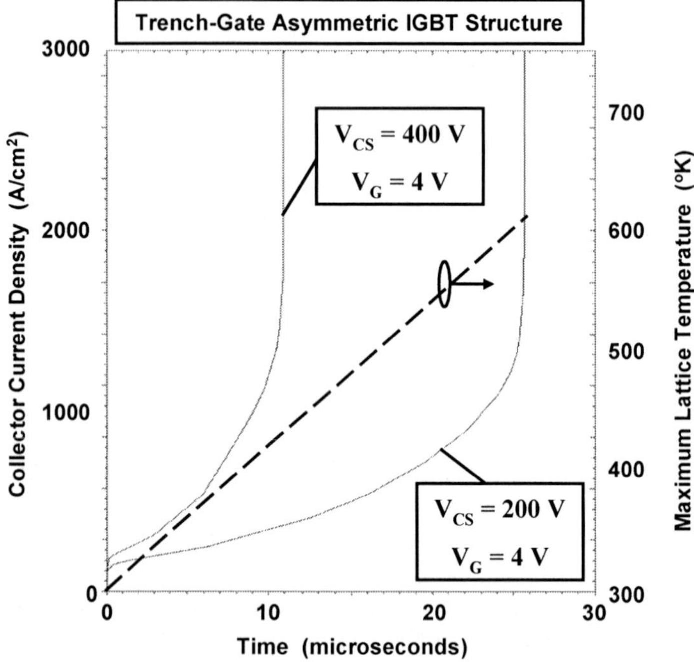

Fig. 9.192 Short-circuit safe operating area for the trench-gate asymmetric IGBT structure

9.12.6 Modified Structures

The trench-gate IGBT structure enables reduction of the on-state voltage drop and improvement of the latch-up current density. Several attempts have been made to improve upon the performance of the basic trench-gate architecture with structural modifications. Two of these cases are discussed in this section.

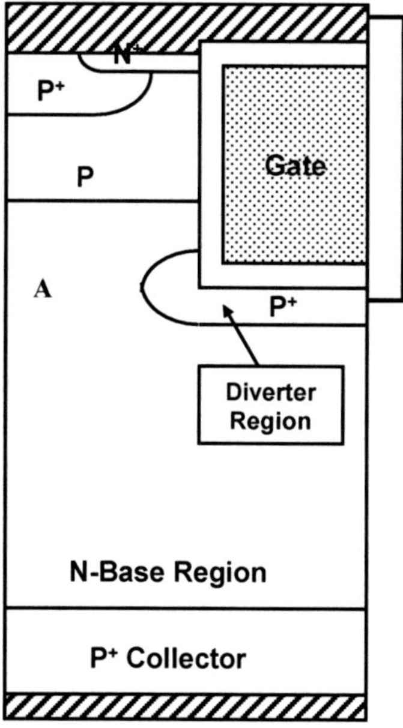

Fig. 9.193 Trench-gate IGBT structure with a diverter region

A cross section of the first modified trench-gate IGBT structure is illustrated in Fig. 9.193. Here, a P^+ diverter region has been added to the basic trench-gate IGBT structure.[54] The diverter is located at the bottom of the trench but is connected to the emitter electrode as indicated on the right-hand side of the structure in the figure. The diverter can be formed after the trench is created by ion implantation of boron orthogonal to the wafer surface. The primary purpose for inclusion of the diverter region is the same as that discussed in Sect. 9.10.5 for the planar IGBT structure. Some of the hole current can be diverted from flowing into the P-base region by the diverter. This allows improving the latch-up current density. The presence of the P^+ region around the trench corner prevents high electric fields from developing at this location during the blocking mode but degrades the on-state voltage drop due to a JFET contribution. The diverter also shields the gate oxide from the high electric fields in the drift region improving the reliability of the device.

Another modified trench-gate IGBT structure that has been proposed is the recessed-gate structure shown in Fig. 9.194.[55,56] The basic concept in the recessed-gate structure is to eliminate the JFET resistance within the planar IGBT structure while maintaining a MOSFET channel on the upper planar surface. With this approach, the difficult planarization step required for the trench-gate structure is avoided. Since the channel is formed on the upper surface, the threshold voltage is controlled by the same process used to fabricate the DMOS planar-gate structure.

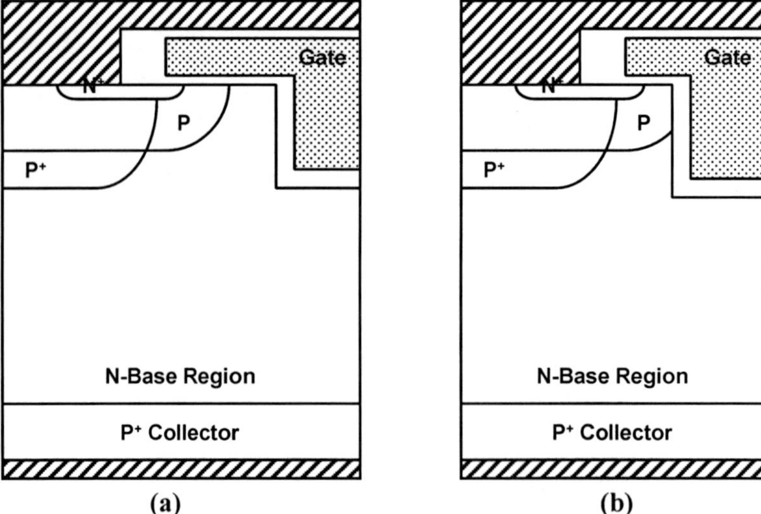

Fig. 9.194 Recessed-gate IGBT structures

Two variants of the recessed-gate structure are shown in Fig. 9.194 based upon the location of the trench in relation to the P-base region. In the case shown in Fig. 9.194a, the entire channel is formed on the upper surface with an accumulation layer formed on the sidewall and bottom of the trench during the on-state. In the case shown in Fig. 9.194b, the channel is partly formed on the upper surface and partly formed along the trench sidewall. An accumulation layer is formed on the rest of the sidewall and bottom of the trench during the on-state. The recessed-gate IGBT structure has been demonstrated to have an on-state voltage drop slightly larger than that of the trench-gate structure. Its latch-up current density is significantly (~3 times) larger than for the planar-gate structure. Due to the larger cell pitch (~20 μm) for the recessed-gate IGBT structure, its saturation current density is smaller than that for the trench-gate structure (with cell pitch of ~7 μm). This produces a superior short-circuit withstand capability. The recessed-gate IGBT structure provides a good compromise in performance between the planar-gate and trench-gate structures while simplifying the fabrication process.

9.13 Blocking Voltage Scaling

Power devices with various blocking voltage ratings are required for different applications. For example, the rectified DC voltage from the 120-V (rms) AC voltage for power distributed in the United States is 170 V, making power devices with blocking voltages of 300 V suitable after allowing for voltage spikes during the switching events. For higher power appliances, it is commonplace to use a 220-V (rms) AC voltage source in the United States, requiring an increase in the blocking voltage rating for the power device to 600 V. In Europe and Asia, it is commonplace to use 440-V (rms) AC voltage sources, requiring a further increase in the blocking voltage rating for the power device to 1,200 V. For high power applications, such as electric locomotives, the voltage rating for the power device escalates even further to over 5,000 V.

9.13.1 N-Base Design

One of the major advantages of the IGBT structure has been the ability to scale its voltage-blocking capability while maintaining good on-state voltage drop and switching performance.[4] The rapid scaling up of the voltage ratings, and hence the power ratings for IGBT devices, that was described in the introduction to this chapter can be attributed to this feature. In the IGBT structure, the wide-base region that supports the blocking voltage operates under high-level injection conditions, resulting in a drastic reduction of its resistance to on-state current flow. If the high-level lifetime is increased as the voltage-blocking capability is increased, a low on-state voltage drop can be maintained in the IGBT structure. This is acceptable from an application's viewpoint because the higher voltage devices are used in systems that operate at lower switching frequencies due to their higher operating power levels.

The scaling of the drift region within n-channel symmetric IGBT devices with different blocking voltage capability is illustrated in Fig. 9.195. The thickness and resistivity of the drift regions shown in the figure are based upon experimental results reported in the literature.[57] Both of these parameters for the N-base region must be increased to achieve the desired blocking voltage capability based upon

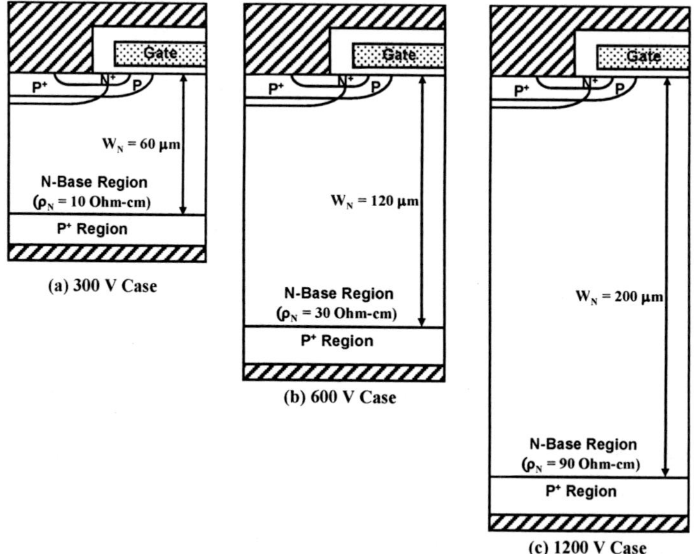

Fig. 9.195 IGBT structures with different blocking voltage ratings

the open-base transistor breakdown analysis for the IGBT structure. The values for the 1,200-V device match those used in the previous sections for the symmetric blocking IGBT structures.

9.13.2 Power MOSFET Baseline

The voltage-blocking capability of the power MOSFET structure can be increased by using a thicker drift region with a reduced doping concentration. In the case of the power MOSFET structure, this produces a very rapid increase in the on-resistance due to unipolar operation as discussed in Chap. 6 (Sect. 6.7.2). For devices with blocking voltage capability in excess of 200 V, the drift region resistance becomes dominant. The one-dimensional drift region resistance (see Sect. 6.1) can be obtained by using

$$R_{\mathrm{ON,SP}}(\mathrm{MOSFET}) = 5.93\times10^{-9}\,\mathrm{BV}^{2.5}. \tag{9.406}$$

In the case of the 300-V power MOSFET, the specific on-resistance for the optimized cell design is about three times the one-dimensional drift region resistance (see Fig. 6.49). When the blocking voltage is increased to 600 and 1,200 V, the specific on-resistance for the optimized cell design of the power MOSFET structure increases to about 2 times and 1.5 times the one-dimensional drift region resistance (see Fig. 6.50). Using these specific on-resistance values, the on-state voltage drop for the power MOSFET can be computed by multiplying the specific on-resistance and the on-state current density. Using an on-state current density of 100 A cm^{-2}, the on-state voltage drops for the power MOSFET structures with blocking voltage capability of 300, 600, and 1,200 V are found to be 2.9, 10.6, and 45 V, respectively. These relatively high on-state voltage drops produce excessive power losses in applications, making the power MOSFET less than ideal as a power switch despite its excellent switching speed.

9.13.3 On-State Characteristics

The change in the on-state characteristics for the IGBT structure can be analyzed by using the analytical model developed earlier in Sect. 9.5.3. As the voltage rating is increased, the increasing width of the N-base region results in an increase in the voltage drop across the N-base region. This produces an increase in the on-state voltage drop for the IGBT structure as shown in Fig. 9.196. The characteristics shown in this figure were obtained by using a high-level lifetime of 20 μs. At an on-state current density of 100 A cm^{-2}, the on-state voltage drop increases from 1.05 V for the 300-V IGBT structure to 1.16 V for the 600-V IGBT structure and 1.34 V for the 1,200-V IGBT structure. These values are substantially smaller than those for the power MOSFET structure with the difference becoming much greater with increasing blocking voltage capability.

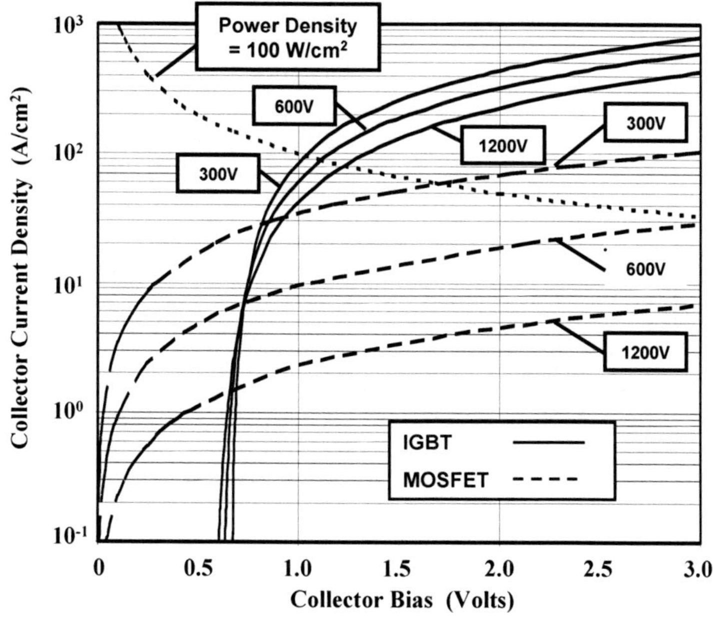

Fig. 9.196 Comparison of the on-state characteristics of IGBT and MOSFET structures with different blocking voltage ratings

The on-state operating point for power devices is usually constrained by the power dissipation as determined by the packaging and heat sink. A typical power dissipation value for power devices is 100 W cm^{-2}. The on-state operating point for the devices can be determined by the intersection point between this power dissipation line (dotted line in Fig. 9.196) and the on-state characteristics. For the case of the IGBT structures, the on-state current density based upon this power dissipation limitation is 95 A cm^{-2} for the 300-V structure, 85 A cm^{-2} for the 600-V structure, and 75 A cm^{-2} for the 1,200-V structure. Consequently, the active area needed for IGBT structures with different voltage ratings does not change drastically. In contrast, the on-state current density for the power MOSFET structures based upon this power dissipation limitation is 60 A cm^{-2} for the 300-V structure, 32Acm^{-2} for the 600-V structure, and 15Acm^{-2} for the 1,200-V structure. Consequently, the active area needed for power MOSFET structures must be greatly increased with increasing voltage rating.

It can be observed in Fig. 9.196 that the on-state *i–v* characteristics for the IGBT structures with different voltage ratings cross over each other. This occurs because the voltage drop across the N-base region is dominant at high-current levels, while at low-current levels, the voltage drop across the P–N junction becomes dominant for all the structures. The voltage drop across the P–N junction is dependent on the maximum injected hole carrier concentration (p_0) at the P$^+$ collector/N-base junction (see (9.77)). At low on-state current levels, the maximum injected hole carrier concentration obtained by using the analytical model is larger for the devices with smaller blocking voltage capability as shown in Fig. 9.197. Consequently, the lower blocking voltage structures exhibit a higher on-state voltage drop at low-current levels.

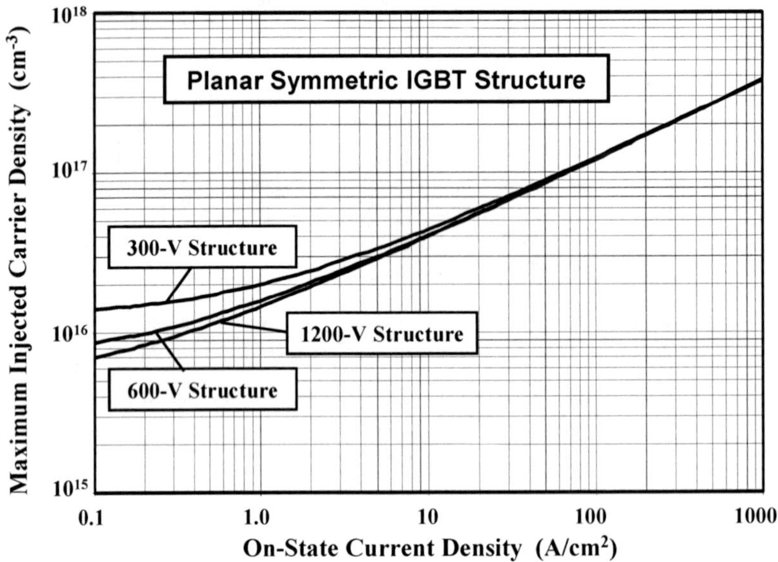

Fig. 9.197 Maximum injected carrier concentration in IGBT structures with different blocking voltage ratings

Simulation Example

To corroborate the analytical model for the on-state characteristics of the IGBT structures with different voltage-blocking capability, the results of two-dimensional numerical simulations for the symmetric structure are discussed here. All the structures had the same planar-gate structure with a cell pitch of 15 μm as described in previous sections. The N-base region parameters were varied to obtain devices with the different blocking voltage ratings. The N-base parameters were a thickness of 60 μm and resistivity of 10 Ω cm for the 300-V structure, a thickness of 120 μm and resistivity of 30 Ω cm for the 600-V structure, and a thickness of 200 μm and resistivity of 90 Ω cm for the 1,200-V structure. A high-level lifetime of 20 μs was used in the N-base region for all three cases.

The on-state characteristics obtained by using the simulations are shown in Fig. 9.198 for the three structures. It can be concluded that the analytical model accurately describes the on-state behavior as the blocking voltage is scaled by comparison of this figure with the IGBT characteristics shown in Fig. 9.196. The on-state voltage drop obtained from the simulations matches that obtained by using the analytical model. In addition, the slope of the *i–v* characteristics changes with blocking voltage rating resulting in a crossover of the *i–v* characteristics as predicted by the analytical model.

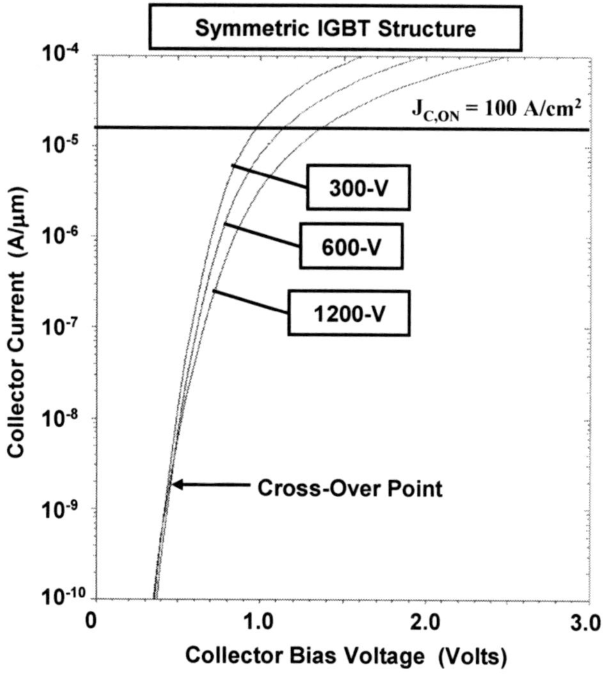

Fig. 9.198 On-state characteristics for symmetric IGBT structures with different blocking voltage ratings

9.13.4 Tradeoff Curve

The impact of reducing the lifetime in the N-base region by electron irradiation has been experimentally studied for structures with different voltage ratings.[57] It was found that the on-state voltage drop for the 300-V structure does not increase as rapidly as that for the 600-V and 1,200-V structures when the lifetime is reduced. This outcome is to be expected because of the smaller thickness of the N-base region for the 300-V structure. The empirically derived tradeoff plot between the turn-off time and the on-state voltage drop for the devices with the three voltage ratings is shown in Fig. 9.199. The on-state voltage drop in this plot was obtained at an on-state current density of 230 A cm^{-2}. Even after enhancing the switching speed of the structures, their on-state voltage drop is far smaller than that of power MOSFET structures. This has made the IGBT structure preferable for all high-voltage applications (above 200 V). Excellent device characteristics have been reported for IGBT devices with a broad range of blocking voltage capabilities ranging from 300[58] to 6,500 V.[59]

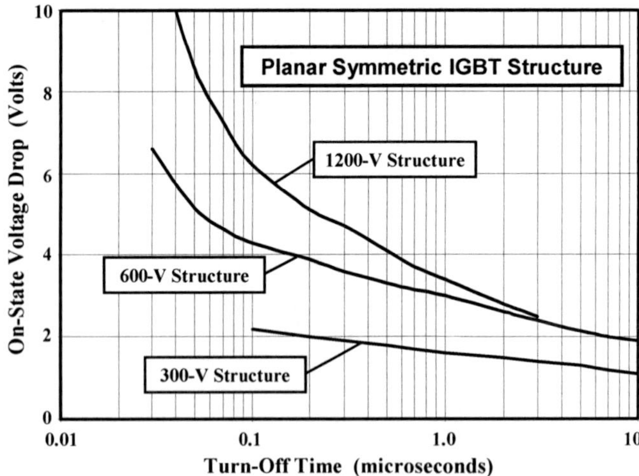

Fig. 9.199 Comparison of the tradeoff characteristics of IGBT structures with different blocking voltage ratings

9.14 High Temperature Operation

Power devices often encounter high ambient temperatures in their applications. Two examples are the placement of IGBT devices within the housing for motors being controlled by adjustable speed drives and the placement of IGBT devices in the base of lamps being controlled by electronic ballasts. In addition, the power dissipation generated within the IGBT structure during normal operation in a circuit produces an increase in the temperature especially at the top of the device structure where the channel is located. The performance of all power devices degrades with increasing operating temperature. As an example, the on-resistance for the power MOSFET structure increases by a factor of 2 times when the temperature increases by 150°C (see Sect. 6.19). This increases the power loss within the device which degrades the efficiency of the power circuit.

9.14.1 On-State Characteristics

One of the attributes of the IGBT structure is that its on-state voltage drop does not increase rapidly with increasing temperature. At the same time, the IGBT structure exhibits a positive temperature coefficient for the on-state voltage drop. This is a desirable feature for power devices because it ensures good current sharing within the chip as well as when chips are paralleled to handle larger currents. To demonstrate this behavior, the on-state *i–v* characteristics obtained for the 1,200-V symmetric IGBT structure discussed in the previous sections are shown in Fig. 9.200 at three ambient temperatures. These characteristics were obtained by using the analytical model developed earlier in Sect. 9.5.3. In computing the change in the on-state characteristics with increasing temperature, it is necessary to account for the change in the intrinsic carrier concentration, the mobility, and the threshold voltage.

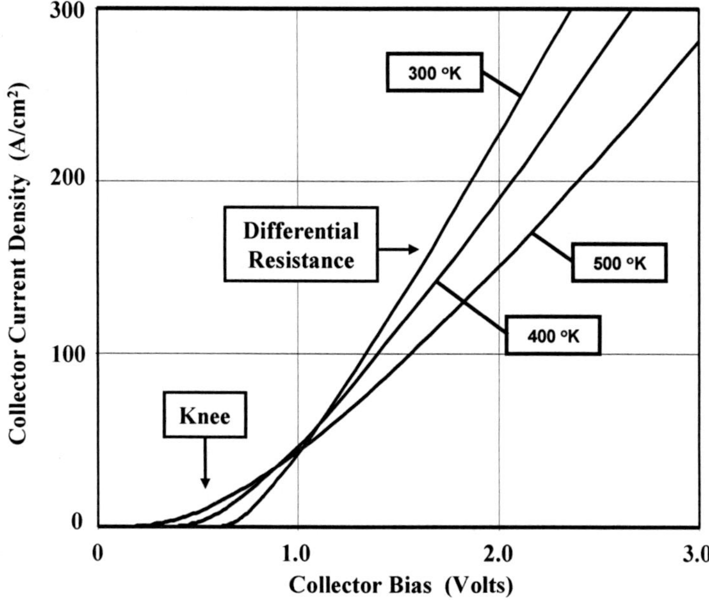

Fig. 9.200 Change in the on-state characteristics of the symmetric IGBT structure with increasing temperature

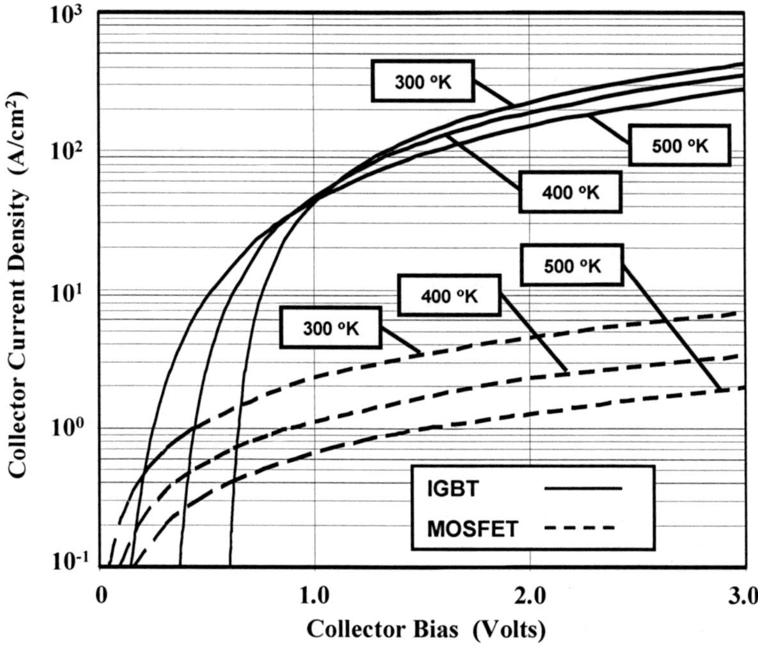

Fig. 9.201 Comparison of the on-state characteristics of the 1,200-V IGBT and MOSFET structures with increasing temperature

As the temperature is increased, the knee voltage for the IGBT structure decreases due to an increase in the thermal voltage while the differential resistance in the portion above the knee becomes larger because of a decrease in the mobility.[60] These phenomena have opposite influence on the on-state voltage drop. Consequently, the on-state voltage drop reduces with increasing temperature at low on-state current density while it increases with increasing temperature at high on-state current density. For the symmetric IGBT structure with 1,200-V blocking voltage capability, the on-state voltage drop is nearly independent of temperature at an on-state current density of about 50 A cm^{-2}. More importantly, the on-state voltage drop has a mild positive temperature coefficient at the typical operating on-state current density of 100 A cm^{-2}. This ensures good current distribution within the IGBT chip and allows paralleling of multiple devices to handle higher power levels in applications.

The changes in the on-state *i–v* characteristics of the 1,200-V symmetric IGBT structure are compared with those for the 1,200-V MOSFET structure in Fig. 9.201. The specific on-resistance for the MOSFET structure increases by a factor of 3.5 times when the temperature increases from 300 to 500 K. This greatly degrades the current-handling capability of the power MOSFET structure at high ambient temperatures. In contrast, the on-state voltage drop for the IGBT structure increases by only a small amount (~20%) allowing its operation at elevated temperatures. Consequently, IGBT devices have been used to manufacture ballasts for compact fluorescent lamps and in heat-sensitive applications such as controllers for space heaters.

Simulation Example

To corroborate the analytical model for the on-state characteristics of the IGBT structure at elevated temperatures, the results of two-dimensional numerical simulations for the symmetric structure are discussed here. This device had the planar-gate structure with a cell pitch of 15 μm as described in previous sections. A high-level lifetime of 20 μs was used in the N-base region for all three cases.

The on-state characteristics obtained by using the simulations are shown in Fig. 9.202 at the three temperatures. The on-state voltage drop increases slightly with temperature in accordance with the analytical model. The on-state voltage drop increases from 1.34 V at 300 K to 1.40 V at 500 K. This positive temperature coefficient for the on-state voltage drop has been an important feature for the IGBT structure because it favors uniform current distribution within chips and allows paralleling of devices to obtain higher power ratings in applications. It can be concluded that the analytical model accurately describes the on-state behavior as a function of the temperature by comparison of this figure with the IGBT characteristics shown in Fig. 9.201. Firstly, the on-state voltage drop obtained from the simulations matches that obtained by using the analytical model at all the three temperatures. In addition, the slope of the *i–v* characteristics changes with temperature resulting in a crossover of the *i–v* characteristics as predicted by the analytical model at a current density of 50 A cm^{-2}.

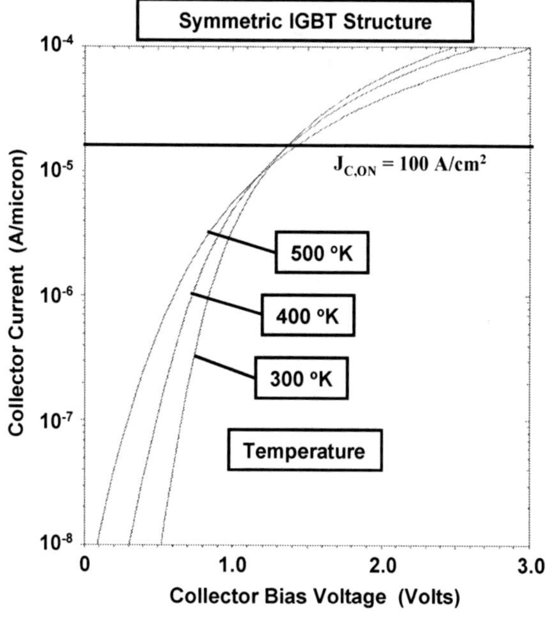

Fig. 9.202 On-state characteristics for the 1,200-V symmetric IGBT structure at various temperatures

9.14.2 Latch-Up Characteristics

One of the problems encountered with high temperature operation of the IGBT structure at elevated temperatures is the reduction of the latch-up current density.[60] This phenomenon can be understood by examination of the temperature dependence of the components of (9.350) that determines the latch-up current density:

$$J_{\mathrm{C,L}} = \frac{V_{\mathrm{bi}}}{\alpha_{\mathrm{PNP,ON}}\rho_{\mathrm{SP}}L_{\mathrm{P}}W_{\mathrm{CELL}}} \propto \frac{V_{\mathrm{bi}}(T)\mu_{\mathrm{pB}}(T)}{\alpha_{\mathrm{PNP,ON}}(T)}. \tag{9.407}$$

The temperature-dependent terms have been highlighted in the above equation for discussion purposes. The built-in potential (V_{bi}) of the N^+ emitter/P-base junction and the mobility of holes (μ_{pB}) in the P-base region decrease with increasing temperature. The current gain of the P–N–P transistor tends to increase slightly with increasing temperature due to an increase in the lifetime. The latch-up current density computed by using the temperature dependence of the built-in potential and the hole mobility is provided in Fig. 9.203. Here, the power coefficient for the temperature dependence of the hole mobility was assumed to be −1.5 due to the high doping concentration of the P-base region. From this plot, it can be observed that these effects conspire to reduce the latch-up current density for the IGBT structure by about a factor of 2–3 times when the temperature increases from

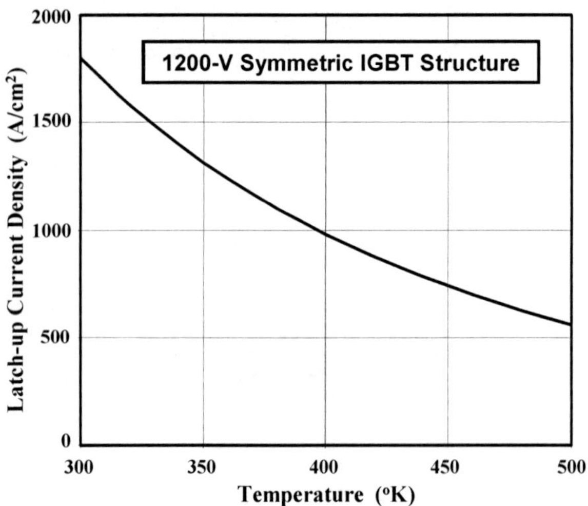

Fig. 9.203 Change in the latch-up current density of the symmetric IGBT structure with increasing temperature

300 to 500 K. Consequently, the various latch-up suppression techniques described earlier in this chapter must be vigorously utilized to suppress latch-up for IGBT structures designed for high temperature applications.

Simulation Example

To corroborate the analytical model for the latch-up characteristics of the IGBT structure at elevated temperatures, the results of two-dimensional numerical simulations for the symmetric structure are discussed here. This device had the

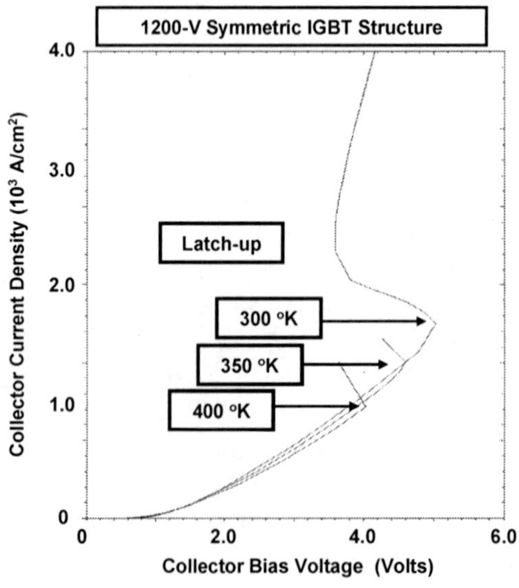

Fig. 9.204 Latch-up characteristics for the 1,200-V symmetric IGBT structure

planar-gate structure with a cell pitch of 15 μm as described in previous sections. A high-level lifetime of 20 μs was used in the N-base region for all three cases. The latch-up characteristics obtained by using the simulations are shown in Fig. 9.204 at three temperatures. The latch-up current density decreases with temperature in good quantitative agreement with the analytical model.

9.15 Lifetime Control Techniques

The switching speed of the IGBT structure is determined by the lifetime in the drift region (wide N-base region for the n-channel device) as previously discussed in Sect. 9.7.4. The two basic methods for reducing lifetime in semiconductor devices are either by introduction of a heavy metal impurity that produces deep levels in the band gap or by particle bombardment to create defects in the crystal.[61] Although both techniques were originally developed and utilized to alter the characteristics of bipolar power devices, only the particle bombardment process is practical for application to MOS-gated devices. Heavy metal impurities segregate to the semiconductor surface during thermal processing. When they accumulate under the gate oxide of devices such as the IGBT structure, the threshold voltage becomes very large and nonuniform across the wafer. In contrast, a uniform distribution of defects can be produced within the wafer by using electron or neutron bombardment with the appropriate energy. However, these energetic particles damage the gate oxide creating significant positive charge that can shift the threshold voltage by many volts. Fortunately, it was discovered that the defects (and charge) produced in the gate oxide can be removed by annealing at low temperature (between 150 and 200°C) while preserving the defects created within the silicon.[62] This breakthrough has allowed the precise and reproducible control of the lifetime in the IGBT structures.[63]

9.15.1 Electron Irradiation

When first introduced as a new device concept, the slow switching speed of the as-fabricated IGBT structure was considered to be a major limitation to its application potential. The high lifetime in the devices when fabricated in a clean manufacturing facility produces long collector current fall times resulting in high switching power loss. This would have relegated the IGBT structure to applications, such as offline appliance controls, at frequencies below 1 kHz. Fortunately, the electron irradiation process was found to provide a simple and precise method for reducing the turn-off time for the IGBT structure allowing its utility for a broad range of applications.

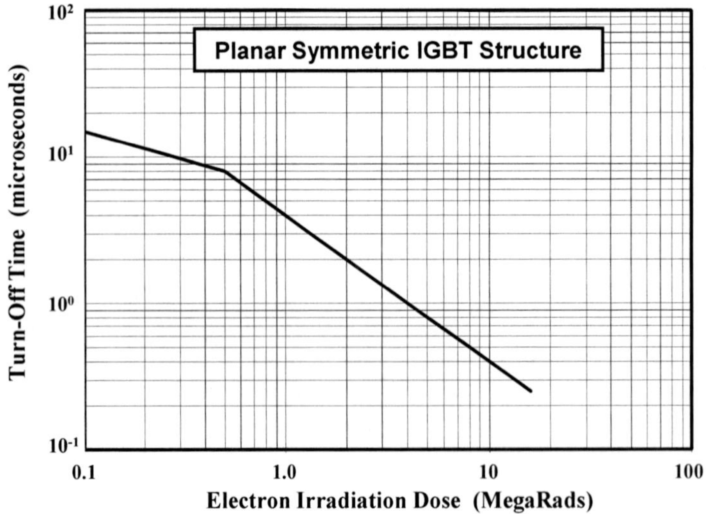

Fig. 9.205 Controlling turn-off time of the symmetric IGBT structure with 3-MeV electron irradiation

The results of measured changes in the turn-off time with increasing electron irradiation dose are shown in Fig. 9.205 for the 600-V symmetric IGBT structure in the case of an electron beam energy of 3 MeV.[64] The turn-off time for the IGBT structure without any electron irradiation was 15 μs. It can be observed from this plot that the turn-off time decreases with increasing electron irradiation dose in accordance with the relationship

$$\frac{1}{t_{\mathrm{OFF}}} = \frac{1}{t_{\mathrm{OFF,i}}} + K_{\mathrm{ER}}\phi_{\mathrm{ER}}, \tag{9.408}$$

where t_{OFF} is the turn-off time after electron irradiation with a dose ϕ_{ER}. In this expression, $t_{\mathrm{OFF,i}}$ is the initial turn-off time before electron irradiation and K_{ER} is the electron irradiation damage coefficient. The damage coefficient extracted from the data is 0.25 microsecond^{-1} Mrad^{-1}. It can be observed that the electron irradiation process allows reducing the turn-off time over a broad range from over 10 μs down to 0.25 μs. The reduction of the lifetime in inverse proportion to the radiation dose, which can be accurately metered, allows precise control of device characteristics.

The reduction of the lifetime resulting from the electron irradiation can be expected to produce an increase in the on-state voltage drop. The measured change in the on-state voltage drop for the 600-V symmetric IGBT structure is shown in Fig. 9.206 for the case of electron irradiation with an energy of 3 MeV.[64] The on-state voltage drop shown in this figure was measured at an on-state current density of 167 A cm^{-2}. Using the data in Figs. 9.205 and 9.206, a tradeoff between the on-state conduction power loss and the turn-off power loss can be performed.

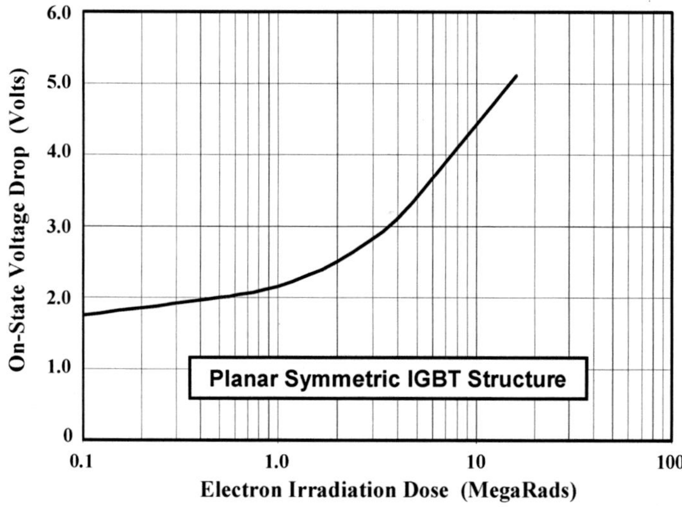

Fig. 9.206 Increase in the on-state voltage drop of the symmetric IGBT structure after 3-MeV electron irradiation

9.15.2 Neutron Irradiation

An alternative process to electron irradiation for controlling the lifetime uniformly within the silicon is by using high energy neutron irradiation.[65] It has been established that the neutron irradiation produces the same tradeoff curve between on-state voltage drop and turn-off time as electron irradiation. The neutron irradiation is performed by exposing the silicon to the neutron flux emanating from the U235 fission reaction. This requires access to a nuclear facility with cadmium shielding of the silicon to avoid exposure to thermal neutrons. The devices cannot be irradiated after mounting on headers because the packages become radioactive. It is therefore not possible to test the devices prior to the irradiation. These restrictions do not apply to electron irradiation which is performed by using a Van de Graff generator. Consequently, electron irradiation is preferred by the industry due to its simplicity and lower cost.

9.15.3 Helium Irradiation

The uniform reduction of the lifetime within the IGBT structure is suitable for the symmetric blocking structure. For the asymmetric structure, where the excess free carriers are remnant only within the N-buffer layer during the current fall time if the collector supply voltage is large, it is beneficial to reduce the lifetime selectively only within the N-buffer layer. A selective reduction of the lifetime deep within the silicon wafer can be obtained by using high-mass ions instead of low-mass electron beams.[66] The concentration of defect centers is about five times greater at the peak when compared with the regions closer to the surface.

Consequently, the impact of the ion bombardment in the N-base region above the N-buffer layer cannot be neglected. Using deep-level transient spectroscopy (DLTS) measurements, it has been found that ion bombardment produces the same deep levels as electron irradiation within the silicon band gap. However, the use of protons (hydrogen ions) can also introduce shallow donor levels within the silicon band gap that can alter the doping profile within the device. As in the case of ion implantation of dopants into semiconductors, the maximum damage after ion bombardment occurs at the range of the particle which is determined by its mass and energy. For the case of a 1,200-V IGBT structure, the N-buffer layer is located at a depth of 100–120 μm below the upper surface. A range of 128 μm with a straggle of about 5 μm can be achieved by using 18-MeV helium ions.[67]

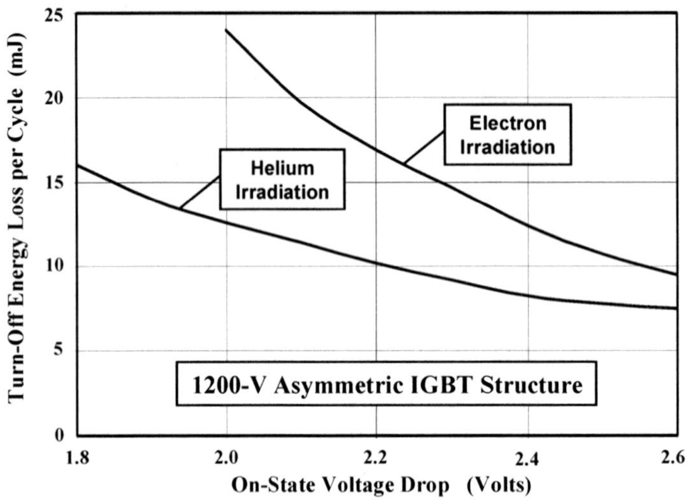

Fig. 9.207 Tradeoff curves for the 1,200-V asymmetric IGBT structure after helium and electron irradiation

The tradeoff curve between the turn-off switching loss per cycle and the on-state voltage drop for the 1,200-V asymmetric IGBT structure has been empirically obtained[68] for the case of helium and electron irradiation. The tradeoff curves are compared in Fig. 9.207 for the two methods of lifetime control. It can be observed that lower losses can be obtained by using the helium bombardment to locate most of the recombination centers in the N-buffer layer.

9.16 Cell Optimization

In Chap. 6, it was demonstrated that the specific on-resistance for the planar-gate power MOSFET structure exhibits a minimum value when the width of the gate electrode is increased. A similar phenomenon is observed for the planar-gate IGBT structure whose on-state voltage drop exhibits a minimum value as the width of the gate electrode is increased.[23,69] It is also important to optimize the cell design of the trench-gate structure by adjusting the mesa width and the trench depth.[52,53,70,71] The impact of cell optimization on the IGBT characteristics is discussed below. In spite of the improved performance that can be derived by using the ALL cell topology, the most commonly used cell topology during the design of planar-gate IGBT structures is a linear cell topology. For this reason, the cell optimization discussed below is based upon a linear cell topology for both the planar-gate and trench-gate cases.

9.16.1 Planar-Gate Structure

For the planar-gate structure, the width of the polysilicon window is determined by process considerations. The optimization of the cell design is performed by varying the width of the gate electrode. The optimum value for the gate width at which the minimum on-state voltage drop occurs can be derived by using the analytical models developed in Sect. 9.5. As an example, the optimization of the gate width for the planar-gate 1,200-V symmetric IGBT structure is shown in Fig. 9.208. The physical parameters for these devices were a polysilicon window of 16 μm, a gate oxide thickness of 500 Å, a channel length of 1.5 μm, and a deep P^+ region junction depth of 5 μm. The gate drive voltage was assumed to be 15 V with a threshold voltage of 5 V. The mobility for electrons in the inversion and accumulation layers was assumed to be 450 and 1,000 $cm^2\ V^{-1}\ s^{-1}$, respectively. The N-base region had a doping concentration of $5 \times 10^{13}\ cm^{-3}$ and a thickness of

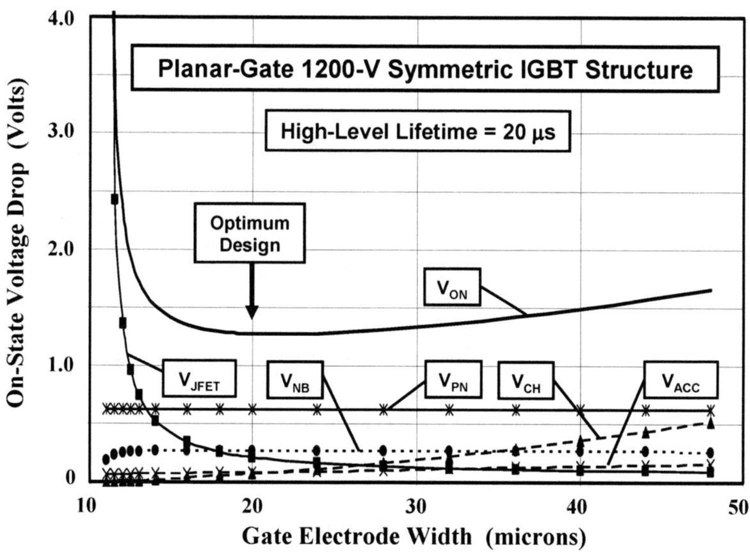

Fig. 9.208 Optimization of the on-state voltage drop of the 1,200-V planar-gate symmetric IGBT structure with a high-level lifetime of 20 μs in the N-base region

200 μm with a high-level lifetime of 20 μs. The JFET region was assumed to have an enhanced effective doping concentration of 5×10^{15} cm^{-3}. All of these parameters are the same as those used in previous sections for the 1,200-V planar-gate symmetric IGBT structure with a gate electrode width of 16 μm. In the figure, the various components of the on-state voltage drop are also shown to provide insight into the optimization of the design. From the figure, it is apparent that the on-state voltage drop exhibits a minimum value at an optimum gate electrode width of 20 μm. At gate widths below 13 μm, the on-state voltage drop increases drastically due to the rapid increase in the JFET component. With a narrow gate electrode width, it is common to see a snapback in the *i–v* characteristics of the IGBT structure, which is undesirable from an application's viewpoint.[67] When the width of the gate electrode is increased beyond the optimum value, the contributions to the on-state voltage drop (V_{CH} and V_{ACC}) from the MOSFET section (shown by the dashed lines in the figure) become larger than that from the JFET region. This produces a gradual increase in the on-state voltage drop with increasing gate width.

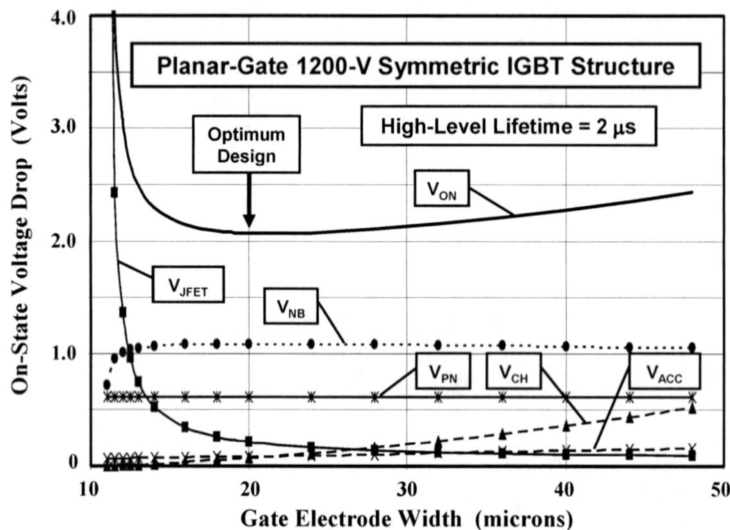

Fig. 9.209 Optimization of the on-state voltage drop of the 1,200-V planar-gate symmetric IGBT structure with a high-level lifetime of 2 μs in the N-base region

The optimization of the gate width for the planar-gate 1,200-V symmetric IGBT structure is shown in Fig. 9.209 when the high-level lifetime in the N-base region is changed to 2 μs, while keeping all the other device parameters the same as defined above. In the figure, the various components of the on-state voltage drop are also shown to provide insight into the optimization of the design. From the figure, it is apparent that the on-state voltage drop exhibits a minimum value at the same optimum gate width of 20 μm. This occurs because the MOSFET and

JFET voltage drop components are not altered by the change in the lifetime in this analytical model. The minimum value for the on-state voltage drop is larger than in the case of a high-level lifetime of 20 μs because the voltage drop (V_{NB}) across the N-base region has increased when the lifetime was reduced.

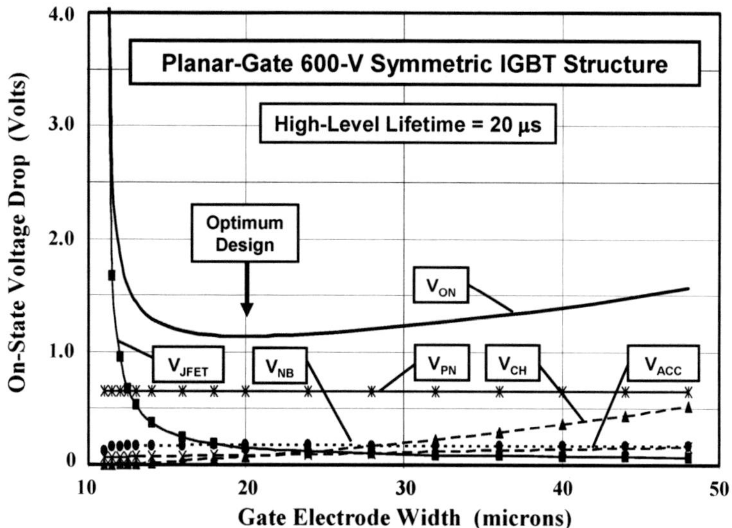

Fig. 9.210 Optimization of the on-state voltage drop of the 600-V planar-gate symmetric IGBT structure with a high-level lifetime of 20 μs in the N-base region

The optimization of the gate width for the planar-gate 600-V symmetric IGBT structure is shown in Fig. 9.210 when the high-level lifetime in the N-base region is 20 μs, while keeping all the other device parameters the same as defined above for the 1,200-V structure. The N-base region had a thickness of 120 μm and a doping concentration of 1.5×10^{14} cm^{-3} for the 600-V structure. In the figure, the various components of the on-state voltage drop are also shown to provide insight into the optimization of the design. From the figure, it is apparent that the on-state voltage drop exhibits a minimum value at the same optimum gate width of 20 μm. This occurs because the MOSFET and JFET voltage drop components are not altered by the changes in the N-base parameters in this analytical model. The minimum value for the on-state voltage drop is smaller than in the case of the 1,200-V structure because the voltage drop (V_{NB}) across the N-base region has decreased for the 600-V structure.

Based upon the above examples, it can be concluded that the same IGBT cell design is adequate for all planar-gate devices even when the voltage rating or switching speed is altered. This is in contrast to the power MOSFET structure where the optimum cell design is a strong function of the blocking voltage capability. In this context, the layout of IGBT structures is much simpler than that for the power MOSFET structure.

For the planar-gate IGBT structure, it is necessary to monitor the change in the latch-up current density during the cell optimization to ensure that it is adequate. For the planar-gate structure, the latch-up current density is given by

$$J_{\mathrm{C,L}} = \frac{V_{\mathrm{bi}}}{\alpha_{\mathrm{PNP,ON}}\rho_{\mathrm{SP}}L_{\mathrm{P}}W_{\mathrm{CELL}}} = \frac{V_{\mathrm{bi}}}{\alpha_{\mathrm{PNP,ON}}\rho_{\mathrm{SP}}L_{\mathrm{P}}(W_{\mathrm{POLY}} + W_{\mathrm{G}})}. \tag{9.409}$$

Based upon this equation, it is apparent that the latch-up current density will reduce as the width of the gate electrode (W_{G}) is increased. This occurs because a larger amount of collector current is collected from under the gate region and fed into the P-base region, producing a larger voltage drop across the N^+ emitter/P-base junction.

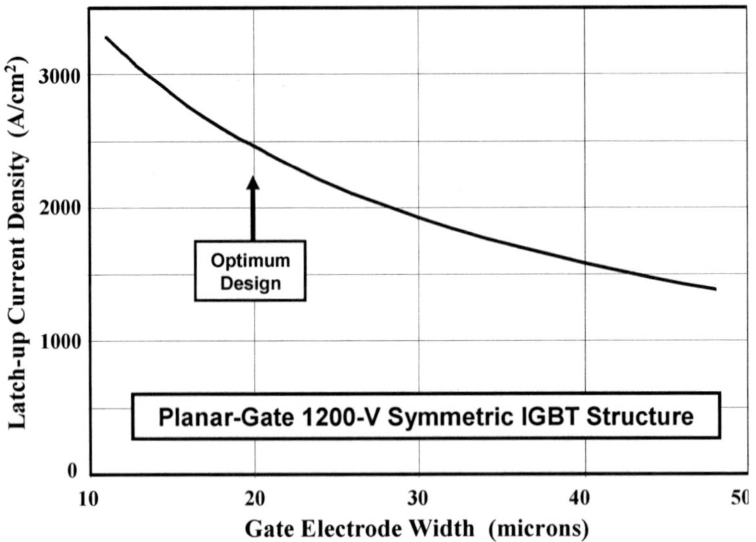

Fig. 9.211 Impact of cell design on the latch-up current density for the 1,200-V symmetric IGBT structure

As an example, the latch-up current density computed by using the above equation is shown in Fig. 9.211 for the case of a 1,200-V IGBT structure. The polysilicon window (W_{POLY}) was assumed to be 16 μm in size and the length (L_{P}) of the N^+ emitter region beyond the edge of the deep P^+ diffusion was assumed to be 2 μm. At the optimum gate electrode width of 20 μm for achieving the lowest on-state voltage drop, the latch-up current density is about 2,500 A cm^{-2}. This value is adequate for suppressing latch-up during device operation even at elevated temperatures.

9.16.2 Trench-Gate Structure

As discussed previously in Sect. 9.12, the trench-gate structure is attractive for the IGBT design because it enables a reduction of the on-state voltage drop when compared with the planar-gate structure. This is especially true for devices with high switching speed because the channel current component is larger. Using a shallow trench depth, it was shown in Sect. 9.12 that the on-state voltage drop for the trench-gate devices is about 0.3 V smaller than for the planar-gate devices due to elimination of the JFET component of the on-state voltage drop. A further reduction of the on-state voltage drop, especially for devices with fast switching speed, can be obtained by optimizing the cell design for the trench-gate IGBT structure.

On-State Operation

In Sect. 9.12.2, it was pointed out that the hole carrier distribution for the n-channel IGBT structure is enhanced under the gate electrode when compared with the hole concentration in the mesa region. The enhancement of the conductivity modulation of the N-base region can be maximized by making the mesa width small and the trench depth large.[71] In this cell design, the hole concentration profile takes a catenary shape from the P^+ collector/N-base junction up to the bottom of the trenches even under the mesa region. The concentration then rapidly falls to zero at the P-base/N-base junction in the mesa region. With this carrier distribution, the voltage drop across the N-base region can be analyzed by using the equations developed for the P–i–N rectifier in Sect. 5.1.3.

The on-state voltage drop across the trench-gate IGBT structure with the enhanced free carrier concentration at the trench is given by

$$V_{ON} = V_{PN} + V_{NB} + V_{ACC} + V_{CH}. \tag{9.410}$$

The voltage drop for the P/N junction in the P–i–N rectifier model is

$$V_{PN} = \frac{kT}{q} \ln\left(\frac{p_0^2}{n_i^2}\right). \tag{9.411}$$

The voltage drop across the N-base region is given by

$$V_{NB} = \frac{4kT}{q}\left(\frac{W_N}{2L_a}\right)^2 \tag{9.412}$$

when the (d/L_a) ratio is less than 2, while it is given by

$$V_{NB} = \frac{6\pi kT}{8q} e^{(W_N/2L_a)} \tag{9.413}$$

when the (d/L_a) ratio is more than 2. Here, the voltage drop in the N-base region is assumed to be twice that for the P–i–N rectifier because the concentration of free carriers near the trench bottom is not as great as near the N$^+$ region in the P–i–N rectifier.

In the deep trench IGBT structure, the accumulation layer extends for the distance between the bottom of the P-base region and the trench bottom surface. The voltage drop across the accumulation layer is given by

$$V_{ACC} = \frac{J_{C,ON}(t_T - x_{JP})W_{CELL}}{2\mu_{nA}C_{OX}(V_G - V_{TH})}, \tag{9.414}$$

where t_T is the trench depth. The voltage drop across the channel is given by

$$V_{CH} = \frac{J_{C,ON}L_{CH}W_{CELL}}{2\mu_{ni}C_{OX}(V_G - V_{TH})}. \tag{9.415}$$

Due to the small cell pitch for this type of trench-gate IGBT structure, the contributions to the on-state voltage drop from the MOSFET portion are smaller than for the planar-gate structure.

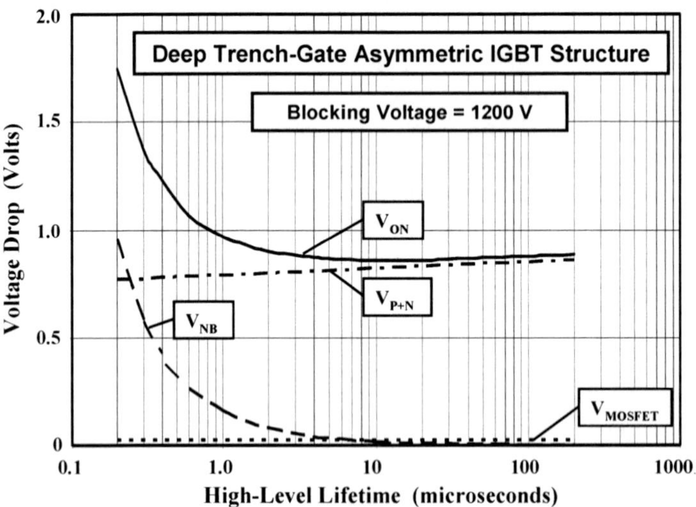

Fig. 9.212 On-state voltage drop for the deep trench asymmetric 1,200-V IGBT structure

The on-state voltage drop obtained by using the above analytical model for the deep (injection-enhanced) trench-gate IGBT structure is shown in Fig. 9.212 for the case of a 1,200-V asymmetric blocking design. It can be observed that the on-state voltage drop is reduced when compared with the shallow trench IGBT structure (see Fig. 9.186) at low lifetime values. The injection enhancement effect has therefore been utilized to fabricate IGBT devices with high (4.5–6.5 kV) blocking voltages with low on-state voltage drops.[71,72]

Simulation Example

To demonstrate the injection enhancement effect in the deep trench IGBT structure, the results of two-dimensional numerical simulations for the 1,200-V asymmetric structure are discussed here. This device had a half-cell pitch of 1.5 μm with a narrow mesa half-width of 1 μm. A trench depth of 10 μm was chosen to produce the injection enhancement effect. The injected carrier distribution with the device at an on-state current density of 100 A cm^{-2} obtained from the simulations is shown in Fig. 9.213 for the case of different high-level lifetime values. It is evident that the injection enhancement effect is not discernable for high lifetime cases. Only when the high-level lifetime is reduced to 0.2 μs, an enhanced concentration of holes is observed at a depth of 10 μm from the surface in the mesa region. This enhanced concentration for holes is not as large as that observed in the case of injection from the N$^+$ cathode region in P–i–N rectifiers. This increases the voltage drop across the N-base region relative to that observed in a P–i–N rectifier.

The on-state voltage drops obtained from the numerical simulations are compared with those predicted by the analytical solution in Fig. 9.214. From this plot, it can be concluded that the analytical model provides an accurate value for the on-state voltage drop for the deep trench IGBT structure because the impact of the enhanced free carrier concentration at the bottom of the trenches has been accounted for by using the modified P–i–N diode model for the voltage drop in the N-base region.

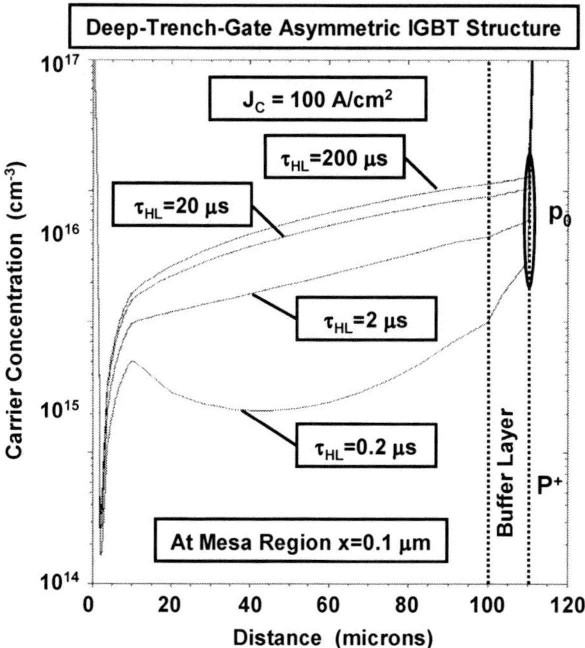

Fig. 9.213 Free carrier concentration in the 1,200-V asymmetric IGBT structure with deep trench-gate structure

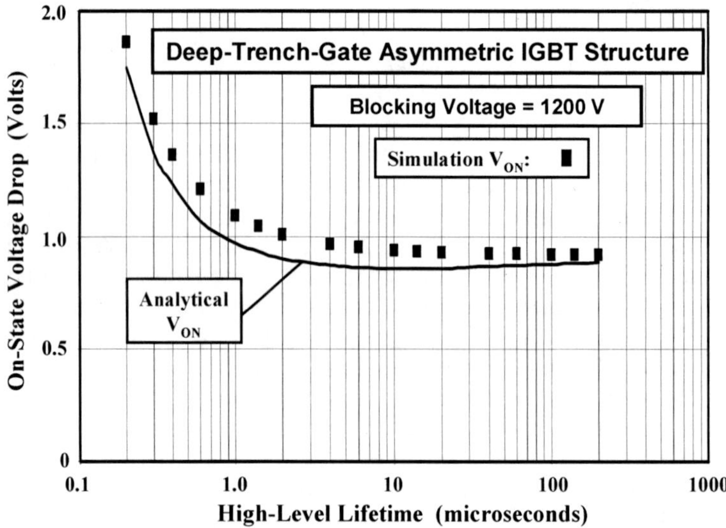

Fig. 9.214 On-state voltage drop for the 1,200-V asymmetric IGBT structure with deep trench-gate structure

Blocking Characteristics

Notwithstanding the lower on-state voltage drop that can be achieved by using the deep trench-gate structure for IGBT structures with high blocking voltage capability, there are several drawbacks to taking this design approach. The first problem is that the process for formation of trenches deeper than 5 μm with a small width suitable for refilling with the polysilicon gate material is complex and expensive. The second problem is that the extension of the trench deep into the N-base region produces an enhancement of the electric field at the trench corners, leading to a degradation of the breakdown voltage as discussed in Sect. 9.12.1. However, the electric field enhancement at the trench corners is ameliorated when the mesa width is made small as required to produce the injection enhancement phenomenon.[70]

Simulation Example

To examine the impact of the deep trenches on the forward-blocking capability of the IGBT structure, the results of two-dimensional numerical simulations for the 1,200-V asymmetric structures are discussed here. The deep trench structure had a half-cell pitch of 1.5 μm with a narrow half-mesa width of 1 μm and a trench depth of 10 μm, while the shallow trench structure had a half-cell pitch of 3.5 μm and a wider half-mesa width of 3 μm with a trench depth of 4 μm. The forward-blocking characteristics for the two structures at 300 K can be compared in Fig. 9.215 for the case of a high-level lifetime of 0.2 μs. The deep trench structure exhibits a smaller leakage current per micron of cell depth orthogonal to the cross section because of its smaller cell pitch. However, the breakdown voltage for the deep trench structure is identical to that for the shallow trench structure. This occurs because the electric field enhancement at the trench corners has been reduced by the small mesa width.

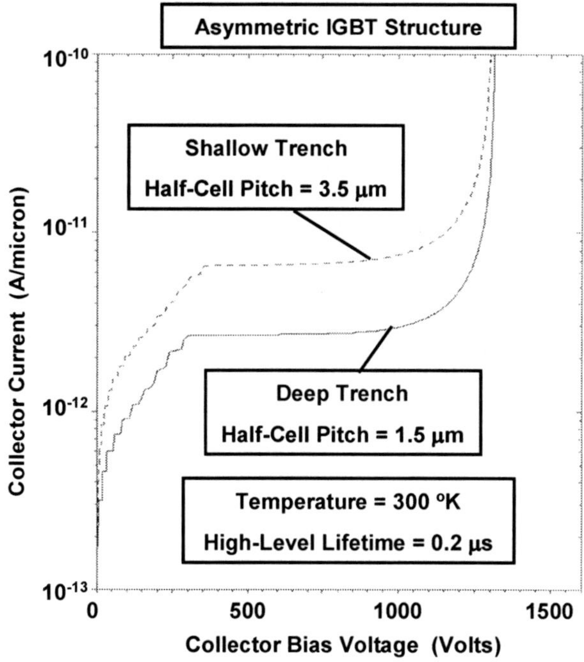

Fig. 9.215 Comparison of the forward-blocking characteristics of the 1,200-V asymmetric IGBT structure with shallow and deep trench-gate structures

The third problem is that the presence of deep trenches in the N-base region can compromise the breakdown voltage at the edges of the chip. Even for devices with shallow trench-gate structures, it is prudent to envelop the edges of the trenches at the ends of the fingers by using a P^+ diffusion that extends deeper than the trenches.[73] This is feasible for trench depths of up to 5 μm. For the much deeper trenches required to produce the injection enhancement effect, the breakdown voltage will be degraded at the edge termination unless very deep P-type diffusions with large thermal budgets are employed during device fabrication.

Switching Performance

The presence of a higher concentration of free carriers in the vicinity of the deep trenches can be expected to have an impact of the switching speed of the IGBT structure. During the first phase of the turn-off process with an inductive load, the stored charge at the P-base/N-base junction must be removed by the advancing space-charge layer. The larger stored charge in the deep trench structure near the bottom of the trenches can be expected to reduce the rate of increase in the collector voltage during the voltage transient. Once the voltage has reached the collector supply voltage, the current decay occurs by recombination in the N-buffer layer. Since this process is the same as for the shallow trench-gate (and planar-gate) IGBT structures, this portion of the transient for deep trench-gate structures can be expected to be similar to the other structures.

Simulation Example

To examine the impact of the deep trenches on the switching behavior of the IGBT structure, the results of two-dimensional numerical simulations for the 1,200-V asymmetric structure are discussed here. The deep trench structure had a half-cell pitch of 1.5 μm with a narrow mesa half-width of 1 μm and a trench depth of 10 μm, while the shallow trench structure had a half-cell pitch of 3.5 μm and a wider mesa half-width of 3 μm with a trench depth of 4 μm. The switching characteristics for the two structures can be compared in Fig. 9.216 for the case of a high-level lifetime of 0.2 μs. The deep trench structure exhibits a longer collector voltage rise time due to the larger stored charge near the trenches. The collector current fall time for the deep trench structure is similar to that for the shallow trench structure because it is controlled by recombination in the N-buffer layer for both devices.

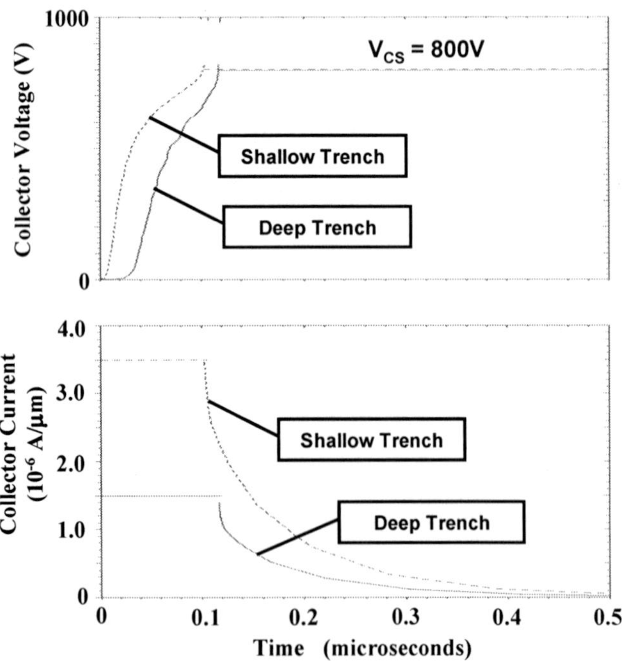

Fig. 9.216 Comparison of the switching characteristics of the 1,200-V asymmetric IGBT devices with shallow and deep trench-gate structures

Short-Circuit Capability

The short-circuit withstand capability for the IGBT structure is dependent upon its saturated current density as discussed in Sect. 9.11.3. The channel density for the deep trench-gate structure with narrow mesa regions is very large leading to a very high saturated current density. This results in a poor short-circuit withstand capability for the deep trench structures.[53] To retain the injection enhancement effect while reducing the saturated current density, it is necessary to reduce the channel density. One method to accomplish this is by increasing the trench width.[53] However, it is difficult to refill and planarized wide trenches making the practical implementation of this approach problematic. A more process compatible approach is to alter the trench-gate cell design to include mesa regions without the N^+ emitter region[74] as shown in Fig. 9.217.

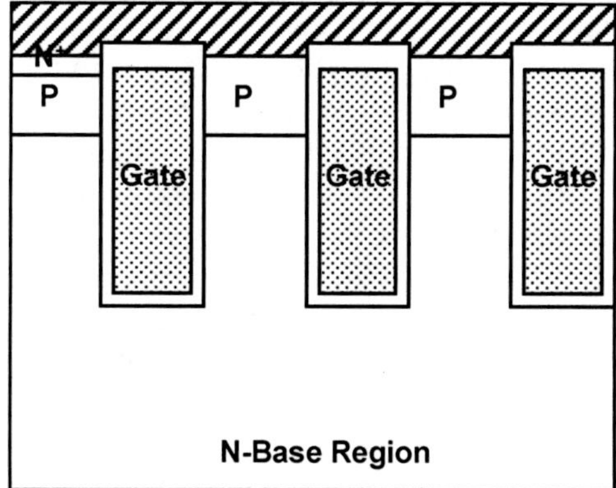

Fig. 9.217 Alternate deep-trench IGBT structure with reduced channel density

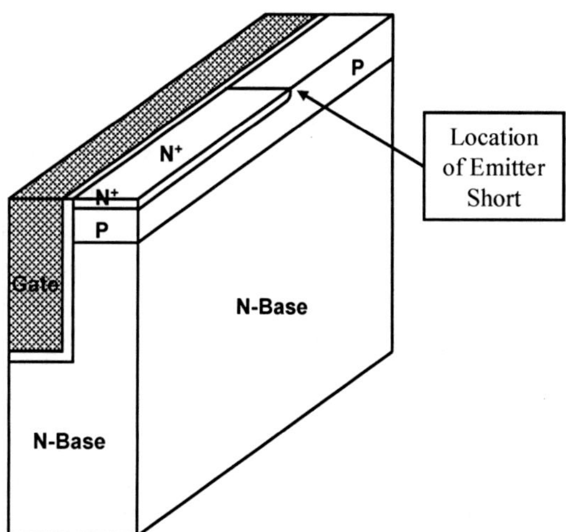

Fig. 9.218 Three-dimensional view of a deep trench IGBT structure with emitter short orthogonal to the cross section

Deep trench-gate IGBT structures with narrow mesa widths require forming the short between the N^+ emitter and the P-base region selectively at locations orthogonal to the cross section as illustrated in Fig. 9.218. Only the upper portion of the device structure is shown in this figure without the gate isolation and the emitter metal layers. This approach to making the emitter short also decreases the channel density. This has a minor impact on the on-state voltage drop and is beneficial for reducing the saturated current for achieving better short-circuit withstand capability.

9.17 Reverse Conducting Structure

The basic IGBT structure contains back-to-back junctions which preclude current flow in the third quadrant of device operation as discussed in Sect. 9.2. In typical motor control circuits from a DC bus power source, the IGBT structure needs only forward-blocking capability. In addition, it is commonplace to connect an antiparallel diode across the IGBT structure, as shown in Fig. 9.28, to carry the load current during a part of the operating cycle. It is advantageous from a system standpoint to integrate the reverse conducting diode within the IGBT structure to eliminate an additional component in the system.

One approach to the integration of a lateral antiparallel diode is by addition of an N^+ region (for an n-channel IGBT structure) around the periphery of the chip outside the edge termination. The N^+ region is connected to the collector terminal while packaging the device. This provides an additional path for current transport that can be created for any of the IGBT structures discussed previously in this chapter. Results have been reported by using this approach for the p-channel IGBT structure.[75] Although simple from an implementation standpoint, this approach is ineffective for high-current devices because the peripheral N^+ region is located at a large distance from the active regions in the middle of the chip. It is preferable that the collector short is placed adjacent to the injecting junction located at the bottom of the IGBT structure.

During the early stages of IGBT device development, the blocking voltage ratings were limited to 1,200 V. Such structures could be most conveniently constructed by using epitaxial growth of the N-buffer and N-base layers on a thick P^+ substrate. The resulting asymmetric blocking structure has weak reverse-blocking capability as discussed in Sect. 9.4.5. The voltage drop across the asymmetric IGBT structure in the third quadrant is too large for reverse conducting operation. One solution[76] for integration of the fly-back diode that is compatible with N-base regions formed by epitaxial growth is illustrated in Fig. 9.219.

The heavily doped N^+ buried layer is first formed by ion implantation of phosphorus at selected locations in the P^+ substrate prior to epitaxial growth of the N-buffer layer. The N^+ buried layer creates a new junction (J_4) within the IGBT structure between two heavily doped regions of opposite conductivity type. Current can be transported across this junction by the tunneling phenomenon producing an ohmic characteristic that allows reverse conducting current to flow through the structure. The reverse conducting diode formed by using this approach is indicated in the figure by the dashed box. The switching speed of the IGBT structure is also improved by the presence of the collector shorts. During the turn-off process, these regions enable the trapped electrons in the N-buffer layer to be transported via the N^+ buried layer into the P^+ collector without undergoing recombination. Experimental validation of the use of an N^+ buried layer to create a faster switching device has been accomplished for IGBT structures capable of supporting 600 V.[77] A reduction in total power losses by 12% was observed at a switching frequency of 15 kHz.

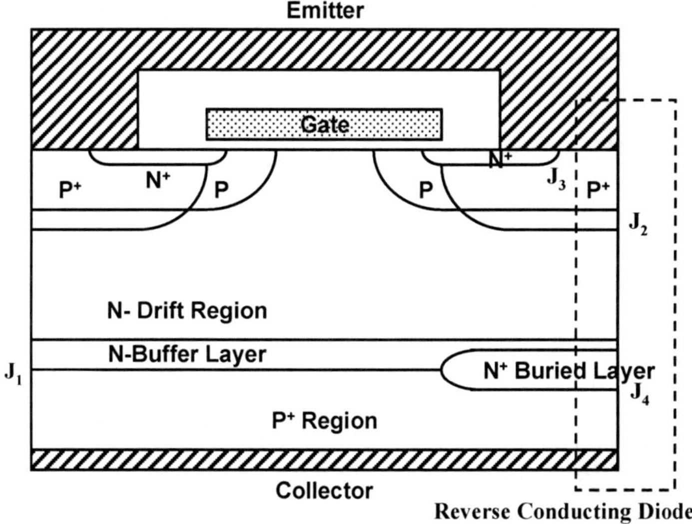

Fig. 9.219 Epitaxial IGBT structure with reverse conduction capability

Device structures based upon using bulk N-type starting material became feasible when the voltage ratings for the IGBT structure were scaled above 1,200 V. The formation of an IGBT structure with reverse conduction capability is relatively straightforward in this case by patterning the P^+ collector diffusion on the backside of the wafer. The zones between the P^+ collector regions are heavily doped with an N-type impurity as shown in Fig. 9.220 to enable forming ohmic contacts to the N^+ region by the backside metal. The reverse conducting diode formed at the collector shorts is indicated in the figure by the dashed box. In this implementation, no additional junctions are formed allowing good current conduction via the diode.

The addition of the collector short introduces a unipolar current conduction path within the IGBT structure during on-state operation. When the gate bias is applied, electron current can be transported from the N^+ collector shorting region through the N-base region and the MOSFET channel. This path has a high resistance due to the low doping concentration and large thickness of the N-base region resulting in relatively low collector current levels. As the collector bias is increased, the lateral unipolar current flow through the resistance (R_{NB}) of the N-base region produces a forward bias across the P^+ collector/N-base junction (J_1) at the middle of the P^+ collector regions. When this forward bias exceeds the built-in potential of the junction, injection of holes begins to occur at the middle of the P^+ collector/N-base junction (J_1) leading to the IGBT mode of operation.

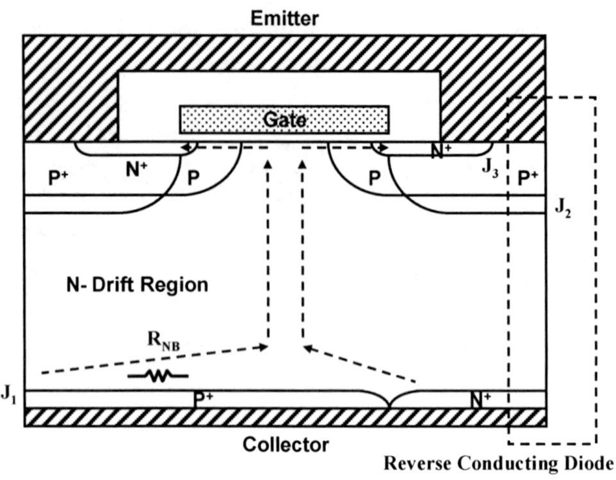

Fig. 9.220 Bulk IGBT structure with reverse conduction capability

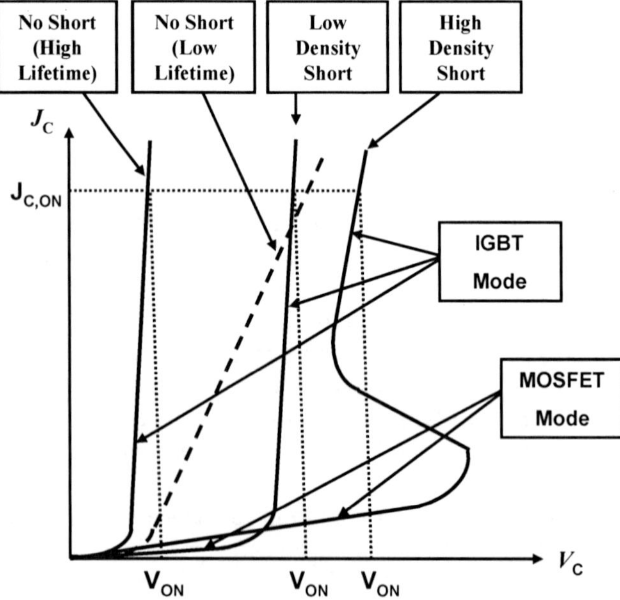

Fig. 9.221 Impact of collector shorting density on on-state characteristics for the IGBT structure

The *i–v* characteristics of the shorted collector IGBT structure can exhibit a snapback,[78] as illustrated in Fig. 9.221, due to the transition from the MOSFET mode to the IGBT mode as the collector current level increases. With a high collector shorting density, the resistance in the initial MOSFET mode is smaller and the device has difficulty entering the IGBT mode until relatively large collector bias voltages are applied. Once the injection from the P^+ collector/N-base junction begins to occur, the conductivity modulation of the N-base region greatly reduces its resistance producing a snapback in the *i–v* characteristics. The low injection efficiency of the P^+ collector/N-base junction at the on-state current density produces reduced hole concentration in the N-base region resulting in a large on-state voltage drop. For a low shorting density, the snapback is not observed but the on-state voltage drop is larger than for an IGBT structure without the collector short and the same lifetime in the N-base region.[79] Since the collector short improves the switching speed, the lifetime in the N-base region for this IGBT structure is usually not reduced. However, to achieve the same switching speed for the unshorted IGBT structure, it is necessary to reduce the lifetime. The on-state characteristic for the IGBT structure with reduced lifetime is shown by the dashed line in the figure. Its on-state voltage drop can be equivalent or even larger than that of the collector shorted structure.

As in the case of other bipolar devices with shorts across the injecting junction (such as the GTO structure), the presence of the collector short in the IGBT structure suppresses the injection of holes from the P^+ collector/N-base junction during on-state operation. This increases the on-state voltage drop for the structure. However, the smaller injected stored charge and the faster removal of free carriers from the N-base region via the collector short reduce the turn-off power loss. A tradeoff curve between the on-state voltage drop and the turn-off energy loss can be generated for the shorted collector IGBT structure by adjusting the density of the shorts. This method for improving the performance of the IGBT structure has been experimentally confirmed.[80] By adjusting the collector short density between 5 and 30%, the turn-off losses could be reduced by 69%. The application of this concept to produce an IGBT structure with an integrated reverse conducting diode has also been experimentally demonstrated[77] more recently.

The forward-blocking characteristics of these structures are also superior to those of the symmetric and asymmetric blocking structures. The presence of the collector short reduces the injection efficiency of the P^+ collector/N-base junction to zero at the relatively low leakage currents that flow during the forward-blocking mode. Consequently, the gain of the P–N–P transistor is essentially zero during the forward-blocking mode making the leakage current equal to the space-charge generation current and the breakdown voltage that of a diode.

Simulation Example

To examine the impact of the collector short on the behavior of the IGBT structure, the results of two-dimensional numerical simulations for the 1,200-V symmetric structure are discussed here. The structural parameters for the device are the same as those described in previous sections. The lifetime (τ_{p0}, τ_{n0}) in the 200-μm thick N-base region for the shorted collector structure was 10 μs. The collector short was formed on the bottom of the N-base region by using N^+ regions with a junction depth of 2 μm. The N^+ regions were located directly below the deep P^+ diffusion in the IGBT cell and had a total width of 2 μm in the 15-μm half-cell used for the simulations. For comparison purposes, the results of numerical simulations for the unshorted IGBT structure are included in this discussion. The lifetime in the N-base region for this structure was adjusted until its on-state voltage drop matched that for the shorted collector structure. The value for the lifetime (τ_{p0}, τ_{n0}) was found to be 0.45 μs.

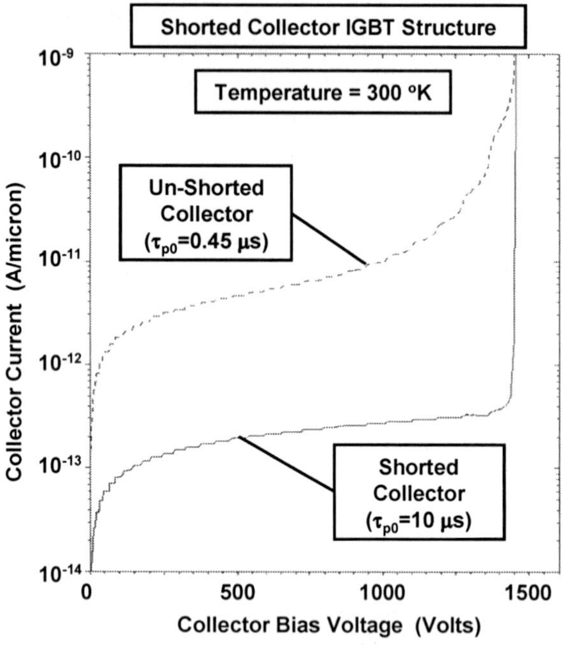

Fig. 9.222 Comparison of the forward-blocking characteristics of the IGBT structures with and without the collector short

The forward-blocking characteristic for the shorted collector IGBT structure is compared with that for the unshorted structure in Fig. 9.222. It can be seen that the leakage current for the shorted collector structure is 40 times smaller than that for the unshorted structure at a collector bias of 800 V. This is partly due to the smaller lifetime (by a factor of 22 times) for the unshorted structure and partly due to the reduction of the gain of the P–N–P transistor to zero in the shorted structure.

The on-state characteristics for the shorted collector IGBT structure are shown in Fig. 9.223 together with those for the unshorted structure. In the case of the unshorted structure, the on-state characteristics are shown for a lifetime equal to that for the unshorted structure (solid line) as well as when the lifetime is reduced to match the on-state voltage drop of the shorted structure (dashed line). It can be observed that the shorted collector structure operates in a MOSFET mode with small current flow until the collector bias voltage reaches 3 V. The device then enters the IGBT mode where the current increases very rapidly due to modulation of the N-base region. In comparison, the IGBT structure without the collector shorts has a smaller off-set voltage and a lower on-state voltage when the lifetime in the N-base region is the same as that for the unshorted structure (10 μs). However, when the lifetime is reduced to 0.45 μs, the on-state voltage drop for the structure without the collector shorts matches that of the shorted collector structure.

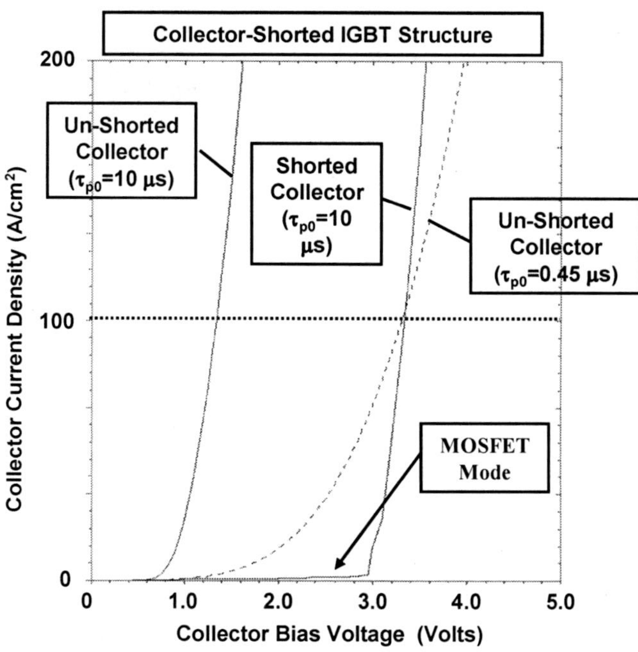

Fig. 9.223 Comparison of the on-state characteristics of the IGBT structures with and without the collector short

The free carrier distribution within the IGBT structure during operation in the on-state provides insight into the operating physics. Three-dimensional views of the on-state hole distribution are shown in Figs. 9.224 and 9.225 for the unshorted and shorted collector IGBT structures, respectively. In the unshorted structure, the hole concentration at the P$^+$ collector/N-base junction is homogeneous across the cell. In contrast, the hole concentration goes to zero at the N$^+$ collector short for the shorted collector IGBT structure. By comparison of the two plots, it can be observed that the hole concentration in the shorted collector structure is five times smaller than in the unshorted structure at the P$^+$ collector/N-base junction. This occurs because of the reduction of the injection efficiency of the P$^+$ collector/N-base junction due to the introduction of the collector short. The turn-off time for the shorted collector structure is reduced due to the smaller stored charge in the N-base region.

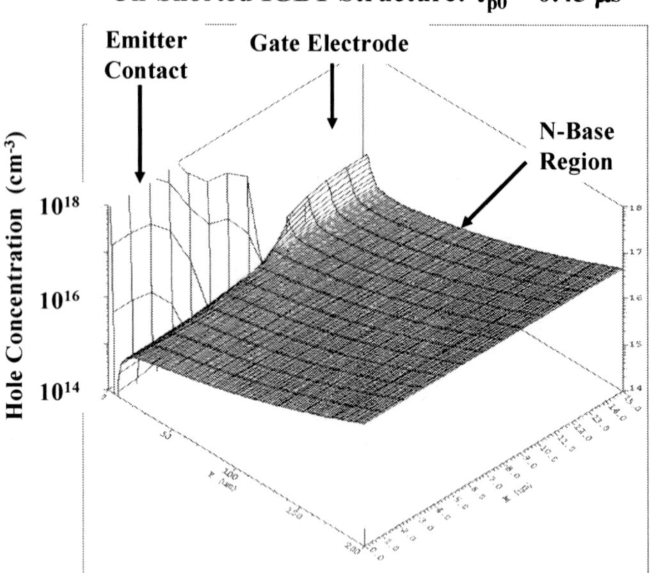

Fig. 9.224 Hole distribution in the on-state for the IGBT structure without the collector short

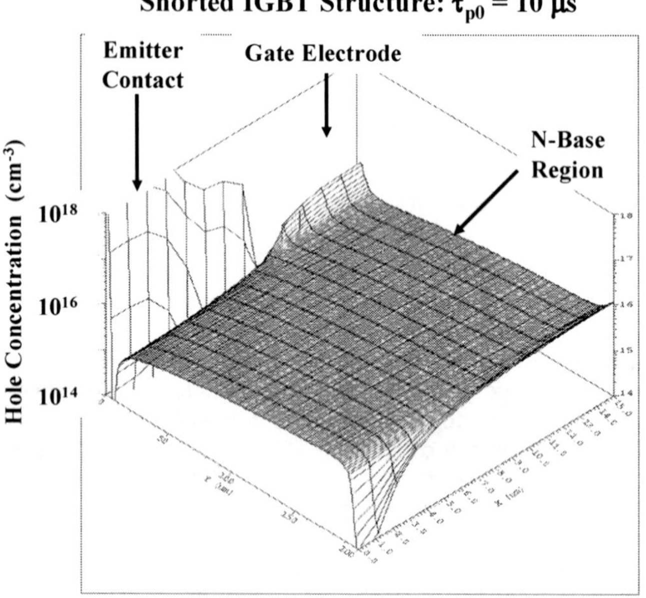

Fig. 9.225 Hole distribution in the on-state for the IGBT structure with the collector short

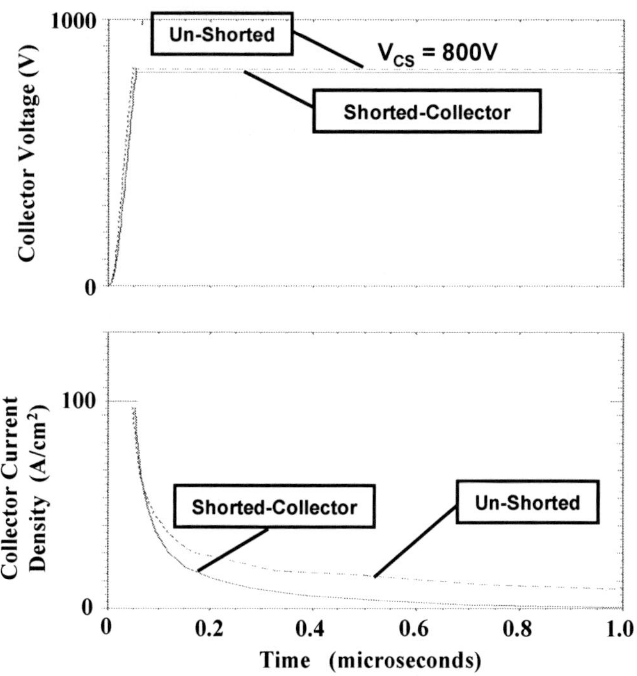

Fig. 9.226 Comparison of the switching characteristics of the IGBT structures with and without the collector short

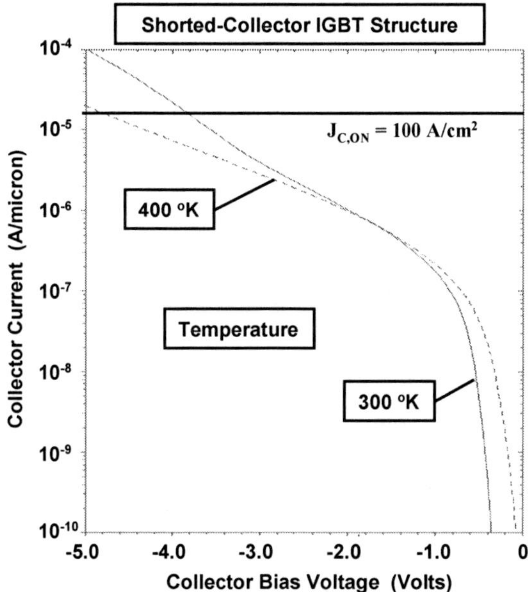

Fig. 9.227 On-state characteristics of the integral diode within the IGBT structures with collector short

The impact of the collector short on the switching characteristics of the IGBT structure can be observed in Fig. 9.226. The characteristics for the shorted collector structure are shown by the solid lines while those for the unshorted structure with a lifetime of 0.45 μs are shown by the dashed lines. The collector voltage rise time for the two structures is nearly identical. The collector current fall time for the shorted collector structure is much smaller (0.27 μs) than for the unshorted structure (0.95 μs) despite the far larger lifetime in the N-base region. This confirms the removal of the stored charge from the N-base region via the short during the turn-off process.

The conduction characteristic of the integral diode in the shorted collector IGBT structure was obtained by using the numerical simulations with negative bias applied to the collector. The gate bias was kept at zero to turn off the path through the MOSFET channel. The integral diode exhibits a relatively large on-state voltage drop as can be observed in the *i–v* characteristics shown in Fig. 9.227 in spite of the large lifetime in the N-base region. This is due to the small area (13%) for the N^+ region used as the collector short. The on-state voltage drop of the diode can be reduced by increasing its relative area within the IGBT structure.

9.18 Summary

The IGBT was originally developed to replace the bipolar transistor in motor control applications. Immediately after the fabrication of the first prototypes, it was apparent that the superior on-state characteristics and simple gate drive requirements for the IGBT structure would greatly reduce the size and cost of the motor control system. The ability to suppress the latch-up of the parasitic thyristor in the IGBT structure and control its switching speed over a broad range were critical breakthrough events to making the device a viable candidate as a power switch. The displacement of bipolar power transistors by the IGBT devices was accomplished in a remarkably short time frame because the IGBT structure could be manufactured using the existing power MOSFET process.

It was initially conjectured that the IGBT structure would be limited to blocking voltages below 2,000 V because MOS-gated thyristors would offer superior tradeoff curves between on-state voltage drop and switching losses.[5] However, the optimization of the IGBT structure, together with the introduction of the trench-gate design, has made the device competitive at much higher voltages. The excellent current saturation capability of the IGBT structure, which cannot be provided by MOS-gated thyristors, is now considered an essential characteristic from an application's perspective. Due to its attractive characteristics, the IGBT structure has been scaled to blocking voltages above 5,000 V and has already replaced the gate turn-off thyristor in traction applications. Due to its excellent overall performance, the IGBT structure is now the predominant power switch technology for applications that operate from power sources with voltages above 200 V (see Fig. 1.2). Until the advent of a cost-effective silicon carbide-based power switch technology, the IGBT structure can be expected to be the dominant device for all medium and high power applications.[81]

Problems

9.1 Determine the doping concentration and width of the N-base region (drift region) for a planar-gate symmetric n-channel IGBT structure to obtain a blocking voltage of 600 V. The lifetime (low-level, high-level, space-charge generation) in the N-base region is 1 μs. Minimize the width of the N-base region within the nearest 5 μm. Provide the values for the depletion width, the base transport factor, and the multiplication factor for the P–N–P transistor at the breakdown condition.

9.2 What is the leakage current density for the planar-gate symmetric n-channel IGBT structure designed in Problem 9.1 at 400 K when it is supporting 100 and 400 V?

9.3 Determine the width of the N-drift region for a planar-gate asymmetric n-channel IGBT structure to obtain a blocking voltage of 600 V if its doping concentration is 1×10^{13} cm^{-3}, assuming that punch-through breakdown voltage conditions are applicable.

9.4 What is the thickness of the N-buffer layer required to prevent open-base transistor breakdown in the planar-gate asymmetric IGBT structure designed in Problem 9.3 if its doping concentration is 1×10^{17} cm^{-3}? Assume that this width is the sum of the depletion region width in the buffer layer and one diffusion length for minority carriers. The lifetime (low-level, high-level, space-charge generation) in the N-drift region is 1 μs. Scale the lifetime for the N-buffer layer based upon its larger doping concentration. Round up your width to the nearest 5 μm.

9.5 What is the leakage current density for the planar-gate asymmetric n-channel IGBT structure designed in Problem 9.4 at 400 K when it is supporting 100 and 400 V if the lifetime (low-level, high-level, space-charge generation) in the N-drift region is 1 μs?

9.6 Determine the on-state voltage drop at an on-state current density of 100 A cm^{-2} for the planar-gate symmetric n-channel IGBT design in Problem 9.1 using the one-dimensional P–i–N/MOSFET model. Use the following parameters: (a) D_a of 15 cm^2 s^{-1}, (b) cell pitch of 16 μm, (c) channel length of 1.5 μm, (d) inversion mobility of 450 cm^2 V^{-1} s^{-1}, (e) accumulation mobility of 1,000 cm^2 V^{-1} s^{-1}, (f) gate oxide thickness of 500 Å, (g) gate bias of 15 V, and (h) threshold voltage of 5 V.

9.7 Determine the on-state voltage drop at an on-state current density of 100 A cm^{-2} for the planar-gate asymmetric n-channel IGBT design in Problem 9.4 using the one-dimensional P–i–N/MOSFET model. Use the following parameters: (a) D_a of 15 cm^2 s^{-1}, (b) cell pitch of 16 μm, (c) channel length of 1.5 μm, (d) inversion mobility of 450 cm^2 V^{-1} s^{-1}, (e) accumulation mobility of 1,000 cm^2 V^{-1} s^{-1}, (f) gate oxide thickness of 500 Å, (g) gate bias of 15 V, and (h) threshold voltage of 5 V.

9.8 Compare the on-state voltage drops for the above IGBT designs with that for a planar-gate power MOSFET structure designed to support 600 V using the same cell parameters.

9.9 Calculate the injected hole concentration at the P$^+$ collector/N-base junction for the planar-gate symmetric IGBT structure under the operating conditions defined in Problem 9.6. Assume an effective doping concentration of 1 × 10^{18} cm^{-3} for the P$^+$ collector region with a diffusion length for electrons of 0.5 μm. What is the injection efficiency at the P$^+$ collector/N-base junction under these operating conditions?

9.10 Determine the on-state voltage drop at an on-state current density of 100 A cm^{-2} for the planar-gate symmetric n-channel IGBT design in Problem 9.1 using the two-dimensional model. Use the following parameters: (a) D_a of 15 cm^2 s^{-1}, (b) cell width of 32 μm, (c) channel length of 1.5 μm, (d) inversion mobility of 450 cm^2 V^{-1} s^{-1}, (e) accumulation mobility of 1,000 cm^2 V^{-1} s^{-1}, (f) gate oxide thickness of 500 Å, (g) gate bias of 15 V, (h) threshold voltage of 5 V, (i) gate electrode width of 16 μm, (j) junction depth of P$^+$ region is 5 μm, and (k) JFET region doping concentration of 5 × 10^{15} cm^{-3}. Provide the values for the voltage drop across the P$^+$/N junction, the N-base region, the JFET region, the accumulation region, and the channel region.

9.11 Calculate the injected hole concentration at the P$^+$ collector/N-base junction for the planar-gate asymmetric IGBT structure under the operating conditions defined in Problem 9.7 with the assumption of low-level injection conditions in the N-buffer layer. Assume an effective doping concentration of 1 × 10^{19} cm^{-3} for the P$^+$ collector region with a diffusion length for electrons of 0.5 μm. What is the injection efficiency at the P$^+$ collector/N-base junction under these operating conditions? What is the hole concentration in the N-buffer layer at the interface between the N-drift and N-buffer regions? What is the hole concentration in the N-drift region at the interface between the N-drift and N-buffer regions?

9.12 Determine the on-state voltage drop at an on-state current density of 100 A cm^{-2} for the planar-gate asymmetric n-channel IGBT design in Problem 9.4 using the two-dimensional model. Use the following parameters: (a) D_a of 15 cm^2 s^{-1}, (b) cell width of 32 μm, (c) channel length of 1.5 μm, (d) inversion mobility of 450 $cm^2 V^{-1} s^{-1}$, (e) accumulation mobility of 1,000 $cm^2 V^{-1} s^{-1}$, (f) gate oxide thickness of 500 Å, (g) gate bias of 15 V, (h) threshold voltage of 5 V, and (i) gate electrode width of 16 μm. Provide the values for the voltage drop across the P^+/N junction, the N-base region, the JFET region, the accumulation region, and the channel region.

9.13 Consider a transparent emitter n-channel IGBT structure having an N-base region with a thickness of 300 μm and a doping concentration of 5×10^{13} cm^{-3}. The lifetime (low-level, high-level, space-charge generation) in the N-base region is 10 μs. The P^+ collector region has a surface concentration of 1×10^{18} cm^{-3} and a depth of 1 μm. Determine the on-state voltage drop at an on-state current density of 100 A cm^{-2} using the two-dimensional model. Use the following parameters: (a) D_a of 15 $cm^2 s^{-1}$, (b) cell width of 32 μm, (c) channel length of 1.5 μm, (d) inversion mobility of 450 $cm^2 V^{-1} s^{-1}$, (e) accumulation mobility of 1,000 $cm^2 V^{-1} s^{-1}$, (f) gate oxide thickness of 500 Å, (g) gate bias of 15 V, (h) threshold voltage of 5 V, and (i) gate electrode width of 16 μm. Provide the values for the voltage drop across the P^+/N junction, the N-base region, the JFET region, the accumulation region, and the channel region.

9.14 Determine the injected hole concentration at the P^+ collector/N-base junction for the planar-gate transparent emitter n-channel IGBT structure under the operating conditions defined in Problem 9.13. What is the injection efficiency at the P^+ collector/N-base junction under these operating conditions?

9.15 Obtain the saturated collector current density for the planar-gate symmetric n-channel IGBT design in Problem 9.1 using the P–N–P transistor/ MOSFET model at a gate bias of 7 V. Use the parameters defined in Problem 9.6.

9.16 Obtain the saturated collector current density for the planar-gate asymmetric n-channel IGBT design in Problem 9.4 using the P–N–P transistor/MOSFET model at a gate bias of 7 V. Use the parameters defined in Problems 9.7 and 9.11.

9.17 Obtain the saturated collector current density for the planar-gate transparent emitter n-channel IGBT design in Problem 9.13 using the P–N–P transistor/MOSFET model at a gate bias of 7 V.

9.18 Calculate the specific output resistance for the planar-gate symmetric n-channel IGBT design in Problem 9.1 at a collector bias of 400 V and a gate bias of 7 V. What is the hole concentration in the space-charge region?

9.19 Calculate the specific output resistance for the planar-gate asymmetric n-channel IGBT design in Problem 9.4 at a collector bias of 400 V and a gate bias of 7 V. What is the hole concentration in the space-charge region?

9.20 Calculate the specific output resistance for the planar-gate transparent emitter n-channel IGBT design in Problem 9.13 at a collector bias of 400 V and a gate bias of 7 V. What is the hole concentration in the space-charge region?

9.21 The planar-gate symmetric n-channel IGBT design in Problem 9.1 is switched off under inductive load conditions from the on-state operating conditions defined in Problem 9.6. Calculate the voltage rise time to reach a collector DC supply voltage of 400 V. What is the current fall time? Obtain the energy loss per cycle.

9.22 The planar-gate asymmetric n-channel IGBT design in Problem 9.4 is switched off under inductive load conditions from the on-state operating conditions defined in Problem 9.7. Calculate the voltage rise time to reach a collector DC supply voltage of 400 V. What is the current fall time? Obtain the energy loss per cycle.

9.23 The planar-gate transparent emitter n-channel IGBT design in Problem 9.13 is switched off under inductive load conditions from the on-state operating conditions. Calculate the voltage rise time to reach a collector DC supply voltage of 2,000 V. What is the current fall time? Obtain the energy loss per cycle.

9.24 Determine the on-state voltage drop at an on-state current density of 100 A cm^{-2} for the planar-gate symmetric p-channel IGBT structure, with the same wide-base parameters as for the n-channel device in Problem 9.10, using the two-dimensional model. Use the same drift region doping concentration and thickness obtained in Problem 9.1. Use the following parameters: (a) D_a of 15 cm^2 s^{-1}, (b) cell width of 32 μm, (c) channel length of 1.5 μm, (d) inversion mobility of 200 cm^2 V^{-1} s^{-1}, (e) accumulation mobility of 400 cm^2 V^{-1} s^{-1}, (f) gate oxide thickness of 500 Å, (g) gate bias of 15 V, (h) threshold voltage of 5 V, (i) gate electrode width of 16 μm, (j) junction depth of N$^+$ region is 5 μm, and (k) JFET region doping concentration of 5 × 10^{15} cm^{-3}. Provide the values for the voltage drop across the N$^+$/P junction, the P-base region, the JFET region, the accumulation region, and the channel region.

9.25 Determine the latch-up current density at 300 K for the planar-gate symmetric n-channel IGBT structure using the following assumptions: (a) common-base current gain of 0.4, (b) average P-base doping concentration of 5×10^{16} cm^{-3}, (c) P-base junction depth of 3 μm, (d) N^{+} emitter junction depth of 1 μm, (e) cell width of 26 μm (pitch of 13 μm), (f) polysilicon window of 10 μm, and (g) N^{+} emitter ion implant mask of 4 μm. Assume that the P^{+} region is not included.

9.26 Determine the improvement in the latch-up current density for the device in Problem 9.25 obtained by the addition of a deep P^{+} region with an average doping concentration of 5×10^{18} cm^{-3} and junction depth of 5 μm. The deep P^{+} region is ion implanted through a photoresist window of 2 μm in the middle of the cell. Use an effective lateral junction depth of 4 μm for the P^{+} region.

9.27 The gate oxide thickness for the planar-gate symmetric n-channel IGBT structure in Problem 9.25 is reduced from 500 to 250 Å while maintaining the same threshold voltage. What is the improvement in the latch-up current density?

9.28 Compare the latch-up current densities for the planar-gate symmetric n-channel IGBT structure in Problem 9.26 if the cell topology is changed from the linear design to the square cell, circular cell, and ALL cell designs. Use an effective lateral junction depth of 4 μm for the P^{+} region.

9.29 A trench-gate asymmetric n-channel IGBT structure is designed with the same parameters for the lightly doped portion of the N-base region and the N-buffer layer as the device in Problem 9.4. Determine the on-state voltage drop at an on-state current density of 100 A cm^{-2} for the device using the two-dimensional model. Use the following parameters: (a) D_a of 15 cm^2 s^{-1}, (b) cell width of 4 μm, (c) channel length of 1.5 μm, (d) inversion mobility of 450 cm^2 V^{-1} s^{-1}, (e) accumulation mobility of 1,000 cm^2 V^{-1} s^{-1}, (f) gate oxide thickness of 500 Å, (g) gate bias of 15 V, (h) threshold voltage of 5 V, (i) trench depth of 4 μm, (j) P-base junction depth of 2.5 μm, and (k) N^{+} emitter junction depth of 1 μm. Assume an effective doping concentration of 1×10^{19} cm^{-3} for the P^{+} collector region with a diffusion length for electrons of 0.5 μm. Provide the values for the voltage drop across the P^{+}/N junction, the N-base region, the accumulation region, and the channel region.

9.30 Calculate the latch-up current density at 300 K for the trench-gate asymmetric n-channel IGBT structure defined in Problem 9.29 using the following assumptions: (a) common-base current gain of 0.4, (b) average

P-base doping concentration of 5×10^{16} cm^{-3}, (c) P-base junction depth of 2.5 μm, (d) N^+ emitter junction depth of 1 μm, and (e) N^+ emitter length of 1 μm.

References

[1] B.J. Baliga, "Enhancement and Depletion Mode Vertical Channel MOS-Gated Thyristors", Electronics Letters, Vol. 15, pp. 645–647, 1979

[2] B.J. Baliga, "How the Super-Transistor Works", Scientific American Magazine, Special Issue on 'The Solid-State-Century', pp. 34–41, January 22, 1998

[3] N. Zommer, "The Monolithic HV BIPMOS", IEEE International Electron Devices Meeting, Abstract 11.5, pp. 263–266, 1981

[4] D.Y. Chen and S. Chin, "Design Considerations for FET-Gated Power Transistors", IEEE Transactions on Electron Devices, Vol. ED-31, pp. 1834–1837, 1984

[5] B.J. Baliga, "Evolution of MOS–Bipolar Power Semiconductor Technology", Proceedings of the IEEE, Vol. 74, pp. 409–418, 1988

[6] B.J. Baliga et al., "The Insulated Gate Rectifier: A New Power Switching Device", IEEE International Electron Devices Meeting, Abstract 10.6, pp. 264–267, 1982

[7] J.P. Russell et al., "The COMFET", IEEE Electron Device Letters, Vol. EDL-4, pp. 63–65, 1983

[8] B.J. Baliga et al., "The Insulated Gate Transistor: A New Three-Terminal MOS-Controlled Bipolar Power Device", IEEE Transactions on Electron Devices, Vol. ED-31, pp. 821–828, 1984

[9] H.R. Chang et al., "Insulated Gate Bipolar Transistor (IGBT) with a Trench Gate Structure", IEEE International Electron Devices Meeting, Abstract 29.5, pp. 674–677, 1987

[10] M. Kitagawa et al., "A 4,500 V Injection Enhanced Insulated Gate Bipolar Transistor (IEGT) Operating in a Mode Similar to a Thyristor", IEEE International Electron Devices Meeting, Abstract 28.3.1, pp. 679–682, 1993

[11] M. Takei, Y. Harada, and K. Ueno, "600 V IGBT with Reverse Blocking Capability", IEEE International Symposium on Power Semiconductor Devices and ICs, Abstract 11.1, pp. 413–416, 2001

[12] T. Naito et al., "1200 V Reverse Blocking IGBT with Low Loss for Matrix Converter", IEEE International Symposium on Power Semiconductor Devices and ICs, Abstract 3.2, pp. 125–128, 2004

[13] H. Takahashi, M. Kaneda, and T. Minato, "1200 V Class Reverse Blocking IGBT for AC Matrix Converter", IEEE International Symposium on Power Semiconductor Devices and ICs, Abstract 3.1, pp. 121–124, 2004

[14] N. Tokuda, M. Kaneda, and T. Minato, "An Ultra-Small Isolation Area for 600 V Class Reverse Blocking IGBT with Deep Trench Isolation Process", IEEE International Symposium on Power Semiconductor Devices and ICs, Abstract 3.3, pp. 129–132, 2004

[15] H. Shigekane, H. Kirihata, and Y. Uchida, "Developments in Modern High Power Semiconductor Devices", IEEE International Symposium on Power Semiconductor Devices and ICs, pp. 16–21, 1993
[16] B.J. Baliga and M.S. Adler, "Measurement of Carrier Lifetime Profiles in Diffused Layers of Semiconductors", IEEE Transactions on Electron Devices, Vol. ED-25, pp. 472–477, 1978
[17] D.J. Roulston, N.D. Arora, and S.G. Chamberlain, "Modeling and Measurement of Minority Carrier Lifetime Versus Doping in Diffused Layers", IEEE Transactions on Electron Devices, Vol. ED-29, pp. 284–291, 1982
[18] Medici User's Manual, Avant! Corporation, Fremont, CA, 2001
[19] B.J. Baliga, "Power Semiconductor Devices for Variable Frequency Drives", Proceedings of the IEEE, Vol. 82, pp. 1112–1122, 1994
[20] A.R. Hefner and D.L. Blackburn, "An Analytical Model for the Steady-State and Transient Characteristics of the Power Insulated Gate Bipolar Transistor", Solid-State Electronics, Vol. 31, pp. 1513–1532, 1988
[21] B.G. Streetman and S.K. Banerjee, "Solid State Electronic Devices", 6th Edition, pp. 190–192, 2006
[22] A.R. Hefner and D.L. Blackburn, "A Performance Trade-Off for the Insulated Gate Bipolar Transistor: Buffer Layer Versus Base Lifetime Reduction", IEEE Transactions on Power Electronics, Vol. PE2, pp. 194–207, 1987
[23] M. Otsuki et al., "The 3rd Generation IGBT Toward a Limitation of IGBT Performance", IEEE International Symposium on Power Semiconductor Devices and ICs, pp. 24–29, 1993
[24] F. Bauer et al., "A Comparison of Emitter Concepts for High Voltage IGBTs", IEEE International Symposium on Power Semiconductor Devices and ICs, pp. 230–235, 1995
[25] K. Mochizuki et al., "Examination of Punch-Through IGBT (PT-IGBT) for High Voltage and High Current Applications", IEEE International Symposium on Power Semiconductor Devices and ICs, pp. 237–240, 1997
[26] B.J. Baliga, "Analysis of the Output Conductance of Insulated Gate Transistors, IEEE Electron Device Letters", Vol. EDL-7, pp. 686–688, 1986
[27] Arthur D. Little, Inc., "Opportunities for Energy Savings in the Residential and Commercial Sectors with High-Efficiency Electric Motors", U.S. Department of Energy Contract No. DE-AC01-90CE23821, December 1999
[28] B.G. Streetman and S.K. Banerjee, "Solid State Electronic Devices", 6th Edition, p. 141, Prentice Hall, Englewood Cliffs, NJ, 2006
[29] B.J. Baliga, "Analysis of Insulated Gate Transistor Turn-Off Characteristics", IEEE Electron Device Letters, Vol. EDL-6, pp. 74–77, 1985
[30] T. Laska et al., "The Field-Stop IGBT (FS-IGBT)", IEEE International Symposium on Power Semiconductor Devices and ICs, pp. 355–358, 2000
[31] T. Takahashi et al., "A Design Concept for the Low Turn-Off Loss 4.5 kV Trench IGBT", IEEE International Symposium on Power Semiconductor Devices and ICs, pp. 51–54, 1998

[32] S. Wang, "Fundamentals of Semiconductor Theory and Device Physics", pp. 312–316, Prentice Hall, Englewood Cliffs, NJ, 1989

[33] M.F. Chang et al., "Comparison of N and P Channel IGTs", IEEE International Electron Devices Meeting, Abstract 10.6, pp. 278–281, 1984

[34] B.J. Baliga, "Gate Enhanced Rectifier", U.S. Patent 4,969,028, Originally Filed December 2, 1980, Issued November 6, 1990

[35] B.J. Baliga et al., "Suppressing Latch-up in Insulated Gate Transistors", IEEE Electron Device Letters, Vol. EDL-5, pp. 323–325, 1984

[36] B.J. Baliga, "Temperature Behavior of Insulated Gate Transistor Characteristics", Solid-State Electronics, Vol. 28, pp. 289–297, 1985

[37] S. Eranen and M. Blomberg, "The Vertical IGBT with Implanted Buried Layer", IEEE International Symposium on Power Semiconductor Devices and ICs, pp. 211–214, 1991

[38] A. Nakagawa et al., "Non-Latch-Up 1200 V, 75A Bipolar-Mode MOSFET with Large SOA", IEEE International Electron Devices Meeting, Abstract 16.8, pp. 860–861, 1984

[39] N. Thapar and B.J. Baliga, "A New IGBT Structure with a Wider Safe Operating Area", International Symposium on Power Semiconductor Devices and ICs, pp. 177–182, 1994

[40] H. Yilmaz, "Cell Geometry Effect on IGT Latch-up", IEEE Electron Device Letters, Vol. EDL-6, pp. 419–421, 1985

[41] B.J. Baliga et al., "New Cell Designs for Improved IGBT Safe-Operating-Area", IEEE International Electron Devices Meeting, Abstract 34.5, pp. 809–812, 1988

[42] V. Parthasarathy et al., "Cell Optimization for 500-V n-Channel IGBTs", IEEE International Symposium on Power Semiconductor Devices and ICs, Abstract 3.5, pp. 69–74, 1994

[43] K. Yoshikawa et al., "A Study on Wide RBSOA of 4.5kV Power-Pack IGBT", IEEE International Symposium on Power Semiconductor Devices and ICs, pp. 117–120, 2001

[44] T. Ogura et al., "Turn-Off Switching Analysis Considering Dynamic Avalanche Effect for Low Turn-Off Loss High-Voltage IGBT", IEEE Transactions on Electron Devices, Vol. ED-51, pp. 629–635, 2004

[45] J.G. Bauer et al., "Investigations of the Ruggedness Limit of 6.5kV IGBT", IEEE International Symposium on Power Semiconductor Devices and ICs, pp. 71–74, 2005

[46] H. Hagino et al., "An Experimental and Numerical Study on the Forward Biased SOA of IGBTs", IEEE Transactions on Electron Devices, Vol. ED-43, pp. 490–499, 1996

[47] N. Iwamuro et al., "Numerical Analysis of Short-Circuit Safe Operating Area for p-Channel and n-Channel IGBTs", IEEE Transactions on Electron Devices, Vol. ED-38, pp. 303–309, 1991

[48] H.R. Chang and B.J. Baliga, "500-V n-Channel Insulated Gate Bipolar Transistor with a Trench Gate Structure", IEEE Transactions on Electron Devices, Vol. ED-36, pp. 1824–1829, 1989
[49] M. Harada et al., "600 V Trench IGBT in Comparison with Planar IGBT", IEEE International Symposium on Power Semiconductor Devices and ICs, pp. 411–416, 1994
[50] A. Bhalla et al., "High Performance Wide Trench IGBTs for Motor Control Applications", IEEE International Symposium on Power Semiconductor Devices and ICs, pp. 41–44, 1999
[51] T. Ogura et al., "4.5-kV Injection-Enhanced Gate Transistors with High Turn-Off Ruggedness", IEEE Transactions on Electron Devices, Vol. ED-51, pp. 636–641, 2004
[52] I. Omura et al., "Carrier Injection Enhancement Effect of High Voltage MOS Devices", IEEE International Symposium on Power Semiconductor Devices and ICs, pp. 217–220, 1997
[53] R. Holtz, F. Bauer, and W. Fichtner, "On-State and Short Circuit Behavior of High Voltage Trench Gate IGBTs in Comparison with Planar IGBTs", IEEE International Symposium on Power Semiconductor Devices and ICs, pp. 224–229, 1995
[54] R. Constapel, J. Korec, and B.J. Baliga, "Trench IGBTs with Integrated Diverter Structures", IEEE International Symposium on Power Semiconductor Devices and ICs, pp. 201–206, 1995
[55] B.J. Baliga, "Methods for Forming Power Semiconductor Devices Having T-Shaped Gate Electrodes", U.S. Patent 6,303,410, Issued October 16, 2001
[56] M. Nemoto and B.J. Baliga, "The Recessed-Gate IGBT Structure", IEEE International Symposium on Power Semiconductor Devices and ICs, pp. 149–152, 1999
[57] T.P. Chow and B.J. Baliga, "Comparison of 300, 600, and 1200 V n-Channel Insulated Gate Transistors", IEEE Electron Device Letters, Vol. EDL-6, pp. 161–163, 1985
[58] P.M. Shenoy, J. Yedniak, and J. Gladish, "High Performance 300 V IGBTs", IEEE International Symposium on Power Semiconductor Devices and ICs, pp. 217–220, 2000
[59] J.G. Bauer et al., "6.5 kV Modules Using IGBTs with Field Stop Technology", IEEE International Symposium on Power Semiconductor Devices and ICs, pp. 121–124, 2001
[60] B.J. Baliga, "Temperature Behavior of Insulated Gate Transistor Characteristics", Solid-State Electronics, Vol. 28, pp. 289–297, 1985
[61] B.J. Baliga and E. Sun, "Comparison of Gold, Platinum, and Electron Irradiation for Controlling Lifetime in Power Rectifiers", IEEE Transactions on Electron Devices, Vol. ED-34, pp. 1103–1108, 1977

[62] B.J. Baliga and J.P. Walden, "Improving the Reverse Recovery of Power MOSFET Integral Diodes by Electron Radiation", Solid-State Electronics, Vol. 26, pp. 1133–1141, 1983
[63] B.J. Baliga, "Fast Switching Insulated Gate Transistors", IEEE Electron Device Letters, Vol. EDL-4, pp. 452–454, 1983
[64] B.J. Baliga, "Switching Speed Enhancement in Insulated Gate Transistors by Electron Irradiation", IEEE Transactions on Electron Devices, Vol. ED-31, pp. 1790–1795, 1984
[65] W.A. Strifler and B.J. Baliga, "Comparison of Neutron and Electron Irradiation for Controlling IGT Switching Speed", IEEE Transactions on Electron Devices, Vol. ED-32, pp. 1629–1632, 1985
[66] D. Silber et al., "Improved Dynamic Properties of GTO-Thyristors and Diodes by Proton Implantation", IEEE International Electron Devices Meeting, Abstract 6.6, pp. 162–165, 1985
[67] H. Akiyama et al., "Partial Lifetime Control in IGBTs with Helium Irradiation Through Mask Patterns", IEEE International Symposium on Power Semiconductor Devices and ICs, pp. 187–191, 1991
[68] Y. Konishi et al., "Optimized Local Lifetime Control for the Superior IGBTs", IEEE International Symposium on Power Semiconductor Devices and ICs, pp. 335–338, 1996
[69] Y. Onishi et al., "Analysis on Device Structures for Next Generation IGBT", IEEE International Symposium on Power Semiconductor Devices and ICs, pp. 85–88, 1998
[70] K. Matsushita, I. Omura, and T. Ogura, "Blocking Voltage Design Considerations for Deep Trench MOS Gate High Power Devices", IEEE International Symposium on Power Semiconductor Devices and ICs, pp. 256–260, 1995
[71] M. Kitagawa et al., "4500 V IEGTs Having Switching Characteristics Superior to GTO", IEEE International Symposium on Power Semiconductor Devices and ICs, pp. 486–491, 1995
[72] J.G. Bauer et al., "Investigations on the Ruggedness Limit of 6.5 kV IGBT", IEEE International Symposium on Power Semiconductor Devices and ICs, pp. 71–74, 2005
[73] N. Thapar and B.J. Baliga, "Influence of the Trench Corner Design on Edge Termination of UMOS Power Devices", Solid-State Electronics, Vol. 41, pp. 1929–1936, 1997
[74] T. Nitta et al., "A Design Concept for the Low Forward Voltage Drop 4500 V Trench IGBT", IEEE International Symposium on Power Semiconductor Devices and ICs, pp. 43–46, 1998
[75] T.P. Chow et al., "P-Channel Vertical Insulated Gate Bipolar Transistors with Collector Short", IEEE International Electron Devices Meeting, Abstract 29.4, pp. 670–673, 1987

[76] B.J. Baliga, "Semiconductor Device Having Rapid Removal of Majority Carriers from an Active Base Region thereof at Device Turn-Off and Method of Fabricating this Device", U.S. Patent 4,782,379, Issued November 1, 1988

[77] H. Takahashi et al., "A High Performance IGBT with New N+ Buffer Structure", IEEE International Symposium on Power Semiconductor Devices and ICs, pp. 474–479, 1995

[78] H. Takahashi et al., "1200-V Reverse Conducting IGBT", IEEE International Symposium on Power Semiconductor Devices and ICs, pp. 133–136, 2004

[79] H. Akiyama et al., "Effects of Shorted Collector on Characteristics of IGBTs", IEEE International Symposium on Power Semiconductor Devices and ICs, pp. 131–136, 1990

[80] N. Iwamuro et al., "Switching Loss Analysis of Shorted Drain Non-Punch-Through and Punch-Through Type IGBTs in Voltage Resonant Circuit", IEEE International Symposium on Power Semiconductor Devices and ICs, pp. 220–225, 1991

[81] B.J. Baliga, "The Future of Power Semiconductor Device Technology", Proceedings of the IEEE, Vol. 89, pp. 822–832, 2001

Chapter 10

Synopsis

Power devices are required for systems that operate over a broad spectrum of power levels and frequencies as discussed in Chap. 1. A variety of power rectifier and transistor structures were discussed in previous chapters for serving these applications. Although the bipolar power transistor and the thyristor were the first technologies developed for this purpose, they have been replaced by power MOSFET and IGBT structures in modern applications due to the resulting simplification of the control circuit and elimination of snubbers. The choice of the optimum device suitable for an application depends upon the device voltage rating and the circuit switching frequency. From the point of view of presenting a unified treatment, it is convenient to analyze a typical pulse-width-modulated (PWM) motor control circuit as an example because it is utilized for both low-voltage applications, such as disk drives in computers, and high-voltage applications, such as the drive train in hybrid electric vehicles and electric locomotives.

10.1 Typical H-Bridge Topology

The control of motors using PWM circuits is typically performed by using the H-bridge configuration shown in Fig. 10.1. In this figure, the circuit has been implemented using four IGBT devices as the switches and four P–i–N rectifiers as the fly-back diodes. This is the commonly used topology for medium and high power motor drives where the DC bus voltage exceeds 200 V. When the H-bridge topology is used for applications that operate from a low DC bus voltage, it is typically implemented using four power MOSFET devices as the switches and four Schottky rectifiers as the fly-back diodes.

The direction of the current flow in the motor winding can be controlled with the H-bridge configuration. If IGBT-1 and IGBT-4 are turned on while

B.J. Baliga, *Fundamentals of Power Semiconductor Devices*, doi: 10.1007/978-0-387-47314-7_10,

maintaining IGBT-2 and IGBT-3 in their blocking mode, the current in the motor will flow from the left side to the right side in the figure. The direction of the current flow can be reversed if IGBT-3 and IGBT-2 are turned on while maintaining IGBT-1 and IGBT-4 in their blocking mode. Alternately, the magnitude of the current flow can be increased or decreased by turning on the IGBT devices in pairs. This method allows synthesis of a sinusoidal waveform across the motor windings with a variable frequency that is dictated by the PWM circuit.[1]

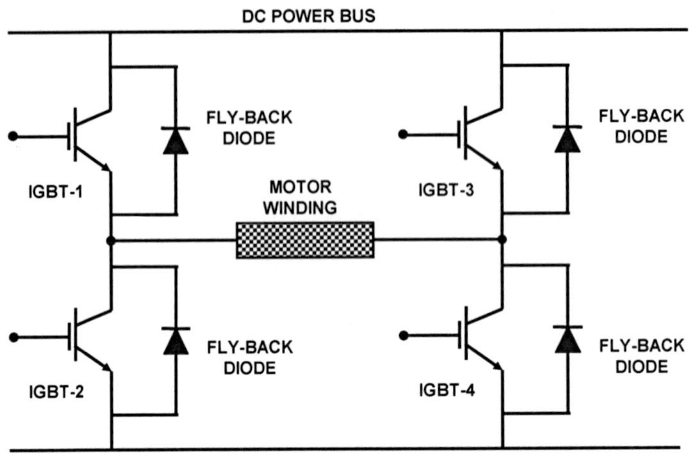

Fig. 10.1 Typical H-bridge topology for motor control

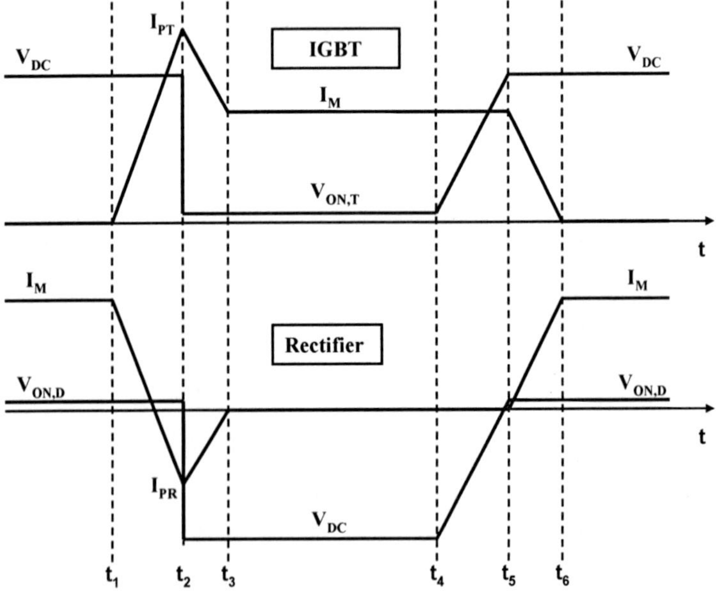

Fig. 10.2 Typical waveforms during PWM operation

The typical waveforms for the current and the voltage across the power transistor and the fly-back diode are illustrated in Fig. 10.2 during just one cycle of the PWM operation. These waveforms have been linearized for simplification of the analysis.[2] The cycle begins at time t_1 when the transistor is turned on by its gate drive voltage. Prior to this time, the transistor is supporting the DC supply voltage and the fly-back diode is assumed to be carrying the motor current. As the transistor turns on, the motor current is transferred from the diode to the transistor during the time interval from t_1 to t_2. In the case of high DC bus voltages, where P–i–N rectifiers are utilized, the fly-back diode will not be able to support voltage until the stored charge in its drift region is removed as discussed in Chap. 5. To achieve this, the P–i–N rectifier must undergo its reverse recovery process. During reverse recovery, substantial reverse current flows through the rectifier with a peak value I_{PR} reached at time t_2. The large reverse recovery current produces significant power dissipation in the diode. Moreover, the current in the IGBT at time t_2 is the sum of the motor winding current I_M and the peak reverse recovery current I_{PR}. This produces substantial power dissipation in the transistor during the turn-on transient. The power dissipation in both the transistor and the diode is therefore governed by the reverse recovery characteristics of the power rectifier.

The power transistor is turned off at time t_4 allowing the motor current to transfer from the transistor to the diode. In the case of an inductive load, such as motor windings, the voltage across the transistor increases before the current is reduced as illustrated in Fig. 10.2 during the time interval from t_4 to t_5. Subsequently, the current in the transistor reduces to zero during the time interval from t_5 to t_6. The turn-off durations are governed by the physics of the transistor structure as discussed in previous chapters. Consequently, the power dissipation in both the transistor and the diode during the turn-off event is determined by the transistor switching characteristics.

In addition to the power losses associated with the two basic switching events within each cycle, power loss is incurred within the diode and the transistor during their respective on-state operation due to a finite on-state voltage drop. It is common practice to trade off a larger on-state voltage drop to obtain a smaller switching loss in the bipolar power devices. Consequently, the on-state power loss cannot be neglected especially if the operating frequency is low. The leakage current for the devices is usually sufficiently small, so that the power loss in the blocking mode can be neglected.

10.2 Power Loss Analysis

The total power loss incurred in the power transistor can be obtained by summing four components:

$$P_{L,T}(\text{total}) = P_{L,T}(\text{on}) + P_{L,T}(\text{off}) + P_{L,T}(\text{turn-on}) + P_{L,T}(\text{turn-off}). \quad (10.1)$$

The power loss incurred in the transistor during the on-state duration from time t_3 to t_4 is given by

$$P_{\mathrm{L,T}}(\mathrm{on}) = \frac{t_4 - t_3}{T} I_{\mathrm{M}} V_{\mathrm{ON,T}}. \tag{10.2}$$

The power loss incurred in the transistor during the off-state duration beyond time t_6 until the next turn-on event is given by

$$P_{\mathrm{L,T}}(\mathrm{off}) = \frac{T - t_6}{T} I_{\mathrm{L,T}} V_{\mathrm{DC}}. \tag{10.3}$$

The leakage current ($I_{\mathrm{L,T}}$) for the transistors is usually very small allowing this term to be neglected during the power dissipation analysis.

The power loss incurred in the transistor during the turn-on event from time t_1 to t_3 can be obtained by analysis of the segments between the time intervals t_1 to t_2 and t_2 to t_3. The power loss incurred during the first segment is given by

$$P_{\mathrm{L,T-1}}(\mathrm{turn\text{-}on}) = \frac{1}{2}\frac{t_2 - t_1}{T} I_{\mathrm{PT}} V_{\mathrm{DC}}, \tag{10.4}$$

where the peak transistor current is dependent on the peak reverse recovery current of the P–i–N rectifier:

$$I_{\mathrm{PT}} = I_{\mathrm{M}} + I_{\mathrm{PR}}. \tag{10.5}$$

In the power loss analysis, it will be assumed that the time duration ($t_2 - t_1$) is determined by the reverse recovery behavior of the P–i–N rectifier and is independent of the operating frequency. The power loss incurred during the second segment is given by

$$P_{\mathrm{L,T-2}}(\mathrm{turn\text{-}on}) = \frac{1}{2}\frac{t_3 - t_2}{T}\left(\frac{I_{\mathrm{PT}} + I_{\mathrm{M}}}{2}\right) V_{\mathrm{DC}}. \tag{10.6}$$

In the power loss analysis, it will be assumed that the time duration ($t_3 - t_2$) is also determined by the reverse recovery behavior of the P–i–N rectifier and is independent of the operating frequency.

The power loss incurred in the transistor during the turn-off event from time t_4 to t_6 can be obtained by analysis of the segments between the time intervals t_4 to t_5 and t_5 to t_6. The power loss incurred during the first segment is given by

$$P_{\mathrm{L,T-1}}(\mathrm{turn\text{-}off}) = \frac{1}{2}\frac{t_5 - t_4}{T} I_{\mathrm{M}} V_{\mathrm{DC}}. \tag{10.7}$$

The time interval ($t_5 - t_4$) is determined by the time taken for the transistor voltage to rise to the DC power supply voltage. This time duration was analyzed for each transistor in the previous chapters. The power loss incurred during the second segment is given by

$$P_{L,T-2}(\text{turn-off}) = \frac{1}{2}\frac{t_6 - t_5}{T} I_M V_{DC}. \tag{10.8}$$

The time interval ($t_6 - t_5$) is determined by the time taken for the transistor current to decay to zero. This time duration was analyzed for each transistor in the previous chapters.

In a similar manner, the total power loss incurred in the power rectifier can be obtained by summing four components:

$$P_{L,R}(\text{total}) = P_{L,R}(\text{on}) + P_{L,R}(\text{off}) + P_{L,R}(\text{turn-on}) + P_{L,R}(\text{turn-off}). \tag{10.9}$$

The power loss incurred in the power rectifier during the on-state duration from time t_6 to the end of the period is given by

$$P_{L,R}(\text{on}) = \frac{T - t_6}{T} I_M V_{ON,R}. \tag{10.10}$$

In writing this expression, it is assumed that the cycle begins at time t_1. The power loss incurred in the power rectifier during the off-state time duration ($t_4 - t_3$) is given by

$$P_{L,R}(\text{off}) = \frac{t_4 - t_3}{T} I_{L,R} V_{DC}. \tag{10.11}$$

The leakage current ($I_{L,R}$) for the power rectifier will be assumed to be very small (even for the silicon Schottky rectifier) allowing this term to be neglected during the power dissipation analysis.

The power loss incurred in the power rectifier during the turn-on event from time t_1 to t_3 can be obtained by analysis of the segments between the time intervals t_1 to t_2 and t_2 to t_3. The power loss incurred during the first segment is much smaller than during the second segment due to the small on-state voltage drop for the power rectifiers. The power loss incurred during the second segment is given by

$$P_{L,R-2}(\text{turn-on}) = \frac{1}{2}\frac{t_3 - t_2}{T} I_{PR} V_{DC}. \tag{10.12}$$

The power loss incurred in the power rectifier during the turn-off event from time t_4 to t_6 can be obtained by analysis of the segments between the time intervals t_4 to t_5 and t_5 to t_6. The power loss incurred during the first segment is negligible due to the low leakage current for the power rectifier. The power loss incurred during the second segment is given by

$$P_{L,R-2}(\text{turn-off}) = \frac{1}{2}\frac{t_6 - t_5}{T} I_M V_{ON,D}. \tag{10.13}$$

This power loss is also small due to the low on-state voltage drop of power rectifiers.

10.3 Low DC Bus Voltage Applications

In this section, the above power loss analysis is applied to a motor control application using a low DC bus voltage with a duty cycle of 50%. The DC bus voltage (V_{DC}) will be assumed to be 20 V as pertains to the backplane power source in desktop computers. In this case, the device blocking voltage rating is typically 30 V. The current being delivered to the motor winding (I_M) will be assumed to be 10 A. Due to the low blocking voltage required for this application, it is commonly implemented using silicon unipolar devices, namely the power MOSFET as the power switch. To reduce the cost and packaging complexity, it is attractive to use the integral body diode within the power MOSFET structure instead of a separate antiparallel fly-back diode. The reverse recovery characteristics of the integral body diode can be optimized by using electron irradiation.[3] Alternately, a Schottky diode has been integrated in the power MOSFET cell for the JBSFET structure[4] allowing suppression of the reverse recovery phenomenon without the packaging complexity of using an external Schottky fly-back diode.

Characteristics	Silicon MOSFET	Silicon IGBT	4H-SiC MOSFET
On-State Voltage Drop (V)	0.05	0.90	0.08
Turn-Off Time ($t_5 - t_4$) (μs)	0.01	0.1	0.01
Turn-Off Time ($t_6 - t_5$) (μs)	0.01	0.1	0.01

Fig. 10.3 Characteristics of transistors with 30-V blocking voltage rating

The characteristics for the transistors that are pertinent to the analysis of the power loss are provided in Fig. 10.3. The on-state voltage drop of 0.05 V for the silicon MOSFET device is based upon using a specific on-resistance of 0.5 mΩ cm^2 and an on-state current density of 100 A cm^{-2}. This specific on-resistance is typical for U-MOSFET devices, as well as the planar-gate SSCFET devices, described in Chap. 6. For comparison purposes, the silicon IGBT device and the silicon carbide power MOSFET device are included in the power analysis. The on-state voltage drop (0.90 V) for the silicon IGBT device with such a low blocking voltage rating is limited by the voltage drop across the P^+ collector/N-base junction. In the case of the 4H-SiC power MOSFET structure, the specific on-resistance becomes limited by the N^+ substrate (0.4 mΩ cm^2) and the channel (0.4 mΩ cm^2) contributions because the drift region contribution is extremely small (see Fig. 6.162 for the shielded trench-gate 4H-SiC MOSFET structure with 1-μm channel length). These devices are also assumed to be operated at an on-state current density of 100 A cm^{-2}.

Characteristics	Silicon Schottky	Silicon P-i-N	4H-SiC Schottky
On-State Voltage Drop (V)	0.5	0.9	1.0
Turn-On Time ($t_2 - t_1$) (μs)	0.01	0.10	0.01
Turn-On Time ($t_3 - t_2$) (μs)	0.01	0.10	0.01
Peak Reverse Recovery Current (A)	0	5	0

Fig. 10.4 Characteristics of rectifiers with 30-V blocking voltage rating

The characteristics for the power rectifiers that are pertinent to the analysis of the power loss are provided in Fig. 10.4. The on-state voltage drop of 0.5 V for the silicon Schottky diode (or the integral diode within the JBSFET structure) is based upon an on-state current density of 100 A cm^{-2} (see Fig. 4.7). For comparison purposes, the silicon P–i–N rectifier (or the integral diode within the MOSFET structure) and the silicon carbide Schottky rectifier are included in the power analysis. The on-state voltage drop (0.90 V) for the silicon P–i–N rectifier (or the integral diode within the MOSFET structure) with such a low blocking voltage rating is limited by the voltage drop across the P/N junction. In the case of the 4H-SiC Schottky rectifier structure, the on-state voltage drop is limited by the barrier height for the metal–semiconductor contact (see Fig. 4.8). These devices are also assumed to be operated at an on-state current density of 100 A cm^{-2}. The reverse recovery current for the Schottky diodes is negligible because of unipolar operation in the on-state. For the P–i–N rectifier (or the integral diode within the MOSFET structure), the peak reverse recovery current is assumed to be equal to half the on-state current. The power loss incurred in the transistor and the fly-back diode can be computed as a function of the operating frequency by using the equations provided in Sect. 10.2 and the numerical values in the above figures.

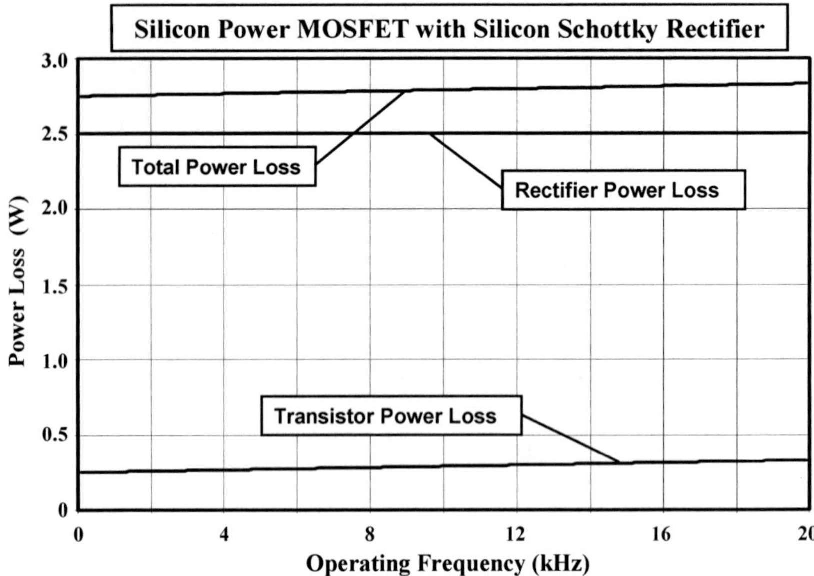

Fig. 10.5 Power losses during motor control with 20-V DC bus: silicon power MOSFET with silicon Schottky rectifier

As an example, the power losses in the case of the silicon power MOSFET as the power switch with a silicon Schottky rectifier as the diode (also representative of the JBSFET structure) are provided in Fig. 10.5 for frequencies ranging to 20 kHz. The power losses during the on-state are dominant in both the transistor and the diode in this case because of the fast switching speeds for the unipolar devices. Consequently, there is only a slight increase in power losses as the frequency increases. The power loss in the rectifier is much larger than that in the transistor due to its much larger on-state voltage drop. The total power loss is only 2.75 W when 200 W of power is delivered to the load.

When the Schottky rectifier is replaced by a P–i–N rectifier representative of the integral body diode in the power MOSFET structure, the power losses increase considerably as shown in Fig. 10.6. A major part of the increase in power loss is due to the larger on-state voltage drop of the P–i–N rectifier. However, the power loss also increases with frequency due to the contribution from the switching losses. The switching loss is associated with the reverse recovery of the P–i–N rectifier. It is worth pointing out that the reverse recovery transient for the power rectifier also produces increased switching losses in the transistor during its turn-on event. The total power loss is increased to about 5 W when 200 W of power is delivered to the load.

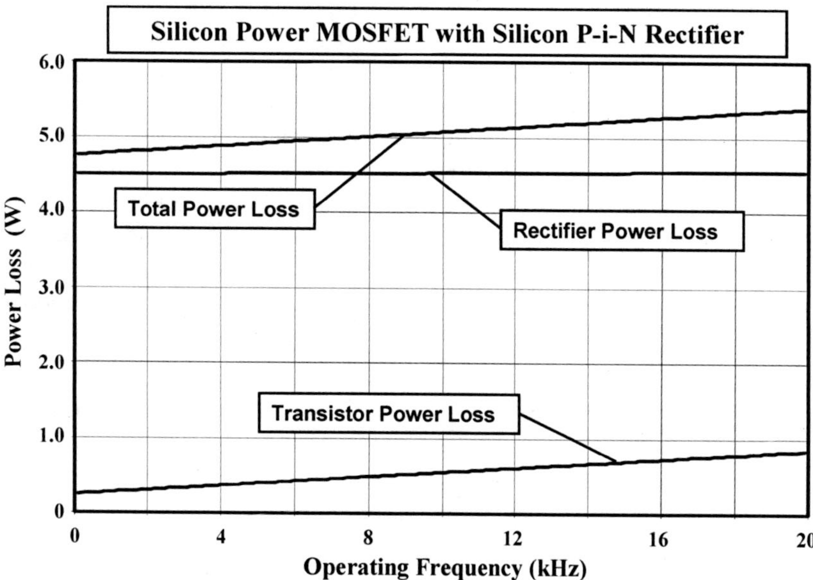

Fig. 10.6 Power losses during motor control with 20-V DC bus: silicon power MOSFET with silicon P-i-N rectifier

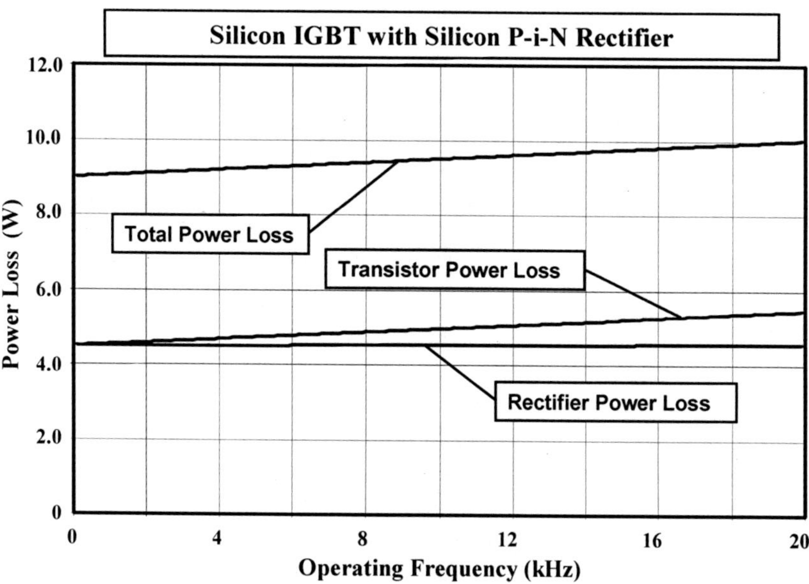

Fig. 10.7 Power losses during motor control with 20-V DC bus: silicon IGBT with silicon P-i-N rectifier

When the silicon power MOSFET is replaced by an IGBT and the silicon Schottky rectifier is replaced by a P–i–N rectifier, the power losses increase by an even greater amount as shown in Fig. 10.7. A major part of the increase in power loss is due to the larger on-state voltage drop of the P–i–N rectifier and the IGBT structure. Moreover, the power loss increases with frequency due to the contribution from the switching losses. The switching loss is associated with the reverse recovery of the P–i–N rectifier producing increased switching losses in the transistor during its turn-on event. The total power loss is increased to about 10 W when 200 W of power is delivered to the load. The above example is provided to illustrate that the silicon power MOSFET structure is more suitable for applications operating from low DC bus voltages than the IGBT structure.

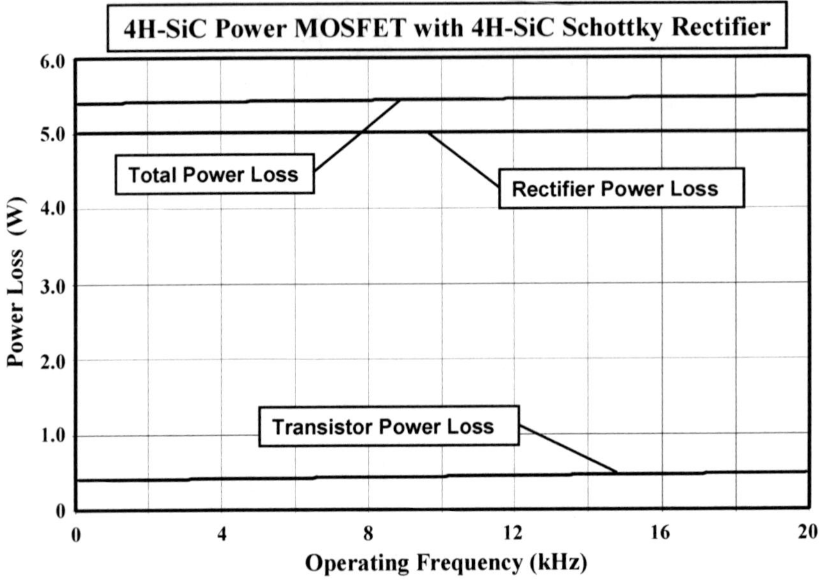

Fig. 10.8 Power losses during motor control with 20-V DC bus: silicon carbide power MOSFET with silicon carbide Schottky rectifier

There is considerable interest in replacing silicon devices with those based upon wide band-gap semiconductors to improve system efficiency. The impact of using a silicon carbide MOSFET device as the transistor and a silicon carbide Schottky rectifier as the fly-back diode is provided in Fig. 10.8. When compared with the silicon unipolar devices, there is a considerable increase in the power dissipation especially due to the high on-state voltage drop for the 4H-SiC Schottky rectifier. The total power loss is increased to about 5.5 W when 200 W of power is delivered to the load. This example is provided to illustrate that silicon unipolar devices are more suitable for applications operating from low DC bus voltages than devices based upon wide band-gap semiconductors.

10.4 Medium DC Bus Voltage Applications

In this section, the above power loss analysis is applied to a motor control application using a medium DC bus voltage with a duty cycle of 50%. The DC bus voltage (V_{DC}) will be assumed to be 400 V as pertains to the power source in a hybrid electric car. In this case, the device blocking voltage rating is typically 600 V. The current being delivered to the motor winding (I_M) will be assumed to be 20 A. Due to the larger blocking voltage required for this application, it is commonly implemented using silicon bipolar devices, namely the IGBT as the power switch and the P–i–N rectifier as the fly-back diode.

Characteristics	Silicon MOSFET	Silicon IGBT	4H-SiC MOSFET
On-State Voltage Drop (V)	10	1.8	0.08
Turn-Off Time ($t_5 - t_4$) (μs)	0.01	0.1	0.01
Turn-Off Time ($t_6 - t_5$) (μs)	0.01	0.2	0.01

Fig. 10.9 Characteristics of transistors with 600-V blocking voltage rating

The characteristics for the transistors that are pertinent to the analysis of the power loss are provided in Fig. 10.9. In the case of the silicon power MOSFET structure, the specific on-resistance is 0.1 Ω cm^2 for a device capable of blocking 600 V as shown in Fig. 6.50. This results in an on-state voltage drop of 10 V for the silicon MOSFET device based upon using an on-state current density of 100 A cm^{-2}. The on-state voltage drop (1.80 V) and switching times for the silicon IGBT device with 600-V blocking voltage rating are based upon scaling the characteristics for the 1,200-V structure modeled in Chap. 9. The on-state voltage drop for the asymmetric IGBT structure shown in Fig. 9.57 and the turn-off waveforms shown in Fig. 9.114 for the 1,200-V structure were used as the basic for the scaling. In the case of the 4H-SiC power MOSFET structure, the drift region contribution increases to 0.03 mΩ cm^2 (see Fig. 3.6), which is still much smaller than the specific on-resistance contributed by the N$^+$ substrate (0.4 mΩ cm^2) and the channel (0.4 mΩ cm^2). Consequently, a specific on-resistance of 0.8 mΩ cm^2 has been used for the silicon carbide MOSFET structure. These devices are also assumed to be operated at an on-state current density of 100 A cm^{-2}.

The characteristics for the power rectifiers that are pertinent to the analysis of the power loss are provided in Fig. 10.10. The on-state voltage drop of 5.4 V for the silicon Schottky diode is based upon an on-state current density of 100 A cm^{-2} (see Fig. 4.7). The high on-state voltage drop for the silicon Schottky diode precludes its use in this application despite its excellent switching behavior. For comparison purposes, the silicon P–i–N rectifier and the silicon carbide Schottky rectifier are included in the power analysis. The on-state voltage drop (2.0 V) for the silicon P–i–N rectifier is typical for fast switching rectifiers. In the case of the 4H-SiC Schottky rectifier structure, the on-state voltage drop is limited by the barrier height for the metal–semiconductor contact due to the low specific on-resistance for the drift region (see Fig. 4.8). These devices are also assumed to be operated at an on-state current density of 100 A cm^{-2}. The reverse recovery current for the Schottky diodes is negligible because of unipolar operation in the on-state. For the P–i–N rectifier, the peak reverse recovery current is assumed to be equal to half the on-state current. Based upon the information provided in Fig. 10.10, the silicon Schottky rectifier is not a viable alternative for power loss analysis in medium voltage applications due to its high on-state voltage drop.

Characteristics	Silicon Schottky	Silicon P-i-N	4H-SiC Schottky
On-State Voltage Drop (V)	5.4	2.0	1.0
Turn-On Time ($t_2 - t_1$) (μs)	0.01	0.20	0.01
Turn-On Time ($t_3 - t_2$) (μs)	0.01	0.20	0.01
Peak Reverse Recovery Current (A)	0	10	0

Fig. 10.10 Characteristics of rectifiers with 600-V blocking voltage rating

The power losses in the case of the silicon power MOSFET as the power switch with a silicon P–i–N rectifier as the diode are provided in Fig. 10.11 for frequencies ranging to 20 kHz. The on-state power losses in the transistor are dominant in this case. The power loss in the transistor increases with increasing frequency due to the increasing turn-on losses associated with the reverse recovery of the P–i–N rectifier. The power loss in the transistor is much larger than in the rectifier due to its large on-state voltage drop. The total power loss is 185 W at 20 kHz when 8,000 W of power is delivered to the load.

When the silicon power MOSFET is replaced by the IGBT structure, the power losses are considerably reduced as shown in Fig. 10.12. A major portion of the reduction in the power loss occurs due to the smaller on-state voltage drop for the IGBT structure. The switching power loss for the IGBT structure increases and becomes much larger than the on-state power loss at 20 kHz. The switching power

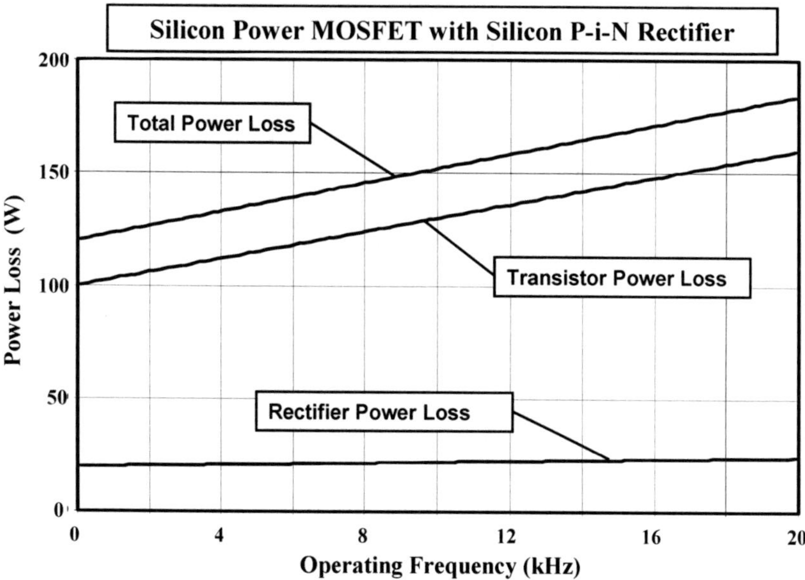

Fig. 10.11 Power losses during motor control with 400-V DC bus: silicon power MOSFET with silicon P-i-N rectifier

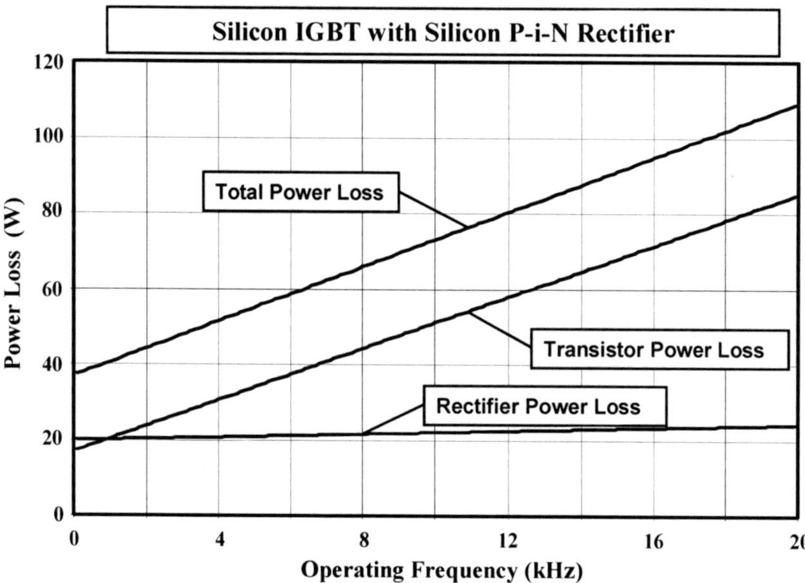

Fig. 10.12 Power losses during motor control with 400-V DC bus: silicon IGBT with silicon P-i-N rectifier

loss in the IGBT during the turn-on event is about twice as large as during the turn-off event. The total power loss is reduced to 110 W when 8,000 W of power is delivered to the load. Due to the greater efficiency in the motor control application, the IGBT-based drives are commonly used for hybrid electric vehicles and other consumer applications, such as air conditioning and ventilation, operated from medium DC bus voltages.

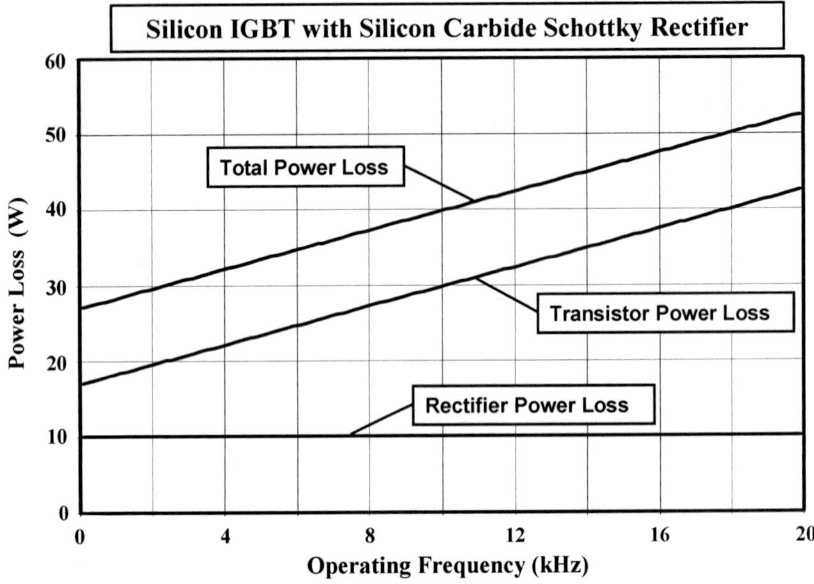

Fig. 10.13 Power losses during motor control with 400-V DC bus: silicon IGBT with silicon carbide Schottky rectifier

Ever since the first demonstration of the low on-state voltage drop and outstanding switching characteristics for the silicon carbide Schottky rectifiers,[5] there has been a considerable interest in replacing the silicon P–i–N rectifiers with this device for improving the performance of motor drives. The reduction in the power losses that can be achieved with this approach is shown in Fig. 10.13. A significant decrease in power loss is achieved due to the smaller on-state voltage drop of the 4H-SiC Schottky rectifier. Moreover, the power loss in the IGBT is reduced due to the elimination of the reverse recovery of the P–i–N rectifier. The total power loss is reduced to 53 W at 20 kHz when 8,000 W of power is delivered to the load. This example represents the earliest potential adoption of silicon carbide-based power devices in commercial applications.

An even superior technical solution can be produced by using the silicon carbide power MOSFET as the switch and the silicon carbide Schottky rectifier as the fly-back diode as illustrated by the power loss provided in Fig. 10.14. In this case, the power loss in the transistor is greatly reduced due to the low on-state voltage drop for the silicon carbide power MOSFET. The largest component of the total power loss is associated with the on-state voltage drop for the silicon carbide Schottky rectifier. The power loss does not increase as rapidly with frequency in this approach because of its implementation with only unipolar devices. The total power loss is reduced to only 14 W at 20 kHz when 8,000 W of power is delivered to the load. This example demonstrates the full potential for improving the efficiency of motor drives by using silicon carbide-based power devices.

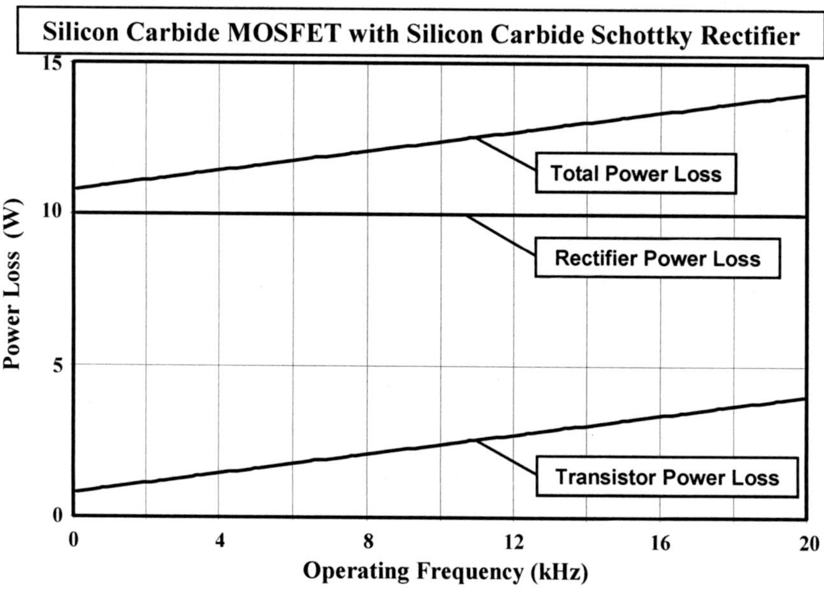

Fig. 10.14 Power losses during motor control with 400-V DC bus: silicon carbide MOSFET with silicon carbide Schottky rectifier

10.5 High DC Bus Voltage Applications

In this section, the above power loss analysis is applied to a motor control application using a high DC bus voltage with a duty cycle of 50%. The DC bus voltage (V_{DC}) will be assumed to be 3,000 V as pertains to the power source for electric locomotive drives such as the Shinkansen bullet train. In this case, the device blocking voltage rating is typically 4,500 V. The current being delivered to the motor winding (I_M) will be assumed to be 1,000 A. Due to the larger blocking voltage required for this application, the motor drive is commonly implemented using silicon bipolar devices as the power switch and the silicon P–i–N rectifier as the fly-back diode. Until the turn of the century, the power switch that was utilized by the traction industry was the gate turn-off thyristor. Since then, it is commonplace to implement the motor drive using IGBT devices.

Characteristics	Silicon MOSFET	Silicon IGBT	4H-SiC MOSFET
On-State Voltage Drop (V)	Very High	3.0	0.38
Turn-Off Time ($t_5 - t_4$) (μs)	-	2.0	0.1
Turn-Off Time ($t_6 - t_5$) (μs)	-	1.0	0.1

Fig. 10.15 Characteristics of transistors with 4,500-V blocking voltage rating

Characteristics	Silicon Schottky	Silicon P-i-N	4H-SiC Schottky
On-State Voltage Drop (V)	Very High	2.5	1.0
Turn-On Time ($t_2 - t_1$) (μs)	-	2.0	0.1
Turn-On Time ($t_3 - t_2$) (μs)	-	2.0	0.1
Peak Reverse Recovery Current (A)	-	1000	0

Fig. 10.16 Characteristics of rectifiers with 4,500-V blocking voltage rating

The characteristics for the transistors that are pertinent to the analysis of the power loss are provided in Fig. 10.15. In the case of the silicon power MOSFET structure, the specific on-resistance increases to a prohibitively large magnitude ($10\ \Omega\ cm^2$) for a device capable of blocking 4,500 V as shown in Fig. 3.6. The silicon MOSFET device is therefore not viable for such high-voltage applications. The on-state voltage drop (3 V) and switching times for the silicon IGBT device with 4,500-V blocking voltage rating are based upon scaling the characteristics for the 1,200-V structure modeled in Chap. 9 as well as the values reported in the literature.[6,7] In the case of the 4H-SiC power MOSFET structure, the drift region contribution increases to $3\ m\Omega\ cm^2$ (see Fig. 3.6), which is comparable to the specific on-resistance contributed by the N^+ substrate ($0.4\ m\Omega\ cm^2$) and the channel ($0.4\ m\Omega\ cm^2$). Consequently, a specific on-resistance of $3.8\ m\Omega\ cm^2$ has been used for the silicon carbide MOSFET structure. These devices are also assumed to be operated at an on-state current density of $100\ A\ cm^{-2}$.

The characteristics for the power rectifiers that are pertinent to the analysis of the power loss are provided in Fig. 10.16. The on-state voltage drop of the silicon Schottky diode is extremely large due to the very high resistance of the drift region (see Fig. 4.7). For comparison purposes, the silicon P–i–N rectifier and the silicon carbide Schottky rectifier are included in the power analysis. The on-state voltage drop (2.5 V) for the silicon P–i–N rectifier and its reverse recovery current are typical for such high-voltage structures.[7] In the case of the 4H-SiC Schottky rectifier structure, the on-state voltage drop is mostly incurred in the drift region due to its high specific on-resistance (see Fig. 4.8). These devices are assumed to be operated at an on-state current density of 100 A cm^{-2}. From the information provided in Fig. 10.16, it can be concluded that the silicon Schottky rectifier is not a viable alternative for such high-voltage applications.

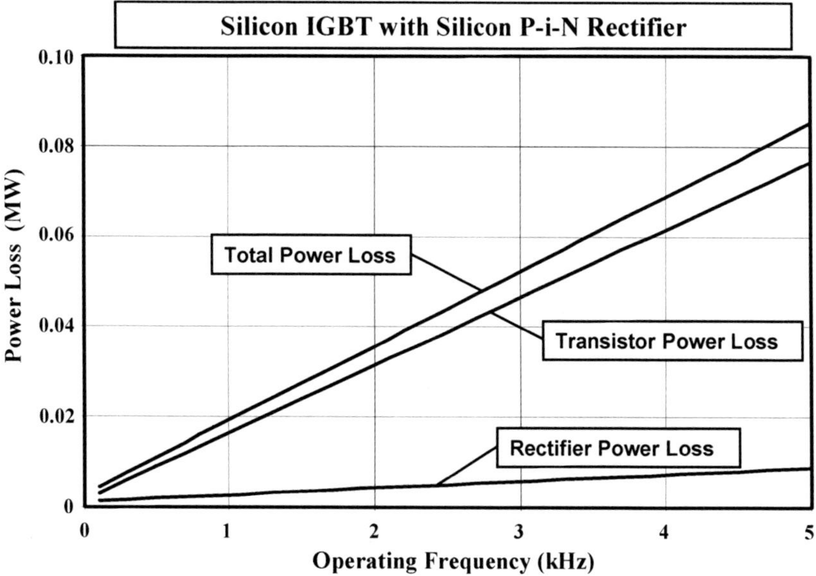

Fig. 10.17 Power losses during motor control with 3,000-V DC bus: silicon IGBT with silicon P-i-N rectifier

As an example, the power losses in the case of the silicon IGBT as the power switch with a silicon P–i–N rectifier as the fly-back diode are provided in Fig. 10.17 for frequencies ranging to 5 kHz. The power losses during switching are dominant in this case. The power loss in the transistor grows with increasing frequency due to the greater turn-on and turn-off losses. The turn-on losses in the silicon IGBT, associated with the reverse recovery of the P–i–N rectifier, are twice as large as the turn-off losses. The total power loss is 0.085 MW at 5 kHz when 3 MW of power is delivered to the load.

There is considerable interest in replacing the silicon P–i–N rectifiers with the silicon carbide-based Schottky rectifier for improving the performance of motor drives even in the case of high DC bus voltage applications. The reduction in the power losses that can be achieved with this approach is shown in Fig. 10.18. A significant decrease in power loss is achieved due to the smaller on-state voltage drop of the 4H-SiC Schottky rectifier. Moreover, the power loss in the IGBT is reduced due to the elimination of the reverse recovery of the P–i–N rectifier. The total power loss is reduced to 0.025 MW at 5 kHz when 3 MW of power is delivered to the load. This example represents another early potential adoption of silicon carbide-based power devices in commercial applications because epitaxial material suitable for delivering blocking voltage capability of over 5,000 V is already available.

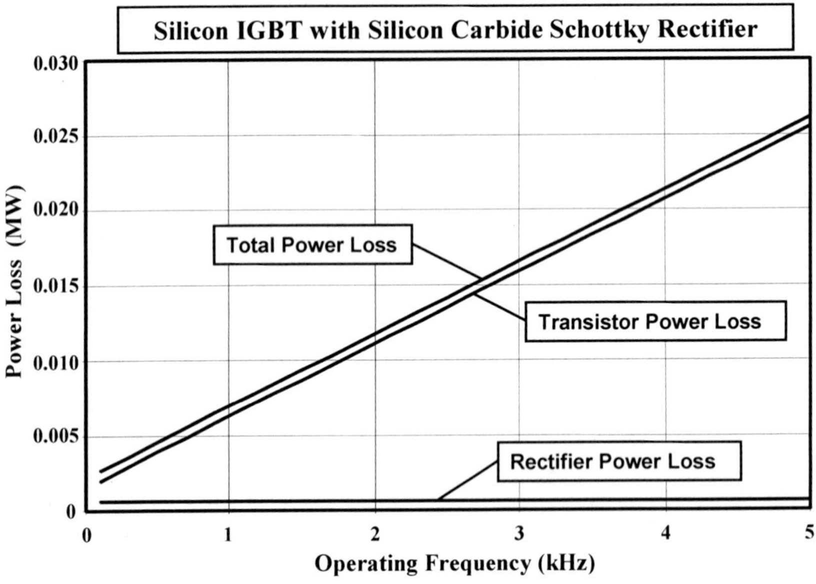

Fig. 10.18 Power losses during motor control with 3,000-V DC bus: silicon IGBT with silicon carbide Schottky rectifier

An even superior technical solution for applications using high DC bus voltages can be produced by using the silicon carbide power MOSFET as the switch and the silicon carbide Schottky rectifier as the fly-back diode as illustrated by the power loss provided in Fig. 10.19. In this case, the power loss in the transistor is greatly reduced due to the low on-state voltage drop for the silicon carbide power MOSFET structure. The turn-on and turn-off losses contribute equally to the increase in power loss with frequency in the silicon carbide power MOSFET structure. The power loss does not increase as rapidly with frequency in this approach because of its implementation with only unipolar devices. The total power loss is reduced to only 0.004 MW at 5 kHz when 3 MW of power is delivered to the load. This example demonstrates the full potential for improving the efficiency of motor drives by using silicon carbide-based power devices.

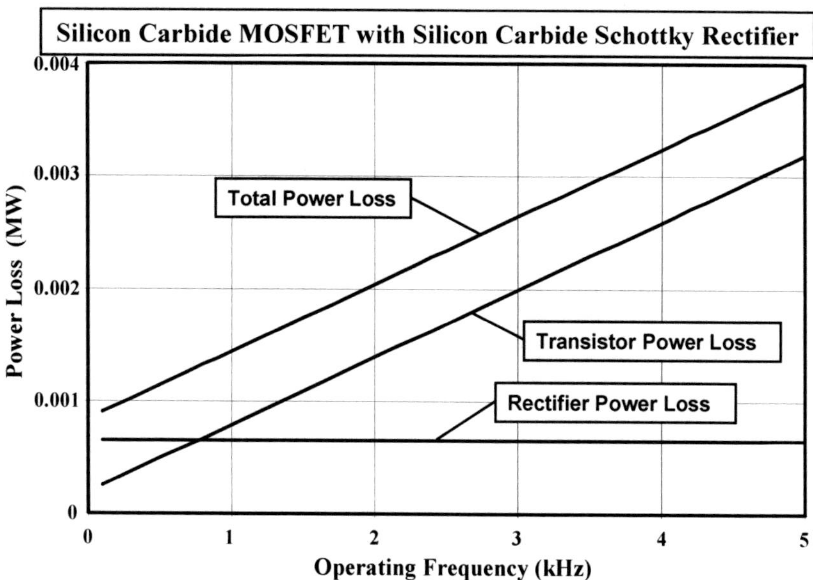

Fig. 10.19 Power losses during motor control with 3,000-V DC bus: silicon carbide MOSFET with silicon carbide Schottky rectifier

10.6 Summary

This chapter allows comparison of the benefits of utilizing various technologies for a broad range of applications. It can be concluded that silicon power MOSFETs and Schottky rectifiers provide the best performance in applications working from a low DC bus voltage. Silicon bipolar devices and those created from wide band-gap semiconductors will not displace this technology in the future. On the other hand, for applications working from a medium DC bus voltage, it is advantageous to utilize the silicon IGBT and P–i–N rectifier as a low cost technology. In these applications, silicon carbide-based Schottky rectifiers greatly reduce power losses when used in conjunction with the silicon IGBT devices. Even further gains in efficiency can be obtained by replacing the silicon IGBT with silicon carbide power MOSFET devices when the cost and manufacturing capability for the silicon carbide switches mature. The same benefits translate to the applications working from a high DC bus voltage as well.

Problems

10.1 Name the four basic components of power loss in transistors and rectifiers used in PWM motor control circuits.

10.2 Consider motor control performed from a 20-V DC bus. What is the best technology for the transistor and rectifier to minimize power losses?

10.3 Calculate the power loss occurring at a PWM operating frequency of 10 kHz for the technology choice in Problem 10.2 if the motor current is 10 A. Use the on-state voltage drop and switching times provided in Figs. 10.3 and 10.4 for the analysis.

10.4 Consider motor control performed from a 400-V DC bus. What is the best silicon technology for the transistor and rectifier to minimize power losses?

10.5 Calculate the power loss occurring at a PWM operating frequency of 10 kHz for the technology choice in Problem 10.4 if the motor current is 20 A. Use the on-state voltage drop and switching times provided in Figs. 10.9 and 10.10 for the analysis.

10.6 Consider motor control performed from a 400-V DC bus. What is the best silicon carbide technology for the transistor and rectifier to minimize power losses?

10.7 Calculate the power loss occurring at a PWM operating frequency of 10 kHz for the technology choice in Problem 10.6 if the motor current is 20 A. Use the on-state voltage drop and switching times provided in Figs. 10.9 and 10.10 for the analysis.

10.8 Consider motor control performed from a 3,000-V DC bus. What is the best silicon technology for the transistor and rectifier to minimize power losses?

10.9 Calculate the power loss occurring at a PWM operating frequency of 3 kHz for the technology choice in Problem 10.8 if the motor current is 1,000 A. Use the on-state voltage drop and switching times provided in Figs. 10.15 and 10.16 for the analysis.

10.10 Consider motor control performed from a 3,000-V DC bus. What is the best silicon carbide technology for the transistor and rectifier to minimize power losses?

10.11 Calculate the power loss occurring at a PWM operating frequency of 3 kHz for the technology choice in Problem 10.10 if the motor current is 1,000 A. Use the on-state voltage drop and switching times provided in Figs. 10.15 and 10.16 for the analysis.

10.12 Based upon the power loss analyses done in the previous problems, estimate the DC bus voltage above which silicon carbide technology will provide an improvement in efficiency for motor control applications.

References

[1] B.K. Bose, *Power Electronics and Variable Frequency Drives* (IEEE, New York, 1997)

[2] B.J. Baliga, Power semiconductor devices for variable-frequency drives, Proceedings of the IEEE, 82, 1112–1122, 1994

[3] B.J. Baliga and J.P. Walden, Improving the reverse recovery of power MOSFET integral diodes by electron irradiation, Solid-State Electronics, 26, 1133–1141, 1983

[4] B.J. Baliga and D.A. Girdhar, Paradigm shift in power MOSFET technology, Power Electronics Technology Magazine, 24–32, 2003

[5] M. Bhatnagar, P.K. McLarty, and B.J. Baliga, Silicon carbide high voltage (400 V) Schottky barrier diodes, IEEE Electron Device Letters, 13, 501–503, 1992

[6] R. Hotz, F. Bauer, and W. Fichtner, On-state and short circuit behavior of high voltage trench gate IGBTs in comparison with planar IGBTs, IEEE International Symposium on Power Semiconductor Devices and ICs, 224–229, 1995

[7] F. Bauer et al., 6.5 kV Modules using IGBTs with field stop technology, IEEE International Symposium on Power Semiconductor Devices and ICs, 121–124, 2001

PROFESSOR B. JAYANT BALIGA'S BIOGRAPHY

Professor Baliga is internationally recognized for his leadership in the area of power semiconductor devices. In addition to over 500 publications in international journals and conference digests, he has authored and edited 15 books (*Power Transistors*, IEEE Press, 1984; *Epitaxial Silicon Technology*, Academic Press, 1986; *Modern Power Devices*, John Wiley, 1987; *High Voltage Integrated Circuits*, IEEE Press, 1988; *Solution Manual: Modern Power Devices*, John Wiley, 1988; *Proceedings of the 3rd Int. Symposium on Power Devices and ICs*, IEEE Press, 1991; *Modern Power Devices*, Krieger Publishing Co., 1992; *Proceedings of the 5th Int. Symposium on Power Devices and ICs*, IEEE Press, 1993; *Power Semiconductor Devices*, PWS Publishing Company, 1995; *Solution Manual: Power Semiconductor Devices*, PWS Publishing Company, 1996; *Cryogenic Operation of Power Devices*, Kluwer Press, 1998; *Silicon RF Power MOSFETs*, World Scientific Publishing Company, 2005; *Silicon Carbide Power Devices*, World Scientific Publishing Company, 2006; *Fundamentals of Power Semiconductor Devices*, Springer Science, 2008; *Solution Manual: Fundamentals of Power Semiconductor Devices*, Springer Science, 2008. In addition, he has contributed chapters to another 20 books. He holds 120 U.S. Patents in the solid-state area. In 1995, one of his inventions was selected for the *B.F. Goodrich Collegiate Inventors Award* presented at the *Inventors Hall of Fame*.

Professor Baliga obtained his Bachelor of Technology degree in 1969 from the Indian Institute of Technology, Madras, India. He was the recipient of the *Philips India Medal* and the *Special Merit Medal (as Valedictorian)* at IIT, Madras. He obtained his Masters and Ph.D. degrees from Rensselaer Polytechnic Institute, Troy NY, in 1971 and 1974, respectively. His thesis work involved Gallium Arsenide diffusion mechanisms and pioneering work on the growth of InAs and GaInAs layers using organometallic CVD techniques. At R.P.I., he was the recipient of the *IBM Fellowship* in 1972 and the *Allen B. Dumont Prize* in 1974.

From 1974 to 1988, Dr. Baliga performed research and directed a group of 40 scientists at the General Electric Research and Development Center in Schenectady, NY, in the area of power semiconductor devices and high voltage integrated circuits. During this time, he pioneered the concept of MOS-Bipolar functional integration to create a new family of discrete devices. He is the *inventor of the IGBT* which is now in production by many international semiconductor companies. This invention is widely used around the globe for air-conditioning, home appliance (washing machines, refrigerators, mixers, etc.) control, factory automation (robotics), medical systems (CAT scanners, uninterruptible power supplies), and electric street-cars/bullet-trains, as well as for the drive-train in electric and hybrid-electric cars under development for reducing urban pollution. The U.S. Department of Energy has released a report that the variable speed motor drives enabled by IGBTs produce an energy savings of 2 quadrillion btus per year (equivalent to 70 GW of power). The widespread adoption of compact fluorescent lamps (CFLs) in place of incandescent lamps, enabled by IGBT ballasts, is producing an additional power savings of 30 GW. The cumulative impact of these energy savings on the environment is a reduction in carbon dioxide emissions from coal-fired power plants by over one trillion pounds per year. Most recently, the IGBT has enabled fabrication of very compact, light-weight, and inexpensive defibrillators used to resuscitate cardiac arrest victims. When installed in fire-trucks, paramedic vans, and on-board airlines, it is projected by the American Medical Association (AMA) to save 100,000 lives per year in the US. For this work, *Scientific American Magazine* named him one of the eight heroes of the semiconductor revolution in its 1997 special issue commemorating the solid-state century.

Dr. Baliga is also the originator of the concept of merging Schottky and p-n junction physics to create a new family of power rectifiers that are commercially available from various companies. In 1979, he theoretically demonstrated that the performance of power MOSFETs could be enhanced by several orders of magnitude by replacing silicon with other materials such as gallium arsenide and silicon carbide. This is forming the basis of a new generation of power devices in the twenty-first century.

In August 1988, Dr. Baliga joined the faculty of the Department of Electrical and Computer Engineering at North Carolina State University, Raleigh, NC, as a Full Professor. At NCSU, in 1991 he established an international center called the *Power Semiconductor Research Center* (PSRC) for research in the area of power semiconductor devices and high voltage integrated circuits, and has served as its founding director. His research interests include the modeling of novel device concepts, device fabrication technology, and the investigation of the impact of new materials, such as GaAs and silicon carbide, on power devices. In 1997, in recognition of his contributions to NCSU, he was given the highest university faculty rank of *Distinguished University Professor of Electrical Engineering*.

Professor Baliga has received numerous awards in recognition for his contributions to semiconductor devices. These include two *IR 100 awards* (1983, 1984), the *Dushman* and *Coolidge Awards* at GE (1983), and being selected among the *100 Brightest Young Scientists in America* by *Science Digest Magazine* (1984). He was elected *Fellow of the IEEE* in 1983 at the age of 35 for his contributions to power semiconductor devices. In 1984, he was given the *Applied Sciences Award* by the world famous sitar maestro Ravi Shankar at the Third Convention of Asians in North America. He received the 1991 *IEEE William E. Newell Award*, the highest honor given by the Power Electronics Society, followed by the 1993 *IEEE Morris E. Liebman Award* for his contributions to the emerging *Smart Power Technology*. In 1992, he was the first recipient of the BSS Society's *Pride of India Award*. At the age of 45, he was elected as foreign affiliate to the prestigious *National Academy of Engineering*, and was one of only four citizens of India to have the honor at that time (converted to regular member in 2000 after taking U.S. citizenship). In 1998, the University of North Carolina system selected him for the *O. Max Gardner Award*, which recognizes the faculty member among the 16 constituent universities who has made the greatest contribution to the welfare of the human race. In December 1998, he received the *J.J. Ebers Award*, the highest recognition given by the IEEE Electron Devices Society for his technical contributions to the solid-state area. In June 1999, he was honored at the Whitehall Palace in London with the *IEEE Lamme Medal*, one of the highest forms of recognition given by the IEEE Board of Governors, for his contributions to development of an apparatus/technology (the IGBT) of benefit to society. In April 2000, he was honored by his Alma Mater, I.I.T.-Madras, as a *Distinguished Alumnus*. In November 2000, he received the *R.J. Reynolds Tobacco Company Award for Excellence in Teaching, Research, and Extension* for his contributions to the College of Engineering at North Carolina State University.

In 1999, Prof. Baliga founded a company, *Giant Semiconductor Corporation*, with seed investment from Centennial Venture Partners, to acquire an exclusive license for his patented technology from North Carolina State University with the goal of bringing his NCSU inventions to the marketplace. A company, *Micro-Ohm Corporation*, subsequently formed by him in 1999, has been successful in licensing the GD-TMBS power rectifier technology to several major semiconductor companies for worldwide distribution. These devices have application in power supplies, battery chargers, and automotive electronics. In June 2000, Prof. Baliga founded another company, *Silicon Wireless Corporation*, to commercialize a novel super-linear silicon RF transistor that he invented for application in cellular base-stations and grew it to 41 employees. This company (renamed *Silicon Semiconductor Corporation*) is located at Research Triangle Park, NC. It received an investment of $10 million from *Fairchild Semiconductor Corporation* in December 2000 to co-develop and market this technology. Based upon his additional inventions, this company has also produced a new generation of Power MOSFETs for delivering power to microprocessors in notebooks and servers. This technology was licensed by his company to Linear Technologies Corporation with transfer of the know-how and manufacturing process. Voltage regulator modules (VRMs) using his transistors are currently available in the market for powering microprocessor and graphics chips in laptops and servers.

Index

CPSIA information can be obtained at www.ICGtesting.com
Printed in the USA
LVOW030211071211

258182LV00003B/1/P